Stetson University

3 4369 00358511 5

W9-BKV-577

262
B78
v. 2

DATE DUE			
WITHDRAWN			

Survey of

ORGANIC SYNTHESES

Volume 2

SURVEY OF

ORGANIC SYNTHESES

VOLUME 2

CALVIN A. BUEHLER
University of Tennessee

DONALD E. PEARSON
Vanderbilt University

Reaction Index prepared by
PAUL F. HUDRLIK
Howard University

A Wiley-Interscience Publication

John Wiley & Sons
New York • London • Sydney • Toronto

Copyright © 1977 by John Wiley & Sons, Inc.

All rights reserved.   Published simultaneously in Canada.

No part of this book may be reproduced by any means,
nor transmitted, nor translated into a machine language
without the written permission of the publisher.

Library of Congress Cataloging in Publication Data *(Revised)*

Buehler, Calvin Adam, 1896–
  Survey of organic syntheses.

  Includes bibliographical references and indexes.
  1. Chemistry, Organic—Synthesis.   I.   Pearson,
Donald Emanual, 1914–   joint author.   II.   Title.

QD262.B78   547'.2   73-112590
ISBN 0-471-11671-8

Printed in the United States of America

10 9 8 7 6 5 4 3 2 1

PREFACE

This book is a continuation of the first volume, published in
1970. The organization is similar in that there are 20 chapters,
each dealing with a functional group as before. Many chapters
include new methods of synthesis or related material not given in
Volume 1. In many cases the treatment is critical, since,
invariably, the most important fact desired is the preferred
method of synthesis at this writing. Again, some difficulty has
been experienced in placing methods, because many reactions
involve a series of steps, of which the key one may be a matter
of opinion.

    An attempt has been made to include all the newer methods of
synthesis. This goal becomes increasingly difficult to achieve
because of the tremendous number of new methods that appear not
only in expected areas but in unexpected ones as well. This
situation has been remedied to some extent in recent years by the
appearance of more sources devoted exclusively to synthesis.
Coverage begins largely in 1969 where Volume 1 ends, and extends
through 1975, with perhaps less thoroughness existing, because of
printing delay, in the last year or two. Addenda have been added
to most chapters to cover syntheses not available when the
chapters were first written.

    Some effort has been expended in making the contents of the
book readily available. Chapter contents appear at the beginning
of each chapter and at the end of the book there are Subject and
Reaction Indexes.

    Frequent reference to Volume 1 is made. This plan offers a
background for the reader and prevents much repetition. Such
references are designated by 1 (for the volume) followed by the
page number.

199842

We are again greatly indebted to Dr. Henry E. Baumgarten, University of Nebraska, for his painstaking examination of the first typed draft of each chapter. His suggestions have added substantially to the value of the book. Our gratitude goes as well to Dr. C. C. Cheng, Midwest Research Institute, for examining Chapters 15, 16, and 17 and to Dr. Richard M. Pagni and Dr. George W. Kabalka, both of the University of Tennessee, for their helpful advice. Mention should also be made of Dr. Marshall G. Frazer of Vanderbilt University, who examined with great care the camera copy of all chapters, and his wife, Virginia, who drew so well the many formulas and equations that appear in the text. Thanks are also due to Dr. L. T. Burka, Vanderbilt University, for proofreading the camera copy of Chapter 11, Ketones, the most difficult of all chapters, and to Dr. Paul F. Hudrlik, Howard University, Washington, D.C., for helpful suggestions while constructing the Reaction Index. Finally, we express our gratitude to our departments for encouragement and secretarial aid.

We will appreciate being informed by readers of any errors or omissions.

Knoxville, Tennessee   Calvin A. Buehler
Nashville, Tennessee   Donald E. Pearson
August, 1976

CONTENTS

ABBREVIATIONS

Ac	Acetyl
Acac	$(CH_3CO)_2CH-$
9-BBN	9-Borabicyclo[3.3.1]nonane
BMS	Borane-methyl sulfide
Bu	Butyl
Dabco	1,4-Diazabicyclo[2.2.2]octane
DBN	1,5-Diazabicyclo[4.3.0]non-5-ene
DBU	1,8-Diazabicyclo[5.4.0]undec-7-ene
DCC	Dicyclohexylcarbodiimide
DDQ	2,3-Dichloro-5,6-dicyano-1,4-benzoquinone
Desyl	$C_6H_5COCHC_6H_5$
Diglyme	Diethylene glycol dimethyl ether
DMA	Dimethylacetamide
DME	Dimethoxyethane
DMF	Dimethyl formamide
DMSO	Dimethyl sulfoxide
DNPH	2,4-Dinitrophenylhydrazone
EDTA	Ethylenediaminetetraacetic acid
Et	Ethyl
Fremy's salt	$(KSO_3)_2NO$
Glyme	1,2-Dimethoxyethane
Hexamine	Hexamethylenetetramine
HMPA	Hexamethylphosphoric triamide, hexamethylphosphoramide
LDIA	Lithium diisopropylamide
LiICA	Lithium N-isopropylcyclohexylamide
Me	Methyl
Ms	Mesyl
MTPI	Methyltriphenoxyphosphonium iodide
MVK	Methyl vinyl ketone
NBS	N-Bromosuccinimide
NCS	N-Chlorosuccinimide
Ph or ϕ	Phenyl
PPA	Polyphosphoric acid
Pr	Propyl
Py	Pyridine
α-SAS	Sodium-9,10-anthraquinone-α-sulfonate

Sulfolane   Tetramethylene sulfone

Sulfurane $(C_6H_5)_2S[OC(CF_3)_2]_2$

TEBA   Triethylbenzylammonium chloride

Thexyl $(CH_3)_2CH-\underset{\underset{CH_3}{|}}{\overset{\overset{CH_3}{|}}{C}}-$

THF   Tetrahydrofuran

TMEDA   N,N,N,N-Tetramethylenediamine

TPE   Tetraphenylethylene

Trityl $(C_6H_5)_3C-$

Ts   Tosyl, p-$CH_3C_6H_4SO_2-$

TTC   Triphenyltetrazolium chloride

Survey of

ORGANIC SYNTHESES

Volume 2

ORGANIC SYNTHESIS

Chapter 1

ALKANES, CYCLOALKANES, AND ARENES

The Wolff-Kishner reduction of ketones to hydrocarbons has
proved to be more reliable than the Clemmensen reduction (A.1).
A reaction supplementary to the W.-K. (Wolff-Kishner) reaction
and of some specificity is tosylhydrazone reduction with sodium
cyanoborohydride (A.1). Many novel methods of reducing benzylic
or allylic alcohols to the hydrocarbon have been introduced (A.2)
and the list of reagents for reducing halides grows longer (A.3).
More discussion of reduction by the use of homogeneous catalysts
is included (A.5), as well as that of sulfur compounds, including
chain-lengthening processes, using thiophene (A.7). Nonaqueous
reductions of diazonium salts are described (A.8).

Although the exact nature of the Wurtz and Grignard-Wurtz
coupling reactions appear to be not completely understood, considerable knowledge has accumulated on what and what not to do in
these preparations (B.2). Silver is best for symmetrical
coupling and the copper anion, for unsymmetrical coupling. However it appears that nickel salts chelated with an ethylene diphosphine may be a better reagent for unsymmetrical coupling
(B.2). While another nickel chelate is of value in Ullmann
coupling (B.3), it seems that the Wurtz coupling at low temperature in the presence of tetraphenylethylene is a superior
method of operation (B.2, Ref. 21). Finally, the oxidation
coupling of lithium diarylcuprates to biaryls is discussed more
fully (B.3).

Among nucleophilic reactions, the addition of carbanions to
alkenes has become a realistic preparation (C.2). In addition,
a better understanding of strong base structure in solution has
advanced our knowledge of the synthesis of carbanions (C.7) and
a method ($S_{RN}I$) has been found to bypass benzyne formation (C.8).

Among electrophilic reactions, the rearrangements to prepare adamantanes are numerous (D.3) and alkylations with trialkylalanes and t-halides to produce hydrocarbons with quaternary carbon atoms are new (D.4).

The discussion on decarboxylation of acids to hydrocarbon
benefits from a recent review (F.1).

In the consideration of free radical reactions, the Pschorr
synthesis, both from stilbene diazonium salts (G.1) and from the
unsaturated stilbene (I_2 + hv, G.3), has been discussed, but the
photochemical coupling of alkenes is so common that examples are

presented more or less in list form (G.3, G.4). It should be
noted, however, that cuprous triflate is a superior catalyst for
alkene coupling because the salt does not deposit on the walls
of the flask during irradiation (G.4). All of the ramifications
of the Simmons-Smith reaction, which are many, diverse, and
capable of adding considerable scope to cyclopropane formation,
are discussed (G.6).

In this chapter reference is frequently made to the Hœuben-
Weyl (H.-W.) books, <u>Methoden der Organischen Chemie</u>, which contain not only comprehensive tables, but also chapters on the
synthesis of stereoisomeric and labeled hydrocarbons.

A. Reduction

1. From Aldehydes or Ketones and Their Derivatives (1, 3)

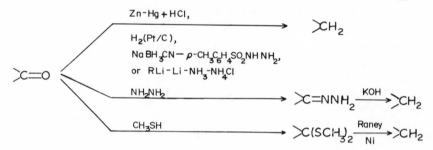

The most recent reviews on the Clemmensen and W.-K.
reductions of carbonyl compounds are in the Augustine [1a,1b]
and H.-W. [2] books. A modified Clemmensen reduction using
active zinc powder in common organic solvents or acetic anhydride saturated with hydrogen chloride has yielded variable
results [2a].

Recent examples confirm the fact that for the preparation
of tri- and tetramethylnaphthalenes the Huang-Minlon modification of the W.-K. reduction is a more reliable procedure than
the Clemmensen reduction [3]. For instance, the W.-K. procedure with β-acetonaphthalene gives an 82% yield of β-ethyl-
naphthalene and no by-products, while the Clemmensen procedure
gives 61% of the hydrocarbon plus 14% of 2-ethyl-1,2-dihydro-
naphthalene plus small amounts of dimers derived from the
pinacol of β-acetonaphthalene. Moreover the Clemmensen procedure does not work well with most diketones; in fact, the
reduction of 1,3-diketones and α,β-unsaturated ketones gives a
useful method for the synthesis of rearranged monoketones while
1,4-diketone reduction leads to unexpected and usually inaccessible compounds [4].

On the other hand, the W.-K. procedure not only has been quite satisfactory, but has been supplemented by the introduction of a modification that can be used in cases in which strong alkalinity cannot be tolerated. This modification capitalizes on the fact that sodium cyanoborohydride is relatively stable to hydrolysis at pH 3 and is fairly specific for the reduction of the imminium group as shown [5,5a]:

$$\text{C=O} \;+\; \text{TsNHNH}_2 \;\xrightarrow{\;H^{\oplus}\;}\; \text{C=N} \overset{H\;H}{\underset{\oplus}{N}} \text{Ts} \;\xrightarrow{\;NaBH_3CN\;}$$

$$\text{CHNHNHTs} \;\longrightarrow\; \text{CH}_2 + C_7H_7SO_2H + N_2$$

The hydrolysis of the $NaBH_3CN$ is so slow in the acid solution that the hydrazone forms preferentially, is protonated, and is then reduced. The hydrocarbon probably results via the diimide (see Ex. b). Here then is a W.-K.-like reaction conducted in dilute acidic rather than basic solution. If no other reducible groups are present in the substrate, tosylhydrazones may be reduced by $NaBH_4$ [2, p. 268].

A comprehensive discussion of Hutchins' method of reduction of carbonyl compounds using both sodium and tetrabutylammonium cyanoborohydride has appeared recently [6]. A noteworthy feature of the method is the fact that aralkyl ketones do not reduce well with sodium cyanoborohydride unless an electron-releasing group is attached to the aryl group.

Direct methods of reduction of carbonyl compounds include catalytic hydrogenation (1, 4) and metal-liquid ammonia reduction. Both methods are restricted to carbonyl compounds that produce benzylic hydrocarbons. For lithium-liquid ammonia reduction, traces of cobalt or aluminum catalyze the reaction [7]:

Tetralin, 96%

The decomposition with sodium benzoate rather than ammonium chloride leads to appreciable amounts of 1-tetralol. The ammonium chloride acts as a proton source to generate tetralol while the reducing agent is still present. The tetralol is then further reduced to tetralin.

Still another modification of the metal-liquid ammonia reducing system consists of alkylation of the ketone followed by reduction of the alkoxide (using again the absolutely essential NH_4Cl) [8].

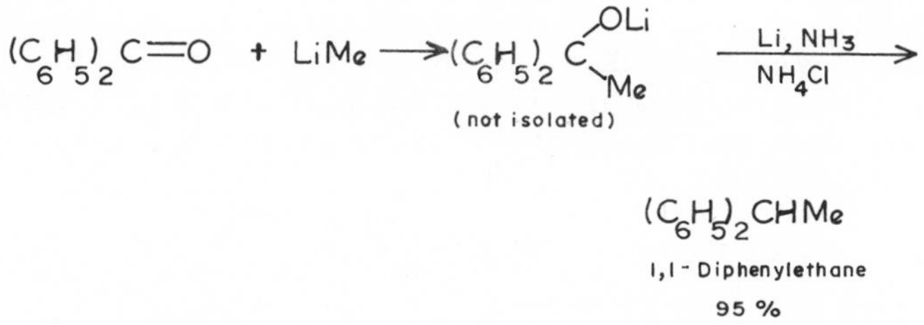
$(C_6H_5)_2C{=}O$ + LiMe → $(C_6H_5)_2C(OLi)(Me)$
(not isolated)
Li, NH_3
NH_4Cl
$(C_6H_5)_2CHMe$
1,1-Diphenylethane
95 %

It is interesting to note that some of the hydrides at elevated temperatures reduce ketones beyond the alcohol stage [9].

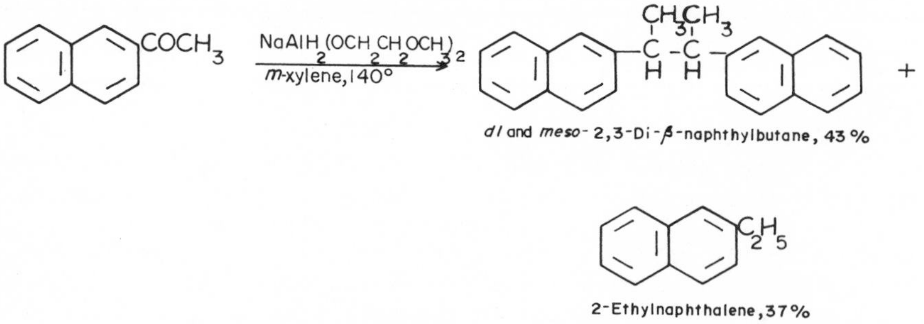
$COCH_3$
$NaAlH_2(OCH_2CH_2OCH_3)_2$
m-xylene, 140°
CH_3CH_3
C - C
H H
+
dl and *meso*-2,3-Di-β-naphthylbutane, 43 %
C_2H_5
2-Ethylnaphthalene, 37 %

Although reduction of ketones with diborane yields the corresponding alcohol, traces of boron trifluoride may carry the reduction further to the hydrocarbon stage. Thus p,p'-dimethoxybenzophenone gives the corresponding diphenylmethane [10].

Triethylsilane is the last reducing agent discussed that is capable of reducing the carbonyl to the methylene group. It is generally used with a strong acid and is restricted in scope to aralkyl ketones or benzaldehydes, exemplified as follows [11]:

$$ArCOR + Et_3SiH \xrightarrow[25^\circ]{CF_3CO_2H} ArCH_2R$$

Yields are often quantitative. Low yields are due to the incursion of some Friedel-Crafts side products.

When for some reason the carbonyl compound is not available or pure, it is possible to reduce the hydrazone (or other derivative) directly to the hydrocarbon (3-4 atm) in acetic acid with Pd on C at 50-60°. Yields were 71-95% (12 ex.) [11a].

a. Preparation of Hydrocarbons by the Wolff-Kishner, Huang-Minlon Modification [12]

The carbonyl compound (0.01 mol) in a mixture of 13.5 g KOH, 10 mol of 85% $NH_2NH_2 \cdot H_2O$, and 100 ml DEG was refluxed for 1 hr, after which the water was removed by distillation until the internal temperature reached 200°. Refluxing was continued for an additional 3 hr and the cooled mixture acidified with HCl and extracted with C_6H_6. Isolation by fractionation yielded diphenylmethane 83% (from benzophenone), propylbenzene 82% (from propiophenone), and cyclohexane, 80 % (from cyclohexanone).

b. Preparation of Hydrocarbons by Hutchins' Modification [5]

The carbonyl compound (1 mmol) and p-$CH_3C_6H_4SO_2OH$ were treated with $NaBH_3CN$ (4 mmol) and the mixture heated to 100-105° for 1-4 hr. Dilution with water, extraction with cyclohexane, evaporation, and crystallization gave yields of hydrocarbons, 62-98%.

c. Preparation of Hydrocarbons by Clemmensen Reduction in the Absence of an Organic Solvent and in the Presence of a Miscible Solvent [1a]

d. Preparation of Hydrocarbons from Thioacetals or Thioketals (1, 5) [1c]

2. From Alcohols (1, 6)

$$ArCH_2OH \longrightarrow ArCH_3$$

The reduction has been reviewed [2, pp. 227-237].
Drastic conditions bring about reduction of the aromatic ring as well. Thus benzyl alcohol, 0.02 mol and hydrazine, 97%, 0.08 mol heated in a bomb at 180° for 8 hr give methylcyclohexane (97%) [2, p. 228].
Our interest here is in milder methods of which there are five of recent origin. By a direct method Shahak and Bergmann [13] reduced diethyl α-hydroxybenzyl phosphonate to toluene (69%) with hydrazine hydrate and aqueous KOH in diethylene glycol.
Of the one-operation methods, Scharf and Döring [14] transformed 3,5-di-t-butylphenyldimethylcarbinol into the methyl ether that was hydrogenated in the presence of Pt to yield 91% of the di-t-butylcumene.
Corey and Achiwa [15] were successful in reducing allylic and benzylic alcohols to the corresponding hydrocarbons by

conversion into the pyridinium sulfate, which mixture was reduced with LiAlH$_4$ in yields of 64-98% (Ex. a):

$$ROH + SO_3 \cdot N \underset{}{\overset{THF}{\longrightarrow}} ROSO_3 \overset{\oplus}{H} \overset{\ominus}{N} \underset{THF}{\overset{LiAlH_4}{\longrightarrow}} RH$$

It will be recalled that benzylic alcohols can be reduced with lithium-liquid ammonia followed by ammonium chloride [A.1: 7,8]. A fourth method is that of Ireland and co-workers [16], who found that the diethyl phosphates and N,N,N',N'-tetramethyl-phosphorodiamidates of alcohols could be reduced in high yield without purification in lithium-ethylamine solution:

$$R_2CHOH \quad \frac{\text{1) BuLi} \quad \text{2) ClPO(NMe}_2)_2}{\text{3) Li/EtNH}_2} \quad R_2CH_2$$

The tetramethylphosphorodiamidate derivative is more suitable for use with primary and secondary alcohols, while the diethyl phosphate grouping appears to be more satisfactory for tertiary alcohols.

A more drastic method of converting alcohols into hydrocarbons was developed by Pines [17]:

$$RCH_2OH \underset{\text{High temp.}}{\overset{H_2, Ni}{\longrightarrow}} RH$$

The reaction consists of a dehydrogenation to the aldehyde followed by a decarbonylation. The decarbonylation step may be stopped by the addition of sulfur:

$$C_6H_5(CH_2)_2CH_2OH \underset{H_2, \triangle}{\overset{Ni-S}{\longrightarrow}} C_6H_5(CH_2)_2CH_3$$

Propylbenzene

In this case there is a tendency for the hydrocarbon chain to isomerize as though the poisoned catalyst were somewhat acidic.

To the many methods of reducing phenols (1, 7) may now be added that of Rossi and Bunnett [18] in which aryl phosphonates (ArOPO(OEt)$_2$) are reduced with sodium and liquid ammonia, possibly via the $S_{RN}I$ mechanism.

a. Preparation of <u>trans</u>-2,6-Dimethyloctadiene [15]

This consisted of 98% from geraniol, $CH_3\overset{CH_3}{C}=CHCH_2CH_2\overset{CH_3}{C}=CHCH_2OH$, and the pyridine-sulfur trioxide complex, the solution of which was treated with $LiAlH_4$.

3. From Halides or Sulfonates (1, 8) [2, pp. 281-290]

$$RX \xrightarrow{\text{2 (H)}} RH + HX$$

The number of reducing agents employed to convert halides into hydrocarbons continues to offer great variety. A listing follows:

1. Lithium aluminum hydride for aromatic bromides and iodides [19] and for alkyl halides [20]. Contrary to previous reports, the reagent works very well with aryl types when it is fresh and used in a ratio of 4 mol:1 of the halide. It is superior to triphenyltin hydride. It reacts with vinyl bridge-head and cyclopropyl halides to give hydrocarbons in high yield. The same hydride has been employed for trifluoromethylbenzenes ($-CF_3 \longrightarrow -CH_3$) with o- and p-hydroxy or amino groups present (80-100%) [21].

2. Sodium (bis-2-methoxyethoxy) aluminum hydride, $NaAlH_2(OC_2H_4OCH_3)_2$, for alkyl and aryl halides [22]. It is claimed that the reduction of the former with this reagent proceeds more rapidly and with less alkene formation than if $LiAlH_4$ is used; in addition aryl halides give higher yields of hydrocarbon than with $LiAlH_4$.

3. Sodium borohydride in DMSO for alkyl and benzyl halides [23] and for t-alkyl halides [24].

4. Sodium cyanoborohydride in HMPA for alkyl bromides and iodides [25]. Being a very mild reducing agent, this hydride is selective in the presence of ester, nitro, or ketone groups. Even neopentyl iodides are reduced.

5. Diborane in CH_3NO_2 for aralkyl halides [26], although benzyl chloride is not reduced. The reagent in THF is less effective. It is interesting to note that diborane may be generated in situ [23], as may be implied in the equation

$$(CH_3O)_2SO_2 + 2NaBH_4 + 6RCH{=\!=}CH_2 \xrightarrow[\text{2) } 6H_2O]{\text{1) DMSO}} 2CH_4 + Na_2SO_4 + 6RCH_2CH_2OH + 2B(OH)_3$$

6.   Lithium iodide-boron trifluoride etherate, specifically for α-haloketones (85-98%) [27]. The mechanism is as shown:

$$F_3B + O{=}CCH_2X + I^{\ominus} \longrightarrow F_3BO{-}C{=}CH_2 + XI$$

with R on the carbonyl carbon and $\downarrow H_2O$

$$O{=}C{-}CH_3 \rightleftharpoons HO{-}C{=}CH_2$$

with R below each carbon.

7.   Sodium in t-butyl alcohol for unsaturated, bicyclic polyhalides [28]. This procedure has been considered the best for the removal of chlorine from an organic molecule.

8.   Zinc and deuterium oxide for α-halogenated esters [29], which is the method of choice for preparing α-deuterated esters, but which is not so specific for preparing α-deuterated ketones.

9.   Formation of the phenyl sulfide followed by reduction with Li-liquid ammonia for tosylates [30]. This method should be considered if elimination is a problem. It was used successfully in the preparation of 9,11-secoprogesterone.

10.   Triethyltin hydrides for aryl bromides [31] (90-97%). Ultraviolet radiation accelerates some reductions.

11.   Tri-n-butyltin hydride for alkyl and aryl halides [32] and with an effective radical initiator for acetylated chloro-deoxy sugars (32b). Tin hydrides with radical initiators have been generally used for the reduction of inert halides and cyclopropyl halides (to avoid ring opening), but it has now been demonstrated that $LiAlH_4$ in refluxing ether reduces inert halides such as bromoadamantane and 1-bromotriptycene to the hydrocarbons in about 80% yield and bromocyclopropane smoothly to the cyclopropane [20].

12.   Dibutylstannic hydride generated by the action of poly-methylhydrosiloxane, $Me_3SiO(MeHSiO)_{\sim35}SiMe_3$, PMHS, on tetrabutyldiacetoxytin oxide [33]. The combination is perhaps the most novel reducing agent of the group listed. Bromides and iodides react exothermically, while chlorides require heating or a free radical initiator. Fluorides do not react. Aromatic bromides do react, but aromatic chlorides do not. Probably the most valuable use of the combined reagent is the stepwise reduction of geminal dihalides (Ex. a).

13.   Vanadous chloride complex $[VCl_2(C_5H_5N)_2]$ for the coupling of benzyl halides to 1,2-diarylethanes [34].

14.   Sodium borodeuteride-palladium chloride for the reduction of aryl halides to deuterated arenes [35]:

$$ArX + NaBD_4 \xrightarrow{PdCl_2} ArD$$

For the coupling of lithium dialkylcopper with halides to form cyclopropanes, see G.6. Photoreduction of chloro- and bromophenols is given under Phenols, C.4.

a. Preparation of trans-1-Bromo-2,3-dimethylcyclopropane [33]

trans-1,1-Dibromo-2,3-dimethylcyclopropane, 0.2 mol, and silicone polymer (G.E. Dri-Film 1040), 18 g, 0.3 eq, were mixed at 0° while bis-tributyltin oxide was added dropwise and the mixture stirred 1.5 hr more after the addition. The total was then allowed to warm to room temperature until it was no longer exothermic or until it had jelled. Distillation at reduced pressure with an efficient dry-ice trap yielded the bromocyclo-propane (85%).

5. From Alkenes and Arenes (1, 11)

$$RCH{=\!=}CH_2 + H_2 \xrightarrow{Pd/C} RCH_2CH_3$$

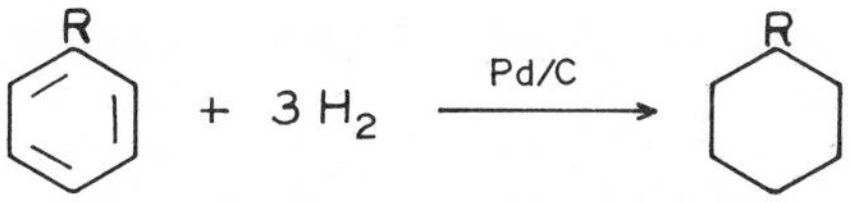

Although the three-step method, involving (a) the Grignard addition to a carbonyl compound, (b) dehydration of the alcohol formed, and (c) reduction of the resulting alkene, has become prosaic in nature, it remains a well-traveled route to alkanes.

A stepwise method, starting with the alkene, has also been employed in the preparation of long-chain hydrocarbons carrying quaternary carbon atoms [35a] such as

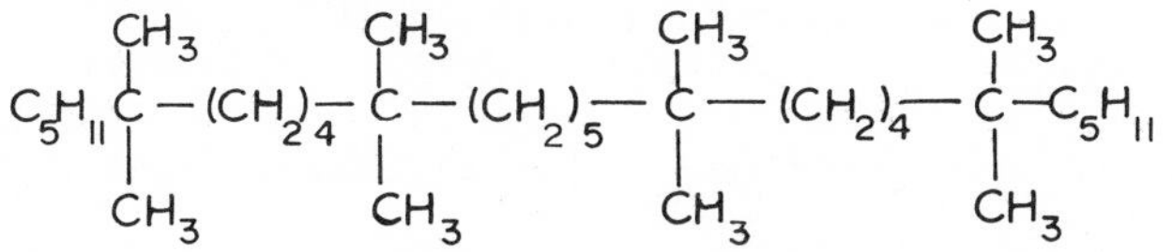

For the reduction of alkenes a tremendous variety [2, pp. 8-226] of chemical reductions and of catalytic reductions have been listed. This section deals only with further advances.

An active form of platinum has been prepared by Eaborn and co-workers [36] from chloroplatinic acid and triethylsilane in

95% ethanol. The product stores well and is more active than the Adams' catalyst in the hydrogenation of alkenes. The investigators have noted as well that a trace amount of acid activates the catalyst made from chloroplatinic acid and sodium borohydride. In addition nickel boride [easily prepared in ethanol from $(NiOAc)_2$ and $NaBH_4$] proved to be an effective catalyst in the Parr bomb hydrogenation of a variety of olefins, mostly in quantitative yield [37]. This catalyst probably breaks down to nearly colloidal nickel [38]. The latter is prepared in non-aqueous solution and is unusually sensitive to steric effects. For example, cyclohexene is reduced very slowly.

Attention has been called to the curious addition of deuterium to Δ-9,10 octalin [39]:

Not only does the product contain three deuterium atoms, but it is a mixture of cis and trans isomers.

Continued interest is being shown in homogeneous catalysis, a theoretical discussion of which has appeared [40]. The following ions with suitable ligands to bring about miscibility are capable of relaying hydrogen to a substrate:

Cu^I, Cu^{II}, Ag^I, Hg^I, Hg^{II}, Co^I, Co^{II}, Pd^{II}, Rh^I, Rh^{III}, Ru^{II},

Ru^{III}, Ir^I.

The hydrogen adds to the ion in one of three ways:

Heterolytic: $Ru^{III}Cl_6^{3\ominus} + H_2 \longrightarrow Ru^{III}Cl_5 H^{3\ominus} + H^\oplus + Cl^\ominus$

Homolytic: $2 Co^{II}(CN)_5^{3\ominus} + H_2 \longrightarrow 2 Co^{III}(CN)_5 H^{3\ominus}$

Dihydride addition: $Ir^I Cl(CO)\left[P(C_6H_5)_3\right]_2 + H_2 \longrightarrow Ir^{III} H_2 Cl(CO)\left[P(C_6H_5)_3\right]_2$

Recent reviews on homogeneous catalysis are those of Schrauzer [41] and Harmon [42]. Of the homogeneous catalysts of Group VII metals those of Rh, Ru, and Ir appear to have been studied to the greatest extent. Common examples are:

(a) RhCl(PPh$_3$)$_3$ ($\underline{1}$, 11), (b) RuCl$_4$(Ph$_3$P)$_3$ ($\underline{1}$, 11), and (c) IrCl (CO)(Ph$_3$P)$_2$.

The homogeneous hydrogenation of alkenes may be accomplished usually at low temperature and pressure. Ordinarily oxygen must be excluded from the reaction mixture. The solvent employed, benzene, toluene, alcohol, and so on, must be deoxygenated before the addition of catalyst.

One advantage of homogeneous catalysis is the high degree of selectivity shown. For example, among the alkenes, terminal double bonds react more readily than internal ones and the rate of hydrogenation of cis alkenes exceeds that of the trans isomers. In addition, many of the problems common in heterogeneous catalysis, such as isomerization, disproportionation, and hydrogenolysis, may often be eliminated.

Chlororhodium tris(triphenylphosphine) has been utilized with hydrogen to convert a series of α,β-olefinic acids, esters, carbonyl compounds, and so on, into saturated types (60-93% yields) in a very specific reduction [43]. For an example of the use of this catalyst, see Example d below.

A homogeneous catalyst of the Ziegler type, referred to as soluble although it may be in a finely suspended form, is the product when alkyl- or aryllithium is added to transition metal 2-ethylhexanoates [44]. In cyclohexane with the lithium:cobalt ratio of 2:9 and under 50 lb pressure of hydrogen, many olefins are hydrogenated satisfactorily.

Many chemical reducing agents have also become available. Chromous sulfate in aqueous DMF is effective in reducing unsaturated acids, esters, or nitriles at 25° (84-100%) [45] (Ex. a). Lithium-liquid ammonia works well with unsaturated acids [46].

Newer reducing agents follow. Sodium-hexamethylphosphoramide in t-butyl alcohol gives quantitative yields in the reduction of a variety of cyclic and acyclic alkenes [47] (Ex. b). 2,4,6-Triisopropylbenzenesulfonyl hydrazide, although less stable than the unsubstituted hydrazide, appears to reduce a series of alkenes, allyl alcohols, and ethyl methacrylate more readily than the latter [48]. Its reductive powers probably stem from its ability to form diimide which itself has been employed [49] (Ex. c).

a.   Preparation of meso-2,3-Dimethylsuccinic Acid [45]

Dimethyl 2,3-dimethylmaleate, 28.4 mmol, in 25 ml of DMF was added to 225 ml of DMF, 100 ml of 0.89 M CrSO$_4$, and 130 ml of water. After 3 days the solution was diluted and extracted with ether. Saponification and acidification of the ethereal residue gave 75% of the free acid.

b. Preparation of Norbornane [47]

Sodium, 0.25 g atom, and 100 ml of HMPA were stirred until a
blue color appeared. t-Butyl alcohol was added and the mixture
was stirred for 5 min, whereupon norbornene, 0.1 mol in 10 ml
HMPA was added over 6 hr at such a rate that the blue color never
quite disappeared. The mixture was then stirred for 12 hr,
poured into water, and extracted with pentane. Distillation gave
73% of norbornane.

c. Preparation of Acenaphthene [49]

Sodium metaperiodate, 2.5 mmol in 5 ml of water, was added
to a mixture of 0.5 mmol of acenaphthylene, 20 mmol of 100%
NH_2NH_2, 2 drops of saturated aqueous $CuSO_4$, and 2 drops of
CH_3COOH in 3 ml of DMSO. Yield of acenaphthene was 91%.

d. Preparation of 5α,6-Dihydroergosterol [50]

Rhodium(III) chloride trihydrate, 2 g, in 70 ml of ethanol
and triphenylphosphine, 12 g in 350 ml of hot ethanol, were
mixed and refluxed for 2 hr. The hot solution was filtered and
the crystalline solid was washed with several portions of ether
(50 ml). The red crystalline solid (85%) was ready for use.
Ergosterol, 5 mmol, benzene, 20-40 ml, and 0.1 g of the
catalyst were placed in a flask that was completely filled with
H_2 at atmospheric pressure. Stirring was begun and continued
until hydrogenation was complete (reduction is rapid in this
case). After proper venting, the solvent was evaporated and from
the residue the sterol was recovered at melting point 173-174°.
By the preceding method cyclohexene is hydrogenated in 15
min; β-nitrostyrene yields 1-phenyl-2-nitroethane in 60-90 hr;
and 1,4-naphthoquinone gives 71% of 1,2,3,4-tetrahydro-1,4-
dioxonaphthalene in 30 min.

7. From Sulfur Compounds (1, 15) [2, pp. 322-331]

$$RSH \xrightarrow{[H]} RH + H_2S$$

$$RSR' \xrightarrow{[H]} RH + R'H + H_2S$$

In a study of the desulfurization of 4,5-diamino-6-hydroxy-
2-mercaptopyrimidine by heating to form hypoxanthine, Nakamizo
and co-workers [51] showed that when Raney alloy per se was used
instead of the usually prepared Raney Ni, it was possible to
reduce the Ni/S ratio from 2.6:1 to 0.2-0.3:1. Fortunately, this

reduction was applicable as well if applied to other sulfur-containing substrates. Besides, the addition of Al or Fe powder was useful in diminishing the Ni/S ratio and combinations of Ni or Co salts and Al or Fe powder were as effective as the Raney alloy. The desulfurization also took place smoothly in aqueous HCl or H_2SO_4.

Lithium aluminum hydride catalyzed by copper chloride and zinc chloride has been used to form the hydrocarbon from the sulfide and the thioketal [52]. Later benzyl sulfides were converted into α-lithio derivatives that were transformed into the lithium quaternary salt. The latter, with cuprous iodide gave the alkylbenzene (69-92%) [53]. In another indirect method [30] a tosylate is converted into a phenyl sulfide which is then reduced by Li and liquid ammonia.

The sulfide reduction with Raney nickel continues to play an important part in the synthesis of m,m'-cyclophanes [54]:

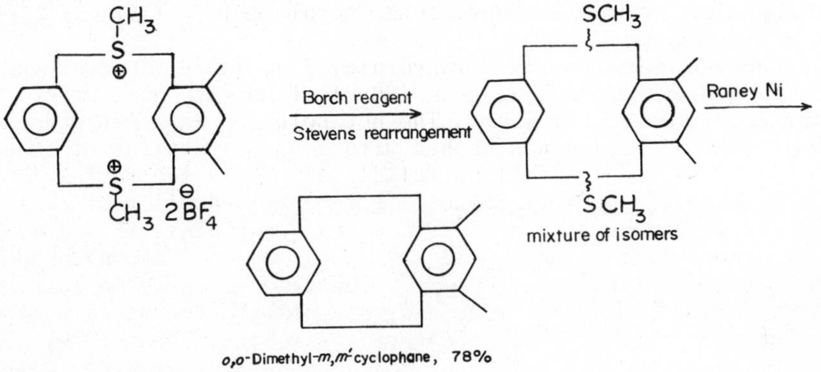

o,o-Dimethyl-m,m'-cyclophane, 78%

The reduction of sulfur compounds has been utilized as well to prepare hydrocarbons with quaternary carbon atoms [2, p. 323).

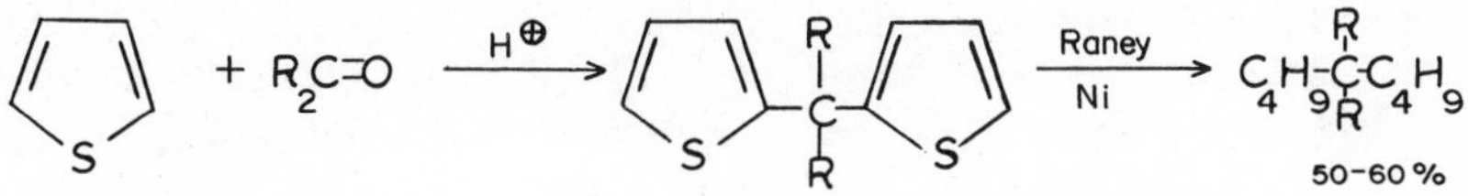

Somewhat similarly, the great condensation ability of formaldehyde has been harnessed to produce cyclononane derivatives [55]:

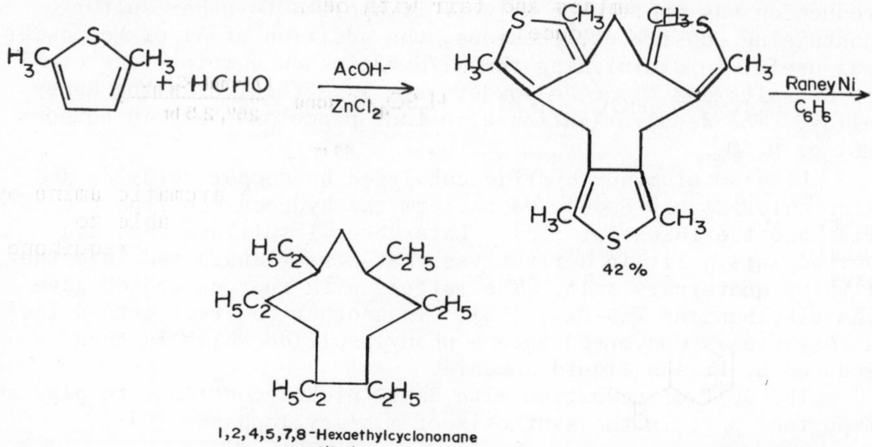

42 %

1,2,4,5,7,8-Hexaethylcyclononane
quantitative

Other active aromatic nuclei, such as veratrole, condense with formaldehyde similarly.

Another series of reactions, which eventually leads to the replacement of the oxygen of a carbonyl group by two alkyl groups, involves the reduction of a sulfone as the final step [56]

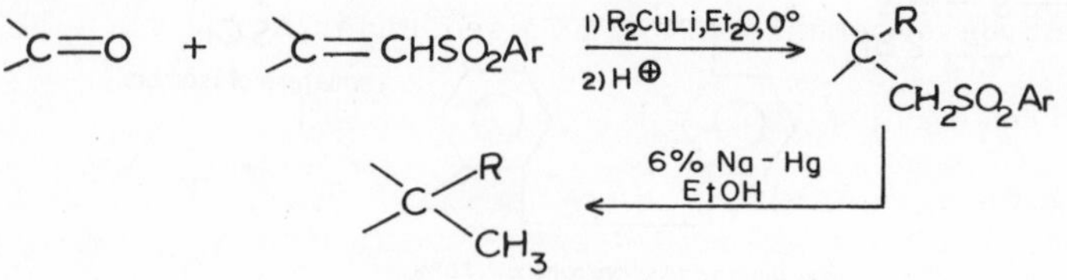

By this procedure, 1-methyl-1-n-butylcyclohexane was prepared in a 38% overall yield.

8.   From Diazonium Salts (1, 16)

$$Ar\overset{\oplus}{N_2} + H_3PO_2 + H_2O \longrightarrow ArH + H_3PO_3 + \overset{\oplus}{H} + N_2$$

Among the reagents that reduce diazonium salts to the corresponding hydrocarbons are alkaline formaldehyde, formamide, stannous salts, glucose in alkaline solution, sulfites under the proper conditions [57], as well as tributylstannane in ether or the more sluggish, but more stable triethylsilane in acetonitrile [57a].

If the reduction of the diazonium salt must be done in nonaqueous solution, McDonald and Richmond [58] recommend that hydroquinone be used as the reducing agent. The yields are quite good with azulene amines and fair with benzidine, a difficult compound at best to reduce:

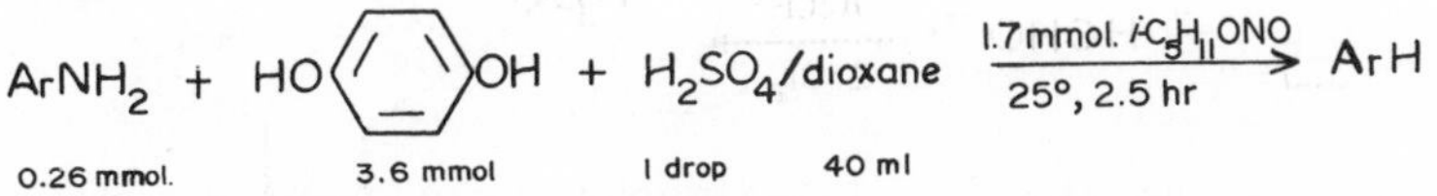

$$\text{ArNH}_2 \;+\; \text{HO}\!\!\left\langle\!\!\bigcirc\!\!\right\rangle\!\!\text{OH} \;+\; \text{H}_2\text{SO}_4/\text{dioxane} \xrightarrow[\;25°,\,2.5\,\text{hr}\;]{\;1.7\,\text{mmol. } i\text{-C}_5\text{H}_{11}\text{ONO}\;} \text{ArH}$$

0.26 mmol.   3.6 mmol   1 drop   40 ml

By the use of aprotic diazotization of the aromatic amine by pentyl nitrite in THF, Cadogan and Molina [59] were able to obtain the corresponding hydrocarbon. Thus 2-aminoanthraquinone gave the quinone as indicated:

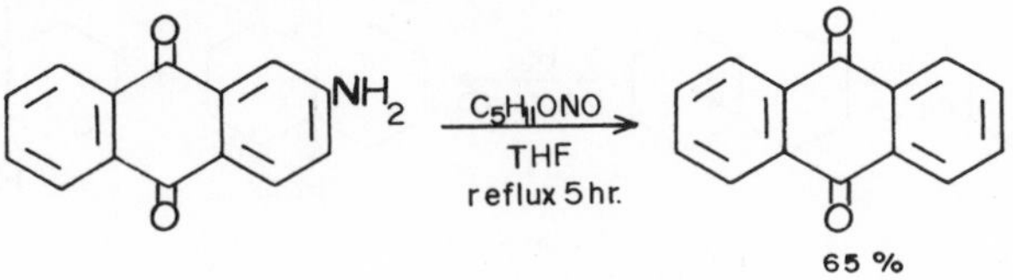

65 %

The method is a simple, one-pot reaction, which does not involve separate diazotization, but the yield may leave something to be desired.

9.   From Alkoxides (1, 17)

Recent developments are discussed under A.2.

10.   From Hydrocarbons (1, 17) [2, p. 62]

The type of tetralin obtained by reduction of alkylnaphthalenes with Li-liquid amine systems without a proton donor depends somewhat on the aryl group [60].

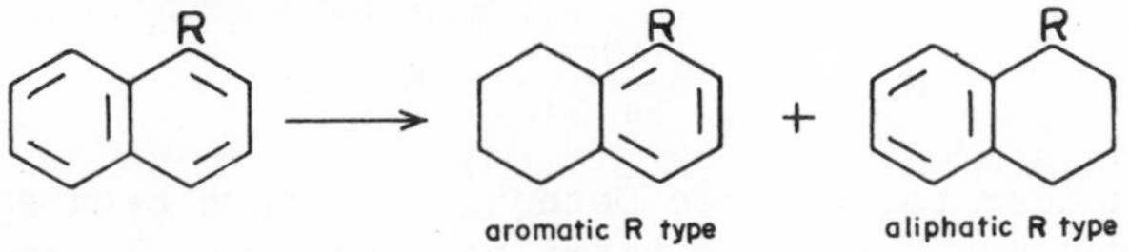

aromatic R type   aliphatic R type

The ratio of aromatic R type to aliphatic R type is 24 when R is methyl and 5.3 when R is t-butyl. The 2-alkylnaphthalenes give ratios close to 1. With a proton donor, naphthalene gives mainly $\Delta^{9,10}$ octalin [61].

The Simmons-Smith reaction has been utilized to construct a quaternary carbon atom in a cycloalkane [2, p. 62]:

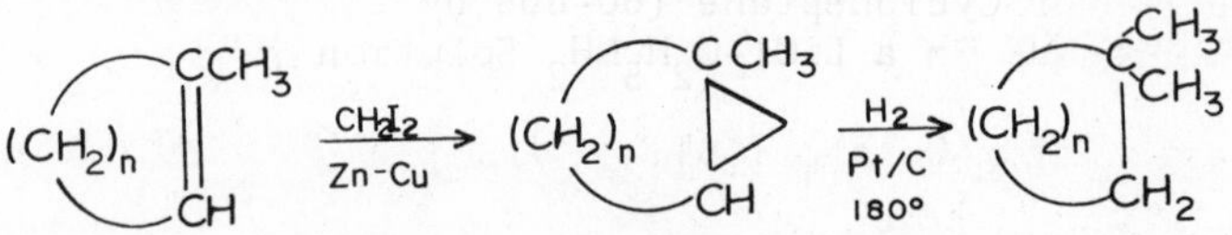

When n = 8, the yield of 1,1-dimethylcyclodecane was 97% and the
1,2-isomer, 3%. Similarly the 9-methyldecalins were obtained
from tricyclo[4.4.1.0]undecane [2, p. 113], which is available
from 9,10-octalin as shown:

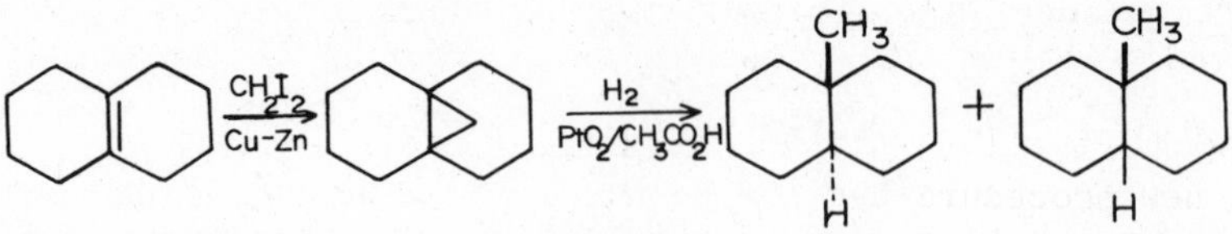

An interesting intraannular cyclization of a metacyclophane
has been found to take place [62]:

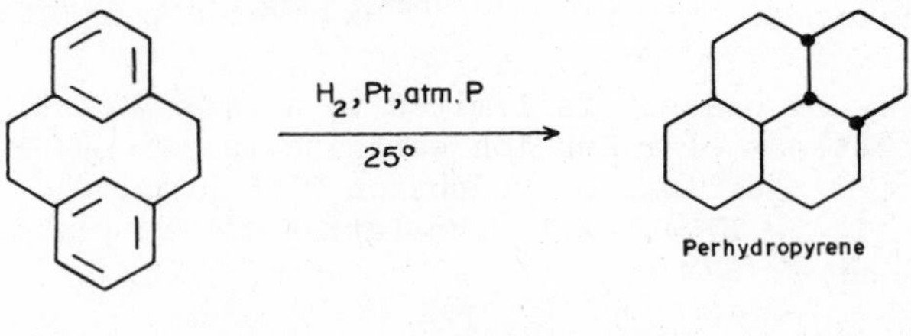

Perhydropyrene

The high melting point of the product (174°) gave rise to the
suspicion that ring closure had occurred.

11. From Nitriles (Reductive Decyanation) [2, p. 345]

$$RCH_2CN \xrightarrow[\text{liq. } NH_3]{Na} RCH_3$$

There appear to be three methods that have been employed in
the reductive decyanation of nitriles. With the exception of (2)
and (3), the decyanation applies best to tertiary cyanides.

1. The Use of an Alkali Metal in Liquid Ammonia or Lithium in an
Alkyl Amine [63] or Potassium in HMPA [64]

2. Electrolysis in a LiCl-Anhydrous Ethylamine Solution [65]
(Ex. a)

3. The Use of Na-Ferric Acetylacetone Fe(acac)$_3$ [66] (Ex. b)

a.   Preparation of Cycloheptane (60-80% by the Electrolysis of Cycloheptyl Cyanide in a $LiCl-C_2H_5NH_2$ Solution [65]

b.   Preparation of n-Octane [66]

Under argon, $Fe(acac)_3$ 10 mmol, octyl cyanide 10 mmol, and sodium sand 20 mg-atoms were stirred in C_6H_6 at 25° for 50-70 hr. After decomposition with 0.5 ml H_2O, analysis for octane by glc indicated 100% yield.

12.   From Carboxylic Acids [2, pp. 279-281]

$$ArCO_2H \longrightarrow ArCH_3$$

A new procedure has now been conceived which, if not one-step, is at least carried to completion without isolation of intermediates [67] (Ex. a).

$$2\ ArCO_2H \xrightarrow{SiHCl_3} (ArCO)_2O \xrightarrow[SiHCl_3]{R_3N} 2ArCH_2SiCl_3 \xrightarrow[MeOH]{KOH-} 2ArCH_3$$

The synthesis at present is limited to aromatic carboxylic acids.
Older methods of reduction entail catalytic hydrogenation under drastic conditions or reduction with hydriodic acid and red phosphorus [2, p. 279]. With the latter method yields of hydro-carbons are about 50%.

a.   Preparation of p-Xylene [67]

Trichlorosilane, 0.6 mol, and 0.1 mol of p-toluic acid in 80 ml of acetonitrile were refluxed for 1 hr. Tri-n-propylamine, 0.264 mol, was added to the cold mixture, the temperature of which was never allowed to rise above 15°. This mixture was refluxed for 16 hr, diluted to 850 ml with ether, and filtered to remove the amine·HCl. The filtrate was concentrated to remove ether and acetonitrile; methanol , 50 ml, was added and the solution was refluxed for 1 hr, followed by the dropwise addition of 1 mol of KOH dissolved in 95 ml methanol and 25 ml H_2O. After 19 hr of further refluxing, the mixture was diluted with 600 ml of H_2O and extracted with pentane. From the pentane solution the p-xylene (74%) was recovered

13.   From Nitro Compounds

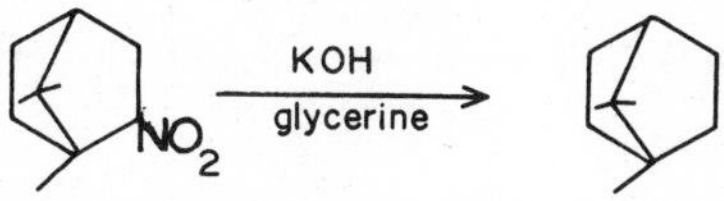

2-Nitrobornanes may be reduced to norbornane by heating in a solution of KOH in glycerol or diethylene glycol [68].

14.   From Esters (see Addenda)

Addenda

A.1.   The Clemmensen reaction has been reviewed [E. Vedejs, Org. React., 22, 401 (1974)].

A.1.   2,3,6,7-Tetramethylanthraquinone was reduced to the corresponding anthracene (85%) by aluminum tricyclohexoxide in boiling cyclohexanol [J. C. Hinshaw, Org. Prep. Proced. Int., 4, 211 (1972)].

A.3.   1-Bromonaphthalene in DMF at 25° was reduced to naphthalene by Cr^{2+} (en)$_x$ [R. S. Wade, C. E. Castro, Org. Syn., 52, 62 (1972)].

A.5.   The reduction of α,α-dialkylalkenes (as well as other substrates) by silanes in trifluoroacetic acid, an ionic type reduction, has been reviewed: chalcones reduce to 1,3-diaryl or 1-aryl-3-alkylarylpropanes, aromatic aldehydes or ketones to hydrocarbons, alcohols that form stable carbonium ions to hydrocarbons, and thiophenes to tetrahydrothiophenes [D. N. Kursanov et al., Synthesis, 633 (1974)].

A.6.   Anthraquinones have been reduced to 9,10-dihydroanthracenes by HI, P, and I_2 [R. N. Renaud, J. C. Stephens, Can. J. Chem., 52, 1229 (1974)].

A.14.   From Esters: Acetates, benzoates, or formates of cholestane-type molecules may be reduced to the hydrocarbon by irradiation in HMPA-H_2O at 254 nm. Sulfonates give the alcohol [J.-P. Pete et al., Chem. Commun., 439 (1975)].

References

1.   R. L. Augustine, Reduction, Dekker, New York, 1968: (a) Clemmensen, pp. 186-194; (b) Wolff-Kishner, p. 171; (c) Raney nickel desulfurization, p. 196.
2.   H.-W., 5, Pt 1a, 1970 (Clemmensen), pp. 244, 450; W.-K., pp. 251, 456.
2a.   S. Yamamura et al., Bull. Chem. Soc. Jap., 45, 264 (1972).
3.   W. P. Duncan, G. W. Keen et al., J. Org. Chem., 37, 142 (1972); P. H. Gore et al., J. Chem. Soc., Perkin Trans. I, 892 (1972).

4. J. G. St. C. Buchanan, P. D. Woodgate, Quart. Rev. (London), 23, 522 (1969).
5. R. O. Hutchins et al., J. Am. Chem. Soc., 93, 1793 (1971).
5a. C. F. Lane, Aldrichimica Acta, 8 (1), 3 (1975).
6. R. O. Hutchins et al., J. Am. Chem. Soc., 95, 3662, 6131 (1973).
7. S. S. Hall et al., J. Org. Chem., 36, 2588 (1971); see also ibid., 37, 760 (1972).
8. S. S. Hall, S. D. Lipsky, Chem. Commun., 1242 (1971); J. Org. Chem., 38, 1735 (1973); 40, 271 (1975).
9. J. Málek, M. Černy, Synthesis, 53 (1973).
10. K. M. Biswas, A. H. Jackson, J. Chem. Soc., C, 1667 (1970).
11. M. P. Doyle et al., J. Org. Chem., 38, 2675 (1973).
11a. J. W. Burnham, E. J. Eisenbraun, J. Org. Chem., 36, 737 (1971).
12. R. S. Monson, Advanced Organic Synthesis, Academic, New York, 1971, p. 55; Ref. 1b.
13. I. Shahak, E. D. Bergmann, Israel J. Chem., 4, 225 (1966).
14. H.-D. Scharf, F. Döring, Chem. Ber., 100, 1761 (1967).
15. E. J. Corey, K. Achiwa, J. Org. Chem., 34, 3667 (1969).
16. R. E. Ireland et al., J. Am. Chem. Soc., 94, 5098 (1972).
17. H. Pines, Intra-Sci. Chem. Rep., 6 (2), 25 (1972).
18. R. A. Rossi, J. F. Bunnett, J. Org. Chem., 38, 2314 (1973).
19. H. C. Brown, S. Krishnamurthy, J. Org. Chem., 34, 3918 (1969).
20. C. W. Jefford et al., J. Am. Chem. Soc., 94, 8905 (1972).
21. N. W. Gilman, L. H. Sternbach, Chem. Commun., 465 (1971).
22. J. Málek, M. Černy, Synthesis, 217 (1972).
23. H. M. Bell et al., J. Org. Chem., 34, 3923 (1969).
24. R. O. Hutchins et al., Tetrahedron Letters, 3495 (1969).
25. R. O. Hutchins et al., Chem. Commun., 1097 (1971).
26. S. Matsumura, N. Tokura, Tetrahedron Letters, 363 (1969).
27. J. M. Townsend, T. A. Spencer, Tetrahedron Letters, 137 (1971).
28. P. G. Gassman, J. L. Marshall, Org. Syn. Coll. Vol. 5, 424 (1973).
29. G. M. Whitesides et al., J. Org. Chem., 37, 3300 (1972).
30. N. S. Crossley, R. Dowell, J. Chem. Soc., C, 2496 (1971).
31. W. P. Neumann, H. Hillgartner, Synthesis, 537 (1971).
32. (a) G. L. Grady, H. G. Kuivila, J. Org. Chem., 34, 2014 (1969); (b) H. Arita et al., Bull. Chem. Soc. Jap., 45, 567 (1972).
33. J. Lipowitz, S. A. Bowman, Aldrichimica Acta, 6 (1), 1 (1973).
34. T. A. Cooper, J. Am. Chem. Soc., 95, 4158 (1973).
35. T. R. Bosin et al., Tetrahedron Letters, 4699 (1973).
35a. R. M. Schisla, W. C. Hammann, J. Org. Chem., 35, 3224 (1970).
36. C. Eaborn et al., J. Chem. Soc., C, 2823 (1969).

37. T. W. Russell, R. C. Hoy, J. Org. Chem., 36, 2018 (1971).
38. C. A. Brown, V. K. Ahuja, J. Org. Chem., 38, 2226 (1973).
39. R. L. Burwell, Intra-Sci. Chem. Rep. 6 (2), 135 (1972).
40. J. Halpern, Disc. Faraday Soc., 46, 7 (1968).
41. G. N. Schrauzer, Transition Metals in Homogeneous
     Catalysis, Dekker, New York, 1971.
42. R. E. Harmon et al., Chem. Rev., 73, 21 (1973).
43. R. E. Harmon et al., J. Org. Chem., 34, 3684 (1969).
44. J. C. Falk, J. Org. Chem., 36, 1445 (1971).
45. C. E. Castro et al., J. Am. Chem. Soc., 88, 4964 (1966).
46. J. E. Shaw, K. K. Knutson, J. Org. Chem., 36, 1151 (1971).
47. G. M. Whitesides, W. J. Ehmann, J. Org. Chem., 35, 3565
     (1970); E. M. Kaiser, Synthesis, 399 (1972).
48. C. B. Reese et al., Chem. Commun., 1132 (1972).
49. J. M. Hoffman, Jr., R. H. Schlessinger, Chem. Commun.,
     1245 (1971).
50. R. S. Monson, Ref. 12, p. 43.
51. N. Nakamizo et al., Bull. Chem. Soc. Jap., 44, 2192 (1971).
52. T. Mukaiyama et al., Bull. Chem. Soc. Jap., 44, 2285
     (1971).
53. T. Mukaiyama et al., Bull. Chem. Soc. Jap., 45, 2244
     (1972).
54. V. Boekelheide, C. M. Tsai, J. Org. Chem., 38, 3931 (1973)
     and previous papers; T. Umemoto et al., Tetrahedron
     Letters, 593 (1973).
55. O. Meth-Cohn, Tetrahedron Letters, 91 (1973).
56. G. H. Posner, D. J. Brunelle, Tetrahedron Letters, 935 (1973).
57. H.-W., 10, Pt. 3, 1965, pp. 113-144.
57a. J. Nakayama et al., Tetrahedron, 26, 4609 (1970).
58. R. N. McDonald, J. M. Richmond, Chem. Commun., 605 (1973).
59. J. I. G. Cadogan, G. A. Molina, J. Chem. Soc., Perkin
     Trans. I, 541 (1973).
60. T. J. Nieuwstad, H. Van Bekkum, Rev. Trav. Chim., 91, 1069
     (1972).
61. E. M. Kaiser, R. A. Benkeser, Org. Syn., 50, 88 (1970).
62. E. Langer, H. Lehner, Tetrahedron Letters, 1143 (1973).
63. P. G. Arapakos et al., J. Am. Chem. Soc., 91, 2059 (1969).
64. H. Normant et al., Compt. Rend., 274C, 797 (1972).
65. P. G. Arapakos, M. K. Scott, Tetrahedron Letters, 1975,
     (1968).
66. E. E. van Tamelen et al., J. Am. Chem. Soc., 93, 7113
     (1971).
67. R. A. Benkeser et al., J. Am. Chem. Soc., 92, 3232 (1970);
     Acc. Chem. Res., 4, 94 (1971).
68. H. Toivonen et al., Tetrahedron Letters, 3203 (1971).

B.   Organometallic and Hydroboration Methods

2. Coupling of Organometallic Compounds or Halides and
    Sodium (Wurtz) (1, 22; H.-W., 5, Pt. 1a, 1970, pp. 485-
    499)

$$RMgX + R'X \longrightarrow RR' + MgX_2$$

$$2\,RX + 2\,Na \longrightarrow RR + 2\,NaX$$

The conventional coupling reaction involving the Grignard
reagent and the Grignard-Wurtz reaction continues to be employed.
Thus Hobbs and Hammann [1] prepared a series of mono-t-alkyl-
benzenes, di-t-alkylbenzenes, and t-alkylphenylalkanes from the
alkyl halide and the Grignard reagent in yields of 11-30%. These
yields, which are relatively high for uncatalyzed Grignard
coupling reactions, were obtained by conducting the coupling
reaction in nonpolar solvents such as hexane. Although not
related to noncatalyzed Grignard coupling, it is of interest to
note that a new method of preparing highly reactive magnesium has
been described [2]. Magnesium halide is reduced with sodium in
diglyme to form a black suspension that reacts with inert
chlorobenzene at 25° for 90 min to form 62% of phenylmagnesium
chloride. In the coupling reaction there is an increasing tendency to use catalysts as well. For example, the Kharasch
coupling

$$2\,Ar\,MgX + RX + CoCl_2 \longrightarrow Ar\,Ar$$

is now believed to proceed via metallic cobalt [3]:

$$2\,ArMgBr + 2\,Co + 2\,C_2H_5Br \longrightarrow 2\,ArCo\,C_2H_5 + Mg\,Br_2 \longrightarrow$$

$$2\,(Ar\cdot) \qquad + \qquad 2\,(C_2H_5\cdot) + 2\,Co$$

$$\downarrow \qquad\qquad\qquad \downarrow$$

$$Ar\,Ar \qquad\qquad CH_2{=}CH_2 + CH_3CH_3$$

Soluble silver is also an effective catalyst [4] as shown, particularly when the alkyl groups are identical:

$$n\text{-}C_4H_9MgBr + n\text{-}C_4H_9Br \xrightarrow{Ag} n\text{-}C_8H_{18}$$
$$79\,\%$$

With soluble copper(I) as a catalyst cross coupling is satisfactory [1] as indicated:

$$n\text{-}C_6H_{13}MgBr \quad + \quad n\text{-}C_4H_9Br \xrightarrow{\text{Cu(I)}} n\text{-}C_6H_{13}C_4H_9n$$

73 %

Copper salts in TMEDA have also been used in coupling aryl
Grignard reagents with primary alkyl or aralkyl halides (33-93%)
[5]. Perhaps the best mixed coupling occurs using lithium dialkyl cuprate (LiCuR$_2$) and R'Br, where R' is primary [6] as
shown:

$$LiCu\left[C(CH_3)_3\right]_2\cdot PBu_3 \quad + \quad nC_5H_1Br \longrightarrow nC_5H_{11}C(CH_3)_3$$

92% (glc)

In fact, the lithium dialkylcuprates now appear to be the
reagent of choice in coupling [6a]. For example, methylcopper
does not react with iodocyclohexane but methylcyclohexane (75%)
is formed with LiCu(CH$_3$)$_2$ [6b]. The cuprates are most conveniently prepared from 2 mol alkyllithium and 1 mol cuprous
halide (usually CuI). An excellent discussion of the structure
and reactions of the organocopper reagents is found in the
literature [6c]. Incidentally, methyllithium reacts with aryl
iodides to form methylarenes without complexation with cuprous
salts. For aralkyl hydrocarbon coupling it would be ideal if the
aryl and alkyl groups were in the same complex so as to break
down in close proximity to each other and thus induce maximum
mixed coupling. On the basis of the high yields obtained (76-
98%) it appears that such a complex has been found [7]:

The diphosphine ligands are much more effective in this reaction
than the monophosphine ones.

Yields are high in either coupling vinyl halides or aromatic
dihalides with Grignard reagents in the presence of nickel
halides (Ex. a). Although primary alkyl groups in the Grignard
reagent do not rearrange in this process, secondary ones do to a
certain extent [8]:

$$\text{NiCl (complex)} + C_6H_5Cl + (CH_3)_2CHMgBr \longrightarrow$$

$$C_6H_5CH(CH_3)_2 + C_6H_5CH_2CH_2CH_3$$

$$\text{A} \qquad\qquad \text{B}$$

Table 1.1.   Yields of Hydrocarbons from Isopropyl Magnesium Bromide

Ligand	Yield (glc)	%A	%B
$(C_6H_5)_2PCH_2CH_2CH_2P(C_6H_5)_2$	89	96	4
$Me_2PCH_2CH_2PMe_2$	84	9	84

Alkyllithiums have been employed at times instead of the Grignard reagent. The preparation of pure benzyl- and 2- and 4-pyridylmethyllithiums, free of nuclear-attached lithium has been reviewed [9]. The purity is ensured by using the least reactive alkyllithium to bring about exchange with toluene or either of the picolines. n-Butyllithium has been recommended for metalation of the 2-methyl group in picolines and $NaNH_2$ in liquid ammonia or lithium diisopropylamide for exclusive metalation of the 4-methyl group [10]. For benzyllithium a combination of n- and sec-butyllithium in THF is employed (no TMEDA). For 2- and 4-pyridylmethyllithiums, 2-thenyllithium is utilized.

Alkyllithiums with optically active secondary halides usually give a racemized coupled product. However, charge-delocalized alkyllithiums such as benzyl-, allyl-, or benzhydryllithium usually give an inverted coupled product [11]. α-Bromo-ketones with lithium cuprates (2RLi + CuI → LiCuR$_2$ + LiI) lead to the formation of hindered ketones as shown [12]:

8-90%
(best with R=CH$_3$)

t-Butyllithium reacts with benzyl chlorides to give neopentyl-benzenes (25-75%) [13]. Again yields are best if a nonpolar solvent is employed. In addition cyclopropyllithium reacts with chlorobenzene to yield 82% of cyclopropylbenzene [14].

At times,the alkylsodium finds use as in the synthesis of 1,1-diphenylpentane [15]:

$$(C_6H_5)_2CH_2 \xrightarrow[\text{liq. NH}_3]{\text{NaNH}_2} (C_6H_5)_2CHNa \xrightarrow{n\text{-}C_4H_9Br} (C_6H_5)_2CHC_4H_9\text{-}n$$

92 %

This method is a general one for the preparation of 1,1-diphenyl-substituted hydrocarbons.

A comparison of conditions for the synthesis of cyclobutane by the Wurtz reaction has been published [16]:

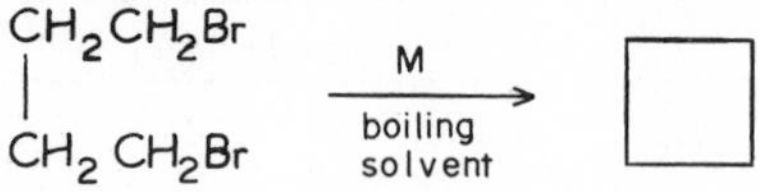

If M = Na in toluene, yield = 7%; if M = Na in xylene, yield = 11%; if M = Li in ether, yield = 20%; and if M = Li/Hg in dioxane, yield = 70%.

Sodium has also been employed in the preparation of bicyclo [1.1.0] butane [17]:

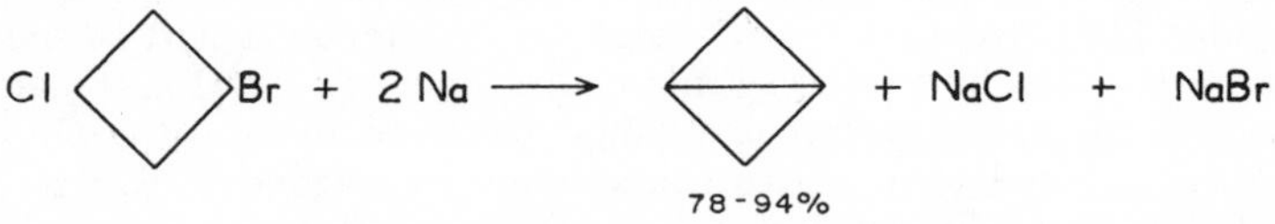

78-94%

The same compound has also been prepared [18] by the use of magnesium. 1-Chloro-3-ethoxycyclobutane was prepared from 1-keto-3-ethoxycyclobutane, a product of ketene and the vinyl ether.

The Wurtz reaction has also proved to be one of the most successful in the formation of [2.2]cyclophanes [19]. Thus Boekelheide and co-workers [20] prepared 8,16-dimethyl[2.2]meta-cyclophane as shown:

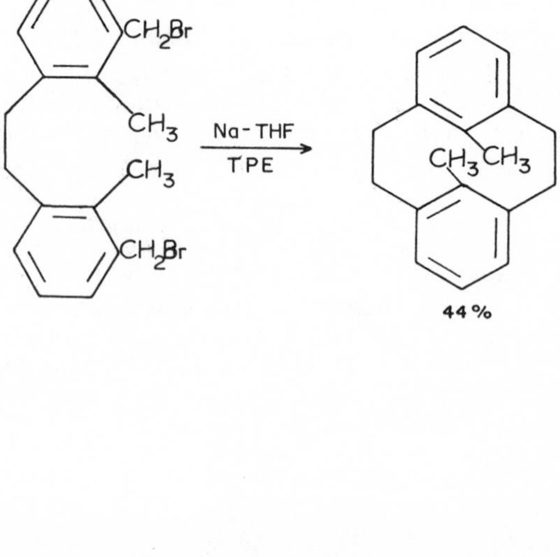

44%

It should be noted that the modern improved method of conducting
the Wurtz reaction consists of allowing the halogen compound in
THF to react at -75° to -80° under nitrogen in the presence of a
catalytic amount of tetraphenylethylene, TPE [21]. In this
manner the sodium dissolves to form

$$(C_6H_5)_2 \underset{Na}{C} - \underset{Na}{C}(C_6H_5)_2$$

so that the reaction may take place in the homogeneous phase.
    Brown [22] has shown that organoboranes with alkaline silver
nitrate offer a general method for the synthesis of bialkyls.
Breuer and Broster [23] modified the procedure of preparing the
organoboranes from the Grignard reagent in the presence of diborane as shown:

$$2 \ RBr \ \xrightarrow[\text{2)} KOH-Ag^{\oplus}]{\text{1)} Mg-BH_3} \ RR$$
$$52\text{-}94 \ \%$$

The procedure is very simple. The organoborane is formed by
reacting the halide, 20 mmol, with magnesium, 20 mmol, in the
presence of BH_3, 30 mmol, in refluxing THF. To achieve the
coupling the product is treated with potassium hydroxide in
methanol and aqueous silver nitrate. Although the coupling
reaction is sensitive to steric effects, it is successful for
aryl and primary and secondary alkyl halides.

a.   Preparation of m-Di-n-butylbenzene [7]

    To $NiCl_2$(dpe), 0.39 mmol, and m-$C_6H_4Cl_2$, 53.9 mmol, in 50 ml
of ether at 0° was added BuMgBr, 120 mmol, in 50 ml of ether in
10 min with stirring. The black mixture was refluxed for 20 hr
to form considerable salt. After decomposition with dilute acid,
the mixture was worked up in the usual way to give 94% of the
product. α-Vinylnaphthalene was prepared similarly in 80% yield.

    3.   Coupling of Aryl Halides (Ullmann) and of Other Arenes
          Including Oxidative Coupling of Cuprates

$$2 \ ArX + Cu \ \longrightarrow \ ArAr + CuX_2$$

$$2 \ ArMgX \ \xrightarrow{MX_n} \ ArAr$$

The Ullmann reaction [23a] continues to occupy an important
place in the synthesis of biaryls [24], but recent work concentrates on reactions at lower temperature. One general type is
the indirect oxidation of an arylmagnesium halide, probably via
a biarylmetallic species:

$$2\ ArMgX + MX_2 \longrightarrow [Ar_2M] \longrightarrow Ar_2$$

M may be either Ni, Co, V, Ti, Cu, Cr, Fe, U, or Tl [25].

For the preparation of terphenyls with 2 mol of the Grignard reagent nickel(II) acetylacetonate proved to be effective [26]:

$$p\text{-}BrC_6H_4Br + 2\ ArMgBr \xrightarrow{Ni(acac)_2} p\text{-}ArC_6H_4Ar \quad >80\%$$

Perhaps the best of the recent methods is that involving the thallium salt [27] (see Ex. a):

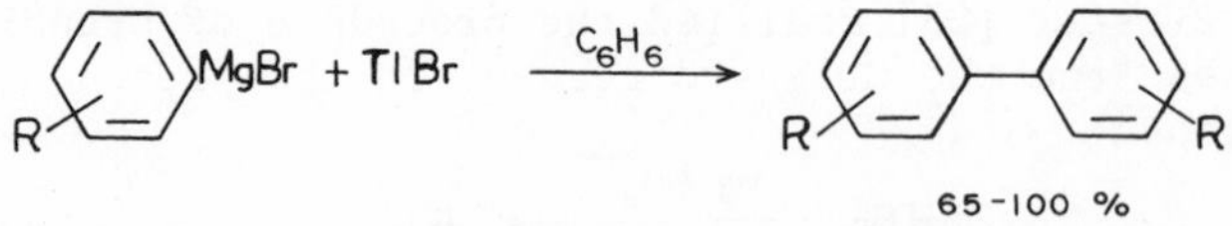

MgBr
+ Tl Br
C_6H_6
R
R
R
65-100 %

However, it is unsuccessful with o-substituted arenes and is limited largely to the preparation of symmetrical biaryls and to those containing substituents unaffected by the Grignard reagent.

Other methods of synthesis begin with the aryl halide rather than the Grignard reagent. One of these involves bis(1,5-cyclooctadiene)nickel(0) [25].

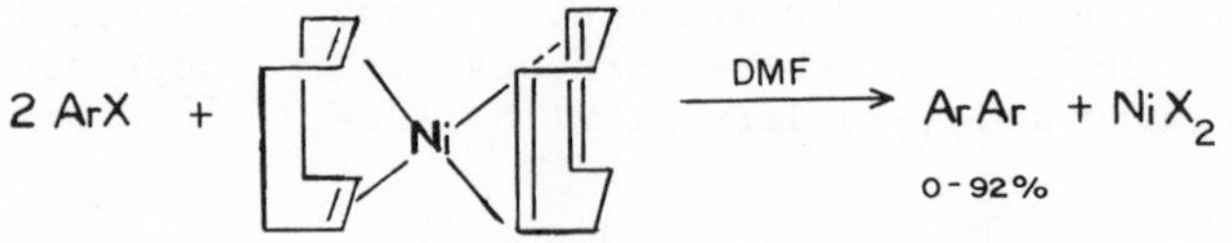

2 ArX +
Ni
DMF
ArAr + NiX_2
0-92%

The reaction, which avoids the Grignard reagent interferences, is retarded by ortho substituents and is stopped by acidic groups.

Two reactions that begin with the arene are not very general, but in certain instances may be useful. One method utilizes a palladium complex and a silver salt [28]:

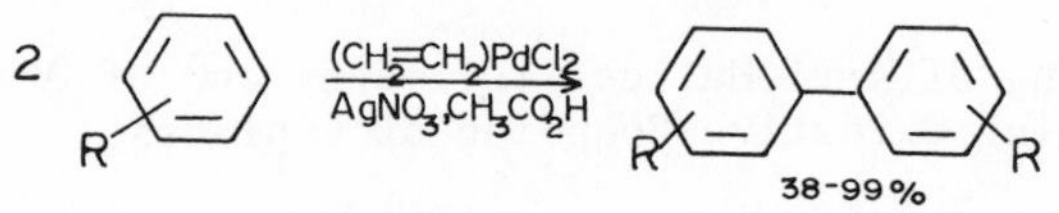

2
R
$(CH_2{=}CH_2)PdCl_2$
$AgNO_3, CH_3CO_2H$
R
R
38-99%

Mixtures of isomeric biaryls were obtained in all cases except in the formation of 3,3'-dinitrobiphenyl. Another method employs a complex oxidation system [29]:

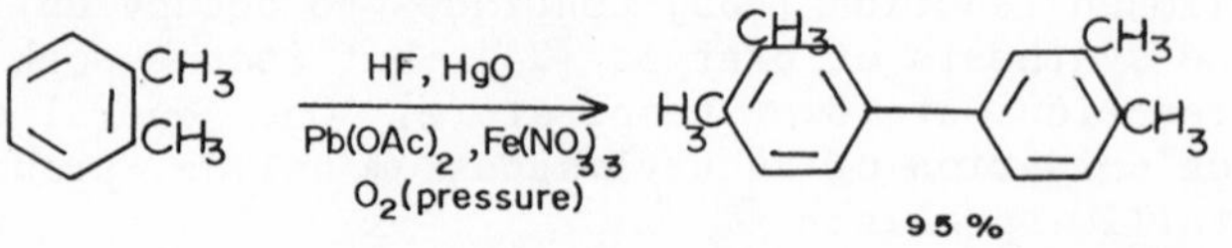

CH_3
CH_3
HF, HgO
$Pb(OAc)_2$, $Fe(NO_3)_3$
O_2(pressure)
CH_3
H_3C
CH_3
CH_3
95 %

In general isomeric mixtures may be expected. Also, such mixtures of biphenyls are obtained in the oxidation of aromatic compounds with palladium(II) catalysts [30]:

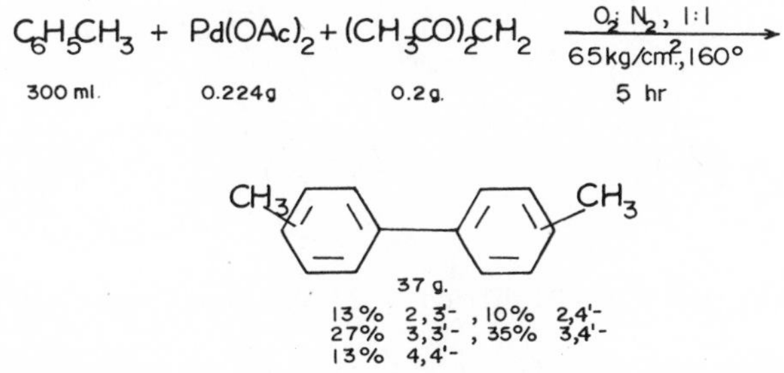

$$C_6H_5CH_3 + Pd(OAc)_2 + (CH_3CO)_2CH_2 \xrightarrow[65kg/cm^2,160°]{O_2; N_2, 1:1}$$

300 ml.   0.224g   0.2g.   5 hr

37 g.

13% 2,3'- , 10% 2,4'-
27% 3,3'- ; 35% 3,4'-
13% 4,4'-

Dimethyl o-phthalate similarly yields a mixture of tetramethyl-biphenylcarboxylates.

Finally, biphenyl formation may occur via a mercurial [31]:

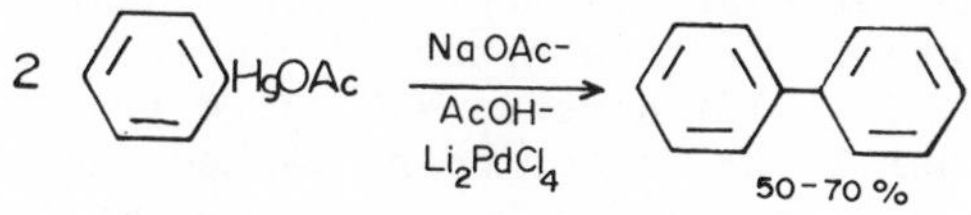

$$2 \underset{}{\bigcirc}\!HgOAc \xrightarrow[\substack{AcOH- \\ Li_2PdCl_4}]{NaOAc^-} \underset{}{\bigcirc}\!-\!\bigcirc$$

50-70 %

The oxidative coupling of cuprates,

$$2 \, LiCuAr_2 \xrightarrow{[O]} Cu^{\oplus} + ArAr + Li^{\oplus}$$

which was first reported by Whitesides [31a], takes place smoothly at 0° and, therefore, is supplementary to the high-temperature Ullmann reaction. For a recent review see the work by Normant [31b]. Many oxidizing agents such as oxygen, cupric salt, or nitrobenzene may be used, but the best in the opinion of the present authors is biacetyl [31c]. The copper atom of the lithium diarylcuprate, which is probably in the form of a cluster, serves as a relay agent to permit the oxidizing agent to form aryl free radicals which then combine. o-Terphenylene has been formed in 60% yield by the method [31b]:

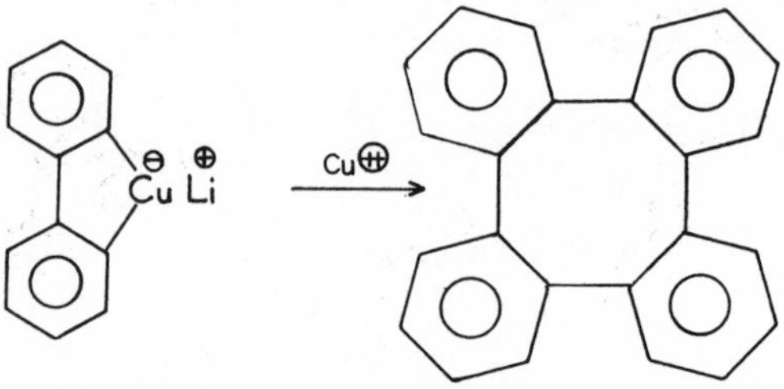

a.   Preparation of 3,3'-Dichloro-2,2'-dimethylbiphenyl [31c]

    To a magnetically stirred solution of 40 mmol of 2-bromo-6-
chlorotoluene in 200 ml of ether under N_2 was added 42 mmol of
commercial sec-BuLi in 2 min (exothermic). After 15 min the
solution of 6-chloro-2-tolyllithium was cooled to 0° and treated
with 8 g (20.4 mequiv) of $(Bu_3PCuI)_4$ (crystalline complex of CuI
and PBu_3). After the yellow color faded to a clear, colorless
solution, 42 mmol of neat biacetyl (syringe) was added. An
orange precipitate formed and after 30 min water was added.
Addition of 100 ml of 10% aqueous NaCN gave two nearly colorless
layers. The ether layer, after being washed thoroughly, dried,
and evaporated, gave a residue which on sublimation yielded
3.75 g (66%), melting point 74.0-74.5°, of the biphenyl.

b.   Preparation of o-,m-,o-,o-,m-,o-, Hexaphenylene [32]

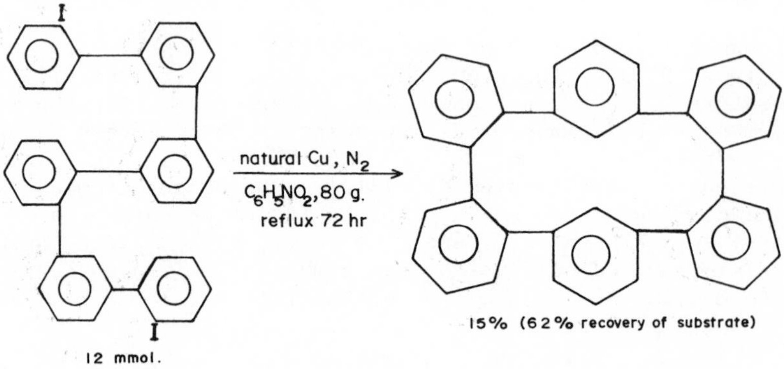

4.   Coupling to Cyclopropanes

    Coupling of 1,3-dichloropropanes to cyclopropanes takes
place primarily by the Wurtz reaction (1, 23), but other

couplings designed similarly to those which produce ethylene from
vic-dihalides, have become available. They are of a reductive
nature. One utilizes Cr(II) [33]:

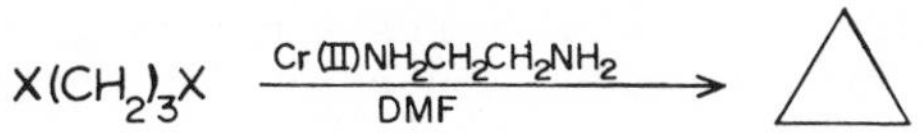

$$X(CH_2)_3X \xrightarrow[\text{DMF}]{\text{Cr(II)}NH_2CH_2CH_2NH_2} \triangle$$

The yields are practically quantitative when X = a halogen or
when one X = Cl and the other OTs. Similar reductions with the
tetramethylene analogs did not produce cyclobutanes.

5. Exhaustive Methylation of Tertiary Alcohols, Ketones,
and Carboxylic Acids by Trimethylaluminum

Tertiary alcohols, ketones, and carboxylic acids may be
methylated exhaustively as indicated [34]:

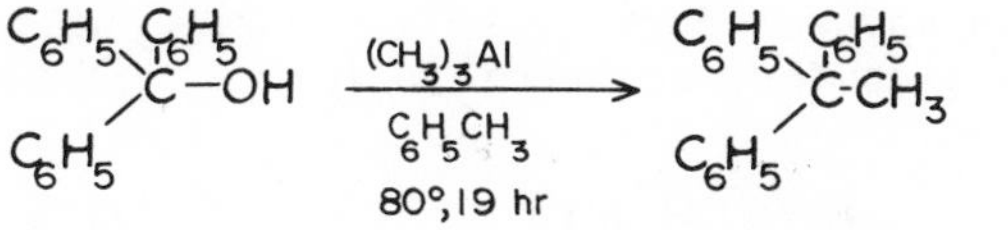

1,1,1-Triphenylethane
86 %

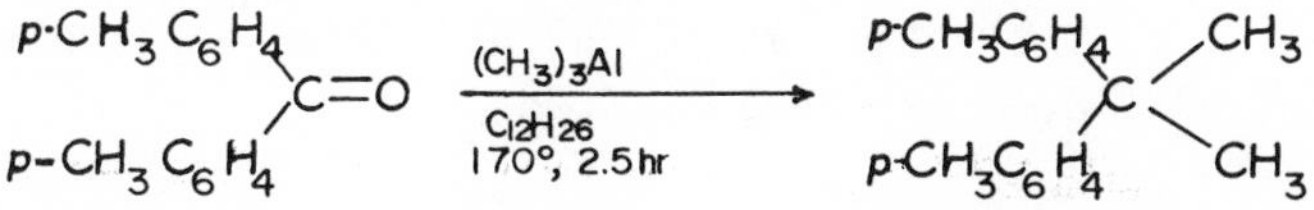

2,2-Di-p-tolylpropane, 63 %

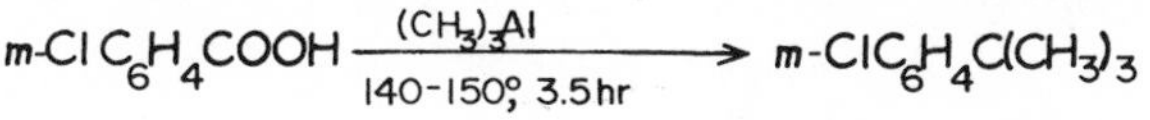

m-Chloro-t-butylbenzene, 62 %

The reaction exhibits carbonium ion character and may be conducted with or without a solvent. An excess of the trimethylaluminum, as was used in the three examples cited, increases the
speed of the reaction.

6. Coupling of Benzylic and Allylic Iodides via Triethyl-
boranes

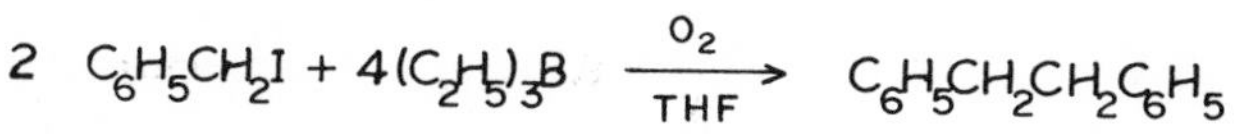

$$2\ C_6H_5CH_2I + 4(C_2H_5)_3B \xrightarrow[\text{THF}]{O_2} C_6H_5CH_2CH_2C_6H_5$$

Benzylic iodides, 1 mol, when treated with 2 mol of tri-ethylborane in THF and a stream of air couple to give bibenzyls (43-87%) [35]. To obtain high yields the triethylborane was employed in 100% excess. The method lends itself to the synthesis of mixed coupling products. Thus 4-phenyl-1-butene, 72%, was obtained from benzyl iodide and allyl iodide (1:4 mol ratio).

7. Cyclization via Hydroboration

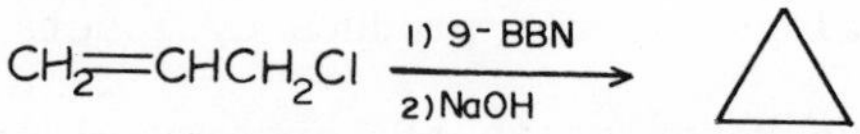

$$CH_2{=}CHCH_2Cl \xrightarrow[\text{2)NaOH}]{\text{1) 9-BBN}} \triangle$$

Allylic chlorides react with 9-borabicyclo [3.3.1] nonane (9-BBN) to give β-(γ-chloropropyl)-9-borabicyclo [3.3.1] nonanes, which with aqueous NaOH form cyclopropanes [36]. The 9-BBN exhibits great selectivity and the boron in the derivative is "unusually open and susceptible to attack by nucleophiles." With allyl chloride the intermediate formation is as shown:

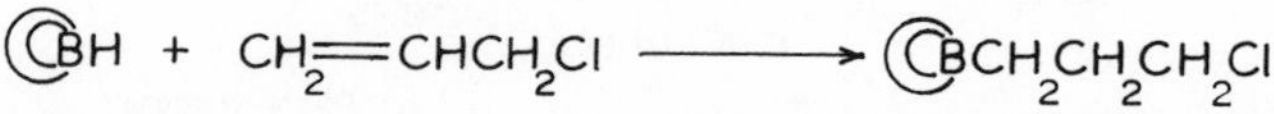

$$\text{\textcircled{C}BH} + CH_2{=}CHCH_2Cl \longrightarrow \text{\textcircled{C}BCH}_2CH_2CH_2Cl$$

Borane, rather than the substituted borane, gives both isomers on addition to allyl chloride. Yields of alkylated cyclopropanes vary from 75 to 92%. The method has also been applied to the cyclobutyl moiety as follows:

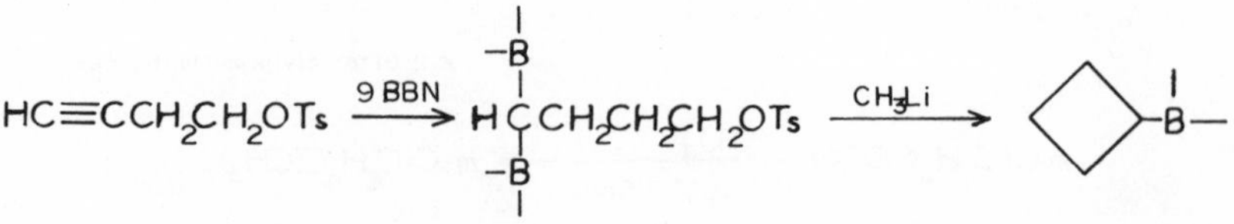

$$HC{\equiv}CCH_2CH_2OTs \xrightarrow{\text{9 BBN}} HC\underset{\overset{|}{-B}}{\overset{\overset{|}{-B}}{|}}CH_2CH_2CH_2OTs \xrightarrow{CH_3Li} \square\!-B{-}$$

8. From 2,2-Benzodioxoles

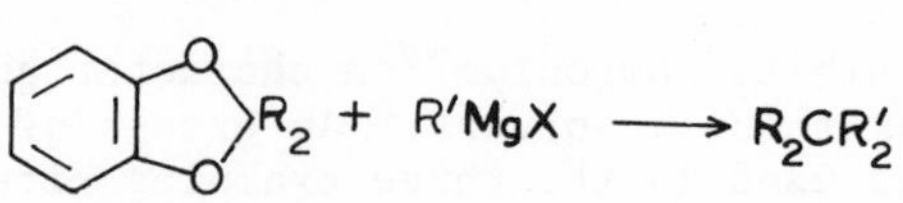

$$\text{(benzodioxole)}R_2 + R'MgX \longrightarrow R_2CR'_2$$

Hydrocarbons containing quaternary carbon atoms have been prepared by the above procedure in 25-81% yields (7 Ex.) [37].

Addenda

B.2, B.3. The oxidative coupling via organocopper compounds has been reviewed [T. Kauffman, Angew. Chem. Intern. Ed. Engl., 13, 291 (1974)].

B.2. For cross coupling of Grignard reagents and alkyl halides lithium dialkylcuprates or Li_2CuCl_4 is effective [M. Tamura, J. Kochi, Synthesis, 303 (1971)].

References

1. C. F. Hobbs, W. C. Hammann, J. Org. Chem., 35, 4188 (1970).
2. R. D. Rieke, P. M. Hudnall, J. Am. Chem. Soc., 94, 7178 (1972).
3. A. McKillop et al., Organomet. Chem. Rev., A8, 135 (1972).
4. M. Tamura, J. Kochi, Synthesis, 303 (1971).
5. K. Onuma, H. Hashimoto, Bull. Chem. Soc. Jap., 45, 2582 (1972).
6. G. M. Whitesides et al., J. Am. Chem. Soc., 91, 4871 (1969).
6a. G. H. Posner, Org. Reactions, 22, 253 (1974).
6b. Ref. 6a, p. 264.
6c. H. O. House, Proceed. R. A. Welsh Foundation, XVII, 1973, p. 101.
7. M. Kumada et al., J. Am. Chem. Soc., 94, 4374 (1972).
8. M. Kumada et al., J. Am. Chem. Soc., 94, 9268 (1972); J. Organomet. Chem., 50, C12 (1973).
9. C. W. Kamienski et al., Chimia, 24, 109 (1970).
10. E. M. Kaiser et al., J. Org. Chem., 38, 71 (1973).
11. L. H. Sommer, W. D. Korte, J. Org. Chem., 35, 22 (1970).
12. J.-E. Dubois et al., Tetrahedron Letters, 177 (1971).
13. C. Eaborn et al., J. Chem. Soc., C, 2505 (1969).
14. W. Kurtz, F. Effenberger, Chem. Ber., 106, 560 (1973).
15. C. R. Hauser et al., Org. Syn., Coll. Vol., 5, 523 (1973).
16. H.-W., 1971, 4, p. 32.
17. G. M. Lampman, J. C. Aumiller, Org. Syn., 51, 55 (1971).
18. J. B. Sieja, J. Am. Chem. Soc., 93, 130 (1971).
19. F. Vögtle, P. Neumann, Synthesis, 85 (1973).
20. V. Boekelheide et al., J. Am. Chem. Soc., 83, 943 (1961).
21. E. Müller, G. Roscheisen, Chem. Ber., 90, 543 (1957).
22. H. C. Brown et al., J. Am. Chem. Soc., 83, 1001, 1002 (1961).
23. S. W. Breuer, F. A. Broster, Tetrahedron Letters, 2193 (1972).
23a. P. A. Fanta, Synthesis, 9 (1974).

24. M. Gosheav et al., Russ. Chem. Rev., 41, 1046 (1972).
25. M. F. Semmelhack et al., J. Am. Chem. Soc., 93, 5908 (1971).
26. R. J. P. Corriu, J. P. Masse, Chem. Commun., 144 (1972).
27. A. McKillop, L. F. Elsom, E. C. Taylor, Tetrahedron, 26, 4041 (1970).
28. Y. Fujiwara et al., Bull. Chem. Soc. Jap., 43, 863 (1970).
29. R. R. Josephson, Ger. Offen., 2,064,301, July 1, 1971; C. A., 75, 63358 (1971).
30. H. Iataaki, H. Yoshimoto, J. Org. Chem., 38, 76 (1973).
31. H. R. Krause et al., Ger. (East), 72, 262, April 12, 1970; C. A., 73, 109510 (1970).
31a. G. M. Whitesides et al., J. Am. Chem. Soc., 89, 5302 (1967).
31b. J. F. Normant, Synthesis, 63 (1972).
31c. J. D. Thoennes, D. E. Pearson, unpublished work.
32. G. Wittig et al., Ann. Chem., 495 (1973).
33. J. K. Kochi, D. M. Singleton, J. Org. Chem., 33, 1027 (1968).
34. A. Meisters, T. Mole, Chem. Commun., 595 (1972).
35. H. C. Brown et al., J. Am. Chem. Soc., 93, 1508 (1971).
36. H. C. Brown, S. P. Rhodes, J. Am. Chem. Soc., 91, 2149, 4306 (1969).
37. G. Bianchetti et al., Ann. Chim. Roma, 60, 483 (1970); S. Cabiddu et al., Ann. Chim. Roma, 60, 580 (1970); Synthesis, 581 (1972).

C. Nucleophilic Reactions (1, 29)

2. From Alkenes and Carbanions (1, 30)

$$RLi(MgBr) + CH_2{=}CH_2 \longrightarrow RCH_2CH_2Li(MgBr) \xrightarrow{H_2O} RCH_2CH_3$$

It now becomes apparent that reactive alkyllithiums, usually lithium with a secondary, tertiary, or highly branched alkyl group, add to ethylene to give an adduct that forms a hydrocarbon on decomposition with water [1]. Although no syntheses were conducted by the investigators of this reaction, there can be no doubt that hydrocarbons can be prepared in this way as shown [2]:

$$C_6H_5C(Et)\!\!=\!\!=\!\!CH_2 + LiCH(CH_3)_2 \xrightarrow[\text{2)}H_2O]{\text{1)}-30°, \text{A atm.}} C_6H_5CH(Et)CH_2CH(CH_3)_2$$

2-Methyl-4-phenylhexane, 46%
(on 10 mmol. scale)

The reactivity of the alkyllithiums in this reaction (and others) apparently depends on the geometry and size of the cluster existing in the alkyllithiums. For example, n-butyllithium forms a cluster that approximates a tetramer in nonpolar solvents. sec-Butyllithium must form a smaller and less stable cluster because it is more reactive. An order of reactivity in exchange with bromobenzene has been listed [3]: menthyllithium > sec-butyllithium > t-butyllithium > n-butyllithium. This order probably applies as well in the addition to ethylene, the n-butyllithium not adding at all. It may well be that n-butyllithium in a solvent that forms a strong ligand bond with the lithium ion may add to alkenes.

The reaction of an alkyllithium with aromatic alkenes such as stilbene is more complex as judged from the variety of products, some of which indicate a free radical character, formed [4]:

$$C_6H_5CH\!\!=\!\!=\!\!CHC_6H_5 + n\text{-BuLi} \xrightarrow[\text{2)}CH_3OH]{\text{1)}THF, 0°, 48 hr}$$

trans

$$C_6H_5CHCH_2C_6H_5 + (C_6H_5CH_2)_2 + C_6H_5CH\!\!=\!\!=\!\!C(C_6H_5)CH_2C_6H_5 + \text{others}$$
n-Bu

1,2-Diphenylhexane, 38%   Bibenzyl, 9%   1,2,3-Triphenylpropene, 19%

Grignard reagents also undergo addition to ethylene, but those reagents that add are restricted to sec, tert, or allylic types. Yields are modest [5]:

$$(CH_3)_2CHMgBr + CH_2\!\!=\!\!=\!\!CH_2 \longrightarrow (CH_3)_2CHCH_2CH_2MgBr \xrightarrow{H_2O}$$

44%

$$(CH_3)_2CHCH_2CH_3$$

Isopentane

In another case under pressure a better yield was obtained [6]:

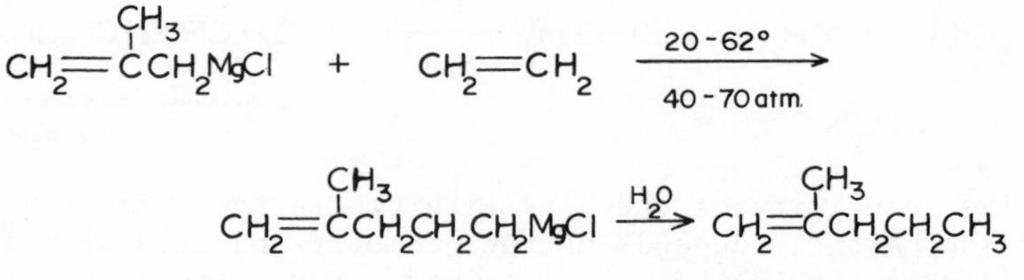

$$CH_2\!\!=\!\!\overset{\overset{\displaystyle CH_3}{|}}{C}CH_2MgCl \quad + \quad CH_2\!\!=\!\!CH_2 \quad \xrightarrow[40-70\,atm.]{20-62°}$$

$$CH_2\!\!=\!\!\overset{\overset{\displaystyle CH_3}{|}}{C}CH_2CH_2CH_2MgCl \xrightarrow{H_2O} CH_2\!\!=\!\!\overset{\overset{\displaystyle CH_3}{|}}{C}CH_2CH_2CH_3$$

2-Methyl-1-pentene, 82%

For Grignard reagent additions to other alkenes see [7].
Dialkyl adducts were obtained with lithium and alkyl bromides [7a]:

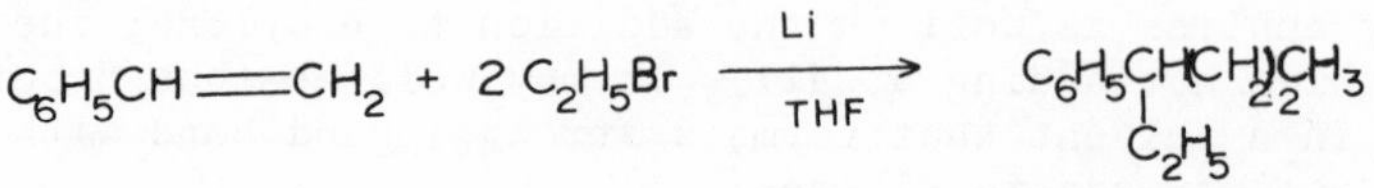

$$C_6H_5CH\!\!=\!\!CH_2 + 2\,C_2H_5Br \xrightarrow[THF]{Li} C_6H_5\overset{\overset{\displaystyle C_2H_5}{|}}{CH}(CH_2)_2CH_3$$

3-Phenylhexane, 86%

To obtain maximum yield it is important that a twofold excess of
the halide be employed. The principal mechanism involved
appears to be:

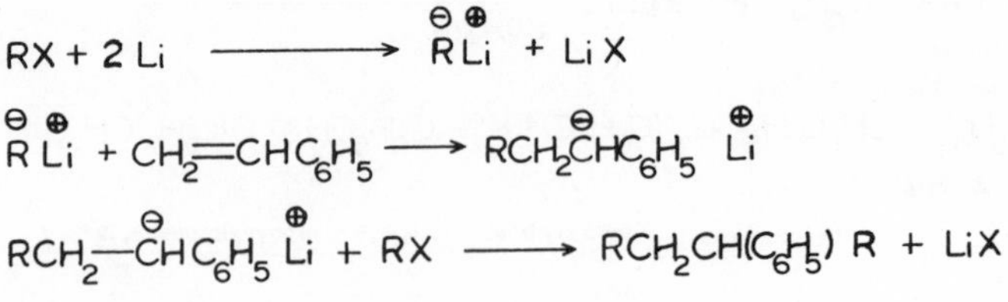

$$RX + 2\,Li \longrightarrow \overset{\ominus}{R}\overset{\oplus}{Li} + LiX$$

$$\overset{\ominus}{R}\overset{\oplus}{Li} + CH_2\!\!=\!\!CHC_6H_5 \longrightarrow RCH_2\overset{\ominus}{CH}C_6H_5\ \overset{\oplus}{Li}$$

$$RCH_2\!\!-\!\!\overset{\ominus}{CH}C_6H_5\ \overset{\oplus}{Li} + RX \longrightarrow RCH_2CH(C_6H_5)\,R + LiX$$

6.   From Multienes or Ynes

Isomerization may be induced in some cases by base
catalysts. When aromatic hydrocarbons or alkanes are formed,
the reactions become a part of this section. One example is the
formation of ethylbenzene [8]:

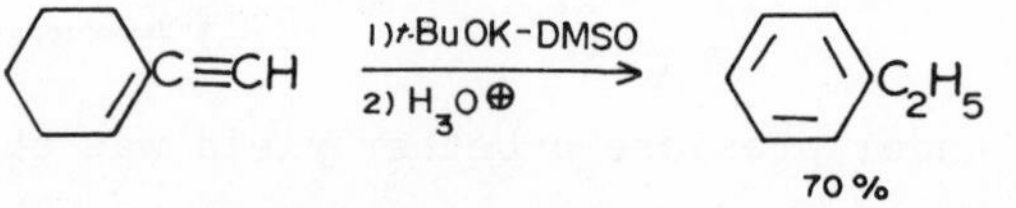

70%

Combined isomerization and dehydrogenation is exemplified in the
formation of p-cymene from limonene [9].

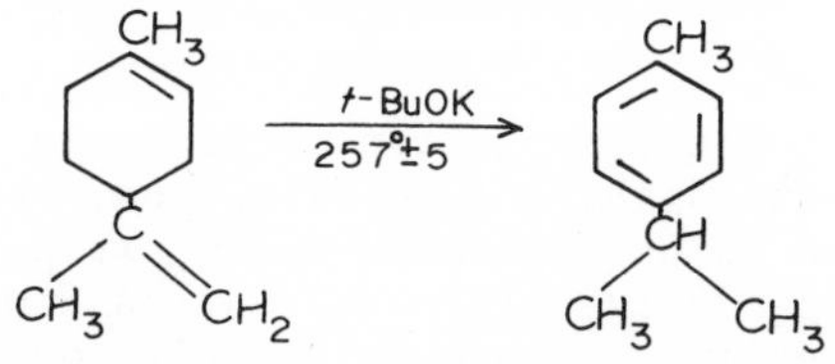

Limonene   p-Cymene, 95-97%

A third example is that of 1,7-diphenyl-1,6-heptadiyne, which with the t-butoxide isomerizes to 4-phenyl-2,3-dihydro-1H-benz(f)indene [$\overline{10}$]:

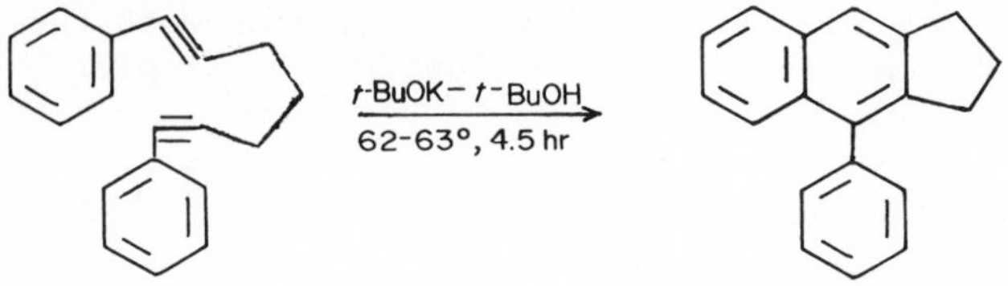

35 %

Finally, a more complex isomerization involving the loss of oxygen in the form of water has been carried out on a large scale [11]:

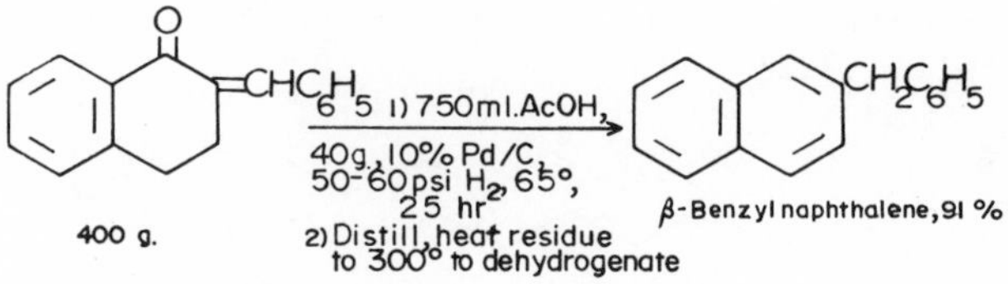

400 g.   β-Benzyl naphthalene, 91 %

7.   From Hydrocarbon Anions and Alkyl Halides or Alkenes

$$Ar_2CH_2 \longrightarrow Ar_2\overset{\ominus}{C}H \xrightarrow{RX} ArCHR$$

A significant step has been made in the understanding of the ability of bases to remove protons from substrates. Whether it be increasing the strength of an acid or that of a base, it is the counterion that must be manipulated in such a way that it becomes independently stable and free from its tendency to form clusters with ions of opposite charge. For increasing acid strength to the point of being superacids, the counterion, fluoroantimonate, [12]

$$HF + SbF_5 \longrightarrow H^{\oplus} + SbF_6^{\ominus}$$

is selected for least association with the proton. For increasing base strength, the opposite ion, the cation, must be stabilized. Such is best done by using, if possible, a large cation and/or ligands that free the cation from association with the anion. For example, the base strength of potassium t-butoxide varies in solvents as shown [13]:

$$DMSO > DMF > C_6H_6 > t\text{-}C_4H_9OH$$

The first two solvents probably form ligands with the potassium cation and thus break down the cluster to a simpler one or to the free anion:

$$(t\text{-BuOK})_n \xrightarrow{DMSO} K(DMSO)_x^{\oplus} + O^{\ominus}\text{-}t\text{-Bu}$$

In the latter form the anion has the greatest tendency to remove protons. Another way to tie up the potassium ion is to use the crown ethers that compare favorably in effect with DMSO as ligands [13]. However, it is now claimed that "cryptates," which are macrocyclic ligands, are better ligands for sodium and potassium ions [14]. The cryptate forms a cation-ligand complex in which the cation is held within a cage [15]. An illustration is the so-called 222:

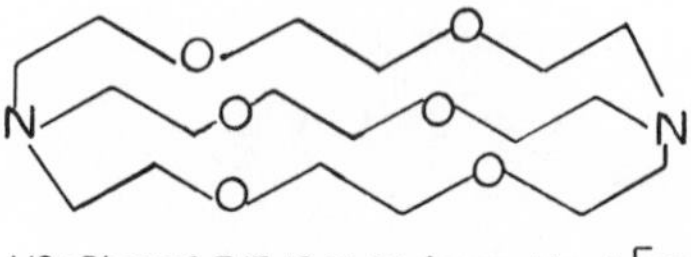

1,10-Diaza-4,7,13,16,21,24-hexaoxabicyclo[8.8.8]-
hexadodecane

Some unusual proton removals by a base and cryptate 222 [14] are:

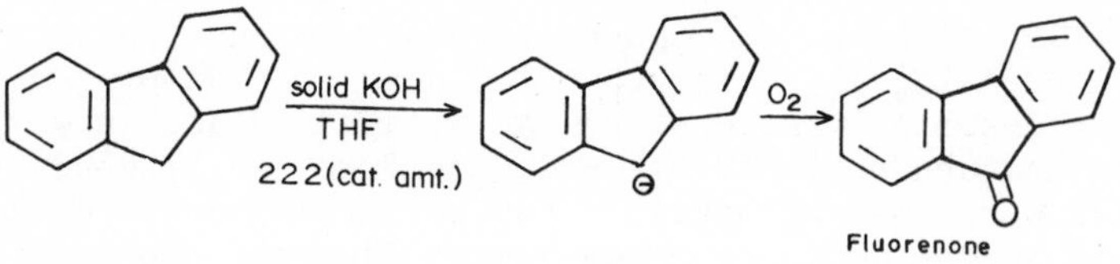

$$(C_6H_5)_2CH_2 \xrightarrow[\text{THF}]{\text{NaNH}_2} (C_6H_5)_2\overset{\ominus}{C}H \xrightarrow{C_6H_5CH_2Cl} (C_6H_5)_2CHCH_2C_6H_5$$

222

Triphenylethane

Conversion of triphenylmethane is similar to that of 1,1,1,2-tetraphenylethane. When 222 is added to sodium t-amylate in benzene, two layers separate. The lower layer is very active and instantly forms the anions of diphenyl- or triphenylmethane from the corresponding hydrocarbons. Butyllithium in hexane does not form anions with diphenyl- or triphenylmethane, but on the addition of 222, anion formation is immediate. No doubt further developments will take place in this interesting field of proton removal.

The impressive work of Pines [16] on the addition of anions to alkenes has now been summarized. These reactions are run at a higher temperature than the Bartlett reaction (C.2) and consequently they are more diverse. Two reactions are illustrated:

$$2\ \underset{C_6H_5}{C}\overset{\overset{CH_2}{\|}}{C}CH_3 \xrightarrow[230°]{\underset{160-}{Na}} 2\ \underset{C_6H_5}{C}\overset{\overset{CH_2}{\|}}{C}\overset{\ominus}{C}H_2 \longrightarrow \underset{C_6H_5}{C}\overset{\overset{CH_2}{\|}}{C}CH_2CH_2\overset{\ominus}{C}\underset{CH_3}{C_6H_5} \xrightarrow{\text{proton source}}$$

1,3-Diphenyl-1-methylcyclopentane
"good yield"

$$C_6H_5CH_2CH_3 \xrightarrow[\text{initiator}]{\underset{C_{14}H_{10}}{Na}} C_6H_5\overset{\ominus}{C}HCH_3 \xrightarrow[\text{2)proton source}]{\text{1)}CH_3CH=CH_2} \underset{CH_3}{C_6H_5\overset{|}{C}H CH(CH_3)_2}$$

α,β-Dimethylpropylbenzene,52%

Many other reactions are described.

A ring closure to produce a naphthalene has recently been achieved by nucleophilic means [17]:

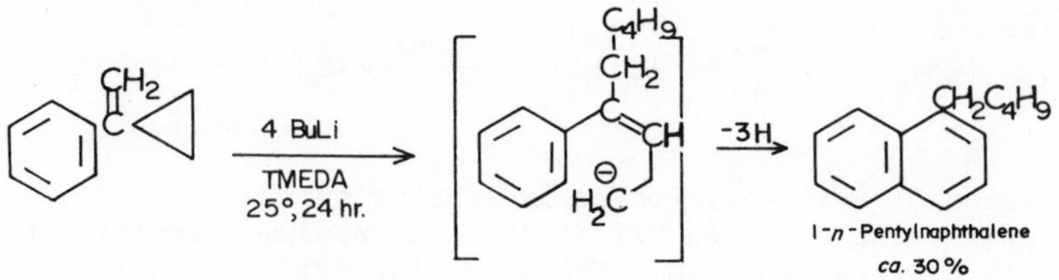

I-n-Pentylnaphthalene
$ca.$ 30%

8. From Hydrocarbon Radical Anions and Alkylating Agents ($S_{RN}I$ Mechanisms)

$$ArX \xrightarrow[R\ominus]{\epsilon \text{ source}} ArR$$

Bunnett and co-workers have found that the benzyne mechanism is bypassed when a displacement reagent is introduced in the presence of an excess of electrons. This condition is arrived at by providing intermittent addition of an alkali metal to liquid ammonia until the blue color is retained.

Among the reactions completed, one is that involved in the formation of 9-phenylfluorene [18]:

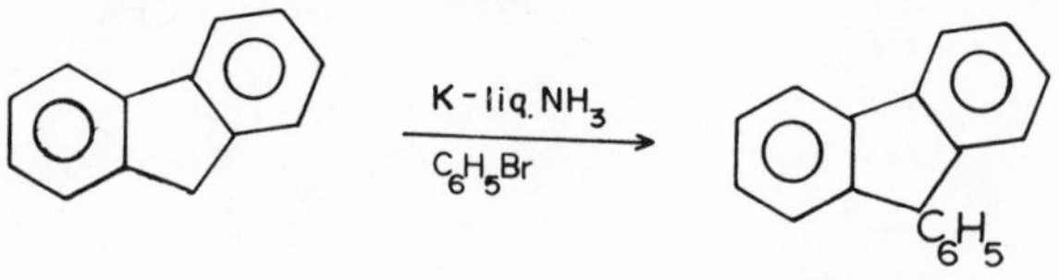

44%
(+5% 9,9-diphenylfluorene)

A possible mechanism has been proposed [19]:

1) ϵ donor + ArX $\longrightarrow$ ArX$^{\ominus}$

2) ArX$^{\ominus} \longrightarrow$ Ar$^{\odot}$ + X$^{\ominus}$

3) Ar$^{\odot}$ + Y$^{\ominus} \longrightarrow$ ArY$^{\ominus}$

4) ArY$^{\ominus}$+ ArX $\longrightarrow$ ArY + ArX$^{\ominus}$

9. From Diphosphoranes

Phenanthrene, as one example, was prepared as indicated in 71% yield [20]. The diylid first formed is oxidized to the diradical and triphenylphosphine oxide and phenanthrene are produced.

References

1. P. D. Bartlett et al., J. Am. Chem. Soc., 75, 1771 (1953); 91, 7425 (1969).
2. J. A. Landgrebe, J. D. Shoemaker, J. Am. Chem. Soc., 89, 4465 (1967).
3. W. H. Glaze, C. H. Freeman, J. Am. Chem. Soc., 91, 7198 (1969).
4. Y. Okamoto et al., Bull. Soc. Chem. Jap., 42, 760 (1969).
5. L. H. Shepherd, U.S. 3,597,488 (1971), Aug. 3, 1971: C. A., 75, 88751 (1971).
6. H. Lehmkuhl, D. Reinehr, J. Organomet. Chem., 25, C47 (1970).
7. H. Lehmkuhl et al., J. Organomet. Chem., 57, 29,39,49 (1973); Ann. Chem., 145 (1975).
7a. N. F. Scilly et al., J. Chem. Soc., Perkin Trans. I, 286 (1972).
8. R. Mantione, Compt. Rend., 264, C 1668 (1967).
9. H. Pines, L. Schaap, J. Am. Chem. Soc., 79, 2956 (1957). For reviews on addition of carbanions see H. Pines, Synthesis, 309 (1974); Acc. Chem. Res., 7, 155 (1974).
10. I. Iwai, J. Ide, Chem. Pharm. Bull. (Tokyo), 12, 1094 (1964); C. A., 61, 14622 (1964).
11. E. J. Eisenbraun et al., Org. Prep. Proced., 2, 37 (1970).
12. G. A. Olah, Angew. Chem., Intern. Ed. Engl., 12, 173 (1973).
13. D. E. Pearson, Calvin A. Buehler, Chem. Rev., 74, 45 (1974).
14. B. Dietrich, J. M. Lehn, Tetrahedron Letters, 1225 (1973).
15. J. M. Lehn et al., Chem. Commun., 1100 (1972); 15 (1973).
16. H. Pines, Intra-Sci. Chem. Rep., 6 (2), 1 (1972).
17. S. Brenner, E. Dunkelblum, Tetrahedron Letters, 2487 (1973).
18. R. A. Rossi, J. F. Bunnett, J. Org. Chem., 38, 3020 (1973).
19. J. F. Bunnett, B. F. Gloor, J. Org. Chem., 38, 4156 (1973).
20. H.-J. Bestmann et al., Chem. Ber., 102, 2259 (1969).

D. Friedel-Crafts Alkylations and Related Reactions

  1. From Arenes and Alkylating or Arylating Agents (1, 33;
      H.-W., 5, Pt. 1a, 1970, pp. 501-540)

Olah [1] continues to provide new insights into the
mechanism of alkylation. Lately, he has shown the variability of
the transition state with the electronegativity of the substituent in the alkylation of arenes with a series of substituted
bezyl halides as shown in two cases:

Table 1.2. Effect of Substituent in Arenes on the Transition
State

Substituent	$k_{Toluene}/k_{Benzene}$ Rate Ratio	o/p Ratio
p-NO$_2$	2.5	1.74
p-OCH$_3$	97	0.41

From figures such as these Olah concludes that benzylic ions with
electron-withdrawing groups form a complex with the arene that
has more π and less σ character, the reverse being true for
electron-releasing groups. Thus the statistical o/p ratio of 2
is decreased with electron-releasing groups and the reaction
becomes more selective for p substitution. Olah has also found a
highly polarized methyl species more closely approximating the
elusive methyl carbonium ion species [2], namely, methyl hexa-
fluoroantimonate,

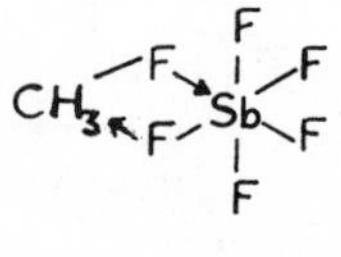

It should be quite reactive in methylation. Later this investigator considered the alkylating power of methyl and ethyl
fluoride-antimony pentafluoride complexes as "unmatched by that
observed for any previously reported methylating or ethylating
agent" [3].
     Roberts has assessed and explained the strange equilibrium

existing in the alkylation of benzene with tertiary halides of
molecular weight greater than the butyl [4]:

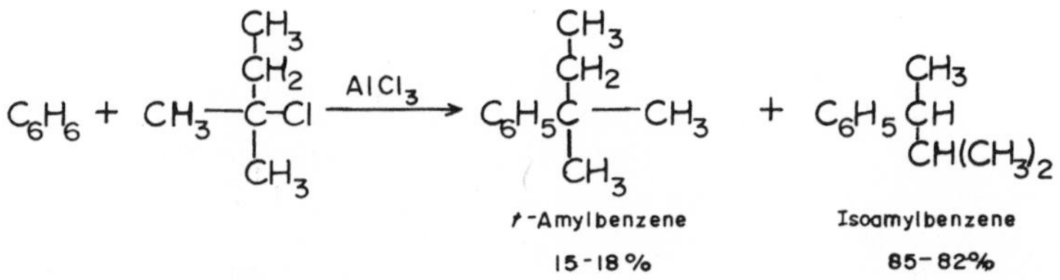

t-Amylbenzene   Isoamylbenzene

15-18%   85-82%

The t-amylbenzene is probably the kinetic product, but since it
tends to reverse more easily than isoamylbenzene, the latter is
the predominant product. If nitrobenzene is used to complex the
catalyst, t-amylbenzene is the sole product.
    There has recently been prepared a series of stable and
soluble fluorophosphate salts, trialkyloxonium and trialkyloxy-
carbenium fluorophosphates, $R_3O^+ PF_6^-$ and $(RO)_3\overset{+}{C}PF_6^-$,

respectively, which give excellent yields in alkylation
reactions [5],
    Phosphate or phosphite esters have been utilized as alky-
lating agents [6]:

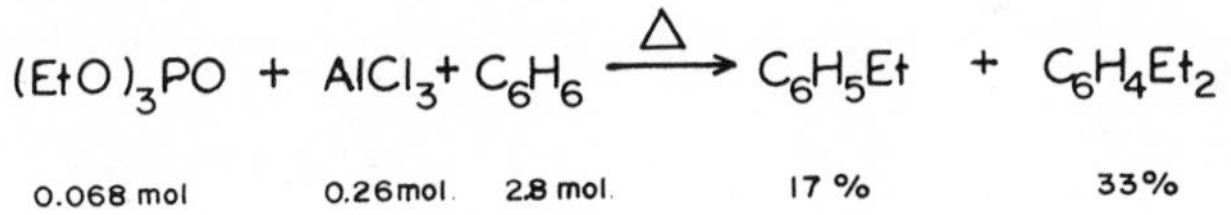

$$(EtO)_3PO + AlCl_3 + C_6H_6 \xrightarrow{\Delta} C_6H_5Et + C_6H_4Et_2$$

0.068 mol   0.26 mol   2.8 mol   17 %   33%

Sulfonic esters have been used similarly (H.-W., 5, Pt. 1a, 1970,
p. 537).
    A homogeneous catalyst, $ArMo(CO)_3$, appears to be quite
similar to $AlCl_3$ in alkylation [7].
    Sulfuric acid led to the formation of a greater amount of
t-alkybenzenes than $AlCl_3$ in alkylation with alkenes [8]. This
acid has been serviceable in alkylation in that it leads to substituted indanes (H.-W., 5, Pt. 1a, 1970, p. 43):

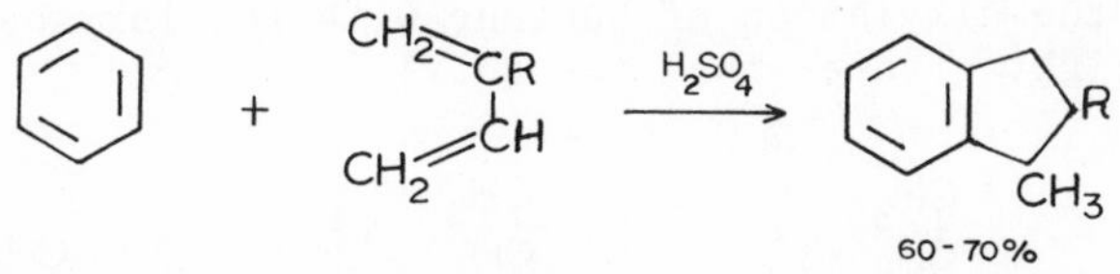

60-70%

The indanes may be further hydrogenated to hexahydroindanes.

An exhaustive methylation process, conducted only on a semimicro stage, has been developed [9]. Not only 5,6,7,8-tetra-methyltetralin (Ex. a), but 2,3,4,5-tetramethylpyrrole (52% from 2,4-dimethyl-5-carbethoxypyrrole at elevated temperatures) and 2,3,4-trimethyl-5-carbethoxypyrrole (64% from the same substrate at 25°) have been produced by the same procedure.

An interesting arylation has been accomplished [10]:

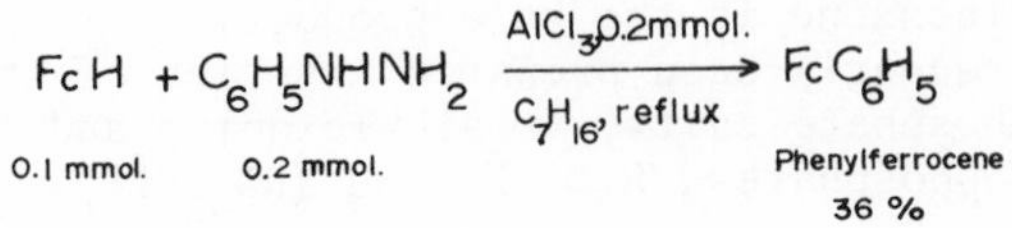

$$FcH + C_6H_5NHNH_2 \xrightarrow[C_7H_{16}, \text{reflux}]{AlCl_3, 0.2 \text{mmol.}} FcC_6H_5$$

0.1 mmol.   0.2 mmol.   Phenylferrocene
36 %

The reaction must depend on the high susceptibility of ferrocene to electrophilic reagents since benzene and anisole are not alkylated in this manner. A free radical arylation of ferrocene by phenylhydrazine and silver oxide or benzoquinone is known [11].

A free radical alkylation has recently been accomplished [12]:

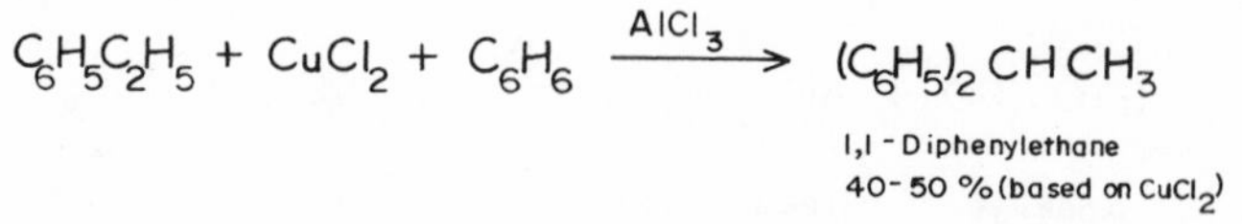

$$C_6H_5C_2H_5 + CuCl_2 + C_6H_6 \xrightarrow{AlCl_3} (C_6H_5)_2CHCH_3$$

1,1-Diphenylethane
40-50 % (based on $CuCl_2$)

a. Preparation of 5,6,7,8-Tetramethyltetralin [9]

Tetralin, 1 ml was added to 2 g of paraformaldehyde, 10 ml of hydriodic acid (d 1.95), and 40 ml of acetic anhydride and the mixture was held at 90-100° for 4 hr and at 116° for 5 hr with stirring. Periodically the mixture was cooled to 60°, de-colorized with hypophosphorous acid (total 7 ml, 50%), and then poured into water. The solid obtained by filtration was refluxed with pyridine and the mixture was poured into dilute acetic acid. After filtration, the precipitate was dried, sublimed, and re-crystallized from methanol, 74%.

b. Preparation of 1,3-Diphenyl- and 1,3,5-Triphenyladamantane
[13]

1-Bromoadamantane, 5 g, t-butyl bromide, 6.3 g in dry (Linde
Type 3A) molecular sieves, benzene, 175 ml, and $AlCl_3$, 250 g,
were refluxed for 20 min. The reaction mixture was poured into
ice water and ether added, and 30 min later 2.7 g of the insoluble material was removed by filtration. Recrystallization from
benzene gave 1,3,5-triphenyladamantane, melting point 230-233°.
From the filtrate 1,3-diphenyladamantane, 3.1 g, melting point
101-104°, was obtained by concentration, filtration, and recrystallization from methanol.

2. From Aromatic Ketones (Cyclization) (1, 39)

Note is made of the condensation of ketones with thiophene
to produce methylene bis-thiophenes (A.7).
1,2,3-Trimethylnaphthalene was formed by the dehydration of
2,3-dimethyl-2-hydroxy-4-oxo-1-phenylpentane [14]:

3. Rearrangement of Hydrocarbons, Including Adamantane
     (1, 41; H.-W., 5, Pt. 1a, 1970, pp. 305-321)

Of all the rearrangements studied it appears that those
involved in the synthesis of adamantane have been pursued the
most vigorously [14a]. Apparently the 11-carbon tricyclic compounds, methyltricyclodecane or tricycloundecane, yield 1-methyl-
adamantane on treatment with $AlCl_3$ or $AlBr_3$, 25% by weight, if
refluxed or held with stirring 2-72 hr at 25°. Any 12-carbon
tricyclic compound, including 1-ethyladamantane, yields 1,3-di-
methyladamantane on similar treatment. Triamantane,

the heptacyclic compound shown, has been synthesized, but as of now such is not the case for tetramantane. A catalyst preferable to the aluminum halides has been used to bring about these re-arrangements [15]. It consists of 0.5% Pt on alumina activated consecutively by H_2, HCl, and $SOCl_2$. The adamantanes are obtained by passing the substrate in dry HCl through a hot tube containing the catalyst at 165-169°. By this procedure exo-2,3-tetramethylenenorbornane yielded 98% of 1-methyladamantane.

More recently an improved, liquid-phase catalyst, fluoro-antimonic acid, for the isomerization of tetrahydrodicyclopenta-diene to adamantane has been reported [16]:

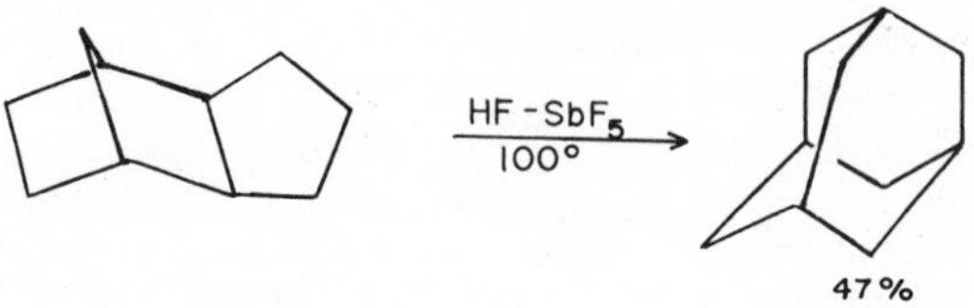

$$\xrightarrow[100°]{HF - SbF_5}$$

47%

a. Preparation of 1-Methylbicyclo(2.2.1)heptane (H.-W., 5, Pt. 1a, 1970, p. 311)

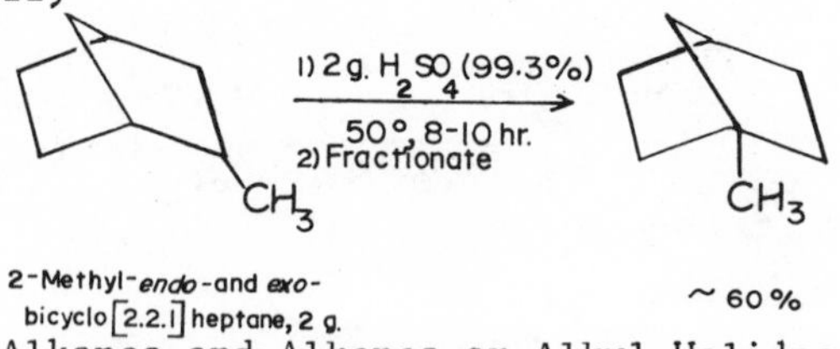

$$\xrightarrow[\substack{50°, 8-10\,hr.\\ 2)\,Fractionate}]{1)\,2g.\,H_2SO_4\,(99.3\%)}$$

2-Methyl-endo-and exo-
bicyclo[2.2.1]heptane, 2 g.   ~ 60%

4.   From Alkanes and Alkenes or Alkyl Halides (1, 42)

Trialkylalanes (R_3Al) are known to give good yields of hy-drocarbons with quaternary carbon atoms from t-alkyl halides [17]:

$$(CH_3)_3CCl \ + \ AlMe_3 \xrightarrow[10\,min.]{-78°} (CH_3)_4C$$

Neopentane
100 %

Alkenes that supposedly form secondary halides do not give good yields:

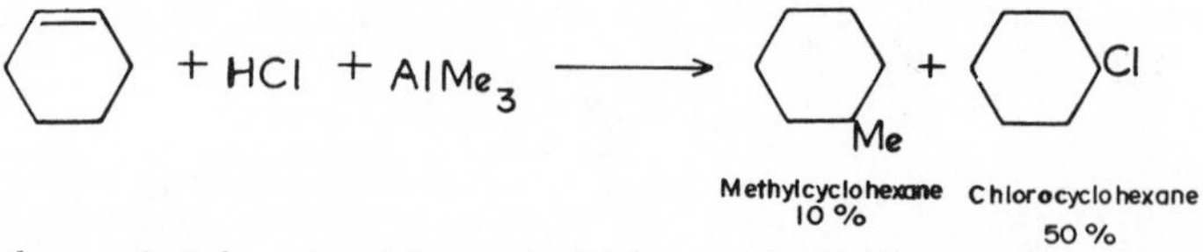

Methylcyclohexane   Chlorocyclohexane
10 %   50 %

Alkenes that yield a tertiary halide work well in the displace-ment [18]:

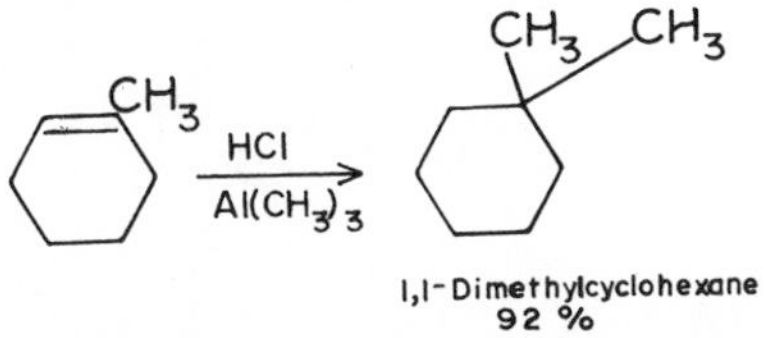

1,1-Dimethylcyclohexane
92 %

Alkylation of diphenylmethane has been accomplished by heating with sodium bis-(2-methoxyethoxy) aluminohydride, $NaAlH_2$ $(OCH_2CH_2OCH_3)_2$ as shown [19]:

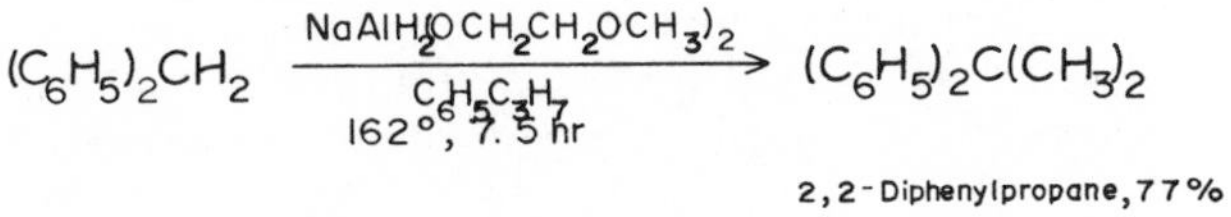

$$(C_6H_5)_2CH_2 \xrightarrow[\substack{C_6H_5C_3H_7 \\ 162°, 7.5 \text{ hr}}]{NaAlH_2OCH_2CH_2OCH_3)_2} (C_6H_5)_2C(CH_3)_2$$

2,2-Diphenylpropane, 77%

The temperature and time must be carefully controlled to prevent the formation of large amounts of other products.

Addenda

D.1. Anhydrous aluminum chloride intercalated in graphite is a milder catalyst than $AlCl_3$ for the alkylation of arenes and gives less polyalkylation [J.-M. Lalancette et al., Can. J. Chem., 52, 589 (1974)].

D.3. Alkylboranes may be used in the synthesis of adamantanes [B. M. Mikhailov, V. N. Smirnov, Izv. Akad. Nauk SSSR (Engl. transl.), 23, 1079 (1974)].

D.3. Norbornadiene may be converted into diamantane by a series of Friedel-Crafts reactions, one of which uses $CuBr_2$ in addition to BF_3 [P. v. R. Schleyer et al., Org. Syn., 52, unchecked procedure 1776 (1972)].

References

1. G. A. Olah, Acc. Chem. Res., 4, 240 (1971).
2. G. A. Olah et al., J. Am. Chem. Soc., 91, 2112 (1969);
    G. A. Olah, P. von R. Schleyer, Carbonium Ions, Wiley-
    Interscience, New York, Vol. II, p. 722.
3. G. A. Olah et al., J. Am. Chem. Soc., 94, 156 (1972).

4. R. M. Roberts et al., J. Org. Chem., 35, 3717 (1970);
    Intra-Sci. Chem. Rep., 6 (2), 89 (1972).
5. G. A. Olah et al., Synthesis, 490 (1973).
6. G. Sosnovsky et al., Synthesis, 142 (1971).
7. M. F. Farona, J. F. White, J. Am. Chem. Soc., 93, 2826
    (1971).
8. J. R. Nooi et al., Rec. Trav. Chim., 88, 398 (1969).
9. S. F. MacDonald et al., Can. J. Chem., 46, 3291 (1968).
10. G. P. Sollott, W. R. Peterson, Jr., J. Org. Chem., 34, 1506
    (1969).
11. A. L. J. Beckwith, R. J. Leydon, Australian J. Chem., 19,
    1381 (1966).
12. L. Schmerling, J. A. Vesely, J. Org. Chem., 38, 312 (1973).
13. H. Newman, Synthesis, 692 (1972).
14. T. S. Chen et al., Synthesis, 620 (1973).
14a. M. A. McKervey, Chem. Soc. Rev., 3, 479 (1974).
15. M. A. McKervey et al., J. Am. Chem. Soc., 93, 2798 (1971).
16. J. A. Olah, G. A. Olah, Synthesis, 488 (1973).
17. J. P. Kennedy, U. S., 3,585,252, June 15, 1971; C. A., 75
    63083 (1971); J. Am. Chem. Soc., 95, 6386 (1973).
18. J. P. Kennedy, S. Sivaram, J. Org. Chem., 38, 2262 (1973).
19. M. Černy, J. Málek, Tetrahedron Letters, 691 (1972).

E. Dehydrogenation of Hydroaromatic Hydrocarbons (1, 49; H.-W.,
5, Pt. 1a, 1970, pp. 445-447)

The discussion in the original text appears to be rather
complete, but representative behavior is illustrated by the conversion of a series of cyclohexanones to 1,2-dialkylbenzenes:

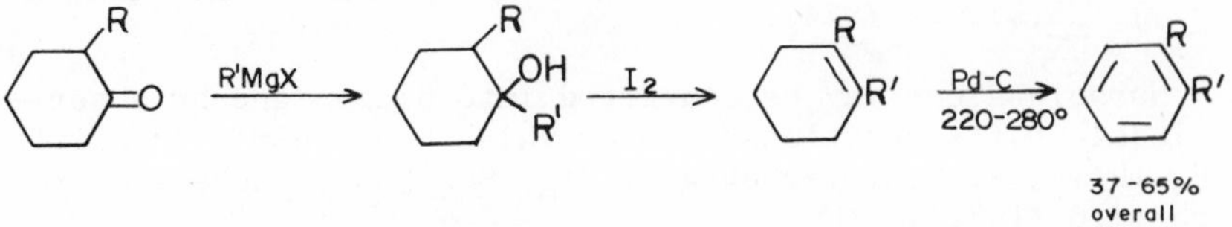

37-65%
overall

Little aromatization occurs in the reaction of cyclooctane
with Pt-C (H.-W., 5, Pt. 1a, 1970, p. 310):

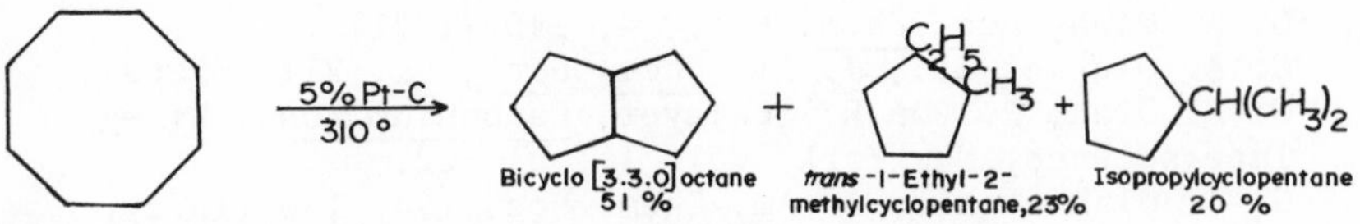

Bicyclo [3.3.0] octane   trans -1-Ethyl-2-   Isopropylcyclopentane
        51 %   methylcyclopentane,23%   20 %

With hydrogen present the cis and trans forms of 1-ethyl-2-
methylcyclopentane are obtained in 52% yield.

At a lower temperature N-lithioethylenediamine is known to effect not only isomerization, but also dehydrogenation [1]:

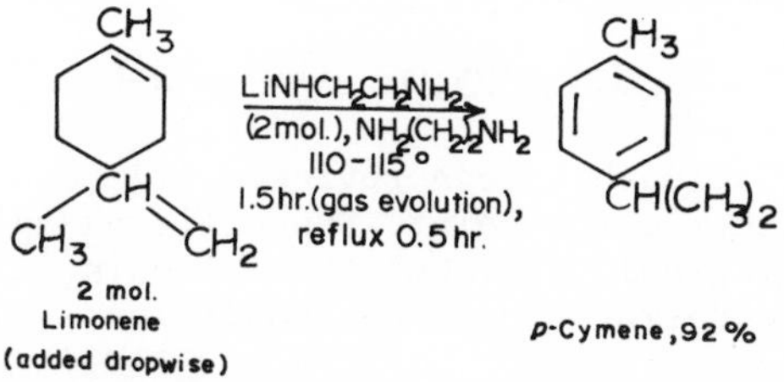

2 mol.
Limonene
(added dropwise)

p-Cymene, 92 %

The reaction does not occur if the limonene concentration is in excess of the reagent concentration. Lithium hydride is no doubt eliminated in the process.

In the dehydrogenation of tetrahydrotoluene serious scrambling of the carbon atoms has been found to take place [2]:

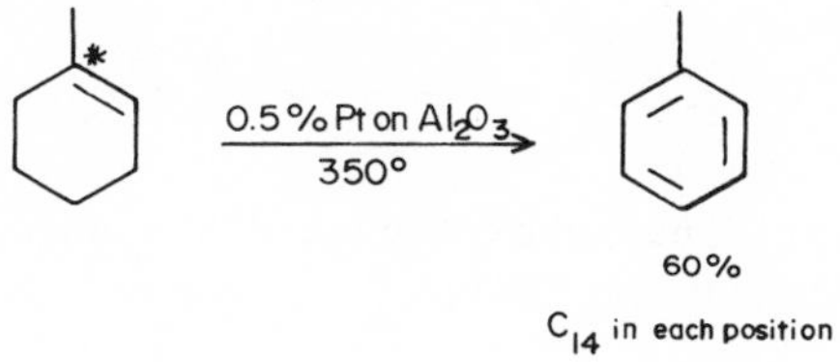

60%

C_{14} in each position

Addenda

E. Limonene has been dehydrogenated and aromatized to p-cymene at 25° in 5 min by $KH-NH_2CH_2CH_2NH_2$ [C. A. Brown, J. Am. Chem. Soc., 95, 982 (1973)].

E. Triphenylcarbinol in refluxing trifluoroacetic acid dehydrogenates tetralin to naphthalene [P. P. Fee, R. G. Harvey, Tetrahedron Letters, 3217 (1974)]. Other reagents for the dehydrogenation of some dihydro compounds are RLi-TMEDA [R. G. Harvey, H. Cho, J. Am. Chem. Soc., 96, 2434 (1974)] and acetone under irradiation [T. Matsuura, Y. Ito, Bull. Chem. Soc. Jap., 47, 1724 (1974)]. 1,6-Methano [10] annulene may be prepared by dehydrogenation of the hexahydrido substrate with the excellent reagent 2,3-dichloro-5,6-dicyano-1,4-benzoquinone, DDQ, in dioxane [E. Vogel et al., Org. Syn., 54, 11 (1974)]. Methods for the synthesis of other annulenes are also given.

References

1. L. Reggel et al., J. Org. Chem., 23, 1136 (1958).
2. J. L. Marshall et al., Tetrahedron Letters, 3491 (1973).

F.   Decarboxylation (<u>1</u>, 52; H.-W., <u>5</u>, Pt. 1a, 1970, p. 332)

1.   From Acids

$$RCO_2H \longrightarrow RH$$

Decarboxylation has been reviewed [1] more from a mechanistic than a synthetic viewpoint. Specific examples are to be found in the original text, but the following classification may be helpful in choosing the proper conditions.

a.   Base Decarboxylation

This is for trichloroacetic acid, α-nitroacetic acid, amino acids, and 2- and 4-pyridinecarboxylic acids. Here the anion formed after decarboxylation is quite stable:

$$CCl_3CO_2Na \longrightarrow \overset{\oplus}{Na}\overset{\ominus}{C}Cl_3 + CO_2$$

Indeed with 2-pyridinecarboxylic acid the anion can be trapped by carbonyl compounds [1, 601]:

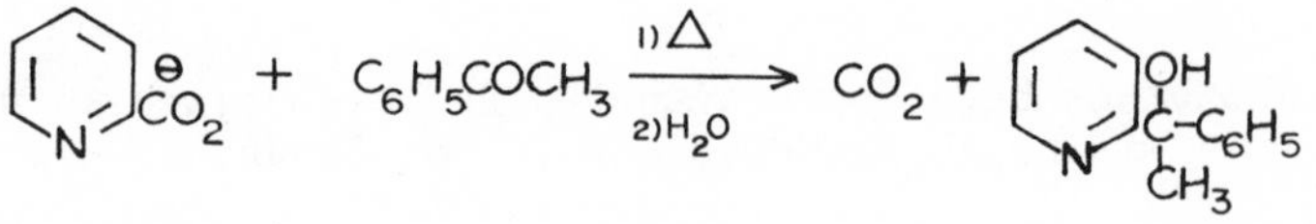

This type of decarboxylation is subject to catalysis by micellar formation with cetyltrimethylammonium bromide [1a].

b.   Acid Decarboxylation

This is for β-keto acids, acids with other electron-withdrawing substituents in the β-position, or acids that form relatively stable carbonium ions on decarboxylation. Here a portion of the driving force arises from the concerted nature of the reaction:

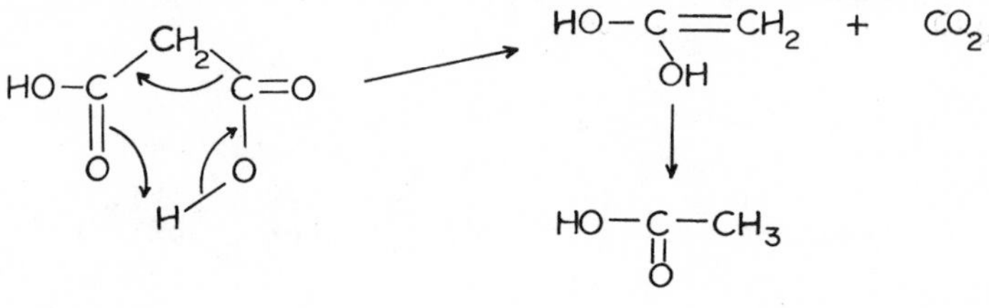

On the other hand, decarbethoxylation of β-keto esters or malonic esters appears to be catalyzed by sodium chloride [2]:

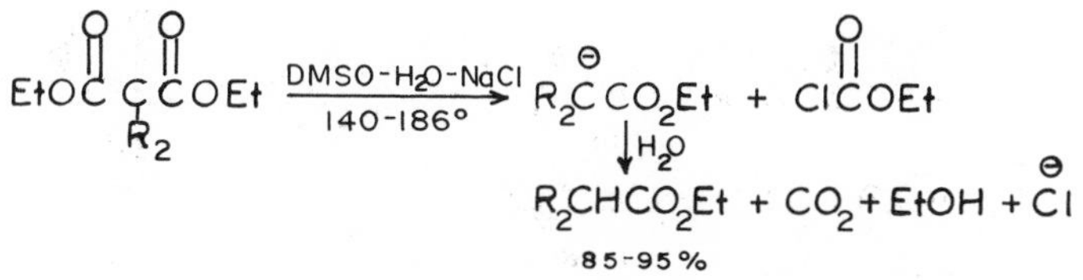

$$EtOC\overset{O}{\underset{}{\|}}\underset{\underset{R_2}{|}}{C}C\overset{O}{\underset{}{\|}}OEt \xrightarrow[140-186°]{DMSO-H_2O-NaCl} R_2\overset{\ominus}{C}CO_2Et + Cl\overset{O}{\underset{}{\|}}COEt$$

$$\downarrow H_2O$$

$$R_2CHCO_2Et + CO_2 + EtOH + \overset{\ominus}{Cl}$$

85-95%

However, for branched-chain acids that form relatively stable cations, the driving force is decarbonylation and formation of a carbonium ion:

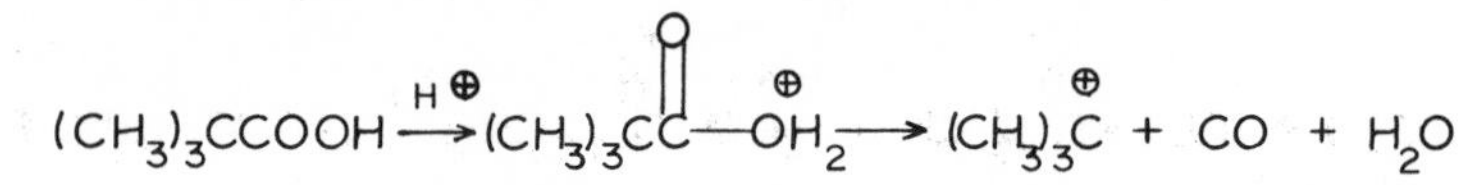

$$(CH_3)_3CCOOH \xrightarrow{H\oplus} (CH_3)_3C\overset{O}{\underset{}{\|}}C\overset{\oplus}{-}OH_2 \longrightarrow (CH_3)_3\overset{\oplus}{C} + CO + H_2O$$

c. Free Radical Decarboxylation

This is for acids that have no destabilizing features and whose decomposition may be represented as shown:

$$ArCO_2Cu\,(quinoline\ ligand) \xrightarrow{\Delta} CO_2 + Cu + [Ar\cdot] \xrightarrow[source]{H} ArH$$

Acyl peroxides derived from acids also decompose to give hydrocarbons (see G.2).

Perhaps the best free radical way of decarboxylation is the oxidative procedure by the use of lead tetraacetate [3]. Lead tetraacetate in chloroform under irradiation gives 50-60% yields of alkanes from secondary acids and 70-80% from primary acids. Alkenes are obtained from tertiary acids and from other acids if the reaction is catalyzed by copper acetate.

The mechanism is as follows:

$$Pb^{IV}O_2CR \xrightarrow[\substack{or\\ h\nu}]{heat} Pb^{II} + [R\cdot] + CO_2 \xrightarrow{HS} RH$$

Thus the reaction proceeds by a free radical mechanism. A few interesting syntheses are listed, all of which include Pb(OAc)$_4$:

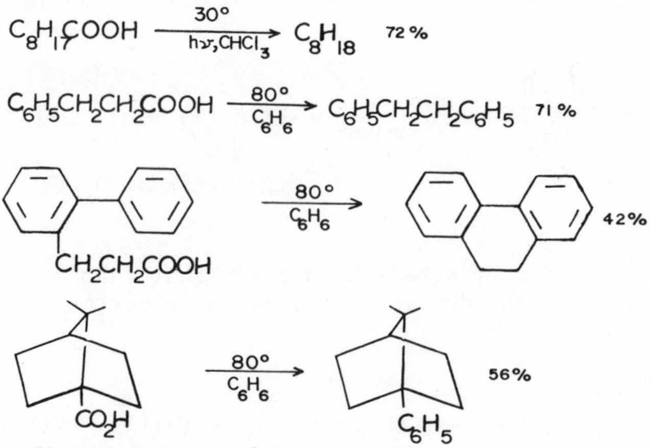

$$C_8H_{17}COOH \xrightarrow[h\nu, CHCl_3]{30°} C_8H_{18} \quad 72\%$$

$$C_6H_5CH_2CH_2COOH \xrightarrow[C_6H_6]{80°} C_6H_5CH_2CH_2C_6H_5 \quad 71\%$$

$$\xrightarrow[C_6H_6]{80°} \quad 42\%$$

$$\xrightarrow[C_6H_6]{80°} \quad 56\%$$

3. From Phenyl Esters of Fatty Acids

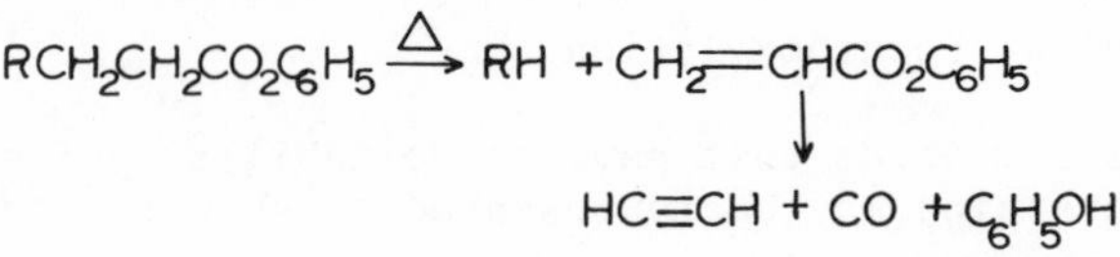

$$RCH_2CH_2CO_2C_6H_5 \xrightarrow{\Delta} RH + CH_2{=}CHCO_2C_6H_5$$

$$\downarrow$$

$$HC{\equiv}CH + CO + C_6H_5OH$$

This high-temperature reaction is carried out best in a bomb [4].

a. Preparation of Pentadecane

Phenyl stearate, 10 g was heated in a bomb at 280-310° for 64 hr. From the products, 3.5 g (59%) of pentadecane and 2 g of stearic acid were isolated.

References

1. L. W. Clark, in S. Patai, The Chemistry of Carboxylic Acids and Esters, Wiley-Interscience, New York, 1969, pp. 589-623.
1a. C. A. Bunton et al., J. Org. Chem., 37, 1388 (1972).
2. A. P. Krapcho, A. J. Lovey, Tetrahedron Letters, 957 (1973).
3. R. A. Sheldon, J. K. Kochi, Org. Reactions, 19, 279 (1972).
4. H.-W., 5, Pt. 1a, 1970, p. 335.

G. Free Radical Reactions

Previous sections (e.g., B, C, E, and F) involve some free

radical reactions, but not as consistently as in this section.

1. From Amines or Diazo Compounds by Coupling with Aromatic
Nuclei (Gomberg-Bachmann and Pschorr) (1, 55)

$$Ar \overset{\oplus}{N_2} X^{\ominus} + C_6H_6 \longrightarrow Ar\,C_6H_5$$

A comparison of yields has been made under various conditions for conducting the reaction [1] with benzene.

Table 1.3.   Yields of Biphenyl in Coupling Reactions

Reactant or Reactants	Biphenyl Yield, %
$C_6H_5-\overset{+}{N_2}$ + NaOH	10
$C_6H_5N(NO)COCH_3$	80
$C_6H_5N=NN(CH_3)_2$	25
$C_6H_5\overset{+}{N_2}$ + CuCl or CuCl(NH_3)	50-80%

It is difficult to maintain the cuprous salt in its monovalent
form.   The complex

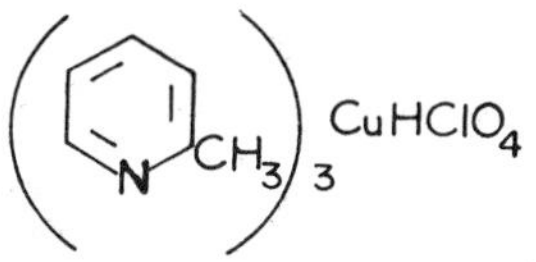

appears to be quite stable and to show promise as a catalyst.   It
has been used to abstract a hydrogen atom from ethanol [2]:

$$Ar\overset{\oplus}{N_2} + CH_3CH_2OH \xrightarrow{\text{Cu complex}} ArH$$

However, rather good yields (50-88%) of symmetrical biaryls,
mostly nitrobiaryls, are obtained by decomposing the diazonium
salt in the presence of cuprous chloride and a strong acid [3] as
shown:

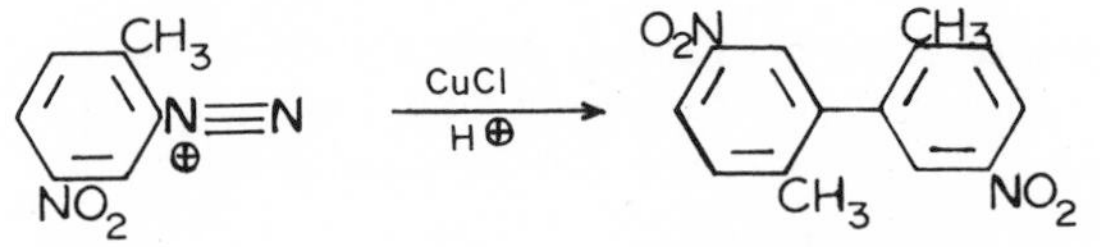

3,3'-Dinitro-6,6'-dimethylbiphenyl
60%

A recent reaction utilizes the tetrafluoroborate with DMSO as a solvent to advantage [4]. It is claimed that yields are better than in the Ullmann reaction (B.3).

There can be no doubt that the Pschorr coupling reaction has been of great benefit in the preparation of phenanthrenes, a considerable number of which have been synthesized in the recent antimalarial program of the U.S. government as described by one contractor [5]. Although copper is usually recommended as a catalyst, sodium iodide appears to be superior in some cyclizations [6]:

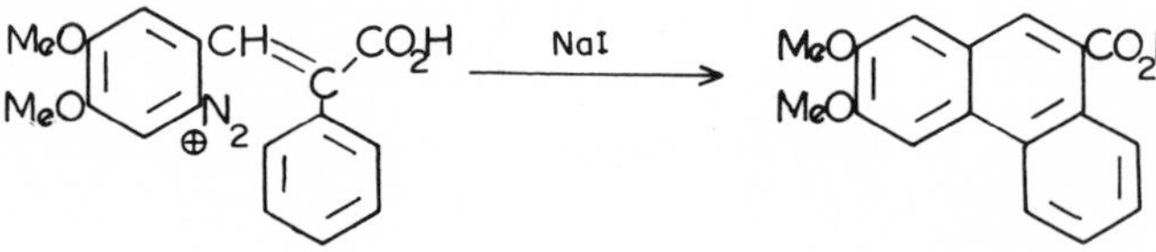

2,3-Dimethoxy-9-phenanthroic acid

72%

Section G.3 includes other Pschorr couplings.

2.   From Peroxides (1, 57)

$$(ArCO_2)_2 + Ar'H \longrightarrow ArCO_2H + ArAr' + CO_2$$

In a recent publication [7] 3,4'-dichlorobiphenyl has been synthesized from bis-3,4-dichlorobenzoyl peroxide and C_6H_6 and a small amount of m-dinitrobenzene in 78-81% yield. On the other hand, acyl peroxides do not couple but instead yield alkanes (H.-W., 5, Pt. 1a, 1970, p. 336):

$$(C_{11}H_{23}CO_2)_2 + C_2H_5OC_2H_5 \xrightarrow[\text{17 days}]{\text{Reflux}} C_{11}H_{24} + C_{11}H_{23}CO_2H + CO_2$$

Undecane
84%

$$+ CH_3CHO + CH_2{=}CH_2$$

Peroxides couple with olefins under special circumstances. An intramolecular reaction of this nature is the cyclization of an unsaturated peroxide [8]:

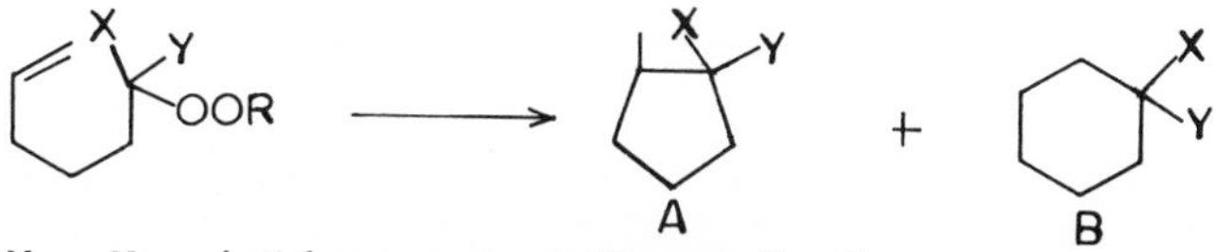

If X and Y = H, yields are A, 100 and B, 0.
If X = CN and Y = COOC$_2$H$_5$ (ω-methyl substituted), yields are A, 0 and B, 100. The 4-(1'-naphthylsubstituted)butyl radical cyclizes to tetrahydrophenanthrene in 10-14% yield [9].

3. From Aryl Compounds by Photochemical Coupling (1, 58)

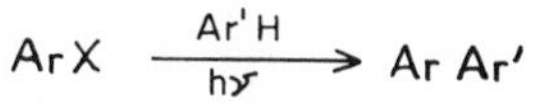

$$ArX \xrightarrow[h\nu]{Ar'H} Ar\,Ar'$$

Reviews on photocyclization of stilbenes and related compounds as illustrated are now available [10,11].

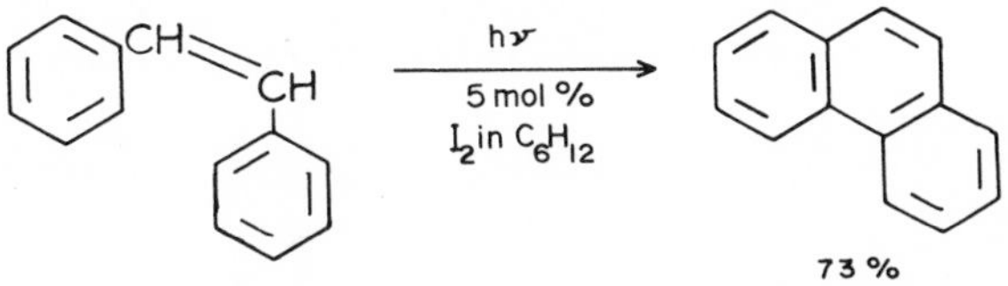

73 %

The iodine serves to dehydrogenate the intermediate cyclohexadiene to phenanthrene. Other papers describe the synthesis of 1-, 3-, or 9-substituted phenanthrenes [12]. Triphenylene has been prepared similarly from o-terphenyl in moderate yield on a 10 mmol scale [13]. More complex ring systems such as hexahelicene have been obtained by the photolysis of 2,7-distyrylnaphthalene and I$_2$ in C$_6$H$_6$ [14].

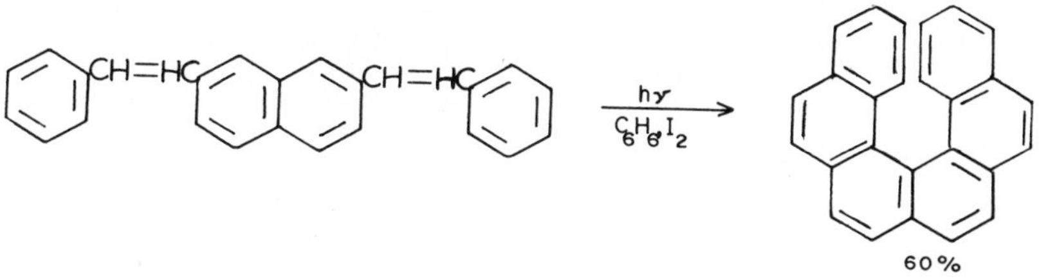

60%

Higher yields of biphenyls are available by the photolysis of arylthallium difluoroacetates in C_6H_6 [15]:

$$ArTl(OCOCF_3)_2 \xrightarrow[\underset{6\ 6}{CH}]{h\nu} C_6H_5Ar$$

78-91%

The reaction possesses specificity and is particularly valuable in the synthesis of unsymmetrical biaryls.

4.   From Arenes by Irradiation or from Alkenes or Azoalkanes by Heat or Irradiation ($\underline{1}$, 59)

Olefins comprise a source of cyclobutanes by coupling:

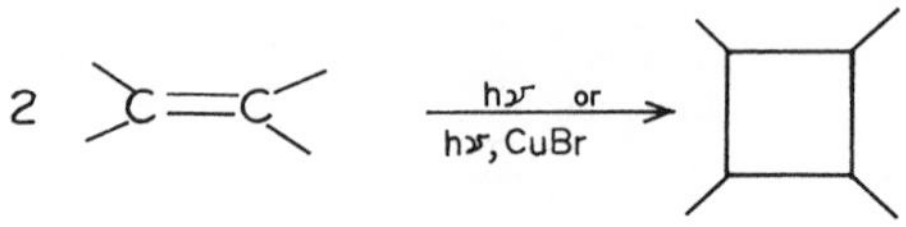

The norbornene dimer, for example, has been prepared in 37% yield by irradiation of 328 g of norbornene for 142 hr with periodic additions of CuBr [16]. Other types of ring closures takes place with alkenes [17]:

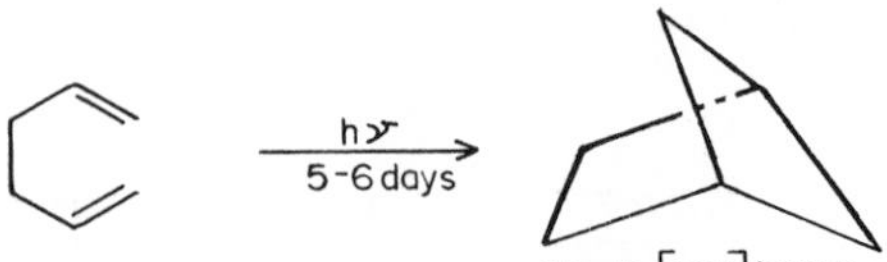

Bicyclo [2.2.1] hexane
10%

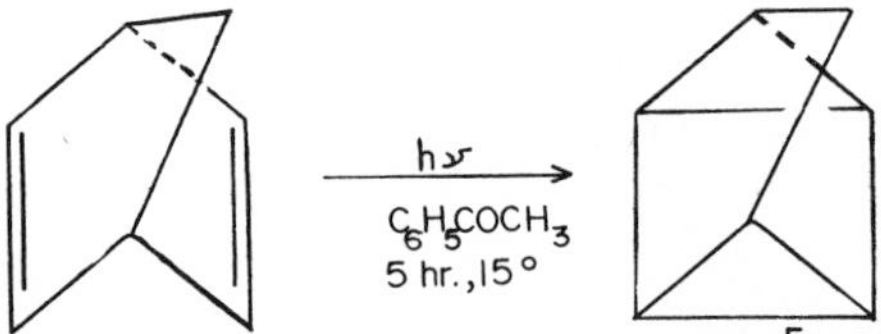

Tetracyclo[3.2.0.0^{2,7}.0^{4,6}]-
heptane, 57 %

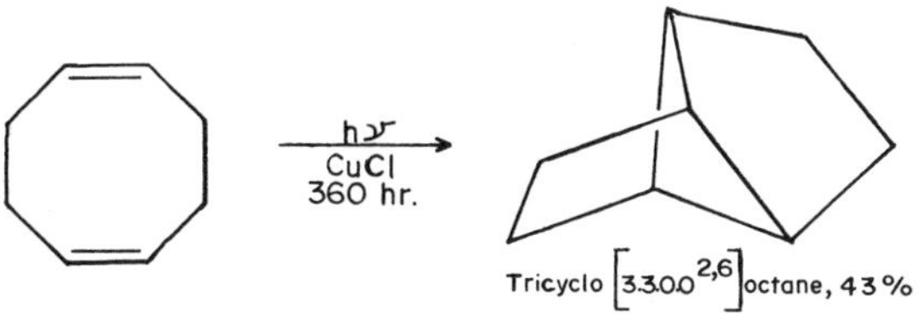
hν
CuCl
360 hr.
Tricyclo [3.3.0.0^2,6] octane, 43%

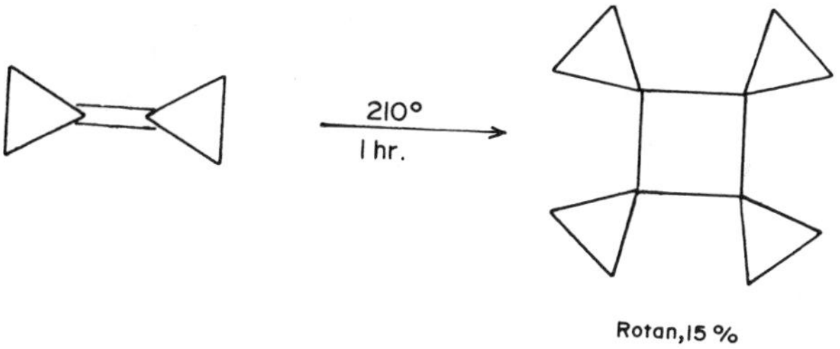
210°
1 hr.
Rotan, 15%

[18]

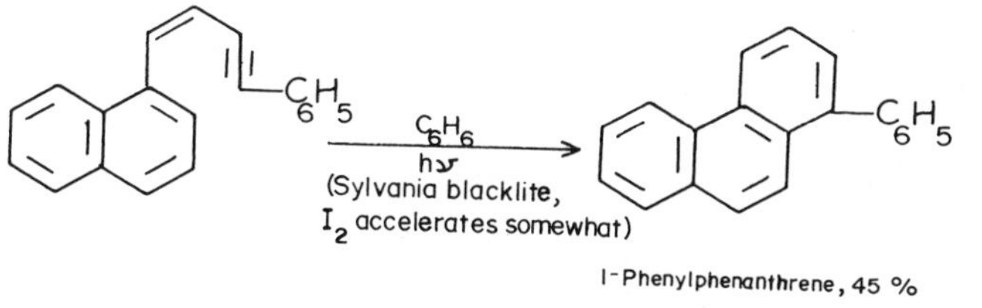
C6H5
C6H6
hν
(Sylvania blacklite,
I2 accelerates somewhat)
C6H5
1-Phenylphenanthrene, 45 %

[19]

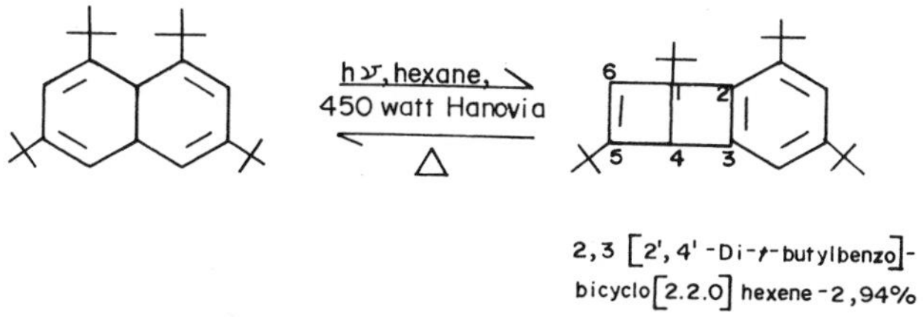
hν, hexane,
450 watt Hanovia
Δ
6
1
2
5
4
3
2,3 [2', 4' -Di-t-butylbenzo]-
bicyclo[2.2.0] hexene-2, 94%

[20]

Cuprous chloride definitely catalyzes the dimerization process, but it has the unfortunate properties of being unstable and of depositing on the walls of the flask. Cuprous triflate, $CuOSO_2CF_3$, is far superior as a catalyst, forms complexes with olefins, and in some cases more than doubles the yield of dimer [21].

In the photodehydrocyclization of stilbenes, Laarhoven and deJong [22] succeeded in preparing a new double helicene as indicated:

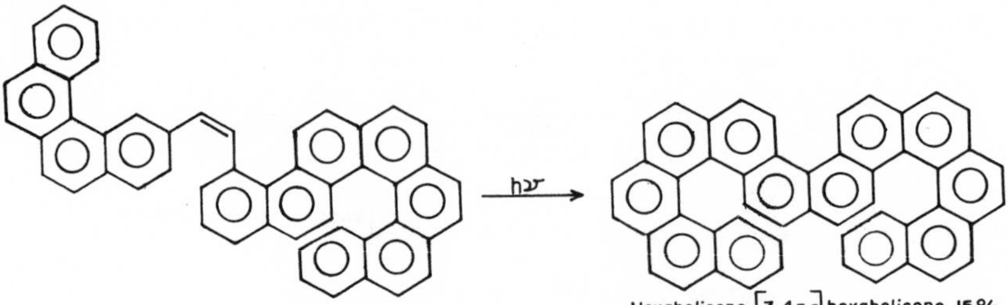

Hexaheliceno [3,4-c] hexahelicene, 15%

Azoalkanes occasionally are excellent sources of coupled hydrocarbons. An example is the formation of a bicyclopentane [22a]:

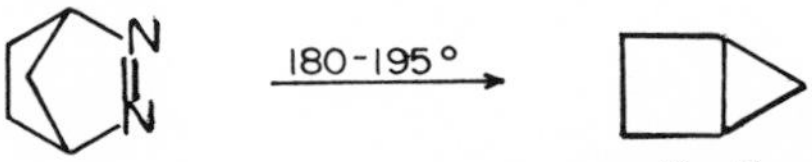

Bicyclo[2.1.0]pentane

The decomposition temperature is usually lower if a stable free radical is formed. Below is given a partial list of azo types with decomposition temperatures [23]:

> Azomethane, 300°; 2,2'-azopropane, 250°; 1,1'-diphenylazoethane, 110°; 1,1,1',1'-tetraphenylazomethane, 64°; azo-bis-isobutyronitrile, 80°

Many cyclobutanes have been synthesized from cyclic azo compounds or from other species that eliminate a stable molecule as shown [24]:

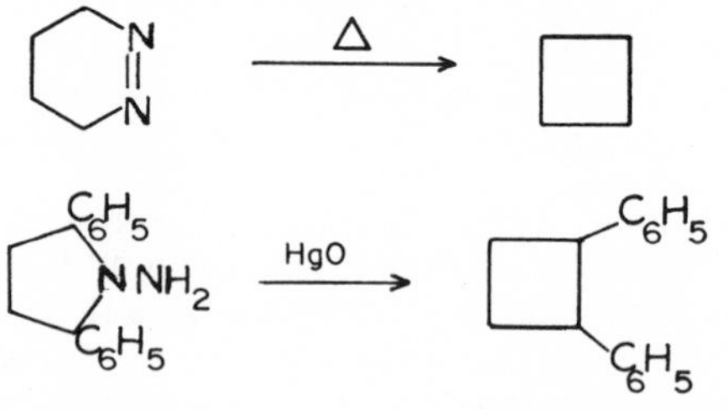

cis and trans

cis and trans-1,2-Diphenyl-cyclobutane, 60-80%

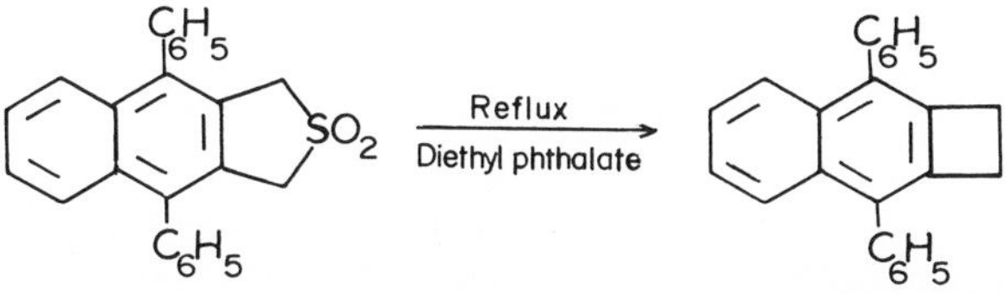

3,8-Diphenylnaphtho[b]-
cyclobutane, 77%   [25]

An azoalkane decomposition has served as an intermediate reaction in the preparation of an azulene [26].

Although aliphatic diazonium salts are not isolable, they must be intermediates in the synthesis of some hydrocarbons [27]:

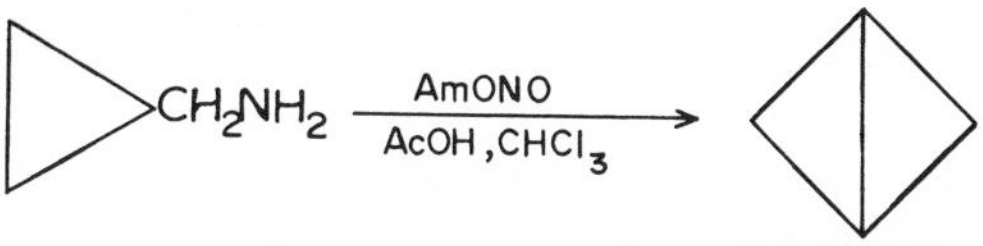

Bicyclobutane, 15%

Diazoalkane decompositions are discussed in G.6.

5. From Decarbonylation of Carbonyl Compounds (1, 61, H.-W., 5, Pt. la, 1970, p. 338)

$$RCHO \longrightarrow CO + [R\cdot] \xrightarrow[\substack{or \\ RCHO}]{\text{Hydrogen source}} RH$$

A recent review of decarbonylation is available [28]. The reaction may be carried out by the addition of a peroxide

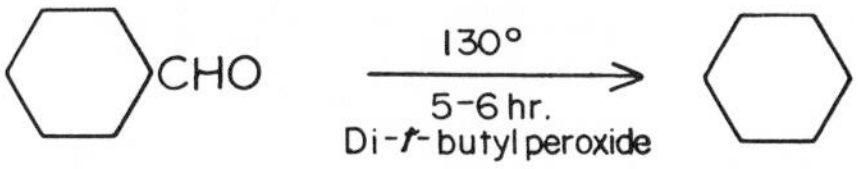

or by irradiation

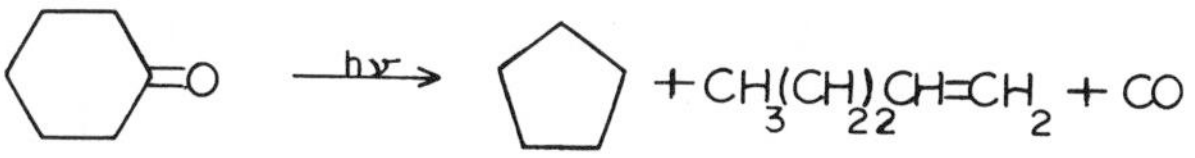

or, most expeditiously, via a palladium or rhodium complex:

$$RCHO + RhCl\left[P(C_6H_5)_3\right]_3 \xrightarrow[C_6H_6]{25°} RH + RhCl(CO)\left[P(C_6H_5)_3\right]_2 + P(C_6H_5)_3$$

A modified complex, $RhCl(CO)[P(C_6H_5)_3]_2$, is effective in the de-carbonylation process [29]. The use of a nitrile as a solvent prevents dimer formation of the rhodium complex and permits de-carbonylation of hindered aldehydes. Some alkene may be formed during decarbonylation.

The reaction proceeds with retention of configuration and the results with deuterated aldehydes indicate that the aldehydo hydrogen becomes attached to the R group to form the product [30]:

$$RCOD \longrightarrow RD + CO$$

Salicylaldehyde is converted into phenol and the cinnamaldehydes into styrenes with no skeletal rearrangements. Metallic palla-dium effects decarbonylation as shown (H.-W., 5, Pt. 1a, 1970, p. 479

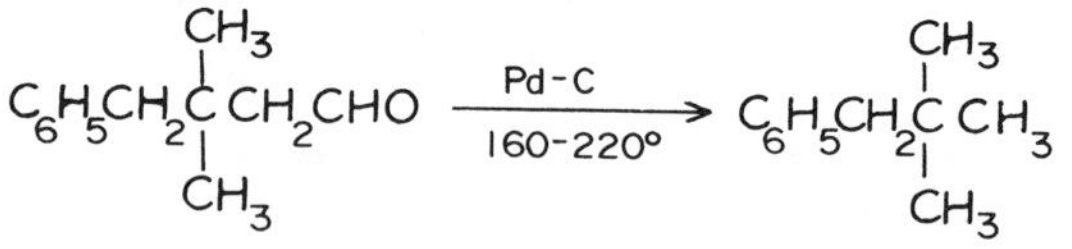

Neopentylbenzene, 90%

Tertiary α-aryl aldehydes as shown decarbonylate readily on irradiation [31]:

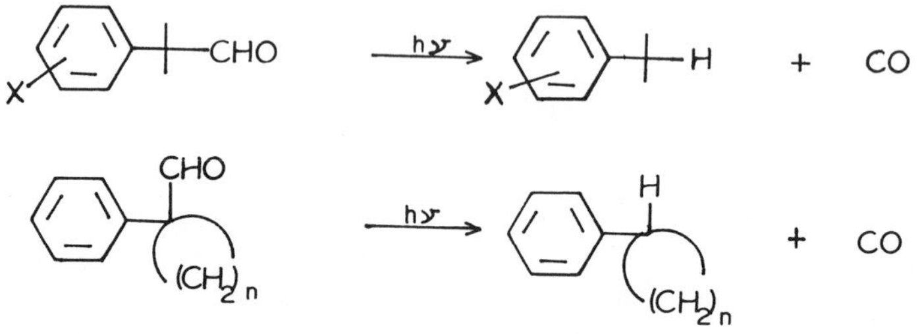

Sodamide occasionally effects decarbonylation perhaps by an ionic mechanism (H.-W., 5, Pt. 1a, 1970, p. 338):

$$(C_6H_5)_2C{=}O + NaNH_2 \xrightarrow{100\text{-}110°} C_6H_6 + C_6H_5CONH_2 \qquad (\underline{1}, 933)$$

NaNH2
Δ

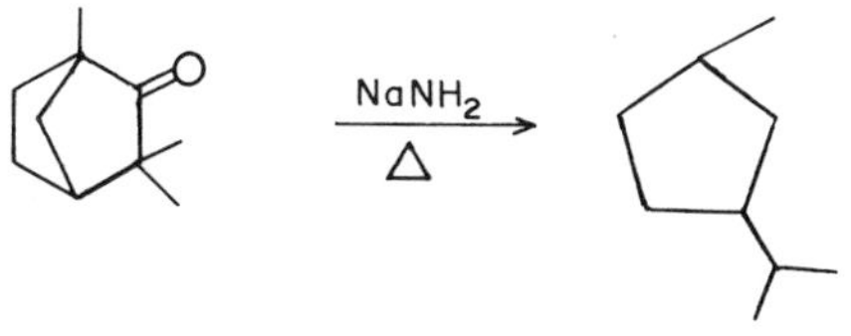

3-Methylisopropyl-
cyclopentane, 36%

a. Preparation of 9,10-Dihydro-9,10-ethanoanthracene [28]

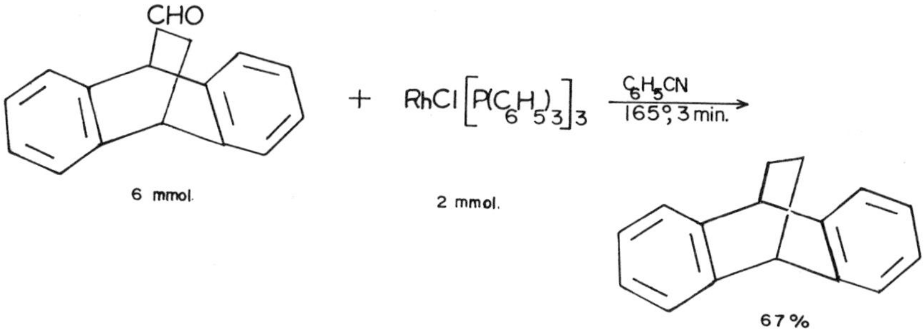

CHO
+ RhCl[P(C6H5)3]3
C6H5CN
165°, 3 min.
6 mmol.
2 mmol.
67%

6. From Carbenes or Carbenoids ($\underline{1}$, 62) [32]

[CH2:]
CH2
XCH2MR
+ RMX
CH2

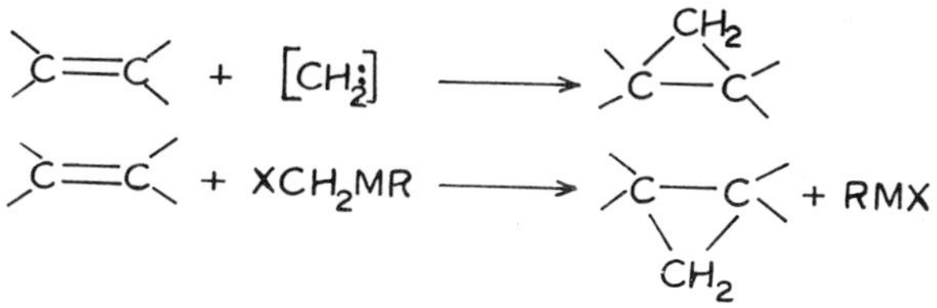

This subject has recently been reviewed [33,34].

The Simmons-Smith reaction, utilizing methylene iodide, a zinc-copper couple, and an alkene to produce a cyclopropane, has been one of the outstanding discoveries of recent times in organic synthesis [35]. Since 1958 few subjects have received greater attention in the literature. The original zinc-copper couple has proved to be a somewhat irreproducible reagent and a zinc dust and cuprous chloride combination has been found to be more effective [36] (see Ex. a). Furukawa [37] has prepared a new carbenoid from cadmium as follows:

$$Et_2Cd + CH_2I_2 \longrightarrow EtCdCH_2I + EtI$$

With this reagent and cyclohexene the yield of norcarane was 86%
using 0.1 mol of the olefin. It was pointed out in this paper
that the addition is not always exclusively <u>syn</u>. In particular,
some <u>anti</u> addition occurred with the carbenoid, EtCdCHICH$_3$. For
other <u>anti</u> additions see the article by Wittig and Jautelåt [38].
Indeed, the carbenoid seems quite sensitive to steric effects.
When one side of an olefinic bond is more hindered than the
other, the attack occurs on the least hindered side [39]. For
example, bicyclo [2.2.1] heptene gave only the single product as
shown:

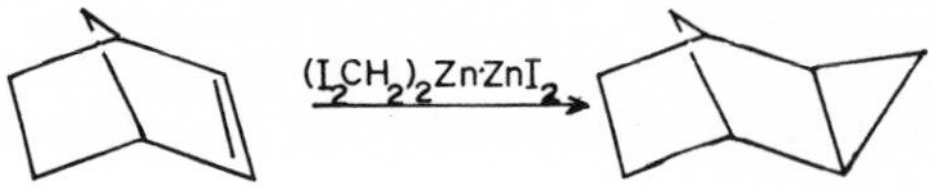

Sawada and Inouye [40] recommend the synthesis of a stock
solution of the carbenoid, EtZnCH$_2$I, prepared as follows: Ethyl
iodide (1 mol) and ZnCu (70 g, 1 g-atom) were stirred overnight
in 90 ml of dry ether. The supernatant liquid was used as stock.
Comparison of yields is given in Table 1.4.

Table 1.4. Comparison of Yields of Cyclopropanes with Three
Reagents

Product	Sawada	Simmons-Smith	Furukawa
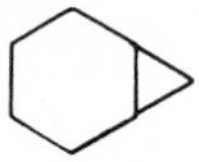	92	61	79
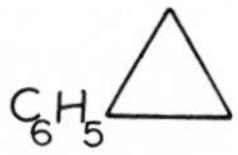	78	32	76
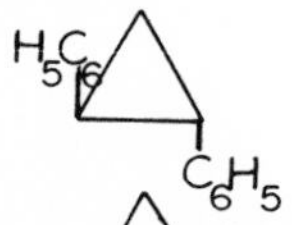	48	18	
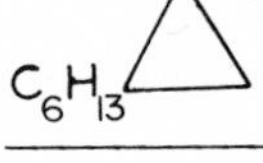	77	70	60

Although the Simmons-Smith reaction failed in the synthesis
of 2-cyclopropylpyridine, the compound (40%) was prepared by the
cyclopropanation of 2-vinylpyridine with dimethylsulfonium
methylide [40a].
Conia and co-workers [41] also have improved the yields of
cyclopropanes by using a Zn-Ag couple and avoiding a hydrolytic

workup.   In fact, the zinc salts are precipitated with pyridine. This couple is particularly effective in forming cyclopropanes from alkyl β-substituted, unsaturated carbonyl compounds as shown:

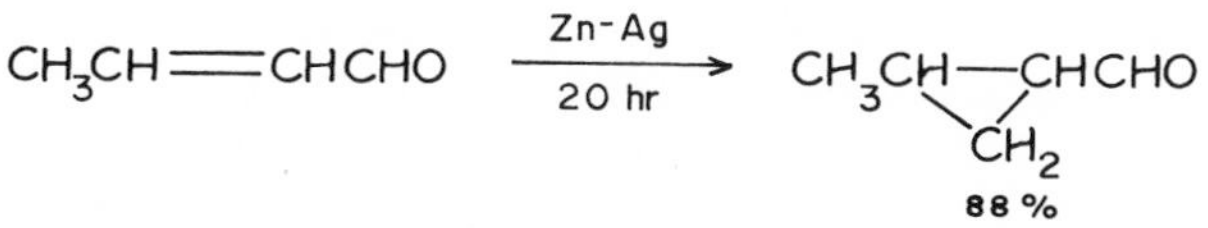

$$CH_3CH = CHCHO \xrightarrow[\text{20 hr}]{\text{Zn-Ag}} CH_3CH-CHCHO$$

$$\underset{CH_2}{\underset{|}{}}$$

88 %

Another modification utilizes the ease of reaction between vinylaluminum alkyls and carbenoids [42]:

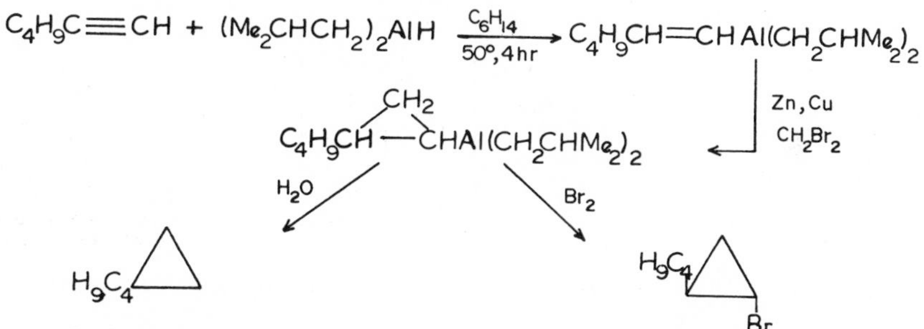

$$C_4H_9C \equiv CH + (Me_2CHCH_2)_2AlH \xrightarrow[\text{50°,4hr}]{C_6H_{14}} C_4H_9CH=CHAl(CH_2CHMe_2)_2$$

$$C_4H_9CH - CHAl(CH_2CHMe_2)_2 \xleftarrow[\text{CH}_2\text{Br}_2]{\text{Zn,Cu}}$$

Butylcyclopropane
73 % (glc)

trans-1-Bromo-2-butyl-
cyclopropane, 64% (glc)

Bis(iodomethyl)mercury and benzyl(iodomethyl)mercury are carbenoids that yield cyclopropanes with alkenes, but their expense and inaccessibility emphasize the advantages of the improved carbenoids previously described [43].   Interestingly, the studies on carbenoids have led to a possible link in mechanism with the Clemmensen reduction (A.1) since a reaction conducted under Clemmensen-like conditions has trapped the inter-mediate as a cyclopropane [44]:

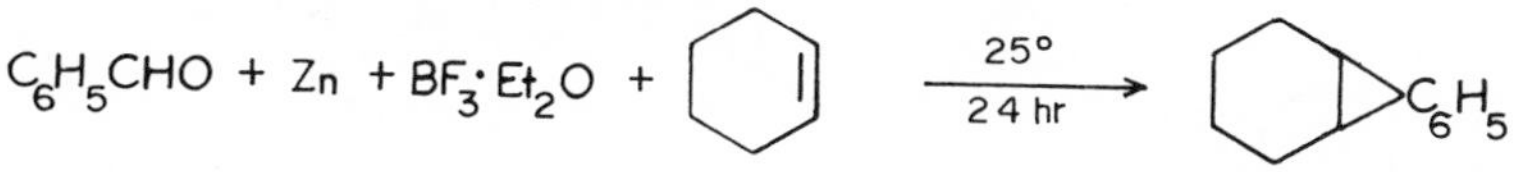

$$C_6H_5CHO + Zn + BF_3 \cdot Et_2O + \underset{}{\bigcirc} \xrightarrow[\text{24 hr}]{25°}$$

7-Phenylnorcarane, 35 %

Cyclopropanes have also been produced from olefins and the stable carbenoid, benzylmercuriodomethane [45]:

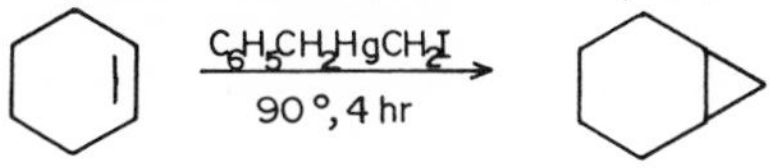

$$\underset{}{\bigcirc} \xrightarrow[\text{90°,4 hr}]{C_6H_5CH_2HgCH_2I}$$

Norcarane, 80%

Diazomethane in the presence of Pd(II) acetate [46] or copper
triflate [47] has been utilized as well in cyclopropane formation. The catalytic activity of the latter is attributed to
coordination of the olefins by the highly electrophilic copper(I)
species.

Less stable carbenoids may be prepared from the alkyllithium
and geminal halides at temperatures from -70 to -120° by using a
solvent comprised of 4 parts THF, 1 part ether, and 1 part petroleum ether [48]. Halocyclopropanes may be prepared from these
carbenoids and alkenes [49].

This type of insertion has led to the formation of naphthobicyclobutane [50] in quite sizeable yields:

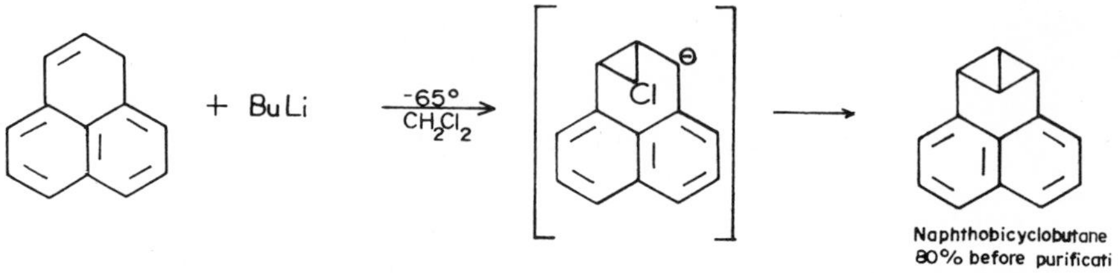

Naphthobicyclobutane
80% before purificati

The compound  made similarly from cyclopentadiene can be
produced on
a rather large scale but is explosive [51].

A carbenoid, made from cuprous oxide and alkylating agents
and stabilized by isocyanides, forms cyclopropanes with alkenes
[52]. However, the carbenoids thus produced usually carry a
functional group such as carbethoxy, and so form cyclopropane-
carbalkoxy esters rather than hydrocarbons.

A typical reaction of lithium carbenoids is as shown [53]:

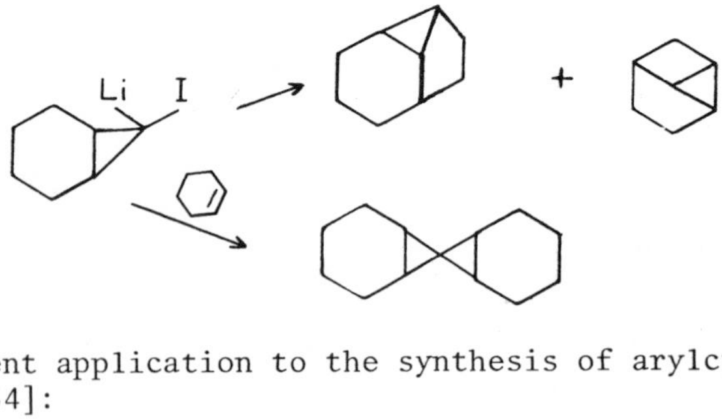

and a recent application to the synthesis of arylcyclopropanes
follows [54]:

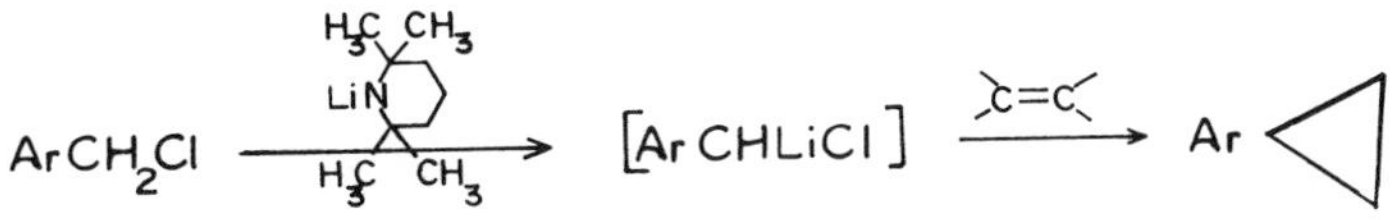

$$ArCH_2Cl \xrightarrow[H_3C \quad CH_3]{\underset{LiN}{H_3C \quad CH_3}} [Ar\,CHLiCl] \xrightarrow{\overset{}{C=C}} Ar\!\!\triangleleft$$

No reaction takes place with cyclohexene, vinylic esters, or vinylic ethers.

With allenes and carbenoids the major products are spiropentanes [55]:

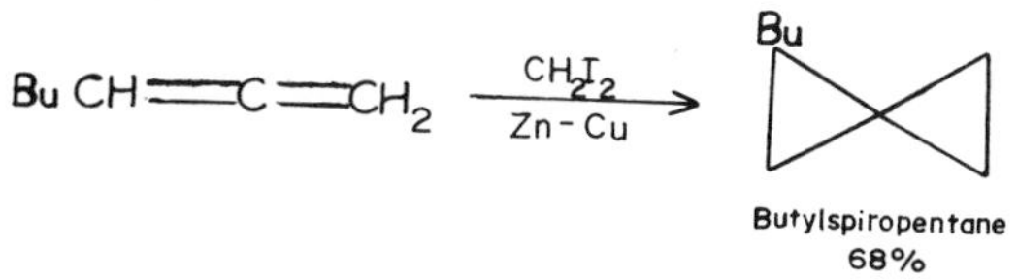

$$Bu\,CH=\!\!=\!\!C=\!\!=CH_2 \xrightarrow[Zn-Cu]{CH_2I_2} \text{(Bu—spiropentane)}$$

Butylspiropentane
68%

In favorable situations for insertion, as with compounds capable of transannular reactions, bicyclic types as shown are formed [67]:

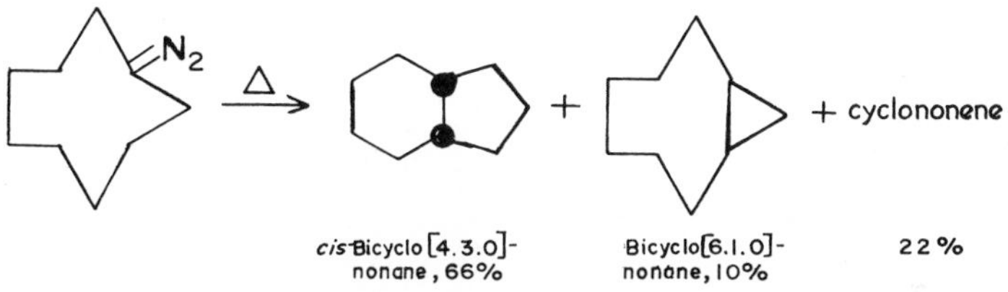

cis-Bicyclo[4.3.0]-
nonane, 66%   Bicyclo[6.1.0]-
nonane, 10%   22%   + cyclononene

With the irradiation of diazomethane and cyclohexene both addition and insertion take place [56]:

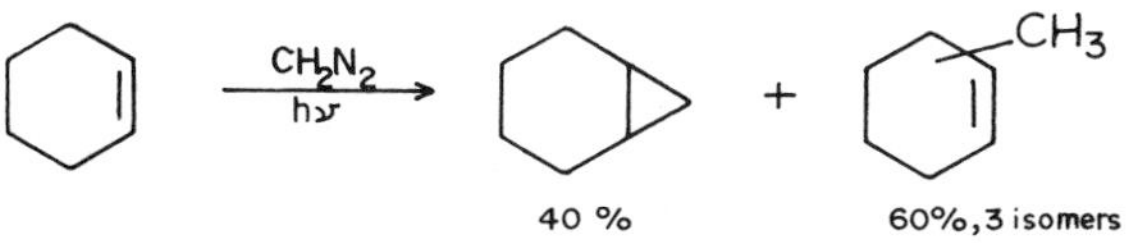

40 %   60%, 3 isomers

Diazo compounds are decomposed by heat, irradiation, or cuprous halides. It has been recommended that 0.14 mmol of CuI.P(OMe)$_3$ be used per 20 mmol of diazo compound [57], although this recommendation was obtained by extrapolation from work on α-diazoacetates. In preparing cyclopropanes from diazomethane and an alkene, great care must be taken to avoid explosions. It is recommended that the H.-W. directions be followed [57a].

Ylids appear to serve as a source of a carbene-like

intermediates. Furthermore the optically active form leads to a
direct asymmetric synthesis [58]:

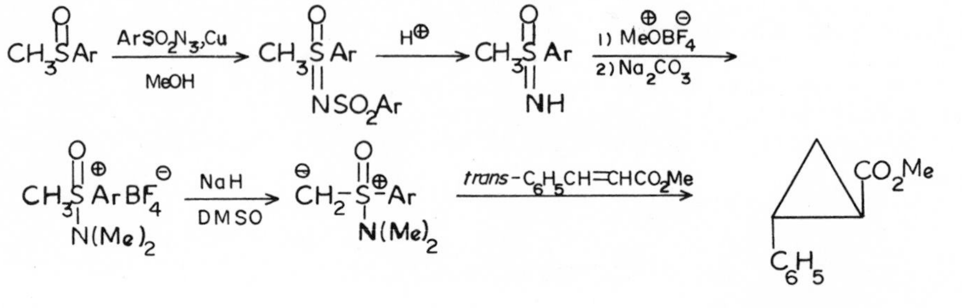

trans-1-Phenyl-2-carbo-
methoxycyclopropane, 76%
α+95°, 30% optically pure

This optical purity is high for a direct, asymmetric synthesis.
    The last example with a cyclopropane to olefin ratio of 55:
45 must involve a carbene or carbenoid of some sort [59]:

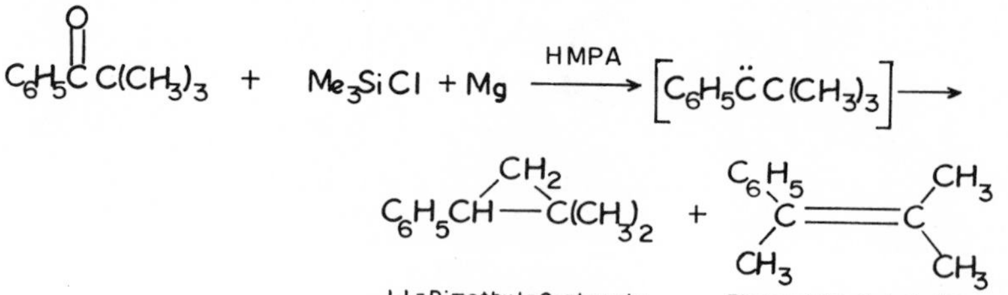

1,1-Dimethyl-2-phenyl-
cyclopropane

Phenyl trimethylethylene

a.   Preparation of Phenylcyclopropane [36]

    Zinc dust, 0.26 g-atom and cuprous chloride, 0.26 mol in 40
ml of ether were stirred and refluxed for 30 min under N_2. Styrene, 0.1 mol and then methylene iodide, 0.13 mol were added and
the mixture was refluxed for 24 hr. The usual workup led to a
69% yield of the cyclopropane.

b.   Preparation of 1,1-Diphenyl-2,2-dichlorocyclopropane [60]

    1,1-Diphenylethylene, 44 mmol cetyltrimethylammonium
chloride, 44 mmol and 11 ml of $CHCl_3$ were stirred at 50° and
NaOH, 337 mmol in 27 ml of water, was added in 15 min. After 2
hr, iced water, 50 ml, was added and after acidification with 10%

H_2SO_4, the mixture was extracted with ether. The residue recovered from the ether extract was crystallized from alcohol to give the dichlorocyclopropane, 94%. Although the product is a halohydrocarbon, its synthesis is included to show the interesting possibilities of cationic micellar cyclopropane formation. Phenanthrene gives the 9,10-dichloromethylene adduct, 79%.

7. Electrolysis of Alkali Metal Carboxylates (Kolbe) and Alkyl Halides (1, 65; H.-W., 5, Pt. 1a, 1970, p. 395)

$$2\ RCOO^{\ominus} \xrightarrow{\ 2e^{\ominus}\ } RR\ +\ 2CO_2$$

Reviews are available, one on mechanism [61], another on synthesis [62], and two others on mechanism and synthesis [63, 63a]. The second review states that optimum conditions include a pH of 3 to 5, low temperature (around 0°), and solution in a medium of high dielectric constant with sufficient alcohol or other miscible solvent to dissolve the substrate. Acids with α substituents and α, β-, and β, γ-unsaturated acids do not couple satisfactorily. Phenylacetic acid gives 50% bibenzyl; β-phenylpropionic acid, 37% of 1,4-diphenylbutane; and phenoxyacetic acid, 45% of 1,2-diphenoxyethane. Crossed Kolbe electrolyses may be conducted as shown:

$$C_6H_5CH_2O\overset{O}{\overset{||}{C}}(CH_2)_3COOH\ +\ MeO\overset{O}{\overset{||}{C}}CH_2CH_2COOH \xrightarrow[0.2\,amp., 24\,volts, 35°]{Pt\ elect., MeOH, C_5H_5N}$$

$$C_6H_5CH_2O\overset{O}{\overset{||}{C}}(CH_2)_3(CH_2)_2CO_2Me \xrightarrow{H_2, Pt} HO\overset{O}{\overset{||}{C}}(CH_2)_5\overset{O}{\overset{||}{C}}OMe$$

Monomethyl ester of
pimelic acid, 24%

The products of the electrolysis depend on the structure of the acid and the experimental conditions employed [62]. Factors such as solvent, composition of electrolyte, anode material, added compounds, current density, temperature, duration of reaction, and potential have been discussed by Wawzonek [63].

Among the best yields reported [63] are those for diesters as indicated:

$$2\ CH_3COO(CH_2)_5COOH \xrightarrow{electrolysis} CH_3COO(CH_2)_{10}OCOCH_3$$

Decamethylene diacetate, 83%

Grignard reagents may be electrolyzed in ether solution to

give approximately a 50% yield of coupled products [64].
Coupling of a Grignard reagent with styrene by electrolysis has
led to a substituted dodecane [63] as shown:

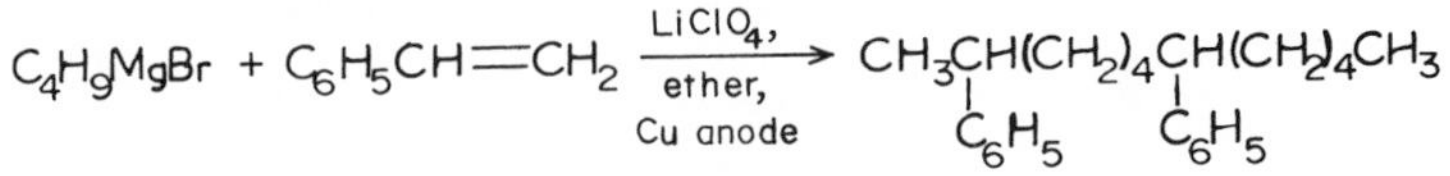

$$C_4H_9MgBr + C_6H_5CH\!=\!CH_2 \xrightarrow[\substack{\text{ether,} \\ \text{Cu anode}}]{\text{LiClO}_4,} CH_3\underset{\underset{C_6H_5}{|}}{CH}(CH_2)_4\underset{\underset{C_6H_5}{|}}{CH}(CH_2)_4CH_3$$

2,7-Diphenyldodecane, current eff. 30%

Some unusual hyarocarbons have been prepared by the electrolysis of alkyl halides. For instance, spiropentane has been
obtained in "high yield" as shown [64a]:

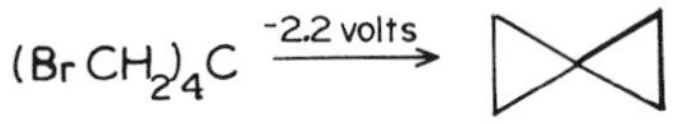

$$(BrCH_2)_4C \xrightarrow{-2.2\,\text{volts}} \bowtie$$

By controlled potential electrolysis, the intermediate cyclopro-

pane, $(BrCH_2)_2C\!\triangleleft$ may be isolated [64b].

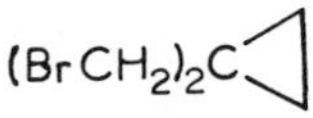

8. From Strained Hydrocarbons by Isomerization with Silver
    Tetrafluoroborate

A remarkable new isomerization reagent has been discovered
for certain strained hydrocarbons [65]. An example is given:

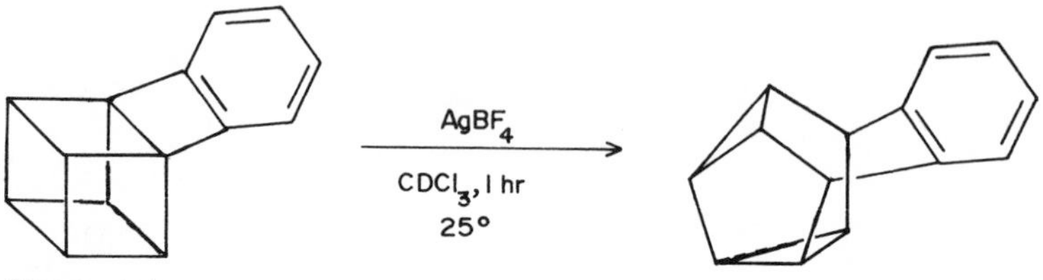

Benzobasketene

$$\xrightarrow[\substack{\text{CDCl}_3,\,1\,\text{hr} \\ 25°}]{\text{AgBF}_4}$$

Benzosnoutene, quantitative

It may be well to keep this reagent in mind for other isomerizations.

The following structures readily form hexamethylbenzene
with 5% silver perchlorate in chloroform [66]:

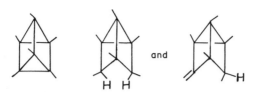

Addenda

G.6.   Oxygen accelerates the formation of the carbenoid in the synthesis from diethylzinc and methylenechloroiodide, but not the addition of the carbenoid to an alkene [S. Miyano et al., Bull. Chem. Soc. Jap., 45, 1946 (1972)].

G.6.   Carbenes prepared from benzyl halides and lithium 2,2,6,6-tetramethylpiperidide give cyclopropanes in good yields [R. A. Olofson, C. M. Dougherty, J. Am. Chem. Soc., 95, 581 (1973)].

References

1.   H.-W., 10, Pt. 3, 1965, pp. 151-163.
2.   A. H. Lewin, R. S. Michl, J. Org. Chem., 38, 1126 (1973).
3.   Ref. 1, p. 162.
4.   M. Kobayashi et al., Bull. Chem. Soc. Jap., 43, 1131 (1970).
5.   E. A. Nodiff et al., J. Med. Chem., 14, 921 (1971).
6.   B. Chauncy, E. Gellert, Australian J. Chem., 22, 993 (1969).
7.   D. H. Hey, M. J. Perkins, Org. Syn., Coll. Vol. 5, 51 (1973).
8.   M. Julia, Acc. Chem. Res., 4, 386 (1971).
9.   J. C. Chottard, M. Julia, Tetrahedron, 28, 5615 (1972).
10.   P. G. Sammes, Synthesis, 636 (1970).
11.   F. R. Stermitz, in O. L. Chapman, Organic Photochemistry, Dekker, New York, 1967, Vol. 1, p. 247.
12.   F. B. Mallory et al., J. Am. Chem. Soc., 86, 3094 (1964); C. S. Wood, F. B. Mallory, J. Org. Chem., 29, 3373 (1964); F. B. Mallory, C. S. Wood, Org. Syn., Coll. Vol. 5, 952 (1973).
13.   T. Sato et al., Bull. Chem. Soc. Jap., 42, 766 (1969).
14.   R. H. Martin et al., Helv. Chim. Acta, 54, 358 (1971).
15.   E. C. Taylor et al., J. Am. Chem. Soc., 92, 6088 (1970).
16.   D. J. Trecker, R. S. Foote, in R. Srinivasan, T. D. Roberts, Organic Photochemical Syntheses, 1, 80 (1971). (The foreword of this book lists useful laboratory practices.

17. Ref. 16, pp. 31-101; H.-W., 4, Pt. 4, 1971, p. 79.
18. P. LePerchec, J. M. Conia, Tetrahedron Letters, 1587 (1970).
19. A. H. A. Tinnemans, W. H. Laarhoven, Tetrahedron Letters, 817 (1973).
20. W. L. Mandella, R. W. Franck, J. Am. Chem. Soc., 95, 971 (1973).
21. R. G. Salmon, J. K. Kochi, Tetrahedron Letters, 2529 (1973).
22. W. H. Laarhoven, M. H. deJong, Rec. Trav. Chem., 92, 651 (1973).
22a. P. G. Gassman, K. T. Mansfield, Org. Syn., Coll., Vol. 5, 96 (1973).
23. H.-W., 10, Pt. 2, 1967, pp. 790-801.
24. H.-W., 4, Pt. 4, 1971, p. 42.
25. M. P. Cava, J. P. van Meter, J. Org. Chem., 34, 538 (1969).
26. P. G. Gassman, W. J. Greenlee, Syn. Commun., 2, 395 (1972).
27. L. Friedman, in Olah and Schleyer, Carbonium Ions II, Wiley-Interscience, New York, 1970, p. 655.
28. J. Tsuji, K. Ohno, Synthesis, 157 (1969).
29. K. Ohno, J. Tsuji, J. Am. Chem. Soc., 90, 99 (1968).
30. H. M. Walborsky, L. E. Allen, J. Am. Chem. Soc., 93, 5465 (1971).
31. H. Küntzel et al., Helv. Chim. Acta, 54, 868 (1971).
32. H.-W., 4, Pt. 3, 1971, pp. 98-150.
33. H. E. Simmons et al., Org. Reactions, 20, 1 (1973).
34. M. Jones, Jr., R. A. Moss, Carbenes, Wiley, New York, 1973; Ref. 32.
35. H. E. Simmons, R. D. Smith, J. Am. Chem. Soc., 80, 5323 (1958).
36. R. J. Rawson, I. T. Harrison, J. Org. Chem., 35, 2057 (1970); E. LeGoff, ibid., 29, 2048 (1964).
37. J. Furukawa et al., Tetrahedron, 26, 243 (1970).
38. G. Wittig, M. Jautelat, Ann. Chem., 702, 24 (1967).
39. H. E. Simmons et al., J. Am. Chem. Soc., 86, 1347 (1964).
40. S. Sawada, Y. Inouye, Bull. Soc. Chem. Jap., 42, 2669 (1969).
40a. R. Levine, G. R. Patrick, J. Org. Chem., 38, 3942 (1973).
41. J. M. Conia et al., Synthesis, 549 (1972).
42. G. Zweifel et al., J. Am. Chem. Soc., 93, 1305 (1971).
43. D. Seyferth et al., J. Am. Chem. Soc., 91, 5027 (1969); J. Organomet. Chem., 39, C41 (1972).
44. I. Elphimoff-Felkin, P. Sarda, Chem. Commun., 1065 (1969).
45. R. Scheffold, U. Michel, Angew. Chem. Intern. Ed. Engl., 11, 231 (1972).
46. R. Paulissen et al., Tetrahedron Letters, 1465 (1972).
47. R. G. Salomon, J. K. Kochi, J. Am. Chem. Soc., 95, 3300 (1973).

48. G. Köbrich, Angew. Chem. Intern. Ed. Engl., 11, 473 (1972).
49. Ref. 32, p. 217.
50. R. M. Pagni, C. R. Watson, Tetrahedron Letters, 59 (1973).
51. T. J. Katz et al., J. Am. Chem. Soc., 93, 3782 (1971).
52. T. Saegusa et al., J. Org. Chem., 38, 2319 (1973).
53. R. A. Moss, Chem. Eng. News, June 30, 1969, p. 50.
54. R. A. Olofson, C. M. Dougherty, J. Am. Chem. Soc., 95, 581
     (1973).
55. P. Battioni et al., Bull. Soc. Chim. France, 3942 (1970)
     and earlier papers.
56. H.-W., 4, Pt. 3, 1971, p. 105.
57. B. W. Peace, D. S. Wulfman, Tetrahedron Letters, 3799
     (1971).
57a. Ref. 56, p. 110.
58. C. R. Johnson, C. W. Schroeck, J. Am. Chem. Soc., 90, 6852
     (1968).
59. J.-P. Pillot et al., Bull. Chem. Soc. France, 3490 (1972).
60. G. C. Joshi et al., Tetrahedron Letters, 1461 (1972).
61. H. K. Vijh, B. E. Conway, Chem. Rev., 67, 623 (1967).
62. G. E. Svadkovskaya, S. A. Voitkevich, Russ. Chem. Rev., 29,
     161 (1960).
63. S. Wawzonek, Synthesis, 285 (1971).
63a. M. M. Baizer, Organic Electrochemistry, Dekker, New York,
     1973, pp. 469-507.
64. H.-W., 5, Pt. 1a, 1970, p. 398.
64a. M. R. Rifi, J. Org. Chem., 36, 2017 (1971).
64b. M. R. Rifi, Org. Syn., 52, 22 (1972).
65. L. A. Paquette, Acc. Chem. Res., 4, 280 (1971); L. A.
     Paquette, G. Zon, J. Am. Chem. Soc., 96, 203 (1974).
66. H. Hogeveen, J. Thio, Tetrahedron Letters, 3463 (1973).
67. H.-W., 5, Pt. 1a, 1970, p. 331.

H. Cycloaddition (1, 67)

The catalyst $Ni(CO_2)_2[P(C_6H_5)_3]_2$ continues to be used for aromatizing acetylenes. Thus dialkynes have been cyclized as shown [1]:

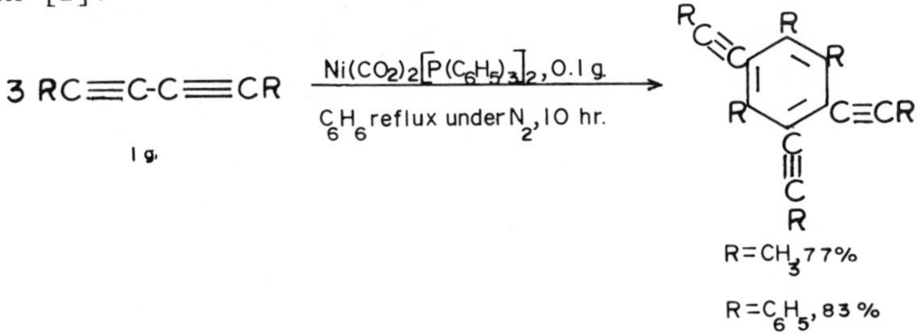

$R=CH_3, 77\%$

$R=C_6H_5, 83\%$

Modified catalysts such as

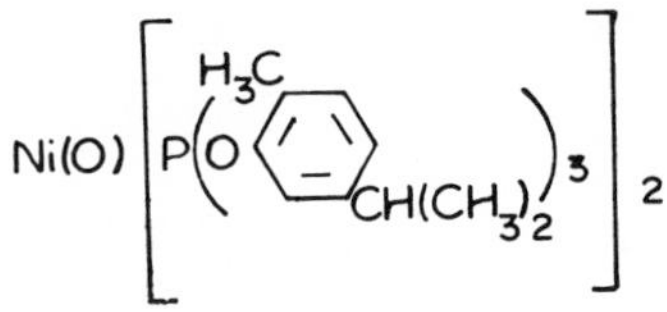

$$\text{Ni(0)} \left[P \left(O - \underset{\underset{CH(CH_3)_2}{\overset{H_3C}{\diagdown}}}{\diagup} \right)_3 \right]_2$$

[2] and $CoCl_2$-Mn in CH_3CN [3] have been recommended. The latter
is simple and usually gives 1,2,4-substituted benzenes in high
yield. Thus phenylacetylene trimerizes as shown:

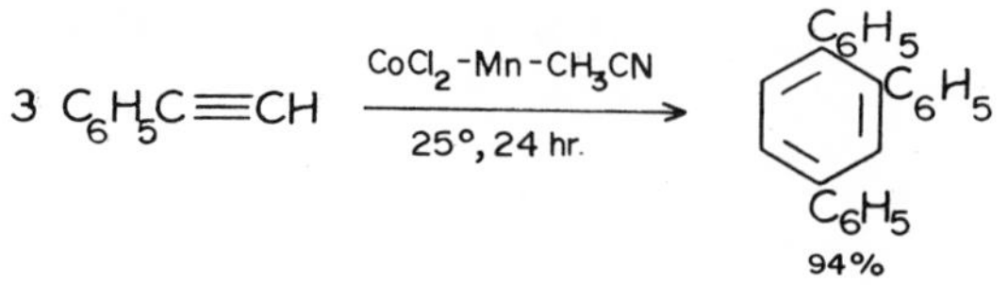

$$3\ C_6H_5C\equiv CH \xrightarrow[25°,24\ hr.]{CoCl_2\text{-Mn-}CH_3CN}$$

94%

Cyclization also has been brought about with 1,8-nonadiyne
[4]:

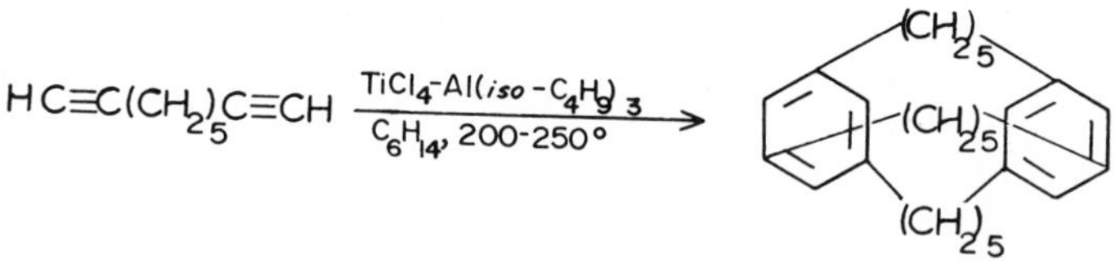

$$HC\equiv C(CH_2)_5C\equiv CH \xrightarrow[C_6H_{14},\ 200\text{-}250°]{TiCl_4\text{-Al}(iso-C_4H_9)_3}$$

1,1',3,3',5,5'- *tris* -(pentamethylene)-
dibenzene

A benzyne procedure has been devised for the synthesis of a
series of triptycenes [5]:

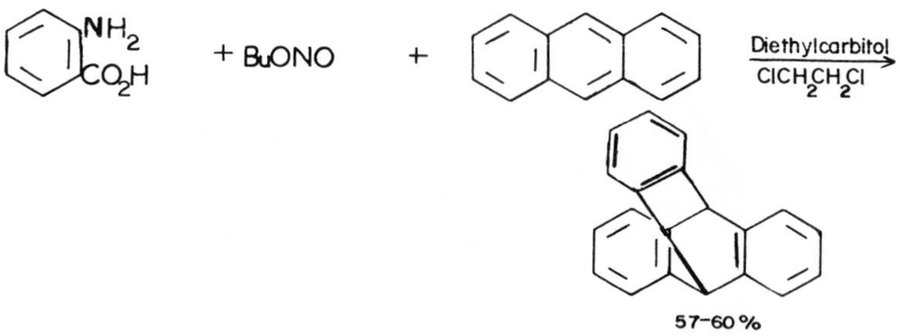

+ BuONO + $\xrightarrow[CICH_2CH_2Cl]{Diethylcarbitol}$

57-60%

Yields are usually somewhat lower if substituted anthranilic acid
is employed.

Hart has used the dienone, readily available from the oxidation of hexamethylbenzene, to condense with benzynes, the adduct of which can be converted into a naphthalene [6].

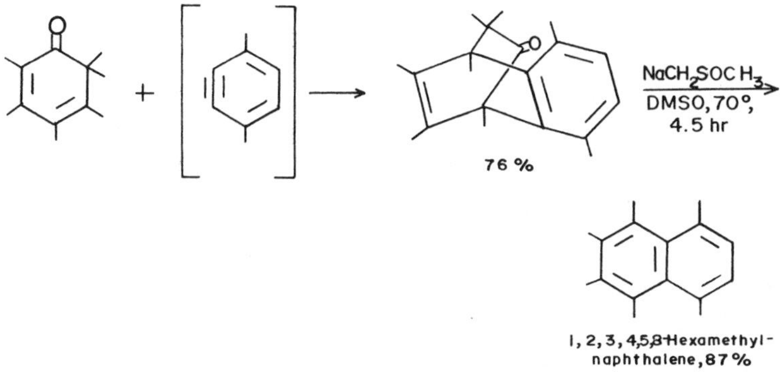

76 %

1,2,3,4,5,8-Hexamethyl-
naphthalene, 87 %

A dimethylketene fragment and two hydrogen atoms are lost in the reverse Diels-Alder, the last step.

The cycloaddition of benzyne with cyclic olefins has recently received attention [7]. Three types of cycloaddition known as 2 + 2, 2 + 4, and ene are possible. Thus with 1,3-cycloheptadiene, the addition product is largely 2 + 4:

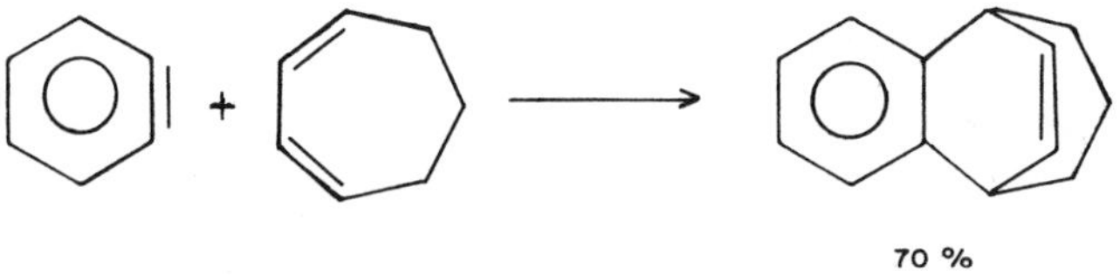

70 %

With 1,3,5-cycloheptatriene, on the other hand, the products are exclusively 2 + 2 and ene:

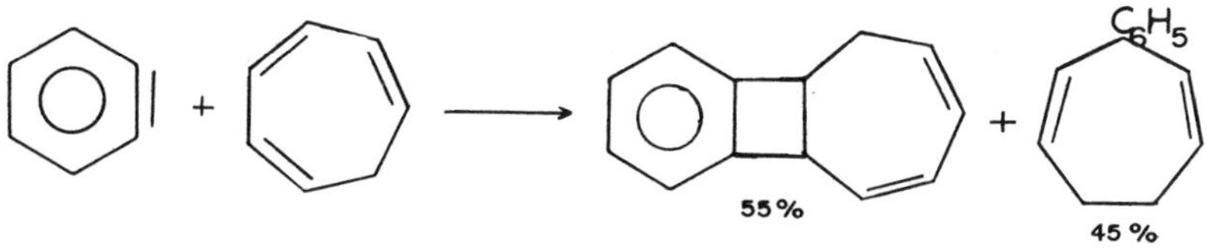

55%

45%

The amounts of the products formed are altered decidedly by the presence of catalytic amounts of Ag^+. In this case cyclic six- and seven-membered di- and triolefins give almost exclusive formation of 2 + 4 type products, although with acyclic systems the catalyst exhibits little or no effect.

Allyl-substituted heterocycles such as benzothiophene or benzofuran may be converted into the benzo derivative of the heterocycle as shown [8]:

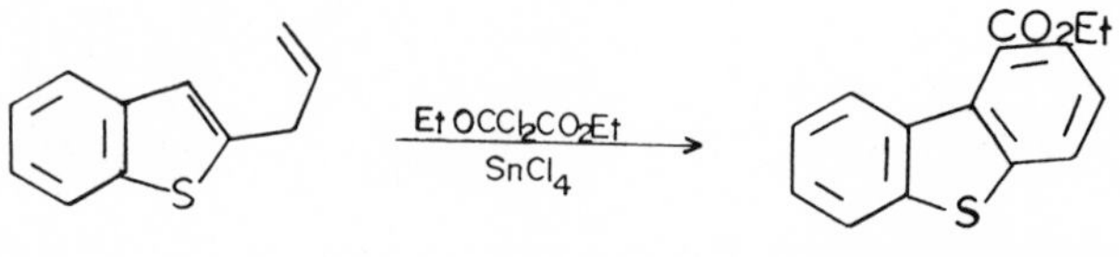

Ethyl dibenzothiophene-
1-carboxylate, 61%

The method represents a one-step process for synthesizing dibenzothiophenes or dibenzofurans and of course requires an aromatic system in which substitution ortho to the allylic group is possible.

Cyclopropanes have been produced by the reaction between diphenylsulfonium isopropylide and diene esters [9]:

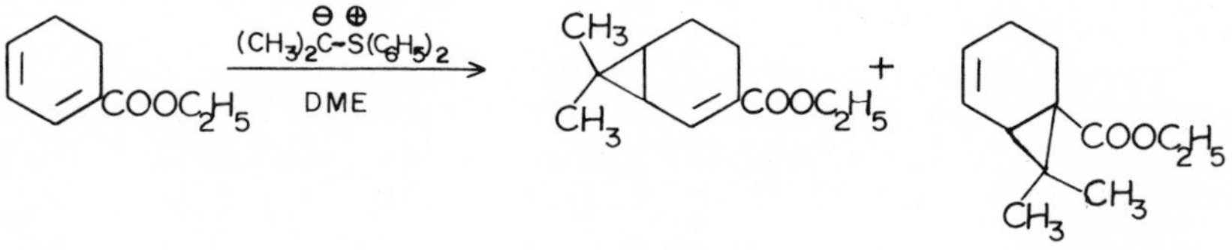

A   B
4   :   1 (total yield, 70%)

The ratio of A to B varies with the solvent used, while in the acylic system addition to the γ,δ double bond occurs regardless of the solvent.

An interesting azulene synthesis has appeared [10]:

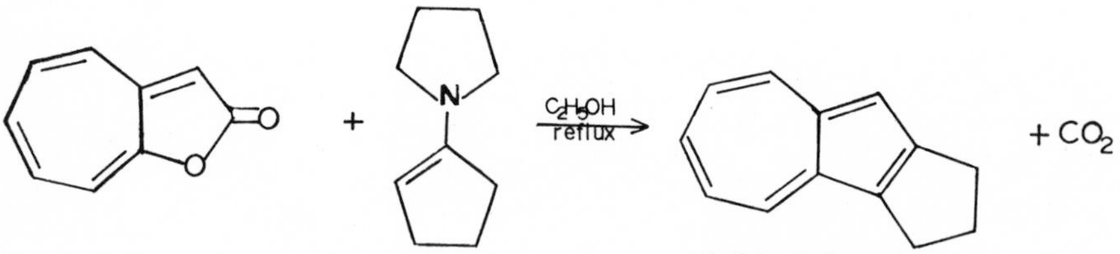

1,2-Trimethyleneazulene
"excellent yield"

The enamine probably adds to the diene system of the lactone and forms the azulene by carbon dioxide elimination.

Addendum

H. The preparation of cyclobutanes by photochemical, metal-
catalyzed, and cation-radical-induced dimerization of monoolefins
has been reviewed [L. J. Kricka, A. Ledwith, Synthesis, 539
(1974)].

References

1. A. J. Chalk, R. A. Jerussi, Tetrahedron Letters, 61 (1972).
2. W. Reppe et al., Angew. Chem. Intern. Ed. Engl., 8, 727
    (1969).
3. G. Agnes, G. Cometti, Organomet. Chem. Syn., 1, 185 (1970/
    1971).
4. V. Yu Vasil'ev, V. O. Reikhsfel'd, Tr. Khim.-Met. Inst.
    Akad. Nauk Kaz. SSR., 18, 29 (1972); C. A., 79, 18461
    (1973).
5. L. Friedman, F. M. Logullo, J. Org. Chem., 34, 3089 (1969).
6. H. Hart et al., J. Am. Chem. Soc., 89, 4554 (1967).
7. P. Crews, J. Beard, J. Org. Chem., 38, 522, 529 (1973).
8. O. Meth-Cohn et al., Chem. Commun., 1251 (1971).
9. C. S. F. Tang, H. Rapoport, J. Org. Chem., 38, 2806 (1973).
10. K. Takase et al., Tetrahedron Letters, 4275 (1971).

Chapter 2

ALKENES, CYCLOALKENES, AND DIENES

One outstanding and comprehensive review on the synthesis of alkenes has been published [1] and is designated here as H.-W.,
followed by volume and page number for many subsections. The
material found in that book is mainly in addition to the discussion in the present work. Other references of interest include the syntheses of cyclopropene [2] and of cyclobutene and
cyclobutadiene [3]. Perhaps of more than average interest is the
article on stereoselective olefin syntheses covering all methods
[4]. Eliminations restricted to alicyclic compounds have been

compiled in a continuing series on alicyclic compounds [5]. Some
of the chapters in the Patai series have a bearing on specific
syntheses, namely, gas phase elimination [6], condensation (to
prepare unsaturated carbonyl compounds) [7], rearrangements [8,9],
alkenes with radicals and carbenes [10], cycloaddition [11], conjugated dienes [12], cumulenes [13], complexes with transition
metals [14], photochemistry [15], olefinic properties of cyclopropanes [16], and polymers with olefin groups [17].

References

1. H.-W., Methoden der Organischen Chemie, 5, Pt. 1b, Verlag,
    Stuttgart, 1972.
2. H.-W., 4, Pt. 3, 1971, p. 679.
3. H.-W., 4, Pt. 4, 1971, pp. 77, 231.
4. A. S. Arora, I. K. Ugi, Ref. 1, p. 728.
5. N. A. LeBel in H. Hart, G. J. Karabatsos, Advances in
    Alicyclic Chemistry, Vol. 3, Academic, New York, 1971, p. 195.
6. S. Patai, Chemistry of Alkenes, Vol. I, Interscience, New
    York, 1964, p. 203.
7. Ref. 6, p. 241.
8. Ref. 6, p. 387.
9. S. Patai, Chemistry of Alkenes, Vol. II, Interscience, New
    York, 1970, p. 115.
10. Ref. 6, p. 585.
11. Ref. 6, p. 739.
12. Ref. 6, p. 955.
13. Ref. 6, p. 1025.
14. Ref. 9, p. 215.
15. Ref. 9, p. 267.
16. Ref. 9, p. 511.
17. Ref. 9, p. 411.

Among new reagents for the dehydration of alcohols to alkenes is
hexamethylphosphoric triamide (HMPA) at high temperature (A.1).
And a surprising new reagent for dehydrohalogenation is the weak,
soft base, tetrabutylammonium bromide in lutidine and acetone
(A.2). It appears to maximize the E_2 character of elimination
so that the alkene product is more trans and more Saytzeff-
oriented than with any other base. The effective bicyclic bases,
DBU and DBN, now have the benefit of a review (A.2). Sodium
thiosulfate is a new reagent for elimination from vic-dihalides
(A.3), whereas an acid halide in pyridine seems a satisfactory
reagent for elimination from iodohydrins (A.5). More satisfactory than xanthates in the Chugaev elimination are the esters of
N,N-dimethylthiocarbamic acid (A.7). Yields in the preparation

of alkenes from quaternary ammonium bases have been improved by carrying out the pyrolysis under vacuum to remove the water quickly (A.8). A surprising elimination is dehydroamidation (A.9), having some characteristics of a reverse Ritter reaction. Either the cis- or the trans-alkene may be obtained from the acetylene compound via boranes or alanes (B.1). Many other unique eliminations to be used in specific instances are described (A.17 to A.25) including, for example, elimination from a vic-dinitro compound with Na_2S in DMF (A.22). An unstable olefin may be stored as its charge-transfer complex and regenerated when desired (A.24).

The reduction of acetylenes to ethylenes by catalytic means benefits from a recent review (B.1). The reduction by chemical means has become more versatile (B.1). In the subsection on reduction of multienes, the unusual but stable catalyst platinum chloride dissolved in tetraethylammonium trichlorostannite (B.2) is cited.

The Woodward-Hoffman rules have been sketched in the Diels-Alder subsection and the "ene" reaction has been more clearly defined (C.2). Organometallic syntheses have expanded considerably in the field of alkylation (C.7), in the use of lithium dialkyl-cuprates (C.9), in arylation through the use of palladium salts (C.10), and in the addition of organometallic compounds to alkenes (C.11).

The isomerization of alkenes from cis to trans can be done stepwise, chemically (D.1.b) or photochemically (D.1.d). The scope of the latter has been been enlarged by conducting the irradiation in the presence of a complexing agent (D.1.d). The Wittig olefination process perhaps may be considered the most useful and general modern method of preparing olefins. Modifications continue to emerge including in situ preparations, the use of phase transfer conditions, and coupling by means of oxidation (E.2). The Simmons-Smith reagent (E.3), the α-metalated isocyanides (E.4), and the anions from boranes and silanes (E.5) may be substituted for the ylid in the Wittig reaction.

A review of stereoselective and stereospecific olefin synthesis is available [1].

A. Elimination (1, 71)

1. From Alcohols (Dehydration) (1, 71)

$$RCH_2CHOHR \longrightarrow RCH{=\!=\!=}CHR + H_2O$$

The splitting of water from alcohols in both vapor [1a] and liquid states [1b] has been reviewed.

New dehydrating agents that have been utilized in the dehydration of alcohols are hexamethylphosphoric triamide [HMPA] [2], methyltriphenoxyphosphonium iodide [MTPI] in HMPA [3], sulfurane

[4], thionyl chloride-pyridine [5], iodine in propionic anhydride [6], methyl(carboxysulfamoyl)triethylammonium hydroxide inner salt [7], zinc chloride [8], lithium aluminum hydride-aluminum chloride [9] or sodium hydride-aluminum chloride [8] followed by hydrolysis, and thorium dioxide at 250° (for sec alcohols) [10]. In addition, reductive dehydration in one operation has been accomplished with sodium-ethanol-liquid ammonia [11], lithium aluminum hydride-ethyl ether [12], and pyridine-sulfur trioxide and lithium aluminum hydride [13].

Monson and Priest [2] were successful in dehydrating primary and secondary alcohols with HMPA at temperatures over 200° to give unrearranged alkenes. Primary alcohols give 1-alkenes (57-87%), while cyclic secondary alcohols lead to cyclic alkenes (95-100%). Of significance in this process is the fact that the reaction does not proceed by a carbonium ion mechanism. Rather, dimethylamine is produced. trans-1(α)Decalol gives 77% $\Delta^{1(9)}$-, 14% trans-Δ^1-, and 9%-Δ^9-octalin.

Hutchins and co-workers [3] found that methyltriphenoxyphosphonium iodide, MTPI, in HMPA was an effective reagent system for the selective dehydration of secondary alcohols under mild conditions. cis-4-t-Butylcyclohexanol gave an 84% yield of the alkene, while with the trans alcohol the yield was 88%. As a rule the Saytzeff is formed in preference to the Hofmann alkene. Primary alcohols stop at the iodide stage, while tertiary ones seem to be inert. Sulfurane, $(C_6H_5)_2S[OC(C_6H_5)(CF_3)_2]_2$, is very effective in the dehydration of secondary and tertiary carbinols at room temperature, generally with good yields [4]. The value of this reagent is shown by the fact that it was possible by its use to obtain the hitherto inaccessible unrearranged alkene from tricyclopropyl carbinol as illustrated:

32%

Lomas and co-workers [5] studied the dehydration of tertiary alcohols by the use of thionyl chloride-pyridine at 0°. Di-t-butylmethylcarbinol gave a 94% yield of 1,1-di-t-butylethylene (Ex. a), whereas di-t-butyl-t-butylmethylcarbinol led largely to the rearranged olefin as shown:

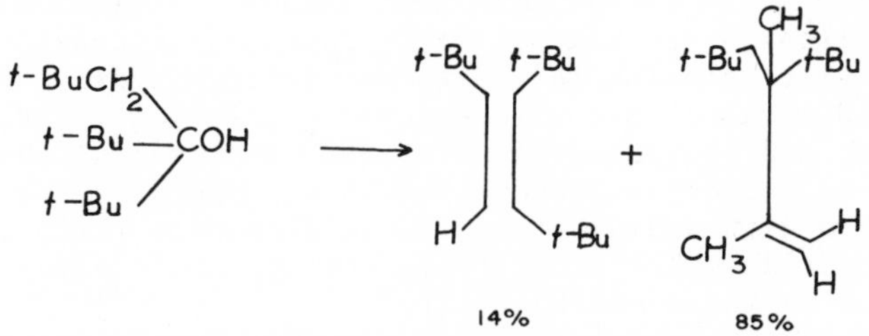

14%   85%

Reeve and Reichel [6] dehydrated 3,4-dimethyl-3,4-hexanediol with I_2 in $(CH_3CH_2CO)_2O$ and C_6H_5NCO as well as with other acidic reagents. The former favors the formation of cis,cis- and cis, trans-3,4-dimethyl-2,4-hexadiene (Saytzeff pathway), while the latter unexpectedly promotes the Hofmann pathway.

Burgess [7] utilized the methyl(carboxysulfamoyl)triethyl-ammonium hydroxide inner salt, $CH_3O\overset{O}{\overset{\|}{C}}\bar{N}SO_2\overset{+}{N}(C_2H_5)_3$, in the dehydration of 1,2-diphenylethanol (Ex. b). Secondary and tertiary alcohols gave olefins but primary ones gave only the urethane.

Billé-Samé and Bergmann [8], in the dehydration of 9,10-dihydro-9-anthracenecarbinols, produced an enlargement of the inner ring:

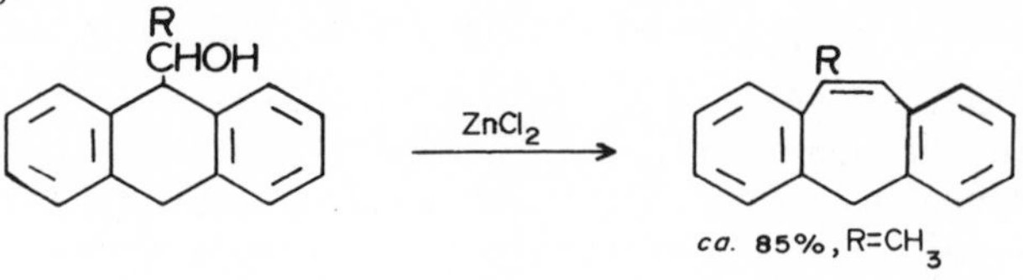

ca. 85%, R=CH$_3$

By the use of $LiAlH_4$-$AlCl_3$ or NaH-$AlCl_3$, Mead [9] obtained superior yields of alkenes from tertiary alcohols:

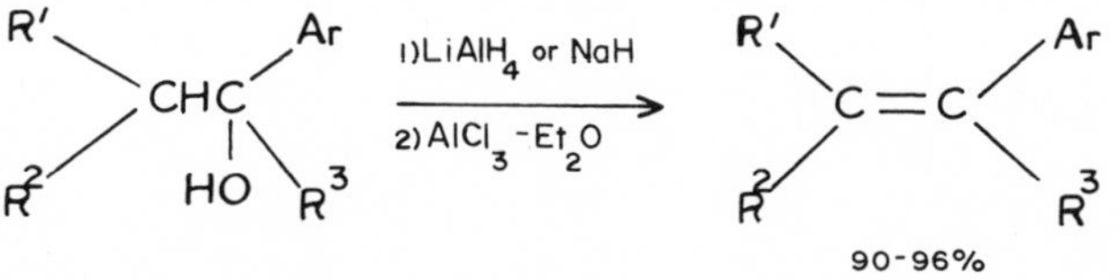

90-96%

In reductive-dehydration Kaiser [11] converted the methylene alcohol into the alkene as shown:

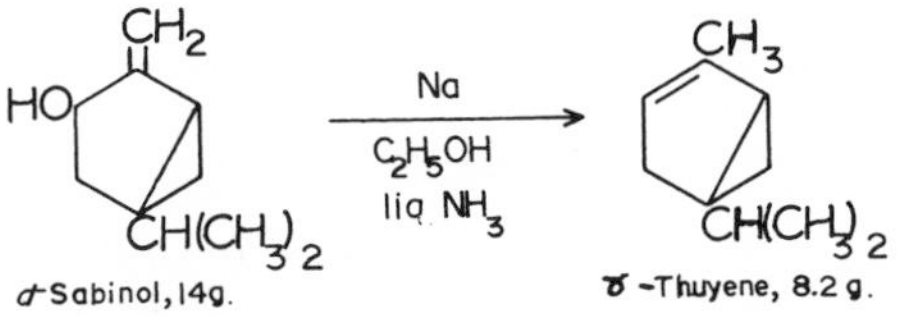

d-Sabinol, 14g.   ở -Thuyene, 8.2 g.

Claesson and Bogentoft [12] produced the 1,3-butadiene:

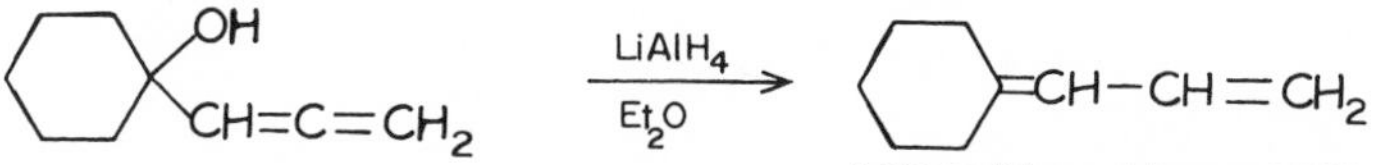

2-Propenylidenecyclohexane, 75 %

In the hands of Corey and Achiwa [13] geraniol, a diene-alcohol, gave a 98% yield of <u>trans</u>-2,6-dimethylocta-2,6-diene by treatment with pyridine-sulfur trioxide, followed, without isolation, by $LiAlH_4$.

a. Preparation of 1,1-Di-t-butylethylene [5] (94% from Di-t-butylmethylcarbinol and $SOCl_2$-C_5H_5N at 0°)

b. Preparation of Symmetrical Diphenylethylene [7] (95% from

1,2-Diphenylethanol and $CH_3O\overset{O}{\overset{\|}{C}}\bar{N}SO_2\overset{+}{N}(C_2H_5)_3$ in C_6H_6 at 50°)

   2. From Alkyl Halides (Dehydrohalogenation) (<u>1</u>, 75) [1a, p. 134]

$$RCH_2CHXR' \longrightarrow RCH = CHR'$$

Scharf and Laux dehydrochlorinated 2,2,3-trichlorobutane with KOH in CH_3OH as shown [14a]:

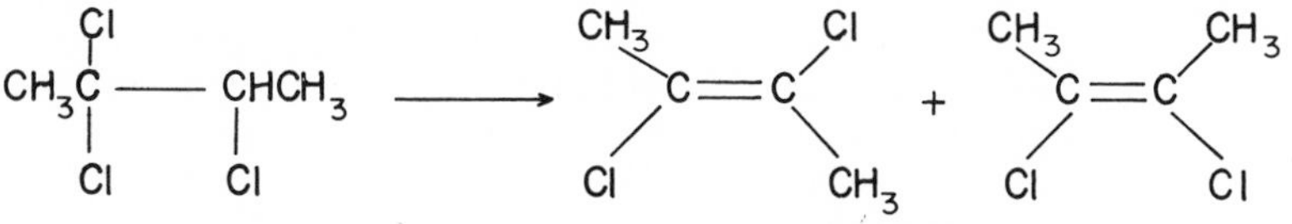

Total yield 70 %
trans:cis = 2:1

With no solvent the <u>trans</u>:<u>cis</u> ratio was 1:1. Said and Tipping [14b] employed the same reagent in the partial dehydrohalogenation of 1,2,3,4-tetrabromo-2,3-dimethylbutane as indicated:

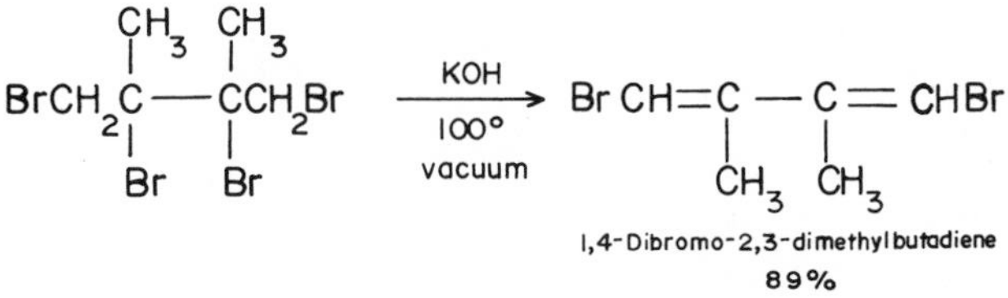

1,4-Dibromo-2,3-dimethylbutadiene
89%

The interesting properties of DBN and DBU now have been reviewed [15]; DBU, 1,8-diazabicyclo[5.4.0]-undec-7-ene at times gives substantially

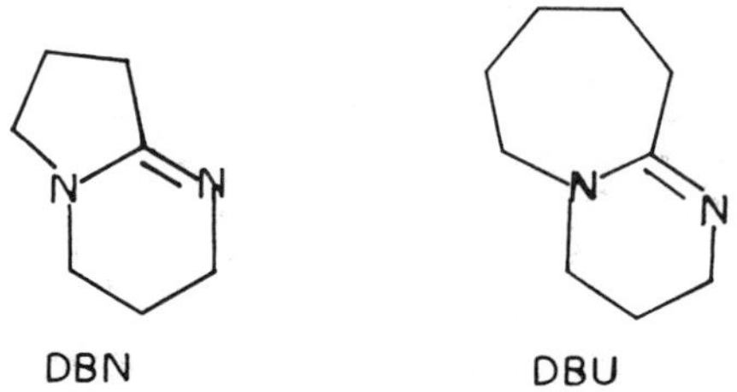

DBN   DBU

twice the yield of alkenes than when DBN, 1,5-diazobicyclo-[4.3.0]nonene-5, is used.   Among dehydrohalogenations possible with DBN and not with other bases is the synthesis of an oxepine as shown and compounds of a similar structure:

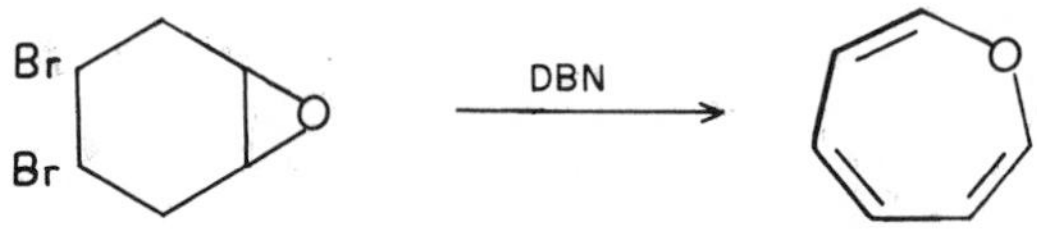

Other reagents gave a phenol; and DBN is also useful as a base in the preparation of the Wittig olefination reagent, $Ph_3P=CHR$.
    A new concept to produce the maximum amount of the true E_2 olefin, a <u>trans</u> olefin of the Saytzeff type, has been proposed. Disregarding the "foolishness" of working with weak bases to produce true E_2 olefins as quoted by one physical organic textbook, Parker has employed the weak (and soft) tetrabutylammonium bromide in acetone-lutidine to increase the amount of <u>trans</u>, Saytzeff-type olefin as shown in Tables 2.1 and 2.2 [16].

Table 2.1

Structure	trans/cis Olefin Ratio Bu$_4$NBr, Lutidine, (CH$_3$)$_2$CO, 75%	t-BuOK-t-BuOH
CH$_3$CH$_2$CHBrCH$_3$(I)	5.9	1.6
Me$_2$CHCHOTsMe(II)	70	2
PhCH$_2$CHBrPh	49	140
PhCH$_2$CHBrMe	14	52
PhCHBrCH$_2$CH$_3$	100	70

Table 2.2

Structure	Saytzeff-Hofmann Olefin Ratio Bu$_4$NBr, Lutidine, (CH$_3$)$_2$CO, 75°	t-BuOK-t-BuOH
I	23.8	0.85
II	333	0.31
Me$_2$CBrCH$_2$CH$_3$	10 (with C$\bar{\text{I}}$)	0.37
Me$_2$CHCH CH$_2$CH$_3$ OTs	12.3	1.04

Thus the products depend on the base and solvent. For example, 2-bromobutane with tetrabutylammonium bromide in acetone containing an excess of 2,6-lutidine gives largely a Saytzeff product with a high proportion of trans. By contrast, t-BuOK-t-BuOH gives largely a Hofmann elimination and a significant amount of the cis olefin by Saytzeff elimination.

Other reactions in addition to dehydrohalogenation may occur with DBU as shown [17]:

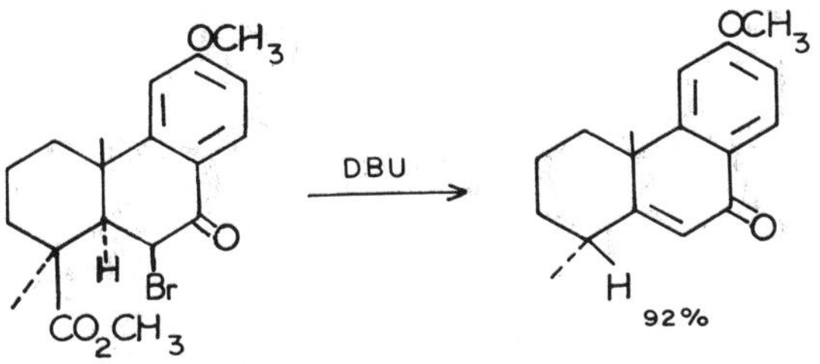

and DBU proved to be useful in the cleaving of other hindered esters.

Bases not previously reported in the elimination are potassium triethyl- or tricyclopentylmethoxide [18], trialkyl phosphites [19], silver and mercuric salts [20], tetramethyl-ammonium dimethyl phosphate [21], lithium amide in dioxane [22], sodium hydride in DMF [23], and lithium diethylamide [24]; in addition heating [25] is sometimes employed.

Potassium triethylmethoxide and tricyclopentylmethoxide in their alcohols were shown to be effective in giving high yields of methylene derivatives (Hofmann elimination) from 2,3-dimethyl-2-chlorobutane and 1-chloro-1-methylcycloalkanes [18].

Hunziker and Müllner [19] employed trimethyl phosphite to produce the 4,6-diene from 7-bromocholesteryl benzoate as shown:

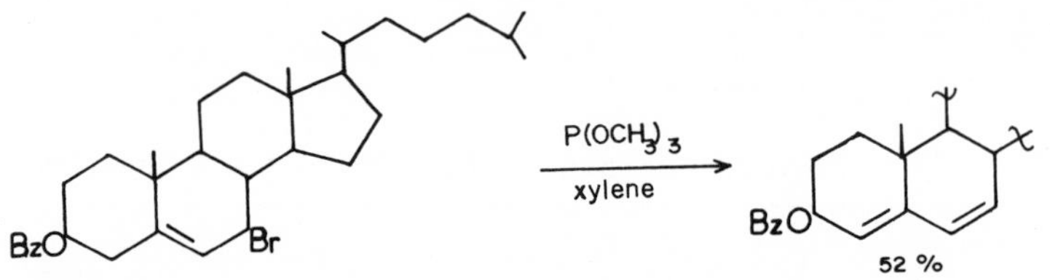

Bartsch and co-workers [20], in the dehydrohalogenation of secondary alkyl bromides and iodides with $AgNO_3$, $AgClO_4$, $AgOCOCH_3$, AgONO, and $Hg(NO_3)_2$ in aprotic and protic solvents, observed a strong preference for the formation of internal olefins with the trans:cis 2-alkene ratio of 1.1-2.8. With 2-iodobutane and an oxyanion base in DMSO at 50°, the ratio of cis:trans olefin varied from 3.0 to 3.8. With an increase in the strength of the base, more of the 1-butene (less stable form) was produced.

Sturtz and co-workers [21] employed tetramethylammonium dimethyl phosphate in the dehydrobromination of α-bromoaldehydes and α-bromoketones in various solvents, at times under pressure. For acyclic types the yields of olefinic aldehydes or ketones varied from 30 to 48%, whereas for cyclic types the variation was 35-80%. The dehydrohalogenation may proceed via an epoxy hemi-

acetal $> C \underset{O}{\overset{O}{\underset{\diagup}{\parallel}}} CHOP(OMe)_2$.

Binger and co-workers [22] utilized lithium amide in dioxane and a trace of water to produce thermally unstable 3-methylcyclopropene (30%) from 1-chloro-2-butene. Earlier the latter had been obtained (∿ 50%) from methallyl chloride and sodamide in THF

[26].   The chloride with potassium amide in THF gives methylenecyclopropane (36%) [27].

McDonald and Reitz [23] succeeded in obtaining the dimethyl
cycloalk-1-ene-1,2-dicarboxylates (21-71%) from the dimethyl
esters of α,α'-dibromoadipic, -pimelic, -suberic, and -azelaic
acids by treatment with NaH in DMF as indicated:

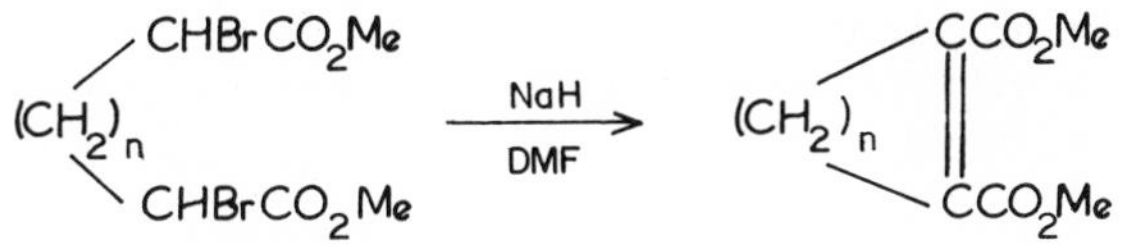

Cyclohexenes as shown have been produced from halocyclo-
hexane or cyclohexyl phenyl sulfonate by treatment with lithium
diethylamide [24]:

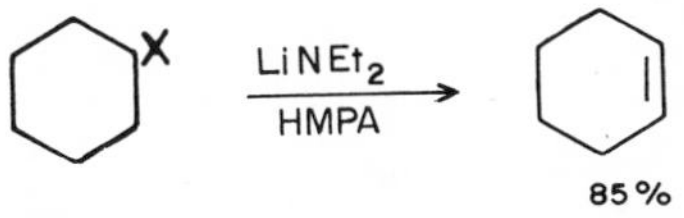

Stilbenes were produced from the arene and α-chloroacetal-
dehyde via the rearrangement of the chloromethyldiarylmethane as
indicated [25] (Ex. a):

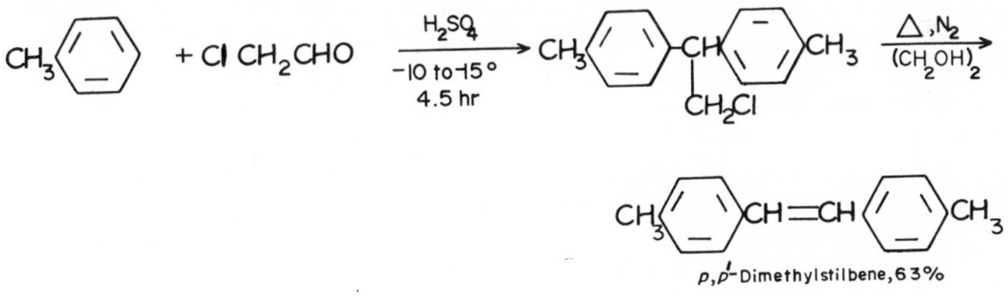

p,p'-Dimethylstilbene, 63%

This one-operation process is applicable as well to phenols and
phenol ethers.

Similarly, by heating to 550-610°, perchloroolefins were
produced [28]:

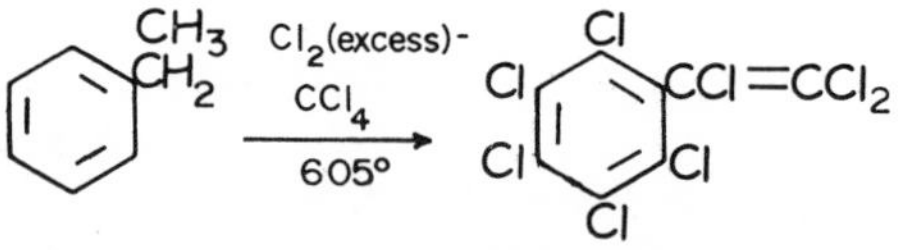

Pentachlorophenyl-1,1,2-trichloroethylene, 84%

This method is applicable to ethylpyridines as well.

Potassium t-butoxide was also employed to give unsaturated orthoesters as shown [29]:

$$RCH_2CHBrC(OEt)_3 \xrightarrow[80-135^\circ]{t\text{-BuOK}} RCH{=}CHC(OEt)_3 \quad 72-75\%$$

With DMSO the t-butoxide served for the synthesis of 3,3-dimethylcyclopropene [30]:

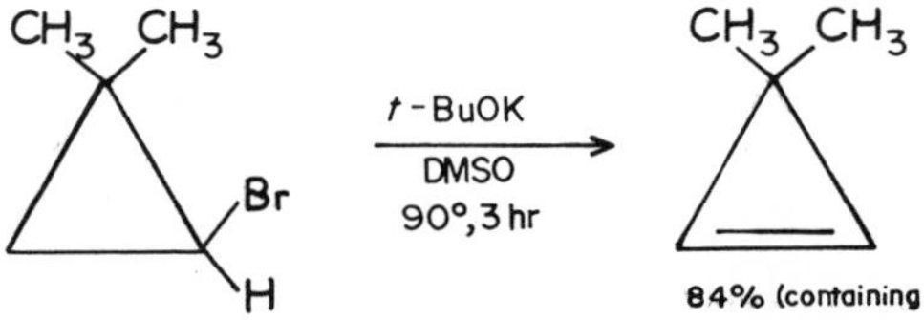

84% (containing 97% alkene)

a. Preparation of 4,4'-Dimethylstilbene [25]

A mixture of 1 mol of toluene and 0.5 mol of chloroacetaldehyde dimethyl acetal at -10° to -15° was stirred for 2.5 hr while 300 ml of concentrated H_2SO_4 was added. Two more hr of stirring and pouring on ice gave an organic phase that when purified was mixed with ethylene glycol and heated under N_2 up to the boiling point of the glycol. Cooling and recrystallization of the solid mass yielded the stilbene, 63%.

3. From Dihalides (Dehalogenation) (1, 80) [1a, p. 180]

$$RCHXCHXR \xrightarrow[95\%\ EtOH]{Zn} RCH{=}CHR$$

$$2\ R_2CX_2 \xrightarrow[ether]{Mg} R_2C{=}CR_2$$

a.   From 1,2-Dihalides

Dehalogenating agents not previously mentioned are KOH [31], mercuric trimethylgeranium [32], sodium dihydronaphthylide [33], titanocene $(C_{10}H_{10}Ti)_2$ [34], ethylene epoxide-tetrabutylammonium bromide [35], dialkylcopperlithium [36], and sodium thiosulfate [37]. Dehalogenation using NaI or KI has been reviewed [1a, p. 195].

Methyl diiodosterculate has been dehalogenated by heating at 50-60° for 36 hr with 10% KOH in C_2H_5OH [31]:

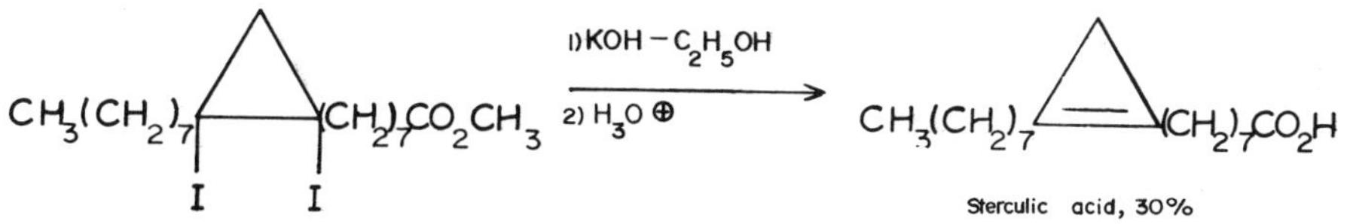

Sterculic acid, 30%

Bennett and co-workers [32] utilized bis-trimethylgermyl-mercury in the debromination of <u>erythro</u> and <u>threo</u> 1,2-dibromides as indicated:

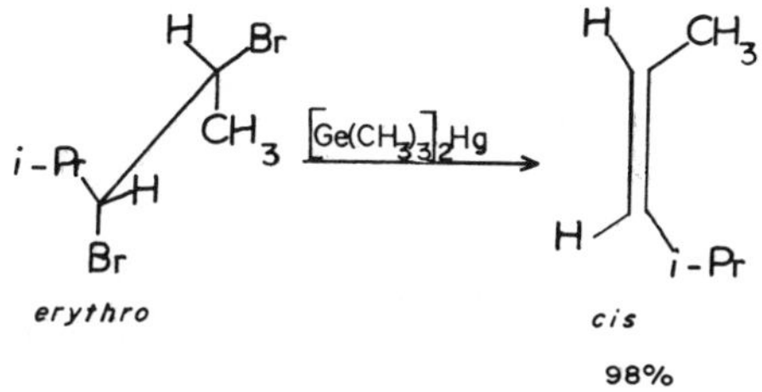

erythro

cis

98%

The <u>threo</u> form gave 96% of the <u>trans</u> isomer.

Garst and co-workers [33] employed sodium dihydronaphthylide to dechlorinate the 1,2-dichlorocyclobutane as indicated:

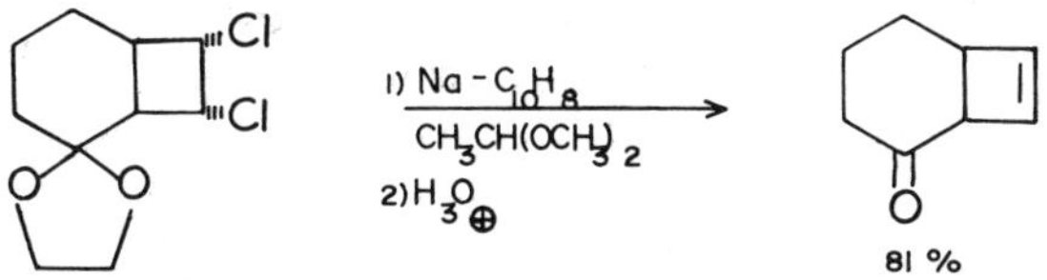

81 %

Merijanian and co-workers [34] found that titanocene, $(C_{10}H_{10}Ti)_2$, abstracts the halogens readily at room temperature from alkyl, allyl, and some vinyl halides, but not from aromatic halides. Unfortunately, the yields leave something to be desired and at times coupling occurs. The best yield reported is that from <u>meso</u>-stilbene dibromide which gave 73.6% of <u>trans</u>-stilbene.

Buddrus and Kimpenhaus [35] obtained vinyl <u>bromide</u> by the treatment of ethylene bromide with ethylene oxide in the presence of halide ions under pressure as shown:

$$BrCH_2CH_2Br \ + \ CH_2{-}CH_2 \xrightarrow[\substack{150° \\ autoclave}]{(n-C_4H_9)_4 N\,Br} CH_2{=\!=}CHBr$$

63%

A more favorable return was achieved by Posner and Ting [36] by the use of a dialkylcopperlithium (Ex. a).

$$R'CHBrCHBrR^2 \xrightarrow[Et_2O]{R_2Cu\,Li} R'CH{=\!=}CHR^2$$

70—98 %

A more available reagent, sodium thiosulfate, was utilized with similar substrates [37]:

$$C_6H_5CHBrCHBrR \xrightarrow[DMSO]{Na_2S_2O_3} \substack{C_6H_5 \\ \diagdown \\ H} C{=\!=}C \substack{H \\ \diagup \\ R}$$

65—99%

Kochi and co-workers (see <u>1</u>, 80) [38] have continued a study of dehalogenation with Cr^{II} complexes at room temperature. From $ClCH_2CH_2Y$, where Y = Cl, OH, NH_2, $OCOCH_3$, OC_6H_5, OTS, and OMe, the yield of ethylene is quantitative; similar results were obtained with $BrCH_2CH_2Y$.

Schroeder and co-workers found that triethyl phosphite [39] (see <u>1</u>, 80) dehalogenates vicinal dihalides satisfactorily if electronegative groups are also attached to the carbon atoms bearing the halogens. Thus $CH_3CHBrCHBrCN$ gives 85% of $CH_3CH=CHCN$ (Ex. b); on the other hand, $CH_3BrCHBrCH_3$ gives no $CH_2=CHCH_3$ but rather traces of phosphonates.

b. From <u>gem</u> Dihalides

A zinc-copper couple was utilized in the coupling of <u>gem</u> dibromoadamantane via dehalogenation [40] as shown to produce adamantylideneadamantane. The reaction occurred via a carbene or carbenoid.

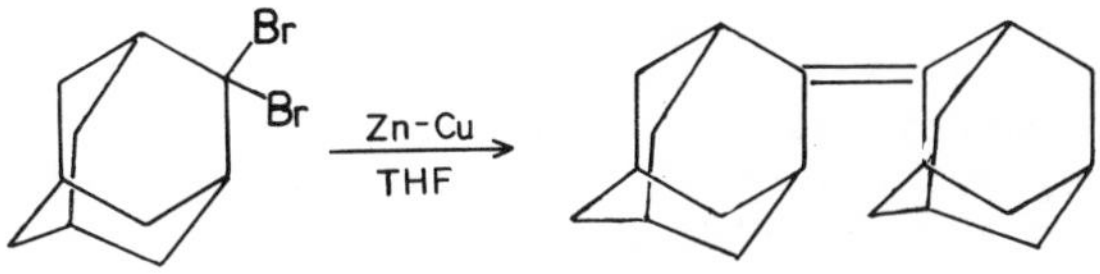

75%

i.   Preparation of Methyl 10-Undecanoate [36]

To a stirred suspension of 1.9 g of CuI in 40 ml of ether at -35° under N_2, 9.9 ml of 2M n-butyllithium was added. Then at -20°, 717 mg of methyl 10,11-dibromoundecanoate in 10 ml of ether was introduced. After stirring for 2 hr, 4 ml of CH_3OH was added and the mixture was poured into a saturated NH_4Cl solution. From the ether extract, 357 mg (90%) of the ester was recovered.

ii.   Preparation of Crotonic Nitrile [39]

This was 85% from 2,3-dibromobutyronitrile and triethyl phosphite by heating to 180-185° for 5-6 hr.

iii. Preparation of Oleic Acid [37]

This was 99% from threo-oleic acid dibromide, 3 equiv of powdered $Na_2S_2O_3$ in DMSO at 60° for 8 hr.

4.   From Halo Ethers (Boord) (1, 82) [1a, p. 212]

$$BrCH_2CH_2OR(Ar) \longrightarrow CH_2{=}CH_2$$

Ethylene has been produced in quantitative yields from the bromoether by elimination of the bromo and alkoxy (aroxy) groups through the use of Cr^{II}en [38].

5.   From Halohydrins (1, 84) [1a, p. 204]

$$ROCH_2CHOHCH_2Cl \xrightarrow{P(OC_2H_5)_3} ROCH_2CH{=}CH_2$$

To eliminate the hypohalous acid, various reagents such as Cr^{II}en [38], triethyl phosphite [41], phosphorus oxychloride in pyridine [42], and methanesulfonyl chloride in pyridine [43] have been employed. Thus Kochi [38] obtained ethylene from β-chloro- and β-bromoethyl alcohol in quantitative yield, and

Henning [41] heated 1-chloro-2-hydroxy-3-methoxy- and 3-n-butoxy-propanes with triethyl phosphite to obtain the corresponding alkoxypropenes (41-55%). The unsaturated ethers were removed as they were formed by using a rotating evaporator. In a similar reaction Corey and Grieco [43] utilized methanesulfonyl chloride in pyridine as the elimination reagent to obtain from the iodo-lactone shown the cyclopentene lactone in superior yield:

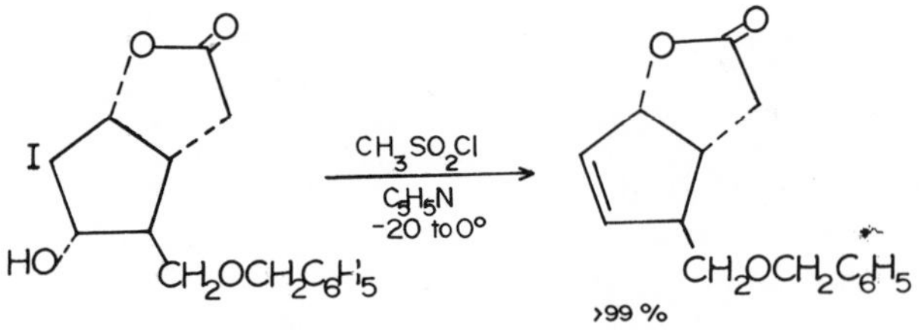

As is seen, this elimination occurs under extremely mild conditions without employing a reducing agent such as Zn or $SnCl_2$. Crabbé and Guzmán [42] accomplished the same elimination with phosphorus oxychloride and pyridine also at a low temperature with a 92% yield. These investigators regard the method to be of general applicability for olefin synthesis.

    6.   From Esters (1, 85) [1a, p. 105]

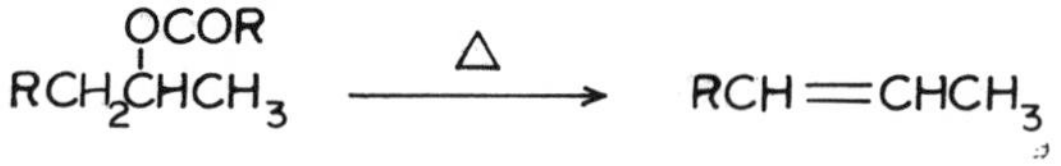

    Wood and Cheng [44] have shown that, with t-BuOK-t-BuOH as the elimination agent, tosylates such as n-octyl- and 24-cholanyl- give low amounts (16-25%) of the alkene as compared to the chloride. However, with secondary tosylates the formation of alkenes is more favorable and the nature of the products depends on the choice of base and solvent [45]. Thus 1-phenyl-3-methyl-2-butyl p-toluenesulfonate may serve for the synthesis of either Hofmann or Saytzeff products (E2H or E2C elimination, respectively) as shown:

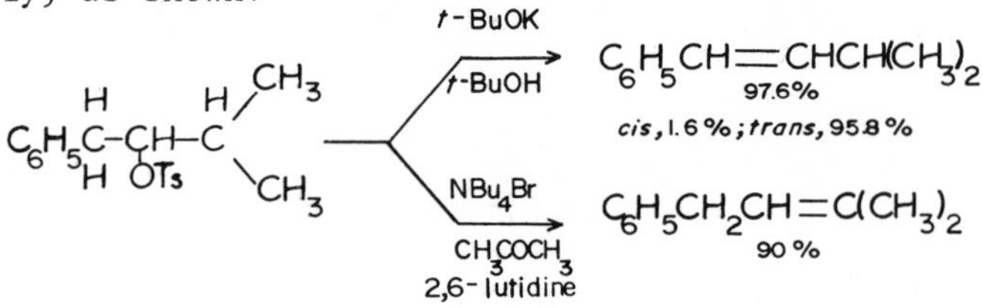

In the case of tosylmethylcyclopropane a partial rearrangement occurs with t-BuOK [46]:

In the case of tosylmethylcyclopropane a partial rearrangement occurs with t-BuOK [46]:

Cyclobutene
50%

Methylenecyclopropane
50%

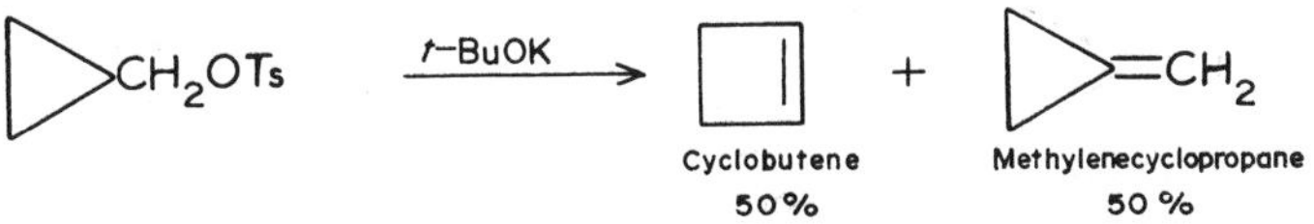

Double elimination from trans-1,2-bis(hydroxymethyl)allyl p-toluenesulfonates leads to vicinal exocyclic dimethylene hydrocarbons [47]. Elimination from the ditosylate is more useful than from the diacetate

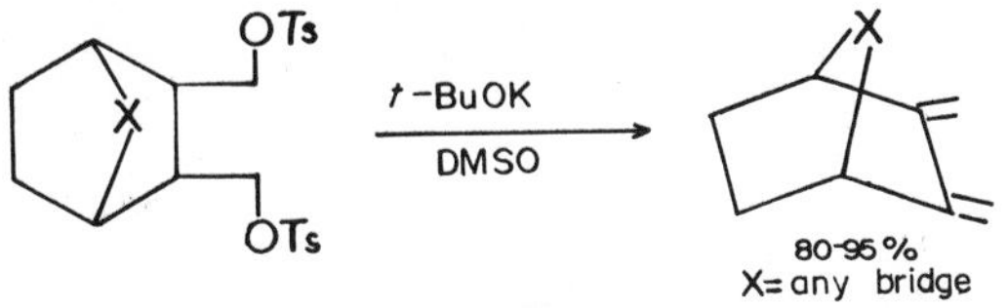

80-95%
X= any bridge

Elimination to form alkenes is also possible from disulfonic esters by treatment with the anion of sodium anthracene or sodium naphthalene [48]. Thus the dimesylate of cis- or trans-1,2-cyclooctanediol gives cyclooctene (Ex. a). The reaction involves reduction of one mesylate group to a carbanion followed by rapid alkene formation.

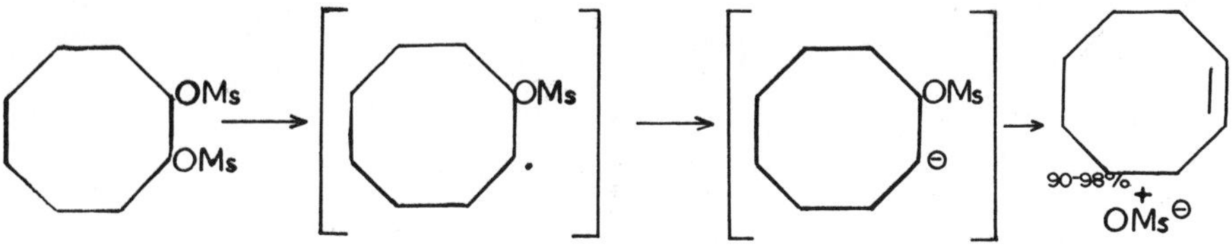

90-98%
+
$OMs^{\ominus}$

Other yields vary within 55-99%. The reaction is nonstereospecific and there appear to be no steric requirements for elimination.

Other substituted esters and related types may undergo elimination to give alkenes as indicated:

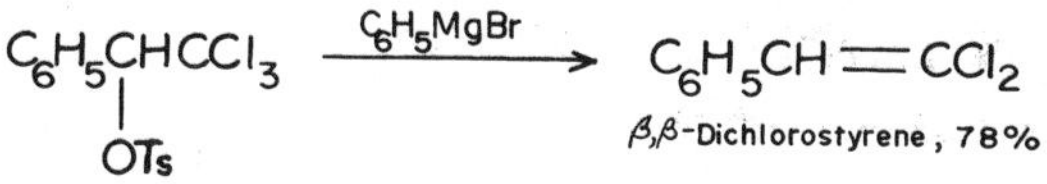

$C_6H_5CHCCl_3$ $\xrightarrow{C_6H_5MgBr}$ $C_6H_5CH=CCl_2$ [49]
|
OTs

β,β-Dichlorostyrene, 78%

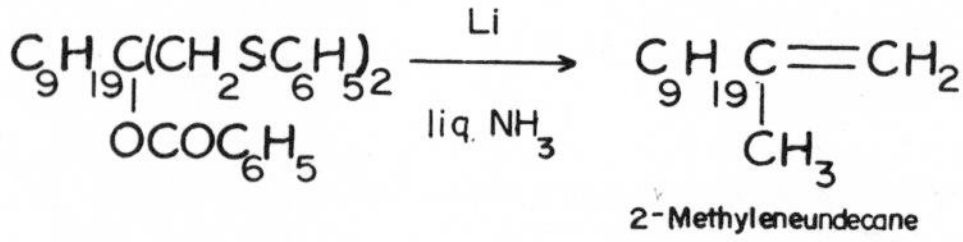

$$C_9H_{19}C(CH_2SC_6H_5)_2 \xrightarrow[\text{liq. NH}_3]{\text{Li}} C_9H_{19}C\!\!=\!\!CH_2$$

with $OCOC_6H_5$ and CH_3 [50]

2-Methyleneundecane

This reaction succeeds while the Wittig olefination for highly hindered ketones fails.

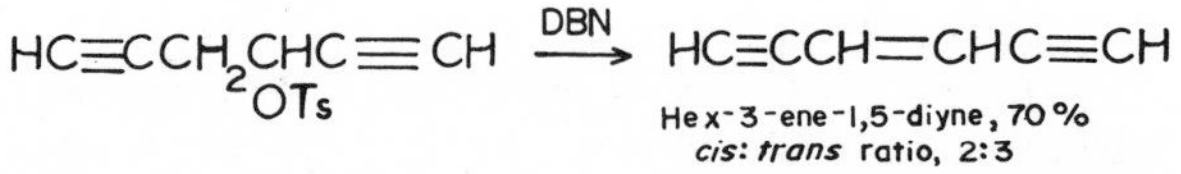

$$HC\!\equiv\!CCH_2CHC\!\equiv\!CH \xrightarrow{\text{DBN}} HC\!\equiv\!CCH\!=\!CHC\!\equiv\!CH$$

with OTs [51]

Hex-3-ene-1,5-diyne, 70%
cis: trans ratio, 2:3

Adam and co-workers [52] developed a method for obtaining alkenes from carbonyl compounds via the β-lactone as shown (for synthesis from other lactones, see A.23):

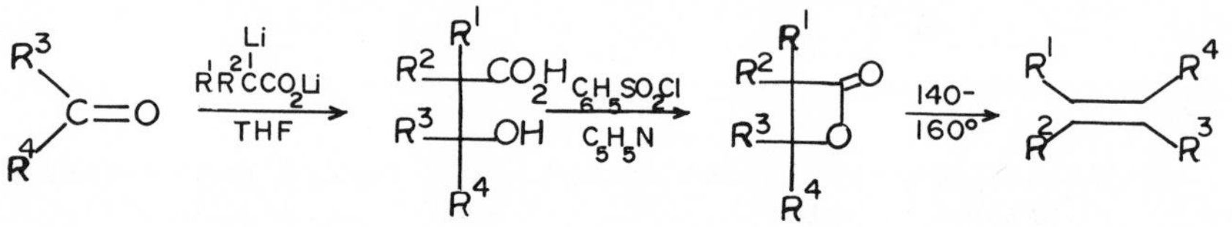

Both the β-lactones and alkenes are obtained in high yield and the latter retain the initial geometry without double bond isomerization. The method constitutes an alternate to the Wittig olefin synthesis (1, 141).

The pyrolysis of isochroman-3-one led to benzocyclobutene [53]:

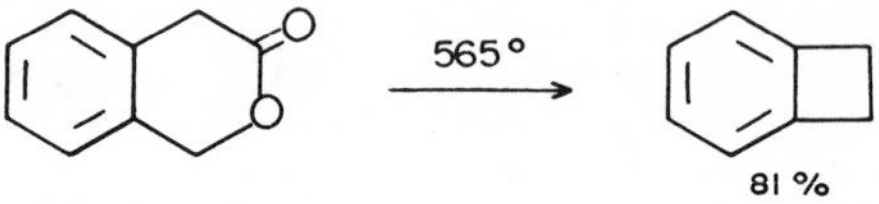

The substrate was prepared by the oxidation of indan-2-one.

The di(4-methylphenyl thioncarbonic ester) of (-)-trans, trans-spiro[4.4]nonan-1,6 diol was converted into the spiro alkene as indicated [54]:

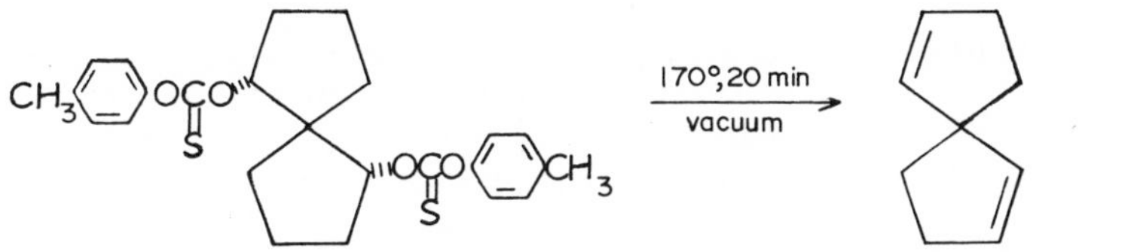

(+)-(5S)-Spiro[4.4]nona-
1,6-diene, 58%

Elimination from lactones is discussed in A.23.

a. Preparation of Cyclooctene [48]

Yield of 98-99% by adding slowly a 0.3M solution of the anthracene anion radical to a degassed, stirred solution of trans- or cis-1,2-cyclooctanediol dimesylate in DMF until the color of the anion radical persisted.

7.   From Xanthates (Chugaev) or O-Alkyl Dimethylthiocarbamates (1, 88) [1a, p. 123]

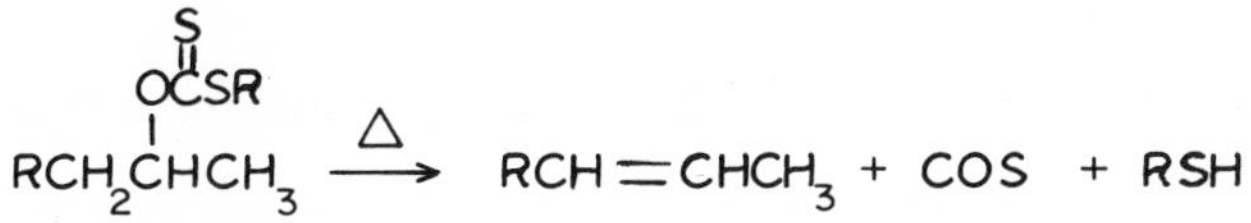

Rutherford and co-workers [55] found that the xanthate salts of some tertiary alcohols gave better yields than the ester. Thus 1-methylcyclopentanol potassium xanthate on being heated led to an 81% yield of 1-methylcyclopentene, while the yield from the ester was 62%. Smaller amounts of isomeric alkenes are usually present as well.

Burke and co-workers [56] studied a modification of the Chugaev reaction in which the xanthate is converted into a derivative (with compounds containing hetero atom unsaturation such as chloroacetic or chloroacetonitrile) that, with dilute aqueous acid or Lewis acids at pH 4 to 7, gave the olefin. This so-called BBT procedure may be represented as:

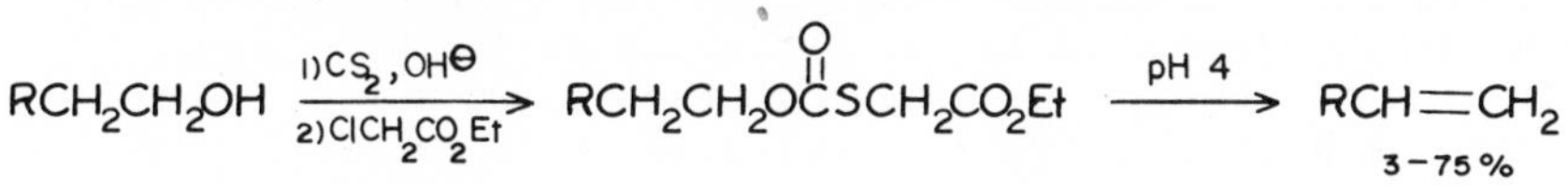

The method is new, ionic as opposed to cyclo, and limited to primary and secondary alcohols.

A failure of the Chugaev reaction has been reported [57]. In the case of S-p-bromophenacyl xanthates of primary alcohols small amounts of alkenes are obtained, but there are formed a great variety of other types not previously encountered in the reaction.

The pyrolysis of O-alkyl dimethylthiocarbamates containing a β hydrogen leads to olefins usually in high yields [58] at a reasonably low temperature. Thus O-α-phenylethyl dimethylthio-carbamate gives the alkene as indicated:

$$2 \ (CH_3)_2 N \underset{\underset{C_6H_5}{S}}{\overset{O}{C}} OCHCH_3 \xrightarrow{180\text{-}200°} \underset{\underset{C_6H_5CCH_3}{\|}}{C_6H_5CCH_3}$$

2,3-Diphenyl-2-butene
81 %

The substrate is readily available from the sodium alkoxide and dimethylthiocarbamyl chloride. The procedure is simpler than the xanthate route and the O-alkyl dimethylthiocarbamates are more readily purified than the xanthates.

8.   From Quaternary Ammonium Hydroxides or Their Salts (Hofmann) (1, 89) [1a, p. 219]

$$RCH\overset{\oplus}{\underset{CH_3}{N(CH_3)_3}} \overset{\ominus}{OH} \xrightarrow{\Delta} RCH = CH_2 + N(CH_3)_3 + H_2O$$

For the decomposition of phosphonium and sulfonium bases see the above H.-W. reference. A recent method for the one-step preparation of quaternary ammonium bases utilizes a structurally hindered organic base of greater strength than the amine involved. In this manner complete alkylation of primary and secondary amines with pK_a values as low as 2.36 and of alicyclic and strong aliphatic amines with pK_a values as high as 11.1 is accomplished [59]. Such a base is 1,2,2,6,6-pentamethylpiperidine.

The well-known phenomenon that is desirable in the Hofmann elimination to heat the quaternary ammnium hydroxide under reduced pressure [60] has been investigated recently by Archer [61], who showed that the removal of water in the quaternary ammonium hydroxide used in the pyrolysis is the key to the problem. To cite one case, N,N-dimethyl-1,2,3,4-tetrahydro-quinolinium hydroxide when concentrated at 40° under 20 mm pressure gave 5% elimination at 160° and 760 mm pressure. On the other hand, the same substrate when dried additionally at 20° under 0.005 mm pressure gave 75% elimination at 60° and 0.005 mm pressure.

Heating the quaternary ammonium hydroxide has been employed to prepare methylene cycloalkenes by a novel 1,4-elimination reaction [62] as shown:

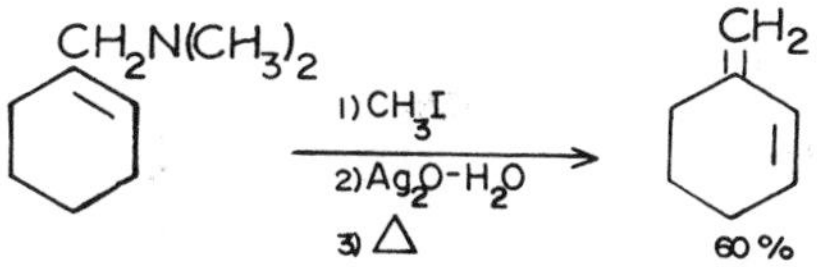

Methylenecycloheptene (40%), methylenecyclooctene (86.5%), and 3,5-dimethyl-methylenecyclohexene (57%) were prepared similarly.

By a like elimination (see 1, 91) a mixture of cis and trans cyclooctenes was obtained from N,N,N-trimethylcyclooctylammonium hydroxide [63] (Ex. a). By extraction with aqueous $AgNO_3$, the trans isomer (40%) is recovered while the cis isomer (30%) remains.

A 1,3,5-triene has been synthesized by Spangler and Woods [64] by the sequence shown:

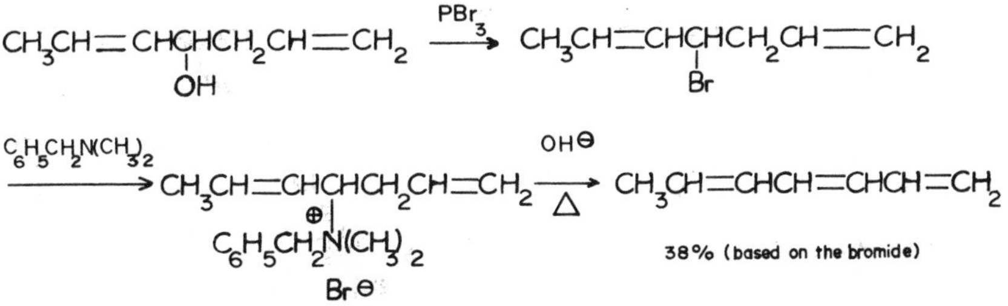

38% (based on the bromide)

It has been shown (1, 91) that alkenes may be obtained directly from the quaternary ammonium bromides. Schmid and Wolkoff [65] achieved this end by the pyrolysis under vacuum of N-methyl-4-alkoxypyridinium iodides as indicated:

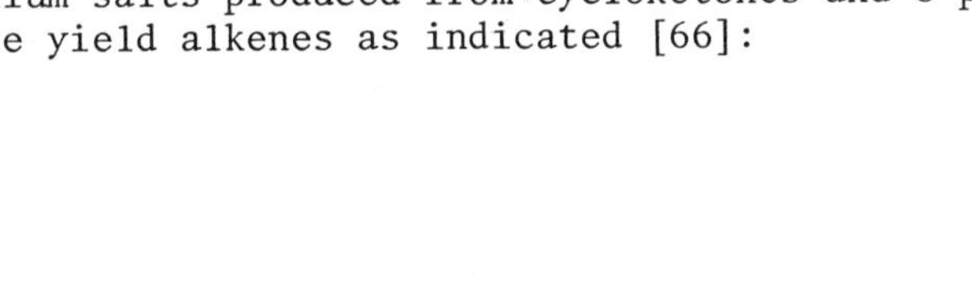

Cyclohexene
90%(21 examples)

Other yields varied from 27 to 100% (21 ex.), the higher yields being usually for cyclenes.

Immonium salts produced from cycloketones and 3-pyrroline perchlorate yield alkenes as indicated [66]:

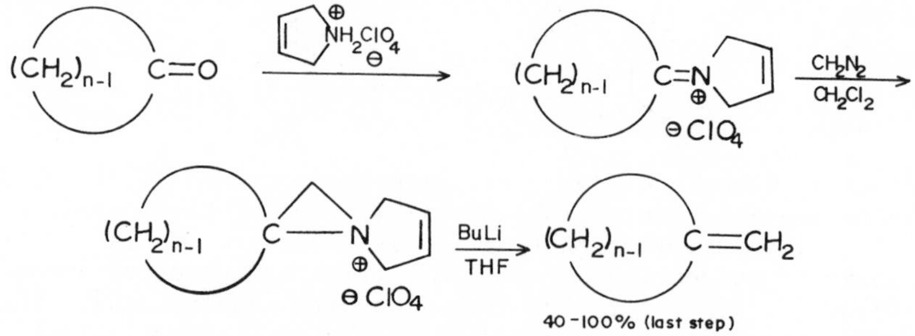

a.   Preparation of trans-Cyclooctene [63]

N,N,N-Trimethylcyclooctylammonium hydroxide solution (prepared from 100 g of the iodide, 76 g of Ag_2O and 35 ml of water)
was added dropwise to a flask heated to 110-125°, whereas
evacuation to a pressure of approximately 10 mm under a constant
sweep of N_2 was carried out. The distillate at 25° was treated
with 200 ml of 5% HCl and the mixture of cis- and trans-
cyclooctenes was extracted in pentane, from which solution the
trans isomer (40%) was recovered by extraction in $AgNO_3$ (present
silver prices detract greatly from methods using it or its compounds).

9.   From β-Dialkylamino Ketones, Amides, and the Like (Dehydroamidation) (1, 92)

A little known reaction is the elimination of the amide
group (dehydroamidation) to form an alkene. Several mechanisms
may apply; in the first a carbonium ion is involved [67]:

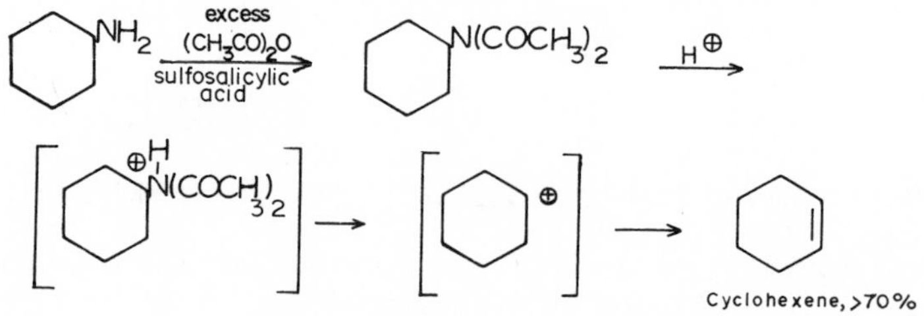

A similar elimination was accomplished earlier [68]. The loss of
the amido group is accomplished best from a tertiary carbon atom,
a fact that reminds one of a reverse Ritter reaction, except that
the amide does not give the nitrile. The second reaction occurs
perhaps by the E_2 mechanism [69].

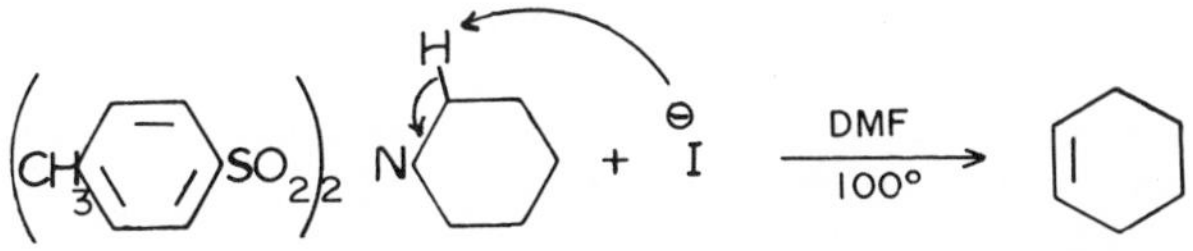

Cyclohexene
90%

trans-Stilbenes have been produced from acetyl and benzoyl derivatives of 1,2-diphenylethylamines by acid hydrolysis as shown [70]:

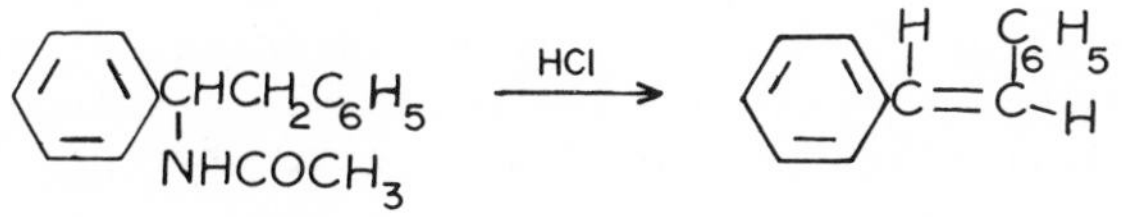

Not all of the types lead to stilbenes. The presence of methoxy groups in the benzene ring, the nature of the acyl group, and the acid concentration are factors affecting yields of 0-97.5%.

Dehydroamidation, particularly of amides of steroids, has been accomplished in fair yield as shown [71]:

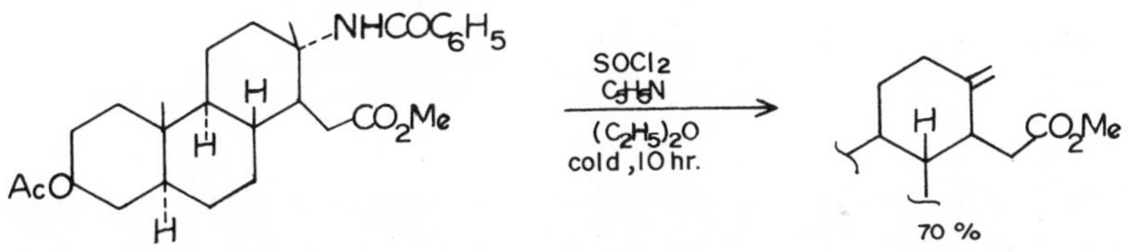

70%

10.   From Amine Oxides (Cope) (1, 93)

$$R_2CHCR_2 \longrightarrow R_2C=CR_2 + Me_2NOH$$
$$\overset{|}{O \leftarrow NMe_2}$$

This cyclo reaction has been reviewed [1a, p. 238].

11.   From Sulfoxides, Sulfones, β-Hydroxysulfinamides, Hydroxythioethers, and β-Hydroxyphosphonamides (1, 93) [1a, p. 250]

$$RCH_2CH_2SOR \overset{\Delta}{\longrightarrow} RCH=CH_2$$

$$(CH_3)_2CHSO_2CH(CH_3)_2 \overset{\Delta}{\longrightarrow} 2\ CH_3CH=CH_2$$

Elimination is a common reaction for compounds of S and P. In particular, photocyclic extrusion to form alkenes is facile [72].

Trost prepared olefinic esters as indicated from esters having the sulfoxide group on the αC [73]:

$$CH_3(CH_2)_6CH_2\overset{H}{\underset{\underset{O \leftarrow SCH_3}{|}}{C}}CO_2C_2H_5 \xrightarrow[\text{reflux}]{C_6H_5CH_3} CH_3(CH_2)_6CH=CHCO_2C_2H_5$$

Ethyl α,β-decenoate,86-88%(overall from substrate)

Good yields of alkenes as a rule were obtained from β-hydroxy sulfoxides [74]. Thus the sulfoxide obtained from t-butyl methyl sulfoxides, as indicated, gave styrene:

$$(CH_3)_3C\underset{O}{\overset{\parallel}{S}}CH_3 \xrightarrow[2)C_6H_5CHO]{1)CH_3Li} (CH_3)_3C\underset{O}{\overset{\parallel}{S}}CH_2\underset{OH}{\overset{}{C}}HC_6H_5 \xrightarrow[\text{or}]{NCS} \xrightarrow{SO_2Cl_2} CH_2=CHC_6H_5$$

82% (last step)

Similarly, β-hydroxysulfoximines, which need not be isolated, serve as a source of alkenes [75].

$$\overset{\diagup}{\underset{\diagup}{}}C=O \; + \; C_6H_5\underset{NCH_3}{\overset{O}{\overset{\parallel}{S}}}CH_2Li \rightarrow -\underset{|}{\overset{|}{C}}CH_2\underset{NCH_3}{\overset{O}{\overset{\parallel}{S}}}C_6H_5 \xrightarrow[\substack{THF \\ AcOH}]{AlHg} \overset{\diagup}{\underset{\diagup}{}}C=CH_2$$

50 - 90 %

Julia and Paris devised a method of elimination from β-hydroxysulfones by the use of $POCl_3$-C_5H_5N [76].

$$C_6H_5SO_2\overset{\ominus}{\underset{\oplus Mg}{C}}H_2 + C_6H_5CHO \xrightarrow[Et_2O]{C_6H_6} CH_2\underset{\underset{C_6H_5}{|}}{\overset{OH}{\underset{SO_2}{|}}}\overset{|}{C}C_6H_5 \xrightarrow[C_5H_5N]{POCl_3} CH_2=CHC_6H_5$$

Styrene,53% overall

Elimination in other sulfones has been reviewed (1, 93) [77].

Kuwajima and co-workers [78] were successful in eliminating the phenylmercapto group from β-hydroxyalkyl phenyl sulfides to form terminal olefins:

$$C_6H_5SCH_2\underset{|}{\overset{OH}{\underset{R_2}{C}}}RR_2 \xrightarrow[\substack{2)(o-C_6H_4O_2)PCl \\ 3) \triangle}]{1) MeLi} CH_2=CR_1R_2$$

64-83%

Olefins have also been obtained from sulfides by first forming
the α-chloro derivative from which elimination is possible by
the use of Ph_3P-$\underline{t}$-BuOK [79]:

$$ArCH_2SCH_2Ar \xrightarrow[CCl_4]{SO_2Cl_2} ArCH_2SCHClAr \xrightarrow[\substack{THF \\ t\text{-BuOK} \\ 24\text{-}36\,hr.}]{(C_6H_5)_3P} ArCH=CHAr \quad 60\text{-}94\%$$

The intermediate is probably the episulfide.

Selenides may also be utilized as a source of alkenes. Thus
Sharpless and co-workers [80] in the preparation of alkyl phenyl
selenides obtained in a 71% conversion a mixture of alkenes:

$$C_6H_5Se \cdots \xrightarrow[25°]{H_2O_2} \text{+} \text{+}$$

l-Butene	trans-2-Butene	cis-2-Butene
61%	28%	11%

The elimination is $\underline{syn}$.

12. From Episulfides or Epoxides ($\underline{1}$, 98)

$$CH_3CH\!-\!CH_2 \xrightarrow{(C_2H_5O)_3P} CH_3CH=CH_2 + (C_2H_5O)_3PO(S)$$
$$\underset{O(S)}{\diagdown\diagup}$$

New reagents that have been used for converting epoxides
into alkenes are Mg/Hg-$MgBr_2$ [81], WCl_6 with Li, LiI, or RLi
[82], Zn-Cu [83], Ph_2PLi-CH_3I [84], sodium (cyclopentadienyl)
dicarbonylferrate with fluoroboric acid and NaI [85], and triphenylphosphine selenide-trifluoroacetic acid [86]. For episulfide elimination the new reagents are diiron nonacarbonyl,
$Fe_2(CO)_9$, triiron dodecacarbonyl, $Fe_3(CO)_{12}$ [87], and the phenyl
radical [88].

Cainelli obtained alkenes (22-80%) from epoxides at 25° by
treatment with Mg/Hg-$MgBr_2$ in THF [81]. Sharpless and co-workers
utilized WCl_6 with Li, LiI, or RLi for the same deoxygenation to
obtain yields of olefins from 37 to 99% [82]. Kupchan and
Maruyama utilized Zn-Cu for the same deoxygenation of the
epoxides of sesquiterpenes, steroids, stilbenes, and octenes (8-
95%) [83].

Vedejs and Fuchs [84] were successful in producing an inversion of an olefin via the epoxide as shown in the case of $\underline{cis}$-
stilbene:

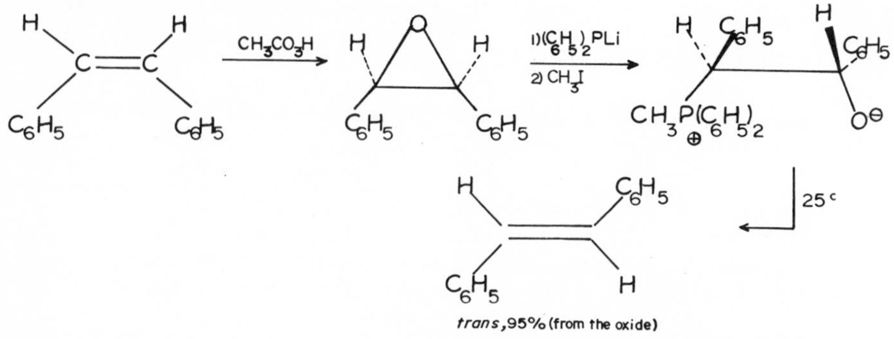

trans,95% (from the oxide)

In this transformation lithium diphenylphosphide opens the epoxide ring stereospecifically while quaternization with CH_3I leads directly to the betaine. The latter generally fragments at 25° to give inversion of the olefin.

Rosenblum and co-workers [85] converted olefins into epoxides and then into olefins with retention of configurations by the use of sodium (cyclopentadienyl) dicarbonylferrate, then fluoroboric acid, and finally sodium iodide. By this procedure the epoxide from cis-2-butene is restored to the original cis-2-butene. Terminal epoxides respond much more rapidly than internal ones, and neither carbonyl nor ester functions appear to interfere with the sequence of reactions.

A new reagent, triphenylphosphine selenide-trifluoroacetic acid, for converting epoxides into olefins, was introduced by Clive and Denyer [86]. The reaction, which proceeds by extrusion of the episelenide, may be represented as:

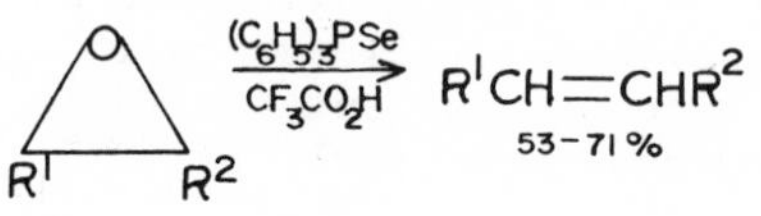

The conversion of cis-cycloalkenes and cis-cycloalkadienes into the corresponding trans forms has been accomplished by Bridges and Whitham [89]. The first intermediate appears to be the epoxide and the second is crystalline and readily purified. The steps in the case of cis-cyclooctene are:

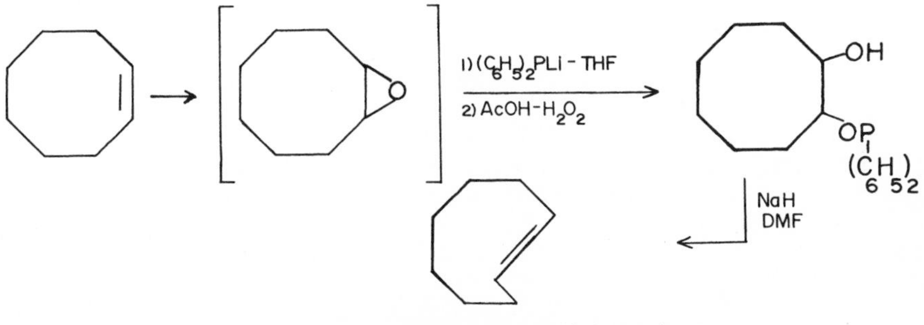

$$\frac{1)(C_6H_5)_2PLi - THF}{2) AcOH-H_2O_2}$$

NaH
DMF

trans-Cyclooctene, 76%(last step)

The method, applicable to di-, tri-, and tetrasubstituted olefins, is convenient and efficient.

Episulfides have been desulfurized by Trost and Ziman with
n-butyllithium, diiron nonacarbonyl, and triiron dodecacarbonyl,
the elimination in each case being stereospecific [87].

It has been proposed that the conversion of substituted
alkyl acetates or similar types with thionyl chloride into alkenes proceeds via the episulfide [90]:

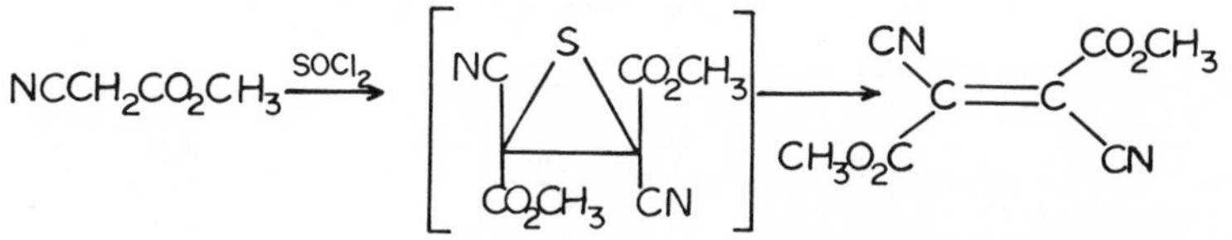

trans-Dimethyl-1,2-dicyanofumarate,65-70%

13. From Ethers, Vinyl Ethers, and Oxetanes (1, 98)

$$CH_3CH = CHCH_2OR \xrightarrow{KOH} CH_2 = CHCH = CH_2$$

It has been shown (1, 98) that alkenes may be obtained from
ethers by the elimination of an alcohol molecule. Mkryan and co-
workers [91], by heating the ether with solid KOH, obtained 1,3-
butadienes as shown above. The yields from a series of 2-butenyl
alkyl ethers varied from 65 to 78%.

Felkin and Tambuté found that butyl allyl ether when treated
with propyllithium in pentane did not undergo the Wittig rearrangement, but instead a displacement occurred as indicated
[92]:

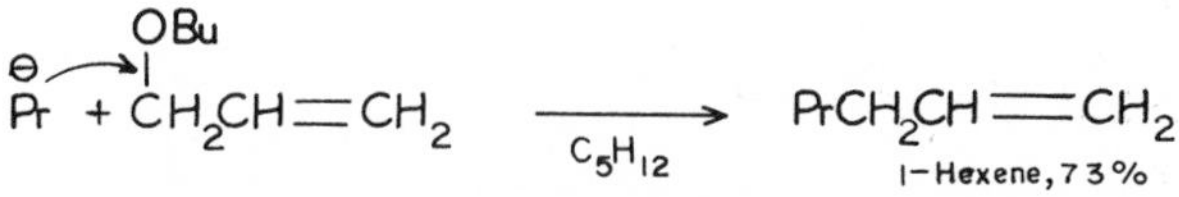

$$\overset{\ominus}{Pr} + \overset{OBu}{CH_2CH{=}CH_2} \xrightarrow[C_5H_{12}]{} Pr\text{-}CH_2CH{=}CH_2$$

1-Hexene, 73%

It is fortunate that the pyrolysis of some oxetanes prepared by the photochemical reaction of acetone and dimethyl maleate gives products different from the original ones [93].

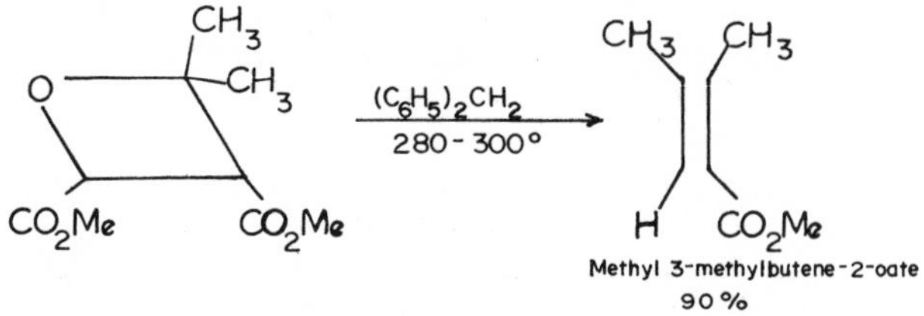

$$\xrightarrow[280-300°]{(C_6H_5)_2CH_2}$$

Methyl 3-methylbutene-2-oate
90%

14.   From Diazo Compounds, Tosylhydrazones, and the Like (1, 98) [1a, pp. 237, 686, 698]

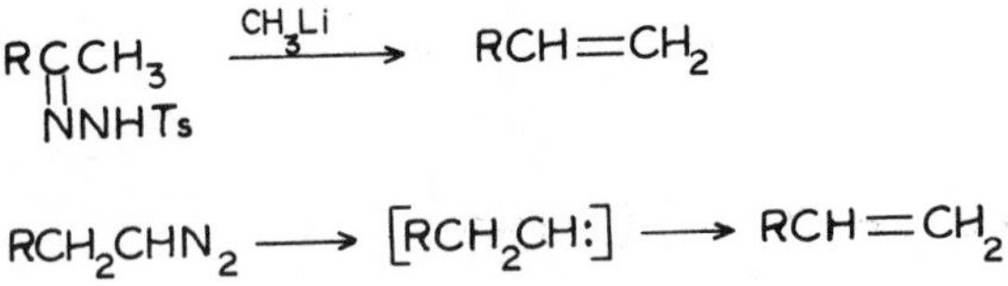

$$\underset{NNHTs}{\overset{R\overset{\text{O}}{C}CH_3}{}} \xrightarrow{CH_3Li} RCH{=}CH_2$$

$$RCH_2CHN_2 \longrightarrow [RCH_2CH{:}] \longrightarrow RCH{=}CH_2$$

The chemistry of carbenes is not necessarily one of inter-molecular insertion (or addition) but preferably, if possible, one of intramolecular hydride shift to form an olefin or more likely a mixture of olefins [94]. Thus for 2-butylidene [CH$_3\ddot{C}$CH$_2$CH$_3$] the following products are obtained: (a) 3.3% 1-butene, (b) 66.6% trans-2-butene, (c) 29.5% cis-2-butene, and (d) 0.5% methylcyclopropene.

Recently diazoalkenes have been converted into diolefins by the use of silver perchlorate or PdCl$_2$-(C$_6$H$_5$CN)$_2$ [95]. Thus diazomethyl-3-cyclohexene responds as indicated

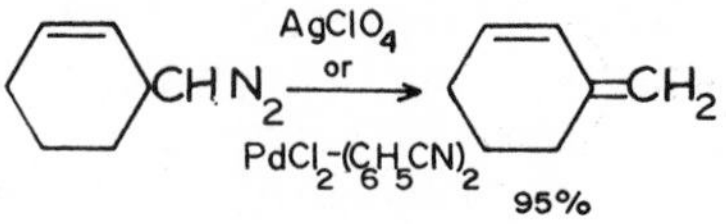

$$\overset{AgClO_4}{\underset{PdCl_2\text{-}(CH_2C_6H_5CN)_2}{\xrightarrow{\text{or}}}}$$

95%

to give methylene-3-cyclohexene.

To produce benzocycloheptatrienes, Moerck and Battiste [96] employed the diazadiene and benzyne and obtained the product without isolation of the intermediate diazo compound:

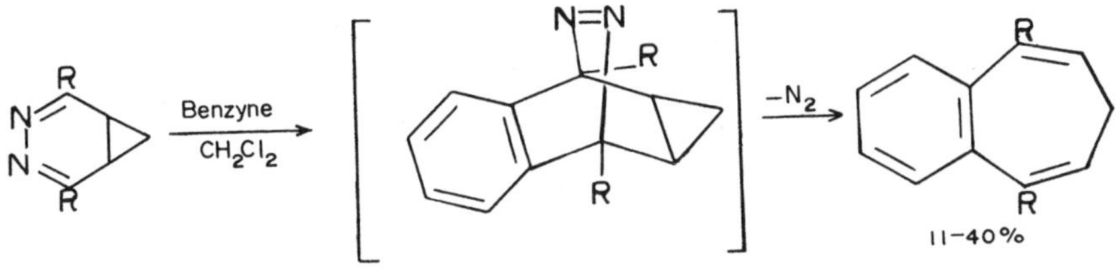

11-40%

Alkyllithiums have been employed rather widely in the elimination of tosylhydrazones. Bose and Steinberg [97] used methyllithium as shown in the synthesis of 3β-hydroxy-Δ⁵,²⁰-pregnadiene from pregnenolone tosylhydrazone:

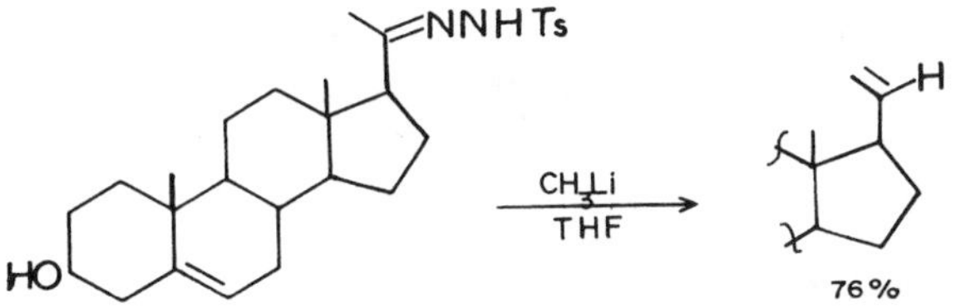

76%

The method has been extended to dienes. Thus Dauben and co-workers [98] synthesized 2-methyl-5-isopropenyl-1,3-cyclohexa-diene from carvone:

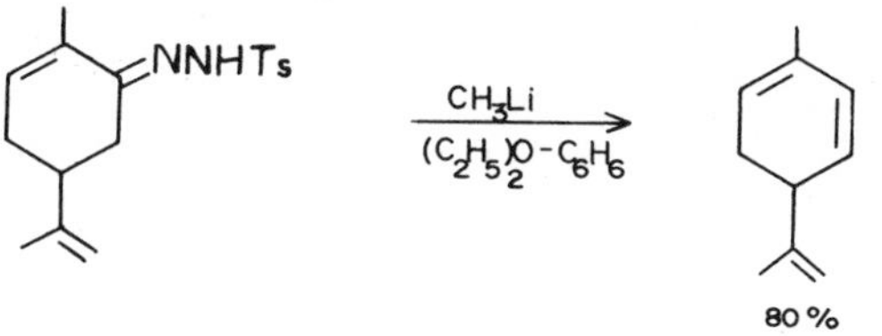

80%

Methyllithium is superior to butyllithium in this reaction, which appears to be restricted to tosylhydrazones with an adjacent methylene or methyl group.

Berson and Olin converted the 1,3-diene into the 1,4-diene via the hydrazo compound in the sequence indicated [99]:

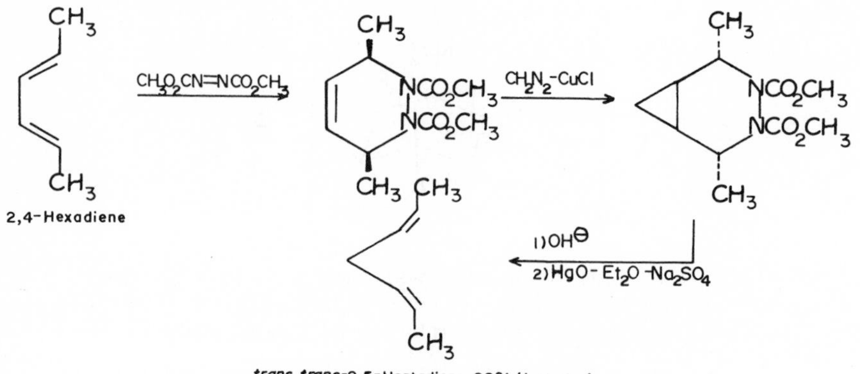

trans,trans-2,5-Heptadiene, 98% (last step)

Newman and ud Din [100] have generated an unsaturated carbene to form methylene cyclopropanes in good yield:

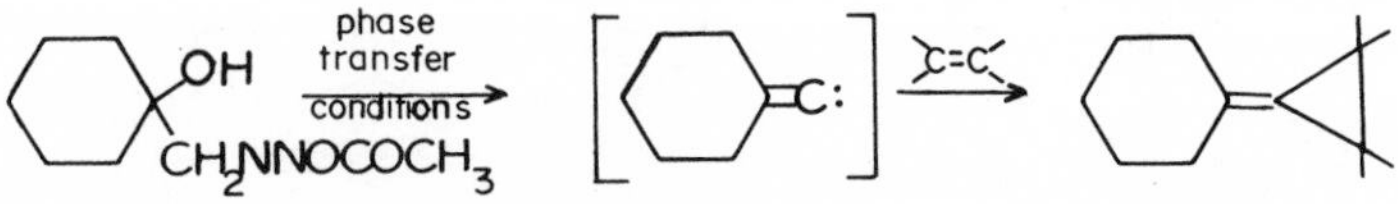

A spiro ring type containing a cyclopropene ring, although quite unstable, has been synthesized by Waali and Jones [101]:

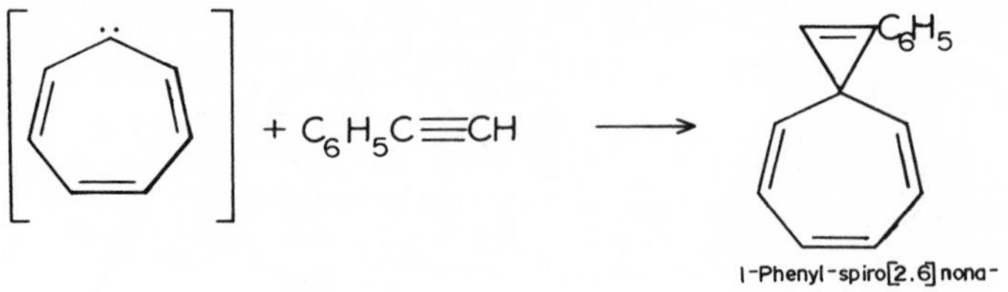

1-Phenyl-spiro[2.6]nona-1,4,6,8-tetraene

In diglyme at 145°, the reaction gave 2-phenylindene. The carbene was available from tropone tosylhydrazone.

16.   From Alkylboron Compounds (1, 100)

For the conversion of alkynes into alkenes via hydroboration, see B.1.

A new method of introducing the methallyl group into the carbon chain of an olefin, via a borane, has been devised [102]:

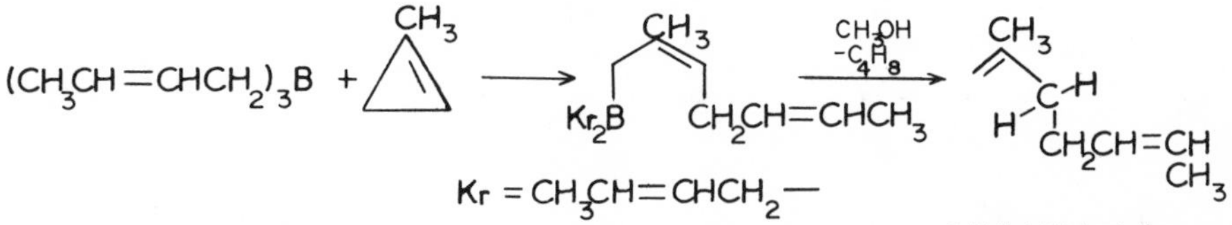

$(CH_3CH=CHCH_2)_3B$ + [triangle with CH₃] ⟶ [structure with Kr₂B, CH₃, $CH_2CH=CHCH_3$] $\xrightarrow[-C_4H_8]{CH_2OH}$ [structure CH₃, C-H, H, $CH_2CH=CH$, CH_3]

$Kr = CH_3CH=CHCH_2-$

2-Methyl-1,5-heptadiene
25% *cis* , 75% *trans*

Heating of boranes with alkenes gives displacement and redistribution reactions [103]. However, in a review of the preparation of organoboranes [104] it is stated that heating leads to more profound changes:

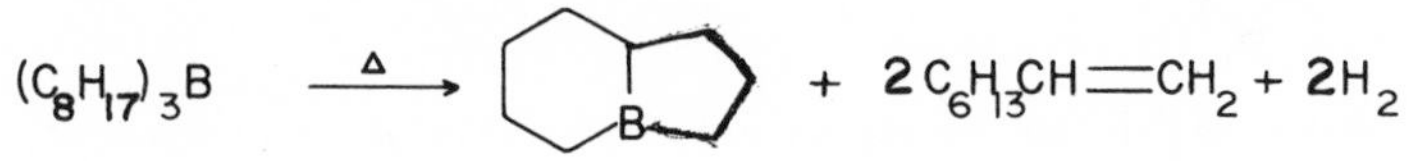

$(C_8H_{17})_3B$ $\xrightarrow{\Delta}$ [cyclic structure with B] + $2 C_6H_{13}CH=CH_2 + 2H_2$

17. From Aziridines

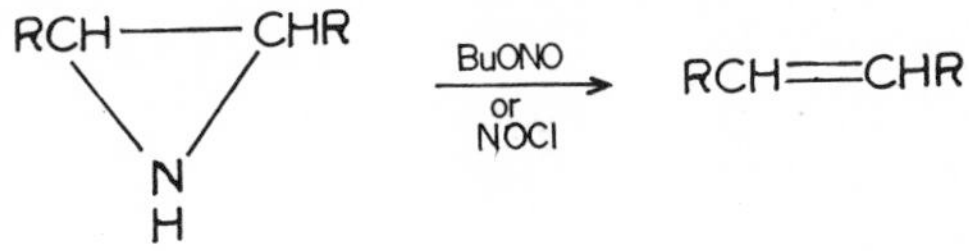

[structure RCH——CHR with N, H] $\xrightarrow[\text{or}]{\text{BuONO}}$ RCH=CHR
 NOCl

Clark has shown that the deamination of aziridines proceeds to the corresponding olefin of retained stereochemistry as indicated [105]:

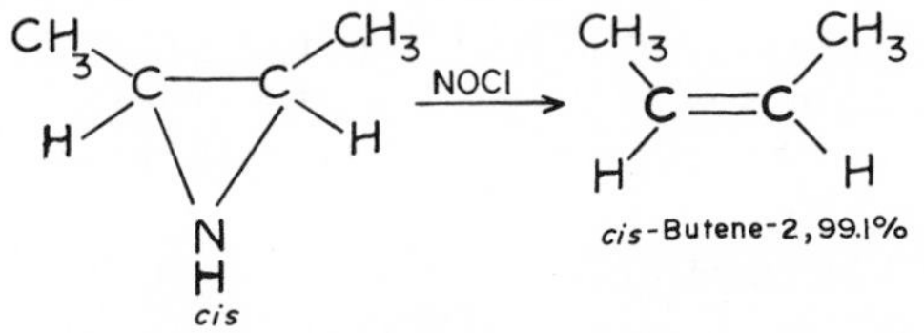

[structure with CH₃, C-C, H, H, N, H, cis] $\xrightarrow{\text{NOCl}}$ [structure CH₃, CH₃, C=C, H, H]
 *cis*-Butene-2,99.1%

The trans aziridine gives 99.7% of the trans alkene.
   The aziridine has been utilized as an intermediate in the conversion of the cis into the trans alkene [106]:

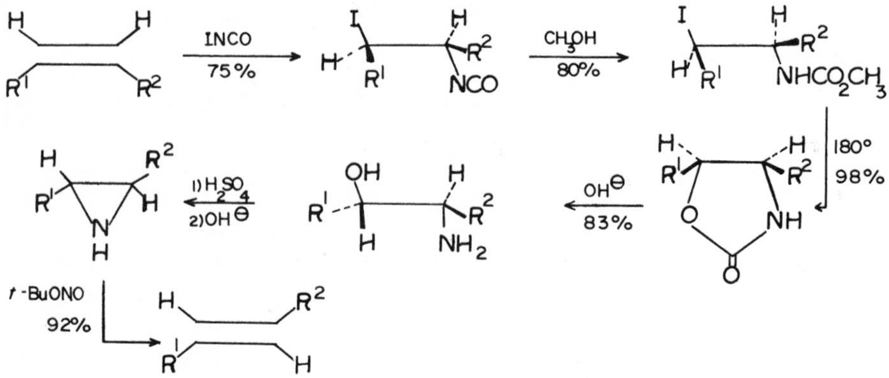

Alkenes may also be obtained from aziridines via the ylid [107]:

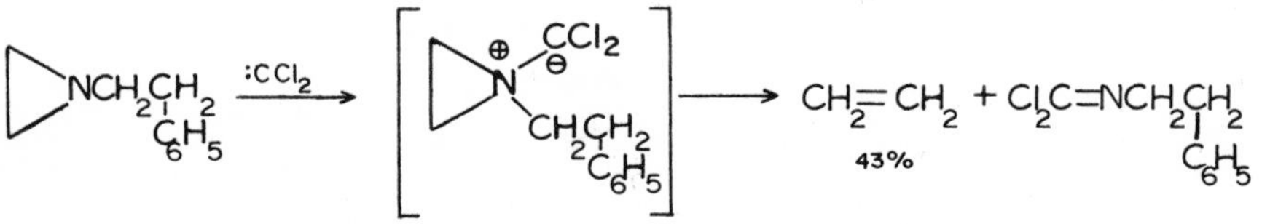

However, this reaction is not successful for four- or five-membered cyclic imines.

18.   From Nitriles [1a, p. 361]

Dehydrocyanation of nitriles with potassium amide in liquid ammonia leads to alkenes in good yield [108]. Thus α,α,β-triphenylpropionitrile gives triphenylethylene:

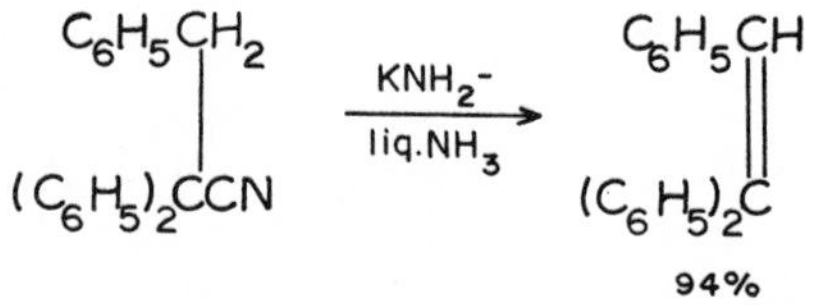

The dehydrocyanation is also applicable to cyanoamino types yielding enamines:

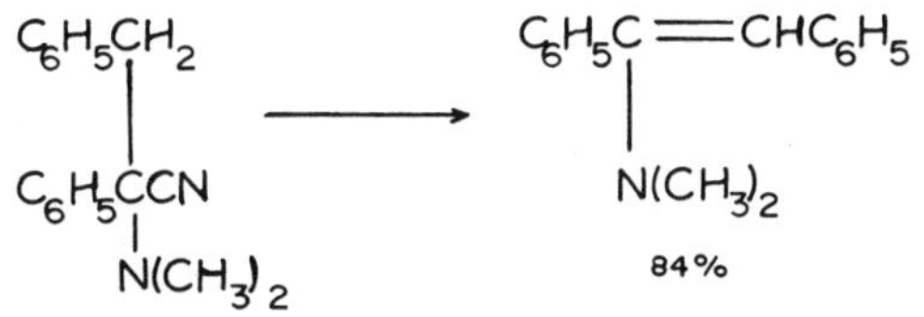

$$C_6H_5CH_2 \atop \underset{\displaystyle N(CH_3)_2}{C_6H_5\overset{|}{C}CN} \longrightarrow \underset{\displaystyle N(CH_3)_2}{C_6H_5\overset{|}{C}=CHC_6H_5}$$

84%

An elimination coupling was achieved with NaH in diglyme [109]:

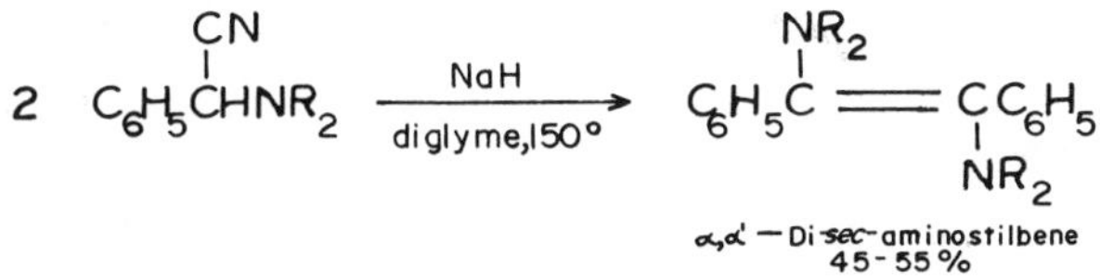

$$2 \; \underset{\displaystyle}{C_6H_5\overset{\displaystyle CN}{\underset{|}{C}}HNR_2} \; \xrightarrow[\text{diglyme,150°}]{\text{NaH}} \; C_6H_5\overset{\displaystyle NR_2}{\underset{\displaystyle NR_2}{C}}=CC_6H_5$$

α,α′—Di sec-aminostilbene
45–55%

19.   From Sulfur Heterocycles and Sulfones

Although episulfides have been discussed under A.12, many larger sulfur-containing ring compounds have served as a source of alkenes. The principal ones of this group are considered here.

The five-membered rings respond as indicated:

1.   2,5-Dihydrothiophenes, on being irradiated followed by heating with triphenylphosphine, lose S and give small amounts of acyclic dienes [110]:

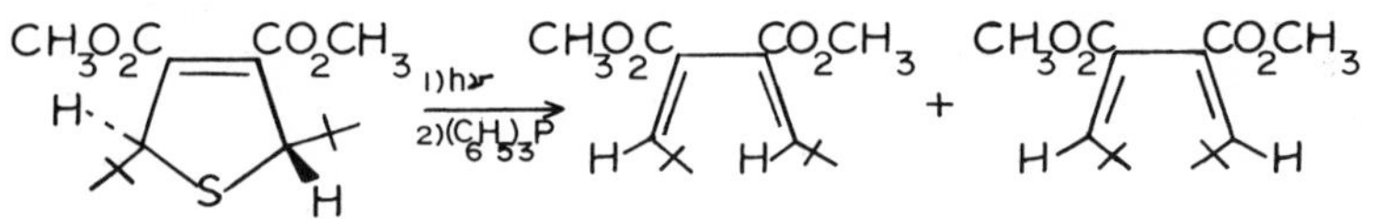

2.   3,5-Diphenylthiophene with nickel chloride and NaBH$_4$ gives a mixture of 1,2-dimethyl-1,2-diphenylethylenes and 2,4-diphenylbutane [111] (reductive elimination):

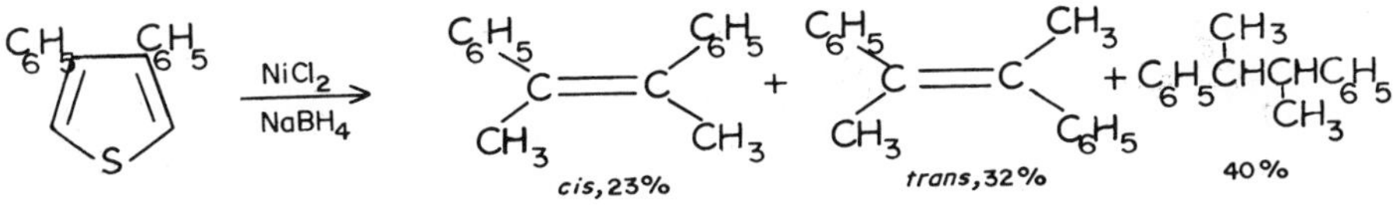

cis,23%   trans,32%   40%

3.   2,5-Dihydrothiophene-S-oxide, tetramethyl diester on irradiation gave the butadiene [112]:

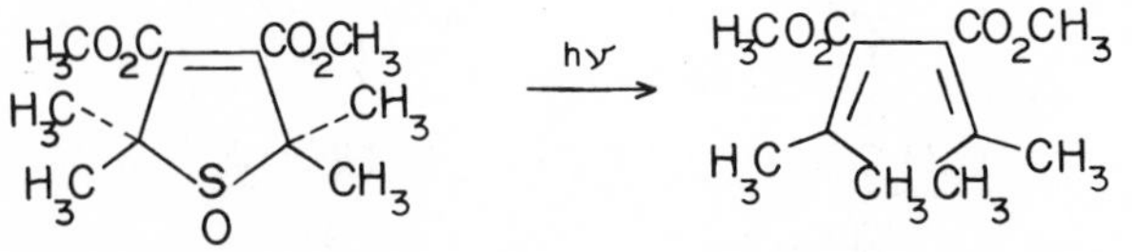

The corresponding S-dioxide gave with heat or irradiation the
tetramethylbutadiene diester as well.

4.   The less-substituted 2,5-dihydrothiophene-S-dioxide upon
being heated in xylene gave the corresponding butadienes usually
trapped as Diels-Alder adducts with dimethyl acetylenedicarboxylate [113].

5.   Sulfolane with dry solutions of $CuCl_2$/LiCl/S at 210-220°
gave the tetrachlorobutadiene [114]:

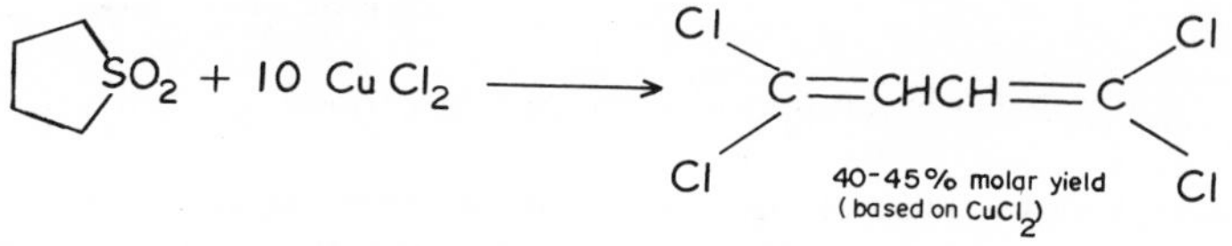

The oxathiolan-5-one with tris(diethylamino)phosphine gave
triphenylethylene [115]:

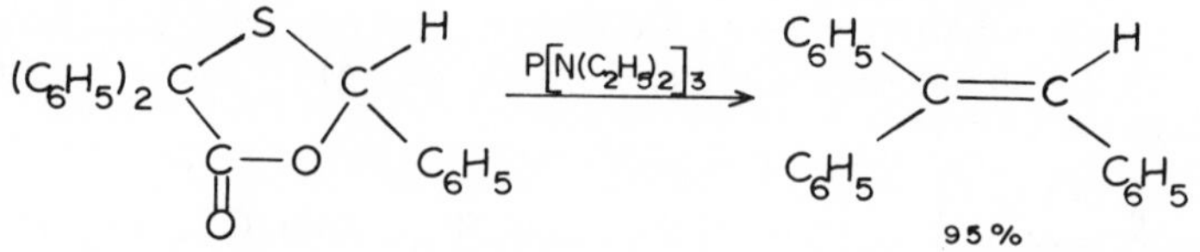

Likewise, the tetrahydrothiadiazole may be oxidized to the azo-
sulfide that in the presence of triphenylphosphine gave bis-
cyclohexylidene [116]:

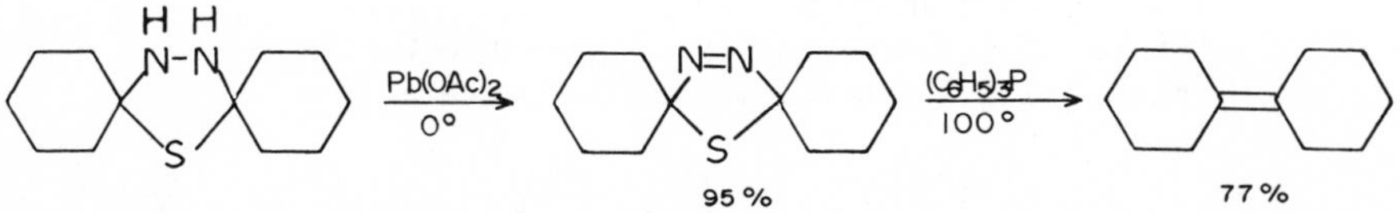

Sulfur has also been eliminated from the 2,11-dithia[3.3]
metacyclophane in a stepwise process involving a Stevens
rearrangement [117]:

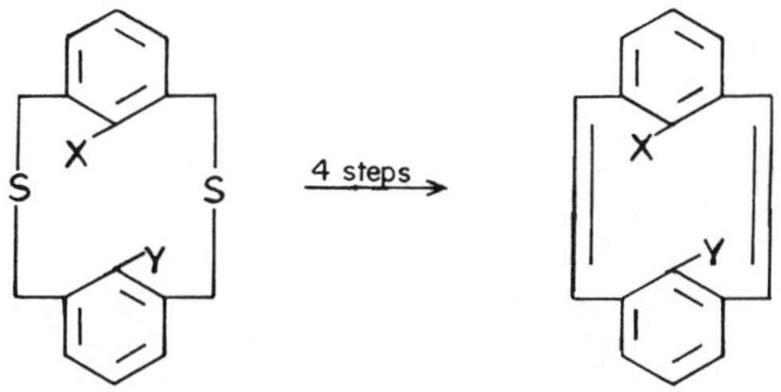

20. From Carbonyl Compounds [1a, p. 364]

Various methods, all of which involve intermediates, have been utilized in synthesizing alkenes from carbonyl compounds. From the aldehyde Carlson and Lee [106] utilized the sequence:

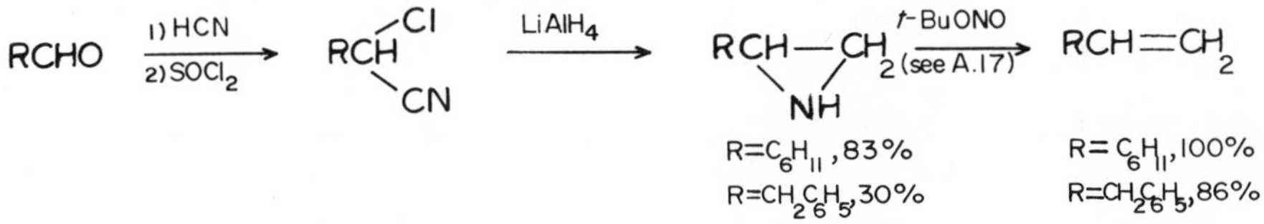

Fetizon and co-workers [118], in starting with a 3-keto-steroid, chose the path as indicated:

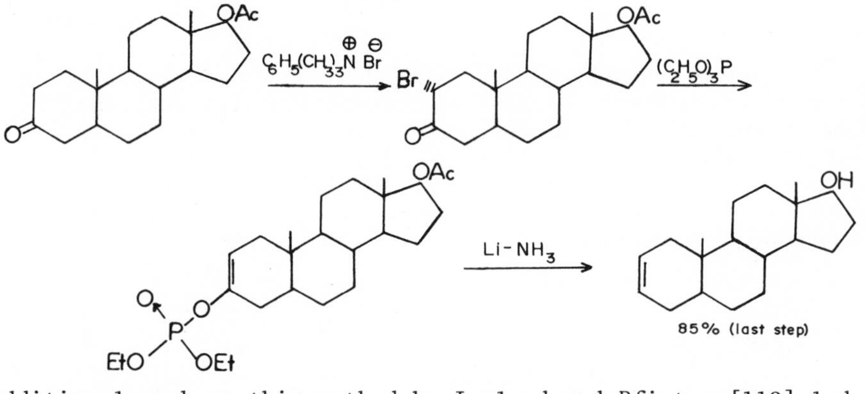

85% (last step)

Additional work on this method by Ireland and Pfister [119] led to a completion of the reaction without isolation of any intermediates.

Sharpless and co-workers [82] have employed the lower valent tungsten halides in the reduction of aldehydes and ketones to alkenes:

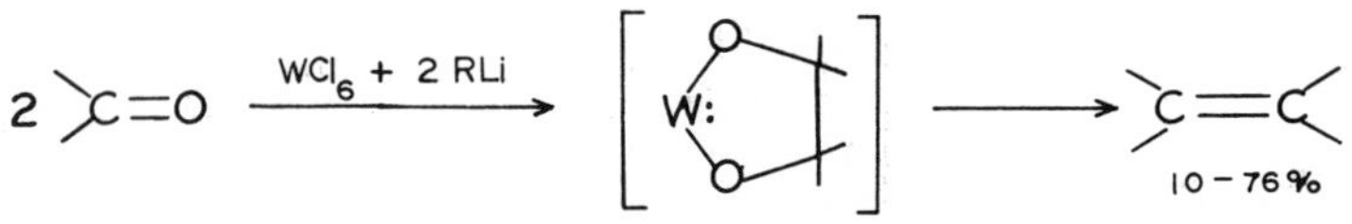

$$2 \; \text{>C=O} \xrightarrow{\text{WCl}_6 + 2\,\text{RLi}} \left[\text{W:} \right] \longrightarrow \text{>C=C<}$$

10 - 76%

Kuwajima and Uchida [120] converted nonenolizable carbonyl compounds efficiently into the corresponding olefin in a one-step scheme, as represented in the syntheses of 1,1-diphenyl-ethylene:

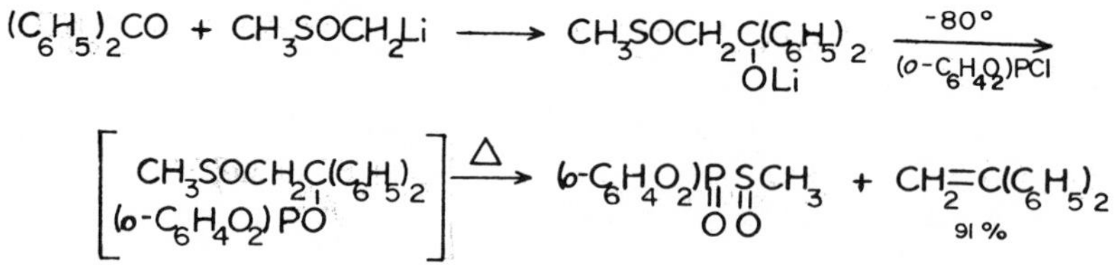

$$(C_6H_5)_2CO + CH_3SOCH_2Li \longrightarrow CH_3SOCH_2\underset{\underset{OLi}{|}}{C}(C_6H_5)_2 \xrightarrow[(o\text{-}C_6H_4O_2)PCl]{-80°}$$

$$\left[\begin{array}{l} CH_3SOCH_2C(C_6H_5)_2 \\ (o\text{-}C_6H_4O_2)PO \end{array} \right] \xrightarrow{\Delta} (o\text{-}C_6H_4O_2)\underset{\underset{O}{||}}{P}\underset{\underset{O}{||}}{S}CH_3 + CH_2{=}C(C_6H_5)_2$$

91%

Hudrlik and Peterson [121] utilized the β-ketosilane and proceeded as shown to synthesize the alkene, 2-methyl-1-heptene:

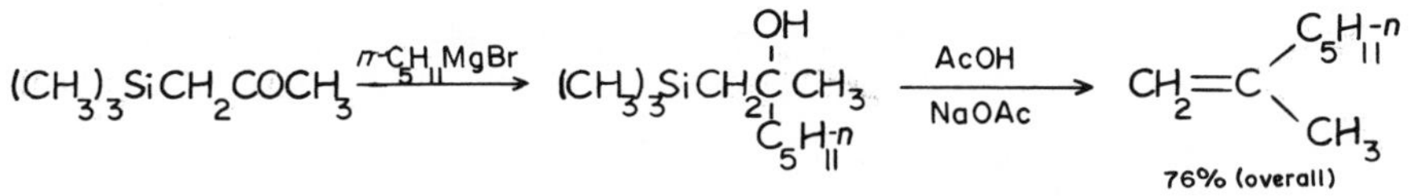

$$(CH_3)_3SiCH_2COCH_3 \xrightarrow{n\text{-}C_5H_{11}MgBr} (CH_3)_3SiCH_2\underset{\underset{C_5H_{11}\text{-}n}{|}}{\overset{\overset{OH}{|}}{C}}CH_3 \xrightarrow[\text{NaOAc}]{\text{AcOH}} CH_2{=}C\underset{\diagdown CH_3}{\overset{\diagup C_5H_{11}\text{-}n}{}}$$

76% (overall)

To convert highly hindered ketones, which are inert to methylenetriphenylphosphorane and proton transfer reactions, Sowerby and Coates [122] developed a method of methylenation involving the formation of an adduct followed by acylation and subsequent reduction. Applied to the highly hindered tricyclic ketone (±)-norizanone, which responds sluggishly to the Wittig reaction, the reaction is:

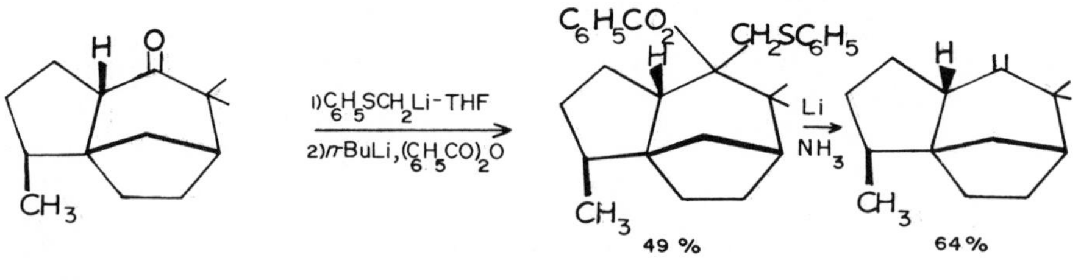

1) $C_6H_5SCH_2Li$-THF
2) n-BuLi,$(C_6H_5CO)_2O$

49%

$\xrightarrow[\text{NH}_3]{\text{Li}}$

64%

Motherwell [123] in a single reaction vessel at 25° converted alicyclic ketones into olefins as shown:

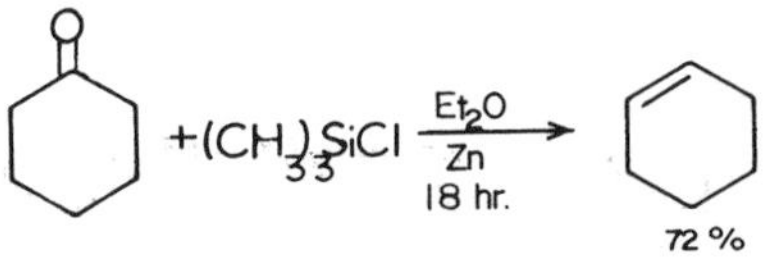

72 %

Yields in other cases were lower. The enol trimethyl silyl ether
is not an intermediate as it gives no alkene under the reaction
conditions.

Ketones have also been converted into alkenes by the use of
the complex formed between $TiCl_3$, THF, and Mg in the presence of
argon [124]. Thus benzophenone leads to tetraphenylethylene:

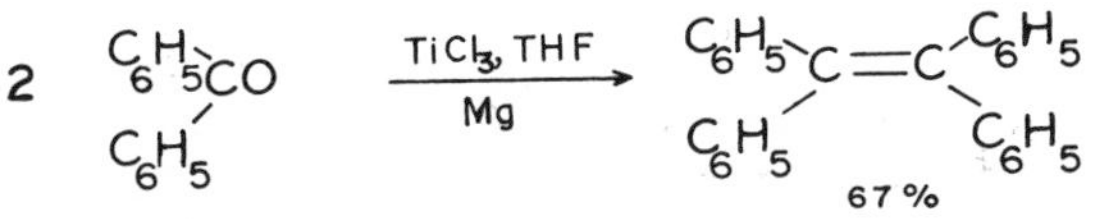

67 %

Other yields are usually lower. This process, which may be
called "reductive pinacolization," may proceed via a free radical
mechanism.

Both unsaturated aldehydes and ketones have been decarbony-
lated [1a, p. 365].

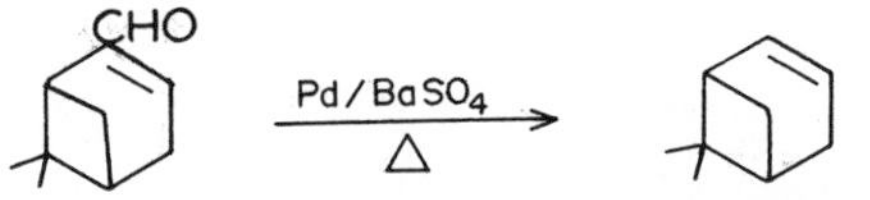

6,6-Dimethylbicyclo-
[3.1.1]heptene-2
73 %

21.   From Hydroxyacetophenones

The reduction of hydroxyacetophenones containing a sufficient
number of hydroxyl groups in the ortho and para positions with
$NaBH_4$ leads to vinyl phenols in good yield [125]. Of the hydroxyacetophenones studied, the best yield was obtained with 2,4,
6-trihydroxyacetophenone:

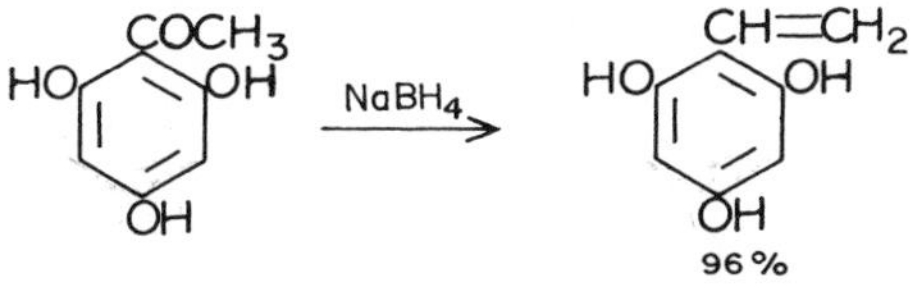

96 %

3-Hydroxyacetophenone gave 100% of the carbinol, which is an intermediate in the formation of the vinyl phenols.

22. From Vicinal Dinitro Compounds

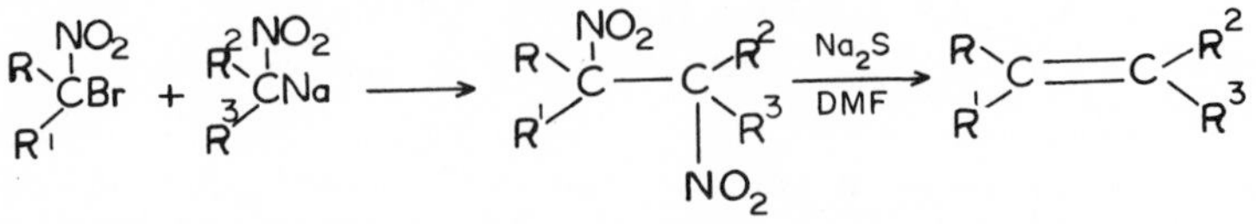

Both symmetrical and a few unsymmetrical acyclic and cyclic olefins are obtained by treating the 1,2-dinitro compound with Na_2S in DMF [126]. Yields vary from 82 to 92% and in five-, six-, and seven-membered ring types the double bond does not isomerize into the ring.

23. From β-Hydroxycarboxylic Acids or β-Lactones [1a, pp. 316, 329]

Sultanbawa [127] obtained a mixture of two diastereoisomeric β-hydroxy acids via the reduction of the corresponding β-keto ester. By treating one diastereoisomer as indicated the <u>cis</u> olefin was obtained:

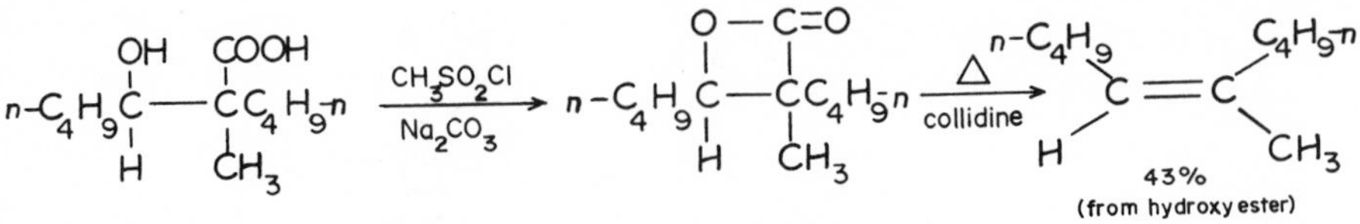

A similar treatment of the second diastereoisomer gave the <u>trans</u> olefin over 95% pure. Thus the elimination is stereospecific.
    For the synthesis of other olefins from β-lactones, see A.6.

24. From Charge-transfer Complexes

Because of the difficulty of storing 1,2-diphenylcyclopropenes as such, it was discovered that the charge-transfer complex formed with 9-dicyanomethylene-2,4,7-trinitrofluorene (DTF) may be stored for prolonged periods at 25° in air [128]. To regenerate the olefin from the charge-transfer complex, a dichloromethane solution of it is treated with alumina, from which the olefin is recovered by elution with pentane.

25. From 4-(Chloromethyl)-4-(trichlorosilyl)cyclohexene

The synthesis of 1,4-exocyclic-endocyclic dienes has been accomplished as shown [129]:

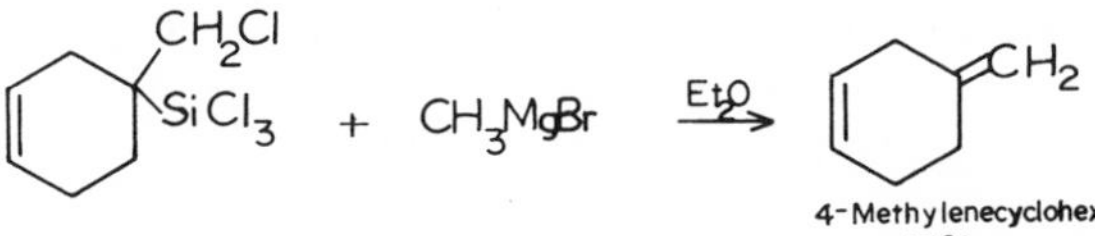

4-Methylenecyclohexene
43%

The substrate was prepared from 1,3-butadiene and 3-chloro-2-(trichlorosilyl)propene (Diels-Alder).

Addenda

General. Methods for synthesizing bridgehead alkenes have been reviewed [R. Keese, Angew. Chem. Intern. Ed. Eng., 14, 528 (1975)]. Bredt's rule has been reviewed [G. L. Buchanan, Chem. Soc. Rev., 3, 41 (1974)].

A.3. Ethylene may be prepared as needed from ethylene bromide and zinc in ethylene glycol [C. A. Brown, J. Org. Chem., 40, 3154 (1975)].

A.9. The pyrolysis of N-alkylacetamides to give olefins is described [H. E. Baumgarten et al., J. Am. Chem. Soc., 80, 4588 (1958)].

A.10. The catalytic transfer hydrogenation reaction, such as the conversion of anthracene into 1,2,3,4-tetrahydroanthracene by tetralin in the presence of palladium on carbon, has been reviewed [G. Brieger, T. J. Nestruck, Chem. Rev., 74, 567 (1974)].

A.12. Epoxides may be converted into olefins by $FeCl_3$-BuLi [T. Fujisawa et al., Chem. Lett., 883 (1974)], by $TiCl_3$-LiAlH$_4$ [J. E. McMurry, M. P. Fleming, J. Org. Chem., 40, 2555 (1975)], by KSeCNO [R. A. W. Johnstone et al., J. Chem. Soc., Perkin Trans., I, 1216 (1975)] or by LDIA followed by acidification [R. G. Riley, J. A. Katzenellenbogen, J. Org. Chem., 39, 1957 (1974)]; cf. J. K. Crandall, L. C. Crawley, Org. Syn., 53, 17 (1973)]. By using the last reagent 3-methyl-3,4-oxido-1-butene gave 2-hydroxymethyl-1,3-butadiene (76% crude).

A.19.

$$(R_2C=CHCH_2)_2SO_2 \xrightarrow[KOH]{CCl_4} R_2C=CHCH=CHCH=CR_2$$

18-89% (5 examples)

Not stereospecific [G. Büchi, R. M. Freidinger, J. Am. Chem. Soc., 96, 3332 (1974)]. See also K. E. Koenig et al., J. Org. Chem., 39, 1539 (1974)]. The mechanism probably involves a dioxythio-epoxide intermediate that eliminates SO_2.

A.19. Fragmentation of 2,5-diphenyl-1,3-oxathiolan with lithium diethylamide gave styrene in high yield [G. H. Witham et al., J. Chem. Soc., Perkin Trans., I, 433 (1974)].

A.23. β-Hydroxyacids, prepared from the lithium α-lithio carboxylates and carbonyl compounds, yield β-hydroxyacids, which cyclize to β-lactones. Thermolysis of the latter gives the olefin [A. P. Krapcho, E. G. E. Jahngen, Jr., J. Org. Chem., 39, 1322, 1650 (1974)].

A.25. A more general elimination occurs from the β-hydroxyethylsilane. Choice of conditions gives either the cis or trans alkene [P. F. Hudrlik, D. Peterson, Tetrahedron Letters, 1133 (1974)].

References

1. J. Reucroft, P. G. Sammes, Quart. Rev. (London), 25, 135 (1971).
1a. H.-W., 5, Pt. 1b, 1972, p. 45.
1b. Ref. 1a, p. 62.
2. R. S. Monson, Tetrahedron Letters, 567 (1971); R. S. Monson, D. N. Priest, J. Org. Chem., 36, 3826 (1971).
3. R. O. Hutchins et al., J. Org. Chem., 37, 4190 (1972).
4. J. C. Martin, R. J. Arhart, J. Am. Chem. Soc., 93, 4327 (1971); 94, 5003 (1972).
5. J. S. Lomas et al., Tetrahedron Letters, 599 (1971).
6. W. Reeve, D. M. Reichel, J. Org. Chem., 37, 68 (1972).
7. E. M. Burgess et al., J. Org. Chem., 38, 26 (1973); P. Crabbé, C. León, J. Org. Chem., 35, 2594 (1970).
8. T. Billé-Samé, E. D. Bergmann, Bull. Soc. Chim. France, 4209 (1972).
9. T. J. Mead et al., Chem. Commun., 679 (1972).
10. J.-M. Bonnier et al., Bull. Soc. Chim. France, 2306 (1972).
11. E. M. Kaiser, Synthesis, 405 (1972).
12. A. Claesson, C. Bogentoft, Acta Chem. Scand., 26, 2540 (1972).
13. E. J. Corey, K. Achiwa, J. Org. Chem., 34, 3667 (1969).
14a. H.-D. Scharf, F. Laux, Synthesis, 582 (1970).
14b. E. Z. Said, A. E. Tipping, J. Chem. Soc., Perkin Trans., I, 1399 (1972).

15. H. Oediger et al., Synthesis, 591 (1972); T. J. Barton, J. Org. Chem., 37, 552 (1972).
16. A. J. Parker, Chem. Tech., 1, 297 (1971); D. J. Lloyd, A. J. Parker, Tetrahedron Letters, 637 (1971).
17. E. J. Parish, D. H. Miles, J. Org. Chem., 38, 1223 (1973).
18. S. P. Acharya, H. C. Brown, Chem. Commun., 305 (1968).
19. F. Hunziker, F. X. Müllner, Helv. Chim. Acta, 41, 70 (1958).
20. R. A. Bartsch et al., J. Org. Chem., 37, 458 (1972); J. Am. Chem. Soc., 95, 3405 (1973).
21. G. Sturtz et al., Bull. Soc. Chim. France, 2962 (1971); 4012 (1971).
22. P. Binger et al., Angew. Chem., Intern. Ed. Engl., 9, 810 (1970).
23. R. N. McDonald, R. R. Reitz, Chem. Commun., 90 (1971).
24. D. Reisdorf, H. Normant, Organomet. Chem. Syn., 1, 375 (1972).
25. R. H. Sieber, Ann. Chem., 730, 31 (1969).
26. F. Fisher, D. E. Applequist, J. Org. Chem., 30, 2089 (1965).
27. P. Binger et al., Angew. Chem. Intern. Ed. Engl., 8, 205 (1969).
28. S. H. Ruetman, Synthesis, 680 (1973).
29. H. Stetter, W. Uerdingen, Synthesis, 207 (1973).
30. P. Binger, Synthesis, 190 (1974).
31. D. A. Rosie, G. G. Shone, Lipids, 6 (8), 623 (1971).
32. S. W. Bennett et al., J. Organomet. Chem., 27, 195 (1971).
33. J. F. Garst et al., Chem. Commun., 78 (1969).
34. A. Merijanian et al., J. Org. Chem., 37, 3945 (1972).
35. J. Buddrus, W. Kimpenhaus, Chem. Ber., 106, 1648 (1973).
36. G. H. Posner, J.-S. Ting, Syn. Commun., 3, 281 (1973).
37. K. M. Ibne-Rasa et al., Chem. Ind. (London), 232 (1973).
38. J. K. Kochi et al., Tetrahedron, 24, 3503 (1968).
39. J. P. Schroeder et al., J. Org. Chem., 35, 3181 (1970).
40. H. W. Geluk, Synthesis, 652 (1970).
41. H. G. Henning, Z. Chem., 6, 463 (1966).
42. P. Crabbé, A. Guzmán, Tetrahedron Letters, 115 (1972).
43. E. J. Corey, P. A. Grieco, Tetrahedron Letters, 107 (1972).
44. N. F. Wood, F. C. Chang, J. Org. Chem., 30, 2054 (1965).
45. A. J. Parker et al., Tetrahedron Letters, 3015 (1971).
46. W. R. Dolbier, Jr., J. H. Alonso, Chem. Commun., 394 (1973).
47. D. N. Butler, R. A. Snow, Can. J. Chem., 50, 795 (1972).
48. J. C. Carnahan, Jr., W. D. Closson, Tetrahedron Letters, 3447 (1972).
49. W. Reeve et al., J. Am. Chem. Soc., 93, 4607 (1971).
50. R. L. Sowerby, R. M. Coates, J. Am. Chem. Soc., 94, 4758 (1972).

51. W. H. Okamura, F. Sondheimer, J. Am. Chem. Soc., 89, 5991
     (1967); H. Oediger et al., Synthesis, 595 (1972).
52. W. Adam et al., J. Am. Chem. Soc., 94, 2000 (1972).
53. R. J. Spangler, J. H. Kim, Synthesis, 107 (1973).
54. H. Gerlach, W. Müller, Helv. Chim. Acta, 55, 2277 (1972).
55. K. G. Rutherford et al., J. Chem. Soc., C582 (1971).
56. N. I. Burke et al., Ind. Eng. Chem. Prod. Res. Dev., 9, 230
     (1970).
57. R. E. Gilman et al., Can. J. Chem., 48, 970 (1970).
58. M. S. Newman, F. W. Hetzel, J. Org. Chem., 34, 3604 (1969).
59. H. Z. Sommer et al., J. Org. Chem., 36, 824 (1971).
60. A. C. Cope, E. R. Trumbull, Org. React., 11, 357, 377
     (1960).
61. D. A. Archer, J. Chem. Soc., C1327 (1971).
62. M. R. Short, J. Org. Chem., 37, 2201 (1972).
63. A. C. Cope, R. D. Bach, Org. Syn., Coll. Vol., 5, 315
     (1973).
64. C. W. Spangler, G. F. Woods, J. Org. Chem., 30, 2218 (1965).
65. G. H. Schmid, A. W. Wolkoff, Can. J. Chem., 50, 1181 (1972).
66. Y. Hata, M. Watanabe, J. Am. Chem. Soc., 95, 8450 (1973).
67. K. Syhora, H. Bocková, Tetrahedron Letters, 2369 (1965).
68. J. W. Cook et al., J. Chem. Soc., 1074 (1949).
69. P. J. DeChristopher et al., J. Am. Chem. Soc., 91, 2384
     (1969).
70. A. Novelli et al., Tetrahedron Letters, 613 (1968).
71. M. Fétizon, N. Moreau, Bull. Soc. Chim. France, 2404 (1966).
72. M. Jones, Jr., R. A. Moss, Carbenes, Vol. 1, Wiley, New
     York, 1973, p. 333.
73. B. M. Trost, T. N. Salzmann, J. Am. Chem. Soc., 95, 6840
     (1973).
74. T. Durst et al., J. Am. Chem. Soc., 95, 3420 (1973).
75. C. R. Johnson et al., J. Am. Chem. Soc., 95, 6462 (1973),
76. M. Julia, J.-M. Paris, Tetrahedron Letters, 4833 (1973).
77. E. R. deWaard et al., Tetrahedron Letters, 1481 (1973);
     E. J. Corey, E. Block, J. Org. Chem., 34, 1233 (1969);
     L. A. Paquette, R. W. Houser, J. Am. Chem. Soc., 91, 3870
     (1969).
78. I. Kuwajima et al., Tetrahedron Letters, 737 (1972).
79. R. H. Mitchell, Tetrahedron Letters, 4395 (1973).
80. K. B. Sharpless et al., Tetrahedron Letters, 1979 (1973).
81. G. Cainelli et al., Chem. Commun., 144 (1970).
82. K. B. Sharpless et al., J. Am. Chem. Soc., 94, 6538 (1972).
83. S. M. Kupchan, M. Maruyama, J. Org. Chem., 36, 1187 (1971).
84. E. Vedejs, P. L. Fuchs, J. Am. Chem. Soc., 93, 4070 (1971);
     J. Org. Chem., 38, 1178 (1973).
85. M. Rosenblum et al., J. Am. Chem. Soc., 94, 7170 (1972).
86. D. L. Clive, C. V. Denyer, Chem. Commun., 253 (1973); T. H.
     Chan, J. R. Finkenbine, Tetrahedron Letters, 2091 (1974).

87. B. M. Trost, S. D. Ziman, J. Org. Chem., 38, 932 (1973).
88. P. B. Shevlin et al., Chem. Commun., 901 (1973).
89. A. J. Bridges, G. H. Whitham, Chem. Commun., 142 (1974).
90. C. J. Ireland, J. S. Pizey, Chem. Commun., 4 (1972).
91. G. M. Mkryan et al., J. Org. Chem. USSR, 3, 1121 (1967).
92. H. Felkin, A. Tambute, Tetrahedron Letters, 821 (1969).
93. G. Jones II et al., Chem. Commun., 374 (1973).
94. Ref. 72, p. 19.
95. M. Sakai, S. Masamune, J. Am. Chem. Soc., 93, 4610 (1971).
96. R. E. Moerck, M. A. Battiste, Chem. Commun., 1171 (1972).
97. A. K. Bose, N. G. Steinberg, Synthesis, 595 (1970).
98. W. G. Dauben et al., J. Am. Chem. Soc., 90, 4762 (1968).
99. J. A. Berson, S. S. Olin, J. Am. Chem. Soc., 91, 777 (1969).
100. M. S. Newman, Z. ud Din, J. Org. Chem., 38, 547 (1973).
101. E. E. Waali, W. M. Jones, J. Org. Chem., 38, 2573 (1973).
102. B. M. Mikhailov et al., Tetrahedron Letters, 4627 (1972).
103. H. C. Brown, Boranes in Organic Chemistry, Cornell U. P.,
     1972, p. 305.
104. R. Koster, Angew. Chem. Intern. Ed. Engl., 3, 174 (1964).
105. R. D. Clark, G. K. Helmkamp, J. Org. Chem., 29, 1316 (1964).
106. R. M. Carlson, S. Y. Lee, Tetrahedron Letters, 4001 (1969).
107. Y. Hata, M. Watanabe, Tetrahedron Letters, 3827, 4659
     (1972).
108. C. R. Hauser et al., J. Am. Chem. Soc., 82, 1786 (1960).
109. J. W. Scheeren, P. E. M. van Helvoort, Syn. Commun., 1, 113
     (1971).
110. R. M. Kellog, J. Am. Chem. Soc., 93, 2344 (1971).
111. H. Wynberg et al., Syn. Commun., 2, 415 (1972).
112. W. L. Prins, R. M. Kellogg, Tetrahedron Letters, 2833
     (1973).
113. J. M. McIntosh, H. B. Goodbrand, Tetrahedron Letters, 3157
     (1973).
114. P. E. Prillwitz, R. Louw, Syn. Commun., 1, 125 (1971).
115. D. H. R. Barton, B. J. Willis, Chem. Commun., 1225 (1970);
     G. H. Witham et al., J. Chem. Soc., Perkin Trans., I, 433
     (1974).
116. D. H. R. Barton et al., Chem. Commun., 1226 (1970).
117. B. H. Mitchell, V. Boekelheide, Tetrahedron Letters, 1197
     (1970).
118. M. Fetizon et al., Chem. Commun., 112 (1969).
119. R. E. Ireland, G. Pfister, Tetrahedron Letters, 2145
     (1969).
120. I. Kuwajima, M. Uchida, Tetrahedron Letters, 649 (1972).
121. P. F. Hudrlik, D. Peterson, Tetrahedron Letters, 1785
     (1972).
122. R. L. Sowerby, R. M. Coates, J. Am. Chem. Soc., 94, 4758
     (1972).

123. W. B. Motherwell, Chem. Commun., 935 (1973).
124. S. Tyrlik, I. Wolochowicz, Bull. Soc. Chim. France, 2147
      (1973).
125. K. H. Bell, Australian J. Chem., 22, 601 (1969).
126. N. Kornblum et al., J. Am. Chem. Soc., 93, 4316 (1971).
127. M. U. S. Sultanbawa, Tetrahedron Letters, 4569 (1968);
      A. P. Krapcho et al., J. Org. Chem., 39, 1322, 1650
      (1974).
128. M. A. Battiste et al., Synthesis, 273 (1974).
129. R. F. Cuneo, E. M. Dexheimer, Organomet. Chem. Syn., 1, 253
      (1971).

B. Reduction

    1. From Acetylenes (Largely Borane Addition) (1, 106) [1]

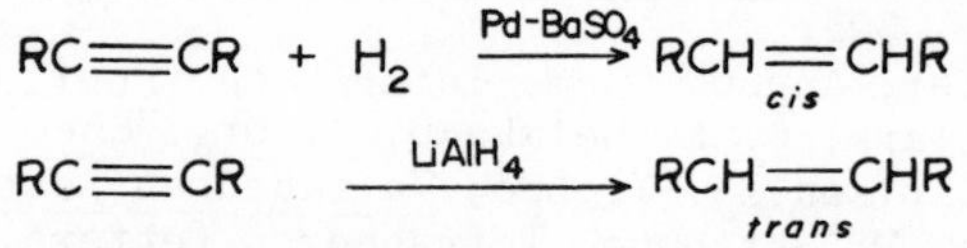

$$RC\equiv CR + H_2 \xrightarrow{\text{Pd-BaSO}_4} \underset{cis}{RCH=CHR}$$

$$RC\equiv CR \xrightarrow{\text{LiAlH}_4} \underset{trans}{RCH=CHR}$$

The alkylborons have proved to be unusually valuable in the synthesis of cis and trans alkenes from acetylenes. The first investigators to report their use in this manner appear to be Cainelli and co-workers [1a], who treated the carbonyl compound with the organometallic intermediate as shown:

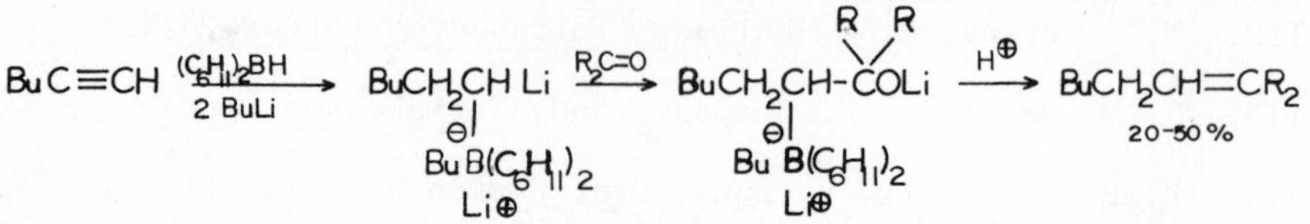

$$BuC\equiv CH \xrightarrow[\text{2 BuLi}]{(C_6H_{11})_2BH} \underset{\substack{\ominus|\\Bu B(C_6H_{11})_2\\Li\oplus}}{BuCH_2CH Li} \xrightarrow{R_2C=O} \underset{\substack{\ominus|\\Bu B(C_6H_{11})_2\\Li\oplus}}{BuCH_2CH-COLi} \xrightarrow{H\oplus} \underset{20-50\%}{BuCH_2CH=CR_2}$$

Zweifel's method which does not involve a carbonyl compound is given in 1, 101. The same investigator has also developed a method [2] for the synthesis of trans alkenes:

$$CH_3(CH_2)_3C \equiv CBr \xrightarrow{(C_6H_{11})_2BH} CH_3(CH_2)_3C = C \begin{smallmatrix} Br \\ \\ \end{smallmatrix} \begin{smallmatrix} C_6H_{11} \\ B \\ C_6H_{11} \end{smallmatrix} \xrightarrow{NaOCH_3}$$

$$\underset{H}{\overset{CH_3(CH_2)_3}{>}} C = C - B \begin{smallmatrix} OCH_3 \\ C_6H_{11} \end{smallmatrix} \xrightarrow{AcOH} \underset{H}{\overset{CH_3(CH_2)_3}{>}} C = C \begin{smallmatrix} H \\ C_6H_{11} \end{smallmatrix}$$

trans-1-Cyclohexyl-1-hexene, 90%

By employing thexyl monoalkylboranes in the Zweifel synthesis,
Brown has utilized an alkylboron that is more readily available
and one in which the total alkyl availability is used [2a].

A method has also been developed for the synthesis of trans
olefins via hydroboration-cyanohalogenation [2b]:

$$RC \equiv CH \xrightarrow{R_2'BH} RCH = CHBR_2' \xrightarrow{BrCN} \underset{R'}{\overset{Br \quad CN}{R CH - CH - BR'}} \xrightarrow[elimination]{cis} \underset{H}{\overset{R}{>}} C = C \begin{smallmatrix} H \\ R' \end{smallmatrix}$$

A method has also been devised, starting with alkynes, for the
introduction of vinyl groups into cycloalkane rings [3].

Because of the instability of the geminal diboroalkanes
toward sodium hydroxide, Zweifel and Steele [4] substituted the
1,1-dialuminoalkane for them. Thus they were able to synthesize
cis and trans hexenes in superior yields:

$$C_2H_5C \equiv CC_2H_5 \xrightarrow{LiAlH_2(CH_2CH \begin{smallmatrix} CH_3 \\ CH_3 \end{smallmatrix})_2} \underset{H}{\overset{C_2H_5}{>}} C = C \underset{Al(CH_2CH \begin{smallmatrix} CH_3 \\ CH_3 \end{smallmatrix})_2}{\overset{C_2H_5}{}} \xrightarrow{CH_3Li}$$

$$\left[\underset{H}{\overset{C_2H_5}{>}} C = C \underset{\underset{CH_3}{\overset{}{\mid}}{Al(CH_2CH \begin{smallmatrix} CH_3 \\ CH_3 \end{smallmatrix})_2}}{\overset{C_2H_5}{}} \right]^{\ominus} Li^{\oplus} \xrightarrow{H_2O} \underset{H}{\overset{C_2H_5}{>}} C = C \begin{smallmatrix} C_2H_5 \\ H \end{smallmatrix}$$

cis-3-Hexene, 90% (glpc)

Similarly, by using lithium diisobutylmethylaluminum hydride, the
trans-3-hexene is obtained in 88% yield (glpc). Thus the first

hydride adds <u>cis</u> to the acetylene with conversion into a <u>cis</u> olefin, but the second hydride, because of its bulk, adds <u>trans</u> with
the eventual formation of the <u>trans</u> olefin. <u>cis,trans</u>-Butadienes
have been synthesized via hydroboration (<u>1</u>, 101). In addition
Zweifel and Polston [5] succeeded in synthesizing <u>cis,cis</u>-conju-
gated dienes from diynes as shown:

$$n\text{-}C_4H_9C\equiv CC\equiv CC_4H_9\text{-}n \quad \xrightarrow[2)\,CH_3CO_2H]{1)\,(C_6H_{11})_2BH}$$

cis,cis —Dodeca-5,7-diene, 79 %

The syntheses of 1,3-dienes [6] and 1,5-dienes have been
reviewed [7].

Numerous reagents have been employed in recent years to
reduce acetylenes to alkenes. Among these are: (a) catecholborane followed by aqueous CH_3COOH [8], (b) diisoamylboron—H_2O_2
[9], (c) borohydride-reduced nickel (P-2Ni]-ethylenediamine—H_2
[10], and (d) lithium in ethylamine [19]. For reductive alkylation to alkenes MeMgBr-(Ph$_3$P)$_2$-NiCl$_2$ [11], C_2H_5Cu-MgBr$_2$ [12],
C_2H_5Cu-MgBr$_2$-n-BuI [13], carbene addition [14], and CH_3MgBr-(Ph
CN)$_2$-PdCl$_2$ [15] have been utilized.

The stereoselectivity of hydrogenation with Pd or P-2 Ni [<u>1</u>,
106] has already been mentioned. With the former as Pd/C, ten
acetylenes were hydrogenated to <u>cis</u> olefins [16]. With P-2 Ni
and NH$_2$CH$_2$CH$_2$NH$_2$, 1-phenylpropyne gave 95% of 1-phenylpropene,
the <u>cis-trans</u> ratio being approximately 200:1 [10a]. In fact,
this catalyst permits the hydrogenation of norbornene double
bonds in the presence of other double bonds [10b]; in addition,
the P-1 Ni catalyst reduces triple bonds to single bonds in the
presence of double bonds [17].

A review of the catalytic semihydrogenation of the triple
bond is available [18]. Fortunately, supported catalysts reduce
terminal triple bonds faster than internal ones and both are
reduced faster than terminal or internal double bonds. For semihydrogenation a lessening in the rate of hydrogenation should
occur after complete conversion of the alkyne into the alkene.
The most successful attempt to accomplish this end experimentally
is that of Lindlar, who employed his lead-poisoned palladium-on-
calcium carbonate catalyst. Nevertheless, palladium is the
catalyst most commonly employed in semihydrogenation to produce,
usually, <u>cis</u>-disubstituted olefins. In the preparation of the
<u>cis</u> form by employing the Lindlar catalyst and very pure
materials a break has been noted in the hydrogen uptake [18].

Although sodium/liquid ammonia/alcohol (Birch reagent) is a common reagent for converting cis alkynes into trans alkenes (1, 107), the same result may be obtained with lithium in ethylamine (Benkeser reagent) [19]. In fact, these reagents are often interchangeable. Thus 5-decyne has been reduced in good yields to trans-5-decene by Na in liquid ammonia at -33° or by lithium in ethyl amine at -78° (see B.3 for a further discussion of these reagents). In the reduction of monoalkylacetylenes with Na in liquid ammonia, the addition of ammonium sulfate led to quantitative yields of the alkene [20].

Lastly, electrocatalytic reductions are possible [21]:

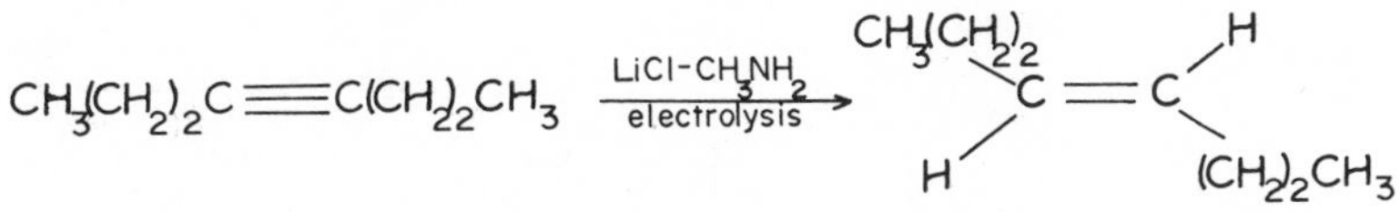

$CH_3(CH_2)_2C{\equiv}C(CH_2)_2CH_3$
LiCl-CH_3NH_2
electrolysis
$CH_3(CH_2)_2$
H
H
$(CH_2)_2CH_3$

trans-Octene, 96%

Nonconjugated aromatic acetylenes were also reduced to trans aromatic olefins in the electrolysis.

2. From Dienes and Trienes (1, 108) [1, p. 599]

Allenes have been reduced to olefins by the use of H_2 in the presence of chlorotris (triphenylphosphine) rhodium [22]:

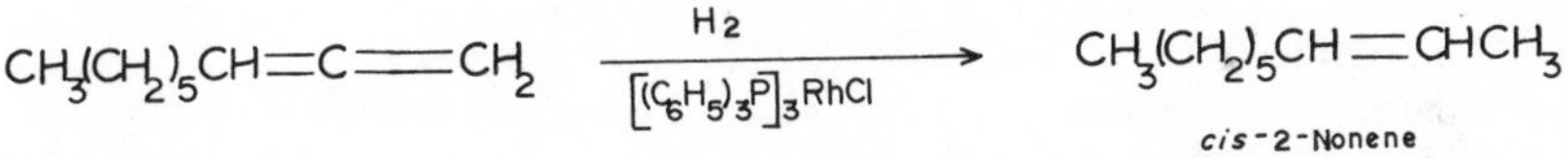

$CH_3(CH_2)_5CH{=}C{=}CH_2$
H_2
$[(C_6H_5)_3P]_3RhCl$
$CH_3(CH_2)_5CH{=}CHCH_3$

cis-2-Nonene
66% conversion (glc)

Ionic hydrogenation occurs when carbonium ions and a molecule that furnishes hydride ions are present [23]. Organosilanes are suitable hydride-ion donors. When present with trifluoroacetic acid, a diene may be partially reduced as shown:

$CH_2CH_2CH{=}CH_2$
CF_3COOH
$(C_2H_5)_3SiH$
H
$(CH_2)_2CH{=}CH_2$

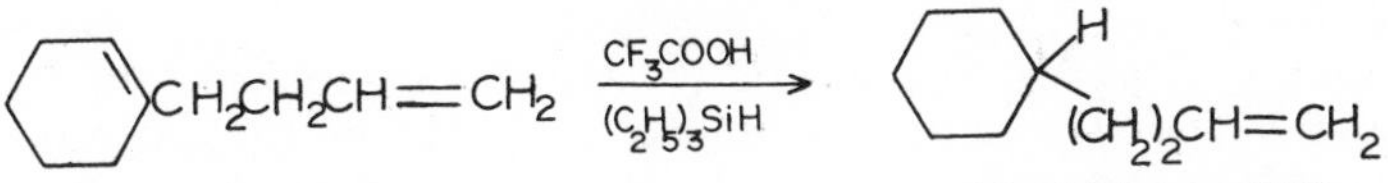

4-Cyclohexyl-1-butene, 65%

The method is not as satisfactory when there is one methylene group between the double bonds; in fact, for conjugated bonds in the ring practically no hydrogenation occurs.

Besides, the expected cyclohexene dimers, in which each ring is partially reduced, form when cyclohexadiene is treated with Na in liquid NH_3 [24]:

2,2'—Biphenylene(main product)

Trienes have also been reduced to alkenes. Thus cis,trans, trans-1,5,9-cyclododecatriene has been converted into cis-cyclododecene [25]:

cis-Cyclodecene, 51–76%

A mixture of cis- and trans-cyclododecenes (99.9%) was obtained by the reduction of 1,5,9-cyclododecatriene with H_2 in the presence of $(Ph_3P)_2NiI_2$-Ph_3P [26].

A rather novel reducing agent for the conversion of multienes into enes is tetraethylammonium trichlorostannite ($Et_4\overset{+}{N}$ $Sn\bar{C}l_3$) mixed with platinum chloride. The salts acts as a molten solvent and ligand for the platinum ion. The bright red solution is an excellent catalyst for carbonylation or for selective hydrogenation of some trienes [27]. For instance, 1,5,9-cyclododecatriene is smoothly hydrogenated to cyclododecene. Moreover, the catalyst is sufficiently stable so that the alkene may be removed by distillation.

3.   From Aromatic Compounds (Birch and Benkeser Reductions) (1, 108) [1, p. 614]

The Birch reduction has been reviewed by other investigators [28-30], and a comparison of the Birch and Benseker reductions may be found in [31].

Both the Birch reduction (sodium and alcohols in liquid ammonia) and the Benkeser reduction (lithium in low-molecular-weight amines) generally yield partially or fully reduced products. As a rule the Benkeser reduction is the more powerful but less selective of the two [31]. However, the selectivity of the former may be increased by the proper choice of solvents. Often there is little choice between the two methods.

In the aromatic series reductions to form cyclohexenes are quite common. Thus t-butylbenzene gives 1-t-butylcyclohexene [32]:

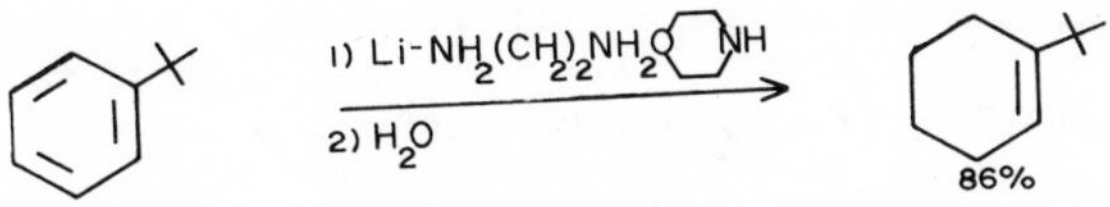

In addition, cyclohexadienes may be obtained by methods shown:

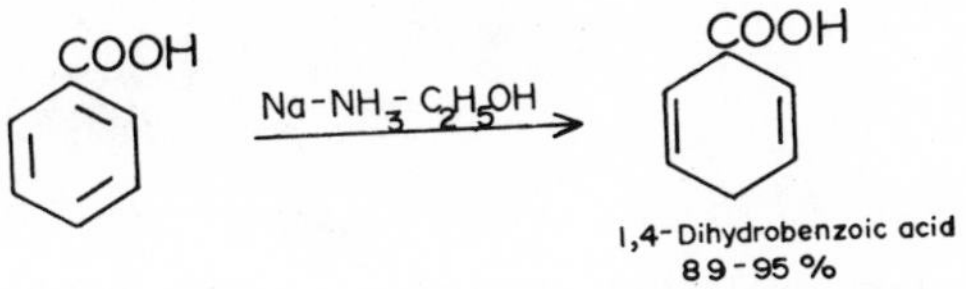

[33]

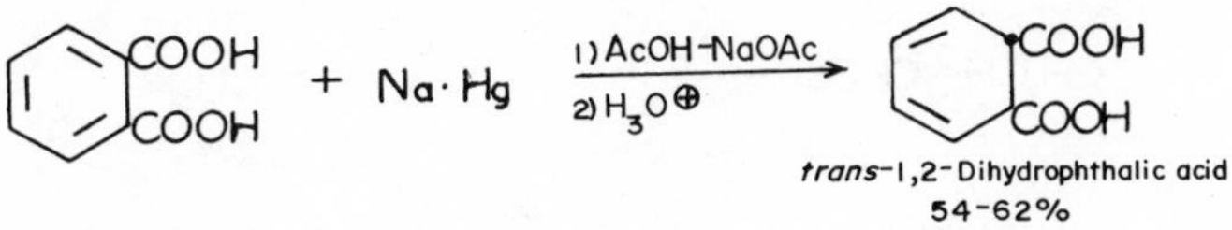

1,4-Dihydrobenzoic acid
89-95 %

[34]

trans-1,2-Dihydrophthalic acid
54-62%

5. From Ketones (Kishner Reduction-Elimination and the Like (1, 110)

A series of tosyl hydrazones of α,β-unsaturated compounds has been reduced to the corresponding alkenes by NaBH$_3$CN refluxing DMF-sulfolane [35]. Yields varied from 36 to 98%.

6. From Enamines (1, 111)

Pyrrolidine enamines of acyclic and cyclic ketones give good yields of olefins on heating with hydrides. For example, the enamines of cyclohexene with AlH$_3$ at 40° for 48 hr gave 80% of cyclohexene [36]. The pyrrolidine enamine of α-substituted cyclohexanones give good yields of 3-substituted cyclohexene as:

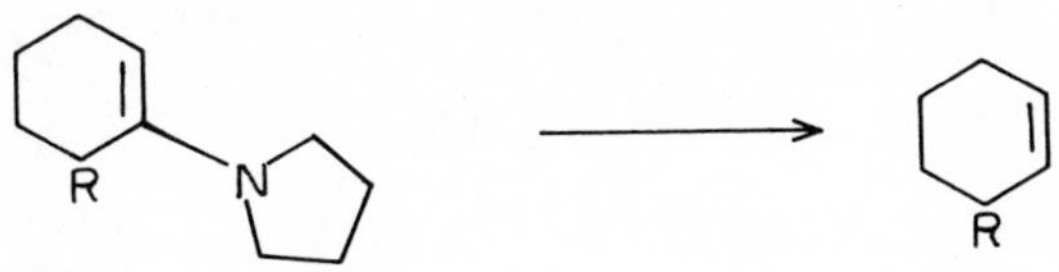

Enamines of straight-chain aldehydes on hydrogenolysis give good yields of olefins, but the yield is much lower for α-branched aldehydes.

7.   From Vinyl Ethers, Vinyl Sulfides, Vinyl Halides, and the Like (Reductive Elimination)

Allyl pyridyl sulfides may be converted into alkenes as indicated [37]:

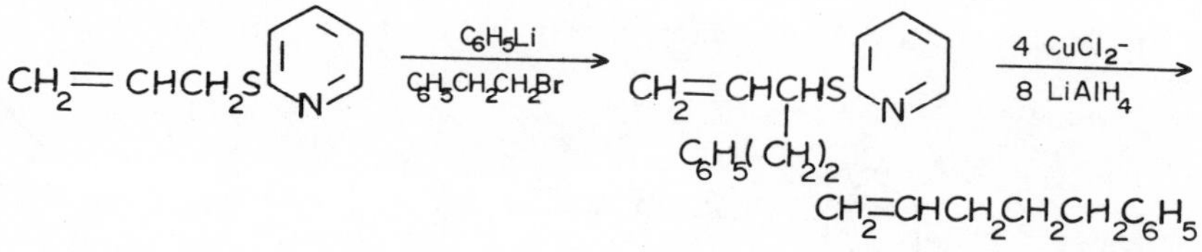

5-Phenyl-1-pentene, 80 %

Pino and Lorenzi [38] subjected vinyl ethers to reductive elimination as indicated:

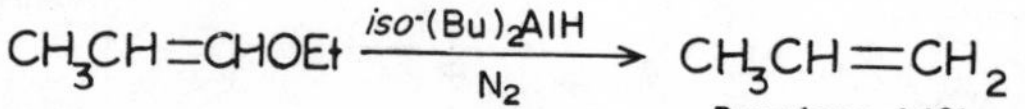

$$CH_3CH=CHOEt \xrightarrow[N_2]{iso\text{-}(Bu)_2AlH} CH_3CH=CH_2$$

Propylene, 44%

Tanabe and co-workers [39] utilized iron pentacarbonyl in the reductive elimination of a series of substituted olefins as shown:

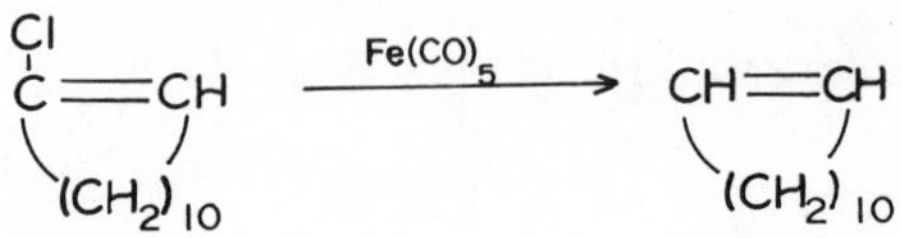

Cyclododecene, 30 %

Evans and co-workers utilized an unsaturated sulfoxide in a method for preparing trisubstituted alkenes [40]:

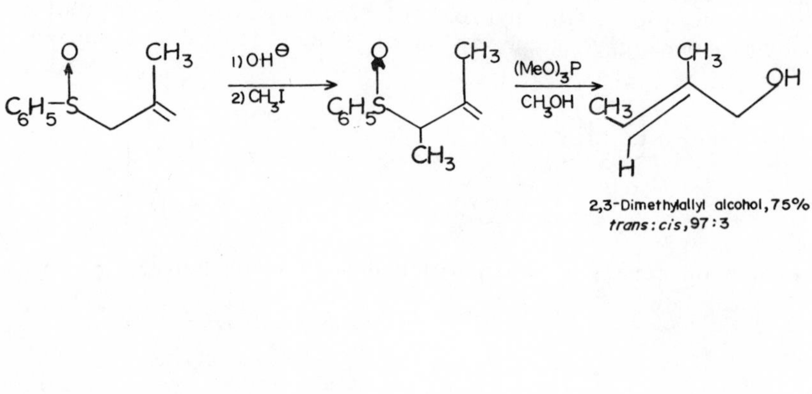

2,3-Dimethylallyl alcohol, 75%
trans:cis, 97:3

Trichloroacetyl chloride has been converted into methylene-cycloalkanes via the dichloromethylenecycloalkane as shown [41]:

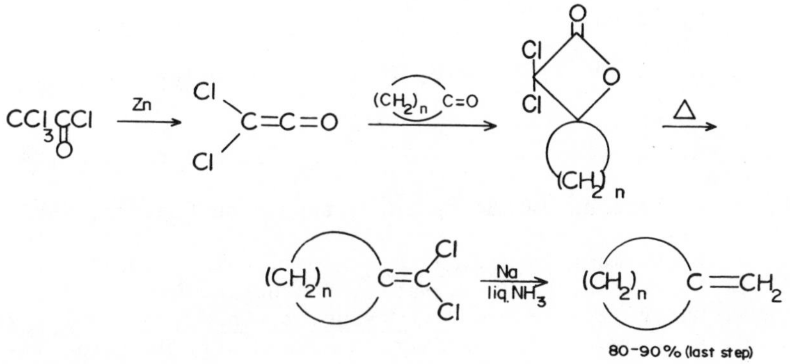

80-90% (last step)

Reductive elimination in α,β-unsaturated phenyl sulfones also leads to alkenes [42]:

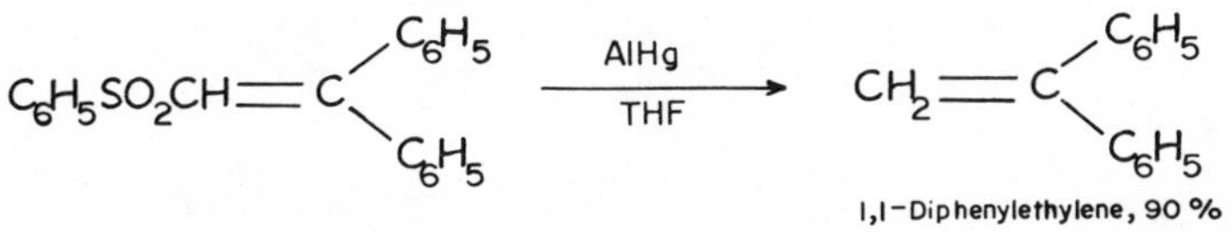

I,I-Diphenylethylene, 90 %

Addendum

B.1.   Diisobutylaluminum hydride gives reasonable stereospecific control in the reduction of acetylenes to alkenes [E. Winterfeldt, Synthesis, 628 (1975)].

References

1.   H.-W., 5, Pt. 1b, 1972, p. 579.
1a.   G. Cainelli et al., Tetrahedron Letters, 4315 (1966).
2.   G. Zweifel, H. Arzoumanian, J. Am. Chem. Soc., 89, 5086 (1967).
2a.   H. C. Brown et al., Synthesis, 555 (1972).
2b.   G. Zweifel et al., J. Am. Chem. Soc., 94, 6560 (1972).
3.   G. Zweifel, R. P. Fisher, Synthesis, 557 (1972); G. Zweifel et al., J. Am. Chem. Soc., 93, 6309 (1971).
4.   G. Zweifel, R. B. Steele, J. Am. Chem. Soc., 89, 2754, 5085 (1967); Tetrahedron Letters, 6021 (1966).
5.   G. Zweifel, N. L. Polston, J. Am. Chem. Soc., 92, 4068 (1970).

6. G. Zweifel, R. L. Miller, J. Am. Chem. Soc., 92, 6678 (1970).

7. J. A. Marshall, Synthesis, 229 (1971).

8. H. C. Brown, S. K. Gupta, J. Am. Chem. Soc., 94, 4370 (1972).

9. G. Holan, D. F. O'Keefe, Tetrahedron Letters, 673 (1973).

10a. C. A. Brown, V. K. Ahuja, Chem. Commun., 553 (1973).

10b. C. A. Brown, Chem. Commun., 952 (1969).

11. J.-G. Duboudin, B. Jousseaume, J. Organomet. Chem., 44, C1 (1972).

12. J. F. Normant, M. Bourgain, Tetrahedron Letters, 2583 (1971).

13. J. F. Normant et al., J. Organomet. Chem., 40, C49 (1972).

14. L. E. Friedrich, R. A. Fiato, Synthesis, 611 (1973).

15. N. Garty, M. Michman, J. Organomet. Chem., 36, 391 (1972).

16. C. A. Brown, A. C. S. Petroleum Chemistry Division Reprints, 12 (4), B-71-B-85 (1967).

17. E. J. Corey et al., Tetrahedron Letters, 1837 (1969); J. Am. Chem. Soc., 91, 4318 (1969).

18. E. N. Marvell, T. Li, Synthesis, 457 (1973).

19. R. A. Benkeser et al., J. Am. Chem. Soc., 77, 3378 (1955); E. M. Kaiser, Synthesis, 397 (1972).

20. A. L. Henne, K. W. Greenlee, J. Am. Chem. Soc., 65, 2020 (1943).

21. R. A. Benkeser, C. A. Tincher, J. Org. Chem., 33, 2727 (1968).

22. M. M. Bhagwat, D. Devaprabhakara, Tetrahedron Letters, 1391 (1972).

23. D. N. Kursanov et al., Synthesis, 633 (1974).

24. D. Y. Myers et al., Tetrahedron Letters, 533 (1973).

25. M. Ohno, M. Okamoto, Org. Syn., 49, 30 (1969).

26. M. T. Musser, U. S. 3,631,210, Dec. 28, 1971; C. A., 76, 72116 (1972).

27. L. W. Gosser, G. W. Parshall, Dupont Innovation, 6 (2), 6 (1975).

28. A. A. Akhrem, I. G. Reshetova, Yu. Titov, Birch Reduction of Aromatic Compounds, IFI, Plenum, New York, 1972.

29. A. J. Birch, G. S. Rao, Advances in Organic Chemistry, Wiley-Interscience, 1972, Vol. 8, p. 1.

30. R. G. Harvey, Synthesis, 161 (1970).

31. E. M. Kaiser, Synthesis, 391 (1972).

32. R. A. Benkeser et al., Tetrahedron Letters (16) 1 (1960).

33. M. E. Kuehne, B. F. Lambert, Org. Syn., Coll. 5, 400 (1973).

34. R. N. McDonald, C. E. Reineke, Org. Syn., 50, 50 (1970).

35. R. O. Hutchins et al., J. Am. Chem. Soc., 95, 3662 (1973).

36. J. M. Coulter et al., Tetrahedron, 24, 4489 (1968).

37. T. Mukaiyama et al., Bull. Chem. Soc. Jap., 44, 2285 (1971).
38. P. Pino, G. P. Lorenzi, J. Org. Chem., 31, 329 (1966).
39. M. Tanabe et al., Tetrahedron Letters, 447 (1973).
40. D. A. Evans et al., Tetrahedron Letters, 1389 (1973).
41. W. T. Brady, A. D. Patel, Synthesis, 565 (1972).
42. V. Pascali, A. Umani-Ronchi, Chem. Commun., 351 (1973).

C.   Addition and Coupling Reactions

1.   From Olefins by Acid or Ziegler Catalysis (1, 113) [1]

The oligomerization of alkenes has been reviewed [1, p. 505] in general and with coordination catalysts in particular [1, p. 537].

α-Alkyl-substituted α,β-unsaturated nitriles or carbonyl compounds give dimers when heated with cyclohexyl isocyanide-copper[I] oxide catalyst [1a]. Thus methyl crotonate responds as shown:

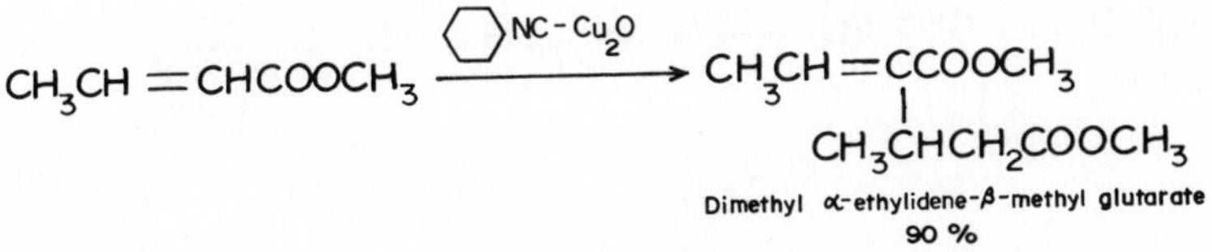

$$CH_3CH=CHCOOCH_3 \xrightarrow{\overset{\bigcirc}{}NC-Cu_2O} CH_3CH=CCOOCH_3$$
$$\underset{\underset{CH_3CHCH_2COOCH_3}{|}}{}$$

Dimethyl α-ethylidene-β-methyl glutarate
90 %

Dimerization of alkenes is also controllable by the use of triethylaluminum and tricrotylboron. Thus 1-butene may be converted into a substituted hexene [2]:

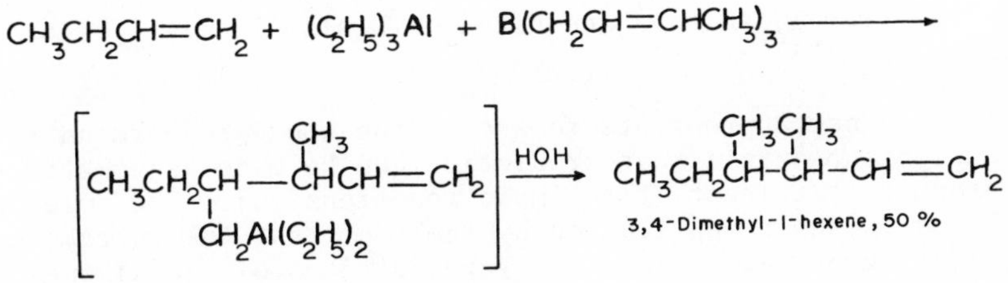

$$CH_3CH_2CH=CH_2 + (C_2H_5)_3Al + B(CH_2CH=CHCH_3)_3 \longrightarrow$$

$$\left[\begin{array}{c} \overset{CH_3}{\underset{|}{}} \\ CH_3CH_2CH-CHCH=CH_2 \\ \underset{CH_2Al(C_2H_5)_2}{|} \end{array} \right] \xrightarrow{HOH} \begin{array}{c} \overset{CH_3}{\underset{|}{}} \overset{CH_3}{\underset{|}{}} \\ CH_3CH_2CH-CH-CH=CH_2 \end{array}$$

3,4-Dimethyl-1-hexene, 50 %

Butadiene with the complex shown gives a good return of the cyclic diene [3]:

$$2 \quad CH_2{=}CHCH{=}CH_2 \xrightarrow[\substack{n\text{-BuLi} \\ CH_3OH}]{(Bu_3P)_2NiBr}$$

2-Methylene vinyl cyclopentane, 90%

The novel arylation, which appears to be quite useful, has been reviewed [4]:

$$C_6H_5CH{=}CH_2 + ArH \xrightarrow[AcONa]{PdCl_2} C_6H_5CH{=}CHAr$$

The complex of the alkene and palladium salt need not be pre-formed, but the concentration of sodium acetate must be high enough to allow the palladium complex to add to the arene. The last step may be the elimination of palladium hydride.

Olefins containing a tertiary carbon atom have been produced from t-butyl chloride, ethylene, and $AlCl_3$ in the sequence shown [5]:

$$CH_3\underset{\underset{CH_3}{|}}{\overset{\overset{CH_3}{|}}{C}}Cl + CH_2{=}CH_2 \xrightarrow[-30°]{AlCl_3} CH_3\underset{\underset{CH_3}{|}}{\overset{\overset{CH_3}{|}}{C}}CH_2CH_2Cl \xrightarrow[\substack{180\text{-}200° \\ 22\ atm}]{AcOK}$$

$$CH_3\underset{\underset{CH_3}{|}}{\overset{\overset{CH_3}{|}}{C}}CH_2CH_2OAc \xrightarrow{\triangle} CH_3\underset{\underset{CH_3}{|}}{\overset{\overset{CH_3}{|}}{C}}CH{=}CH_2$$

3,3-Dimethyl-1-butene

2. From 1,3-Dienes and Alkenes (Diels-Alder) (1, 116) [1, p. 433]

It is now appropriate to sketch the Woodward-Hofmann rules in sufficient detail that prediction can be made as to structures of products, at least, for simple reactions [6]. The rules are weakened in their application by the fact that the intramolecular cyclization or intermolecular cycloadditions of the alkenes or polyene at hand must proceed via a concerted reaction path when in reality a polar or radical mechanism may prevail under some circumstances. While the W.-H. rules must therefore be applied with some caution, a large body of concerted reactions has been rationalized through their application. For a 4π-electron system, the following possibilities for a concerted, electrocyclic reaction arise (it is assumed that the butadiene is in the <u>cis</u> form shown if cyclization occurs):

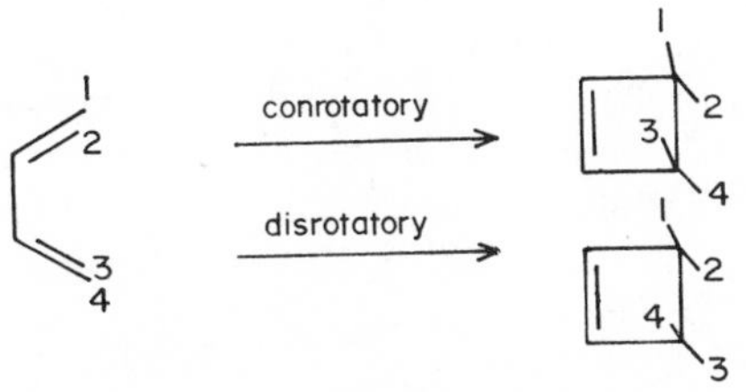

The above must be true because the groups 1, 2, 3, and 4 are nearly in the same plane in the butadienes. As the cyclobutene forms, the groups must rotate out of the plane of the ring. They do so in two ways. They can rotate in the same direction (conrotatory) or the opposite (disrotatory) as shown above. The W.-H. rules for an electrocyclic reaction of substrate with $q\pi$ electrons are shown in Table 2.3 [7]:

Table 2.3

q	Thermal	Photochemical
4n	Conrotatory	Disrotatory
4n + 2	Disrotatory	Conrotatory

These rules derive from observations concerning the conservation of symmetry in the new orbitals formed in the concerted reaction. Space does not permit this derivation, but the original source may be consulted [6]. As an example of the $q=4n\,\pi$ electron system, the cyclic heptadiene illustrated yields the <u>cis</u>-bicyclic cyclobutene by thermal means:

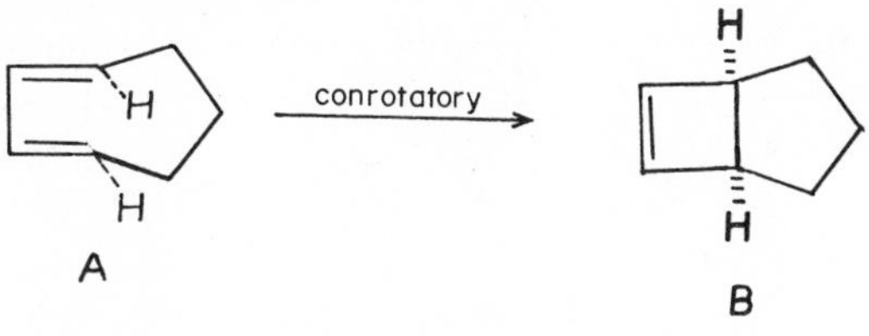

By photochemical means the heptadiene should yield the <u>trans</u>-bicyclic cyclobutene. A number of similar photochemical cycloadditions has been reviewed by Coyle [8].

Moreover, the principle of microscopic reversibility states that the pathway, either backward or forward, for a reversible reaction should be the same.   Although all Diels-Alder reactions are not truly reversible, those that are should follow W.-H. rules.   Thus B should yield A as it does by thermal means.   As an example of the 6π (4n+2) electron system, both isomers of the following triene have been shown to be disrotatory in their concerted thermal reactions:

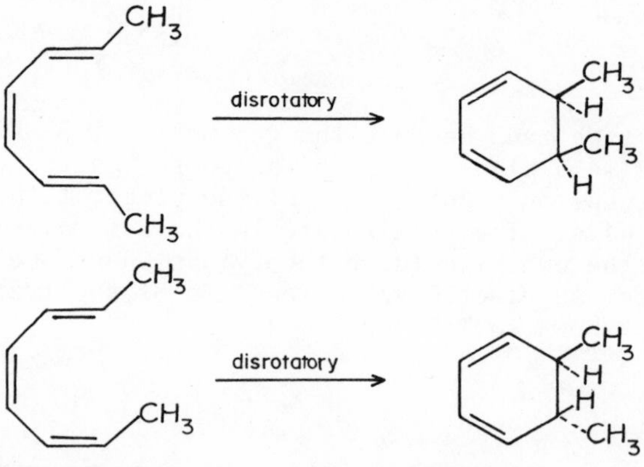

For a two-component, cycloaddition reaction, what Woodward calls the supra,supra pathway is conceptionally simple.   The two π bonds, for instance, of two ethylene systems approach each other with coalescence of the π orbitals to form sigma orbitals as shown:

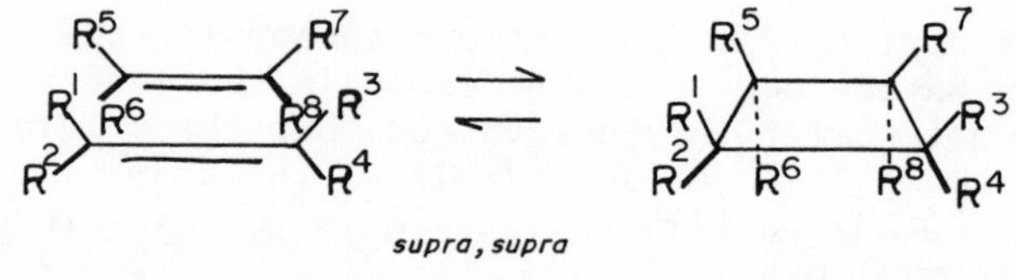

supra, supra

On the other hand, the supra,antara process involves severe twisting of one of the ethylene molecules as shown:

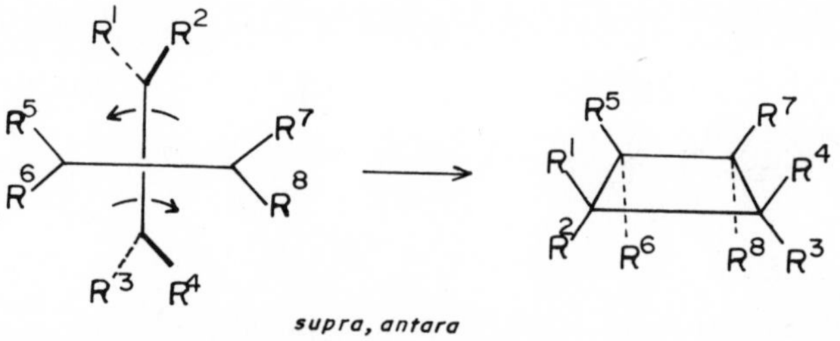

supra, antara

An example of <u>supra</u>,<u>antara</u> dimerization, if the concerted
reaction holds, is the following [9]:

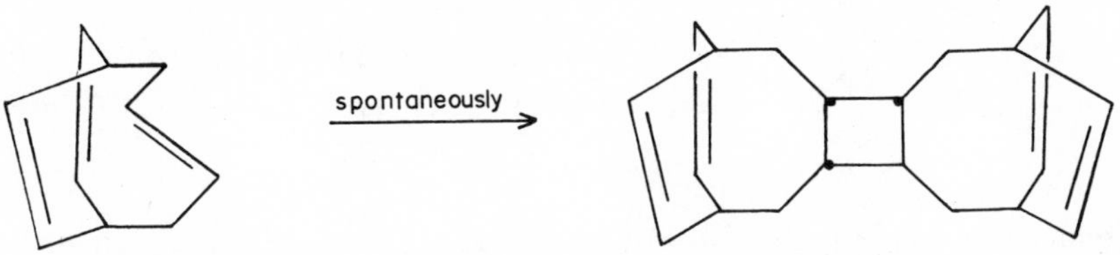

Recent examples of Diels-Alder reactions are countless.
Therefore, only a few are given to illustrate trends. Differ-
ences in catalyzed and uncatalyzed Diels-Alder reactions are
still found [10]:

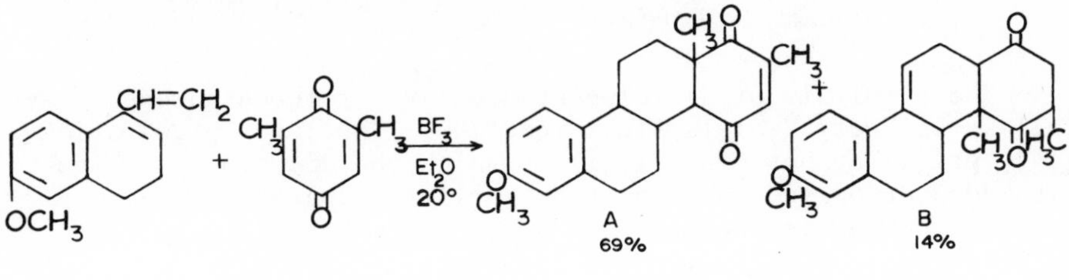

Compound B is the major product of an uncatalyzed Diels-Alder
reaction. Mannich bases occasionally are used as sources of
dienes as shown [11]:

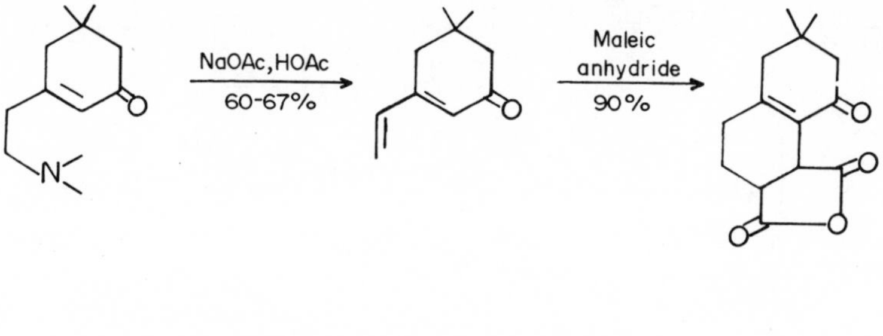

Cyclopropene addition to dienes is facile [12] as shown in a review.

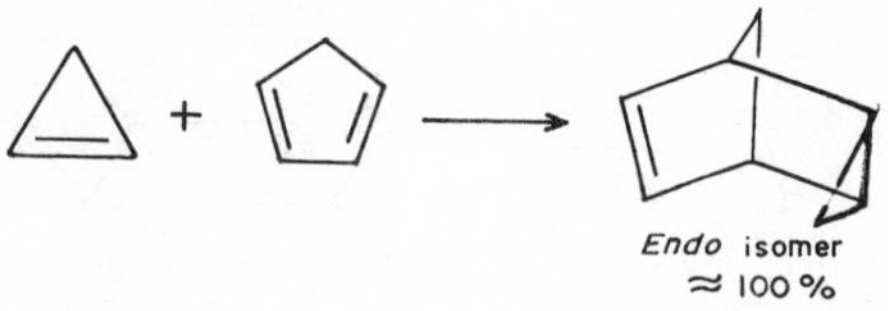

Endo isomer
≈ 100 %

The "ene" reaction (1, 118) now has been reviewed so that its scope may be realized [13]. The "ene" reaction is like the Diels-Alder except that one olefin group in this diene is replaced by a hydrogen atom:

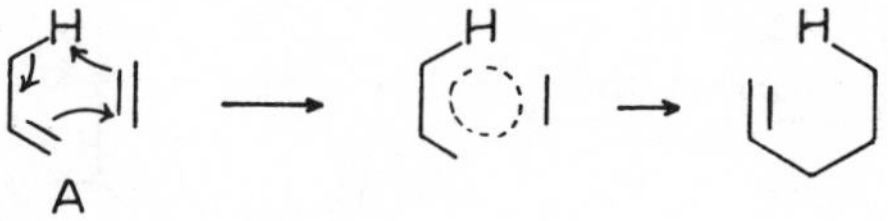

A

The enophile, A, may be maleic anhydride, an acetylene, an azo, or a carbonyl type. For instance, the following occurs:

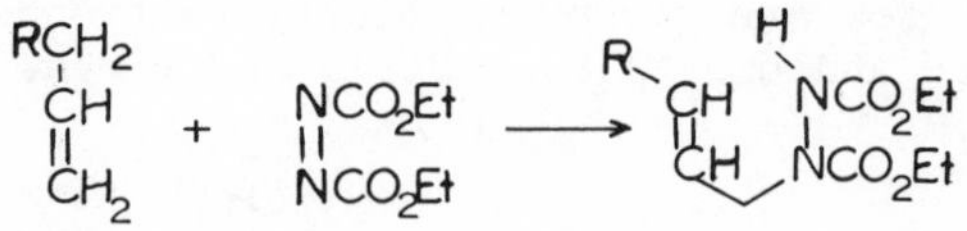

The pyrolysis of dicyclopentadiene to cyclopentadiene is a well-known "retro" Diels-Alder reaction. The later preparation [14], utilizing hot mineral oil to which the dicyclopentadiene was added dropwise, supplants the earlier preparation [15] because foaming is avoided.

3.   From Olefins and Acetylenes (1, 124)

2,3-Diphenylbutadiene has been produced from diphenylacetylene (Ex. a) by a method not highly recommended [16]:

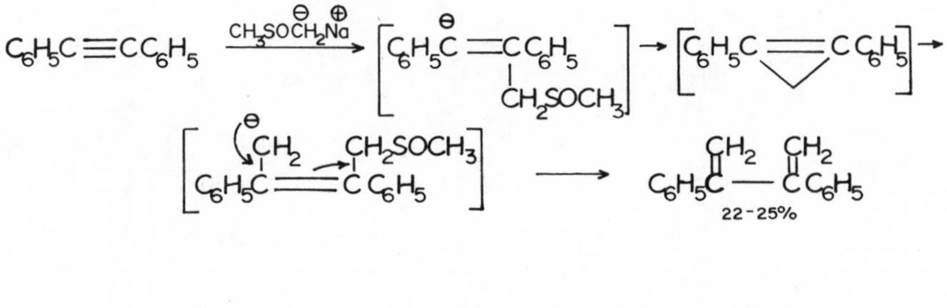

22-25%

This unconventional method is simple operatively.

An alkene has also been obtained from an alkyne by intramolecular cyclization [17]:

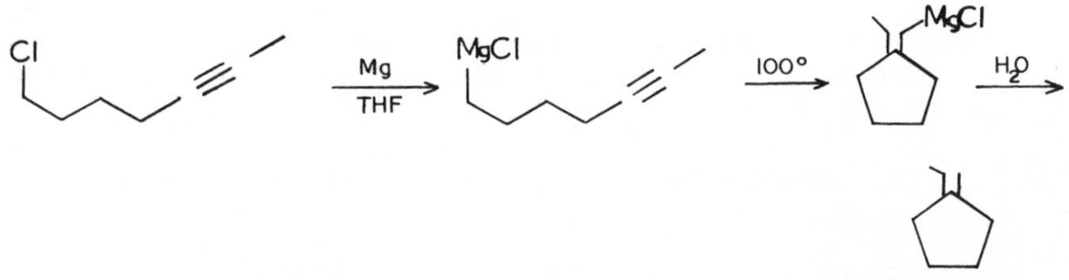

Ethylidene cyclopentane
ca90%

a.   Preparation of 2,3-Diphenyl-1,3-butadiene [16]

Yield was 22-25% from DMSO and sodium hydride under N_2 at 30° to which diphenylacetylene in DMSO was added followed by heating at 65° for 2 hr.

4.   From Unsaturated Systems (Anionic Addition) (1, 127)

Alkylated cyclopropenes have been produced by the action of alkali metal amide on methallyl chloride [18]:

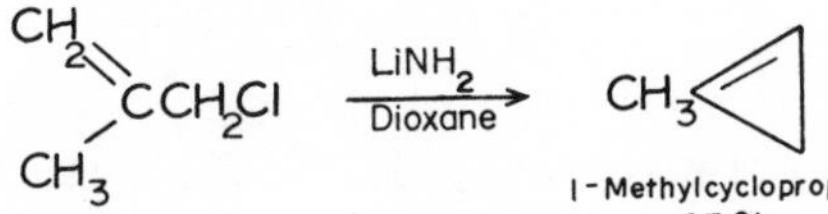

1-Methylcyclopropene
43%

Sodamide also produces some methylenecyclopropane.

The dianionic naphthalene addition to ethylene has been accomplished as shown [19]:

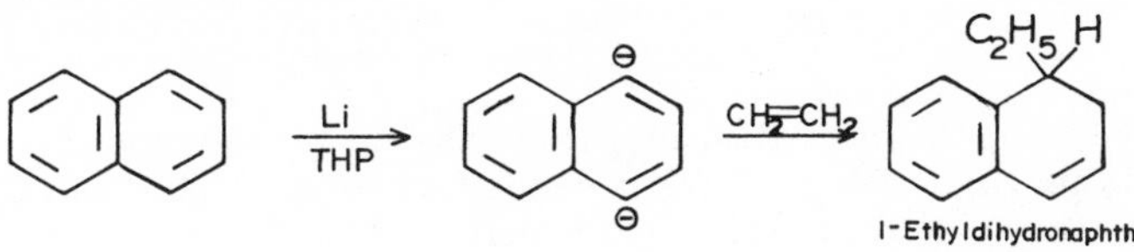

1-Ethyldihydronaphthalene
45%

Cyclopropenes may be alkylated as indicated [20]:

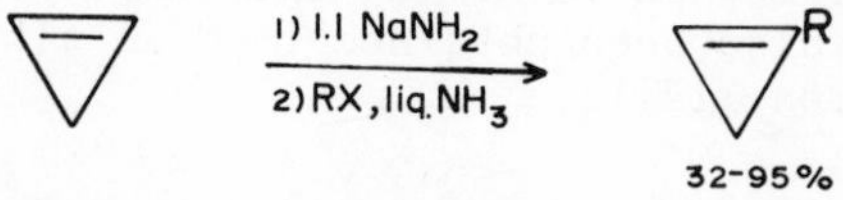

32-95%

The monoalkyl derivative is unstable but may be stored as a 1M solution in CCl_4 at -20° for prolonged periods. Two alkyl groups may also be introduced.

   7.   From Organometallic Compounds (Alkylation, Arylation ($\underline{1}$, 129)

$$RMgX + CH_2{=}CHCH_2X \longrightarrow RCH_2CH{=}CH_2$$

$$RMgX + R'CH{=}CHX \longrightarrow R'CH{=}CHR$$

   This section utilizes some π-allyl metal derivatives whose role in organic synthesis has been reviewed [21].
   Various organometallic compounds and unsaturated compounds have been employed in this transformation. Heck and Nolley [22] used the arylpalladium iodide as shown under C.10.
   Corriu and Masse [23] conducted the reaction with the conventional Grignard reagent and vinyl bromide:

$$C_6H_5CH{=}CHBr + RMgX \xrightarrow[Ni(acac)_2]{Et_2O,25°} C_6H_5CH{=}CHR$$
50-75 %

Nickel(II)   acetylacetonate, 0.1-0.5%, was the most effective catalyst.
   Tamura and Kochi [24] achieved success with the Grignard reagent and ferric chloride (Ex. a):

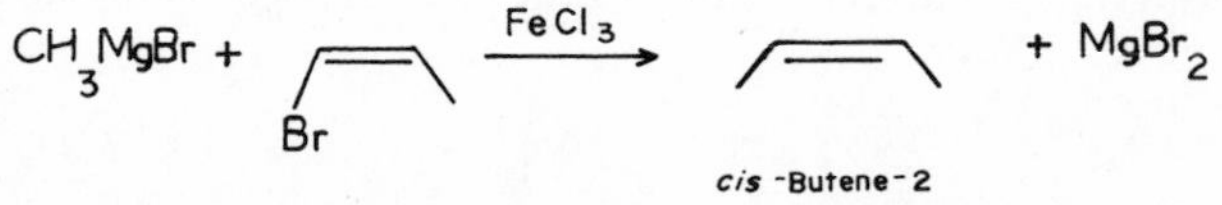

cis-Butene-2

Similarly, the <u>trans</u> bromide gives the <u>trans</u>-butene-2.
   Akiyama and Hooz [25] proceeded through the coupling of the lithio methallyl ion with an organic halide:

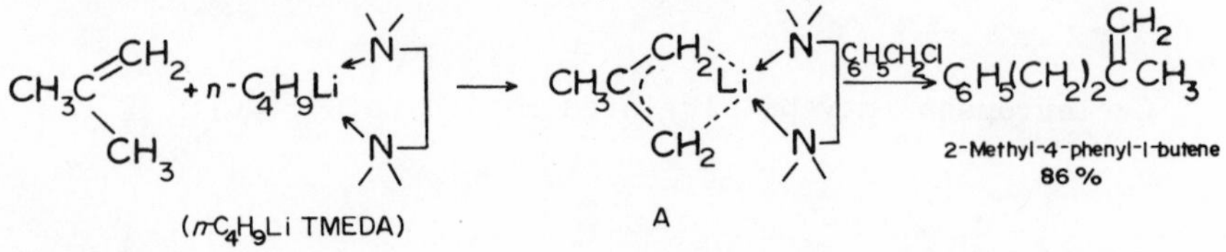

(n-C_4H_9Li TMEDA)   A

The complex A is also effective in addition to carbonyl compounds to produce methallylcarbinols. In general, the procedure is attractive for accomplishing either allylation or methallylation. Dihaloalkenes have been alkylated as indicated [26]:

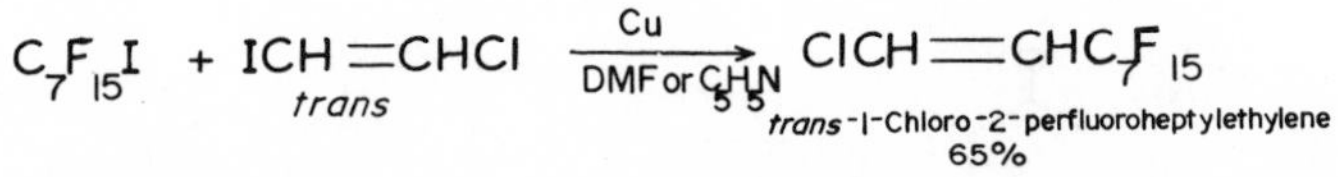

$$C_7F_{15}I + ICH\!=\!CHCl \xrightarrow[\text{DMF or } C_5H_5N]{Cu} ClCH\!=\!CHC_7F_{15}$$

$\textit{trans}$

trans-1-Chloro-2-perfluoroheptylethylene

65%

The methylation of styrene was accomplished as shown [27]:

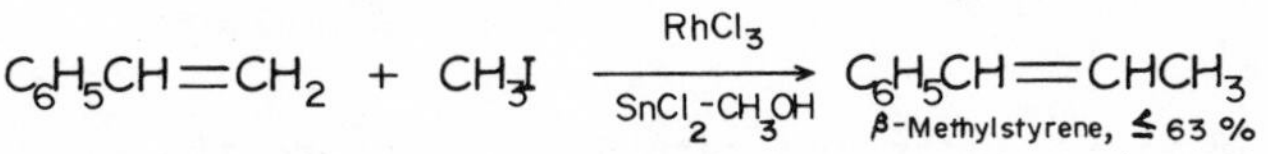

$$C_6H_5CH\!=\!CH_2 + CH_3I \xrightarrow[\text{SnCl}_2\text{-CH}_3\text{OH}]{RhCl_3} C_6H_5CH\!=\!CHCH_3$$

β-Methylstyrene, $\leqq$ 63 %

πAllylnickel halides of value in alkylation and arylation may be obtained through the use of nickel carbonyl as indicated [28] in the synthesis of methallylbenzene:

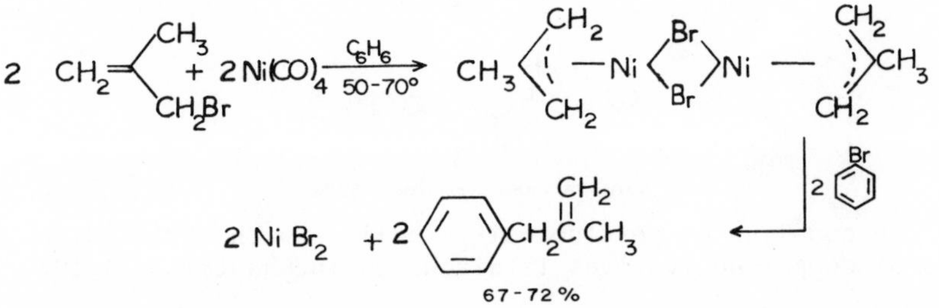

67-72 %

The π-allylnickel halide method leads to a yield better than that obtained from phenylmagnesium bromide and the methallyl halide and the π-allyl complex has the advantage of being nonnucleo-philic and nonbasic in character.

Lehmkuhl and Reinehr [29] employed ethylene with the unsaturated Grignard reagent under pressure to obtain the rearranged product as shown:

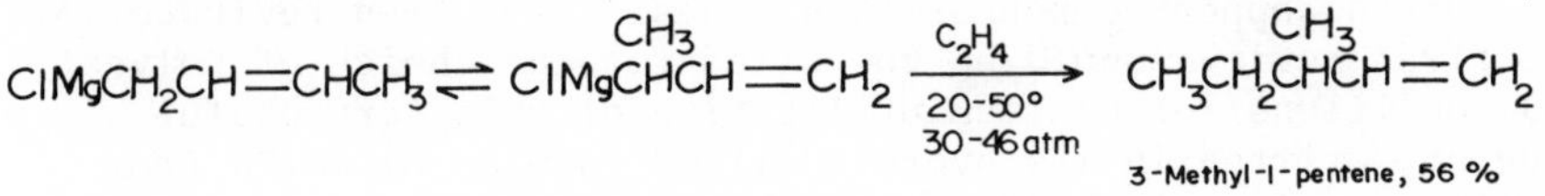

$$ClMgCH_2CH\!=\!CHCH_3 \rightleftharpoons ClMgCHCH\!=\!CH_2 \xrightarrow[\substack{20\text{-}50° \\ 30\text{-}46\,\text{atm}}]{C_2H_4} CH_3CH_2CHCH\!=\!CH_2$$

3-Methyl-1-pentene, 56 %

8.   From Vinyl Grignard Reagents or the Like (<u>1</u>, 131)

Vinyl organometallic compounds have been utilized in forming

dienes or tetrasubstituted olefins, or in coupling to form butadienes.

Corey and Chen [30] treated methyl trans-3-methyl-2,4-
dienoate with vinylcopper, followed by quenching with methanol,
to obtain the olefin as shown:

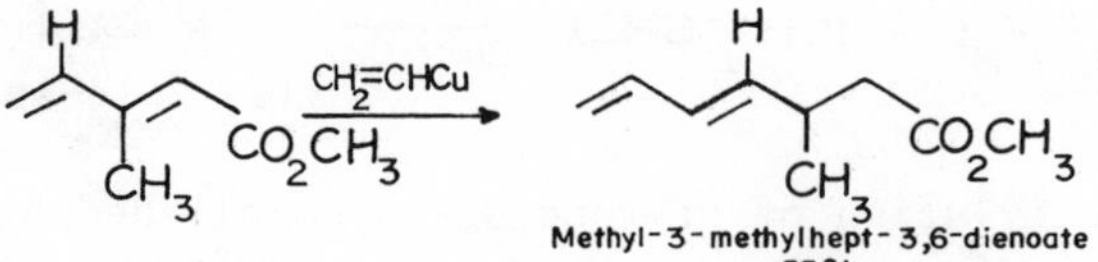

Methyl-3-methylhept-3,6-dienoate
55%
(stereochemical purity 95%)

By the use of vinylcopper and methyl-2-butynoate the stereoselective synthesis of trisubstituted olefins is possible:

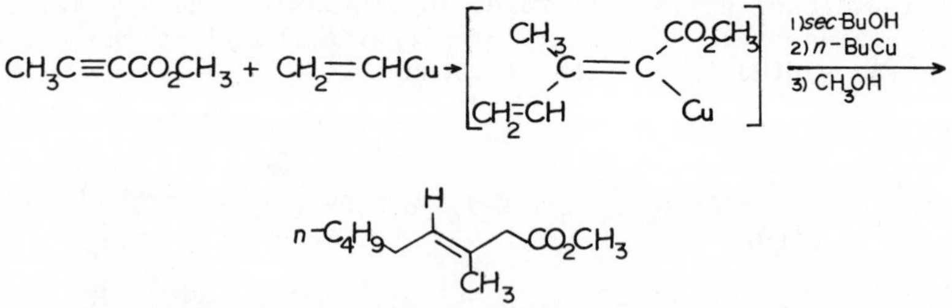

Methyl 3-methyl-3-nonenoate, 50%

The coupling of perfluoroalkyl halides with aryl iodides by
means of copper derivatives is discussed in Chapter 7, A.19.

9.   From Lithium Dialkylcuprates and the Like

$$LiCu(CH_2CH=CH_2)_2 \xrightarrow{[o]} CH_2=CHCH_2CH_2CH=CH_2$$

$$LiCuR_2 + CH_2=CHCH_2Cl \longrightarrow RCH_2CH=CH_2$$

Organocopper compounds in synthesis have been reviewed [31].
The first review should be consulted for the choice of solvent
and conditions for maximum utilization of this very useful
reagent, written in its oversimplified form as LiCuR$_2$. Frequently, it is complexed with tributylphosphine and on occasion
may be used in catalytic amounts. Its greatest asset is its
propensity for 1,4 addition to α,β-unsaturated carbonyl compounds.
Not only is 1,4 addition enhanced by this reagent, but in one
example the rate has been accelerated from 10- to 100-fold by a

catalytic amount [32]. In addition to its value in 1,4 addition, oxidative coupling with a variety of mild oxidizing agents is an attractive synthesis, and alkylation with reactive halides is a feasible process [32].

Vig and co-workers [33] employed lithium diisopropenyl copper (prepared from lithium, 2-bromopropene, and cuprous iodide) in the propenylation of 3-methyl-1-bromo Δ^2-cyclohexene:

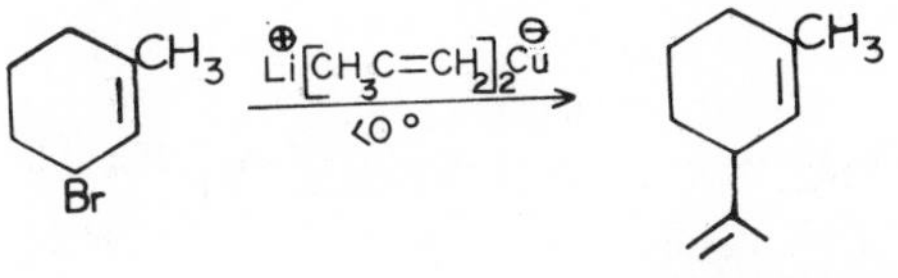

3-Methyl-1-*isopropenyl*-Δ^2- cyclohexene, 76%

For a reductive alkylation with a lithium dialkylcuprate, see C.11.

10.   From Alkenes and Arenes with Palladium Salts (Arylation)

Heck and Nolley [22] utilized the aryl palladium iodide as shown:

$$C_6H_5PdI + C_6H_5CH=CH_2 \xrightarrow{(C_4H_9)_3N} C_6H_5CH=CHC_6H_5$$

trans-Stilbene, 75%

The reaction is a one-operation process in which the palladium is introduced as palladium acetate. No solvent is employed, steam-bath temperature in an open flask is utilized and tri-n-butyl-amine serves to remove the hydrogen halide formed. The reaction takes place with aryl iodides, benzyl chloride, and β-bromo-styrene.

Sodium p-toluenesulfinate has also been employed with palladium(II) chloride in the arylation [34]:

$$C_6H_5CH=CH_2 + p-CH_3C_6H_4SO_2Na + PdCl_2 \xrightarrow{CH_3CN} p-CH_3C_6H_4CH=CHC_6H_5$$

trans-*p*-Methylstilbene, 63%

The Meerwein arylating procedure (1, 146) is a related method for arylating unsaturated acids to give arylalkenes [1, p. 356]:

$$ArCH=CHCOOH \xrightarrow{Ar'N_2^{\oplus}} ArCH=CHAr'$$

11.   From Unsaturated Organometallic Compounds and Alkenes, and the Like

This synthesis is similar to the addition of Grignard reagent or carbanions to alkenes to form alkanes (Chapter 1, C.2).

The allylation of an unsaturated amine may be achieved as indicated [35]:

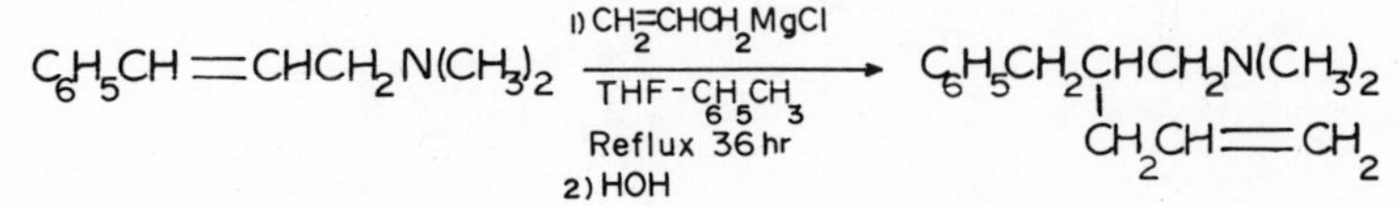

$$C_6H_5CH=CHCH_2N(CH_3)_2 \xrightarrow[\substack{THF-CH_5CH_3 \\ Reflux\ 36hr \\ 2)HOH}]{1)CH_2=CHCH_2MgCl} C_6H_5CH_2CHCH_2N(CH_3)_2$$

$$\underset{CH_2CH=CH_2}{|}$$

I-Dimethylamino-2-benzyl-4-pentene,44%

Alkynes have been subjected to reductive alkylation by means of lithium dimethylcopper [36]:

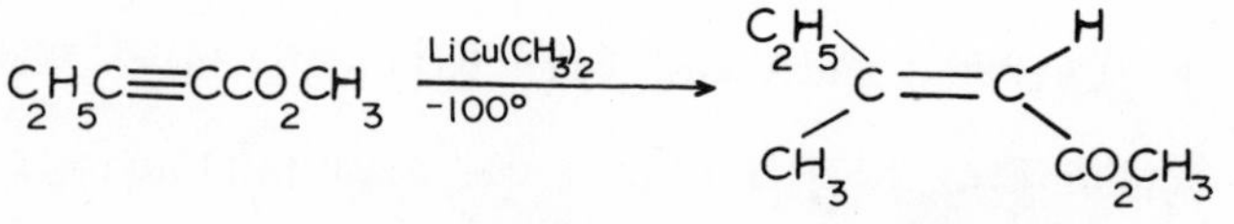

$$C_2H_5C\equiv CCO_2CH_3 \xrightarrow[-100°]{LiCu(CH_3)_2}$$

$$\underset{\substack{CH_3 \qquad\qquad CO_2CH_3}}{\overset{\substack{C_2H_5 \qquad\qquad H}}{C=C}}$$

cis + trans, 95%

The dienic alcohol with the allyl Grignard reagent responded as shown [37]:

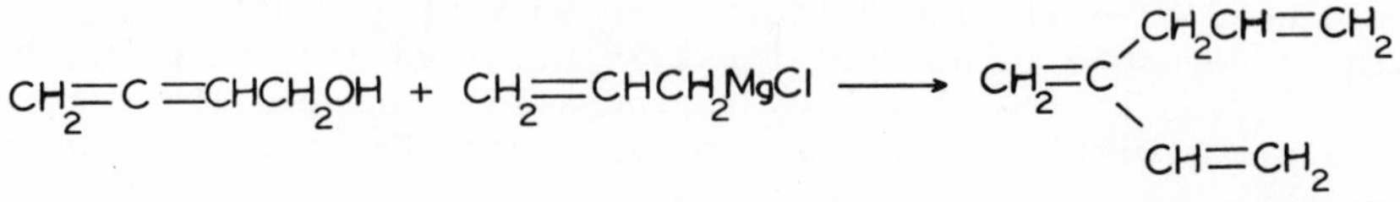

$$CH_2=C=CHCH_2OH + CH_2=CHCH_2MgCl \longrightarrow CH_2=C\underset{CH=CH_2}{\overset{CH_2CH=CH_2}{\big<}}$$

2-Vinylpenta-I,4-diene, 81%

Cyclization occurred to form 1,2-benzocyclohepta-1,3-diene when o-allylstyrene was treated with sodium in isopropylamine [38]:

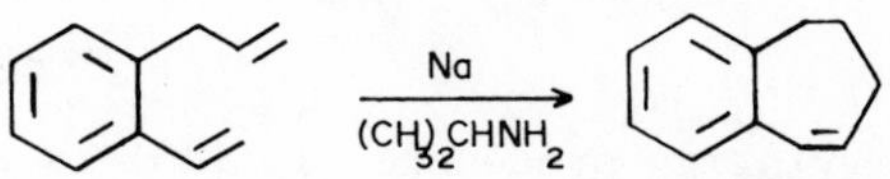

55%

The substrate is available from homophthaldehyde in a Wittig reaction.

12.   Coupling of Halides and of Carbonyl Compounds (Mostly via Carbenes or Free Radicals) [1, p. 427]

The various halides were coupled as indicated:

$$2\ C_6H_5CH_2Cl \xrightarrow[\substack{2)\ H_3O^{\oplus}}]{\substack{1)\ LiN(C_2H_5)_2 \\ Et_2O}} C_6H_5CH=CHC_6H_5 \qquad [39]$$

trans-Stilbene, 80%

$$2\ C_6H_5CH=CHMgBr \xrightarrow[\text{SOCl}_2]{\text{THF}} C_6H_5CH=CHCH=CHC_6H_5 \qquad [40]$$

1,4-Diphenyl-1,3-butadiene, 89%

Coupling also occurs with phenylethynylmagnesium bromide.

$$2\ RCH=CHBr \xrightarrow[\substack{DMF\ or \\ Et_2O,\ 25°}]{Ni(COD)_2} RCH=CHCH=CHR \qquad [41]$$

48-70% (mixture of *trans,trans* and *cis,trans*)

$$2\ CH_2=CHCH_2I \xrightarrow[(C_2H_5)_3B]{O_2} (CH_2=CHCH_2)_2 \qquad [42]$$

1,5-Hexadiene, 97%

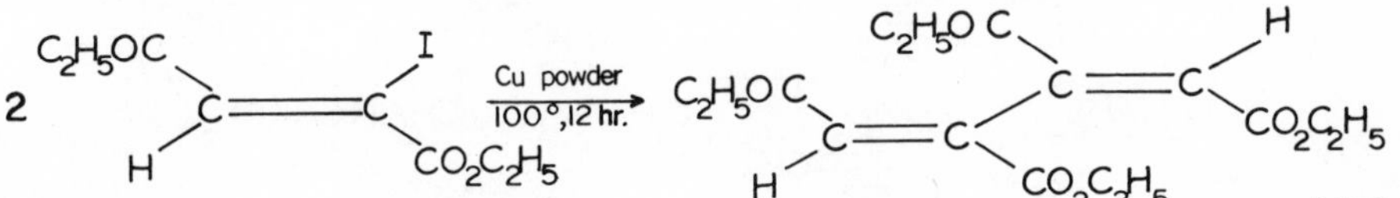

[43]

trans,trans-1,2,3,4-Tetracarboxy-1,3-butadiene
96%

The iodomaleate at 100° in 48 hr coupled to give 87% of <u>cis,cis</u>- and 13% of <u>trans,trans</u>-butadiene tetraester.

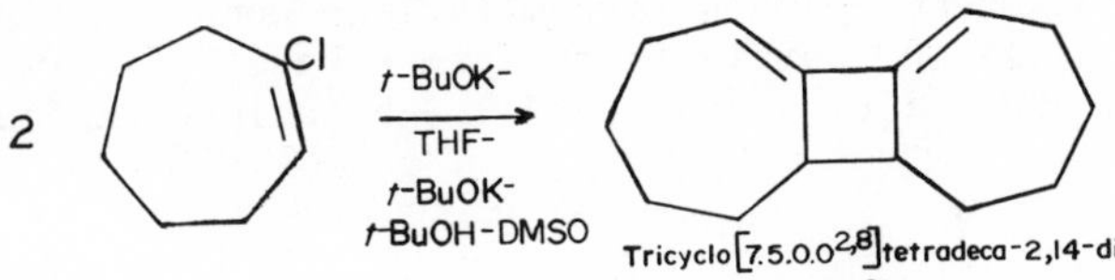

[44]

Tricyclo[7.5.0.0^{2,8}]tetradeca-2,14-diene
54%

Several aldehydes were coupled to give alkenes as shown:

$$2\ C_6H_5CHO \xrightarrow[\text{THF}]{WCl_6-BuLi^-} C_6H_5CH=CHC_6H_5 \qquad [45]$$

Stilbene, 76%

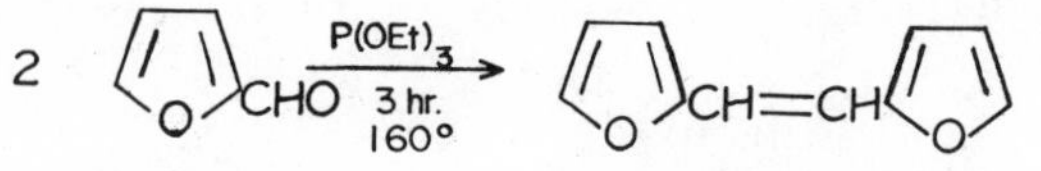

Difurylethylene, no yield given   [46]

Other types which couple [1, p. 418] are:

$$2 \ R_2C{=}S \xrightarrow{\Delta} R_2C{=}CR_2 + 2\,S$$

$$2 \ R_2CN_2 \longrightarrow R_2C{=}CR_2 + 2\,N_2$$

13. From Allyl Alcohols and Orthoesters (Orthoester Claisen)

Ethyl orthoacetate and an allyl alcohol combine as shown to give a diene ester [47]:

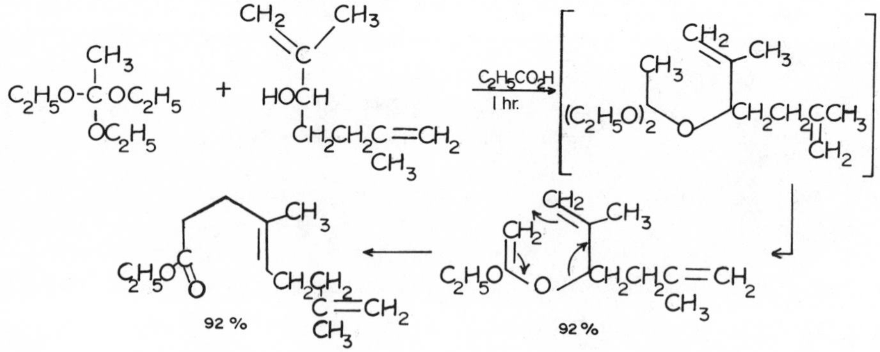

The Claisen rearrangement has been discussed [47a]

14. From Alkenes and Active Methylene Compounds

1,3-Dienes react with active methylene compounds in the presence of palladium catalysts to form alka-2,7-dienyl deriva-tives. With ethyl acetoacetate and 1,3-butadiene the equation is [48]:

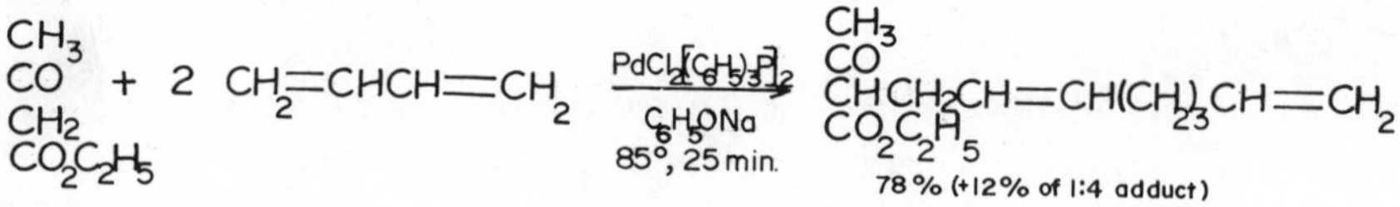

78% (+12% of 1:4 adduct)

Platinum catalysts may also be used but the products are different.

Allylic alkylation was achieved by the use of 2-n-propyl-1-pentene and a π-alkyl-palladium complex as shown [49]:

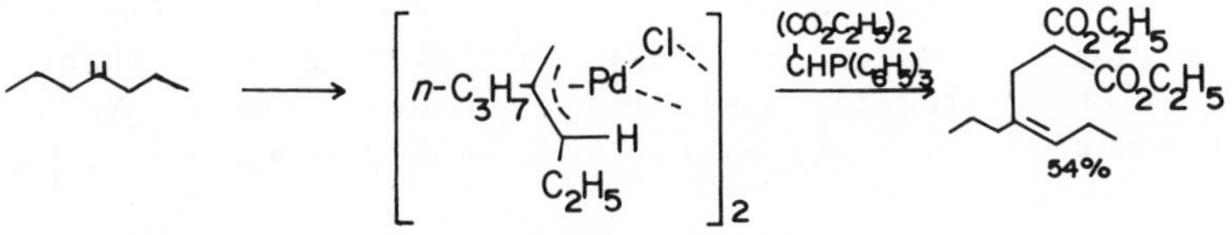

15.   From Vinyl Ethers and Triallylboranes

1,4-Dienes may be prepared by the reaction of vinyl ethers and triallylboranes. Thus 1,4-pentadiene may be produced as shown [50] from triallylboron and vinyl butyl ether.

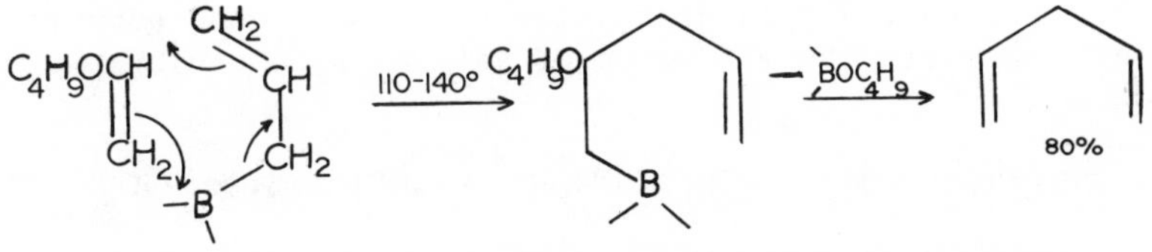

16.   From Pyridazine-N-Oxide and the Grignard Reagent

Pyridazine-N-oxide reacts with the Grignard reagent to give different products depending on the solvent employed as indicated [51]:

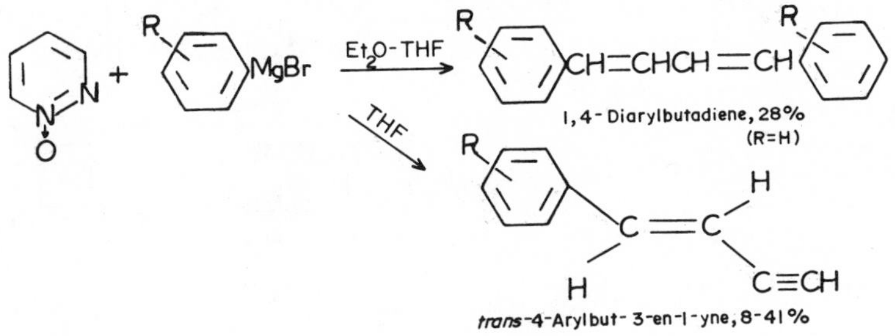

1,4-Diarylbutadiene, 28%
(R=H)

trans-4-Arylbut-3-en-1-yne, 8-41%

17.   From Vinyl Halides (Coupling) See Addenda.

Addenda

C.1; C.7. The product from cyclodimerization by nickel catalysis is very much dependent on the ligand attached to the nickel
[P. Heimbach, Angew. Chem. Intern. Ed. Engl., 12, 975 (1973)].

C.2. Dienes with maleic anhydride in the presence of chloranil
give completely aromatic Diels-Alder adducts. Yields are higher
as the adduct becomes polynuclear [M. Zander, Ann. Chem., 723,
27 (1969)].

C.2. Silver perchlorate catalyzes the Diels-Alder reaction
between benzyne and 1,3-cyclohexadiene [P. Crews et al., J. Org.
Chem., 38, 522 (1973); Tetrahedron Letters, 4697 (1971)].

C.2. The ene reaction between methylenecyclohexane and methyl
acrylate has been found to be catalyzed by anhydrous $AlCl_3$
[B. B. Snider, J. Org. Chem., 39, 255 (1974)]. Vinyl trichlorosilane is not only a good dienophile but also a good enophile.
As such it permits the extension of a terminal chain by two carbon atoms:

$$RCH_2CH=CH_2 + CH_2=CHSiCl_3 \longrightarrow RCH=CH(CH_2)_3SiCl_3$$

[A. Laporterie et al., Compt. Rend., 278, C375 (1974)].

C.2. A Diels-Alder reaction followed by the reverse thereof
follows:

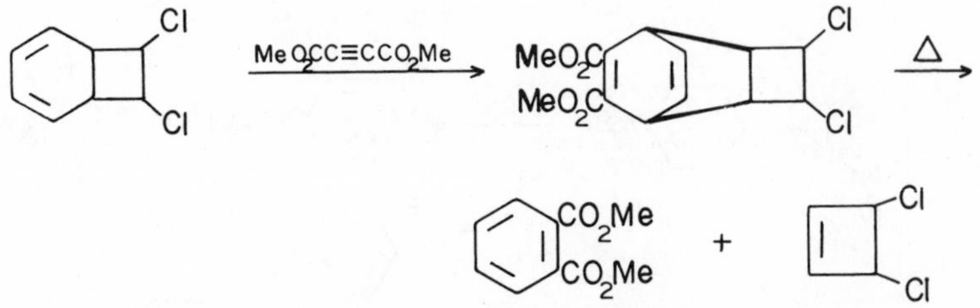

[R. Pettit, J. Henery, Org. Syn., 50, 36 (1970)].

C.3. Acetylene alternates for Diels-Alder reactions are 2-
phenyl- or 2-thiono-1,3-dioxol-4-ene [W. K. Anderson, R. H.
Dewey, J. Am. Chem. Soc., 95, 7161 (1973)].

C.3. (E,E)-1,3-Dienes may be produced in good yields from
acetylenes:

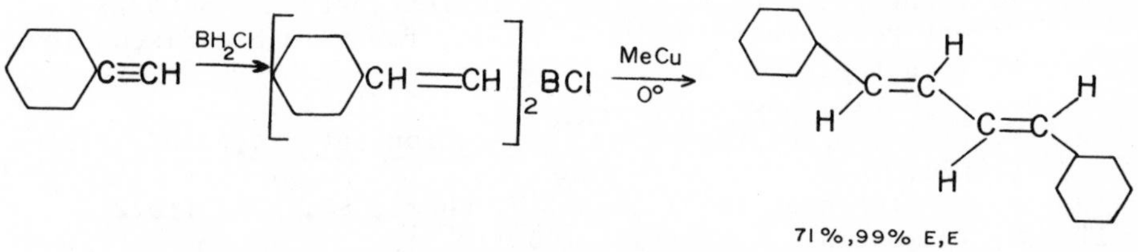

71%,99% E,E

[Y. Yamamoto et al., J. Am. Chem. Soc., 97, 5606 (1975)].

C.7. The product from the reaction of $[(CH_3)_2CH]_2$ CuMgBr and butadiene, $[(CH_3)_2CHCH_2CH=CHCH_2]$ Cu, may be alkylated with allyl bromide to give $(CH_3)_2CHCH_2CH=CHCH_2CH_2CH=CH_2$ [J. Normant et al., J. Organomet. Chem., 92, C28 (1975)].

C.9. Two excellent reviews on organic copper reagents are H. O. House, Proceedings of the R. A. Welsh Foundation, XVII, 1974, p. 101 and J. P. Marino, Ann. Rep. Med. Chem., 10, 327 (1975)].

C.9. For the best method for the preparation of lithium dialkylcuprates see Ketones, Addenda, G.3.

C.11. The addition of allyl Grignard reagents to olefins has been summarized [H. Lehmkuhl et al., Ann. Chem., 103, 119 (1975)].

C.17. Vinyl halides, particularly those containing electron-withdrawing groups, are coupled in high yield to butadiene by bis(1,5-cyclooctadiene) nickel(0) in DMF [M. F. Semmelhack et al., J. Am. Chem. Soc., 94, 9234 (1972)].

References

1. H.-W., 5, Pt. 1b, 1972, p. 490.
1a. T. Saegusa et al., J. Org. Chem., 35, 670 (1970); Synthesis, 548 (1970).
2. A. Stefani, Helv. Chim. Acta, 56, 1192 (1973).
3. K. Masui et al., Bull. Chem. Soc. Jap., 44, 1956 (1971).
4. I. Moritani, Y. Fujiwara, Synthesis, 524 (1973).
5. V. A. Soldatova et al., Neftekhimiya, 13, 272 (1973); C. A., 79, 41801 (1973).
6. R. B. Woodward, R. Hoffmann, The Conservation of Orbital Symmetry, Verlag Chemie GmbH, Academic, New York, 1970.

7. We thank Dr. Patrick Coffey, Nicolet Instruments, 5225
    Verona Road, Madison, Wis., 57311, for valuable discussions
    of this topic.
8. J. D. Coyle, Chem. Soc. Rev., 3, 329 (1974).
9. K. Kraft, G. Koltzenburg, Tetrahedron Letters, 4357, 4723
    (1967).
10. R. A. Dickinson et al., Can. J. Chem., 50, 2377 (1972).
11. M. Tramontini, Synthesis, 769 (1973).
12. M. L. Deem, Synthesis, 675 (1972).
13. H. M. R. Hoffmann, Angew. Chem. Intern. Ed. Engl., 8, 556
    (1969).
14. M. Korach et al., Org. Syn., Coll. Vol., 5, 414 (1973).
15. R. B. Moffatt, Org. Syn., Col. Vol., 4, 238 (1963).
16. I. Iwai, J. Ide, Org. Syn., 50, 62 (1970).
17. H. G. Richey, Jr., A. M. Rothman, Tetrahedron Letters,
    1457 (1968)
18. R. Köster et al., Ann. Chem., 1219 (1973).
19. J. C. Carnahan, Jr., W. D. Closson, J. Org. Chem., 37,
    4469 (1972).
20. A. J. Schipperijn, P. Smael, Rec. Trav. Chim., 92, 1159
    (1973).
21. R. Baker, Chem. Rev., 73, 487 (1973).
22. R. F. Heck, J. P. Nolley, Jr., J. Org. Chem., 37, 2320
    (1972).
23. R. J. P. Corriu, J. P. Masse, Chem. Commun., 144 (1972).
24. M. Tamura, J. Kochi, Synthesis, 303 (1971); S. M. Neumann,
    J. K. Kochi, J. Org. Chem., 40, 599 (1975).
25. S. Akiyama, J. Hooz, Tetrahedron Letters, 4115 (1973).
26. P. L. Coe et al., J. Chem. Soc., Perkin Trans., I, 639
    (1972).
27. I. Yu Levitin et al., Izv. Akad. Nauk SSSR, Ser. Khim.,
    1188 (1973); C. A., 79, 52884 (1973).
28. M. F. Semmelhack, P. M. Helquist, Org. Syn., 52, 115
    (1972).
29. H. Lehmkuhl, D. Reinehr, J. Organomet. Chem., 25, C47
    (1970).
30. E. J. Corey, R. H. K. Chen, Tetrahedron Letters, 1611
    (1973).
31. (a) J. F. Normant, Synthesis, 63 (1972); (b) T. Saegusa
    et al., Kogyo Kagaku Zasshi, 72 (8), 1627 (1969); C. A.,
    72, 11709 (1970); G. H. Posner, J. S. Ting, Tetrahedron
    Letters, 683 (1974).
32. H. O. House, W. F. Fischer, Jr., J. Org. Chem., 33, 949
    (1968).
33. O. P. Vig et al., J. Indian Chem. Soc., 45, 1026 (1968).
34. R. Selke, W. Thiele, J. Prakt. Chem., 313, 875 (1971).
35. H. G. Richey, Jr., et al., Tetrahedron Letters, 2183
    (1971).

36. J. B. Siddall et al., J. Am. Chem. Soc., 91, 1853 (1969).
37. H. G. Richey, Jr., S. S. Szucs, Tetrahedron Letters, 3785
     (1971).
38. A. Kergomard et al., Synthesis, 149 (1973).
39. D. Reisdorf, H. Normant, Organomet. Chem. Syn., 1, 375
     (1972); Ref. 1, p. 429.
40. A. Uchida et al., J. Org. Chem., 37, 3749 (1972).
41. M. F. Semmelhack et al., J. Am. Chem. Soc., 94, 9234
     (1972).
42. H. C. Brown et al., J. Am. Chem. Soc., 93, 1508 (1971).
43. T. Cohen, T. Poeth, J. Am. Chem. Soc., 94, 4363 (1972).
44. A. T. Bottini et al., Tetrahedron, 29, 1975 (1973).
45. K. B. Sharpless et al., J. Am. Chem. Soc., 94, 6538 (1972).
46. B. A. Arbuzov, V. M. Zoroastrova, Izvest. Akad. Nauk SSSR
     Otdel Khim. Nauk, 1030 (1960); C. A., 54, 24627 (1960).
47. W. S. Johnson, D. J. Faulkner et al., J. Am. Chem. Soc.,
     92, 741 (1970).
47a. R. I. Trust, R. E. Ireland, Org. Syn., 53, 116 (1973).
48. G. Hata et al., Chem. Ind. (London), 1836 (1969).
49. B. M. Trost, T. J. Fullerton, J. Am. Chem. Soc., 95, 292
     (1973).
50. B. M. Mikhailov, Yu. N. Bubnov, Tetrahedron Letters, 2127
     (1971).
51. G. Okusa et al., Chem. Pharm. Bull., 17, 2502 (1969);
     Synthesis, 548 (1970).

D. Isomerization and Thermal Reactions

1. From Alkenes, Polyalkenes, or Other Compounds (Base,
    Acid, Thermal, Photochemical, or via π-Complex Iso-
    merization or Rearrangement (1, 133) [1]

a. Base Isomerization (1, 133)

b. Acid Isomerization

    trans-Cyclononene has been converted into the cis form as
indicated [2]:

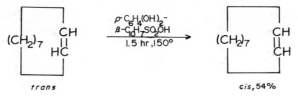

trans   cis, 54%

c. Thermal Isomerization

3-Cyclohexenylmagnesium bromide in ether is in equilibrium with 2-cyclopentenylmagnesium bromide [3]:

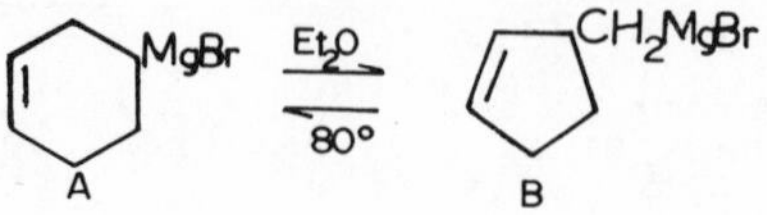

The equilibrium constant, B/A $\approx$ 8 for 1.3M solution in ether. B is more stable than A in this solution by 1.46 kcal/mol.

d. Photochemical Isomerization

This topic has been reviewed recently [4]. A premise for the photochemical isomerization of a cis to a trans isomer or vice versa is that one of the isomers will not be as easily activated as the other. The equilibrium then will be in favor of the isomer least affected. To promote a shift one may either select a wavelength of light more readily absorbed by one isomer than the other or add a sensitizer that reacts more readily with one isomer than the other. A third way is to add a complexing agent to stabilize one isomer more than the other. Such was the case in the isomerization of cis- to trans-cyclooctene [5]. If the cis form is photolyzed in the presence of cuprous chloride, the equilibrium obtained is such that 19% of trans-cyclooctene may be isolated. This isolation is made easy by the fact that the cis may be removed from the trans isomer by its solubility in aqueous silver nitrate. On the other hand, irradiation of trans-β-ionol in the presence of 2-acetophenone gave a quantitative yield of the cis isomer [6]:

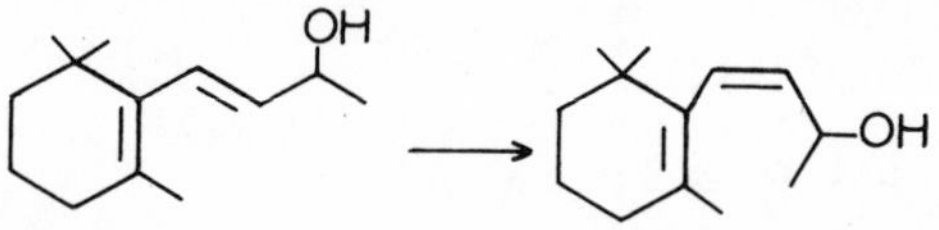

It now becomes clear why cyclohexenes (and a few other types) are unique in giving methyl ethers when the cyclohexene is irradiated in methanol. Irradiation gives some of the trans form that is so unstable that it reacts with methanol to relieve the strain [4].

Irradiation of cis-2-butene in the presence of Na-Al$_2$O$_3$ gave largely the trans form [7]:

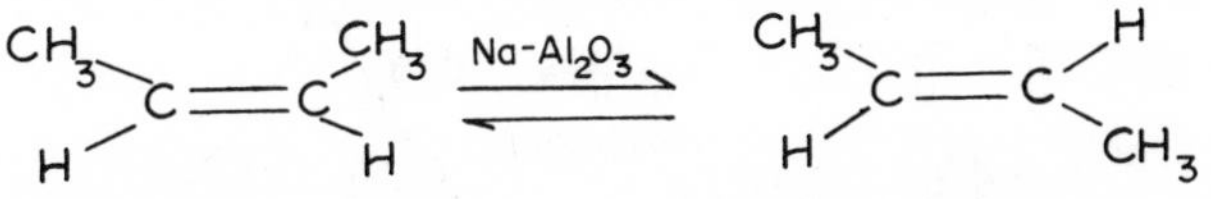

$$CH_3-C=C-CH_3 \quad (cis) \xrightarrow{\text{Na-Al}_2O_3} \quad CH_3-C=C-H, CH_3 \quad (trans)$$

cis

trans
73 %

In the case of cyclotetradeca-1,8-diene, irradiation in the presence of acetophenone converted the cis,cis largely into the trans,trans isomer [7]:

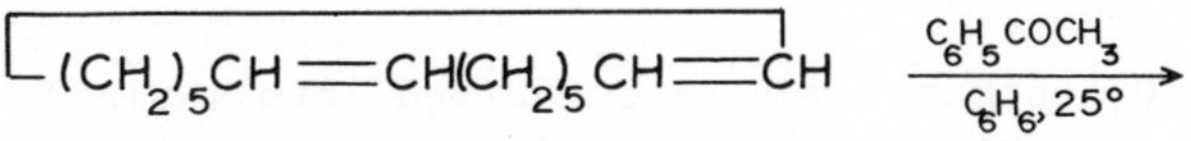

$$(CH_2)_5 CH=CH(CH_2)_5 CH=CH \xrightarrow[C_6H_6, 25°]{C_6H_5COCH_3}$$

cis, cis   trans, trans, 79 %

(-)-Caryophyllen was converted into the iso form quantitatively by irradiation [8]:

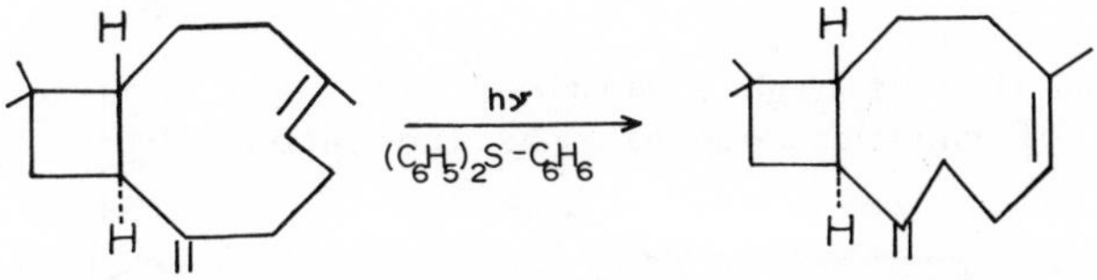

$$\xrightarrow[(C_6H_5)_2S-C_6H_6]{h\nu}$$

(-)-Caryophyllen

(-)- Isocaryophyllen
ca. 100 %

Other isomerizations, some protound, are illustrated as described by Coyle [4]:

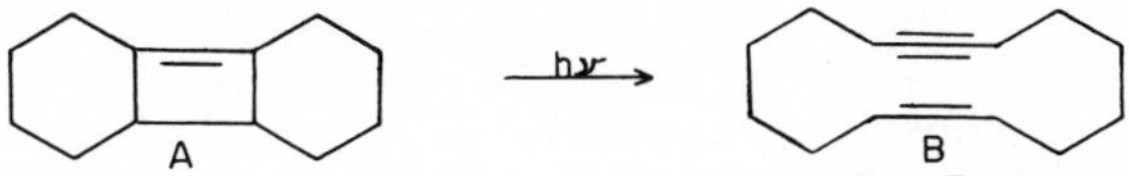

A

$\xrightarrow{h\nu}$

B
Dodeca-7-enyne

One noteworthy point is that each isomer of A gives a specific isomer of B.

$$\xrightarrow{h\nu}$$

Quadricyclene

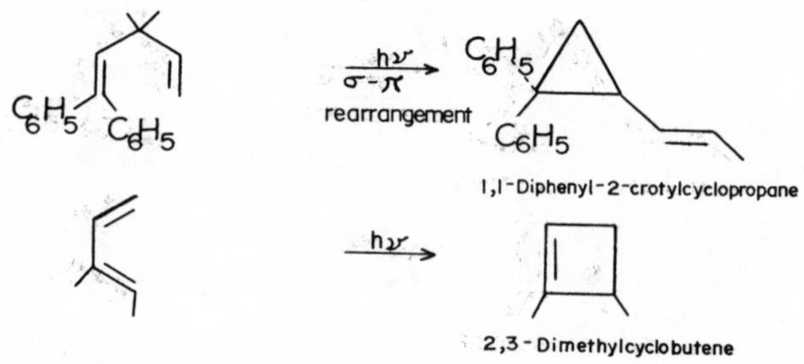
hν
σ-π
rearrangement
1,1-Diphenyl-2-crotylcyclopropane
hν
2,3-Dimethylcyclobutene

e. Isomerization via π-Complexes

Bicyclo [4.2.0] octene-7 may be converted into 1-vinylcyclohexene by the use of $(Ph_3P)_3RhCl$ [9]:

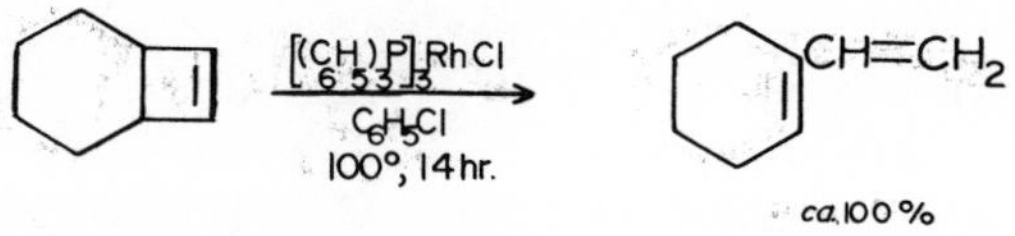
CH=CH2
100°, 14 hr.
ca. 100 %

The opening of highly strained rings has been accomplished by the use of rhodium complexes as indicated [10]:

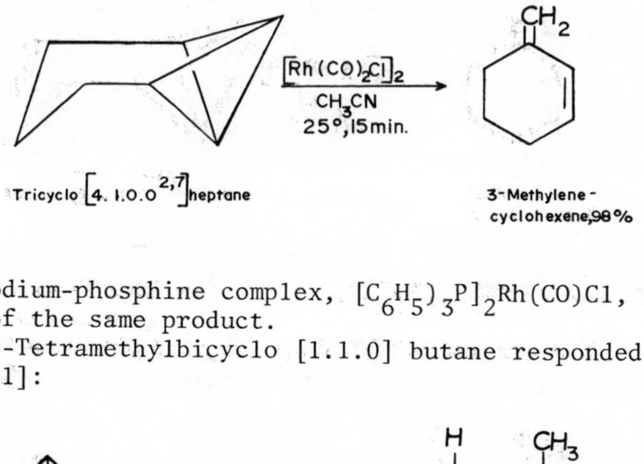
CH3CN
25°, 15 min.
Tricyclo [4.1.0.0^2,7] heptane
3-Methylene-cyclohexene, 98 %

The rhodium-phosphine complex, $[(C_6H_5)_3P]_2Rh(CO)Cl$, gave a 92% return of the same product.

2,2,4,4-Tetramethylbicyclo [1.1.0] butane responded as indicated [11]:

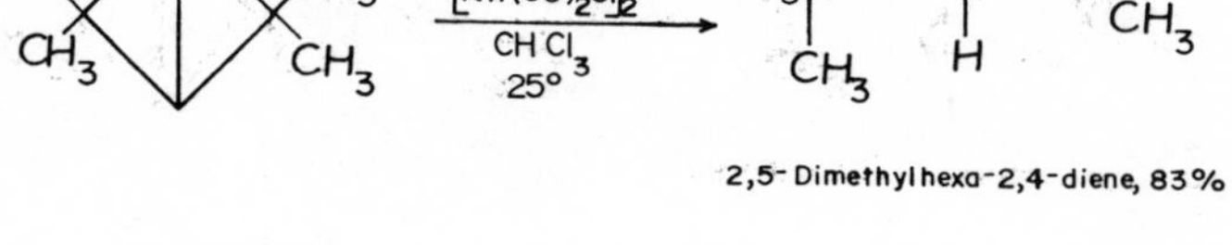
25°
2,5-Dimethylhexa-2,4-diene, 83%

Treatment of 1,1'-bishomocubane with bis(1,5-cyclooctadiene) nickel led to the formation of the dienes shown [12]:

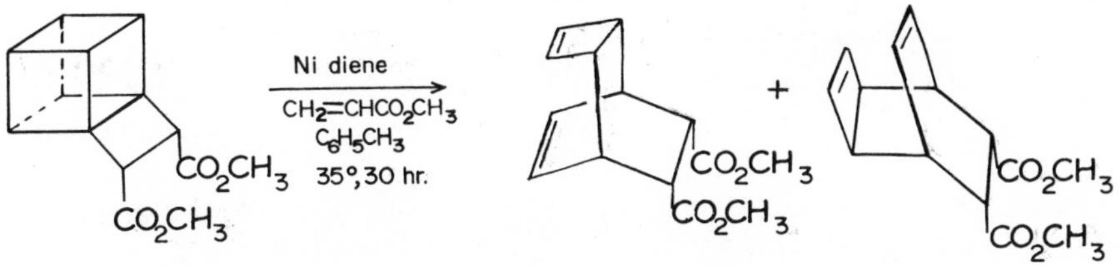

3.   From Small-ring Compounds (<u>1</u>, 138)

For the formation of small-ring compounds, see D.2.
1-(2-Methylprop-1-enylidene)-2-phenylcyclopropane rearranges on being heated to 1-isopropylidene-2-methylene-3-phenylcyclopropane [13]:

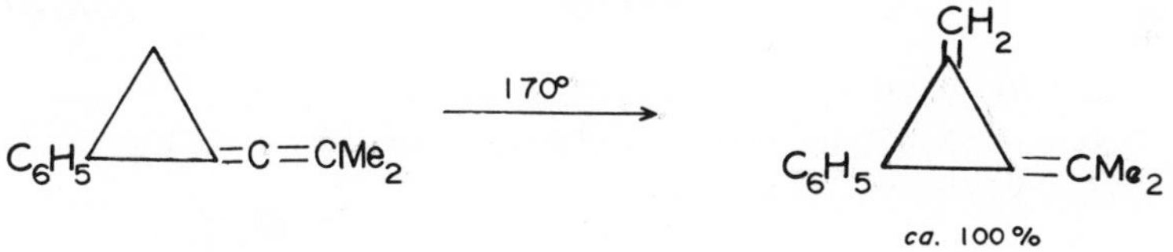

Carbethoxycyclopropenes rearrange under reducing conditions to give cyclobutenes [14].   Thus 3-carbethoxy-1,2-dipropylcyclopropene which is available from 4-octyne and ethyl diazoacetate, leads to 1,2-dipropylcyclo-1-butene:

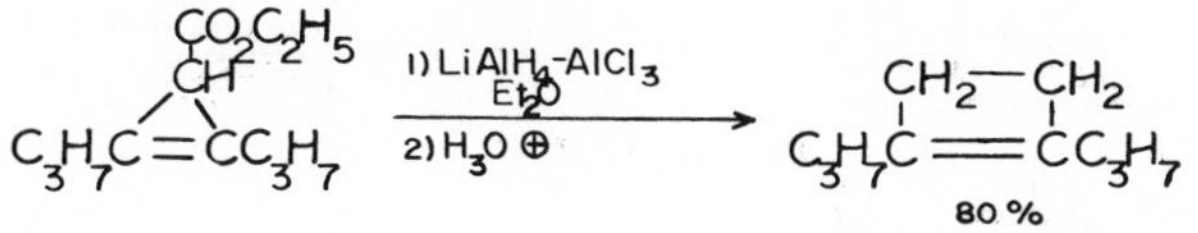

The method is probably applicable to the synthesis of 1,2-dialkyl- 1,2,3-trisubstituted, and -1,2,3,3-tetrasubstituted cyclobutenes.
Bicyclo[6.1.0]enes may be converted into cyclononadienes by the use of a palladium(II) salt complex [15] as indicated:

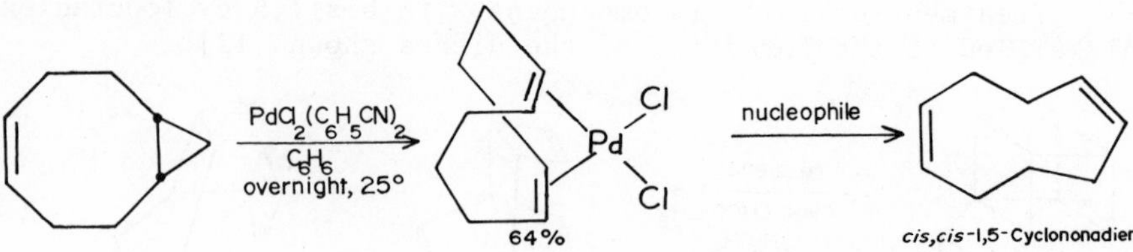

64%   cis,cis-1,5-Cyclononadiene

Bicyclic undecadienes have been produced from the cyclopropane adduct of acrolein in a Wittig reaction [16]:

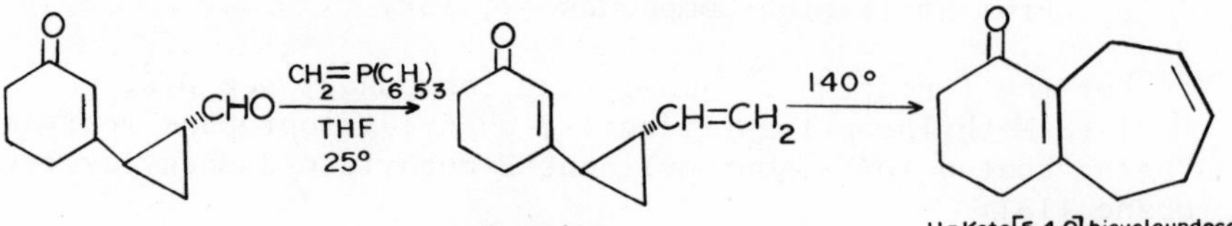

main product   11-Keto[5.4.0] bicycloundeca-1(7), 3-diene

For other thermal rearrangements, see D.1.

Cycloheptadienes are available from vinylcyclopropanealdehyde as indicated [17]:

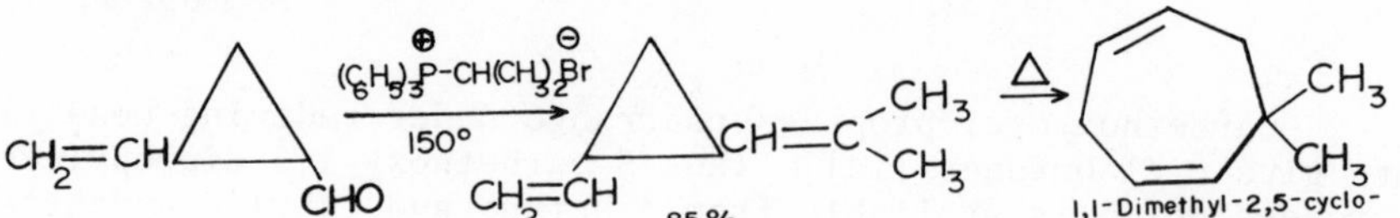

1,1-Dimethyl-2,5-cyclo-heptadiene,38%

In cis,trans isomers isomerization occurs below 25°.

Surprisingly, Breslow and co-workers found that a facile 1,5-alkyl shift occurs when bicyclo [3.2.0] hepta-1,3-diene is heated mildly [18]:

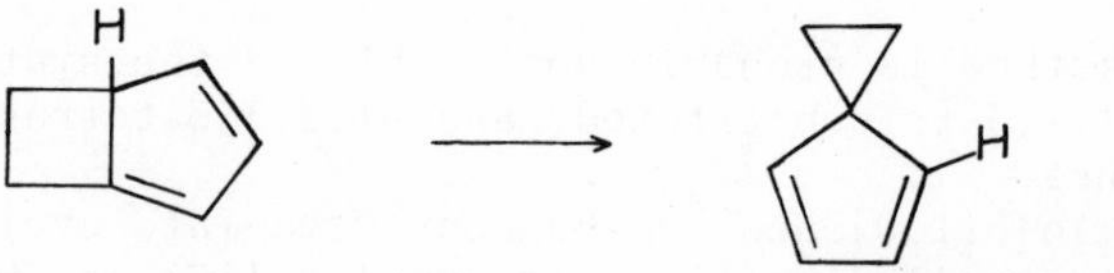

Spiro[2,4]hepta-4,6-diene

Perhaps the spirodiene is somewhat aromatic and thus the cyclopropane ring electrons are partially delocalized in the cyclopentadiene ring.

5. From Hydrocarbons (Dehydrogenation) (1, 139) [1,
    p. 376)

A recent review on the catalytic dehydrogenation of hydrocarbons is available [19]. The dehydrogenation products of
straight-chain paraffins are aromatic hydrocarbons and olefins.
In the case of n-butane and n-pentane, some diolefins are produced as well. Dehydrogenation catalysts are extremely numerous
but usually involve Group IV-VI elements. As has been mentioned
previously (1, 139), the process is best adapted to industrial
use.

From cyclohexane at 500° one obtains butadiene, ethylene,
hydrogen, and a small amount of propylene. From aromatic hydrocarbons demethanation tends to occur [1, p. 265].

1-Butene has been converted into 1,3-butadiene, 92%, by
passing the gas over Type G84 catalyst in the presence of an
electrical glowing discharge [20].

A common reagent for the dehydrogenation of various organic
compounds is 2,3-dichloro-5,6-dicyano-1,4-benzoquinone (DDQ).
It has been utilized to produce an alkene as shown [21]:

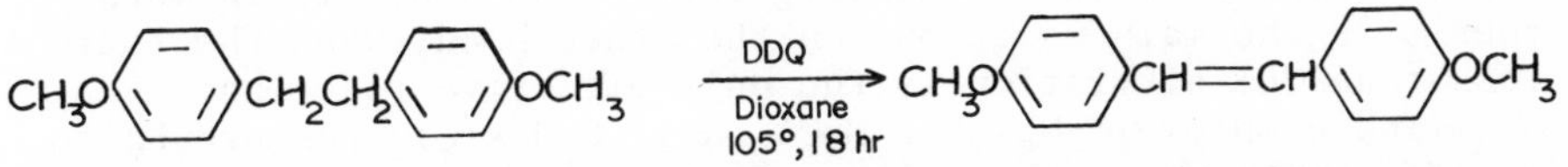

trans-4,4'-Dimethoxystilbene, 83-85%

Rhodium complexes are also effective in opening some tri-
and tetracyclic alkenes to form olefins [22]:

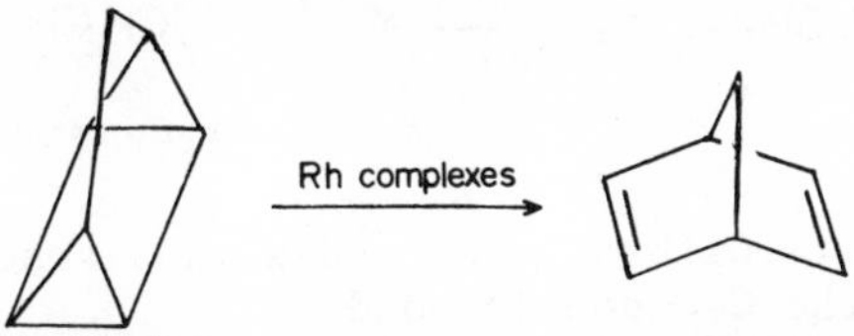

Quadricyclene   Norbornadiene,100%

For cyclanones Pd(II) metal complexes with a Cu(II) co-
catalyst with O_2 or air give cyclenones [23]:

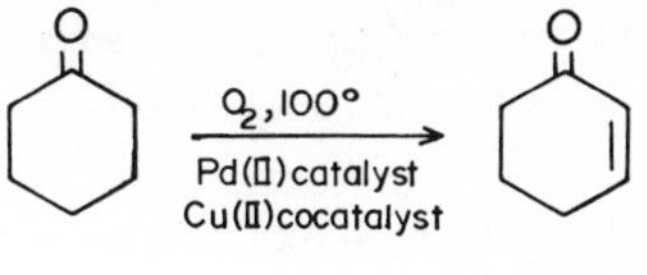

2-Cyclohexenone
15-30% conversion
selectivity 95%-90%

6.   Disproportionation of Alkenes [1, p. 285]

$$2 \; CH_3CH = CHCH_3 \xrightarrow[\text{EtAlCl}_2]{\text{WCl}_6} CH_3CH = CH_2 + CH_3CH_2CH = CHCH_3$$

The topic was discussed in 1, 115 but seems deserving of a separate section. Treating alkenes with WCl_6 and cocatalysts such as $LiAlH_4$ [24,25], $NaBH_4$ [25], $EtAlCl_2$-EtOH [24], BuLi [24] or C_3H_7MgBr [26] results in hastening the time at which the disproportion equilibrium is reached. Thus hept-3-ene gives the products as indicated [24]:

$$\text{Hept-3-ene} \xrightarrow[\substack{N_2 \\ C_6H_5Cl, \, 2hr}]{WCl_6 - LiAlH_4} \text{Hex-3-ene} + \text{Hept-3-ene} + \text{Oct-4-ene}$$

$$\qquad\qquad\qquad\qquad\qquad\qquad\qquad\quad 18.5\% \qquad\quad 39\% \qquad\quad 23\%$$

As a matter of fact, tungsten is not the only catalyst which accomplishes disproportionation [1, p. 285].

The Lewis acid, $WCl_6 - C_2H_5AlCl_2$, is also an effective catalyst for olefin metathesis and alkylation [27]. When the olefin:catalyst molar ratio is greater than 400, olefin metathesis is the main reaction; on the other hand, when the ratio is less than 70, alkylation is the only effective reaction.

The complex $[C_6H_5)_3P]_2 \; MoCl_2(NO)_2 \; C_2H_5AlCl_2$ has proved to be an efficient catalyst for the metathesis of linear terminal and nonterminal ω-arylalkenes [28]. Thus 3-phenylpropene-1 gives the product shown:

cis,trans-1,4-Diphenyl-2-butene, 90%

The reaction fails with styrene and with alkenes tri- or tetra-substituted at the C=C double bond.

In starting with the primary alcohol, n-amyl alcohol, Nefedov and co-workers [29] obtained largely hexenes as shown:

$$CH_3(CH_2)_3CH_2OH \xrightarrow[\substack{5-80atm.CO \text{ or } N_2}]{\substack{CuCl_2 \\ 140-200°}} \text{Pentenes} + \text{Hexenes}$$

$$\qquad\qquad\qquad\qquad\qquad\qquad\qquad\qquad \text{largely } cis, trans\text{-2-Hexenes} \\ \text{up to } 76\%$$

7.   From 5,6-Dihydro-4H-1,3-oxazines

Thermolysis of 5,6-dihydro-4H-1,3-oxazines leads to cleavage to amidomethylation products [30]:

2-Phenyl-5,6-dihydro-
4H-1,3-oxazine

trans-N-($\triangle^2$-Alkenyl)benzamide

Addendum

D.1. The photochemistry of alkenes, including isomerization, ring opening or closing, fragmentation, photo Diels-Alder, and cycloaddition, has been reviewed [J. D. Coyle, Chem. Soc. Rev., 3, 329 (1974)]. It must be remembered that a short ultraviolet wavelength is needed for alkene activation.

References

1. H.-W., 5, Pt. 1b, 1972, pp. 565-577.
2. A. T. Blomquist et al., J. Am. Chem. Soc., 74, 3643 (1952).
3. A. Maercker, R. Geuss, Angew. Chem. Intern. Ed. Engl., 9, 909 (1970).
4. J. D. Coyle, Chem. Soc. Rev., 3, 329 (1974).
5. J. A. Deyrup, M. Betkouski, J. Org. Chem., 37, 3561 (1972).
6. R. S. H. Liu et al., J. Org. Chem., 38, 1247 (1973).
7. C. Moussebois, J. Dale, J. Chem. Soc., C260 (1966).
8. K. H. Schulte-Elte, G. Ohloff, Helv. Chim. Acta, 51, 548 (1968).
9. J. C. Duggan, H. Weingarten, Syn. Commun., 1, 161 (1971).
10. P. G. Gassman, T. J. Atkins, J. Am. Chem. Soc., 93, 1042 (1971).
11. P. G. Gassman et al., J. Am. Chem. Soc., 93, 1812 (1971).
12. H. Takaya et al., Chem. Lett., 781 (1973).
13. T. B. Patrick et al., Tetrahedron Letters, 423 (1971).
14. W. J. Gensler et al., J. Am. Chem. Soc., 93, 3828 (1971).
15. G. Albelo, M. F. Rettig, J. Organomet. Chem., 42, 183 (1972).
16. J. P. Marino, T. Kaneko, Tetrahedron Letters, 3975 (1973).
17. L. Jaenicke et al., Ann. Chem., 1252 (1973).
18. M. Oda, R. Breslow, Tetrahedron Letters, 2537 (1973).
19. V. K. Skarchenko, Russ. Chem. Rev., 40, 997 (1971).
20. R. N. Schindler et al., Synthesis, 581 (1974).
21. J. W. A. Findlay, A. B. Turner, Org. Syn., 49, 53 (1969).
22. H. Hogeveen et al., J. Am. Chem. Soc., 89, 2486 (1967); Tetrahedron Letters, 3667 (1973).
23. R. J. Theissen, J. Org. Chem., 36, 752 (1971).

24. S. A. Matlin, P. G. Sammes, Chem. Commun., 174 (1973).
25. G. J. Leigh et al., Chem. Commun., 1202 (1972).
26. P. A. Raven, E. J. Wharton, Chem. Ind. (London), 282 (1972).
27. L. Hocks et al., Tetrahedron Letters, 2719 (1973).
28. G. Descotes et al., Synthesis, 364 (1974).
29. B. K. Nefedov et al., Izv. Akad. Nauk SSSR, Ser. Khim, 1886
     (1973); C. A., 79, 136387 (1973).
30. R. R. Schmidt, Synthesis, 335 (1972).

E. Condensation Reactions

   2. From Carbonyl Compounds and Phosphoranes (Wittig) ($\underline{1}$,
        141) [1]

$$RCHO + (C_6H_5)_3P{=}CHR' \longrightarrow RCH{=}CHR' + (C_6H_5)_3PO$$

The Wittig reaction, involving ylids, is now widely used in
the synthesis of alkenes. A new synthesis of ylids has been
described [1a] and their reactions reviewed [2]. Advances made
include their utilization by the phase transfer method [3]:

$$(EtO)_2\underset{\underset{O}{\|}}{P}CH_2R' \xrightarrow[\substack{\text{phase transfer} \\ \text{conditions}}]{R_2C=O} R_2C{=}CHR'$$

$$51-77\%$$

Similarly, phase transfer conditions have been used to synthesize
stilbene [4]. In the case of the synthesis of o-nitrostyrene
from the o-nitrophosphorane, $o\text{-}NO_2C_6H_4CH{=}PPh_3$, and aqueous formaldehyde, phase transfer conditions are not necessary because of
the activity of formaldehyde and its sluggish activity toward
alkali. Yield of o-nitrostyrene was 90% [5]. A base other than
aqueous alkali utilized to prepare the ylid is that obtained by
dissolving potassium in HMPA to form $(OP(\bar{N}Me_2)\bar{N}(Me)_2\bar{K}$ [1a].
Alkenes from an aldehyde and this ylid were synthesized in 55-
84% yields. Carbon tetrahalides with tridimethylaminophosphine,
TDAP, respond similarly [6]. Symmetrical alkenes are made by
simply oxidizing this solution of the ylid with air [1a].

$$(C_6H_5)_3P{=}CHR \xrightarrow{[o]} RCH{=}CHR$$

$$48-70\%$$

Another oxidizing agent is sulfur, which may be utilized with the
ylid prepared by the use of BuLi [7].

The recent literature contains numerous other examples of the use of the phosphorane in synthesis [8].

Many other modifications of the Wittig olefin synthesis have appeared. In some cases the ylid has been prepared in situ. Thus to prepare difluoromethylene olefins the preferred method, which avoids any fluoride ion isomerization or addition reactions, is as shown [9]:

$$RCOR_f + CF_2Br_2 \xrightarrow[\text{Diglyme, 24 hr}]{\overset{(C_6H_5)_3P}{70°}} F_2C = C(R)(R_f)$$

82-87%

In other cases the β-oxido ylid, generated by an alkyl-lithium such as butyllithium, is employed since this intermediate permits the overall reaction to become stereospecific. Thus 2-methyl-cis-2-nonen-1-ol was synthesized as indicated [10]:

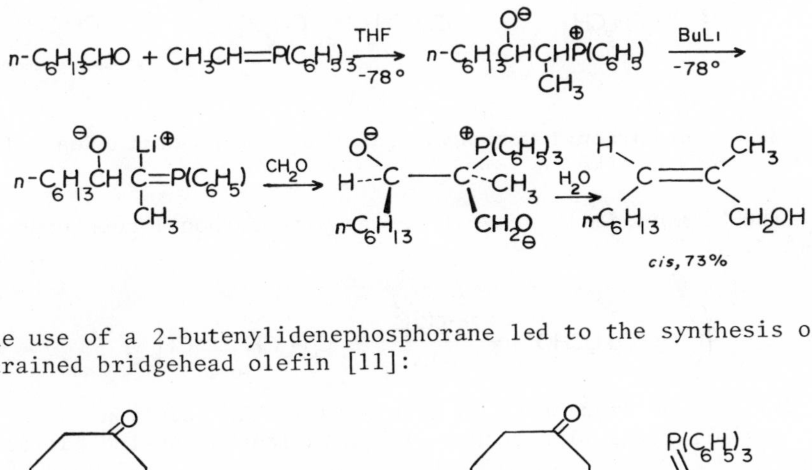

The use of a 2-butenylidenephosphorane led to the synthesis of a strained bridgehead olefin [11]:

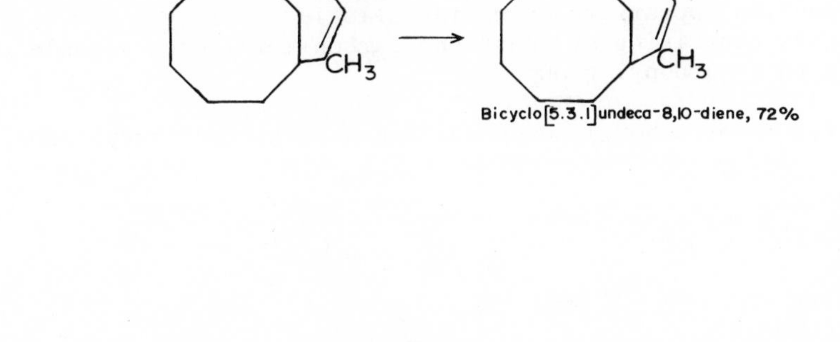

Bicyclo[5.3.1]undeca-8,10-diene, 72%

Alkenes may also be obtained from sulfonium ylids [12].   If not the ylid, the dianion, which reacts with benzophenone to yield the alkene, may be formed [13]:

$$C_6H_5SO_2CLiH + (C_6H_5)_2CO \longrightarrow C_6H_5SO_2CH=C(C_6H_5)_2 \xrightarrow{\text{AlHg}} CH_2=C(C_6H_5)_2$$

1,1-Diphenylethylene
90%

The reaction appears to be stereoselective in that E isomers are obtained when possible.   For uses of E and Z designation for cis,trans isomers, see the work by Blackwood [14].

Methylenation may also be accomplished with a methyl sulfide and a ketone [15]:

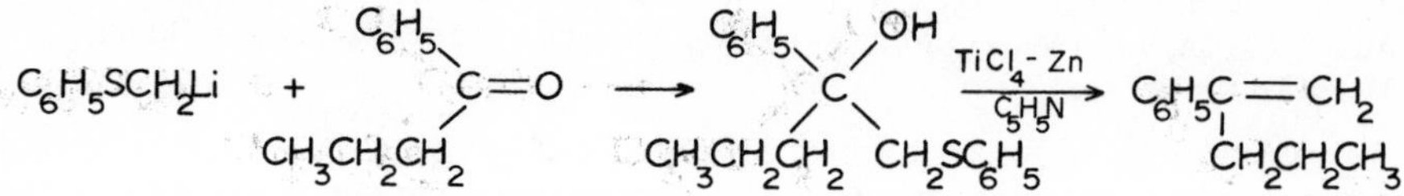

$$C_6H_5SCH_2Li \quad + \quad \begin{matrix} C_6H_5 \\ \diagdown \\ CH_3CH_2CH_2 \end{matrix} C=O \longrightarrow \begin{matrix} C_6H_5 \\ \diagdown \quad \diagup \\ CH_3CH_2CH_2 \quad CH_2SC_6H_5 \end{matrix} \xrightarrow[C_5H_5N]{TiCl_4-Zn} \begin{matrix} C_6H_5C=CH_2 \\ | \\ CH_2CH_2CH_3 \end{matrix}$$

2-Phenyl-1-pentene
85%

3. From Carbonyl Compounds and the Simmons-Smith Reagent or the Like

The Simmons-Smith reagent reacts with carbonyl compounds as shown [16]:

$$C_6H_5CHO \xrightarrow[Zn]{Zn-ClCH_2I} C_6H_5CH=CH_2$$

80%

The presence of zinc metal is necessary in the reaction.

A similar reaction occurs with methylene magnesium halides [17]:

$$R_2CO \quad + \quad CH_2(MgI)_2 \longrightarrow R_2C=CH_2$$

30-80%

The reaction appears to be characteristic of all gem dimetallic compounds containing at least one carbon-metal bond capable of adding to a carbonyl group.

4. From Carbonyl Compounds and α-Metalated Isocyanides

As with phosphoranes (E.2) and the Simmons-Smith reagent (E.3), the α-metalated isocyanides may be used to convert carbonyl compounds into alkenes [18]:

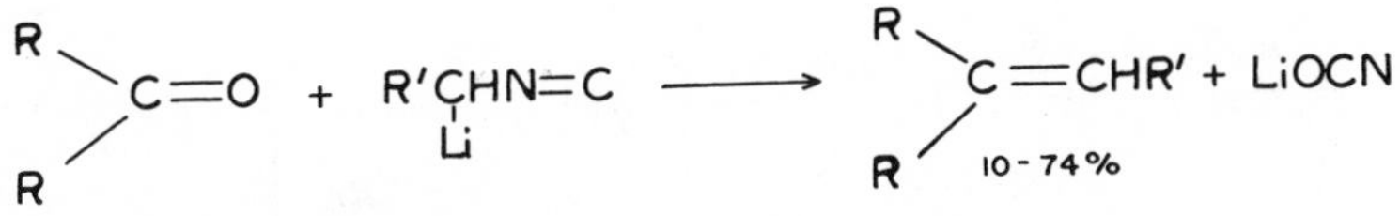

$$\underset{R}{\overset{R}{>}}C=O \;+\; R'\underset{Li}{CH}N{=}C \longrightarrow \underset{R}{\overset{R}{>}}C{=}CHR' + LiOCN$$

10 - 74%

The metalated isocyanides are more reactive than the alkylidene-triphenylphosphoranes.

5. From Carbonyl Compounds and Boranes or Silanes

Syntheses have already been cited involving silanes [A.20, 121, 123].

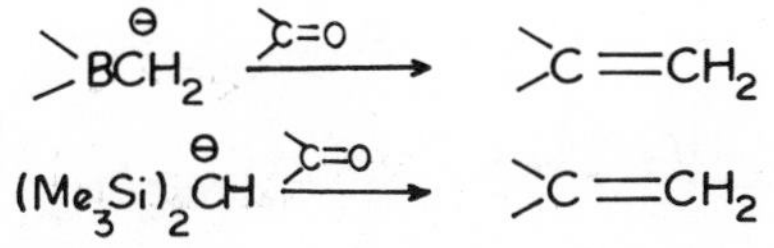

$$>BCH_2^{\ominus} \xrightarrow{\;>C=O\;} \;>C{=}CH_2$$

$$(Me_3Si)_2\overset{\ominus}{CH} \xrightarrow{\;>C=O\;} \;>C{=}CH_2$$

In other words, these anions behave like ylids. The problem with the borane lies in the formation of the anion. Only highly hindered bases, such as lithium 2,2,6,6-tetramethyl-piperidide, probably because they fail to complex with the boron atom, make this conversion possible [19].
    In this manner methylenecyclohexane has been produced by the action of β-methyl-9-borabicyclononane on cyclohexanone [19] in the presence of a strong, hindered base:

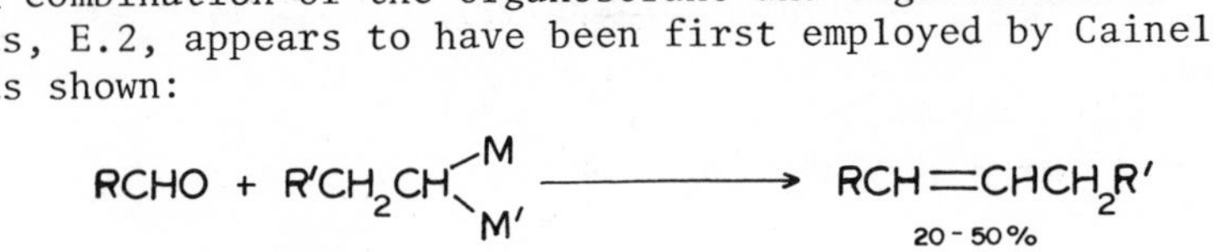

55-65%

A combination of the organoborane and organometallic methods, E.2, appears to have been first employed by Cainelli [20] as shown:

$$RCHO \;+\; R'CH_2CH\underset{M'}{\overset{M}{<}} \longrightarrow RCH{=}CHCH_2R'$$

20 - 50%

$$M = Li, \; M' = BR_2$$

Silyl-substituted anions, prepared from the silane and NaOMe-HMPA, react with carbonyl compounds as indicated to give the alkene [21]:

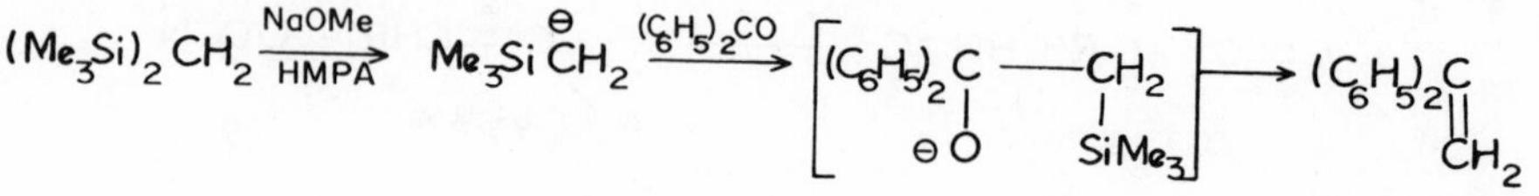

$$(Me_3Si)_2CH_2 \xrightarrow[HMPA]{NaOMe} Me_3Si\overset{\ominus}{C}H_2 \xrightarrow{(C_6H_5)_2CO} \left[(C_6H_5)_2\underset{\underset{\ominus O}{|}}{C} - \underset{\underset{SiMe_3}{|}}{CH_2} \right] \rightarrow (C_6H_5)_2\underset{\underset{CH_2}{||}}{C}$$

1,1-Diphenylethylene
53%(from ketone)

6. From a Schiff Base and an Alkyl Arene (Siegrist Reaction)

2,7-Distyrylnaphthalene, of interest in the synthesis of [6] helicene, has been prepared as indicated [22]:

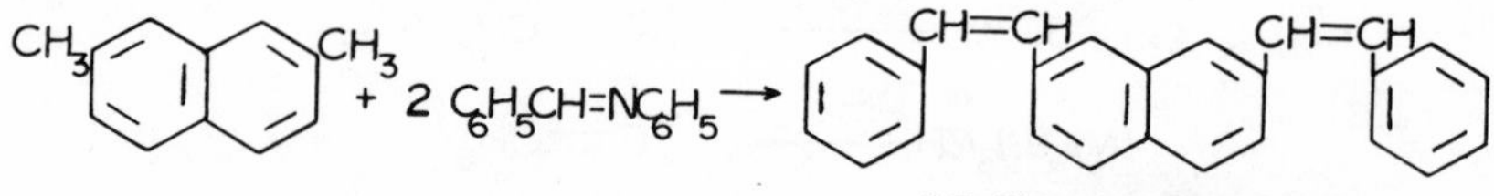

2,7-Distyrylnaphthalene, 89%

[6] Helicene
60%

Addenda

E.2. The advantages of phosphonate over phosphorane carbanions (ylids) are: (a) the former are more nucleophilic, (b) they may be elaborated by alkylation, (c) they are cheaper, and (d) the water-soluble phosphate ion formed from phosphonates permits easier separation of the olefin [J. Boutagy, R. Thomas, Chem. Rev., 74, 87 (1974)].

E.2. The Bestman oxidation procedure is a ring closure of a diylid:

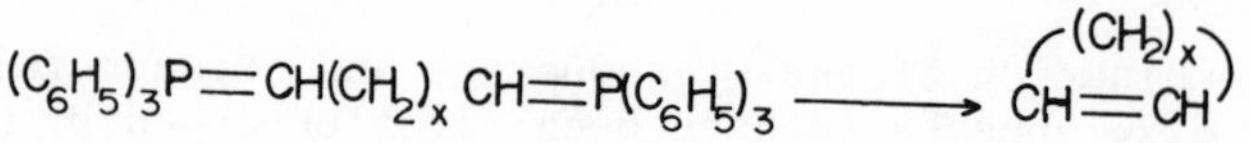

$$(C_6H_5)_3P=CH(CH_2)_x\,CH=P(C_6H_5)_3 \longrightarrow \underset{CH=CH}{\overset{(CH_2)_x}{\frown}}$$

[Example in J. A. Deyrup, M. F. Bethouski, J. Org. Chem., 40, 284 (1975)].

E.2. cis-Olefins from Wittig ylids and alkyl halides may be formed by using DMSO or HMPA in the mixture with THF as the solvent [P. E. Sonnet, Org. Prep. Proced. Int., 6, 269 (1974)].

E.2.   The conversion, $>C=O \rightarrow >C=CMe_2$, may be performed by triphenylphosphine and carbon tetrabromide followed by methylation with $LiCuMe_2$ [G. H. Posner et al., Tetrahedron Letters, 1373 (1975)].

E.3.   Aluminum metal and methylene bromide form a stable substance, $Br_2AlCH_2AlBr_2$, which reacts with ketones to replace the carbonyl oxygen with the methylene group [A. Bongini et al., J. Organomet. Chem., 72, C4 (1974)].

E.5.   The carbanion of "methane triborane," $HC[(BOMe)_2]_3$, reacts with benzaldehyde to form $C_6H_5CH=CHB(OH)_2$, which with I_2 and alkali gives styrene [D. S. Matteson, Synthesis, 147 (1975)].

E.12.   The evolutionary process by which ylids and their derivatives were discovered has been discussed [G. Wittig, Acc. Chem. Res., 7, 6 (1974)].

References

1.   H.-W., 5, Pt. 1b, 1972, p. 383.
1a. H. J. Bestmann, W. Stransky, Synthesis, 798 (1974).
2.   P. A. Lowe, Chem. Ind. (London), 1070 (1970).
3.   C. Piechucki, Synthesis, 869 (1974).
4.   G. Märkl, A. Merz, Synthesis, 295 (1973).
5.   M. Butcher et al., Australian J. Chem., 26, 2067 (1973).
6.   J. C. Combret et al., Tetrahedron Letters, 1035 (1971).
7.   H. Mägerlein, G. Meyer, Chem. Ber., 103, 2995 (1970).
8.   H. Wagner, Chem. Ber., 102, 2259 (1969); E. J. Corey et al.,
     J. Am. Chem. Soc., 92, 6635 (1970); S. V. McKinley, J. W.
     Rakshys, Jr., Chem. Commun., 134 (1972); G. Wittig et al.,
     Chem. Ber., 95, 2514 (1962); H. Oediger et al., Synthesis,
     596 (1972); G. Büchi, H. Wüest, Helv. Chim. Acta, 54, 1767
     (1971); G. Koszmehl, B. Bohn, Angew. Chem. Intern. Ed.
     Engl., 12, 237 (1973); E. J. Corey, R. A. Ruden, Tetra-
     hedron Letters, 1495 (1973).
9.   D. G. Naae, D. J. Burton, J. Fl. Chem., 1, 123 (1971/72);
     Syn. Commun., 3, 197 (1973).
10. E. J. Corey, H. Yamamoto, J. Am. Chem. Soc., 92, 226 (1970);
     M. Schlosser et al., Synthesis, 29 (1971).
11. W. G. Dauben, J. Ipaktschi, J. Am. Chem. Soc., 95, 5088
     (1973).
12. H. Nozaki et al., Tetrahedron Letters, 2303 (1967).
13. V. Pascali, A. Umani-Ronchi, Chem. Commun., 351 (1973).
14. J. E. Blackwood et al., J. Am. Chem. Soc., 90, 509 (1968).
15. T. Mukaiyama et al., Chem. Lett., 1523 (1974).

16.   S. Miyano et al., J. Organomet. Chem., 12, 263 (1968); 10, 518 (1967); Nippon Kagaku Kaishi, 1760 (1972); C. A., 77, 151,556 (1972).
17.   F. Bertini, G. Cainelli et al., Tetrahedron, 26, 1281 (1970).
18.   U. Schöllkopf, F. Gerhart, Angew. Chem. Intern. Ed. Engl., 7, 805 (1968); F. Kienzle, Helv. Chim. Acta, 56, 1671 (1973).
19.   M. W. Rathke, R. Kow, J. Am. Chem. Soc., 94, 6854 (1972).
20.   G. Cainelli et al., Tetrahedron Letters, 4315 (1966).
21.   H. Sakurai et al., Tetrahedron Letters, 4193 (1973).
22.   R. H. Martin et al., Helv. Chim. Acta, 54, 358 (1971).

F.   Decarboxylation, Decarbonylation, and Dehydroxylation

1.   From Unsaturated Acids ($\underline{1}$, 144) [1]

β-Anthracene-9-acrylic acid has been decarboxylated to give 68% of 9-vinylanthracene by heating with quinoline and a trace of Cu at 100° under vacuum [1a].

3.   From Succinic Acids or Their Peresters ($\underline{1}$, 146) [1, p. 351]

The dicarboxylic acids were converted, via the diacyl chlorides, into their di-t-butyl peresters, which were decomposed either thermally or photochemically [2]:

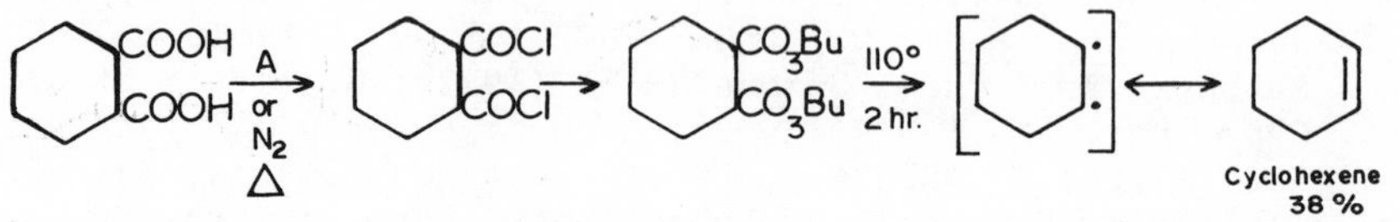

The equation illustrates the general principle that simultaneous (or near simultaneous) generation of free radicals on vicinal carbon atoms is an accelerated, facilitated process, since two free radicals on adjacent carbon atoms provide the double bond of the alkene.

β-Truxinic acid was electrolyzed, as shown, to give cis-3,4-diphenylcyclobutene [3]:

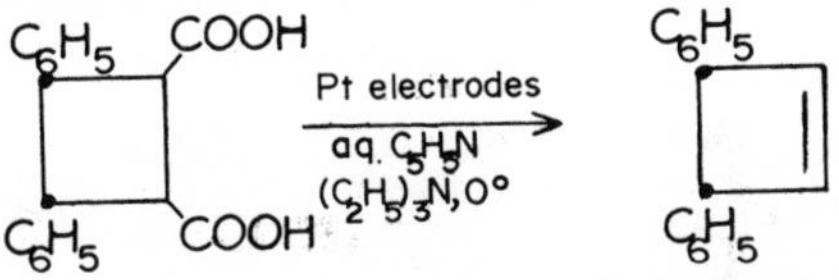

cis-3,4-Diphenylcyclobutene
24%

6. From Diols Directly and via Thionocarbonates and Dioxalanes (1, 148)

Reagents such as thiocarbonyldiimidazole (1, 148), trimethyl phosphite [4], bis(1,5-cyclooctadiene) nickel [5], and isopropyl iodide followed by Zn-EtOH [6] have been employed in forming alkenes from thionocarbonates. Thus the thionocarbonate of cis-1,2-dihydroxycyclobutene responded as indicated [4]:

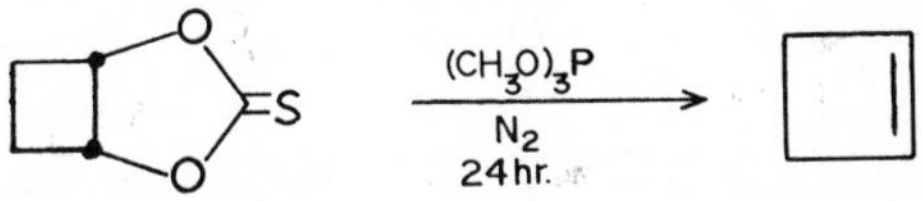

Cyclobutene
91%

Other intermediates have been utilized as well in the conversion of glycols into alkenes. Eastwood and co-workers [7] used the 1,3-dioxalane as indicated:

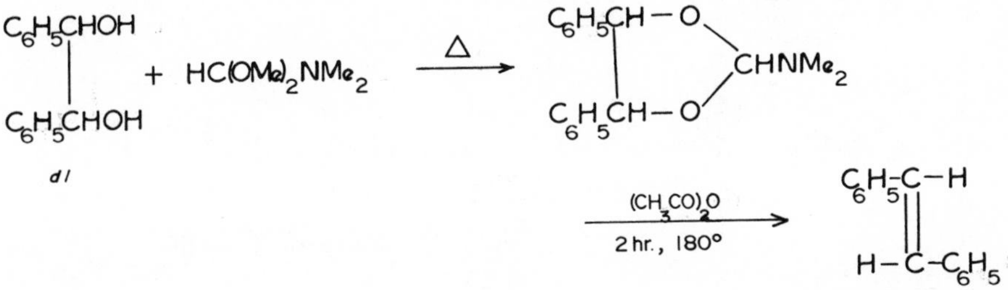

trans-Diphenylethylene
80%

An early intermediate was the disulfonic acid ester which with NaI and acetone gave the alkene [8]:

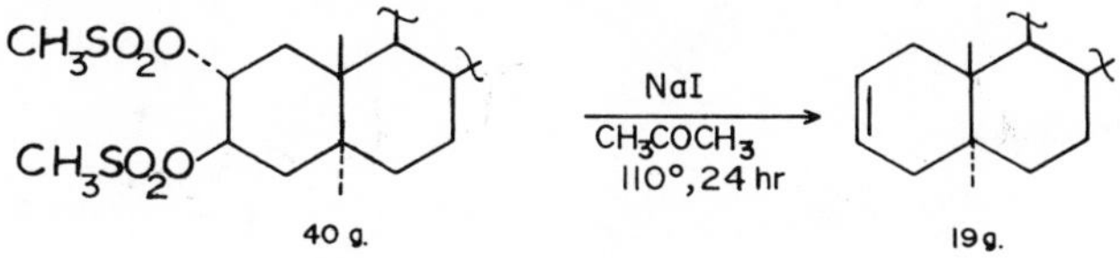

Direct elimination from the pinacol is also possible. Thus dl-2,3-diphenylbutane-2,3-diol led to trans-2,3-diphenyl-2-butene as shown [9]:

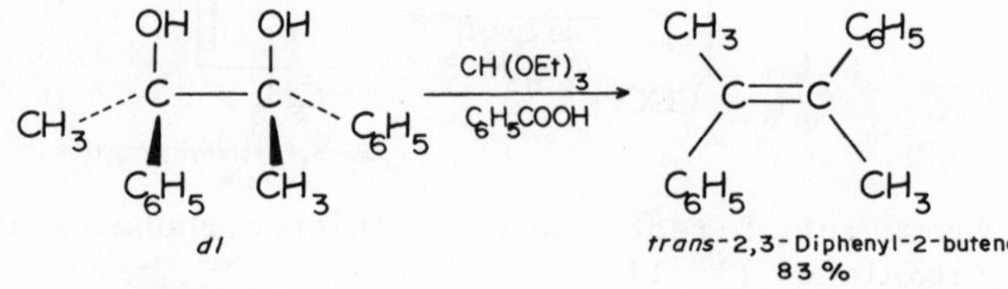

trans-2,3-Diphenyl-2-butene
83%

The reaction is quite useful in preparing olefins. It proceeds via the dioxolane:

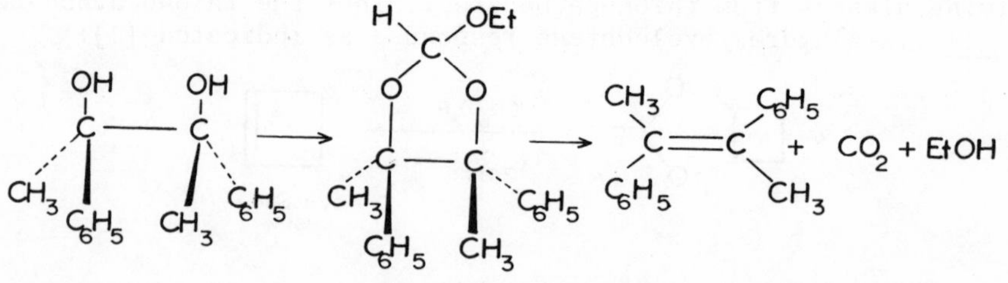

In a one-operation process Sharpless and Flood [10] converted the 1,2-diol into the lithium salt which, without isolation, was treated with K_2WCl_6 as indicated:

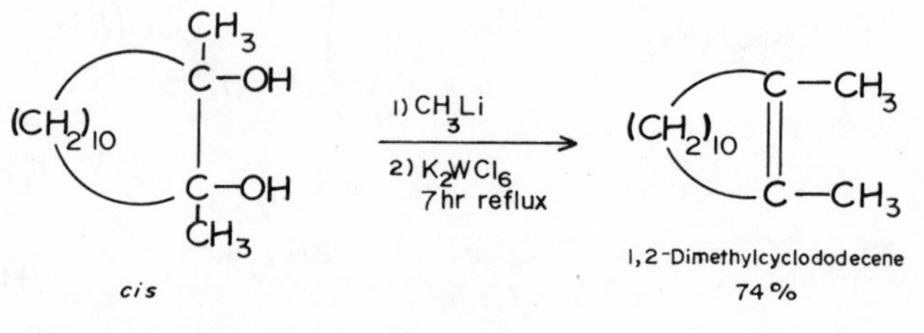

1,2-Dimethylcyclododecene
74%

An elimination to yield the alkene has also been accomplished with benzaldehyde acetals as shown [11]:

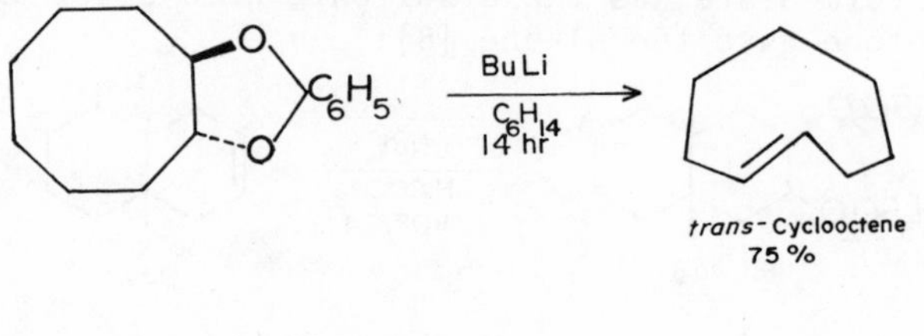

trans-Cyclooctene
75%

7. From Saturated Carboxylic Acids (Oxidative Decarboxylation) [1, p. 343]

Reviews of oxidative decarboxylation are available [12,12a].
The oxidation of an acid such as cyclohexanecarboxylic acid with
$Pb(OCOCH_3)_4$ gives a mixture of cyclohexyl acetate and cyclohexene. It is possible to increase the amount of cyclohexene by
using DMF as a solvent or by adding cupric acetate in catalytic
amounts. The method has been of synthetic value, particularly
when the radical attached to the COOH group is a primary or
secondary alkyl or a cycloalkyl group.
The oxidation of nonanoic acid gave octene-1 as indicated
[13]:

$$CH_3(CH_2)_7CO_2H \; + \; Pb(OAc)_4 \; \xrightarrow[\text{LiOAc}]{\text{Cu(OAc)}_2} \; CH_3(CH_2)_5CH{=}CH_2$$

Octene-1, 91 %

An improved oxidation procedure has been described [14,15]. The
latter uses dicarboxylic acids and states that the lead tetraacetate should be added gradually for maximum yields of unsaturated acids.
A similar elimination was achieved with sodium 2-chloro-
cyclohex-4-ene-1-carboxylate [16]:

Cyclohexa-1,4-diene
70 %

Addenda

F.6. meso-1,5-Hexadiyne-3,4-diol may be completely dehydroxy-
lated by treatment with p-toluenesulfonyl chloride-C_5H_5N followed
by NaI to give the hexadiyne-3-ene (68%) [H. P. Figeys, M.
Gelbcke, Tetrahedron Letters, 5139 (1970)].

F.7. Electrolytic oxidation of N-carbethoxy-7-azanorbornane-2,
3-dicarboxylic acid results in the formation of the 2,3-diene
[A. P. Marchand, R. W. Allen, J. Org. Chem., 40, 2551 (1975)].

F.7. Additional examples of oxidative decarboxylation may be
found in Fieser and Fieser, Reagents for Organic Synthesis,
Wiley, New York, 1967, Vol. I, p. 554; 1969, Vol. II, p. 235.

F.7.   An old method of preparing trans-cinnamyl bromide consists
of treating cis-cinnamic acid dibromide with $NaHCO_3$ or alcoholic
KOAc [E. Grovenstein, Jr., S. P. Theophilou, J. Am. Chem. Soc.,
77, 3795 (1955)].

References

1.   H.-W., 5, Pt. 1b, 1972, p. 287.
1a. S. D. Paul et al., Indian J. Chem., 10, 321 (1972).
2.   S. Masamune et al., Chem. Commun., 98 (1969).
3.   J. I. Brauman, W. C. Archie, Jr., Tetrahedron, 27, 1275
     (1971).
4.   W. Hartmann et al., Tetrahedron Letters, 853 (1972).
5.   M. P. Semmelhack, R. D. Stauffer, Tetrahedron Letters,
     2667 (1973).
6.   E. Vedejs, E. S. C. Wu, Tetrahedron Letters, 3793 (1973).
7.   F. W. Eastwood et al., Tetrahedron Letters, 5223 (1970).
8.   H. L. Slates, N. L. Wendler, J. Am. Chem. Soc., 78, 3749
     (1956).
9.   T. Hiyama, H. Nozaki, Bull. Chem. Soc. Jap., 46, 2248
     (1973).
10. K. B. Sharpless, T. C. Flood, Chem. Commun., 370 (1972).
11. G. H. Whitham et al., J. Chem. Soc., Perkin Trans., I,
     2332 (1973).
12. R. A. Sheldon, J. K. Kochi, Org. Reactions, 19, 279 (1972).
12a. H. O. House, Modern Synthetic Reactions, Benjamin, Reading,
     Mass., 1972, p. 373.
13. J. D. Bacha, J. K. Kochi, Tetrahedron, 24, 2215 (1968).
14. Ref. 1, p. 305.
15. Yu. N. Ogibin et al., Synthesis, 889 (1974).
16. W. P. Norris, J. Org. Chem., 33, 4540 (1968).

Chapter 3

ALKYNES, ALLENES, AND CUMULENES

A new, comprehensive, and definitive treatment of acetylenes has
become available [1], as well as a superior book on synthesis
[2] and an excellent review on the synthesis of stable cyclo-
alkynes of medium ring size [3]. The best way for preparing
terminal acetylenes from 1,2-dibromides is elimination with
$NaNH_2$ or NaH in DMSO at 65-70° for 9 hr as a maximum (A.1). A
new elimination for the preparation of arylacetylenes looks
promising; it involves the loss of chloro and carboxaldehyde
groups from arylchloroacroleins, ArCCl=CHCHO, readily available
by the Vilsmeir reaction from acetophenone (A.1). Monohydrazones as well as diphenylhydrazones may be converted into
acetylenes by oxidation (A.4).
     A host of mixed diaryl acetylenes has become available via
coupling of cuprous arylacetylides (B.1). An inverse correlation exists in the addition of acetylides to ketones; the more
enolic the ketone, the smaller the yield of acetylenic alcohol
(B.2). Terminal diacetylenes are available by coupling the
cuprous acetylide with $BrC \equiv CSiEt_3$ (C.1). In general, for
coupling of acetylenes to form diacetylenes the best catalyst is
cuprous chloride complexed with tetramethylethylenediamine
(C.1). Some generalities are given for the conversion of acetylenes into isomeric acetylenes or dienes, including a description
of the "conducted tour" mechanism (D.1). As in the alkene

family, a disproportionation of acetylenes, which probably occurs
via a cyclobutadiene intermediate, has been found to occur at
high temperature with a tungsten catalyst (D.5).

An improved survey of allene preparations was made possible
by the publication of another definitive text listed in E.1.
This book, together with the two previously mentioned, has con-
tributed greatly to advance our knowledge of the syntheses of
acetylenes and allenes.

A. Elimination

1. From Dihalides and Vinyl Halides (Dehydrohalogenation)
    (1, 151)

$$RCHXCH_2X$$
$$RCX_2CH_3 \longrightarrow RC{\equiv}CH$$
$$RCX{=}CH_2$$

For the preparation of terminal acetylenes from higher
alkyl halides and sodium acetylide in liquid ammonia by the
usual method, insignificant yields are obtained. By a second
method, the bromination-dehydrobromination procedure, rearrangements occur during the dehydrobromination step. The rearrangement may be avoided by the use of $NaNH_2$ in liquid ammonia, but
this solvent is inconvenient to handle. To overcome these
difficulties, Klein and Gurfinkel [4] in the dehydrobromination
step employed either $NaNH_2$ or NaH in DMSO at 65-70° for 9 hr.
Yields of alkynes, based on the dibromide, from 1-decyne to 1-
hexadecyne varied within 72-94% (Ex. a). In this reaction more
prolonged heating (21-30 hr) converts the 1-alkyne into the 2-
alkyne.

Other bases in DMSO have been employed in an improved
procedure for the synthesis of t-butylacetylene. Thus Collier
and Macomber [4a] utilized the following steps:

$$+CH{=}CH_2 + Br_2 \xrightarrow{-78°} +\underset{\underset{Br}{|}}{C}H-\underset{\underset{Br}{|}}{C}H_2 \xrightarrow[DMSO]{t\text{-}BuOK} +C{\equiv}CH$$
$$\text{90\%} \qquad\qquad\qquad\qquad \text{91\%}$$

Kocienski [4b] began with pinacolone dichloride as
indicated:

$$\underset{\substack{| \\ CH_3}}{\overset{CH_3\ CH_3}{CH_3CCCl_2}} \xrightarrow[\substack{DMSO \\ <40°}]{t\text{-BuOK}} \underset{\substack{| \\ CH_3}}{\overset{CH_3}{CH_3CC}} \equiv CH$$

95%

The substrate is readily available from pinacolone and PCl_5 and thus the method is inexpensive, requiring no low temperature.

Elimination from vinyl halides or 1,2-dihalides with lithium diethylamide has been reviewed [5]. With this reagent 1-phenyl-1-chloroethylene in ether gave an 84% yield of phenylacetylene. Similarly, meso-1,2-diphenyl-1,2-dichloroethane gave a 61% yield of tolane.

Elimination may also be accompanied by decarboxylation. For instance, β-bromocinnamic acids readily give phenylacetylenes when heated with an aqueous base [6]. Dehydrohalogenation may occur first followed by the decarboxylation of the phenyl propiolate anion:

$$\underset{\substack{Br \\ |}}{C_6H_5C}=CHCOOH \xrightarrow{\overset{\ominus}{OH}} C_6H_5C\equiv C\overset{\ominus}{CO_2} \xrightarrow{H_2O} C_6H_5C\equiv CH + H\overset{\ominus}{CO_3}$$

In addition, the carboxyl group becomes a rather unusual leaving group in the transitory formation of benzyne:

$$\underset{N=N\ \oplus}{\overset{CO_2H}{\bigcirc}} \longrightarrow \left[\bigcirc | \right] + N_2 + CO_2 + H^{\oplus}$$

The benzyne formation has been reviewed [7].

With an electron-withdrawing group other than carboxyl, a smooth dehydrohalogenation occurs. Thus in the preparation of phenylpropargyl aldehyde acetal [8], the acetylene is formed:

$$C_6H_5CH=\underset{\substack{Br \\ |}}{C}CH(OEt)_2 \xrightarrow[EtOH]{KOH} C_6H_5C\equiv CCH(OEt)_2$$

80−86%

In addition, the closely related β-chlorovinyl aldehydes prepared by the Vilsmeier reaction may be converted into the corresponding acetylenes, with the loss of the aldehyde group, by a base [9]:

$$ArCOCH_3 \xrightarrow[\text{DMF}]{\text{POCl}_3} Ar\overset{\overset{\displaystyle Cl}{|}}{C}=CHCHO \xrightarrow[\text{Dioxane-H}_2O]{\text{NaOH}}$$

$$ArC\equiv CH + HCO_2H + NaCl$$
$$52-98\%$$

Ferrocenylacetylene has been prepared (88%) by elimination from ferrocenyl chloroacrolein [10].

In the dehydrohalogenation of vicinal dihalides, the acetylene generated may add to the base, particularly if electron-attracting groups are present in the acetylene:

$$C_6H_5\overset{\overset{\displaystyle Br}{|}}{C}H\overset{\overset{\displaystyle Br}{|}}{C}HCO_2Et \xrightarrow[\text{EtOH}]{\text{NaOEt}} C_6H_5CH=C\overset{\displaystyle CO_2Et}{\underset{\displaystyle OEt}{}}$$

Newman and Merrill [11] overcame this tendency by using NaH in benzene as the base with only catalytic amounts of ethyl alcohol. Sodium ethoxide then is regenerated in small amounts and the acetylene rather than the vinyl ether is obtained (Ex. b).

Wolinsky [12] prepared 2-butyne from the vicinal dibromide through the use of the base, t-BuOK in p-cymene, as shown:

$$CH_3\overset{\overset{\displaystyle CH_3}{|}}{\underset{\underset{\displaystyle Br}{|}}{C}}CH_2Br \xrightarrow[\text{p-Cymene}]{\text{t-BuOK}} CH_3C\equiv CCH_3$$
$$65\%$$

The mechanism appears to involve a carbene as indicated:

$$CH_3\overset{\overset{\displaystyle CH_3}{|}}{\underset{\underset{\displaystyle Br}{|}}{C}}CH_2Br \xrightarrow{-HBr} \left[CH_3\overset{\overset{\displaystyle CH_3}{|}}{C}=CHBr\right] \xrightarrow{\text{t-BuOK}}$$

$$\left[CH_3\overset{\overset{\displaystyle CH_3}{|}}{C}=\overset{\displaystyle \ominus}{C}Br\right] \xrightarrow{-Br^{\ominus}} \left[CH_3\overset{\overset{\displaystyle CH_3}{|}}{C}=C:\right] \rightarrow CH_3C\equiv CCH_3$$

Potassium t-butoxide was also used by Mitchell and Sondheimer
[13] to prepare 1,8-diethynylnaphthalene by elimination from the
tetrabromide:

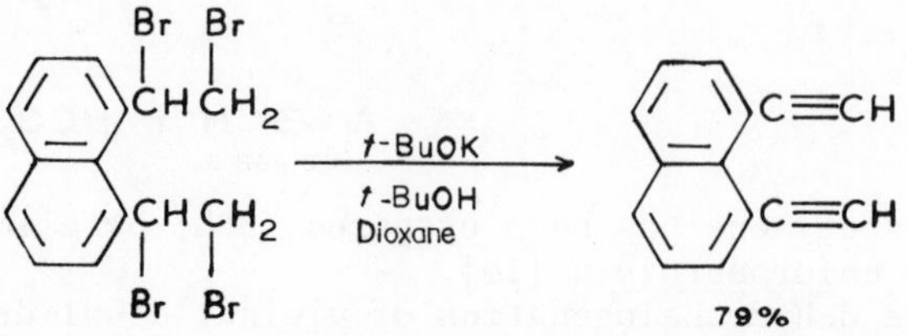

79%

Not only olefins with hydrogen and halogen in vicinal
positions, but also those containing these elements at the same
carbon atom may be converted into alkynes. The latter elimination, known as the Fritsch-Buttenberg-Wiechell rearrangement,
may be accomplished with bases such as lithiumorganic reagents
[14] or potassium t-butoxide in diglyme [15]. Bender and co-
workers [15] prepared a series of alkynes by the treatment of
1,1-diaryl-2-bromoethenes with the t-butoxide as shown:

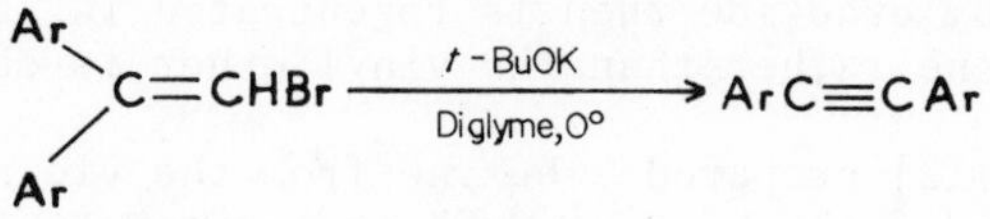

When Ar=C_6H_5, m-$CH_3C_6H_4$, p-$CH_3C_6H_4$, p-ClC_6H_4, p-FC_6H_4, and
p-$CH_3OC_6H_4$, satisfactory amounts of the alkyne were obtained.
The formation of the alkyne failed when Ar=m- or p-$CF_3C_6H_4$ in
that the product was the bis(trifluoromethylphenyl) vinyl t-
butyl ether. The mechanism of the FBW rearrangement is as
follows:

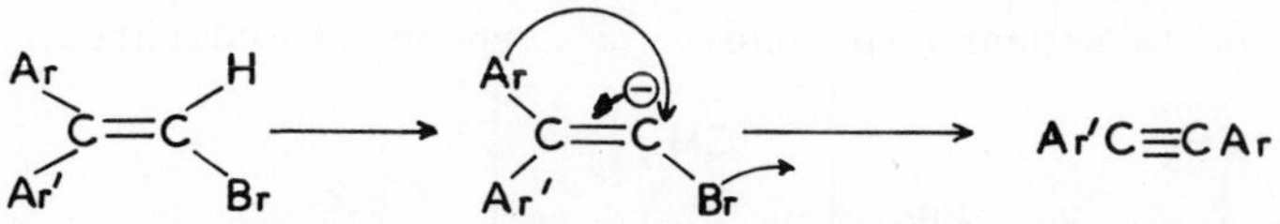

A safe and convenient synthesis of dichloroacetylene, DCA, has
been devised by Siegel and co-workers [16]. By this method trichloroethylene in a liquid medium containing ether and KOH is
heated at 140° under nitrogen (Ex. c):

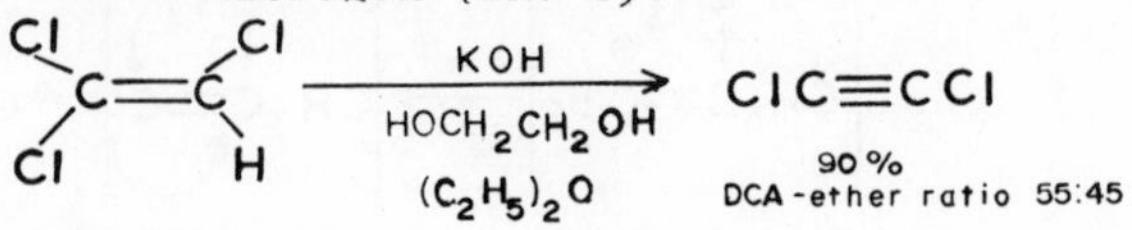

90%
DCA-ether ratio 55:45

The presence of ether reduces the hazard of explosions and keeps the side reactions of the product at a minimum.

To prepare ynamines by a convenient, general laboratory rule, Strobach [17] treated the 1,1-difluoroalkene or the β,β-difluorostyrene with lithium diethylamide as shown:

$$RCH{=}CF_2 \; + \quad 2\,LiN(C_2H_5)_2 \longrightarrow RC{\equiv}CN(C_2H_5)_2$$

$$45{-}84\%$$

The difluoroalkenes may be obtained readily from aliphatic or aromatic aldehydes and sodium chlorodifluoroacetate [18].

Monohaloacetylenes may be produced by elimination from dihaloalkenes [19] such as:

$$RCH{=}CX_2 \xrightarrow[\text{C_2H_5OH}]{\text{KOH}} RC{\equiv}CX$$

However, the interesting and useful synthesis of Corey and Fuchs [20] points the way to the conversion of RCH=CX$_2$ into RC≡CH without the isolation of the haloacetylene. In the process the transformation of an aldehyde into an acetylene with one more carbon atom results:

$$RCHO + \; CBr_4 + \; (C_6H_5)_3P + \; Zn \longrightarrow$$

$$RCH{=}CBr_2 \xrightarrow[\text{1.5\% LiHg-THF}]{\substack{\text{1) BuLi, THF} \\ \text{2) H}_2\text{O} \\ \text{or}}} RC{\equiv}CH \qquad R = C_7H_9, \; \text{Nonyne, 94\%}$$

Other methods of preparation are given in B.3.

a.   Preparation of 1-Decyne [4]

Sodamide, 8 g, was pulverized under paraffin oil and then added to DMSO, 80 g, after which the mixture was heated under N$_2$ at 65-70° for 1 hr. 1,2-Dibromodecane was added to 8 molar equiv of the sodamide mixture and the total was heated at 65-70° for 9 hr. Extraction first in ice water, then in ether followed by washing and distillation gave 1-decyne, 94%

b.   Preparation of Ethyl Phenylpropiolate [11]

Ethyl dibromocinnamate, the amount prepared from 0.65 mol of ethyl cinnamate and bromine, in dry benzene was treated with 1.33 mol of NaH, and the suspension was heated to boiling. Heating was discontinued and 2 ml of absolute ethanol was added

with stirring. Additional 2-ml portions of ethanol were added
from time to time to maintain the reaction for 3 hr. Ethanol, 20
ml, was finally added to decompose the remaining hydride and the
mixture was cooled in ice. Ethyl ether, 5% HCl, and water were
added and the crude ester, 65 g, was recovered.

c. Preparation of Dichloroacetylene [16]

Trichloroethylene and ether (1:1-mol mixture) were added
dropwise at 3-5 ml/min to a mixture of 400 g of technical flake
KOH and 350 ml of ethylene glycol while nitrogen was passed
through the system at 100 ml/min at 140°. The contents of the
receiver (attached to the reaction flask through an upright condenser at -10°) and trap (immersed in a dry ice bath) were mixed,
the water layer was decanted, and from the ether solution there
was recovered by distillation a 90% yield of a 55:45 DCA-ether
mixture.

d. Preparation of Cyclodecyne [3]

A mixture of 200 g of 1,1-dichlorocyclodecane, 100 g of
NaOH, and 250 g of methanol was heated at 150° for 6 hr in an
autoclave. Fractionation of the product gave 64% of the alkyne.
If the reaction mixture was heated to 180°, an equilibrium
mixture of the alkyne and the allene formed.

2. From β-Chloroacetals, β-Chloroethers, or Vinyl Ethers
(1, 154)

$$RCH{=}CHOR' \longrightarrow RCH{\equiv}CH$$

Since metal alkoxide elimination is more difficult to achieve
than hydrohalide elimination, stronger bases such as alkyllithium
or sodamide are used. The elimination beginning with an acetal
is given [21]:

$$\underset{\text{liquid }NH_3}{\overset{\text{Cl}}{\underset{}{CH_2CH(OEt)_2}}} \xrightarrow[\text{liquid } NH_3]{3NaNH_2} NaC{\equiv}COEt \xrightarrow{H_2O} HC{\equiv}COEt$$

57-60%

Ethyl ethynyl ether

Moreover, the alkynyl ether may be converted into an alkyne [22]:

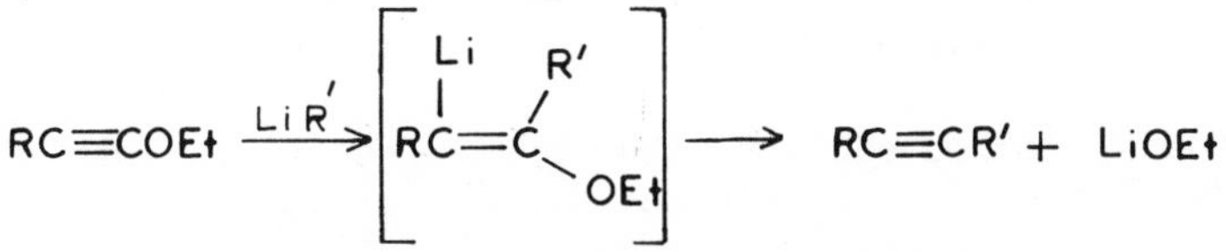

$$RC\equiv COEt \xrightarrow{LiR'} \left[\begin{array}{c} \overset{Li}{|} \quad R' \\ RC=C \\ \qquad \diagdown OEt \end{array} \right] \longrightarrow RC\equiv CR' + LiOEt$$

A combination of alkylation and debromination was achieved by Gelin and co-workers [23] in the synthesis of 1-hydroxy-2-alkynes as indicated:

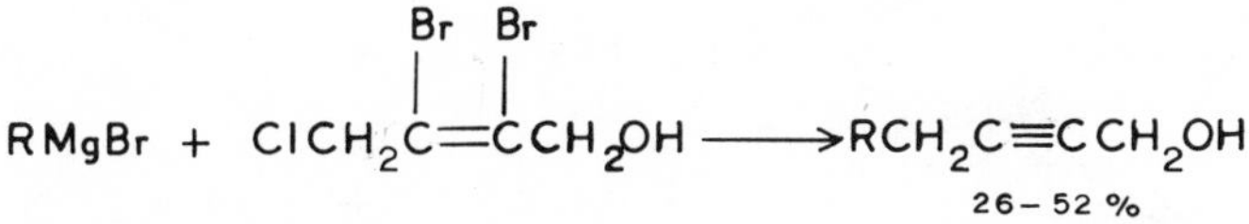

$$RMgBr + ClCH_2\overset{\overset{Br}{|}}{C}=\overset{\overset{Br}{|}}{C}CH_2OH \longrightarrow RCH_2C\equiv CCH_2OH$$

26–52 %

1,1-Dichloroalkenes, when treated with methyllithium or sodium, undergo dehalogenation with rearrangement similar to that exhibited in the Fritsch-Buttenberg-Wiechell rearrangement [24] (A.1).

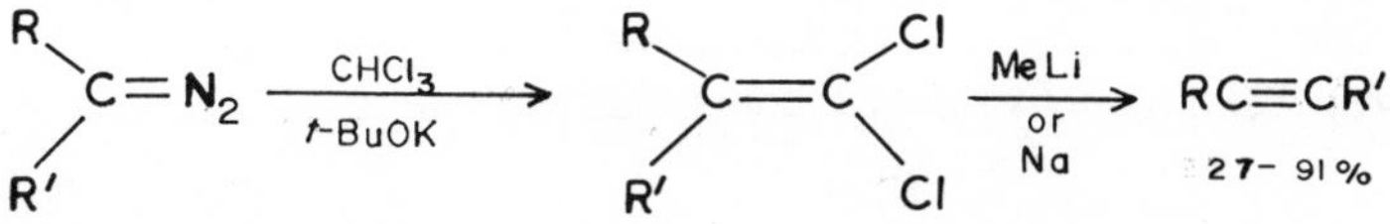

$$\overset{R}{\underset{R'}{\diagup}}C=N_2 \xrightarrow[t\text{-BuOK}]{CHCl_3} \overset{R}{\underset{R'}{\diagup}}C=C\overset{\diagup Cl}{\diagdown Cl} \xrightarrow[Na]{MeLi \atop or} RC\equiv CR'$$

27–91 %

4.   From Dihydrazones and Related Compounds (1, 156)

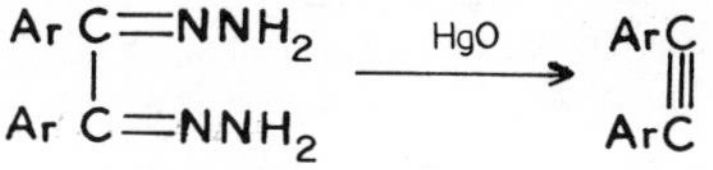

$$\begin{array}{c} Ar\ C\!=\!NNH_2 \\ | \\ Ar\ C\!=\!NNH_2 \end{array} \xrightarrow{HgO} \begin{array}{c} Ar\underset{\displaystyle \|\|\|}{C} \\ Ar C \end{array}$$

Besides mercuric oxide, another satisfactory reagent for the oxidation of dihydrazones to acetylenes is oxygen catalyzed by cuprous chloride in pyridine [25]:

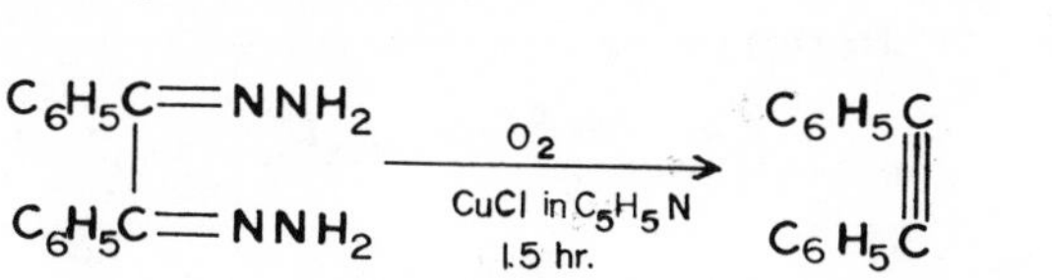

$$\begin{array}{c} C_6H_5C\!=\!NNH_2 \\ | \\ C_6H_5C\!=\!NNH_2 \end{array} \xrightarrow[\substack{CuCl\ in\ C_5H_5N \\ 1.5\ hr.}]{O_2} \begin{array}{c} C_6H_5C \\ \underset{\displaystyle \|\|\|}{} \\ C_6H_5C \end{array}$$

Tolane, 96.6%

Theis and Dessy [26] were successful in oxidizing substituted benzyl ketone hydrazones to acetylenes with mercurous trifluoroacetate in refluxing ether or dioxane:

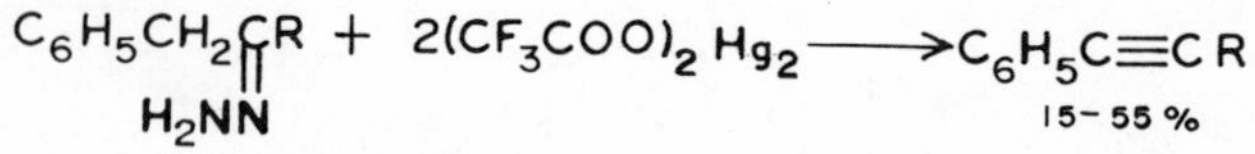

$$C_6H_5CH_2\underset{\underset{H_2NN}{\|}}{C}R \ + \ 2(CF_3COO)_2 \ Hg_2 \longrightarrow C_6H_5C\equiv CR$$

15-55%

Oxygenated solvents, such as those indicated, which form addition compounds with trifluoroacetic acid, must be employed as prevent the addition of the CF_3COOH formed to the acetylene. Also, azine formation is minimized by adding the hydrazone dropwise to the refluxing reaction medium.

The α,β-epoxyketone fragmentation studied by Eschenmoser [27] involves a monohydrazone and may be generalized as follows:

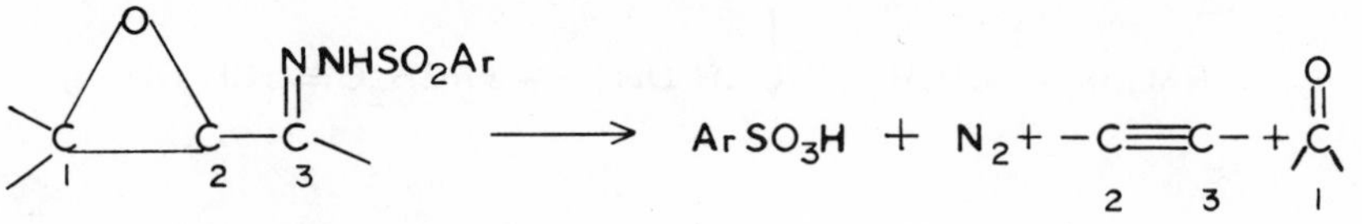

A specific example follows:

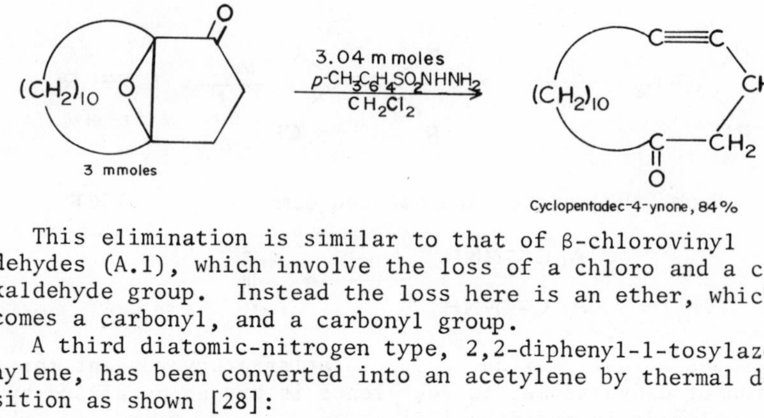

3 mmoles

3.04 m moles
p-$CH_3C_6H_4SO_2NHNH_2$
CH_2Cl_2

Cyclopentadec-4-ynone, 84%

This elimination is similar to that of β-chlorovinyl aldehydes (A.1), which involve the loss of a chloro and a carboxaldehyde group. Instead the loss here is an ether, which becomes a carbonyl, and a carbonyl group.

A third diatomic-nitrogen type, 2,2-diphenyl-1-tosylazo-ethylene, has been converted into an acetylene by thermal decomposition as shown [28]:

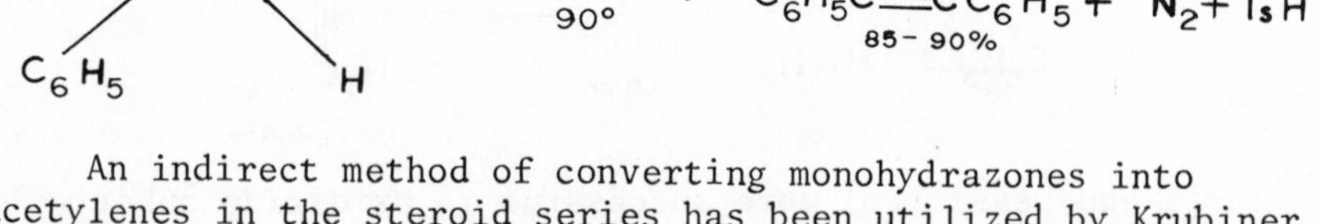

$$C_6H_5C\equiv CC_6H_5 + \ N_2 + Ts\,H$$

85-90%

An indirect method of converting monohydrazones into acetylenes in the steroid series has been utilized by Krubiner and co-workers [29]. These investigators treated pregnenolone hydrazone with iodine and triethylamine to form

20-iodopregna-5,20-dien-3β-ol, which with alkali gave 3β-hydroxy-pregn-5-en-20-yne:

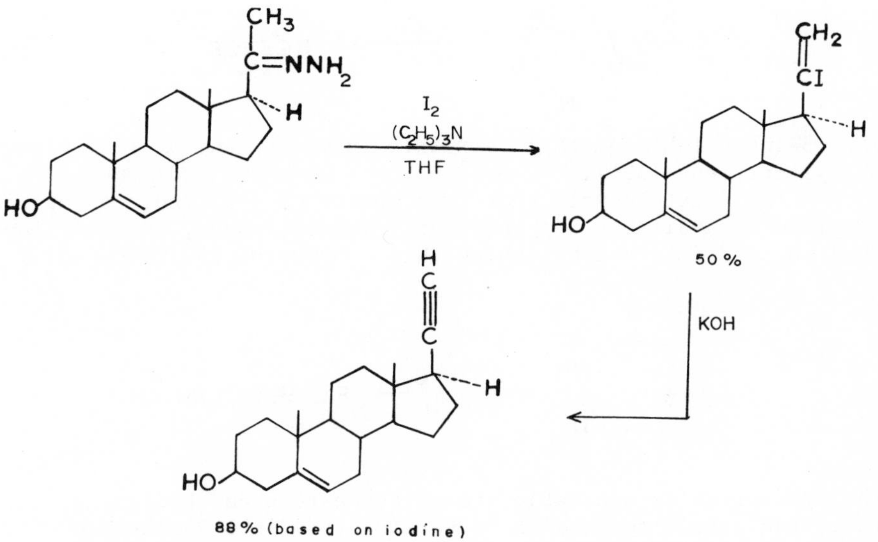

50 %

88 % (based on iodine)

A heterocycle which gives acetylenes on heating is N-nitrosooxazolidone [30]:

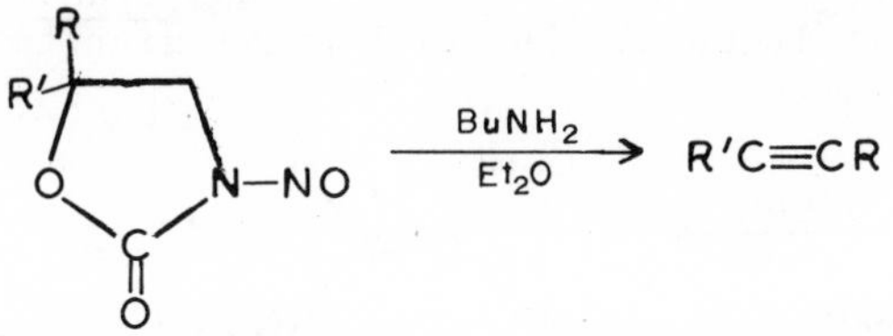

$$R'C{\equiv}CR$$

Yields are quantitative when R is a phenyl group and R' is a hydrogen, methyl, or phenyl group. Absence of a phenyl group in position 5 leads to very low yields. The features of the reaction resemble those of the FBW rearrangement (see A.1).

Another heterocycle that serves as a source of alkynes is the substituted 5-methyl-1H-tetrazole. Pyrolysis at temperatures varying from 110 to 200° leads to alkynes [31], as shown with R and R' = C_6H_5 or H and X = halogen, OH, NH_2 or N_3:

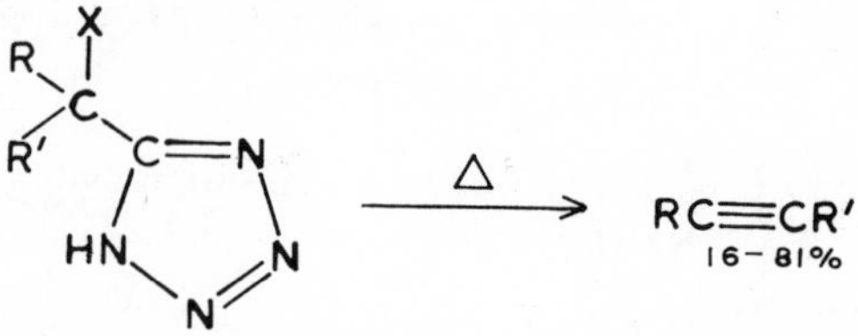

Thus 5-(phenylchloromethyl)tetrazole in mesitylene at 176° gives phenylacetylene (59%).

A third heterocycle that is a source of 2-alkynoic esters is the readily available 5-pyrazolone. Taylor, Robey, and McKillop [32] succeeded in accomplishing this conversion as indicated:

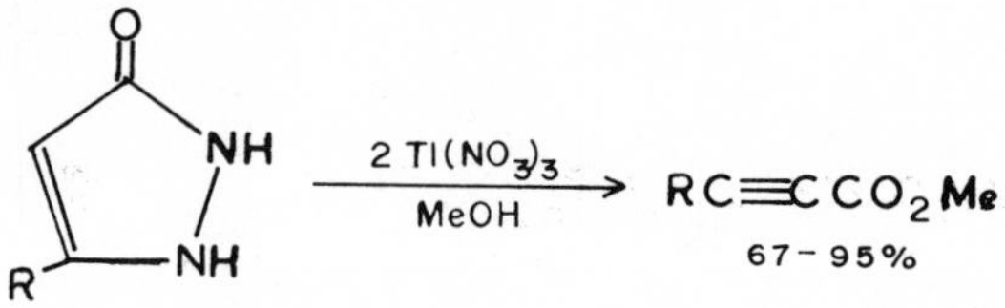

The pyrazolone is available from β-ketoesters and hydrazine; in fact, the direct conversion of β-ketoesters into 2-alkynoic esters is possible without isolation of the intermediate, 5-pyrazolone.

The decomposition of vinylamines (1, 157) has been extended to include acetylenes containing electron-withdrawing groups such as CN, CHO, COCH$_3$, and COOC$_2$H$_5$ [33]. Thus trans-3-amino-2-phenylacrylonitrile on heating with butyl nitrite gives phenyl-propiolic nitrile:

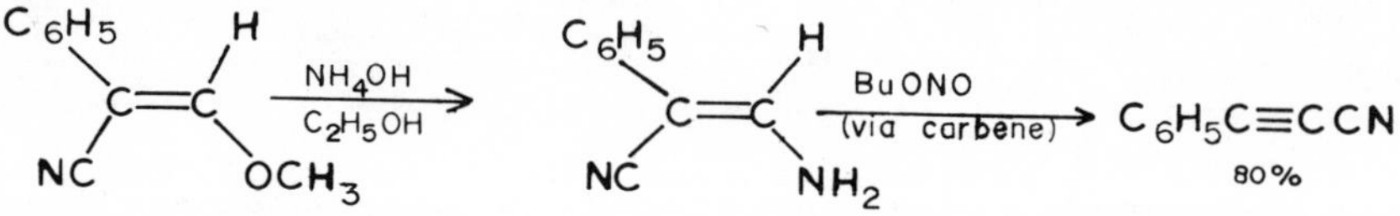

Similarly, the CN group may be replaced by CHO, COCH$_3$, or COOC$_2$H$_5$ and the C$_6$H$_5$ may be replaced by C$_{10}$H$_7$. The intermediate in the reaction is undoubtedly the unsaturated carbene

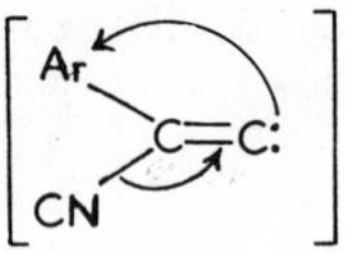

from which the acetylene is formed as indicated in

A.1.

5. From Quaternary Ammonium Bases ($\underline{1}$, 158)

The yield of tolane via the quaternary ammonium salt of a vinylamine has been improved [34] over that given in the original text as indicated:

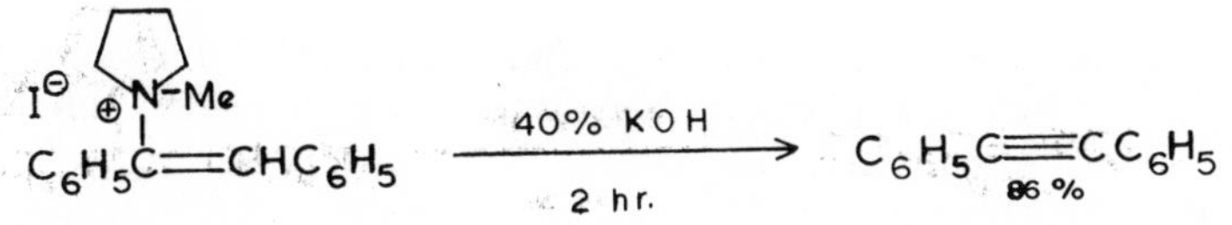

$$C_6H_5\overset{\displaystyle \overset{I^{\ominus}\ \ \overset{\oplus}{N}-Me}{|}}{C}=CHC_6H_5 \quad \xrightarrow[\text{2 hr.}]{\text{40\% KOH}} \quad C_6H_5C\equiv CC_6H_5 \ \ \text{86\%}$$

However, the reaction remains of limited usefulness.

6. From Olefins ($\underline{1}$, 159)

The β elimination of cis- and trans-β-chlorovinyltrimethyl-silane has been accomplished as shown [35]:

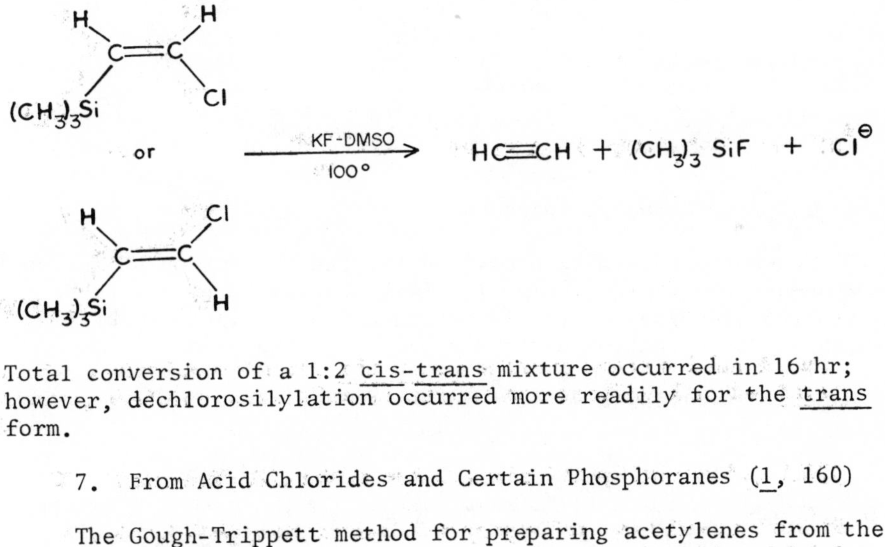

$$
\begin{array}{c}
\underset{(CH_3)_3Si}{\overset{H}{\diagdown}}C=C\underset{Cl}{\overset{H}{\diagup}} \\
\text{or} \\
\underset{(CH_3)_3Si}{\overset{H}{\diagdown}}C=C\underset{H}{\overset{Cl}{\diagup}}
\end{array}
\quad \xrightarrow[\text{100°}]{\text{KF-DMSO}} \quad HC\equiv CH \ + \ (CH_3)_3\,SiF \ + \ Cl^{\ominus}
$$

Total conversion of a 1:2 cis-trans mixture occurred in 16 hr; however, dechlorosilylation occurred more readily for the trans form.

7. From Acid Chlorides and Certain Phosphoranes ($\underline{1}$, 160)

The Gough-Trippett method for preparing acetylenes from the acid chlorides and the phosphoranes has been applied to the formation of diacetylenes by the use of α,β-acetylenic acid chlorides [36]:

$$R'C\equiv CCOCI + (C_6H_5)_3P\!\!=\!\!CHR^2 \xrightarrow{Et_3N}$$

$$(C_6H_5)_3P\!\!=\!\!CR^2COC\equiv CR' \xrightarrow[vacuum]{280-300°} (C_6H_5)_3PO + \underset{9-30\%}{R'C\equiv C\!-\!C\equiv CR^2}$$

In a later method Imaev and Shakirova [37] prepared phenyl-acetylene by heating phenacyl bromide with triphenyl phosphite:

$$C_6H_5COCH_2Br \xrightarrow[\substack{120-132°\\vacuum}]{(C_6H_5O)_3P} \underset{50\%}{C_6H_5C\equiv CH}$$

8. From Some Sulfur Compounds (1, 160)

Carpino and co-workers [38] in a study of 2,3-diphenylthi-irene-1,1-dioxide found that decomposition occurs readily on heating to give diphenylacetylene:

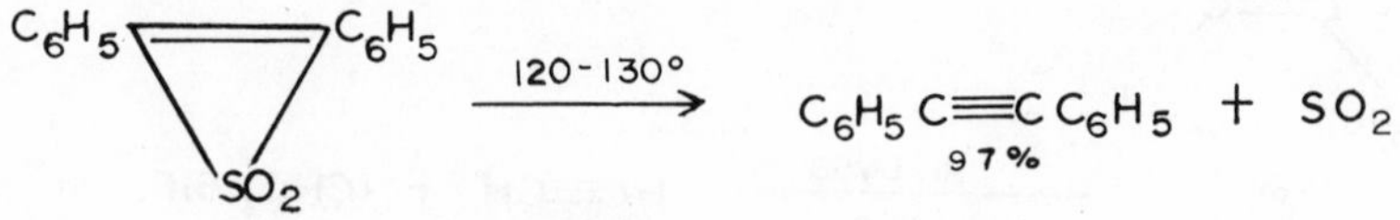

$$\xrightarrow{120-130°} \underset{97\%}{C_6H_5C\equiv CC_6H_5} + SO_2$$

9. From Carbonyl Compounds

Six methods have been devised for the conversion of carbonyl compounds into alkynes. The first is that of Corey and Fuchs where the aldehyde is transformed into the ethylidene dibromide which on debromination gives the alkyne as shown in A.1.

The second method of Colvin and Hamill [39] is a one-step synthesis involving a diazo compound as indicated:

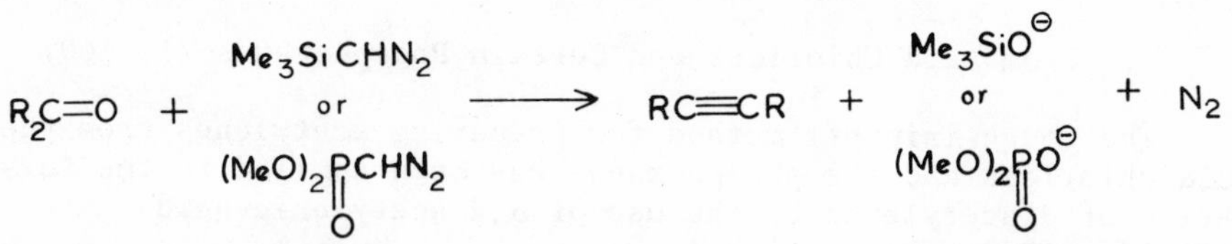

$$R_2C\!\!=\!\!O + \begin{matrix}Me_3SiCHN_2\\or\\(MeO)_2\underset{O}{\overset{\|}{P}}CHN_2\end{matrix} \longrightarrow RC\equiv CR + \begin{matrix}Me_3SiO^{\ominus}\\or\\(MeO)_2\underset{O}{\overset{\|}{P}}O^{\ominus}\end{matrix} + N_2$$

However the scope of this reaction is severely limited due to the fact that enolizable carbonyl compounds give greatly reduced yields. Phenylacetaldehyde, for example, gives a 30% yield of 3-phenylpropyne.

The third method involves the condensation of a carbonyl compound with an olefinic halide or dihalide in the presence of lithium diethylamide [5]. The reaction may be illustrated with the preparation of 1-chloro-2(1'-hydroxy-1'-cyclohexyl) ethyne.

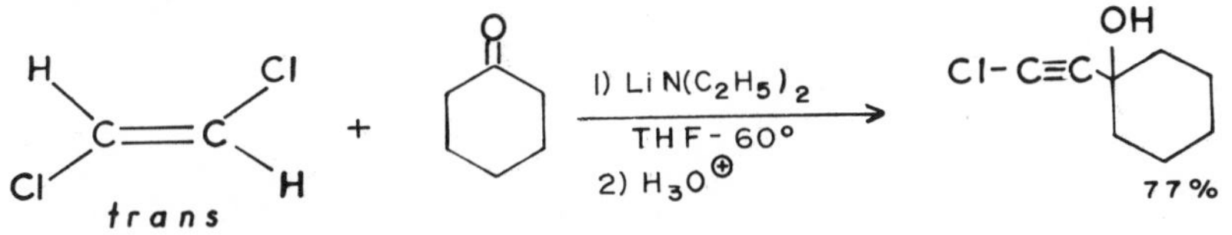

A fourth method of elimination occurs with α-diazo-β-hydroxycarbonyl compounds [40]. Thus α-diazo-β-hydroxyketones and esters lead to substituted acetylenes as indicated:

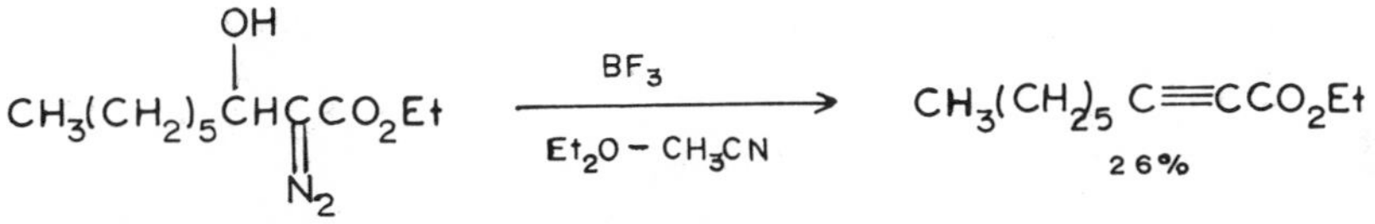

In a similar manner α-diazo-β-hydroxy-2,4-dichlorobenzyl phenyl ketone gave 2,4-dichlorophenylethynyl phenyl ketone as shown:

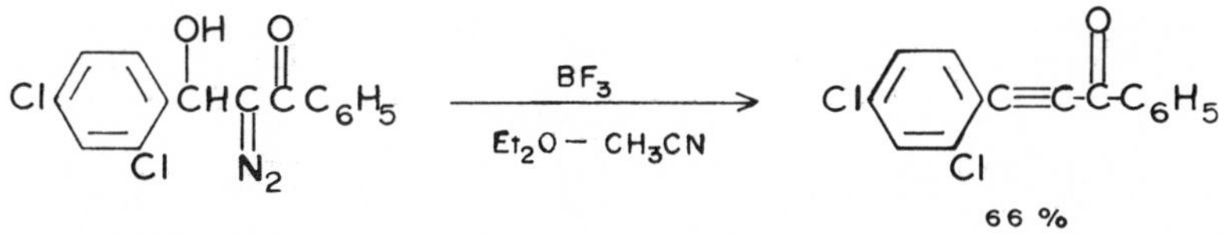

A fifth method involves elimination from an α,α'-dibromo cyclic ketone to the cyclopropenone that readily loses carbon monoxide [3]:

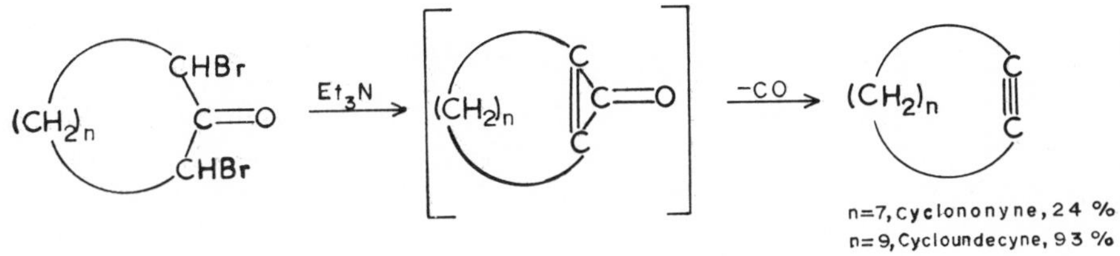

n=7, cyclononyne, 24 %
n=9, Cycloundecyne, 93 %

A sixth method consists of the oxidation of a cyclic semi-carbazone to a diazoselenium intermediate that readily loses selenium and nitrogen to form the acetylene [3]:

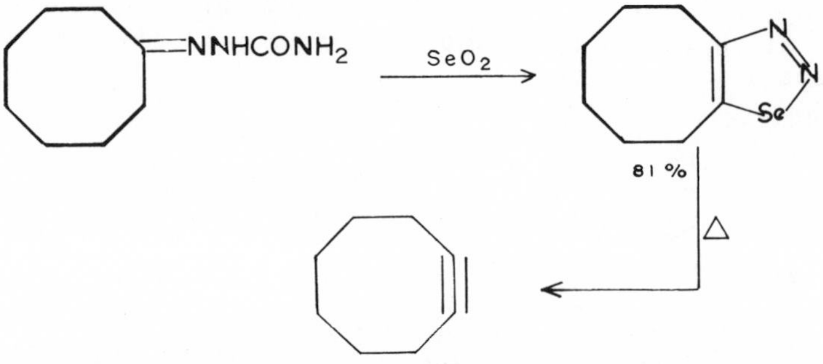

Cyclooctyne, 49%

Indeed this reaction appears to give the highest yield of cyclooctyne obtained. Elimination from 1-chloro-2-bromocyclo-octene with lithium amalgam gave 3% cyclooctyne and from 1-bromocyclooctene with $NaNH_2$ at 200° the return was 17%. In the latter case the product was purified by the addition of silver nitrate, with which it forms a stable complex.

a.   Preparation of 1-Nonyne [20]

Octanal, 1 equiv, was added to a reagent prepared from Zn dust, 2 equiv, triphenylphosphine, 2 equiv, and carbon tetra-bromide, 2 equiv, in methylene chloride at 23° for 24-30 hr. A solution of 0.875 g of the product, 1,1,-dibromonon-1-ene, in 17 ml of THF at -78° under N_2 was then treated with 5.6 ml of 1.16M solution of n-butyllithium in pentane. The reaction mix-ture was stirred for 1 hr at -78°, warmed to 25°, and maintained for 1 hr at that temperature. Addition of water, extraction with pentane, and distillation gave 0.35 g (95%) of the alkyne.

b.   Preparation of Diphenylacetylene [39]

Dimethyldiazomethyl phosphite, 1.1 equiv, in THF at -78° was treated with 1.1 equiv of butyllithium and the mixture was allowed to react for 5 min. Benzophenone, 1 equiv, in THF was added slowly, and the mixture was permitted to warm to 25° during 20 hr. After the usual work-up, tolane was isolated in 80% yield.

10.   From Nitrosooxazolidones (see Addenda)

References

1. H. G. Viehe, Chemistry of Acetylenes, Dekker, New York, 1969.
2. L. Brandsma, Preparative Acetylenic Chemistry, American Elsevier, New York, 1971.
3. H. Meier, Synthesis, 235 (1972).
4. J. Klein and E. Gurfinkel, Tetrahedron, 26, 2127 (1970).
4a. W. L. Collier, R. S. Macomber, J. Org. Chem., 38, 1367 (1973).
4b. P. J. Kocienski, J. Org. Chem., 39, 3285 (1974).
5. D. Reisdorf and H. Normant, Organomet. Chem. Syn., 1, 393 (1972).
6. Ref 1, p. 143.
7. Ibid,, p. 1063.
8. C. F. H. Allen, C. O. Edens, Jr., Org. Syn. Col. Vol., 3, 731 (1955).
9. Ref. 1, p. 147.
10. M. Rosenblum et al., J. Organomet. Chem., 6, 173 (1966).
11. M. S. Newman, S. H. Merrill, J. Am. Chem. Soc., 77, 5549 (1955).
12. J. Wolinsky, J. Org. Chem., 26, 704 (1961); K. L. Erickson, J. Wolinsky, J. Am. Chem. Soc., 87, 1142 (1965).
13. R. H. Mitchell, F. Sondheimer, Tetrahedron, 24, 1397 (1968).
14. D. Y. Curtin, E. W. Flynn, J. Am. Chem. Soc., 81, 4714 (1959).
15. D. F. Bender et al., J. Org. Chem., 35, 939 (1970).
16. J. Siegel et al., J. Org. Chem., 35, 3199 (1970).
17. D. R. Strobach, J. Org. Chem., 36, 1438 (1971).
18. S. A. Fuqua et al., J. Org. Chem., 30, 1027 (1965).
19. Ref. 1, p. 667.
20. E. J. Corey, P. L. Fuchs, Tetrahedron Letters, 3769 (1972).
21. E. R. H. Jones et al., Org. Syn., Coll. Vol., 4, 404 (1963).
22. J. G. A. Kooyman et al., Rec. Trav. Chim., 87, 69 (1968).
23. S. Gelin et al., Bull. Soc. Chim. France, 4513 (1969).
24. H. Reimlinger, Chem. Ind. (London), 1306 (1969).
25. J. Tsuji et al., Tetrahedron Letters, 4573 (1973).
26. R. J. Theis, R. E. Dessy, J. Org. Chem., 31, 624 (1966).
27. A. Eschenmoser et al., Helv. Chim. Acta, 54, 2896 (1971).

28. G. Rosini, S. Cacchi, J. Org. Chem., 37, 1856 (1972).
29. A. M. Krubiner et al., J. Org. Chem., 34, 3502 (1969).
30. H. P. Hogan, J. Seehafer, J. Org. Chem., 37, 4466 (1972);
     M. S. Newman, L. F. Lee, ibid., 37, 4468 (1972).
31. H. Behringer, M. Matner, Tetrahedron Letters, 1663 (1966).
32. E. C. Taylor, R. L. Robey, A. McKillop, Angew. Chem. Intern.
     Ed. Engl., 11, 48 (1972).
33. M. Cariou, Bull. Soc. Chim. France, 210 (1969).
34. J. B. Hendrickson, J. R. Sufrin, Tetrahedron Letters, 1513
     (1973).
35. R. F. Cunico, E. M. Dexheimer, J. Am. Chem. Soc., 94, 2868
     (1972).
36. S. T. D. Gough, S. Trippett, J. Chem. Soc., 543 (1964).
37. M. G. Imaev, A. M. Shakirova, Dokl. Akad. Nauk SSSR, 163(3),
     656 (1965); C. A., 63, 11338 (1965).
38. L. A. Carpino et al., J. Am. Chem. Soc., 93, 476 (1971).
39. E. W. Colvin, B. J. Hamill, Chem. Commun., 151 (1973).
40. E. Wenkert, C. A. McPherson, Syn. Commun., 2, 331 (1972).

B. Nucleophilic Reactions

    1. From Acetylenic Salts and Alkylating or Arylating
        Agents (1, 162)

$$RC\equiv CNa + R'X \longrightarrow RC\equiv CR'$$

    The metal acetylide appears to give better yields if the
metathesis with a primary halide is carried out in DMF (see
Ex. a). A simple means of preparing the cuprous acetylide is
now available [1]:

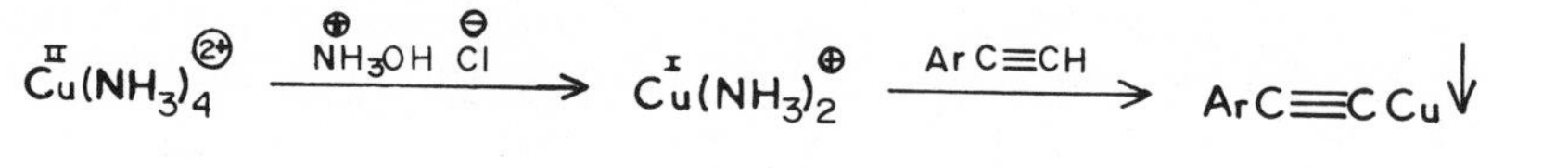

$$\overset{\text{II}}{Cu}(NH_3)_4^{2\oplus} \xrightarrow{\overset{\oplus}{NH_3}OH \ \overset{\ominus}{Cl}} \overset{\text{I}}{Cu}(NH_3)_2^{\oplus} \xrightarrow{ArC\equiv CH} ArC\equiv C\,Cu\downarrow$$

The compound couples with active aryl iodides to form diarylacetylenes and with either ortho hydroxy or amino ring types to
form furans, indoles, or other heterocycles, as for example:

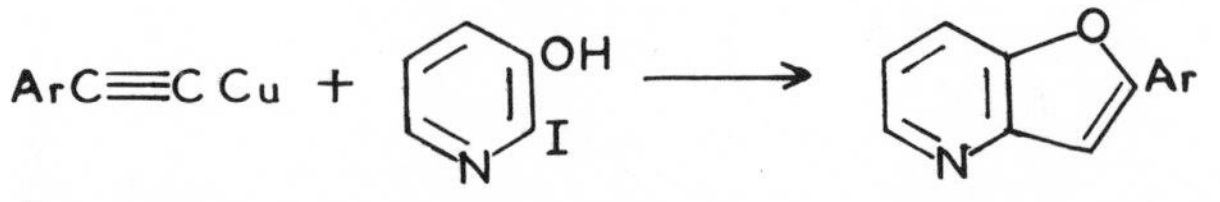

Recently, the lithium acetylide of phenyl acetylene was prepared
in THF at room temperature, although for other acetylenes refluxing was necessary [1a].

A list of the coupling products of the cuprous acetylide with aryl iodides is available [2].

The lithium acetylides are more soluble in organic solvents but much less reactive. In fact, temperatures as high as 150° in an autoclave are necessary to bring about maximum metathesis. However, HMPA as a solvent increases the yields to as high as 73-90% [3]. For maximum yields in the reaction of relatively reactive allyl halides with sodium acetylides, a cuprous salt as a catalyst is necessary. Benzyl halides inexplicably give poor yields of coupled products [4].

Although copper acetylides couple fairly well with aryl halides, a new method has been developed to produce the coupling products [5] (see Ex. b).

$$ArCu + IC\equiv CSiMe_3 \longrightarrow ArC\equiv CH$$

In the synthesis of alkadiynoic acids [6], acidic groups in the alkyl halide are tolerated provided sufficient lithium amide is added to form the salt and provided the salt is soluble. In this case the more soluble lithium salt may be preferred:

$$HC\equiv C(CH_2)_n C\equiv CH + Br(CH_2)_n CO_2H \xrightarrow[THF]{LiNH_2}$$

$$HC\equiv C(CH_2)_n C\equiv C(CH_2)_n CO_2H$$

1-Phenyl-1,4-pentadiyne has been synthesized by the action of phenylethynylmagnesium bromide on propargyl bromide in the presence of Cu_2Cl_2 in THF [7]. This method is the "best and often the sole route to the 1,4-diynes." By treatment with sodium hydroxide the conjugated diyne, 1-phenyl-1,3-pentadiyne, becomes available without heating. This is the "method of choice if the corresponding 'skipped' diyne or allene is available" (see Ex. c). The equation for the two-step process follows:

$$C_6H_5C\equiv CMgBr + BrCH_2C\equiv CH \xrightarrow[THF]{Cu_2Cl_2}$$

$$C_6H_5C\equiv CCH_2C\equiv CH \xrightarrow[C_2H_5OH]{NaOH} C_6H_5C\equiv C-C\equiv CCH_3$$

$$45-57\% \qquad\qquad 54-75\%$$

In the case of <u>tert</u>-propargyl bromides, the Grignard reagent reacts to form allenic products [8]:

$$\underset{\underset{MgBr}{\overset{\displaystyle R}{|}}}{CH_3C}\text{-}C\equiv CH \;\rightleftharpoons\; \underset{\overset{\displaystyle R}{|}}{CH_3C}=C=CHMgBr \;\xrightarrow{R'X}\; \underset{\overset{\displaystyle R}{|}}{CH_3C}=C=CHR'$$

a. Preparation of 1-Eicosyne [9]

Sodium, 110 g was dissolved in 3 1 of liquid ammonia while acetylene was introduced. When the blue color disappeared, the acetylene flow was interrupted, 2 1 of dry DMF was added, and the ammonia was allowed to evaporate. Stearyl bromide, 800 g was added slowly and the mixture was stirred at 70° for 3 hr; a liter of water was added with cooling and the dark brown mixture was stirred. The ethereal extract was washed in the usual way and evaporated. On distillation the residue gave the colorless, solid acetylene, 75%, boiling point 132-140°/0.2 mm.

b. Preparation of 2-Trifluoromethylphenylacetylene [5]

2-Trifluoromethylphenylmagnesium bromide, 1 equiv was added to a <u>vigorously stirred</u> suspension of freshly prepared cuprous bromide, 1.1 equiv in ether. Iodotrimethylsilylacetylene (0.8 mol from $Me_3SiC\equiv CSiCMe_3$ + ICl $\rightarrow$ $Me_3SiC\equiv CI$) was added and the mixture was stirred at 0° for 3 hr and at 20° for 6 hr. The mixture was evaporated, the residue was taken up in methanol, and the methanolic solution was treated with dilute aqueous alkali to liberate the arylacetylene, 51% when purified. The reaction appears to be general provided reactive groups are protected.

c. Preparation of 1-Phenyl-1,4- and 1-Phenyl-1,3-Pentadiyne [7]

Anhydrous Cu_2Cl_2, 2 g was added to phenylethynylmagnesium bromide (prepared from Mg, 0.81 mol, C_2H_5Br, 1 mol, and $C_6H_5C\equiv CH$, 1 mol, in THF) and after heating under reflux for 20 min in a N_2 atmosphere, $BrCH_2C\equiv CH$, 0.81 mol, in THF was added slowly so as to maintain gentle refluxing. After heating for an additional 30-40 min the cooled mixture was poured into an ice-water slush of H_2SO_4. Extraction with ether and distillation of the dried extract under N_2 gave 51-64 g (45-57%) of the 1,4-diyne.

The 1,4-diyne, 0.07 mol, was stirred with 2 g of NaOH in 50 ml of C_2H_5OH under N_2 for approximately 2 hr. After pouring the mixture into water and extracting with ether, the dried extract was distilled as before to give 5.4-7.5 g (54-75%) of the 1,3-diyne.

d. Preparation of 5-Propynyl-2-(3,3-dimethoxypropynyl) thiophene [10]

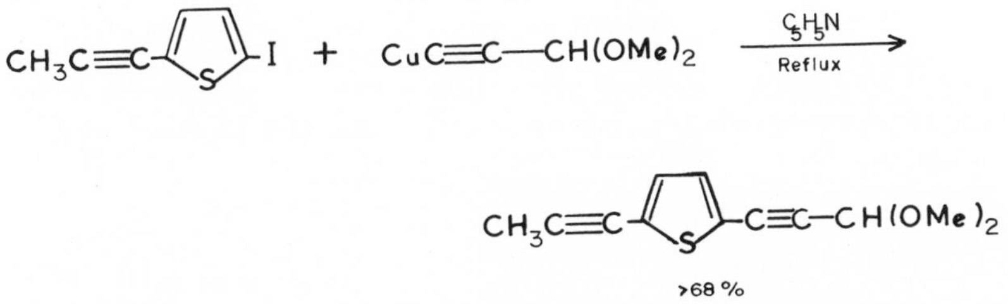

>68 %

2. From Acetylenic Salts and Carbonyl Compounds (1, 164)

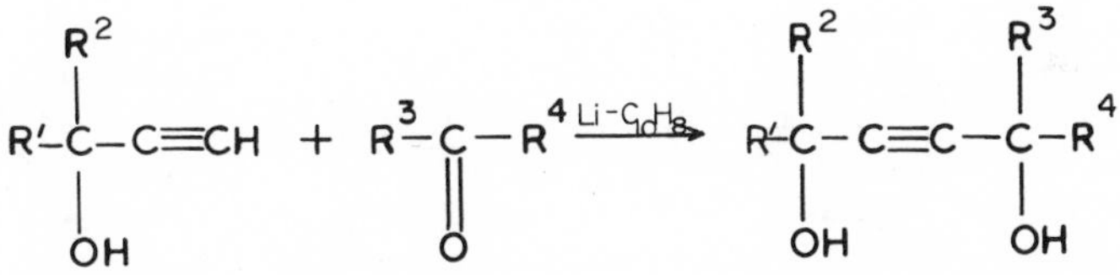

The addition of acetylenic salts to carbonyl compounds has been surveyed critically [11]. Both acetylenic Grignard reagents and lithium acetylides appear to be the most common agents. An inverse correlation of the enolization constant of the ketone and the yield has been found, that is, acetophenone with a relatively large enolization constant gives a 55% yield of the ethynyl derivative while diisopropyl ketone with an enolization constant some 4000 times smaller gives a 100% yield of the ethynyl derivative. Although acetylene and potassium hydroxide do not form an acetylene salt, they evidently form a complex sufficiently ionic to add to a carbonyl compound. The reaction using the simple reagents noted is known as the Favorskii addition [12]

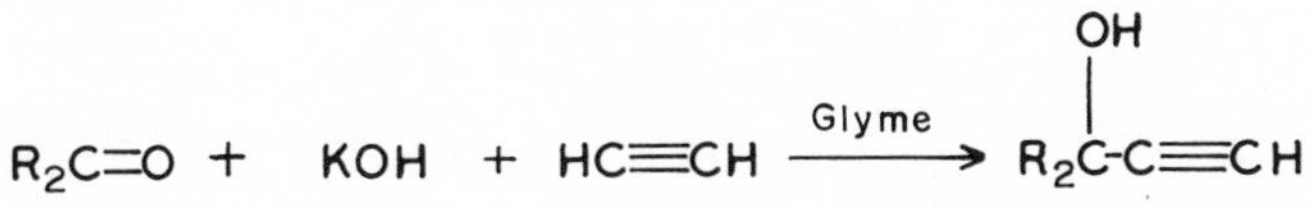

(see Ex. a). Industrially, the reaction may be carried out at high pressure by using only catalytic amounts of base. Copper acetylide is used to prepare propargylic alcohol from acetylene and aqueous formaldehyde in an autoclave [13].

Although the use of the acetylenic Grignard reagent is more common in the carbonyl-addition reaction, a convenient one-step process using the acetylenic alcohol with $Li-C_{10}H_8$ has been developed by Watanabe and co-workers [14] for the preparation of acetylene diols. It was shown that the dilithio derivative of the alcohol is an intermediate and that $Li-C_{10}H_8$ led to better yields in the one-step synthesis of 1-(1'-hydroxycyclohexyl)-3-methyl-1-pentyn-3-ol than other metalating agents such as n-BuMgBr. The method may be modified, without isolation of the intermediate, to give acetylenic diols from two ketones and acetylene as shown:

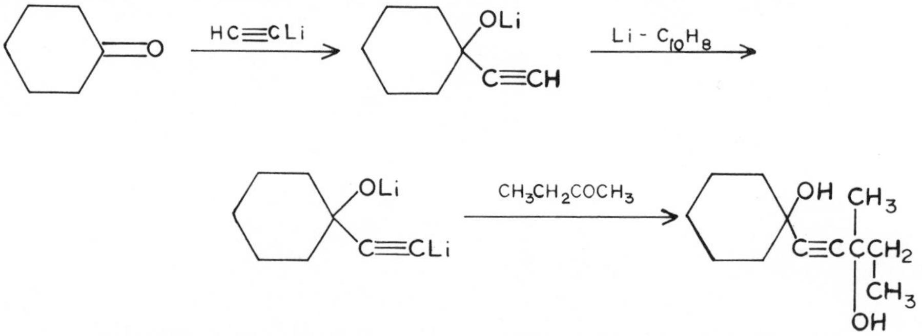

Yields for a series of diols vary within 20-80% [see Ex. b).

It is of interest to know that a double bond can be hydrated in an enyne, via the cobalt carbonyl complex, without affecting the triple bond as indicated [15]:

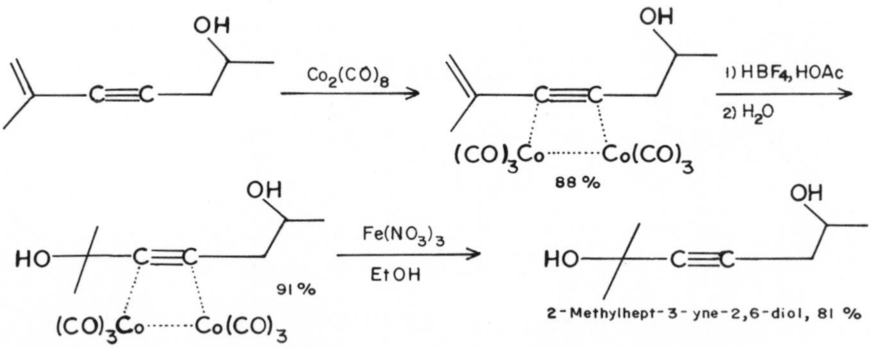

Without protection the acetylenic group would be hydrated. Reduction of the enyne complex may also be accomplished with di-imide to yield an acetylene.

It is of interest to note that the Favorskii addition reaction may be reversed as indicated [16]:

$$\text{HOCH}_2\text{C} \equiv \text{CCH}_2\text{OH} \xrightarrow[160-185°]{\text{K}_2\text{CO}_3} \text{HOCH}_2\text{C} \equiv \text{CH} + \text{CH}_2\text{O}$$

Formerly (1, 165), it was suggested that traces of powdered sodium hydroxide be used.

a. Preparation of 3-Hydroxy-4-methylpentyne [12]

A finely divided suspension of equal weights of KOH and glyme was diluted with solvent to give 6.67 mol of KOH and 1600 g of glyme. After 11 g of ethanol was added, the solution was stirred and saturated with acetylene at -10° to 0°. Then 3.33 mol of isobutyraldehyde containing 11 g of ethanol was added in 2 hr, after which the mixture was kept at -10° to 0° while an excess of acetylene was introduced. On treatment with ice water, extraction and purification gave 87% of the pentyne.

b. Preparation of 1-(1'-Hydroxycyclohexyl)-3-methyl-1-pentyn-3-ol [14]

To naphthalene, 0.04 mol in THF, Li, 0.08 mol was added and the mixture was agitated at 25° in a N_2 atmosphere. After 1 hr, ethynylcyclohexanol, 0.04 mol in THF was added slowly. Methyl ethyl ketone, 0.04 mol was then added gradually and stirring was continued as before for an additional 3 hr. Saturation with NH_4Cl solution followed and extraction with isopropyl ether gave an organic layer that was dried and distilled to give 4.5 g (50%) of the acetylenic diol.

3. From Acetylenic Salts and Halogens, Ortho Esters, or Isocyanates (1, 166)

The preceding methods have been discussed in depth [17]. The simplest method to obtain a chloro- or bromoacetylene is to treat the acetylene with sodium hypochlorite or bromine water. In addition, the metal acetylide may be treated with the benzene-sulfonyl halide [18]:

$$\text{RC} \equiv \text{CM} + \text{C}_6\text{H}_5\text{SO}_2\text{X} \longrightarrow \text{RC} \equiv \text{CX} + \text{C}_6\text{H}_5\text{SO}_2\text{M}$$

The reaction of an acetylide with ortho esters is as follows:
$$\text{RC} \equiv \text{CMgX} + \text{HC(OR')}_3 \longrightarrow \text{RC} \equiv \text{CCH(OR')}_2 \xrightarrow{\text{H}^{\oplus}} \text{RC} \equiv \text{CCHO}$$

Thus ortho esters are sources of acetylenic aldehydes. Isocyanates, on the other hand, are sources of propargylic acid
derivatives (1, 166):

$$RC\equiv CMgX \ + \ C_6H_5-N\!=\!C\!=\!O \longrightarrow RC\equiv C\overset{O}{\overset{\|}{C}}NHC_6H_5$$

5.   From Acetylenic Salts and Epoxides or Acid Derivatives

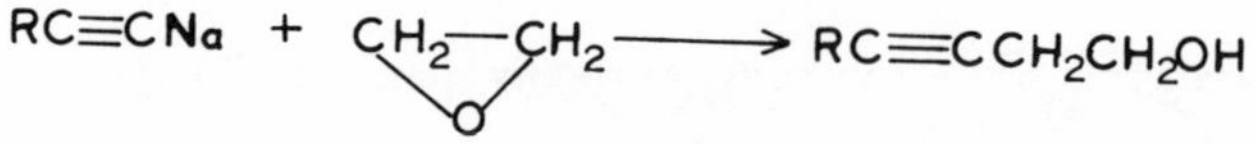

$$RC\equiv CNa \ + \ CH_2\!\!-\!\!CH_2 \longrightarrow RC\equiv CCH_2CH_2OH$$

Yields on this addition are good but decrease as the chain length
of the epoxide increases [19]. For unsymmetrical epoxides the
ring opening involves the bond expected in a nucleophilic attack:

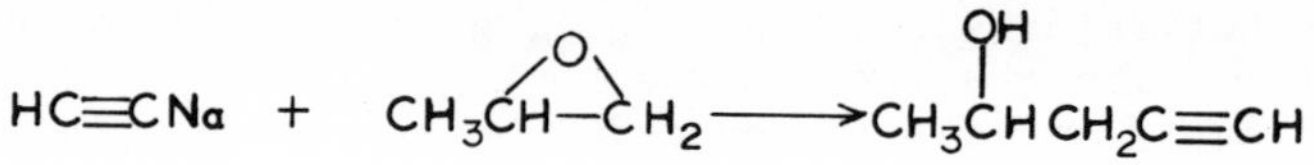

$$HC\equiv CNa \ + \ CH_3CH\!\!-\!\!CH_2 \longrightarrow CH_3\overset{OH}{\underset{}{C}}HCH_2C\equiv CH$$

4-Hydroxy-1-pentyne, 38%

For acyl halides the less nucleophilic silver acetylide appears
to give the best yields of acetylenic ketones [20]:

$$HC\equiv CAg \ + \ RCOCl \ \xrightarrow{CCl_4} \ HC\equiv CCOR + AgCl$$

Esters do not react as well with acetylides to form the acetylenic ketone.
    Another way to synthesize acetylenic ketones is via the tri-
methylsilane derivative treated with an aroyl chloride and AlCl$_3$
as shown [21]:

$$Me_3SiC\equiv CSiMe_3 \ + \ ArCOCl \ \xrightarrow{AlCl_3}$$

$$ArCOC\equiv CSiMe_3 \ \xrightarrow[aq.C\,H_3OH]{Borax} \ ArCOC\equiv CH$$

40-94%

Care must be taken to avoid Michael addition of the base to the
product.
    Acetylenic aldehydes have been synthesized by Gorgues [22]
as shown:

$$C_6H_5(C\equiv C)_2CH(OEt)_2 \xrightarrow[25°]{2\ HCOOH} C_6H_5(C\equiv C)_2CHO + 2HCOOC_2H_5$$

5-Phenylpent−2,4-diynal

The aldehyde decomposes at room temperature to give the relatively stable dimer:

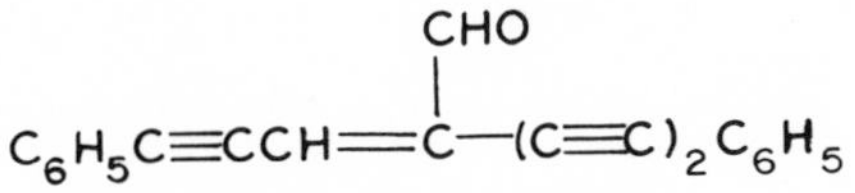

$$C_6H_5C\equiv CCH=\overset{\displaystyle CHO}{\overset{\displaystyle |}{C}}-(C\equiv C)_2C_6H_5$$

Addenda

B.2. Good yields of acetylenic carbinols are obtained by the reaction of butyllithium with acetylene followed by the addition of a carbonyl compound with both reactions at -78° [M. M. Midland, J. Org. Chem., 40, 2250 (1975)].

B.2. Acetylenic carbinols may be degraded to form the ketones from which they originated by refluxing in toluene with Ag_2CO_3 on celite [G. R. Lenz, Chem. Commun., 468 (1972)].

References

1.   D. C. Owsley, C. E. Castro, Org. Syn., 52, 128 (1972).
1a. H. Ogura, H. Takahashi, Syn. Commun., 3, 135 (1973).
2.   H. G. Viehe, Chemistry of Acetylenes, Dekker, New York, 1969, p. 632.
3.   D. N. Brattesani, C. H. Heathcock, Syn. Commun., 3, 245 (1973).
4.   Ref. 2, pp. 189-192.
5.   R. Oliver, D. R. M. Walton, Tetrahedron Letters, 5209 (1972).
6.   D. E. Ames et al., J. Chem. Soc., 4373 (1965).
7.   H. Taniguchi et al., Org. Syn., 50, 97 (1970).
8.   Y. Pasternak, J. C. Traynard, Bull. Soc. Chim. France, 356 (1966).
9.   E. F. Jenny, K. D. Meier, Angew. Chem., 71, 245 (1959).
10. F. Bohlmann, W. Skuballa, Chem. Ber., 106, 497 (1973).
11. Ref. 2, pp. 207-211.
12. H. A. Stansbury, Jr., W. R. Proops, J. Org. Chem., 27, 279 (1962).
13. Ref. 2, p. 240.
14. S. Watanabe et al., Can. J. Chem., 47, 2343 (1969; Chem. Ind. (London), 1489 (1969).

15. K. M. Nicholas, R. Pettit, Tetrahedron Letters, 3475 (1971).
16. Z. A. Navrezova, A. V. Shchelkunov, Tr. Khim. Met. Inst., Akad. Nauk Kaz. SSR, 18, 9 (1972); C. A., 79, 4903 (1973).
17. Ref. 2, p. 651.
18. Ibid., p. 672.
19. Ibid., p. 241.
20. Ibid., p. 248.
21. D. R. M. Walton, F. Waugh, J. Organomet. Chem., 37, 45 (1972).
22. A. Gorgues, Ann. Chim. (Paris), 7, 373 (1972).

C. Free Radical and Cyclo Reactions

1. From Acetylenes (Oxidative Coupling) (1, 168)

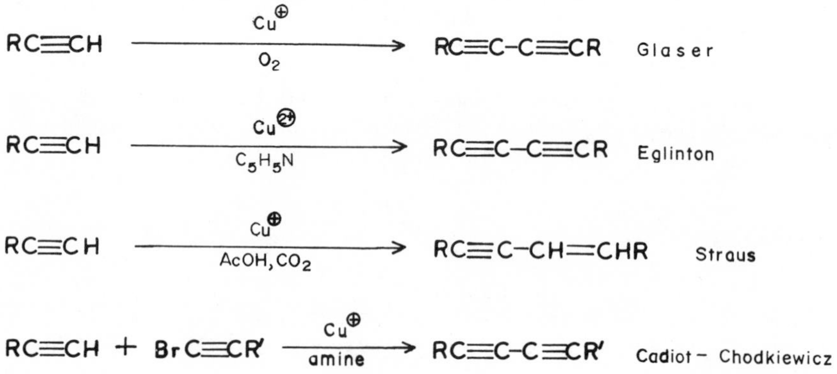

$$RC\equiv CH \xrightarrow[\;O_2\;]{Cu^{\oplus}} RC\equiv C-C\equiv CR \quad \text{Glaser}$$

$$RC\equiv CH \xrightarrow[\;C_5H_5N\;]{Cu^{\ominus}} RC\equiv C-C\equiv CR \quad \text{Eglinton}$$

$$RC\equiv CH \xrightarrow[\;AcOH,CO_2\;]{Cu^{\oplus}} RC\equiv C-CH\equiv CHR \quad \text{Straus}$$

$$RC\equiv CH + BrC\equiv CR' \xrightarrow[\;amine\;]{Cu^{\oplus}} RC\equiv C-C\equiv CR' \quad \text{Cadiot-Chodkiewicz}$$

The preceding couplings have been reviewed critically [1]. The very specific nature of the copper ion in the reaction indicates that the free radicals are being generated within a cluster so that they are in close enough proximity to couple:

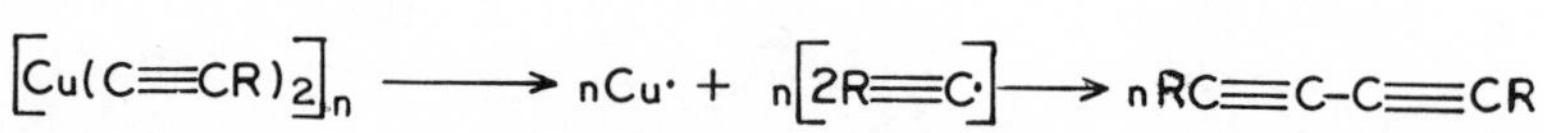

$$\left[Cu(C\equiv CR)_2\right]_n \longrightarrow nCu\cdot + n\left[2R\equiv C\cdot\right] \longrightarrow nRC\equiv C-C\equiv CR$$

It may be noted that the Cadiot-Chodkiewicz coupling may lead to unsymmetrical diacetylenes, which makes it a more versatile method. The following groups have been situated on either side of the diacetylenic product: OH, OR, Ar, R, $O=\bigcirc=O$, COOH, COOR, $CONH_2$, S, SiR_3. The latter is important in synthesizing diacetylenes with a terminal acetylenic group [2]:

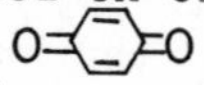

$$C_6H_5C\equiv CH \ + \ BrC\equiv C\,SiEt_3 \ \xrightarrow[\substack{Cu_2Cl_2 \\ EtNH_2}]{NH_2OH-HCl}$$

$$\underset{50\%}{C_6H_5C\equiv C-C\equiv C\,SiEt_3} \ \xrightarrow[2)\,H^{\oplus}]{1)\,NaOH-CH_3OH} \ \underset{\substack{1-Phenylbuta-1,3-diyne \\ 100\%}}{C_6H_5C\equiv C-C\equiv CH}$$

Formerly, the alkylcarbinol (R$_2$C(OH)-) or carboxyl groups were used for protection in synthesizing polyacetylenes [3].

Polyacetylenes have been synthesized in the following manner [4]:

$$Et_3SiC\equiv CH \ \xrightarrow[\substack{TMEDA \\ 2)\,H_3O^{\oplus}}]{1)\,CuCl-air} \ \underset{86\%}{Et_3SiC\equiv C-C\equiv C\,SiEt_3}$$

$$\xrightarrow[2)\,H_3O^{\oplus}]{\substack{1)\,Controlled\ cleavage \\ by\ aq.\ NaOH-CH_3OH}} \ Et_3Si\,C\equiv C-C\equiv CH$$

The product may be subjected to further coupling. Care must be used in the workup since the compounds become more unstable as the acetylenic chain increases.

Directions for the preparation of a superior catalyst are as follows [4]: About 1 g of freshly precipitated cuprous chloride is added to a stirred solution of 0.5 g of TMEDA in 20 ml of acetone. The green supernatant liquid serves as the catalyst and may be stored over nitrogen.

a. Preparation of Di-α-pyridylbutadiyne, α-NC$_5$H$_4$C≡C-C≡CC$_5$H$_4$N-α [5]

A solution of α-pyridylacetylene, 50 mmol, in 10-20 ml of glyme was added in 10 min to a stirred mixture of cuprous chloride, 10 mmol, and tetramethylethylenediamine, 13 mmol, in 40-50 ml of glyme held at 30-35°. Into this strongly agitated solution, oxygen was passed for 20-60 min. The solid from the reaction mixture was separated under vacuum and placed on neutral Al$_2$O$_3$ in a column where it was eluted with benzene. Recrystallization gave the butadiyne, melting point 122-123°, 79%.

5. From Acetylenes and Boranes (Unsymmetrical Coupling)

$$RC\equiv CLi \ + \ BR'_3 \longrightarrow \overset{\oplus}{Li} \ \overset{\oplus}{BR'_3}C\equiv CR \xrightarrow{\ I_2\ } R'C\equiv CR$$

Brown and co-workers have shown that coupling occurs involving different R groups when the Li acetylide reacts with an alkylboron [6]. For instance, when R = butyl and R' = cyclopentyl, 1-cyclopentyl-1-hexyne is obtained (100%, glpc). Yields in general varied within 91-100% (glpc).

In place of iodine, methanesulfinyl chloride, MeSOCl, may be used to bring about coupling. With this reagent when R = phenyl and R' = sec-butyl, 1-phenyl-3-methyl-1-pentyne was isolated in 62% yield [7].

References

1. H. G. Viehe, Chemistry of Acetylides, Dekker, New York, 1969, p. 597.
2. R. Eastmond, D. R. M. Walton, Tetrahedron, 28, 4591 (1972).
3. Ref. 1, p. 622.
4. D. R. M. Walton et al., Tetrahedron, 28, 4601 (1972).
5. U. Fritzsche, S. Hunig, Tetrahedron Letters, 4831 (1972).
6. H. C. Brown et al., J. Am. Chem. Soc., 95, 3080 (1973).
7. M. Naruse et al., Tetrahedron Letters, 1847 (1973).

D.   Isomerization and Disproportionation

   1.   Acetylenes to Allenes

Base-catalyzed isomerization is the chemistry of the anionic species as shown in its most simplified form:

$$H-\overset{|}{\underset{|}{C}}-C\equiv \overset{\ominus}{C} \ \rightleftharpoons \ \overset{|}{\underset{|}{C}}=C=\overset{\ominus}{C}H$$

Generalities will follow [1]. Potassium alkoxides generally isomerize 1- to 2-alkynes, while sodium metal or sodamide isomerizes 2- to 1-alkynes. The equilibrium between allenes and alkynes in base catalysis lies in the direction of the alkyne, but other products are formed, as will become evident. If carbanions are not formed in base isomerization, a "conducted tour" mechanism might apply, as already shown in its simplified form at the beginning of the section or as illustrated for a second type of acetylene [2]:

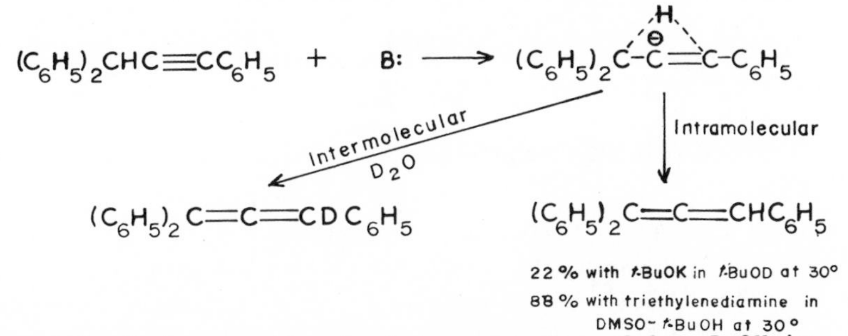

$(C_6H_5)_2CHC\equiv CC_6H_5$ + B: $\longrightarrow$ $(C_6H_5)_2\overset{\ominus}{C}-C\equiv C-C_6H_5$ (with H above)

Intramolecular

Intermolecular
D_2O

$(C_6H_5)_2C\!=\!C\!=\!CDC_6H_5$ $(C_6H_5)_2C\!=\!C\!=\!CHC_6H_5$

22% with t-BuOK in t-BuOD at 30°

88% with triethylenediamine in
DMSO-t-BuOH at 30°

About 22% of the reaction is intramolecular with t-BuOK in
t-BuOD at 30°, but with triethylenediamine in DMSO-t-BuOH at 30°
the amount increased to 88%. The products from the isomerization of 1-heptyne by treatment with t-BuOK-t-BuOH at 196° were
2-heptyne, 2,3-heptadiene, 3-heptyne, and 3,4-heptadiene [3],
whereas from 1-pentyne with alcoholic KOH at 175° the main
product (95%) was 2-pentyne. However, t-BuOK in DMSO seems to
favor dienes as shown [4]:

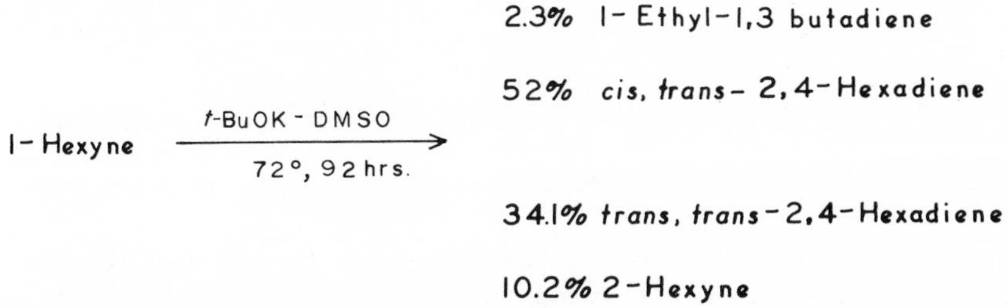

2.3% 1-Ethyl-1,3 butadiene

52% cis, trans- 2,4-Hexadiene

1-Hexyne $\xrightarrow[72°, 92\,hrs.]{t\text{-}BuOK-DMSO}$

34.1% trans, trans-2,4-Hexadiene

10.2% 2-Hexyne

Since strong bases in favorable solvents seem to give the same
equilibrium, aqueous alkali appears to be as good a reagent as
any to obtain at least some of the allene in isomerization [5].
    The extent of the hexyne-hexallene rearrangement has been
investigated recently by Carr and co-workers [6]. Although, as
shown here, there is disagreement among investigators using
different bases, on whether isomerization proceeds to C-3 and
beyond, these investigators in studying the hexyne-hexallene
system in the presence of t-BuOK-t-BuOH conclude that the isomerization involves more than C-1 and C-2. The actual rates of
isomerization as determined by the Carr group (two numbers refer
to the allene, one to the alkyne) are in the sequence:

1,2->1- >> 2,3- >> 3- > 2-

    When functionalized allenes or acetylenes are synthesized,
the equilibrium may favor one or the other. For instance, α-
bromoallenes with primary or secondary amines undergo the
allene-acetylene rearrangement as shown [7]:

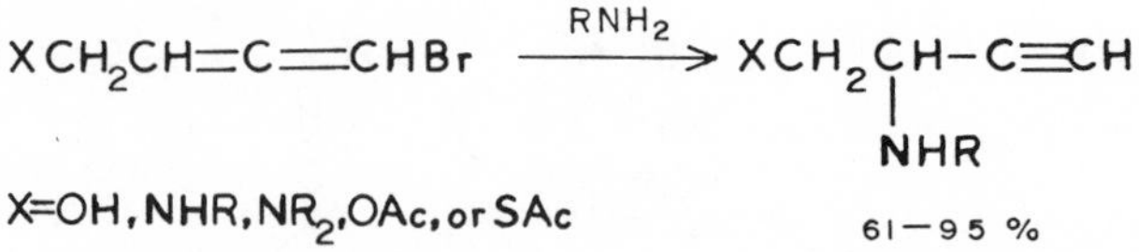

$$XCH_2CH=C=CHBr \xrightarrow{RNH_2} XCH_2\underset{\underset{NHR}{|}}{CH}-C\equiv CH$$

X=OH, NHR, NR$_2$, OAc, or SAc   61−95 %

X = OH, NHR, NR$_2$, OAc, or SAc. In addition, SAc may be substituted for NHR. A similar rearrangement occurs in the conversion of phenylallene into 3-phenyl-3-alkyl-1-propyne via the steps as indicated [8]:

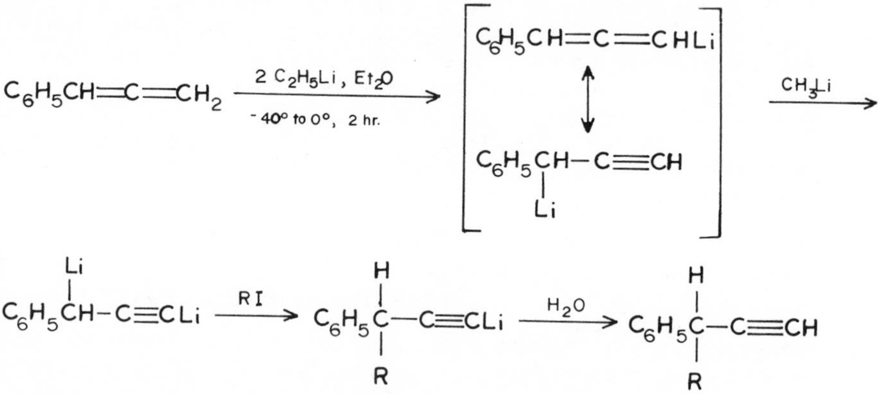

$$C_6H_5CH=C=CH_2 \xrightarrow[-40° \text{ to } 0°, \ 2 \text{ hr.}]{2 \ C_2H_5Li, \ Et_2O} \left[\begin{array}{c} C_6H_5CH=C=CHLi \\ \updownarrow \\ C_6H_5\underset{\underset{Li}{|}}{CH}-C\equiv CH \end{array} \right] \xrightarrow{CH_3Li}$$

$$C_6H_5\underset{\overset{|}{Li}}{CH}-C\equiv CLi \xrightarrow{RI} C_6H_5\underset{\underset{R}{|}}{\overset{\overset{H}{|}}{C}}-C\equiv CLi \xrightarrow{H_2O} C_6H_5\underset{\underset{R}{|}}{\overset{\overset{H}{|}}{C}}-C\equiv CH$$

To obtain clean alkylation it was necessary to proceed via the dilithio compound. The reverse rearrangement occurs in good yield when diethylaminoalkoxyalkynes are treated with an alkyllithium followed by hydrolysis [9] as indicated:

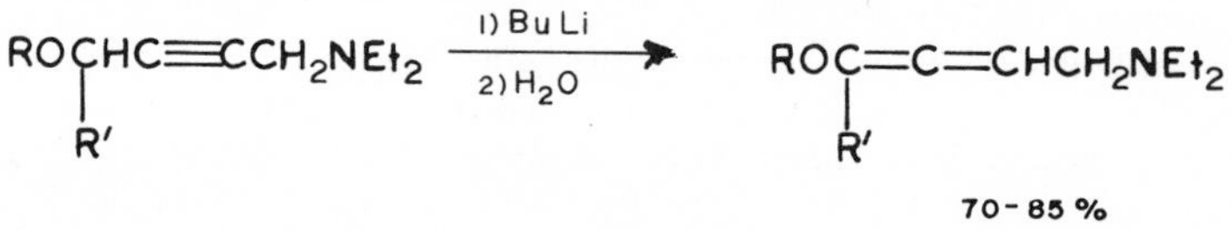

$$RO\underset{\underset{R'}{|}}{CH}C\equiv CCH_2NEt_2 \xrightarrow[2)H_2O]{1) BuLi} RO\underset{\underset{R'}{|}}{C}=C=CHCH_2NEt_2$$

70−85 %

The Grignard reagent poses interesting relationships in which acetylenic, allenic, or 1,3-dienic bromides are converted into an equilibrium mixture of Grignards, the major portion of which is frequently not an allenic Grignard, but one from which an allenic product may be derived. This result arises because the addition reaction of the Grignard and a carbonyl compound is often of a concerted type as illustrated [10]:

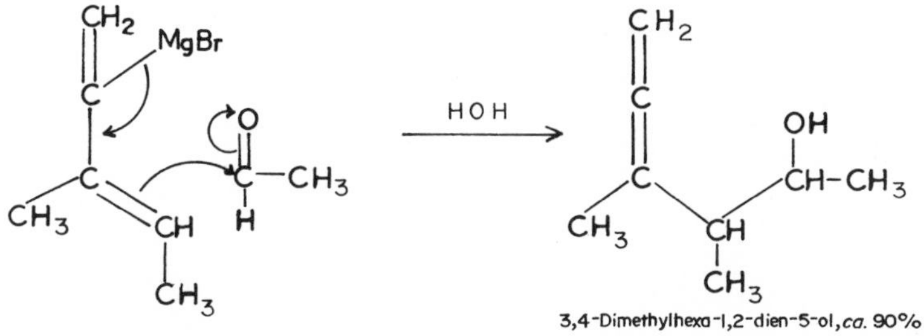

3,4-Dimethylhexa-1,2-dien-5-ol, *ca.* 90%

In addition, the major Grignard reagent from propargylic halides is the allenic Grignard [11]:

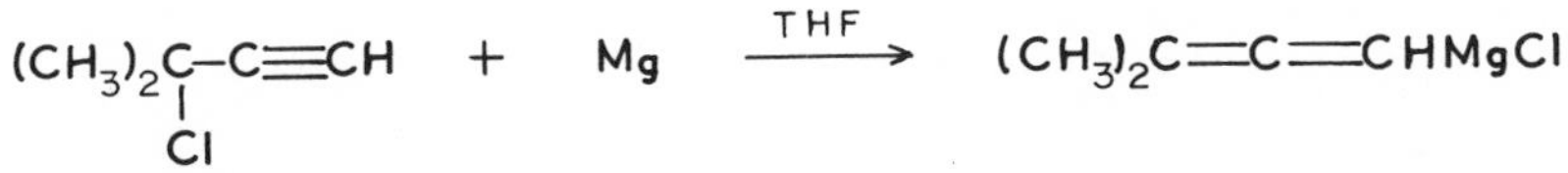

$$(CH_3)_2\underset{\underset{Cl}{|}}{C}-C\equiv CH \;+\; Mg \;\xrightarrow{\;THF\;}\; (CH_3)_2C=C=CHMgCl$$

The isomerization of 1- to 2-alkynes by lithium acetylide has been reviewed [11a].

a. Preparation of 3-Methylamino-4-hydroxy-1-butyne [7]

1-Bromo-4-acetoxybutadiene-1,2, 0.1 mol was added slowly to an aqueous solution of methylamine, 0.04-0.05 mol, at 25°. After the mixture stood for 15 hr, it was treated with 50 ml of 2% aqueous HCl, saturated with NaCl, and the methylaminobutyne was recovered from the ethereal extract, 93%.

2. Allenes to Conjugated Dienes

Acid-catalyzed isomerization tends to isomerize allenes to conjugated dienes [12]:

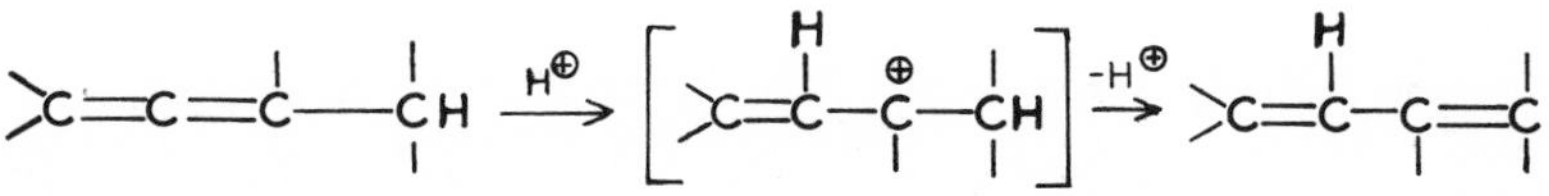

The best yields are obtained when the allenes are highly substituted. However, some allenes retain their structure in the product as shown:

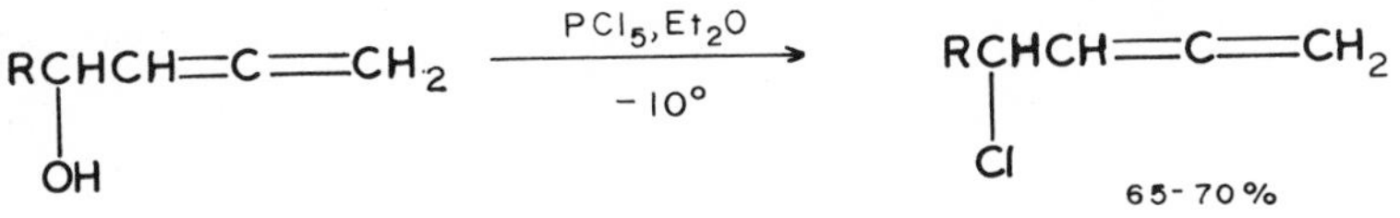

65-70%

3.   Enynes to Ene-allenes and Alkyl Benzenes

1,4-Enynes (usually as part of an acid structure) rearrange to conjugated ene-allenes at room temperature and then to conjugated trienes on heating [13]:

$$CH_3(CH_2)_4 C \equiv C CH_2 CH = CH(CH_2)_7 CO_2H \xrightarrow{\textit{t}-BuOK}$$

cis-9-Octadecen-12-ynoic acid

$$CH_3(CH_2)_4 CH = C = CH - CH = CH(CH_2)_7 CO_2H \xrightarrow{\Delta}$$

cis-Octadecentrienoic acid

$$CH_3(CH_2)_4 CH = CH - CH = CH - CH = CH(CH_2)_6 CO_2H$$

8,10,12-Octadecentrienoic acid, 70%

On the other hand, cyclohexenyl acetylenes, obtainable from acetylenic carbinols, give high yields of aromatic hydrocarbons [14]:

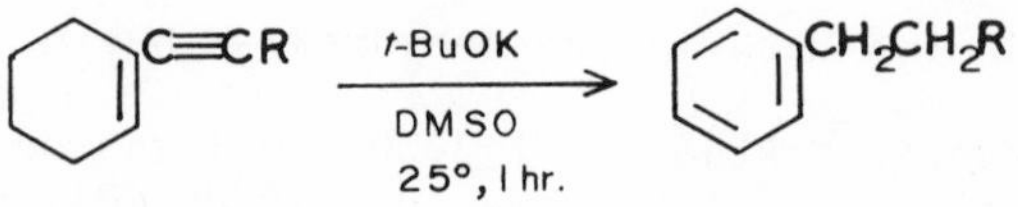

4.   Diynes to Dienynes, Aromatic Hydrocarbons, and Conjugated Octadienes

1,5-Hexadiyne rearranges to 33% 1,3-hexadiene-5-yne in the presence of t-BuOK-t-BuOH, while 1,6-heptadiyne in t-BuOK-diglyme gives mainly toluene [11]. A fairly selective base isomerization is as follows [15]:

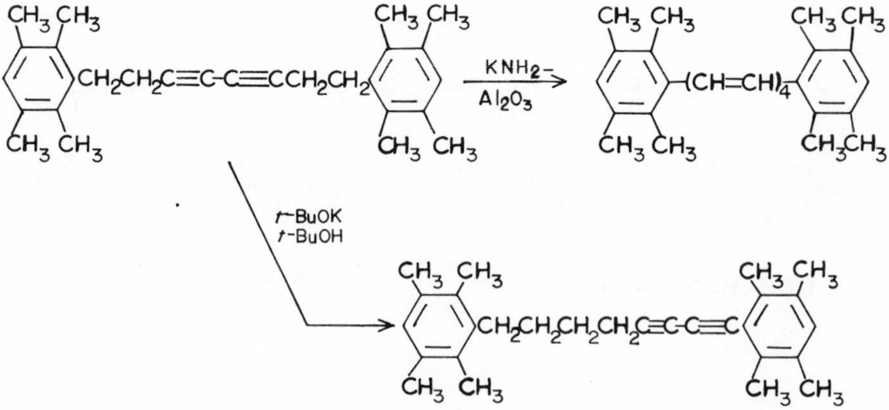

5. Disproportionation

Pennella and co-workers [16] found that 2-pentyne (mixed with cyclohexene as an internal standard and diluent) when heated at 200-450° at atmospheric pressure in the presence of activated W_2O_3 on silica gave largely 2-butyne and 3-hexyne in approximately 1:1 mol ratio as shown:

$$2\ CH_3CH_2C{\equiv}C\,CH_3 \xrightarrow{\ W_2O_3\ } CH_3C{\equiv}CCH_3\ +\ CH_3CH_2C{\equiv}CCH_2CH_3$$

The investigators suggest that this result is consistent with the formation of a four-membered intermediate that cleaves to give the two symmetrical alkynes.

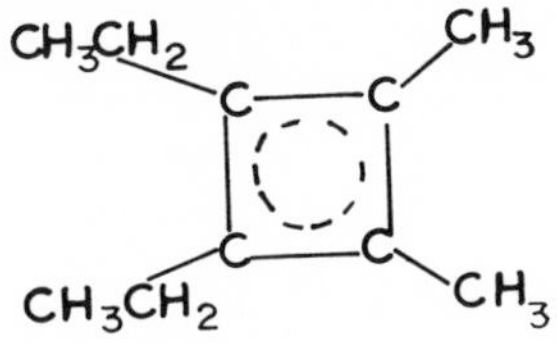

Addenda

D.1. The reagent of choice for the rearrangement of internal to terminal acetylenes is the monopotassium salt of trimethylene-diamine. Rearrangement occurs in seconds at 0°, but a blockage occurs if a branched chain is present. [C. A. Brown, A. Yamashita, J. Am. Chem. Soc., 97, 891 (1975)].

D.1. The rate of rearrangement of 3-hexyne is accelerated if the mixed catalyst, $NaNH_2$-$NH_2CH_2CH_2NH_2$, is aged. This result probably occurs because of the increased amount of the sodium salt of ethylenediamine formed [J. H. Wotig et al., J. Org. Chem., 38, 489 (1973)].

References

1. T. F. Rutledge, Acetylenes and Allenes, Reinhard, New York, 1969, p. 35.
2. D. J. Cram, Fundamentals of Carbanion Chemistry, Academic, New York, 1965, p. 189.
3. R. J. Busby, Quart, Rev. (London), 24, 585 (1970).
4. M. L. Farmer et al., J. Org. Chem., 31, 2885 (1966).
5. W. Smadja, Ann. Chim. (Paris), 10, 105 (1965); C. A., 63, 6834 (1965).

6. M. D. Carr et al., J. Chem. Soc., Perkin Trans., II, 668 (1973).

7. M. V. Mavrov et al., Tetrahedron, 25, 3277 (1969).

8. L. Brandsma, E. Mugge, Rec. Trav. Chim., 92, 628 (1973).

9. R. Mantione, B. Kirschleger, Compt. Rend., 272, C786 (1971).

10. Y. Pasternak and J.-C. Traynard, Bull. Soc. Chim. France, 356 (1966).

11. G. F. Hennion, C. V. DiGiovanna, J. Org. Chem., 31, 970 (1966).

11a. J. H. P. Tyman, Syn. Commun., 5, 21 (1975).

12. Ref. 1, p. 47.

13. Ibid., pp. 42-45.

14. R. Mantione, Compt. Rend., 264, C1668 (1967).

15. J. Cousin, A. J. Hubert, J. Chem. Soc., Perkin Trans., I, 1653 (1972).

16. F. Pennella et al., Chem. Commun., 1548 (1968).

E. Allenes and Cumulenes

1. Allenes (1, 170)

An excellent review of the preparation of allenes has been published [1].

a. Elimination

Recent publications emphasize the importance of elimination in the preparation of allenes. The methods are as follows.

1. Dehalogenation of 2,3-dihalopropenes [1, 2].

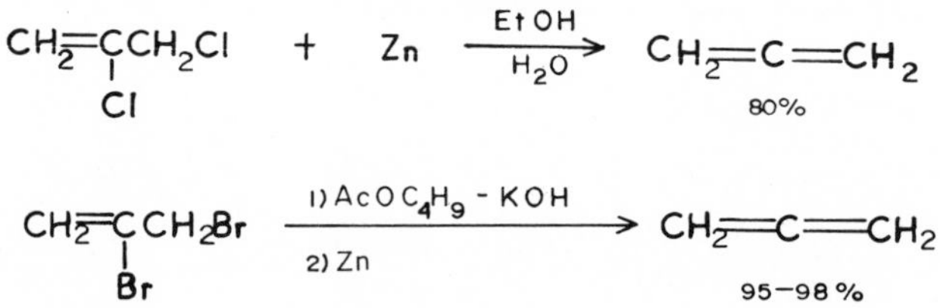

$$CH_2{=}CCH_2Cl \ (\text{with } Cl) \ + \ Zn \ \xrightarrow[H_2O]{EtOH} \ CH_2{=}C{=}CH_2 \quad 80\%$$

$$CH_2{=}CCH_2Br \ (\text{with } Br) \ \xrightarrow[2) Zn]{1) AcOC_4H_9 - KOH} \ CH_2{=}C{=}CH_2 \quad 95{-}98\%$$

2. Hydroboration of propargyl chlorides followed by hydrolysis [3].

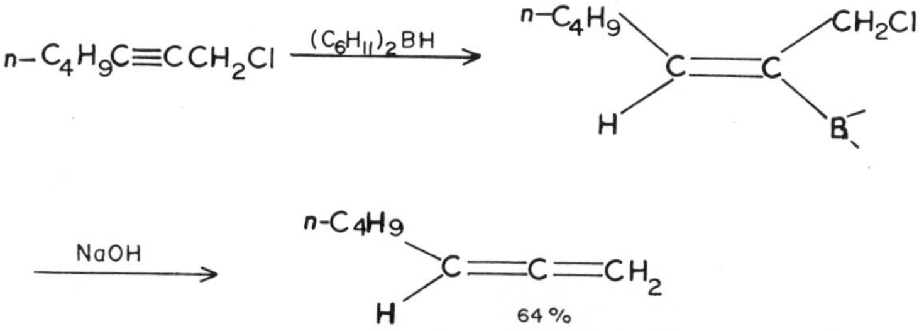

$$n-C_4H_9C\equiv CCH_2Cl \xrightarrow{(C_6H_{11})_2BH}$$

$$\xrightarrow{NaOH}$$

64%

3. Pyrolysis of halogenated propiolactones [4a] and 1-acetoxymethyl-2,2,3,3-tetrafluorocyclobutane [4b].

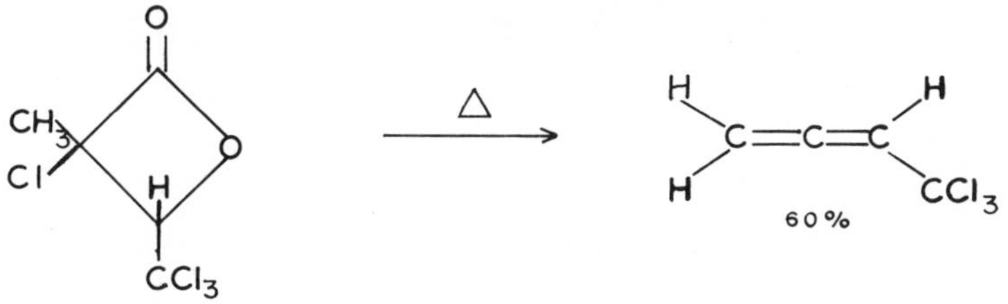

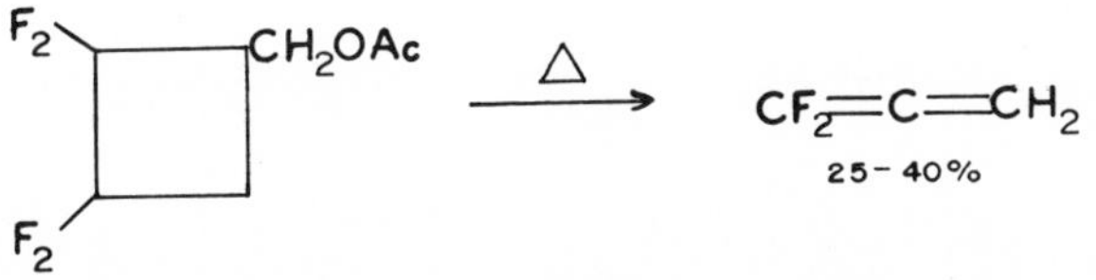

60%

25-40%

4. Treatment of ethynylcarbinol acetates with lithium dialkylcopper reagents [5].

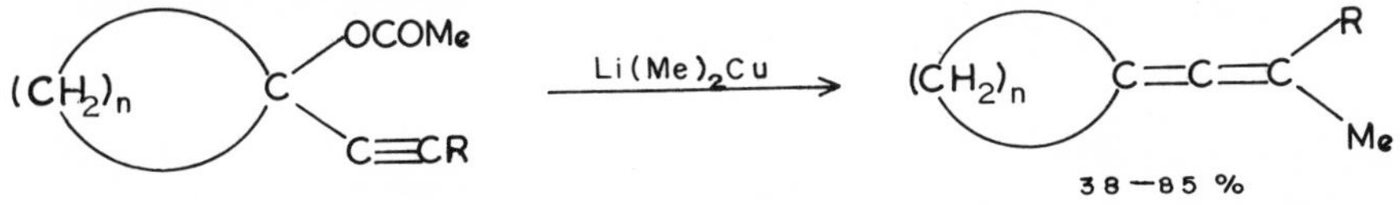

38—85 %

5. Conversion of cyclooctene to an allene (1,2-cyclo-nonadiyne).

a. Direct method [6].

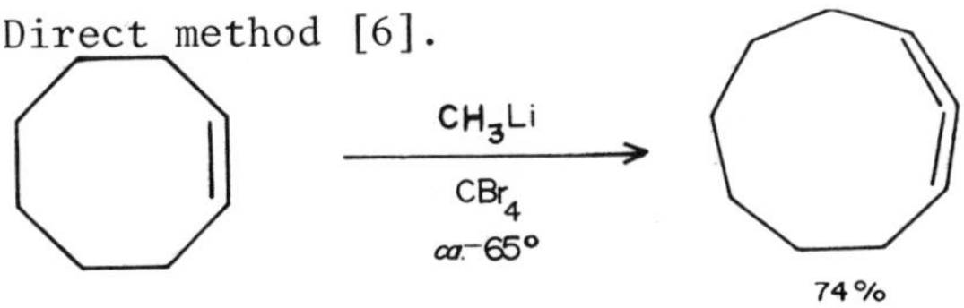

74%

b.   Via 9,9-dibromobicyclo (6.1.0) nonane [7].

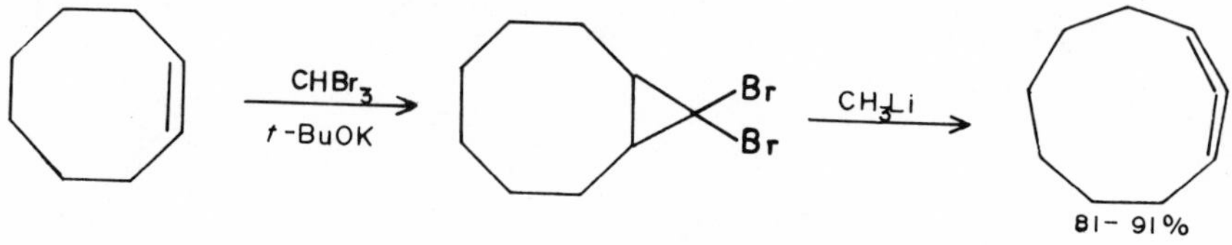

Methyl- or butyllithium brings about the same elimination from 1,1-dibromocyclopropanes [8].

6.   Pyrolysis of diketene [9].

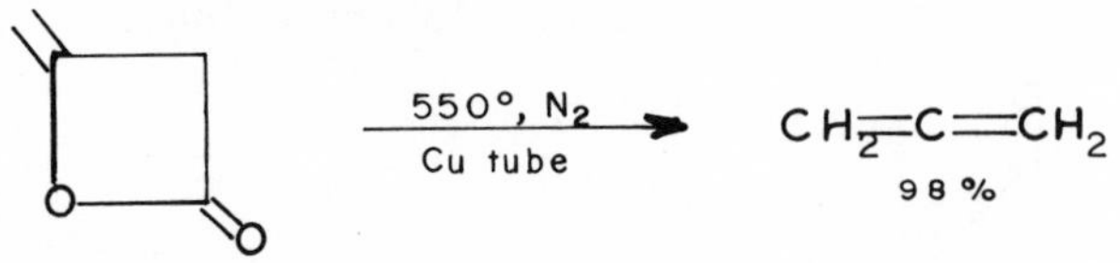

$$CH_2 = C = CH_2$$

98 %

7.   Dehydration of tetrasubstituted allyl alcohols, 3-alkyl-2,4-pentadiones, or ethyl α-alkylacetoacetates. The first of these methods [10] is illustrated as shown:

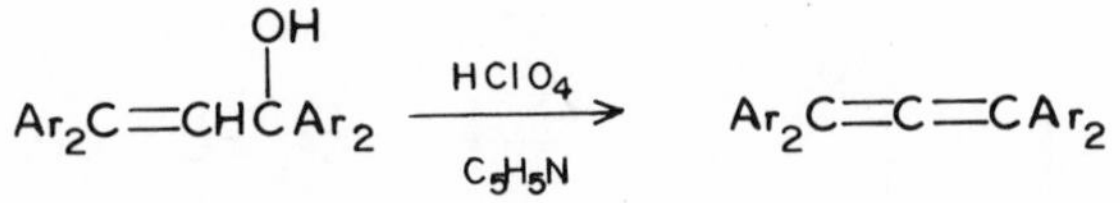

t-Butyl may be substituted for one of the aryl groups. The tetraarylallenes may cyclize with acid to form triarylindenes [11].

The second method, which employs triphenylphosphine di-bromide as the dehydrating agent, may be illustrated with the preparation of 1-methyl-1-acetylallene [12].

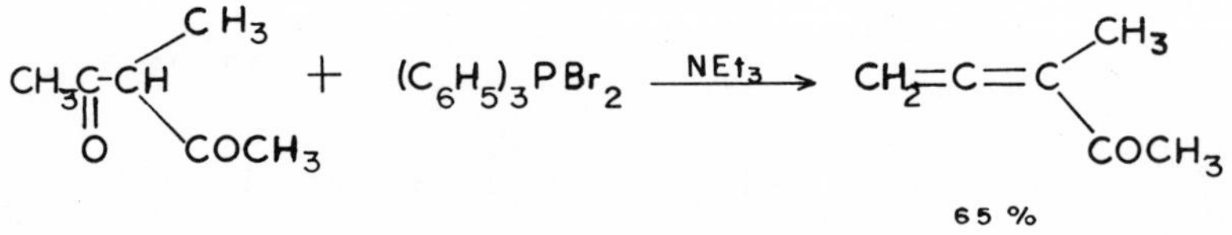

65 %

8.   By the Wittig reaction [13].

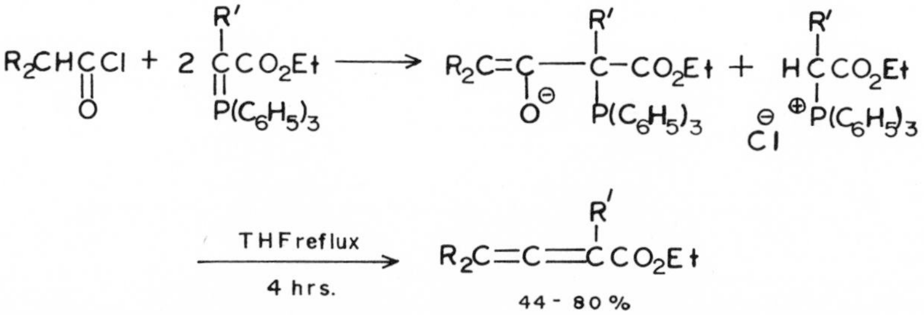

$$R_2CHCCl + 2 \underset{\underset{P(C_6H_5)_3}{|}}{\overset{R'}{\underset{|}{C}}CO_2Et} \longrightarrow R_2C=\underset{\underset{P(C_6H_5)_3}{|}}{\overset{R'}{\underset{|}{C}}-\overset{R'}{\underset{|}{C}}-CO_2Et} + \underset{\overset{\ominus}{\underset{Cl}{}}}{H\overset{R'}{\underset{|}{C}}CO_2Et}$$

$$\xrightarrow[\text{4 hrs.}]{\text{THF reflux}} \quad R_2C=C=\overset{R'}{\underset{|}{C}}CO_2Et$$

44 – 80 %

The allenic esters do not rearrange to acetylenic esters during the reaction, but they do polymerize on standing in light.

9. Dehydrohalogenation of 1,1,3,3-tetrafluoro-3-bromopropene [14].

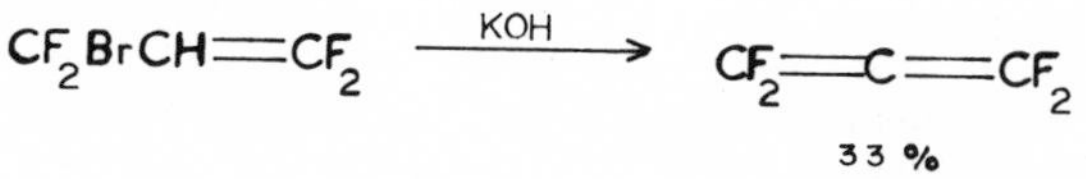

$$CF_2BrCH=CF_2 \xrightarrow{KOH} CF_2=C=CF_2$$

33 %

10. Dehydration of N-propargylamides [15].

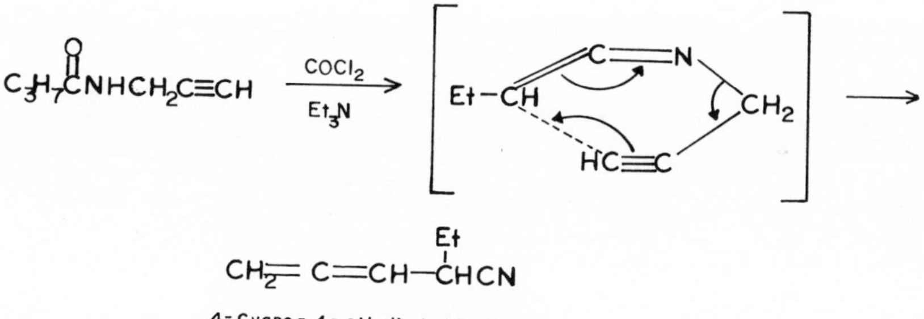

$$C_3H_7\overset{O}{\underset{||}{C}}NHCH_2C\equiv CH \xrightarrow[Et_3N]{COCl_2} \left[\text{ Et–CH} \overset{C \equiv N}{\underset{HC\equiv C}{\diagdown}} CH_2 \right] \longrightarrow$$

$$CH_2=C=CH-\overset{Et}{\underset{|}{C}}HCN$$

4-Cyano-4-ethylbut-1,2-diene
28 %

11. Alkylation of 1-haloallenes with lithium dialkylcuprate reagents [16].

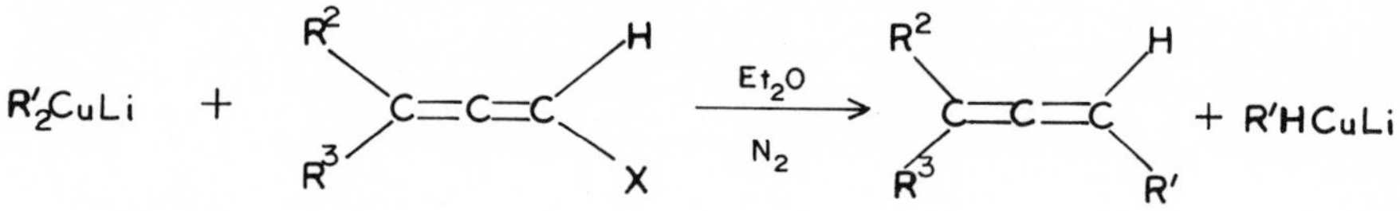

$$R'_2CuLi + \quad \overset{R^2}{\underset{R^3}{>}}C=C=C\overset{H}{\underset{X}{<}} \xrightarrow[N_2]{Et_2O} \quad \overset{R^2}{\underset{R^3}{>}}C=C=C\overset{H}{\underset{R'}{<}} + R'HCuLi$$

b. Via Carbene Addition

A carbene addition to alkenes has become rather useful for the preparation of allenes. The carbene generated from the easily available t-acetylenic chlorides is utilized as shown [17]:

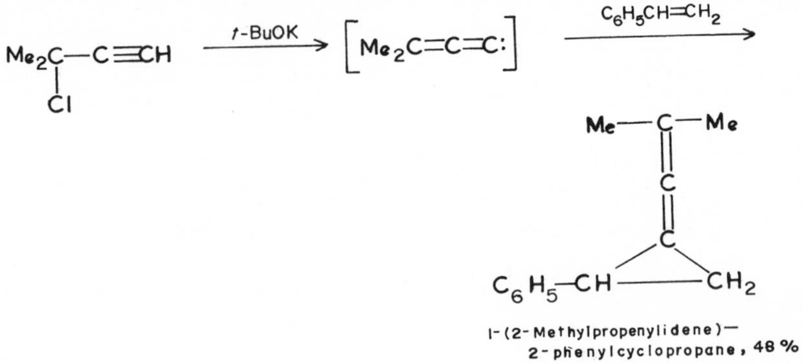

$$Me_2C-C\equiv CH \xrightarrow{t-BuOK} \left[Me_2C=C=C: \right] \xrightarrow{C_6H_5CH=CH_2}$$

(with CI below the Me_2C carbon)

Me—C—Me
‖
C
‖
C
C_6H_5—CH ——— CH_2

l-(2-Methylpropenylidene)—
2-phenylcyclopropane, 48%

Apparently, the same intermediate is involved in the coupling of acetylenic carbinols (1, 173):

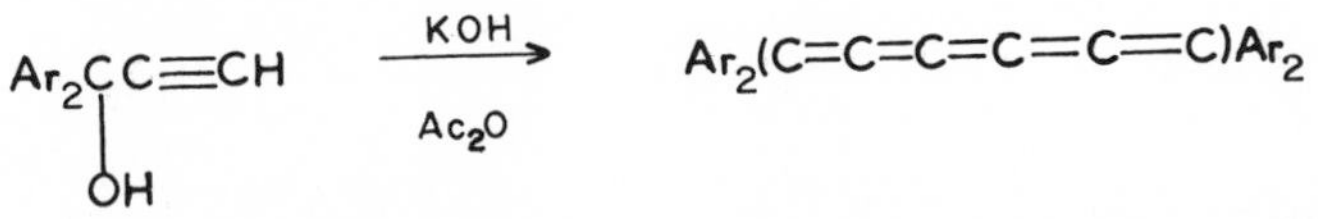

$$Ar_2CC\equiv CH \xrightarrow[Ac_2O]{KOH} Ar_2(C=C=C=C=C=C)Ar_2$$

(with OH below the Ar_2C carbon)

A carbonium ion mechanism was originally assigned to this reaction.

c. Concerted Addition of Ynamines and Carbon Dioxide [18]

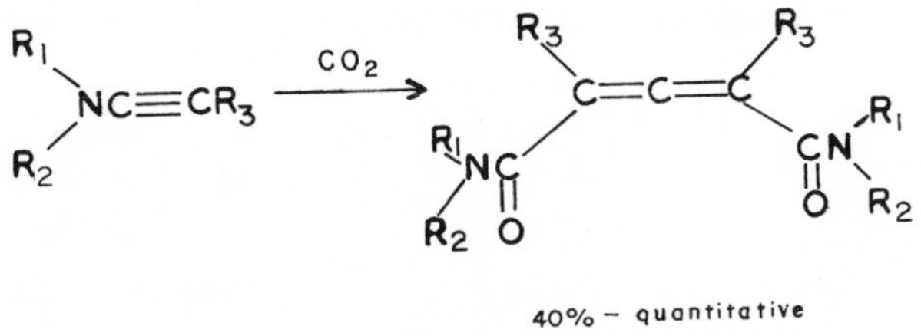

40% — quantitative

d. Oxidation of Pyrazolones by Thallium Nitrate

The pyrazolones, which may be oxidized to allenes, are readily available from β-ketoesters as indicated [19]:

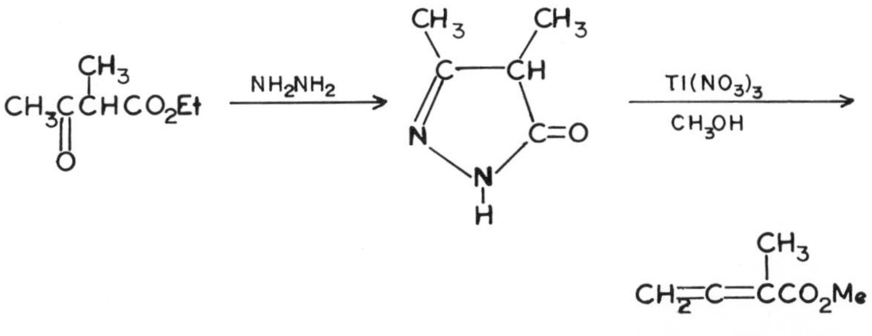

$$CH_3\overset{CH_3}{\underset{O}{\underset{||}{C}}}CHCO_2Et \xrightarrow{NH_2NH_2}$$

[structure of pyrazolone with CH₃ groups]

$$\xrightarrow[CH_3OH]{Tl(NO_3)_3}$$

$$CH_2{=}C{=}\overset{CH_3}{\underset{}{C}}CO_2Me$$

Methyl¯2¯methylbuta¯
dienoate, 50%

e. Isomerization

For a discussion of the formation of allenes from acetylenes, see D.

2. Cumulenes (1, 173)

Newer methods for the synthesis of cumulenes follow:

a. Reaction of Carbenes with Diazo Compounds [20]

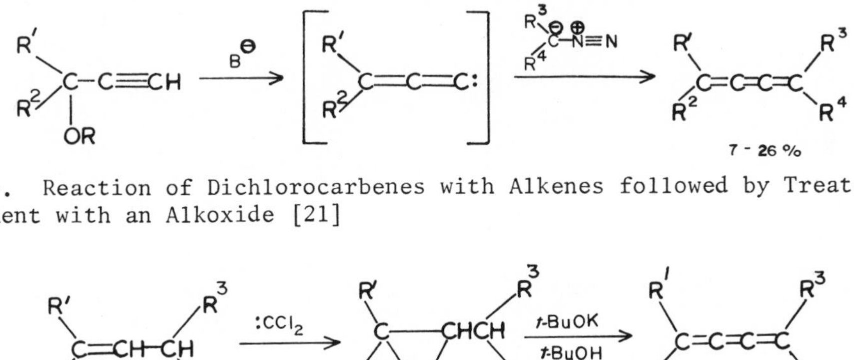

$$\overset{R'}{\underset{R^2}{C}}\underset{OR}{\overset{}{C}}{-}C{\equiv}CH \xrightarrow{B^\ominus} \left[\overset{R'}{\underset{R^2}{C}}{=}C{=}C{:} \right] \xrightarrow{\overset{R^3}{\underset{R^4}{C}}{-}\overset{\ominus}{N}{\equiv}N} \overset{R'}{\underset{R^2}{C}}{=}C{=}C{=}\overset{R^3}{\underset{R^4}{C}}$$

7 - 26 %

b. Reaction of Dichlorocarbenes with Alkenes followed by Treatment with an Alkoxide [21]

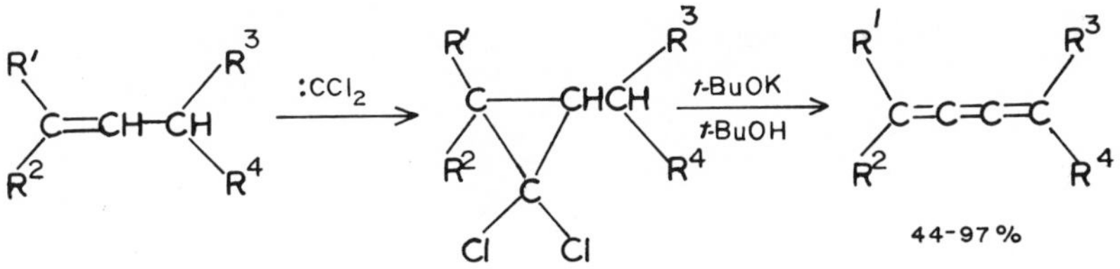

$$\overset{R'}{\underset{R^2}{C}}{=}CH{-}\overset{R^3}{\underset{R^4}{C}}H \xrightarrow{:CCl_2} \overset{R'}{\underset{R^2}{C}}\underset{\underset{Cl\ \ \ Cl}{C}}{\overset{}{}}CH\overset{R^3}{\underset{R^4}{C}}H \xrightarrow[t\text{-BuOH}]{t\text{-BuOK}} \overset{R'}{\underset{R^2}{C}}{=}C{=}C{=}\overset{R^3}{\underset{R^4}{C}}$$

44-97 %

Addenda

E.1. 1,1-Dimethylpropargyl chloride may be converted under
phase transfer conditions to the carbene of dimethylallene,
$(CH_3)_2C{=}C{=}C{:}$, which gave dimethylvinylidenecyclopropanes with

alkenes [S. Julia et al., Compt. Rend., 278, C1523 (1974)].

E.1. Allene may be alkylated with butyllithium and RX. A second
alkyl group, R^1, may be introduced similarly to give $R^1CH=C=CHR$
[G. Linstrumelle, D. Michelot, Chem. Commun., 561 (1975)].

E.1. Divinylallene was prepared (44% overall) by treatment of
vinylethynylmagnesium bromide with allyl bromide (CuCl catalyst)
followed by rearrangement with KOH [W. Mödlhammer, H. Hopf,
Angew. Chem. Inter. Ed. Engl., 14, 501 (1975)].

E.1. A review on the synthesis of chiral allenes has been published. One illustration consists of the conversion of the
optically active cyclopropanecarboxylic acid first to the isocyanate, then to the ethyl carbamate, and finally to the chiral
allene by treatment with N_2O_4 followed by a base [R. Rossi, P.
Diversi, Synthesis, 25 (1973)].

References

1. T. F. Rutledge, Acetylenes and Allenes, Reinhold, New York,
    1969, p. 9.
2. H. N. Cripps, E. F. Keifer, Org. Syn., 42, 12 (1962).
3. G. Zweifel et al., J. Am. Chem. Soc., 92, 1427 (1970).
4a. W. T. Brady, A. D. Patel, Chem. Commun., 1642 (1971).
4b. Ref. 1, p. 11.
5. P. Rona, P. Crabbé, J. Am. Chem. Soc., 90, 4733 (1968).
6. K. G. Untch et al., J. Org. Chem., 30, 3572 (1965).
7. L. Skattebøl, S. Soloman, Org. Syn., 49, 35 (1969).
8. W. R. Moore, H. R. Ward, J. Org. Chem., 25, 2073 (1960);
    27, 4179 (1962); L. Skattebøl, Tetrahedron Letters, 167
    (1961).
9. R. T. Conley, T. F. Rutledge, U. S. Patent, 2,818,456,
    December 31, 1957; C. A., 52, 6391 (1958); Ref. 1.
10. K. Ziegler, C. Ochs, Chem. Ber., 55, 2257 (1922); Ref. 1.
11. H. Fischer, in S. Patai, The Chemistry of Alkenes, Inter-
    science, New York, 1964, Vol. 1, p. 1036.
12. G. Buono, Tetrahedron Letters, 3257 (1972).
13. H. F. Bestmann, H. Hartung, Chem. Ber., 99, 1198 (1966);
    Ref. 1, p. 10.
14. Ref. 1, p. 11.
15. Ibid., p. 12.
16. S. R. Landor et al., Chem. Commun., 593 (1972).
17. H. D. Hartzler, J. Am. Chem. Soc., 83, 4990, 4997 (1961).
18. J. Ficini, J. Pouliquen, J. Am. Chem. Soc., 93, 3295 (1971).
19. E. C. Taylor, A. McKillop et al., J. Org. Chem., 37, 2797
    (1972).

20. H. Reimlinger, R. Paulissen, _Tetrahedron Letters_, 3143 (1970).
21. S. Kajigaeshi et al., _Tetrahedron Letters_, 4887 (1971).

Chapter 4

ALCOHOLS

No general review of alcohol syntheses has come to our attention. Patai's publication on the chemistry of the hydroxyl group [1] offers much in this regard in having chapters on such topics as free-radical and electrophilic hydroxylation (p. 133), oxymetallation (p. 193), biological formation (p. 755), synthesis of O_{18}-labeled compounds (p. 797), and protection of the hydroxyl group (p. 1001).

A curious method of forming the hydroxyl from the methoxy group using iodine, $NaBH_4$ in carbon tetrachloride and methanol is described (A.9). It would be convenient if it were general, but it is probably restricted to β-methoxyesters. An interesting trend in hydrolysis is to first convert the ethyl into either a vinyl ether or a β-haloethyl ether. The first is then hydrolyzed by acid and the second into the alkoxide by a divalent metal (A.9).

All the facets of Brown's hydroboration process are described (B.2) and in addition an in situ hydroboration process that does not depend on using preprepared diborane or diborane complexes (B.2, Ex. b).

Other recent trends and improvements are given. The use of the enol of trimethylsilyl ethers in condensations to prepare alcohols (B.4, F.2, H), especially acyloins (C.8), is one example. Another is the ozonide reduction to alcohols, which has been improved by carrying out the hydrogenation in two stages (D.3). It may be too early to so state, but the addition of organometallic compounds to ketones to form tertiary alcohols may have undergone a profound improvement simply by adding a mixture of the alkyl halide and the ketone to a lithium suspension (E.1).

Finally, these points may be of interest: (a) the superiority of the alkyllithium addition over Grignard addition has been noted, (b) the Reformatsky reaction has been improved by the presence of trimethyl borate, which acts as a buffer (E.3), and (c) more directed aldol reactions other than those of Wittig have been described.

A. Solvolysis

1. From Esters ($\underline{1}$, 176)

$$RCOOCH_2R' \xrightleftharpoons{HOH} RCOOH + R'CH_2OH$$

β-Substituted alcohols are readily prepared by reaction of ethylene carbonates with nucleophilic reagents [2]. The substrate and the reagent in equivalent amounts are simply heated in diethylene glycol:

$$\xrightarrow{LiCl} ClCH_2CH_2OH \quad 75-80\,\%$$

$$\xrightarrow{NaCN} NCCH_2CH_2OH \quad 55-60\,\%$$

$$\xrightarrow{C_6H_5NH_2} C_6H_5NHCH_2CH_2OH \quad 85\,\%$$

For success at least one of the methylene groups of the carbonate must be unsubstituted. To prevent reaction between the product and the reagent, prompt removal of the former is essential.

A superior method for the solvolysis of tosylates consists of treating the substrate in THF under nitrogen with sodium naphthalene, usually at 25° [3]:

$$ROSO_2\!\!\left\langle\!\!\!\!\begin{array}{c}\\ \\\end{array}\!\!\!\!\right\rangle\!\!CH_3 \xrightarrow[C_{10}H_8]{Na} ROH + SO_2 + \left\langle\!\!\!\!\begin{array}{c}\\ \\\end{array}\!\!\!\!\right\rangle\!\!CH_3$$

$$> 90\,\%\ \text{(glc)}$$

No epimerization occurs in the process. More recently tosylates, mostly of steroids, have been solvolyzed by irradiation in ether or ether-methanol with yields as high as 78% [4].

5. From Amines (1, 180)

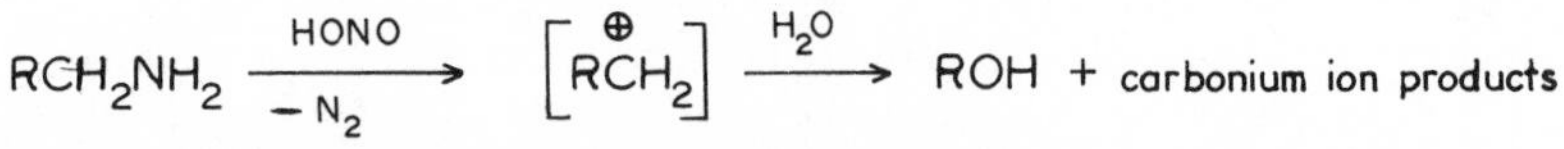

$$RCH_2NH_2 \xrightarrow[-N_2]{HONO} \left[RCH_2^{\oplus} \right] \xrightarrow{H_2O} ROH + \text{carbonium ion products}$$

A recent discussion on the mechanism of the reaction has appeared in the literature [5]. In a modified, indirect method Tsirkel and co-workers [6] have converted 1,6-diaminohexane into the 1,6-hexanediol:

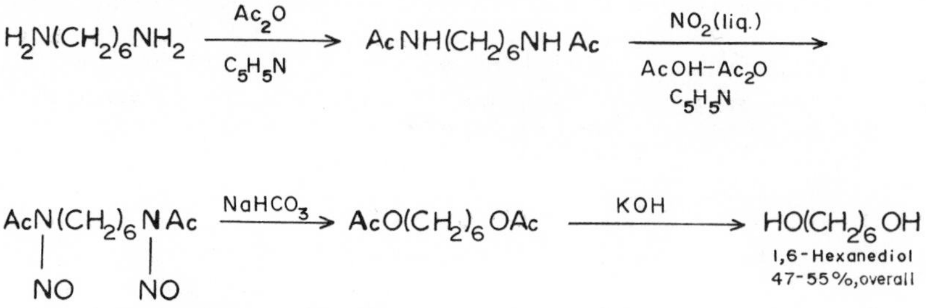

$$H_2N(CH_2)_6NH_2 \xrightarrow[C_5H_5N]{Ac_2O} AcNH(CH_2)_6NHAc \xrightarrow[\substack{AcOH-Ac_2O \\ C_5H_5N}]{NO_2(liq.)}$$

$$\underset{\substack{| \\ NO}}{AcN}(CH_2)_6\underset{\substack{| \\ NO}}{NAc} \xrightarrow{NaHCO_3} AcO(CH_2)_6OAc \xrightarrow{KOH} \underset{\substack{1,6\text{-Hexanediol} \\ 47\text{-}55\%,\text{overall}}}{HO(CH_2)_6OH}$$

7. From Ethers, Mostly Cyclic (1, 182)

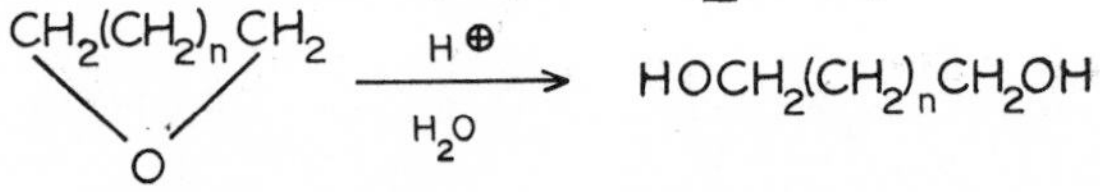

$$\underset{O}{\overset{CH_2(CH_2)_n CH_2}{\diagup \diagdown}} \xrightarrow[H_2O]{H^{\oplus}} HOCH_2(CH_2)_n CH_2OH$$

A recent review on the base-catalyzed rearrangements of epoxides, many of which produce alcohols, is available [7].

The opening of epoxide rings is at times subject to control [8]. Thus for unsymmetrically substituted epoxides acids give the alcohol expected, while the methoxide ion leads to the other possible alcohol:

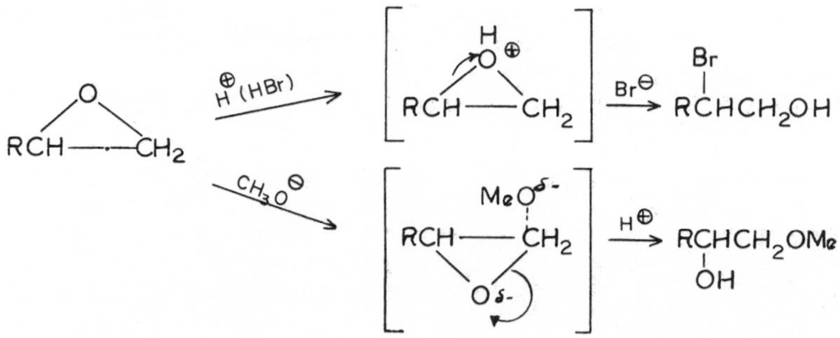

Thus the H^+ of the HBr apparently attacks the ring O while $\overline{MeO}$ attacks the unsubstituted ring C.

In the presence of a strong acid and DMSO, the glycol obtained from the cis epoxide is the threo form, while from the trans isomer the erythro form is obtained [9].

The importance of the solvent is illustrated in the

treatment of styrene oxide with HCl [8]. In a nonpolar solvent
inversion occurs while retention of configuration results with a
polar one:

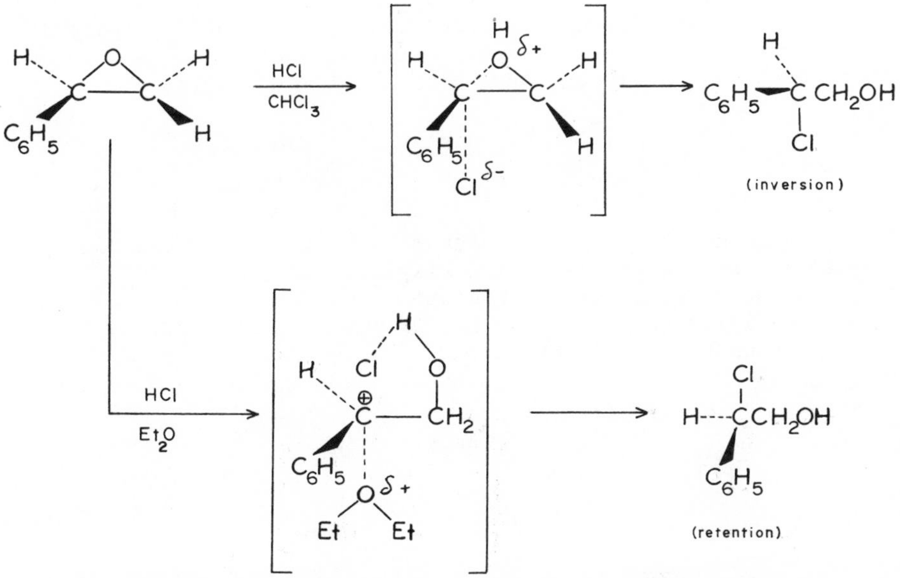

By contrast, treatment of propylene oxide with lithium
phosphate leads by isomerization to allyl alcohols [10].

A series of 3-fluoro-2-propanols has been prepared by the
ring opening of 3-fluoro-1,2-propylene oxide [11]. Tosylhydrazones of larger ring structure have been cleaved as shown [12]:

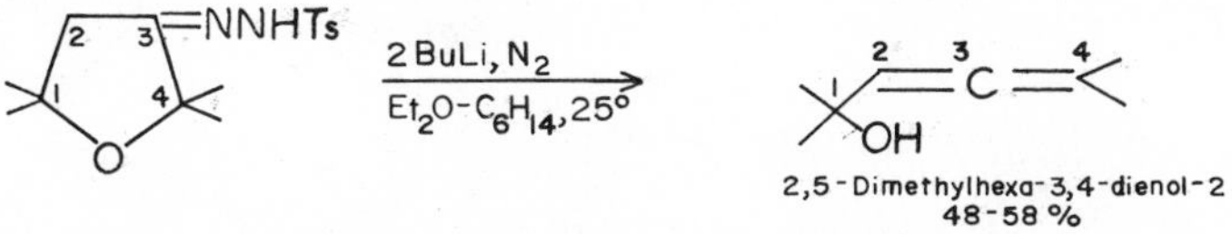

2,5-Dimethylhexa-3,4-dienol-2
48-58 %

Epoxides of cycloalkanes may also be cleaved to alcohols.
Thus cyclohexene oxide with lithium alkylamides gives a series of
alcohols, the principal two of which are 2-cyclohexenol and 2-
aminocyclohexanol [13]:

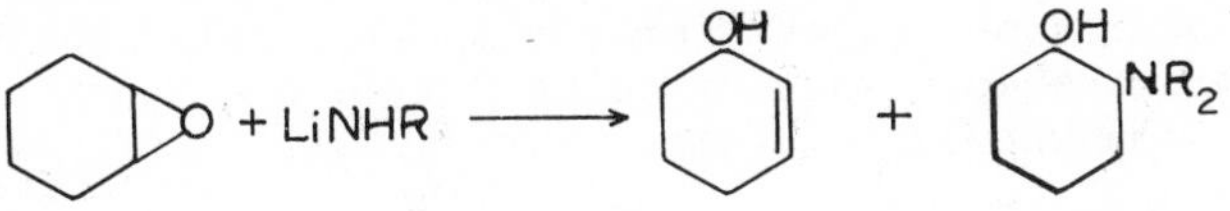

Quantitative yields of the 2-cyclohexenol are obtained by using $[(n-C_3H_7)_2N]_2NLi$ or $[(n-C_4H_9)_2N]_2NLi$, while with $LiNHC_3H_7-n$ the yield of the product is 77%.

In a previous study Johnson and co-workers cleaved cyclohexene oxide with organocopper reagents [14]. Their principal product was the 2-alkyl or 2-aryl cyclohexanol, which was obtained in 81% yield by the use of lithium diphenylcuprate:

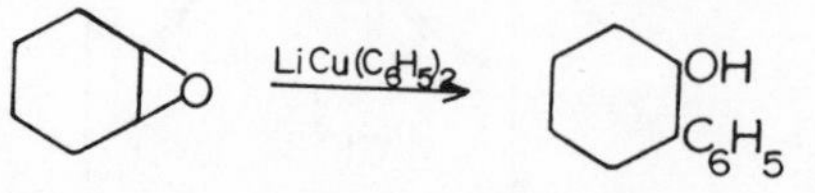

The lithium dialkyl cuprates proved to be more reactive toward epoxides than the corresponding alkyllithiums.

In the cleavage of cyclooctene oxide the allylic alcohol is obtained exclusively with the use of potassium t-butoxide or lithium phosphate [15] (Ex. a).

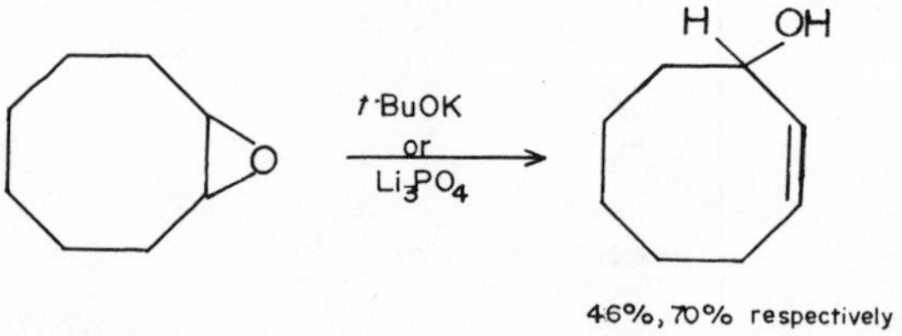

46%, 70% respectively

This same conversion was accomplished by Sharpless and Lauer [16] by proceeding via the phenyl selenide as shown:

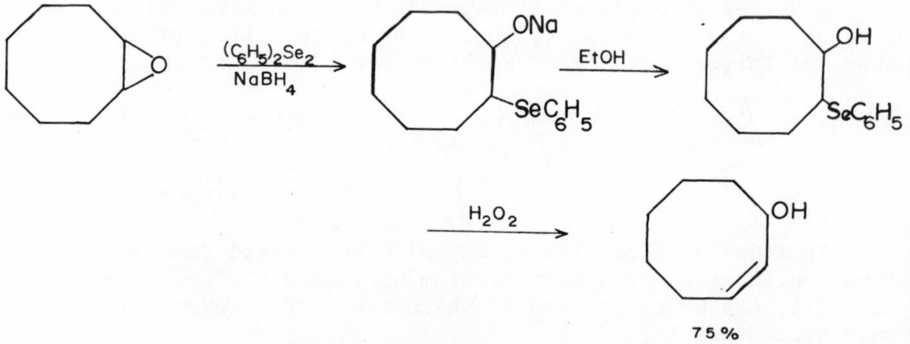

75%

This one-operation process for the epoxide-allylic alcohol conversion has the advantage of being a mild one, particularly in the last step where room temperature usually suffices.

Carbohydrates have been debenzylated by treatment with bromine in sulfolane followed by irradiation [17]. Thus methyl

2,3,4,6-tetra-O-benzyl-α-D-glucopyranoside gives 73% of methyl-α-
D-glucopyranoside:

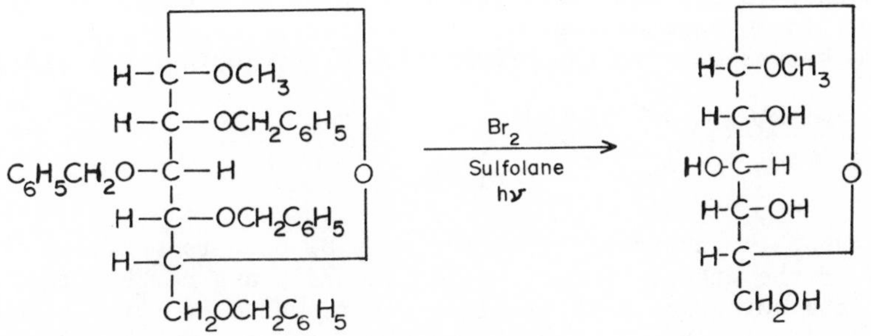

Here the α-bromobenzyl ether, which can be readily hydrolyzed,
probably is formed. The method, which appears to be specific for
debenzylation, has been applied to oligosaccharides and a poly-
saccharide as well.

8. From Sulfoxides (Sulfoxide-Sulfenate Rearrangement)

α-Substituted methallyl p-tolyl sulfoxides rearrange to
allylic sulfenate esters, which with the proper nucleophiles give
allyl alcohols [18].

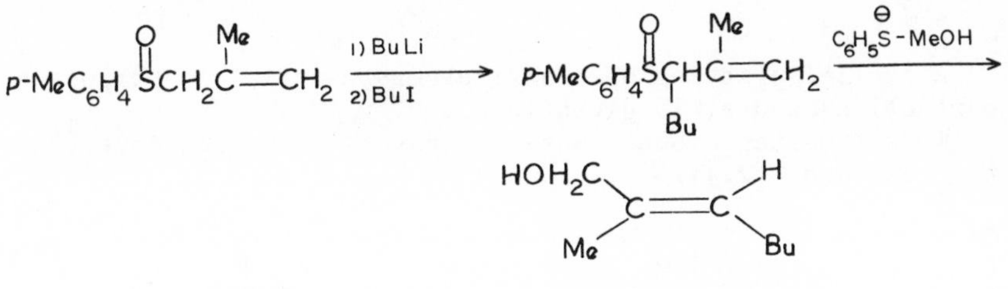

trans-2-Methyl-2-heptenyl alcohol
90%

The mechanism involved appears to be:

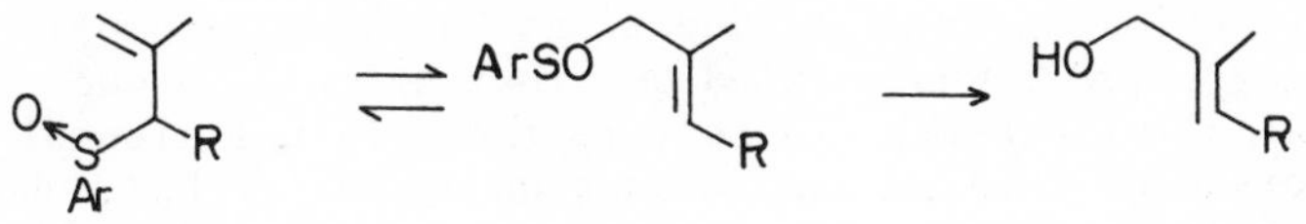

9. From β-Haloethyl Ethers, Vinyl Ethers, and the Like

$$ROCH_2CH_2X + M \longrightarrow ROMX + CH_2{=}CH_2 \xrightarrow{H^{\oplus}} ROH$$

Sometimes a combination of an alkyllithium and lithium halide is used in place of a divalent metal. One advantage of the method is the fact that the ethers may be cleaved to the alcohol in nonaqueous medium, if necessary. Apparently, the method serves best in the preparation of cyclopropanols [19].

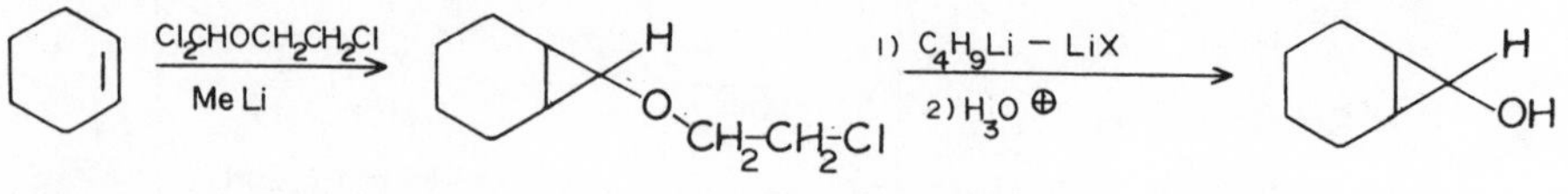

Unfortunately an excess of alkene must be used to maximize the yield of the ether. This reaction occurs via a carbene that forms the norcarane β-chloroethyl ether, which in turn is converted into the easily hydrolyzed vinyl ether. Other cyclopropanol syntheses, including those of Cottle and DePuy, are described in detail [20] and discussed in E.1.

A related method has been employed in prostaglandin synthesis [21]:

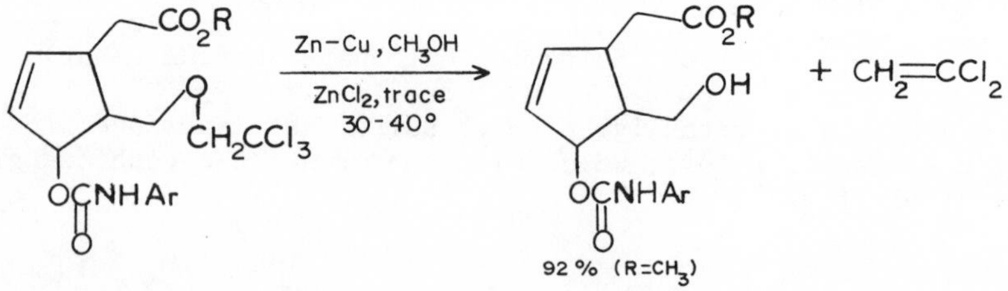

A facile splitting of benzyl ethers, similar in purpose to the reactions above, is given in D.9.

In a two-step process 3-allyloxyoxetane may be converted into 3-oxetanol [22]:

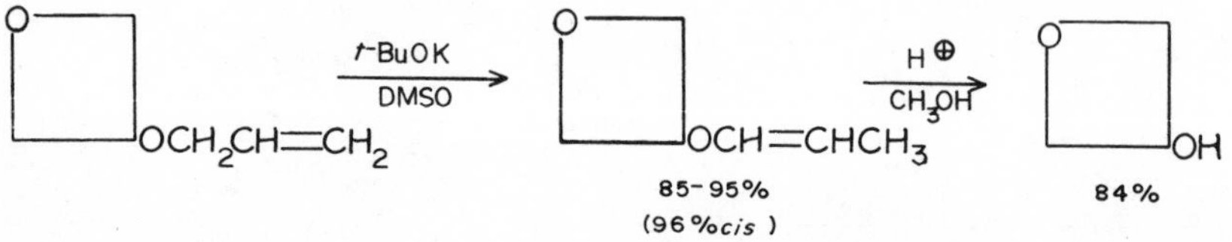

A similar reaction under mild conditions has been accomplished by Corey and Suggs [23], who pointed out that hydroxyl groups may be protected as allyl ethers. To restore the original alcohol these investigators used rhodium complexes such as $RhCl(PPh_3)_3$ under neutral aprotic conditions in the first step and acid at pH 2 in the second step:

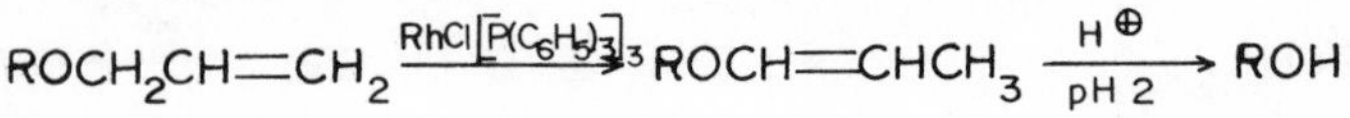

$$ROCH_2CH=CH_2 \xrightarrow{RhCl[P(C_6H_5)_3]_3} ROCH=CHCH_3 \xrightarrow[pH\ 2]{H^{\oplus}} ROH$$

Allyl ethers of methanol, 1-decanol, and cholesterol were all converted into the corresponding alcohol in >90% yield.

The methyl ether of methyl D-3-hydroxynonanoate was cleaved by sodium borohydride and iodine as indicated [24].

$$CH_3(CH_2)_5 \underset{\underset{OCH_3}{|}}{C} DCH_2COOCH_3 \xrightarrow[\underset{45°}{CCl_4 - CH_3OH}]{NaBH_4 - I_2 -} CH_3(CH_2)_5 \underset{\underset{OH}{|}}{C} DCH_2COOCH_3$$

$[\alpha]_D^{25} -2.8°$

Methyl 3D- 3-hydroxynonanoate, 77%

$[\alpha]_D^{25} -20.0°$

In the cleavage methanol was present to convert the borate ester formed into the methyl ester by transesterification. Note that optical activity was retained in the solvolysis.

The equation for the reaction in more detail is:

$$3 \ ROCH_3 + 2 \ I_2 + MBH_4 \longrightarrow B(OR)_3 + 3 \ CH_3I + MI + 2 \ H_2$$
$$\downarrow 3 \ CH_3OH$$
$$3 \ ROH + B(OCH_3)_3$$

The fact that the reaction is not general (in our opinion) may suggest a complex mechanism.

   10. From Alcohols (O Deuteration)

$$ROH \longrightarrow ROD$$

In a recent review [25] the·deuteration appears to proceed more favorably by treating the intermediate magnesium alkoxide or borate ester with deuterium oxide.

Addendum

Finely ground $CaCl_2$ suspended in a mixture of 33% of cis- and 67% of trans-4-t-butylcyclohexanol gives a complex from which 83% of the trans isomer (1% cis) could be recovered [K. B. Sharpless et al., J. Org. Chem., 40, 1252 (1975)].

References

1. S. Patai, The Chemistry of the Hydroxyl Group, Pts. 1 and 2, Interscience, New York, 1971.
2. E. D. Bergmann, I. Shahak, J. Chem. Soc., C899 (1966).
3. W. D. Closson et al., J. Am. Chem. Soc., 88, 1581 (1966).

4. J. P. Péte, Tetrahedron Letters, 4555 (1971).
5. C. J. Collins, Acc. Chem. Res., 4, 315 (1971).
6. T. M. Tsirkel et al., Tr. Vses. Nauch.-Issled. Inst. Sin. Natur. Dushistykh. Veshchestv., 8, 311 (1968); C. A., 71, 38249 (1969).
7. V. N. Yandovskii, B. A. Ershov, Russian Chem. Rev., 41, 403 (1972).
8. D. N. Kirk, Chem. Ind. (London), 109 (1973).
9. M. A. Khuddus, D. Dwern, Tetrahedron Letters, 411 (1971).
10. A. G. Polkovnikova et al., Khim. Prom. (Moscow), 49, 170 (1973); C. A., 79, 4908 (1973).
11. E. D. Bergmann et al., Synthesis, 646 (1971).
12. A. M. Foster, W. C. Agosta, J. Org. Chem., 37, 61 (1972).
13. C. L. Kissel, B. Rickborn, J. Org. Chem., 37, 2060 (1972).
14. C. R. Johnson et al., J. Am. Chem. Soc., 92, 3813 (1970).
15. M. N. Sheng, Synthesis, 194 (1972).
16. K. B. Sharpless, R. F. Lauer, J. Am. Chem. Soc., 95, 2697 (1973).
17. C. Y. Meyers et al., J. Org. Chem., 33, 4292 (1968).
18. P. A. Grieco, Chem. Commun., 702 (1972).
19. U. Schöllkopf et al., Org. Syn., Coll. Vol. 5, 859 (1973).
20. C. H. DePuy, Acc. Chem. Res., 1, 33 (1968).
21. E. J. Corey, G. Moinet, J. Am. Chem. Soc., 95, 6831 (1973).
22. J. A. Wojtowicz, R. J. Polak, J. Org. Chem., 38, 2061 (1973).
23. E. J. Corey, J. W. Suggs, J. Org. Chem., 38, 3224 (1973).
24. G. Odham, B. Samuelsen, Acta Chem. Scand., 24, 468 (1970).
25. L. Verbit, Synthesis, 254 (1972).

B. Addition and Substitution (Friedel-Crafts)

1. From Alkenes (1, 185)

In the presence of Lewis acids such as stannic or aluminum chloride, 2-methylpropene, 2-methyl-1-pentene, and cyclohexene react with chloral at low temperature to form β,γ-ethylenic trichloromethylcarbinols [1]. Thus 2-methylpropene responds as indicated:

$$CH_3C{=}CH_2 \ + \ CCl_3CHO \xrightarrow{SnCl_4} CH_3C{-}CH_2CHCCl_3$$

with CH_3 on the first carbon and CH_2, OH on the product.

2-Methylallyltrichloromethyl carbinol

85%

Thallium sulfate adds to cyclic olefins to give <u>trans</u>-diols (46-67%, 2 ex.) [2], whereas chlorine in the presence of DMF gives the chlorohydrin [3]:

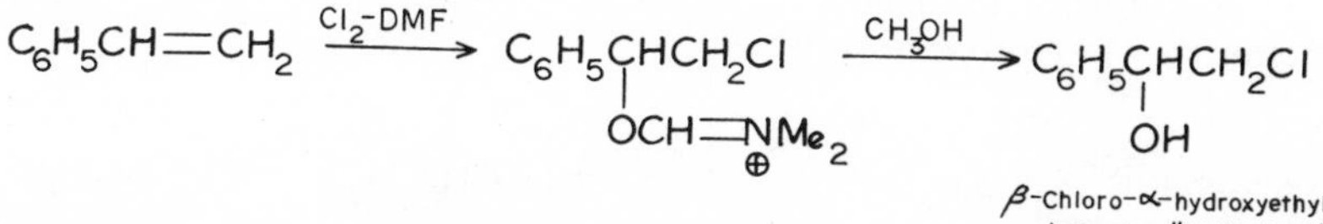

$$C_6H_5CH{=}CH_2 \xrightarrow{Cl_2\text{-}DMF} C_6H_5\underset{\underset{OCH{=}\overset{\oplus}{N}Me_2}{|}}{C}HCH_2Cl \xrightarrow{CH_3OH} C_6H_5\underset{\underset{OH}{|}}{C}HCH_2Cl$$

β-Chloro-α-hydroxyethyl-
benzene, "high yield"

Apparently, carbonium ions can add to an alkene carrying a group capable of forming an alcoholic group as shown [4].

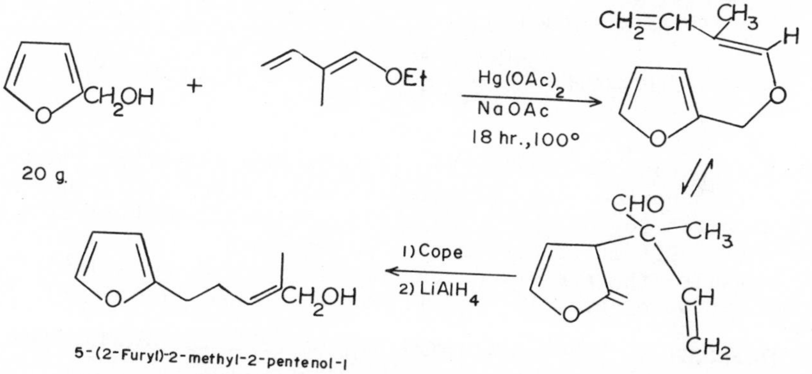

20 g.

5-(2-Furyl)-2-methyl-2-pentenol-1

ca. 5 g.

2. From Alkenes or Alkynes via Boranes and Boric Acid Esters (1, 186)

The hydroboration process of H. C. Brown has extended greatly the methods for preparing various types of alcohols. For our purpose we shall apply the process to the synthesis of: (a) primary and secondary alcohols, (b) tertiary alcohols, (c) diols, (d) cycloalkanols, (e) alkenols, and (f) cyclohexanediols. In its customary form hydroboration of the olefin or alkyne is followed by oxidation with alkaline peroxide.

a. Primary and Secondary Alcohols

Primary alcohols may now be produced by controlled oxidation of the trialkylboron followed by hydrolysis of the intermediate boron derivative [5] (Ex. a):

$$R_3B + 1.5\ O_2 \longrightarrow \left[(RO)_3 B\right] \xrightarrow{3\ H_2O} 3\ ROH + (HO)_3B$$

Pure oxygen is used with THF as the solvent at 0° and the

reaction is stopped at exactly 100% oxygen absorption.   Yields
from a series of trialkylboranes exceed 90% (glpc).

More common methods for preparing primary or secondary
alcohols involve hydroboration of alkynes [6] or alkenes [7]
followed by peroxide oxidation:

$$Bu C \equiv CH \xrightarrow{2BH_3} \left[BuCH_2CH \begin{smallmatrix} B< \\ \\ B< \end{smallmatrix} \right] \xrightarrow{NaOH} \left[BuCH_2CH_2B< \right] \xrightarrow{H_2O_2} BuCH_2CH_2OH$$

n-hexyl alcohol
80%

Small amounts of the hexaldehyde, 1,2-diols, 2-hexanol, and 2-
hexanone are also formed.

$$RCH \!\!=\!\! CH_2 \xrightarrow[\text{2) } C_2H_5OH-NaOH-H_2O_2]{\text{1) } BH_2Cl -Et_2O} RCH_2CH_2OH$$

Monochloroborane in ethyl ether is much more directive in its
effect than borane. For example, 1-hexene gives >99% 1-hexanol;
styrene gives 96% of the primary alcohol; and 1-methylcyclopen-
tene gives >99% of trans-2-methylcyclopentanol.

Finally, the preferred reagent for the hydroboration of
alkenes is 9-borabicyclo [3.3.1] nonane, 9-BBN, in THF [8].

$$RCH = CH_2 \xrightarrow[\text{THF}]{\bigcirc B-H} \left[\begin{smallmatrix} RCH_2CH_2 \\ | \\ B \end{smallmatrix} \right] \xrightarrow[\text{NaOH -}H_2O_2]{C_2H_5OH-} RCH_2CH_2OH$$

>99 %

The regioselectivity here surpasses that of other hydroborating
agents, particularly in the case of internal olefins. An in
situ preparation of diborane to prepare normal alcohols has
become available (see Ex. b). Yields are comparable to those
obtained starting with pure diborane.

b.   Tertiary Alcohols

In this synthesis Lane and Brown [9] brominated the trialkylborane under irradiation and then oxidized the reaction
mixture in the usual manner:

$$(C_2H_5)_3B \xrightarrow[\substack{H_2O \\ h\nu}]{Br_2-} \begin{smallmatrix} CH_3CH_2 \\ | \\ CH_3CB(OH)_2 \\ | \\ CH_3CH_2 \end{smallmatrix} \xrightarrow[\substack{OH \\ \ominus}]{H_2O_2} \begin{smallmatrix} CH_3CH_2 \\ | \\ CH_3C \, OH \\ | \\ CH_3CH_2 \end{smallmatrix}$$

3-Methyl-3-pentanol
85%

In the bromination-irradiation step the rearrangement of alkyl groups from boron to carbon probably occurs as follows:

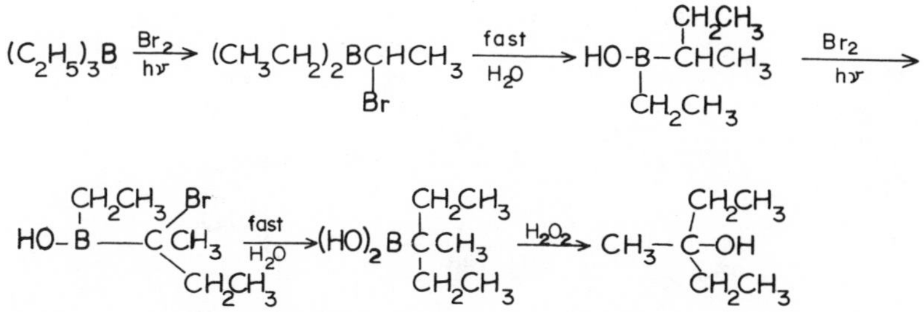

$$(C_2H_5)_3B \xrightarrow[h\nu]{Br_2} (CH_3CH_2)_2BCHCH_3 \xrightarrow[H_2O]{fast} HO-B-CHCH_3 \xrightarrow[h\nu]{Br_2}$$

(with CH₂CH₃ branch on the BCHCH₃ and CH₂CH₃ groups on the HO-B-CHCH₃ center)

$$HO-B-CCH_3 \xrightarrow[H_2O]{fast} (HO)_2B\,CCH_3 \xrightarrow{H_2O_2} CH_3-C-OH$$

The method represents a procedure for combining terminal or internal olefins to produce highly substituted tertiary alcohols.

Further elaboration of the method for the synthesis of other highly substituted tertiary alcohols [10], of tertiary alcohols containing functional substituents [11], and of cis-2-alkylcyclohexanols [12] has been achieved. Furthermore, NBS in the presence of water has been used successfully rather than photochemical bromination in the formation of tertiary alcohols [13].

c. Diols

1,3-Diols have been prepared by the hydroboration of allyl-lithium derivatives [14]. Thus allylbenzene was converted into 1-phenyl-1,3-propanediol in the steps indicated:

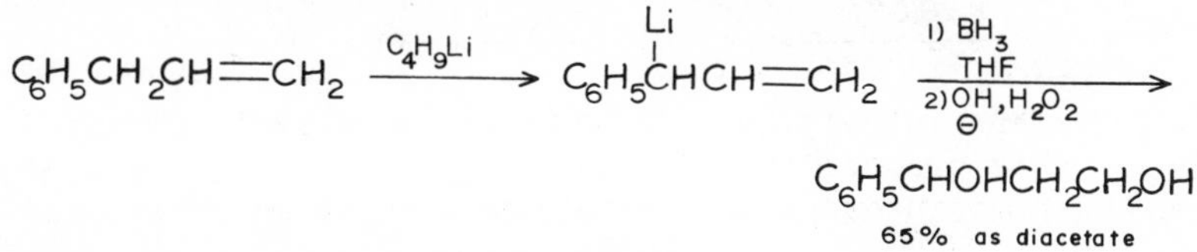

$$C_6H_5CH_2CH=CH_2 \xrightarrow{C_4H_9Li} C_6H_5\overset{Li}{C}HCH=CH_2 \xrightarrow[2)OH,H_2O_2]{1) BH_3 \; THF}$$

$$C_6H_5CHOHCH_2CH_2OH$$

65% as diacetate

d. Cycloalkanols

The synthesis of cis-2-alkylcyclohexanols has already been given. Other cyclohexanols have been produced from the tosylate of 3-butyn-1-ol [15] and from bicyclo [4.1.0] heptane [16] as shown:

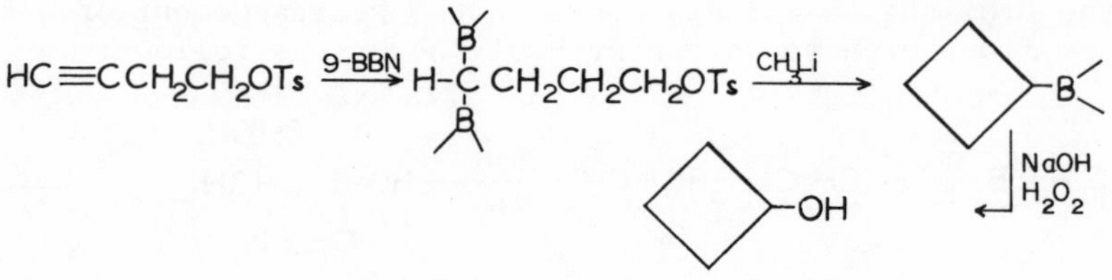

HC≡CCH₂CH₂OTs $\xrightarrow{\text{9-BBN}}$ H-C CH₂CH₂CH₂OTs $\xrightarrow{\text{CH}_3\text{Li}}$

Cyclobutanol, 65 %

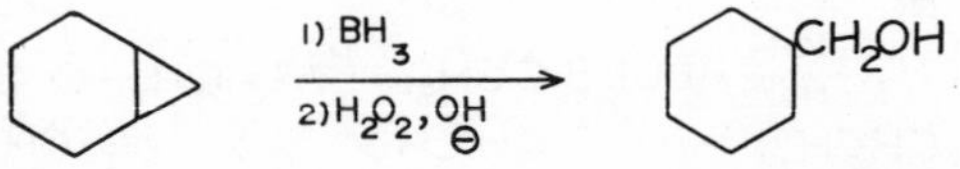

$$\xrightarrow[\text{2)H}_2\text{O}_2,\text{OH}^{\ominus}]{\text{1) BH}_3}$$ CH₂OH

Cyclohexylmethanol
98 %

e.   Alkenols

2-Cyclohexenol has been synthesized from cyclohexanone via the enamine [17].

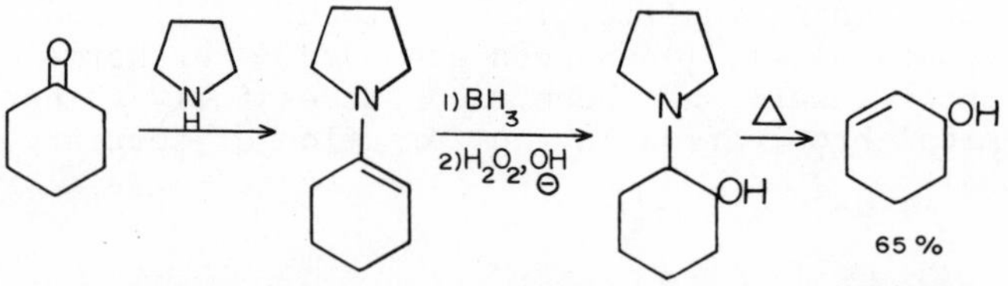

$$\xrightarrow[\text{2)H}_2\text{O}_2,\text{OH}^{\ominus}]{\text{1)BH}_3}$$

65 %

A 3-cyclopentenol has become available from cyclopentadiene [18] as indicated:

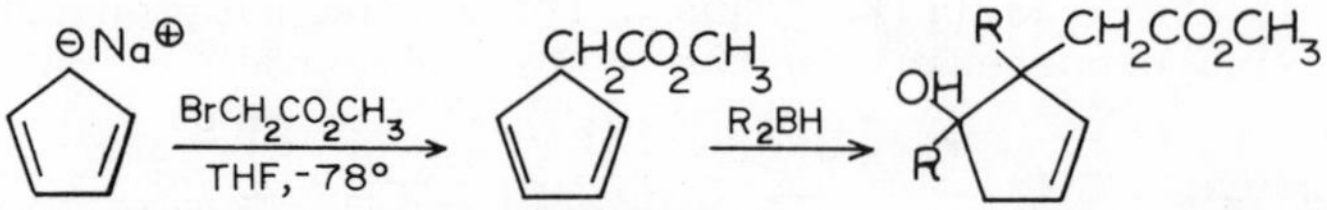

$$\xrightarrow[\text{THF,-78°}]{\text{BrCH}_2\text{CO}_2\text{CH}_3}$$ $\xrightarrow{\text{R}_2\text{BH}}$

With R=(+)-di-3-pinanyl, an optically active 3-cyclopentenol is obtained. This product has been utilized in the synthesis of prostaglandin of interest in the preparation of an artificial gene.

Unsaturated aliphatic alcohols have been produced by opening the ring of 1,3-butadiene monoxide by treatment with a trialkyl-boron in the presence of catalytic amounts of oxygen or a free radical initiator [19]:

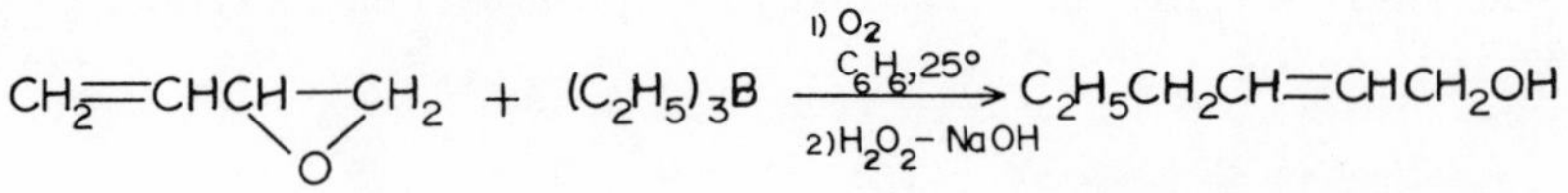

CH₂=CHCH—CH₂ + (C₂H₅)₃B $\xrightarrow[\text{2)H}_2\text{O}_2-\text{NaOH}]{\substack{\text{1) O}_2 \\ \text{C}_6\text{H}_6,25°}}$ C₂H₅CH₂CH=CHCH₂OH

2-Hexenol,75% (89% trans)

f. Cyclohexanediols

Cyclohexanediols have been produced by the hydroboration of
2-cyclohexenone [20] or of the sodium enolate of cyclohexanone
[21]:

trans—Cyclohexane-1,2-diol

76 %

g. Preparation of 2,4,4-Trimethyl-1-pentanol [5]

2,4,4-Trimethyl-1-pentane, 16.9 g, in 50 ml of THF at 0°
was hydroborated by the dropwise addition of 18 ml of a 2.68 M
solution of borane in THF at 0°, followed by stirring at 25° for
1 hr. The solution at 0° in a properly flushed flask was treated
with oxygen during stirring until a theoretical amount was
absorbed. From the reaction mixture treated with 18 ml of 3 N
NaOH there was recovered 17.2 g (885) of the alcohol.

h. In situ Preparation of Primary Alcohols [22]

All parts of the apparatus as well as the $NaBH_4$ were oven
dried. The acetic acid was made up to 100% acid with acetic
anhydride. To the mixture of the olefin (0.1 mol), $NaBH_4$
(0.037 mol) and 250 ml of dry THF, cooled to 10-20° and purged
with N_2, a solution of 0.037 mol of acetic acid in 50 ml of THF
was added dropwise over 1 hr and the mixture was stirred for an
additional 2 hr. Then sodium hydroxide, 0.1 mol in 20 ml of
water, was added followed by 14 ml of 30% H_2O_2 and the mixture
was again stirred for 2 hr. After the layers were separated,
the organic layer was worked up in the usual manner to give
76-92% of straight-chain primary alcohols or cycloalkanols.

3. From Alkenes (via Mercurials) (<u>1</u>, 188)

This subject has been reviewed [23]. In contrast to the
hydroboration procedure (B.2) the mercuration procedure of
obtaining alcohols from alkenes consists of a Markownikoff
hydration. No intermediates need be isolated and the method as
a rule gives high yields of alcohols from mono-, di-, tri-, and
tetraalkyl, as well as from phenyl-substituted olefins [24].

Carlson and Funk [25] have carried out the oxymercuration-demercuration procedure by employing a chiral mercury(II) carboxylate and thus obtained optically active alcohols:

$$RCH = CH_2 \xrightarrow[2)NaBH_4 - \overset{\ominus}{O}H]{1)Hg(O\overset{*}{C}R)_2} R\overset{*}{C}H - CH_3$$

To cite one example, 1-decene was converted, by using Hg(II)-(+)-tartrate, into 2-decanol in which the excess of (-) isomer over the (+) isomer was 17.4%.

Diols have been produced by the oxymercuration-demercuration method from both acyclic unsaturated alcohols and dienes [26]. Thus allyl alcohol gives a quantitative yield of 1,2-propanediol, whereas 1,4-pentadiene forms 2,4-pentanediol with a 83% yield (see Ex. a).

a. Preparation of 2,4-Pentanediol [26]

To mercuric acetate, 20 mmol, in 10 ml of water, 10 ml of the THF was added to produce a deep yellow color. 1,4-Pentadiene, 10 mmol was added, after which the yellow color disappeared. The mixture was then stirred for 45 min and 20 ml of 3M NaOH was added followed by 20 ml of 0.5M NaBH$_4$ in 3M NaOH. Stirring was continued until almost all of the mercury had coagulated and then the aqueous layer was saturated with K$_2$CO$_3$. The diol was recovered (83%) from the organic layer to which two THF extractions had been added.

4. From Alkenes and Carbonyl Compounds (Prins) ($\underline{1}$, 188)

$$R'CH = O + \overset{\oplus}{H} \rightleftharpoons R'\overset{\oplus}{C}HOH \xrightarrow[H_2O]{RCH=CH_2} RCHOHCH_2CHOHR'$$

Stapp [27] found that by carrying out the Prins reaction between 1-olefins and aqueous formaldehyde at elevated temperatures under pressure two principal products, 4-alkyl-1, 3-dioxane and cis,trans-3-alkyltetrahydropyran-4-ol, were obtained. Thus 1-pentene gave with 100% conversion the products shown:

$$C_2H_5CH_2CH = CH_2 \xrightarrow[H_2SO_4]{2 CH_2O - H_2O}$$

46 %   42 %

A new aldol type reaction was discovered when trimethylsilyl enol ethers of ketones react with aldehydes or ketones in the presence of TiCl$_4$ [28]. Thus the trimethylsilyl enol ether of acetophenone reacts with phenylacetaldehyde as indicated:

$$C_6H_5\overset{\overset{\displaystyle OSi(CH_3)_3}{|}}{C}=CH_2 + C_6H_5CH_2CHO \xrightarrow[2)H_2O]{1)TiCl_4} C_6H_5\overset{\overset{\displaystyle O}{||}}{C}CH_2\overset{\overset{\displaystyle OH}{|}}{\underset{\underset{\displaystyle H}{|}}{C}}-CH_2C_6H_5$$

1,4-Diphenyl-3-hydroxybutan-1-one

78 %

It is of interest to note that the cross-aldol addition products and not the self-addition or condensation product of the aldehyde or ketone, are obtained by this method.

6. From Boranes and Carbon Monoxide (1, 191)

$$R_3B + CO \xrightarrow{125°} R_3CBO \xrightarrow[NaOH]{H_2O_2} R_3COH$$

An excellent review of the organoborane-carbon monoxide reaction is available [29]. It may be employed not only to synthesize tertiary alcohols, but primary and secondary ones as well. The former, in which a single alkyl group has been transferred from boron to carbon, may be obtained by carrying out the first step with lithium trimethoxyaluminohydride and the second with aqueous NaOH rather than NaOH and H_2O_2. The secondary alcohol may be obtained by simply using an equimolar quantity of water in the first step.

The reaction has been utilized in the synthesis of acyclic and cyclic mixed carbinols [30] and highly branched carbinols [31].

The organoborane-carbon monoxide reaction has been modified by substituting chlorodifluoromethane for carbon monoxide [32]:

$$Bu_3B + HCClF_2 \xrightarrow[2)AcONa-H_2O_2]{1)LiOEt_3-THF} Bu_3COH$$

98%

The method has the advantage for readily isomerized organoboranes in that it proceeds more readily at lower temperatures than when carbon monoxide is employed.

A second modification to the organoborane-carbon monoxide reaction is that of Pelter and co-workers [33], who first formed the trialkylcyanoborate, which with trifluoroacetic anhydride followed by alkaline peroxide gave the trialkylcarbinol as shown:

$$R_3B \xrightarrow{NaCN} R_3\overset{\ominus}{B}CN\overset{\oplus}{Na} \xrightarrow[2)H_2O_2-\overset{\ominus}{OH}]{1)(CF_3CO)_2O} R_3COH$$

73-79%

(based on olefin)

This method is a convenient one which proceeds at low temperatures under homogeneous conditions. The probable mechanism follows:

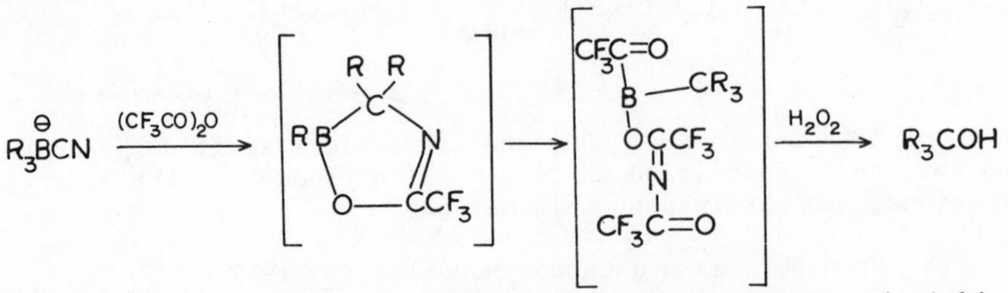

Finally, α,α-dichloromethyl methyl ether, DCME, and lithium triethylcarboxide were introduced as a reagent to convert organoboranes into tertiary carbinols [34] (Ex. a).

$$R_3B \ + \ CHCl_2OCH_3 \ + \ LiOCEt_3 \ \xrightarrow[\text{2)NaOH-H}_2\text{O}_2]{\text{1) THF,25}°} R_3COH$$

This method is a low-temperature one giving yields (glpc) over 90% when the alkyl groups in the tertiary carbinol are straight chain. In addition, it gives yields more satisfactory than the chlorodifluoromethane method for tertiary carbinols containing highly-branched alkyl groups.

a. Preparation of 2,3,3-Trimethyl-4-(n-pentyl)-4-nonanol

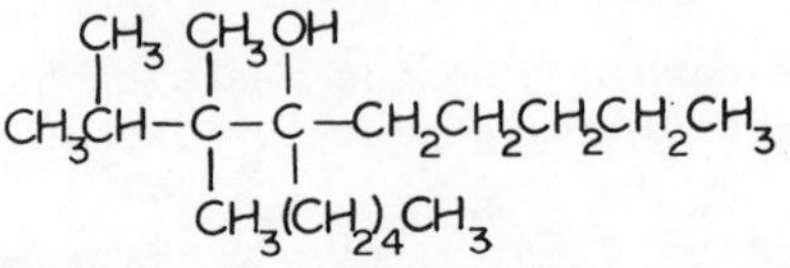

[34]

Thexyl borane, 25.2 ml of a 1.98M solution in THF at 0° was treated with 100 mmol of 1-pentene. To ensure completion of the hydroboration, the mixture was stirred an additional hour at 0°. Then over a 10-min period, DCME, 5.0 ml, 55 mmol, was added, followed by 50 mmol of lithium triethylcarboxide and 27 ml of a 1.84M solution in hexane. At room temperatue 50 ml of 95% ethanol was added and then 12 g of NaOH. Slow addition of 40 ml of 30% hydrogen peroxide at 0° followed by warming to 50-60° for 1 hr completed the oxidation. After salting the aqueous phase with NaCl, the organic phase separated. On evaporation of solvents, the alcohol, 80%, boiling point 114°/0.2 mm, was recovered by distillation under reduced pressure.

7. From Cyclic Ethers or Some Aldehydes and Arenes
    (Friedel-Crafts) (1, 191)

$$ArH + CH_2\!\!-\!\!CH_2 \ \xrightarrow{\ AlCl_3\ }\ ArCH_2CH_2OH$$

In a reinvestigation of the benzene-propylene oxide reaction under strictly anhydrous conditions Milstein [35] found, in contrast to previous investigations in which 1-phenyl-2-propanol was reported, that only 2-phenylpropanol was formed plus a trace of a high-boiling compound:

$$C_6H_6 + CH_2\!\!-\!\!CHCH_3 \ \xrightarrow{\ AlCl_3\ }\ C_6H_5\underset{\underset{CH_3}{|}}{C}HCH_2OH$$

64%

In a more extensive study Nakamoto and co-workers [36] examined the Friedel-Crafts alkylation of benzene with some oxiranes and oxetanes. Mixtures of products were obtained in most cases and benzylethylene oxide and cyclohexene oxide did not give Friedel-Crafts reaction products. For unsymmetrical cyclic ethers, the bond cleavage occurred exclusively between the secondary (or tertiary) C and the O atom.

The Friedel-Crafts reaction of benzene and chloral has now been shown to yield reliably 60% β,β,β-trichloro-α-hydroxy-ethylbenzene [37].

Addendum

B.2. When the strongly alkaline conditions for oxidizing boranes to alcohols cannot be used, it is recommended that trimethyl-amine-N-oxide be employed [G. W. Kabalka, H. C. Hedgecock, Jr., J. Org. Chem., 40, 1776 (1975)].

References

1. E. I. Klimova et al., J. Org. Chem. USSR, 5, 1308, 1315 (1969).
2. C. Freppel et al., Can. J. Chem., 49, 2586 (1971).
3. A. DeRoocker, P. de Radzitzky, Bull. Soc. Chim. Belges, 79, 531 (1971).
4. A. F. Thomas, M. Ozainne, J. Chem. Soc., C220 (1970).
5. H. C. Brown et al., J. Am. Chem. Soc., 93, 1024 (1971).
6. G. Zweifel, H. Arzoumanian, J. Am. Chem. Soc., 89, 291 (1967).
7. H. C. Brown, N. Ravindran, J. Org. Chem., 38, 182 (1973).
8. C. G. Scouten, H. C. Brown, J. Org. Chem., 38, 4092 (1973).

9. C. F. Lane, H. C. Brown, J. Am. Chem. Soc., 93, 1025 (1971).
10. H. C. Brown, C. F. Lane, Synthesis, 303 (1972).
11. H. C. Brown et al., Synthesis, 304 (1972).
12. N. Miyamoto et al., Tetrahedron Letters, 4597 (1971).
13. H. C. Brown, Y. Yamamoto, Synthesis, 699 (1972).
14. J. Klein, A. Medik, J. Am. Chem. Soc., 93, 6313 (1971).
15. H. C. Brown, S. P. Rhodes, J. Am. Chem. Soc., 91, 4306 (1969).
16. B. Rickborn, S. E. Wood, J. Am. Chem. Soc., 93, 3940 (1971).
17. J.-J. Barieux, J. Gore, Bull. Soc. Chim. France, 1649 (1971).
18. H. G. Khorana, Chem. Eng. News (Sept. 17, 1973), p. 18.
19. H. C. Brown et al., J. Am. Chem. Soc., 93, 2792 (1971).
20. M. Zaidlewicz, I. Uzarewicz, Rocz. Chem., 43, 937 (1969); C. A., 71, 70167 (1969).
21. J. Klein et al., Tetrahedron Letters, 2845 (1972).
22. V. Hach, Synthesis, 340 (1974).
23. H. Arzoumanian, J. Metzger, Synthesis, 527 (1971).
24. H. C. Brown, P. J. Geoghegan, Jr., J. Org. Chem., 35, 1844 (1970).
25. R. M. Carlson, A. H. Funk, Tetrahedron Letters, 3661 (1971).
26. H. C. Brown et al., Organomet. Chem. Syn., 1, 7 (1970/1971).
27. P. R. Stapp, J. Org. Chem., 35, 2419 (1970).
28. T. Mukaiyama et al., Chem. Letters, 1011 (1973).
29. H. C. Brown, Acc. Chem. Res., 2, 65 (1969).
30. H. C. Brown et al., J. Am. Chem. Soc., 92, 6648 (1970).
31. E. Negishi, H. C. Brown, Synthesis, 197 (1972).
32. H. C. Brown et al., J. Am. Chem. Soc., 93, 2070 (1971).
33. A. Pelter et al., Chem. Commun., 1048 (1971).
34. H. C. Brown et al., J. Org. Chem., 38, 2422, 3968 (1973).
35. N. Milstein, J. Heterocyclic Chem., 5, 337 (1968).
36. Y. Nakamoto et al., Kogyo Kagaku Zasshi, 72, 2594 (1969); C. A., 72, 100192 (1970).
37. W. Reeve, Synthesis, 131 (1971).

C. Reduction

1. From Organic Oxygen Compounds and Metal Hydrides (1, 193)

The most complete general reference on metal hydride reduction is that of Augustine [1]. Of the solid reducing agents, the metal hydrides are by far the most common. Although lithium aluminum hydride and sodium borohydride are the most widely used, many other hydrides are now available. For our

purpose we shall discuss them on the basis of the substrate
attacked, in other words, under the headings: (a) aldehydes and
ketones, (b) unsaturated ketones, and (c) acids, esters, and
salts.

In recent years aldehydes and ketones have been reduced to
alcohols by the use of comparatively new reagents such as lithium
and sodium cyanohydridoborate, $LiBH_3CN$ and $NaBH_3CN$ [2], sodium
bis-(2-methoxyethoxy) aluminum hydride, $NaAlH_2(OCH_2CH_2OCH_3)_2$ [3],
and sulfurated sodium borohydride, $NaBH_2S_3$ [4]. Of these four
types, $NaBH_3CN$ or $LiBH_3CN$ is effective in reducing not only
aldehydes and ketones but many other types as well (Ex. a). For
best results the pH is held at 3-4, preferably in a protic solvent such as methanol at 25°. Yields of alcohol cover the range
of 67-93%. Sodium bis-(2-methoxyethoxy) aluminum hydride differs
from most others in being soluble in aromatic hydrocarbons and in
being very stable in air. Yields of alcohols from the carbonyl
compounds tested vary within 90-97% (glc) (Ex. b) if the
sterically hindered 2,4,6-trimethylacetophenone is omitted from
consideration. As a rule the double bond in α,β-unsaturated
aldehydes and ketones are not affected by this reducing agent.
In the reduction of alkyl cyclohexanones the trans alcohol is
usually formed in excess, although the ratio between the cis and
trans forms varies with the temperature. The sulfurated sodium
borohydride leads to quantitative yields of alcohols only at low
temperatures (disulfides are produced at high temperatures) and
with nonhindered ketones the mol ratio of reagent to substrate
must be 1:1. Even so, very hindered ketones are not affected.
By using a chiral aluminum hydride and a ketone Giongo and co-
workers [5] achieved an asymmetric synthesis of a secondary
alcohol:

Phenylmethylcarbinol, 95%
$[\alpha]_{589m\mu}^{25}$ $-12.04°$

The reducing agent was prepared from $LiAlH_4$ and N-methyl-N-
phenethylamine hydrochloride. Lesser optical activity of the
alcohol was noted by using the α-phenethylamine-borane complex in
the reduction [6].

Hydrosilylation is preferable to catalytic hydrogenation if
the alcohol is susceptible to hydrogenolysis [7]:

$$R_2CO \; + \; R_2'SiH_2 \; \xrightarrow{\text{RhCl}\left[P(C_6H_5)_3\right]_3} \; R_2CHOSiHR_2' \; \xrightarrow{H_3O^\oplus} R_2CHOH$$

If a chiral ligand is employed, an optically active alcohol is obtained.

Ketones are considered here alone not because aldehydes would not be affected as well under the conditions stated, but because many experiments have been conducted on the selective reduction of unsaturated ketones. In one of these Brown and Hess [8] were successful in reducing unsaturated ketones with AlH$_3$ without affecting the doubly bound carbon atoms. Thus 3-methyl-2-cyclopentenone gives 3-methyl-2-cyclopentenol (Ex. c).

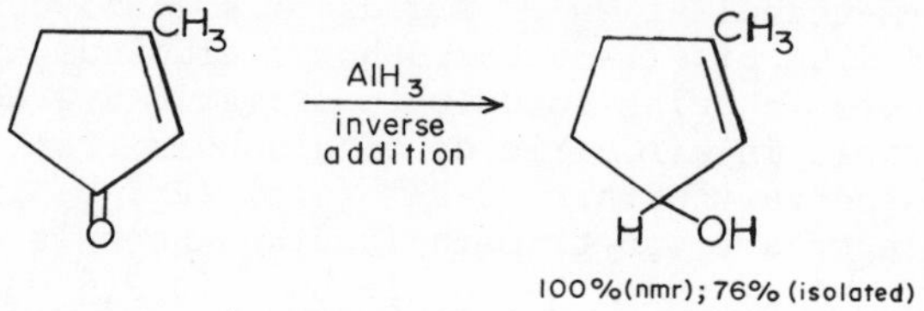

100%(nmr); 76% (isolated)

Wilson and co-workers recommend di-isobutylaluminum hydride [9] for the selective reduction of 2-ene-1,4-diones and 2-enones. They found this hydride to be superior to AlH$_3$ in the reduction of 2-cyclopentenone to the unsaturated alcohol (99% vs 90%) (glc).

Alumina treated with NaBH$_4$ is effective in reducing ketones containing alkali-sensitive functional groups in nonpolar solvents or in the solid state [10]. Thus 4-androstene-3, 17-dione, 4 g was converted stepwise into testosterone, 450 mg, as indicated:

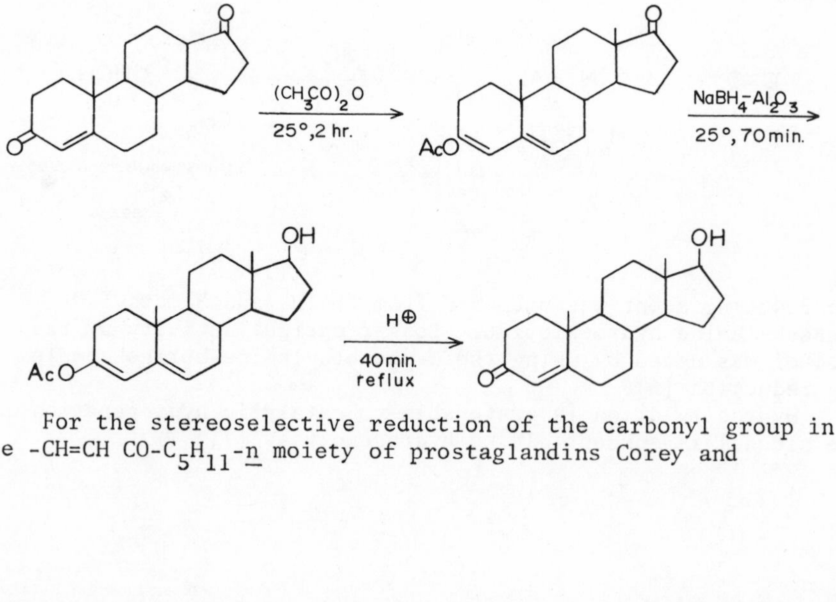

For the stereoselective reduction of the carbonyl group in the -CH=CH CO-C$_5$H$_{11}$-$\underline{n}$ moiety of prostaglandins Corey and co-workers [11] employed the borohydride ion obtained by treating racemic or (+)-limonene and thexyl borane as shown:

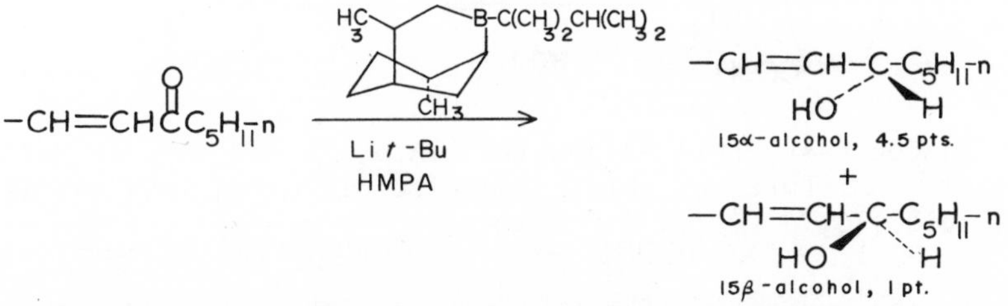

Lithium tri-sec-butylborohydride offers enzyme-like specificity in the reduction of ketones [12]. With this reagent 2-methylcyclopentanone gives cis-2-methylcyclopentanol 98% (0°), 99.3% (-78°) (Ex. d). In contrast, lithium trimethoxyaluminohydride at 0° gives only 69% of the cis alcohol. Fortunately the new reagent may be prepared in satisfactory form by simply mixing LiAlH (OMe)$_3$ and sec-Bu$_3$ B, both of which are available commercially.

To reduce cyclanones stereospecifically potassium triisopropoxyborohydride appears to be the reagent of choice [13]. For example, with 2-methylcyclohexanone it gives the less stable cis-2-methylcyclohexanol (92%):

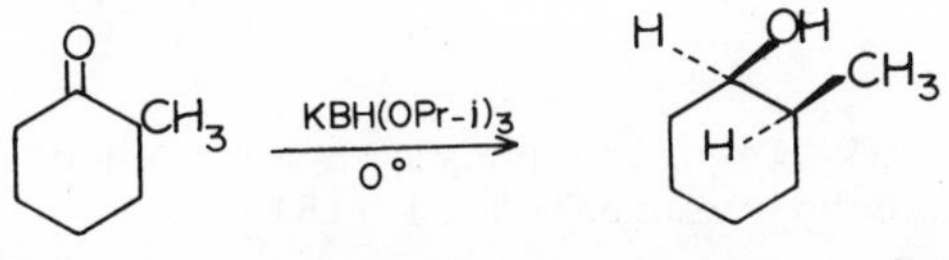

By contrast, NaBH$_4$ gives 31% and LiAlH(OBu-t)$_3$, 27% of the same geometrical isomer at the same temperature.

Aliphatic or aromatic carboxylic acids may be reduced rapidly and quantitatively to the corresponding alcohols by borane in THF [14]. Thus adipic acid leads to 1,6-hexanediol (100%) (Ex. e):

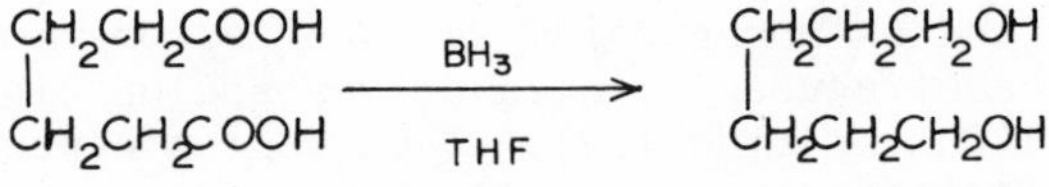

The reagent offers these advantages over LiAlH$_4$ or AlH$_3$:   (a) the reaction is rapid and quantitative; (b) the stoichiometric quantity of borane brings the reaction to completion under mild conditions; and (c) the presence of functional groups such as nitro, halogen (alkyl and aryl), nitrile, ester, epoxide, and tosylate is tolerated.

At about the same time a process was patented for converting camphoric acid into 1,2,2-trimethyl-1,3-bis(methylol)cyclopentane by passing borane through a methyl borate-THF solution of the acid [15].

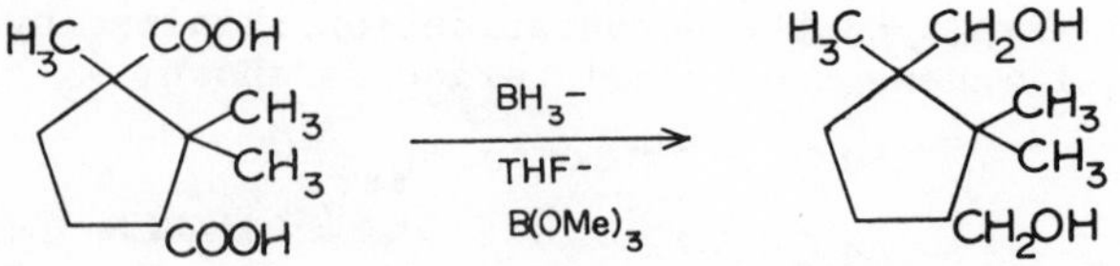

92 %

In contrast, thio acids with LiAlH$_4$, LiAlH$_4$-BF, or NaBH$_4$-AlCl$_3$ give mixtures of thiols and alcohols often in a nearly 1:1 mol ratio [16]:

$$2 \; RCOSH \xrightarrow{LiAlH_4} RCH_2OH \; + \; RCH_2SH$$

In the reduction of methyl esters with sodium trimethoxyborohydride Bell and Gravestock [17] found that the rate of reduction decreases in the order primary > secondary > tertiary. Thus it is possible to reduce one type of ester group in the presence of another. For example, the diester-acid A with sodium trimethoxyborohydride gave the ester-acid-alcohol B as shown:

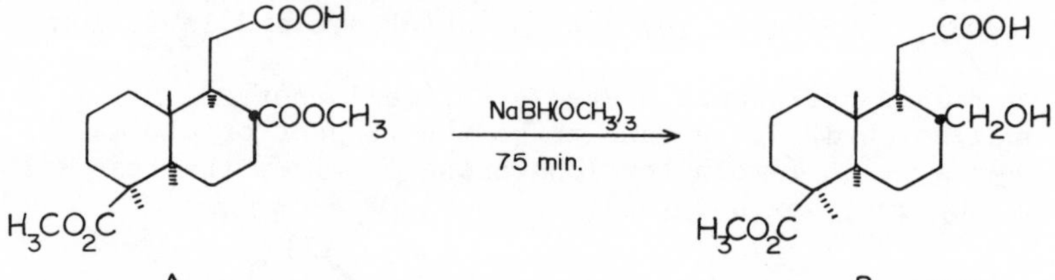

A   B   >95 %

1-Cyclopentene-1,3-diethylcarboxylate was reduced to the unsaturated dialcohol as indicated [18]:

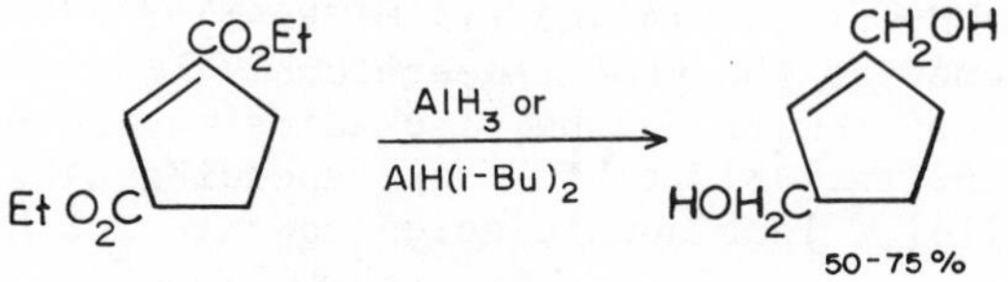

50-75 %

Sodium and bromomagnesium salts of aliphatic and aromatic acids have been reduced to alcohols by sodium bis (2-methoxyethoxy) aluminum hydride [19]. Thus sodium benzoate gives benzyl alcohol as indicated:

$$C_6H_5COONa \xrightarrow[2) H_3O \oplus]{1) NaAl(OCH_2CH_2OCH_3)_2H_2} C_6H_5CH_2OH$$

100 % (glc)

The results compare favorably with those obtained with the free acid. As the authors have shown, in two cases LiAlH$_4$ works equally well.

a. Preparation of Cyclohexanol [2]

Cyclohexanone, 4 mmol, and a trace of bromocresol green in 6 ml of methanol was treated with 2 mmol of NaBH$_3$CN. The solution turned deep blue and 2 N HCl was added carefully to restore the yellow color. The solution was then stirred for about 1 hr while 1 drop of acid was added occasionally to restore the yellow color. From the methanol there was recovered 350 mg (88%) of the alcohol.

b. Preparation of 1-Octanol [3]

Octanal, 50 mmol was stirred and cooled while adding over 2-4 min 21 mmol of sodium bis-(2-methoxyethoxy) aluminum hydride from a 25% solution in benzene. After stirring 1 hr, the mixture was decomposed with 10% HCl and the aqueous layer was washed with ether, from which solution the alcohol (97%) was recovered.

c. Preparation of 3-Methyl-2-cyclopentenol [8]

3-Methyl-2-cyclopentenone, 10 g in THF under N$_2$ at 0° was treated with 90 ml of a 0.77M AlH$_3$ solution in THF over a 15-min period. The reaction mixture was then stirred at 0° for an additional 30 min, after which it was hydrolyzed with aqueous NaOH. From the combined organic layer and ethereal extract of the aqueous layer there was recovered 7.8 g (76%) of the carbinol.

d. Preparation of cis-2-Methylcyclopentanol [12]

To 5.0 ml of 1.0M LiAl(OCH$_3$)$_3$H in THF under N$_2$ was added 1.25 ml of tri-sec-butylborane (5.0 mmol). After 30 min the mixture was treated at -78° with 1.25 ml of a 2.0M solution of 2-methylcyclopentanone (2.5 mmol). Stirring for 3 hr was followed by hydrolysis and oxidation to give 99.3% of the alcohol.

e. Preparation of 1,6-Hexanediol [14]

Adipic acid, 3.65 g, in 15 ml of THF was cooled to 0° and to the mixture was added 27 ml of 2.39M borane solution in THF gradually. After stirring for 6 hr at 25°, the excess hydride was destroyed carefully with 15 ml of a 1:1 mixture of THF and water. From the organic and aqueous phases there was recovered 3.0 g (100%) of pure 1,6-hexanediol.

2. From Cyclic Ethers ($\underline{1}$, 196)

$$RCH\!-\!CH_2 \xrightarrow{\text{LiAlH}_4} RCH_2CH_2OH \quad + \quad RCHOHCH_3$$
$$\diagdown\!\!_{O}\!\!\diagup$$

Cyclic ethers have been cleaved to give alcohols by metal hydrides, borane organometallic compounds, hydrogenation, and sodium in liquid ammonia. Lithium tri-tert-butoxyaluminohydride-triethylborane proved to be unusually effective. With cyclohexene oxide, for example, it gives cyclohexanol (100%) [20].

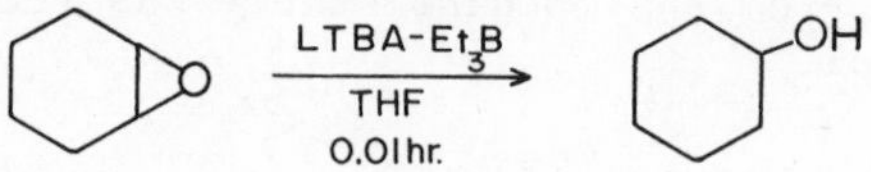

Lithium-ethylenediamine was found to be preferable to lithium-ethylamine, particularly in the cleavage of bicyclic epoxides. Thus norbornene oxide gives norborneol in 87% yield using the former as compared to 64% by the use of the latter [21]:

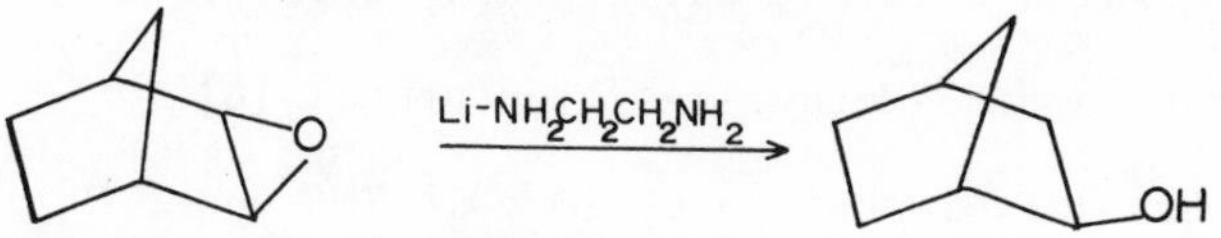

Trimethylaluminum has been utilized to cleave the epoxy ring in cis- and trans-ethyl 3-phenylglycidates [22]. Two products, threo- and erythro-butyrates, were obtained in each case, the percentages of which were different with the cis and trans.

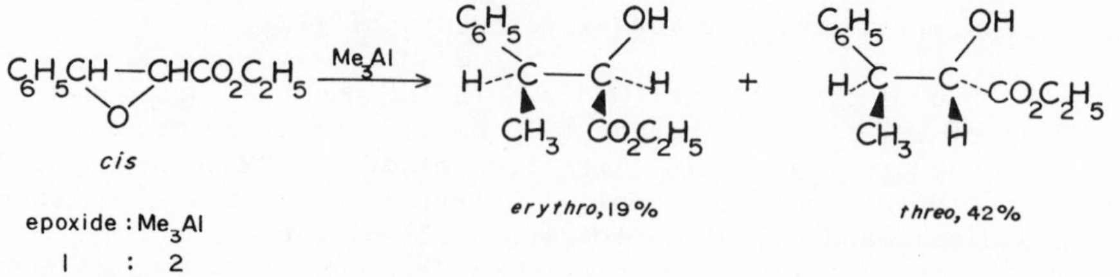

The trans form on the other hand gave 66% of the erythro and 22% of the threo form.

Both aliphatic and aromatic epoxides may be cleaved to alcohols by the use of alkali metals in liquid ammonia [23]. For example, styrene oxide gives 2-phenylethanol in 83% yield:

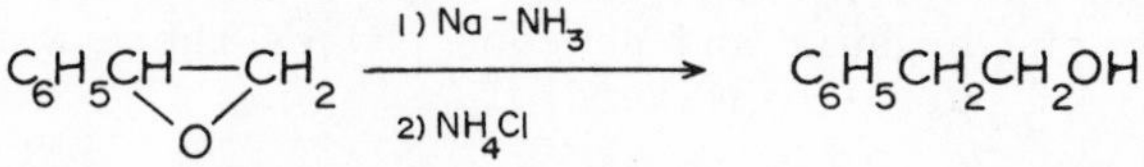

It is of interest to note that $LiAlH_4$ yields 1-phenylethanol from the same epoxide. The first observation suggests that the reduction here proceeds via the 1,3-dianion, $C_6H_5\overline{C}HCH_2\overline{O}$.

Sodium in liquid ammonia was also employed in the synthesis

of <u>cis</u>-4β, 10β-dimethyl-6-isopropylidenedecal-7β-ol from the epoxynitrile as shown [24]:

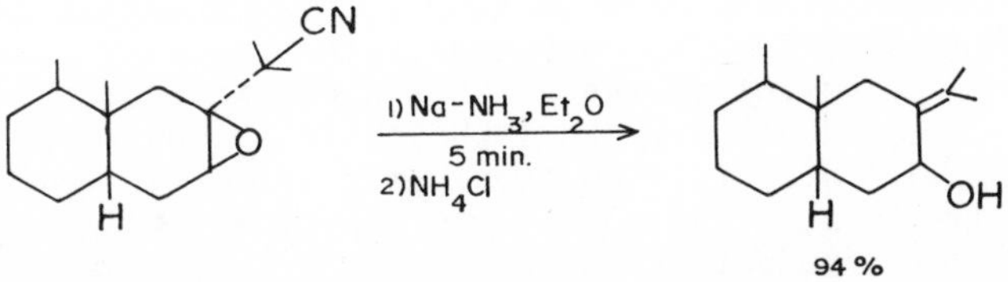

94 %

1-Heptene has been converted into 1-heptanol via the epoxide as indicated [25]:

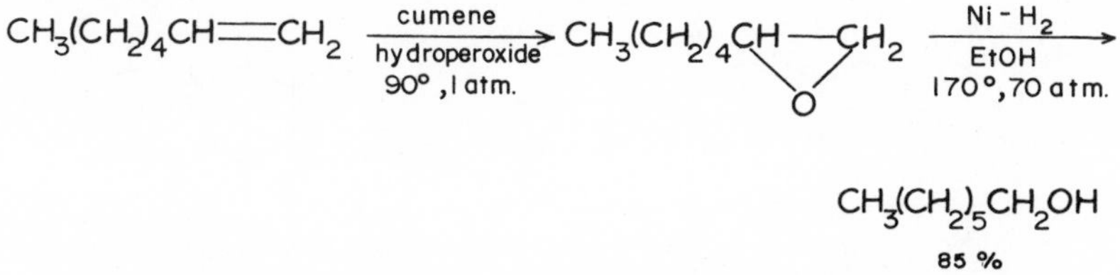

$$CH_3(CH_2)_4CH=\!\!=\!\!CH_2 \xrightarrow[\substack{hydroperoxide \\ 90°,1\,atm.}]{cumene} CH_3(CH_2)_4CH\!-\!CH_2 \xrightarrow[\substack{EtOH \\ 170°,70\,atm.}]{Ni-H_2}$$

$$CH_3(CH_2)_5CH_2OH$$

85 %

3. From Carbonyl Compounds and Other Reducing Agents (1, 197)

Reducing agents other than the metal hydrides (C.1) and hydrogen, in the presence of a catalyst (C.6), have been utilized in great variety. A limited number is discussed here.

Among the metals, nickel, zinc, and potassium on graphite have found use recently. Thus precipitated nickel was employed to reduce benzaldehyde to benzyl alcohol (>90% glc) [26], zinc dust served for converting benzil into benzoin (93%) [27], and potassium on graphite gave cycloheptanol (92%) from cyclohepta-none [28]. Hydrazine, diimide and thiourea dioxide have been of value. 7-Methylenebicyclo [3.3.1] nonan-3-one was reduced to the hydroxynoradamantane:

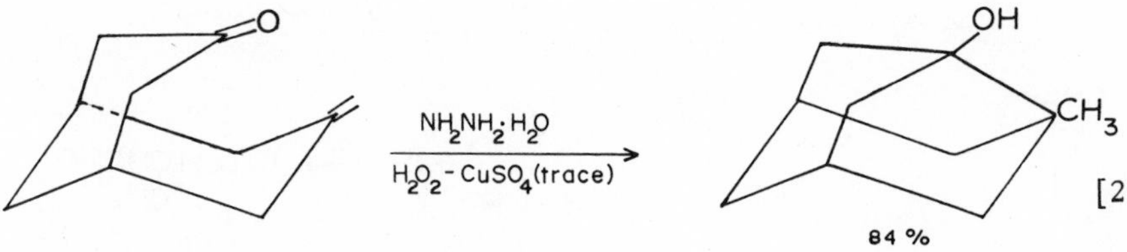

84 %

[29]

Similarly, nitrofluorenones were reduced to aminofluorenols [30]. Aromatic aldehydes with diimide from potassium azodicarboxylate gave alcohols usually in 62-84% (glc) yields [31]. Ketones, when reduced with thiourea dioxide, give yields from 74.2 to 100% (glc).

$$\text{RCOR} \xrightarrow[\underset{\ominus}{\overset{OH}{}}]{(NH_2)_2C=SO_2} \text{RCHOHR}$$

[32]

Complexes of iridium compounds with trimethyl phosphite have been extremely useful in the synthesis of axial alcohols. Thus chloroiridic acid and trimethyl phosphite in 90% aqueous isopropyl alcohol (Henbert reduction) give selective reduction of 3-oxo groups in steroids [33]. Oxo functions at C-6, C-11, C-12, C-17, and C-20 are not affected. Thus 5α-androstane-3,17-dione led to the 3α-ol.

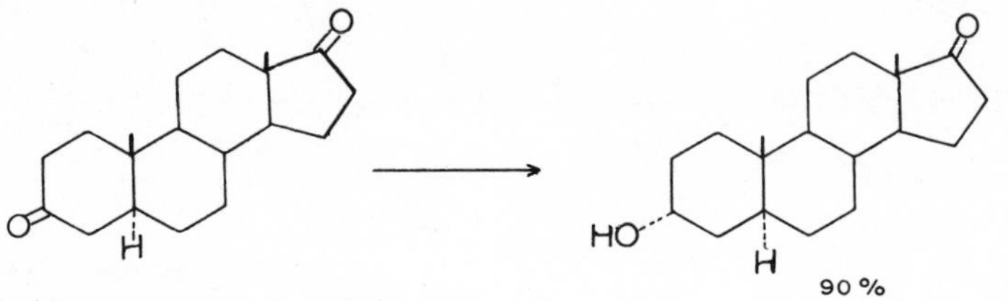

90%

In a similar manner 4-t-butylcyclohexanone gives cis-4-t-butyl-cyclohexanol [34] (Ex. a).

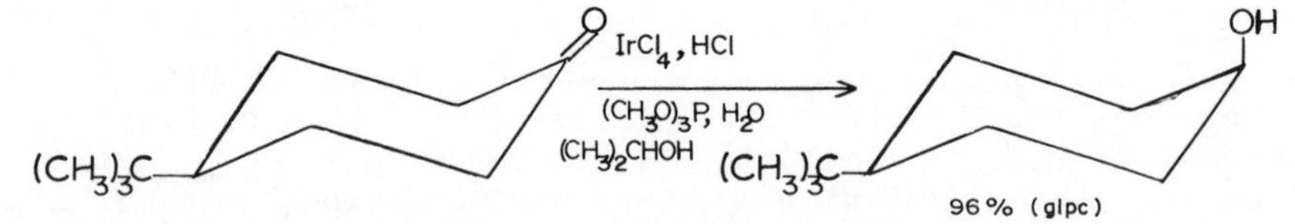

96% (glpc)

And diethyl phosphite reduces ketones to form phosphoric acid esters that may be saponified to the alcohols [35].

Although the use of the Meerwein-Ponndorf-Verley reduction of aldehydes and ketones has declined in recent years, it may be used to prepare optically active alcohols as shown [36]:

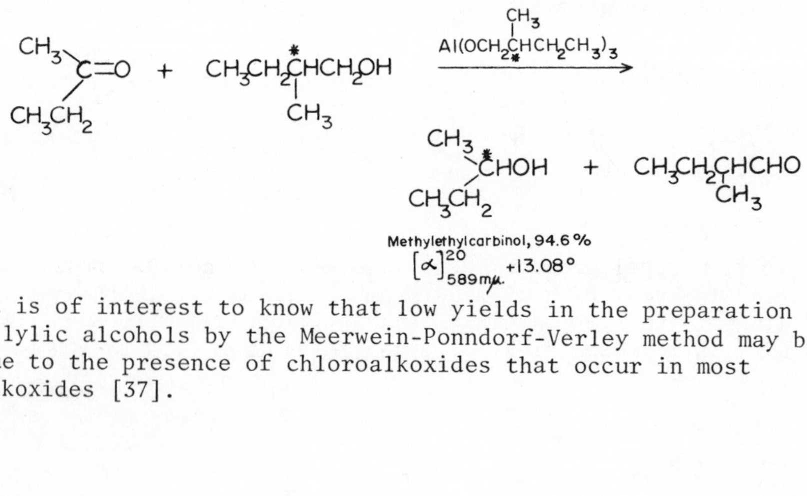

Methylethylcarbinol, 94.6%
$[\alpha]_{589m\mu}^{20} +13.08°$

It is of interest to know that low yields in the preparation of allylic alcohols by the Meerwein-Ponndorf-Verley method may be due to the presence of chloroalkoxides that occur in most alkoxides [37].

A new method for the reduction of quinones to hydroquinones and α-diketones to α-hydroxyketones consists of heating with benzpinacol [38]. Thus benzil responds to give benzoin (85%) plus benzophenone (Ex. b). The method is extremely simple and as a rule yields are good. Above 100° the benzpinacol forms the benzhydryl free radical, which serves as the reducing agent.

Although the Birch reduction of acetophenone gives ethylbenzene, lithium-methylamine reduction gives mainly methylcyclohexenylcarbinol,

or an isomer [39].

a. Preparation of cis-4-t-Butylcyclohexanol [34]

To a solution of 4 g of iridium tetrachloride in 4.5 ml of concentrated HCl were added 180 ml of water and 50 ml of trimethyl phosphite. The solution was added to a solution of 30.8 g of 4-t-butylcyclohexanone in 635 ml of isopropyl alcohol, after which the mixture was refluxed for 48 hr. On removal of the alcohol by evaporation the remaining solution was diluted with water and extracted with ether, from which extract there was obtained 29-31 g (93-99%) of the cis alcohol.

b. Preparation of Benzoin [38]

A mixture of 210 mg of benzil, 700 mg of benzpinacol, and 5 ml of decalin was immersed in an oil bath at 165°. After 15 min the solution was cooled; filtration gave 176 mg (85%) of benzoin.

6. From Organic Oxygen Compounds and Hydrogen with Catalysts (1, 201)

Although catalytic hydrogenation has certain advantages, as pointed out in 1, 201, it is no longer used as widely as formerly. However there are cases as shown in which the method is a desired one:

1. To reduce unsaturated aldehydes without affecting the doubly bonded carbon atoms [40].

$$C_6H_5CH=CHCHO + H_2 \xrightarrow[\substack{750-1000\,psig.\\100°\\CH_3CHOHCH_3}]{5\% \; Os/C} C_6H_5CH=CHCH_2OH$$

Cinnamyl alcohol, 95%

2.   To convert nitriles into alcohols in one operation [41].

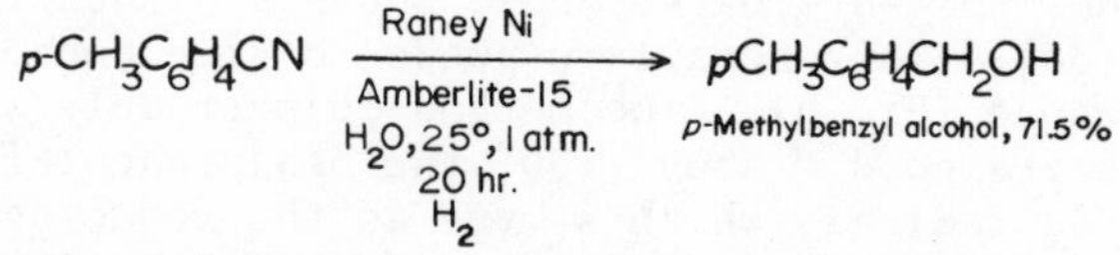

$$p\text{-}CH_3C_6H_4CN \xrightarrow[\substack{\text{Amberlite-15} \\ \text{H}_2\text{O},25°,1\,\text{atm.} \\ 20\,\text{hr.} \\ \text{H}_2}]{\text{Raney Ni}} p\text{-}CH_3C_6H_4CH_2OH$$

p-Methylbenzyl alcohol, 71.5%

No amine is reported in this case, but unfortunately it is produced in the other examples studied.

3.   To reduce aldehydes by homogeneous hydrogenation under
mild conditions [42].

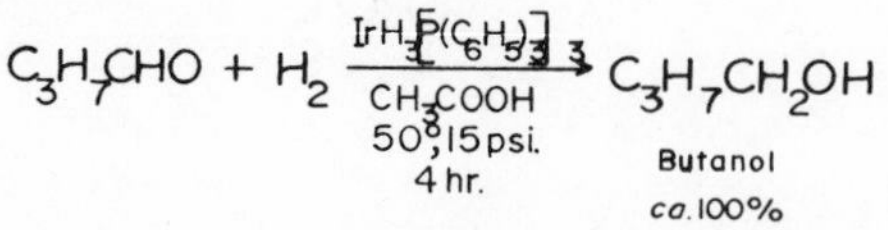

$$C_3H_7CHO + H_2 \xrightarrow[\substack{\text{CH}_3\text{COOH} \\ 50°,15\,\text{psi.} \\ 4\,\text{hr.}}]{[IrH_3P(C_6H_5)_3]_3} C_3H_7CH_2OH$$

Butanol
ca. 100%

In connection with the preparation of Raney nickel
catalysts, it should be noted that the demineralization may be
hastened by the use of an ion-exchange resin such as amberlite
IR 120 [43].

8.   From Carbonyl Compounds, Esters, or Acids (Mostly
      Bimolecular Reduction (1, 204)

Pinacol Reaction

It is claimed that aliphatic and cycloaliphatic ketones were
reduced under mild conditions in yields superior to those
obtained previously. The method consists essentially of adding
HgCl$_2$ to aluminum foil, washing the metal with dichloromethane or
THF, then adding the solvent and the ketone, and bringing the
mixture to reflux [44]. Thus cyclohexanone gives the corresponding pinacol:

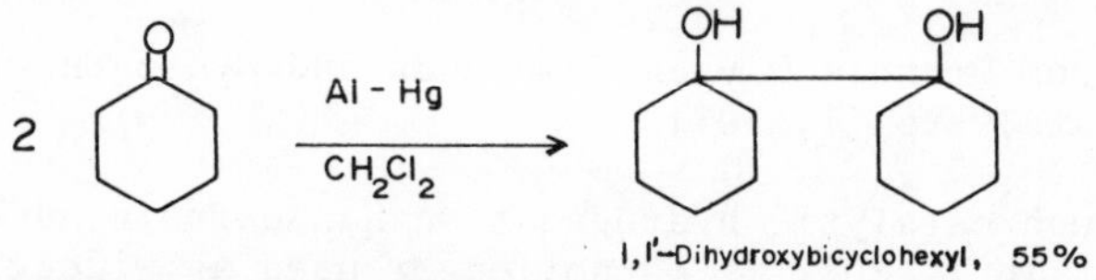

1,1'-Dihydroxybicyclohexyl, 55%

Yields from other ketones were somewhat lower. In all cases the
dihydric alcohol formed was largely the dl form.

In a novel pinacol formation Chan and Vinokur [45] formed
the pinacol via the trimethylsilane as shown:

$$C_6H_5\overset{O}{\underset{}{C}}CH_3 \xrightarrow[\text{HMPA}]{\text{(CH}_3)_3\text{SiCl-Mg}} C_6H_5\overset{CH_3}{\underset{(CH_3)_3SiO}{C}}-\overset{OSi(CH_3)_3}{\underset{CH_3}{C}}C_6H_5 \xrightarrow{\text{aq NH}_4\text{Cl}} C_6H_5\overset{CH_3}{\underset{OH}{C}}-\overset{OH}{\underset{CH_3}{C}}C_6H_5$$

2,3-Diphenyl-2,3-butanediol

88%

dl,meso ratio 1:1

Benzaldehyde with similar treatment gave a 3:1 mixture of dl- and meso 1,2-diphenyl-1,2-ethanediols in 90% yield. Corey and Carney [46], who had previously employed magnesium and dimethyldichlorosilane in the synthesis of cis- and trans bicyclic diols, point out the importance of the chlorosilane in eliminating undesirable side reactions.

Pinacols may also be obtained from carboxylic acids by treatment with alkyllithium compounds and TiCl$_3$ in 1,2-dimethoxyethane [47]. Thus benzoic acid in the molar amounts shown responds to give three principal products:

$$2\ C_6H_5COOH + TiCl_3 + 5MeLi \longrightarrow$$

$$C_6H_5COCH_3 + C_6H_5\overset{CH_3\ CH_3}{\underset{OH\ OH}{C}-C}C_6H_5 + C_6H_5\overset{O}{\underset{CH_3}{CH}C}C_6H_5$$

2,3-Diphenylbutane-2,3-

diol,33%, *dl,meso* ratio 2.4:1

Although the yield is low, separation of the diol may be accomplished by chromatography on silica gel.

Benzoin Condensation

The organic-inorganic contact in the benzoin condensation may be improved by using tetrabutylammonium cyanide rather than sodium cyanide as the catalyst [48]. Thus sodium cyanide is effective at 90° in 50% CH$_3$OH, but it is rather ineffective in water alone at room temperature. On the other hand, tetrabutylammonium cyanide under the latter conditions gives a 70% yield of benzoin from benzaldehyde:

$$2\ C_6H_5CHO \xrightarrow[\text{H}_2\text{O, 25°}]{\text{(C}_4\text{H}_9)_4\text{NCN}} C_6H_5COCHOHC_6H_5$$

Acyloin Condensation

An excellent review on the reactions of esters with sodium in the presence of trimethylchlorosilane is now available [49]. In the presence of the silane the reaction mixture is kept neutral and thus base-catalyzed side reactions, such as β-elimination and Claisen or Dieckmann condensations, are prevented. However trimethylchlorosilane has no effect on side reactions proceeding via a free radical mechanism.

By trapping the enediol as a bis(siloxy)alkene, yields of the acyloin are increased and the scope of the original method may be extended considerably. The bis(siloxy)alkenes may be isolated and stored satisfactorily under nonhydrolytic conditions or converted into acyloins by acid hydrolysis or alcoholysis. The two-step reaction may be represented as follows (Ex. a):

$$2\ R'COOR^2 \xrightarrow[4\,ClSi(CH_3)_3]{4\,Na\,+} \begin{matrix} R'COSi(CH_3)_3 \\ \| \\ R'COSi(CH_3)_3 \\ \text{56-91\%} \end{matrix} \xrightarrow{H_3O^{\oplus}} \underset{\text{58-99\%}}{R'\overset{O}{\overset{\|}{C}}CHOHR'}$$

$$+$$
$$2\ R^2OSi(CH_3)_3 + 4\ NaCl$$

See Ketones, H.8, for an alternate preparation of acyloins. Mixed acyloins are formed by the reaction of ketones with the salt of a cyanoacetal [G. L. A. Maldonado, Rev. Soc. Quim. Mex., 16, 200 (1972)]; C. A., 78, 15437 (1973)].

a.   Preparation of Propionoin [49]

Toluene, 600 ml, and sodium, 46 g under N_2 were heated and the metal was broken up by stirring vigorously. After cooling the toluene was decanted off and the sodium sand was washed with ether (300 ml). Then 600-800 ml of ether and 217.3 g of trimethylchlorosilane were added and 1 mol of ethyl propionate was dropped into the mixture while stirring. The mixture was permitted to reflux until the sodium had disappeared, after which the filtrate was fractionally distilled to give a 91% yield of the bis-(siloxy)alkene. The latter on hydrolysis with N HCl and THF gave a 90% yield of the acyloin.

10.   From Trityl Ethers (W.-K. Reduction)

The trityl group can be easily introduced into most primary and some secondary alcohols to form ethers, stable to many reagents. If the trityl group contains a keto group as in 9-phenyl-9-hydroxyanthrone, its ether becomes stable to most base-catalyzed reactions but is cleaved by the W.-K. reduction. Thus alcohols may be recovered from these particular ethers, as indicated in the case of n-hexadecanyl tritylone ether [50]:

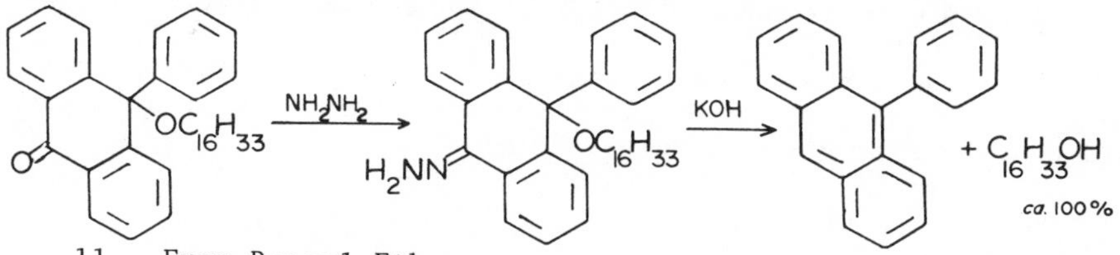

11.   From Benzyl Ethers

$$C_6H_5CH_2OR \xrightarrow{[H]} C_6H_5CH_3 + ROH$$

The hydrogenolysis of benzyl ethers has been reviewed [51]. The process has found two distinct uses. It has been employed in removing the benzyl group introduced at times to protect the alcoholic group and in synthesizing various compounds containing the benzyl group. The hydrogenolysis, which is usually conducted with hydrogen and a metal catalyst, proceeds smoothly often in quantitative yield. Nickel or platinum may be used as the catalyst, although palladium is preferred if hydrogenation of the nucleus is to be prevented.

In an early experiment butyl benzyl ether was converted into butanol and toluene [52]:

$$C_4H_9OCH_2C_6H_5 \xrightarrow[\substack{150-250atm. \\ 175°}]{\substack{H_2 \\ Raney\ Ni}} C_4H_9OH + C_6H_5CH_3$$
$$\qquad\qquad\qquad\qquad\qquad\qquad 92\% \qquad 92\%$$

More recently Brown and Gallivan [53] carried out the hydrogenolysis of benzyl 2-hydroxybutyl ether to obtain 1,2-butanediol:

$$\underset{\underset{OH}{|}}{CH_3CH_2CHCH_2OCH_2C_6H_5} \xrightarrow[\substack{(Brown\ and\ Brown \\ apparatus)}]{\substack{H_2 \\ Pd/C}} CH_3CH_2CHOHCH_2OH$$
$$\qquad\qquad\qquad\qquad\qquad\qquad\qquad 83.7\%$$

The yield would have been higher since the product was contaminated by some 1,3-butanediol, produced not by the reaction, but as a result of the BH_3 addition utilized in forming the original alcohol.

Van Bekkum and co-workers [54] showed that the rate of hydrogenolysis of benzyl alkyl ethers to alcohols is enhanced by carrying out the reaction in an acid medium.

12.   From Amides (New)

$$RCONH_2 \xrightarrow{electroreduction} RCH_2OH$$

Amides were reduced to alcohols in an electrolytic cell with platinum electrodes and LiCl in monomethylamine as an electrolyte [55]. Yields from straight-chain aliphatic amides varied from 4 to 97%. In the presence of a proton source such as ethanol aldehydes are produced as well.

Addenda

C.1. The reduction of acids with diborane in methyl sulfide, BMS, which compares favorably with the use of BH_3-THF [14], is described [C. F. Lane et al., J. Org. Chem., 39, 3052 (1974)]. The yields of benzyl alcohols formed are enhanced by the addition of trimethyl borate, which probably forms the more easily reduced mixed anhydride as an intermediate.

C.1a. Lithium tri-sec-butylborohydride is said to be the most stereoselective reducing agent. For instance, with 2-methyl-cyclohexanone it gives the less stable cis alcohol (99.3%) [H. C. Brown, S. Krishnamurthy, J. Am. Chem. Soc., 94, 7159 (1972)].

C.1a. Crown ethers appear to enhance the reduction of ketones by sodium borohydride, although the study was not directed towards a synthetic approach [T. Matsuda, K. Koida, Bull. Chem. Soc. Jap., 46, 2259 (1973)].

C.1a. Substituted 1,3-propanediols may be prepared by the for-mylation of an appropriate enolic ketone with ethyl formate and reduction of the product with aluminum hydride, made from $LiAlH_4$ and an equivalent of 100% H_2SO_4 [E. J. Corey, D. E. Cane, J. Org. Chem., 36, 3070 (1971)].

C.1c. Substituted 1,3 propanediols may be made by the reduction of malonic esters with $LiAlH_4$ [J. A. Marshall et al., J. Org. Chem., 32, 113 (1967)].

C.3. The reduction of cyclic ketones with dicyclohexylborane in diglyme at 0° gives consistently the less stable, cis alcohol [H. C. Brown, V. Varmer, J. Org. Chem., 39, 1631 (1974)]. In addition, with regard to regioselectivity, the reagent $MeS_2BH_2)Li\cdot2(CH_2OCH_3)_2$ has been studied [J. Hooz et al., J. Am. Chem. Soc., 96, 274 (1974)].

C.6. High pressure and high temperature hydrogenation of phenols leads to cyclohexanols. Homogeneous catalysts such as a nickel salt-$(C_2H_5)_3$ Al may be used [R. E. Harmon et al., Chem. Rev., 73,

50 (1973)]. It is pointed out that low-pressure hydrogenation (Parr) may be accomplished with 5% Rh on alumina in ethanol [A. I. Meyers et al., J. Org. Chem., 29, 3427 (1964); Org. Syn., 51, 103 (1971)].

C.8. The pinacols of aldehydes and ketones may be prepared by Zn + TiCl$_4$ at 0° [T. Mukaiyama et al., Chem. Lett., 1041 (1973)].

C.8. The benzoin condensation may be carried out rapidly at room temperature by the phase transfer method [J. Dockx, Synthesis, 451 (1973)].

References

1.   R. L. Augustine, Reduction, Dekker, New York, 1968, pp. 1-94; E. R. H. Wilson, Chem. Soc. Rev., 5, 23 (1976).
2.   R. F. Borch et al., J. Am. Chem. Soc., 91, 3996 (1969); 93, 2897 (1971).
3.   M. Capka et al., Tetrahedron Letters, 3303 (1968); Coll. Czech. Chem. Commun., 34, 118 (1969).
4.   J. M. Lalancette et al., Can. J. Chem., 47, 739 (1969); Synthesis, 526 (1972).
5.   G. Giongo et al., Tetrahedron Letters, 3195 (1973).
6.   R. F. Borch, S. R. Levitan, J. Org. Chem., 37, 2347 (1972).
7.   R. J. P. Corriu, J. J. E. Moreau, Chem. Commun., 38 (1973); J. Organomet. Chem., 64, C51 (1974).
8.   H. C. Brown, H. M. Hess, J. Org. Chem., 34, 2206 (1969).
9.   K. E. Wilson et al., Chem. Commun., 213 (1970).
10. F. Hodosan, N. Serban, Rev. Roumaine Chem., 14, 121 (1969).
11. E. J. Corey et al., J. Am. Chem. Soc., 93, 1492 (1971).
12. H. C. Brown, S. Krishnamurthy, J. Am. Chem. Soc., 94, 7159 (1972).
13. C. A. Brown et al., Chem. Commun., 391 (1973).
14. H. C. Brown et al., J. Org. Chem., 38, 2786 (1973).
15. J. Plesek et al., Czech. 149, 279, June 15, 1973; C. A., 79, 146061 (1973).
16. G. E. Heasley, J. Org. Chem., 36, 3235 (1971).
17. R. A. Bell, M. B. Gravestock, Can. J. Chem., 47, 2099 (1969).
18. E. J. Corey, R. L. Danheiser, Tetrahedron Letters, 4477 (1973).

19. M. Cerny, J. Málek, Coll. Czech. Chem. Commun., 36, 2394 (1971).
20. H. C. Brown et al., J. Am. Chem. Soc., 94, 1750 (1972).
21. H. C. Brown et al., J. Org. Chem., 35, 3243 (1970).
22. D. Abenhaim, J.-L. Namy, Tetrahedron Letters, 1001 (1972).
23. E. M. Kaiser et al., J. Org. Chem., 36, 330 (1971).
24. J. A. Marshall, G. M. Cohen, J. Org. Chem., 36, 877 (1971).
25. Y. M. Paushkin et al., Neftekhimiya, 9, 758 (1969); C. A., 72, 31135 (1970).
26. K. Sakai et al., Bull. Chem. Soc. Jap., 40, 1548 (1967); 41, 1902 (1968); 43, 1172 (1970).
27. W. Kreiser, Ann. Chem., 745, 164 (1971).
28. J.-M. Lalancette et al., Can. J. Chem., 50, 3058 (1972).
29. K. Kimoto, M. Kawanisi, Chem. Ind. (London), 1174 (1971).
30. H.-L. Pan, T. L. Fletcher, Synthesis, 192 (1972).
31. D. C. Curry et al., J. Chem. Soc., C1120 (1967).
32. K. Nakagawa, K. Minami, Tetrahedron Letters, 343 (1972).
33. P. A. Browne, D. N. Kirk, J. Chem. Soc., C1653 (1969).
34. E. L. Eliel et al., Org. Syn., 50, 13 (1970).
35. H. Timmler, J. Kurz, Chem. Ber., 104, 3740 (1971).
36. S. Yamkshita, J. Organomet. Chem., 11, 377 (1968).
37. J. H. P. Tyman, J. Appl. Chem. Biotechnol., 22, 465 (1972); C. A., 77, 34066 (1972).
38. M. B. Rubin, J. M. Ben-Bassat, Tetrahedron Letters, 3403 (1971).
39. E. M. Kaiser, Synthesis, 391 (1972).
40. P. N. Rylander, D. R. Steele, Tetrahedron Letters, 1579 (1969).
41. A. Gauvreau et al., Bull. Soc. Chim. France, 126 (1969).
42. R. S. Coffey, Chem. Commun., 923 (1967).
43. H. Wynberg et al., Syn. Commun., 1, 25 (1971).
44. A. A. P. Schreibmann, Tetrahedron Letters, 4271 (1970).
45. T. H. Chan, E. Vinokur, Tetrahedron Letters, 75 (1972).
46. E. J. Corey, R. L. Carney, J. Am. Chem. Soc., 93, 7318 (1971).
47. E. H. Axelrod, Chem. Commun., 451 (1970).
48. J. Solodar, Tetrahedron Letters, 287 (1971).
49. K. Rühlmann, Synthesis, 236 (1971).
50. W. E. Barnett, L. L. Needham, Chem. Commun., 170 (1971).
51. W. H. Hartung, R. Simonoff, Org. Reactions, 7, 263 (1953).
52. E. M. Van Duzee, H. Adkins, J. Am. Chem. Soc., 57, 147 (1935).
53. H. C. Brown, R. M. Gallivan, Jr., J. Am. Chem. Soc., 90, 2906 (1968).
54. H. Van Bekkum et al., J. Catalysis, 20, 58 (1971).
55. R. A. Benkeser et al., J. Org. Chem., 35, 1210 (1970).

D. Oxidation

The most recent monographs on oxidation are those of Augustine [1].

1. From Saturated Hydrocarbons (1, 212)

Little success has been achieved in oxidizing saturated hydrocarbons to alcohols except in the case of those containing tertiary carbon atoms. The most commonly used oxidizing agents are the permanganate ion and chromium(VI) compounds in aqueous acetic acid or aqueous pyridine [1a]. In either case yields are low and dehydration or oxidation beyond the alcohol stage may occur.

Microbiological oxidation continues to be an important mode of oxidation, as witness the conversion of 3-α-hydroxy-5α-andro-stan-17-one into 5-α-androstan-15α-ol [2]. The subject has been reviewed [3].

Wiberg and Foster [4] succeeded in oxidizing (+)-3-methyl-heptane to 3-methyl-3-heptanol as shown:

$$CH_3CH_2CH(CH_2)_3CH_3 \xrightarrow[HClO_4,CH_3CO_2H]{Na_2Cr_2O_7} CH_3CH_2C(CH_2)_3CH_3$$

with substituent CH_3 on the starting material and OH, CH_3 on the product.

(+)

(+) 10%

75-85% retention of configuration

The same authors also oxidized cis-decalin to cis-9-decalol:

$$\xrightarrow[CH_3COOH]{\substack{Na_2Cr_2O_7 \\ HClO_4}}$$

18 %

The mechanism of oxidation of a tertiary carbon atom with per-manganate probably involves hydrogen atom abstraction [5].

More satisfactory yields were obtained in oxidizing adamantane with fluorosulfonic acid [6]:

1) F SO$_2$OH unpurified, 25°
2) H$_2$O

48%

Pure fluorosulfonic acid is not effective; the crude acid, prepared from a mixture of potassium bifluoride and oleum, was employed in the reaction. However, chromic acid oxidation of adamantane gave 71% of 1-adamantanol and 9% of 2-adamantanone [7].

2. From Organometallic Compounds ($\underline{1}$, 214)

Hydroxyl groups have been introduced into carboxylic acids containing α-hydrogen atoms by aeration of the lithiated carboxylic acids [8] (Ex. a):

$$\begin{array}{c} R' \\ \diagdown \\ \diagup \\ R^2 \end{array} CHCOOH \xrightarrow[\substack{2)air \\ 3)H_2O}]{1)C_4H_9Li-THF} \begin{array}{c} R' \\ \diagdown \\ \diagup \\ R^2 \end{array} \overset{\displaystyle |}{\underset{\displaystyle OH}{C}} COOH$$

32-90 %

a. Preparation of Benzilic Acid [8]

Diisopropylamine, 2 molar equiv, and THF in a nitrogen atmosphere at 0-5° were treated with butyllithium, 2.1 molar equiv, in heptane in a fine stream. The mixture was stirred for 15-30 min at 0-5° and then diphenylacetic acid, 1 molar equiv, in THF was dropped into the mixture with continued stirring and cooling. After stirring for an additional 30 min at 0-5° and then at 40-50° for 1-1.5 hr, the reaction was cooled to 25° and air was bubbled into the solution for ∿18 hr. Water, ∿2 vol, was added and the mixture was extracted with ether, from which 90% of the hydroxy acid was recovered.

3. From Olefins (Ozonization Followed by Reduction) ($\underline{1}$, 215)

An excellent discussion of the reaction of ozone with olefins is available [9]. In addition, a recent discussion of the mechanism appears in [10]. The ozonides have been the subject of extensive investigations and to account for their diverse reactions it does not seem possible to assign a single formula to them. At present there is evidence for at least three structures, which Story has interrelated as shown:

Story's peroxyepoxide   Staudinger's molozonide   Criegee's zwitterion

Although it is suggested that chemical reduction is preferable to catalytic hydrogenation in converting ozonides into alcohols [11], White and co-workers [12] obtained excellent yields by controlled catalytic hydrogenation as indicated:

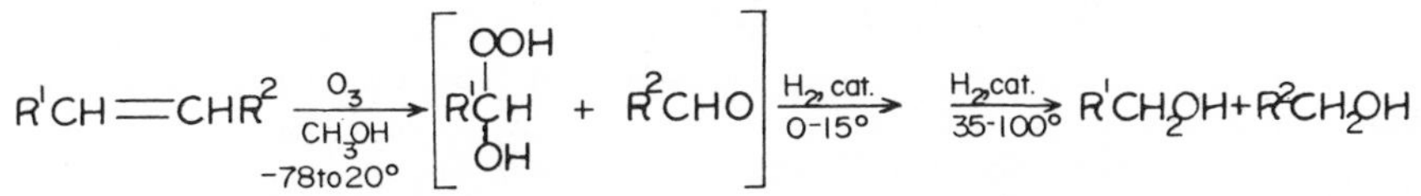

$$R'CH\!=\!CHR^2 \xrightarrow[\substack{CH_3OH \\ -78to20°}]{O_3} \left[\underset{\overset{|}{OH}}{\overset{\overset{|}{OOH}}{R\overset{2}{C}H}} + R^2CHO \right] \xrightarrow[0-15°]{H_2\,cat.} \xrightarrow[35-100°]{H_2cat.} R'CH_2OH + R^2CH_2OH$$

The first stage of the reduction with Pt at 15-50 psi reduces the hydroperoxide, while the final stage with Pt at 50-150 psi completes the reduction to the alcohols. Cyclenes give glycols in 77-95% yield by the method.

4. From Olefins (Allylic or Similar Oxidation) (1, 216)

a. By Selenium Dioxide

The oxidation of 2-methyl-2-heptene by SeO_2, which is stereospecific, was accomplished by Bhalerao and Rapoport [13]:

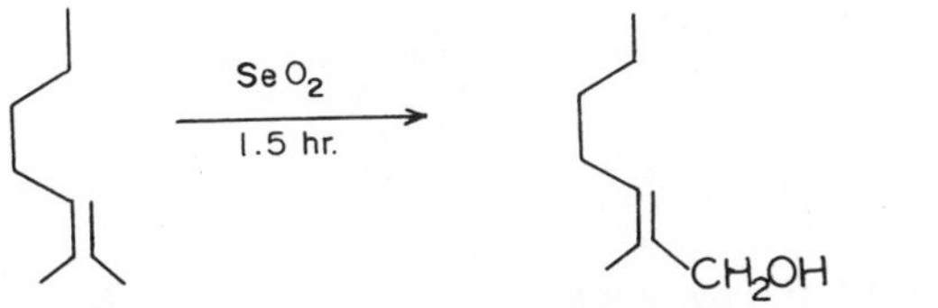

trans -2-Methylhept-2-enol, 98%

About 2% of the cis alcohol was also formed; longer times gave the aldehyde.

5. From Alkenes (cis or trans Addition) (1, 219)

The reagents employed for the conversion of alkenes into cis or trans glycols have been enumerated (1, 219). o-Sulfoperbenzoic acid, prepared from o-sulfobenzoic anhydride and hydrogen peroxide, is a new reagent that may be used without isolation in the preparation of the epoxide or trans glycol [14] (Ex. a):

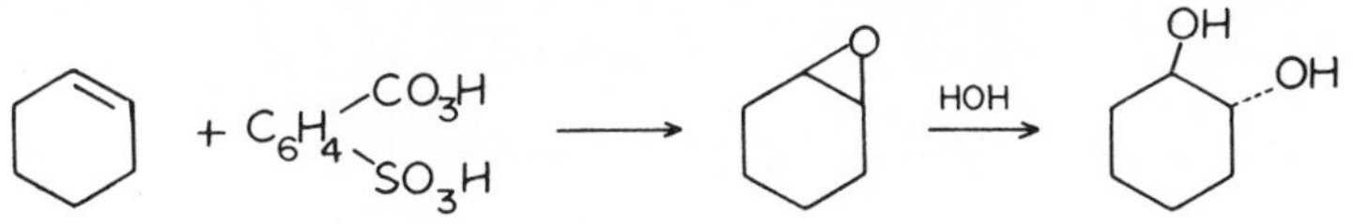

trans -1,2-Cyclohexanediol
88 %

The reagent is usually employed in an aqueous-acetone solution, which is reasonably stable at room temperature. Other yields vary from 78-89% and no esterification of the glycol occurs.

To eliminate the toxicity hazard associated with the use of

osmium tetroxide in the formation of glycols from alkenes, Lloyd
and co-workers [15] utilized potassium osmate. This reagent is
employed either with hydrogen peroxide or sodium chlorate as the
catalyst and acetic acid to neutralize the potassium osmate
solution. Of the two catalysts, sodium chlorate is the most
convenient, although it should not be used, because of an
explosion hazard, if the product is to be recovered from the
reaction mixture by distillation. This reagent gave cis-
cyclohexanediol from cyclohexene (76%). Likewise, glycerol was
obtained with H_2O_2 as the catalyst in 67% yield from allyl
alcohol.

The use of potassium permanganate as the oxidant in the
conversion of alkenes into cis-1,2 glycols has been improved by
employing phase-transfer catalysis [16]. Thus cis-cyclooctene
may be transformed into cis-1,2-cyclooctanediol:

50 %

Previous yield using aqueous, basic $KMnO_4$ was 7%.

The Woodward method of using iodine, silver acetate, and
moist acetic acid in the conversion of olefins into cis diols
has been modified in that the expensive silver salt may be
eliminated. Thus Mangoni and co-workers [17] formed cis-1,2-
cyclohexanediol from cyclohexene (86%) (Ex. b).

a. Preparation of trans-1,2-Cyclohexanediol [14]

Cyclohexene, 0.1 mol in a minimum amount of acetone was
treated with 0.11 mol of the aqueous acetone peracid solution
(o-sulfobenzoic anhydride, 1.0 mol, and 30% H_2O_2, 1.3 mol, in
acetone at 0-4°). After 0.5-1.5 hr, most of the acetone was
removed under reduced pressure and the diol was extracted with
ether, from which solution it was isolated in 88% yield.

b. Preparation of cis-1,2-Cyclohexanediol [17]

Cyclohexene, 1.35 mmol, 20 ml of glacial acetic acid, KIO_3,
0.34 mmol, and I_2, 0.675 mmol, were stirred at 25° for 5 hr and
then refluxed with potassium acetate, 1.35 mmol, for 7 hr. Upon
hydrolysis with alkali the cis diol, 86%, was recovered.

7. From Ketones

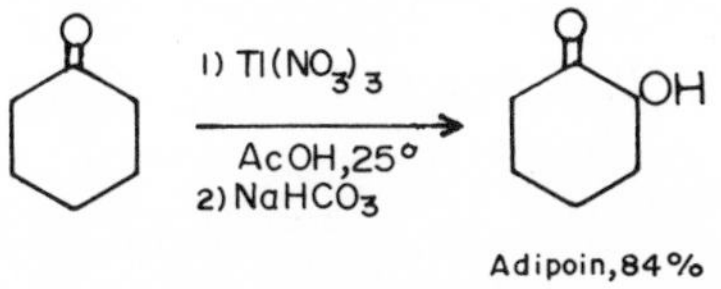

Adipoin, 84%

The oxidation of cyclohexanones with thallium nitrate in acetic acid has been investigated by Taylor and McKillop [18]. Adipoin was obtained in good yield as shown under the conditions stated (Ex. a). It is interesting to note that if the filtrate obtained after the removal of thallium nitrate is heated above about 40° for a few minutes no adipoin is obtained, but instead the product is cyclopentanecarboxylic acid, also in 84% yield.

It is suggested that the mechanism is as follows:

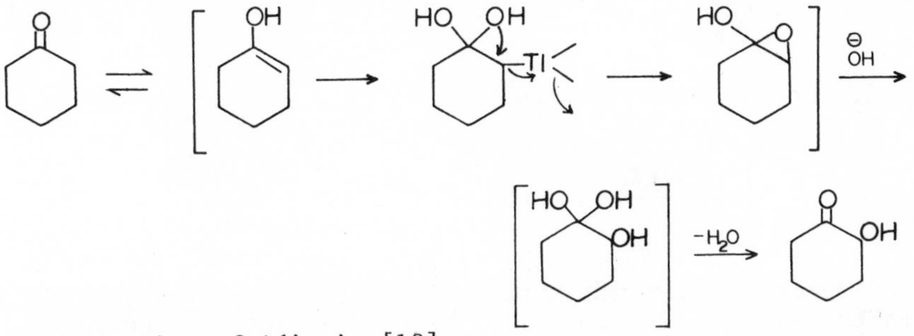

a. Preparation of Adipoin [18]

Thallium(III) nitrate, 18 g, was added to a solution of 4 g of cyclohexanone in 40 ml of acetic acid. Thallium(I) nitrate, which precipitated almost immediately, was removed by filtration. After the filtrate was neutralized with NaHCO$_3$, the solution was allowed to stand overnight. From the chloroform extract adipoin dimer, 3.43 g (84%) was recovered.

8.   From β,γ-Unsaturated Acids (Decarboxylation)

Still and co-workers oxidized the dioxolane [19] of the β,γ-carboxylic acid A with m-chloroperbenzoic acid to obtain the dioxolane epoxycarboxylic acid B, which on heating gave the corresponding methylene alcohol C:

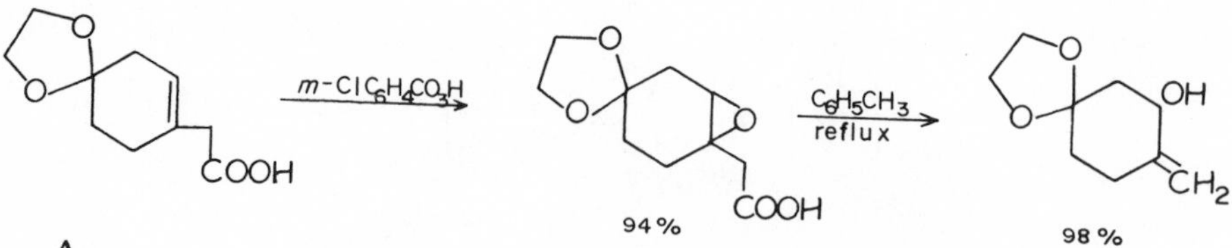

A   B   C

The reaction appears to be general. The path taken from B to C
may be represented as:

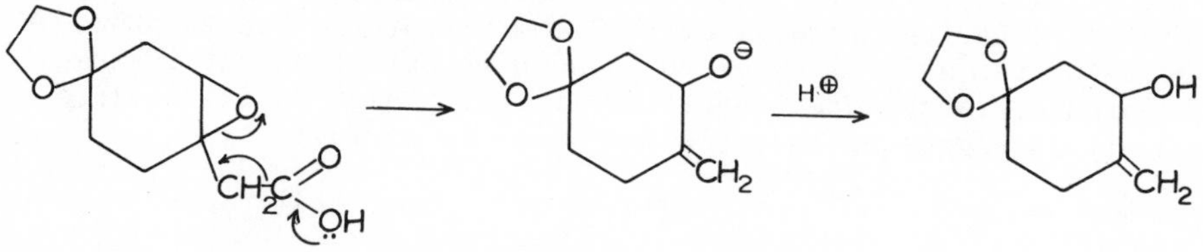

9.   From Benzylic Ethers

$$C_6H_5CH_2OR \xrightarrow[\text{2) } H_2O]{\text{1) }(C_6H_5)_3C^{\oplus}} ROH + C_6H_5CHO$$

In the steroid family the benzylic ether protective group
may be removed by hydride abstraction with triphenylcarbonium
tetrafluoroborate at 0-20° [20]. Five examples gave 60-90%
yields, whereas one did not respond.

10.   From Aromatic Structures

Tetraphenylcyclopentadienone has been oxidized to the cis-
glycol [21]:

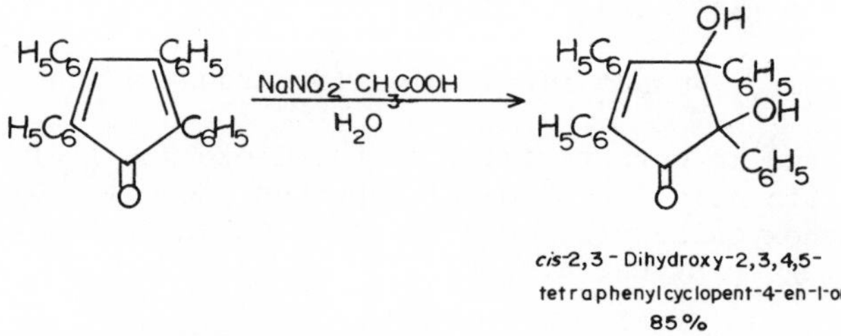

cis-2,3 - Dihydroxy-2,3,4,5-
tetraphenylcyclopent-4-en-1-one
85%

11.   From Silyl Vinyl Ethers (see Addenda).

12.   From Amides or Esters (see Addenda).

Addenda

D.7. Although conversion of a ketone α-methylene group to the acyloin has been accomplished previously (p. 249), such is also possible by preparing the ketone enolate at -70° and adding molybdenum peroxide-pyridine-HMPA [E. Vedejs, J. Am. Chem. Soc., 96, 5944 (1974)].

D.11. Trimethylsilyl vinyl ethers add diborane to give predominantly the β-borane which upon oxidation followed by hydrolysis gives 1,2-glycols [G. L. Larson et al., J. Organomet. Chem., 76, 9 (1974)].

D.12. If an N,N-dialkylamide or an ester is treated with LDIA, oxidized with oxygen, and the peroxide reduced with NaHSO$_3$, α-hydroxy amides or α-hydroxyesters, respectively, are obtained in good yield [H. H. Wasserman, B. H. Lipshutz, Tetrahedron Letters, 1731 (1975)].

References

1.   R. L. Augustine, Oxidation, Vol. 1, 1969; Vol. 2, 1971, Dekker, New York.
1a. Ref. 1, Vol. 1, pp. 2-6.
2.   G. D. Meakins et al., J. Chem. Soc., C1136 (1971).
3.   S. K. Erickson, in S. Patai, The Chemistry of the Hydroxyl Group, Pt. 2, Interscience, New York, 1971, p. 755.
4.   K. B. Wiberg, G. Foster, J. Am. Chem. Soc., 83, 423 (1961).
5.   J. I. Brauman, A. J. Pandell, J. Am. Chem. Soc., 92, 329 (1970).
6.   B. M. Lerman et al., J. Org. Chem. (USSR), 7, 1110 (1971).
7.   R. C. Bingham, P. v. R. Schleyer, J. Org. Chem., 36, 1198 (1971).
8.   G. W. Moersch, M. L. Zwiesler, Synthesis, 647 (1971).
9.   J. S. Belew, in Ref. 1, Vol. 1, p. 259.
10. P. R. Story et al., J. Am. Chem. Soc., 93, 3042, 3044 (1971).
11. J. A. Sousa, A. L. Bluhm, J. Org. Chem., 25, 108 (1960); A. J. Hubert, J. Chem. Soc., 4088 (1963); D. G. M. Diaper, D. L. Mitchell, Can. J. Chem., 38, 1976 (1960).
12. R. W. White et al., Tetrahedron Letters, 3587 (1971).
13. U. T. Bhalerao, H. Rapoport, J. Am. Chem. Soc., 93, 4835 (1971).
14. J. M. Bachhawat, N. K. Mathur, Tetrahedron Letters, 691 (1971).
15. W. D. Lloyd et al., Synthesis, 610 (1972).

16. W. P. Weber, J. P. Shepherd, Tetrahedron Letters, 4907 (1972).
17. L. Mangoni et al., Tetrahedron Letters, 4485 (1973).
18. E. C. Taylor, A. McKillop et al., J. Org. Chem., 37, 3381 (1972).
19. W. C. Still, Jr. et al., Tetrahedron Letters, 1421 (1971).
20. D. H. R. Barton et al., Chem. Commun., 1109 (1971).
21. S. Ranganathan, S. K. Kar, J. Org. Chem., 35, 3962 (1970).

E. Organometallic Reactions

1. From Carbonyl Compounds, Esters, and Carbonates (1, 224)

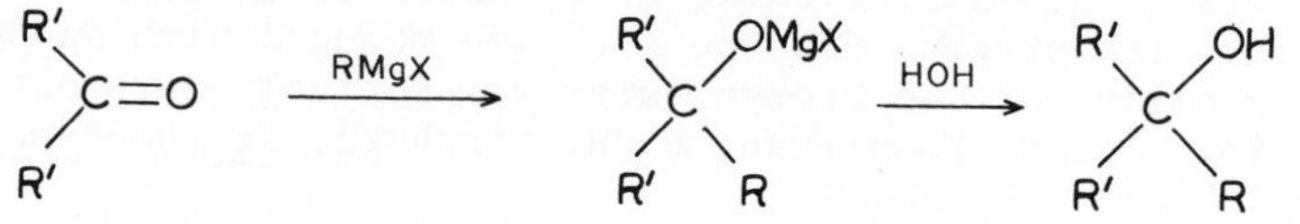

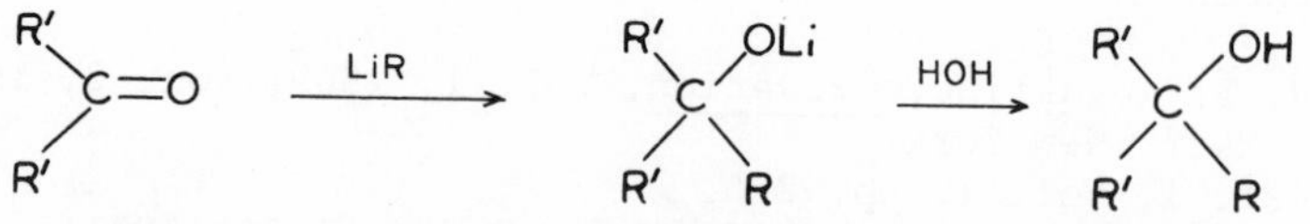

These reactions have been widely used in the synthesis of alcohols. The most common method consists of starting with the carbonyl compound, and there is an increasing tendency to use the lithium rather than the Grignard reagent. In particular, the method has been used extensively in the synthesis of secondary and tertiary alcohols. Its use will be illustrated in the preparation of secondary, tertiary, and unsaturated alcohols.

One of the better methods of preparing secondary or tertiary alcohols via a lithium reagent is that of Rathke [1] (Ex. a). Thus benzaldehyde reacts with lithio ethyl acetate, prepared from ethyl acetate and lithium bis(trimethylsilyl) amide in THF at -78°, at low temperature almost instantly to give after hydrolysis ethyl β-phenyl-β-hydroxypropionate.

$$C_6H_5CHO + LiCH_2CO_2C_2H_5 \xrightarrow{HOH} C_6H_5CHOHCH_2CO_2C_2H_5$$
$$80\%$$

Yields exceed those of the usual Reformatsky procedure.

The β-hydroxy type may also be prepared from the dilithium salt, the Ivanov reagent [2]:

$$\underset{R}{\overset{R}{>}}C=O \quad \xrightarrow[\text{2)}H_2O]{1)\ \underset{R}{\overset{R}{>}}\underset{Li}{C}-COOLi} \quad \begin{array}{c} R \\ R-\!\!\!\!\overset{|}{\underset{|}{C}}\!\!\!\!-COOH \\ R-\!\!\!\!\overset{|}{\underset{|}{C}}\!\!\!\!-OH \\ R \end{array}$$

This method offers an advantage if the β-hydroxy acid is desired, but yields are not as satisfactory as in the lithio ethyl acetate method.

In a modified procedure in which the alkyllithium is not prepared in advance, Pearce and co-workers [3] simply added a mixture of the carbonyl compound and the alkyl halide dropwise to a suspension of lithium in THF. The temperature was held at 0° by the rate of addition and the alcohol was recovered after hydrolysis. Thus propionaldehyde and ethyl bromide gave the secondary alcohol:

$$CH_3CH_2CHO + C_2H_5Br \xrightarrow[\text{2)}H_2O]{\text{1)}Li-THF} CH_3CH_2CHOHC_2H_5$$
Pentan-3-ol, 90 %

In the preparation of diphenylcarbinol, the yield was 95%. With sodium instead of lithium the yield of the same alcohol was 44%; with the Grignard reagent it was 55%. In fact, yields of secondary as well as tertiary (from ketones or esters) alcohols are invariably better than by the Grignard procedure.

The introduction of large alkyl groups through the lithium derivative has been possible. Thus Bartlett and co-workers [4] utilized isopentyllithium with benzophenone to prepare diphenyl-isopentylcarbinol:

$$\underset{C_6H_5}{\overset{C_6H_5}{>}}C=O + (CH_3)_2CH(CH_2)_2Li \xrightarrow{H_2O} \underset{C_6H_5}{\overset{C_6H_5}{>}}\underset{(CH_2)_2CH(CH_3)_2}{\overset{OH}{C}}$$
(unstated yield)

In a similar manner Eaton and co-workers [5] introduced the hydroxypropyl group (Ex. b)

1-(3-Hydroxypropyl)-cyclohexanol
96 %

The parent of this reagent is 3-bromopropanol, which is treated with ethyl vinyl ether to form the acetal.

Other interesting applications are the addition of a Schiff base anion to form a series of aminoethanols [6]:

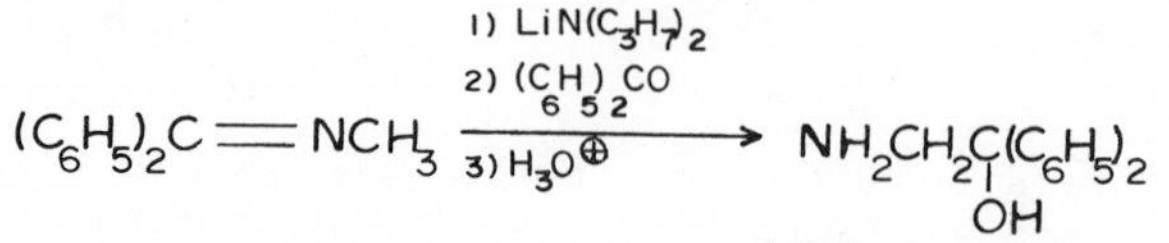

$$(C_6H_5)_2C = NCH_3 \xrightarrow[\substack{2)~(CH_3)_2CO \\ 3)~H_3O^\oplus}]{1)~LiN(C_3H_7)_2} NH_2CH_2C(C_6H_5)_2$$
$$\qquad\qquad\qquad\qquad\qquad\qquad\qquad OH$$

β-Amino-α,α-diphenylethanol, 76%

Similar products may be obtained in a different way [7]:

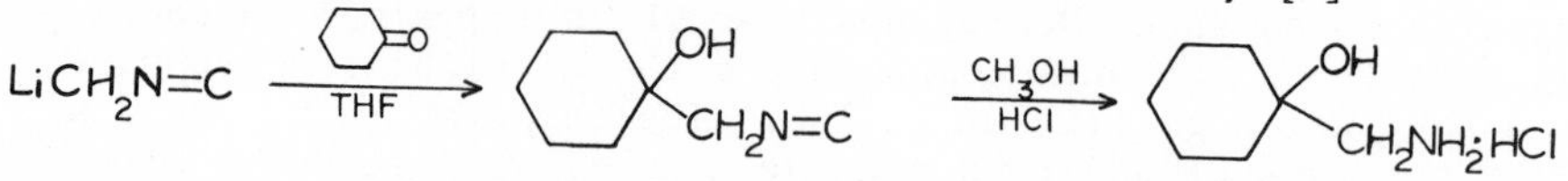

The ring of the product may be expanded to form cycloheptanone.

Various glycols have been prepared from oxalic esters and Grignard reagents [8a], while fluorinated tertiary alcohols are available from fluoroketones and Grignard reagents [8b].

The formation of a three-membered ring by the Cottle synthesis from the Grignard adduct followed by hydrolysis has been utilized in the synthesis of 1-aryl cyclopropanols [9]:

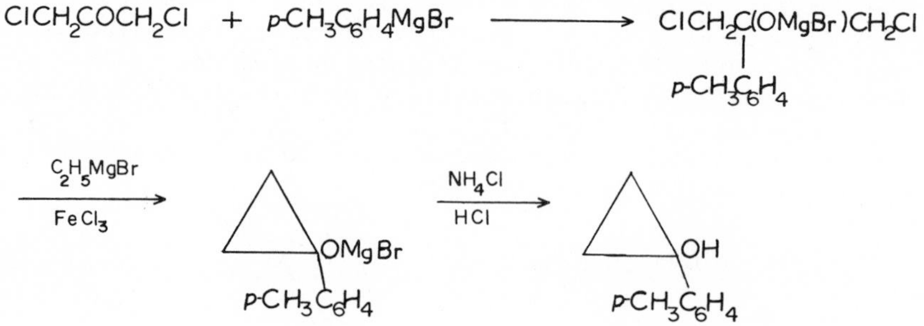

$$ClCH_2COCH_2Cl ~+~ p\text{-}CH_3C_6H_4MgBr \longrightarrow ClCH_2C(OMgBr)CH_2Cl$$
$$\qquad\qquad\qquad\qquad\qquad\qquad\qquad\qquad\qquad\qquad p\text{-}CH_3C_6H_4$$

p-Tolylcyclopropanol, 51-57%

The method offers a general procedure for the preparation of 1-alkyl or 1-aryl cyclopropanols.

When benzophenones are treated with a hindered Grignard reagent, such as t-butylmagnesium chloride, 1,2-, 1,4-, and 1,6-addition products, as well as the pinacol, may be formed [10]. Thus unsubstituted benzophenone gives largely the 1,2- and 1,6-addition products:

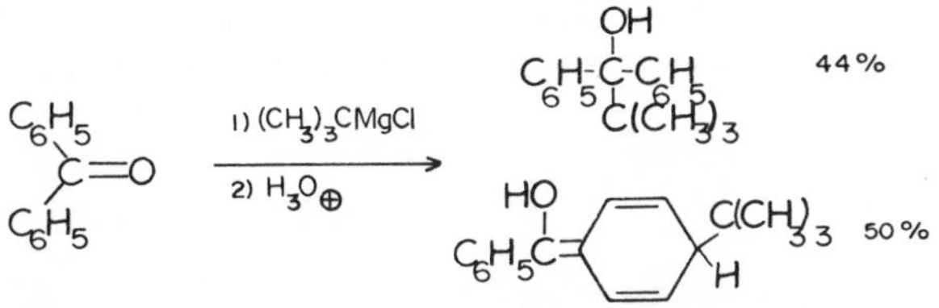

The ratios between the types of products formed are dependent largely on the steric factors involved. The mechanism is thought to be a free radical one in which the 1,2-addition product results from a combination of the t-butyl radical and the benzo-phenone ketyl radical anion:

$$(C_6H_5)_2C{=}O \; + \; (CH_3)_3CMgCl \longrightarrow (C_6H_5)_2\overset{\cdot}{C}{-}\overset{\ominus}{O} \; + \; Mg\overset{\oplus}{C}l + (CH_3)_3C\cdot$$

The reaction of organometallic compounds with a cyclic ketone such as 4-t-butylcyclohexanone, followed by hydrolysis, gives both the axial and equatorial alcohols in which the ratio between the two products varies with the ratio between the sub-strate and with the solvent employed [11] as indicated:

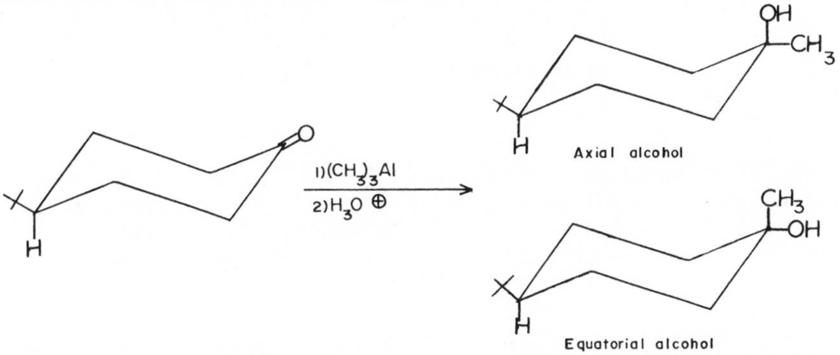

When the $(CH_3)_3Al$:ketone ratio in benzene is 0.5, the ratio of axial:equatorial alcohol is 80:20. On the other hand, when the $(CH_3)_3Al$:ketone ratio in benzene is 3.0, the ratio of axial: equatorial alcohol is 12:88. These effects are explained on the basis of two factors, the so-called compression effect and the steric approach factor. The latter tends to an approach from the least hindered side to give the axial alcohol, while the com-pression effect, which predominates when high ratios of the tri-methylaluminum are present, leads to the equatorial alcohol.

As has already been indicated, reduction of the ketone is at times a competing reaction in the alkylation of ketones by means of an organometallic reagent. Chastrette and Amouroux [12] found

that the percentage of addition of a Grignard reagent to a
hindered ketone can be increased at the expense of reduction by
complexing with lithium perchlorate or tetrabutylammonium
bromide. For example, diisopropyl ketone with propylmagnesium
bromide in ether responds as follows:

$$\begin{array}{c} i\text{-Pr} \\ i\text{-Pr} \end{array}\!\!\!C\!=\!O \xrightarrow[\text{2)H}_3O^\oplus]{\text{1)2 PrMgBr}} \begin{array}{c} i\text{-Pr} \quad OH \\ \diagdown \; C \diagup \\ i\text{-Pr} \quad Pr \end{array} \; + \; \begin{array}{c} i\text{-Pr} \\ i\text{-Pr} \end{array}\!\!\!CHOH$$

                                    36 %   62%

Premixing the Grignard reagent with 1.5 mol $LiClO_4$ or 1 mol of
Bu_4NBr led to an increase in yield of the tertiary alcohol to
70%. It is thought that the effect is due to complexation of
the salt with the Grignard reagent, a process which would increase its ionic character and thus make the transition state
more polar than that without a salt. A similar increase
in the yield of the tertiary alcohol may be obtained with
$MgBr_2$ or MgI_2.

The asymmetric synthesis of tertiary alcohols has been
accomplished as shown [13]:

$$\begin{array}{c} R \\ R^1 \end{array}\!\!\!C\!=\!O + R^2_2Cd \xrightarrow[\text{(+)(Me}_2CHO_2C(CHOMe)_2COCHMe_2]{\text{Et}_2O} \begin{array}{c} R \quad OH \\ \diagdown \; C \diagup \\ R^1 \quad R^2 \end{array}$$

                                          0.9 to 2.5% optical yield

In the four examples investigated, R, R^1, R^2 were C_2H_5, H, C_6H_5,
C_6H_5, H, C_2H_5, $C_6H_5, H, i\text{-}C_3H_7$, and C_6H_5, CH_3, C_2H_5, respectively.
Olefinic carbinols may be obtained by the so-called Scoopy
reaction [14]. In this reaction ethyltriphenylphosphonium bromide was converted into the ylid with butyllithium, after which
the ketone was added. The white precipitate formed was then
treated with sec-butyllithium and then formaldehyde was added.
The overall reaction with hexanal follows:

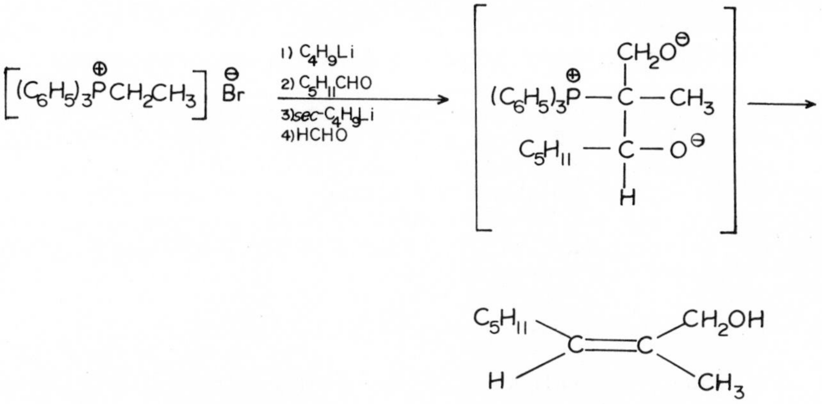

$$\left[(C_6H_5)_3 \overset{\oplus}{P} CH_2CH_3 \right] \overset{\ominus}{Br} \quad \xrightarrow[\substack{3) sec\text{-}C_4H_9\text{-}i \\ 4) HCHO}]{\substack{1) C_4H_9Li \\ 2) C_5H_{11}CHO}} \quad \left[\begin{array}{c} CH_2O^{\ominus} \\ | \\ (C_6H_5)_3 P - C - CH_3 \\ | \\ C_5H_{11} - C - O^{\ominus} \\ | \\ H \end{array} \right] \longrightarrow$$

$$C_5H_{11} \diagdown \quad \diagup CH_2OH$$
$$C = C$$
$$H \diagup \quad \diagdown CH_3$$

71 %

cis: trans   >99:1

Thus in this particular case only the primary alcohol is obtained. Usually, a secondary or tertiary alcohol is produced as well. At any rate the method possesses limitations as a means of synthesizing a particular unsaturated alcohol.

Unsaturated alcohols also have been prepared as follows [15a]:

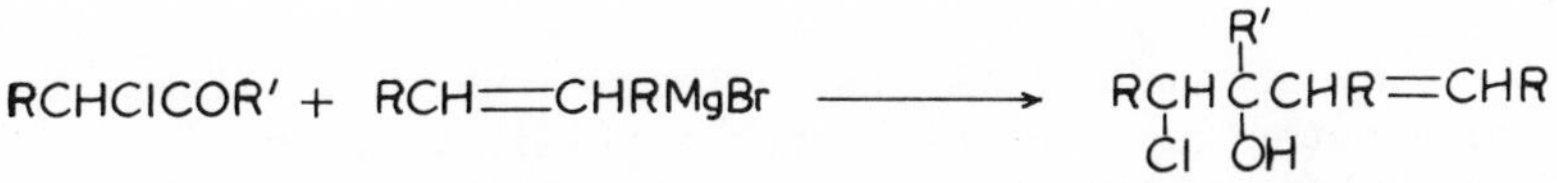

$$RCHClCOR' + RCH=CHRMgBr \longrightarrow \underset{\substack{| \quad | \\ Cl \quad OH}}{RCH\overset{R'}{C}CHR=CHR}$$

A second method involves an acetylene and diisobutylalane [15b]:

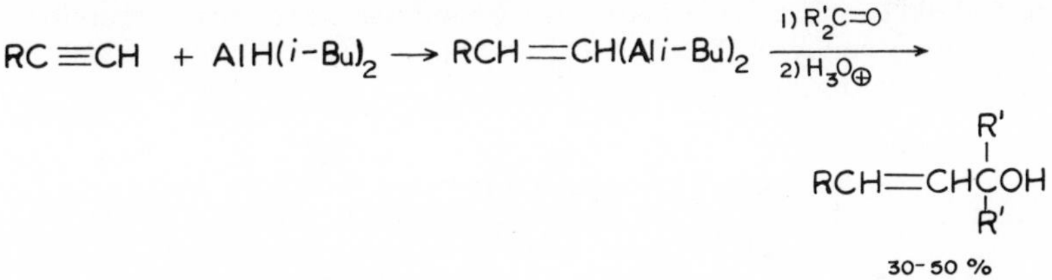

$$RC\equiv CH + AlH(i\text{-}Bu)_2 \longrightarrow RCH=CH(Ali\text{-}Bu)_2 \xrightarrow[2)H_3O^{\oplus}]{1)R_2'C=O}$$

$$RCH=CH\overset{R'}{\underset{R'}{C}}OH$$

30-50 %
(4 examples)

A third method involves a carbonyl compound and allylzinc or allylcadmium [16] [see 13]:

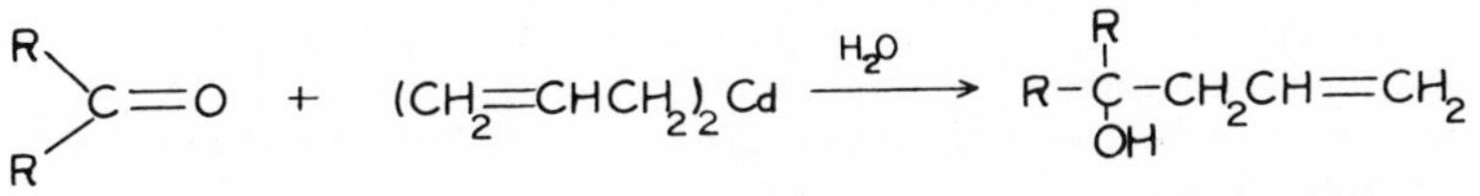

$$\underset{R}{\overset{R}{\diagdown}} C=O + (CH_2=CHCH_2)_2Cd \xrightarrow{H_2O} R-\underset{OH}{\overset{R}{\underset{|}{C}}}-CH_2CH=CH_2$$

A recent acetylenic salt addition to carbonyl compounds is that of propynyllithium or propynylsodium to cyclohexanone to give 1-propynyl-1-cyclohexanol (91%) [17]. The reaction has been discussed more fully elsewhere (1, 164).

a.   Preparation of Ethyl 3-Hydroxy-5-phenyl-4-pentenoate [1]

    Lithio ethyl acetate, prepared from lithium bis(trimethyl-silyl)amide in THF, 25 ml of a 1.0M solution, was treated with ethyl acetate at dry ice-acetone temperature in a nitrogen atmos-phere dropwise for 2 min and the mixture was stirred for an additional 15 min. Cinnamaldehyde, 25 mmol, was added and after 5 min the solution was hydrolyzed with HCl. There was recovered 5.17 g (94%) of the hydroxyester.

b.   Preparation of 1-(3-Hydroxypropyl) Cyclohexanol [5]

    The adduct of cyclohexanone and 1-ethoxyethyl 3-lithiopropyl ether, 24 g, was stirred into 100 ml of a 60:40 mixture of water and ethanol and 4 ml of concentrated HCl. After 15 min, the neutralized solution was reduced to a small volume, and the organic material in the residue was taken up in chloroform, from which solution there was recovered 15.8 g (96%) of the diol.

    2.   From Epoxides, Oxetanes, and the Like (1, 228)

$$CH_2{-}CH_2 \xrightarrow{CH_3Li} CH_3CH_2CH_2OLi \xrightarrow{H_2O} CH_3CH_2CH_2OH$$

    A study of the use of dimethylmagnesium, methyllithium, and lithium dimethylcuprate in the nucleophilic ring opening of 1,2-epoxybutane has shown that these reagents are superior to the Grignard reagent [18]. Lithium dimethyl cuprate in particular is recommended in that it gives the expected secondary alcohols as predominant products [19]. For example, cyclohexene oxide responds as shown:

trans-2-Methylcyclohexanol
76 %

This reagent is selective in that it does not affect the carbonyl group of esters.
    The radical anion of naphthalene at 65° results in the cleavage of the THF moiety to form 1-(4-hydroxybutyl)-1,4-dihydronaphthalene, the α-isomer,

50%, containing a small amount of the β-isomer [20].

3. From Carbonyl Compounds and α-Halo Esters (Reformatsky)
    (1, 229)

A recent review of the Reformatsky reaction has been published [21].

In attempts to increase yields, the Reformatsky reaction in
recent years has been the subject of numerous investigations.
Frankenfeld and Werner [22] examined each step in the procedure
when using an aliphatic aldehyde and ethyl bromoacetate. They
succeeded in improving yields (55-80%) over those previously
obtained (29-40%) when operating on a fairly large scale (3-5
mol) in the laboratory. These investigators, after conducting
many runs, recommend: (a) Use of a 5-10% excess of ethyl bromoacetate and a 20-30% excess of zinc, (b) use of neutral, dry,
activated zinc, (c) temperature of 80-85°, and (d) addition of
50% sulfuric acid to the cooled reaction mixture ($\leq$35°) in the
hydrolysis.

Some investigators have conducted the reaction on esters
other than the ethyl ester with the hope that side reactions
would be less prevalent. For example, Cornforth and co-workers
[23], in using t-butyl esters in THF obtained yields of β-hydroxy
esters varying from 18 to 87%. The free acids were obtained by
solvolysis with HCl in CH_2Cl_2 at -10° for 48 hr or with CF_3CO_2H
at 25° for 1 hr. Horeau [24] similarly employed the trimethylsilyl α-bromoesters and obtained yields of the β-hydroxy acids,
without isolation of the corresponding ester, of 30-80%.

There have been attempts as well to improve yields and the
quality of the product by isolating the zinc bromoester complex,
which is readily prepared from the bromoester and zinc in
methylal [25], and treating it with the carbonyl compound.   In
this manner many β-hydroxy esters were prepared in yields
superior (31-87%) to the usual method. A similar method was
employed in the synthesis of β-hydroxy α,α-dichloroesters [26]
via the stable dichlorozinc enolate (Ex. a).

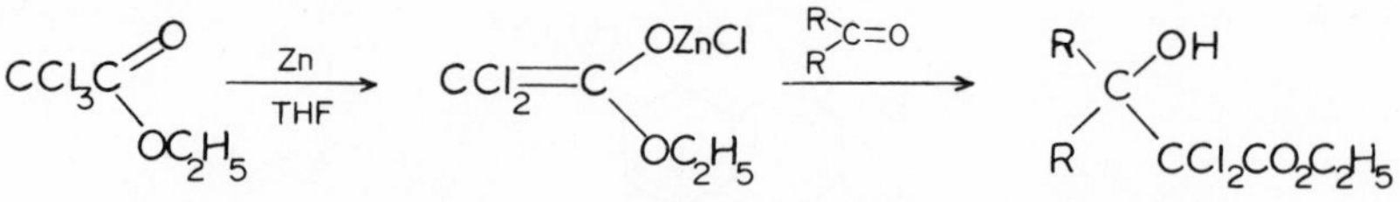

51 - 81 %

The best procedure for conducting the Reformatsky reaction consists of using tetrahydrofuran-trimethyl borate or benzene as a solvent at room temperature [27]. Yields with the former range from 85-98% (Ex. b). With benzene as a solvent they are slightly lower, particularly with the more reactive carbonyl compounds. The trimethyl borate appears to buffer the reaction at near neutrality or on the slightly acidic side and thus the products are more stable.

Some success has been achieved in the condensation by starting with the α-bromo acid and proceeding by first blocking the carboxyl group as shown [28]:

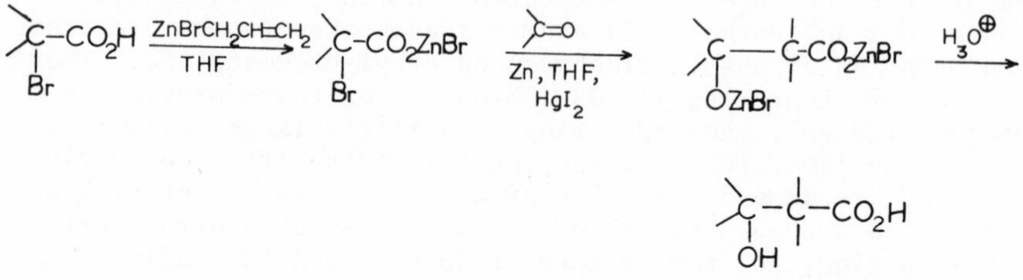

26-97 %

The Reformatsky reaction has been utilized as well in the synthesis of olefinic β-hydroxynitriles [29], esters of tertiary α-hydroxyacids [30], and 1-hydroxy-2-naphthoates [31].

a. Preparation of Ethyl 2,2-Dichloro-3-hydroxy-4-methyl-pentanoate [26]

Zinc, 0.1 g atom, powdered by heating at 200° in a current of N_2, was covered by 30 ml of THF and at 30° 0.1 mol ethyl trichloroacetate in 90 ml THF was added carefully to the cooled mixture. After 4 hr, the zinc was completely dissolved and 0.08 mol of isobutyraldehyde was added. On stirring for 2 hr, hydrolyzing with NH_4Cl, and acidifying with HCl, 11.5 g (51%) of the hydroxydichloroester was recovered from the ethereal extract.

b. Preparation of Ethyl 3-Phenyl-3-hydroxypropionate [27]

Zinc, 6.54 g under static N_2 pressure at 25° was treated with a solution of 10.6 g of benzaldehyde in 25 ml of THF and 25 ml of trimethyl borate. Ethyl bromoacetate, 11.1 ml with

stirring was introduced all at once and the total was stirred for
12 hr. After hydrolysis with a solution of NH$_4$Cl and glycerine,
the ethereal extract yielded 18.5 g (95%) of the β-hydroxyester.

4. From Allyl Alcohols

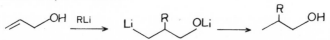

This reaction occurs in a hydrocarbon solvent after several
hours [32]. Yields with various alkyllithiums vary from 4 to
60%. The yield is improved at times with the addition of TMEDA.
The lithium adduct, as might be expected, is a potential intermediate for various other transformations; for example, in dry
ice the butyllithium adduct with acid gives β-butyl-γ-butyrolac-
tone.

5. From Mercurials (see Addenda).

6. From Some Vinyl Ethers (see Addenda).

Addenda

E.1. If 2.9 equiv of trimethylaluminum is added to 1 equiv of
t-butylcyclohexanone, the tertiary alcohol obtained is 75%
equatorial [E. C. Ashby, J. T. Laemmle, J. Org. Chem., 40, 1469
(1975)]. The stereochemistry of this addition with various
organometallic compounds has been reviewed [E. C. Ashby, J. T.
Laemmle, Chem. Rev., 75, 521 (1975)].

E.1. The addition of optically active LiCH$_2$SOC$_7$H$_7$ to benzaldehyde gives diastereoisomers that were separated from each other
in approximately 16% yields. Reduction of each gave the
enantiomers of α-phenylethanol in nearly 100% optical purity
[G. Tsuchihashi et al., Tetrahedron Letters, 4605 (1972)].

E.1. Sodium acetylide addition to carbonyl compounds in liquid
ammonia is enhanced by the addition of magnesium salts [J. N.
Gardner, Can. J. Chem., 53, 2157 (1975)].

E.1. Silylated hydrocarbons having an activated Si-C bond
(allyl-, aryl-, vinyl-, or propargylsilanes) add to chloral and
after hydrolysis give the corresponding α-trichloromethylated
carbinols [R. Calas et al., J. Organomet. Chem., 93, 43 (1975)].

E.1. The carbenoid of methylene chloride may be prepared in situ
in the presence of cyclohexanone of 0° with lithium

dicyclohexylamide to give 1-dichloromethylcyclohexanol (89%) [H. Yamamoto et al., J. Am. Chem. Soc., 96, 3010 (1974)]. If the ketone acceptor is not present, the carbenoid must be generated at -120° to -70° [G. Köbrich, Angew. Chem. Intern. Ed. Engl., 11, 473 (1972)].

E.1. Reduction and enolization are minimized if butyllithium is added at -78° to the carbonyl compound [J. D. Buhler, J. Org. Chem., 38, 904 (1973)].

E.4.

$$RC\equiv CCH_2OH \; + \; CH_2=CHCH_2MgCl \xrightarrow{CuI} \underset{\underset{\underset{RCH=CCH_2CH=CH_2}{\overset{|}{CH_2OH}}}{\overset{|}{CH_2OH}}}{RC(MgCl)=CCH_2CH=CH_2} \; \overset{CH_2OMgCl}{\xrightarrow{H_2O}}$$

[B. Jousseaume, J.-G. Duboudin, J. Organomet. Chem., 91, C1 (1975)].

E.5. RHgBr may be treated with NaBH$_4$ in DMF saturated with oxygen to yield the alcohol, ROH. The free radical, R·, evidently is a precursor [C. L. Hill, G. M. Whitesides, J. Am. Chem. Soc., 96, 870 (1974)].

E.6. Vinyl benzyl ether with butyllithium and TMEDA at -27° gives α-phenylallyl alcohol (90%). The substrate with butyllithium perhaps forms benzaldehyde and vinyllithium, which add to give the product. If metalation does not occur first, yields are high [G. Büchi et al., J. Am. Chem. Soc., 96, 2576 (1974)].

References

1. M. W. Rathke, J. Am. Chem. Soc., 92, 3222 (1970).
2. G. W. Moersch, A. R. Burkett, J. Org. Chem., 36, 1149 (1971).
3. P. J. Pearce et al., Chem. Commun., 1160 (1970); J. Chem. Soc., Perkin Trans. I, 1655 (1972).
4. P. D. Bartlett et al., J. Am. Chem. Soc., 91, 6362 (1969).
5. P. E. Eaton et al., J. Org. Chem., 37, 1947 (1972).
6. Th. Kaufmann et al., Angew. Chem., 82, 138 (1970).
7. U. Schöllkopf, P. Böhme, Angew. Chem., 83, 490 (1971).
8a. I. I. Lapkin et al., J. Org. Chem. USSR, 5, 868 (1969).
8b. C. J. Willis, Can. J. Chem., 50, 512 (1972).
9. C. H. DePuy, R. A. Klein, Org. Syn., Coll. Vol., 5, 1058 (1973).
10. T. Holm, I. Crossland, Acta Chem. Scand., 25, 59 (1971).
11. E. C. Ashby et al., J. Org. Chem., 38, 2526 (1973).

12. M. Chastrette, R. Amouroux, Chem. Commun., 470 (1970); Bull. Soc. Chim. France, 4348 (1970).
13. H.-J. Bruer, R. Haller, Tetrahedron Letters, 5227 (1972).
14. M. Schlosser, D. Coffinet, Synthesis, 575 (1972).
15a. J. Huet et al., Compt. Rend., 271, C430 (1970).
15b. H. Newman, Tetrahedron Letters, 4571 (1971).
16. D. Abenheim et al., Bull. Soc. Chim. France, 4038 (1969).
17. W. N. Smith, E. D. Kuehn, J. Org. Chem., 38, 3588 (1973).
18. R. W. Herr, C. R. Johnson, J. Am. Chem. Soc., 92, 4979 (1970).
19. C. R. Johnson et al., J. Am. Chem. Soc., 92, 3813 (1970).
20. T. Fujita et al., Synthesis, 630 (1972).
21. M. Gaudemar, Organomet. Chem. Review, 8A, 183 (1972).
22. J. W. Frankenfeld, J. J. Werner, J. Org. Chem., 34, 3689 (1969).
23. D. A. Cornforth et al., J. Chem. Soc., C2799 (1969).
24. A. Horeau, Tetrahedron Letters, 3227 (1971).
25. J. Curé, M. Gaudemar, Bull. Soc. Chim. France, 2471 (1969).
26. B. Castro et al., Bull. Soc. Chim. France, 3521 (1969).
27. M. W. Rathke, A. Lindert, J. Org. Chem., 35, 3966 (1970).
28. M. Gaudemar et al., J. Organomet. Chem., 36, C33 (1972).
29. N. Goasdoue, M. Gaudemar, J. Organomet. Chem., 28, C9 (1971).
30. I. I. Lapkin, Y. V. Ionov, J. Org. Chem., USSR, 6, 2428 (1970).
31. M. Pailer, O. Vostrowsky, Monatsh. Chem., 102, 951 (1971).
32. J. K. Crandall, A. C. Clark, Tetrahedron Letters, 325 (1969).

F. Addition of Simple Anions or Nucleophilic Molecules to Carbonyl Compounds

2. From Alkali Cyanides or Hydrogen Cyanide (1, 233)

$$\text{>C=O} \ + \ HCN \ \longrightarrow \ \text{>C}\begin{smallmatrix}OH\\CN\end{smallmatrix}$$

The reaction has now been adapted to the preparation of β-aminoalcohols [1]:

$$RCH_2COR \xrightarrow[\text{Stork method}]{\text{House or}} RCH=CR \ (OSiMe_3) \xrightarrow[H_2SO_4]{HCN} RCH_2CR \ (OSiMe_3)(CN) \xrightarrow{LiAlH_4} RCHCR \ (OH)(CH_2NH_2)$$

37–47%

(3 examples)

Reference

1. W. H. Parham, C. S. Roosevelt, Tetrahedron Letters, 923
    (1971).

G. Addition of Carbanions

1. From Carbonyl Compounds (Aldol or Ketol) or Carbonyl
    Compounds and Acid Derivatives (Claisen-Schmidt)
    (1, 235)

Methylcalcium iodide has been recommended as a catalyst in
the aldol condensation [1]. With acetone it forms first
diacetone alcohol, which is then converted into mesityl oxide:

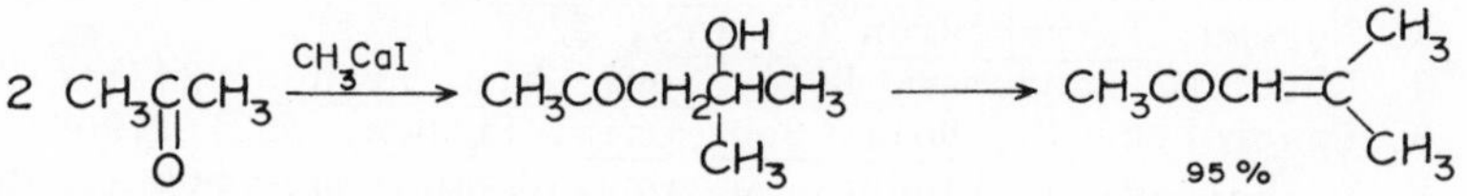

Aluminum chloride in pyridine was utilized by Takahashi and
Schmid [2] in the synthesis of aldols from long-chain aldehydes.
Thus decanal gave 2-octyl-3-hydroxydodecanal (Ex. a).

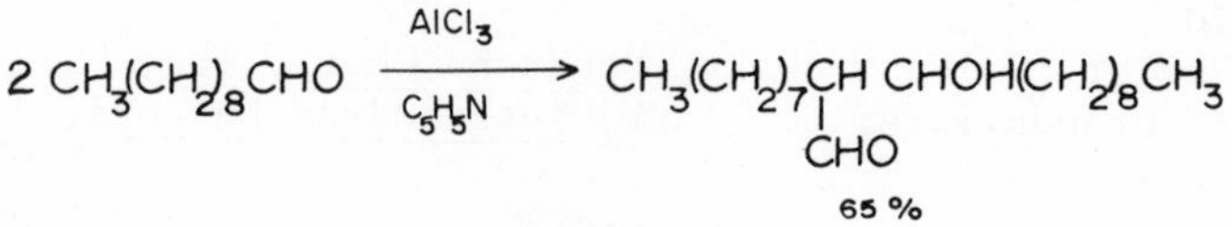

The aldol reaction of the dianion of β-keto esters was
accomplished in steps as shown [3]:

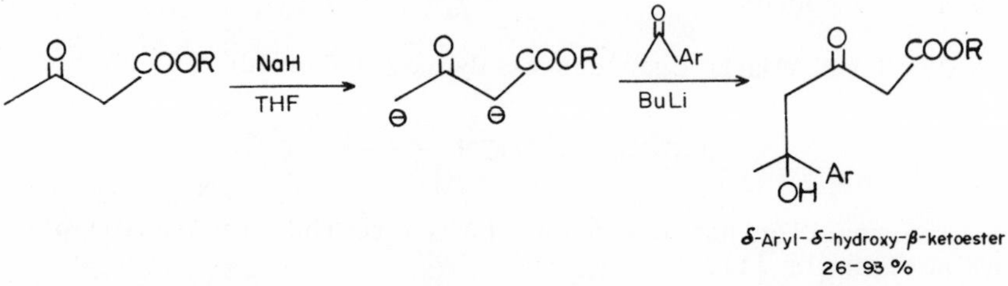

δ-Aryl-δ-hydroxy-β-ketoester
26-93 %

The process leads to alkylation of β-keto esters at the γ carbon.
In one example methyl acetoacetate condensed with pivaldehyde to
give methyl 2-acetyl-3-hydroxy-5,5-dimethylpentanoate (82%).
    The difficulty of controlling the aldol condensation has
been pointed out (1, 235). One method of value for this purpose
is the directed procedure of Wittig in which success is due to
the formation of a bidentate chelate. House and co-workers [4]

applied the same principle to the conventional aldol condensation in that they intercepted the intermediate keto alkoxide in the series of equilibria as a metal chelate:

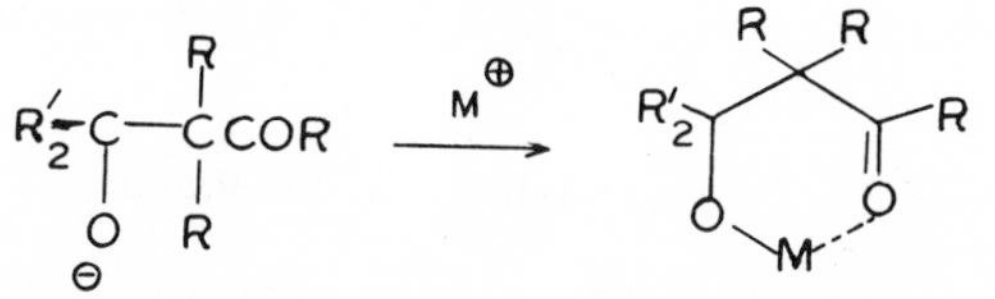

The conditions necessary for success in starting with the pre-formed lithium enolate are given in great detail by the authors. Usually, $ZnCl_2$ or $MgBr_2$ in Et_2O or Et_2O-DME are employed at low temperature and the reaction is of short duration. To cite one of the many syntheses studied, the lithium enolate of phenyl methyl ketone and benzaldehyde gave the aldol as shown (Ex. b):

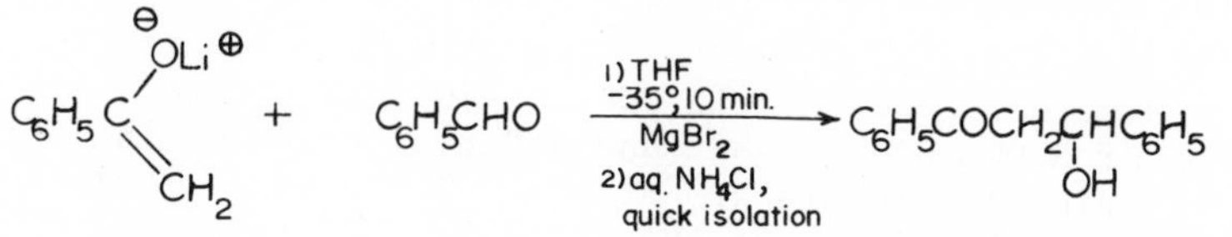

1,3-Diphenyl-1-hydroxy-3-keto-propane, 81 %

The investigators emphasize the instability of some of these aldol products.

The hydroxymethylation of ketones has been accomplished as shown in the case of t-butylcyclohexanone [5]:

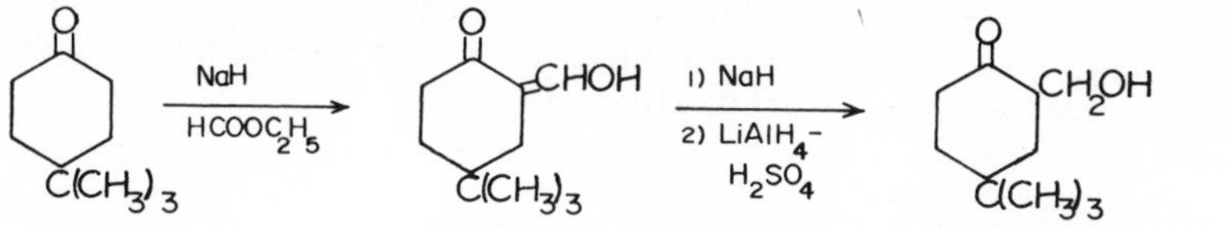

2-Hydroxymethyl-4-t-butylcyclohexanone, 90%

The aldol condensation occurs without a catalyst in the temperature range of 200-300°. If electron-withdrawing groups are attached to the α position of the aldehyde, the reaction occurs at temperatures as low as 100°. Thus at this temperature glyoxalic acid adds to acetone to give 4-oxo-2-pentenoic acid, $CH_3COCH=CHCO_2H$, in high yield [6].

Aldol products of chloral and fluoral have been described [7]. In addition, an interesting intramolecular aldol condensation has given rise to a partially optically active product by using quinine as the base [8]:

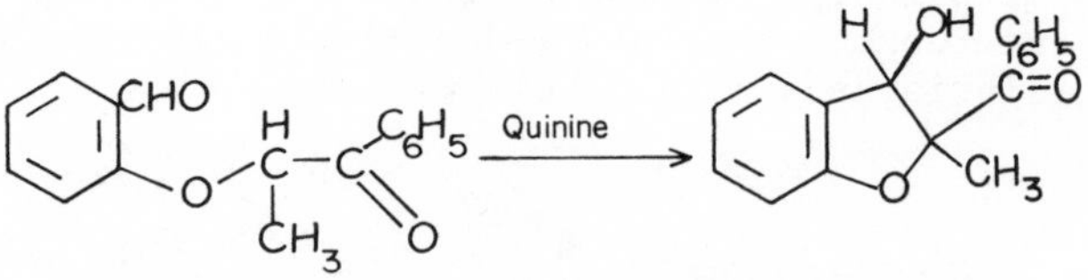

a.   Preparation of 2-Octyl-3-hydroxydodecanal [2]

Aluminum chloride, 12 mmol, in absolute C_5H_5N was cooled under N_2 and decanal, 2 mmol in C_5H_5N, was added at 0-3°, after which the mixture was stirred for 45 min. From the ethereal extract the aldol (65%) was recovered.

b.   Preparation of 1,3-Diphenyl-1-hydroxy-3-ketopropane [4]

The lithium enolate from 5 mmol of MeLi and 0.73 g of the silyl enol ether, $\overset{\displaystyle OSiMe_3}{\underset{\displaystyle |}{PhC}}=CH_2$, in THF at -40° was treated with a solution of 5.0 mmol of $MgBr_2$ and 5.0 mmol of PhCHO in THF. After stirring for 10 min at -35 to -50° the usual isolation procedure gave 0.77 g (81%) of the ketol.

2.   From Formaldehyde (Tollens) (1, 239)

The addition of formaldehyde to any compound with an active hydrogen atom has been reviewed [9].
The factors affecting rates in the condensation of formaldehyde and aliphatic aldehydes have been discussed by Vik [10]. Cyclohexane -1,1,4,4-tetramethanol has been prepared in 81% yield from 1,4-cyclohexanedicarboxaldehyde [11].

3.   From Alcohols (Guerbet)

The condensation of butanol with sodium butoxide was studied in the presence of phosphine complexes of salts of Rh, Ir, Pt, and Ru [12]. Highest yields (90%) were obtained with $Rh(CO)_2+8P(C_2H_5)_3$, $RuCl_3 \cdot 3H_2O+4P(C_4H_9)_3$, and $Ir(CO)Cl(PPh)_3$ in the equation:

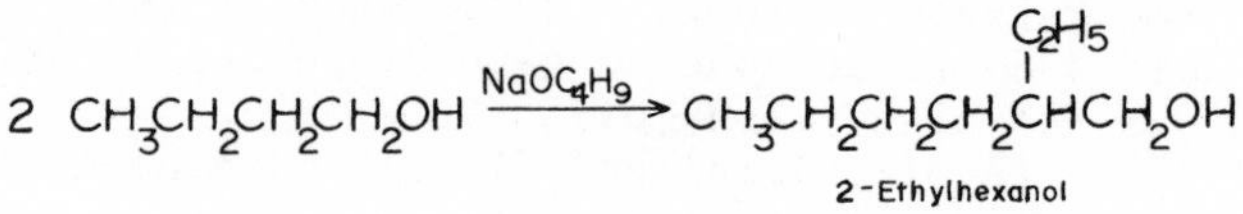

$$2 \ CH_3CH_2CH_2CH_2OH \xrightarrow{NaOC_4H_9} CH_3CH_2CH_2CH_2\overset{\displaystyle C_2H_5}{\underset{\displaystyle |}{CH}}CH_2OH$$

2-Ethylhexanol

7.   From Ethers (Wittig Rearrangement) (1, 243)

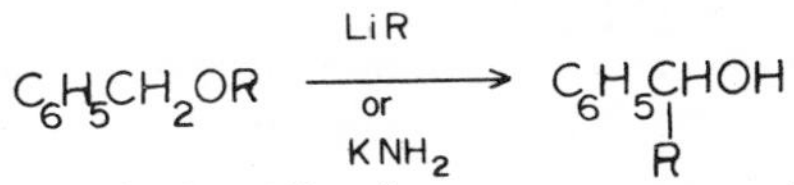
LiR
$C_6H_5CH_2OR$
or
KNH_2
C_6H_5CHOH
R

The mechanism of the Wittig rearrangement has been discussed [13].

9. From Allylic Sulfoxides (see Addendum)

Addendum

G.9. From Allylic Sulfoxides. Allylic sulfoxides may be alkylated and rearranged as follows [D. A. Evans, G. C. Andrews, Acc. Chem. Res., 7, 147 (1974)]:

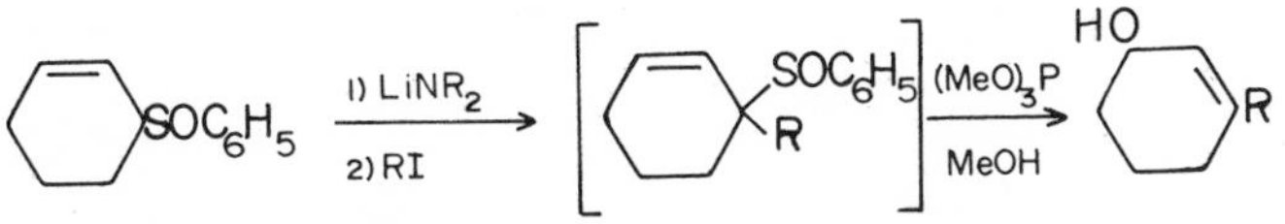
SOC_6H_5
1) $LiNR_2$
2) RI
SOC_6H_5
R
$(MeO)_3P$
MeOH
HO
R

For mechanism see p. 215.

References

1. A. V. Bogatskii et al., J. Gen. Chem. USSR, 40, 1167 (1970).
2. T. Takahashi, H. H. O. Schmid, Chem. Phys. Lipids, 3, 185 (1969); C. A., 71, 70037 (1969).
3. S. N. Huckin, L. Weiler, Tetrahedron Letters, 4835 (1971).
4. H. O. House et al., J. Am. Chem. Soc., 95, 3310 (1973).
5. E. J. Corey, D. E. Cane, J. Org. Chem., 36, 3070 (1971).
6. H. Müller et al., Angew. Chem. Intern. Ed. Engl., 10, 846 (1971).
7. E. Kiehlmann, Angew. Chem. Intern. Ed. Engl., 10, 839 (1971).
8. U. Obenius, G. Bergson, Chem. Acta Scand., 26, 2546 (1972).
9. J. Mathieu, J. Weill-Raynal, Formation of the C-C Bonds, Vol. 1, Georg Thieme, Stuttgart, 1973, pp. 18-33.
10. J.-E. Vik, Acta Chem. Scand., 26, 3165 (1972).
11. C. F. Beam, W. J. Bailey, J. Chem. Soc., C2730 (1971).
12. G. Gregorio, G. F. Pregaglia, J. Organomet. Chem., 37, 385 (1972).
13. U. Schöllkopf, Angew. Chem. Intern. Ed. Engl., 9, 763 (1970).

H. From Free Radicals, Cyclo Processes, Electrolytic Reactions, and the Like

The degradation of acids to alcohols occurs via the acid chloride by treatment with m-chloroperbenzoic acid in pyridine [1]. Thus 2-ethylhexanoic acid responds as indicated:

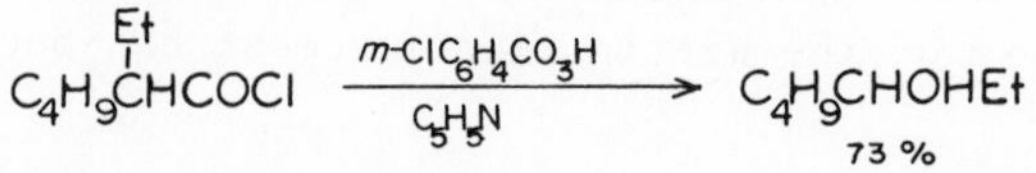

$$C_4H_9\overset{\overset{\displaystyle Et}{|}}{C}HCOCl \xrightarrow[C_5H_5N]{m\text{-}ClC_6H_4CO_3H} C_4H_9CHOHEt$$

73 %

Yields with other alkyl groups are somewhat lower.

Trimethylsilyl enol ethers with the Simmons-Smith reagent form trimethylsilyl cyclopropyl ethers which may be degraded to cyclopropanols [2]. The steps are shown with the trimethylsilyl enol ether of acetophenone:

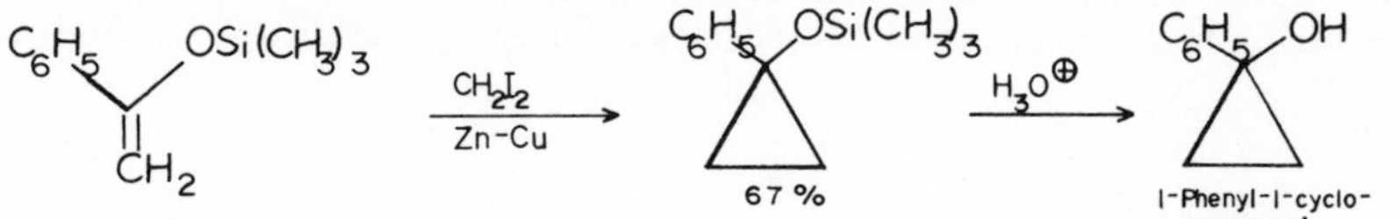

67 %

l-Phenyl-l-cyclo-propanol
73%
(without isolation of intermediate)

An electrolytic reaction of unsaturated ketones gives a number of cyclic tertiary alcohols [3]:

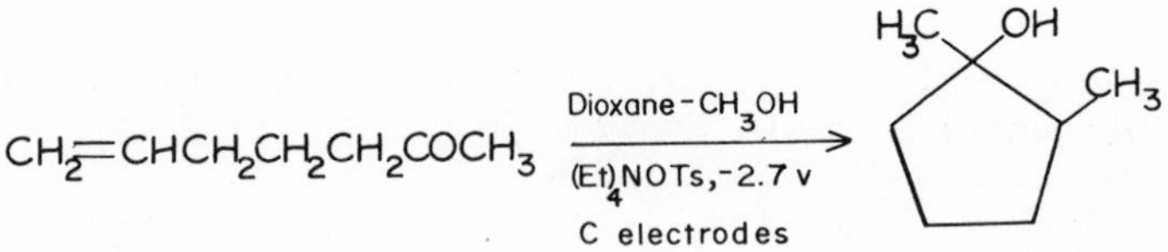

$$CH_2{=}CHCH_2CH_2CH_2COCH_3 \xrightarrow[\substack{(Et)_4NOTs,-2.7v \\ C\ electrodes}]{Dioxane-CH_3OH}$$

1,2- Dimethylcyclopentanol
66 %
(other yields, 35-47 %)

Addenda

H. The methylol free radical, ·CH$_2$OH, generated from methanol and t-butyl peroxide, attacks heterocycles such as quinaldine to give the methylol derivative, 4-hydroxymethylquinaldine (86%) [R. O. C. Norman, Chem. Ind. (London), 974 (1973)].

H. The Simmons-Smith reagent reacts with the trimethylsilyl vinyl ether of cyclohexanone in high concentration in ether to give after hydrolysis 2-methylidenecyclohexanol (68%). At low concentration in ether the normal cyclopropanation product was obtained [S. Murai et al., J. Org. Chem., 39, 858 (1974)].

H.   Methylolation is carried out on heterocycles with ammonium
peroxysulfate in methyl alcohol, water, and sulfuric acid.   The
reaction is a free-radical one involving attack of $\cdot CH_2OH$ [F.
Minisci, Synthesis, 18 (1973)].

References

1.   D. B. Denney, N. Sherman, J. Org. Chem., 30, 3760 (1965).
2.   G. M. Rubottom, M. I. Lopez, J. Org. Chem., 38, 2097 (1973).
3.   T. Shono, M. Mitani, J. Am. Chem. Soc., 93, 5284 (1971).

Chapter 5

PHENOLS

The synthesis of phenols by the hydrolysis of aryl halides shows
increasing use of Cu-Cu$_2$O (A.2) and that by the hydrolysis of

diazonium tetrafluoroborates leads to superior yields when trifluoroacetic acid and its salts are employed as a solvolytic
medium [A.6]. 4- and 5-Alkylresorcinols have become more readily
available of late by capitalizing on the concomitant benzyne-
direct replacement mechanisms during the fusion of m-chlorocre-
sols with alkali (A.2).

The oxidation of arenes to phenols by oxygen or peroxides
has been accomplished to a limited degree (B.1). A more
promising procedure from the arene is to utilize a one-pot
oxidation with thallium trifluoroacetate and lead tetraacetate
(B.10). In the direct oxidation of arenes mixtures of ortho and
para phenols are obtained when both are possible, a fact that
calls attention to the electrophilic character of the attacking
species, be it free radical or positive ion. For specific
orientation control the conversion of the aromatic Grignard reagent to the organoborane followed by peroxide oxidation should
be considered (B.2). Photochemical hydroxylation of aromatic
compounds has been reviewed recently (B.1). The scope is quite
broad, but yields are disappointingly low in most cases. Oxidation coupling of phenols also has been reviewed of late and this
article has led to a more extensive discussion (B.9).

In electrophilic reactions the great activity of the phenol
nucleus has been demonstrated by alkylations carried out in
aqueous solution (D.1). Rare 2,2'-dihydroxytriphenylmethanes
have become available by condensation of aryloxymagnesium bromides with benzaldehyde (D.1). Chlorination of a particular
phenol with sulfuryl chloride has been found to depend on the
presence of thiolacetic acid (D.2).

As is true in so many cases, for aromatic compounds the
syntheses of phenolic compounds from straight-chain substrates
strike the eye. Two are to be found in E.3: 5-Alkylresorcinols
from the Michael condensation of acetoacetic and acrylic esters
and a naphthol from the photochemical ring closure of a β-
phenylisopropylidene malonic ester. A third is in F.3, the
cyclo reaction of methallyl chloride, acetylene, carbon monoxide,
and catalyst to form m-cresol.

A.   Solvolysis

    2. From Halides (1, 247)

$$\text{Ar Cl} \xrightarrow{\text{NaOH}} \text{ArOH}$$

The activity of aromatic halides has been reviewed recently
[1].

3-Hydroxypyrene has been synthesized from 3-bromopyrene by
heating under pressure as shown [2].

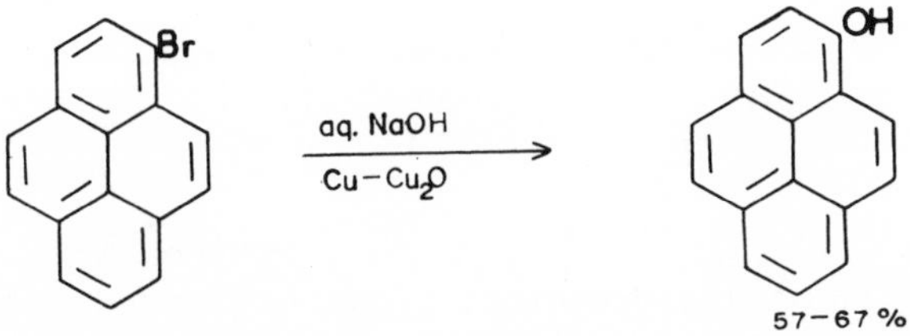

aq. NaOH
Cu—Cu$_2$O

57—67 %

The method is applicable to the preparation of other phenols such as 2,3,6-trimethylphenol. Since two mechanisms, elimination-addition (benzyne) and S$_N$2 displacement, are operative, the copper-cuprous oxide system is introduced to suppress the benzyne route, which leads to the formation of mixtures of isomers.

In another reaction in which the benzyne is desired, DMSO essentially adds to the benzyne to give some unusual phenols [3]:

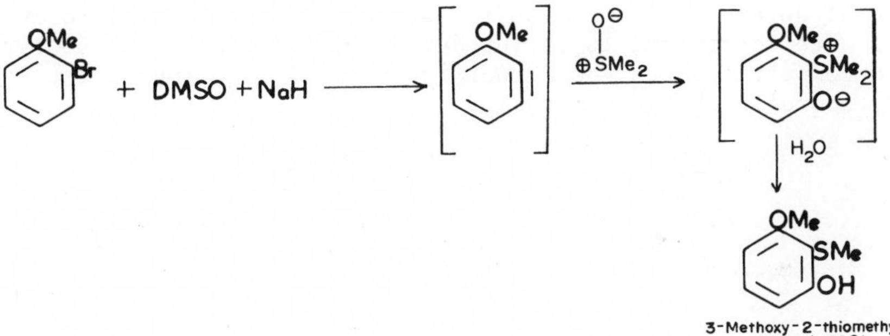

3-Methoxy-2-thiomethyl-phenol, 30%

In displacement in perhalophenols the reaction may be selective [4]:

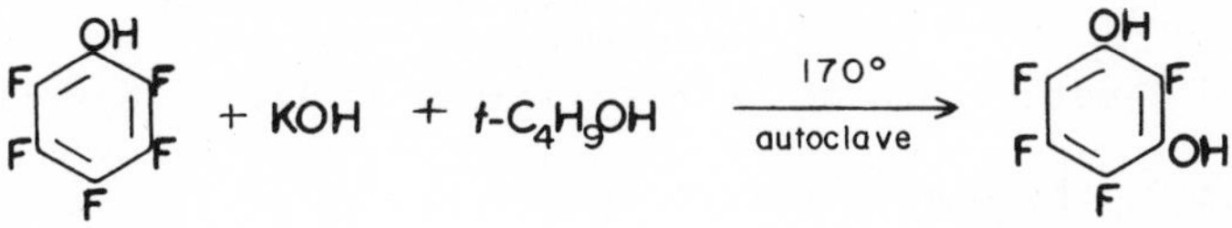

2.2 g.

Tetrafluororesorcinol
1.5 g. as the dimethyl ether

It is noteworthy that displacement in 2-nitro-4-trifluoro-methylchlorobenzene shows no hydrolytic effect on the trifluoro-methyl group [5]:

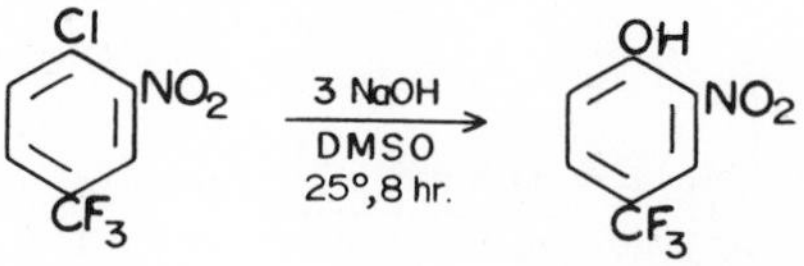

2- Nitro - 4 - trifluoro-
methylphenol, 96 %

However, if the trifluoromethyl group is <u>ortho</u> to the nitro group, as in 4-nitro-3-trifluoromethylchlorobenzene, it is lost while the chloro group is unaffected:

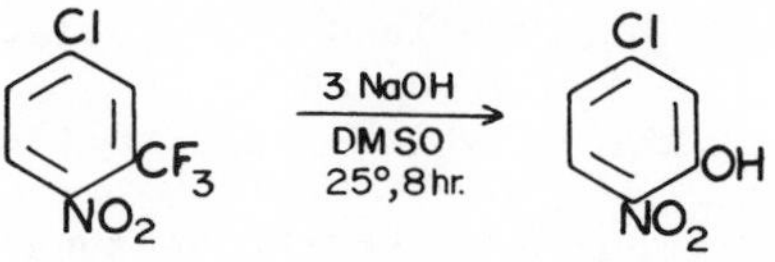

2 - Nitro-5-chloro-
phenol, 93%

Orcinol is a relatively expensive compound, but recent commercial developments promise to make both 4- and 5-alkylresorcinols more economical [6]. The development depends on a benzyne or a competitive benzyne-direct replacement reaction of the corresponding chloro-n-cresols:

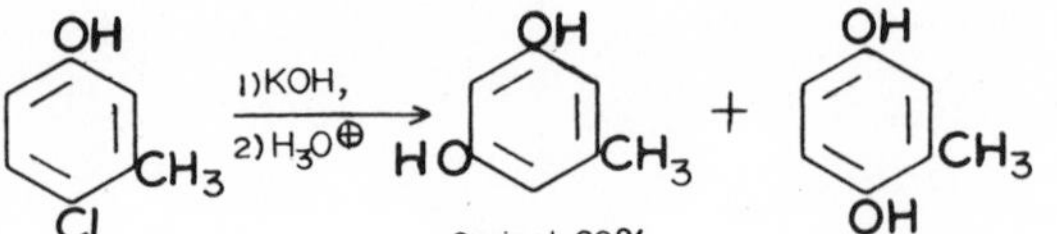

Orcinol, 60%   Methylhydroquinone, 3-6 %
(via benzyne)

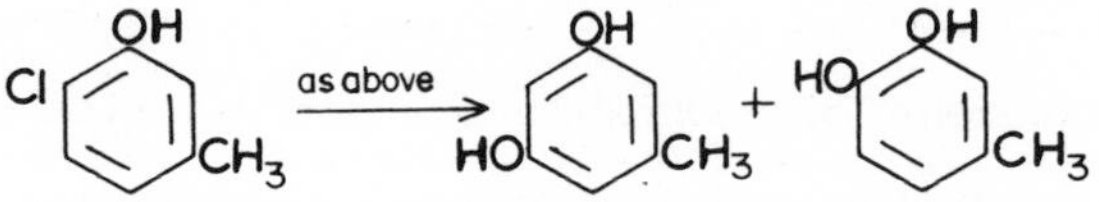

Orcinol, 60%   4-Methylcatechol, 40%
(via benzyne)

Orcinol and 4-methylcatechol are separated easily by distillation.

Recently, the photochemical hydrolysis of aryl halides has been accomplished [7] (Ex. a).

a. Preparation of Resorcinol [7]

(Seventy-four percent by irradiation for 5 hr of 17.6 mmol of m-chlorophenol in a solution of 10 g of KOH in 200 ml of water housed in a Vycor glass flask.)

4.   From Ethers and Basic Reagents (1, 253)

$$ArOR + R'O^{\ominus} \longrightarrow ArO^{\ominus} + ROR'$$

Additional basic reagents which have been used in de-
methylating methyl ethers by alkyl fission are :   (a) aqueous
methylamine (40%) under pressure [8], (b) lithium iodide-
collidine [9], (c) the thioethoxide ion in DMF [10], (d) potassium thiophenoxide in diethylene glycol [11], and (e) methylmagnesium iodide in boiling xylene [11a].   The LiI-collidine
reagent in 10 hr gives an almost quantitative yield, after
acidification, of β-naphthol from the methyl ether.   The thioethoxide ion, a preferred reagent, which is effective at low
temperatures and a short reaction time, gives, also after
acidification, yields of phenols from substituted anisoles of
94-98%.   With this reagent it is of interest to note that the
bromine in 4-bromo-3-methylanisole is not affected and a hindered ether, 2-t-butylanisole, responds without difficulty.   In
addition, the dimethyl ethers of dihydric phenols are monode-
methylated.   The potassium thiophenoxide ion produces minimum
isomerization of the $\Delta^{9(11)}$ double bond in Δ-$^{9(11)}$ trans-
tetrahydrocannabinol to the Δ^8 and Δ^9 positions [11].
     A cleavage of limited scope is the splitting of the methyl-
enedioxy ring as shown [12]:

o-Ethoxyphenol, 40 %

a.   Preparation of m-Cresol [10]

One gram of 3-methylanisole, 2.5 equiv of sodium thioethoxide, and 40 ml of DMF were refluxed under nitrogen for 3
hr.   After cooling, the reaction mixture was acidified and
extracted with ether.   The phenol recovered from the alkaline
washing of the ether layer amounted to 94%.

5.   From Ethers by Acid Cleavage (1, 254)

$$3 \ ArOMe + BBr_3 \longrightarrow (ArO)_3B + 3 \ MeBr$$
$$\downarrow H^{\oplus}$$
$$3 \ ArOH$$

Boron tribromide in methylene chloride has been utilized in
the demethylation of 3,3'-dimethoxybiphenyl to produce 3,3'-
dihydroxybiphenyl in 77-86% yield [13].

In the demethylation of 1,7-bis (p-methoxyphenyl)-heptanone-4 with hydrogen bromide cyclization occurs in addition as shown [14]:

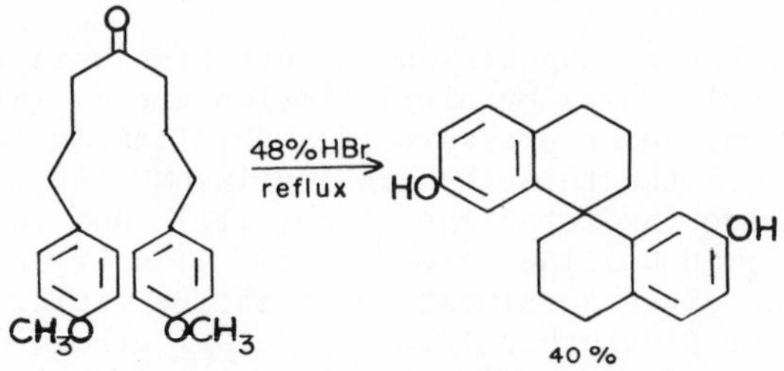

If in place of refluxing HBr, 100 ml of 48% HBr in 100 ml of acetic acid is refluxed with 3 g of the substrate, the diphenol is obtained (85%) with no indication of any cyclization.

The phenacyl group has been recommended as a protective group for phenols, indicating that the group is both easily attached and removed [15]. Such has been accomplished as shown:

$$Ar\,OH + C_6H_5COCH_2Br \xrightarrow{K_2CO_3} Ar\,OCH_2COC_6H_5 \xrightarrow{Zn-AcOH} Ar\,OH$$

It may be assumed that the phenacyl group is cleaved easily for the same reason that the trifluoroethyl group is cleaved in esters.

6. From Diazonium Salts (1, 255)

$$Ar\,N_2^{\oplus}X^{\ominus} \xrightarrow{H_2O} ArOH + N_2 + HX$$

The thermal decomposition of each of the fluoroborates of the o, m-, and p-hydroxybenzenediazonium salts gives 37, 35, and 69%, respectively, of o, m-, and p-fluorophenols [16].

In a good general method [17] the tetrafluoroborate salt is isolated and then decomposed by boiling in trifluoroacetic acid and potassium trifluoroacetate:

$$Ar\,N_2BF_4 \xrightarrow[\substack{CF_3CO_2K \\ 2)\,H_2O}]{1)\,CF_3CO_2H} Ar\,OH$$

Yields are 75% or better with some heterocyclic amines.

7. From Certain Esters by Photolysis

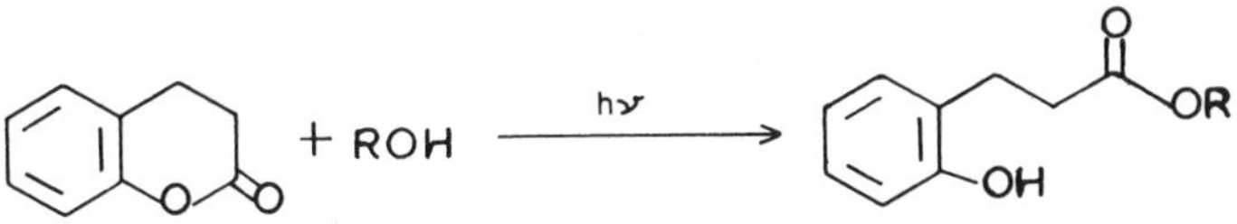

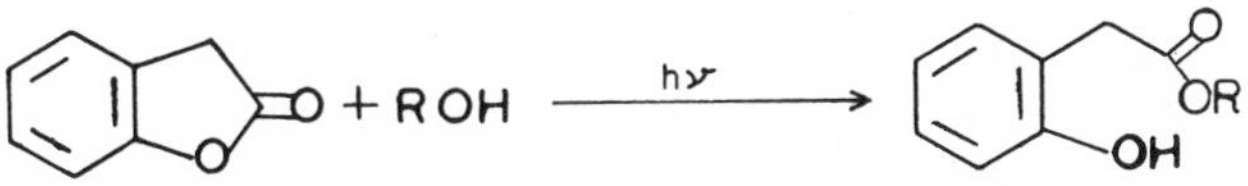

Gutsche and co-workers [18] have discovered that dihydro-coumarin and 2-keto-2,3-dihydrobenzofurans are cleaved by irradiation in alcoholic solution to phenols. It has also been found that other products reminiscent of a carbonium ion pre-cursor may be formed if there are substituents at the 3-position of the benzofuran:

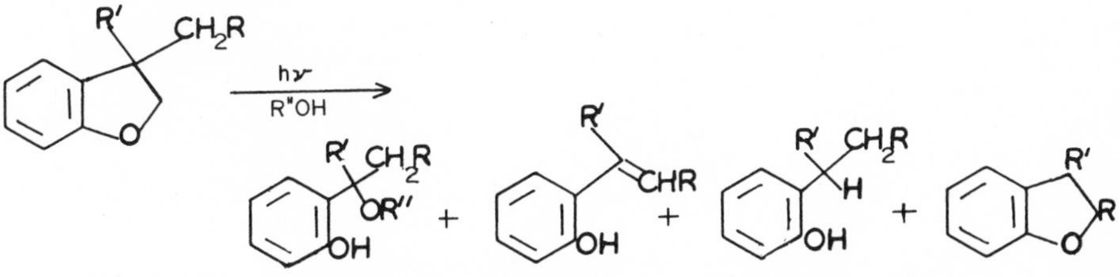

8.   From Phenols (O-Deuteration)

$$ArOH \xrightarrow{ClSi(CH_3)_3} ArOSi(CH_3)_3 \xrightarrow{CH_3OD} ArOD$$
$$\qquad\qquad\quad 90\% \qquad\qquad\qquad\quad 95\%$$

The best way to deuterate phenols is first to convert the phenol into the o-trimethylsilyl derivative with chlorotrimethyl-silane and then to reflux with deuterated methanol for 1 hr [19].

Addendum

A.4. Monodemethylation of a resorcinol dimethyl ether may be brought about with NaSEt in DMF [R. N. Mirrington, G. I. Fentrill, Org. Syn., 52, Unchecked Procedure 1820 (1972)], while demethylation of some phenolic ethers occurs by treatment with anhydrous Na$_2$S in N-methylpyrrolidone at 140-150° for 2-4 hr [M. S. Newman et al., private communication].

References

1. C. A. Fyfe, in S. Patai, The Chemistry of the Hydroxyl Group, Pt. 1, Wiley-Interscience, New York, 1971, p. 83.
2. W. H. Gumprecht, Org. Syn., 48, 94 (1968).
3. A. J. Birch et al., Austral. J. Chem., 24, 2179 (1971).
4. J. Burdon et al., J. Chem. Soc., 5152 (1965).
5. R. L. Jacobs, J. Org. Chem., 36, 242 (1971).
6. W. M. Carmichael et al., Chem. Ind. (London), 685 (1973).
7. T. Matsuura, K. Omura, Synthesis, 173 (1974).
8. I. M. Lockhart, N. E. Webb, Chem. Ind. (London), 1230 (1970).
9. I. T. Harrison, Chem. Commun., 616 (1969).
10. G. I. Feutrill, R. N. Mirrington, Tetrahedron Letters, 1327 (1970); Austral. J. Chem., 25, 1719, 1731 (1972).
11. C. G. Pitt et al., J. Org. Chem., 36, 721 (1971).
11a. S. Cabiddu et al., Gazz. Chim. Ital., 99, 771, 1095 (1969).
12. S. Cabiddu et al., Gazz. Chim. Ital., 98, 800 (1968).
13. J. F. W. McOmie, D. E. West, Org. Syn., 49, 50 (1969).
14. I. Kawasaki et al., Bull. Chem. Soc. Jap., 44, 1986 (1971).
15. J. B. Hendrickson, C. Kandall, Tetrahedron Letters, 343 (1970).
16. O. Daněk et al., Collection Czech. Chem. Commun., 32, 1642 (1967).
17. J. M. Muchowski et al., Can. J. Chem., 51, 2347 (1973).
18. C. D. Gutsche et al., J. Am. Chem. Soc., 90, 5855 (1968); J. Org. Chem., 38, 1993 (1973).
19. D. R. M. Walton et al., J. Chem. Soc., C, 1577 (1970).

B. Oxidation

A review, including reactions which introduce hydroxyl groups into aromatic compounds by oxidation, has been published [1].

1. From Aromatic Compounds and a Reagent Supplying Oxygen (1, 259)

A review on a portion of this subject has been published [2]. After many years of indoctrination with the belief that the oxidation of hydrocarbons to phenols is not a good method of synthesis because the phenol is more easily oxidized than the hydrocarbon, three papers challenging this doctrine have been published recently.

Veseley and Schmerling [3] have shown that liquid H_2F_2

and CO_2 under pressure (the latter effects solubility of the hydrocarbon in H_2F_2) so modify the oxidation of hydrocarbons that phenols are obtained. A typical oxidation scheme is shown:

$$C_6H_5CH_3 + H_2O_2(30\%) + H_2F_2 + CO_2 \xrightarrow[\text{Pressure}]{0°} CH_3C_6H_4OH$$

4 mol   0.39 mol   15.0 mol   49 mol %
                                                   (o:p ratio 2.3:1)

As is seen, the process gives an isomeric mixture of cresols. Naphthalene gives 30% 1-, 9% of 2-naphthol, and 17% of 1,5-naphthalenediol. Phenols give rather high yields of dihydroxybenzenes with no CO_2 present to affect solubility.

Kurz and Johnson [4] utilized anhydrous aluminum chloride as a catalyst as shown in a typical procedure:

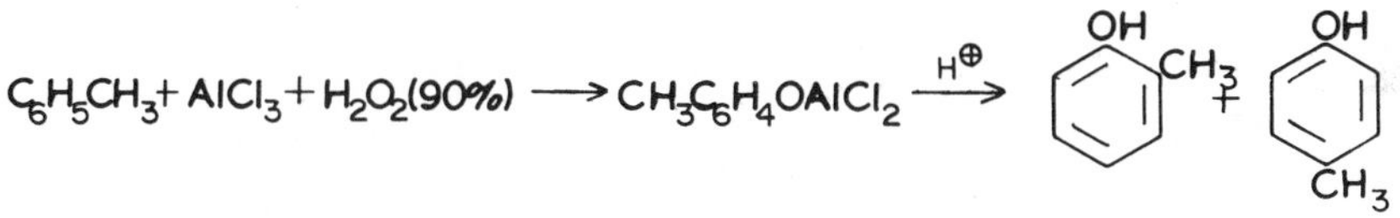

40% (o:p ratio 2:1)

Very little of higher oxidation products is formed in the process. Nuclear chlorination was a minor competing reaction with toluene.

Hart [5] employed trifluoroperacetic acid and boron trifluoride (see 1, 259). This mixture at -40° with mesitylene gives mesitol in an essentially quantitative yield, but with other substrates it is less discriminate in its attack in that it produces cyclohexadienones as well by an attack at a substituted position:

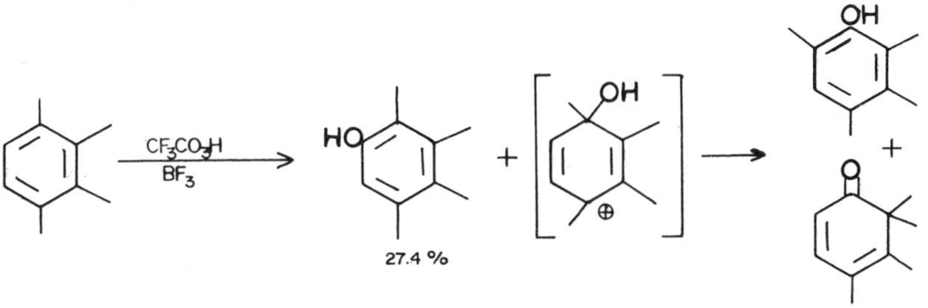

27.4 %   small amounts

Obviously, the dienones are valuable in their own right (1, 68). In considering these three methods the work of Kovacic

(1, 260) using diisopropyl peroxydicarbonate under electrophilic conditions should not be overlooked.

The problem of orientation did not arise in the introduction of the hydroxyl group by Barton and co-workers [6] as indicated:

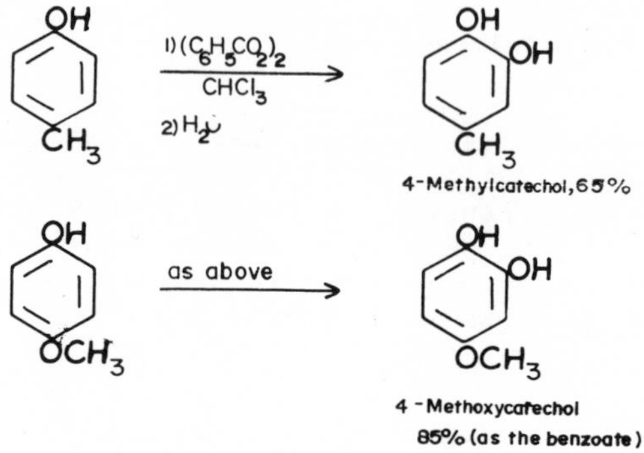

4-Methylcatechol, 65%

4 -Methoxycatechol
85% (as the benzoate)

Lead tetraacetate in glacial acetic acid has been used in the synthesis of 3-fluoranthenol [7] as shown:

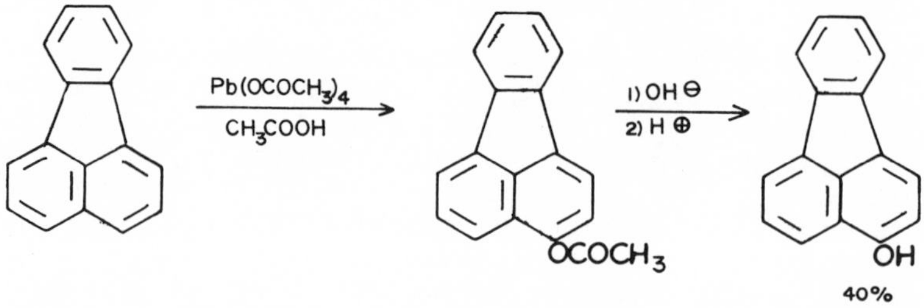

40%

By contrast, lead dioxide in acetic or formic acid gives with phenols p-benzoquinones and diphenoquinones [8].

In addition to chemical hydroxylation just described, photochemical hydroxylation has now been reviewed [9]. Phenol and hydrogen peroxide in water, irradiated at 2537 Å for 6 hr, yield 26% catechol and 14% hydroquinone. The o,p-orientation in photochemical hydroxylation indicates the electrophilic character of the hydroxyl free radical. Water-insoluble phenols were best irradiated in acetonitrile, again in rather low yields (see Ex. a). Anisole in the presence of AlCl₃ yields approximately 11% of o-hydroxyanisole. Heterocyclic-N-oxides on irradiation appear to yield atomic oxygen, which is capable of attacking hydrocarbons. For example, pyridazine N-oxide and mesitylene together yield 23% of mesitol. Azoxybenzenes on irradiation yield 5-15% of o-hydroxyazobenzenes, and hindered nitroarenes appear to yield phenols via an oxirane,

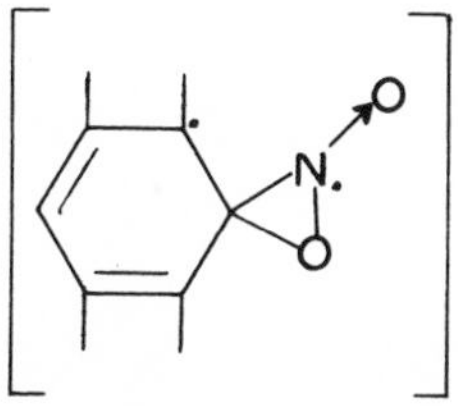

for instance, nitrodurene yields 91% durenol.

a. Preparation of 4-Phenylcatechol [9]

p-Phenylphenol, 29 mmol, 35% hydrogen peroxide, 30 ml, and 200 ml of acetonitrile were irradiated with a 10 watt, low-pressure mercury lamp at 40° for 2.5 hr. After removing the hydrogen peroxide with aqueous $NaHSO_3$, and after chromatographing on silica gel, the residue from the eluate gave 0.6 g (14%) of the catechol.

2. From Organometallic compounds (1, 260)

$$ArMgX \xrightarrow{B(OCH_3)_3} ArB(OCH_3)_2 \xrightarrow[2)H_2O_2]{1)H^\oplus} ArOH$$

Breuer and Broster [10] devised a method of synthesizing phenols from aryl bromides without the isolation of any intermediate as shown:

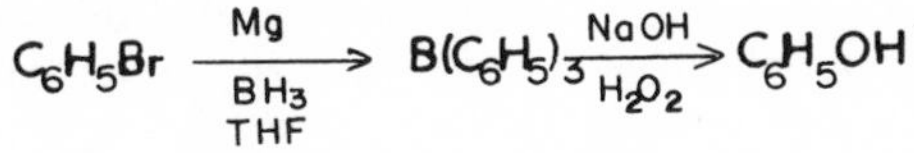

$$C_6H_5Br \xrightarrow[\substack{BH_3 \\ THF}]{Mg} B(C_6H_5)_3 \xrightarrow[H_2O_2]{NaOH} C_6H_5OH$$

Yields of phenols varied from 61 to 77%. The use of the borane offers a simple, mild, and convenient method for the synthesis of phenols via the Grignard reagent. This procedure minimizes the coupling side reactions of the Grignard reagent.

Mention has been made previously of oxidizing the aryl-lithium to the phenol with nitrobenzene (1, 261), but Byck and Dawson [11] carried out the oxidation by first converting the lithium compound into a Grignard reagent and then oxidizing with oxygen:

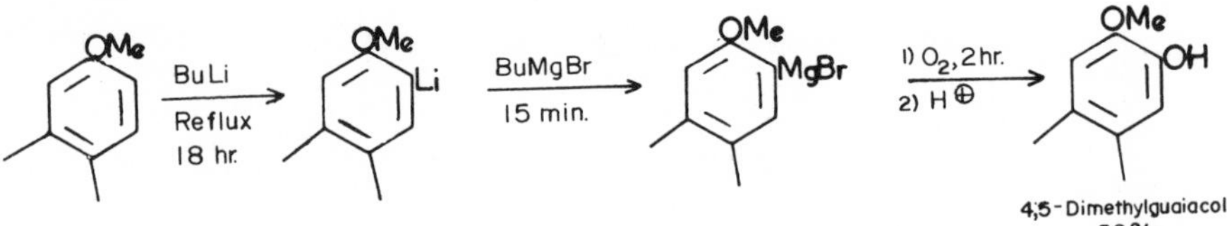

4,5-Dimethylguaiacol
59%

Phenols are also available from the Grignard reagent and 1,3-benzodioxoles [12] as shown:

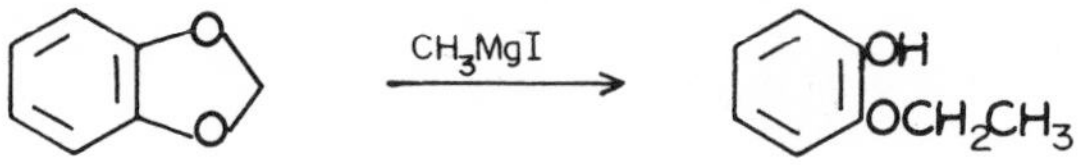

2-Ethoxyphenol, 40%

Other Grignard reagents give yields varying from 22-52%.

5.   From Carbonyl Compounds (Including Dakin) (1, 263)

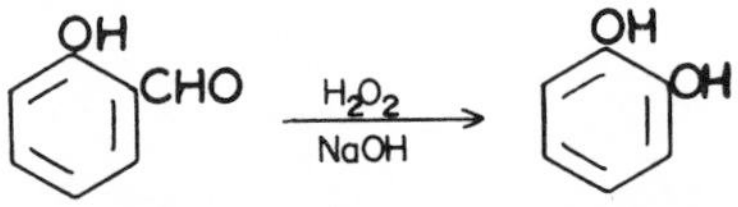

Dihydric phenols have been prepared from p-hydroxybenzaldehyde, salicylaldehyde, 5-bromosalicylaldehyde, and vanillin by treatment with alkaline peroxide [13] in yields of 72-85%. Nitro- and aminobenzaldehydes gave no dihydric phenols.   3-Formylindole gave a 70% yield of indigo.

Occasionally, insoluble sodium or potassium salts, which resist conversion, are formed in the Dakin reaction.   The use of tetramethylammonium hydroxide as the base may prevent precipitation in such cases and improve the yield [14]:

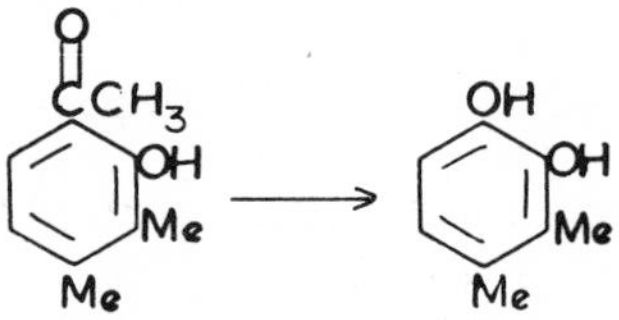

3,4-Dimethylcatechol;
2.5% (using $KOH-H_2O_2$)
25 % (using $Me_4NOH-H_2O_2$)

6. From Cyclic Ketones or Epoxides (Dehydrogenation)
    (1, 264)

A cyclic ether, 1,4-cyclohexadiene dioxide, has been converted into catechol as shown [15]:

70-75 %

The formation of the o-dihydric phenol is accounted for as
follows:

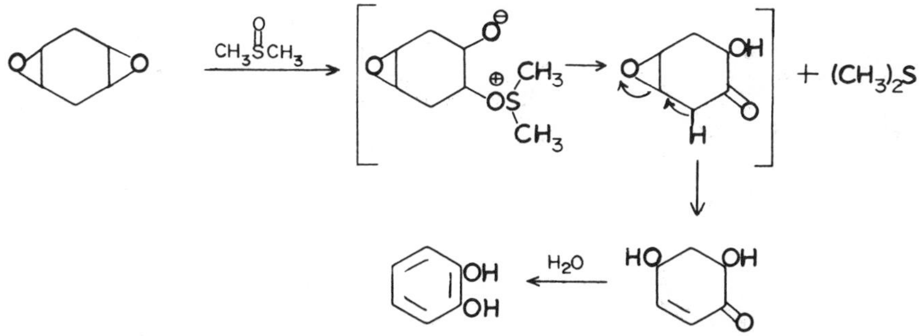

Cyclohexanediones are readily and quantitatively dehydrogenated to their respective phenols; resorcinol is obtained from
1,3-, catechol from 1,2-, and hydroquinone from 1,4-cyclohexane-
dione [16]:

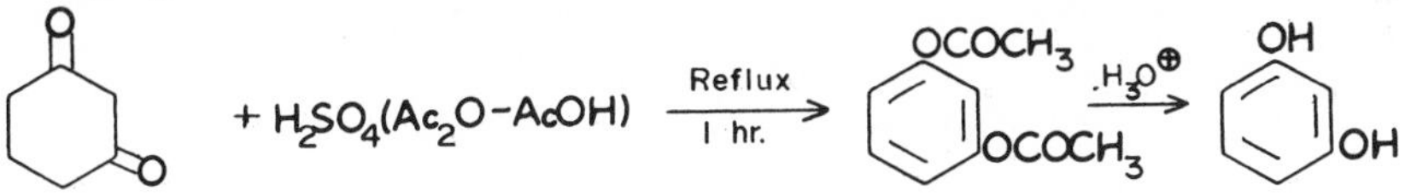

Aromatization of 5,5-dimethyl-1,3-cyclohexanedione gives 4,5-
dimethylresorcinol, a product of methyl migration as well.
A study of the dehydrogenation of methyltetralones by Pd-C
showed that some methylnaphthols were obtained, but the main
product was the methylnaphthalene [17].

9. From Phenols (Oxidative Coupling) (1, 266)

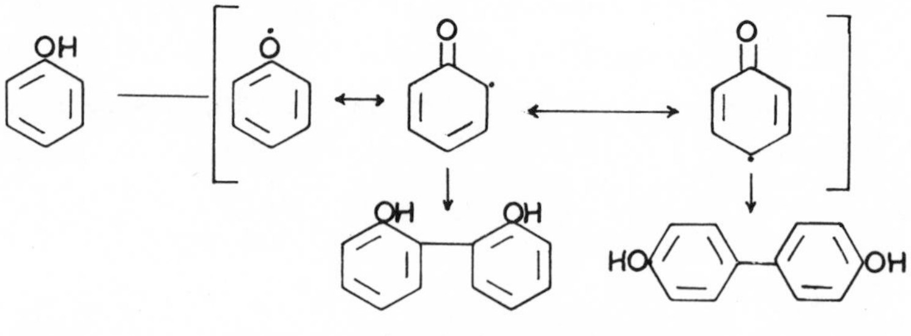

(or mixed *o*, *p* coupling)

Oxidative coupling has been reviewed recently [18]. The phenoxy free radical has a longer life than an alkyl or aryl free radical, does not abstract hydrogen atoms from solvents, and frequently forms a dimer on coupling. As more alkyl or aryl groups are situated on the ring, the more stable the free radical becomes and the higher the yield of the coupling product results, provided no steric problems arise. In such cases disproportionation ensues:

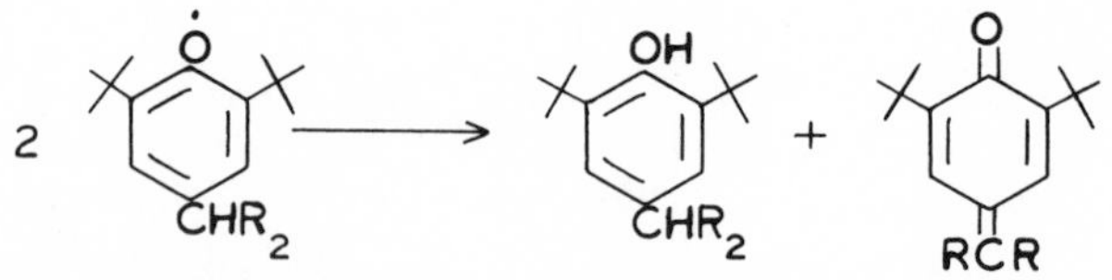

With the short-lived phenoxy radicals such as from phenol itself, the final products are quite complex. Indeed, even from the radical formed from p-cresol, the dimer, trimer, and Pummerer's ketone may be isolated. The latter is derived from ring closure of an o-,p-coupling:

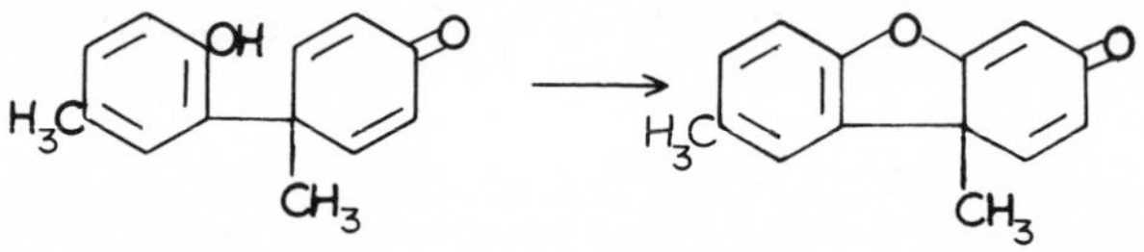

Diphenoquinones may result if the oxidizing conditions are too strenuous:

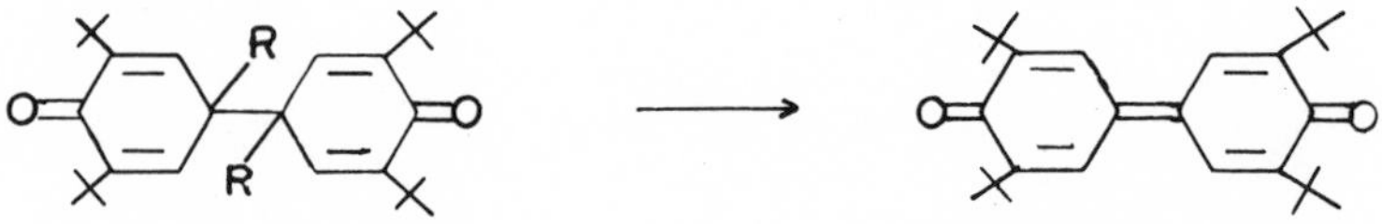

2,6,2,'6' - Tetra-*t*-butyldipheno-quinone

However, the quinones may be reduced back to the bisphenol (see Ex. a). In the equation above the identical tetra-t-butyldi-phenoquinone is obtained when R=COOH, CHO, or halogen.

Another type of coupling occurs at the site of a methyl group:

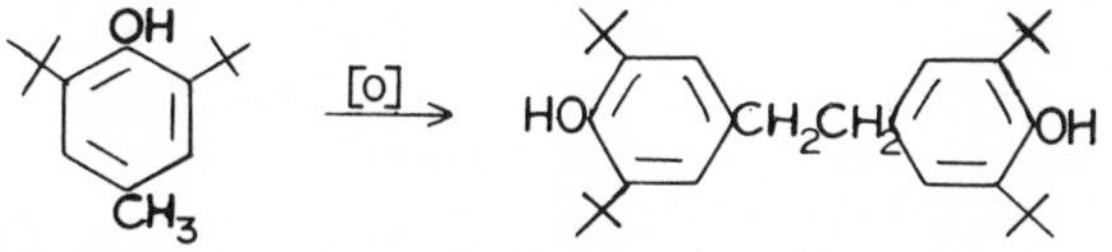

The product may be oxidized further to the stilbenequinone.

The structure of the product also depends on the reagent since the same phenol may be oxidized to the quinone (with the loss of methyl) by β-manganese dioxide (pyrolusite) under the conditions indicated

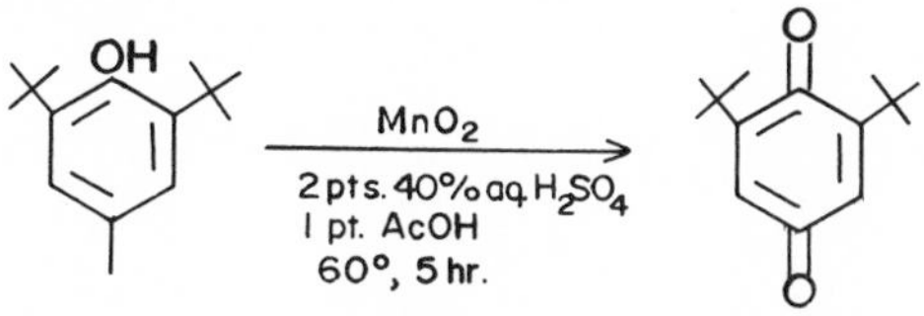

74 %

or if sufficient methanol is present the adduct is isolated in 60% yield [19].

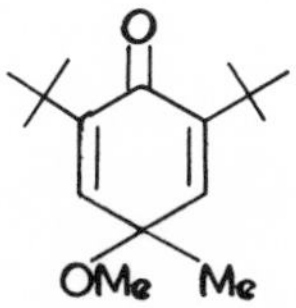

In addition to the variations described pre-viously, the presence of Ti(IV) compounds enables one to isolate the hydroperoxide [20]:

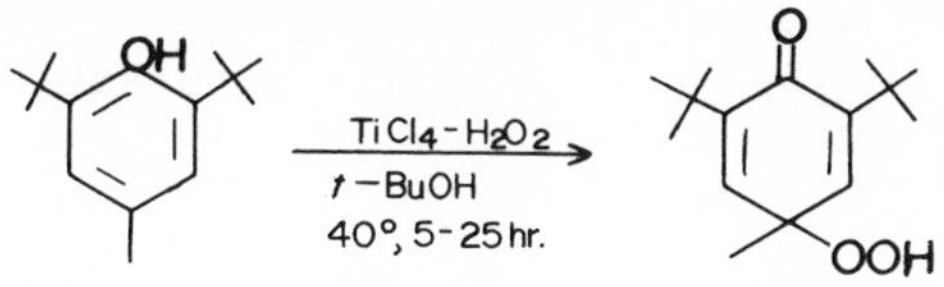

60-90 %

Halogen in a position para to the hydroxyl group may be replaced on oxidation to give the diphenyl ether [21]:

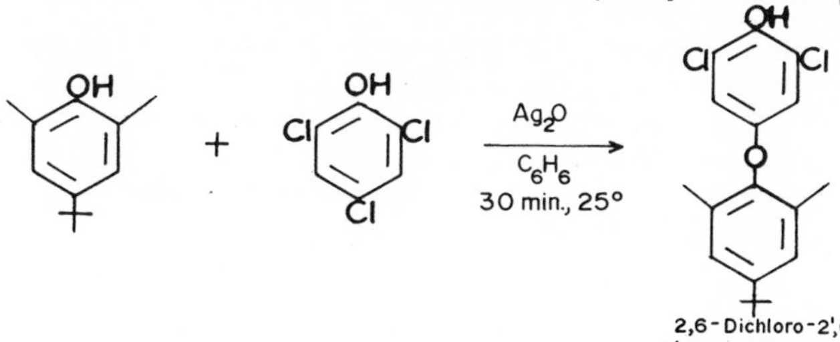

2,6-Dichloro-2',6'-dimethyl-
4'-t-butyldiphenyl ether, 80 % (crude)

The most widely used reagent in coupling is potassium ferricyanide [22] in alkaline solution, although ferric chloride sometimes leads to dimeric phenols not obtainable with the use of the ferricyanide. For example, 1-naphthol is oxidized to all the three possible dimeric phenols and 2-naphthol to the 1,1'-bisnaphthol in good yield with ferric chloride, while the ferricyanide gives mostly polymeric material. Lead tetraacetate in benzene with phenols affords small amounts of dimers as does oxygen in alkaline solution:

Manganese sulfate with the primary oxidant may increase the yield of the coupling products [23]:

a. Preparation of 2,2'-Dihydroxy-3,3',4,4'-tetramethyl-5,5'-dimethoxydiphenyl [24]

One gram of the phenol in 20 ml of methanol was treated with 10 ml of 1N KOH and stirred rapidly with 5 g of $K_3Fe(CN)_6$ in 50 ml of 0.5N KOH. A thick, flocculent, blue-violet precipitate formed; it was washed with water, CH_3OH, and petroleum ether to give 0.9 g (90%) of the diphenoquinone.

The diphenoquinone was reduced with dithionate in alcohol and the crude product, recrystallized from methanol (charcoal), gave 0.6 g (66%) of coupled phenol.

10. From Arenes via the Arylthallium Difluoroacetates

$$ArH \xrightarrow{TTFA} ArTl(O_2CCF_3)_2 \xrightarrow[\text{2) P(C}_6\text{H}_5\text{)}]{\text{1) Pb(OCOCH}_3\text{)}_4} ArO_2CCF_3 \xrightarrow{OH^\ominus} ArOH$$

Taylor and McKillop applied their well-known thallation synthesis to the preparation of phenols without the isolation of intermediates [25]. In the three-step process as represented above, the yields of phenols from seven arenes varied within 39-78% (based on hydroxylation of the isolated arylthallium difluoroacetate). No activating groups are necessary in the arene and the isomer orientation is subject to control.

References

1. S. Patai, The Chemistry of the Hydroxyl Group, Pt. 1, Wiley-
    Interscience, New York, 1971.
2. D. F. Sangster, Ref. 1, p. 133.
3. J. A. Veseley, L. Schmerling, J. Org. Chem., 35, 4028
    (1970).
4. M. E. Kurz, G. J. Johnson, J. Org. Chem., 36, 3184 (1971).
5. H. Hart, Acc. Chem. Res., 4, 337 (1971); H. Hart, C. A.
    Buehler, J. Org. Chem., 29, 2397 (1964).
6. D. H. R. Barton et al., Chem. Commun., 550 (1969).
7. M. I. Shenbor, G. A. Cheban, J. Org. Chem. USSR, 5, 140
    (1969).
8. C. R. H. I. deJonge et al., Tetrahedron Letters, 1881
    (1970).
9. T. Matsuura, K. Omura, Synthesis, 173 (1974).
10. S. W. Breuer, F. A. Broster, J. Organomet. Chem., 35, C5
    (1972); S. W. Breuer et al., Chem. Commun., 1475 (1971);
    G. M. Pickles, F. G. Thorpe, J. Organomet. Chem., 76, C23
    (1974).
11. J. S. Byck, C. R. Dawson, J. Org. Chem., 32, 1084 (1967).
12. S. Cabiddu et al., Gazz. Chim. Ital., 99, 771 (1969);
    C. A., 72, 12621 (1970).
13. A. Chatterjee et al., J. Indian Chem. Soc., 46, 429 (1969).
14. W. Baker et al., J. Chem. Soc., 1615 (1953).
15. B. McKague, Can. J. Chem., 49, 2447 (1971).
16. M. S. Kablaoui, Organic Paper 29, Abstracts 166th ACS
    National Meeting, August 26-31, 1973.
17. E. J. Eisenbraun et al., J. Org. Chem., 36, 686 (1971).
18. Ref. 1, M. L. Mihailović, Z. Cekovic, pp. 505-530; W. I.
    Taylor, A. R. Battersby, Oxidative Coupling of Phenols,
    Dekker, New York, 1967.
19. H. Dietl, H. S. Young, J. Org. Chem., 37, 1672 (1972).
20. R. G. R. Bacon, L. C. Kuan, Tetrahedron Letters, 3397
    (1971); I. Saito et al., ibid., 239 (1970).
21. D. A. Bolon, J. Org. Chem., 38, 1741 (1973).
22. M. L. Mihailović, Z. Cekovic, Ref. 1, pp. 516-567.
23. H. Tanaka et al., Bull. Chem. Soc. Jap., 43, 212 (1970).
24. D. Schulte-Frohlinde, F. Erhardt, Ann. Chem., 671, 92
    (1964).
25. E. C. Taylor, A. McKillop, J. Am. Chem. Soc., 92, 3520
    (1970).

C. Reduction

    1. From Quinones (1, 267)

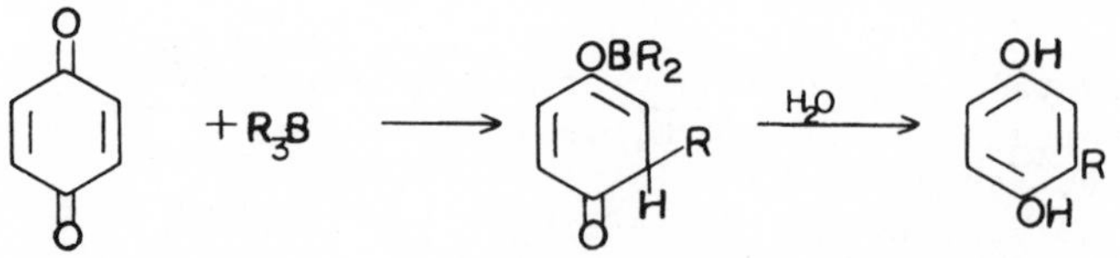

A comparatively new method of preparing alkylated dihydric phenols is the hydrolysis of the product formed when a quinone is treated with a trialkylborane [1]. Crude yields of alkyl-hydroquinones vary from 86 to 99%. To date the reaction appears to have been conducted only with p-benzoquinone and 1,4-naphthoquinone, and in the latter case, if the quinone is very pure, oxygen introduction may be necessary. The synthesis represents an excellent method for preparing 2-alkylhydro-quinones and 2-alkyl-1,4-naphthalenediols.

Alkylation and reduction have also been accomplished by the use of π-allylnickel bromide [1a]. Thus p-benzoquinone gives allylhydroquinone:

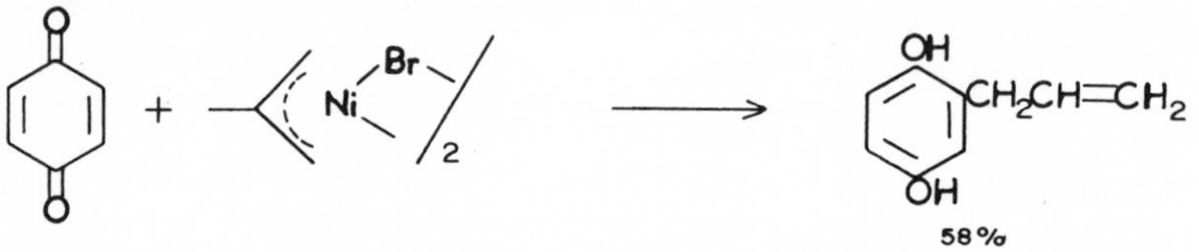

Direct reduction of quinones to the hydroquinone is described (1, 267).

A recent reagent described to reduce diphenoquinones is diphenylsilane [1b].

a. Preparation of 2-Cyclohexyl-1,4-naphthalenediol [1]

Borane, 50 mmol, in 25 ml of THF was mixed with 150 mmol of cyclohexene in 25 ml of THF and the mixture was stirred at 50° for 3 hr. Water, 60 mmol, was added followed by 50 mmol of 1,4-naphthoquinone in 50 ml of THF. Air was then passed into the reaction which was complete in 20 min. Steam distillation removed the solvent, borinic acid, and boronic acid, and the naphthalenediol by glc analysis amounted to 90%.

3. From Aromatic Ethers (1, 269)

$$ArOR \xrightarrow{[H]} ArOH + RH$$

The original text describes the debenzylation of benzyl ethers with hydrogen and Raney nickel. A recent synthesis employs Pd and a small amount of acid [2]:

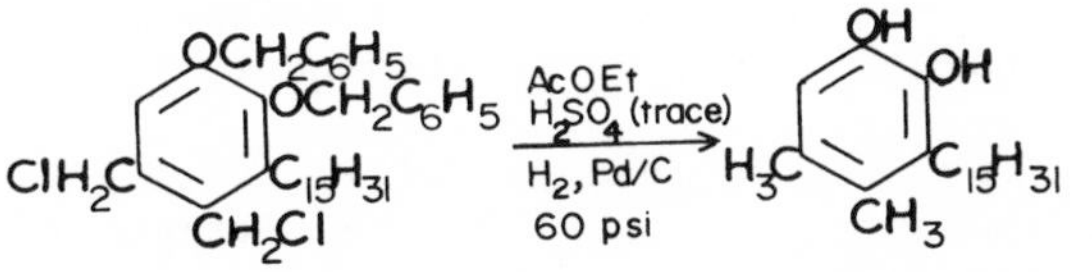

4,5-Dimethyl-3-penta-
decylcatechol, 91%

The cleavage of diphenyl ethers by sodium in pyridine was also described in the original text. This reaction appears to be general [3].

4. Photoreduction of Halophenols (New)

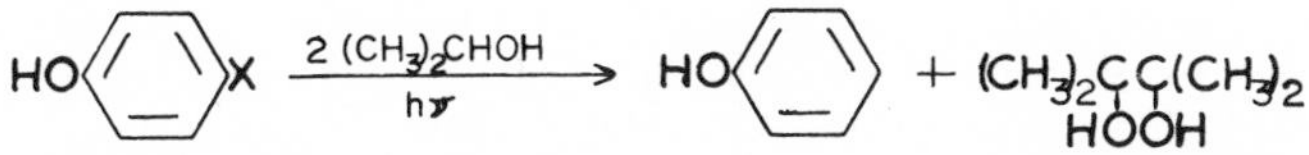

Pinhey and Rigby [4] found that o- and p-chloro-, o- and p-bromo-, and o- and p-iodophenols in isopropanol under nitrogen when irradiated were converted into phenols with yields varying from 86-98% (glc). The method is less satisfactory in preparing unsubstituted phenols from m-halophenols since there is a tendency for the halo group to be replaced by the alkoxy group of the alcohol. m-Chlorophenol, for example, with isopropanol on irradiation gives 62% (glc) of m-isopropoxyphenol.

References

1. G. W. Kabalka, J. Organomet. Chem., 33, C25 (1971); M. F. Hawthorne, M. Reintjes, J. Am. Chem. Soc., 87, 4585 (1965).
1a. L. S. Hegedus et al., J. Am. Chem. Soc., 94, 7155 (1972).
1b. H.-D. Becker, J. Org. Chem., 34, 2469 (1969).
2. D. I. Lerner, C. R. Dawson, J. Org. Chem., 38, 2096 (1973).
3. J. Gripenberg, T. Hase, Acta. Chem. Scand., 20, 1561 (1966).
4. J. T. Pinhey, R. D. G. Rigby, Tetrahedron Letters, 1267, 1271 (1969).

D. Electrophilic Reactions

   1. From Phenols by Alkylation or Arylation or by Rearrangement (1, 270)

The great versatility of the alkylation of phenols is brought out in the original text and in a recent review [1]. Both kinetic (ortho and para) and thermodynamic (meta) alkylated products may be isolated by the proper choice of conditions. A cyclo process, using aluminum phenoxide, is available to give ortho-alkylated products. Noncatalyzed alkylation of phenols with alkenes at 320° gives the o-alkylated phenols as the principal product.

Phenol has been tritylated as shown [2]:

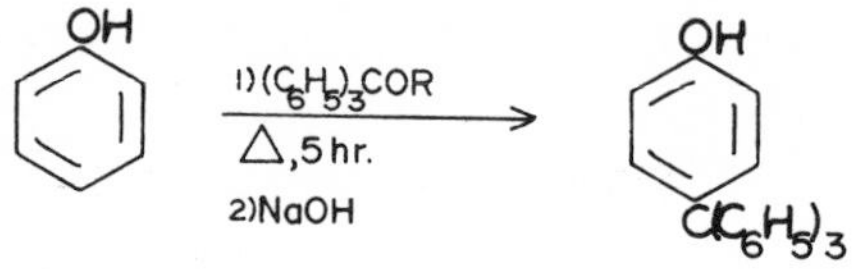

When R=CH$_3$, C$_2$H$_5$, C$_3$H$_7$, i-C$_3$H$_7$, C$_4$H$_9$, C$_5$H$_{11}$, C$_6$H$_5$CH$_2$, C$_6$H$_5$, the yield of the trityl phenol varied within 95-100%.

Resorcinol has been alkylated in the 2-position by treatment with linalool in the presence of the dineopentyl acetal of HCON(CH$_3$)$_2$ as shown [3]:

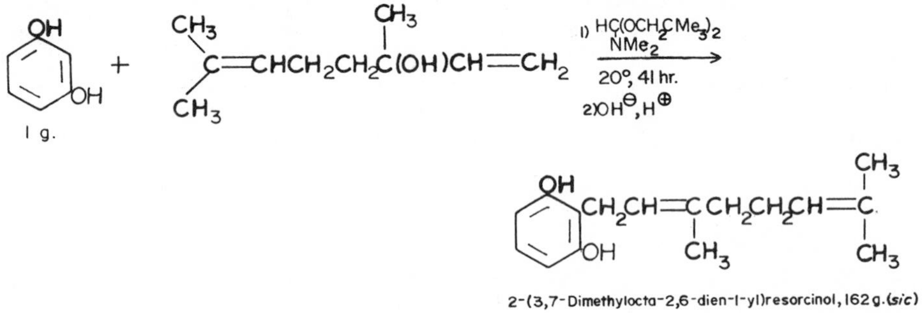

2-(3,7-Dimethylocta-2,6-dien-1-yl)resorcinol, 162 g.(sic)

Phenol with 1,5-hexadiene and a catalyst primarily forms the tetrahydronaphthols [4]:

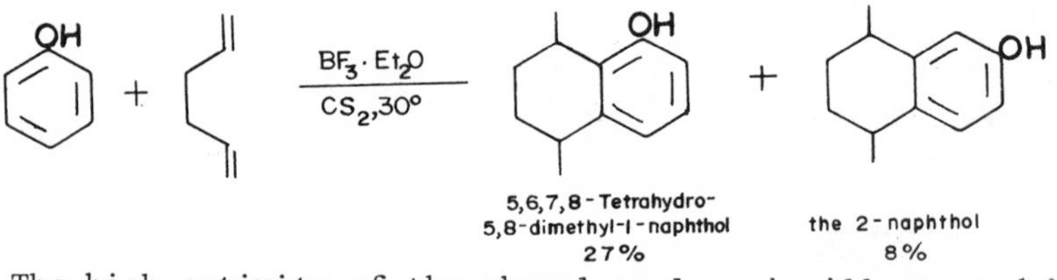

5,6,7,8-Tetrahydro-
5,8-dimethyl-1-naphthol
27%

the 2-naphthol
8%

The high activity of the phenol nucleus is illustrated by the following alkylations carried out in aqueous solution [5]:

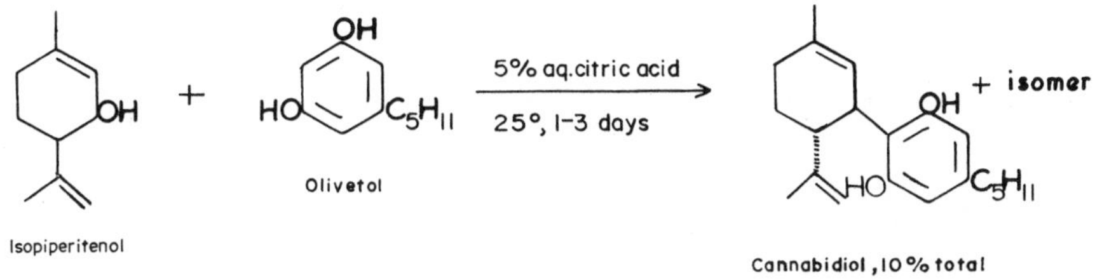

Isopiperitenol

Olivetol

5% aq.citric acid
25°, 1-3 days

+ isomer

Cannabidiol, 10% total

The product is an intermediate in the synthesis of tetrahydro-cannabidiol.

A second example is as shown [6]:

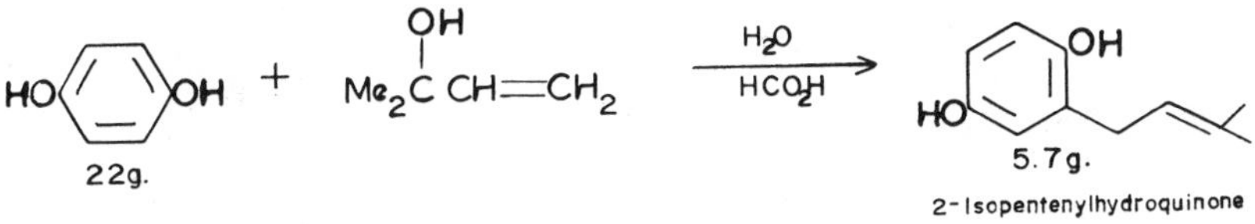

22g.

Me_2C CH=CH_2

H_2O
HCO_2H

5.7g.

2-Isopentenylhydroquinone

An unusual alkylation occurs with the Schiff base derived from pyridinecarboxaldehyde [7]:

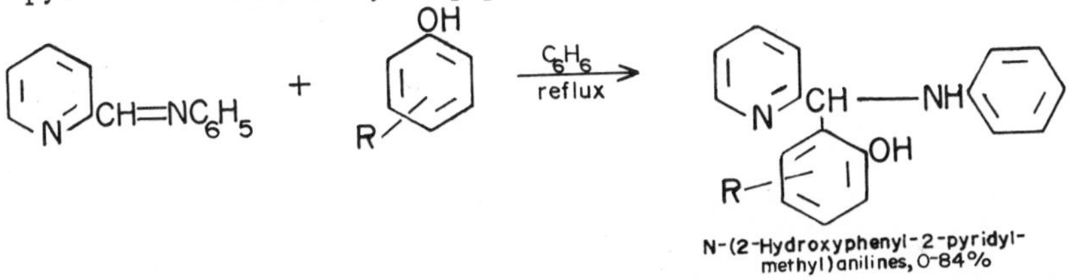

CH=NC_6H_5

C_6H_6
reflux

N-(2-Hydroxyphenyl-2-pyridyl-methyl)anilines, 0-84%

The Fries reaction also gives rise to phenols by rearrangement. It is discussed in Chapter 11, Ketones, 1, 655 and in the present text, Chapter 11, C.3.

A review of the reaction of carbonyl compounds, usually acetone, on phenols is available [8]. The products from acetone and phenol vary with the condensing agent and other experimental conditions as shown:

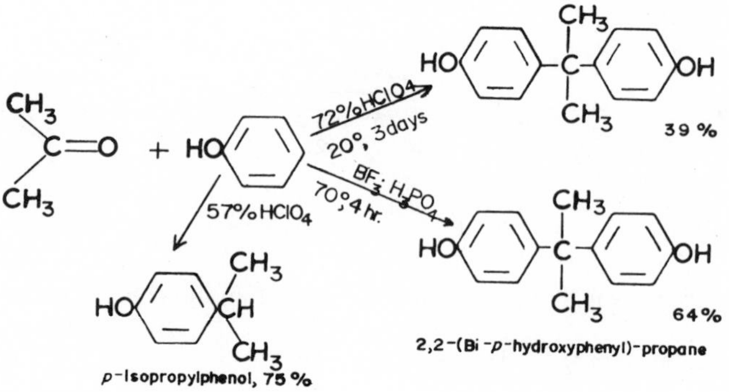

p-Isopropylphenol, 75 %

2,2-(Bi-*p*-hydroxyphenyl)-propane

Many more complex products, some of which are phenols, may be produced in this reaction.

The almost completely unknown 2,2'-dihydroxy derivatives of triphenylmethane were synthesized by the reaction of benzaldehyde and the phenoxymagnesium halide as indicated [9]:

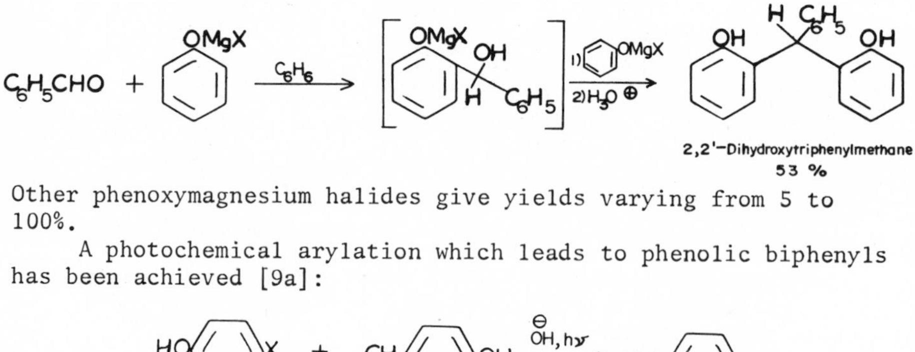

2,2'-Dihydroxytriphenylmethane
53 %

Other phenoxymagnesium halides give yields varying from 5 to 100%.

A photochemical arylation which leads to phenolic biphenyls has been achieved [9a]:

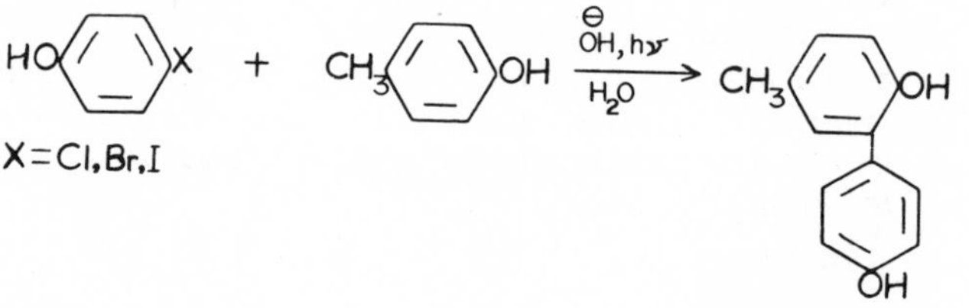

2-(4'-Hydroxyphenyl)-
4-cresol, 41%

2. From Phenols and Halogenating Agents (1, 273)

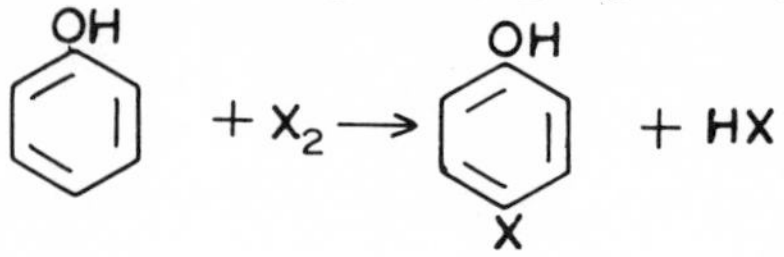

Thiols have been found to effect chlorination of bisphenols [10]:

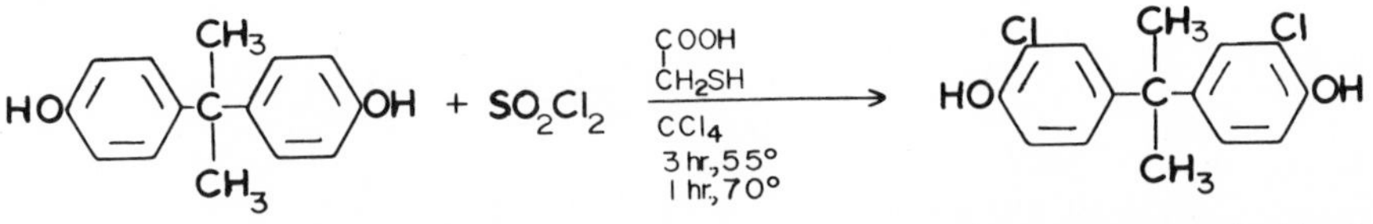

4,4'-Isopropylidene-bis-(2-chlorophenol)
83-89%

In the absence of the sulfur-containing catalyst no halogenation occurred.

Tetrabromocyclohexadienone, 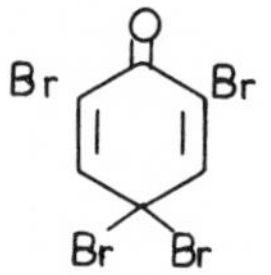 , appears to be a selective brominating agent. In nonpolar solvents such as carbon tetrachloride substitution is ortho. As the polarity of the solvent increases the p:o ratio becomes greater [10a]. However, it is doubtful that the ortho halogenation is as selective as that described in the use of the halogen in t-butylamine and toluene at dry-ice temperature [10b].

6. From Cyclodienones (1, 275)

It appears that the variability in the path of rearrangement of dienones to phenols is as much a structural as a medium effect [11]:

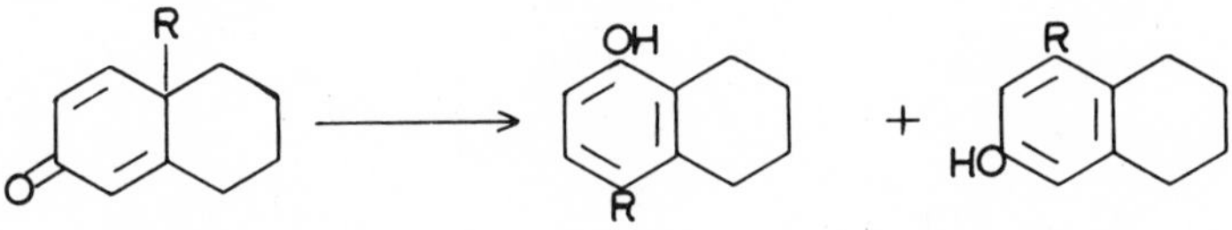

When R=Me, with 30-50% aqueous H+, the products are 20% A and 80% B; with Ac$_2$O-H$_2$SO$_4$, the products are 80% A and 20% B (as the acetate); and when R=Et, Pr, or Bu, the main product is B regardless of the medium.

7. From Cyclic Glycols (Pinacol Rearrangement) (1, 277)

It is conceivable that the following rearrangement of an epoxide proceeds via the pinacol rearrangement. At least Bruice and co-workers [12] have offered proof for the presence of a precursor of the pinacol. In the equation

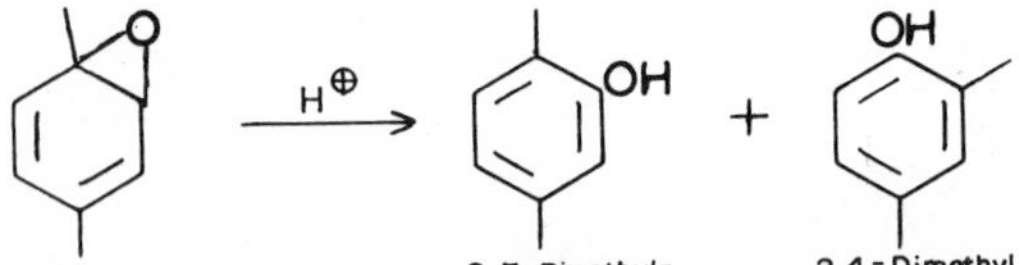

2,5-Dimethyl-
phenol   2,4-Dimethyl-
phenol

these investigators suggest that ... and possibly

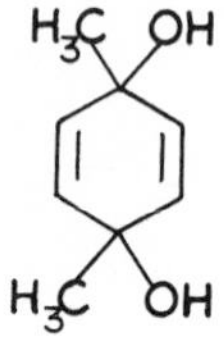

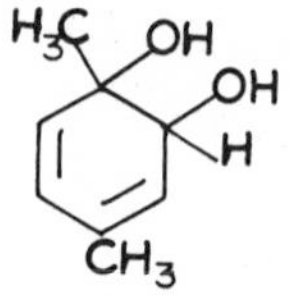

are intermediates. Conversion to the
phenol may proceed, in the latter case,
as follows:

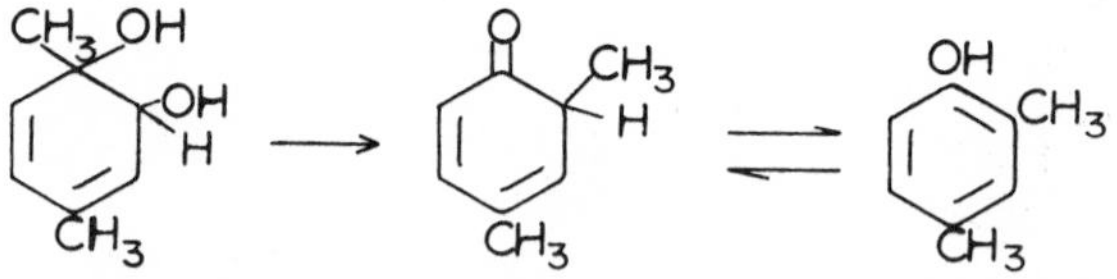

8. From Arenes and Some Acylating Agents (1, 278)

The Fries reaction also gives rise to phenols by rearrangement. It is discussed in Chapter 11, Ketones (1, 655) and in the present text, Chapter 11, C.3.

9. From Phenols and N-Chloroamines

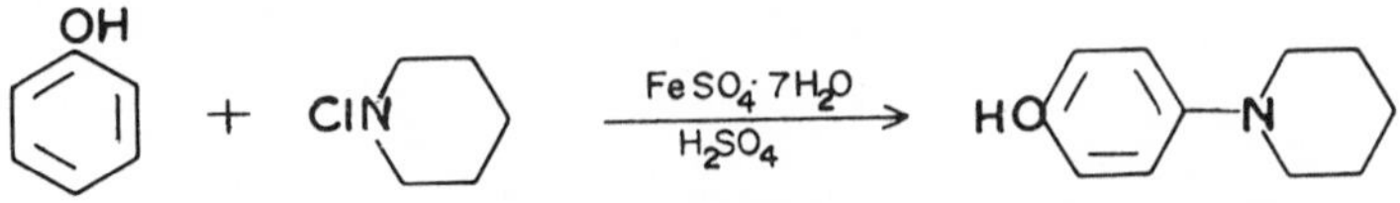

N,N-Dialkylaminophenols have been prepared as indicated [13]. The yield of N-(p-hydroxyphenyl) piperidine obtained is 74% (based on the piperidine); from the mother liquid of benzene recrystallization, the o-hydroxy isomer may be isolated in small amounts.

10. Isomerization

A new reagent for the quantitative isomerization of sterically hindered benzoquinone methides is neutral aluminum oxide. The isomerization, as indicated, results at the same time in the isomerization of the alkenyl group [14].

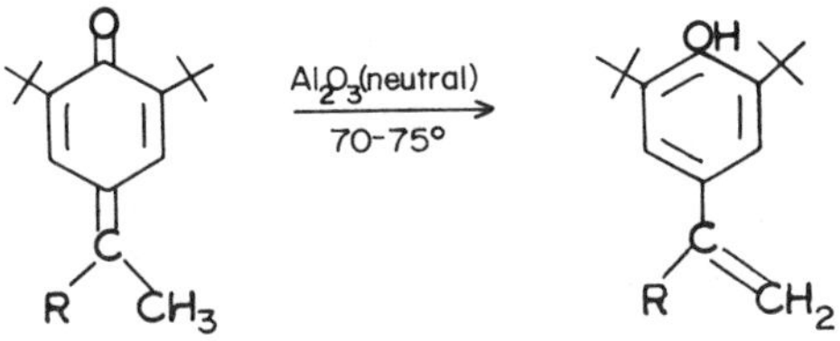

quantitative

The isomerization of t-butylphenols in liquid hydrogen fluoride has been studied recently [14a].

11. From N,O-Diarylhydroxylamines (Cox Rearrangement)

$$C_6H_5ONHC_6H_5 \xrightarrow{H^{\oplus}} HO\text{-}\langle\langle\ \rangle\rangle\text{-}\langle\langle\ \rangle\rangle\text{-}NH_2$$

This rearrangement is similar to that of benzidine but more limited in scope. Carbamates appear to respond well [15]:

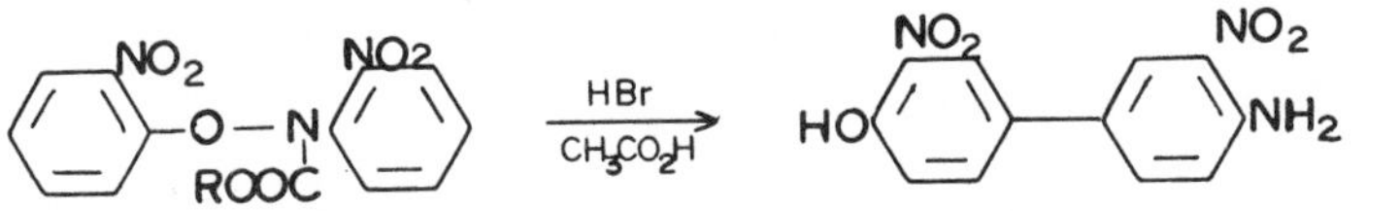

4-Amino-3,3'-dinitro-4'-hydroxybiphenyl
(unstated yield)

References

1. D. A. R. Happer, J. Vaughan, in S. Patai, The Chemistry of the Hydroxyl Group, Pt. 1, Wiley-Interscience, New York, 1971, pp. 418-431.
2. P. F. Butakus, R. Yu. Sabonene, J. Org. Chem. USSR, 5, 521 (1969).
3. A. Eschenmoser et al., French 1,543,647, October 25, 1968; C. A., 72, 31435 (1970).
4. J. M. Balquist, E. R. Degginger, J. Org. Chem., 36, 3345 (1971).
5. B. Cardillo et al., Tetrahedron Letters, 945 (1972).
6. L. Jurd et al., Tetrahedron Letters, 2275 (1971).
7. S. Miyano, N. Abe, Tetrahedron Letters, 1909 (1970).
8. J. Kahovec, Chem. Listy, 65, 397 (1971).
9. G. Casiraghi et al., Tetrahedron Letters, 3969 (1971).
9a. K. Omura, T. Matsuura, Synthesis, 28 (1971).
10. I. M. Bilik et al., J. Org. Chem. USSR, 5, 334 (1969).
10a. Y. Calo et al., Chimica e Industria, 53, 467 (1971);
10b. D. E. Pearson et al., J. Org. Chem., 32, 2358 (1967).
11. H. J. Shine, C. E. Schoening, J. Org. Chem., 37, 2899 (1972).

12. T. C. Bruice et al., J. Am. Chem. Soc., 94, 7876 (1972).
13. F. Minisci et al., Org. Prep. Proced., 1, 87 (1969).
14. D. Braun, B. Meier, Angew. Chem. Intern. Ed. Engl., 10, 566 (1971).
14a. J. R. Norell, J. Org. Chem., 38, 1929 (1973).
15. T. Sheradsky, G. Salemnick, Tetrahedron Letters, 645 (1971).

E.   Nucleophilic Reactions

   1.   From Carbonyl Compounds (1, 280)

   The cyclization of triketo acids to resorcyclic acids has been given (1, 280) [1]. It is of interest that ring closures of this type are variable, depending on the pH as in the following example [2]:

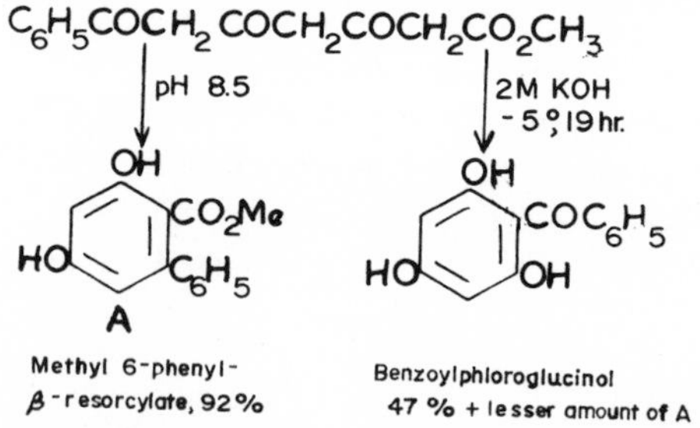

   Methyl 6-phenyl-
   β-resorcylate, 92%

   Benzoylphloroglucinol
   47 % + lesser amount of A

In a similar manner 9-phenyl-3,5,7,9-tetraketononanoic acid underwent an aldol-type cyclization as indicated [3]:

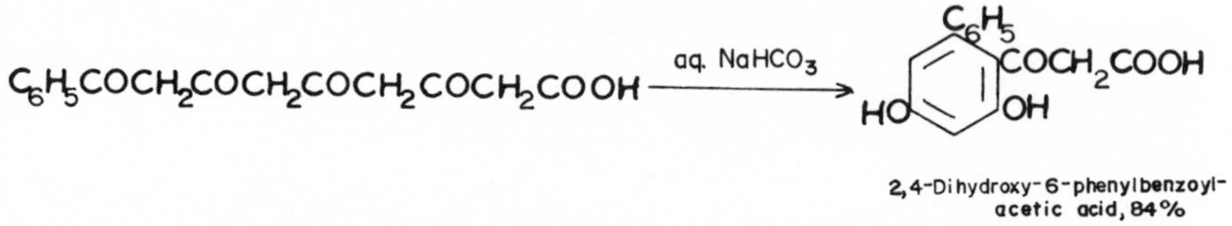

   2,4-Dihydroxy-6-phenylbenzoyl-
   acetic acid, 84%

   2.   Rearrangement of o-Acyloxyacetophenones

   o-Acyloxyacetophenones undergo an intramolecular Claisen reaction in the presence of a base and in a solvent such as benzene, toluene, ether, or pyridine to give o-hydroxy-β-diketones, the so-called Baker-Venkataraman rearrangement [4]:

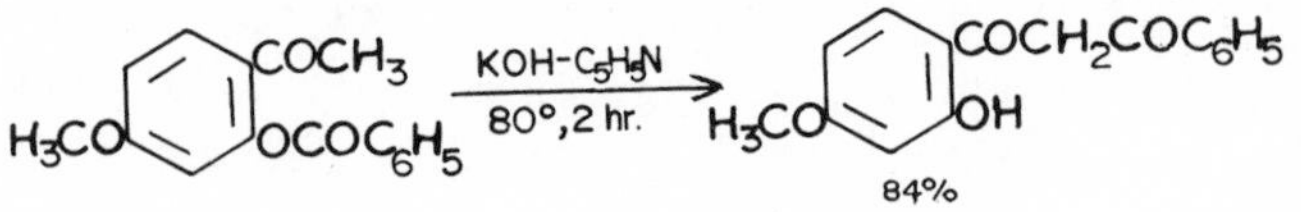

   84%

The rearrangement is an intramolecular one in which the product is of importance in chromone synthesis.

3. From Esters (Michael)

A general source of substituted resorcinols is provided by the Michael addition of acetoacetic ester to acrylic esters [6] as shown:

$$RCH{=}CHCO_2Et \ + \ CH_3COCH(Na)CO_2Et \longrightarrow$$

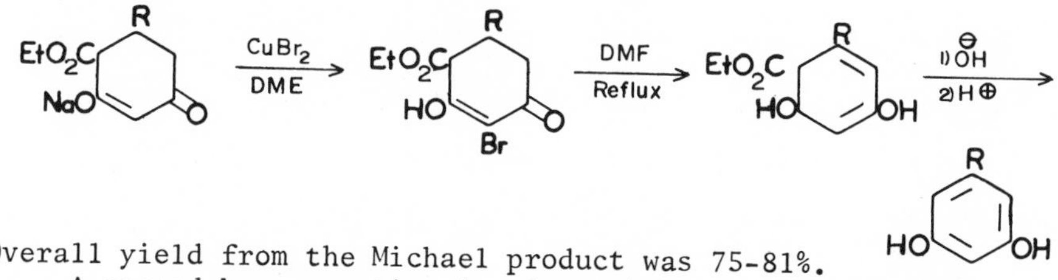

Overall yield from the Michael product was 75-81%.
A second base reaction involves the ring closure of a malonic ester derivative [7]:

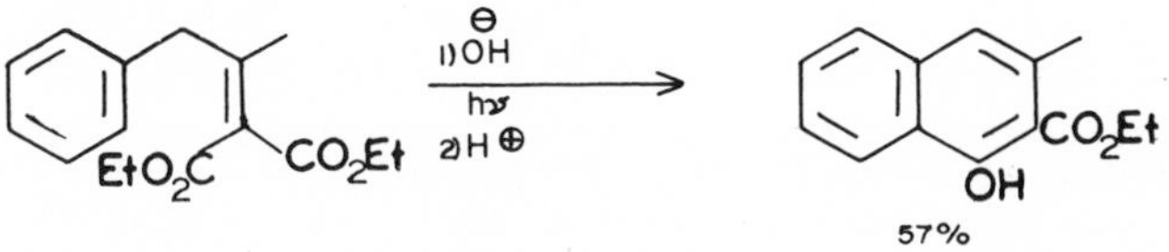

57%

References

1. T. T. Howarth, T. M. Harris, J. Am. Chem. Soc., 93, 2506 (1971).
2. T. M. Harris, R. L. Carney, J. Am. Chem. Soc., 89, 6734 (1967).
3. T. M. Harris, G. P. Murphy, J. Am. Chem. Soc., 93, 6708 (1971).
4. W. Baker, J. Chem. Soc., 1381 (1933), 1953 (1934); K. Venkataraman et al., J. Chem. Soc., 1767 (1934), 868 (1935).
5. H. Schmid, K. Banholzer, Helv. Chim. Acta, 37, 1706 (1954).
6. R. S. Marmor, J. Org. Chem., 37, 2901 (1972).
7. N. C. Yang et al., J. Org. Chem., 34, 1845 (1969).

F. Cyclo Reactions

1. From Allyl Aryl Ethers (Claisen Rearrangement) and Diaryl Ethers (1, 282)

Diaryl ethers rearrange to give phenols in a manner similar to the Claisen rearrangement. Thus the o-methyldiaryl ethers rearrange thermally, the maximum yield having been obtained in the case of 2,6-dimethylphenyl phenyl ether [1]:

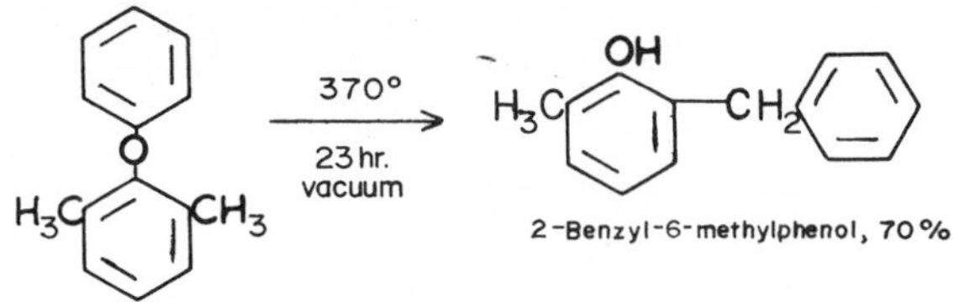

2-Benzyl-6-methylphenol, 70%

The free radical mechanism appears to be:

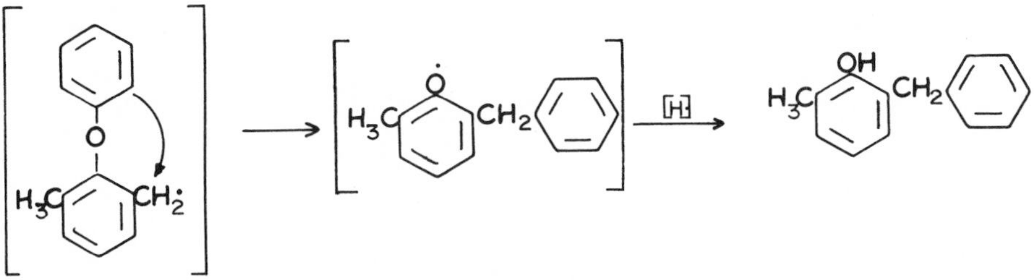

3.   From Acetylenes and Carbon Monoxide

Reppe appears to have been the first to synthesize hydroquinone from acetylene, carbon monoxide, and water or hydrogen [2]. The method follows:

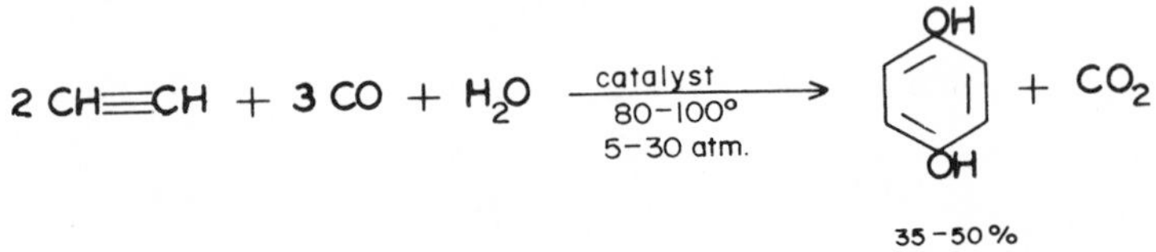

$$2\ CH\equiv CH + 3\ CO + H_2O \xrightarrow[\substack{80-100° \\ 5-30\ atm.}]{catalyst}$$

35-50%

Catalysts employed were complex salts such as $[Fe(NH_3)_6][Co(CO)_4]_2$ or $[Co(NH_3)_6][Co(CO)_4]_2$ and yields as high as 70% were obtained by increasing pressures to 600-700 atm, the partial pressure of C_2H_2 to about 70 atm, the temperature to above 170°, the use of suitable solvents such as dioxane, and the introduction of the proper amount of water.

Pino and co-workers [3] synthesized hydroquinone from C_2H_2, CO, and ruthenium tetracarbonyl with H_2 and with H_2O. The highest yield ($\leq$ 65%) was obtained with water under relatively low partial pressure of CO at 150-250°.

Cassar and co-workers [4] improved the cyclization by conducting the reaction at room temperature and atmospheric pressure on a catalytic system formed in situ from tetracarbonyl nickel and iodide ions. The synthesis of m-cresol from methallyl chloride, acetylene, and carbon monoxide is shown (Ex. a):

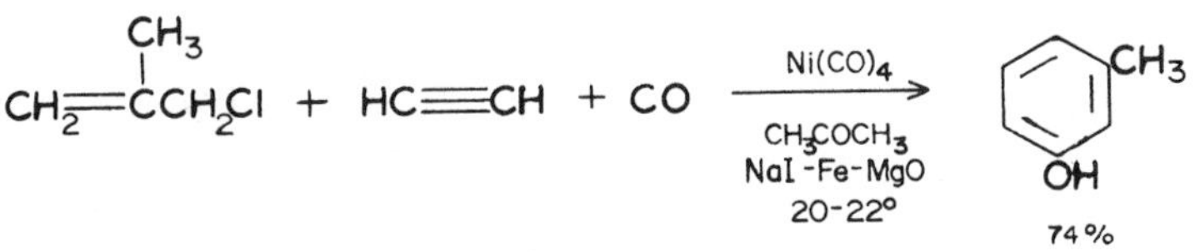

$$CH_2{=}CCH_2Cl + HC{\equiv}CH + CO \xrightarrow[\substack{CH_3COCH_3 \\ NaI\text{-}Fe\text{-}MgO \\ 20\text{-}22°}]{Ni(CO)_4}$$

(structure with CH_3 and OH) 74%

a. Preparation of m-Cresol [4]

Carbon monoxide-acetylene, 50:50 mixture, was bubbled into 800 ml of acetone and 0.9 g of NaI, 27.5 g of powdered Fe, 12 g of MgO, and 13 g of Ni(CO)$_4$ were added to the solution. Methallyl chloride, 95 g, was then added in 5 hr at 20-22°. After 3 additional hr, the flow of acetylene-carbon monoxide was stopped. The acetone and volatile products, 730 ml, were removed by distillation at ordinary pressure and the residue was acidified and extracted with ether. After removing the ether, the phenol was recovered (74%) by distillation under reduced pressure.

4. From Azoniapolycyclic Salts

This unusual and striking cycloreaction makes available a number of substituted naphthols, phenanthrols, and other poly-cyclic phenols in good to excellent yield. One example is shown to illustrate this versatile reaction [5]:

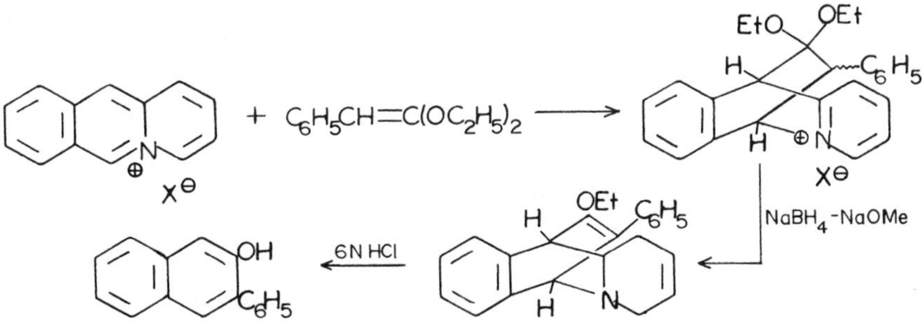

3-Phenyl-2-naphthol, 92 %   (partially reduced)

When it is considered that the structure of the azonia salt and the ketene acetal may be varied, the scope of the reaction becomes extensive. It has been applied most frequently to poly-nuclear hydrocarbons in which case the azonia salt structure is varied. These salts are available by the Bradsher ring closure of the quaternary salts of benzylpyridine.

Addendum

F.1. The rate of Claisen rearrangement of a phenyl allyl ether
is accelerated by as much as 10^5 times by using trifluoroacetic
acid as the solvent [U. Svanholm, V. D. Parker, Chem. Commun.,
645 (1972)].

References

1. A. Factor et al., J. Org. Chem., 35, 57 (1970).
2. A. Magin et al., Angew. Chem. Intern. Ed. Engl., 8, 727
    (1969); W. Reppe, H. Vetter, Ann. Chem., 582, 133 (1953).
3. P. Pino et al., Chem. Ind. (London), 1732 (1968).
4. L. Cassar et al., Organomet. Chem. Syn., 1, 302 (1970/1971).
5. D. L. Fields, J. Org. Chem., 36, 3002 (1971).

G. Irradiation Reactions

    1. From Phenols (Isomerization)

    Although the irradiation of phenols has been of limited
value to date, a few phenols such as 2,6-di-t-butyl-4-methyl-
phenol have been isomerized to the 2-methyl-3,6-di-t-butyl isomer
[1]:

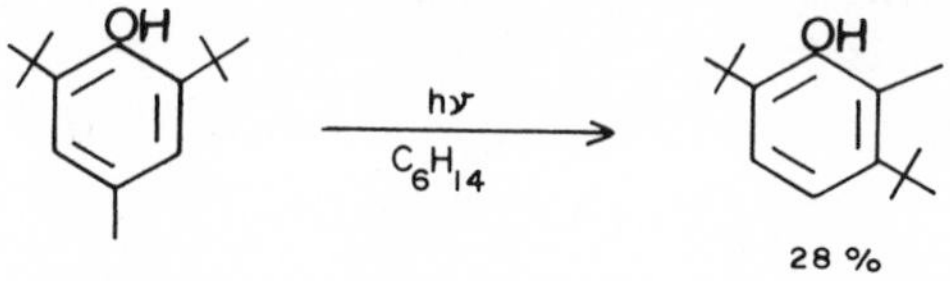

28 %

    Irradiations in this chapter have already been discussed
under A.2 (Ex. a), A,7, B.1, and C.4

Reference

1. T. Matsuura et al., Tetrahedron Letters, 3727 (1970).

Chapter 6

ETHERS

In addition to the H.-W. compendium on the preparation of ethers [1], the Patai monograph [2] describes a number of preparations.

The versatile preparation of epoxides from carbonyl compounds and sulfur ylids now has been published in Organic Syntheses (C.4). Moreover, epoxide preparations from an alkene and a peroxide have been improved by adding a stabilizer to prevent useless decomposition of the peroxide (D.1) or by using a two-phase system to protect sensitive epoxides from further reaction. The metathesis of an alkoxide and an alkyl halide in DMSO to form aliphatic ethers has been studied (A.1), and that between an ester and alcohol has been found to benefit by using 2,6-di-t-butylpyridine as the acid acceptor (A.2). Also within the metathetical group of reactions the synthesis of a crown ether is described (A.3). An interesting template effect has been found in the synthesis of a macrocyclic ether by a Friedel-Crafts reaction (B.9). Among electrophilic preparations of ethers, the addition of mercuric acetate in alcoholic solution to an alkene followed by reductive removal of the mercury group (B.10) and what appears to be a superior synthesis of dioxane (B.11) are described.

The synthesis of vinyl ethers has been collected in a small section at the end of the chapter.

References

1. H.-W., Methoden der Organischen Chemie, 6, Georg Thieme
    Verlag, Stuttgart, 1965, Pt. 3, p. 1.

2. H. Feuer, J. Hooz, in S. Patai, <u>The Chemistry of the Ether Linkage</u>, Wiley-Interscience, New York, 1967, p. 445.

A. Metathesis

   1. From Halides (Williamson) (<u>1</u>, 286)

$$\text{RONa} + \text{R'X} \longrightarrow \text{ROR'} + \text{NaX}$$

     Further study of the use of DMSO in the Williamson reaction was completed by Smith and co-workers [1]. While the conventional synthesis of butyl ether from NaOH, butyl chloride, and an excess of butyl alcohol led to a 60% yield of butyl ether in 14 hr, by replacing the excess of alcohol by DMSO the yield of ether rose to 95% in 9.5 hr. However the formation of the alkoxide from NaOH and alcohol in DMSO was slow, but could be followed by the change in the appearance of the precipitate from a somewhat slimy to a granular form. It required as much as 3 hr for the alkoxide formation and an equal amount of time for metathesis of the alkoxide.

     Other primary alkyl chlorides responded similarly, but for secondary alkyl chlorides and primary alkyl bromides elimination was the major reaction. Unreactive halides such as vinyl chloride, phenyl bromide, and 2,4-dinitrobromobenzene were not etherified in DMSO.

     The synthesis of a number of haloethers such as 4-chloropentyl methyl ether has been discussed by Peterson and Slama [2] and the alkyl ethers of α-alkylbenzoins, ArCOCOHAr, with an R group on the central carbon, have been described by Heine [3]. The magnesium halide salt of a tertiary alcohol reacts with alkyl halides to give the ethers in 30-60% yields [4].

   2. From Esters and Related Types (<u>1</u>, 290)

     A series of secondary and tertiary alcohols has been alkylated by titration with a standardized solution of sodium methyl sulfinyl carbanion to form the anion of the alcohol, which was then treated with an alkylating agent [5]. By this method alkylation occurred readily at room temperature to give ethers often difficult, if not impossible, to obtain satisfactorily by the usual alkylation procedure. t-Butyl and t-amyl alcohols, for example, gave the methyl ethers in 78 and 81% yields, respectively.

     The standardized solution of sodium methyl sulfinyl carbanion was prepared by dispersing 15 g of NaH (50%) in 200 ml of DMSO by means of ultrasound waves while the mixture was protected from the atmosphere by a 1-cm layer of mineral oil on the surface. The preparation of an ether with the reagent is given (Ex. a).

Macrocyclic ethers were prepared from polyethylene glycols and the tosylates of the same by adding a mixture of the two dropwise during 5 hr to a heated mixture of t-BuOK in t-BuOH and C_6H_6 [6]. Similarly dineopentyl ether was prepared from the sodium alkoxide and tosylate by refluxing one week [7]. Likewise, phenols with a small amount of NaH and ethyl oxalate in DMF (essential) gave the ethyl ethers [8].

By the use of the hindered pyridine, 2,6-di-t-butylpyridine, which is a good acceptor of protons and a poor acceptor of alkyl groups, Jensen and Neese [9] were able to effect a satisfactory synthesis of 1,4-dioxane as indicated:

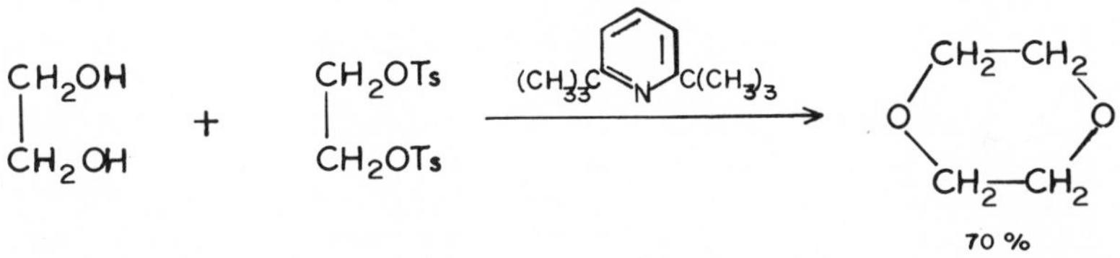

2,6-Di-t-butylpyridine acts as a normal base toward free protons but does not act as a nucleophile, and thus base-catalyzed elimination is held at a minimum. Similarly, pentabromoanisole has been prepared from hexabromobenzene and sodium methoxide in pyridine [10].

Butyrolactone is another ester that has been used in the synthesis of ethers [11]:

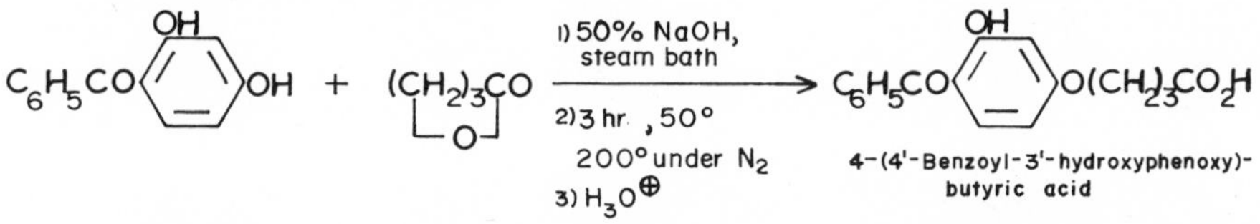

An aluminum salt has been treated with dimethyl sulfate to give an ether [12]:

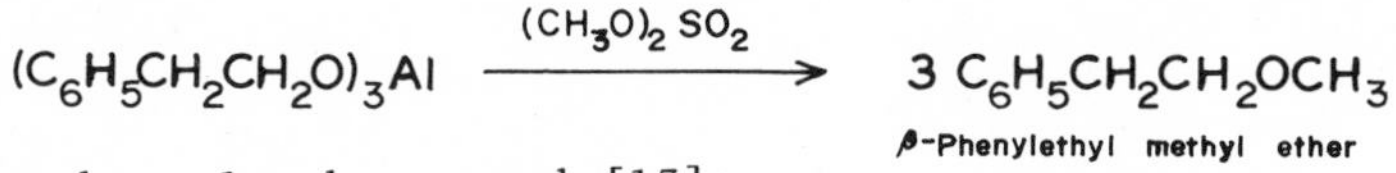

A perester has also been used [13]:

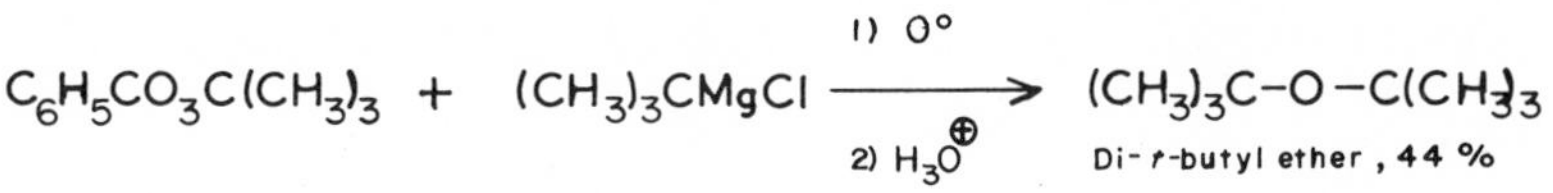

A similar preparation gave t-butyl phenyl ether in 78-84% yield [13a].

Since trimethylsilyl ethers are rather sensitive to both acid and base, it would be desirable to have a more stable protective group for hydroxyl groups. Corey and Venkateswarlu [13b] suggested the use of the more promising t-butyldimethyl group as a substitute. It was introduced as shown by treating the alcohol with t-butyldimethylsilyl chloride with imidazole as a catalyst:

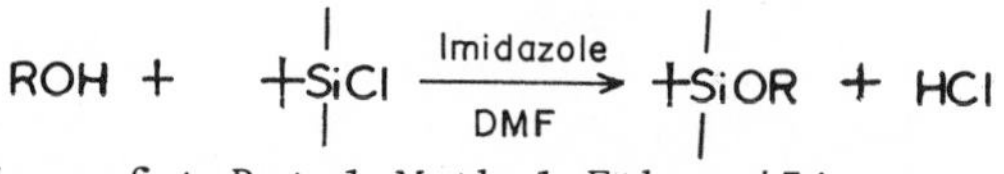

$$\text{ROH} + \overset{|}{\underset{|}{+}}\text{SiCl} \xrightarrow[\text{DMF}]{\text{Imidazole}} \overset{|}{\underset{|}{+}}\text{SiOR} + \text{HCl}$$

a.   Preparation of t-Butyl Methyl Ether [5]

t-Butyl alcohol in a small amount of dry DMSO containing a trace of triphenylmethane was treated with an equivalent amount of $CH_3SO_2CH_2Na$ with stirring and cooling. The bright red solution formed was treated with 10% excess of $(CH_3)_2SO_4$ while cooling and stirring for 10 min. If the ether did not separate as an upper layer, methylene chloride and water were added and the product was recovered by extraction from the aqueous DMSO layer with methylene chloride. Purification in the usual manner gave a 78% yield.

3.   From Aromatic Halides (Ullmann) (1, 292)

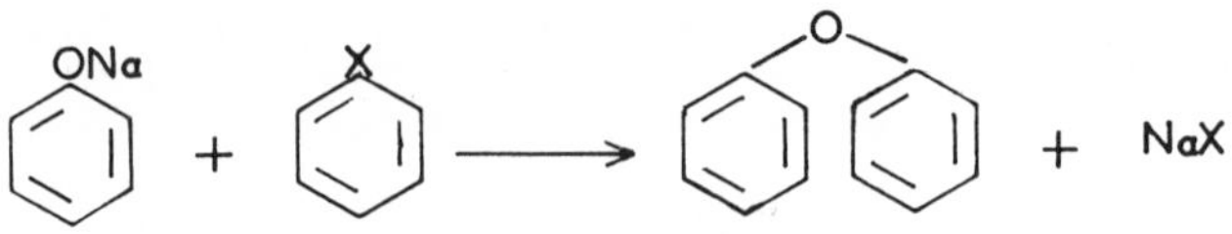

Seven new polyphenyl ethers have been synthesized by a modified Ullmann reaction as indicated [14]:

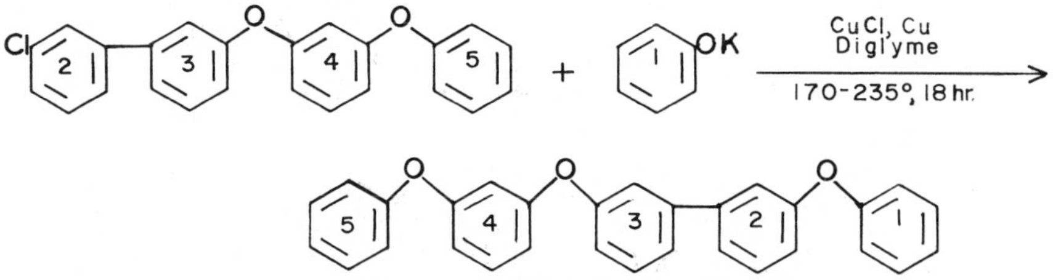

3,3'-(Phenoxy-m-phenoxyphenoxy)-biphenyl, 56%

It appears that the 4,4'-fluoro atoms in perfluorobiphenyl are reactive under the conditions of the Ullmann reaction as shown [15]:

$$C_6F_5C_6F_5 + KO\langle\!\!\!\!\bigcirc\!\!\!\!\rangle CH_3 \xrightarrow{\text{DMF}} CH_3\langle\!\!\!\!\bigcirc\!\!\!\!\rangle OC_6F_4C_6F_4O\langle\!\!\!\!\bigcirc\!\!\!\!\rangle CH_3$$

4,4'-Bis(*p*-methylphenoxy)-perfluorobiphenyl

96 %

Hindered diphenyl ethers by a modified Hems synthesis have been prepared by Neville and Moir [16] by treating the phenol with N(4'-carbomethoxy-2,6'-dinitro) phenylpyridinium chloride as shown:

2',6'- Dinitro-4'-carbomethoxy-
phenyl- 2- methylphenyl ether, 67 %

The diphenyl ethers are recovered more easily by passing the reaction mixture into acidified ice water and chloroform and shaking for 30 min. Although a diphenyl ether containing nitro groups in the 2,2',6-positions has been synthesized by this pro-cedure, one containing nitro groups in the 2,2',6,6' positions could not be formed. The pyridine hydrochloride formed evidently split these highly hindered ethers. In addition, some O-phenylsalicylates have been prepared recently by the Hems method [16a].

The question of the alkylation of oxygen or carbon in phenols, β-ketoesters, β-diketones, and other potential ambident anions has been discussed (1, 288). A recent review [17] has stated the case more clearly. To encourage oxygen alkylation (in order to synthesize ethers), the oxygen atom anion should be made as free and unfettered as possible. Such can be done by conducting the metathesis in a polar, aprotic solvent (HMPA, DMSO, or DMF) with a large counter ion (R_4^+N or K^+), a low concen-tration of anion, a hard leaving group, and an alkylating agent of low S_N2 reactivity [17]. Another way to making the counter ion large is to conduct the metathesis in a polyether solvent. Allylation of phenols gives rise to 2-, 4-, and O-substitution as well [18].

A preparation of a macrocyclic ether, such as dibenzo-18-crown-6-polyether, which forms complexes with alkali cations, is described (Ex. a). The syntheses of 18-crown-6 [18a][18b], 21-crown-7 [18b], 24-crown-8 [18b], and dicyclohexyl-18-crown-6

polyethers [21] have also been published. Similar compounds
called "cryptates" (e.g., polyoxadiamines) are of current
interest as ligands for cations [19] and are available commercially [18c].

Flavonoids are formed rather readily from chalcones [20]:

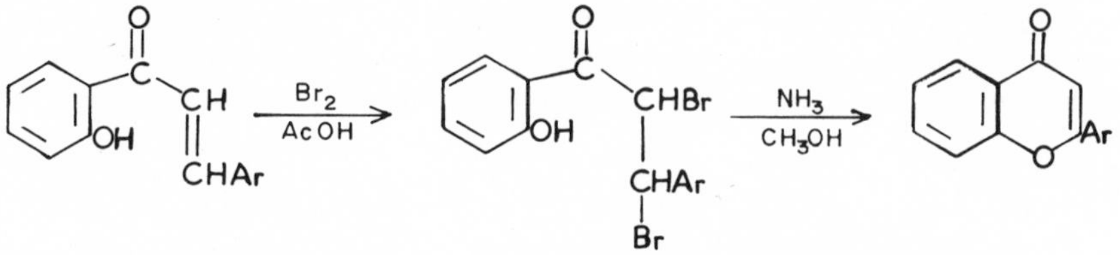

a. Preparation of Dibenzo-18-crown-6-polyether [21]

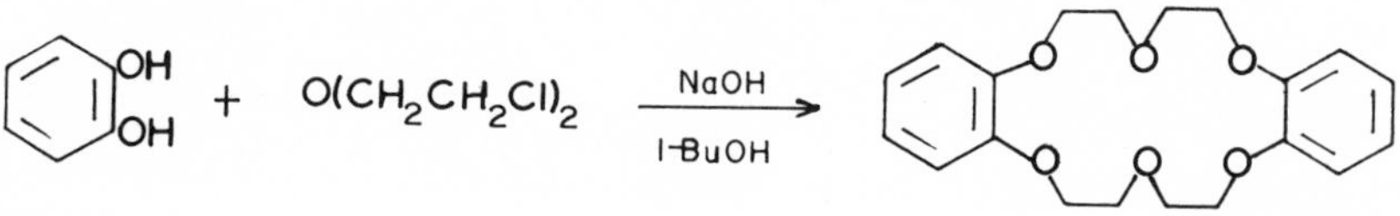

Catechol, 300 mol and 2 1 of commercial 1-butanol in a N_2 atmosphere were stirred while 3.05 mol of NaOH pellets were added.
After heating to reflux, 1.55 mol of bis(2-chloroethyl) ether in
150 ml of 1-butanol was added dropwise with stirring and heating
for 2 hr. After additional refluxing with stirring for another
hr and then cooling to 90°, the mixture was treated with another
3.05 mol of NaOH pellets. Then again after refluxing and stirring for 30 min, another 1.55 mol of bis(2-chloroethyl) ether in
150 ml of 1-butanol was added with stirring and heating for 2 hr.
Further refluxing and stirring for 16 hr was followed by acidification with HCl. Approximately 700 ml of 1-butanol was distilled
from the mixture while water was added to maintain a constant
volume. After the temperature exceeded 99°, the slurry was
cooled to 30-40°, diluted with 500 ml of acetone, stirred, and
filtered. The residual crude product was purified by treatment
with water and acetone to give 221-260 g (39-48%) of the ether,
melting point 161-162°.

   4. From Halohydrins and Related Types (Intramolecular
       Displacement) (1, 294)

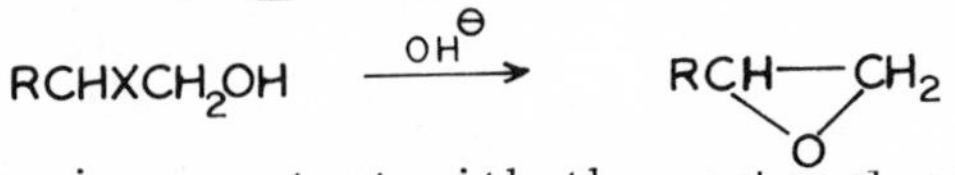

   This synthesis may start with the carbonyl compound and
methylene bromide, which with Li and THF gives the ethylene
oxide, via the lithium halohydrin as shown [22]:

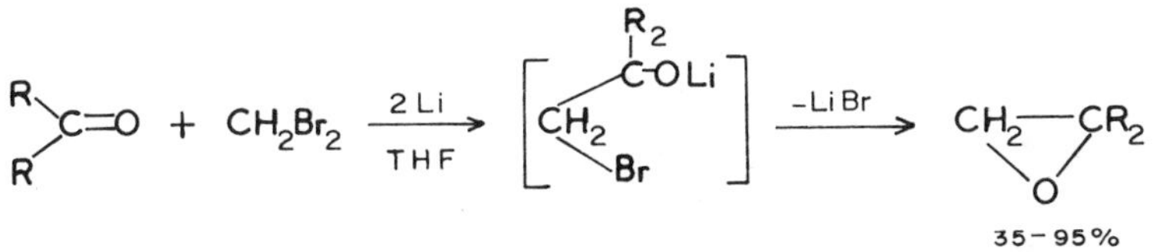

$$35\text{-}95\%$$

Aldehydes and ketones lead to epoxides usually in satisfactory yields, but α,β-unsaturated carbonyl compounds do not respond. If magnesium is used in place of lithium, an alkene is formed (1, 129).

A recent paper has given the method for obtaining either the cis or trans epoxide by starting with a single glycol, meso-2,3-butanediol [23]:

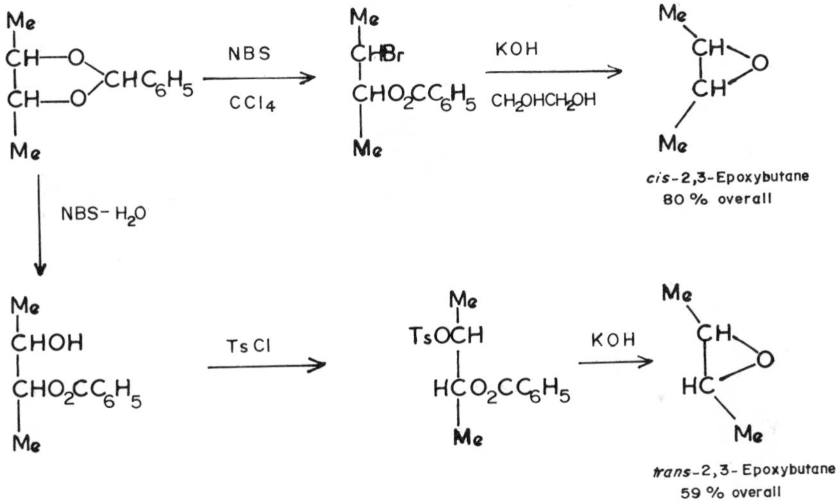

cis-2,3-Epoxybutane
80 % overall

trans-2,3-Epoxybutane
59 % overall

Epifluorohydrin has been prepared as follows [24]:

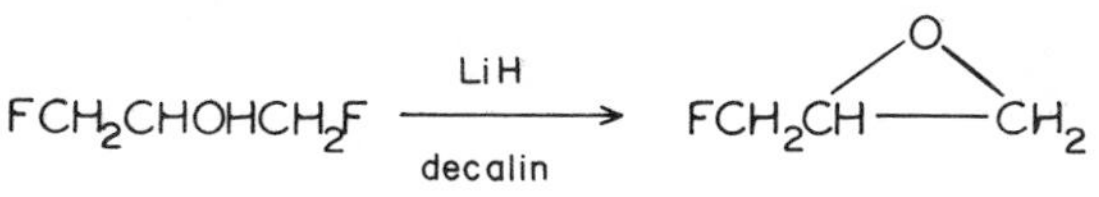

5.   From Alcohols or Phenols and Onium Salts or Bases
     (1, 296)

$$R\overset{\ominus}{O} \ + \ R'_4\overset{\oplus}{N} \longrightarrow ROR' \ + \ R'_3N$$

Although in the original text the powerful alkylating ability of triethyloxonium tetrafluoroborate was mentioned (1, 297), it was not stated that this reagent has a greater tendency than any other to give O-alkylation of an ambident anion. An example of

its ability to alkylate the enol rather than the keto form is
shown [25]:

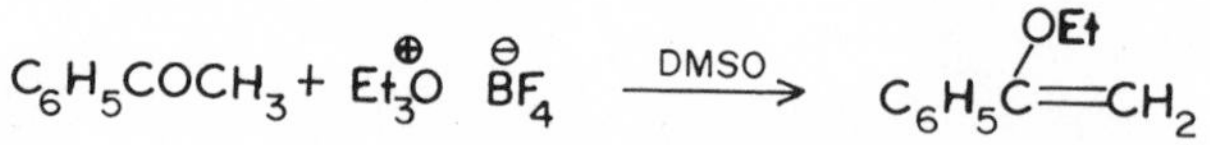

$$C_6H_5COCH_3 + Et_3\overset{\oplus}{O}\ \overset{\ominus}{BF_4} \xrightarrow{DMSO} C_6H_5\overset{OEt}{\underset{}{C}}=CH_2$$

Ethyl 1-phenylvinyl ether
68%

Oxetanes have been synthesized as indicated by treating
alkyloxyphosphonium salts with sodium methoxide in methanol [26]:

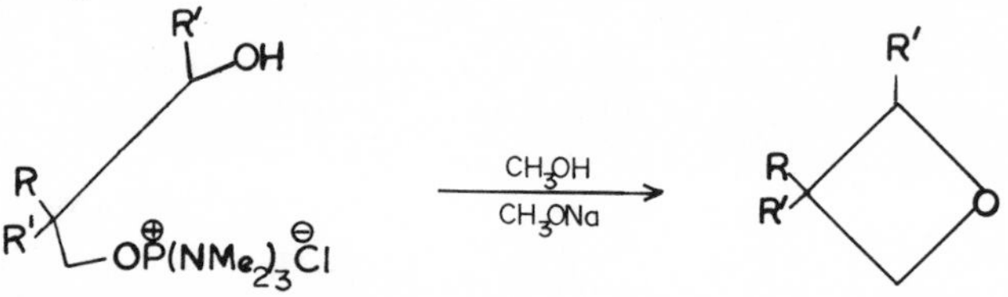

$$\xrightarrow[CH_3ONa]{CH_3OH}$$

From a series of phosphonium salts the yields varied from
10 to 70%.

6. From Grignard Reagents and α-Chloroethers, and the Like
    (1, 298)

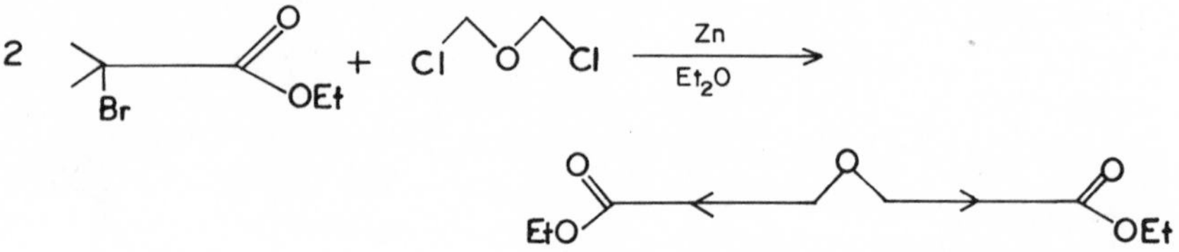

1,5-Dicarbethoxy-1,1,5,5-tetramethyl-3-oxapentane, 60%

It is well known that zinc organometallics do not add to the
carbonyl group of esters, as is well illustrated in the
Reformatsky reaction. Instead it has been found that coupling
takes place between active halogen compounds and the preformed
Reformatsky reagent, as illustrated above, to give ethers [27].

8. From Alcohols and Electronegatively Substituted Anilines

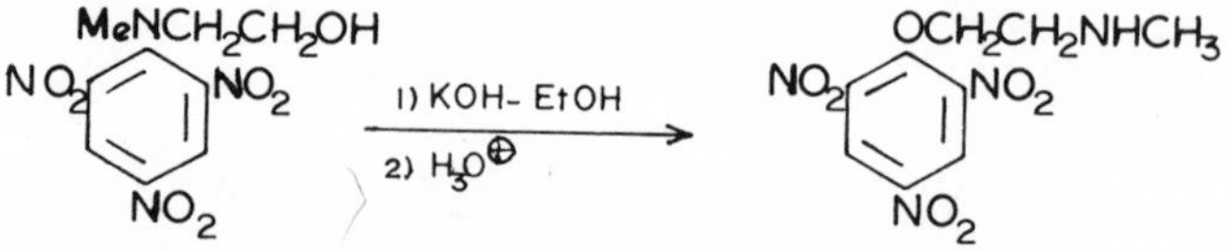

$$\xrightarrow[2)\ H_3O^{\oplus}]{1)\ KOH-\ EtOH}$$

Although the Smiles rearrangement occurs readily, the
reverse occurs only with electronegative anilines such as is
shown. The Meisenheimer complex,

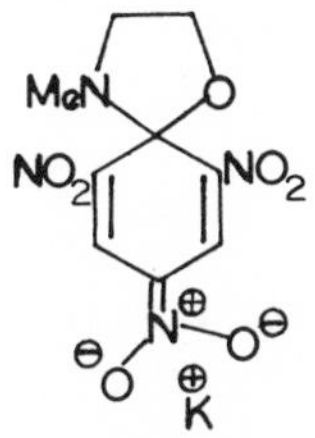

can be isolated and quickly acidified to give a 70% overall yield of N-methyl-β-aminoethyl picryl ether [28].

Addenda

General. Trimethylsilylacetamide reacts with alcohols to form the trimethylsilyl ether as shown:

$$ROH + (CH_3)_3SiNHCOCH_3 \longrightarrow ROSi(CH_3)_3 + CH_3CONH_2$$

[L. Birkofer, A. Ritter, Newer Meth. Prep. Org. Chem., 5, 211 (1968)].

A.1. The macrocyclic ethers, 12-crown-4 and 15-crown-5 have been prepared by a modified Williamson reaction. Of these ethers 18-crown-6 best forms a ligand with the potassium ion, while 15-crown-5 is a good complexing agent for sodium and potassium ions as is 12-crown-4 for the lithium ion [C. L. Liotta et al., Tetrahedron Letters, 4029 (1974)].

A.2. A review on macrocyclic polyethers has been published [C. J. Pederson, H. K. Frensdorf, Angew. Chem. Intern. Ed. Engl., 11, 16 (1972)] and another on crown ether applications in synthesis [G. W. Gokel, H. D. Durst, Synthesis, 168 (1976)]; see also Aldrichimica Acta, 9, 3 (1976).

A.2. Superior methylation of alcohols with dimethyl sulfate is carried out by a phase-transfer method [A. Merz. Angew. Chem. Intern. Ed. Engl., 12, 846 (1973)]. A similar method has been utilized for simple and hindered phenols [A. McKillop et al, Tetrahedron, 30, 1379 (1974)].

A.6. 2,4-Dichlorophenyl acetals react with alkyl Grignard reagents to form unsymmetrical ethers [H. Ishikawa, T. Mukaiyama, Chem. Lett., 305 (1975)].

A.9.   A thionthiazole has been used to convert an epoxide into a thioepoxide [V. Calo, Chem. Commun., 621 (1975)].

References

1. R. G. Smith et al., Can. J. Chem., 47, 2015 (1969).
2. P. E. Peterson, F. J. Slama, J. Org. Chem., 35, 529 (1970).
3. H.-G. Heine, Ann. Chem., 735, 56 (1970).
4. J.-C. Combret, Y. Leroux, Compt. Rend., 266, C 1178 (1968).
5. B. and K. Sjöberg, Acta Chem. Scand., 26, 275 (1972).
6. J. Dale, P. O. Kristiansen, Acta Chem. Scand., 26, 1471 (1972).
7. V. W. Gash, J. Org. Chem., 37, 2197 (1972).
8. E. E. Smissman et al., J. Org. Chem., 37, 3944 (1972).
9. F. R. Jensen, R. A. Neese, J. Org. Chem., 37, 3037 (1972).
10. I. Collins, H. Suschitzky, J. Chem. Soc., C 2337 (1969).
11. T. Haga et al., Japan Kokai 7,318,266, March 7, 1973; C. A. 79, 18,411 (1973).
12. Y. Chaturvedi, U. S. 3,734,970, May 22, 1973; C. A., 79, 18,356 (1973).
13. S.-O. Lawesson, N. C. Yang, J. Am. Chem. Soc., 81, 4230 (1959).
13a. C. Frisell, S.-O. Lawesson, Org. Syn., Coll. Vol., 5, 924 (1973).
13b. E. J. Corey, A. Venkateswarlu, J. Am. Chem. Soc., 94, 6190 (1972).
14. W. C. Hammann et al., J. Chem. Eng. Data, 15, 352 (1970).
15. B. F. Malichenko et al., J. Org. Chem. USSR (Engl. transl.), 9, 342 (1973).
16. G. A. Neville, R. Y. Moir, Can. J. Chem., 47, 2787 (1969).
16a. S. Swaminathan et al., Chem. Commun., 358 (1972).
17. W. J. Le Noble, Synthesis, 1 (1970).
18. W. J. Le Noble et al., J. Org. Chem., 36, 193 (1971).
18a. D. J. Cram et al., J. Org. Chem., 39, 2445 (1974).
18b. R. N. Greene, Tetrahedron Letters, 1793 (1972).
18c. E. M. Laboratories, Inc., Elmsford, N.Y., 10523.
19. B. Dietrich et al., Tetrahedron, 29, 1629, 1647 (1973).
20. G. Litkei et al., Acta Chim. (Budapest), 76, 95 (1973); C. A., 79, 18,533 (1973).
21. C. J. Pedersen, Org. Syn., 52, 66 (1972).
22. G. Cainelli et al., Chem. Commun., 1047 (1969).
23. D. A. Seeley, J. McElwee, J. Org. Chem., 38, 1691 (1973).
24. E. D. Bergmann et al., Synthesis, 646 (1971).
25. G. J. Heiszwolf, H. Kloosterziel, Chem. Commun., 51 (1966).
26. B. Castro, C. Selve, Tetrahedron Letters, 4459 (1973).
27. P. Y. Johnson, J. Zitsman, J. Org. Chem., 38, 2346 (1973).

28. C. F. Bernasconi et al., J. Org. Chem., 38, 2838 (1973).

B. Electrophilic Type Preparations

1. From Alcohols or Phenols (1, 301)

$$2\ RCH_2OH\ \xrightarrow{H_2SO_4}\ RCH_2OCH_2R\ +\ H_2O$$

Methyl (α,α-dimethyl-3,5-di-t-butylbenzyl) ether has been produced from the alcohol by treatment with CH_2OH-HCl [1]. 1,2-Dihydric alcohols form epoxides with dimethyl formamide dimethyl acetal as shown [2]:

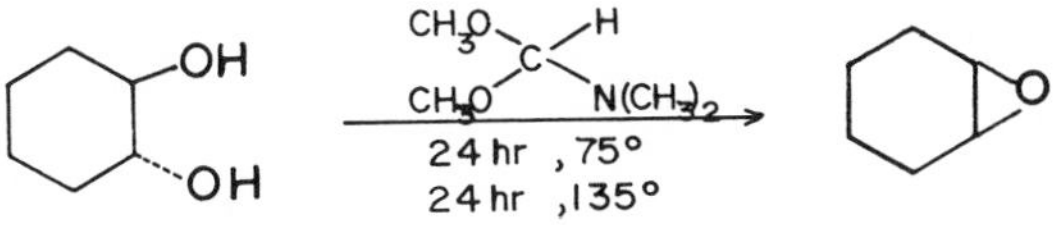

trans— Cyclohexanediol   cis—Cyclohexene oxide
                                              88 %

The cis isomer does not give an epoxide but rather the acetal,

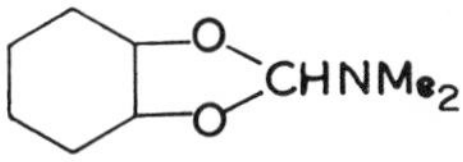

It has been stated (1, 303) that the treatment of an alcohol with diazomethane to form the ether is catalyzed by boron trifluoride etherate. It now appears that the same reactions may be catalyzed by rhodium salts [3]:

$$(CH_3)_3COH\ +\ N_2CHCO_2Et\ \xrightarrow[25°]{Rh_2(OAc)_4\,(trace)}\ (CH_3)_3COCH_2CO_2Et$$

Ethyl t-butoxyacetate, 82 %

Stereospecificity was noted in ring closure of meso and dl-hexanediol-2,5, respectively, to trans- and cis-2,5-dimethyl-furans by refluxing with a trace of phosphoric acid [4].

A homoallylic type of cation may be formed in the reaction of steroidal alcohols with dimethyl phosphite to form methyl ethers [5]. Phenyl ethers may be formed similarly with diphenyl phosphite.

A rather unusual alkylation of phenols may be brought about with certain β-aminoalcohols [6].

$$ArOH\ +\ \underset{\underset{Me}{|}}{\overset{\overset{Me}{|}}{HOCH_2CNMe_2}}\ \xrightarrow{DCC}\ \underset{\underset{Me}{|}}{\overset{\overset{Me}{|}}{ArOCH_2CNMe_2}}$$

The ethyleniminium ion, $CH_2 \overset{\oplus}{\underset{\diagup \diagdown}{N Me_2}} CMe_2$, as an intermediate is
suggested, but the structure of the product, which would be

expected to be $Me_2NCH_2\text{-}\overset{\overset{Me}{|}}{\underset{\underset{Me}{|}}{C}}OAr$, is unexpected.

2.   From Halides or Esters (SN_1 Process) (1, 302)

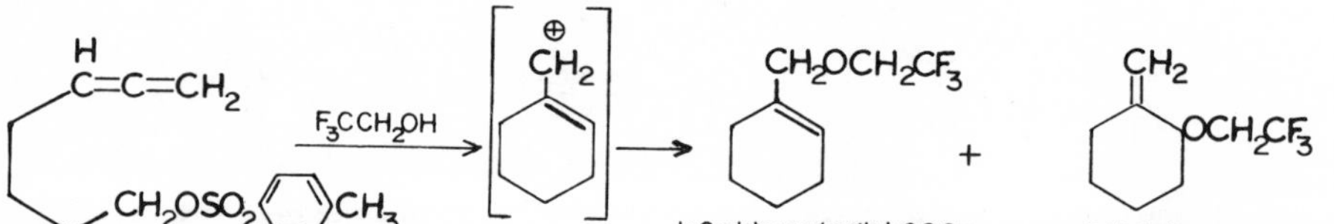

1-Cyclohexenyl methyl-2,2,2-
trifluoroethyl ether

2-Methylenecyclohexyl-
2,2,2-trifluoroethyl ether

(Total 65%, containing 45% of each)

Solvolysis of hepta-5,6-dienyl toluene-p-sulfonate in tri-
fluoroethanol leads to the 2-methylenecyclohexyl cation which
gives the ethers as indicated [7].

3.   From Vinyl Esters or Ethers (1, 305)

$$(C_6H_5)_3COR + R'OH \xrightarrow{AcOH} (C_6H_5)_3COR' + R OH$$

R' may be
$CH_2{=}CH{-}CH_2{-\!-}$

This exchange occurs in high yield if an excess of R'OH is
used [8].   The acidic nature of nitroform permits additions to
vinyl acetate [9]:

$$CH_2{=}CHO_2CCH_3 + ROH + HC(NO_2)_3 \longrightarrow CH_3\overset{}{\underset{\underset{OR}{|}}{C}}HC(NO_2)_3$$

2-Alkoxy-1,1,1-trinitropropanes
28-76%

6.   From Acetals (or Orthoformates), Alkenes, and the Like
(1, 308)

$$CH_3COCH{=\!=}CH_2 + HC(OMe)_3 \xrightarrow{H^{\oplus}} CH_3\overset{\overset{(OMe)_2}{|}}{C}CH_2CH_2OMe$$

This reaction is serviceable in that the product may be converted into 2-methoxybutadiene in 45% yield by $KHSO_4$ at 150° [10].

9. From Aryl Ethers (Friedel-Crafts)

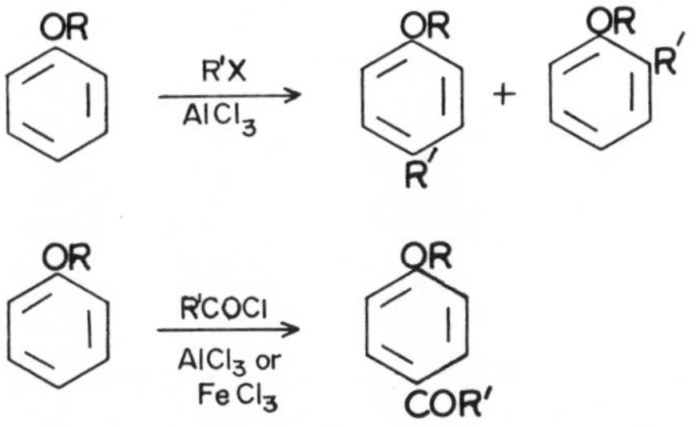

Since the ether group activates the aromatic ring, electrophilic substitution of aromatic ethers allows for easy entry to a wide variety of ethers. The substitution usually takes place in the para position, but in certain cases bonding (e.g., hydrogen bonding) may occur first with the ether group, in which case the substitution occurs predominantly in the ortho position [11].

Of interest is the Friedel-Crafts acylation of aromatic ethers (or other substrates with active nuclei) with little or no catalyst [12], the small amount of catalyst being ferric chloride, iodine, zinc chloride, or iron. Obviously, the isolation of the pure product was made easier by the presence of so little catalyst. One example in which no catalyst is used is given (Ex. a).

In the alkylation of anisole with a series of olefins and with γ-valerolactone, the substitution occurs primarily in the ortho position [13]. The extent of ortho alkylation is a function of the solvent and of the basic functionality of the alkylating agent. In some cases the ortho percentage was as high as 97%.

A template effect has been noted in the preparation of a macrocyclic ether by a Friedel-Crafts type of reaction [14]:

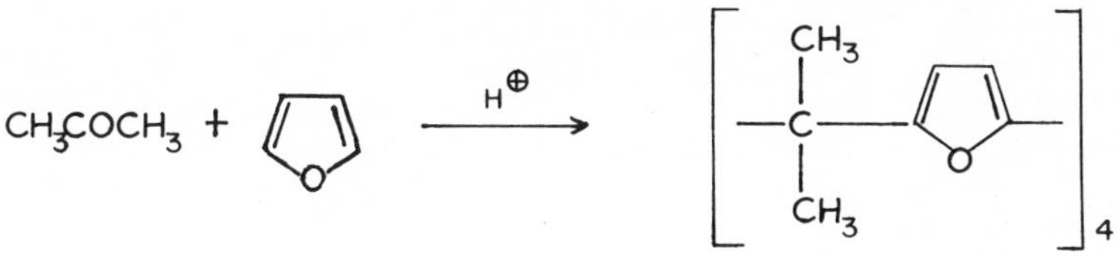

The yield of tetrameric cyclic ether is 18% in the absence of salts but is increased to about 40% when Li, Ca, Zn, or Mg perchlorates are added. The cations of these salts appear to be loci for clustering of the furan nuclei, a phenomenon that increases the possibilities for cyclization.

The reaction of phloroglucinol and citral in refluxing pyridine yields a tetracyclic compound [14a]:

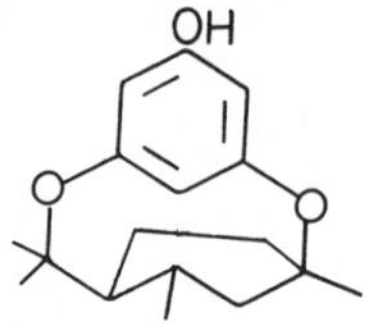

35 %

a.   Preparation of 4-Methoxybenzophenone [12]

Anisole and benzoyl chloride (0.21 mol of each) were heated for 4 days, the internal temperature being 176° the first day and 190°, 220°, and 255° on the following 3 days. The cooled mixture was then poured into 10% aqueous NaOH, warmed, and stirred overnight. The solid, formed on cooling the tan oil, was crystallized from hexane to give 19.7 g (45%) of the methoxy ketone.

10.   From Alkenes and Mercury Salts in Alcohol

The solvomercuration-demercuration procedure originally applied to the synthesis of alcohols (1, 188) has now been modified, by using an alcohol rather than water in the first step, to prepare ethers by Markownikoff addition [15] as indicated:

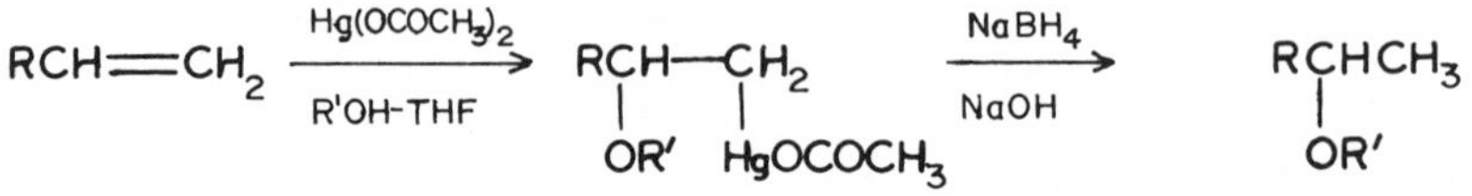

$$RCH\!\!=\!\!CH_2 \xrightarrow[R'OH-THF]{Hg(OCOCH_3)_2} \underset{\underset{OR'\ \ HgOCOCH_3}{|\ \ \ \ \ |}}{RCH\!-\!CH_2} \xrightarrow[NaOH]{NaBH_4} \underset{\underset{OR'}{|}}{RCHCH_3}$$

By this procedure methoxyethers are obtained in 83-100% yield and even t-butoxyethers were synthesized satisfactorily, except in cases in which the steric strain was too great. The substitution of $Hg(OCOCF_3)_2$ for $Hg(OCOCH_3)_2$ increased the rate and led to better yields in most cases.

Although the mode of addition is immaterial in the preparation of simple cyclic ethers, such is not true for substituted ones. The C_4-C_7 cis cycloalkenes give trans addition of the $HgOCOCH_3$ and OR groups exclusively. On the other hand, trans-cyclooctene and trans-cyclononene give cis addition exclusively [16].

Another way of accomplishing ether formation by an addition-reduction series of reactions is shown [17]:

NBS
$(CH_3)_3COH$
Bu_3SnH

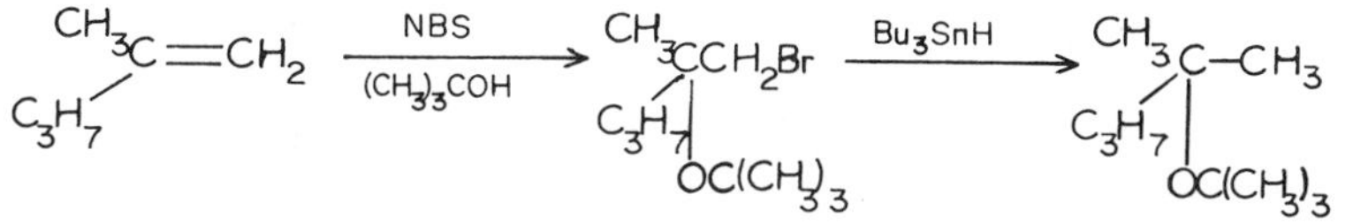

2-Methyl-2-pentyl-*t*-butyl ether, 73 %

Strained hydrocarbons also give ethers by catalysis with mercury salts [18]:

C_6H_5HgCl
CH_3OH
OCH_3

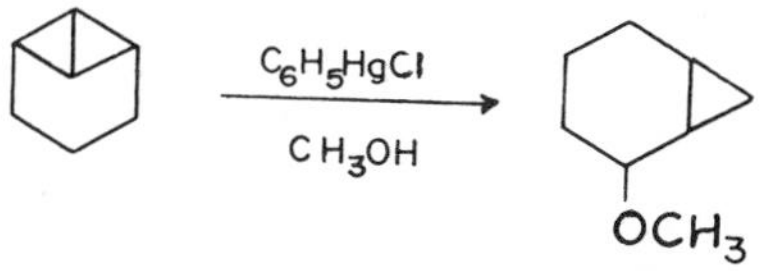

2-Methoxynorcarane (No yield given)

11. From Alkenes and Glycols (Prins) or the Like

The Prins reaction has been discussed (1, 188). It often yields a variety of products, such as allyl and other unsaturated alcohols, m-dioxanes, and 4-hydroxytetrahydropyrans. If the reaction is conducted with 1-alkenes, good yields of monoalkyl-1, 4-dioxanes may be obtained as shown [19]:

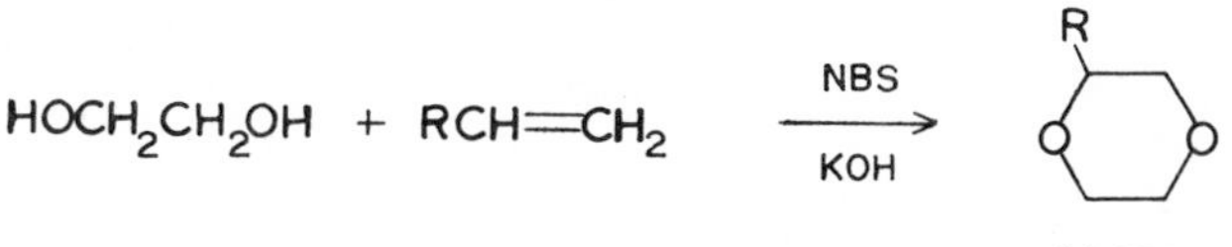

$HOCH_2CH_2OH + RCH{=}CH_2$
NBS
KOH
R

86-97% (crude)

The mechanism is as follows:

NBS
$HOCH_2CH_2OH$
$-H^{\oplus}$
KOH

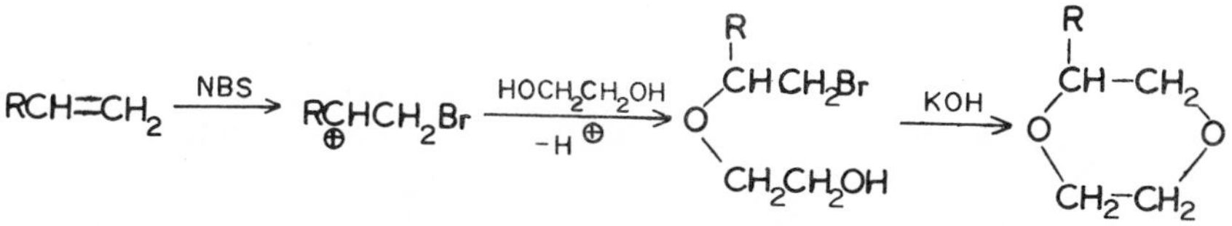

Chloromethyl ether has been added to an alkene [20]:

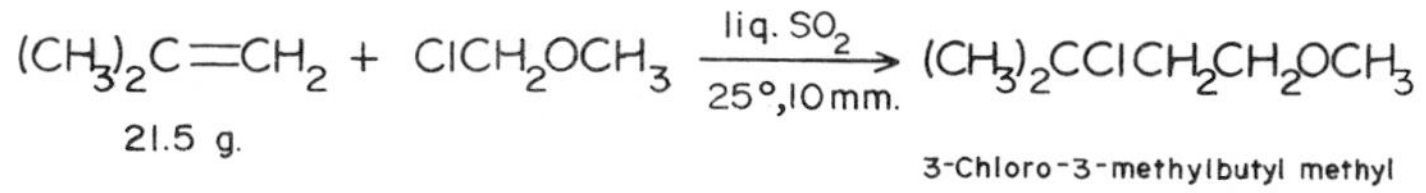

$(CH_3)_2C{=}CH_2 + ClCH_2OCH_3$
liq. SO_2
25°,10mm.
$(CH_3)_2CClCH_2CH_2OCH_3$
21.5 g.

3-Chloro-3-methylbutyl methyl ether, 7.8 g.

a. Preparation of n-Butyl-1,4-dioxane [19]

A mixture of 0.11 mol of $CH_3CH_2CH{=}CH_2$, 0.1 mol of NBS, and

1 mol of ethylene glycol at 50° was stirred rapidly for 1 hr. Then 0.6 mol of KOH was added and the reaction mixture was heated to 150° for 2 hr. After pouring the mixture into one liter of H_2O and neutralizing with 10% HCl, the product was extracted with ether. Recovery in the usual manner led to 86% of the crude alkyl 1,4-dioxane.

12. From Glycols and Sulfuranes (to Form Epoxides)

$$-\underset{\underset{OH}{|}}{C}-\underset{\underset{OH}{|}}{C}- \; + \; (C_6H_5)_2S\left[OC(CF_3)_2C_6H_5\right]_2 \; \xrightarrow{25°}$$

"A"

$$-\overset{\overset{\displaystyle O}{\triangle}}{\underset{|}{C}-\underset{|}{C}}- \; + \; (C_6H_5)_2SO \; + \; 2\,C_6H_5(CF_3)_2COH$$

The sulfurane, A, has been found to be an efficient and unique dehydrating agent [21]. Its synthesis has been reported [22]. trans-1,2-Cyclohexanediol yields the epoxide, while the cis isomer yields only cyclohexanone. The intermediate in the reaction is the mixed sulfurane:

$$\text{"A"} \; + \; ROH \; \longrightarrow \; (C_6H_5)_2\,S\!\!\underset{OC(CF_3)_2C_6H_5}{\overset{OR}{\diagup}}$$

which gives considerable carbonium ion character or potential to R. 1,3-Propanediol does not form the oxetane, but 2,2-dimethyl-1,3-diol does in good yield.

Addenda

B.1. Dicyclopropylcarbinol has been converted into the ether (81%) by using polymer-protected aluminum chloride, P -AlCl$_3$ [D. C. Neckers et al., J. Am. Chem. Soc., 94, 9284 (1972)].

B.5. Formation of the methyl ether from phenols by allowing the latter to stand in methanol and 0.1 N HCl at room temperature for seven hours depends on the amount of ketonization of the enolic form, the phenol. However, the following summary suggests other factors exist. Percent formation of methyl ether from the phenol under the above conditions: phenol, 8%; 1-naphthol, 14%; 2-naphthol, 16%; 2-phenanthrol, 29%; anthrone, 0% (a steric factor?); 6-OH-benz[a]pyrene, 0%; 7,12-dimethyl-6-OH-benz[a]-anthracene, 84% (M. S. Newman, J. Am. Chem. Soc., 98, 3237 (1976). Of course, refluxing increases the yield of the methyl ether.

References

1. H.-D. Scharf, F. Döring, Chem. Ber., 100, 1761 (1967).
2. H. Neumann, Chimia, 23, 267 (1969).
3. H. Reimlinger et al., Tetrahedron Letters, 2233 (1973).
4. M. Lj. Mihailovic et al., J. Chem. Soc., Perkin Trans., I, 2460 (1972).
5. Y. Kashman, J. Org. Chem., 37, 912 (1972).
6. F. L. Bach et al., J. Med. Chem., 11, 987 (1968).
7. R. G. Bergman et al., Chem. Commun., 679 (1973).
8. P. F. Butakus, R. Y. Sabonene, J. Org. Chem. USSR, 5, 521 (1969).
9. V. I. Grigos, L. T. Eremenko, Izv. Akad. SSSR, Ser. Khim., 2566 (1969); Synthesis, 597 (1971).
10. L. J. Dolby, K. S. Marshall, Org. Prep. Proced., 1, 229 (1969).
11. D. E. Pearson, C. A. Buehler, Synthesis, 460 (1971).
12. D. E. Pearson, C. A. Buehler, Synthesis, 533 (1972).
13. R. A. Kretchmer, M. B. McCloskey, J. Org. Chem., 37, 1989 (1972).
14. M. and F. Chastrette, Chem. Commun., 534 (1973).
14a. L. Crombie, R. Ponsford, J. Chem. Soc., C 788 (1971).
15. H. C. Brown, M.-H. Rei, J. Am. Chem. Soc., 91, 5646 (1969).
16. T. G. Traylor et al., J. Org. Chem., 38, 2306 (1973).
17. G. L. Grady, S. K. Chokshi, Synthesis, 483 (1972).
18. E. Müller, Tetrahedron Letters, 1201 (1973).
19. A. Jovtscheff et al., Monatsh., 102, 114 (1971).
20. N. Tokura, Synthesis, 641 (1971).
21. J. C. Martin et al., University of Illinois, Urbana, Private Communication.
22. R. J. Arhart, J. C. Martin, J. Am. Chem. Soc., 94, 4997 (1972).

C. Nucleophilic Type Reactions

3. From Carbonyl Compounds and Certain Halides (Darzens) (1, 316)

$$C_6H_5CHO + ClCH_2CO_2C_2H_5 \xrightarrow{NaOR} C_6H_5CH\underset{O}{\overset{}{-}}CHCO_2C_2H_5$$

Although the Darzens condensation is usually carried out in an alkaline medium, Sipos and co-workers [1] showed that, similar to the aldol condensation, success may be achieved in an acid medium. By using absolute ethanol containing 10% HCl, a series of epoxyketones was prepared from benzaldehydes and phenacyl chlorides or bromides, as shown, with the maximum yield being 77%:

$$C_6H_5CHO \ + \ C_6H_5COCH_2X \longrightarrow C_6H_5COCH\!-\!CHC_6H_5$$

No epoxyketone formation occurred when p-methyl- or p-methoxy-benzaldehyde was the aldehyde employed in the reaction.

Villieras and co-workers [2] synthesized glycidic acid esters and amides from aldehydes and trichloroacetic acid esters and amides by the use of tris (dimethylamino) phosphine as shown:

$$\begin{array}{c}CH_3 \\ \diagdown \\ CH_3 \end{array}\!\!CHCHO \ + \ CCl_3CO_2Me \xrightarrow[\text{THF, 0-5}^\circ]{P[N(CH_3)_2]_3}$$

$$\begin{array}{c}CH_3 \qquad\qquad Cl \\ \diagdown \qquad\quad | \\ CH\!-\!CH\!-\!C \\ \diagup \quad \diagdown O \diagup \ \diagdown CO_2Me \\ CH_3 \end{array} \ + \ [(CH_3)_2N]_3PCl_2$$

Methyl-α-chloro-β-isopropylglycidate
80%

The formation of glycidic esters does not occur satisfactorily with straight-chain aliphatic aldehydes, benzaldehyde, or ketones. In addition to α-chloroamides, α-bromolactones may be employed in the condensation.

The Darzens condensation also occurs between a geminal dibromide, butyllithium or lithium itself in THF, and a carbonyl compound [3] as shown:

$$\begin{array}{c}R \\ \diagdown \\ \ \ \ C \ Br_2 \\ \diagup \\ R' \end{array} \xrightarrow[\text{THF}]{\text{BuLi}} \left[\begin{array}{c}R \diagdown \quad \diagup Li \\ C \\ R' \diagup \ \diagdown Br \end{array}\right] \xrightarrow{\text{R''COR'''}}$$

$$\begin{array}{ccc} R'' \ OLi \qquad R \\ \diagdown \ | \qquad \diagup \\ C\!-\!C \\ \diagup \qquad | \ \diagdown R' \\ R''' \qquad Br \end{array} \longrightarrow \begin{array}{c} R'' \qquad O \qquad R \\ \diagdown \ \diagup \diagdown \ \diagup \\ C\!-\!C \\ \diagup \qquad\quad \diagdown R' \\ R''' \end{array}$$

45-95%(1:1 *cis-trans* mixture)

Aliphatic, alicyclic, and aromatic aldehydes and ketones respond satisfactorily.

Tavares and co-workers found that the Darzens condensation, using potassium t-butoxide as a base, was satisfactory for preparing α,β-epoxysulfones [4] and α,β-epoxysulfoxides [5]. However, in applying the method to the synthesis of the corresponding sulfides, it was unsatisfactory. The use of "Dabco"

(1,4-diazabicyclo[2.2.2.]octane) as a catalyst overcame this
difficulty [6] as shown:

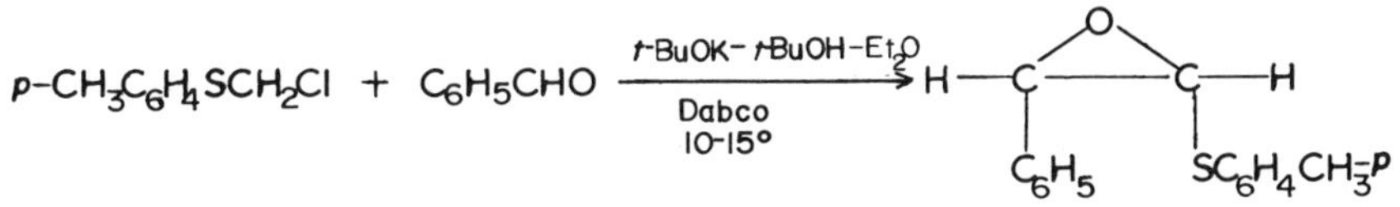

"Good yield"
trans—cis ratio 1 : 4.5

A most unusual epoxide formation is as indicated [7]:

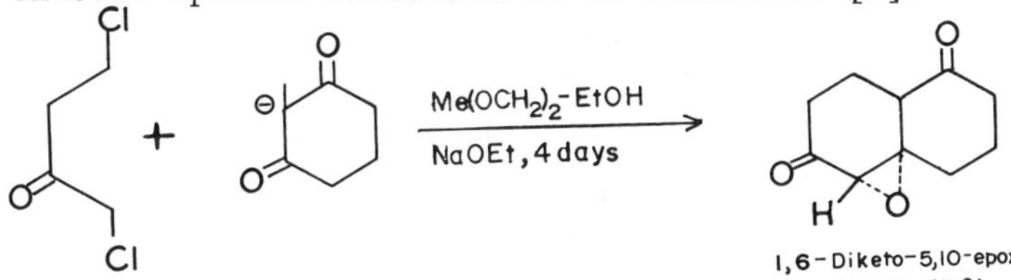

1,6 – Diketo-5,10-epoxy-
decalin, 43 %

4. From Carbonyl Compounds and Dimethylsulfonium Methylide
    or Dimethyloxosulfonium Methylide (1, 317)

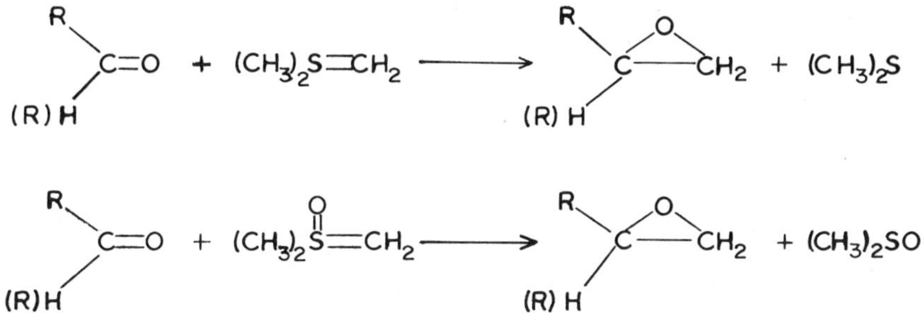

In place of a sulfonium methylide, a sulfonium cyclopropyl-
ide leads to an interesting type of epoxide [8]:

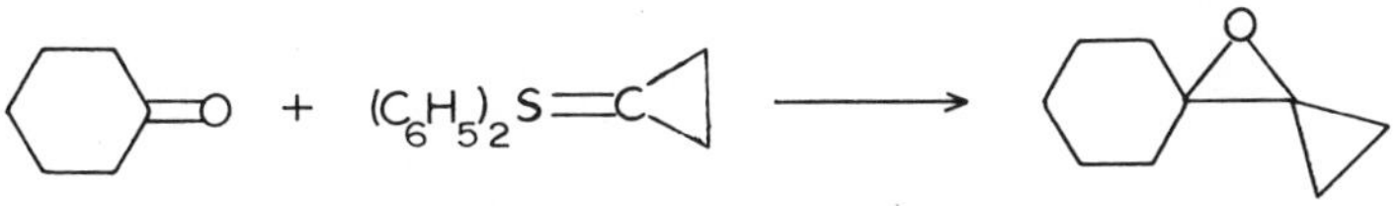

An oxaspiropentane, 97 %

The general method for the conversion of aldehydes and
ketones into oxiranes by the use of dimethyloxosulfonium methylide has now appeared in the literature [9]. It has been shown
that ylids by reaction with an alkene form cyclopropanes [10].
In a similar manner carbonyl compounds form oxiranes as shown:

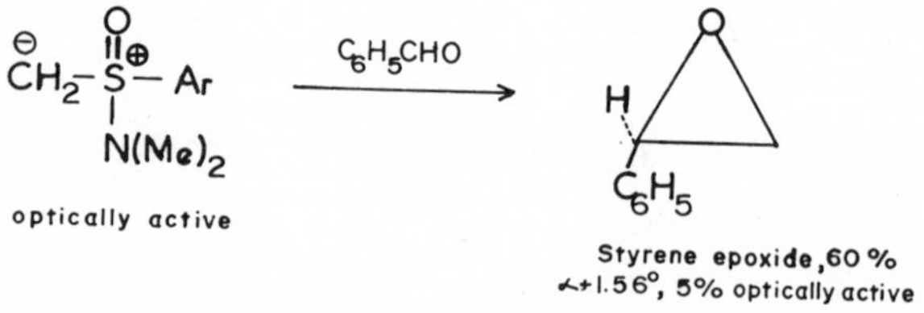

Styrene epoxide, 60 %
∠+1.56°, 5% optically active

The preparation of epoxides from certain diaryl ketones, such as dibenzosuberone, as shown is not satisfactory by the usual procedure of employing dimethylsulfonium methylide or dimethyloxosulfonium methylide. To overcome this difficulty Muchowski and co-workers [11] found that by the addition of 50% excess of trimethylsulfonium iodide at 25° to a slurry of NaH in DMSO containing the ketone, the epoxide was obtained in 90% yield:

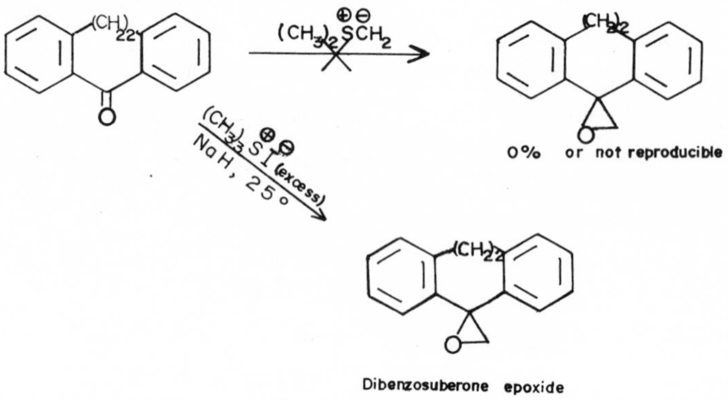

0%   or not reproducible

Dibenzosuberone epoxide

An elegant synthesis of 3- and 3,4-substituted furans has now become available via the formylation of ketones [12]:

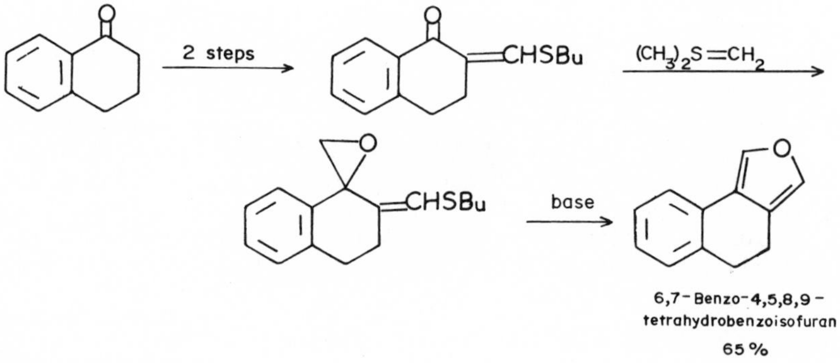

6,7-Benzo-4,5,8,9-
tetrahydrobenzoisofuran

65 %

6. Isomerization of Epoxides

5β,6β-Epoxysteroids may be isomerized with potassium cyanate in 50% aqueous ethanol to the 5α,6α-forms [13]:

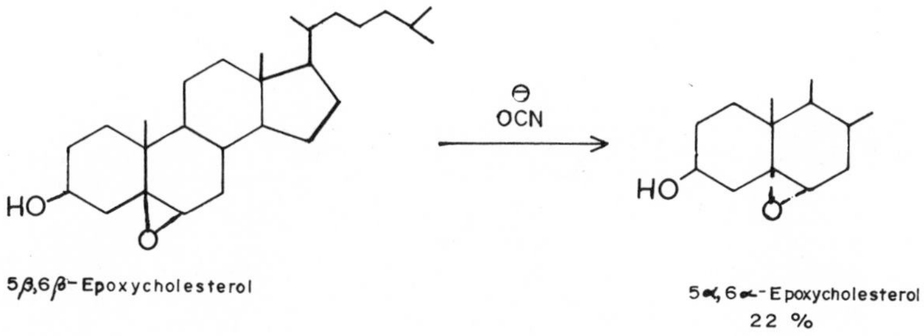

5β,6β-Epoxycholesterol

5α,6α-Epoxycholesterol
22 %

It is suggested that the isomerization occurs via the trans-equatorial ring opening:

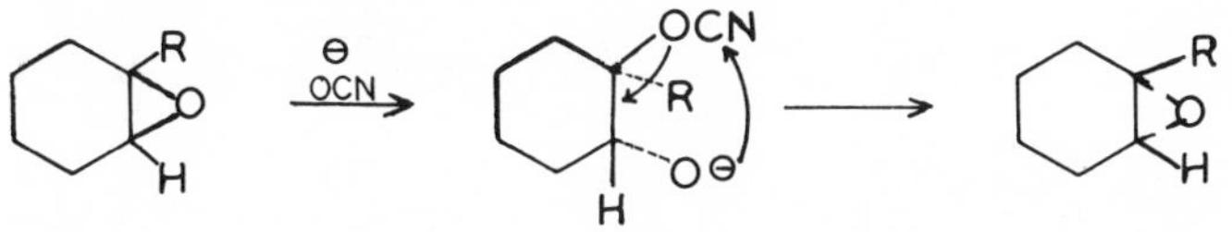

The isomerization does not occur with 5α,6α-epoxysteroids.

Addenda

C.3. A glycidic thiol ester has been prepared for the first time by the Darzens reaction [D. J. Dagli, J. Wemple, J. Org. Chem., 39, 2938 (1974)].

C.3. A Darzens reaction has been achieved with lithium dicyclo-hexylamide [H. Yamamoto et al., J. Am. Chem. Soc., 96, 3010 (1974)].

C.4. In place of dimethylsulfonium methylide to furnish the methylene group for the preparation of epoxides from carbonyl compounds, methylene bromide and BuLi at -80° may be used. A number of diepoxides were prepared in this way [P. Weyerstahl et al., Chem. Ber., 108, 2391 (1975)].

C.4. Epoxides may be prepared from aldehydes by the phase-
transfer method by using trimethylsulfonium iodide and tetrabutylammonium iodide [A. Merz, G. Markl, Angew. Chem. Intern. Ed.
Engl., 12, 845 (1973)].

References

1. Gy. Sipos et al., J. Chem. Soc., C 1154 (1970).
2. J. Villieras et al., Bull. Soc. Chim. France, 898 (1971);
    J. D. White et al., J. Am. Chem. Soc., 93, 281 (1971).
3. G. Cainelli et al., Tetrahedron, 28, 3009 (1972).
4. P. F. Vogt, D. F. Tavares, Can. J. Chem., 47, 2875 (1969).
5. D. F. Tavares et al., Tetrahedron Letters, 2373 (1970).
6. D. F. Tavares, R. E. Estep, Tetrahedron Letters, 1229
    (1973).
7. S. Danishefsky, G. A. Koppel, Chem. Commun., 367 (1971).
8. B. M. Trost et al., Tetrahedron Letters, 923 (1973).
9. E. J. Corey, M. Chaykovsky, Org. Syn., 49, 78 (1969).
10. C. R. Johnson et al., J. Am. Chem. Soc., 90, 3890, 6852
    (1968).
11. J. M. Muchowski et al., Can. J. Chem., 47, 4327 (1969).
12. M. E. Garst, T. A. Spencer, J. Am. Chem. Soc., 95, 250
    (1973).
13. K. Jankowski, J.-Y. Daigle, Synthesis, 32 (1971).

D. Oxidation or Free Radical Reactions

   1. From Olefins (1, 320)

$$RCH{=\!=}CH_2 \xrightarrow{\;C_6H_5CO_2OH\;} RCH{-\!-}CH_2$$
$$\underset{O}{\diagdown\diagup}$$

     Discussions of epoxidation have been published [1, 2, 2a].
Although the reagent is determined largely by the stability of
the epoxide under the reaction conditions, m-chloroperbenzoic
acid is perhaps the preferred reagent for the oxidation. A
general procedure is given (Ex. a).
     It is interesting to note that the decomposition of
m-chloroperbenzoic acid may be prevented even at 90° by 4,4'-
thio-bis-(6-t-butyl-3-methylphenol) [3]. Epoxides that form with
difficulty can now be made in the presence of the inhibitor. At
present the reaction has been conducted only on a milligram
scale with 1-octene, 1-dodecene, and methyl methacrylate. The
substrate, 100 mg, in 5 ml of ethylene dichloride containing
1.2 equiv of m-chloroperbenzoic acid and 1-2 mg of the

thiobisphenol was heated at 90° for 1 hr to give the epoxide in almost quantitative yield.

For epoxides which are sensitive to further conversions, a biphase system consisting of methylene chloride and aqueous sodium bicarbonate has led to increased yields [4]:

5,6-Epoxy-6-methyl-
heptan-2-one, 83-85%

Ordinarily, this epoxide undergoes facile rearrangement to 1,3,3-trimethyl-2,7-dioxabicyclo[2.2.1] heptane.

However an even more selective agent for epoxide formation is a mixture of vanadyl acetyl acetonate and t-butyl hydroperoxide. With this mixture in refluxing benzene, geraniol was oxidized to 2,3-epoxygeraniol, isolated as the acetate, in 93% yield [5].

Other catalysts employed in the hydroperoxide oxidation are molybdenum or other vanadium compounds [6]. The catalyst may be soluble or insoluble and the hydroperoxide may be generated in situ. The olefin may be used as the solvent, temperatures are in the 50-130° range, and the time varies from minutes to several days. Yields are sometimes almost quantitative. Recently molybdenum hexacarbonyl has been employed satisfactorily with t-butyl hydroperoxide in the epoxidation of 1-hexene [7].

Epoxidation by the use of o-sulfoperbenzoic acid, freely soluble in water or aqueous acetone, in acetone has been carried out with a series of olefins [8]. Yields vary from 78 to 89%. In contrast to certain other peracids, no esterification of the diol has been observed in the reaction.

Other oxidizing agents have been used to convert olefins into epoxides. Klinot and co-workers [9] converted 19β,28-epoxy-18α-olean-2-ene into the formate, which was hydrolyzed as shown:

Of interest is the fact that the epoxide possessed a configuration opposite to that obtained by the use of peracids.

Carlson and Ardon [10] found that unhindered methylenecyclohexanes may be converted via bromohydrins into epoxides with an axial methylene group as indicated:

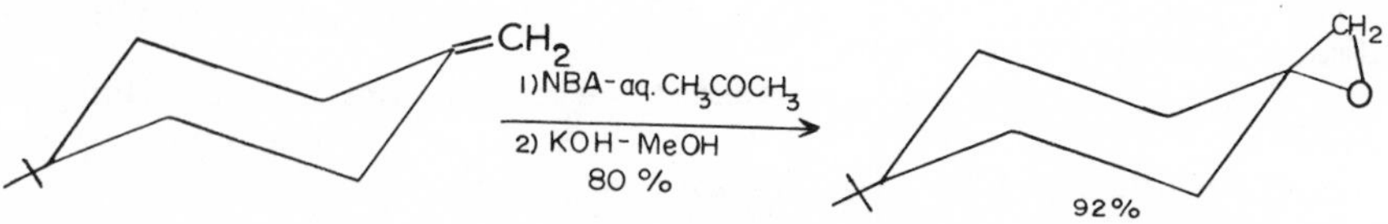

Finally epoxides have been prepared in the laboratory from alkenes by the of molecular oxygen in the presence of chelates of MoO_2^{+2} as shown [11]:

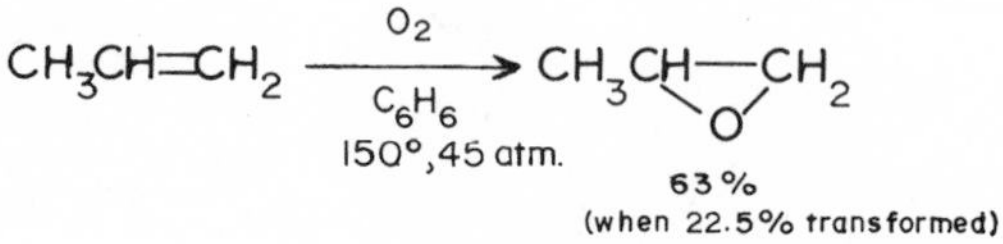

Epoxidation of an alkene (general procedure) [1]. To 0.1 mol of the olefin in 100 ml of methylene chloride was added, with stirring, a solution of 23.0 g (0.113 mol) of 85% m-chloroperbenzoic acid in 250 ml of methylene chloride with cooling if necessary. The mixture was kept at room temperature (0-5 hr) to complete the reaction, and the m-chlorobenzoic acid was removed by washing with a 10% solution of NaOH.

2. Miscellaneous Preparations (1, 322)

a. Electrolysis

The introduction of methoxy groups in hydroquinone dimethyl ether by electrolysis has been described (Ex. b, 1, 322) as well as the introduction of the oxymethyl group in dimethylaniline (Ex. c, 1, 322). Decarboxylation in methanol by electrolysis has now been accomplished [12]:

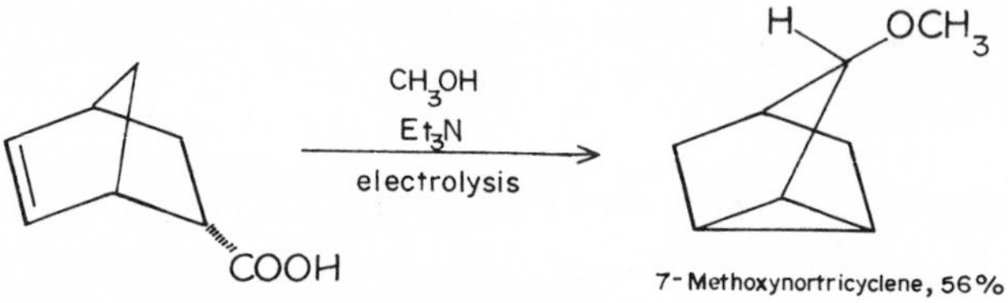

7-Methoxynortricyclene, 56%

b. From Peroxides (1, Ex. a, p. 322)

2-Acetyl-2,3-dihydrobenzofuran has been synthesized by the m-chloroperbenzoic acid oxidation of the chlorobutenylbenzene [13]:

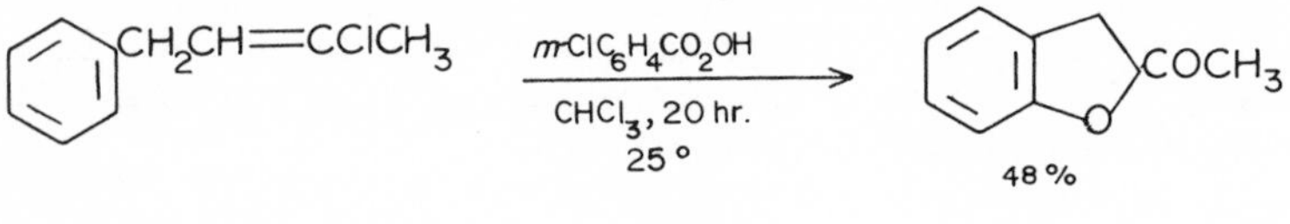

c. From Oxy Free Radicals

2,4,6-Triphenylphenoxyl has been prepared as shown [14]:

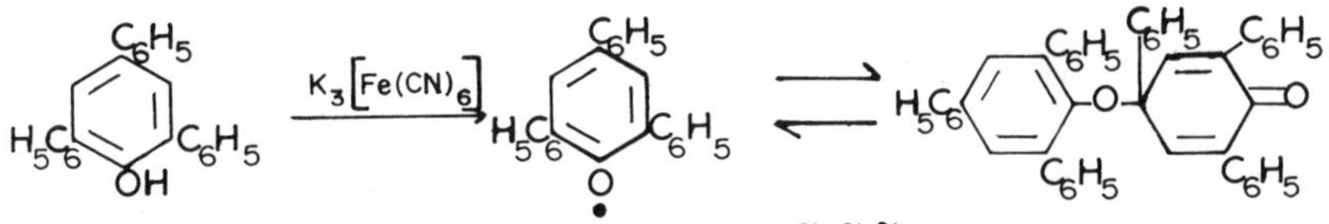

81-91 %

As seen, the monomer, an aroxyl radical, is in equilibrium with the dimer, an ether. This is an example of a radical rather stable toward oxygen.

Tetrahydrofurans have been produced via oxy free radicals by the lead tetraacetate oxidation of alcohols [15]. Alcohols with at least one hydrogen on the Δ carbon atom undergo intramolecular oxidative cyclization to give ethers in modest yield. Two examples are given:

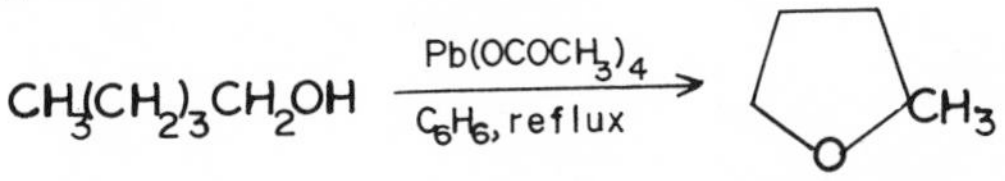

2-Methyltetrahydrofuran, 43%

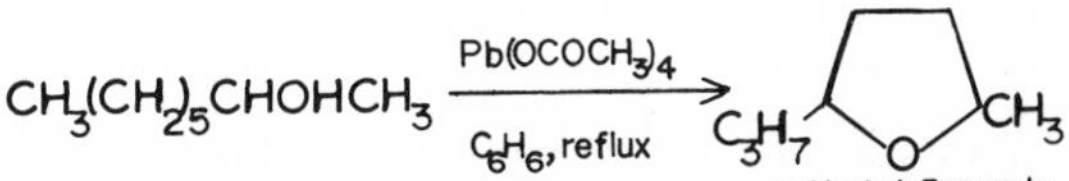

2-Methyl-5-propyl-tetrahydrofuran, 40%

The best results occur with 0.1 mol of the alcohol and 0.1+ mol of lead tetraacetate in refluxing benzene. Products other than furans are sometimes formed:

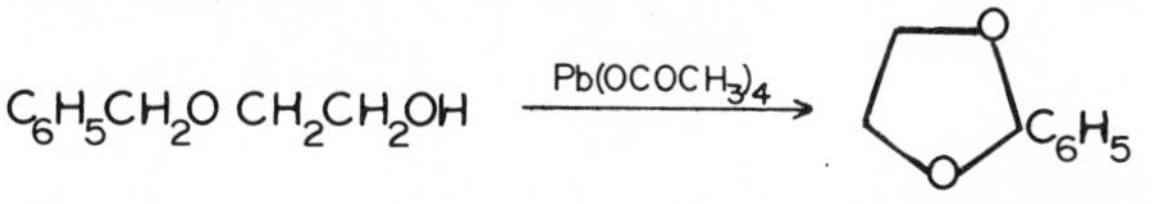

2-Phenyl-1,3-dioxolane
53%

Large-ring unsaturated alcohols yield bicyclic ether acetates:

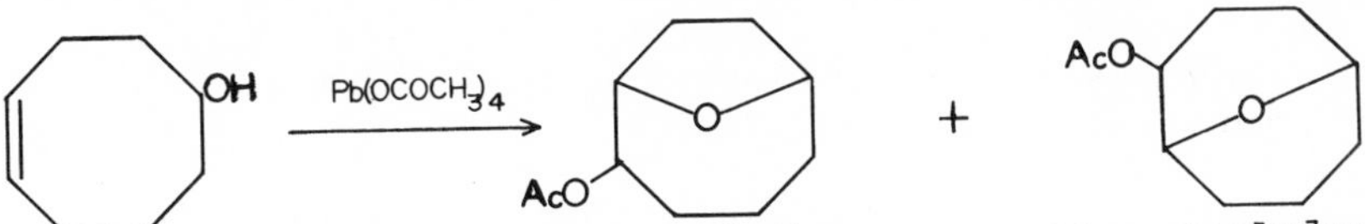

+

2-Acetoxy-9-oxa[4.2.1]nonane
28%

2-Acetoxy-9-oxa[3.3.1]nonane
42%

The hypobromite method of the synthesis of tetrahydrofurans has been somewhat erratic until it was found that particular salts of the silver ion (the reagent for generating positive bromine) produced a profound difference in yields [16].   Silver acetate, sulfate, and carbonate give much better yields of tetrahydrofurans than the nitrate, oxide, or trifluoroacetate.   An example follows:

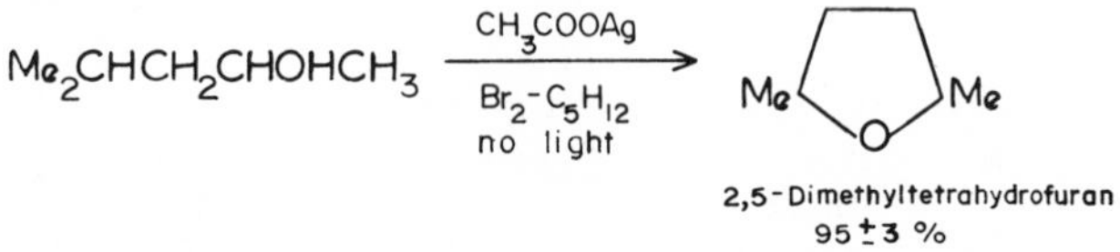

$$Me_2CHCH_2CHOHCH_3 \xrightarrow[\substack{Br_2-C_5H_{12} \\ no\ light}]{CH_3COOAg}$$

2,5-Dimethyltetrahydrofuran
95 ± 3 %

The reaction is not stereospecific and is liable to form ketones as by-products.   The hypoiodite has also been used to form tetrahydrofurans [17].

d.   From Carbenes

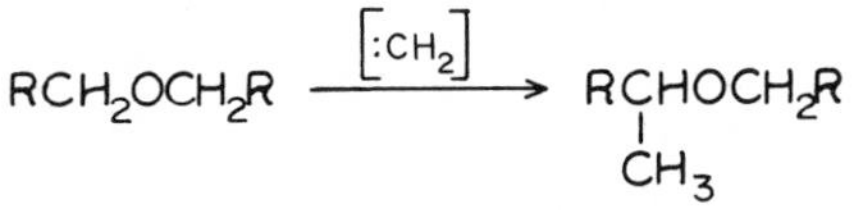

$$RCH_2OCH_2R \xrightarrow{[:CH_2]} \underset{\overset{|}{CH_3}}{RCHOCH_2R}$$

Random insertion tends to occur, but an example of selective insertion is given [18]:

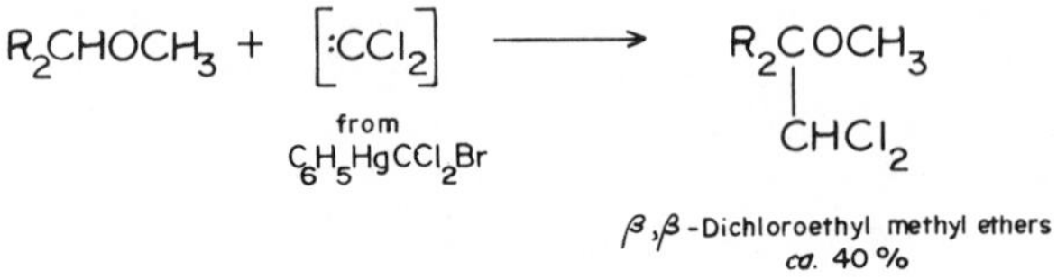

$$R_2CHOCH_3 + [:CCl_2] \xrightarrow{\substack{from \\ C_6H_5HgCCl_2Br}} \underset{\overset{|}{CHCl_2}}{R_2COCH_3}$$

β,β-Dichloroethyl methyl ethers
ca. 40 %

Addenda

D.1.   Cyclohexen-2-enyl hydroperoxide and its methyl derivatives in the presence of vanadyl acetyl acetonate catalyst give largely the cis-2,3-epoxycyclohexanol [T. Itoh et al., Bull. Chem. Soc. Jap., 48, 1337 (1975)].

D.1.   If two olefinic groups are situated in the same alcoholic molecule, the olefinic group nearest the hydroxy group (usually as part of an allylic alcohol system) may be selectively converted into the epoxide by t-BuOOH and VO(acac)$_2$ in a benzene reflux

[K. B. Sharpless, R. C. Michaelson, J. Am. Chem. Soc., 95, 6136 (1973)].

D.1. Epoxidation over metal ketenides has been reviewed [D. Bryce-Smith, Chem. Ind. (London), 154 (1975)].

D.1. Dicyclohexenyl-1,1^1 by oxidation with p-nitroperbenzoic acid may be converted into a 72% yield of the diepoxide (83% dl, 17% meso) [H. Christol et al., Compt. Rend., 278, C 883 (1974)].

References

1. S. N. Lewis, in Augustine's Oxidation, Vol. 1, Dekker, New York, 1969, p. 223.
2. R. Hiatt, in Augustine and Trecker's Oxidation, Vol. 2, Dekker, New York, 1971, p. 113.
2a. D. Swern, Organic Peroxides, Vol. II, Wiley-Interscience, New York, 1971, p. 355.
3. Y. Kishi et al., Chem. Commun., 64 (1972).
4. W. K. Anderson, T. Veysoglu, J. Org. Chem., 38, 2267 (1973).
5. K. B. Sharpless, R. C. Michaelson, J. Am. Chem. Soc., 95, 6136 (1973).
6. Ref. 2, p. 117.
7. V. S. Markevich, N. K. Shtivel, Neftekhimiya, 13, 240 (1973); C. A., 79, 41669 (1973).
8. J. M. Bachhawat, N. K. Mathur, Tetrahedron Letters, 691 (1971).
9. J. Klinot et al., Coll. Czech. Chem. Commun., 35, 3610 (1970).
10. R. G. Carlson, R. Ardon, J. Org. Chem., 36, 216 (1971).
11. J. Rouchaud et al., Bull. Soc. Chim. Belges, 80, 365, 453 (1971).
12. S. Wawzonek, Synthesis, 285 (1971).
13. T.-L. Ho, C. M. Wong, Org. Prep. Proced. Intern., 4, 265 (1972).
14. K. Dimroth et al., Org. Syn., 49, 116 (1969).
15. M. L. Mihailović, Z. Ceković, Synthesis, 209 (1970).
16. N. M. Roscher, Chem. Commun., 474 (1971).
17. J. Kalvoda, K. Heusler, Synthesis, 501 (1971).
18. D. Seyferth et al., J. Org. Chem., 35, 1993 (1970).

E. Reduction

1. From Acetals or Ketals (1, 324)

$$R_2C(OR')_2 \xrightarrow{[H]} R_2CHOR'$$

In addition to the reducing agents mentioned in the original text, acyclic acetals and ketals have been reduced by 50-100% excess of a reagent containing sodium borohydride and boron trifluoride at 25-30° [1].

Also β-chloroethyl acetals have been reduced by $LiAlH_4$ in THF [2]:

$$R_2C = CHCH \underset{OCH_2CH_2X}{\overset{OEt}{\diagup}} \xrightarrow[THF]{LiAlH_4} R_2C = CHCH_2OEt$$

Allyl ethyl ether, 30-65%

To reduce simple acetals to ethers trichlorosilane under γ irradiation has been employed [3].

2.   From Aldehydes or Ketones (1, 325)

Ketones may be converted into ethers via the secondary alcohol [4], which need not be isolated:

$$\triangleright\!\!-\!\!\underset{O}{\overset{||}{C}}C_6H_5 \xrightarrow[2) MeI]{1) NaH \atop DME} \left[\triangleright\!\!-\!\!\underset{OH}{\overset{|}{C}}HC_6H_5\right] \longrightarrow \triangleright\!\!-\!\!\underset{OCH_3}{\overset{|}{C}}HC_6H_5$$

α-Methoxybenzyl-cyclopropane, 60%

A new general method of reducing aldehydes or ketones to ethers employs a trialkylsilane in alcoholic, acid media as shown [5]:

$$\underset{}{>}\!C{=}O + R_3SiH + R'OH \xrightarrow{H^\oplus} \underset{}{>}\!CHOR' + R_3SiOH$$

Isolated yields vary from 26-88%. The use of sulfuric, trifluoroacetic, or trichloroacetic acid is satisfactory and the time required is usually less than one hour.

a.   Preparation of Benzyl Methyl Ether [5]

Sulfuric acid, 1 ml of 97% was added dropwise to a stirred solution of 2.5 ml of methanol, 5.0 mmol of benzaldehyde, and 5.5 mmol of triethylsilane at 0°. After warming to room temperature, the mixture was stirred for 1 hr. The workup procedure, which involved pentane and saturated squeous NaCl, gave 87% of the ether.

3.   From Esters or Lactones (1, 325)

$$-COOR \xrightarrow[BF_3]{NaBH_4} -CH_2OR$$

Alkyl aliphatic esters have been reduced with trichlorosilane under γ irradiation [6] as shown:

$$RCOOR' \xrightarrow[\gamma\text{-irradiation}]{Cl_3SiH} RCH_2OR'$$
$$77\text{-}100\%$$

The degassed mixture of the two components was subjected to irradiation of 9.6 MR at a dose rate of 0.6 MR/hr at 25°.

In a similar manner γ, δ, and ε lactones have been reduced to cyclic ethers [7] as indicated:

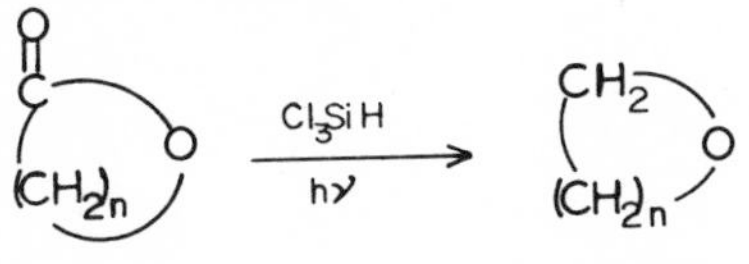

62 - 100%

5.   From Peroxides

$$ROOR \xrightarrow{[H]} ROR$$

Ascaridole has been converted into an epoxide by reduction [8]:

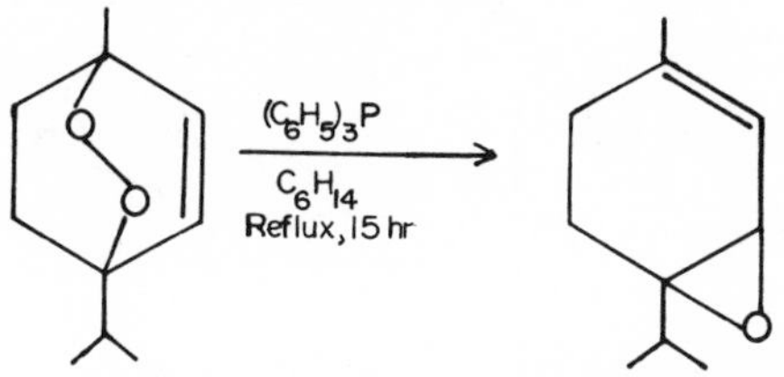

1 - Methyl-4-isopropyl-
3,4-epoxycyclohexene ,(no yield)

A more straightforward, if sluggish, reduction is that of Horner and Jurgeleit [9]:

$$(CH_3)_3C-O-O-C(CH_3)_3 \xrightarrow[30\,hr.\,in\,bomb]{(C_6H_5)_3P} (CH_3)_3C-O-C(CH_3)_3$$

Di-t-butyl ether, 81 %

References

1. B. C. Subba Rao et al., Indian J. Chem., 3, 123 (1965).
2. F. Nerdel et al., Chem. Ber., 102, 3102 (1969).
3. J. Tsurugi et al., J. Org. Chem., 37, 4349 (1972).
4. J.-L. Pierre, P. Arnaud, Bull. Soc. Chim. France, 2107 (1967).
5. M. P. Doyle et al., J. Am. Chem. Soc., 94, 3659 (1972).
6. J. Tsurugi et al., J. Am. Chem. Soc., 91, 4587 (1969).
7. R. Nakao et al., J. Org. Chem., 37, 76 (1972).
8. G. O. Pierson, O. A. Runquist, J. Org. Chem., 34, 3654 (1969).
9. L. Horner, W. Jurgeleit, Ann. Chem., 591, 138 (1955).

G. Photochemical Reactions

1. Ionic Addition to Cyclohexenes

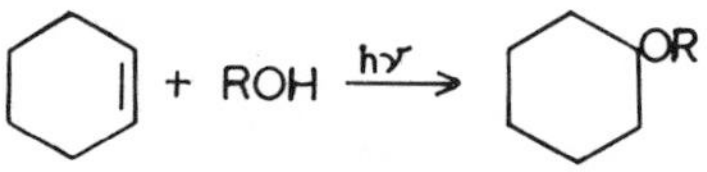
+ ROH
hν
OR

This Markownikoff addition occurs with water, alcohols, and carboxylic acids, but only in the case of cyclohexenes and cycloheptenes [1]. The reaction of methanol with 1-menthene, thought to occur via the carbonium ion, is as follows:

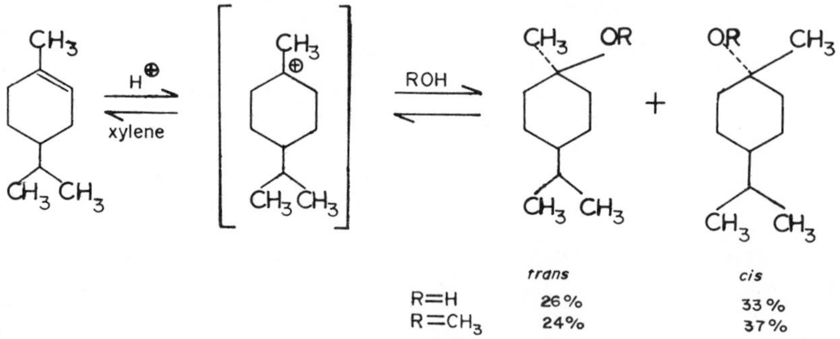
CH3
H⊕
xylene
CH3 CH3
ROH
OR
+
trans
cis
R=H 26% 33%
R=CH3 24% 37%

2. Methoxylation of Aromatic Compounds

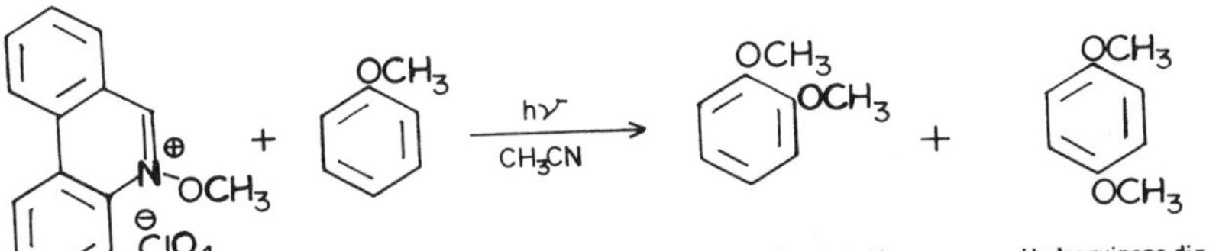

Catechol dimethyl ether 17%   Hydroquinone di- methyl ether, 5%

As seen, the methoxylation of anisole was accomplished by irradiation with N-methoxyphenanthridinium perchlorate in solution in acetonitrile [2]. In a similar manner benzene, toluene, and benzonitrile were methoxylated in low yield.

The nitro group in aromatic compounds may be replaced by the methoxy group in methanol-sodium methoxide solution under irradiation. The latest of such transformation is shown [3]:

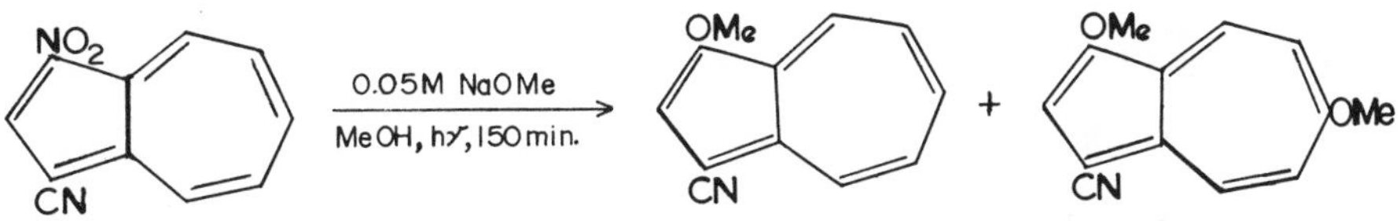

25 mg.   9 mg.   2mg.

I-Methoxy-3-cyanoazulene   1,6-Dimethoxy-3-cyanoazulene

3. Decarboxylation of Acids

$$ArOCH_2COOH \xrightarrow{h\nu} ArOCH_3 + CO_2$$

This reaction has been conducted in the presence of an aromatic ketone activator with thiophenol as the solvent. Yields are usually moderate [4].

References

1. J. A. Marshall, Acc. Chem. Res., 2, 33 (1969); H. Kato, M. Kawanisi, Tetrahedron Letters, 865 (1970).
2. J. D. Mee et al., J. Am. Chem. Soc., 92, 5814 (1970).
3. E. Havinga et al., Tetrahedron, 29, 867 (1973); R. L. Letsinger, R. R. Hautala, Tetrahedron Letters, 4205 (1969).
4. R. S. Davidson et al., J. Chem. Soc., C 3480 (1971).

H. Elimination

1. From Carbonates

Anisyl phenyl carbonates have been decarboxylated by heating in hexamethylphosphoramide or toluene [1] as shown:

$$X\text{—}\langle\bigcirc\rangle\text{—O}\overset{O}{\overset{\|}{C}}\text{-O-CH}_2\langle\bigcirc\rangle\text{OCH}_3 \xrightarrow[\text{HMPA}]{\triangle} X\text{—}\langle\bigcirc\rangle\text{OCH}_2\langle\bigcirc\rangle\text{OCH}_3$$

60-97 %

X= H,Cl,NO$_2$,CN,CH$_3$,OCH$_3$

A rather difficult elimination occurs with sodium methyl carbonate [2]:

$$2\ \text{MeOCO}_2\text{Na} \xrightarrow{332°} \text{MeOMe} + \text{Na}_2\text{CO}_3 + \text{CO}_2$$

58%

Lastly, diaryl ethers have been prepared in 17-90% yield by heating diaryl carbonates at 180-260° with potassium carbonate [3].

2.   From Specific Haloketals

Feugeas and Normant [4] describe, without experimental details, an unusual elimination reaction:

$$\text{Me}\text{—}\overbrace{\underset{O\quad O}{\bigcirc}}^{\text{—(CH}_2)_n\text{Br}} \xrightarrow[\text{2)H}_3\text{O}^{\oplus}]{\text{1)Mg-(C}_2\text{H}_5)_2\text{O}} \text{HOCH}_2\text{CH}_2\text{O}\overset{\text{Me}}{\underset{\|}{C}}\overbrace{\bigcirc}^{\text{(CH}_2)_n}$$

When n = 1, a vinyl ether is obtained. Apparently, the attack of a Grignard reagent on a ketal has taken place:

$$\text{Me}\text{—}\overbrace{\underset{O\quad O}{\bigcirc}}^{\text{—(CH}_2)_n\text{MgBr}} \longrightarrow \text{Me}\overset{\text{(CH}_2)_n}{\underset{O\quad OMgBr}{\bigcirc}}$$

3.   From Alkyl or Aryl Sulfites

$$\text{RO}\overset{O}{\overset{\|}{S}}\text{OR} \xrightarrow{\triangle} \text{ROR} + \text{SO}_2$$

This neglected reaction does not appear to be general. When R is a long alkyl group, pyrolysis of the sulfite ester gives an alkene and an alcohol rather than the ether [5]. However, with R = benzyl or phenethyl "excellent yields" of the respective ethers were obtained. With dibenzyl sulfite the decomposition begins at 130°, but with diphenethyl sulfite it takes place at 290°.

4.   From Other Sulfur Compounds

$$\text{ArOCCl}_2\text{SCl} \xrightarrow{\text{Cl}_2} \text{ArOCCl}_3 + \text{SCl}_2$$

Chlorination of the substrate at 40-45° gave the trichloro-methyl ether in 95% yield [6]. Desulfurization may also be accomplished by another method [7]:

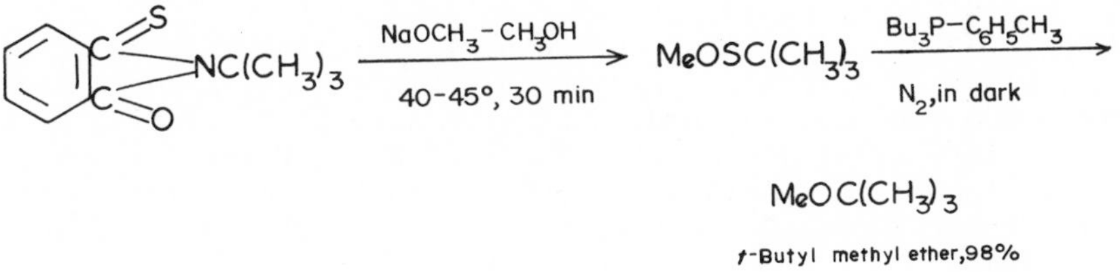

f-Butyl methyl ether, 98%

References

1. J. M. Prokipcak, T. H. Breckles, Can. J. Chem., 49, 914 (1971).
2. J. M. Criscione, W. H. Bernauer, U.S. 2,860,170, November 11, 1958; C. A., 53, 7012 (1959).
3. E. Müller et al., Angew. Chem., 82, 79 (1970).
4. C. Feugeas, H. Normant, Bull. Soc. Chim. France, 1441 (1963).
5. P. Carré, D. Libermann, Bull. Soc. Chim. France, 1248 (1934).
6. E. Kuhle, Synthesis, 563 (1971).
7. D. H. R. Barton et al., Chem. Commun., 1466 (1970).

I. Vinyl Ethers

A vinyl ether is a hybrid of an alkene and ether and there-fore differs from an ether not only in chemical behavior, but also in some of its means of preparation. It is related to an acetal structure, as demonstrated by its easy acid hydrolysis to an aldehyde. Indeed the chemistry of vinyl ethers has been described as that intermediate between alkenes and enamines [1]. Hence the preparations of vinyl ethers have been considered separately in this section. Two recent reviews are available on these syntheses [2,3].

1. The Reppe or Favorskii-Shostakovskii Condensation (1, 313)

$$\text{HC} \equiv \text{CH} + \text{ROH} \xrightarrow{\text{base}} \text{ROCH} = \text{CH}_2$$

There can be no doubt that yields of vinyl ethers are higher if the reaction is conducted under pressure. Shostakovskii [2] states that the danger of pressurized acetylene is overrated since both the vapors of the alcohol and vinyl ether stabilize the acetylene. A typical vinylation is given (Ex. a). Unsaturated alcohols do not give good yields of vinyl ethers, but a number of other structured features are tolerated. Phenol adds well to acetylene, while a thiol reacts in preference to an alcohol. Water accelerates the rate of addition of phenols but retards that of alcohols. If pressure vessels are to be avoided, acetylene may be passed through a 5-10% alcoholic solution of KOH and the vinyl ether may be distilled out as it forms [3]. From stearyl alcohol the yield of the vinyl ether is 85% after heating for 1.5 hr at 180°. Pyridones add rather readily to give vinyl ethers [4]:

2-Vinyloxypyridine, *ca* 60%

a.   Preparation of the Monovinyl Ether of Diethylene Glycol [5]

An autoclave containing diethylene glycol and solid KOH was freed from air by successively purging with N_2 and C_2H_2. Acetylene was then passed in under a pressure of 15.5 atm, after which the autoclave was heated at 140-150° for 2-2.5 hr (maximum pressure, 23 atm). On cooling, followed by vacuum distillation, treatment with K_2CO_3 and a second vacuum distillation gave 87% of the vinyl ether which polymerized on long standing.

    2.   Dehydrohalogenation of Haloethers [2]

This elimination proceeds quite smoothly. The α-chloro atom in ethers is much more reactive than the β-chloro atom. Successful dehydrochlorination of the β-chloroethyl ethers requires distillation from the solid base. Nevertheless the process should not be ignored as a possible method of synthesis. Even the following elimination has been accomplished [6]:

$$CH_2=CHOCH=CHBr \quad + \quad \underset{\text{1-2 parts}}{KOH\text{(powdered)}} \quad \xrightarrow{110°} \quad CH_2=CHOC\equiv CH$$

Vinyl ethynyl ether, 40%

Dombroski and Hallensleben [6a] have recommended that p-nitrophenyl vinyl ether be prepared in a two-step reaction from p-nitrophenol and ethylene bromide in aqueous sodium hydroxide. The β-bromoethyl ether formed is dehydrohalogenated with tBuOK in t-BuOH.

3. Vinyl Exchange (1, 305)

This exchange is unique for vinyl ethers. The ease of exchange is in the order: vinyl ester > vinyl aryl ether > vinyl alkyl ether [7]. The mechanism probably consists of a series of additions of the elements of HgOR' followed by elimination to give the original or new vinyl ether. Thus ROH should be eliminated or removed or R'OH should be used in large excess. Vinyl phenyl ether has been used to prepare vinyl β-naphthyl ether (75%) [7]. Also divinyl mercury has been employed to prepare a series of vinyl aryl ethers [8].

4. Elimination from Acetals and Ketals

$$\underset{}{RCH_2\overset{R'}{\underset{|}{C}}(OR'')_2} \quad \xrightarrow{\text{catalyst}} \quad RCH=\overset{R'}{\underset{|}{C}}OR''$$

The elimination is sometimes quite difficult, is accompanied by low yields, and would seem more adaptable to a flash type of pyrolysis. One simple example is shown [9]:

$$CH_3CH(O\overset{CH_3}{\underset{|}{C}}HC\equiv CH)_2 \quad \xrightarrow[N_2]{\text{,} P_2O_5} \quad CH_2=CHO\overset{}{\underset{CH_3}{C}}HC\equiv CH$$

α-Methylpropargyl vinyl ether, 30%

2-Methoxypropene, of value in the protection of alcohol functions, has been synthesized by an improved method from acetone dimethyl ketal as shown [10].

$$\underset{CH_3}{\overset{CH_3}{\diagdown}}\underset{OCH_3}{\overset{OCH_3}{\diagup}}C \quad \xrightarrow[\substack{C_6H_5N\text{, Diglyme} \\ 110-120°}]{\substack{CH_2=CO \\ CH_2=CO \text{,} C_6H_5CO_2H}} \quad \underset{CH_3O}{\overset{CH_3}{\diagdown}}C=CH_2$$

90%

Another example is elimination from an α-methoxyacetal [11]:

$$(MeO)_2CHCH_2OMe \xrightarrow[165°,bomb]{t-BuOK} MeOCH{=\!=}CHOMe$$

196 mmoles

cis and *trans*
ca.6 mmoles,with recovery
of 180 mmoles

The elimination may be accomplished easily at times with cyclic ketals by refluxing in an apparatus fitted to remove the alcohol as it is formed and then changing to total reflux to remove the last traces of alcohol [11a]:

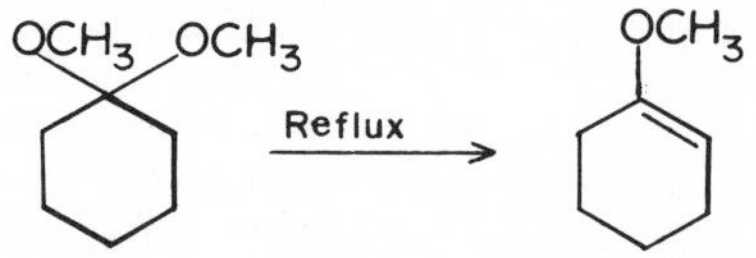

I-Methoxycyclohexene
86%

A more complex example shows an interaction between aldehyde and acetal moieties followed by loss of R'OH [12].

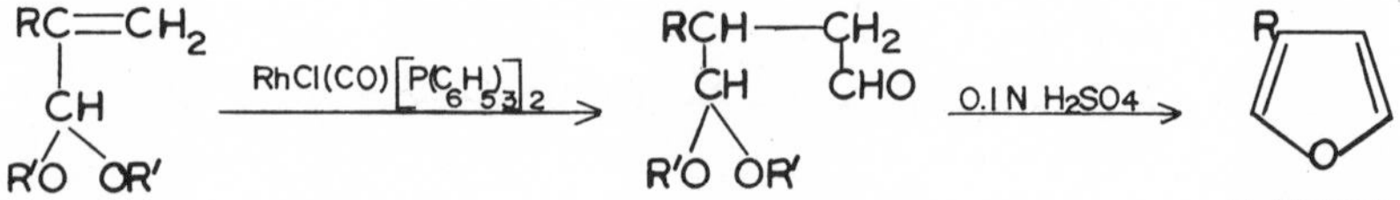

a succinaldehyde monoacetal
60-85%

3-Alkylfuran
40 - 80%

5.   From Vinyl Chloride and an Alkoxide (<u>1</u>, 287)

$$CH_2{=\!=}CHCl + RONa \longrightarrow CH_2{=\!=}CHOR$$

The mechanism involves a fast dehydrohalogenation and a slower addition of the alkoxide [13]. Thus conditions used for the addition of alkoxides to acetylenes are applicable to vinyl chloride and a typical Reppe-Favorskii-Shostakovskii reaction results [14].

As is expected, the method may result in the formation of acetylenes, vinyl ethers, or a mixture of the two. For example, α-bromopentachlorostyrene gave each of the expected products depending on the experimental conditions [15]:

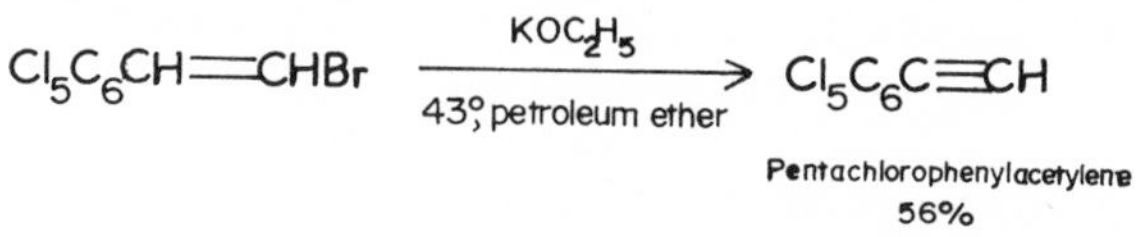

$$Cl_5C_6CH\!=\!CHBr \xrightarrow[\text{43° petroleum ether}]{KOC_2H_5} Cl_5C_6C\!\equiv\!CH$$

Pentachlorophenylacetylene
56%

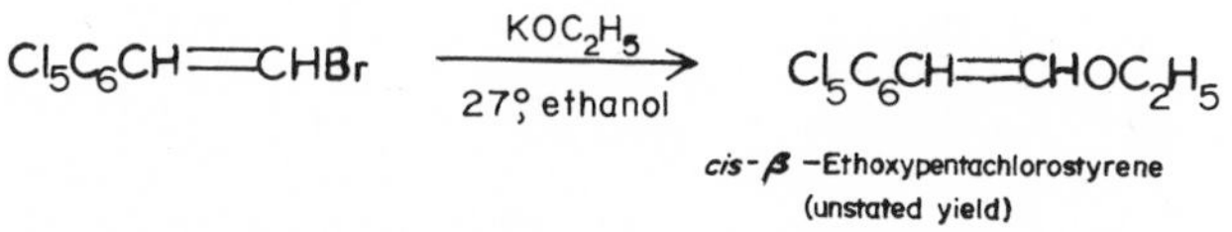

$$Cl_5C_6CH\!=\!CHBr \xrightarrow[\text{27° ethanol}]{KOC_2H_5} Cl_5C_6CH\!=\!CHOC_2H_5$$

cis-β −Ethoxypentachlorostyrene
(unstated yield)

To prevent acetylene formation in a similar reaction Newman and Merrill [16] used sodium hydride in benzene with a catalytic amount of ethanol.

6.   From Carbonyl Compounds, Acid Chlorides, and the Like

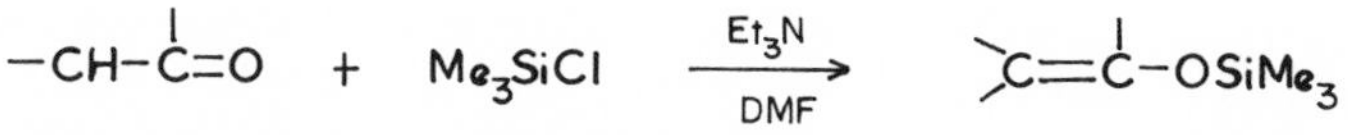

$$-CH\!-\!\overset{|}{C}\!=\!O \ + \ Me_3SiCl \xrightarrow[DMF]{Et_3N} \quad \underset{}{>}\!C\!=\!\overset{|}{C}\!-\!OSiMe_3$$

This subject has been reviewed [17].

Tosychloride with α-cyanoaldehydes yields first the vinyl tosylate and then the divinyl ether [17a]:

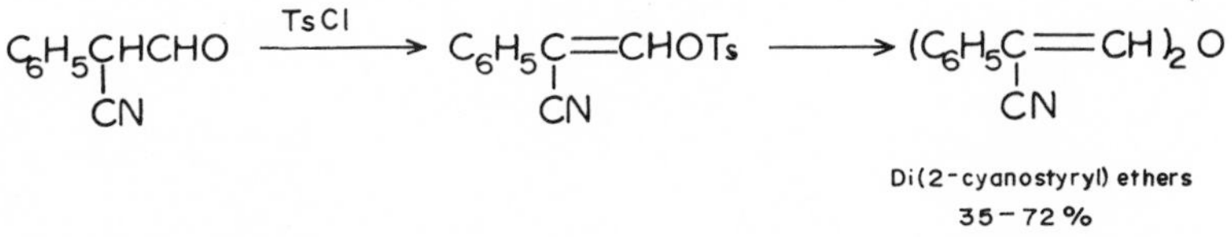

$$\underset{CN}{C_6H_5\overset{|}{C}HCHO} \xrightarrow{TsCl} \underset{CN}{C_6H_5\overset{|}{C}\!=\!CHOTs} \longrightarrow (\underset{CN}{C_6H_5\overset{|}{C}\!=\!CH})_2O$$

Di(2-cyanostyryl) ethers
35−72 %

The thallium salt of 1,3-cyclohexadione or dimedone when treated with an alkyl iodide gives the vinyl ether,

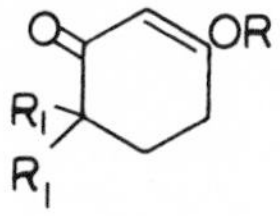

in 35-100% yield [17b]

(R₁= H or CH₃)

7.   From Carbenes and Cyclopropanes

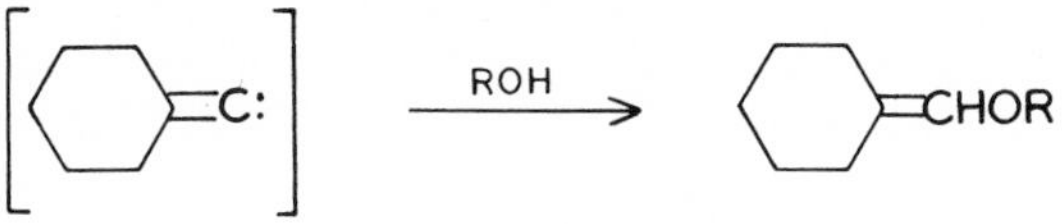

When 3-nitroso-1-oxa-3-azaspiro [4,5]decan-2-ene is suspended in an alcohol containing the sodium alkoxide, the
evolution of nitrogen is rapid and the vinyl ether of the alcohol
employed may be isolated from the reaction mixture [18].   These
facts may be accounted for as indicated:

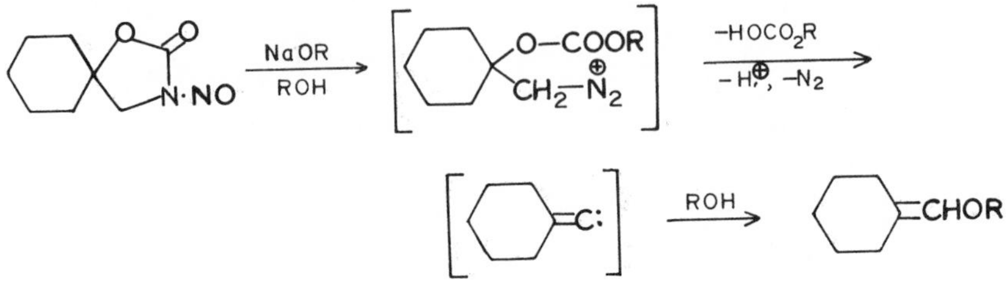

In this manner a series of vinyl ethers have been prepared with
yields varying from 54 to 90%.

The carbene may be generated in an alternate method by dissolving the nitrosooxaazolidone in pentane containing methyl-
tricaprylammonium chloride and the substance to be attacked,
followed by the addition of aqueous NaOH dropwise at -10° [19].
In the presence of ketones such as cyclohexanone, substituted
divinyl ethers are formed:

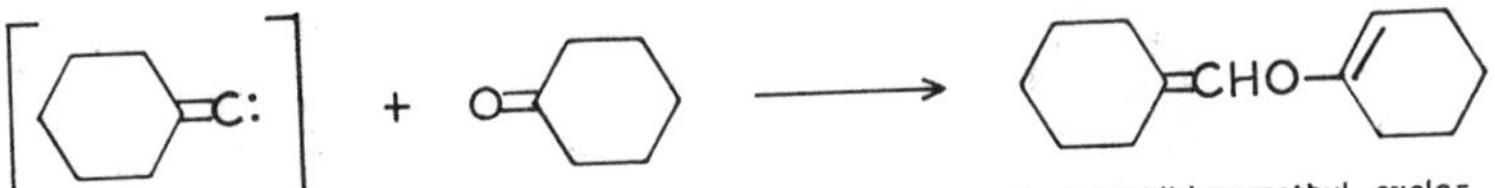

Cyclohexylidenemethyl cyclohexenyl ether, 32%

The reaction is general but yields may be low.   Ethers are not
obtained with either aldehydes, which form α,β-unsaturated
ketones, halides, which form vinyl halides, azides, which form
vinyl azides, phosphites, which form vinyl phosphonates, or
isothiocyanates, which form vinyl isothiocyanates.

Casey and co-workers [20] have utilized the reaction of the
metal-carbene complexes with Wittig reagents to produce vinyl
ethers.

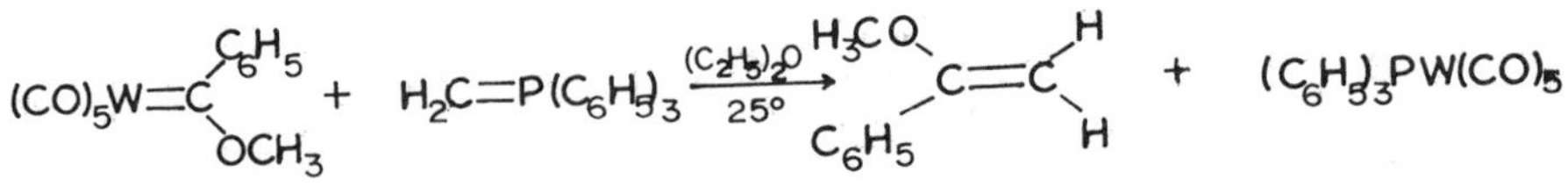

82% as acetophenone

However, the reaction failed for alkylmethoxycarbene complexes. To overcome this difficulty these investigators substituted a diazoalkane for the Wittig reagent as shown:

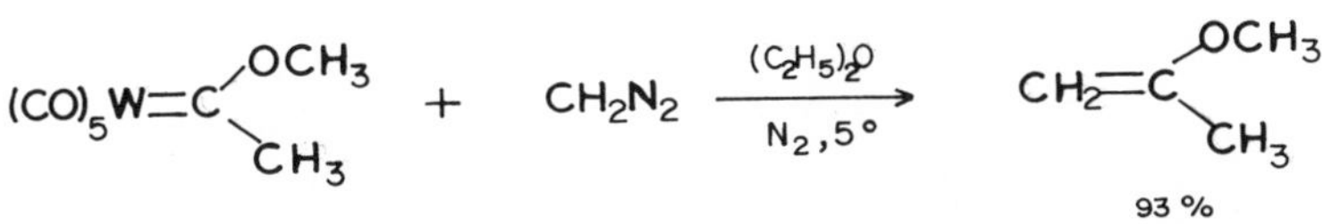

93 %

A possible displacement on a cyclopropane generates an ether [21]:

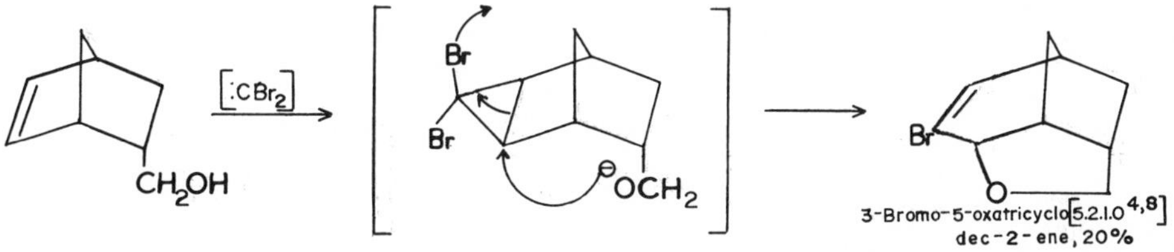

3-Bromo-5-oxatricyclo[5.2.1.0^{4,8}] dec-2-ene, 20%

The reaction is conducted in a two-phase system in which bromo-form is added slowly to a benzene-50% KOH mixture. The yield of the cyclic ether was about 20% after chromatography

8.   From Active Methylene Compounds and Ethyl Chloroformate (Vilsmeier-Haack Reaction)

Although the Vilsmeier-Haack reaction is a method for the introduction of an aldehyde group into a ring (1, 586), Ikawa and co-workers [22] discovered that the reaction may be used to synthesize ethers by employing an active methylene compound and ethyl chloroformate in the presence of DMF as shown:

The steps occurring here probably are

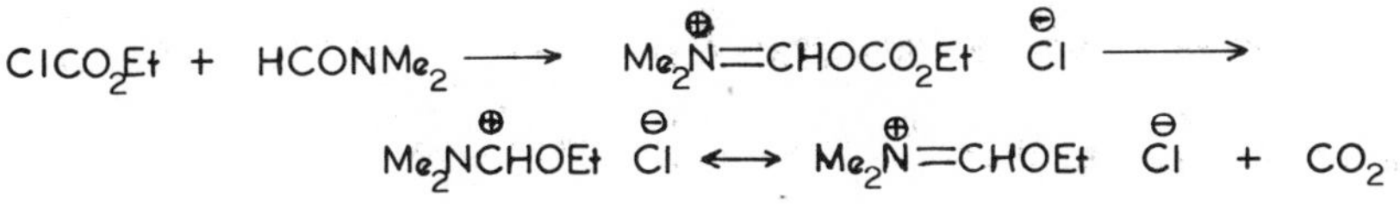

The latter attacks the enimino form of malonitrile, NCCH=C=NH, to form EtOCH=C$\langle^{CN}_{CN}$. This product is an intermediate in the synthesis of thiamine.

9.   From Isomerization of Propargyl Ethers

$$RC{\equiv}CCH_2OEt \xrightarrow{\textit{t}-BuOK} RCH{=}C{=}COEt$$

The reaction appears to be quite general for the synthesis
of allenic or cumulenic ethers [23]. The corresponding thio
ethers are more stable [24]. Moreover, both the thio ethers
[24] and the oxygen ethers may be metallated and then alkylated
to give a wide range of cumulenic ethers [25]:

$$\underset{\overset{|}{R}}{Me}OCHC{\equiv}CCH_2OEt \xrightarrow[Et_2O,-28°]{BuLi} \underset{\overset{|}{Li}}{RCH}{=}C{=}C{=}COEt \xrightarrow{R'Br}$$

$$RCH{=}C{=}C{=}\underset{\overset{|}{R'}}{C}OEt$$

Cumulenic ether, 54-76 %

10.   From Phosphoranes and Certain Esters and the Like

$$RCH{=}P(C_6H_5)_3 + CF_3CO_2Et \xrightarrow[Reflux]{C_6H_6} RCH{=}\underset{CF_3}{\overset{OEt}{C}}$$

The vinyl ether is formed in 41 to 78% yield where R is an
alkyl or aryl group [26]. Ethyl α-fluoroacetate may be substituted for the trifluoroacetate.

11.   From π-(2-Methoxyallyl)nickel Dibromide Complex and Ketones (see Addenda)

Addenda

I.4. A series of cyclic vinyl ethers may be prepared conveniently by refluxing the ketal, formed from the ketone, trimethyl orthoformate, and acid without isolation, with an arrangement for removing the alcohol as the vinyl ether is formed [R. A.
Wohl, Synthesis, 38 (1974)].

I.10. A phosphorus ylid reacts ,with PhC(OCH$_3$)=W(CO)$_5$ to give
methyl phenylvinyl ethers [C. P. Casey, T. J. Burkhardt, J. Am.
Chem. Soc., 94, 6543 (1972)].

I.11. π-(2-Methoxyallyl) nickel dibromide complex may be treated
with a ketone to form β-hydroxylvinyl ethers [L. S. Hegedus,
R. K. Steverson, J. Am. Chem. Soc., 96, 3250 (1974)].

References

1. F. Effenberger, Angew. Chem. Intern. Ed. Engl., **8**, 295 (1969).

2. M. F. Shostakovskii et al., Russ. Chem. Rev. (Engl. transl.) **37**, 907 (1968).

3. T. F. Rutledge, Acetylenes and Allenes, Reinhold, New York, 1969, p. 284.

4. M. F. Shostakovskii et al., Khim. Atsetilena, Tr. Vses. Konf., 3rd, 1968, p. 90; C. A., **79**, 18536 (1973).

5. M. F. Shostakovskii et al., J. Gen. Chem. USSR (Engl.), **34**, 2125 (1964).

6. L. Brandsma, J. F. Arens, Rec. Trav. Chim., **79**, 1307 (1960).

6a. J. R. Dombroski, M. L. Hallensleben, Synthesis, 693 (1972).

7. H. Lüssi, Helv. Chim. Acta, **49**, 1681 (1966).

8. D. J. Foster, E. Tobler, J. Am. Chem. Soc., **83**, 851 (1961).

9. M. F. Shostakovskii et al., J. Org. Chem. USSR, **1**, 1532 (1965).

10. M. S. Newman, M. C. Vander Zwan, J. Org. Chem., **38**, 2910 (1973).

11. J. T. Waldron, W. H. Snyder, J. Org. Chem., **38**, 3059 (1973).

11a. R. A. Wohl, Synthesis, 38 (1974).

12. C. Botteghi et al., J. Org. Chem., **38**, 2361 (1973).

13. H. G. Viehe, Chemistry of Acetylenes, Dekker, New York, 1969, p. 105.

14. M. F. Shostakovskii et al., Russ. Chem. Rev., **33**, 66 (1964).

15. G. Huett, S. I. Miller, J. Am. Chem. Soc., **83**, 408 (1961).

16. M. S. Newman, S. H. Merrill, J. Am. Chem. Soc., **77**, 5549 (1955).

17. Y. I. Baukov, I. F. Lutsenko, Organomet. Chem. Rev., **6A**, 355 (1970).

17a. M. Davis, J. A. Fisher, Australian J. Chem., **23**, 205 (1970).

17b. M. T. Pizzorno, S. M. Albonico, Chem. Ind. (London), 425 (1972).

18. M. S. Newman, A. O. M. Okorodudu, J. Org. Chem., **34**, 1220 (1969).

19. M. S. Newman, W. C. Liang, J. Org. Chem., **38**, 2438 (1973).

20. C. P. Casey et al., J. Am. Chem. Soc., **94**, 6543 (1972); Tetrahedron Letters, 1421 (1973).

21. T. Sasaki et al., J. Org. Chem., **38**, 2230 (1973).

22. K. Ikawa et al., Tetrahedron Letters, 3279 (1969).

23. L. Brandsma et al., Rec. Trav. Chim., **87**, 916 (1968).

24. R. Mantione, Bull. Soc. Chim. France, 4514 (1969).

25. R. Mantione, L. Brandsma et al., Rec. Trav. Chim., 89, 97
     (1970).
26. H. J. Bestmann et al., Chem. Ber., 103, 2011 (1970).

Chapter 7

HALIDES

Rydon reagents (R_3P or $(RO)_3P$ and a halogen source) (see A.2) have become of age as reagents for the conversion of alcohols into halides. Molybdenum hexafluoride appears to be a milder agent than usual for the conversion of carbonyl compounds into gem-difluorides (A.3). Phase transfer conditions have aided in displacement of mesylate by halide (A.5, Addenda) and macro-cyclic ether, 18 crown 6, has accelerated and increased the yield in the exchange of bromide or chloride with fluoride (A.6, Addenda). The Schiemann reaction has been improved by irradiation rather than by the heating of the tetrafluoroborate (A.9). In the same section in which the convenient conversion of a hydrazone into a gem-diiodide is described the iodine serves both as the oxidizing and displacing agent. The Cristol modification of the Hunsdiecker reaction is recommended for the transformation of an acid into the halide, although the new thallium salt method is competitive (A.11). The interesting conversion of carboxyl into the trifluoromethyl group by SF_4 has been improved by adding an excess of liquid HF (A.12). The bromination of trialkylborons to alkyl bromides has been made more feasible by the addition of sodium methoxide to the bro-mination reaction (A.18). A study of the coupling of trifluoro-methyl iodide with aryl iodides has led to an alternate route to trifluoromethylarenes (A.19).

   In an extensive subsection on the addition of halocarbenes to alkenes, the Seyferth method has deservedly been given more prominence, and the phase transfer method has been improved (B.5). The carbenoid preparation from diethylzinc and bromoform has been shown to be a free radical chain reaction (B.5, Addenda). Perhaps the most novel of all new halogenation pro-cedures is that using N-haloamines in strong acid with Fe(II) to give halogenation in preponderance at the ω-1 position in alkyl halides, esters, ethers, and alcohols (C.1). Optically active 2-bromohexanone may be produced by the bromination of the optically active enamine (C.5, Addenda). Thallation has proved to be a quite versatile method of aromatic substitution (D.1).

For instance, from $C_6H_5CH_2CH_2OH$ the o-, m-, or p-iodo derivatives
may be obtained depending on the conditions. Throughout this
section on aromatic substitutions are to be found other rather
unique methods of obtaining aromatic halides, even if most of
them are limited in scope.

The H.-W. series [1,2] was consulted at times. In this
connection H.-W. will be used in the references. Reviews on
fluorine compounds are now available [3-5].

References

1. H.-W., Fluorine and Chlorine Compounds, 5, Pt. 3, 1962.
2. H.-W., Bromine and Iodine Compounds, 5, Pt. 4, 1960.
3. Fluorine Chemistry Reviews, 5, Dekker, New York, 1971.
4. R. E. Banks, M. G. Barlow, Fluorocarbons and Related
   Chemistry, 1, 2, The Chemical Society, London, 1971, 1974.
5. M. Hudlicky, Organic Fluorine Chemistry, Plenum, New York,
   1971.

A.   Displacement

1.   From Alcohols and Hydrogen Halides (1, 330)

$$ROH \xrightarrow{\text{HX}} RX$$

Interest in developments in this field have waned or have
been diverted to A.2. A large-scale preparation of t-butyl
bromide has been described, however (Ex. a). Only methanol of
the aliphatic alcohols reacts with hydrogen fluoride to form an
alkyl fluoride [1]. A maximum yield of methyl fluoride was
obtained at 375° under pressure (78%).

A.   Preparation of t-Butyl Bromide [2]

To a cold, stirred solution of 2.5 mol of sodium bromide,
2.5 mol of t-butyl alcohol, and 100 ml of water was added dropwise 2.45 mol, 135.5 ml, of concentrated H_2SO_4 and the mixture
was heated on a steam-bath with collection of a fraction boiling
within 65-80°. After washing and drying, the product was re-
distilled at boiling point, 71-73°, 180-190 g, 53-55%.

2.   From Alcohols and Phosphorus Halides (1, 332)

$$ROH + PX_3 \longrightarrow ROPX_2 + HX$$

$$\overset{\ominus}{X} \cdots\cdots \overset{\delta+}{R} \cdots\cdots \overset{\delta-}{OPX_2} \longrightarrow XR + \overset{\ominus}{O}PX_2$$

The method using triphenylphosphine and carbon tetrachloride as the reagent has been described (1, 332). It seems to have proved its worth since good yields have been obtained in the syn-thesis of compounds such as 2-acetoxyethyl chloride, 4-acetoxy-butyl chloride,α-chloroacetaldehyde diethyl acetal, and neo-pentyl chloride [3]. The reagent minimizes isomerization and produces pure inversion under mild conditions (Ex. a). In addition, carbon tetrabromide may be used to prepare bromides [4].

The Rydon reagents have of late been modified to suit the occasion. Downie [5] claims that hexamethylphosphorous tri-amide, $P(NMe)_3$, obviates the more difficult separation to be found with triphenylphosphine. Castro [6], with this reagent and carbon tetrachloride, has prepared the chlorohydrin of 2,2-dimethyl-1,3-propanediol. He also has made the perchlorate of the intermediate, $RO\overset{+}{P}(NMe_2)_3$ $\bar{C}lO_4$, which can be used to make bro-mides or iodides. Rydon [7] recommends a mixture of triphenyl phosphite and methyl iodide at 80° and then at 130° to synthesize iodides from hindered alcohols and a premixing of triphenyl phos-phite and methyl iodide with heating prior to reaction with a sensitive alcohol at 25° to form the same halide. Verheyden and Moffatt [8] have used the same reagent to form iodides of the 5' hydroxyl group in nucleosides and found that DMF increases the reaction rate. With phenyltetrafluorophosphorane ($C_6H_5PF_4$), the trimethylsilyl ether must be formed first in order to get good yields of alkyl fluorides [9]. Some olefin was formed with a few alcohols while ethylene bromohydrin gave more ethylene bromide (40%) than 1-bromo-2-fluoroethane (10%). Benzyl alcohol was polymerized by the same reagent. The alkyl diphenylphosphinite, $(C_6H_5)_2POR$, gave the alkyl halide with either HX or X_2 [10]. Although complete inversion was noted in many cases, partial racemization of optically active 2-bromo- or iodooctane was detected merely on being stored. However, for converting optically active alcohols into the inverted optically active alkyl halides, the reagent, $ROP(OC_6H_5)_2$, should be treated with one equivalent of hydrogen halide and the excess should be removed under vacuum [11].

Some phenolic compounds are easily converted into the bromoarene by triphenylphosphine and bromine in acetonitrile at 60-70° [12]. α-Naphthol, 2- and 3-hydroxypyridine, 8-hydroxy-quinoline, and p-nitrophenol are among the phenols transformed into the corresponding bromides in this manner. 6-p-Fluoro-phenyl-3-cyano-2(1H)-pyridone has been converted into the 2-chloro derivative (91%) by phenylphosphorous oxychloride ($C_6H_5POCl_2$) [13]. λ-Phenylallyl alcohol has also been converted into its bromide [14]. In this connection, the isomerization of allyl alcohols is minimized by Rydon reagents [14a].

3. From Carbonyl Compounds (1, 335)

$$R_2C{=}O \xrightarrow{PCl_5} R_2CCl_2$$

1,1-Dichloro-1-cyclopropylethane has been prepared in 64% yield as shown above when methyl cyclopropyl ketone was added dropwise to phosphorus pentachloride stirred in carbon tetrachloride under reflux [15]. Although sulfur tetrafluoride has been used to make the gem-difluoroalkanes (from which vinyl fluorides may be made with Al_2O_3) [15a] and trifluoromethyl-aminoacids [16], molydenum hexafluoride may be the reagent of choice for this conversion because the reaction avoids pressure equipment and tolerates glass apparatus [17]. Yields of gem-difluorides ranged from 17-55% (9 ex.). Phenylsulfur trifluoride ($C_6H_5SF_3$) also may be used without pressure equipment [18] (see A.12).

A new reagent for the synthesis of gem-dichlorides from aldehydes is trichloromethylisocyanide dichloride [19]:

$$RCH{=}O \;+\; CCl_3N{=}CCl_2 \xrightarrow[120°,\,3\,hr.]{FeCl_3(cat.\,amt.)} RCHCl_2 \;+\; Cl\underset{O}{C}N{=}CCl_2$$

Yields ranged in two preparations from 55% for dichloropropane to 98% for o-chlorobenzalchloride.

A novel reaction involves the displacement of the trifluoroacetyl group in some heterocycles [20]:

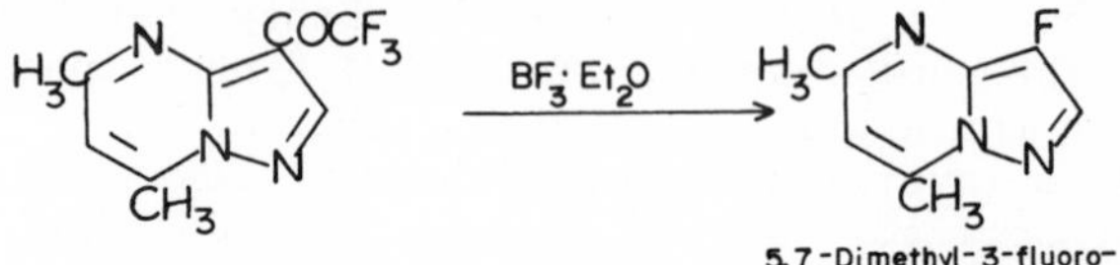

5,7-Dimethyl-3-fluoro-
pyrazolo(1,5-a)pyrimidine, 10%

4. From Alcohols and Thionyl Chloride (1, 336)

$$ROH \xrightarrow{SOCl_2} RCl + SO_2 + HCl$$

An example of using a very small amount of DMF as a catalyst, rather than pyridine, has been published [21]. But the more recent reagent recommended for use with thionyl chloride and an alcohol is one equivalent of HMPA [22]. Yields of chlorides ranged within 43-90% (7 ex.). Inversion occurs in this displacement [23].

6. From Halides or Sulfonates (Finkelstein Halide Interchange) (1, 339)

$$RCl + NaI \longrightarrow RI + NaCl$$

The Finkelstein exchange has been carried out with methylene chloride in dry DMF at 90-95° to prepare methylene iodide in 66% yield [24].

To convert aromatic iodides into the corresponding chlorides, the iodides are irradiated with 3000-Å light in carbon tetrachloride for 5 hr [25]. Eleven examples gave yields of 51-96%, 4-iodoaniline being an exception (0% yield). Yields were based on the fact that most conversions were on the order of 75% with 25% return of starting material. A heterocyclic exchange may be illustrated in the case of 2,5-dibromopyridine, which yielded 95% of 2-fluoro-5-bromopyridine on heating with potassium hydrogen fluoride in an autoclave at 280° for 4 hr [26]. The bridgehead halide exchange of Pincock may have broader applications than noted [27]. For example, in the presence of aluminum bromide, made from aluminum and bromine, and methylene iodide, 1-bromoadamantane gave 1-iodoadamantane in 47% yield with some 30% adamantane; with chloroform rather than methylene iodide 1-chloroadamantane was obtained in 89% yield.

Although yields are not the best, some scrambling reactions appear to give specific products which can be separated. Below are two examples [28]:

$$CF_2BrCBrFCl \xrightarrow[\text{25°}]{\text{AlCl}_3, 3 \text{g.}} CF_3CClBr_2$$

176 g.

1,1-Dibromo-1-chloro-trifluoroethane, 146 g.

$$CF_2BrCF_2Br \xrightarrow[\text{22 hr. reflux}]{\text{AlCl}_3, 10 \text{g.}} CF_3CBr_3$$

400 g.

1,1,1-Tribromotrifluoro-ethane, 148 g.

The substrates are made by the addition of bromine to the proper alkene under irradiation. Other rearrangements of this nature are discussed (C.1).

For the rare conversion of a bromide into a chloride, the reagent, silver difluorochloroacetate in refluxing acetonitrile (12 hr), has been recommended [29]. Five examples gave 70-95% yields.

If the conversion of a chloride into an iodide is difficult, it may be done via the sulfonate ester such as is shown [30]:

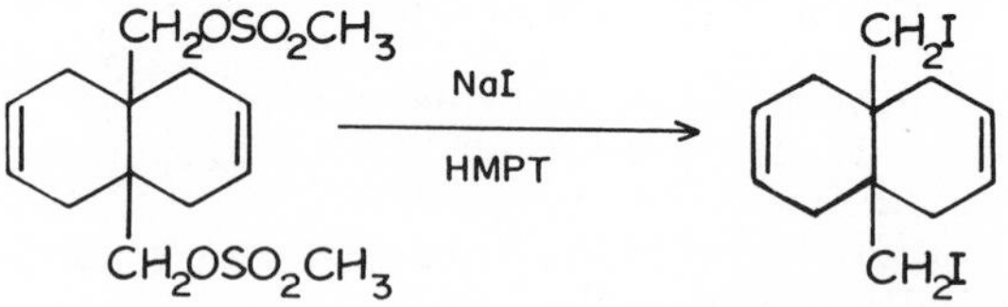

9,10-*bis*-Iodomethyl-
$\triangle^{2,6}$- hexalin, 76 %

7. From Iminoesters or Thiothiazolines (<u>1</u>, 342)

The yields in the above reaction are quite good [31]. Moreover when R is allyl, the substrate can be further alkylated and treated with methyl iodide to give allyl iodides as shown:

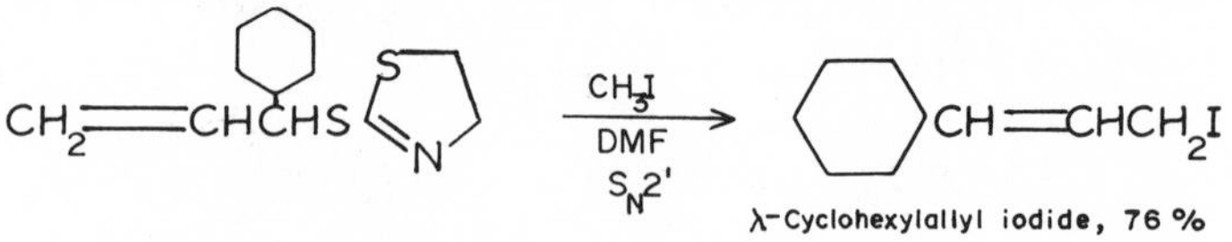

λ-Cyclohexylallyl iodide, 76 %

8. From Ethers (<u>1</u>, 342)

$$ROR \longrightarrow 2\,RX$$

The use of triphenylphosphine dibromide has been extended to cover a series of dialkyl or aryl alkyl ether cleavages to the alkyl bromide in 18-77% yields (7 ex.) [32]. Alkyl t-butyl ethers gave mostly isobutene. The same reagent in acetonitrile was used to synthesize 7-bromonorbornane from 7-t-butoxynorbornane in 74-83% yield [33]. It, or the dichloride, was used also to prepare a series of vicinal dihalides (8 ex.) (32-86%) [34]. With cis-cyclohexene oxide both cis and trans-1,2-dichlorocyclohexanes were obtained unless the reactions were run in benzene, in which case the cis isomer only was obtained. But another report states that refluxing the same oxide (0.1 equiv) with triphenylphosphine (0.15 equiv) in 50 ml of carbon tetrachloride gave the cis-dichloride with only a trace of the trans isomer [35]. Other dichlorides were prepared as well (50-80%) (5 ex.).

Forcing reactions on 2- or 4-alkoxyheterocycles or on alkoxypropionitriles bring about splitting to the halide, as shown in the following two examples:

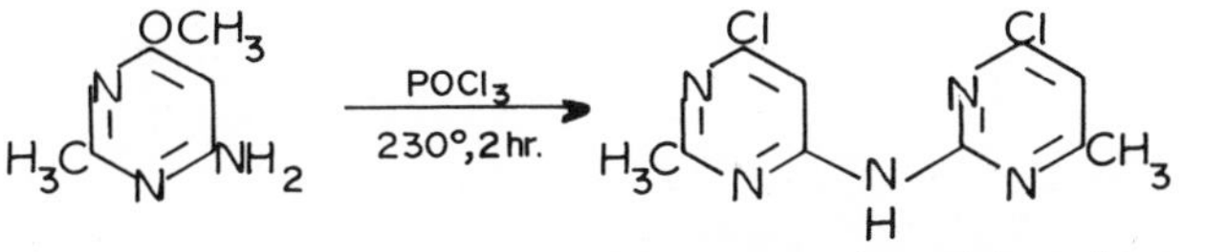

$$\text{6,6'-Dichloro-2,2'-dimethyl—4,4'-imino-}$$
$$\text{dipyrimidine, 90 \%}$$

[36]

$$\text{MeOCH}_2\text{CH}_2\text{CN} \xrightarrow[\substack{2)\text{remove in}\\ \text{vacuo}}]{\substack{1)\text{PCl}_5\\ \text{reflux1hr}}} \text{ClCH}_2\text{CH}_2\text{CN}$$

[37]

$$\beta\text{-Chloropropionitrile,80\%}$$

A new reagent for iodide formation is boron triiodide [38]. Together, THF and this reagent in equivalent quantities at 25° gave 1,4-diiodobutane in 79% yield.

ɣ.   From Diazonium Salts (Sandmeyer) or Diazoalkanes
     (1, 344)

$$\text{ArN}_2\text{X} \longrightarrow \text{ArX} + \text{N}_2$$

Two references have appeared on possible improvement of the Schiemann reaction (1, 345). Both utilize irradiation of the tetrafluoroborate rather than thermolysis. With one, improvement of the yields of fluoroanilines are obtained [39] and with the other, fluoroimidazoles and fluorohistidines have been prepared by this method for the first time [40].

An interesting isomerization has been detected in the decomposition of diazomethyl ketones [41]:

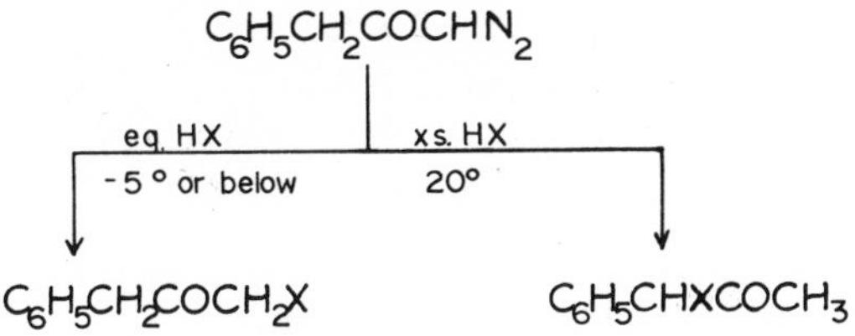

$$\text{C}_6\text{H}_5\text{CH}_2\text{COCHN}_2$$

| eq. HX | xs. HX |
| -5° or below | 20° |

$$\text{C}_6\text{H}_5\text{CH}_2\text{COCH}_2\text{X} \qquad\qquad \text{C}_6\text{H}_5\text{CHXCOCH}_3$$

Yields of either isomer were about 60-70%. A rather simple reaction occurs on the irradiation of diazomethane in carbon tetrachloride to form 60% bis-(chloromethyl)dichloromethane, $\text{Cl}_2\text{C(CH}_2\text{Cl)}_2$ (H.-W., 5, Pt. 3, 1962, p. 991).

Another halide preparation consists of the oxidation of the carbazate [42]:

$$ROH \xrightarrow[\text{2) NH}_2\text{NH}_2]{\text{1) COCl}_2} \underset{\text{1 eq.}}{ROCONHNH_2} \xrightarrow[\substack{\text{CH}_2\text{Cl}_2, \text{C}_5\text{H}_5\text{N(1 eq.)} \\ 20 \text{min., R.T.}}]{\text{NBS(1 eq.)}} RBr$$

This reaction is best suited for those alcohols such as adamantanol that do not easily form alkenes.

The hydrazones of aldehydes and ketones are oxidized by iodine to give _gem_ and vinyl halides as shown [43]:

1,1-Diiodocyclohexane	1-Iodocyclohexene
24 %	41 %

The diazoalkane is no doubt an intermediate and many more examples exist.

Ethylidene iodide has been prepared recently in 30% yield based on acetaldehyde [44]. If ethyl diazoacetate is treated with NBS in HF and C_5H_5N, ethyl bromofluoroacetate is obtained in 50% yield [45].

10. From Amides (von Braun) (<u>1</u>, 347)

$$RNHCOR' \xrightarrow{PCl_5} RCl + R'CN + POCl_3 + HCl$$

Unlike the von Braun reaction, thioformates yield the dichloromethyl sulfide [46]:

$$RSCH{=}O \xrightarrow[35\text{-}45°]{PCl_5} \underset{100\text{-}110°}{RSCHCl_2} \longrightarrow RSCCl_3$$

Yields of dichloromethyl sulfides for 11 examples were 76-87%. The PCl_5 should be added to a slight excess of the thioformate; otherwise, the trichloromethyl sulfide is formed. Yields for the latter were 75-95% for 14 examples.

11. From Carboxlic Acids or Their Salts (Hunsdiecker, Kochi, and Barton) (<u>1</u>, 348)

$$RCO_2Ag \xrightarrow{Br_2} RBr + CO_2 + AgBr$$

The Hunsdiecker reaction should be conducted under anhydrous conditions. The Cristol modification, using HgO and the acid, in

place of the silver salt, is much less sensitive to the presence
of water due to the solubility of the mercuric salt in the nonaqueous phase [47]. This modification is recommended because it
is simpler and yields are good, provided the mercuric salt is
soluble in the nonaqueous solvent. Attempts to make tertiary
halide acids other than bridgehead types fail, as do those to
produce iodides.

The Kochi reaction, involving conversion of the acid into
the alkyl chloride by means of lead tetraacetate and lithium
chloride, is supplementary to the Hunsdiecker one because it
works best with acids that give tertiary chlorides and least
with acids that give primary halides. An improvement of the
Kochi method occurs with the use of a DMF-HOAc (5-1) solvent
with N-chlorosuccinimide as the chlorine source [48]. For
instance, 1-chlorobicyclo[2,2,2]-octane was obtained in 95%
yield from the corresponding acid. Benzoic acid gave no
chlorobenzene. Both cis- and trans-4-t-butylcyclohexanecar-
boxylic acids form the identical product in the Kochi reaction,
67% cis- and 33% trans-4-t-butylchlorocyclohexane, a fact
indicating a common intermediate, which is perhaps the 4-t-
butylcyclohexyl free radical [49].

At present the thallium-salt method looks quite promising
as a substitute for the Hunsdiecker, provided the following
stoichiometry is used [50]:

$$2 \; RCO_2Tl \; + \; 3 \; Br_2 \longrightarrow 2 \; RBr \; + \; 2 \; CO_2 + \; Tl_2Br_4$$

Yields of bromides were 83-98% (7 ex.).

12. From Acids, Acid Chlorides, Some Sulfonyl Chlorides,
and Related Compounds (1, 350)

It was stated in the previous volume, p. 350, that no
satisfactory procedure for the decomposition of sulfonyl
chlorides to chlorides was known. Omitted was the work of Blum
[51], whereby benzenesulfonyl chloride was converted into
chlorobenzene in 79% yield by heating it with a catalytic amount
of $[(C_6H_5)_3P]_3RhCl$.

The excellent reaction $-COOH \xrightarrow{\;\;SF_4\;\;} -CF_3$ (1, 351) has been
improved by using in addition to the above reagent an excess of
hydrogen fluoride [52]. Temperatures of this reaction could be
lowered by 50° in this manner. 4-Nitro-1-trifluoronaphthalene
was prepared in 83% yield by heating the nitro acid with SF_4 and
H_2F_2 at 40° under pressure for 10 hr. Other examples were prepared at higher temperatures.

13. From Organic Carbonyl, Related Compounds, and Certain
Fluorides (1, 351)

This section now has been discussed (A.3, A.12)

14.   From Alcohols and Dialkylaminotetrafluoroethanes
      (1, 351)

N(2-Chloro-1,1,2-trifluoroethyl)diethylamine, as indicated
in 1, 351, continues to be employed in replacing the hydroxyl
group of steroids with fluorine [53]. If N,N-diethyl-1,2,2-
trichlorovinylamine is substituted for the trifluoro reagent,
the chloride of the steroid may be obtained.

16.   From Sulfonium Salts (1, 352)

$$Me_2\overset{\oplus}{S}Br \ \overset{\ominus}{Br} \ + \ ROH(xs.) \ \xrightarrow[\text{4-5 hr.}]{80°} \ RBr$$

The reaction takes place with inversion [54]. High yields
ranging from 37% for t-butyl to 78% for dodecyl bromide were
obtained.

17.   From Alcohols and Selected Acid Chlorides or Cyanuric
      Chloride

$$ROH \ \xrightarrow[\text{DMF}]{Me\ SO_2Cl} \ RCl$$

This reaction probably involves first the formation of the
ester that is converted by the chloride ion into RCl. Unactivated alcohols are not affected by this reagent but allylic
alcohols are converted into the chlorides without rearrangement
at 0° in the presence of lithium chloride and collidine [55].
The 6-hydroxyl group of methyl glycosides is also converted into
the chloride group [56].

Cyanuric chloride, added portionwise to an
alcohol held 10-20° below its boiling point
converts the alcohol into the chloride, which
can be removed by distillation [57]. Pri-
mary, secondary, and tertiary alcohols give 29-92% yields of
the chloride. If the reaction mixture is held at 80-100° for
3-5 hr in the presence of sodium iodide, 30-76% yields of
iodides are obtained with the exception of t-butyl alcohol,
which is converted into isobutylene [58].

18.   From Boranes

$$R_3B \ \xrightarrow{X_2} \ RX \ + \ R_2BX$$

The boranes may be exchanged with mercuric oxide to give dialkylmercurials, which with bromine give alkyl bromides [59]. But the more direct bromination of trialkylborons was made possible by the discovery that sodium methoxide in methanol greatly accelerates the bromination step [60]. Methyl ω-bromoundecanoate was made in 85% yield in this manner. Yields were lower in the preparation of <u>sec</u> alkyl bromides. They were increased, however, when 9-borobicyclo [3.3.1] nonane was used to react with some internal olefin [61]. The acceleration of bromination with sodium methoxide and the fact that the alkyl bromides are formed by an inversion process suggests displacement from an anionic form [62].

$$\overset{-\delta\ +\delta}{Br \sim Br}\ \ R - \overset{\overset{R}{|}}{\underset{\underset{R}{|}}{B}} - \overset{\ominus}{O}CH_3 \longrightarrow Br\ \underset{\text{Inverted}}{R}\ +\ R_2BOCH_3 + \overset{\ominus}{Br}$$

Alkyl iodides also may be made by treatment of the borane with iodine in sodium methoxide solution [63]. Secondary alkyl iodide yields, however, drop to about 30%. An alternate way to prepare iodides from the borane is by treatment with an alkyl iodide, and air, a process probably involving free radicals [64]. Of all the iodides used with the borane, <u>t</u>-butyl iodide appears to work best. A method of making <u>trans</u>-vinyl iodides consists of treating catecholborane with an acetylene, hydrolyzing to remove the catechol, and finally adding iodine as shown [65]:

$$RC \equiv CH\ +\ HB\overset{O}{\underset{O}{\diagdown}}\diagdown\!\!\bigcirc \longrightarrow RCH = CHB\overset{O}{\underset{O}{\diagdown}}\diagdown\!\!\bigcirc \overset{H_2O}{\longrightarrow}$$

$$RCH = CHB(OH)_2 \xrightarrow[Et_2O, 0°, 15\ min.]{I_2, NaOH} RCH = CHI \quad R = C_6H_{13},\ 71\%$$

19.   From Perfluoroalkyl Halides and Perfluoroalkyl Copper

$$CF_3I\ +\ ArI \xrightarrow{Cu} Ar\,CF_3$$

The preceding reaction was studied in considerable detail by McLoughlin and Thrower [66]. About 40 examples were carried out in DMSO, DMF, or pyridine at 110-130° with yields of 17-70%. Apparently, the reaction succeeds because of ready formation of a perfluoroalkyl copper intermediate. Although the above authors used higher homologues of a perfluoroalkyl iodide, Kobayashi [67] has found that trifluoromethyl iodide works as well using a stainless-steel pressure vessel. Moreover, yields were higher using copper generated from copper sulfate and zinc [68]. Although yields were lower, trifluoromethyl bromide, rather than the iodide in HMPA, could be used in this process. The economy of this method, despite the lower yield, may make it attractive [69]. Since the perfluoroalkyl group is a useful group to control lipid-water miscibilities in medicinal chemistry, a new way of synthesizing the compounds containing it is always welcome. Others have accomplished this end by the reaction of (perfluoroheptyl) copper on arenes [70].

20.   From Chloroformates and Metal Halides

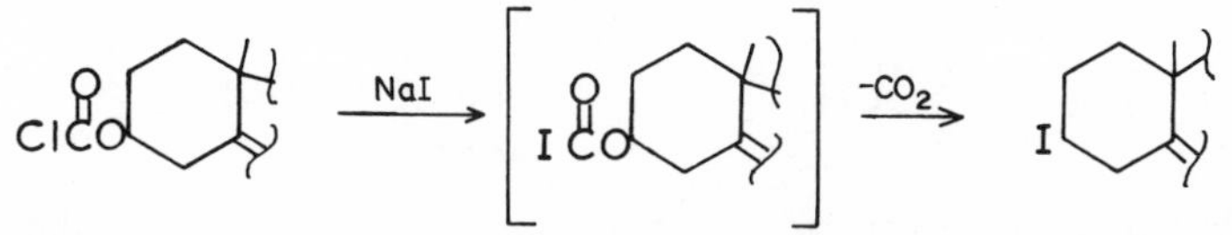

Cholesteryl iodide was prepared as above in an acetone-ether mixture held at 35° for 70 hr [71]. The product decomposes easily.

21.   From Alcohols and the Methiodide of Carbodiimides

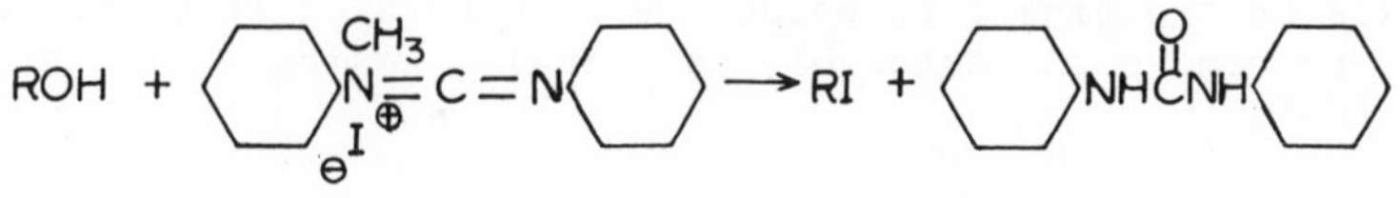

This reaction was conducted in THF, C_6H_6, or C_6H_{14} at 35-50° for 2-62 hr to give 18-89% yields of primary and secondary alkyl iodides [72]. The crude product was filtered through silica to remove the urea that did not crystallize.

22.   From Nitronaphthalene by Irradiation

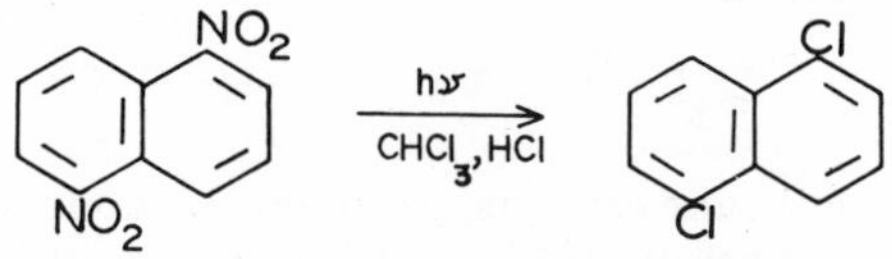

The preceding reaction gave 85% 1,5-dichloronaphthalene [73]. With the 1,8-dinitro isomer it gave only 50% of the 1,8-dichloride plus 50% of a trichloronaphthalene.

23.   From Hydrazides and Phosphorus Pentachloride (see Addenda)

Addenda

A.1. Phase transfer reactions may be conducted even under acidic conditions. Thus water-insoluble primary alcohols may be converted into alkyl halides by concentrated aqueous HCl with a phase transfer catalyst. Yields are usually 94% or better after 45 hr [D. Landini et al., Synthesis, 37 (1974)].

A.1. Acetic acid catalyzes the reaction between glycols and hydrohalic acids, perhaps because of an orthoacetic intermediate [J. H. Exner, Tetrahedron Letters, 1197 (1975)].

A.2. In the sterol family the Rydon reagent with NBS usually gave high yields of bromides [A. K. Bose, B. Lal, Tetrahedron Letters, 3937 (1973)].

A.3. Carbonyl compounds may be converted into gem difluorides by R_2NSF_3 [W. J. Middleton, J. Org. Chem., 40, 574 (1975)].

A.3. Triphenylphosphine dihalides are satisfactory for the conversion of cyclic β-diketones into β-halo-α,β-unsaturated ketones. Thus triphenylphosphine dibromide and 1,3-cyclohexanedione give 3-bromo-2-cyclohexene-1-one (97%) [E. Piers, I. Nagakura, Syn. Commun., 5, 193 (1975)].

A.3. The use of a Rydon-like reagent, Ph_3P-Cl_2, on benzaldehyde led to the formation of benzylidene chloride (59%) [L. Horner et al., Ann. Chem., 626, 26 (1959)].

A.5. Phase transfer conditions at 100° may be used to convert optically active methanesulfonates into halides with inversion, except with sodium iodide, where the racemic iodide was obtained [D. Landini et al., Synthesis, 430 (1975)].

A.6. Halide exchange with the fluoride ion has been carried out by mixing potassium fluoride with the halide in the presence of catalytic amounts of the macrocyclic ether, 18-crown-6 in acetonitrile. The reagent is referred to as the "naked fluoride ion" [C. L. Liotta, H. P. Harris, J. Am. Chem. Soc., 96, 2250 (1974)].

A.11. The Cristol modification of the Hunsdiecker reaction gives 1-bromo-3-chlorocyclobutane (35-40%) from 3-chlorocyclobutanecarboxylic acid [G. M. Lampman, J. C. Aumiller, Org. Syn., 51, 106 (1971)].

A.12. Acid chlorides may be converted into trifluoromethyl compounds by MoF$_6$ in fair to poor yields [F. Mathey, J. Bensoam, Compt. Rend., 276, C 1569 (1973)].

A.13. Selenium tetrafluoride, boiling point 106°, may be used to convert carbonyl compounds into gem difluorides without the use of pressure equipment, as is the case with SF$_4$ [G. Olah et al., J. Am. Chem. Soc., 96, 925 (1974)]. Also, dialkylaminosulfur trifluoride, R$_2$NSF$_3$, may be used similarly [L. N. Markovskii et al., Synthesis, 787 (1973)].

A.23. Phenyl acetohydrazide may be converted into 2,2-dichloro-1-phenylethane (60%) and adipic hydrazide into 1,1,6,6-tetra-chlorohexane (55%), each with PCl$_5$ [V. S. Mikhailov et al., J. Org. Chem. USSR, 9, 1847 (1973)].

References

1. S. F. Politanskii et al., J. Org. Chem. USSR (Eng. transl.), 10, 697 (1974).
2. M. Mihalic et al., Synthesis, 417 (1972).
3. J. B. Lee, T. J. Nolan, Can. J. Chem., 44, 1331 (1966).
4. J. Hooz, S. S. H. Gilani, Can. J. Chem., 46, 86 (1968).
5. I. M. Downie et al., Chem. Commun., 1350 (1968).
6. B. Castro et al., Tetrahedron Letters, 4455 (1973).
7. H. N. Rydon, Org. Syn., 51, 44 (1971).
8. J. P. H. Verheyden, J. G. Moffatt, J. Org. Chem., 35, 2319 (1970).
9. D. U. Robert, J. G. Riess, Tetrahedron Letters, 847 (1972).
10. H. R. Hudson et al., J. Chem. Soc., Perkin Trans., I, 1595 (1972).
11. H. R. Hudson, Synthesis, 112 (1969).
12. J. P. Schaefer et al., Org. Syn., 49, 6 (1969).
13. G. L. Walford et al., J. Med. Chem., 14, 339 (1971).
14. J. P. Schaefer et al., Org. Syn., 48, 51 (1968).
14a. E. I. Snyder, J. Org. Chem., 37, 1466 (1972).
15. W. Schoberth, M. Hanack, Synthesis, 703 (1972).
15a. D. R. Strobach, G. A. Boswell, Jr., J. Org. Chem., 36, 818 (1971).
16. R. M. Babb, F. W. Bollinger, J. Org. Chem., 35, 1438 (1970).
17. F. Mathey, J. Bensoam, Tetrahedron, 27, 3965 (1971).
18. L. Field, Synthesis, 110 (1972).
19. K. Findeisen et al., Synthesis, 601 (1972).
20. D. E. O'Brien et al., Tetrahedron Letters, 3149 (1973).

21. Y. I. Chumakov, Z. N. Murzinova, Metody Polucheniya Khim. Reaktivov i Preparatov, 11, 105 (1964); C. A., 64, 19549 (1966).
22. J. F. Normant et al., Compt. Rend., 269, C 1325 (1969).
23. J. F. Normant, H. Deshayes, Bull. Soc. Chim. France, 2854 (1972).
24. N. Altabev et al., Chem. Ind. (London), 331 (1973).
25. F. Kienzle, E. C. Taylor, J. Org. Chem., 35, 528 (1970).
26. J. Wielgat, Rocz. Chem., 45, 931 (1971); C. A., 75, 98411 (1971).
27. R. E. Pincock et al., J. Am. Chem. Soc., 95, 2030 (1973).
28. D. J. Burton, L. J. Kehoe, J. Org. Chem., 35, 1339 (1970).
29. J. A. Vida, Tetrahedron Letters, 3447 (1970).
30. L. A. Paquette, J. C. Philips, Tetrahedron Letters, 4645 (1967).
31. K. Hirai, Y. Kishida, Tetrahedron Letters, 2743 (1972).
32. A. G. Anderson, Jr., F. J. Freenor, J. Org. Chem., 37, 626 (1972).
33. A. P. Marchand, W. R. Weimar, Jr., Chem. Ind. (London), 200 (1969).
34. A. N. Thakore et al., Tetrahedron, 27, 2617 (1971).
35. N. S. Isaacs, D. Kirkpatrick, Tetrahedron Letters, 3869 (1972).
36. S. Nishigaki et al., Tetrahedron Letters, 539 (1969).
37. V. I. Shevchenko et al., J. Gen. Chem. USSR (Eng. transl.), 36, 485 (1966).
38. T. P. Povlock, Tetrahedron Letters, 4131 (1967).
39. R. C. Petterson et al., J. Org. Chem., 36, 631 (1971).
40. K. L. Kirk, L. A. Cohen, J. Am. Chem. Soc., 93, 3060 (1971).
41. K. Brewster, R. M. Pinder, Synthesis, 307 (1971).
42. D. L. J. Clive, C. V. Denyer, Chem. Commun., 1112 (1971).
43. A. Pross, S. Sternhell, Australian J. Chem., 23, 989 (1970).
44. E. C. Friedrich et al., Syn. Commun., 5, 33 (1975).
45. G. A. Olah, J. Welch, Synthesis, 896 (1974).
46. D. H. Holsboer, A. P. M. Van der Veek, Rec. Trav. Chim., 91, 349 (1972).
47. N. J. Bunce, J. Org. Chem., 37, 664 (1972).
48. C. A. Grob et al., Synthesis, 493 (1973).
49. R. D. Stolow, T. W. Giants, Tetrahedron Letters, 695 (1971).
50. A. McKillop, E. C. Taylor et al., J. Org. Chem., 34, 1172 (1969).
51. J. Blum et al., Tetrahedron Letters, 3041 (1966); J. Org. Chem., 35, 1895 (1970).

52. B. V. Kunshenko et al., J. Org. Chem., USSR (Eng. transl.), 10, 896 (1974).
53. G. B. Spero et al., Steroids, 11, 769 (1968).
54. S. Oae et al., Chem. Commun., 212 (1973).
55. E. W. Collington, A. I. Meyers, J. Org. Chem., 36, 3044 (1971).
56. M. E. Evans et al., J. Org. Chem., 33, 1074 (1968).
57. S. R. Sandler, J. Org. Chem., 35, 3967 (1970).
58. S. R. Sandler, Chem. Ind. (London), 1416 (1971).
59. J. J. Tufariello, M. M. Hovey, Chem. Commun., 372 (1970).
60. H. C. Brown, C. F. Lane, J. Am. Chem. Soc., 92, 6660 (1970).
61. C. F. Lane, H. C. Brown, J. Organomet. Chem., 26, C51 (1971).
62. H. C. Brown, C. F. Lane, Chem. Commun., 521 (1971).
63. H. C. Brown et al., J. Am. Chem. Soc., 90, 5038 (1968).
64. A. Suzuki, H. C. Brown et al., J. Am. Chem. Soc., 93, 1508 (1971).
65. H. C. Brown et al., J. Am. Chem. Soc., 95, 5786 (1973).
66. V. C. R. McLoughlin, J. Thrower, Tetrahedron, 25, 5921 (1969).
67. Y. Kobayashi et al., Chem. Pharm. Bull., 18, 2334 (1970).
68. Y. Kobayashi et al., Tetrahedron Letters, 4095 (1969).
69. Y. Kobayashi et al., Chem. Pharm. Bull., 20, 1839 (1972).
70. P. L. Coe, N. E. Milner, J. Fl. Chem., 2, 167 (1972/73).
71. D. N. Kevill, F. L. Weitl, J. Org. Chem., 32, 2633 (1967).
72. R. Scheffold, E. Saladin, Angew. Chem. Intern. Ed. Engl., 11, 229 (1972).
73. G. Frater, E. Havinga, Rec. Trav. Chem., 89, 273 (1970).

B. Addition to Unsaturated Compounds and Epoxides

1. Hydrogen Halides (1, 356)

$$RCH{=\!=\!=}CH_2 \xrightarrow{\text{HX}} RCHXCH_3$$

The addition of hydrogen fluoride to isopropenyl chloride has been moderated by using nitrobenzene as a solvent and by conducting the reaction for 10 min at 14° to give 75% of 2-chloro-2-fluoropropane [1].

Anti-Markownikoff addition to strained olefins, usually bicyclic, has been shown to give mostly cis addition, in contrast to the usual trans addition [2]. Moreover, cis addition has been found to occur with 1,2-dimethylcyclohexene [3]. With hydrogen chloride in methylene chloride at -98°, the

cis:trans ratio was 88/12, but in ether at 0° the ratio was 25/
75. Molecular sieves with hydrogen bromide also give a preponderance of anti-Markownikoff addition with some alkenes [4].
However, this addition may take place by a free radical
mechanism. Thallium(I) acetate or other thallium organic salts
add to alkenes to form the 2-acetoxyalkylthallium adducts [5].
When iodine in acetic acid is added to the adducts, β-iodoethyl
acetates are formed in 30-98% yields (8 ex.).

2. Halogens (1, 359)

$$RCH = CH_2 \xrightarrow{X_2} RCHXCH_2X$$

If additions with halogens are too slow, as in the case of
5,5,6,6-tetrafluoronorbornene, the radical pathway, namely,
illumination of the halogen solution, is suggested [6]. Cupric
bromide or chloride (the latter either hydrated or anhydrous)
are recommended as reagents in the addition of a halogen to an
alkene [7]. In acetonitrile cyclohexene with cupric bromide
gave 73% of 1,2-dibromocyclohexane. A better ligand than
acetonitrile is ethylene-bis-diphenylphosphine. As little as 5
mole % of this ligand can be used in bromination with cupric
bromide. Moreover, mixed halides can be prepared by this procedure; for instance, cupric chloride in iodine with cyclohexene gave 95% of 1,2-chloroiodocyclohexane [8]. An interesting
mixed addition is as follows [9]:

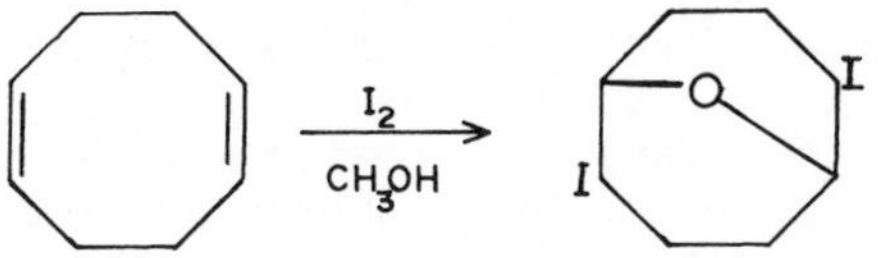

2,6-Diiodo-9-oxabicyclo[3.3.1]nonane
30%

A second mixed addition occurs with chlorine in DMF to give
crystalline adducts which may be utilized to form vic-dichlorides
or chlorohydrins as shown [10]:

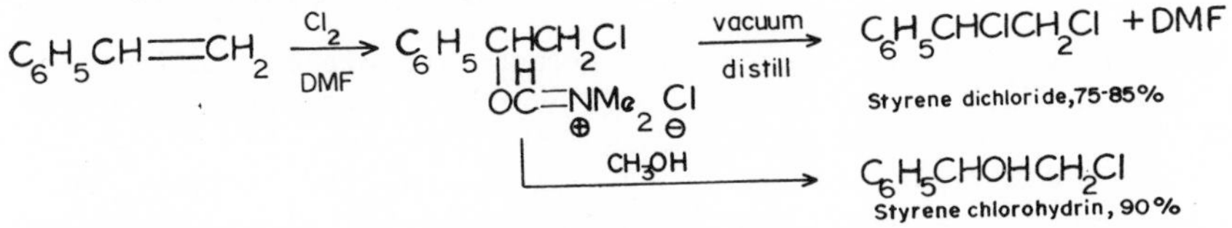

This is the best yield of dichloride reported. Other mixed
additions are to be found in B.3.

Direct addition of fluorine is not well controlled and lead tetrafluoride does not necessarily give the desired product as witness [11]:

$$(C_6H_5)_2C{=\!\!=}CH_2 \Big\langle \quad \xrightarrow{PbF_4} \quad C_6H_5CF_2CH_2C_6H_5$$

$$\xrightarrow[\substack{CCl_3F \\ -78°}]{F_2} \quad (C_6H_5)_2C{=\!\!=}CHF$$
major product

It has been reported that cis addition to cyclohexene, 100 mmol, takes place with antimony pentachloride in carbon tetrachloride at 78° for 5 min [13.9 mmol cis, 2.8 mmol trans] [12]. The double bond in adamantylideneadamantane does not add bromine, but a unique, stable bromonium bromide [13] is formed.

3. Compounds Containing a Halogen Attached to a Hetero Atom (1, 363)

$$RCH{=\!\!=}CH_2 \xrightarrow{HOX} RCHOHCH_2X$$

The addition of N-bromoacetamide in liquid hydrogen fluoride to an alkene is a means of synthesizing vic-bromo-fluorides [14]. Furthermore, the bromofluorides are reduced with tributylstannane (Bu$_3$SnH) at 50° for 1 hr to the alkyl fluoride, as evidenced by the formation of 73% of fluorocyclo-hexane from 1-bromo-2-fluorocyclohexane. Other reducing reagents failed.

The addition of methyl hypobromite (from N-bromophthalimide in methanol) to vinyl ethers gave stereospecific addition, that is, a trans ether gave the erythro α bromoether and a cis ether gave the threo form [15].

In moist DMSO, NBS gave the following bromohydrin with a conjugated diene [16]:

$$CH_3CH{=\!\!=}CHCH{=\!\!=}CHCH_3 \xrightarrow[\text{Moist DMSO}]{NBS} CH_3CHBrCHOHCH{=\!\!=}CHCH_3$$
erythro-5-Bromo-4-hydroxy-2-hexene, 93%

Five other conjugated dienes gave comparable yields, but one did not react. Nitrogen trichloride (2 equiv) is reported to give good yields of 1,2-dichloroalkanes from alkenes [17].

The addition of iodoazide to α,β-unsaturated ketones, such as cyclopentenone or cyclohexenone, gave 30-44% of α-iodocyclo-pentenone or -cyclohexenone [18].

4. Free Radicals (1, 365)

$$RCH{=}CH_2 + CCl_4 \longrightarrow RCHClCH_2CCl_3$$

Burton and Kehoe continue to exploit this addition with cupric chloride-ethanolamine at reflux [19]. The advantage lies in the fact that the perhalide, in this case carbon tetrachloride, need not be in large excess to prevent telomerization. Among other catalysts that may be used are Fe(II), Co(II) and, in the following example, benzoyl peroxide [20]:

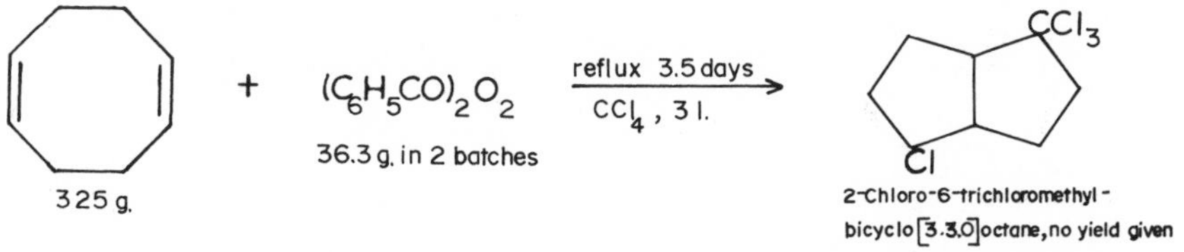

325 g.

+

$(C_6H_5CO)_2O_2$

36.3 g. in 2 batches

reflux 3.5 days
$\xrightarrow{\hspace{2cm}}$
CCl_4, 3 l.

2-Chloro-6-trichloromethyl-
bicyclo[3.3.0]octane, no yield given

A new catalyst apparently applicable at lower temperature is $(Ph_3P)_3RuCl_2$; it, 0.05 g, converts 10 mmol of the alkene into the adduct as shown [21]:

$$C_6H_{13}CH{=}CH_2 + CCl_4 \xrightarrow[\substack{\text{sealed tube}\\\text{catalyst}}]{80°} C_6H_{13}CHClCH_2CCl_3$$

1,1,1,3-Tetrachorononane
97%, 75% conversion

Irradiation also can be used to promote the free radical reaction [22]:

$$CF_3CHICF_3 + CH_2{=}CHF \xrightarrow{h\nu} (CF_3)_2CFCH_2CHFI$$

1,1,1,2,4-Pentafluoro-4-iodo-
2-trifluoromethylbutane
98%, containing 1% of an isomer

Rearrangements of these perhalo compounds are discussed in A.6.
Iron carbonyl in isopropyl alcohol is a catalyst for the addition of 1,1,1-trichloroalkanes to alkenes [23]:

$$Cl(CH_2)_2CCl_3 + CH_2{=}CH_2 \xrightarrow[\text{isopropyl alcohol}]{Fe(CO)_5, 135°} Cl(CH_2)_2\overset{\displaystyle Cl}{\underset{\displaystyle Cl}{C}}CH_2CH_2Cl$$

62 atm.
falling to 43 atm.

1,3,3,5-Tetrachloropentane
56%

Without catalysts, telomerization tends to occur [24].   With peroxides haloforms may be added [25]:

$$CHCl_3 + Cl_2CH(CH_2)_3CH = CH_2 \xrightarrow[\text{peroxycarbonate}]{\text{Dicyclohexyl}} Cl_2CH(CH_2)_5CCl_3$$

1,1,1,7,7-Pentachloroheptane
40%

The product is hydrolyzed to 7,7-dichloroheptanoic acid with 94% sulfuric acid. Dicobalt octacarbonyl also has been used in the addition of carbon tetrachloride to alkenes [26]:

$$CCl_4 + RCH = CH_2 \xrightarrow[\substack{20-160° \\ \text{autogenous P}}]{Co_2(CO)_8} RCHClCH_2CCl_3$$

If the preceding reaction is run in the presence of carbon monoxide at 60-200 atm, 4,4,4-trichlorobutyryl chlorides, $CCl_3CH_2CHRCOCl$, are obtained.

Electrophilic alkylation or arylation is discussed in B.6.

5.   Halocarbenes or Halocarbenoids ($\underline{1}$, 368)

$$[Cl_2C:] + RCH = CH_2 \longrightarrow RCH \overset{\displaystyle CCl_2}{\underset{\displaystyle \diagup \diagdown}{}} CH_2$$

More references have appeared on this subject than on any other in the chapter on halides. These include several improvements in the generation and addition of the carbene to the alkene. The two-phase, strong alkali, quaternary base system, called "phase transfer" system, is one improvement that apparently brings the anion into the organic phase, where the carbene is generated in close contact with the alkene. Benzyltrimethylammonium chloride in 50% sodium hydroxide is used most frequently, but even quaternary phosphonium salts serve as well [27]. In making dibromocyclopropanes by the two-phase system, yields were 26-73% using a rather large excess of bromoform [28]. However a modification that improves these yields by some 10-30% and uses less bromoform (or chloroform) consists of the inclusion of small amounts of ethanol [29]. A second outstanding improvement in halocarbene synthesis is the use of lithium triethylmethoxide in hexane as the base to react with the haloform [30]. With this heterogeneous system, dichlorocyclopropane synthesis may be carried out at room temperature, although a temperature of 65° completes the reaction in several hours. The strong base, lithium triethylmethoxide, is superior to the commercial potassium $\underline{t}$-butoxide. A third outstanding preparation of

dihalocarbene involves the smooth decomposition of phenyltri-
halomethyl mercury in inert solvents to the dihalocarbene and
phenylmercuric halide [31]. It was not so at the time of writing
the previous volume, but the discovery that the preparation of
the reagent takes place readily in THF with commercial potassium
t-butoxide and that phenylmercuric bromide is exceptionally
selective in formation has made this reaction quite useful.
These points are illustrated:

$$C_6H_5HgBr + HCCl_3 + KO\ t\text{-}Bu \xrightarrow{THF} C_6H_5Hg\,CBrCl_2 \xrightarrow[80°,\ stilbene]{hexane}$$

may be isolated

$$\begin{array}{c} C_6H_5CH\!-\!CHC_6H_5 \\ \diagdown\,/ \\ C \\ Cl_2 \end{array}$$

3,3-Dichloro-1,2-diphenyl-
cyclopropane, 90%

Perfluorocyclohexene does not react and basic amino groups in the
alkene interfere with the carbene addition. An amide or di-
phenylamino group, however, does not interfere. In other words,
as the basicity is decreased the reaction becomes feasible. Tri-
fluoromethylmercuric iodide with sodium iodide and cyclohexene
gave 7,7-difluoronorcarane [32].

At least one example has been published, namely, the
preparation of 2-butyl-1,1-dichloro-3-trimethylsilylcyclopro-
pane, in which the phase transfer method is superior to the
Seyferth method [33].

The monobromocyclopropane seems best made from di(trimethylsilyl)sodamide, $(Me_3Si)_2NNa$, and methylene bromide with the
alkene in pentane at 25° (10 ex.) (5-54%) [34]. Another method
is used to make gem-difluorocyclopropanes. It employs the
phosphonium salt, Ph_3PCF_2BrBr, with cesium or potassium fluoride
in the presence of the alkene to give good yields [35]. The
monofluorocyclopropane is best made from the carbenoid derived
from diiodofluoromethane and diethyl zinc, which with cyclohexene
gives the monofluoronorcarane in 91% yield [36].

Other reactions are brought about by halocarbenes. For
instance, alcohols are converted into chlorides:

$$ROH + \left[:CCl_2\right] \longrightarrow \left[RO\overset{\ominus}{C}Cl_2\right] \longrightarrow RCl$$

I

The other products presumably are carbon monoxide and chloride
ion. The reaction is best carried out by the two-phase method
using strong alkali, benzyltriethylammonium chloride with
chloroform added dropwise to a vigorously stirred solution [37].

Six examples, including the conversion of adamantyl alcohol to
1-chloroadamantane, are described with yields of 40-94%. Just as
with indene (1, 369), so can indole be ring-enlarged with dichlorocarbene to form 3-chloroquinoline in 36% yield, although
the carbene is generated by thermal means (550°) from chloroform [38]. At -100°, carbon tetrachloride reacts with butyllithium to form trichloromethyllithium. At about -65°, it decomposes to the carbene and at this temperature adds to haloni-
trobenzenes as follows [39]:

$$NO_2\text{-benzene}(F) + Li\,CCl_3 \xrightarrow{-65°} NO_2\text{-benzene}(CHCl_2, F)$$

5-Fluoro-2-nitrobenzaldichloride
60 %

Ketones are converted into α-chloroketones as shown [40]:

$$R_2C=O \xrightarrow{Cl_2CHLi} R_2\overset{O^{\ominus}}{\underset{|}{C}}CHCl_2 \xrightarrow{BuLi} \left[R_2\overset{O^{\ominus}}{\underset{|}{C}}\overset{\ominus}{C}Cl_2\right] \xrightarrow{-LiCl}$$

$$R\overset{O^{\ominus}}{\underset{|}{C}}=CClR \xrightarrow{H^{\oplus}} R\overset{O}{\underset{\parallel}{C}}CHClR$$

Benzaldehyde gave phenacyl chloride in 72% yield, and cyclopentanone gave 64% of 2-chlorocyclohexanone by the above method.
    The following conversion $\supset C=O \rightarrow \supset C=CX_2$ has been
accomplished by the reagent hexamethyltriaminophosphine:

$$CF_2Br_2 + 2\,(Me_2N)_3P + {\textstyle>}C=O \xrightarrow[25°]{Triglyme} {\textstyle>}C=CF_2 \qquad [41]$$

The intermediate formed with an aldehyde appears to be as
follows [42]:

$$RCHO + CCl_4 + (Me_2N)_3P \xrightarrow[THF]{-78°} RCHO\overset{CCl_3}{\underset{\overset{|}{O}}{P}(NMe_2)_3} \xrightarrow{H_2O} RCHOHCCl_3$$

$$\overset{\overset{|}{CHR}}{\underset{CCl_3}{}} \xrightarrow{CCl_4} RCH=CCl_2$$

50-70%
10 examples

    The decomposition of 5,5-dialkyl-3-nitrosooxazolidones
yields vinyl carbenes which react with lithium halides or sodium
iodide to form vinyl halides [43]:

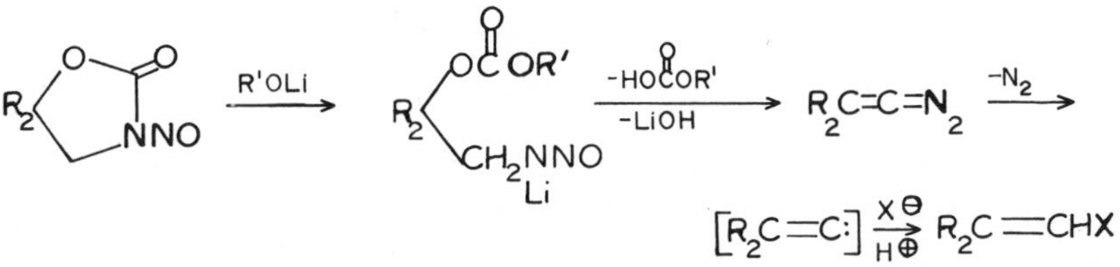

$$\left[R_2C=C\ddot{:}\right] \xrightarrow[H^\oplus]{X^\ominus} R_2C=CHX$$

A helpful review has been published on the conversion of the readily available 1,1-dihalocyclopropanes to other compounds [44]. Only the transformations which lead to other halogen compounds are mentioned here. The reduction of the gem dihalides to mono-halides may be accomplished with tributylstannane, methylmagnesium bromide, or sodium dimsyl. The latter gives considerable trans monohalide. Ring expansion is brought about even by simple distillation such as:

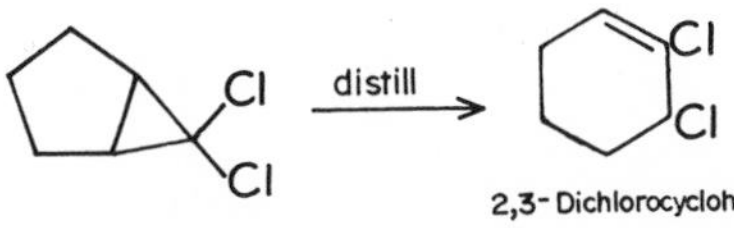

See E.4 for other ring expansions.

With quinoline at high temperature dehydrohalogenation also accompanies ring expansion:

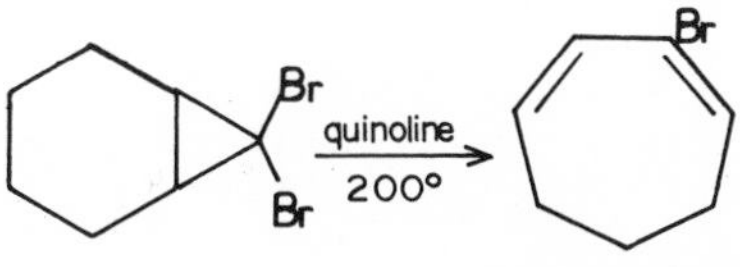

The gem-dichlorocyclopropane from cyclopentadiene has not been isolated but yields chlorobenzene.

In some cases, such as the gem-bromochlorocycloindene, ring expansion occurs to give α-bromo- and α-chloronaphthalene [44]:

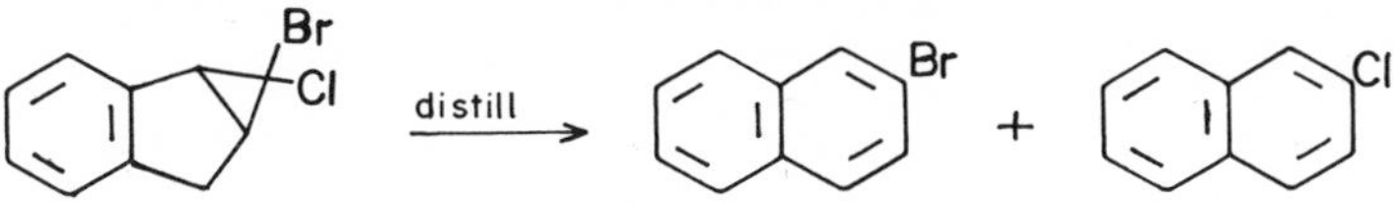

Aqueous silver nitrate appears generally to expand the ring and form the halohydrin:

6. Alkylation or Acylation Reagents (Mostly Electrophilic)
    (1, 370)

$$R'CH = CH_2 + R_3CX \xrightarrow{AlCl_3} R'CHXCH_2CR_3$$

t-Butyl, benzyl, and benzhydryl chlorides have been added to
phenylacetylene using zinc chloride and methylene chloride
saturated with hydrogen chloride to give β-alkyl-α-chlorostyrenes
in high yields [45]. β,β'-Dichlorodiethyl ketone has been made
in 95% yield from a mixture of β-chloropropionyl chloride and
aluminum chloride in methylene chloride through which ethylene
was bubbled for 3 hr at 20° [46]. The workup should be held
below 20° to avoid decomposition of the ketone. It is preferable
to use the crude ketone rather than to distill it. Free radical
modes of alkylation are discussed in B.4.

7. From Epoxides and Hydrogen Halides (1, 372)

The complex of hexamethyltriaminophosphine, $(Me_2N)_3P$, with
chlorine reacts with cyclohexene oxide to give about 75% of
1,2-dichlorocyclohexane, 95% trans [47]. The combination of
chlorine with triphenylphosphine gave nearly a 50:50 mixture of
cis- and trans isomers, although it is reported that triphenylphosphine and carbon tetrachloride gave 80% of 1,2-dichloro-
cyclohexane, 95% cis [48]. A complex of triethylamine and
potassium hydrogen fluoride (KF·2HF) made in situ to which
cyclohexene oxide was added gave 60% of 2-fluorocyclohexanol
[49]. A number of 3-fluoro-1-halo-2-propanols have been made by
the addition of the appropriate acid to the epifluorohydrin
[50]. As is well known, the addition of an epoxide to a brominating medium is an excellent method of scavenging hydrogen
bromide [51].

8. From Haloalkenes (Cycloaddition)

Diels-Alder syntheses from halodienes have been described
elsewhere (H.-W., 5, Pt. 3, 1962, p. 982).
The addition of trifluoroethylenes to cyclopentadiene at
155° for 72 hr gave a telomer as well as monomers [52]:

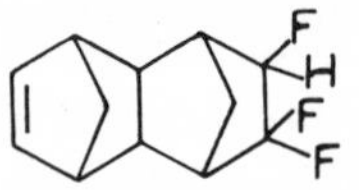

Dibromoketene has been generated as shown and condensed with cyclopentadiene [53]:

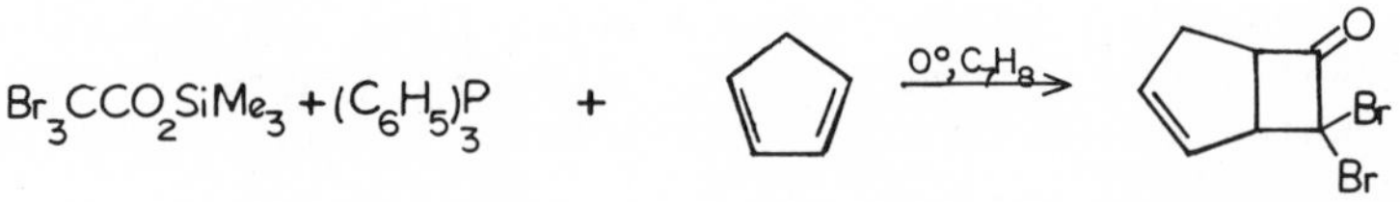

3,3-Dibromobicyclo[3.2.0]-
hept-5-enone-2, 89%

A most useful retrocyclo reaction has been developed as shown [54]:

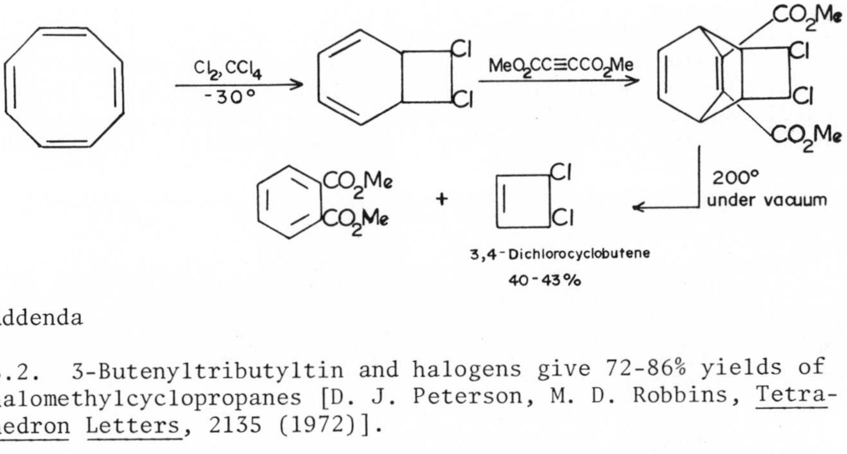

3,4-Dichlorocyclobutene
40-43%

Addenda

B.2.   3-Butenyltributyltin and halogens give 72-86% yields of halomethylcyclopropanes [D. J. Peterson, M. D. Robbins, Tetrahedron Letters, 2135 (1972)].

B.2.   Molybdenum pentachloride and carbon tetrachloride with olefins give 1,2-dichlorides in which the cis:trans addition ratio is 5-12:1 [S. Uemura et al., Bull. Chem. Soc. Jap., 47, 3121 (1974)].

B.3.   t-Butyl hypochlorite catalyzed by boron trifluoride has been employed to convert cyclohexene into trans-2-chlorocyclohexyl t-butyl ether (49%) [C. Walling, R. T. Clark, J. Org. Chem., 39, 1962 (1974)].

B.5.   Phase transfer reactions, especially their use in generating halocarbons, have been reviewed [E. V. Dehmlow, Angew. Chem. Intern. Ed. Engl., 13, 170 (1974)]. High concentration of reagents, chloroform or methylene chloride as the organic solvent, and excess of the anionic reagent are keys to success. Novelties by phase transfer: 1,1-Diiodocyclopropane generation; insertion of a carbene into cis-, but not trans-decalin; amine salt separation (tert more organic soluble), and isocyanide preparation improvement.

B.5.   The carbenoid preparation from diethylzinc and bromoform (or other halides) has been shown to be a free-radical chain reaction, which is accelerated to give better yields either by passing air over, or by irradiation of, the reaction [S. Miyano et al., Bull. Chem. Soc. Jap., 46, 892 (1973) and previous papers[.

B.5.   1-Dichloromethyladamantane has been prepared in 54% yield from dichlorocarbene [I. Tabushi, J. Am. Chem. Soc., 92, 6670 (1970)].

B.6.   Aliphatic acylation via automatic gasimetry has been used to prepare chloroketones. Thus ethylene prepared as shown in Addenda, Alkenes, A.3, was passed into propanoyl chloride-AlCl$_3$ in CH$_2$Cl$_2$ and 1-chloro-3-pentanone (96%) was obtained on hydrolysis [C. A. Brown, J. Org. Chem., 40, 3154 (1975)].

References

1.   J. L. Webb, J. E. Corn, J. Org. Chem., 38, 2091 (1973).
2.   T. G. Traylor, Acct. Chem. Res., 2, 152 (1969).
3.   K. B. Becker, C. A. Grob, Synthesis, 789 (1973).
4.   L. C. Fetterly et al., Mol. Sieves Pap. Conf., 102 (1967, publ. 1968); C. A., 71, 60339 (1969).
5.   R. C. Cambie et al., Chem. Commun., 359 (1971).
6.   B. E. Smart, J. Org. Chem., 38, 2027 (1973).
7.   W. C. Baird, Jr. et al., J. Org. Chem., 36, 3324 (1971).
8.   W. C. Baird, Jr. et al., J. Org. Chem., 36, 2088 (1971).
9.   J. N. Labows, Jr., D. Swern, J. Org. Chem., 37, 3004 (1972).
10. A. De Roocker, P. deRadzitzky, Bull. Soc. Chim. Belges, 79, 531 (1970).
11. R. F. Merritt, J. Org. Chem., 31, 3871 (1966).
12. S. Uemura et al., Chem. Commun., 1064 (1971).
13. H. Wynberg et al., Tetrahedron Letters, 4579 (1970).
14. G. L. Grady, Synthesis, 255 (1971).
15. E. M. Gaydou, Tetrahedron Letters, 4055 (1972).
16. D. R. Dalton, R. M. Davis, Tetrahedron Letters, 1057 (1972).
17. K. W. Field, P. Kovacic, Org. Syn., 50, D69-8 (1970).
18. J. M. McIntosh, Can. J. Chem., 49, 3045 (1971).

19. D. J. Burton, L. J. Kehoe, J. Org. Chem., 35, 1339 (1970).
20. R. Dowbenko, U. S. 3,585,244, June 15, 1971; C. A., 75, 63252 (1971).
21. H. Matsumoto et al., Tetrahedron Letters, 5147 (1973); Tetrahedron Letters, 899 (1975).
22. R. N. Haszeldine et al., J. Chem. Soc., Perkin Trans., I, 574 (1973); ibid., 649 (1973).
23. R. Kh. Freidlina et al., Izv. Akad. Nauk SSSR (Engl. transl.), 1110 (1969).
24. Ref. 23, p. 1701.
25. Ref. 23, p. 2220 (1973).
26. T. Susuki, J. Tsuji, J. Org. Chem., 35, 2982 (1970).
27. C. M. Starks, J. Am. Chem. Soc., 93, 195 (1971).
28. L. Skattebøl et al., Tetrahedron Letters, 1367 (1973).
29. M. Makosza, M. Fedorynski, Syn. Commun., 3, 305 (1973).
30. R. H. Prager, H. C. Brown, Synthesis, 736 (1974).
31. D. Seyferth, Acc. Chem. Res., 5, 65 (1972).
32. D. Seyferth, S. P. Hopper, J. Organometal. Chem., 26, C62 (1971).
33. R. B. Miller, Syn. Commun., 4, 341 (1974).
34. B. Martel, J. M. Hirat, Synthesis, 201 (1972).
35. D. J. Burton, D. G. Naae, J. Am. Chem. Soc., 95, 8467 (1973).
36. J. Nishimura, J. Furukawa, Chem. Commun., 1375 (1971).
37. I. Tabushi et al., J. Am. Chem. Soc., 93, 1820 (1971).
38. J. M. Patterson et al., J. Org. Chem., 37, 1849 (1972).
39. E. T. McBee et al., J. Org. Chem., 36, 2907 (1971).
40. H. Taguchi et al., Tetrahedron Letters 4661 (1972).
41. D. G. Naae, D. J. Burton, Syn. Commun., 3, 197 (1973).
42. J. C. Combret et al., Tetrahedron Letters, 1035 (1971).
43. M. S. Newman, C. D. Beard, J. Am. Chem. Soc., 91 5677 (1969); 92, 4309 (1970).
44. R. Barlet, Y. Vo-Quang, Bull. Soc. Chim. France, 3729 (1969).
45. G. Melloni et al., J. Chem. Soc., Perkin Trans., I, 2491 (1973).
46. G. R. Owen, C. B. Reese, J. Chem. Soc., C2401 (1970).
47. A. C. Oehlschlager et al., Tetrahedron, 27, 2617 (1971).
48. N. S. Isaacs, D. Kirkpatrick, Tetrahedron Letters, 3869 (1972).
49. G. Farges, A. Kergomard, Bull. Soc. Chim. France, 3647 (1969).
50. E. D. Bergmann et al., Synthesis, 646 (1971).
51. S. J. Cristol, H. W. Mueller, J. Am. Chem. Soc., 95, 8489 (1973).
52. B. E. Smart, J. Org. Chem., 38, 2027 (1973).

53. T. Okada, R. Okawara, Tetrahedron Letters, 2801 (1971).
54. R. Pettit, J. Henery, Org. Syn., 50, 36 (1970).

C. Aliphatic Substitution

1. From Aliphatic Hydrocarbons, Alkylarenes, or Alkyl Heterocycles (1, 376)

A very definite trend in orientation has been noted in the halogenation of alkyl halides, esters, acids, or other aliphatic compounds using as a reagent a N-haloamine in sulfuric acid. The orientation is called the ω-1 rule [1] and may be generalized as follows:

$$CH_3CH_2(CH_2)_n \, X' \xrightarrow[H_2SO_4, \, FeSO_4]{\overset{\oplus}{R_2NHX}} CH_3CHX(CH_2)_n \, X'$$

X' = Halogen, COOH, ester, hydroxyl

Of course other isomers are produced, but the preponderant product is the ω-1 one as shown and it is present to the extent of 71-79%. The reaction no doubt is of a free radical nature, favoring secondary over primary substitution, and the reagent is bulky or of such an electronic nature that the functional group is avoided as much as possible, the overall result being the ω-1 orientation. See C.6 for other applications of the ω-1 rule. The reaction is helpful also in giving less di- when monosubstitution is desired (Ex. a). On the other hand, a high degree of selectivity in ω-substitution has been obtained with chlorine in 90% sulfuric acid, namely, as much as 79% 4-chloro-butyric acid in mixture with the isomers from the unsubstituted acid [2]. The highest amount of primary substitution in iso-pentane has been obtained by photochemical chlorination with N-chloro-di-t-butylamine in 30% H_2SO_4 at 15° [3]. The amount of secondary chlorination is still three times that of the primary. Silver tetrafluoroantimonate has been found to be an electro-philic catalyst for aliphatic halogenation [4]. Isobutylene dibromide has been synthesized in 49% yield from t-butyl bromide, the silver catalyst, and bromine in methylene chloride, while silver bromide precipitates during the reaction.
  2-Methyladamantane refluxed in neat bromine for 4 hr gave ax- and eq-4-bromo-2-dibromomethyleneadamantane [5]. It has been pointed out that ionic halogenation of adamantane gives high ratios of 1- to 2-haloadamantane as compared to free radical halogenation [6].

Two reagents used in the halogenation of toluenes to form benzyl halides are phosphorus pentachloride at 200° in a sealed tube [7] and the N-bromo derivatives of caprolactam [8]. To prepare benzyl halides with deactivating groups on the aromatic ring, see D.6.

With 10 mol % of aluminum chloride for 4 hr at 25° any of the dichloropropenes were converted into an equilibrium mixture of about 90% of the 1,2-, 8% of the 1,3-, 1% of the 1,1-, and 0.3% of the 2,2-isomer [9]. Other isomerizations of this nature are to be found in A.6.

a. Preparation of Bromocyclohexane [1]

To a stirred mixture of 0.2 mol of N-bromodimethylamine, 0.4 mol of cyclohexane, 100 ml of sulfuric acid, and 15.5 ml of acetic acid, 14 g of powdered iron(II) sulfate heptahydrate was added. The temperature was 5° during addition and rose to 40° afterward. After workup, 17.9 g of   bromo- and 2.3 g of mixed dibromocyclohexanes were obtained.

2. From Alkenes (Allylic Halogenation) (Wohl-Ziegler)
    (1, 380)

Of interest is the bromination of 1,3-cyclooctadiene with NBS in carbon tetrachloride to form a mixture of 5-bromo-1,3- and 3-bromo-1,4-cyclooctadiene [10].

3. From Acetylenes (1, 382)

$$RC \equiv CH \longrightarrow RC \equiv CX$$

Two room-temperature substitutions have been proposed. One involves the treatment of the terminal acetylene in $CHCl_3$ and C_5H_5N with iodinium nitrate to give 40-60% of the iodoacetylene (3 ex.) [11]. The second is as indicated [12]:

$$Me_3Si \, C \equiv CSiMe_3 \xrightarrow{\text{ICl}} Me_3Si \, C \equiv CI + Me_3SiCl$$
$$95\%$$

The two-phase system has been used to convert acetylenic alcohols into iodoallenes. The appropriate alk-1-yn-3-ol in petroleum is added to copper (I) iodide, ammonium iodide, copper powder, and concentrated hydriodic acid [13]. The product as shown is recovered from the petroleum layer:

$$R_2COHC \equiv CH \longrightarrow R_2C = C = CHI$$

A

31 - 79% , 6 examples

If $R_2COHC\equiv CX(X=Cl$ or Br) is the substrate, the product with 60%
HBr is $R_2C=C=CXBr$ (9 ex.). If the acetate of A is treated with
methylmagnesium iodide and magnesium iodide, the same type of
compound, $R_2=C=CHI$, is one of the products [14].

4. From Ethers, Alcohols, and Sulfides (1, 382)

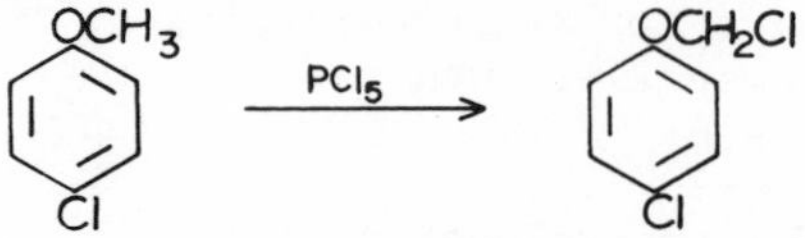

The preceding reaction is quite simple, consisting of
distillation of the phosphorus trichloride and then the chloromethyl p-chlorophenyl ether, 68-80% [15].
Surprisingly, unsaturated alcohols and KI + PPA give poor
yields of saturated iodides [16]. For instance, allyl alcohol
gave 32% of 2-iodopropane. See C.6 for an application of the
ω-1 rule to aliphatic ethers and alcohols.

5. From Carbonyl and Nitro Compounds (1, 383)

$$RCOCH_3 \longrightarrow RCOCH_2X$$

A selective pair of sequences has been found as illustrated
[17]:

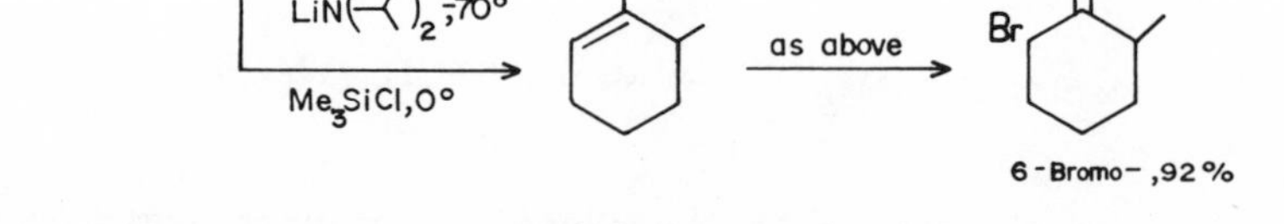

Reuss and Hassner have exploited the trimethylsilylenol ether
route by performing the halogenation directly on the enol ether
at 25° (8 ex.), 42-95% [18]. With the enol acetate of crotonaldehyde and chlorine in CCl_4 λ-chlorocrotonaldehyde was prepared
in 62% yield [19]. For another enolate substitution, see C.9.
Also, a direct chlorination of crotonaldehyde has been carried
out with cupric chloride in DMF to give the same product, 15% at
14% conversion [20]. This method may be changed to a catalytic

one (with respect to $CuCl_2$) by adding hydrochloric acid and introducing air during the chlorination [20]. Periodic acid has been used to iodinate highly enolic ketones such as dibenzoyl-methane (43-46% yields) [21]. Other enolic carbonyl compounds, such as $(NC)_2C=CHOH$, have been converted to vinyl chlorides with phosphorus pentachloride at 25° [22]. Among fluorination reagents for vinyl acetate are CF_3OF, SF_5OF, and $CF_2(OF)_2$ [23].

Rather than by direct halogenation of the sodium salts of nitroalkanes (1, 384), chlorination of the nitroalkane may be brought about by t-butyl hypochlorite in the presence of 1-hexene (or other olefins) [24]. The alkene is essential perhaps to form a Π complex with the hypochlorite, the complex being the active agent. Another way is to use NCS or related N-chloro deriva-tives [25]. Shechter's method (1, 385) using perchloryl fluoride, $FClO_3$, has been used to fluorinate dinitroalkanes [26].

6. From Acid Derivatives, Including Nitriles (1, 386)

$$RCH_2CO_2R \longrightarrow RCHXCO_2R + HX$$

The ω-1 rule has been applied to acids, alcohols, and ethers, in which for C_6 and C_8 compounds the yields of ω-1 chloro derivatives of acids, alcohols, or ethers were on the order of 60-80% with 90-92% orientation in the above positions [27]. A way to increase the yield of the ω- (not ω-1) chlorination product is to adsorb the acid on alumina (about 3-7% acid). With the tails of the acids protruding from the alumina, the amount of ω-chlorooctanoic acid is increased from 17 to 33% using chlorine in CCl_4 with irradiation [28]. For esters reluctant to halo-genate in the α-position, a very fast iodination may be carried out on the enolate of the ester by adding it to iodine in THF at -78° [29]. The enolate is made from the ester and lithium cyclohexylisopropylamide, and it must be added to the iodine in THF, not the reverse addition. Yields were 80-97% (7 ex.). Acid halides have been halogenated with NBS and a trace of strong acid (58-75%, 5 ex.) [30].

The N-bromination of imides seems to work well with acetohypobromite (CH_3CO_2Br) [31], but the preparation would be more valuable if the acetohypobromite were made from a salt other than silver acetate. Maleinimide and maleic anhydride react as follows with thionyl chloride (the anhydride gives the dichloro-anhydride) [32]:

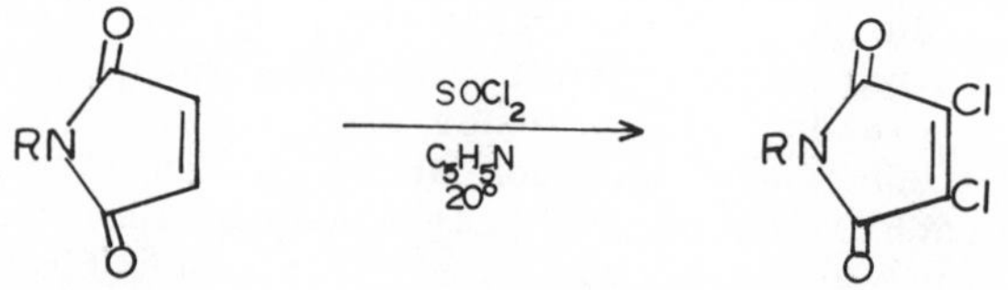

A method of forming an iodomethyl sulfoxide is now available
[33]:

$$RSOCI + CH_2N_2 + KI \longrightarrow RSOCH_2I$$

and for the synthesis of an α-haloketone, from a valuable acid,
the classical malonic ester procedure should be considered [34]:

$$ArCOCI + CH_2(CO_2Et)_2 \xrightarrow[THF,EtOH]{Mg} ArCOCH(CO_2Et)_2 \xrightarrow[2)\,48\%\,HBr]{1)\,Br_2,CHCl_3} Ar\,COCH_2Br$$

Some nitriles may be chlorinated by a unique method [35].

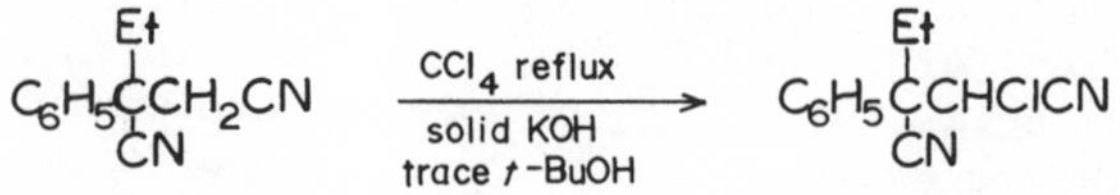

α-Chloro-β-phenyl-β-ethyl-
succinonitrile, 57%

8. From Sulfur Compounds

Sulfur monochloride has been prepared from molten sulfur at
165° and chlorine gas [36]. Sulfides with N-chlorosuccinimide
are usually oxidized to the halosulfonium halide, $R_2S^+ClCl^-$,
which is solvolyzed to the sulfoxide or for aromatic sulfides
with more reagent to the sulfone [37]. The amount of
chlorinating agent determines the stage at which the chlorination
stops. With an excess of the reagent, thiols, sulfides, disulfides, or xanthates are oxidized to the corresponding sulfonyl
halide [38]. In the presence of potassium fluoride and hydrogen
fluoride, the thiol is oxidized by chlorine to the sulfonyl
fluoride [39].

9. From Organometallic Compounds and Boranes (1, 388)

$$RM + X_2 \longrightarrow RX + MX$$

Alkyl fluorides have been prepared by the reaction of the
alkyllithium with perchloryl fluoride ($FClO_3$) [40]. In addition,

enolates of ketones have been converted into α-fluoroketones by the same reagent.

Boranes are capable of halogenation if the carbon attachment is not primary [41]:

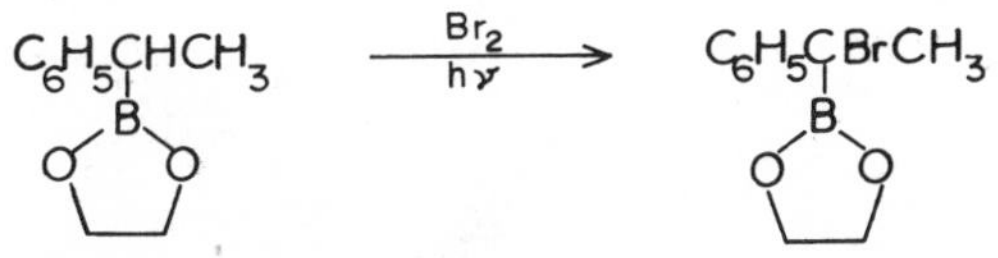

$$C_6H_5CHCH_3 \xrightarrow[h\gamma]{Br_2} C_6H_5CBrCH_3$$

Another type of substitution is shown [41]:

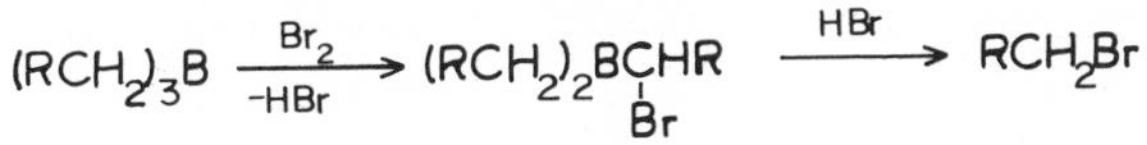

$$(RCH_2)_3B \xrightarrow[-HBr]{Br_2} (RCH_2)_2BCHR \xrightarrow{HBr} RCH_2Br$$

Addenda

C.5. Tetrabromocyclohexadienone is a useful reagent for the bromination of methyl α,β-unsaturated ketones at the methyl position [V. Calo, Tetrahedron, 29, 1625 (1973)].

C.5. Partly optically active 2-bromocyclohexanone was prepared by the bromination of the enamine from cyclohexanone L-proline ester [K. Hiroi, S.-I. Yamada, Chem. Pharm. Bull., 21, 54 (1973)].

C.5. The β-hydroxyethylimino derivative of a methyl steroidal ketone may be halogenated by NCS or NBS at the methyl position [J. F. W. Keana, R. R. Schumaker, Tetrahedron, 26, 5191 (1970)].

References

1. F. Minisci, Synthesis, 1 (1973).
2. N. C. Deno et al., J. Am. Chem. Soc., 92, 5274 (1970).
3. N. C. Deno et al., J. Am. Chem. Soc., 93, 2065 (1971).
4. G. A. Olah, P. Schilling, J. Am. Chem. Soc., 95, 7680 (1973).
5. J. R. Alford et al., J. Chem. Soc., C, 880 (1971).
6. P. Kovacic, J.-H. C. Chang, Chem. Commun., 1460 (1970).
7. R. D. Kimbrough, Jr., R. N. Bramlett, J. Org. Chem., 34, 3655 (1969).
8. J. Kőrősi, Monatsh., 100, 1222 (1969)
9. W. E. Billups et al., Tetrahedron, 26, 1095 (1970).
10. S. Moon, C. R. Ganz, J. Org. Chem., 34, 465 (1969).
11. U. E. Diner, J. W. Lown, Can. J. Chem., 49, 403 (1971).

12. D. R. M. Walton, M. J. Webb, J. Organomet. Chem., 37, 41
     (1972).
13. P. M. Greaves et al., J. Chem. Soc., C, 667 (1971).
14. J. Gore, M. -L. Roumestant, Tetrahedron Letters, 1027
     (1971).
15. H. Gross, W. Bürger, Org. Syn., 49, 16 (1969).
16. R. Jones, J. B. Pattison, J. Chem. Soc., C, 1046 (1969).
17. P. L. Stotter, K. A. Hill, J. Org. Chem., 38, 2576 (1973).
18. R. H. Reuss, A. Hassner, J. Org. Chem., 39, 1785 (1974).
19. J. Castells et al., Tetrahedron Letters, 493 (1971).
20. H. K. Dietl et al., Tetrahedron Letters, 1719 (1973).
21. A. J. Fatiadi, Chem. Commun., 11 (1970).
22. K. Friedrich, H. K. Thieme, Synthesis, 111 (1973).
23. R. H. Hesse et al., Chem. Commun., 122 (1972).
24. V. L., G. E. Heasley et al., Tetrahedron Letters, 4819
     (1971).
25. A. L. Fridman et al., Izv. Akad. Nauk, SSSR, Ser. Khim (Eng.
     transl.), 2465 (1969).
26. L. V. Okhlobystina, V. M. Khutoretskii, Izv. Akad. Nauk,
     SSSR, Ser. Khim. (Eng. transl.), 1095 (1969).
27. N. C. Deno et al., J. Am. Chem. Soc., 93, 438 (1971).
28. N. C. Deno et al., J. Am. Chem. Soc., 92, 1451 (1970).
29. M. W. Rathke, A. Lindert, Tetrahedron Letters, 3995 (1971).
30. J. G. Gleason, D. N. Harpp, Tetrahedron Letters, 3431
     (1970).
31. T. R. Beebe, J. W. Wolfe, J. Org. Chem., 35, 2056 (1970).
32. H. M. Relles, J. Org. Chem., 37, 3630 (1972).
33. C. G. Venier, H. J. Barager, III, Chem. Commun., 319 (1973).
34. R. E. Olsen, U.S. 3,714,168, Jan. 30, 1973; C. A., 78,
     124,463 (1973).
35. G. Morel et al., Tetrahedron Letters, 1031 (1971).
36. F. Lautenschlaeger, N. V. Schwartz, J. Org. Chem., 34, 3991
37. M. Fieser, L. F. Fieser, Reagents for Org. Syn., 3, 36
     (1972).
38. H.-W., 9, 1955, p. 392.
39. A. G. Beaman, R. K. Robins, J. Am. Chem. Soc., 83, 4038
     (1961).
40. M. Schlosser, G. Heinz, Chem. Ber., 102, 1944 (1969).
41. D. J. Pasto, K. McReynolds, Tetrahedron Letters, 801 (1971).

D. Aromatic Substitution ($\underline{1}$, 392)

   1. From Aromatic Hydrocarbons

$$ArH \xrightarrow{X_2} ArX + HX$$

A significant advance in the halogenation of arenes and other compounds consists of thallation followed by treatment with bromine in the case of bromides and with KI in the case of iodides. These reactions have been reviewed [1]. In the preparation of bromides, thallium triacetate and an arene equal in reactivity to or more reactive than benzene are first heated together. The intermediate, arylthallium diacetate, may be isolated or immediately reacted with bromine [2]. Over 36 examples have been studied. Thallium tribromide tetrahydrate may also be used to brominate arenes [3]. In the preparation of iodides the arylthallium compound is treated with aqueous potassium iodide or with iodine, if anhydrous conditions are desired [4].

Thallium(III) (trifluoroacetate) is used to extend the scope of thallation to some deactivated aromatic compounds. Typical results are as follows: toluene gave 98% iodotoluenes (91% p, 9% o); chlorobenzene gave 80% chloroiodobenzene (77% p, 23% o). The orientation may be manipulated to a certain extent as indicated for phenylethyl alcohol [5]

With benzoic and phenylacetic acids in the thallation process, considerable ortho substitution is obtained. In the preparation of chlorobenzenes, cupric or cuprous chloride has been used to decompose the arylthallium dichloride [6].

A second advance consists of the iodination of arenes, more active than benzene, by copper chloride in the presence of an iodide salt, preferably hydrated ferrous iodide [7]. Typical yields are 65% for the preparation of iodobenzene, 81% for iodotoluene (53% o, 47% p). Interestingly, anthracene gave 9-chloro- rather than 9-iodoanthracene.

A number of rather specialized but ingenious halogenations have been described. To obtain pure 4-chloroxylene (free from 3-chloro), it is best to form the sulfone from o-xylene first

and chlorinate it in the pure form by irradiation in chloroform
[8]. This technique may have broader applications. Side-chain
chlorination is avoided if N-chloropolymaleinimide, rather than
NCS, is used [9]. For instance, cumene gave 88% of 4-chloro-
cumene with no trace of α-chlorocumene. The backbone of the
polymer no doubt furnishes a polar environment for ionic
chlorination to occur. Any halocyclooctatetraene is available
by forming the 1,2-dihalo adduct at -60° with either bromine or
chlorine and treating this with t-BuOK at -45° [10]. The iodo
compound is made from cyclooctatetraenelithium and iodine [8] and
the fluoro compound, from bromocyclooctatetraene and silver
fluoride in pyridine [11]. Benzene may be chlorinated in aqueous
solution by stirring 200 ml of the arene with 500 ml of 6N HCl
and a small volume of acetic acid, while 0.5 mol of hydrogen
peroxide is added at a rate of 1 ml/hr. After continuous
stirring for 8 hr and then workup, 50-55% of chlorobenzene is
obtained based on the peroxide [12]. Another peroxide oxidation
is conducted with anhydrous aluminum chloride at 0° to give 10%
phenol and 28% chlorobenzene [13].

Other reagents used for halogenation are α,α,α-tribromo-
acetophenone and AlCl$_3$ [14], bromine with tetramethylammonium
bromide in acetic acid (to brominate 1-methylnaphthalene) [15],
and bromine in liquid sulfur dioxide with a catalytic amount of
aluminum chloride [16]. p-Bromofluorobenzene was obtained in
high yield by this method. Other reagents are trichloroisocyanuric acid with H$_2$SO$_4$ or FeCl$_3$ catalyst suitable for phenols,
arylamines [17], and tetrafluoroxenon [18]. With benzene there
was spontaneous gas evolution with the latter and isolation of
13% fluorobenzene with a trace of biphenyl.

1,6- and 1,8-Dibromopyrene are best made from 1-bromopyrene
and limited amounts of bromine [19]. They both tend to give
1,3,6-tribromopyrene. The isomerization of halobenzenes with
aluminum chloride (1, 393) continues to be studied [20].

2.   From Phenols and Phenol Ethers (1, 396)

Accomplishments here appear to consist of finding greater
selectivity for certain positions in the phenolic ring by choice
of catalysts or conditions. For salicylic acids, the titanium
tetrachloride present in bromination tends to increase the
amount of 3-bromosalicylic acid [21]. Without the catalyst 5-
bromosalicylic acid is the predominant product. These results in
all probability indicate that the attack at the 3-position is
favored by a cyclic mechanism involving bromine and the complex
of the acid with titanium tetrachloride. Thallium salts do not
seem as subject to steric hindrance effects in phenol or anisole
substitution such as in the bromination of

OCH$_3$ With thallium salts and bromine, as much as 40% of 4-
bromo-3-t-butylanisole is obtained. Without thallium
salts, the 6-bromoisomer is predominant [22].
   The yield is increased to 73% for the dibromination of
the dimethyl ether of hydroquinone by adding acetic acid and
potassium acetate [23]. Without potassium acetate the yield is
40%. The yield of 3-bromoresorcinol dimethyl ether was high
when the ether was treated with the complex of bromine and
dioxane at -20° [24]. The chlorination of bisphenols by sulfuryl
chloride is catalyzed by methyl mercaptan or other thiol compounds [25]. Apparently, CH$_3$SCl is the chlorinating agent. The
isomerization of 4-bromo-m-cresol by hydrogen bromide has been
shown to give an equilibrium mixture of about 20% 2-, 20% 4-, 20%
6-bromo-m-cresol, m-cresol, and a mixture of dibromo-m-cresols
[26].

3. From Anilines or Anilides (1, 398)

   Dialkylanilines have been brominated in the para position
in high yield by tetrabromocyclohexadienone in methylene
chloride at -20° [27]. 2-Aminophenanthrene has been brominated
quantitatively by hydrogen bromide, DMSO, and water to 2-amino-
1-bromophenanthrene [28]. 4-Nitroaniline has been iodinated to
2-iodo-4-nitroaniline in 96% yield by heating the substrate and
iodine in chlorobenzene and water at 90-94° for 35 hr while
hydrogen peroxide was added [29].
   The hydroxylamine rearrangement has been examined in detail
to maximize yields of p-fluoroanilines with the following
results [30]:

For success R must not be in the para position; in addition, no
ortho-fluoronation should be observed.

4. From Selected Heterocyclic Compounds (1, 400)

   The preparation of 3-bromoquinoline and 4-bromoisoquinoline
has been facilitated by brominating each of the respective substrates, as hydrochlorides, in nitrobenzene at 180° for 4-5 hr
[31]. The rearrangement of 2-bromo- to 3-bromothiophenes has
been brought about by reaction of the former with sodamide in
liquid ammonia for 10 min [32]. Yields for a series of bromo-
thiophenes ranged from 64-87%. Iodothiophenes gave poorer

rearrangement yields, and chlorothiophenes did not rearrange. Tetraiodothiophene is reduced to appreciable and separable quantities of 3-iodo- and 3,4-diiodothiophene by aluminum amalgam in alcohol [33]. To prepare 3-iodofuran, the following sequence is used [34]:

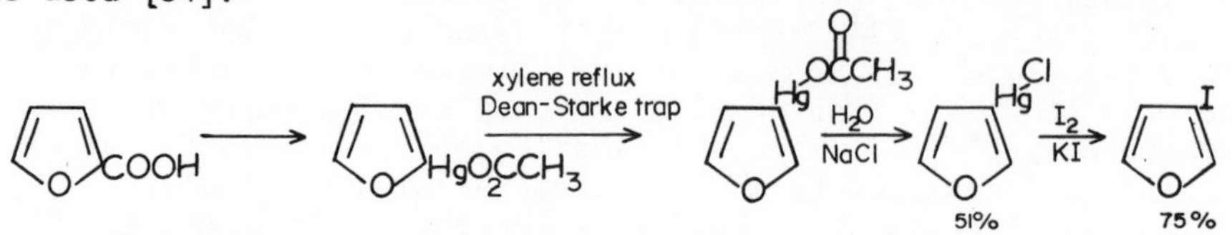

The 3-iodofuran contains about 5% of the 2-isomer.

The best reagent for chlorinating carbazoles appears to be 1-chlorobenzotriazole [35]. Yields with this reagent in methylene chloride were 79% for 3-chloro-, 64% for 3,6-dichloro-, and 61% for 1,3,6,8-tetrachlorocarbazole. Indoles behaved as follows [36]:

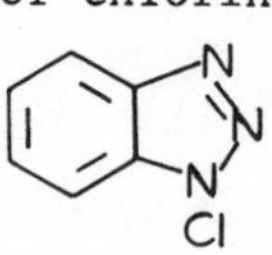

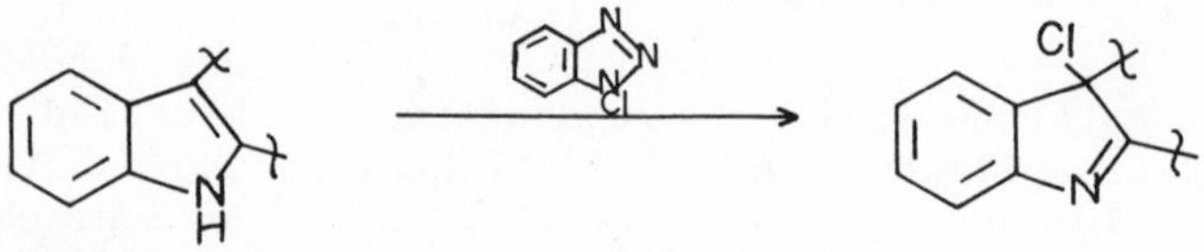

Uracil was fluorinated to 5-fluorouracil (85%) with fluoroxytrifluoromethane, CF_3OF [37].

5. From Aromatic Compounds with Electron-withdrawing Groups (1, 402)

Best conditions for iodinating nitrobenzenes have been devised [38]. Iodine is first stirred with 20% oleum and the nitrobenzene added to the stirred mixture. At 25° in 20 hr nitrobenzene gave only m-nitroiodobenzene (52%), and under the same conditions p-nitrotoluene with 2 equiv of iodine gave 2,6-diiodo-4-nitrotoluene (76%). No iodine atom could be substituted ortho to the nitro group. The attacking species is probably I_3^+. Another way to iodinate deactivated aromatic compounds is to add a mixture of nitric and sulfuric acids to the substrate and iodine in acetic acid [39]. 3,3'-Diiodobenzophenone and 2,7-diiodophenathraquinone were synthesized in this way.

A selective ortho chlorination of azobenzene has been carried out as follows [40]:

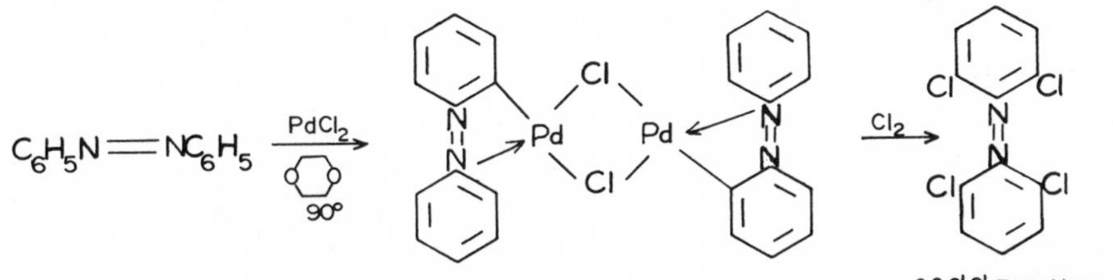

$$C_6H_5N = NC_6H_5 \xrightarrow[\underset{90°}{\bigcirc}]{PdCl_2}$$

2,6,2',6'-Tetrachloro-
azobenzene, 39%

Yields were higher where hydrogen chloride was swept out.
2-Chloroazobenzene was made in 68% yield by making the chlorine
the limiting reagent.

6. From Aromatic Compounds by Haloalkylation and the Like
    (1, 404)

$$ArH + CH_2O + HCl \longrightarrow ArCH_2Cl + H_2O$$

For deactivated arenes, conversion is good if fuming
sulfuric acid (60%) is added dropwise to the substrate dissolved
in an excess of chloromethyl methyl ether (danger, a carcinogen!)
[41]. The yield of 3,5-bis(chloromethyl)-1-nitromesitylene was
88%. Zinc chloride has been suggested as the catalyst for
chloromethylation of nitrophenols [42]. Mono- or dichloro-
methylation of the dibenzyl ether of a catechol has been carried
out with paraformaldehyde and dry HCl in acetic acid [43].
Arenes have been halomethylated with 1,4-bis[halomethoxy]-butane
or 1-chloro-4-halomethoxybutane in the presence of ZnBr$_2$ or
SnCl$_4$ at 60-65° for 3 hr (41-96%) [44].
Chlorovinylation of benzene has been accomplished as
follows [45]:

$$C_6H_6 + Cl_2 = CHPd\left[P(C_6H_5)_3\right]_2 Cl + AgOAc \xrightarrow[\text{Reflux 1.5 hr}]{\text{HOAc}} C_6H_5CH = CCl_2$$

excess

β,β-Dichlorostyrene, 80%

The silver acetate removes the chloride ligand and permits the
benzene to coordinate with the complex.
Chloromethylation of alkenes also occurs as shown (H.-W., 5,
Pt. 3, 1962, p. 1004):

$$CH_3CH = CH_2 + HCl + H_2C = O \longrightarrow CH_3CHClCH_2CH_2Cl$$

1,3-Dichlorobutane

7.   From Aromatic Compounds by Friedel-Crafts

A Friedel-Crafts alkylation has been conducted as shown [46]:

$$C_6H_5CHOHCCl_3 + C_6H_5Br \xrightarrow[\substack{30\,min.\ at\ 10° \\ 5\ hr.\ at\ 25°}]{H_2SO_4\text{-}SO_3} C_6H_5CHC_6H_4Br-p$$

l-*p*-Bromophenyl-l-phenyl-2,2,2-
trichloroethane, 59-74%

The extensive work of Yakobson in the perfluoroarene field is represented by a Friedel-Crafts type of condensation [47]:

$$\xrightarrow[SbF_5,\ 90°,2\,hr.]{S}$$

Perfluorodiphenylsulfide, 95%

References

1. E. C. Taylor, A. McKillop, Acc. Chem. Res., 3, 338 (1970).
2. A. McKillop, E. C. Taylor et al., Tetrahedron Letters, 1623 (1969); J. Org. Chem., 37, 88 (1972).
3. K. Ichikawa et al., Bull. Chem. Soc. Jap., 44, 2490 (1971).
4. A. McKillop et al., Tetrahedron Letters, 2427 (1969); J. Am. Chem. Soc., 93, 4841 (1971).
5. E. C. Taylor, A. McKillop et al., J. Am. Chem. Soc., 93, 4845 (1971).
6. K. Ichikawa et al., Chem. Commun., 169 (1971).
7. W. C. Baird, Jr., J. H. Surridge, J. Org. Chem., 35, 3436 (1970).
8. B. Miller, J. Org. Chem., 38, 1243 (1973).
9. C. Yaroslavsky, E. Katchalski, Tetrahedron Letters, 5173 (1972).
10. R. Huisgen et al., Chem. Ber., 104, 2412 (1971).
11. R. Huisgen et al., ibid., 2405 (1971).
12. H. I. X. Mager, W. Berends, Rec. Trav. Chem., 91, 630 (1972).
13. M. E. Kurz, G. J. Johnson, J. Org. Chem., 36, 3184 (1971).
14. B. I. Stepanov, V. F. Traven, J. Org. Chem. USSR (Engl. transl.), 5, 374 (1969).
15. J.-M. Bonnier, J. Rinaudo, Bull. Soc. Chim. France, 2092 (1971).
16. J.-P. Canselier, Bull. Soc. Chim. France, 1785 (1971).
17. E. C. Juenge et al., J. Org. Chem., 35, 719 (1970).
18. N. C. Yang et al., J. Org. Chem., 35, 4020 (1970).

19. J. Grimshaw, J. Trocha-Grimshaw, J. Chem. Soc., Perkin Trans., I, 1622 (1972).
20. A. A. Spryskov et al., Izv. Vyssh. Ucheb. Zaved. Khim. Tekhnol., 14, 79 (1971); C. A., 75, 5368 (1971).
21. M. V. Sargent et al., Chem. Commun., 214 (1972); J. Chem. Soc., Perkin Trans., I, 340 (1973).
22. K. L. Erickson, H. W. Barowsky, Chem. Commun., 1596 (1971).
23. G. C. Misra et al., Labdev, Part A8, 217 (1970); C. A., 74, 111681 (1971).
24. K. L. Rinehart, Jr. et al., J. Org. Chem., 35, 849 (1970).
25. I. M. Bilik et al., J. Org. Chem. USSR, 5, 334 (1969).
26. E. J. O'Bara et al., J. Org. Chem., 35, 16 (1970).
27. V. Calo et al., J. Chem. Soc., C, 3652 (1971).
28. H.-L. Pan, T. L. Fletcher, Synthesis, 610 (1973).
29. I. Toth, Helv. Chim. Acta, 54, 1486 (1971).
30. T. B. Patrick et al., J. Org. Chem., 39, 1758 (1974).
31. T. J. Kress, S. M. Costantino, J. Heterocyl. Chem., 10, 409 (1973).
32. M. G. Reinecke et al., J. Org. Chem., 36, 2690 (1971).
33. H.-W., 5, 1960, Pt. 4, p. 775.
34. R. A. Bell et al., Can. J. Chem., 50, 3749 (1972).
35. A. Ledwith et al., J. Chem. Soc., C, 2775 (1971).
36. K. V. Lichman et al., J. Chem. Soc., C, 2539 (1971).
37. R. H. Hesse et al., J. Org. Chem., 37, 329 (1972).
38. A. C. Darby et al., J. Chem. Soc., C, 1480 (1970).
39. V. T. Slyusarchuk, A. N. Novikov, Izv. Tomsk. Politekh. Inst., 196, 138 (1969); C. A., 73, 98672 (1970).
40. D. R. Fahey, J. Organometal. Chem., 27, 283 (1971).
41. H. Suzuki, Bull. Chem. Soc. Japan, 43, 3299 (1970).
42. V. Böhmer, J. Deveaux, OPPI, 4, 283 (1972).
43. D. I. Lerner, C. R. Dawson, J. Org. Chem., 38, 2096 (1973).
44. G. H. Olah et al., Synthesis, 560 (1974).
45. I. Moritani et al., J. Organomet. Chem., 27, 279 (1971).
46. A. B. Galun. A. Kalir, Org. Syn., 48, 27 (1968).
47. G. G. Yakobson et al., J. Org. Chem. USSR (Engl. transl.), 10, 802 (1974).

E. Miscellaneous

1. From Polyhalides (Reduction or Elimination) (1, 408)
   (See E.1 Addendum)

A number of useful, miscellaneous reductions of polyhalides have been published. Benzotrichloride was reduced very rapidly at 25° to benzal dichloride with hexaethyltriaminophosphine in the presence of ethanol [1]. gem-Dibromocyclopropanes were reduced in good yield to the monobromo derivative by Cr(II) acetate in DMSO [2] and 1,1,1-trichloroalkanes to 1,1-dichloro-alkanes by a mercaptan and iron pentacarbonyl [3]. A chlorine atom α to a carbonyl group was selectively replaced by hydrogen in the steroid family [4]:

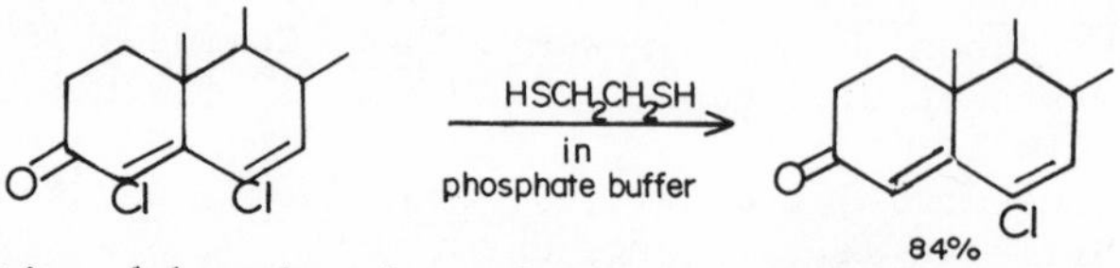

Selective debromination of tetrabromothiophene to give the 3,4-dibromothiophene in 75% yield was accomplished by Mg in Et_2O to which ethylene bromide was added [5]. Pentabromobenzene was made in 95% yield from hexabromobenzene by treating a slurry of the latter in THF at 0° with phenylmagnesium bromide and then decomposing the C_6Br_5MgBr with water [6]. 1,2,4,5-Tetrabromo-benzene was prepared similarly in 51% yield from hexabromoben-zene and 2 equiv of ethylmagnesium bromide [6]. 1,2,3,4-Tetra-bromobenzene was made in 55% yield from hexabromobenzene by refluxing with hydrazine in ethanol for 24 hr [7]. A rather interesting reduction which accompanies condensation is illus-trated [8]:

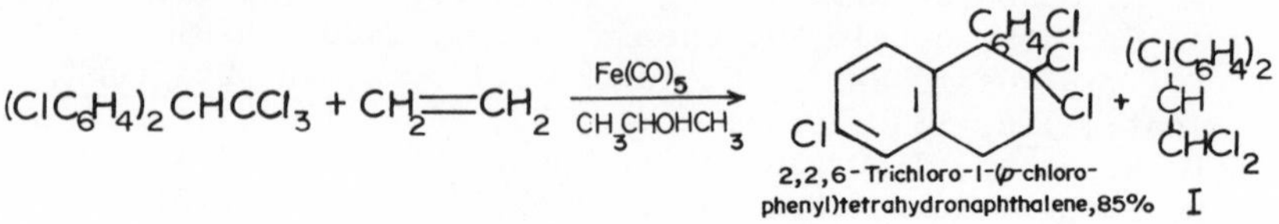

Dichlorodiphenyltrichloroethane has also been reduced to I (92%) by nickel tetracarbonyl in THF at 40° [9].

The ease of dehydrofluorination is indicated by the follow-ing results [10]:

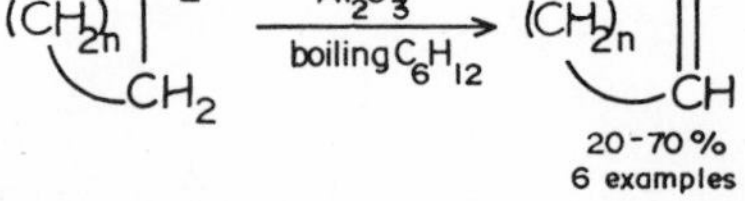

Coupling of some haloalkenes may be brought about as shown (H.-W., 5, Pt. 3, 1962, p. 998):

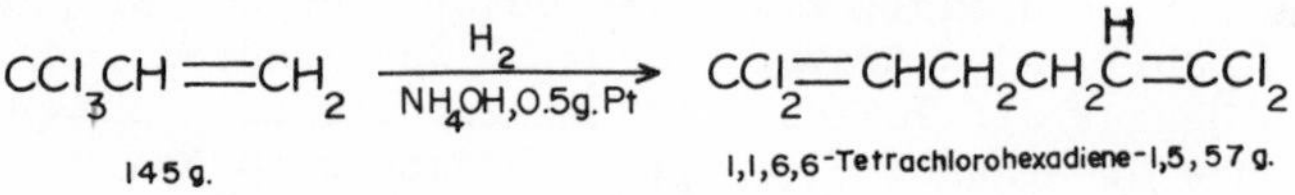

4. From Halides (Rearrangement) or Halocyclopropanes (Ring Opening)

The rings in cyclobutylidenemethyl halides have been expanded by adding the substrate dropwise to solid potassium t-butoxide at 100° [11]:

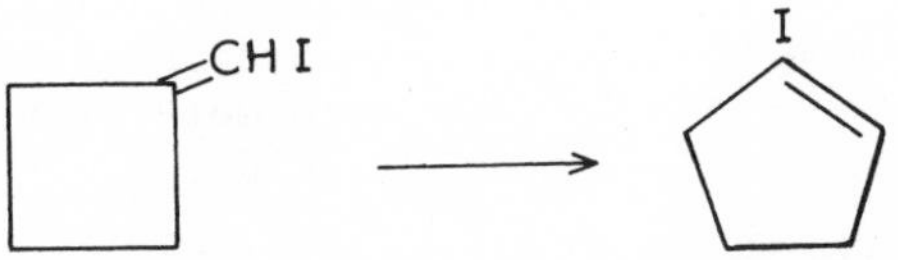

1-Iodocyclopentene
64-70%

The corresponding chloride gave 48-52% and the bromide, 45-55% of the halocyclopentene. Other intramolecular rearrangements are to be found in the literature (H.-W., 5, Pt. 3, 1962, p. 351) and [B.5, 44].

Rather than utilizing nucleophilic reagents, Sandler has cleaved the rings of dibromopropanes by thermal means at 195° for 5 hr [12]:

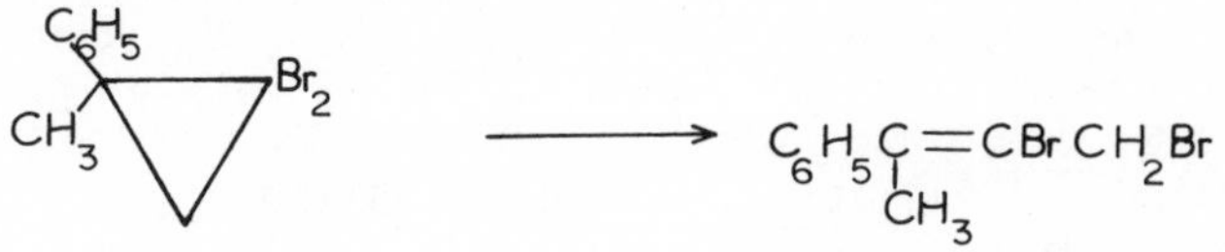

$C_6H_5C=CBrCH_2Br$
$\quad\quad |$
$\quad\quad CH_3$

2-Bromo-3-phenylmethallyl bromide
87%

Ring opening of fluorinated cyclobutanes is described elsewhere (H.-W., 5, Pt. 3, 1962, p. 348).

5. From Halides (Oxidation)

Occasionally, oxidation is used to prepare halides as shown [13]:

$$7\ CF_2ICF_2I + 7\ HF + H_5IO_6 \xrightarrow[\text{pressure}]{150°} 7\ CF_3CF_2I + 4\ I_2 + 6\ H_2O$$

Perfluoroacids frequently are prepared by oxidation of the corresponding perfluoroalkene (H.-W., 5, Pt. 3, 1962, p. 363).

Addenda

E.1. Elimination of hydrogen bromide by a strong base from several 1,2-dibromocycloalkanes gave high yields of seven-membered 1-bromocycloalkenes; the yields were lower in the case of smaller 1-bromocycloalkenes [P. Caubere et al., Tetrahedron, 30, 1289 (1974)].

E.4. The removal of bromine or iodine from polyhaloaromatic compounds with t-BuOK in DMSO-t-BuOH takes place by a direct attack of the dimsyl anion on the halogen, preferably one adjacent to another. The equilibrium between 1,2,4- and 1,3,5-tribromobenzene in the presence of potassium anilide in liquid ammonia is accelerated by 1,2,4,5-tetrabromobenzene. Halogen exchange and disproportionation are much more frenzied in HMPA-t-BuOK [J. F. Bunnett, Acc. Chem. Res., 5, 139 (1972)].

References

1.   I. M. Downie, J. B. Lee, Tetrahedron Letters, 4951 (1968).
2.   T. Shirafuji et al., Bull. Chem. Soc. Jap., 44, 3161 (1971).
3.   R. G. Petrova, R. Kh. Freidlina et al., Izv. Akad. Nauk, SSSR, Ser. Khim (Engl. transl.), 1483 (1970).
4.   R. A. Le Mahieu et al., J. Am. Chem. Soc., 93, 1664 (1971).
5.   M. Janda et al., Synthesis, 545 (1972).
6.   C. F. Smith et al., J. Organomet. Chem., 33, C21 (1971).
7.   I. Collins, H. Suschitzky, J. Chem. Soc., C, 2337 (1969).
8.   R. Kh. Freidlina et al., Dokl. Akad. Nauk, SSSR (Engl. transl.), 212, 744 (1973).
9.   T. Kunieda et al., Chem. Commun., 885 (1972).
10. D. R. Strobach, G. A. Boswell, Jr., J. Org. Chem., 36, 818 (1971).
11. K. L. Erickson et al., J. Org. Chem., 36, 1024 (1971).
12. S. R. Sandler, Chem. Ind. (London), 1481 (1968).
13. H. Millauer, Angew. Chem. Intern. Ed. Engl., 12, 929 (1973).

Chapter 8

AMINES

Such a diverse subject as the synthesis of amines is bound to humble the chronicler who attempts a complete coverage. Selections had to be made based on the authors' experiences and their basic understanding and judgment, all of which have limitations.

For the reduction of nitro compounds (A.1) and nitriles (A.3) to amines considerable effort has been expended in modifications to produce a more effective reduction. Sulfurated sodium borohydride is one of the new reagents (A.1) and another (A.6), particularly for the reductive alkylation of carbonyl compounds, is sodium cyanoborohydride. A third, a combination of sodium borohydride and Raney nickel, has proved to be effective in the reduction of nitriles (A.3). Of all of these diborane appears to be the most general. It is the best for amides (A.4) and fortunately it may now be prepared in situ (Chapter 1, A.3, Ref. 23; Chapter 4, B.2, Ex. b). A rather unique trap in neutral solution for an amine is hydrochloric acid generated by hydrogenation of chloroform (A.3, Ref. 32). And the reduction of a cyanohydrin, which is difficult to form, is accomplished via the trimethylsilyoxy derivative (A.3, Ref. 33).

A Gabriel reaction has been conducted with an alcohol rather than the halide as the substrate (B.2). A mixture of acetic and sulfuric acids appears to be capable of hydrolyzing a number of sulfonamides to amines (B.4). This is a significant advance, particularly if it turns out to be general.

Since 1,2,2,6,6-pentamethylpiperidine forms an acid salt readily and resists alkylation, it serves admirably in removing acid during the exhaustive alkylation of amines (C.1). Evidence is now available that direct replacement of the halogen by an amino group on the less aromatic heterocycles is quite devious in that addition, ring-opening, and ring-closure to the amino-heterocycle are involved (C.1). Displacement of a phosphate group in an aryl phosphate by an amino group in blue liquid ammonia is doubtless a free radical reaction of synthetic merit (C.4). The scope of the alkylation of 2-, 3-, or 4-haloquinolines by Grignard reagents has been broadened by the discovery that the reaction is nickel-salt catalyzed (C.9).

A new, perhaps more searching, Mannich reagent, $Me_2\overset{+}{N}=CH_2$, has been reported (D.3). Myosmine has been synthesized from nicotinylpyrrolidone by an operatively simple reaction (D.8), and ortho-alkylated anilines have been produced (D.10) by a cycloaddition of a sulfur ylid. The addition of an optically active sulfoxide to a nitrile followed by reduction represents an important method for the synthesis of optically active amines (E.2).

Amination of aliphatic hydrocarbons and alkyl halides with nitrogen trichloride is described (F.2). The reactions of nitrenes (F.4) and nitrenium ion-free radicals (F.8) have been delineated largely as a result of the publication of reviews.

Among molecular rearrangements attempts have been made to cite at least one recent application (G.2, G.3, G.4, G.5). A new, quite mild synthesis of indoles via cyclization of a sulfur ylid is described (G.7).

The chapter concludes with the addition of a new section on the use of enamines in amine synthesis (I). The chemistry of this type is being investigated actively at the present time.

A.   Reduction

1.   From Nitro Compounds ($\underline{1}$, 413)

$$ArNO_2 \longrightarrow ArNH_2$$

Typical reducing agents and conditions were discussed previously ($\underline{1}$, 413). An electrolytic reaction was not included then, but since the nitro group is one of the most easily reduced of all groups electrochemically, one is given now (Ex. a).

Hydride reduction of the nitro-to-amino group has not been noticeably prominent since basic conditions lead to intermediate reduction products. Also, sodium borohydride does not reduce nitro compounds, and $LiAlH(OMe)_3$ and $NaAlH_2(OC_2H_4OMe)_2$ react only slowly [1]. Nitroarenes and sodium borohydride in DMSO at 85° give azo- and azoxybenzenes as well as the aniline [2]. With other modified hydrides, some success has been achieved. Sulfurated sodium borohydride, $NaBH_2S_3$, is claimed to be a selective reducing agent of the nitro group [3]:

$$\text{NaBH}_4 + 3\text{ S} \xrightarrow[25°]{\text{THF}} \underset{\substack{\text{I eq.} \\ + \text{H}_2}}{\text{NaBH}_2\text{S}_3} \xrightarrow{\text{ArNO}_2, \text{I eq.}} \text{ArNH}_2$$

Yields were around 80%, with no interference from halogen, ester, nitrile, or an olefin group. A combination of sodium borohydride and cobaltous chloride also reduces nitro compounds in low yields [4], which have been traced to the fact that azoxy compounds are the main products [5]. For reduction to the amine, cupric chloride is superior to cobaltous chloride.

The catalytic hydrogenation method (1, 414) is a very fine one (see Ex. b), but recent progress concentrates more on modified procedures. For instance, it is recommended that the best general synthesis of β-arylethylamines is via the α,β-unsaturated nitro compounds [6]:

$$\text{ArCH} = \text{CHNO}_2 \xrightarrow[\text{H}_2(800\text{ psi}), \text{Pd/C}]{\text{aq. HCl}} \text{ArCH}_2\text{CH}_2\text{NH}_2\cdot\text{HCl}$$

Platinum sulfide on carbon with hydrogen at high pressure is a sluggish reductive system for nitro compounds, but has the advantage of low susceptibility to poisons [7]. With this catalyst some nitroarylsulfides have been reduced to the corresponding aminoarylsulfide. Colloidal iridium with hydrogen at atmospheric pressure and room temperature appears to give fair-to-good yields of hydroxylamines (20-90%) if the reduction is stopped when 2 equiv of hydrogen have been absorbed [8]. Another reaction which apparently forms the intermediate hydroxylamine is the following [9]:

$$\underset{\text{271 g.}}{\text{C}_6\text{H}_5\text{NO}_2} + \underset{\text{205 g.}}{\text{H}_2\text{SO}_4} + \underset{\text{1700 ml.}}{\text{H}_2\text{O}} \xrightarrow[\overset{\oplus}{\text{C}_{12}\text{H}_{25}}\overset{\ominus}{\text{NMe}_3\text{Cl}}]{\text{Pt/C}, 1.4\text{g.}, \text{H}_2, 5\text{hr.}} \underset{p\text{-Aminophenol}, 154\text{ g.}}{\text{NH}_2\text{---}\langle\ \rangle\text{---OH}}$$

The intermediate hydroxylamine is rearranged to the final product by the acid.

A number of chemical reducing agents have been described. Hydriodic acid, 57%, at 140-150° gave amines in 22-79% yields [10]. Sodium in methanol and liquid ammonia or lithium in methylamine occasionally gave anilines, but more often gave di- or tetrahydroanilines with substituted nitrobenzenes [11]. 2-Aminoethanol was found to be an interesting reducing agent [12]:

6-β-Hydroxyethylamino-2-amino-
phenanthridine, 90 %

Sodium hydrogen sulfide in boiling metnanol-toluene has been found to be selective in reducing dinitrobiphenyls to nitro-aminobiphenyls [13].

Two metal carbonyl systems using quite contrasting conditions have been described: (a) carbon monoxide and water with a rhodium salt and pyridine at 120 atm pressure and 150° [14] and (b) $Fe_3(CO)_{12}$ + methanol refluxed 10-17 hr in C_6H_6 [15]. However the most versatile of chemical reducing systems is the Cadogan reaction consisting of refluxing the nitro compound with some trivalent species of phosphorus [16]. The reaction intermediate is more than likely the nitrene (1, 487), but it may be variable. When 2-nitrophenylmesitylene is refluxed in trimethyl phosphite, the aniline or its complex with trimethylphosphite is the major product. If the reaction is carried out in cumene, where reagents are more dilute, 8,10-dimethylphenanthridine, an insertion product, is found to the extent of 13%. With $(EtO)_2PMe$ and nitrobenzene in diethylamine, 2-diethylamino-3-phenyl-3H-azepine was formed in 83% yield. The Cadogan method is discussed further in F.4. Reducing activity of phosphorus compounds is in the order: $(EtO)_2PMe > (Et_2N)_3P > (EtO)_3P > PCl_3$ (PCl_3 is inactive)

a. Preparation of p-Aminobenzoic Acid Hydrochloride [17]

The anode compartment consisted of a porous cup with Pt gauze anode (5 x 5 cm), and an anolyte of 35.6% H_2SO_4 (30 ml). The cathode cell consisted of lead with a thin layer of spongy lead and 500 ml of 8.7% hydrochloric acid in a one-liter beaker. p-Nitrobenzoic acid (15 g) was added to the catholyte at 70° and current density of 6 A/dm^2 was applied until more than the theoretical amount of current had reduced the acid. The catholyte solution was evaporated to give 92% of the product.

b. Preparation of cis-1,3,5-Triaminocyclohexane [18]

This was accomplished by catalytic Pd reduction of 1,3,5-trinitrobenzene in propionic anhydride to the tripropionylamino-benzene (79%) and a second reduction with hydrogen and Nishimura catalyst to tripropionylaminocyclohexane (75-80%); separation of cis-trans isomers as hydrochlorides and saponification.

2. From Nitroso, Amine Oxide, Azo, Hydrazino, Azido, Imino, and Related Compounds (1, 417)

a. From Nitroso Compounds

$$ArNO \longrightarrow ArNH_2$$

p-Dinitrosobenzene has been prepared in over 90% yield by the oxidation of p-quinone dioxime in water either with $FeCl_3$ or Cl_2 [19].

Although aromatic nitroso compounds are usually reduced by hydrogen and a catalyst, they may be reduced by diborane as well [20]. From a number of nitrosoarenes and diborane, aryl amines were produced in yields of 62 to 90%. Diborane with aliphatic gem-nitrosonitro compounds produced the hydroxylamine [20]:

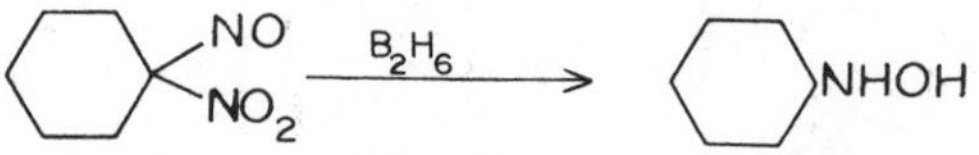

N-Cyclohexylhydroxylamine, 71 %

Aromatic nitrosoamines have been reduced with iron pentacarbonyl [21]:

$$ArN(NO)R \xrightarrow[\text{Refluxing } (C_4H_9)_2O]{Fe(CO)_5} ArNHR + Fe(CO)_4 + CO_2$$

b. From Amine Oxides

$$R_3NO \longrightarrow R_3N$$

First, to synthesize the amine oxide a convenient preparation, which requires no basification or extraction and which offers excellent yields, is now available. In it one utilizes the tertiary amine with m-chloroperbenzoic acid in chloroform at 0° and passes the reaction product through alkaline aluminum oxide [22]. In a second convenient preparation, in which the reaction time is short and isolation of the anhydrous amine often simple, the tertiary amine is oxidized with t-butyl hydroperoxide at 65° using vanadium oxyacetylacetonate as a catalyst [23].

Second, to reduce amine oxides to amines, one reagent, iron pentacarbonyl, compares favorably with triphenylphosphine

(1, 418) [21]. Another milder reducing agent, dithiophosphoric acid ester, $(RO)_2PS_2H$, is also effective in the reduction [24] and perhaps possesses some of the selective powers of sulfur dioxide mentioned in 1, 418. A third reduction of amine oxides, at least for those of pyridine and quinoline, is an electrolytic process using a mercury cathode (78-96%) [25].

c. From Hydrazones

$$\text{>C=NNH}_2 \xrightarrow{\text{B}_2\text{H}_6} \text{>CHNHNH}_2$$

The preparation of 1,2-disubstituted hydrazines from the corresponding hydrazones is carried out with diborane (Ex. a).

d. From Azides

$$\text{R N}_3 \xrightarrow{[\text{H}]} \text{RNH}_2$$

A very nice synthesis of polyamines utilizes the azide route [26]:

$$\text{ROH} \longrightarrow \text{ROSO}_2\text{C}_6\text{H}_5 \longrightarrow \text{RN}_3 \xrightarrow{\text{LiAlH}_4} \text{RNH}_2$$
$$50\text{-}76\%$$
$$(4 \text{ examples})$$

For example, pentaerythritol gave 62% of tetraaminomethylmethane as the hydrochloride. Steroid azides are reduced to the amines in high yield by sodium borohydride catalyzed by a complex of $CoBr_2$ and α,α'-dipyridyl [27]. For the reaction of alkyl azides with trialkylborons or dialkylchloroboranes, see F.4.

e. From Imines

Imines of benzophenone are reduced photochemically to substituted benzhydrylamines (60-90%) [28]:

$$(\text{C}_6\text{H}_5)_2\text{CCl}_2 + \text{RNH}_2 \longrightarrow (\text{C}_6\text{H}_5)_2\text{C=NR} \xrightarrow[\text{h}\nu]{\text{CH}_3\text{CHOHCH}_3} (\text{C}_6\text{H}_5)_2\text{CHNHR}$$

a. Preparation of 1-Benzyl-2-isopropylhydrazine

Benzylhydrazine, 2.4 g, acetone, 1.2 g, and acetic acid, 2 drops in 200 ml of dry bis-2-methoxyethyl ether were allowed to stand for 40 min to produce the hydrazone. Diborane [29], 2 mol with N_2 was passed into the above solution at ambient temperature. Then dry hydrogen chloride was passed through to give a mixture of mono- and dihydrochloride, which was filtered and basicified (2.6 g) [30].

3. From Nitriles ($\underline{1}$, 419)

$$(R) \, ArCN \longrightarrow (R) \, Ar \, CH_2NH_2$$

Until recently, sodium borohydride has been ineffective in reducing nitriles to amines. Two new procedures using catalysts with the borohydride are proposed. The first employs Raney nickel (Ex. a) and the second, more than 1 equiv of a transition metal salt (Ex. b). Because of the large amount of expensive catalyst employed in the latter, the former preparation (Ex. a) appears to be preferable. Acrylonitrile gave allylamine in 70% yield using the transition metal salt method.

In reduction of nitriles with sodium and liquid ammonia, it has been pointed out that primary and secondary nitriles produce a mixture of the products of both reduction and decyanation [31],

$$RC \equiv N \xrightarrow[\text{liq. NH}_3]{\text{Na}} \underset{\text{reduction}}{RCH_2NH_2} + \underset{\text{decyanation}}{RH}$$

while tertiary nitriles, R_3CCN, give almost exclusively the decyanation product. The decyanation process can be eliminated to the extent of obtaining 30-100% yields of amines from tertiary nitriles by using calcium and liquid ammonia rather than sodium and the liquid.

Ingenious modifications of the hydrogenation and metal hydride methods give wider scope to the preparation of amines from nitriles. One modification is to reduce the nitrile catalytically in the presence of chloroform to form the hydrochloride salt in situ in a neutral medium [32]:

$$\underset{\text{2 mmol.}}{RCN} + \underset{\text{1 ml.}}{CHCl_3} + \underset{\text{3 atm.}}{H_2} \xrightarrow[\text{100 mg. Pd/C}]{\text{50 ml. EtOH}} \underset{\text{High yield}}{RCH_2NH \cdot HCl}$$

A minimum amount of chloroform should be used. The preceding conditions work well for the reduction of substituted ribofuranosides containing acid labile isopropylidene and acetal groups. A second modification is essentially the reduction of a trimethylsilyl cyanohydrin for those ketones that form cyanohydrins with difficulty [33]:

l-Aminomethyl-l-hydroxy-
cyclohexane, 74%

A third modification is described in 1, 420-421.

a. Preparation of Benzylamine (NaBH$_4$—Raney Ni Method) [34]

A solution of 7.6 g of NaBH$_4$ in 25 ml of 8N sodium hydroxide
was added dropwise to a strongly stirred solution of 20.6 g of
benzonitrile in 90 ml of methanol containing 10 g of 50% aqueous
Raney Ni held at 50°. After addition the mixture was stirred
several minutes, the nickel filtered, and the methanol distilled
from the filtrate. The residue was treated with 15 g of solid
KOH and the benzylamine layer was separated, dried, and distilled; yield was 17 g (80%) containing 0.4% benzyl alcohol.

b. Preparation of Benzylamine (NaBH$_4$—Transition Metal Salt
Method [4]

Sodium borohydride (0.5 mmol) was added portionwise to a
stirred solution of 0.05 mol of benzonitrile and 0.1 mol of
cobaltous chloride hexahydrate in 300 ml of methanol at 20°.
After addition, the mixture was stirred for 1 hr and 100 ml of
3 N HCl was added to dissolve the black precipitate. The methanol
was removed by distillation and the residue was made alkaline
with concentrated NH$_4$OH and the amine layer was then extracted
with ether. Distillation of the extract gave 3.6 g (72%) of the
amine.

4. From Amides, Hydrazides, or Isocyanates (1, 421)

$$RCONR'_2 \longrightarrow RCH_2NR'_2$$

Although lithium aluminum hydride is a desirable reducing
agent (1, 421), diborane appears to be equally, if not more,
effective as a general reducing agent for amides. In addition,
diborane in THF has been used to reduce isatins and oxindoles to
indoles in 45-81% yields [35].
Sodium borohydride may be utilized in reducing t-amides to
amines [36] by the addition of C$_5$H$_5$N, but, as with nitriles,
A.3, the reduction appears to become more effective by carrying
it out in the presence of about 2 equiv of a transition metal

salt [4]. Yields of amines ranged from 30-70% and groups such as vinyl, sulfur, or halogen are unaffected. In that connection the in situ preparation of diborane should be considered (Chapter 4, B.2, Ex. b) (Chapter 1, A.3, Ref. 23).

If an amide is recalcitrant to reduction, it may be reduced by converting it first into the imino ether fluoborate [37].

$$RCONR_2 \xrightarrow[CH_2Cl_2]{Et_3O^{\oplus} BF_4^{\ominus}} RC\overset{OEt}{=\!\!=\!\!=}\overset{\oplus}{N}R_2\ BF_4^{\ominus} \xrightarrow[EtOH]{NaBH_4} RCH_2NR_2$$

High yield

5.   From Oximes or Hydrazones (1, 423)

$$(R)ArC\overset{R'}{=\!\!=\!\!=}NOH \longrightarrow (R)Ar\overset{R'}{C}HNH_2$$

A number of reagents accomplish this reduction (1, 423), but it appears that none do so with such versatility as diborane. Diborane in THF at 25° is reported to give the corresponding N-alkyl or aryl substituted hydroxylamines from oximes, but at 105-110° in THF-diglyme it gives the corresponding amines in yields of 62-84% [38]. Sulfurated sodium borohydride, $NaBH_2S_3$, is also reported to be a good reagent for reducing oximes [39a], while magnesium in methanol saturated with ammonium acetate at 50° is less satisfactory because nitrile formation and rearrangement are more frequent [40]. A series of α-aminoketones has been prepared by catalytic reduction of the α-oximino ketones [41]. An oxime of an erythromycin failed to be reduced satisfactorily by catalytic hydrogenation. When $TiCl_3$ was used, this oxime was reduced to the imine, which could be reduced further to the amine [39]:

$$R'C\overset{O}{\diagup}\diagdown_{R}C=NOH \xrightarrow{TiCl_3} R'C\overset{O}{\diagup}\diagdown_{R}C=NH \xrightarrow[MeOH]{NaBH_4} R'C\overset{O}{\diagup}\diagdown_{R}CHNH_2$$

The authors found that the imine could be isolated only in cases where the oxime was quite hindered. The $TiCl_3$ solution acted as its own indicator for reduction (dark → colorless). Reductive alkylation of oximes could be carried out on nonenolizable ketones as shown for benzophenone [42]:

$$(C_6H_5)_2C{=}NOH \xrightarrow[\text{liq.NH}_3]{\text{Na,}} (C_6H_5)_2\overset{\overset{\text{Na}}{|}}{C}{-}NHNa \xrightarrow[\text{2)NH}_4\text{Cl}]{\text{1)Me}_2\text{NCH}_2\text{CH}_2\text{Cl}} (C_6H_5)_2\overset{\overset{\text{NH}_2}{|}}{C}CH_2CH_2N(CH_3)_2$$

blue soln.

3-Amino-3,3-diphenylpropyldimethyl-
amine, 60%

It is of interest to note that hydroxylamine and hydroxyl-amine-O-sulfonic acids form diimide, HN=NH, in situ which can be used to reduce the double bond in unsaturated acids [43].

6.   From Carbonyl Compounds and Amines (Reductive
       Alkylation) (1, 424)

$$\underset{}{>}C{=}O \;+\; RNH_2 \;\rightleftharpoons\; \underset{A}{>C{=}NR} \;\longrightarrow\; >CHNHR$$

Among the metal hydrides used for this reduction, lithium or sodium cyanoborohydride has been found to be ideal. Carbonyl compounds themselves are reduced by this reagent at pH 3-4, but hardly at all at pH 7. At the lower pH the imino intermediate A is preferentially reduced, leading to rather good yields of amines. Borch has carried out over 30 reductive alkylations in 21-90% isolated yields [44]. An organic synthesis preparation is also reported [45]:

$$\bigcirc{=}O \;+\; Me_2NH{\cdot}HCl \;+\; KOH \xrightarrow{\text{Na BH}_3\text{CN}} \bigcirc NMe_2$$

N,N-Dimethylcyclohexylamine, 52-55%

The reaction is experimentally simple and adaptable to the synthesis of primary, secondary, and tertiary amines and α-amino acids.

Reductive alkylation, using cyclohexanone and methylamine (liquid) to form methylcyclohexylamine, has been carried out electrolytically by allowing the two substrates to stand for 6 hr (to form the Schiff base), adding LiCl, and passing a 2-A current through the solution containing platinum electrodes for several hr [46]. A very simple operation to give reductive alkylation is the heating of substituted methylamines in the presence of palladium black [47]:

$$RCH_2NH_2 \xrightarrow[\substack{0.2-0.5\text{g.}\\25-200°\\\text{stir 3-20 hr.}}]{\text{Pd black}} [RCH{=}NH] \xrightarrow{RCH_2NH_2} \underset{\underset{RCH_2NH}{|}}{RCHNH_2} \xrightarrow[+H_2]{-NH_3} (RCH_2)_2NH$$

5-10 g.

Benzylamine in the presence of palladium black at 80° gave 45% dibenzylamine; N-ethylaniline at 150° gave 98% of diethylaniline. N-Heterocycles are also available by this synthesis [47]:

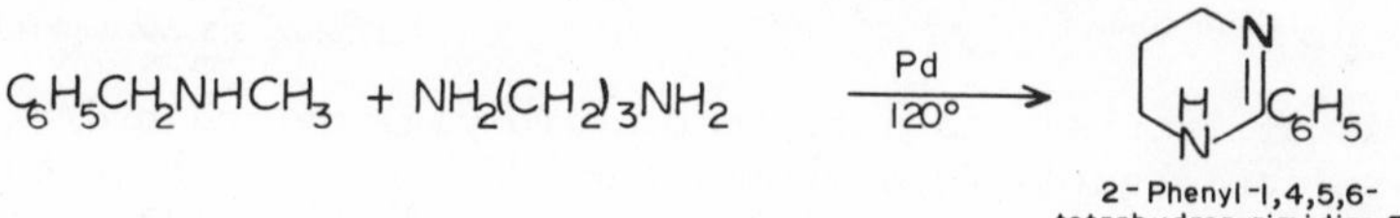

$$C_6H_5CH_2NHCH_3 + NH_2(CH_2)_3NH_2 \xrightarrow[120°]{Pd}$$

2-Phenyl-1,4,5,6-
tetrahydropyrimidine, 75%

The last new reductive alkylation modification described here is essentially a synthesis of carbonyl compounds by the ozonization of olefins, addition of an amine to the carbonyl compounds formed, and then reduction [48]. The key to the success of the synthesis is that at a low temperature (0-15°), partial hydrogenation must be used to reduce the hydroperoxides to the carbonyl compounds. The low temperature avoids free radical decomposition of the hydroperoxide. The ozonization of α-pinene, followed by reductive alkylation, gives a most unique diamine [48]:

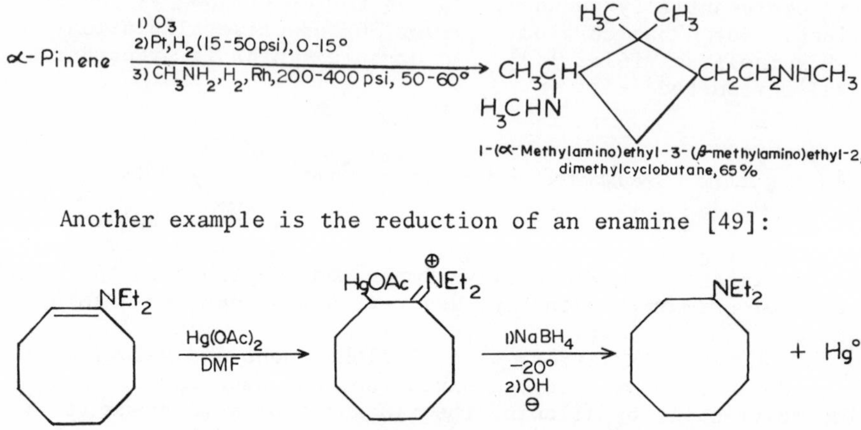

$$\alpha\text{-Pinene} \xrightarrow[\substack{2)\,Pt,H_2\,(15-50\,psi),0-15° \\ 3)\,CH_3NH_2,H_2,Rh,200-400\,psi,\,50-60°}]{1)\,O_3}$$

1-(α-Methylamino)ethyl-3-(β-methylamino)ethyl-2,2-
dimethylcyclobutane, 65%

Another example is the reduction of an enamine [49]:

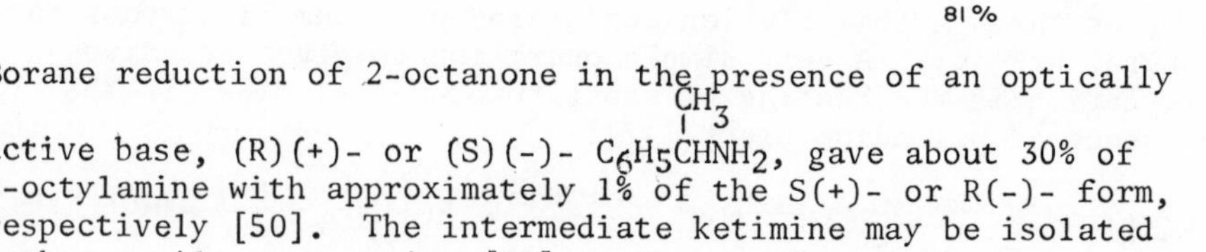

Cyclooctyldiethylamine
81%

Borane reduction of 2-octanone in the presence of an optically active base, (R)(+)- or (S)(-)- C$_6$H$_5$CHNH$_2$ with CH$_3$, gave about 30% of 2-octylamine with approximately 1% of the S(+)- or R(-)- form, respectively [50]. The intermediate ketimine may be isolated rather easily on occasion [51] as the enamine:

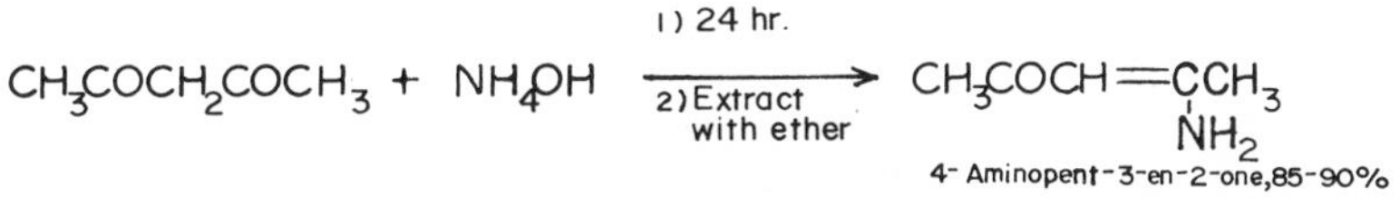

$$CH_3COCH_2COCH_3 + NH_4OH \xrightarrow[\substack{2) \text{Extract} \\ \text{with ether}}]{1)\ 24\ hr.} CH_3COCH = CCH_3$$
$$\underset{NH_2}{|}$$

4-Aminopent-3-en-2-one, 85-90%

One synthesis takes advantage of a favorable cyclodecarboxylation to bring about reductive alkylation [52]:

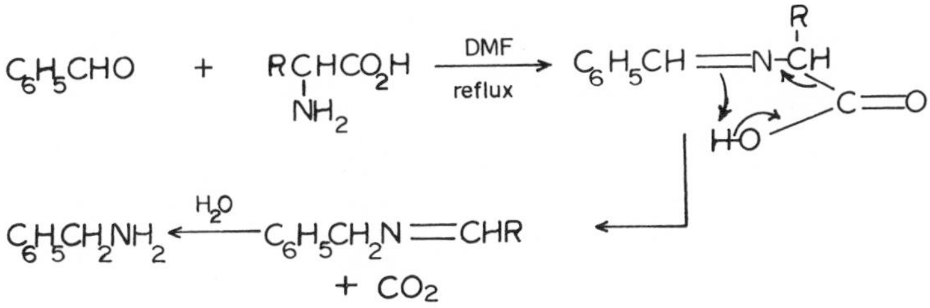

The synthesis works best using isovaline as the amino acid. Although most syntheses in this section start with the carbonyl compound, one may begin with the acid and proceed via the aldehyde (see A.12).

7. From Carbonyl Compounds and Amines (Leuckart-Wallach and Eschweiler-Clarke Reactions)

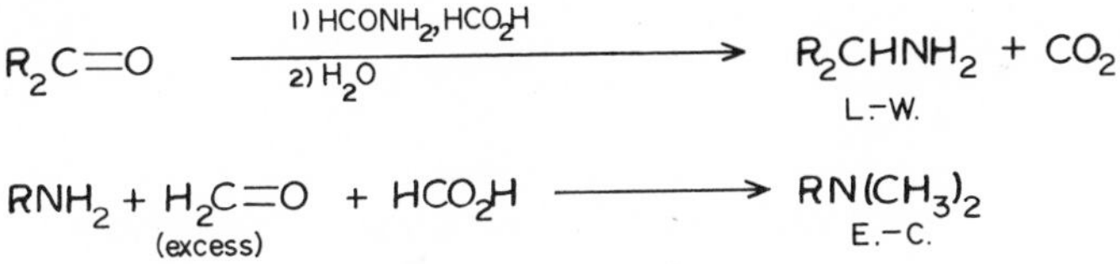

$$R_2C = O \xrightarrow[2)\ H_2O]{1)\ HCONH_2, HCO_2H} R_2CHNH_2 + CO_2$$

L.-W.

$$RNH_2 + H_2C = O + HCO_2H \longrightarrow RN(CH_3)_2$$
$$\underset{(\text{excess})}{}$$

E.-C.

For the Leuckart-Wallach reaction with aliphatic ketones of 15 carbons or more, it has been shown that the addition of nitrobenzene (to give homogeneity), an increase in the amount of formamide, and rapid stirring increased the yield of 9-aminoheptadecane from 40 to 73% [53]. It was also found that the formamide was easily hydrolyzed to the amino salt by stirring and heating with 5-10% HCl [53].

In the Eschweiler-Clarke reaction, a number of experiments are reported in which the reducing agent, formic acid, is replaced by other reducing agents. The sodium cyanoborohydride reagent of Borch [54] comes to mind immediately, and it has been found to be an efficient reducing agent (see Ex. a and b).

Another replacement for the formic acid is sodium borohydride [55].

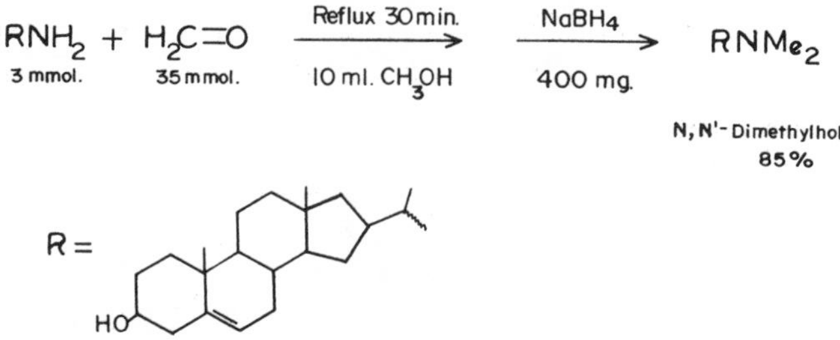

$$RNH_2 + H_2C{=}O \xrightarrow[\text{10 ml. CH}_3\text{OH}]{\text{Reflux 30 min.}} \xrightarrow[\text{400 mg.}]{\text{NaBH}_4} RNMe_2$$

3 mmol.   35 mmol.

N,N'-Dimethylholafebrine
85%

R =

HO

A third replacement is formaldehyde itself, but conditions are so drastically changed that the reaction is designed for only those amines that will survive the treatment [56]:

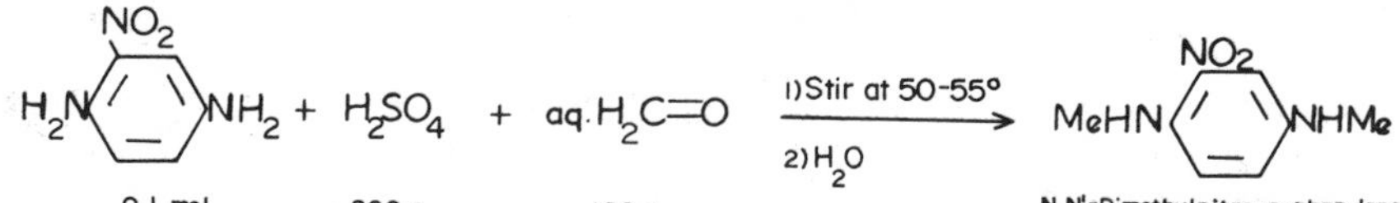

$$H_2N{-}C_6H_3(NO_2){-}NH_2 + H_2SO_4 + \text{aq.}H_2C{=}O \xrightarrow[\text{2)H}_2\text{O}]{\text{1)Stir at 50-55°}} MeHN{-}C_6H_3(NO_2){-}NHMe$$

0.1 mol.   200 g.   100 g.

N,N'-Dimethylnitro-p-phenylene-
diamine, 83%

Only a single methylation occurs on each nitrogen. With tri-oxymethylene the reaction is more rapid and convenient and also gives a purer product. Other nitroamines respond similarly.

Another Clarke-Eschweiler reaction begins with the nitro compound (in place of the amine) and proceeds via catalytic hydrogenation [57]:

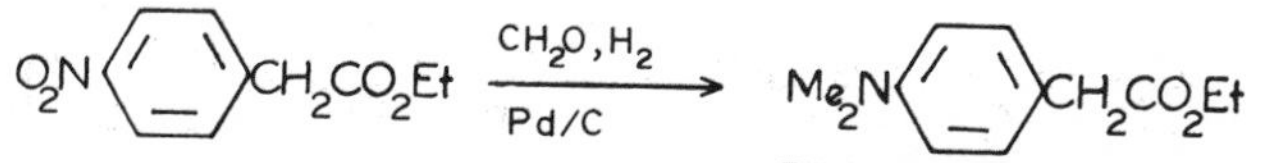

$$O_2N{-}C_6H_4{-}CH_2CO_2Et \xrightarrow[\text{Pd/C}]{\text{CH}_2\text{O,H}_2} Me_2N{-}C_6H_4{-}CH_2CO_2Et$$

Ethyl 4-dimethylaminophenylacetate
67-77 %

More precise studies on the Eschweiler-Clarke reaction show that a large excess of formaldehyde is wasteful and that the addition of sodium formate favors the methylation step [58].

In certain unsaturated amines ring closure may take place as shown [59]:

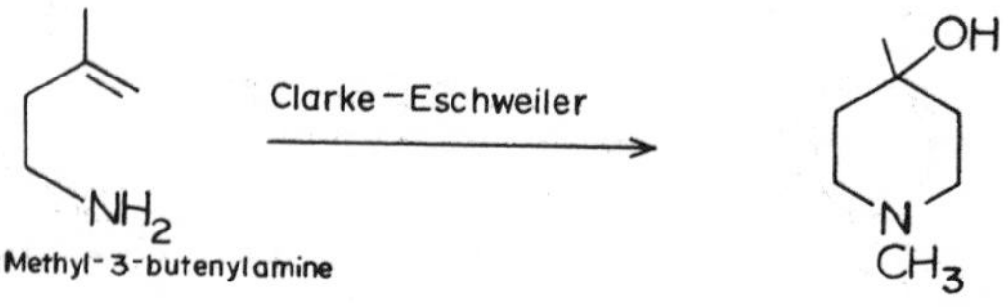

3-Methyl-3-butenylamine

Clarke—Eschweiler

N,4-Dimethyl-4-
hydroxypiperidine, 46%

A type of Prins may be involved, although the reaction is conducted in hydrazine alone.

With ketones and secondary amines, enamines are produced
[60].

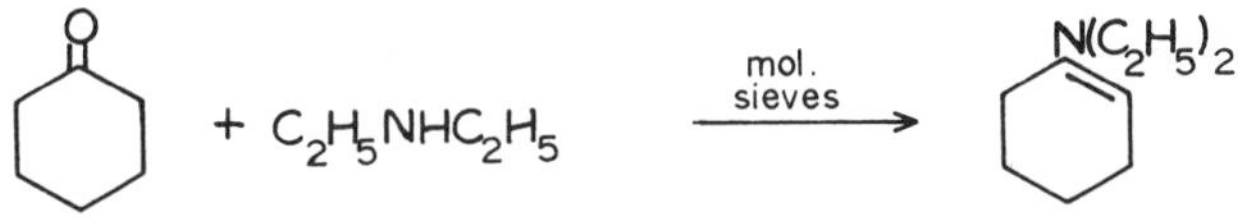

I-Diethylamine-I-cyclohexene
78%

Primary amines lead to ketimines.

a. Preparation of N-Methyl-N-ethylbenzylamine [54] (for
"Reactive" Amines)

To a stirred solution of 5 mmol of N-ethylbenzylamine and
2 ml (25 mmol) of 37% aqueous $H_2C=O$ in 15 ml of acetonitrile was
added 8 mmol of $NaBH_3CN$. The exothermic reaction was stirred
15 min, acetic acid was added dropwise until neutral, and the
mixture was stirred for an additional 45 min. The solvent was
evaporated and the residue, made basic, was extracted with
ether, then with HCl, made basic again, and extracted. Yield of
amine was 735 mg (98%).

b. Preparation of m-Nitrodimethylaniline [54] (for "Unreactive"
Amines Such as m-Nitroaniline)

The same procedure as above with an increase of the $H_2C=O$
to 50 mmol and the $NaBH_3CN$ to 15 mmol. Then after stirring, 0.5
ml of acetic acid was added followed by stirring for 2 hr. Again
0.5 ml acetic acid was added followed by stirring for 30 min.
Yield was 68%.

8. From Aromatic Amines or Azaaromatic Compounds ($\underline{1}$, 429)

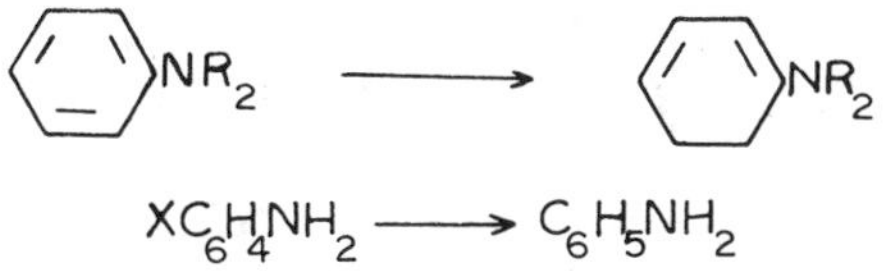

The best way to obtain dialkyldihydroanilines
is to reduce the corresponding dialkylaniline with
lithium and t-amyl alcohol [61]. To obtain the tetrahydroaniline,
, lithium in liquid ammonia and t-butyl alcohol seems 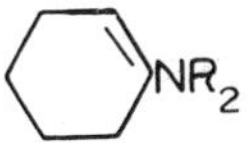
the best reagent [62]. The complete reduction by
ruthenium catalyzed hydrogenation (70-100 kg/cm²)

to the hexahydro compound has been shown to be improved by the presence of lithium hydroxide [63].

The replacement of <u>ortho</u> or <u>para</u> bromo or iodo groups by hydrogen in anilines may be accomplished with thiophenol [64]:

$$p\text{-Br } C_6H_5NH_2 \xrightarrow[\text{reflux}]{C_6H_5SH} C_6H_5NH_2$$
$$\text{Aniline,88\%}$$

<u>m</u>-Bromoaniline is not reduced and <u>p</u>-chloroaniline, only to the extent of 16%.

10.   From Quaternary Salts (Hydrogenolysis) (<u>1</u>, 431)

The chemical reduction of tetrahydroquinoline quaternary salts now has been carried out electrolytically [25]:

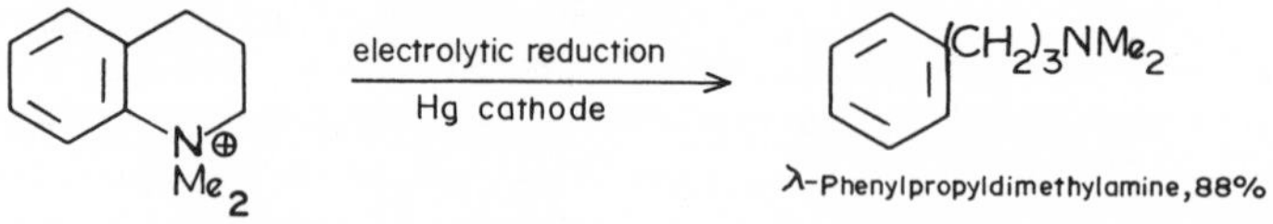

λ-Phenylpropyldimethylamine,88%

11.   From Nitrogen (Fixation)

Studies on the fixation of nitrogen continue to show advancement, but not sufficiently to say that they are of practical use. The system of van Tamelen seems to be the best [65]:

$$\underset{\substack{0.01\,\text{mol.}}}{Cp_2TiCl_2} + \underset{\substack{1\,g.}}{Mg \text{ powder}} + \underset{\substack{60\,ml. \\ \text{absorbed}}}{N_2} \xrightarrow[\text{(dry)}]{\text{THF}} \text{a nitride} \xrightarrow{Et_2C=O}$$

$$\text{3-Aminopentane } + \text{Di(3-pentyl)amine}$$
$$\text{25-50\% based on fixed nitrogen}$$

Naphthalene, lithium, and titanium tetrachloride in THF at a pressure of 100 atm N_2 gave 2% 1-naphthylamine [66]. Another report from Russia [67] indicates that titanium tetrachloride, aluminum powder, and aluminum bromide in benzene with 90 atm of N_2 gave 300 mol of ammonia per mol of titanium tetrachloride.

12.   From Acids

$$RCO_2H + CH_3NH_2 \xrightarrow{M^\circ} RCH=NCH_3 \xrightarrow{H_2,\,cat.} RCH_2NHCH_3$$

Rather than to begin with the carbonyl compound, as in A.6, one may begin with the acid and reduce it to the Schiff base with lithium in methylamine [68] (Ex. a). Moreover, the NCH_3 in the Schiff base may be replaced with other NR groups to give upon hydrogenation secondary amines containing a group other than methyl:

$$RCH\!=\!NCH_3 + BuNH_2 \longrightarrow RCH\!=\!NBu \xrightarrow{\;H_2\;} RCH_2NHBu$$
$$+$$
$$CH_3NH_2$$

a.   Preparation of N-Methylpentylamine [68]

Pentanoic acid, 0.25 mol in 500 ml of methylamine with lithium wire (0.255 g-atom) for 10-20 min gave a blue solution. An additional 0.5 g-atom of lithium was added. After 4.5 hr the mixture was decomposed with 250 ml of water saturated with ammonium chloride. The pentane extract was dried, concentrated to 300 ml, and the imine hydrogenated (3 g of 10% Pd/C) in a low pressure Parr apparatus to give 52% of the product. Obviously, the aldehyde, pentanal could be obtained (66% yield) if the imine was hydrolyzed rather than reduced.

Addenda

General. For the synthesis of heterocycles, not all of which are N-heterocycles, see A. R. Katritzky, A. J. Boulton, Advances in Heterocyclic Chemistry, Vol. 18, Academic, New York, 1975, and earlier volumes; A. Weissberger, E. C. Taylor, The Chemistry of Heterocyclic Compounds, Vol. 29, 1974, and previous volumes, and a narrower review, E. S. Krongauz et al., Russ. Chem. Rev. (Engl. transl.), 39, 747 (1970).

A.2. Sodium cyanoborohydride has been used to reduce an azide to an amine [H. C. J. Ottenheijm et al., Rec. Trav. Chim., 94, 138 (1975)].

A.4. The properties of diisobutylaluminum hydride and triiso-butylaluminum as reducing agents have been reviewed [E. Winterfeldt, Synthesis, 617 (1975)]. The first reagent is not hazardous as a 20-25% solution in toluene. It is superior to $LiAlH_4$ in the reduction of amides to amines and does not reduce carbon-carbon double bonds in the same molecule. The second, which is stereospecific in its effect, may be used as a bulky reducing agent of Meerwein-Ponndorf characteristics (Alcohols, C.3, Ref. 36).

A.6.   The use of sodium cyanoborohydride as a reducing agent has been reviewed [C. F. Lane, Synthesis, 135 (1975)].

A.9.   Arylaminomethylsuccinimides may be hydrogenolyzed to N-methylarylamines by sodium borohydride in DMSO (34-84%, 12 ex.) [S. B. Kadin, J. Org. Chem., 38, 1348 (1973)].

References

1.   J. Malek, M. Cerný, Synthesis, 217 (1972).
2.   R. O. Hutchins et al., J. Org. Chem., 36, 803 (1971).
3.   J. M. LaLancette, J. R. Brindle, Can. J. Chem., 49, 2990 (1971).
4.   T. Satoh, Y. Suzuki et al., Tetrahedron Letters, 4555 (1969).
5.   T. Satoh et al., Chem. Ind. (London), 1626 (1970).
6.   D. P. Wagner et al., Syn. Commun., 1, 47 (1971).
7.   F. S. Dovell, H. Greenfield, J. Am. Chem. Soc., 87, 2767 (1965).
8.   K. Taya, Chem. Commun., 464 (1966).
9.   R. G. Benner, Belg., 706,591, April 1, 1968; C. A., 70, 28618 (1969).
10. L. Krasnec, Z. Chem., 11, 110 (1971); Synthesis, 53 (1972).
11. E. M. Kaiser, Synthesis, 408 (1972).
12. H.-L. Pan, T. L. Fletcher, Synthesis, 592 (1971).
13. J. P. Idoux, J. Chem. Soc., C, 435 (1970).
14. A. F. M. Iqbal, Tetrahedron Letters, 3385 (1971).
15. J. M. Landesberg et al., J. Org. Chem., 37, 930 (1972).
16. J. I. G. Cadogan, M. J. Todd, J. Chem. Soc., C, 2808 (1969).
17. S. Wawzonek, Synthesis, 285 (1971).
18. H. Stetter et al., Chem. Ber., 103, 200 (1970).
19. Y. S. Khishchenko et al., Zh. Prikl. Khim. (Leningrad), 42, 2384 (1969); C. A., 72, 31347 (1970).
20. H. Feuer, D. M. Braunstein, J. Org. Chem., 34, 2024 (1969).
21. H. Alper, J. T. Edward, Can. J. Chem., 48, 1543 (1970).
22. J. Cymerman-Craig, K. K. Purushothaman, J. Org. Chem., 35, 1721 (1970).
23. M. N. Sheng, J. G. Zajacek, Org. Syn., 50, 56 (1970).
24. S. Oae et al., Tetrahedron, 28, 2981 (1972).
25. L. Horner, H. Röder, Chem. Ber., 101, 4179 (1968).
26. E. B. Fleischer et al., J. Org. Chem., 36, 3042 (1971).
27. K. Ponsold, J. Prakt. Chem., 36, 148 (1967).
28. M. Fischer, Chem. Ber., 100, 3599 (1967).
29. G. Zweifel, H. C. Brown, Org. Reactions, 13, 1 (1963).
30. J. A. Blair, R. J. Gardner, J. Chem. Soc., C, 1714 (1970).

31. A. R. Doumaux, Jr., J. Org. Chem., 37, 508 (1972).
32. J. A. Secrist, III, M. W. Logue, J. Org. Chem., 37, 335 (1972).
33. W. E. Parham, C. S. Roosevelt, Tetrahedron Letters, 923 (1971).
34. R. A. Egli, Helv. Chim. Acta, 53, 47 (1970).
35. H. Plieninger et al., Synthesis, 84 (1972).
36. Y. Kikukawa et al., Jap., 6,925,761, October 30, 1969; C. A., 72, 12588 (1970); I. Saito et al., Chem. Pharm. Bull., 18, 1731 (1970); C. A., 73, 110093 (1970).
37. R. F. Borch, Tetrahedron Letters, 61 (1968).
38. H. Feuer, D. M. Braunstein, J. Org. Chem., 34, 1817 (1969).
39. G. H. Timms, E. Wildsmith, Tetrahedron Letters, 195 (1971).
39a. J. M. LaLancette, J. R. Brindle, Can. J. Chem., 48, 735 (1970).
40. J. K. Sugden, Chem. Ind. (London), 260 (1969).
41. H. E. Smith, A. A. Hicks, J. Org. Chem., 36, 3659 (1971).
42. J. A. Gautier et al., Bull. Soc. Chim. France, 2916 (1968); Compt. Rend., 263, C1164 (1966).
43. W. Dürckheimer, Ann. Chem., 721, 240 (1969).
44. R. F. Borch et al., J. Am. Chem. Soc., 93, 2897 (1971); 91, 3996 (1969).
45. R. F. Borch, Org. Syn., 52, 124 (1972).
46. R. A. Benkeser, S. J. Mels, J. Org. Chem., 35, 261 (1970).
47. S.-I. Murahashi et al., J. Am. Chem. Soc., 95, 3038 (1973).
48. R. W. White et al., Tetrahedron Letters, 3591 (1971).
49. R. D. Bach, D. K. Mitra, Chem. Commun., 1433 (1971).
50. R. F. Borch, S. R. Levitan, J. Org. Chem., 37, 2347 (1972).
51. M. J. Lacey, Australian J. Chem., 23, 841 (1970).
52. G. P. Rizzi, J. Org. Chem., 36, 1710 (1971).
53. V. A. Semenov, D. I. Skorovarov, J. Org. Chem., USSR, 5, 39 (1969).
54. R. F. Borch, A. I. Hassid, J. Org. Chem., 37, 1673 (1972).
55. B. L. Sondengam et al., Tetrahedron Letters, 261 (1973).
56. A. Halasz, Chem. Ind. (London), 1701 (1969).
57. M. G. Romanelli, E. I. Becker, Org. Syn., Coll. Vol. 5, 552 (1973).
58. S. H. Pine, B. L. Sanchez, J. Org. Chem., 36, 829 (1971).
59. A. C. Cope, W. D. Burrows, J. Org. Chem., 31, 3099 (1966).
60. K. Taguchi, F. H. Westheimer, J. Org. Chem., 36, 1570 (1971); E. P. Kyba, Org. Prep. Proc., 2, 149 (1970).
61. A. J. Birch et al., J. Chem. Soc., C, 637 (1971).
62. A. J. Birch et al., ibid., C, 2409 (1971).
63. S. Nishimura et al., Bull. Chem. Soc. Japan, 44, 240 (1971).
64. M. W. Barker et al., J. Org. Chem., 37, 3555 (1972).

65. E. E. van Tamelen, H. Rudler, J. Am. Chem. Soc., 92, 5253
     (1970); for a general review see M. M. T. Khan, A. E.
     Martell, Homogeneous Catalysis by Metal Compounds, Vol. I,
     Academic, New York, 1974, p. 266.
66. M. E. Vol'pin et al., Izv. Akad. Nauk, SSSR, Ser. Khim.,
     2858 (1969); C. A., 72, 78745 (1970).
67. A. Brownstein, Chem. Tech., 70 (1971).
68. A. O., J. H. Bedenbaugh et al., J. Am. Chem. Soc., 92,
     5774 (1970).

B. Hydrolysis or Solvolysis

With the exception of B.2, the examples in B could be placed
in C since ammonia, $NH_3 \rightarrow RNH_2$, or the carbon adjacent to the
amino group, $CHNH_2 \rightarrow C(R)NH_2$, is alkylated. The examples are
placed here to emphasize the solvolytic aspect of the synthesis.

   1. From Ureas, Urethanes, Isocyanates, Isothiocyanates,
       Isocyanides, and Hexahydrotriazines (1, 436)

The preparation of meso-2,4-diaminopentane posed a problem
since the sodium-alcohol reduction of the related dioxime gave a
mixture of isomers inseparable except with considerable loss.
The problem was solved by preparing a derivative that held the
groups in the proper orientation to give the meso diamine on
hydrolysis as shown [1]:

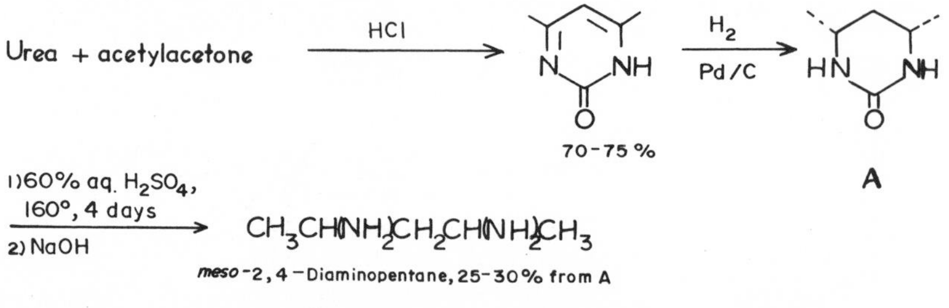

Urea + acetylacetone $\xrightarrow{\text{HCl}}$ (70-75%) $\xrightarrow[\text{Pd/C}]{H_2}$ **A**

$\xrightarrow[\text{2) NaOH}]{\substack{\text{1)60\% aq. } H_2SO_4, \\ 160°, 4 \text{ days}}}$ $CH_3CH(NH_2)CH_2CH(NH_2)CH_3$

meso-2,4-Diaminopentane, 25-30% from A

Since the solvolysis of an α-isocyanoester proved to be
remarkably easy, a way was open to synthesize a number of
alkylated amino esters in the following manner [2]:

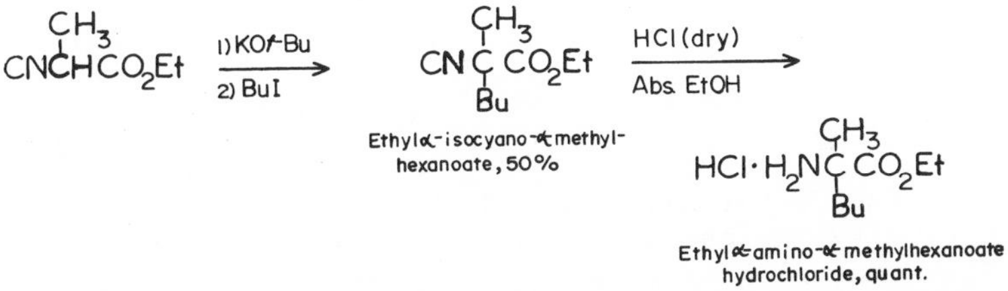

1) KO*t*-Bu
2) BuI
Ethyl α-isocyano-α-methylhexanoate, 50%
HCl (dry)
Abs. EtOH
Ethyl α-amino-α-methylhexanoate hydrochloride, quant.

The hexahydrotriazines are capable of undergoing a simultaneous alkylation, hydrolysis, and reduction. The conditions for accomplishing this combination of events are as follows [3]:

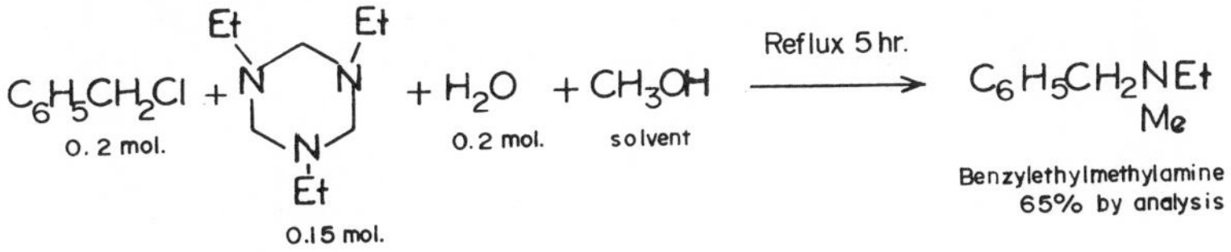

0. 2 mol.
0.15 mol.
0.2 mol.
solvent
Reflux 5 hr.
Benzylethylmethylamine
65% by analysis

Where water was omitted, the yield dropped to 41%.

2. From Amides or Imides (Including Gabriel) (1, 437)

$$RCONH_2 + R'I \longrightarrow RCONHR' \longrightarrow R'NH_2$$

The Gabriel synthesis has now been performed using an alcohol rather than an alkyl halide! Conditions are mild in that one mmol of each component is held at 25° [4]:

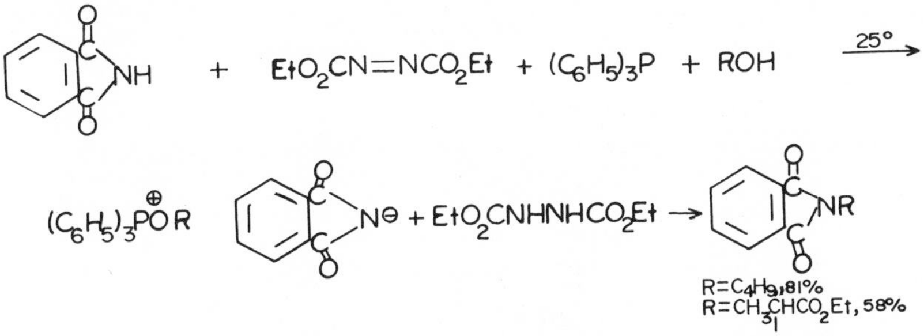

25°
R=C4H9, 81%
R=CH3CHCO2Et, 58%

The usual hydrazinolysis was carried out to give the amine (1, 438). Moreover, optically active 2-octanol was converted as above to 2-aminooctane, 48% yield, of nearly 100% inverted amine.

The alkylation of trifluoroacetamides is facilitated by the presence of the trifluoroacetyl group [5]:

$$\mathrm{RNHCOCF_3 + CH_3I\,(xs.)} \xrightarrow[\text{2) } H_2O,\ \Delta]{\text{1) powdered KOH in acetone}} \mathrm{RNHCH_3}\quad 77\text{-}98\%$$

Propyl iodide is less active in this alkylation, which, however, may be accomplished by using propyl methanesulfonate. Guanidine also serves well in alkylation with alkyl halides. When mixed with a primary alkyl halide in alcohol, refluxed, and saponified, yields of alkyl amines are of the order of 13-79% [6].

The symmetrical alkylation of hydrazine has proved to be quite a problem, but can be accomplished by starting with di-*t*-butyl hydrazoformate [7]. A synthesis is shown [8]:

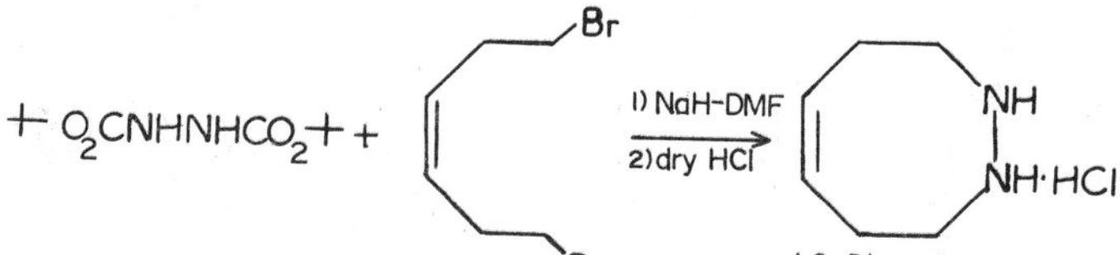

1,2-Diaza-5-cyclooctene hydrochloride, 70%

3. From Cyanamides (1, 439)

$$\mathrm{R_2NC{\equiv}N \longrightarrow R_2NH}$$

Improvements have been made in both the alkylation of cyanamide to dialkylcyanamide and the hydrolysis of the latter. The first step employs sodium dimsyl in DMSO and the second proceeds via an isourea [9]:

$$\mathrm{2\ RBr + H_2NCN} \xrightarrow[\text{DMSO, 25°}]{\mathrm{NaCH_2SOCH_3}} \underset{\text{satisfactory yields}}{\mathrm{R_2NCN}} \xrightarrow{\mathrm{KCN\text{-}CH_3OH}} \underset{\text{O-Methylisourea}}{\mathrm{R_2NC(=NH)OCH_3}} \xrightarrow[\text{18 hr. 2) OH}^{\ominus}]{\text{1) 80\% aq. HOAc}} \underset{55\text{-}81\%}{\mathrm{R_2NH}}$$

No allylic rearrangement takes place if one of the R groups is a substituted allyl, and the potassium cyanide mainly serves as a source of basic ions since sodium methoxide may be used in its place.

4. From N-Substituted Amides and Sulfonamides (1, 440)

$$\mathrm{ArSO_2NHR \longrightarrow RNH_2}$$

This synthesis has always been a problem. The best of the untidy procedures still survives [10] (1, 440):

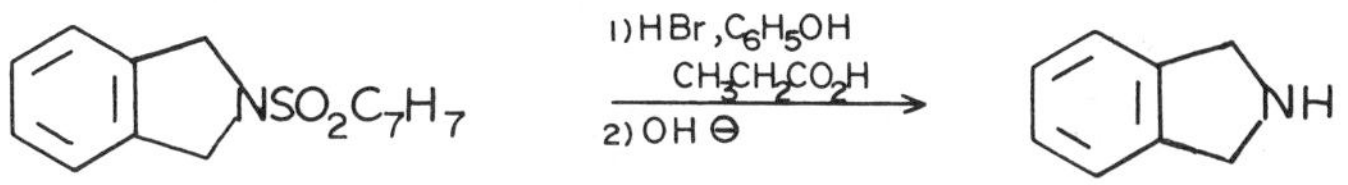

1,3-Dihydroisoindole, 63-71%

But now two procedures, one acidic and the other basic, make the situation with regard to the hydrolysis of sulfonamides look much brighter. The first takes advantage of the fact that sulfuric acid in acetic acid is much stronger than in water [11]:

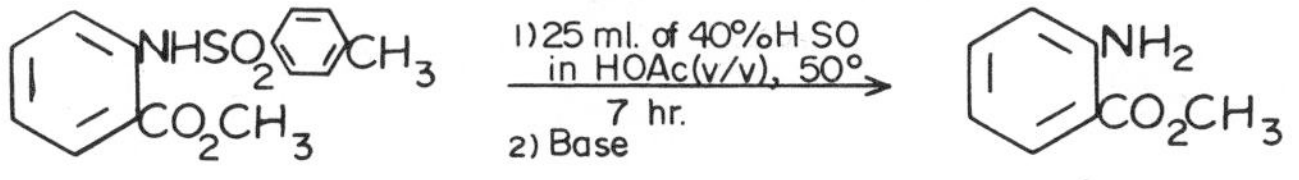

2 g.

Methyl anthranilate, 84 %

In this paper [11] several cases are discussed in which the HBr method failed, whereas the 40% H_2SO_4 in acetic acid worked. The second method reduces, rather than hydrolyzes, the sulfonamide [12]:

$$\text{N-Tosyldeoxyephedrine} \xrightarrow[C_6H_6 \text{ or glyme}]{NaAlH_2(OCH_2CH_2OMe)_2,\ 4\ eq.} \underset{64\ \%}{\text{Deoxyephedrine}}$$

Other methods, which are of interest, serve special purposes. The best of these describes a sulfonamide derivative which is unstable to reduction [13]:

$$RNH_2 + C_6H_5COCH_2SO_2Cl \xrightarrow[CH_2Cl_2,\ 0°]{C_5H_5N} \underset{ca.\ 90\ \%}{C_6H_5COCH_2SO_2NHR} \xrightarrow{R'X,\ K_2CO_3}$$

$$\underset{ca.\ 80\ \%}{C_6H_5COCR'HSO_2N(R)(R')} \xrightarrow[-C_6H_5COCH_2R',\ -SO_2]{Zn,\ HOAc} \underset{\text{Butylmethylamine, 75 \%}}{RNR'H}$$

However, the method is not recommended as a good general one [14]. Tosylamides in some cases may be cleaved by irradiation [15].

$$TsNEt_2 \xrightarrow[6\text{-}20 hr.]{Et_2O,\ h\nu} Et_2\overset{\oplus}{N}H_2 \cdot \overset{\ominus}{O}Ts$$

5. From Quaternary Salts (<u>1</u>, 441)

$$R_4NX \longrightarrow R_3N$$

Recently, it has been shown that a number of nucleophilic groups are capable of dealkylating the quaternary salt (provided it does not form an alkene) [16]:

$$C_6H_5\overset{\oplus}{N}Me_3\overset{\ominus}{I} + (C_6H_5)_3P \xrightarrow{DMF} C_6H_5NMe_2$$
$$96\%$$

Thiourea, sodium thiosulfate, or sodium azide may be substituted for triphenylphosphine to give yields of dimethylaniline in the range of 78-98%. Another reagent is mentioned in C.1 as well as another tertiary amine preparation from the quaternary salt.

References

1. R. O. Hutchins, B. E. Maryanoff, J. Org. Chem., 37, 1829 (1972).
2. U. Schöllkopf et al., Angew. Chem. Intern. Ed. Engl., 10, 331 (1971).
3. Y. Ohshiro et al., Synthesis, 89 (1971).
4. O. Mitsunobu et al., J. Am. Chem. Soc., 94, 679 (1972).
5. R. A. W. Johnstone et al., J. Chem. Soc., C 2223 (1969).
6. P. Hébrard et al., Bull. Soc. Chim. France, 1938 (1970); Tetrahedron Letters, 13 (1969).
7. L. Carpino, Acc. Chem. Research, 6, 191 (1973).
8. L. Carpino, J. P. Masaracchia, J. Org. Chem., 37, 1851 (1972).
9. A. Donetti et al., Tetrahedron Letters, 3327 (1969); J. Org. Chem., 37, 3352 (1972).
10. J. Bornstein et al., Org. Syn., Coll. Vol., 5, 406 (1973).
11. P. D. Carpenter, M. Lennon, Chem. Commun., 664 (1973).
12. E. H. Gold, E. Babad, J. Org. Chem., 37, 2208 (1972).
13. J. B. Hendrickson, R. Bergeron, Tetrahedron Letters, 345 (1970).
14. J. B. Hendrickson, R. Bergeron, Tetrahedron Letters, 3839 (1973).
15. A. Abad et al., Tetrahedron Letters, 4555 (1971).
16. T.-L. Ho, Syn. Commun., 3, 99 (1973).

C. Metathetical or Other Reactions Leading to Metathetical Products

1. From Halides and Amines (Including Ammonia) (1, 444)

$$RX + R'NH_2 \longrightarrow RNHR' \cdot HX$$

1,2,2,6,6-Pentamethylpiperidine has been found to be a useful amine to carry out exhaustive methylation to the quaternary
salt [1] (Ex. a). It does not tend to add methyl iodide and yet
it is basic enough to remove hydrogen iodide from the hydrogen
iodide salt of a secondary or tertiary amine. Although many ways
are available for the synthesis of the tertiary amine from the
quaternary salt, another proved method utilizes high temperature
degradation with ethanolamine [2]:

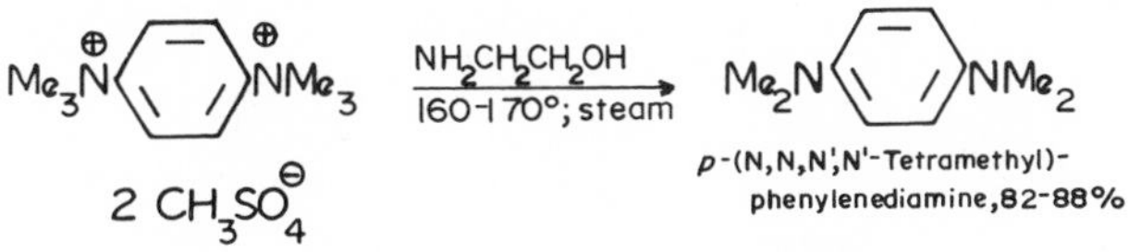

p-(N,N,N',N'-Tetramethyl)-
phenylenediamine,82-88%

A similar compound is prepared in C.3. Other methods are found
in H.3.

The displacement of chlorine by ethylenimine has been shown
to differ from that by diethylamine [3]:

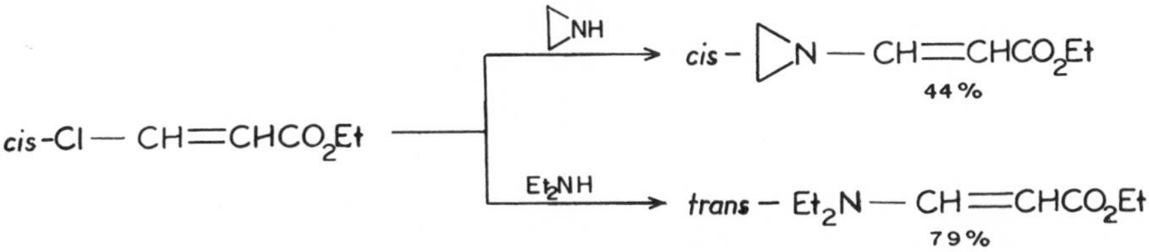

Caution: it should be noted that ethylenimine is a possible
carcinogen.

The fact that p-nitrocumyl chloride (p-O$_2$NC$_6$H$_4$CC1Me$_2$)
alkylates piperidine to give 91% of N-p-nitrocymylpiperidine,
while cumyl chloride does not alkylate the amine whatsoever,
indicates intrusion of a new mechanism [4]. Since the former
reaction is accelerated by light and completely interfered with
by oxygen, Kornblum [4] has concluded that the nitrocumyl
reaction is a free radical one rather than simple S$_N$2 displacement. The great advantage of the use of free radical reactions
lies in the fact that steric hindrance of the substrate or
reagent does not play as important a role as in S$_N$2 reactions.

Displacement in unsaturated halides is complicated by
allylic or similar-type rearrangements [5]. A study now has been
made on some substituted allenic bromides [6]:

3-Methylamino-4-hydroxybutyne
93%

Other amines give comparable yields.

The quaternization of amines sometimes is complicated by liberation of HX, which unites with the amine. This is especially true with allylic halides. In this case ethylene oxide should be added in a room temperature quaternization to remove the HX [7]. However, the pentamethylpiperidine method looks more attractive [1].

The displacement of halogen from aromatic and heterocyclic halides by amines has been outlined in 1, 444. Already mentioned here is the fact that some displacements are copper-catalyzed. The first is a method of making 2,3-diaminopyridine by means of a reagent consisting of 550 ml of 35% NH_4OH containing about 10.5 g of cuprous chloride (reagent A) [8]:

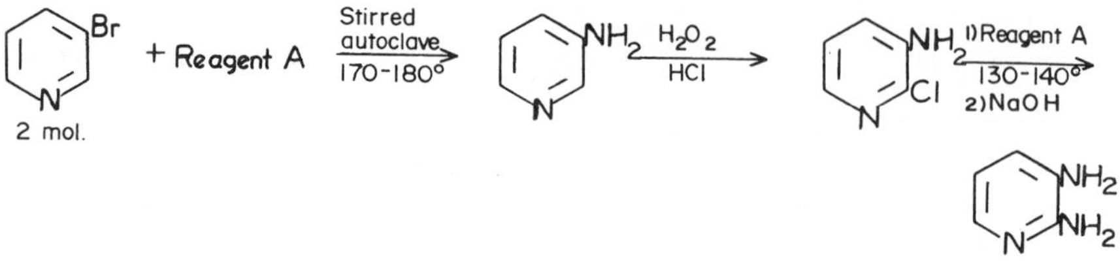

A cuprous oxide-catalyzed condensation was found to be superior to a cuprous chloride-catalyzed one [9]:

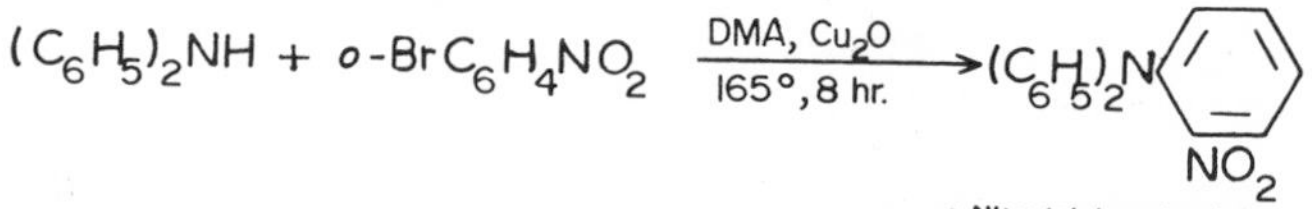

$$(C_6H_5)_2NH + o\text{-}BrC_6H_4NO_2 \xrightarrow[165°, 8\ hr.]{DMA,\ Cu_2O} (C_6H_5)_2N\text{—}$$

o-Nitrotriphenylamine, 51%
after chromatography

A copper bronze catalyst was used in a third example [10]:

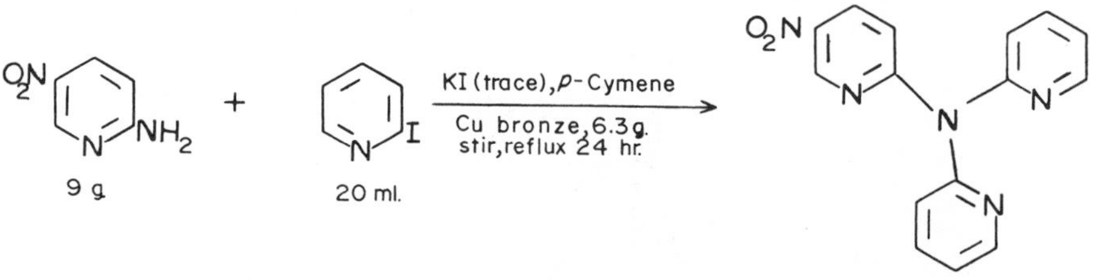

3-Nitro-tri-α-pyridylamine, 10.1 g.

It is also important to realize that displacements from heterocyclic, active halides may be extremely complicated and in certain cases lead to unusual products. It all happens because the nucleophile may not displace halogen, but rather add to the ring and produce ring opening. For instance, it appears that the following reaction is quite simple [11]:

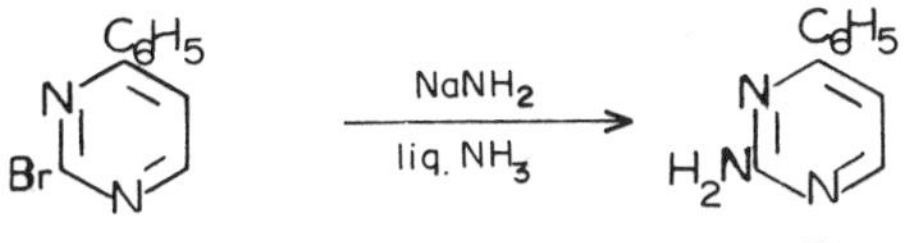
NaNH2
liq. NH3

I

In reality, direct displacement occurs to the extent of about 12% and the remainder of the reaction gives the identical product via ring opening:

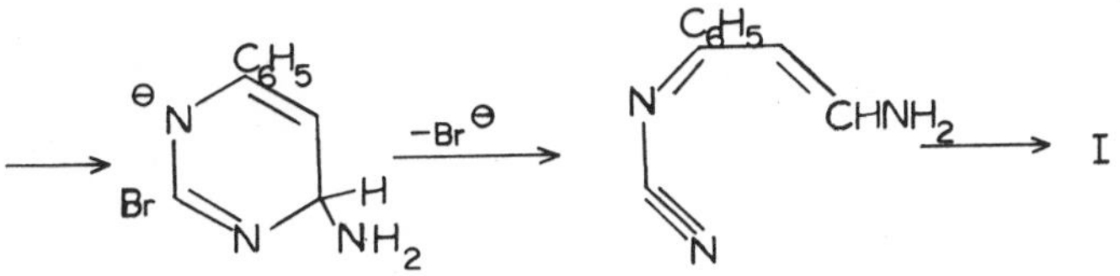

These additions are to be expected more with strong bases or with heterocycles that are less aromatic, such as those containing more than one heteroatom or that are polynuclear. Another example is given in which the course of the reaction is changed [12]:

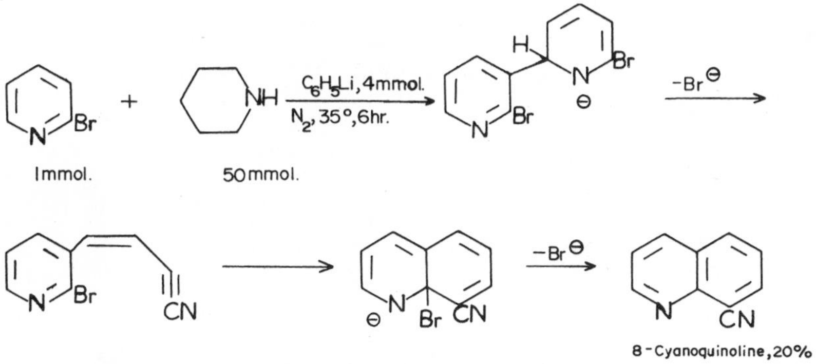
C6H5Li, 4mmol.
N2, 35°, 6hr.
1mmol.
50mmol.
8-Cyanoquinoline, 20%

Alkylations have also been conducted with guanidine [13], hexahydrotriazines [14], and triflamides [15].

a. Preparation of 1-Trimethylamino-2-(4-trimethylaminophenyl)-ethane Dimethiodide [1]

$$p\text{-}H_2NC_6H_4CH_2CH_2NH_2 \longrightarrow p\text{-}(CH_3)_3\overset{\oplus}{N}C_6H_4CH_2CH_2\overset{\oplus}{N}(CH_3)_3 \; I^{\ominus} \; I^{\ominus}$$

β-(p-Aminophenyl)ethylamine, 7.3 mmol, 1,2,2,6,6-pentamethylpiperidine, 29.2 mmol, and methyl iodide, 88 mmol, in 10 ml of DMF evolved heat and deposited a precipitate. After standing overnight, 250 ml of a 6% DMF in acetone solution was added to dissolve the precipitate that was a mixture of the dimethiodide and hydrogen iodide salts. The mixture was refluxed 20 min and the remaining solids were filtered, dissolved in methanol, and reprecipitated with ethyl acetate to give 3.4 g (97%) of product.

2. From Halides and Alkali Metal Amides (Including Aryne Additions) (<u>1</u>, 446)

$$RX + R'NHNa \longrightarrow R'NHR$$

A simple procedure for alkylating amines consists of making the amide with lithium naphthalene and then treating with an alkyl halide [16]:

$$(C_6H_5CH_2)_2NH + C_{10}H_8 \cdot Li \xrightarrow[2) C_6H_{13}Br]{1)THF} (C_6H_5CH_2)_2NC_6H_{13}$$

Dibenzylhexylamine, 90%

Complete methylation of some aromatic amines has been carried out with sodium hydride in THF [17]:

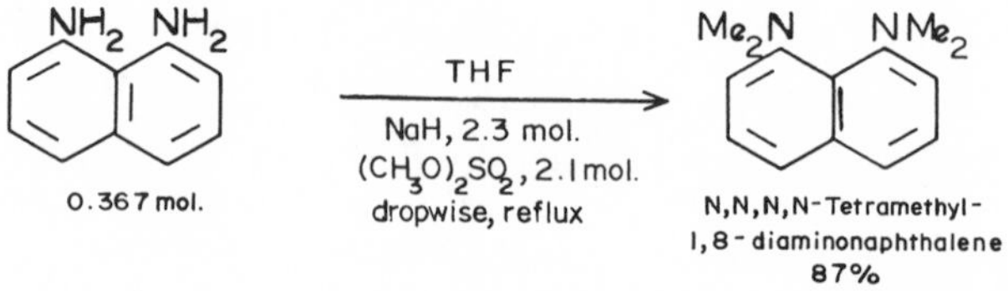

0.367 mol.

THF

NaH, 2.3 mol.
(CH$_3$O)$_2$SO$_2$, 2.1mol.
dropwise, reflux

N,N,N,N-Tetramethyl-
1,8-diaminonaphthalene
87%

The next three procedures may be considered as substitutes for the Gabriel synthesis of primary amines. The amino group in such a synthesis is doubly protected. The first utilizes imines or N-methylimines [18]:

$$(C_6H_5)_2C=NH + LiNEt_2 \xrightarrow[HMPA]{C_6H_6} (C_6H_5)_2C=NLi \xrightarrow{C_3H_7Br}$$

$$(C_6H_5)_2C=NC_3H_7 \xrightarrow{H^\oplus} C_3H_7NH_2$$

82%

Moreover, the N-methylimine is acidic enough to be alkylated, a fact which serves to add scope to the reaction [18]:

$$(C_6H_5)_2C=NCH_3 \xrightarrow[\substack{C_6H_6, \\ HMPA}]{LiNEt_2} (C_6H_5)_2C=NCH_2Li \xrightarrow[2)H_2O]{1)(C_6H_5)_2CO}$$

$$(C_6H_5)_2C=NCH_2C(C_6H_5)_2 \xrightarrow[2)base]{1)aq.HCl} H_2NCH_2C(C_6H_5)_2$$
$$\underset{62\%}{} \quad \underset{OH}{} \qquad \underset{OH}{}$$

2-Amino-1,1-diphenylethanol
90 %

Sulfenimides also may be alkylated [19]:

$$(ArS)_2NH \xrightarrow[\substack{THF \\ -20° \, or \, lower}]{BuLi} (ArS)_2NLi \xrightarrow{RX} (ArS)_2NR \xrightarrow[\substack{or \\ RSH}]{\overset{\oplus}{H};OH^{\ominus}} RNH_2$$

Primary alkyl halides gave about a 60% yield of amines, the
corresponding tosylates about 70%, and one secondary alkyl halide
gave a 17% yield. The sulfenimide could be alkylated with
acrylonitrile and benzyltrimethylammonium hydroxide as the base.
The third method involves the alkylation of N-benzyltriflamide
[20]:

$$C_6H_5CH_2\overset{\ominus}{N}SO_2CF_3 \xrightarrow{RI} \underset{\underset{CH_2C_6H_5}{|}}{RNSO_2CF_3} \xrightarrow[\substack{DMF, \\ 100°, 3hr.}]{NaH} RN=CHC_6H_5 \xrightarrow[\substack{2)OH^{\ominus}}]{1) \, THF \\ HCl} RNH_2$$

R= heptyl,
80%

N-Benzyltriflamide is used since the anion of triflamide itself
gives mostly the dialkylation product $CF_3SO_2NR_2$.
    An ingenious alkylation method, which makes available a
number of alkylated pyridines, has been developed for pyridine.
The method takes advantage of the structure of the first nucleophilic addition product of the amine [21]:

$$C_5H_5N + C_6H_5Li \longrightarrow \text{[pyridine adduct]} \xrightarrow[THF,0°]{RX} \text{[5-Methyl-2-phenylpyridine]}$$

Best to isolate

5-Methyl-2-phenyl-
pyridine, 34%
R=CH_3

Moreover, simple <u>meta</u> alkylated pyridines may be obtained as
follows [22]:

$$C_5H_5N(xs.) + LiAlH_4 \xrightarrow[24 \, hr.]{25°} \underset{\substack{| \\ Li}}{\text{[enamine]}} \xrightarrow[\substack{stir, cool}]{C_6H_5CH_2Cl} \text{[3-Benzylpyridine]}$$

I

An enamine

3-Benzylpyridine, 63%

Alkylation takes place in the 3-position because I is a metalated enamine with powerful nucleophilic properties at the 3-position.

The mechanism of the arylation of amides has always been complicated since either or both direct replacement or intermediate benzyne formation may take place (1, 447). The presence of an intermediate benzyne is indicated by the formation of the minor product in the reaction [23]:

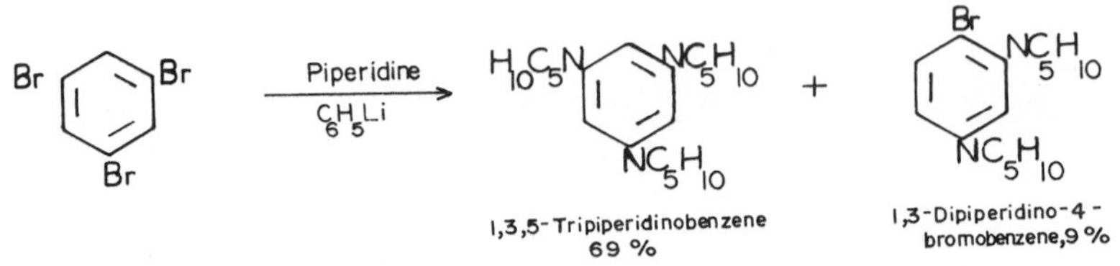
Piperidine
C6H5Li
1,3,5-Tripiperidinobenzene 69 %
1,3-Dipiperidino-4-bromobenzene, 9 %

Some influence may be exerted on controlling metathesis versus benzyne addition [24]:

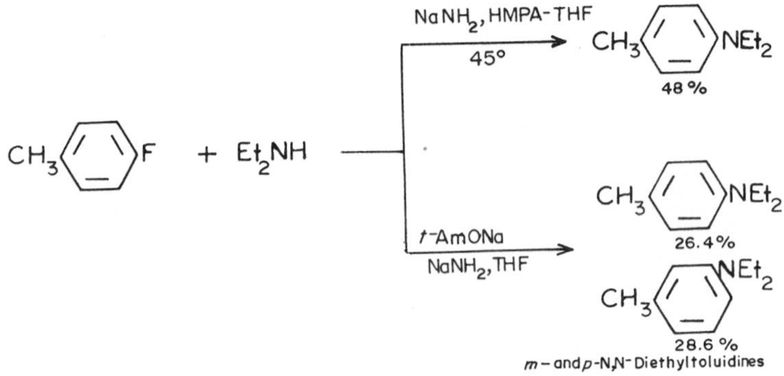
NaNH2, HMPA-THF
45°
48 %
t-AmONa
NaNH2, THF
26.4 %
28.6 %
m- and p-N,N-Diethyltoluidines

The best interpretation is that the amide by itself effects a direct replacement, but when mixed with sodium t-amylate gives almost 100% benzyne formation. The mixed amide-t-amylate base is probably less associated and therefore more basic and capable of removing a proton from the substrate [25]. Another way of modifying the course of reaction is by adding an almost equal molar amount of some inorganic salt, such as sodium thiocyanate or sodium cyanamide, which has resonance possibilities in its anion [26].

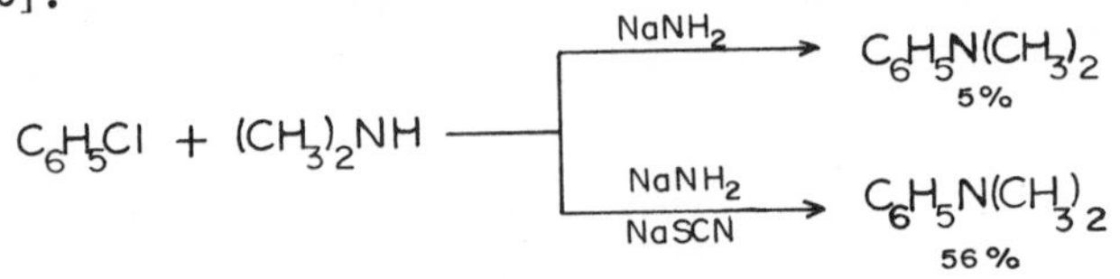
NaNH2
5%
NaNH2
NaSCN
56 %

In some cases the arylation of amides is the method of choice for
preparing alkylated arylamines (see Ex. a).

     Yields of meta-substituted anilines from o-bromoaromatics
(o-BrC$_6$H$_4$OCH$_3$, o-ClC$_6$H$_4$Cl, o-ClC$_6$H$_4$-NMe$_2$) via benzynes (NaNH$_2$,
RNH$_2$, reflux) were about 85%, but contaminated with 2-5% of
o-isomers [27].

a. Preparation of N-t-Butylaniline [28]

     Bromobenzene 0.1 mol and 300 ml of t-butylamine were mixed
with 0.3 mol of sodamide and the mixture was stirred at 25° for
6 hr under nitrogen. Pouring into water followed by the workup
for an amine gave the desired product in 72% yield, the best
yield compared to that of any other method of alkylation.

     3. From Alkyl Sulfonates, Phosphates, or Phosphites and
         Amines (1, 448)

              RNH$_2$ +   R'OSO$_2$Ar $\longrightarrow$ R NHR'

     The methylation of aromatic amines, such as m-trifluoro-
methylaniline, to the tertiary amine (55-58%) has been carried
out with trimethyl phosphate at 150° for 2 hr [29]. The author
states that no quaternization takes place. That may be true for
this amine, but not for others. The reader's attention is
directed to 1, 450, Ex. b.4, where not only quaternization,
which is overcome, takes place, but also a mixture of PPA and
methanol is used rather than the more expensive trimethyl
phosphate.

     If it is desirable to employ a more forceful alkylation, the
amine first is treated with sodium hydride to form the sodamide
as in C.2 and then dimethyl sulfate is added. In this way,
N,N,N,N-tetramethyl-1,8-diaminonaphthalene was prepared in 87%
yield from 1,8-diaminonaphthalene (C.2). Similar examples are
to be found in C.7 and C.8.

     In substituting a phosphite for the more usual phosphate,
Amos and Gillis [30] prepared diethylaniline as shown:

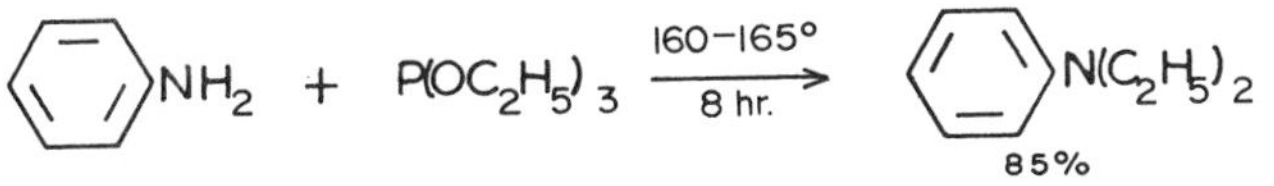

                                  160-165°
     ⟨‾⟩NH$_2$ + P(OC$_2$H$_5$)$_3$ $\xrightarrow{\text{8 hr.}}$ ⟨/‾⟩N(C$_2$H$_5$)$_2$
                                                              85%

     4. From Phenols and Amines or Ammonia (Including Bucherer)
         (1, 450)

                     ArOH $\longrightarrow$ ArNH$_2$

Direct replacement can be carried out on 2-hydroxypyridine [31]:

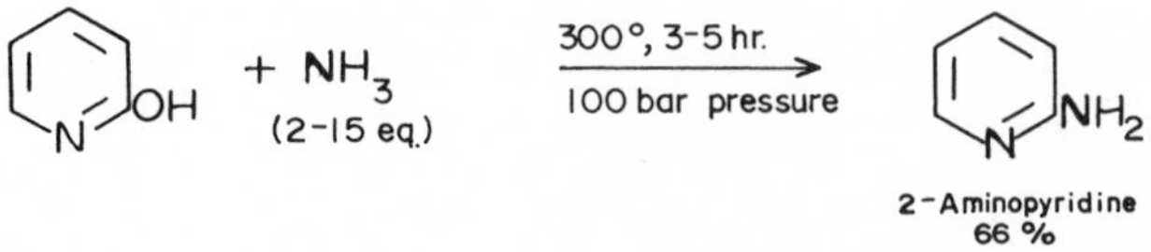
+ NH3
(2-15 eq.)
300°, 3-5 hr.
100 bar pressure
2-Aminopyridine
66 %

Also a novel free radical type displacement has been found [32]:

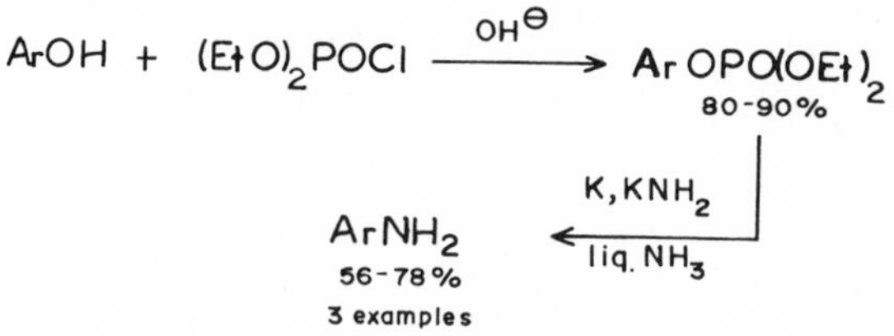
ArOH + (EtO)2POCl
OH⊖
ArOPO(OEt)2
80-90%
K, KNH2
liq. NH3
ArNH2
56-78%
3 examples

The electrons are furnished to the ester by the solvated electrons from K-liq NH_3 to produce the free radical, which breaks down to the aryl free radical. The latter possibly combines with the amide anion to form an aryl anion-free radical,

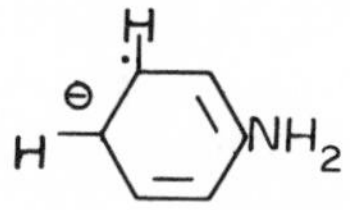

which loses an electron to form aniline.

Another method of forming an aniline from a phenol is to synthesize a phenoxyquinazoline, which rearranges to N-arylquinazolone, which is then hydrolyzed [33]:

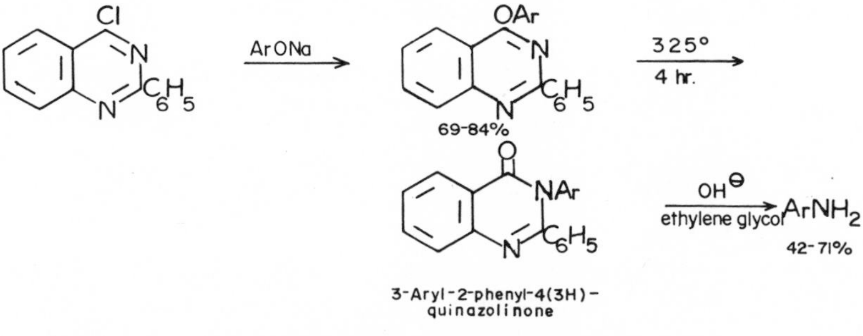
ArONa
325°
4 hr.
69-84%
OH⊖
ethylene glycol
ArNH2
42-71%
3-Aryl-2-phenyl-4(3H)-quinazolinone

7. From Alcohols and Ammonia or Amines (1, 455)

$$ROH + NH_3 \longrightarrow RNH_2$$

Sulfamates of alcohols which form relatively stable
carbonium ions rearrange to form the amine. A specific example
is given [34]:

$$C_6H_5CH=CHCH_2OH \xrightarrow{Me_2NSO_2Cl} C_6H_5CH=CHCH_2OSO_2N(CH_3)_2 \xrightarrow[60°]{MeOCH_2CH_2OMe}$$

$$C_6H_5\underset{\underset{SO_3^{\ominus}}{\overset{\oplus}{N(CH_3)_2}}}{CHCH}=CH_2 \xrightarrow[2)OH^{\ominus}]{1)H^{\oplus}} C_6H_5\underset{N(CH_3)_2}{CHCH}=CH_2 + C_6H_5CH=CHCH_2N(CH_3)_2$$

+ isomer

trans-N,N-Dimethyl-α-phenyl-
allylamine, 81 %

trans-N,N-Dimethylcinnamylamine, 19%

N-Carboalkoxysulfamate esters may also be employed in the
conversion of alcohols into urethanes or amines [35]:

$$CH_3O_2C\overset{\oplus}{N}SO_2\overset{\oplus}{N}Et_3 + CH_3(CH_2)_5OH \longrightarrow CH_3(CH_2)_5OSO_2\overset{\ominus}{N}CO_2CH_3 + \overset{\oplus}{N}HEt_3$$

$$\downarrow 95°$$

$$CH_3(CH_2)_5NHCO_2CH_3$$
>90%

The inner salt, methyl (carboxysulfamoyl) triethylammonium
hydroxide is readily available.

A third method of converting an alcohol into an amine is
shown in the sequence [36]:

$$ROH \xrightarrow[CCl_4]{P(NMe_2)_3} RO\overset{\oplus}{P}(NMe_2)_3\overset{\ominus}{Cl} \xrightarrow{NH_4ClO_4} RO\overset{\oplus}{P}N(Me_2)_3\overset{\ominus}{ClO_4} \xrightarrow[DMF]{R'NHR''} \underset{R''}{R'NR}$$

20-83%

The method has been employed largely in converting secondary into
tertiary amines.

9. From Organometallics or Carbanions and Amine Derivatives
    (1, 456)

$$RMgX \quad + \quad NH_2Cl \longrightarrow RNH_2 + MgClX$$

A number of modifications and variations have been studied.
The tosylates of hydroxylamines have been used in place of
the chloramines [37]:

$$Me_2NOTs + C_6H_5MgBr \xrightarrow[25°]{10\ min.} C_6H_5N(Me)_2$$
295 mg.   Dimethylaniline, 82 mg.

N,N-Dimethylcyclohexylamine was prepared similarly.

A Mannich-like substrate has been utilized to obtain a series of tertiary amines as represented by one example [38]:

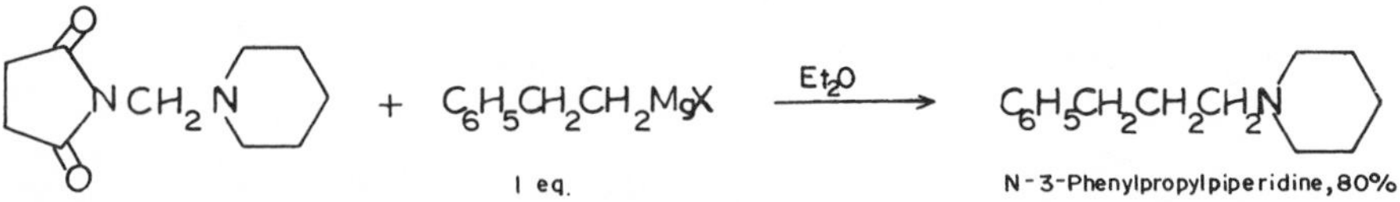

N-3-Phenylpropylpiperidine, 80%

Substituting the succinyl group with other groups gave the following yields: phthalyl 70%, tosyl 35%, benzoyl 0%. When the secondary amino group in the Mannich base was varied, yields were 28-83% (11 ex.). Or when the Grignard reagent was changed, yields varied from 38-80% (12 ex.).

Trichloromethylamines were found to be capable of alkylation [39]:

$$R_2NCCl_3 + 3\ R'MgX \xrightarrow[\text{Reflux 2 hr.}]{Et_2O} R_2NCR'_3$$

31 − 60%
(6 examples)

If aminomethylation (D.3) is not specific or difficult to accomplish, the intervention of an organolithium compound [40] may serve well:

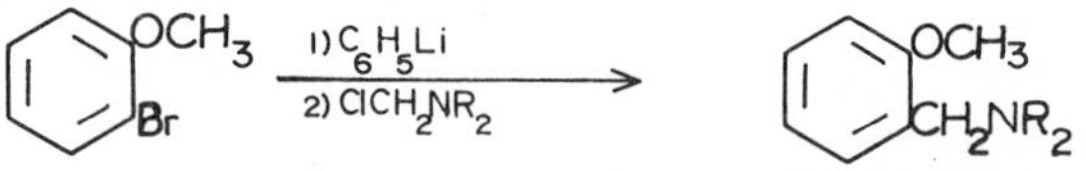

o-Dialkylaminomethylanisoles

Yields on this type and others, such as the resorcinol dimethyl ether, are 52-82%. This reaction succeeds in cases in which direct aminomethylation fails.

Impetus has been given to the alkylation of certain heterocycles by the discovery that the nickel chloride-triphenylphosphine complex catalyzes the reaction [41]:

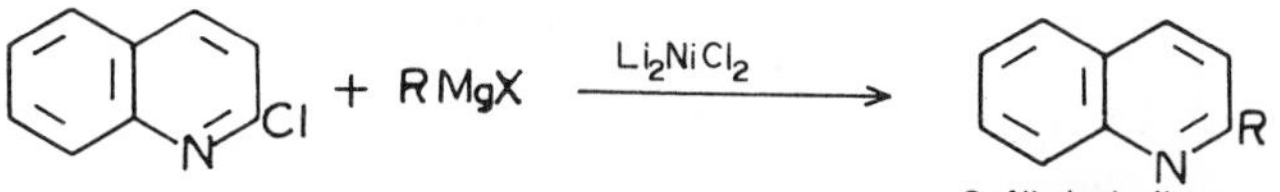

2-Alkylquinolines
0-90%, 9 examples

Without catalyst intractable mixtures were obtained. 3-Bromo-gave 3-methylquinoline in 65% yield and 4,7-dichloro- gave 7-chloro-4-methylquinoline in 77% yield. This reaction sometimes can be brought about by the direct reaction of the unsubstituted heterocycle and alkyllithium reagents (1, 476). This preparation involves a free radical mechanism (see F.7).

10. From Isocyanides (see Addenda).

Addenda

C.2. A mixture of sodamide and sodium t-butoxide seems to be a
more basic reagent than either component by itself. For example,
the reaction of bromobenzene with diethylamine via a benzyne is
accelerated remarkably by the mixed reagent [P. Caubere, Acc.
Chem. Res., 7, 301 (1974)].

C.3. Quaternization of 2,6-di-t-butyl pyridine with methyl
fluorosulfonate gives 80% of the quaternary salt in contrast to
methyl iodide, which gives 20% [Y. Okamoto, K. I. Lee, J. Am.
Chem. Soc., 97, 4015 (1975)].

C.10. An optically active cyclopropylisocyanide may be converted
into its anion   , which may be alkylated without any

racemization whatsoever [H. M. Walborsky, M. P. Periasamy, J. Am.
Chem. Soc., 96, 3711 (1974); 97, 5930 (1975)].

References

1. H. Z. Sommer et al., J. Org. Chem., 36, 824 (1971); 35, 1558
    (1970).
2. S. Hünig et al., Org. Syn., Coll. Vol., 5, 1018 (1973).
3. W. E. Truce, M. L. Gorbaty, J. Org. Chem., 35, 2113 (1970).
4. N. Kornblum, F. W. Stuchal, J. Am. Chem. Soc., 92, 1804
    (1970).
5. H.-W., Methoden der Organischen Chemie, 4th ed., Vol. 11,
    Pt. 1, G. Thieme Verlag, Stuttgart, 1957, p. 47.
6. M. V. Mavrov et al., Tetrahedron, 25, 3277 (1969).
7. A. Donetti, E. Bellora, Tetrahedron Letters, 3573 (1973).
8. W. Brooks, A. R. Day, J. Heterocycl. Chem., 6, 759 (1969).
9. R. G. R. Bacon, D. J. Maitland, J. Chem. Soc., C, 1973
    (1970).
10. J. C. Lancaster, W. R. McWhinnie, J. Chem. Soc., C, 2435
    (1970).
11. A. P. Kroon, H. C. van der Plas, Rec. Trav. Chim., 92, 1020
    (1973).
12. H. N. M. van der Lans, H. J. den Hertog, Tetrahedron
    Letters, 1887 (1973).
13. M. Olomucki, P. Hébrard, Tetrahedron Letters, 13 (1969).
14. Y. Ohshiro et al., Synthesis, 89 (1971).

15. R. A. W. Johnstone et al., J. Chem. Soc., C, 2223 (1969).
16. K. Suga et al., Chem. Ind. (London), 78 (1969).
17. H. Quast et al., Synthesis, 558 (1972).
18. T. Cuvigny, P. Hullot, Compt. Rend., 272, C, 862 (1971).
19. T. Mukaiyama et al., Bull. Chem. Soc. Jap., 44, 2797 (1971);
     Tetrahedron Letters, 3411 (1970).
20. J. B. Hendrickson, R. Bergeron, Tetrahedron Letters, 3839
     (1973).
21. C. S. Giam, J. L. Stout, Chem. Commun., 478 (1970).
22. C. S. Giam, S. D. Abbott, J. Am. Chem. Soc., 93, 1294
     (1971).
23. F. Effenberger et al., Chem. Ber., 103, 1440 (1970).
24. P. Caubère, M.-F. Hochu, Bull. Soc. Chim. France, 2854
     (1969).
25. D. E. Pearson, C. A. Buehler, Chem. Rev., 74, 45 (1974).
26. E. R. Biehl et al., J. Org. Chem., 35, 2454 (1970).
27. E. R. Biehl et al., J. Org. Chem., 36, 3252 (1971).
28. E. R. Biehl et al., J. Org. Chem., 36, 1841 (1971).
29. W. A. Sheppard, Org. Syn., Coll. Vol., 5, 1085 (1973).
30. D. Amos, R. G. Gillis, Australian J. Chem., 22, 1555
     (1969).
31. P. Aviron-Violet, A. Blind, French Patent 2,050,940, May 7,
     1971; C. A., 76, 85705 (1972).
32. R. A. Rossi, J. F. Bunnett, J. Org. Chem., 37, 3570 (1972).
33. R. A. Scherrer, H. R. Beatty, J. Org. Chem., 37, 1681
     (1972).
34. E. H. White, C. A. Elliger, J. Am. Chem. Soc., 87, 5261
     (1965).
35. E. M. Burgess et al., J. Am. Chem. Soc., 92, 5224 (1970).
36. B. Castro, C. Selve, Bull. Soc. Chim. France, 4368 (1971).
37. D. H. R. Barton et al., J. Chem. Soc., C, 2204 (1971).
38. M. Sekiya, Y. Terao, Chem. Pharm. Bull., 18, 947 (1970).
39. V. P. Kukhar, V. I. Pasternak, Synthesis, 611 (1972).
40. H. Böhme, U. Bomke, Arch. Pharm., 303, 779 (1970).
41. E. D. Thorsett, F. R. Stermitz, J. Heterocycl. Chem., 10,
     243 (1973).

D. Additions (Mainly Nucleophilic)

   1. From Unsaturated Compounds and Amines or N-Haloamines or
      Carbon Monoxide and Amines (1, 460)

$$\text{>C=C<} \; + \; RNH_2 \longrightarrow \text{HC—CNHR}$$

About 5% butyllithium has been found to be an effective catalyst for the addition of secondary amines to styrene in cyclohexane at 50°. The yields of β-dialkylaminoethylbenzenes ranged within 41-88% (7 ex.) [1]. In a similar addition to butadiene to give $R_2NCH_2CH=CHCH_3$, cis and trans forms with yields of 48-81% (5 ex.) were obtained. Other amines give similar yields [2]. In still another addition of secondary amines to butadienes it was found that the reaction may be carried out with a catalytic amount of nickel(II) laurate complexed with a dialkyl phenylphosphinite, $ArP(OR)_2$ [3].

The addition of N-chloroamines to unsaturated compounds follows the scheme shown [4]:

$$R_2NCl + R'CH=CH_2 \xrightarrow[AcOH]{H_2SO_4} R'CHClCH_2NR_2$$
$$16-49\%$$

These reactions usually are conducted in acetic-sulfuric acid mixtures. Yields are much better for addition of N-chloroamines to dienes or electronegatively substituted alkenes. On occasion irradiation assists the process. Substitutions with N-chloro-amines are to be found in F.8.

An interesting addition of amines to alkenes takes place with carbon monoxide in the presence of catalysts [5]:

0.25 mol.	0.5 mol.	0.5 mol.	140 atm.		Dimethylaminomethylcyclohexane 91%

The combination catalyst is superior to each single catalyst.

Tertiary amines may also be produced from secondary ones [6] as indicated:

$$(C_2H_5)_2NH + CH_2=CH_2 + C_2H_5Li \xrightarrow[\substack{140°,70atm.\\15\ hr.}]{TMEDA} (C_2H_5)_3N$$
$$82.5\%$$

2. From Carbonyl Compounds or Acid Derivatives and Amines or Hydrazines (1, 461)

$$2R_2NH + CH_2O \longrightarrow R_2NCH_2NR_2 + H_2O$$

Weingarten and co-workers continue to exploit the addition of amine types to carbonyl compounds. With amides amidinium salts [7] are obtained:

$$\text{RCONMe}_2 + \text{Me}_2\text{NH} + \text{TiCl}_4 \xrightarrow[\text{C}_5\text{H}_{12}]{\text{12 hr.,}25°} \overset{\oplus}{\text{RC}}(\text{NMe}_2)_2\overset{\ominus}{\text{Cl}}$$

37-70%,3 examples

The addition is subject to steric hindrance to judge from the fact that substituted pivalamides do not respond.

Cyclization resulted in the treatment of cyclohexanone with N,N'-dimethylhydrazine [8]:

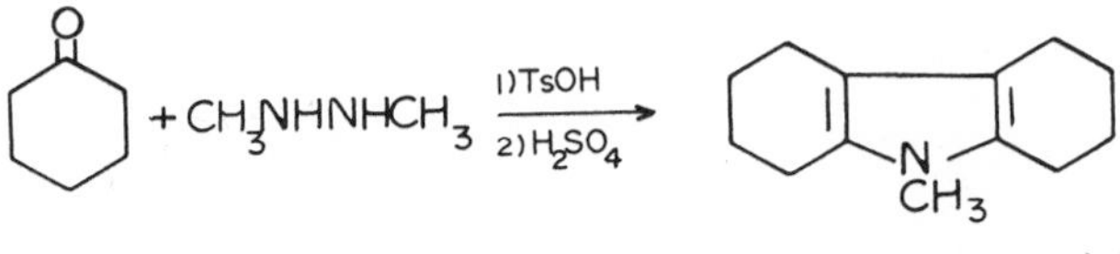

N-Methyloctahydrocarbazole, 80%

3. From Active Hydrogen Compounds, Formaldehyde, and Amines (Mannich) (1, 463)

Recalcitrant Mannich reactions, such as with 1,1-diphenyl-acetone, may be carried out better in a high boiling solvent such as ethylene glycol [9]. Yields are nevertheless low. Both sec- and tert-toluidines have been studied to examine the orientation. With sec-toluidines, both N- and nuclear aminomethylation are observed, the latter by prolonged boiling on a steam bath with acetic acid as a catalyst [10]. Cyclodecanone with methylamine in a Mannich reaction leads to a mixture of bi- and tricyclic compounds [11]:

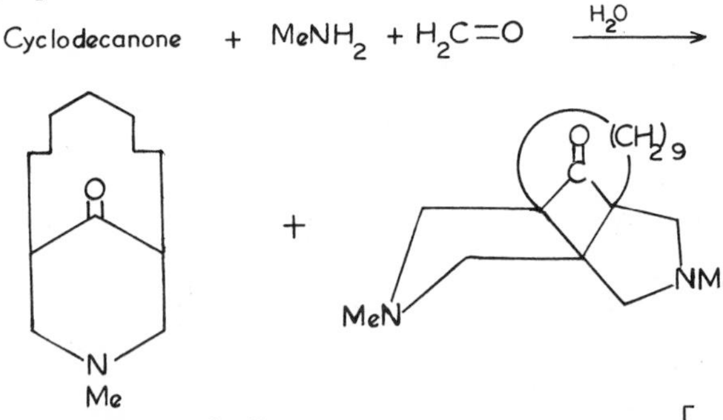

$$\text{Cyclodecanone} \;+\; \text{MeNH}_2 + \text{H}_2\text{C}{=}\text{O} \xrightarrow{\text{H}_2\text{O}}$$

3-Methyl-3-azabicyclo[7.3.1]-tridecane

13,16-Dimethyl-13,16-diazatricyclo[9.3.1]^{1,11}-octadecane-18-one

The next example is truly as though a substituted formalde-hyde were utilized in the Mannich reaction; to be more specific, the fragment $-\underset{\underset{NR_2}{|}}{CH}-CONR_2'$ is utilized as the active hydrogen compound. It is formed from methyl dichloroacetate as follows [12]:

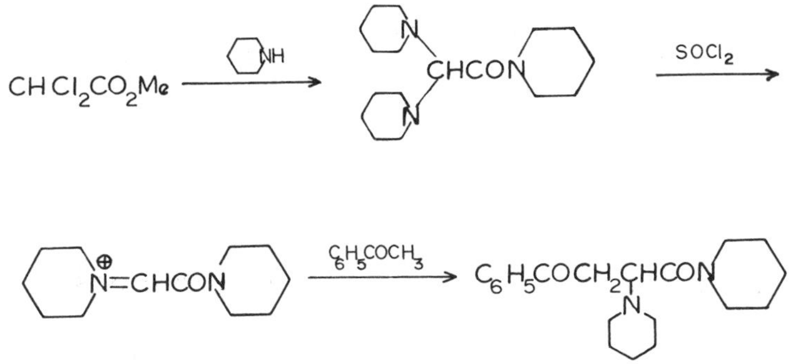

2-Piperidino-3-benzoylpropiopiperidide
93%

Many other examples that utilize different active hydrogen compounds and secondary amines are described.

The last example describes a new highly efficient Mannich reagent made as follows [13]:

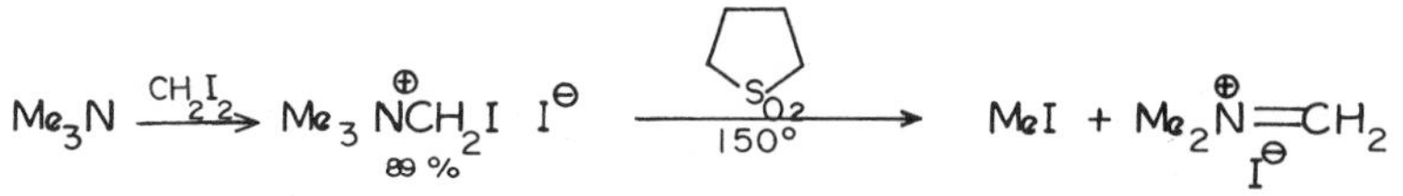

This reagent is efficient in adding to complex systems such as substituted cobalt corrin:

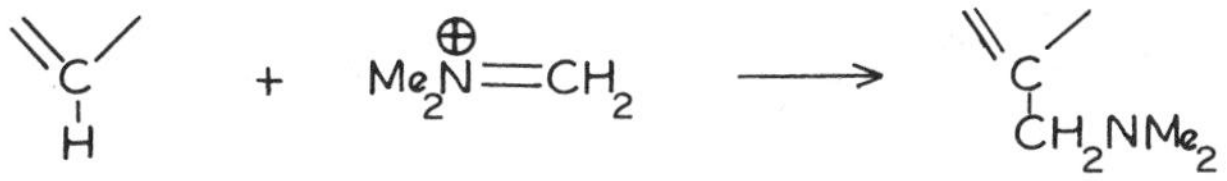

8.   From N-Aroylpyrrolidones or Piperidones

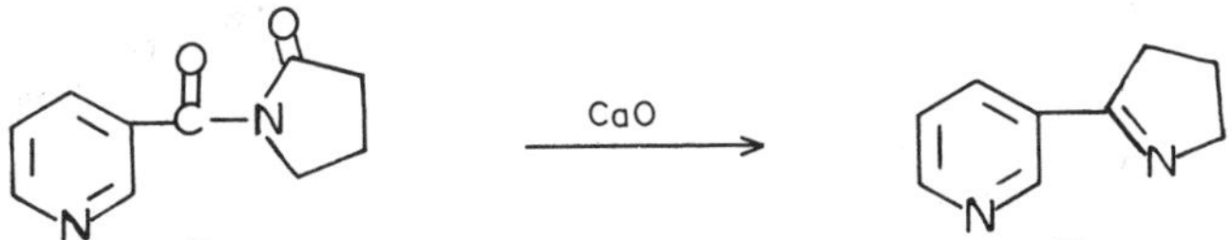

What can only be considered a ring opening and subsequent Dieckmann-like condensation has led to an operatively very simple and useful reaction, in which a free flame distillation is involved [14]:

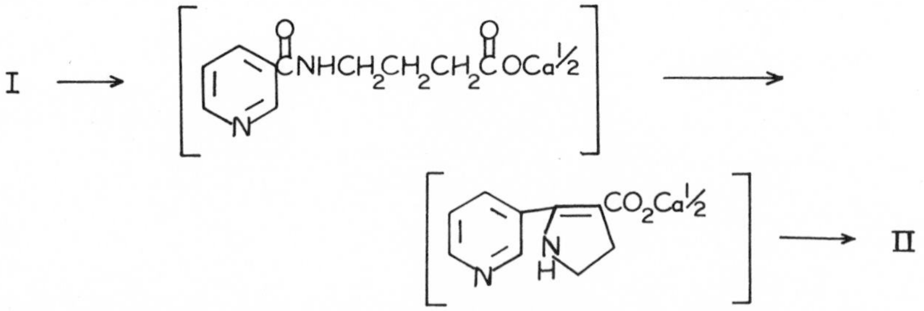

$$I \longrightarrow \left[\begin{array}{c} \text{(pyridine)}\overset{O}{\overset{\|}{C}}NHCH_2CH_2CH_2\overset{O}{\overset{\|}{C}}OCa\tfrac{1}{2} \end{array} \right] \longrightarrow$$

$$\left[\begin{array}{c} \text{(pyridine-pyrroline)} CO_2Ca\tfrac{1}{2} \end{array} \right] \longrightarrow II$$

The yield of myosmine, II, was 67%, the by-product being 2-pyrrolidone. Benzoylpyrrolidone gives 2-phenylpyrroline under the above conditions.

A similar condensation and decarboxylation of caprolactam leads to the rather interesting aminoketone, 1,11-diamino-6-undecanone [15]:

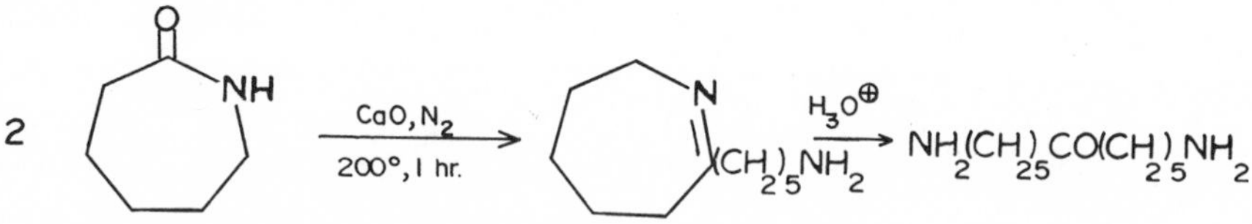

$$2 \;\; \text{(caprolactam)} \;\; \xrightarrow[200°, 1\,hr.]{CaO, N_2} \;\; \text{(imine-}(CH_2)_5NH_2) \;\; \xrightarrow{H_3O^{\oplus}} \;\; NH_2(CH_2)_5CO(CH_2)_5NH_2$$

9. From Carbenes and Amines

$$[CH_2{:}] + R_2NH \longrightarrow R_2NCH_3$$

This subject has been reviewed [16,17]. The reaction is occasionally catalyzed by cuprous cyanide [18]:

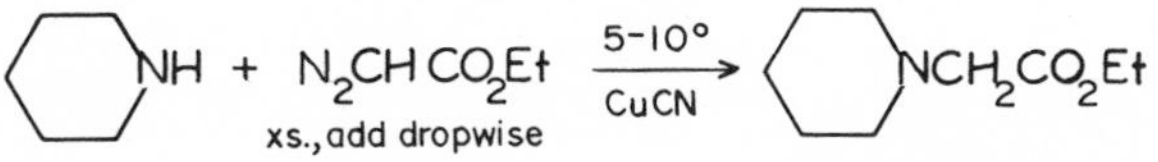

$$\text{(piperidine)}NH + N_2CHCO_2Et \xrightarrow[CuCN]{5-10°} \text{(piperidine)}NCH_2CO_2Et$$

xs., add dropwise

N-Carboethoxymethylpiperidine
72 %

On the other hand, tertiary amines give poor yields of insertion products [16]:

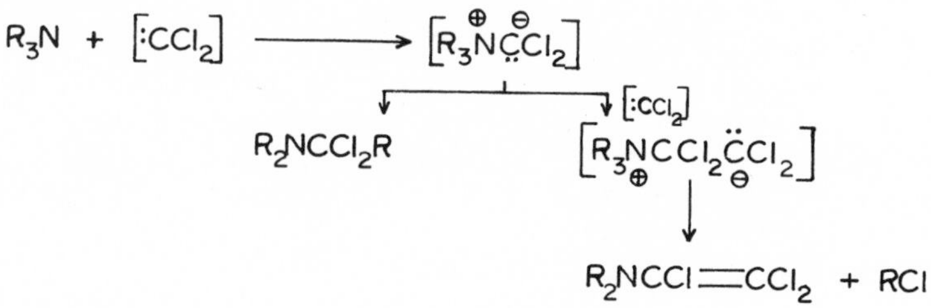

$$R_3N + [{:}CCl_2] \longrightarrow \left[R_3\overset{\oplus}{N}\overset{\ominus}{C}Cl_2 \right]$$

$$R_2NCCl_2R \qquad \left[R_3\overset{\oplus}{N}CCl_2\overset{\ddot{}}{\underset{\ominus}{C}}Cl_2 \right]$$

$$R_2NCCl{=\!=\!=}CCl_2 + RCl$$

10. From Sulfur Ylids

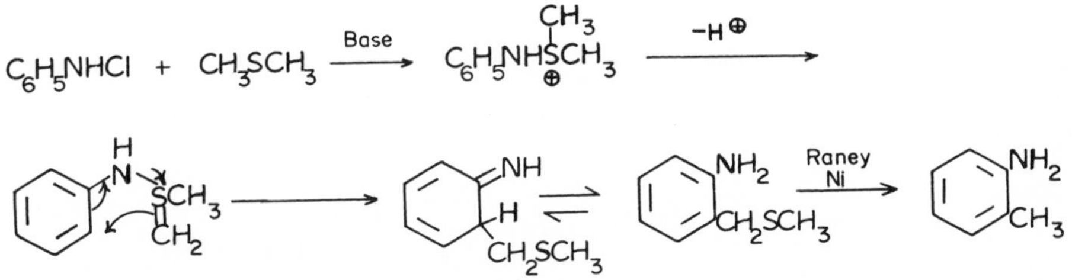

C_6H_5NHCl + CH_3SCH_3 $\xrightarrow{\text{Base}}$ $C_6H_5NH\overset{+}{S}CH_3$ $\xrightarrow{-H^{\oplus}}$

The ortho alkylation is a very mild reaction, not confined to methylation [19]. For instance, tetrahydrothiophene gives the ortho product, which is desulfurized to o-butylaniline in 40% overall yield. Similarly, 3-methyl-2-aminopyridine has been obtained in 50% overall yield [20].

11. From Nitriles and a Strong Base

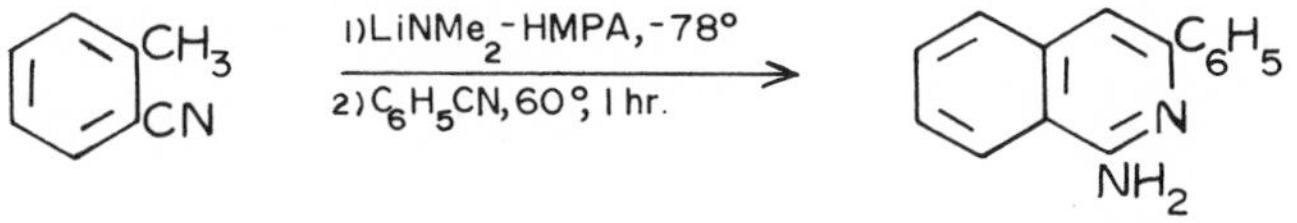

1-Amino-3-phenylisoquinoline
70%

Many other aryl nitriles may be added in the second step to produce isoquinolines [21].

References

1. J. C. Falk et al., J. Org. Chem., 37, 4243 (1972).
2. N. Imai et al., Tetrahedron Letters, 3517 (1971).
3. D. Rose, Tetrahedron Letters, 4197 (1972).
4. R. S. Neale, Synthesis, 1 (1971).
5. A. F. M. Iqbal, Helv. Chim. Acta, 54, 1440 (1971).
6. H. Lehmkuhl, D. Reinehr, J. Organomet. Chem., 55, 215 (1973).
7. H. Weingarten et al., J. Org. Chem., 35, 1542 (1970).
8. E. Schmitz, H. Fechner, Org. Prep. Proc., 1, 253 (1969).
9. J. V. Greenhill, M. D. Mehta, J. Chem. Soc., C, 1549 (1970).
10. M. Miocque, J.-M. Vierfond, Bull. Soc. Chim. France, 1907 (1970).
11. C. W. Thornber, Chem. Commun., 238 (1973).
12. J. Gloede et al., Arch. Pharm., 302 354 (1969).

13. A. Eschenmoser et al., Angew. Chem. Intern. Ed. Engl., 10, 330 (1971).
14. B. P. Mundy et al., J. Org. Chem., 37, 1635 (1972).
15. R. Harada, K. Teramoto, Japan 6,926,661, Nov. 7, 1969; C. A., 72, 12605 (1970).
16. W. Kirmse, Carbene Chemistry, Academic, New York, 1971, p. 409.
17. T. L. Gilchrist, C. W. Rees, Carbenes, Nitrenes, and Arynes, Appleton-Century-Crofts, New York, 1969, p. 81.
18. T. Saegusa et al., Tetrahedron Letters, 6131 (1966).
19. P. G. Gassman, G. Gruetzmacher, J. Am. Chem. Soc., 95, 588 (1973).
20. P. G. Gassman, C. T. Huang, J. Am. Chem. Soc., 95, 4453 (1973).
21. E. M. Kaiser et al., Synthesis, 805 (1974).

E. Addition (Mostly Organometallic Methods)

2. From Nitriles Followed by Reduction (1, 474)

$$RMgX + R'CN \longrightarrow RC{=}NMgXR' \xrightarrow{[H]} RCHNH_2R'$$

Rather than perform the reduction step as shown above, the nitrile addition step has been forced to give the tert-carbinamine [1]:

$$2C_6H_5CN + 3\ EtMgBr \xrightarrow[\substack{2)\ H,OH \\ \oplus\ \ominus}]{1)\ C_6H_5CH_3} C_6H_5COCH_2CH_3 + C_6H_5\overset{Et}{\underset{Et}{C}}NH_2$$

Propiophenone, 40%

α,α-Diethylbenzylamine, 60%

Different nitriles and Grignard reagents give yields of amines ranging from 10-60%. This type of preparation has been accomplished before with allyl Grignard reagents [2].

An optically active amine of rather high optical purity has been synthesized by the typical addition-reduction procedure [3]:

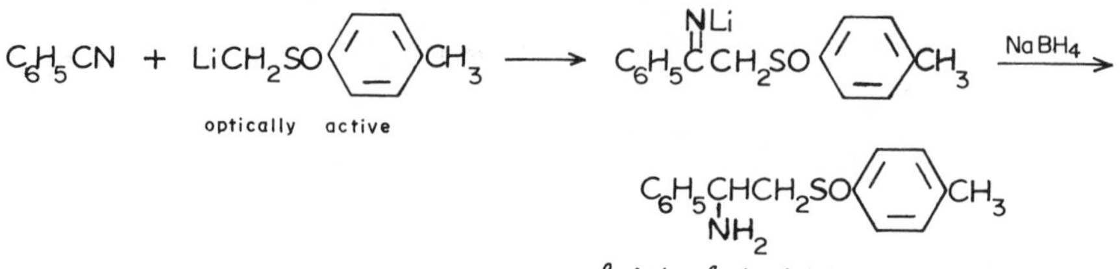

$$C_6H_5CN + LiCH_2SO\text{—}C_6H_4\text{—}CH_3 \longrightarrow C_6H_5\overset{NLi}{\underset{}{C}}CH_2SO\text{—}C_6H_4\text{—}CH_3 \xrightarrow{NaBH_4}$$

optically active

$$C_6H_5CHCH_2SO\text{—}C_6H_4\text{—}CH_3$$
$$\underset{NH_2}{}$$

β-Amino-β-phenylethyl-p-tolylsulfoxide, 82.5%

The product was reduced to an optically active α-phenethylamine by Raney nickel.

   3.   From Azomethines or Like Compounds and Oxonium Salts or Ethynyl Amines (1, 474)

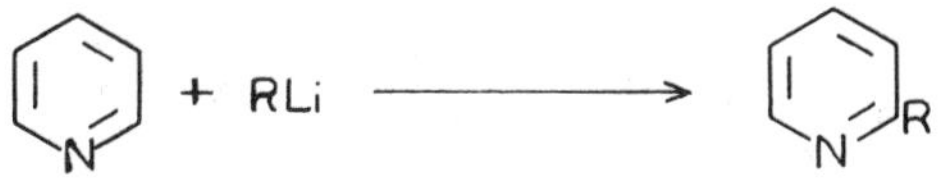

Two papers have appeared on the addition of excess alkyllithium to pyridines to give 2,6-dialkylpyridines and in the case of t-butyllithium to give also 2,4,6-tri-t-butylpyridine [4,5].

The same results can be obtained by conducting the reaction in a free radical manner. For instance, the ethyl free radical can be generated from propionic acid and allowed to react with quinaldine to give 4-ethyl-2-methylquinoline in 98% yield. However, the free radical does not distinguish well between the 2- and 4-positions of pyridine or quinoline [6].

The phenyl group has been substituted in quinolines to form 2-phenylquinolines (1, 476) [7]:

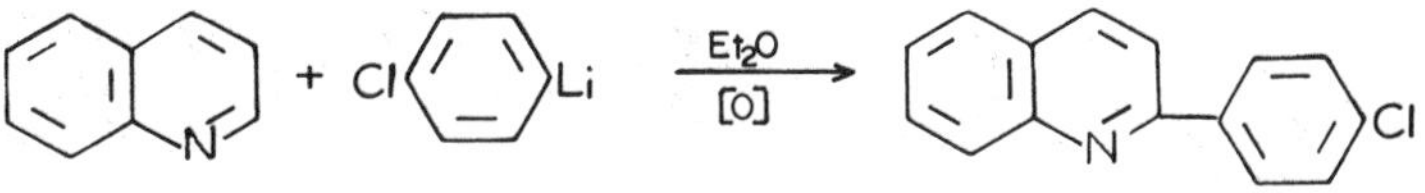

However, p-fluorophenyllithium did not react with 3-sec-butoxypyridine [8].

Not only may an optically active sulfoxide add to nitriles to give optically active amines (E.2) but also it adds to Schiff bases [3]:

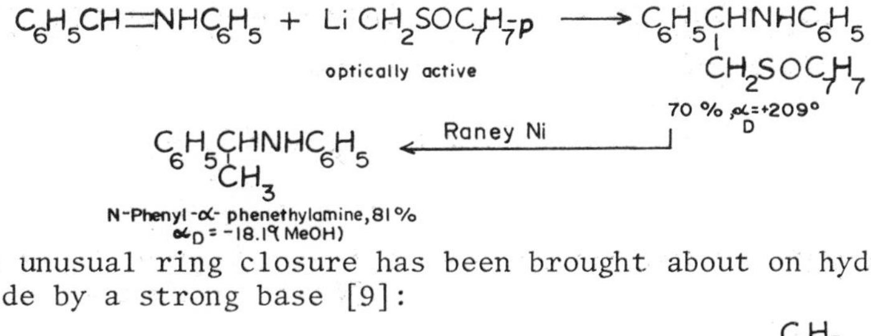

An unusual ring closure has been brought about on hydrobenzamide by a strong base [9]:

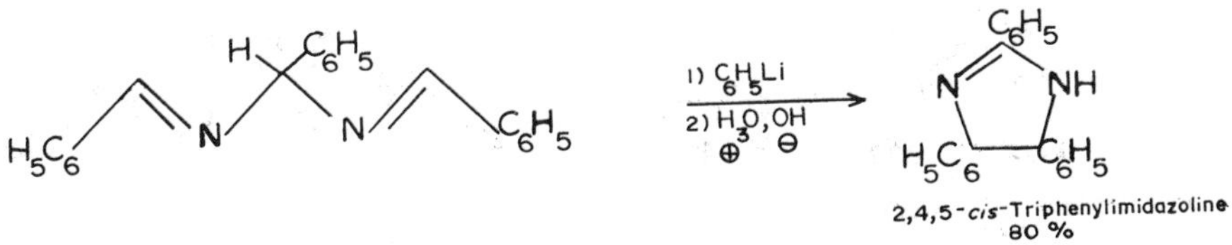

2,4,5-cis-Triphenylimidazoline
80 %

Trichloromethylalkylamines are formed by reactions of Schiff bases with trichloroacetic acid [10]:

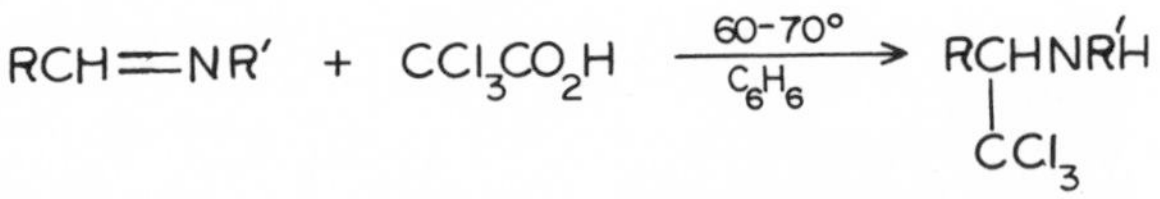

$$RCH{=}NR' \ + \ CCl_3CO_2H \ \xrightarrow[\substack{C_6H_6}]{60\text{-}70°} \ RCHNR'H$$
$$\underset{\displaystyle CCl_3}{|}$$

An interesting chemistry of ethynylamines is shown to be capable of leading to substituted allylamines by the following transformations [11]:

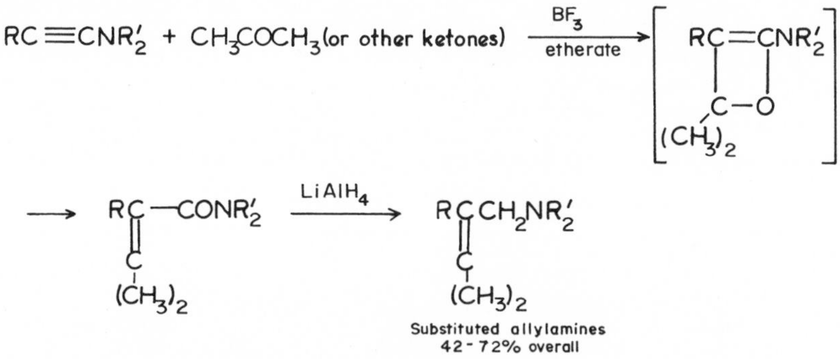

$$RC{\equiv}CNR'_2 \ + \ CH_3COCH_3 \text{(or other ketones)} \ \xrightarrow[\text{etherate}]{BF_3} \ \left[\begin{array}{c} RC{=}CNR'_2 \\ | \quad\ | \\ \underset{(CH_3)_2}{C-O} \end{array} \right]$$

$$\longrightarrow \ \underset{\displaystyle \underset{(CH_3)_2}{\overset{|}{C}}}{RC-CONR'_2} \ \xrightarrow{LiAlH_4} \ \underset{\displaystyle \underset{(CH_3)_2}{\overset{|}{C}}}{RC CH_2NR'_2}$$

Substituted allylamines
42-72% overall

The formation of pyridine compounds can be brought about from aromatic oxonium salts as follows [12]:

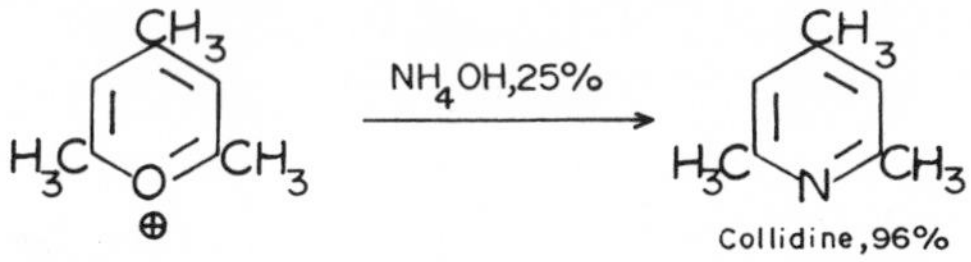

Collidine, 96%

5. From Allyl- and Propargylamines and Allyl Grignard Reagents

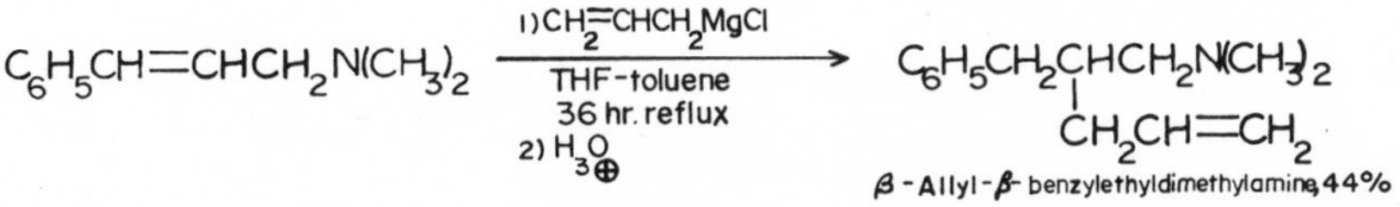

$$C_6H_5CH{=}CHCH_2N(CH_3)_2 \ \xrightarrow[\substack{\text{THF-toluene}\\ \text{36 hr. reflux}\\ 2)\ H_3O^{\oplus}}]{1)CH_2{=}CHCH_2MgCl} \ \underset{\displaystyle \underset{CH_2CH{=}CH_2}{|}}{C_6H_5CH_2CHCH_2N(CH_3)_2}$$

β-Allyl-β-benzylethyldimethylamine,44%

The preceding reaction has been carried out by Richey and co-workers [13]. The yield of 2-benzyl-4-pentenylamine was 44%. With the primary in place of the tertiary amine, the yield was 23%. The corresponding acetylenic tertiary amine gave 51% of the product.

References

1. G. Alvernhe, A. Laurent, Tetrahedron Letters, 1057 (1973).
2. H. R. Henze et al., J. Am. Chem. Soc., 73, 4915 (1951).
3. G. Tsuchihashi et al., Tetrahedron Letters, 3389 (1973).
4. F. V. Scalzi, N. F. Golob, J. Org. Chem., 36, 2541 (1971).
5. R. F. Francis et al., Chem. Commun., 1420 (1971).
6. F. Minisci, Synthesis, 1 (1973).
7. D. E. Pearson et al., J. Med. Chem., 14, 1218 (1971); 13, 383 (1970).
8. G. L. Walford et al., J. Med. Chem., 14, 339 (1971).
9. D. H. Hunter, S. K. Sim, Can. J. Chem., 50, 669 (1972).
10. A. Lukasiewicz et al., Rocz. Chem., 46, 2321 (1972); C. A., 79, 4889 (1973).
11. R. Fuks, H. G. Viehe, Chem. Ber., 103, 564 (1970).
12. G. N. Dorofenko, S. V. Krivau, Metody Poluch. Khim. Reactivov Prep., 149 (1967); C. A., 71, 61161 (1969).
13. H. G. Richey et al., Tetrahedron Letters, 2183 (1971).

F. Other Additions and Substitutions

1. From Olefins and Nitriles (Ritter) (1, 479)

$$\overset{\oplus}{R} + R'CN \longrightarrow R'\overset{\oplus}{C}=NR \xrightarrow{H_2O} R'\overset{O}{\overset{\|}{C}}NHR \xrightarrow{OH^{\ominus}} RNH_2$$

This reaction is discussed more thoroughly in reference to carboxylic acid amides, Chapter 18, D.4, but since it is used to synthesize amines, a recent procedure is given (Ex. a).

a. Preparation of 2-Amino-2,3,3-trimethylbutane (t-Heptylamine) [1]

To 1 mol of sulfuric acid in 500 ml of acetic acid 1.1 mol of acetonitrile was added carefully at 20°. 2,3,3-Trimethylbutan-2-ol (1 mol) was added and the mixture was stirred at 25° for 24 hr. The diluted mixture gave the crystalline amide, which was saponified with potassium hydroxide in ethylene glycol to give 61% of the amine, boiling point 120°.

2. From Hydrocarbons, Haloamines, and the Like (1, 480)

$$RH + NCl_3 \xrightarrow{AlCl_3} RNH_2$$

Kovacic has added some interesting contributions to the synthesis of amines by this method. It will be recalled that amination of toluene with trichloroamine in the presence of aluminum chloride gave 39-43% of m-toluidine (1, 480). More details have now been published [2]. Trichloroamine always gave better yields than monochloroamine. Adamantane with the former gave 85% of 1-aminoadamantane. Alkyl halides with trichloroamine and aluminum chloride gave tert-carbinamines. For instance, t-butyl chloride gave 90% of t-butylamine [3]. Hydrocarbons also give the amine from the stablest carbonium ion formed but the product yield is somewhat controllable [4].

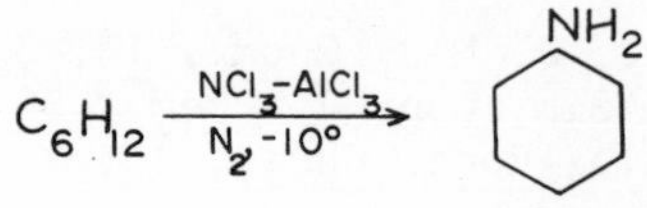

$$C_6H_{12} \xrightarrow[\text{N}_2, -10°]{\text{NCl}_3-\text{AlCl}_3}$$

Cyclohexylamine
45% of 80% purity

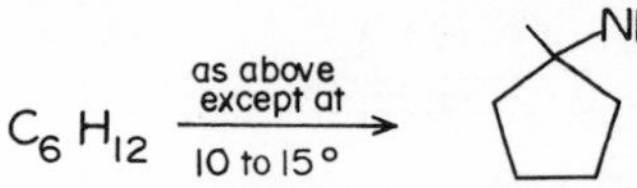

$$C_6H_{12} \xrightarrow[\text{10 to 15°}]{\text{as above}\atop\text{except at}}$$

1-Methylcyclopentylamine
46%

Decalin gave a fair yield of cis-1-aminodecalin and cyclooctane, a mixture of 1-amino-1,3- and 1,4-dimethylcyclohexanes.

Hydroxylamine hydrochloride and aluminum chloride can also be used as an aminating agent, but orientation, for aromatic hydrocarbons, is not well controlled [5]:

$$\text{HONH}_2\cdot\text{HCl} + \text{AlCl}_3 + \text{C}_6\text{H}_5\text{CH}_3 \xrightarrow{100°} 66.8\% \text{ Toluidines}$$
0.4 eq.   0.8 eq.   o, m, p : 40 : 15 : 45

Hydroxylamine also can be used to add to alkenes via a free radical reaction [6]:

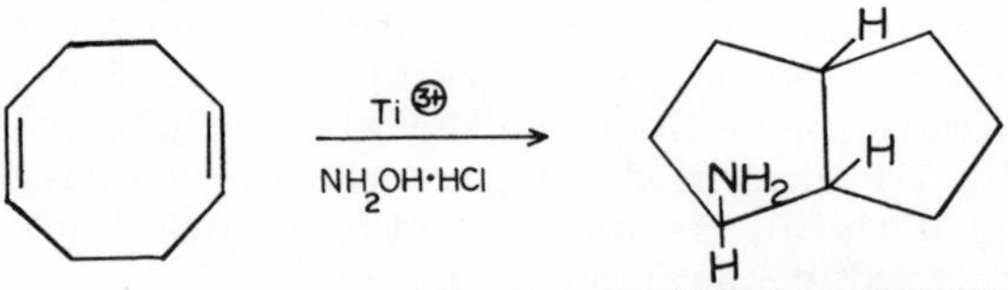

$$\xrightarrow[\text{NH}_2\text{OH}\cdot\text{HCl}]{\text{Ti}^{3\oplus}}$$

exo-1-Amino-cis-octahydropentalene
40-50%

The free radical amination is discussed more fully in F.7.

3. From Aromatic Compounds (Ring Closures, Friedel-Crafts, Skraup, etc.) (1, 482)

The Skraup reaction was discussed in some detail ($\underline{1}$, 482). Now other, selected reactions are considered. A field this broad cannot be covered completely, but certain syntheses appear to be of sufficient significance to be noted.

Friedel-Crafts reactions on 1,3-, 1,4-, and 1,5-aminoalcohols have been carried out to give substituted amines. For success the alcohol must be a tertiary one [7]. It is possible also to acylate phenethylamines [8]:

$$C_6H_5CH_2CH_2NMe_2 + AlCl_3 \xrightarrow[\text{2) OH}^{\ominus}]{\text{1) } C_6H_5SO_2Cl} C_6H_5SO_2 \langle\underline{\quad}\rangle CH_2CH_2NMe_2$$

2 eq.

p-Benzenesulfonylphenethyldimethylamine , *ca.* 50%
m p. 62-64°

A similar reaction has been carried out with benzoyl chloride ($\underline{1}$, 481). A nice ring closure of a phenethylamine type has been accomplished with super-PPA [9]:

$$p\text{-CH}_3C_6H_4CH_2CH_2NHCHO \xrightarrow[\substack{\text{PPA heated to 170°} \\ \text{and then the formamide} \\ \text{is added in a slow stream} \\ \text{with super PPA at 150°}}]{\text{65 g. }P_2O_5\text{ in 325 g.}}$$

7-Methyl-1,2,3,4-
tetrahydroisoquinoline, 93%

The synthesis of indole is discussed in G.7.

4.   From Nitrene Intermediates and the Like ($\underline{1}$, 485) (R N:
       or Other Reactive Intermediates)

The discussion concerning the various reactive intermediates of nitrogen compounds in synthesis continues to be applicable ($\underline{1}$, 412; $\underline{1}$, 485). Since that time a monograph has been published on nitrenes [10] and several reviews will be mentioned throughout this section. The nitrenes are prepared most frequently from the thermolysis or irradiation of azides. The alkyl azides are highly reactive and poor synthetic intermediates [10,11]. Vinyl, carbonylsulfonyl-, acyl-, cyano-, and aminonitrenes are less reactive and in a limited way undergo some of the reactions of the aryl azides that have been more thoroughly studied [10,11]. The reactions of the aryl azides are illustrated:

Hydrogen abstraction $[ArN]$ + RH $\longrightarrow$ ArNH$_2$

Insertion $[ArN]$ + RH $\longrightarrow$ ArNHR

Coupling $2 [ArN]$ $\longrightarrow$ ArN$\equiv$NAr

The above are competing reactions. In hydrogen abstraction,
which tends to be prevalent, the hydrogen atoms are abstracted
one at a time leading to free radicals and then followed by
characteristic reactions of the latter.
Other possible reactions are:

Intramolecular insertion $[C_6H_5N]$ $\xrightarrow{R_2NH}$

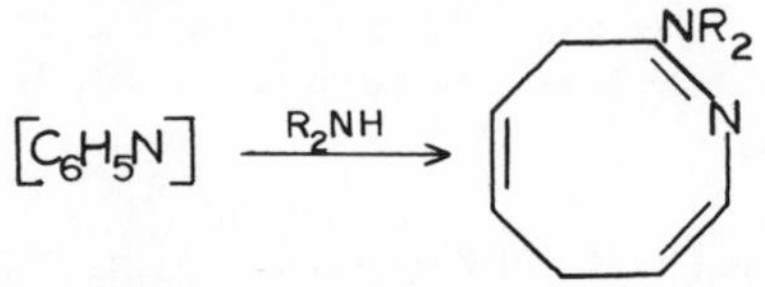

This azepine formation has been described (1, 488).

Complex formation $[Ar\ N]$ $\xrightarrow{(EtO)_3P}$ $ArN=P(OEt)_3$

Aziridine $[ArN]$ + $\overset{\diagdown}{\diagup}C=C\overset{\diagup}{\diagdown}$ $\longrightarrow$
formation

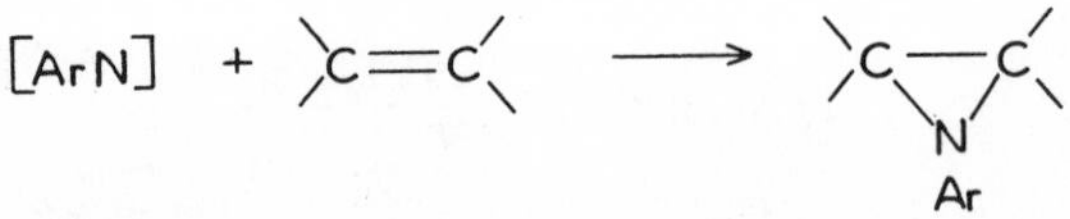

The latter reaction does not occur frequently with aromatic
azides but does with more reactive nitrenes.
    Nitrenes, or rather complexed nitrenes, are formed by the
Cadogan reaction, later developments of which have been reviewed
[12]:

$ArNO_2$ + xs. $(EtO)_3P$ $\longrightarrow$ $\left[ArN = P(OEt)_3\right]$ → $\left[Ar\ddot{N}:\right]$

And this reaction has led to the synthesis of many, useful
cyclization products [13]. Also it has been carried out
thermally [14].
    Of late the blocked ortho effect has been studied in detail;
in cases in which straight ring closure cannot take place, the
resultant convolutions of the intermediate are observed [12,15].

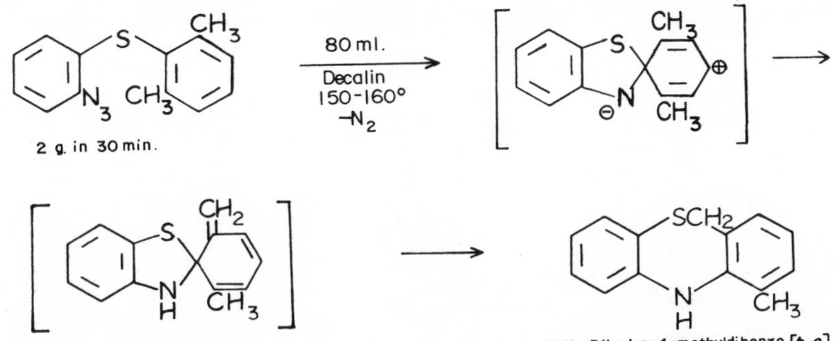

5,11-Dihydro-4-methyldibenzo [b,e],
[1,4]-thiazepine, 73%

The Cadogan method for preparation of amines by reduction of nitro compounds is discussed in A.1.

The decomposition of the azide in a protic solvent gave more of the insertion product [16]

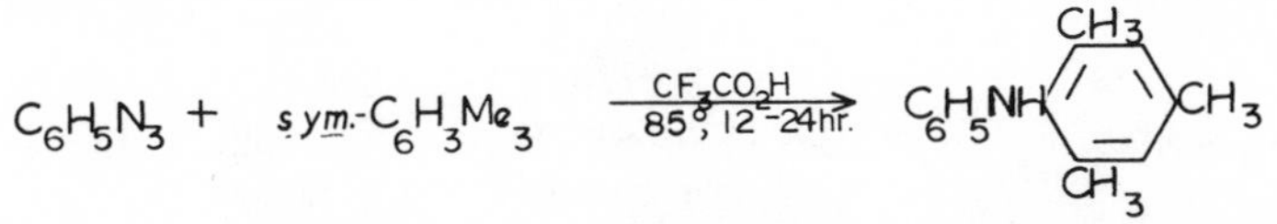

Phenylmesitylamine, 55%

Usually, more product is obtained thermally. In place of mesitylene, toluene and anisole, but not benzene, may be used as the insertion host. A two-step reaction may be carried out [17]:

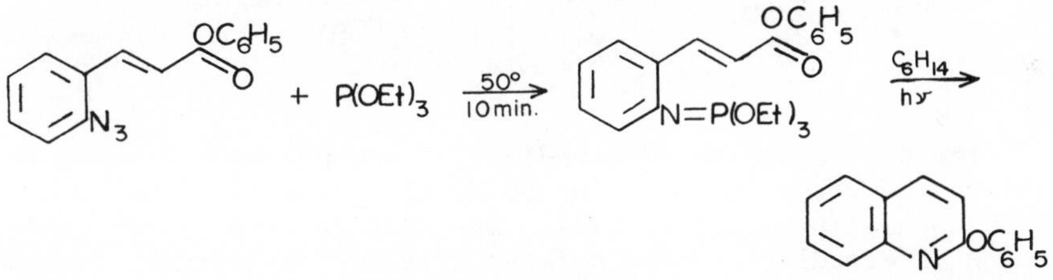

2-Phenoxyquinoline, 95%

It has been observed that irradiated nitrene products are dependent on the sensitizer in some cases [18]:

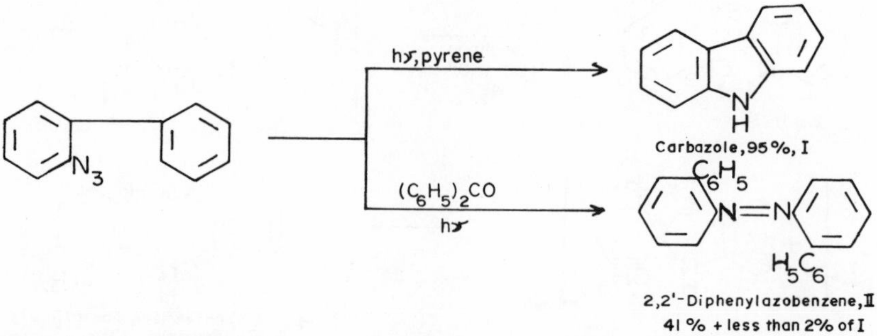

Carbazole, 95%, I

2,2'-Diphenylazobenzene, II
41% + less than 2% of I

Compound I is thought to be formed by a singlet and II, by a triplet nitrene.

The remainder of the examples occur by different nitrogen-reactive intermediates. Alkyl azides react with trialkylborons to form secondary amines by a reductive alkylation process [19]:

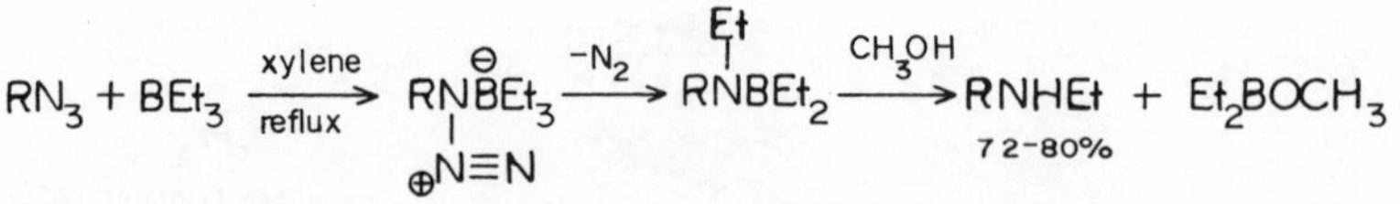

$$RN_3 + BEt_3 \xrightarrow[\text{reflux}]{\text{xylene}} \underset{\underset{\oplus}{N\equiv N}}{RNBEt_3^\ominus} \xrightarrow{-N_2} RN\overset{Et}{\underset{|}{B}}Et_2 \xrightarrow{CH_3OH} \underset{72-80\%}{RNHEt + Et_2BOCH_3}$$

By substituting a dialkylchloroborane for triethylboron yields are similar but the reaction is more rapid [20]:

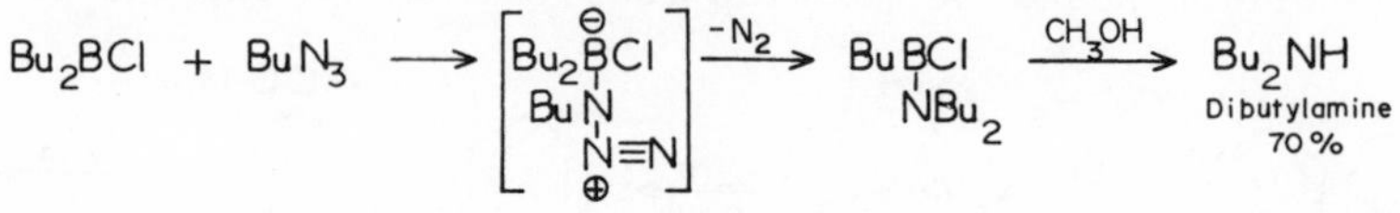

$$Bu_2BCl + BuN_3 \longrightarrow \begin{bmatrix} Bu_2\overset{\ominus}{B}Cl \\ Bu\,N \\ \underset{\oplus}{N\equiv N} \end{bmatrix} \xrightarrow{-N_2} \underset{NBu_2}{BuBCl} \xrightarrow{CH_3OH} \underset{\substack{\text{Dibutylamine} \\ 70\%}}{Bu_2NH}$$

The nitrenium ion free radical, $R_2\overset{+}{N}H$, has been invoked to explain the formation of 3,10'-biphenothiazine from the oxidation of phenothiazine (1, 486) [21]. The same particle is presumed to be the intermediate in the Hofmann-Löffler transformation (1, 485). In a related reaction, other products besides ring-closure ones are possible [22]:

$$Bu_2NCl \xrightarrow[\text{2) aq NaBr}]{\text{1) } CF_3CO_2H, CCl_4} HO(CH_2)_4NHBu$$

4-Hydroxydibutylamine, 76%

The hydrolysis as above is essential.

The nitrenium ion, $R_2\overset{\cdot\cdot}{N}{}^+$, has been studied of late [10,23] and interesting rearrangements have been discovered as shown for one example [23]:

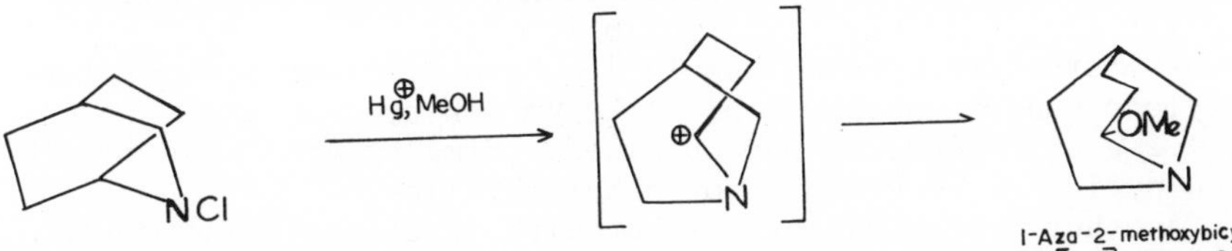

1-Aza-2-methoxybicyclo-
[3.2.1] octane, 60%

Further discussion of the nitrenium ion is to be found in ($\underline{1}$, 485). Other reactions involve more uncertain intermediates, including the Dewar pyridine detected from the irradiation of pyridine [24]. Although the intermediate cannot be isolated, it has been reduced to 2-azabicyclo [2.2.0]hex-5-ene by sodium borohydride.

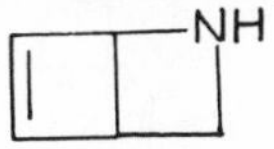

A series of N-oxo Schiff bases has been made by the elimination of what appears to be a product of a nitrene [25]:

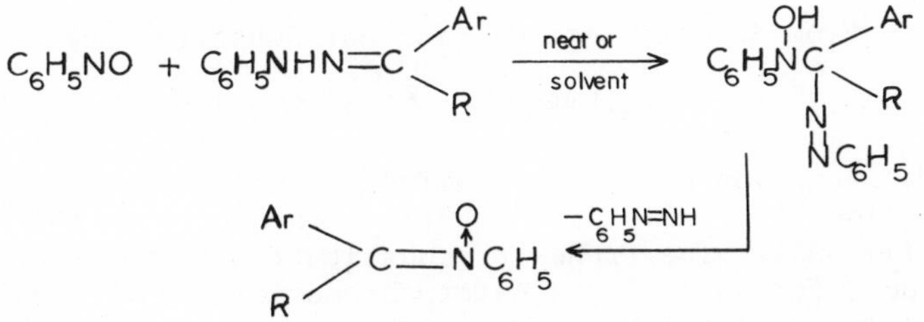

There were four examples using three aromatic aldehydes and fluorenone.

5. From 1,3- or 1,4-Dipolar Compounds ($\underline{1}$, 488)

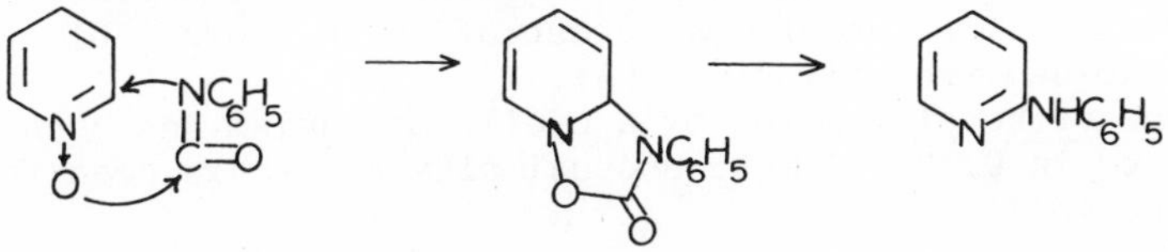

2-Anilinopyridine was prepared in 77% yield by heating the N-oxide and phenylisocyanate in DMF at 110° for 7 hr. Vacuum distillation gave the product [26].

6. From Azo, Azomethine, Nitroso, or Nitrile Compounds (Mostly Cycloaddition) (1, 491)

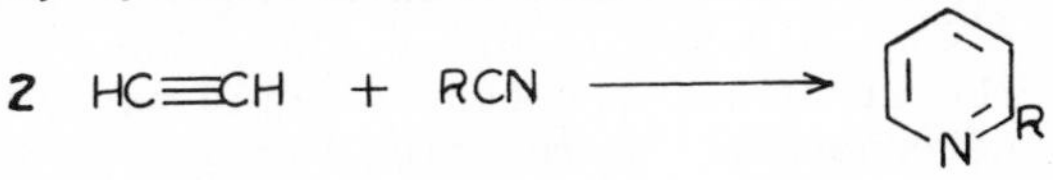

This reaction has been accomplished by means of a cobalt catalyst complexed with triphenylphosphine, tetraphenyl- and the unsubstituted cyclopentadiene [27]. From 2.3 g of acetonitrile, acetylene under a pressure of 11 kg/cm^2 in benzene at 70° for 7 hr, 1.2 g of 2-picoline was obtained. Similar yields of substituted pyridines were found for other nitriles.

Another cycloaddition is that with an ethynylamine and a dimer of dimethylketene to give a λ-pyrene [28]:

Other cycloadditions with the ethynylamine are described.

7. From Aromatic Amines (Nuclear Substitution)

$$C_6H_5NH_2 \xrightarrow{X_2} XC_6H_4NH_2$$

Halogenation of aromatic amines, usually carried out on the acetanilide, is facile, but substitution occurs on the amino group (1, 398). The trend for substituting in the free aniline continues, for which 2,4,4,6-tetrabromocyclohexa-2,5-dienone has been utilized as the halogenating agent, to give p-bromoaniline in 87% yield [29]. Iodination of deactivated anilines, such as p-nitroaniline, has been brought about by stirring a mixture of the aniline, iodine, water, and chlorobenzene at 90-93° for 35 hr. As the pH approaches 2 (from HI release), hydrogen peroxide is added to increase the pH. The chlorobenzene serves to suppress sublimation of iodine out of the mixture [30]. Yield of 2-iodo-4-nitroaniline was 95%.

New ortho alkylation by a cyclic mechanism has been discussed in D.10. A most unusual alkylation has been found for acridine [31]:

9-Phenacylacridine, 68 %

Employing dimethylaniline rather than acetophenone,9-(p-dimethylphenyl)acridine was obtained in 62% yield. The former reaction may occur via the attack of the benzoylacridine complex on the enol form of acetophenone. A reaction similar to the latter example has been brought about photochemically [32]:

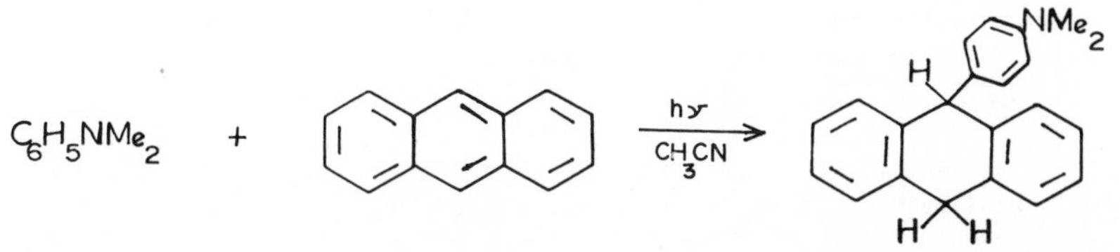

$C_6H_5NMe_2$ +   [anthracene structure] $\xrightarrow[\text{CH}_3\text{CN}]{h\nu}$ [9-(p-dimethylaminophenyl)-9,10-dihydroanthracene structure with NMe₂]

9 (*p*-Dimethylaminophenyl)-9,10-
dihydroanthracene, 60-65 %

Of course, this mechanism is free radical. If benzene is used as a solvent in place of acetonitrile, only the photodimer of anthracene is obtained.

   8.   From N-Chloroamines and Aromatic or Heterocyclic Compounds (Free Radical Reactions)

$$R_2\overset{\oplus}{N}HCl \xrightarrow{M^{\oplus}} \left[R_2\overset{\oplus}{N}H\cdot\right] \xrightarrow{ArH} Ar\,NR_2$$

The above ion free radical reaction has been reviewed [33]. Among the metal ions that may be used are Fe^{2+}, Ti^{3+}, Cu^+, Cr^{2+}, and, indeed, sodium sulfate has served to catalyze one amination (although in a manner differing from oxidation). Conversions with benzene and various N-chloroamines, of which one example is given (Ex. a), ranged within 15-79%. The chloroamine in acid solution can be reduced photochemically as well. Unfortunately with substituted benzenes, a mixture of o and p-isomers is obtained.

a.   Preparation of Dimethylaniline [33]

   To a mixture of N-chlorodimethylamine (4.3 g), acetic acid (50 ml), C_6H_6 (30 ml), and H_2SO_4 (83 ml) was added 1 equiv of finely ground ferrous sulfate. The temperature rose from 15° to 27° in 5 min and stirring was continued for an additional 15 min before pouring the mixture into ice and making it basic. Yield of the amine was 5 g (76%).

   9.   From Copper Benzoates and Ammonia

$$2\,Cu(O\overset{O}{\overset{\|}{C}}C_6H_5)_2 + 2\,NH_3 \xrightarrow[\substack{220°,\text{bomb} \\ 35\ atm.}]{C_6H_5CO\,NH_4} 2\,Cu\overset{O}{\overset{\|}{O}}CC_6H_5 + CO_2 + C_6H_5NH_2 \ [34]$$

72%

It will be interesting to see if this reaction is more general.

Addenda

F.4.   α-Arylvinylazides, $ArCN_z=CH_2$, in refluxing toluene give arylazirines (54-63%) [A. G. Hortmann et al., J. Org. Chem., 37, 322 (1972)].

F.6.   The exothermic cyclo addition of the enamine of cyclohexanone and morpholine to isobenzofuroxan, , gives 1,2,3,4-tetrahydrophenazine-N,N-dioxide (48%). The reaction has proved to be valuable in the preparation of heterocycles [M. J. Haddadin, C. H. Issidorides, Tetrahedron Letters, 3253 (1965)].

F.6.   Benzofuroxan in the preparation of difficulty accessible heterocyclic compounds has been reviewed [K. Ley, F. Seng, Synthesis, 415 (1975)].   Similarly, the use of 1-azirines to form heterocycles by cycloaddition has been reviewed [D. J. Anderson, A. Hassner, Synthesis, 483 (1975)].

References

1.   J. W. Timberlake et al., Synthesis, 632 (1972).
2.   J. W. Strand, P. Kovacic, J. Am. Chem. Soc., 95, 2977 (1973).
3.   P. Kovacic, M. K. Lowery, J. Org. Chem., 34, 911 (1969).
4.   P. Kovacic et al., J. Org. Chem., 35, 2146 (1970).
5.   P. I. Fedorova et al., Zh. Prikl. Khim. (Leningrad), 46, 1079 (1973); C. A., 79, 42066 (1973).
6.   R. P. A. Sneeden, Synthesis, 259 (1971).
7.   A. C. Cope, W. D. Burrows, J. Org. Chem., 31, 3093 (1966).
8.   D. E. Pearson, M. Y. Moss, unpublished work.
9.   D. M. Bailey et al., J. Med. Chem., 16, 151 (1973).
10. W. Lwowski, Nitrenes, Interscience, New York, 1970.
11. R. K. Smalley, H. Suschitzky, Chem. Ind. (London), 1338 (1970); S. Hünig, Helv. Chim. Acta, 54, 1721 (1971).
12. J. I. G. Cadogan, Acc. Chem. Res., 5, 303 (1972).
13. J. I. G. Cadogan, R. K. Mackie, Chem. Soc. Rev., 3, 87 (1974).
14. R. J. Sundberg et al., J. Am. Chem. Soc., 91, 658 (1969).
15. J. I. G. Cadogan, S. Kulik, J. Chem. Soc., C, 2621 (1971).
16. R. J. Sundberg, K. B. Sloan, J. Org. Chem., 38, 2052 (1973).
17. D. G. Saunders et al., Chem. Commun., 29 (1973).
18. J. S. Swenton et al., Chem. Commun., 1263 (1969).
19. H. C. Brown et al., J. Am. Chem. Soc., 93, 4329 (1971).
20. H. C. Brown et al., J. Am. Chem. Soc., 94, 2114 (1972).
21. Y. Tsujino, Tetrahedron Letters, 763 (1969).

22. R. S. Neale, Synthesis, 3 (1971).
23. P. G. Gassman, Acc. Chem. Res., 3, 26 (1970).
24. K. E. Wilzbach, D. J. Rausch, J. Am. Chem. Soc., 92, 2178 (1970).
25. R. G. Landolt et al., J. Org. Chem., 35, 845 (1970).
26. R. Huisgen et al., Chem. Ber., 102, 926 (1969).
27. Y. Wakatsuki, H. Yamazaki, Tetrahedron Letters, 3383 (1973).
28. R. Gompper, J. Stetter, Tetrahedron Letters, 233 (1973).
29. V. Caló et al., J. Chem. Soc., C, 3652 (1971).
30. I. Toth, Helv. Chim. Acta, 54, 1486 (1971).
31. S. G. Potashnikova et al., Metody Poluch. Khim. Reaktivov Prep., 137 (1971); C. A., 79, 91960 (1973).
32. C. Pac, H. Sakurai, Tetrahedron Letters, 3829 (1969).
33. F. Minisci, Synthesis, 1 (1973).
34. G. G. Arzoumanidis, F. C. Rauch, Chem. Commun., 666 (1973).

G. Molecular Rearrangements

In 2, 3, 4, and 5 the isocyanate is invariably the intermediate.

2. From Amides via the Haloamide (Hofmann) (1, 496)

$$RCONH_2 \xrightarrow{NaOX} RNH_2 + CO_2$$

An example of a fairly recent application is given [1]:

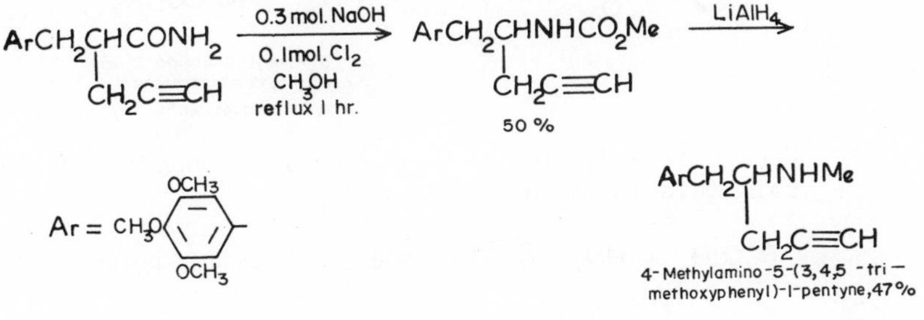

Symmetrical hydrazines have been made from sulfamides [2]:

$$RNHSO_2NHR \xrightarrow{NaOCl} RNHNHR$$

Dibutylhydrazine,62%
R= Bu

Excess NaOCl produced the azoalkane.

3. From Hydrazides or Acid Chlorides via the Azide
    (Curtius) (1, 498)

$$RCON_3 \longrightarrow RN=C=O \ + \ N_2$$

The intermediate isocyanate has been shown to insert into
the thiophene ring [3]:

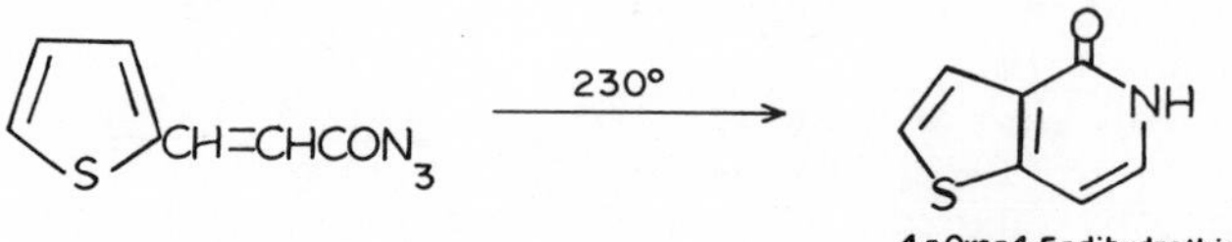

4 - Oxo-4,5-dihydrothieno-
[3,2-c]pyridine, 70%

4. From Hydroxamic Acids (Lossen) (1, 500)

$$RCONHOH \longrightarrow RN=C=O \longrightarrow RNH_2$$

In alkaline degradation of the above, the 2,4-dinitrophenyl
ether, C_6H_5CONHO-(NO2)(NO2) has been shown to rearrange many
times faster than the p-nitro ether [4].
The benzenesulfonate ester in the
following example rearranges instantly [5]:

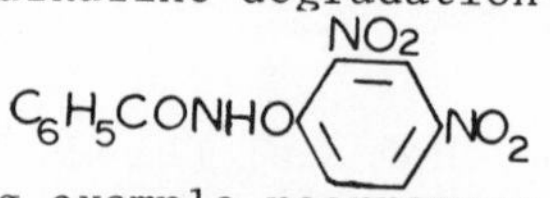

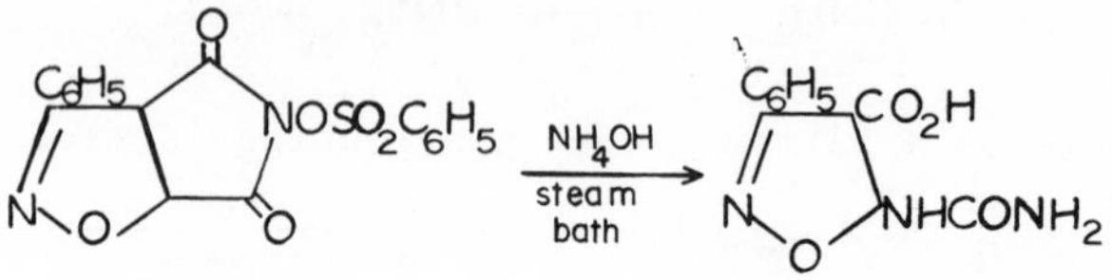

cis-3-Phenyl-5-ureido-2-iso-
oxazoline-4-carboxylic acid
51% + the amide, 28%

5. From Carboxylic Acids or Carbonyl Compounds and Hydra-
    zoic Acid (Schmidt) (1, 501)

$$RCO_2H \ + \ HN_3 \xrightarrow{H_2SO_4} RN=C=O \longrightarrow RNH_2$$

Tertiary carboxylic acids fragment to the carbonium ion
which combines with the hydrazoic acid to form a product that
rearranges to give aniline [6]:

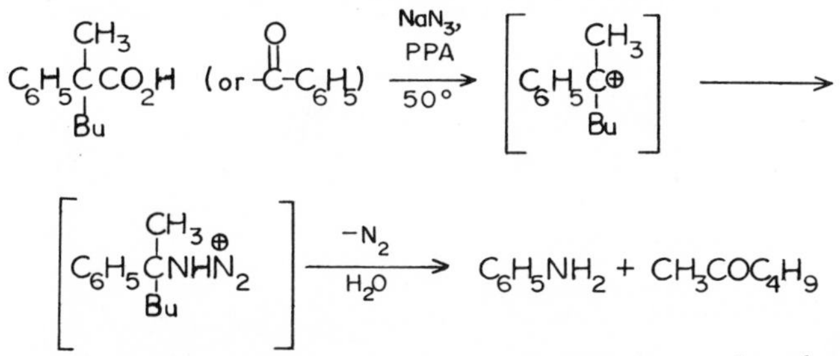

$$C_6H_5\overset{CH_3}{\underset{Bu}{C}}CO_2H \ \ (\text{or} \ \overset{O}{C}-C_6H_5) \ \xrightarrow[50°]{\underset{PPA}{NaN_3,}} \ \left[C_6H_5\overset{CH_3}{\underset{Bu}{C}\oplus} \right] \longrightarrow$$

$$\left[C_6H_5\overset{CH_3}{\underset{Bu}{C}}NHN_2 \right]^{\oplus} \ \xrightarrow[H_2O]{-N_2} \ C_6H_5NH_2 + CH_3COC_4H_9$$

Ketones in the Schmidt reaction rearrange as in the Beckmann
[7]:

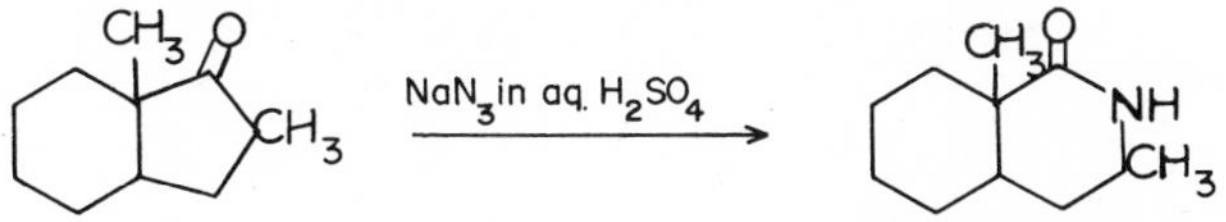

cis-3,8a-Dimethyl-3,4,4a,5,6,7,8,8a-
octahydroisocarbostyril, 27-31%

6.   From Hydrazobenzenes (Benzidine Rearrangement) (1, 503)

$$C_6H_5NHNHC_6H_5 \ \xrightarrow{H\oplus} \ H_2N\langle\underline{\ \ }\rangle\langle\underline{\ \ }\rangle NH_2$$

The benzidine rearrangement references have dropped off
noticeably. The ion free radical of tetraphenylhydrazine has
been found to give the oxidized product of a benzidine [8]:

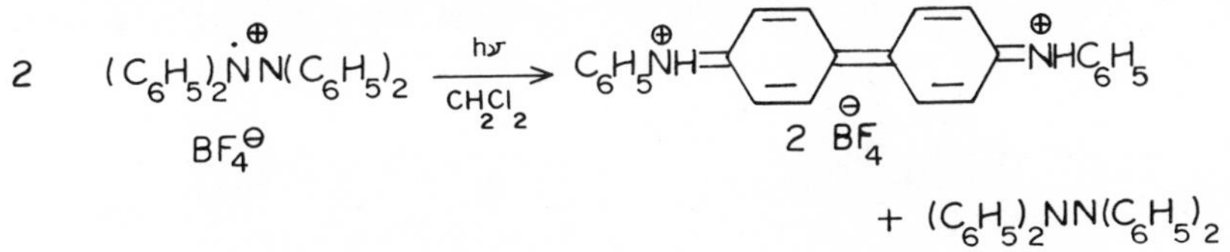

$$2 \ \ (C_6H_5)_2\overset{\cdot\oplus}{N}N(C_6H_5)_2 \ \xrightarrow[CHCl_2]{h\nu} \ C_6H_5\overset{\oplus}{N}H=\langle\underline{\ \ }\rangle=\langle\underline{\ \ }\rangle=\overset{\oplus}{N}HC_6H_5$$

$$BF_4^{\ominus} \qquad\qquad 2\ B\overset{\ominus}{F}_4$$

$$+ \ (C_6H_5)_2NN(C_6H_5)_2$$

7.   From Arylhydrazones (Fischer Indole Synthesis) (1, 504)

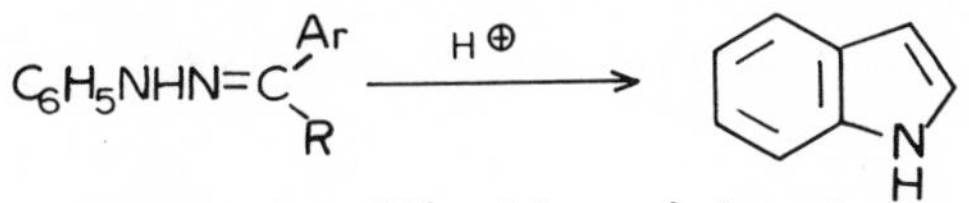

$$C_6H_5NHN=C\overset{Ar}{\underset{R}{\diagdown}} \ \xrightarrow{H\oplus} \ \text{indole}$$

A series of indoles (14-79%, 12 ex.) has been prepared from
phenylhydrazones in PPA at 180° [9]. A facile indole synthesis
has been carried out with a dienehydrazine [10]:

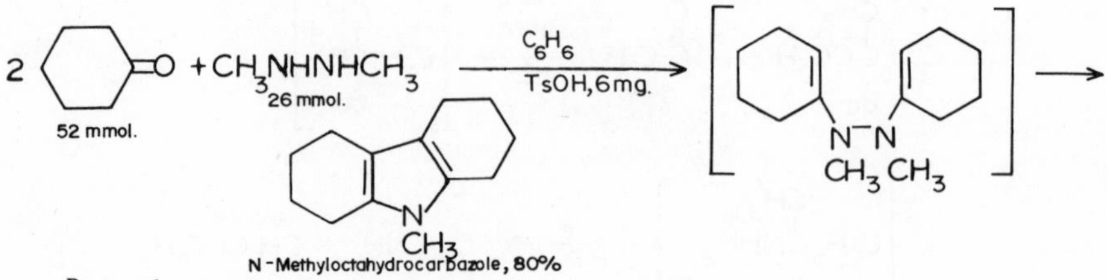

N-Methyloctahydrocarbazole, 80%

But the most important advance, based on novelty and mildness of reaction, is an indole synthesis via an ylid mechanism [11]:

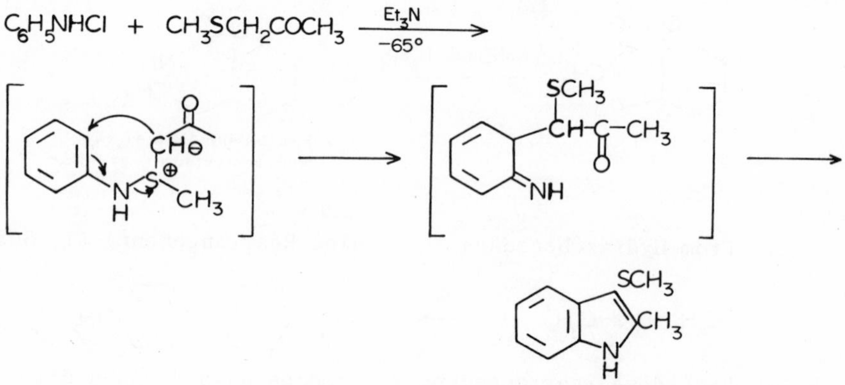

2-Methyl-3-thiomethylindole
69%

The thiomethyl group could be removed by Raney Ni at 25° in at least 70% yield.

Indole formation also has been found to occur with β-bromoallylamines [12]:

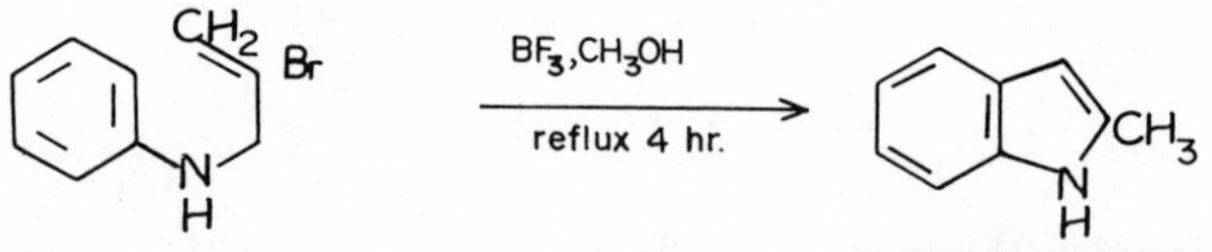

2-Methylindole, 90% crude

8. From Benzyltrialkylammonium Halides (Stevens, Sommelet-
    Hauser) (1, 505)

Stevens:

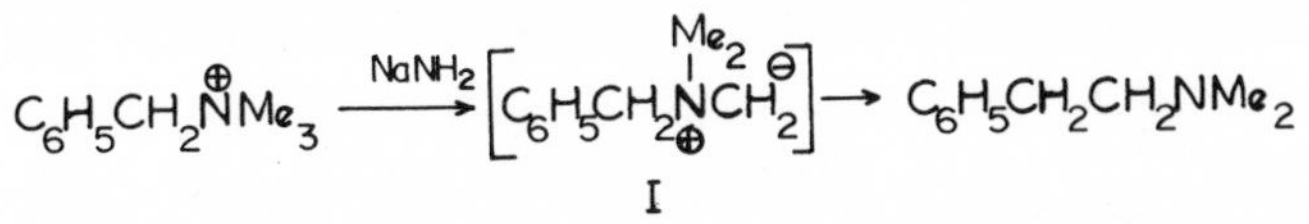

$$C_6H_5CH_2\overset{\oplus}{N}Me_3 \xrightarrow{NaNH_2} \left[C_6H_5\overset{\ominus}{C}H\overset{\overset{Me_2}{|}}{\underset{\oplus}{N}}CH_2 \right] \rightarrow C_6H_5CH_2CH_2NMe_2$$

I

Sommelet Hauser:

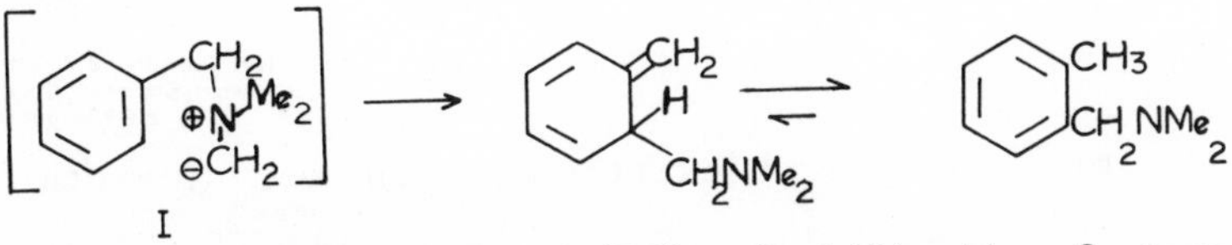

This subject has been reviewed [13]. Stabilization factors for the intermediate ylid determine the course of the reactions, but there is some control by the conditions utilized [14]:

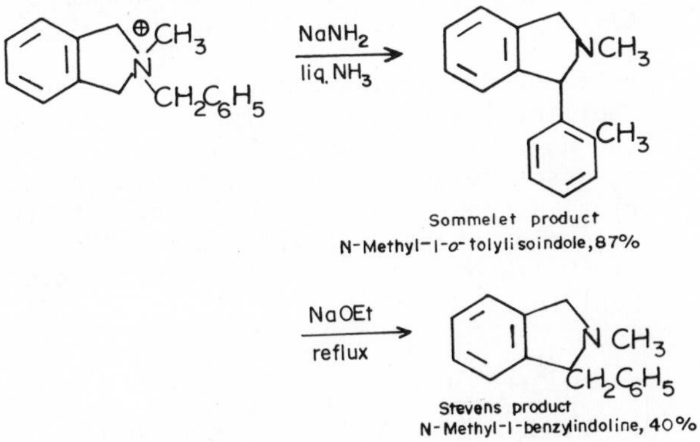

Sommelet product
N-Methyl-1-o-tolylisoindole, 87%

Stevens product
N-Methyl-1-benzylindoline, 40%

Control by stabilization of the ylid is evident in the rearrangement of A, which gives B or benzyl migration exclusively [13]:

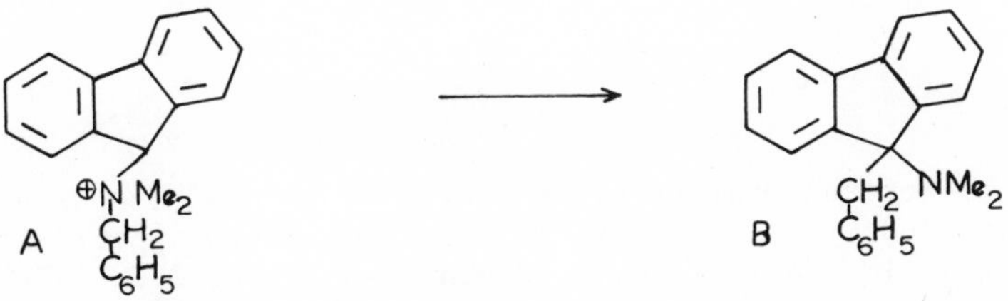

The Sommelet product was obtained from a diallylic quaternary salt [15]:

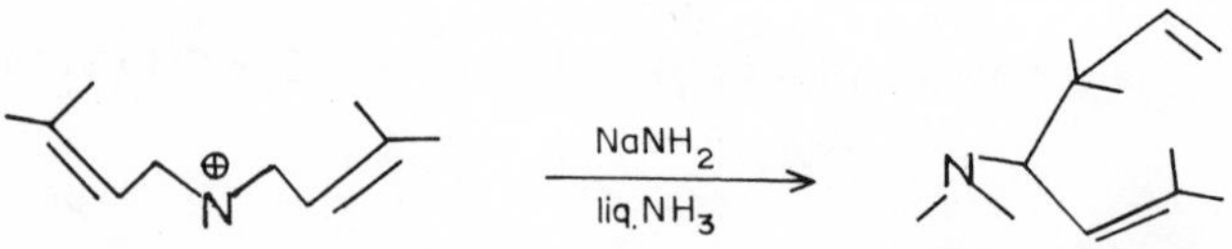

N,N-Dimethyl-3,3,6-trimethyl-
hepta-1,5-diene-4-ylamine
80% of 87% purity

10. Nitrogen-to-nitrogen Rearrangement (Dimroth)

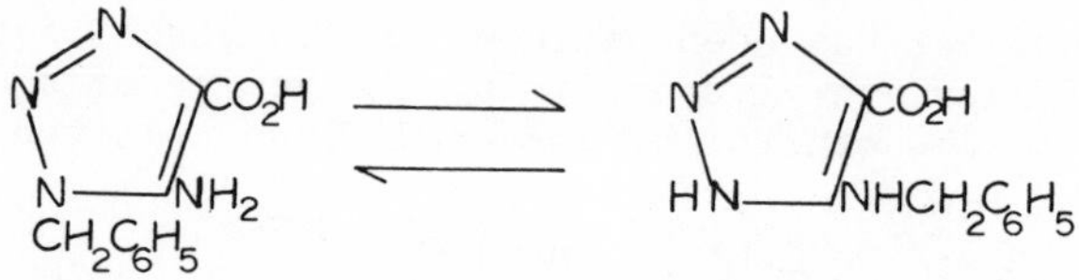

In the preceding 3-benzyl-4-amino-1,2,3-triazole-5-car-boxylic acid, boiling 1 N potassium hydroxide for 6 hr brought the substrate to an equilibrium of 40% product, 55% substrate. The product regresses to substrate in hot, neutral ethanol [16].

Unsymmetrical diphenylhydrazine rearranges to the symmetrical isomer, using excess methyllithium to form the dianion [17]. With N-methyl-N-phenyl-o-phenylenediamine prolonged refluxing with MeLi in Et$_2$O gave phenyl migration [17].

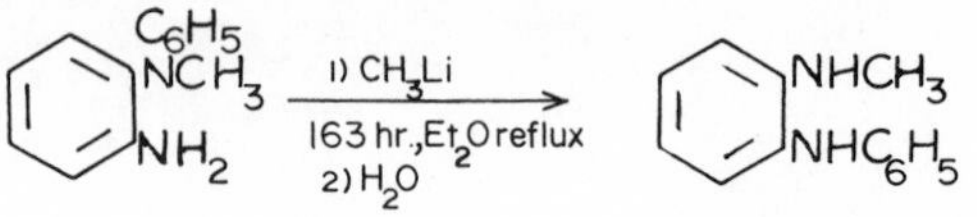

N-Methyl-N'-phenyl-o-phenylene-
diamine, 89 %

11. Claisen Rearrangement of N-Allylheterocycles (see Addenda)

Addenda

G.2. The Hofmann rearrangement yields of amines (as carbamates) may be raised by using methyl rather than sodium hypobromite as the reagent [P. Radlick, L. R. Brown, Synthesis, 290 (1974)].

G.2. A series of amides was rearranged by treatment with lead tetraacetate to the carbamate, which without isolation may be converted into the amine hydrochloride (72-90%) with HCl [H. E. Baumgarten et al., J. Org. Chem., 40, 3554 (1975)].

G.11.   N-Crotylindole may be rearranged to 3-α-methylallylindole
(low yield) at 450-470° [J. M. Patterson et al., J. Org. Chem.,
39, 486 (1974)].

References

1.   F. J. McCarty et al., J. Med. Chem., 13, 814 (1970).
2.   R. Ohme, H. Preuschhof, Ann. Chem., 713, 74 (1968).
3.   F. Eloy, A. Deryckere, Bull. Soc. Chim. Belges, 79, 301
     (1970).
4.   J. S. Swenson et al., J. Org. Chem., 38, 3956 (1973).
5.   W. J. Tuman, L. Bauer, J. Org. Chem., 37, 2983 (1972).
6.   R. T. Conley et al., J. Org. Chem., 37, 4095 (1972).
7.   E. J. Moriconi, M. A. Stemniski, J. Org. Chem., 37, 2035
     (1972).
8.   U. Svanholm, V. D. Parker, J. Am. Chem. Soc., 94, 5507
     (1972).
9.   J. Schmitt et al., Bull. Soc. Chim. France, 1227 (1969).
10. E. Schmitz, H. Fechner, Org. Prep. Proced., 1, 253 (1969).
11. P. G. Gassman, T. J. van Bergen, J. Am. Chem. Soc., 95, 590
     (1973).
12. E. W. Gill et al., J. Chem. Soc., C, 74 (1970).
13. S. H. Pine, Org. Reactions, 18, 403 (1970).
14. G. Wittig, Bull. Soc. Chim. France, 1921 (1971).
15. V. Rautenstrauch, Helv. Chim. Acta, 55, 2233 (1972).
16. A. Albert, J. Chem. Soc., C, 230 (1970).
17. R. West, H. F. Stewart, J. Am. Chem. Soc., 92, 853 (1970).

H.   Degradation of Amines

   1.   From Amines (1, 510)

$$R_3N \longrightarrow R_2NH$$

   An interesting degradation occurs via the peroxide of 2-
nitropropane [1]:

$$Et_3N + CH_3CHCH_3 \xrightarrow[\substack{C_5H_5N \\ 50°}]{CuCl,O_2} Et_2NNO + CH_3CH{=}O$$
$$\substack{NO_2} \qquad \substack{50\%}$$

The oxygen uptake was measured while the mixture was shaken and
the nitrosoamine was reduced to diethylamine. Hindered amines
worked well, but not those in which the t-amino group is a part
of a five-membered ring.

The hydrogenolysis of bicycloaza compounds has also been studied of late. The hydrogenolysis of the readily available indolizidines has been carried out in three ways, of which two are indicated [2]: (a) debenzylation with alkyl or aryl chloroformates and (b) reduction with $LiAlH_4$ and $NiCl_2$. This type of cleavage has been accomplished with lithium in liquid ammonia [3].

β-Hydroxyethylamines have a tendency to cleave to the amine and carbonyl compound in strong acid [4]:

$$C_6H_5CHOHCH_2NR_2 \xrightarrow{H^\oplus} C_6H_5CH_2CHO + R_2\overset{\oplus}{N}H_2$$

3.   From Quaternary Salts (1, 512)

$$R_4\overset{\oplus}{N}\overset{\ominus}{X} \longrightarrow R_3N$$

Formerly (1, 512), a synthesis of ethynyl dialkylamines was described. This section is now made more general to include the degradation of any quaternary salt. Aromatic quaternary salts are rather easily degraded to tertiary amines by steam distilling from a basic solution (1, 450, Ex. b.4) or by refluxing in DMF for prolonged periods (yields of 6-96%). However, by the latter procedure methylpyridinium iodide was demethylated only 70% in 96 hr [5]. The acetates of aromatic quaternary salts (made by passing through an acetate ion-exchange resin) may be dealkylated by refluxing in benzene-acetonitrile for 1 hr [6]. Aliphatic quaternary salts require a higher reflux temperature (xylene and acetonitrile).

For aliphatic quaternary salts that are difficult to dealkylate, a new procedure has been worked out using lithium propanethiolate ($LiSC_3H_7$) in HMPA at 0° for 30 min [7]. This reagent has a high propensity for demethylation as opposed to dealkylation.

Another degradation of quaternary salts is found in the literature [C.1, 2].

References

1.   P. Misbach et al., Angew. Chem. Intern. Ed. Engl., 9, 892 (1970).
2.   M. G. Reinecke, R. G. Daubert, J. Org. Chem., 38, 3281 (1973).
3.   J. P. Yardley et al., J. Med. Chem., 10, 1088 (1967).
4.   S. A. Fine et al., J. Org. Chem., 38, 2089 (1973).

5. D. Aumann, L. W. Deady, <u>Chem</u>. <u>Commun</u>., 32 (1973).
6. N. D. V. Wilson, J. A. Joule, <u>Tetrahedron</u>, <u>24</u>, 5493 (1968).
7. R. O. Hutchins, F. J. Dux, <u>J</u>. <u>Org</u>. <u>Chem</u>., <u>38</u>, 1961 (1973).

I. Amine Synthesis from Enamines

Three reviews are available [1-3]. The first is repetitious in sections, but Chapters 4-6 and 8 contain an enormous amount of information on this expanding field. It is not the intent to summarize these reviews, but to outline those parts devoted solely to the synthesis of amines. Other uses are described in <u>1</u>, 703.

The synthesis of enamines has been described (<u>1</u>, 461; <u>1</u>, 703). A room temperature synthesis with 40 g of Linde molecular sieves 5A per 0.1 mol of ketone has been recorded more recently [4]. Schiff bases also are made in this manner. Reactions of enamines to prepare amines are given as follows.

1. Alkylation of Enamines

Only allyl, benzyl, or propargyl iodides, or α-halogenated types, such as ethers, ketones, esters, or nitriles, on occasion give satisfactory yields [1]. The lack of generality may be accounted for by initial N-alkylation followed by less facile migration to carbon. A case in point is the N-alkylation and isolation of this product from the enamine of cyclopentan-1,2-dione [5]. The second alkyl group will alkylate the least sub-stituted position to give

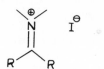

The quaternized salt may lose HI to give unsaturated amines such as

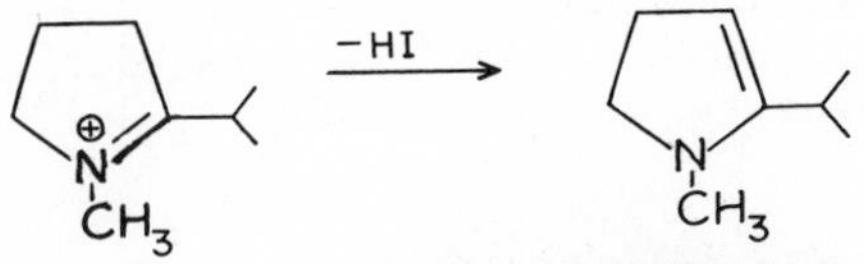

N-Methyl-2-isopropyl-
2-pyrroline

Interestingly enough, the enamine is resistant to reduction by lithium aluminum hydride under conditions in which an ester group is reduced to an alcohol [2]. But the enamine can be reduced by catalytic hydrogenation by sodium borohydride or under the conditions of the Leuckart-Wallach reaction to give tertiary amines [1]; another reduction of enamines is given in the literature [A.6, 49].

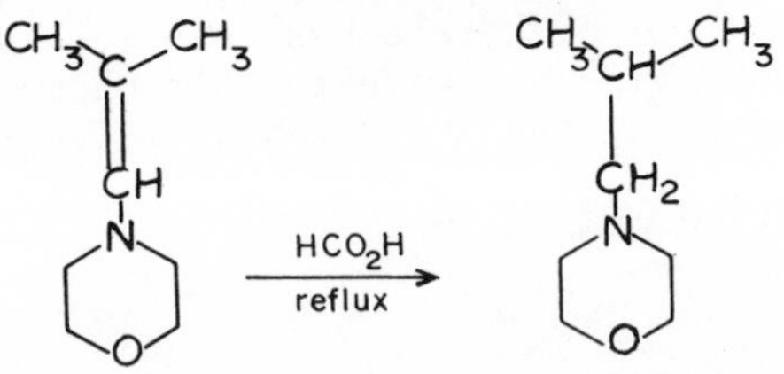

N-Isobutylmorpholine

It appears that some enamines can be alkylated in the allylic position of the enamine, a fact that broadens the scope of alkylation [6]:

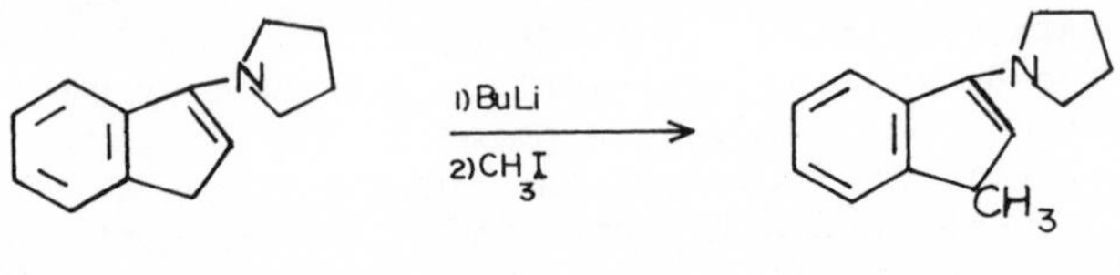

2. Cyclobutane Formation as an Intermediate to Unsaturated Aminocycloalkanones

Acrylonitrile, methyl vinyl sulfone, and α,β-nitrostyrene are expected and do give carbon alkylation, but from an enamine derived from an aldehyde they form cyclobutanes preferentially. Acrolein, however, with a ketone enamine does give a cyclobutane (as the ketone) as well [2]:

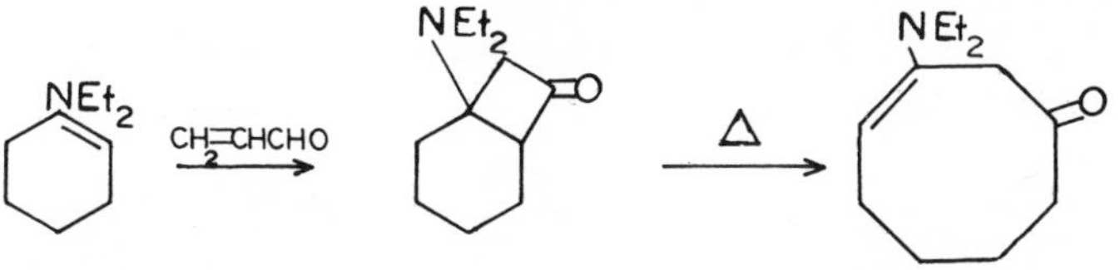

3-Diethylamino-3-cyclooctenone

The cyclobutanone cannot be isolated except in the case of seven-, eight-, or twelve-membered ketones.

On the other hand, an acetylene dicarboxylic ester with an enamine usually gives the expanded ring product, which involves initially an unstable cyclobutane intermediate [7]:

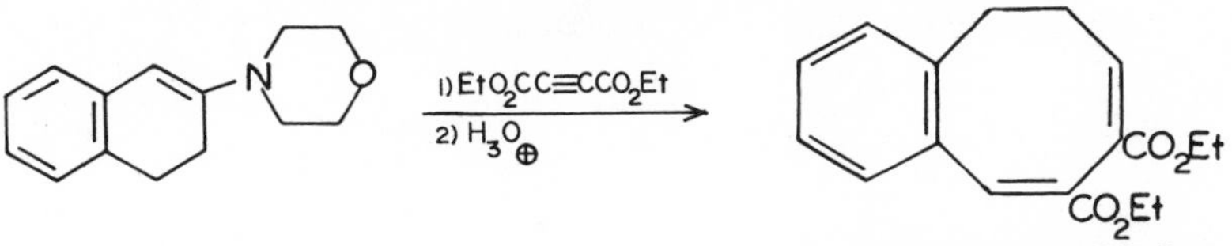

Diethyl I,2-benzocyclobuta-3,5-diene-4,5-dicarboxylate

3. Diacylation [2]

Forcing acylation produces the diacyl product:

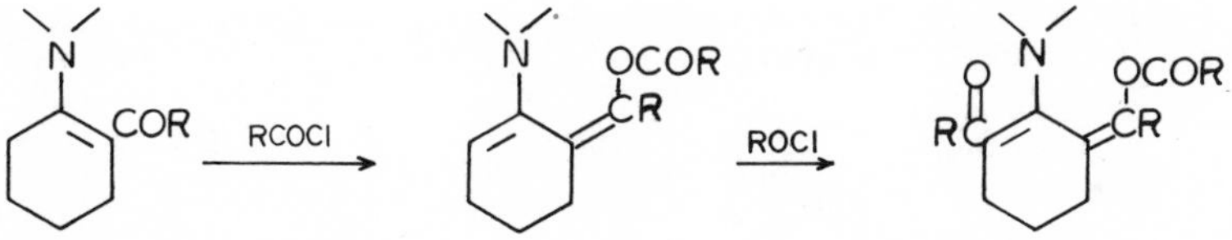

One unsymmetrical diacylation product has been reported ($\underline{1}$, 691, Ref. 36).

4. ω-Aminoacid Formation [2]

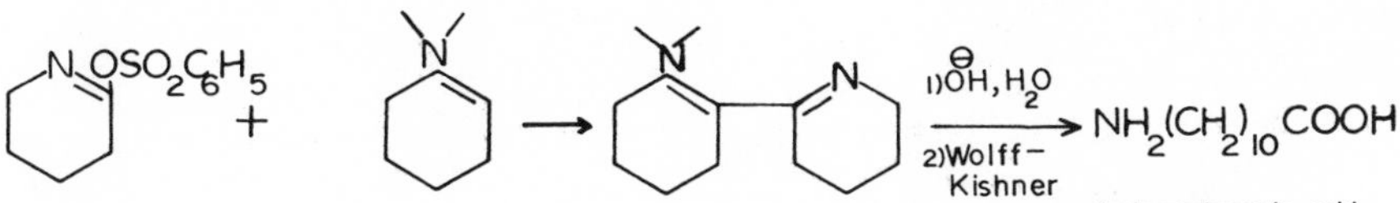

ω-Aminoundecanoic acid

Compound 1 was obtained from the Beckmann rearrangement of cyclopentanone oxime benzenesulfonate ester.

5. Reactions of Carbenes [8] and Diazoalkanes [9] in the Formation of Cyclopropanes and Aziridinium Salts

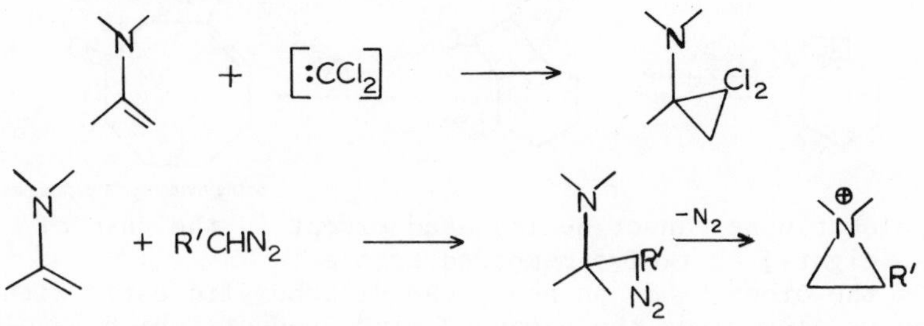

A recent carbene addition is shown [10]:

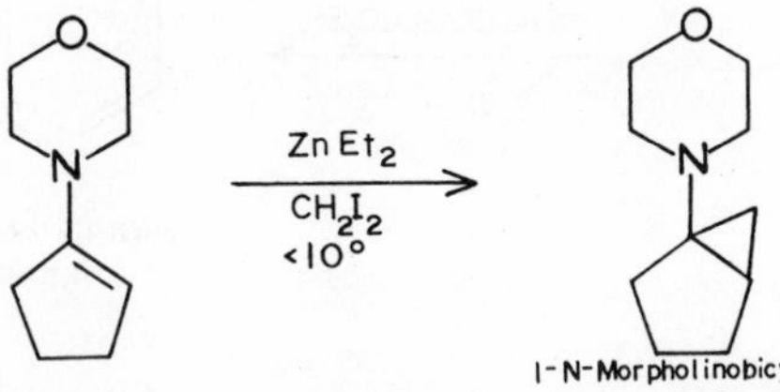

1-N-Morpholinobicyclo[3.1.0]hexane
68%

Diazomethane and cuprous chloride give the same product.

6.   Iminium Salts [11]

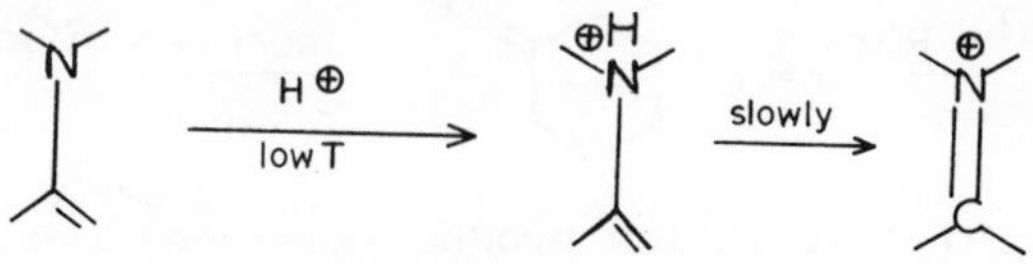

The structure is the same as that of a quinolinium salt and the two types have somewhat similar reactions. Grignard, or aryl- or alkyllithium reagents react to form addition products, as for instance [12]:

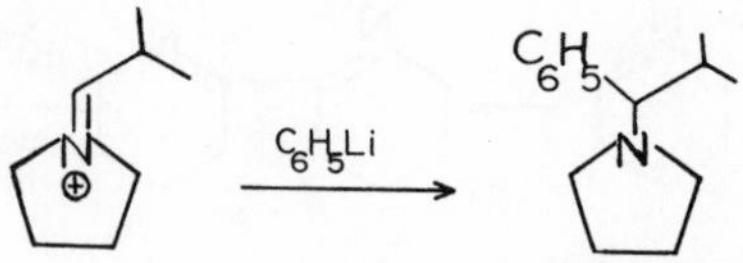

N-(α-Phenylisobutyl)-
pyrrolidine, 70%

The iminium salts also are reduced to tertiary amines.

7.   Heterocyclic Formation

The number of heterocycles formed by enamines is almost infinite. Examples are given of four-, five-, and six-membered heterocycles [2]. Phenylisocyanate and enamines with no β-hydrogen atoms give azetidones:

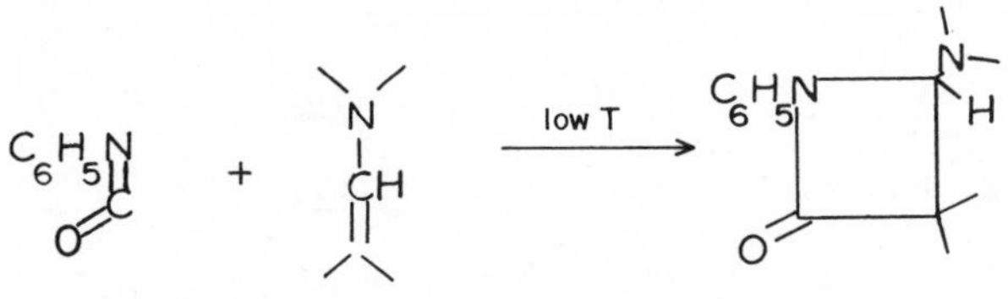

but with sulfur present at higher temperatures, a 1,3-thiazolidin-2-one is formed:

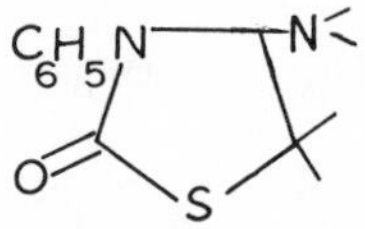

1,3-Dipolar compounds also give five-membered rings. Although cyclohexanone enamines usually give cyclo products, in numerous others straightforward Michael adducts [13] are formed.

An example of six-membered ring formation is [2]:

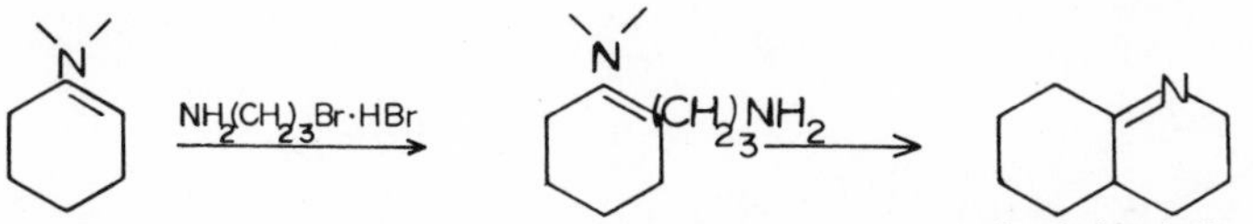

Octahydroquinoline

The reaction of primary and secondary enamines with electro-philic alkenes, alkynes, or diketenes frequently gives pyridines or α- or λ-pyridones [2]. Even substituted adamantanes may be obtained via enamines [14].

8.   5-Hydroxyindoles via the Nenitzescu Reaction

This synthesis, which has been reviewed [15], is illustrated:

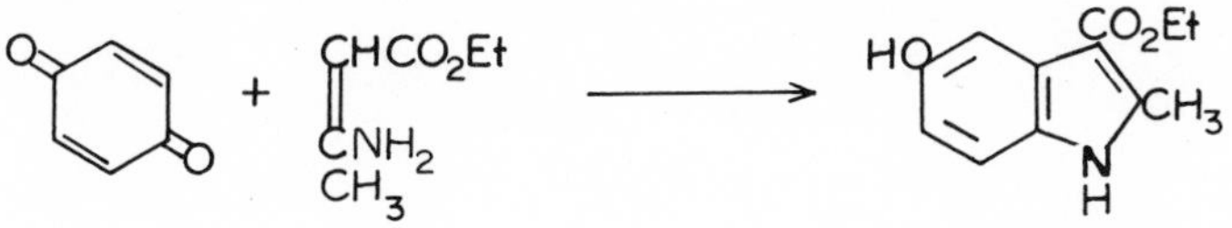

References

1. A. G. Cook, Enamines: Synthesis, Structure, and Reactions, Dekker, New York, 1969.
2. P. W. Hickmott, H. Suschitzky, Chem. Ind. (London), 1188 (1970).
3. S. F. Dyke, The Chemistry of Enamines, Cambridge U. P., 1973.
4. K. Taguchi, F. H. Westheimer, J. Org. Chem., 36, 1570 (1971).
5. K. Sato et al., J. Org. Chem., 38, 551 (1973).
6. H. W. Thompson, B. S. Huegi, Chem. Commun., 636 (1973).
7. Ref. 1, p. 131.
8. Ref. 1, p. 161.
9. Ref. 1, p. 192.
10. M. E. Kuehne, J. C. King, J. Org. Chem., 38, 304 (1973).
11. Ref. 1, p. 171.
12. Ref. 1. p. 184.
13. A. P. Gray et al., J. Org. Chem., 36, 1449 (1971).
14. P. W. Hickmott et al., J. Chem. Soc., Perkin Trans., I, 2063 (1973).
15. G. R. Allen, Jr., Org. Reactions, 20, 337 (1973).

Chapter 9

ACETALS
AND KETALS

Methods have been found to both synthesize acetals and ketals in nonacidic media and cleave them to the corresponding carbonyl compounds in nonacidic media (A.1). Another method forms a ketal that is relatively more resistant to acid hydrolysis than the typical ketal (A.6). A specific catalyst, other than the hydronium ion, has been found for the synthesis of acetals, namely, benzenesulfonylhydroxamic acid (A.1). Best of all reagents for forming ketals from alcohols are substituted tri-methyl silyl ethers (B.1). The metathesis of the Grignard reagent with an orthoformate has been improved (C.2). Two rather unique new methods have been published and described, one utilizing the mercuric addition product of a terminal alkene (D.1) and the other oxidizing DMSO in the presence of an alcohol (D.6). A summary of the methods for the synthesis of acetals and ketals is available [1].

To show that ketal formation is sometimes important in controlling the course of the reaction, the following synthesis is given [1a]:

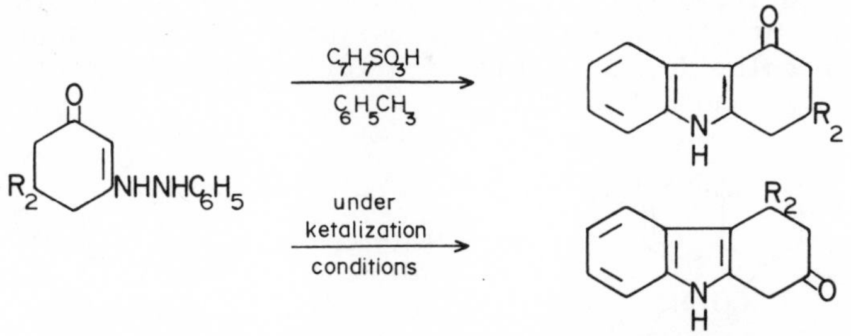

2-Ketotetrahydrocarbazole

A. Nucleophilic Addition of Alcohols to Carbonyl Groups or Exchange

1. From Carbonyl Compounds and Alcohols (1, 514)

The 2,3-ketals of uridine were prepared by utilizing the ketone as solvent, di-p-nitrophenyl phosphate as catalyst, and the dimethyl ketal of acetone as dehydrating agent as shown [2].

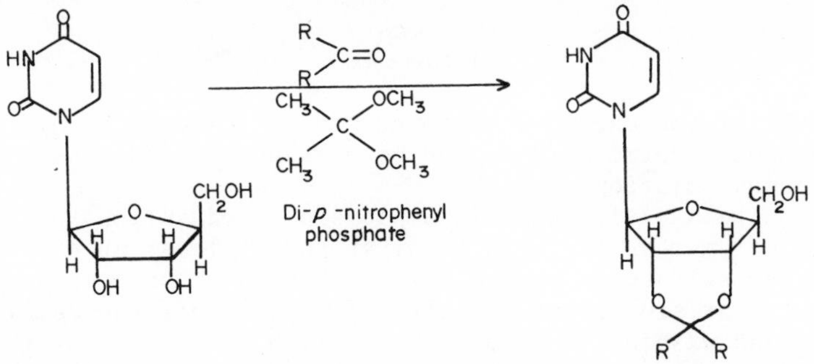

The method requires only room temperature, low concentration of acid catalyst, and it gave high yields with a variety of nucleosides.

To prepare di-sec-alkyl ketals by a direct method the use of molecular sieves is recommended as indicated [3]:

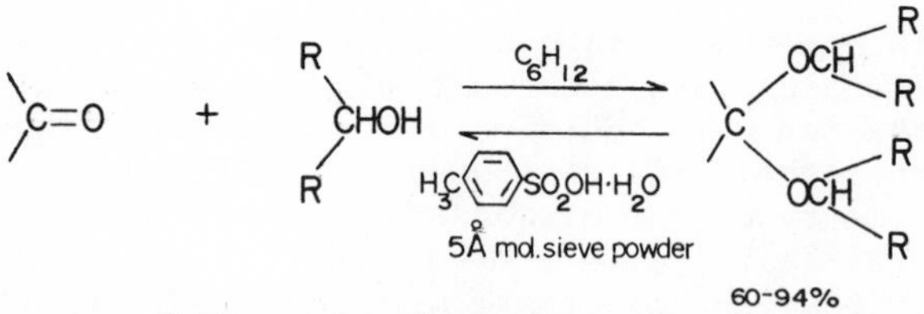

60-94%

The effectiveness of the molecular sieves is shown by the fact that the conversion in their absence was less than 3% while with them it was always better than 80%.

Recently, Welch [4] devised a method for converting the aldehyde into the acetal by the use of an alcohol, a trace of hydrogen chloride, and a bit of Drierite at refrigerator temperature as shown (Ex. c):

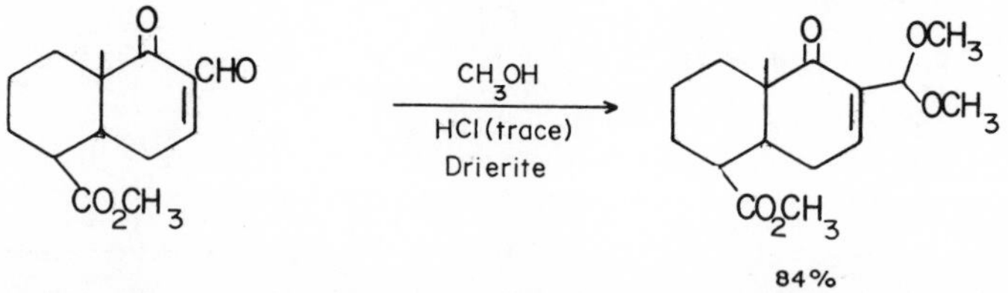

84%

If side reactions such as ring closure with olefinic aldehydes tend to occur during acetal formation, a modification as given is desirable [5]:

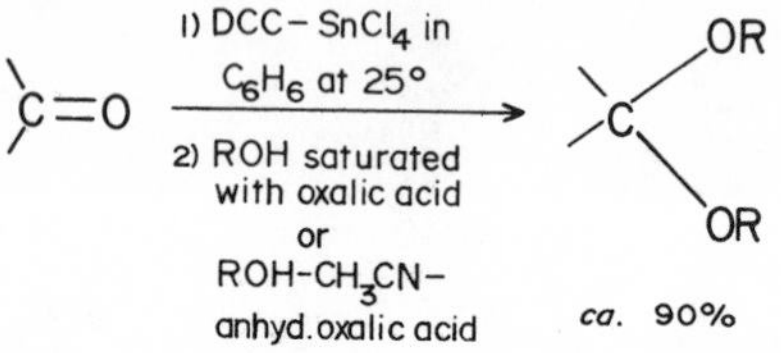

ca. 90%

Retrieval of the carbonyl compound can be achieved by hydrolysis with aqueous acetic acid in THF or dioxane to produce a homogeneous solution.

A quite useful synthesis employs a dried sulfonated polystyrene polymer as the acid catalyst and at least an equivalent of calcium sulfate to remove water. Yields of acetals and ketals in about 6 hr are around 90% after filtering off the calcium sulfate and washing it alternatively with ether and water [5a].

Terephthaldehyde and other similar dialdehydes have been converted into monoacetals on an insoluble polymer support system incorporating a diol functional group. In this way the free aldehyde group could be transformed as with the Wittig reagent [6]. Lastly, o-nitrophenylethylene glycol with aldehydes and ketones forms acetals and ketals that are photosensitive [7].

Irradiation of the acetal or ketal for 6 hr at 25° in benzene
returned the carbonyl compound in yields of 31-90%.

Voelter and Djerassi acetalized steroid ketones by the use
of the alcohol and a homogeneous catalyst in the presence of H_2
as shown [8]:

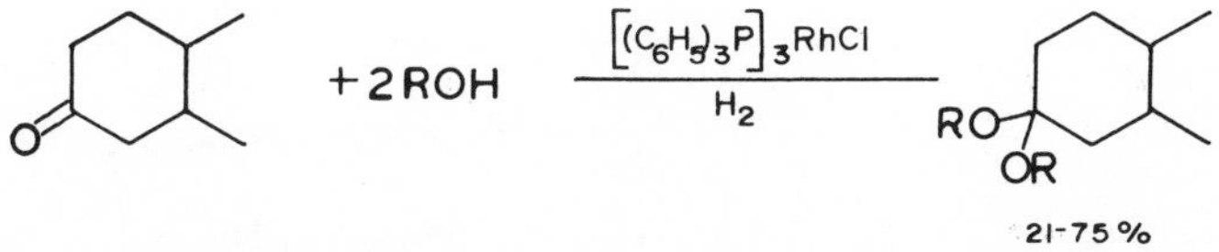

21-75%

The method is specific only for 2- and 3- oxo groups. α,β-
Unsaturated ketones are reduced first to the saturated ketones,
which then ketalize.

A general method for preparing thioacetals or thioketals is
given in Ex. b, but another method utilizes a neutral medium [9].

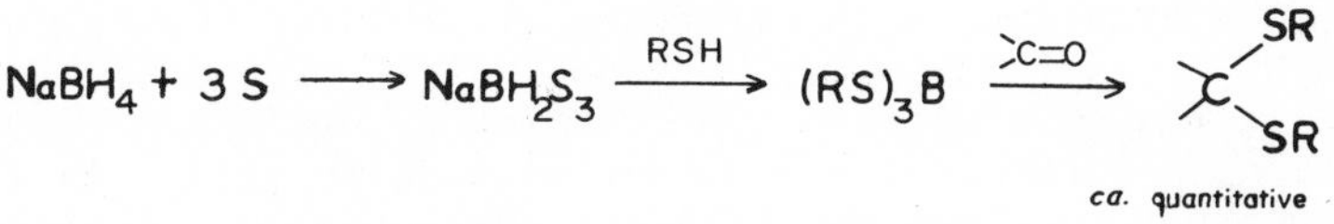

ca. quantitative

The ketone is retrieved from the thioketal by treatment with
silver oxide in aqueous methanol [10].

A second nonacidic method consists of refluxing the ketone
in 2-chlorothanol containing a suspension of lithium carbonate.
The example illustrates a case in which acid cannot be tolerated
[11].

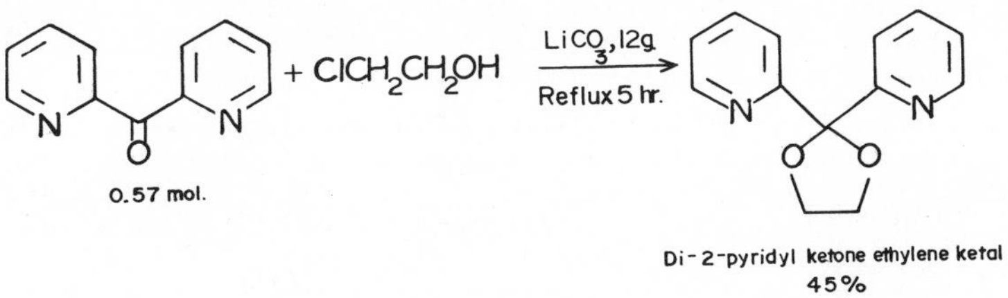

0.57 mol.

Di-2-pyridyl ketone ethylene ketal
45%

Another example of the use of ethylene chlorohydrin has been
mentioned previously (1, 517).

Focusing on a nonacidic medium for hydrolyzing the ketal
rather than a nonacidic medium for synthesizing one, Corey and
Ruden [12] recommend the following synthesis:

$$R_2C{=}O \ + \ HOCH_2CHOHCH_2Br \ \xrightarrow{\ p\text{-}C_6H_4CH_3SO_3H\ } \ R_2C\underset{O}{\overset{O}{\big<}}{-}CH_2Br \ \xrightarrow[CH_3OH]{Zn}$$

A
93-98%

$$BrZnOCH_2CH{=\!=}CH_2 \ + \ R_2C{=}O$$
89-96%

The deketalization is remindful of the preparation of alkenes from β-haloethyl ethers (Boord synthesis, 1, 82). Since the ketal, A, is resistant to change by either liquid NH_3, $NaBH_4$ in ethanol, m-chloroperbenzoic acid, MeLi in ether, or Jones reagent, it is possible to alter the R group before regeneration with zinc.

Formerly, α-bromoacetals were prepared (one-pot) by bromination of an aldehyde in the presence of dibenzoyl peroxide followed by acetalization with ethanol. This procedure has been improved in the bromination step by using t-butyl peroxide and irradiation to generate the bromination species [13]. In a similar one-pot reaction the following has been accomplished [14]:

$$CH_2{=}CHCHO \ \xrightarrow[\ 2)\,EtOH\text{-}HCl\ ]{1)\,Br_2\,\text{-}Et_2O,\,0°} \ EtOCH_2CHBrCH(OEt)_2$$

α-Bromo-β-ethoxypropionaldehyde
diethyl acetal

Ketalization for some ketones, particularly aryloxyacetones, is accelerated by irradiation, but the conversion is low [15].

Lastly, a specific catalyst, benzenesulfonylhydroxamic acid has been found for acetalization. In the mechanism, as indicated, the hydroxamic acid probably acts as a stable leaving group [16]:

$$\underset{C_6H_5SO_2NOH}{RCHOH} \ \underset{H_2O}{\overset{MeOH}{\rightleftharpoons}} \ \underset{C_6H_5SO_2NOH}{RCHOCH_3} \ \rightleftharpoons \ \Big[RCH{=}\overset{\oplus}{O}CH_3\Big] + C_6H_5SO_2\overset{\ominus}{N}OH$$

$$\downarrow MeOH$$

$$C_6H_5SO_2NHOH + RCH(OCH_3)_2$$

Although the hydroxamic acid is a true catalyst, quantities equivalent to that of the aldehyde were used.

a.   Preparation of Cyclohexanone Diisopropyl Ketal [3]

A mixture of cyclohexanone, 0.10 mol, isopropyl alcohol,
0.40 mol, cyclohexane, 150 ml, and p-toluenesulfonic acid mono-
hydrate, 3.8 g, was stirred at 0° and molecular sieve powder type
5A, 40 g, was added. After stirring for 2 hours, triethylamine,
6 ml, was added to quench the reaction. The mixture was then
filtered and the product, 82%, was recovered in the usual manner
from the organic layer.

b. Preparation of Dibutylthioacetals or Ketals [17]

Two equiv of butylmercaptan and 1 equiv of carbonyl compound
were heated at 50-60° while a slow stream of HCl was passed
through the mixture. Steam distillation followed and the remaining oil was dried over CaCl$_2$ and then distilled under reduced
pressure. Yields were about 50-60%.

c. Preparation of 1-Keto-5-Carbomethoxy-9-methyl-$\Delta^{2,3}$-2-
octalincarboxaldehyde Dimethyl Acetal [4]

The unsaturated ketoaldehyde, 0.360 g was treated with
anhydrous methanol containing a few drops of methanolic hydrogen
chloride. A few chips of Drierite, 8 mesh were added and the
mixture was refrigerated for 48 hr. The acid mixture was
neutralized with solid NaHCO$_3$ and then 20 ml of a saturated
NaHCO$_3$ solution was added. The ethereal extract, obtained by
four 25-ml extractions, was washed with 100 ml of water, 25 ml of
saturated aqueous NaCl, dried with Na$_2$SO$_4$, and filtered through
MgSO$_4$ to give a solution, which, when concentrated in vacuo,
gave 0.35 g (84%) of the acetal.

d. Preparation of Cyclohexane-1,4-dione-bis-ethylene Ketal [18]

4. From Carbonyl Compounds and Orthoesters or
    Orthosilicates (1, 520)

Schank has prepared the monoketals of 1,2-cyclic diketones,
as intermediates for the synthesis of reductones, by the use of
orthoesters [19].

6. From Acetals or Ketals and Alcohols (Transacetalization or Transketalization) (1, 523)

$$\text{>C(OC}_2\text{H}_5)_2 + \text{CCl}_3\text{CH}_2\text{OH} \xrightarrow[\underset{1.5\ eq.}{C_6H_6}]{H^\oplus} \overset{\text{OCH}_2\text{CCl}_3}{\underset{\text{OC}_2\text{H}_5}{C<}} \xrightarrow[2.5\ eq.]{\text{CCl}_3\text{CH}_2\text{OH}} \text{>C(OCH}_2\text{CCl}_3)_2$$

This exchange [20] occurs readily. The di-2,2,2-trichloroethyl acetals are more stable toward acid than the diethyl acetals, but they are readily removed in neutral solution as the original carbonyl compound by zinc in ethyl acetate.

8. From gem-Dihalides, gem-Disulfates, or α-Chloroethers (1, 525)

$$\text{RCHClR}' \xrightarrow[C_5H_5N]{R'OH} \text{RCH(OR}')_2$$

In the reaction of an α-chloroether with oxetane to form the acetal, pyridine is not needed [21]:

$$\text{RCHClR}' + \underset{O}{\diamond} \longrightarrow \overset{\text{OR}'}{\underset{|}{\text{RCHOCH}_2\text{CH}_2\text{CH}_2\text{Cl}}}$$

ca. 60%

9. From Acetals or Ketals and Chloral

Dietrich and co-workers [22] discovered that chloral may be inserted, as shown, into the C-O-C bonds of cyclic, linear, substituted or unsubstituted acetals to form novel coacetals:

$$\text{CCl}_3\text{CHO} + \text{H}_2\text{C(OCH}_3)_2 \xrightarrow[(Et)_2O]{BF_3} \overset{}{\underset{\underset{\text{CCl}_3}{|}}{\text{CH}_3\text{OCHOCH}_2\text{OCH}_3}}$$

3-Trichloromethyl-2,4,6-trioxo-
heptane, 85%

The reaction did not occur with aldehydes other than chloral nor with ketals susceptible of undergoing an aldol condensation. With cyclic acetals ring enlargement occurs, Catalysts employed, besides BF_3, were PF_5 and $SbCl_5$. The insertion is of interest although the structure of the product formed is not always predictable.

10. From Paraformaldehyde and Hydrochloric Acid

Formaldehyde dichloromethyl acetal was produced from paraformaldehyde and hydrochloric acid in the presence of an alkene such as allyl chloride [23]:

$$\text{HO-(CH}_2\text{O)}_n\text{H} \quad \xrightarrow[\text{CH}_2\text{=CHCH}_2\text{Cl}]{\text{HCl}} \quad \text{CH}_2\text{(OCH}_2\text{Cl)}_2$$

55 %

The olefin (ethylene and sulfolene may also be used) serves as a catalyst. If the olefin is not used, the product is predominantly bis-chloromethylether. The method is the best available (at publication) for the synthesis of this particular acetal.

11. From Formals and Alcohols (see Addenda)

Addenda

A.1. Polymer-protected aluminum chloride, Ⓟ -AlCl$_3$, as a catalyst improves the yield of acetal formation between benzaldehydes and alcohols. The effect is most pronounced when electron-withdrawing groups are present in the aldehydes [D. C. Neckers et al., J. Org. Chem., 40, 959 (1975)].

A.4. Diethylene orthocarbonate with a trace of acid at 25° is very effective in an exchange with carbonyl compounds, with the exception of benzophenone [D. H. R. Barton et al., Chem. Commun., 432 (1975)].

A.11. Alcohols exchange with formals to give mixed formals by using P$_2$O$_5$ in CHCl$_3$ as a catalyst [K. Fuji et al., Synthesis, 276 (1975)].

References

 1. E. Schmitz, I. Eichhorn, in S. Patai, Chemistry of the Ether Linkage, Interscience, New York, 1967, p. 309.
1a. R. F. Borch, R. G. Newell, J. Org. Chem., 38, 2729 (1973).
 2. A. Hampton et al., J. Am. Chem. Soc., 87, 5481 (1965).
 3. D. P. Roelofsen, H. van Bekkum, Synthesis, 419 (1972); D. P. Roelofsen et al., Rec. Trav. Chim., 90, 1141 (1971).
 4. S. C. Welch of the University of Houston was kind enough to supply us with this unpublished method.
 5. N. H. Andersen, H. Uh, Syn. Commun., 3, 125 (1973)].
5a. V. I. Stenberg et al., J. Org. Chem., 36, 2550 (1971); 39, 2815 (1974).

6. C. C. Leznoff, J. Y. Wong, Can. J. Chem., 51, 3756 (1973).
7. J. Hébert, D. Gravel. Can. J. Chem., 52, 187 (1974).
8. W. Voelter, C. Djerassi, Chem. Ber., 101, 1154 (1968); Synthesis, 550 (1970).
9. J. M. Lalancette, A. Lachance, Can. J. Chem., 47, 859 (1969).
10. D. Gravel et al., Chem. Commun., 1323 (1972).
11. G. R. Newkome et al., Tetrahedron Letters, 1599 (1973).
12. E. J. Corey, R. A. Ruden, J. Org. Chem., 38, 834 (1973).
13. R. Pallaud, L. N. Lang, Compt. Rend., 273, C, 418 (1971).
14. J. N. Wells, M. S. Strahl, J. Pharm. Sci., 60, 533 (1971); Synthesis, 141 (1972).
15. M. K. M. Dirania, J. Hill, J. Chem. Soc., C, 1213 (1971).
16. A. Hassner et al., J. Org. Chem., 35, 1962 (1970).
17. E. Müller in H.-W., Methoden der Organischen Chemie, 9, 1955, p. 200.
18. N. Bensel, P. Weyerstahl, Synthesis, 665 (1973).
19. K. Schank, Synthesis, 176 (1972).
20. J. L. Isidor, R. M. Carlson, J. Org. Chem., 38, 554 (1973).
21. E. Vilsmaier, Ann. Chem., 742, 135 (1970).
22. H. J. Dietrich et al., J. Heterocycl. Chem., 7, 27 (1970).
23. P. R. Staff, J. Org. Chem., 34, 1143 (1969).

B. Addition to Unsaturated Linkages

1. From Vinyl Ethers, Halides, or Acetylenes and an Alcohol or Phenol (1, 528)

$$CH_2{=}CHOR \underset{\longleftarrow}{\overset{H^{\oplus}}{\longrightarrow}} CH_3\overset{\oplus}{C}HOR \underset{\longleftarrow}{\overset{R'OH}{\longrightarrow}} CH_3\underset{OR'}{C}HOR + H^{\oplus}$$

To prepare ketene acetals, Kuryla and Hyre [1] utilized the sodium alkoxide and vinylidene chloride as shown:

$$2\ CH_3OCH_2CH_2ONa + CH_2{=}CCl_2 \xrightarrow[N_2]{Xylene} CH_2{=}C\begin{smallmatrix}OCH_2CH_2OCH_3\\[1mm]OCH_2CH_2OCH_3\end{smallmatrix} + 2\ NaCl$$

56-75%

This general method is a one-step, simple, time-saving procedure, which involves readily available, inexpensive reagents and gives satisfactory yields. The reaction probably proceeds via a series of dehydrochlorinations and additions involving an intermediate acetylene compound.

The most versatile of the vinyl reagents for the preparation of ketals are the vinyl trimethylsilyl ethers. They are quite reactive, as is evidenced by their exothermic addition to alcohols, an example of which is shown [2]:

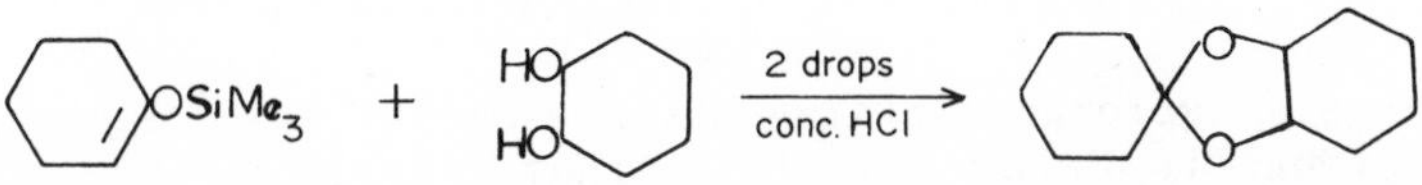

20 ml.   20 ml.   trans-Cyclohexanediol ketal of
                          trans   cyclohexanone, 80 %

Other complicated additions have been accomplished as in the conversion of a vinyl acetylene into a ketoacetal [3]:

$$MeOCH{=\!=\!=}CHC{\equiv}CR \xrightarrow[Hg SO_4,]{CH_3OH-H_2O} (MeO)_2CHCH_2\overset{O}{\overset{\|}{C}}CH_2R$$

31–70 %

Although mercuric acetate has been recommended as a catalyst for the above solvations, mercuric oxide milled in ethanol works satisfactorily as well [4]. Acetals were produced by Gaydou [5] from the addition of an alcohol to the vinyl ether as indicated:

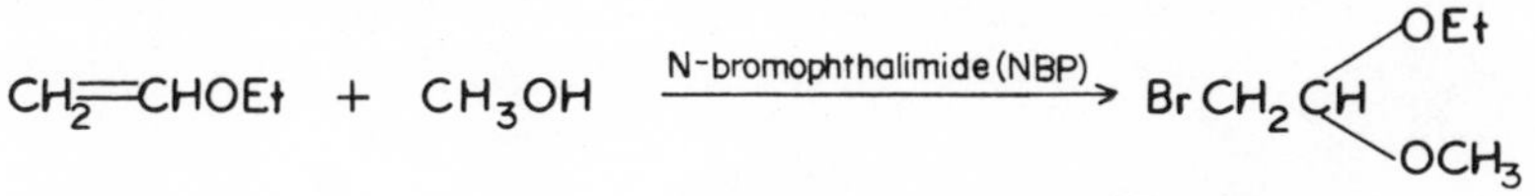

$$CH_2{=}CHOEt + CH_3OH \xrightarrow{\text{N-bromophthalimide (NBP)}} BrCH_2\overset{OEt}{\underset{OCH_3}{CH}}$$

Bromoacetaldehyde methyl
ethyl acetal, 82 %

The reaction is stereospecific in that cis-1-ethoxy-1-butene gave the threo acetal

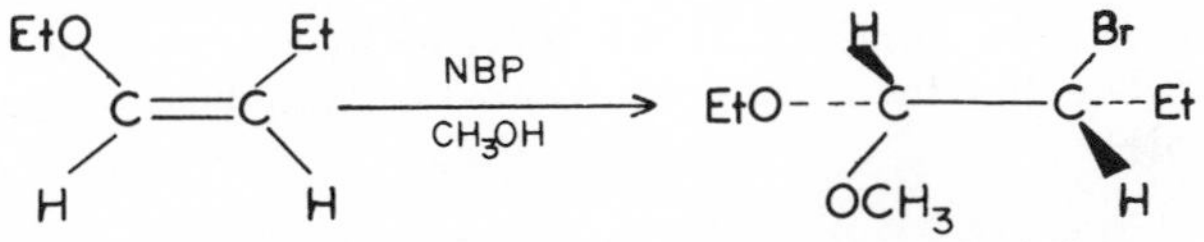

cis   threo-α-Bromobutyraldehyde
                          methyl ethyl acetal

while the trans isomer gave the erythro acetal

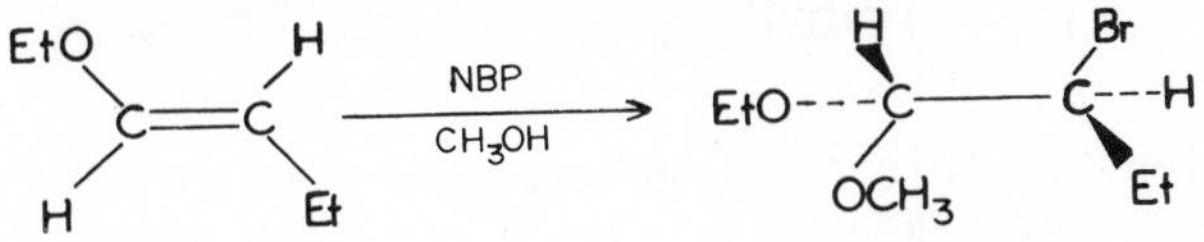

trans   erythro-α-Bromobutyraldehyde
                          methyl ethyl acetal

Vinyl ethers, as is well known, add to alcohols in the

presence of acid catalysts (1, 528). For instance, 2,3-dihydro-pyran with a trace of acid appears to form about seven ketal groups of the following general structure with sucrose [6]:

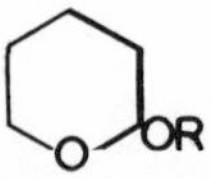

Since a cyclopropane ring often exhibits some degree of unsaturation, it is not surprising that a cyclopropane ether may be converted into an acetal [6a]:

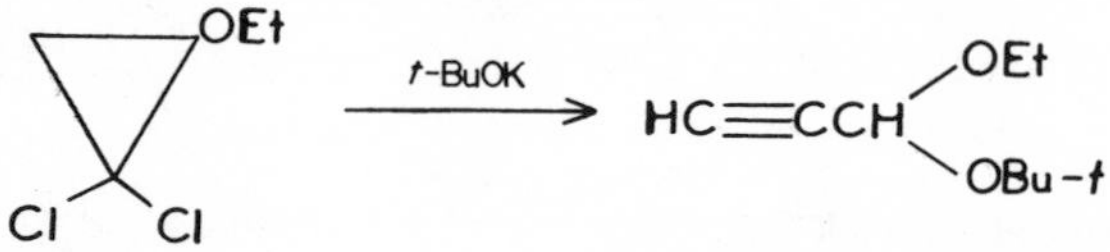

Ethyl *t*-butyl propynal acetal, 73%

The synthesis of ketene acetals usually falls under this section, but it will be referred to only by a general reference [7].

3.   From Vinyl Esters and Alcohols (1, 530)

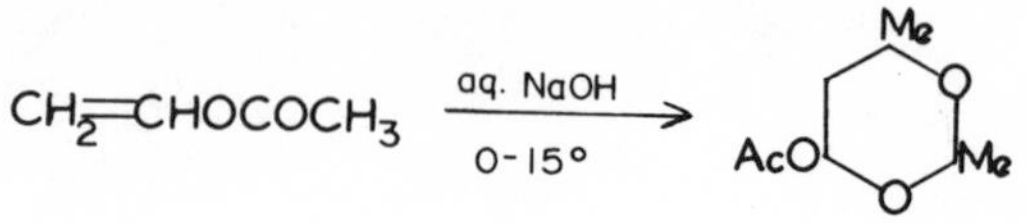

2,4-Dimethyl-6-acetoxy-
1,3-dioxane,(acetomethoxane),
52%

The above synthesis appears to be an isolated example [8].   The structure of the product is such that it probably originates from crotonaldehyde.

6.   From Acetylenes, Ethyl Orthoformate, and the Like (1, 532)

$$RC \equiv CH + HC(OEt)_3 \longrightarrow RC \equiv CCH(OEt)_2$$

Variations of this electrophilic addition have now been accomplished. For example, an acetal has been added to a vinyl ether [9]:

$$\text{BrCH}\!=\!\text{CHOEt} \quad + \quad \text{RCH(OEt)}_2 \xrightarrow[\text{neat, }\triangle]{\text{BF}_3\cdot\text{Et}_2\text{O}} \underset{\substack{\text{EtO} \quad \text{Br}}}{\text{RCH CHCH(OEt)}_2}$$

2 - Bromo - 3 - ethoxy - 3 - alkylpro-
pionaldehyde diethyl ketal, 37-66%

And the behavior of the reactive dialkylaminoorthoformates, or the quaternary salt, indicate that a carbonium ion of an acetal is generated [10] and attacks the acinitromethane:

$$(\text{EtO})_2\text{CH}\overset{\oplus}{\text{N}}\text{Et}_3 \quad + \quad \text{CH}_3\text{NO}_2 \longrightarrow (\text{EtO})_2\text{CHCH}_2\text{NO}_2$$

Nitroacetaldehyde diethyl ketal
44%

The product eliminates ethanol readily to form a vinyl ether.

9.   From β-Alkoxyacrylonitriles or Vinyl Nitro Compounds

The addition of an alkoxide to a β-alkoxyacrylonitrile leads to the cyanoacetal:

$$\text{ROCH}\!=\!\text{CHCN} \xrightarrow{\text{R'ONa}} \underset{\substack{|}}{\overset{\text{OR'}}{\text{ROCHCH}_2\text{CN}}}$$

However, in a substituted acrylonitrile, such as α-methoxymethyl-cinnamonononitrile, the addition occurs as shown [11]:

$$\underset{\text{CN}}{\text{C}_6\text{H}_5\text{CH}\!=\!\text{CCH}_2\text{OCH}_3} \xrightarrow[\text{MeONa}]{\text{MeOH}} \underset{\text{CN}}{\text{C}_6\text{H}_5\text{CH}_2\text{CHCH(OMe)}_2}$$

α-Cyano-β-phenylpropionaldehyde
dimethyl acetal, 74 %

With α-bromovinyl nitro compounds the addition proceeds via the acetylene derivative [12]:

$$\text{C}_6\text{H}_5\text{CH}\!=\!\text{CBrNO}_2 \xrightarrow[\text{2)H}_3\text{O}^\oplus]{\text{1)RONa - THF}} \underset{\substack{\text{OR}}}{\overset{\text{OR}}{\text{C}_6\text{H}_5\text{CCH}_2\text{NO}_2}}$$

α-Nitroacetophenone ketal
78-94%

10.   From Carbenes, Ethynyl Ethers, and Alkoxides

This reaction has become possible because the vinyl carbenes to be utilized are available from nitrosooxazolidones [13]:

$$\left[Me_2 \, C{=}\!{=}\!C{:} \right] \;+\; HC{\equiv}COEt \;\xrightarrow{\;\text{LiOEt-EtOH}\;}\; Me_2C{=}C{=}CHCH(OEt)_2$$

4-Methyl-2,3-pentadienal diethyl
acetal, 35%

Addendum

B.2. The use of dihydropyran to make ketals has two disadvantages: it introduces an asymmetric carbon into the substrate and its ketal is invariably a liquid. On the other hand, the use of 2-methoxypropene overcomes these disadvantages [C. B. Reese et al., _J. Am. Chem. Soc._, _89_, 3366 (1967)]. The synthesis of the latter has been described recently [M. S. Newman, M. C. Vander Zwan, _J. Org. Chem._, _38_, 2910 (1973)].

References

1. W. C. Kuryla, J. E. Hyre, _Org. Syn._, _47_, 78 (1967).
2. G. L. Larson, A. Hernandez, J. Org. Chem., 38, 3935 (1973).
3. A. N. Belyaeva et al., J. Org. Chem. USSR, 6, 1545 (1970).
4. S. S. Pizey, _Synthetic Reagents_, Vol. 1, Wiley, London, 1974, p. 307.
5. E. M. Gaydou, _Tetrahedron Letters_, 4055 (1972).
6. S. A. Barker et al., J. Chem. Soc., 3158 (1962).
6a. L. Skattebøl, _J. Org. Chem._, _31_, 1554 (1966).
7. E. Müller, H.-W., _Methoden der Organischen Chemie_, _7_, 1968, Pt. 4, p. 340.
8. N. P. Solov'eva et al., _Maslo-Zhir. Prom._ (8), 25 (1973); _C. A._, _79_, 115, 509 (1973).
9. S. M. Makin et al., _J. Org. Chem. USSR_, _5_, 2049 (1969).
10. S. Kabusz, W. Tritschler, _Synthesis_, 312 (1971).
11. A. Brossi et al., _J. Med. Chem._, _14_, 462 (1971).
12. M. Shiga et al., _Bull. Chem. Soc. Jap._, _43_, 841 (1970).
13. M. S. Newman, C. D. Beard, _J. Org. Chem._, _35_, 2412 (1970).

C. Displacement by Alkoxides or Complexes

1. From Ketones, Alkoxides, Alkylating Agents, and the Like (_1_, 535)

$$R_2C{=}O \;\xrightarrow[\;R'X\;]{\;R'ONa\;}\; R_2C(OR')_2$$

This limited reaction has been extended to the alkylation of the α-carbon in the S-oxide of thioformals [1], after which the monooxodithioacetals may be converted easily into the acetals:

$$\underset{\substack{\displaystyle CH_2\overset{\displaystyle O}{\overset{\|}{S}}Me \\ \diagdown SMe}}{} \xrightarrow[\text{2)RX}]{\text{1)NaH-THF}} \underset{\substack{\displaystyle RCH\overset{\displaystyle O}{\overset{\|}{S}}Me \\ \diagdown SMe}}{} \xrightarrow[\text{H}^\oplus\text{(trace)}]{HC(OEt)_3} RCH(OEt)_2$$

37-92%   85-88%

2. From Grignard Reagents and Orthoformates (1, 536)

By substituting mono- or diphenylalkylorthoformates for the usual ethyl orthoformate in the reaction with Grignard reagents, Stetter and Reske [2] found that improved yields of acetals were obtained. Thus phenyldiethyl orthoformate and butylmagnesium bromide gave pentanal diethyl acetal as shown:

$$n\text{-}C_4H_9MgBr \;+\; C_6H_5OCH\underset{OC_2H_5}{\overset{OC_2H_5}{\diagup}} \longrightarrow n\text{-}C_4H_9CH\underset{OC_2H_5}{\overset{OC_2H_5}{\diagup}}$$

90.5%

This yield is higher than when ethyl orthoformate is used.

Another Grignard reaction involves the ring opening of an orthoester to give the δ-ketoketal [3]:

$$R'\underset{O}{\overset{}{\bigcirc}}\underset{OEt}{\overset{OEt}{}} \;+\; RMgX \xrightarrow[\text{2)aq.NH}_4\text{Cl}]{\text{1)}\triangle} R'\overset{O}{\overset{\|}{C}}CH_2CH_2CH_2\overset{R}{\underset{|}{C}}(OEt)_2$$

19-48%

References

1. K. Ogura, G. Tsuchihashi, Tetrahedron Letters, 3151 (1971).
2. H. Stetter, E. Reske, Chem. Ber., 103, 643 (1970).
3. V. M. Thuy, Bull. Soc. Chim. France, 4429 (1970).

D. Oxidation or Reduction

1. From Ethylene Derivatives by Oxidation (1, 536)

Alkyl ethylenes have been converted into ethylene ketals by reaction with mercuric acetate and p-toluenesulfonic acid, after which the product was treated with $LiPdCl_2$-Li_2CO_3 as shown [1]:

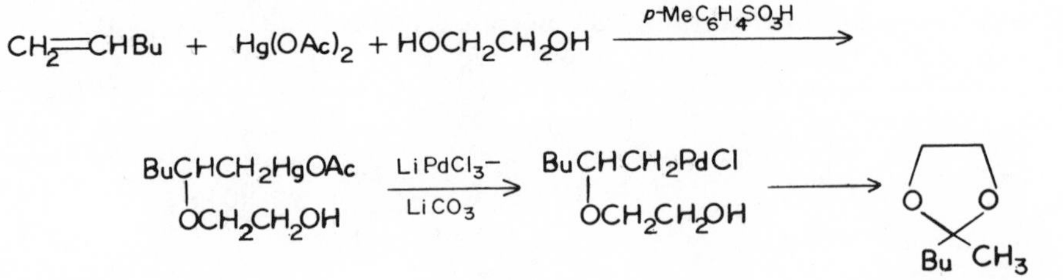

$$CH_2{=}CHBu \ + \ Hg(OAc)_2 \ + HOCH_2CH_2OH \ \xrightarrow{\ p\text{-}MeC_6H_4SO_3H\ }$$

$$\underset{\overset{|}{OCH_2CH_2OH}}{BuCHCH_2HgOAc} \ \xrightarrow[\text{LiCO}_3]{\text{LiPdCl}_3{}^-} \ \underset{\overset{|}{OCH_2CH_2OH}}{BuCHCH_2PdCl} \ \longrightarrow$$

2-Butyl-2-methyl-
1,3-dioxalane, 82%

Another example consists of the oxidation of 1-butene in the presence of palladous and cupric chlorides with ethylene glycol to form butanone ethylene ketal (65%) [2].

Furan derivatives may also be converted into ketals by singlet oxygen [3].

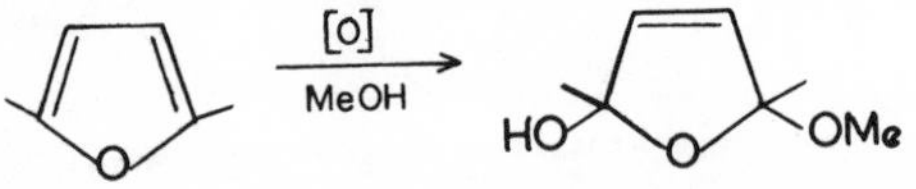

$$\xrightarrow[\text{MeOH}]{\text{[O]}}$$

2-Hydroxy-5-methoxy-2,5-
dimethyldihydrofuran, 48 %

The singlet oxygen for this reaction was generated from CO_2 exposed to microwaves.

3. From Diazoalkanes or α-Diazocarbonyl Compounds (1, 538)

$$R\overset{O}{\overset{\|}{C}}CHN_2 \ \xrightarrow[\text{EtOH}]{(CH_3)_3COCl} \ R\overset{O}{\overset{\|}{C}}CH(OEt)_2$$

Zwanenberg has recently applied the t-butyl hypochlorite oxidation of diazoalkanes to diazoketones [4]. Thus diazo-acetophenone with t-butyl hypochlorite in ethyl alcohol gives phenylglyoxal diethyl acetal (51%). The α-chloroether,

O
‖
RCCHOR', is an intermediate in this reaction.
|
Cl

5. From Orthoesters by Reduction (1, 538)

$$HC(OEt)_3 \ \longrightarrow \ CH_2(OEt)_2$$

This reduction has now been accomplished with carbohydrate ketals containing an orthoester group by the use of lithium

aluminum hydride and aluminum chloride in methylene chloride-
ether under reflux [5].

6. From Alcohols, Dimethyl Sulfoxide, and N-Bromo-
    succinimide

$$CH_3SOCH_3 \xrightarrow[ROH, 50°]{NBS, 2eq.} CH_2(OR)_2$$

This new reaction appears to take place by a series of steps
[6]:

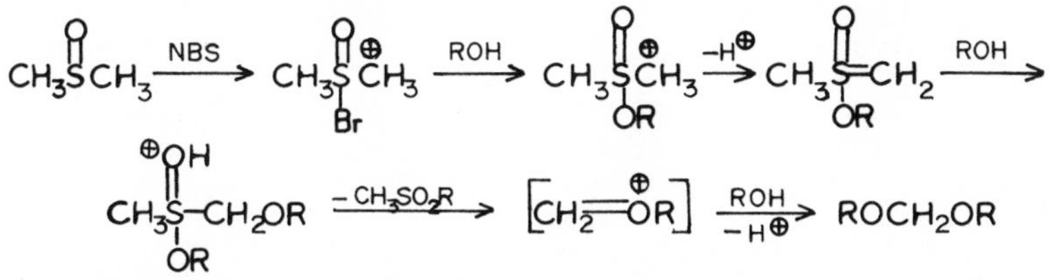

Cyclohexanol as the alcohol gives 86% of dicyclohexyl formal,
boiling point 75-76°, 0.25 mm.

Another rather strange reaction that appears to be simple
displacement [7] follows:

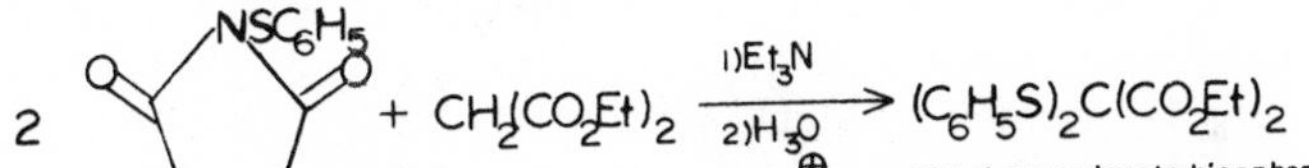

Ethyl oxomalonate bis-phenyl-
thio ketal, 93%

This reaction has been added to emphasize the oxidation
state of the central carbon atom in malonic ester.

7. From β-Halopropionic Esters (Reduction)

The intermediate in this interesting reduction is apparently
trapped since a cyclopropane is formed [8]:

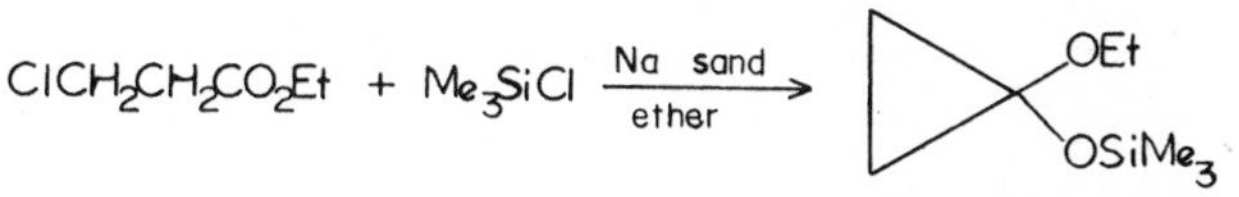

Cyclopropanone ethyl trimethylsilyl ketal 78%

References

1. D. F. Hunt, G. T. Rodeheaver, Tetrahedron Letters, 3595 (1972).
2. W. G. Lloyd, B. J. Luberoff, J. Org. Chem., 34, 3949 (1969).
3. K. Gollnick, G. Schade, Tetrahedron Letters, 857 (1973).
4. B. Zwanenburg et al., Rec. Trav. Chim., 90, 429 (1971).
5. S. S. Bhattacharjee, P. A. Gorin, Can. J. Chem., 47, 1195 (1969).
6. S. Hanessian et al., J. Am. Chem. Soc., 94, 8929 (1972).
7. T. Mukaiyama et al., Tetrahedron Letters, 5115 (1970).
8. K. Rühlman, Synthesis, 236 (1971).

E.   Electrolytic Alkoxylation

   4.   Electrolysis of Vinyl Ethers in Alcohol

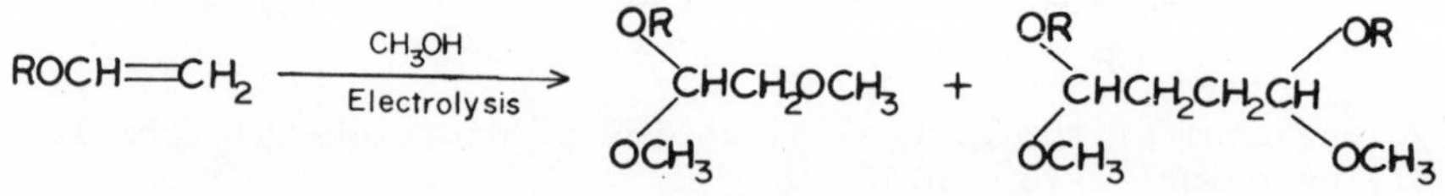

$$ROCH{=}CH_2 \xrightarrow[\text{Electrolysis}]{CH_3OH} \underset{OCH_3}{\overset{OR}{CHCH_2OCH_3}} + \underset{OCH_3}{\overset{OR}{CHCH_2CH_2CH}}\overset{OR}{\underset{OCH_3}{}}$$

   A recent bulletin, describing techniques for electroorganic synthesis, with examples including ketal preparations, is now available [1].   The electrolytic oxidation of vinyl ethers in methanol gives the dimethoxy product (usually 60-75%) and succindialdehyde acetal (usually 15-24%) [2].   The cell employed contained two concentric platinum gauze anodes, a nickel cathode, and the solvent was 0.5-1% KOH in methanol.   A current of 4 A and 7 V was applied for 24 hr at 5°.

References

1. G. Popp, Eastman Organic Chemical Bulletin, 45, No. 3, 1 (1973).
2. B. Belleau, Y. K. Au-Young, Can. J. Chem., 47, 2117 (1969).

F.   Cyclo Reactions

   The inherent instability of cyclopentadienone precludes its use as a diene in a Diels-Alder reaction.   However, it is possible to circumvent the formation of the unstable ketone and obtain its Diels-Alder product as follows [1]:

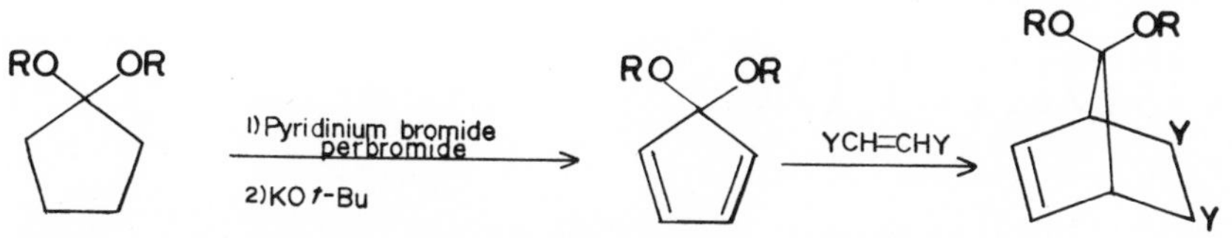

A similar result has been achieved by using 5,5'-dimethoxy-1,2,3, 4-tetrachlorocyclopentadiene [1a].

One of the products of the cyclo addition of vinyl ethers to ketones under irradiation is a ketal [2]:

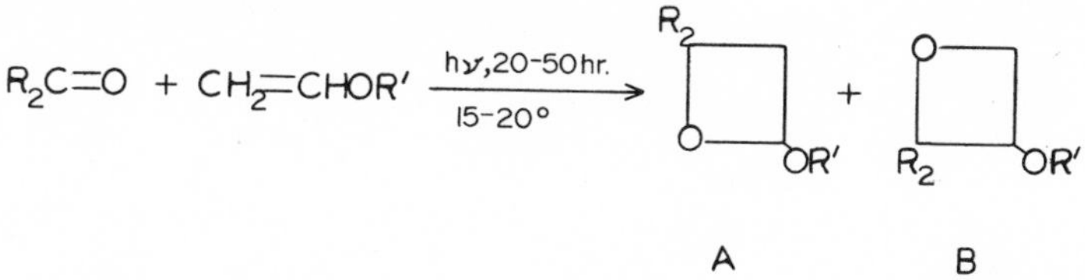

The first product A is an acetal. The total yield of 60-70% consisted of 30% of A, which may be solvolyzed to an open-chain acetal:

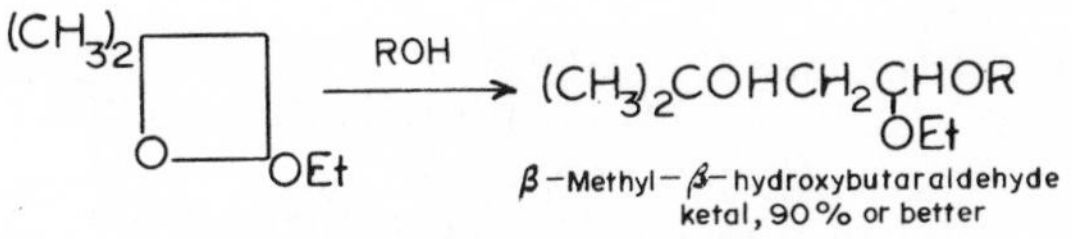

Addendum

F.   Methyl acrylate and 1,1-diethoxyethene refluxed 8 days in acetonitrile gives methyl 2,2-diethoxycyclobutanecarboxylate 60%) [Ph. Amice, J. M. Conia, Bull. Soc. Chim. France, 1015 (1974)].

References

1.   P. E. Eaton, R. A. Hudson, J. Am. Chem. Soc., 87, 2769 (1965).
1a. P. G. Gassman, J. L. Marshall, Org. Syn., Coll. Vol., 5, 424 (1973).
2.   S. H. Schroeter, C. M. Orlando, Jr., J. Org. Chem., 34, 1181, 1188 (1969).

ALDEHYDES

Numerous good aldehyde preparations are found in a single annual issue of Organic Synthesis [1]. Although the Sarett reagent, $CrO_3 \cdot 2C_5H_5N$ has proved its worth in the oxidation of primary alcohols to aldehydes (A.1), the Corey reagent, $C_5H_5NHCrO_3Cl$ (Addenda A.1) not only appears to be more efficient, but has less tendency to oxidize the aldehyde to the acid. Hot dilute nitric acid has its place in performing the same function (A.1). Etard's reagent CrO_2Cl_2, has been extended to oxidize alkenes to aldehydes of the same number of carbon atoms (A.12). The conversion $-CH_2NO_2 \rightarrow -CH=O$ now seems more applicable to synthesis (A.18).

It appears that lithium aluminum tri-tert-butoxyhydride at low temperature is the method of choice for reduction of acid chlorides to aldehydes (B.3), but the Rosenmund reduction has been improved in two ways (B.3). Another elegant reduction of acid chlorides is carried out in the presence of glycol mesylate, which traps the aldehyde as the acetal (B.3).

An unexpected substitute for the cyanide in the Gattermann synthesis of aldehydes is s-triazine (C.1). However, in this Friedel-Crafts section (C.6) the Vilsmeier reaction stands out in its phenomenal expansion of variations. Replacements are proposed for the Lewis acid, $POCl_3$, and even for the formylating agent, DMF (C.6). The Lewis or protonic acid and the formylating agent are employed in an attack on active hydrogen compounds including acetylene, on enamines, and on alkenes as well as aromatics and heterocycles (C.6). Among the new reagents to attach formyl groups to aromatics are hexamethylenetetramine (C.12) and trioxane (C.13).

The versatile Meyers synthesis of aldehydes from masked carbonyl groups is discussed (D.2, D.3) and the scope of hydration of acetylenes to aldehydes is expanded beyond the synthesis of acetaldehyde (D.6).

Ketenethioacetal monoxides have been utilized in the synthesis of both aldehydes and ketones (D.7). Some α-alkylations of aldehydes are described not only via the enamine, but via the aldehyde itself (F.4). Even the Wittig reagent of special

$$Ph_2\overset{\overset{O}{\|}}{P}-CHNMe_2$$

structure, $Ph_2\overset{O}{\overset{\|}{P}}-CHNMe_2$, has been used to prepare aldehydes (F.1).

Alkylation, arylation, or alkylolation of any formic acid derivative, the product of which on hydrolysis leads to aldehydes, is discussed in G. The compound DMF remains outstanding in this connection, but other formic acid derivatives of interest are 1,1,3,3-tetramethylbutylisocyanide of acceptable, if not pleasant, odor and phenyl diethyl orthoformate (G). In addition, the potential aldehydo group may be carried as a part of the reagent such as the Schiff base.

The Claisen-Cope rearrangement of vinyl alkyl ethers is undergoing an expansion of its scope as a means for preparing aldehydes (H.1). A one-pot method starting with the allyl alcohol and ethyl vinyl ether has been suggested. For the latter, the allyl vinyl sulfide offers some advantages. There is a trend to vary the electronic centers as well, particularly the vinyl center, all of which suggests other extensions to the synthesis of aldehydes by the Claisen-Cope rearrangement.

Reference

1. Org. Syn., 51 (1971).

A. Oxidation

1. From Primary Alcohols (1, 545)

$$RCH_2OH \longrightarrow RCHO$$

The number of oxidizing agents utilized in this transformation is legion. Those not previously mentioned, which may serve for both the formation of aldehydes and ketones, are: (a) nitric acid-DME [1], (b) the complex of dimethyl sulfide with Cl_2 or NCS [2] or dimethyl sulfoxide with Cl_2 [2], (c) pyridine dichromate, $(C_5H_5NH)_2Cr_2O_7)$ [3], (d) CrO_3-HMPA [4], (e) CrO_3-graphite [5], (f) Ag_2CO_3 on celite [6], (g) Ag_2O in acid [7], (h) chloramine [8], (i) neutral aqueous $Na_2Cr_2O_7$ [9], (j) CuO (vapor phase oxidation) [10], (k) silver(II) picolinate [11], (l) lead(II) acetate-C_5H_5N [12], (m) DMSO-Ac_2O [13], (n) polymer

$\langle \bigcirc \rangle$CH$_2$N=C=NCH(CH$_3$)$_2$ [14], (o) active cobalt oxide-O_2 [15], and

(p) potassium ferrate [16].

New reagents utilized only in oxidizing primary alcohols to aldehydes are: (a) nickel peroxide [17] and (b) nickel oxide (special) [18].

Reviews of the sulfoxide-carbodiimide oxidation of alcohols (1,546) are available [19]. The general procedure suggested is to dissolve the alcohol in a 10-50% mixture of DMSO in C_6H_6 or $CH_3COOC_2H_5$ containing 3 molar equiv of DCC. Following the addition of 0.5 molar equiv of H_3PO_4, dichloroacetic acid, or pyridinium triacetate (the latter prevents isomerization of the substrate), the reaction mixture is kept at room temperature for a time determined by the acid introduced. The recovery of the carbonyl compound has been reviewed [19a, pp. 11-12]. The DMSO-DCC reagent, which has been applied to the oxidation of a great variety of alcohols, is a powerful oxidizing medium capable of oxidizing inert alcohols under mild conditions and is suitable for use with sensitive alcohols. Another way to use DMSO without the acidic reagent is to prepare the 4-nitrobenzene-sulfonyl ester and then oxidize with DMSO [20]. The intermediate in both cases appears to be the cation, $RCH_2OS\overset{+}{\underset{CH_3}{\overset{CH_3}{<}}}$, which with the base eliminates a proton and then dimethyl sulfide.

A second oxidation reagent rather widely used is the complex of the oxide of chromium and pyridine. The most common form appears to be the Sarett reagent, $CrO_3 \cdot 2C_5H_5N$ [21], which is now available commercially. (Caution: the complex formed between CrO_3 and C_5H_5N is highly exothermic!) By conducting the oxidation in methylene chloride, the rate of aldehyde formation is increased at least 20-fold over the rate observed in pyridine solution. Reaction conditions are mild and the isolation of products is easy (Ex. a).

A third reagent of more recent origin is the complex formed between dimethyl sulfide and chlorine, $(CH_3)_2\overset{+}{S}Cl\overset{-}{Cl}$ or N-chloro-succinimide, NCS,

$$\text{(succinimide ring)} \quad N\overset{\oplus}{S}(CH_3)_2\overset{\ominus}{Cl} \qquad [2]$$

The steps involved in the oxidation are:

$$(CH_3)_2S + Cl_2 \xrightarrow[0°]{CCl_4} (CH_3)_2\overset{\oplus}{S}Cl \overset{\ominus}{Cl} \xrightarrow[-25°]{\text{>CHOH}}$$

$$(CH_3)_2\overset{\oplus}{S}-OC\underset{H}{<} \overset{\ominus}{Cl} \xrightarrow{Et_3N} \text{>C=O} + (CH_3)_2S + Et_3NHCl$$

The method is simple, highly selective, and efficient. The use of the NCS complex is the more convenient of the two since hydrogen chloride formation is avoided and the reaction products are somewhat cleaner. By the method benzyl alcohol gave benzaldehyde in 98% and 1-octanol gave octanal in 95% yield. The dimethyl sulfoxide-chlorine complex has an advantage over the dimethyl sulfide complexes in that halide formation is reduced, but the former is unsatisfactory for unsaturated alcohols [2].

A fourth reagent, also of recent origin, is hot HNO_3 in aqueous glyme [1]. For monoarylcarbinols it is the reagent of choice in that the procedure is simple and gives isolated yields of aldehydes from 86-96% (Ex. b).

One of the older oxidizing agents employed in the oxidation of primary alcohols to aldehydes is manganese dioxide (1, 546). Some attention has been given recently to the preparation of this reagent. Belew and Tek-Ling [22] ozonized manganese(II) nitrate to obtain an active oxide. Goldman [23] devised a reproducible procedure involving the azeotropic distillation of water. Carpino [24] prepared an active form of the oxide by treatment of $KMnO_4$ with decolorizing carbon, whereas Vereshchagen and co-workers [25] found that γ-MnO_2, prepared from an aqueous solution of manganese sulfate, was more active in the oxidation of alcohols than the Attenburrow product (1, 546).

a. Preparation of 1-Heptanal [21b]

To 650 ml of methylene chloride, during stirring, 77.5 g of $CrO_3 \cdot 2C_5H_5N$ was added at 25° followed by 5.8 g of 1-heptanol. After 20 min, the usual workup resulted in the recovery of 4.0-4.8 g (70-84%) of the aldehyde, boiling point 80-84°/65 mm.

b. Preparation of 4-Bromobenzaldehyde [1]

A solution of 0.01 mol of 4-bromobenzyl alcohol in a mixture of 20 ml of 1,2-dimethoxyethane and 2 ml of aqueous HNO_3 was heated under reflux for 1 hr. After extraction with C_6H_6, the organic layer was washed with saturated $NaHCO_3$ solution, then water, and dried (Na_2SO_4). From the filtered solution 96% of the aldehyde was recovered by distillation.

5. From Benzylic (or Allylic) Alcohols and Halides with
    Nitrogen-Oxygen Reagents (1, 551)

p-Nitrosodimethylaniline has been utilized in the conversion of the dipyridinium bromide into the dialdehyde (Kröhnke
reaction, A.13) as shown [26]:

$$Br^{\ominus}C_5H_5\overset{\oplus}{N}CH_2\langle\underline{\quad}\rangle COCO\langle\underline{\quad}\rangle CH_2\overset{\oplus}{N}C_5H_5Br^{\ominus} \xrightarrow[2)H_3O^{\oplus}]{1)p\text{-}NO\,C_6H_4N(CH_3)_2\text{-}NaOH}$$

$$O=C\overset{H}{\langle\underline{\quad}\rangle}COCO\langle\underline{\quad}\rangle\overset{H}{C}=O$$

4,4'-Diformylbenzil

The yield has been improved by starting with the 4,4'-bis(tri-
phenylphosphoniomethyl) benzil dibromide which gave the dialdehyde in 60% yield (overall). o-Nitrobenzaldehyde has been
prepared in an overall yield of 47-53% by the Kröhnke reaction
[27].

6. From Ethylenic Compounds and Related Reactions (1, 553)

$$RCH=CHR \xrightarrow{O_3} RCH\overset{O}{\underset{O-O}{\diagdown}}CHR \xrightarrow{H_2} 2\ RCHO$$

Nonanal has been produced from 1,1-diphenyl-1-decene by the
ozonolysis of a CH_2Cl_2-CH_3OH solution as indicated [28]:

$$CH_3(CH_2)_7CH=C\overset{C_6H_5}{\underset{C_6H_5}{\diagup}} \xrightarrow[2)KI\text{-}CH_3CO_2H]{1)O_3} CH_3(CH_2)_7CHO$$

The use of tris(dimethylamino)phosphine in converting the
ozonide into the aldehyde permitted ready separation of the
product [29].

In the steroid series Shalon and Elliott [30] converted the

$$-CH=C\overset{C_6H_5}{\underset{C_6H_5}{\diagup}}$$ group into the aldehyde group (Barbier-Wieland

degradation) by the use of RuO_4 in a neutral medium, while
Giacopello and co-workers [31] used V_2O_5-H_2O_2 for the conversion
of the propenyl group.

7. From Glycols (<u>1</u>, 555)

$$RCHOHCHOHR \longrightarrow 2\ RCHO$$

Although periodic acid and lead tetraacetate appear to be the most widely used reagents for the oxidation of glycols to aldehydes or ketones, other reagents have been used for this purpose. Glycols, for example, were oxidized to carbonyl compounds with a mixture corresponding of 2 parts of $Pb(H_2PO_4)_2$ and 1 part of $H_2[Pb(H_2PO_4)_2(HPO_4)_2]$ prepared from Pb_3O_4 and 85% H_3PO_4. Thus octene-4 gave butyraldehyde as indicated [32]:

$$CH_3CH_2CH_2CH = CHCH_2CH_2CH_3 \longrightarrow 2\ CH_3(CH_2)_2CHO$$

83%

as the 2,4-dinitrophenylhydrazone

McKillop and Taylor [33] used thallium(III) nitrate or thallium(I) ethoxide in this oxidation to produce aldehydes or ketones. Thus hydrobenzoin gave benzaldehyde as shown:

$$C_6H_5CHOHCHOHC_6H_5 \xrightarrow[CH_3CO_2H]{Tl(NO_3)_3 \cdot 3H_2O} 2\ C_6H_5CHO$$

61 %

By the use of active precipitated manganese(IV) oxide (available from E. Merck, Darmstadt), Ohloff and Giersch [34] were able to convert 1,2-<u>cis</u> diols and analogous <u>trans</u> types with a flexible arrangement of their hydroxyl groups <u>into</u> aldehydes or ketones usually in high yield. Thus <u>cis</u> 1,2-cyclo-dodecanediol gave 1,12-dodecanedialdehyde as indicated:

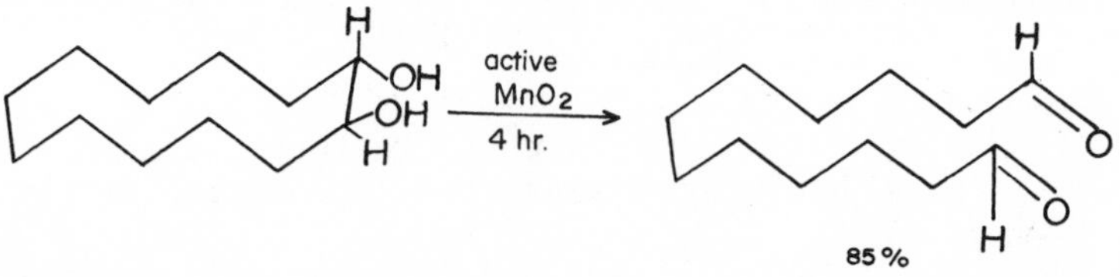

85%

On the other hand, 9,10-trans-decalindiol remained unchanged and the <u>cis</u> isomer gave a high yield of the 1,2-diketone.

10. From Halides and Miscellaneous Oxidizing Agents (Dimethyl Sulfoxide, t-Butyl Hydroperoxide, Sodium Dichromate, Trimethylamine Oxide, or Disodium Tetracarbonylferrate(-II))

Recently, disodium tetracarbonylferrate(-II) has been proven of value in the synthesis of aldehydes from alkyl halides [35]:

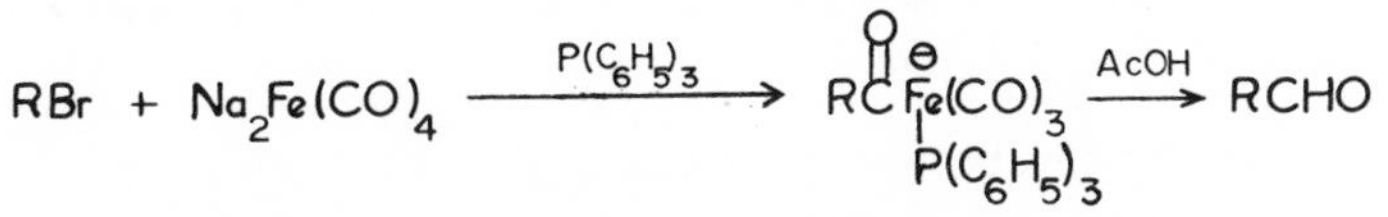

$$RBr + Na_2Fe(CO)_4 \xrightarrow{P(C_6H_5)_3} RC\overset{O}{\overset{\|}{\underset{P(C_6H_5)_3}{Fe(CO)_3}}} \xrightarrow{AcOH} RCHO$$

Yields of aldehydes from a series of alkyl bromides vary within 50-99% (glpc). Primary bromides respond more satisfactorily than secondary ones. Later the alkyl and acyl tetracarbonylferrate anions, which are intermediates in the reaction, were isolated [36]. Thus from these aldehydes or ketones may be obtained:

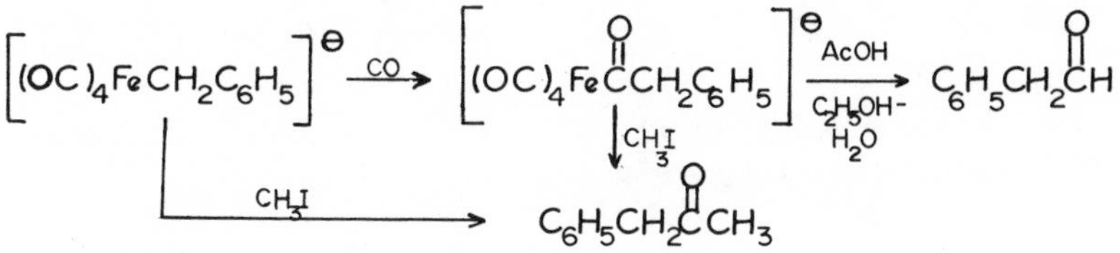

$$\left[(OC)_4FeCH_2C_6H_5\right]^{\ominus} \xrightarrow{CO} \left[(OC)_4Fe\overset{O}{\overset{\|}{C}}CH_2C_6H_5\right]^{\ominus} \xrightarrow[\substack{C_2H_5OH \\ H_2O}]{AcOH} C_6H_5CH_2\overset{O}{\overset{\|}{C}}H$$

$$\downarrow CH_3I$$

$$\xrightarrow{CH_3I} C_6H_5CH_2\overset{O}{\overset{\|}{C}}CH_3$$

11. From Methylarenes (1, 560)

Ceric ammonium nitrate in acid solution has been proven effective in oxidizing methylarenes to the corresponding aldehyde as indicated [37]:

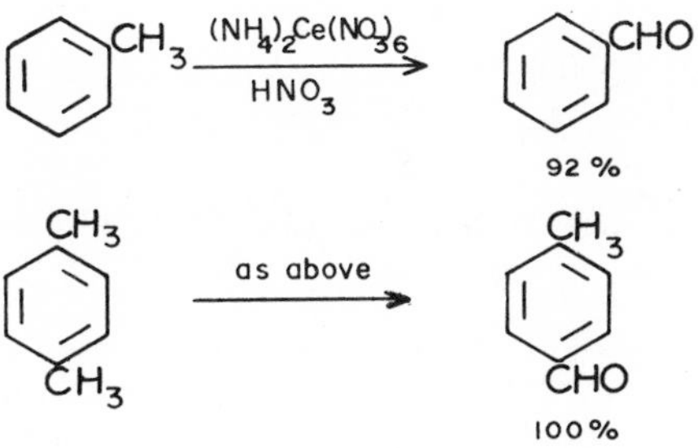

92 %

100%

12. From Methylarenes or Alkenes and Chromyl Chloride (Etard Oxidation) (1, 560)

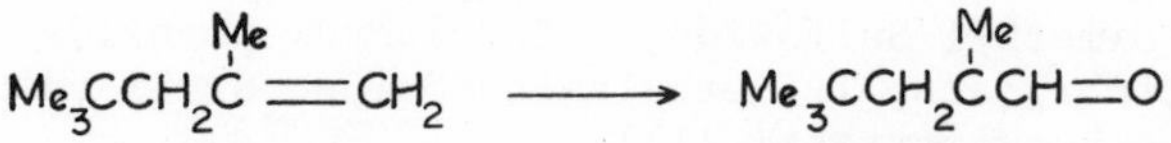

$$Me_3CCH_2\overset{\overset{\displaystyle Me}{|}}{C}=CH_2 \longrightarrow Me_3CCH_2\overset{\overset{\displaystyle Me}{|}}{C}CH=O$$

Etard's reagent, CrO_2Cl_2, oxidizes terminal olefins to aldehydes of the same number of carbon atoms [38]. It was added dropwise to 2,4,4-trimethyl-1-pentene at 0-5° and the mixture was then treated with zinc to prevent overoxidation as the temperature increased. The aldehyde was obtained in approximately 75% yield.

13. From Methyl Heterocycles (1, 561)

Ultraviolet irradiation of 2- and 3-methylindoles in acetic acid produces the corresponding indolecarboxaldehydes in low yield [39]:

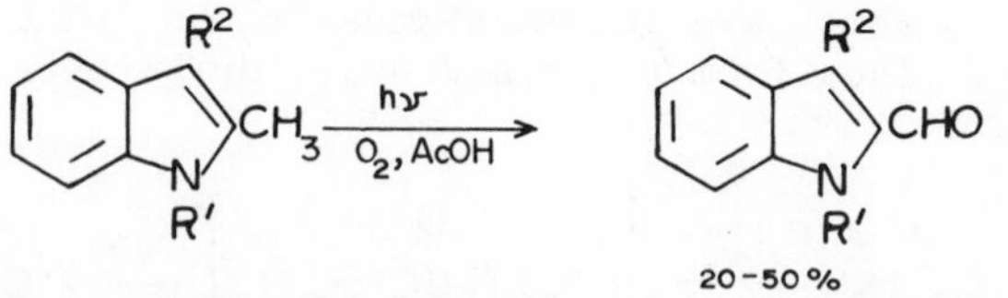

20-50%

The Kröhnke reaction (1, 551, A.5) has been applied to methyl heterocycles to give aldehydes in 43-72% overall yields [40]:

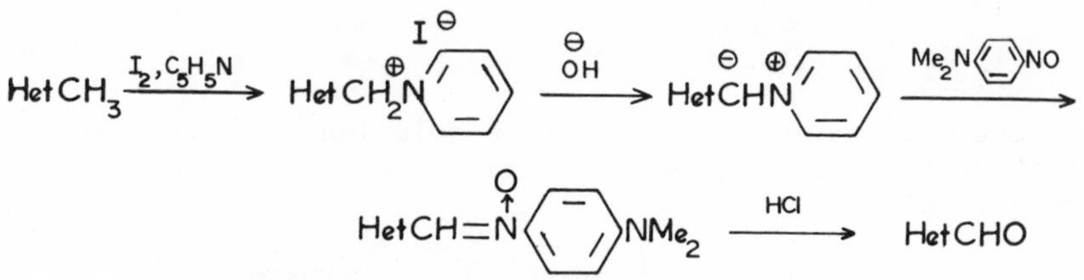

15. From Active Hydrogen Compounds, the Nitroso Ion, and the Like (1, 562,672)

$$YC\overset{\ominus}{H_2} \xrightarrow{NO^{\oplus}} Y\overset{|}{C}HNO \longrightarrow Y\overset{|}{C}=NOH \longrightarrow Y\overset{|}{C}=O$$

Both aldehydes and ketones have been prepared by this procedure, which has been reviewed [41]. To nitrosate hydrocarbons irradiation or other free radical initiation is necessary. The most important of these substitutions is the irradiation of cyclohexane and nitrosyl chloride to form cyclohexanone oxime [42]. If the initial nitroso compound cannot tautomerize to the oxime, it tends to dimerize. Substitution in active hydrogen compounds may be brought about by both acid and base. Several examples illustrate this type of substitution:

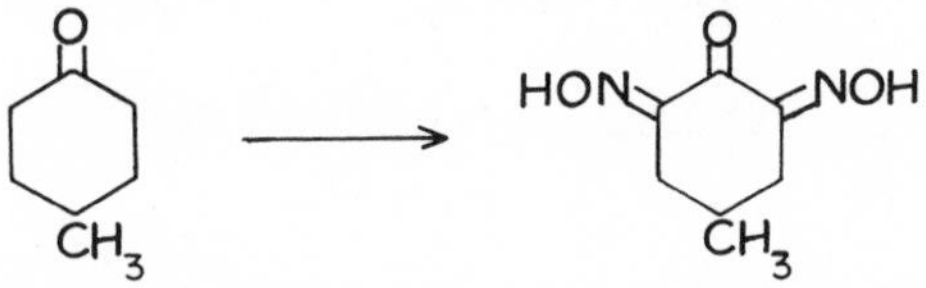

5-Methyl-1,2,3-cyclohexanetrione-1,3-dioxime was obtained in 70% yield by treating the ketone with isoamyl nitrite and acid at -5° [43]. A second example illustrates the base-catalyzed procedure via nitrosation [44]:

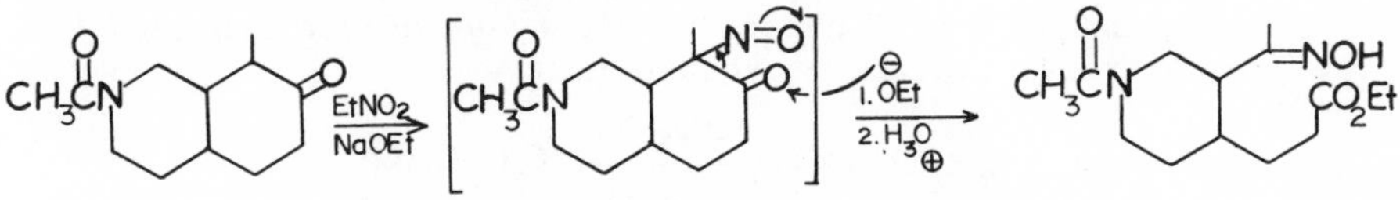

This method is used to degrade ketones to acids [45].

Irradiation of cycloheptyl nitrite leads to profound changes with the isolation of some 7-nitrosoheptanal [46]. Irradiation of nitroanthracene

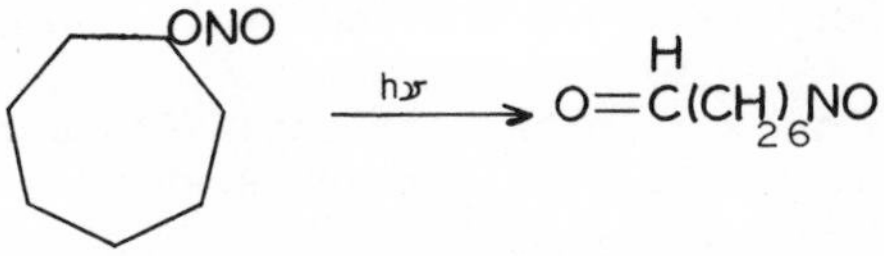

gave anthraquinone monoxime [46].

Additions of nitrosyl chloride to alkenes give α-chloro-β-nitroso adducts that can be hydrolyzed to ketones. Similar additions, which permit carbonyl formation, occur with other nitrosyl compounds such as:

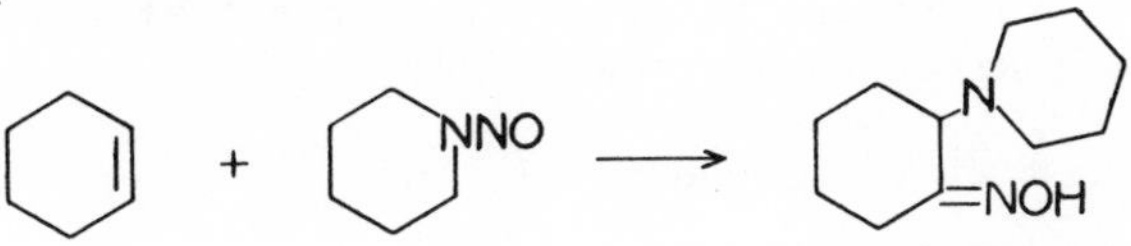

[47]

2-Piperidinocyclohexanone oxime

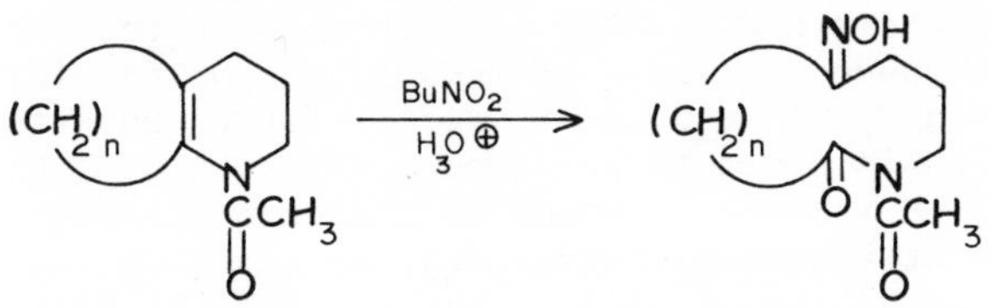

[48]

The yields of oximinoimides were 80-100% when n = 4, 5, and 10.

The nitrosation of β-keto sulfoxides was investigated by Otsuji and co-workers [49] as shown:

$$C_6H_5COCHSOCH_3 \xrightarrow[\text{HCl}]{\text{NaNO}_2} C_6H_5COC=NOH + C_6H_5COCOCH_3$$
$$CH_3 CH_3$$

After 15 min both products were isolated, but after 2 hr only 1-phenylpropane-1,2-dione (70%) was obtained.

16.   From Ethers by Peroxidation or Treatment with Trityl Tetrafluoroborate (1, 563)

Oxidation of benzyl ethers to aldehydes may be achieved by the use of trityl tetrafluoroborate [50]:

$$C_6H_5CH_2OCH_3 \xrightarrow[20°, 2 hr.]{(C_6H_5)_3CBF_4} C_6H_5CHO$$
$$ 75\%$$

17.   From Phenylcarbinol Oxidative Splitting (1, 564)

The checked synthesis of o-anisaldehyde via the phenylcarbinol by the method described previously (1, 564) is now available [51].

Oxidation of an alkylphenylcarbinol with ceric ammonium nitrate gives the aldehyde and ketone as shown [52]:

$$C_6H_5CHOHC(CH_3)_3 \xrightarrow[CH_3CN]{(NH_4)_2Ce(NO_3)_6} C_6H_5CHO + C_6H_5COC(CH_3)_3$$
$$ 95\% 0.5\%$$

The amount of aldehyde formed increases substantially as the alkyl radical formed in the process becomes more stable, in other words, the return is greatest with a tertiary carbon radical and least with a primary one.

18.   From Nitro Compounds (1, 565)

$$RCH_2NO_2 \longrightarrow RCHO$$

A recent method for the conversion of nitro compounds into carbonyl compounds is that of McMurry and Melton [53]. The nitro compound, if acid sensitive, is first converted into the anion, which is added to an aqueous solution of $TiCl_3$-NH_4OAc, ratio 1:3, pH ∿ 5-6. Reaction to the carbonyl compound occurs in minutes at room temperature (Ex. a). Yields vary from 45 to 90%. The reduction is milder than the Nef reaction (see Ketones, D.8) and more tolerant of other functional groups.

Nonsensitive nitro compounds may be treated with an unbuffered aqueous solution of $TiCl_z$. The mechanism is probably as shown:

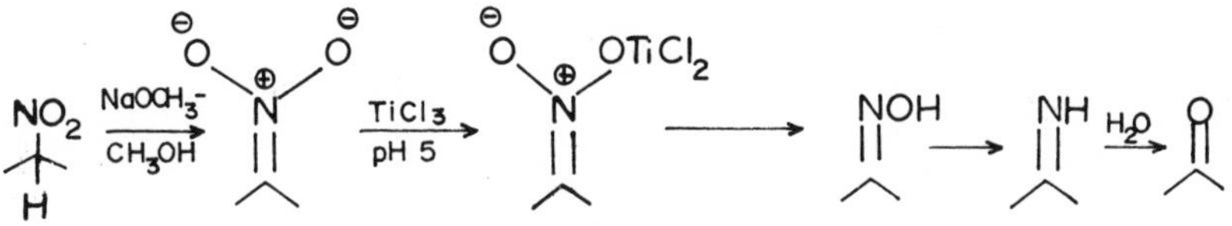

a. Preparation of Phenylacetaldehyde [53]

β-Nitroethylbenzene was dissolved in CH_3OH (0.5 M) and treated with 1 equiv of $NaOCH_3$. A buffered $TiCl_3$-NH_4OAc solution (NH_4OAc, 0.06 mol in 15 ml of H_2O and 20% of $TiCl_3$, 0.01 mol) was then added at 25° under N_2. Recovery in the usual manner gave 70% of the aldehyde.

19. From β-Ketosulfoxides (Pummerer Rearrangement) (<u>1</u>, 566)

The synthesis of phenylglyoxal from the β-ketosulfoxide is now available in <u>Organic Synthesis</u> [54]. It appears to be a method suitable for the synthesis of other glyoxals. The Pummerer rearrangement involved has been the subject of numerous recently discussed investigations [55]. The mechanism involves the oxidation of the methylene carbon by the oxygen of the sulfoxide, a process that may occur in a number of ways, one of which is:

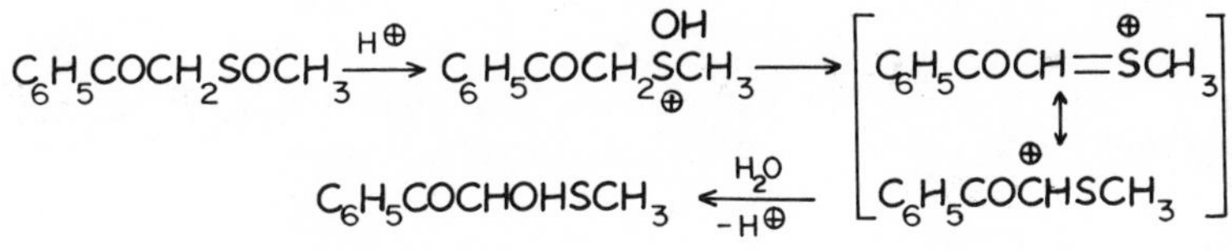

The sulfur is reduced and the active methylene group oxidized to produce the hemimercaptal.

21. From Amines

$$ArCH_2NH_2 \longrightarrow ArCHO$$

The methods for converting amines into aldehydes or ketones may be classified as direct or indirect. Illustrations follow.

a. Direct Methods

Benzaldehyde has been obtained from benzylamine by

oxidation with neutral $KMnO_4$ [56] or K_2FeO_4 [57] and from dimethylbenzylamine by the use of MnO_2 [58].

b. Indirect Methods

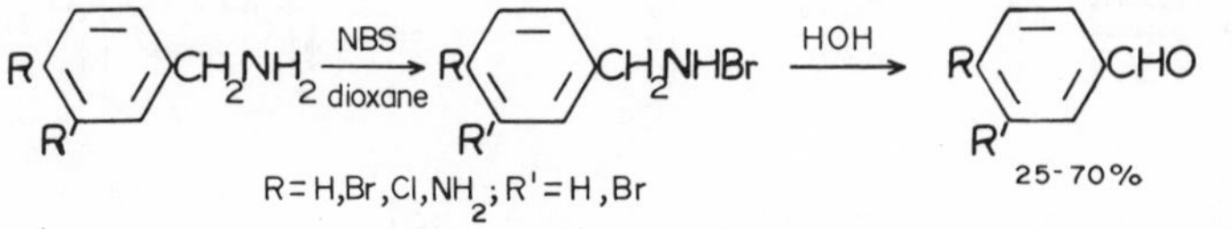

$$R = H, Br, Cl, NH_2; R' = H, Br$$

25-70% [59]

1) Tautomerization
2) Oxalic acid

C_6H_5CHO
69 %

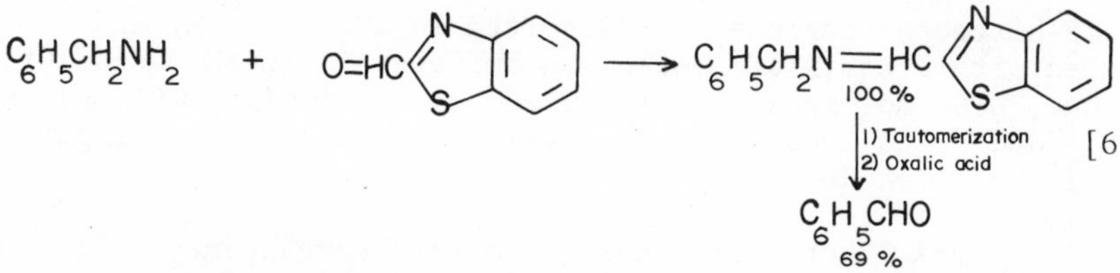

[60]

The method is better for ketones and with both carbonyl types is applicable to the degradation of the proper alkaloids.

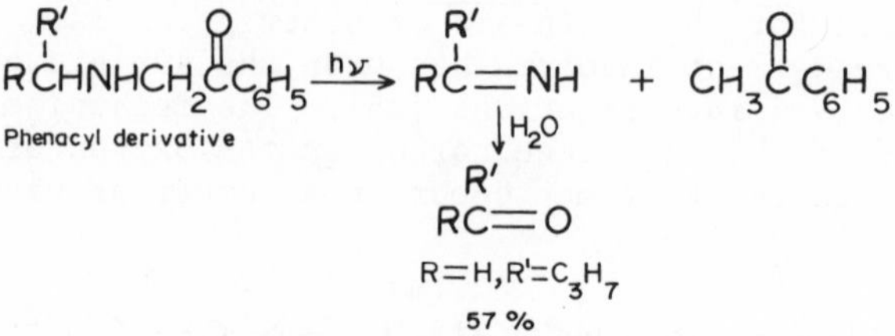

$R = H, R' = C_3H_7$
57 %

[61]

The preceding is a convenient method for conversion of amines into carbonyl compounds under mild, nonoxidative conditions.

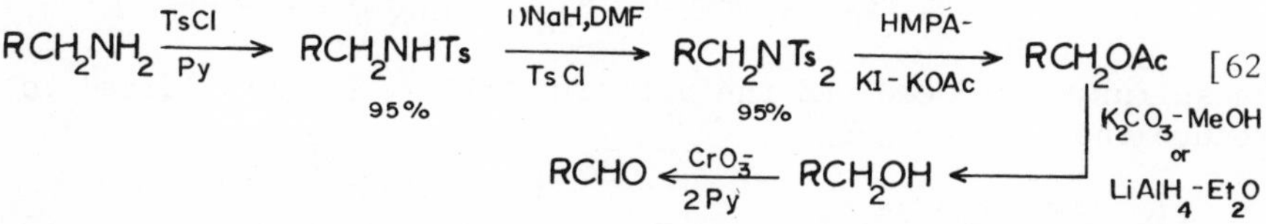

[62]

This sequence is considered usable for the conversion of all but the most hindered primary amines into aldehydes

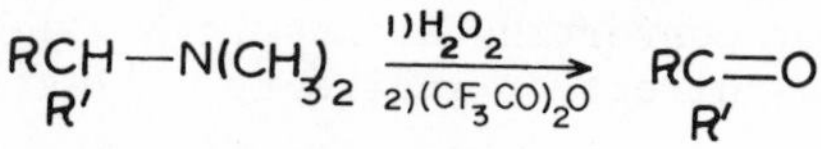

[63]

This method of proceeding via the amine oxide has been utilized by Corey in the bergamotene synthesis, the yield of ketone being 57%.

22. Oxidative Coupling of Aldehydes

Aldehydes couple to form C-C and C-O products on refluxing them over activated MnO_2, nickel peroxide, or PbO_2. Thus iso-butyraldehyde responds as shown [64]:

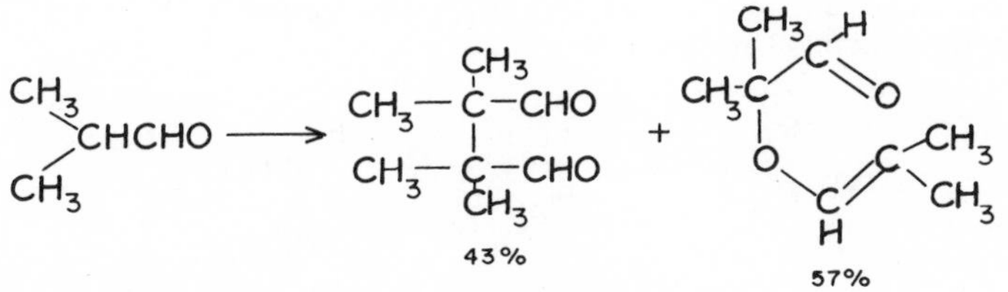

43%

57%

23. From Specific Azo Compounds

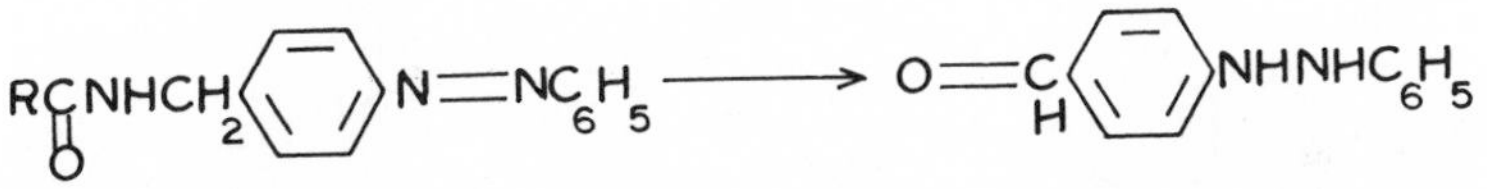

This unusual reaction was carried out by heating the substrate in 0.36N KOH in alcohol under nitrogen for 3 hr [65], after which the hydrazo usually was oxidized to the azo compound by H_5IO_6. With $R=CH_3$, the overall yield of p-phenylhydrazobenzaldehyde was 85%. The ortho phenylazobenzylamides gave 2-phenylindazoles:

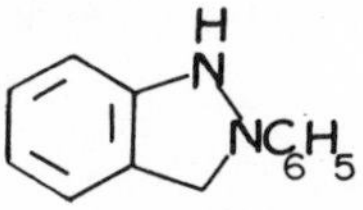

The meta derivatives gave no product, a fact indicating the tautomeric nature of the oxidation involving quinone structures.

24. From Alkenes via Hydroboration

For a discussion of these methods see Aldehydes, E.8 and F.6.

25. From Epoxides

An unexpected cleavage occurs when cyclohexene epoxide is treated with periodic acid in that 1,6-hexanedial is obtained [66]:

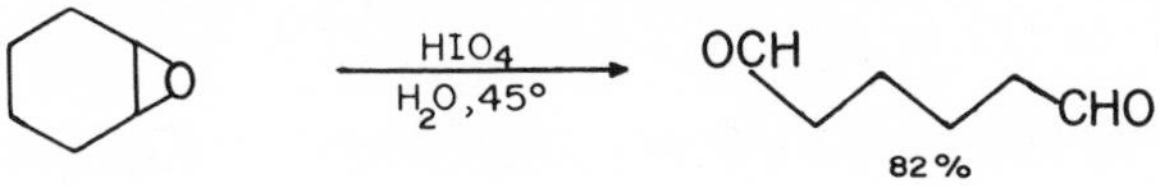

82%

26.   Acetoxylation of Aldehydes and Ketones

A recent review [67] on the acetoxylation of carbonyl compounds is available. The reaction has usually been conducted with lead tetraacetate alone or in the presence of boron trifluoride. Yields are rarely high. One example each of aldehyde and ketone acetoxylation are given:

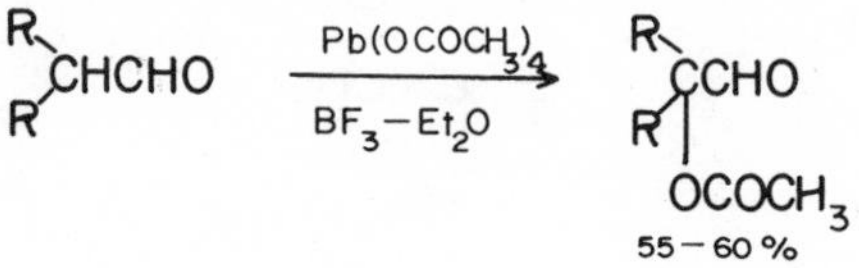

$$\underset{R}{\overset{R}{\diagdown}}CHCHO \xrightarrow[BF_3-Et_2O]{Pb(OCOCH_3)_4} \underset{\underset{55-60\%}{\overset{|}{OCOCH_3}}}{\overset{R}{\underset{R}{\diagdown}}}CCHO \qquad [68]$$

68%
[69] 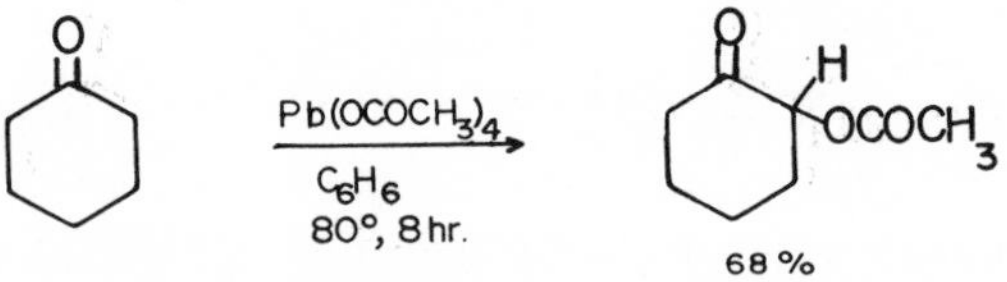

Addenda

A.1. Pyridinium chlorochromate, $C_5H_5NHCrO_3Cl$, is more readily made than the Sarett reagent, is safer, and stores better. Moreover, less reagent (1.5 equiv) is needed to oxidize the alcohol to the aldehyde, and there appears to be no further oxidation. It is used in nonaqueous media. To date the reagent appears to be the reagent of choice for the preparation of aldehydes [E. J. Corey, J. W. Suggs, Tetrahedron Letters, 2647 (1975)].

A.1. Another procedure that appears to produce no oxidation product beyond the aldehyde is the use of triethyltin methoxide followed by bromine [T. Mukaiyama et al., Chem. Lett., 145 (1975)].

A.1. A polymer with a sulfonium salt attached to a number of units reacts with a primary alcohol to form an alkoxysulfonium salt, $\text{(P)}\overset{+}{S}MeOCH_2R\ \bar{C}l$, which with triethylamine gives good yields of RCH=O [G. A. Crosby et al., J. Am. Chem. Soc., 97, 2232 (1975)].

References

1.   A. McKillop, M. E. Ford, Syn. Commun., 2, 307 (1972).

2. E. J. Corey, C. U. Kim, J. Am. Chem. Soc., 94, 7586 (1972); Tetrahedron Letters, 919 (1973); J. Org. Chem., 38, 1233 (1973).
3. W. M. Coates, J. R. Corrigan, Chem. Ind. (London), 1594 (1969).
4. R. Bengelmans, Bull. Soc. Chim. France, 335 (1969).
5. J. M. LaLancette et al., Can. J. Chem., 50, 3058 (1972).
6. M. Fetizon et al., Tetrahedron, 29, 1011 (1973); Compt. Rend., 267C, 900 (1968).
7. L. Syper, Tetrahedron Letters, 4193 (1967).
8. G. A. Jaffari, A. J. Nunn, J. Chem. Soc., C, 823 (1971).
9. D. G. Lee, W. A. Spitzer, J. Org. Chem., 35, 3589 (1970).
10. M. Y. Sheikh, C. Eadon, Tetrahedron Letters, 257 (1972).
11. J. B. Lee et al., Can. J. Chem., 47, 1649 (1969).
12. R. E. Partch, Tetrahedron Letters, 3071 (1974).
13. J. D. Albright, L. Goldman, J. Am. Chem. Soc., 87, 4214 (1965).
14. O. M. Shen, Tetrahedron Letters, 3285 (1972).
15. J. S. Belew et al., Chem. Commun., 634 (1970).
16. R. J. Audette et al., Tetrahedron Letters, 279 (1971).
17. R. E. Atkinson et al., Chem. Commun., 718 (1967).
18. J. S. Belew, C. Tek-Ling, Chem. Commun., 1100 (1967).
19a. J. G. Moffatt, in R. L. Augustine-D. J. Trecker, Oxidation, Vol. 2, Dekker, New York, 1971, p. 1.
19b. T. Durst, Advances in Organic Chemistry, Vol. 6, Interscience, New York, 1969, p. 345.
20. C. H. Snyder et al., Synthesis, 655 (1971).
21. a. R. Ratcliffe, R. Rodehurst, J. Org. Chem., 35, 4000 (1970); b. J. C. Collins, W. N. Hess, Org. Syn., 52, 5 (1972); c. K.-E. Stensio, Acta Chem. Scand., 25, 1125 (1971).
22. J. S. Belew, C. Tek-Ling, Chem. Ind. (London), 1958 (1967).
23. I. M. Goldman, J. Org. Chem., 34, 1979 (1969).
24. L. A. Carpino, J. Org. Chem., 35, 3971 (1970).
25. L. I. Vereshchagen et al., J. Org. Chem. USSR, 8, 1143 (1972).
26. B. Krieg, Chem. Ber., 102, 371 (1969).
27. A. Kalir, Org. Syn., Coll. Vol. 5, 825 (1973).
28. M. Fétizon et al., J. Org. Chem., 38, 1732 (1973).
29. A. Furlenmere et al., Rec. Trav. Chim., 50, 2387 (1967).
30. Y. Shalon, W. H. Elliott, Syn. Commun., 3, 287 (1973).
31. D. Giacopello et al., Org. Prep. Proced. Int., 4, 13 (1972).
32. F. Huber, M. S. A. El-Meligy, Chem. Ber., 102, 872 (1969).
33. A. McKillop, E. C. Taylor et al., J. Org. Chem., 37, 4204 (1972).

34. G. Ohloff, W. Giersch, Angew. Chem. Intern. Ed. Engl., 12, 401 (1973).
35. M. P. Cooke, Jr., J. Am. Chem. Soc., 92, 6080 (1971).
36. W. O. Siegl, J. P. Coleman, J. Am. Chem. Soc., 94, 2516 (1972).
37. L. Syper, Tetrahedron Letters, 4493 (1966).
38. F. Freeman et al., J. Org. Chem., 33, 3970 (1968).
39. C. A. Mudry, A. R. Frasca, Tetrahedron, 29, 603 (1973).
40. W. Ried, R. M. Gross, Chem. Ber., 90, 2646 (1957).
41. J. H. Boyer, in H. Feuer, The Chemistry of the Nitro and Nitroso Groups, Interscience, New York, 1969, Pt. 1, p. 215.
42. Ref. 41, p. 218.
43. D. C. Batesky, N. S. Moon, J. Org. Chem., 24, 1694 (1959).
44. Ref. 41, p. 228.
45. W. Dieckmann, Chem. Ber., 33, 579 (1900).
46. Ref. 41, p. 221.
47. Ref. 41, p. 230.
48. J. R. Mahagan et al., Synthesis, 313 (1973).
49. Y. Otsuji et al., Bull. Chem. Soc., Jap., 44, 219 (1971).
50. D. H. R. Barton et al., Chem. Commun., 1109 (1971).
51. A. J. Sisti, Org. Syn., Coll. Vol., 5, 49 (1973).
52. W. S. Trahanovsky, J. Cramer, J. Org. Chem., 36, 1890 (1971).
53. J. E. McMurry, J. Melton, J. Org. Chem., 38, 4367 (1973); J. Am. Chem. Soc., 93, 5309 (1971).
54. G. J. Mikol, G. A. Russell, Org. Syn., Coll., Vol., 5, 937 (1973).
55. C. R. Johnson, W. G. Phillips, J. Am. Chem. Soc., 91, 682 (1969); T. Durst, Adv. Org. Chem., 6, 356 (1969).
56. S. S. Rawalay, H. Shechter, J. Org. Chem., 32, 3129 (1967).
57. P. J. Smith et al., Tetrahedron Letters, 279 (1971).
58. K. Schlögel, M. Walser, Tetrahedron Letters, 5885 (1968); O. Meth-Cohn, H. Suschitzky, Chem. Ind. (London), 443 (1969).
59. S. H. Dandegaonker, P. D. Pagar, Curr. Sci., 42, 539 (1973); C. A., 79, 91724 (1973).
60. V. Calo et al., J. Chem. Soc., Perkin Trans., I, 1652 (1972).
61. J. H. Hyatt, J. Org. Chem., 37, 1254 (1972).
62. N. H. Anderson, H. Uh, Syn. Commun., 2, 297 (1972).
63. E. J. Corey et al., J. Am. Chem. Soc., 93, 7016 (1971); A. Cavé, Tetrahedron, 23, 4681 (1967).
64. J. C. Leffingwell, Chem. Commun., 357 (1970).
65. F. H. Stodola, J. Org. Chem., 37, 178 (1972).
66. J. P. Nagarkatti, K. R. Ashley, Tetrahedron Letters, 4599 (1973).

67. D. J. Rawlinson, G. Sosnovsky, Synthesis, 577 (1973).
68. J.-J. Riehl, A. Fougerousse, Bull. Soc. Chim. France, 4083
     (1968).
69. G. W. K. Cavill, D. H. Solomon, J. Chem. Soc., 4426 (1955).

B. Reduction

3. From Acid Chlorides (Rosenmund and Brown) or Anhydrides (1, 571)

$$RCOCl \xrightarrow[\text{Pd(BaSO}_4)]{\text{H}_2} RCHO$$

Rosenmund

Improvements in the Rosenmund reduction continue to appear.
To eliminate disadvantages such as long reaction time at high
temperatures, some hydrogen hazards, and high catalyst to substrate ratios, Rachlin and co-workers [1] conduct the reaction in
a closed system under low pressure. From 3,4,5-trimethoxybenzoyl
chloride, Pd/C, and quinoline S, they obtained, at 50 p.s.i. of
H_2 and 25° in 1 hr, a 64-83% yield of the aldehyde. Peters and
Ván Bekkum [2] have shown that ethyldiisopropylamine is an
effective hydrogen chloride acceptor which with H_2 and Pd/C in
acetone is very satisfactory in the reduction at room temperature
and atmospheric pressure. With 1-methylcyclohexanecarbonyl
chloride the aldehyde was obtained in 78% yield.

With the acid chlorides of saturated fatty acids, White and
co-workers [3] obtained excellent results using an acetone
solution with Pd/BaSO$_4$, dimethylaniline, as the hydrogen chloride
acceptor, and hydrogen under slight pressure at 25° for 30 min.
By this procedure palmitaldehyde was obtained in 96.3% yield.

The Brown reduction of acid chlorides using lithium aluminum
tri-tert-butoxyhydride at low temperature is the method of choice.
A model preparation by this method is the synthesis of 3,5-di-
nitrobenzaldehyde in 60-63% crude yield at -78° [4]. About 30
aldehydes both aromatic and aliphatic have been synthesized in
this manner.

An elegant reduction of an acid chloride may be
illustrated by that of cyclobutanecarbonyl chloride with a glycol
mesylate and sodium borohydride [5]. The mesylate serves to trap
the aldehyde as an acetal and thus prevents further reduction.

Other related reductions have been reported. Watanabe and
co-workers used sodium tetracarbonylferrate followed by
acetic acid as indicated [6]:

$$RCOCl + Na_2Fe(CO)_4 \xrightarrow{\text{THF}} \overset{\oplus}{Na} \left[Fe(CO)_4COR\right]^{\ominus} \xrightarrow{\text{AcOH}} RCHO$$

65-95%

Citron employed triethylsilane as the reducing agent as shown
[7]:

$$RCOCl + (C_2H_5)_3SiH \xrightarrow[25°]{Pd/C} RCHO + (C_2H_5)_3SiCl$$
$$5-65\%$$

Aromatic acyl bromides may be reduced to aldehydes (58-80%)
by irradiation in ether solution [8].

4. From Nitriles, Amides, Acylpyrazoles, or Imidazoles
    (1, 572)

$$RCN \text{ or } RCONH_2 \xrightarrow[2) H^\oplus]{1) LiAlH(OC_2H_5)_3} RCHO$$

Lithium triethoxyaluminohydride continues to be used in the
partial reduction of nitriles to aldehydes [9]. It reduces both
aromatic and aliphatic nitriles in 1 hr to the corresponding
aldehydes in 68-96% yields. Likewise, diisobutylaluminum hydride has served for a similar reduction [10]. In a similar
manner the dimethylamides of aliphatic, alicyclic, aromatic, and
heterocyclic acids may be usually converted into the corresponding aldehydes in yields of 60-90% with $LiAlH_2(OC_2H_5)_2$ or
$LiAlH(OC_2H_5)_3$ [8]. Sodium aluminum hydride in THF has also
served well in the reduction of dimethylamides to aldehydes, the
yield range being 70-90% [11]. Thus these reducing agents appear
to be the reagents of choice in the reduction of nitriles and
amides to aldehydes.

7. From Nitriles, Raney Nickel, and Reducing Agents
    (1, 576)

$$RCN \xrightarrow[HCOOH]{Raney Ni} RCHO$$

The use of formic acid alone or with formaldehyde and Raney
nickel has come to be rather common. It has been used in the
synthesis of p-formylbenzenesulfonamide [12], o-methylbenzalde-
hyde [13], and 3,4,5-trimethoxybenzaldehyde [14]. Yields using
formic acid (75% preferred) were from 93 to 100% for four
aromatic aldehydes [15]. For the reduction of hindered nitriles
moist, preformed Raney nickel is recommended [16].
     A study of the reaction of nitriles with hydrazine hydrate
and Raney nickel emphasizes the limitations of this method for
the synthesis of aldehydes. Only benzonitriles form aldazines,
the necessary intermediates in the aldehyde route [17].

9.   From Esters, Lactones, or Ortho Esters (1, 578)

$$RCOOR' \xrightarrow{(i\text{-}C_4H_9)_2AlH} RCHO$$

Deuterated aldehydes have been prepared as indicated [18]:

$$RCOOCH_3 \xrightarrow{LiAlD_4} RCD_2OH \xrightarrow[\text{on celite}]{Ag_2CO_3} R\underset{\underset{O}{\|}}{C}D$$

ca. 80%

(last step)

Diisobutylaluminum hydride has been used in the reduction of N-alkyl-2-carbomethoxyazetidines, -pyrrolidines, and -piperidines [19] in ether at 72° to the corresponding aldehyde:

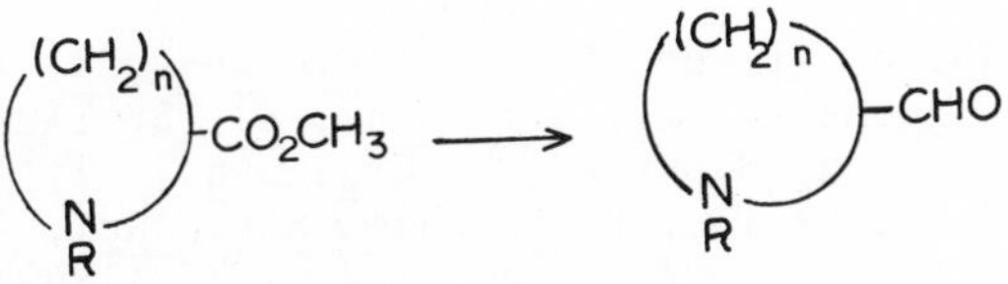

Thus N-isobutyl-2-carbomethoxyazetidine gave 89% of the aldehyde.

11.   From Acids or Acid Anhydrides (1, 580)

Carboxylic acids have been reduced to aldehydes by the use of diisobutylaluminum hydride [20]. Thus caproic acid in ether at -70° to -75° gave caproaldehyde as indicated:

$$CH_3(CH_2)_4COOH \longrightarrow CH_3(CH_2)_4CH\left[OAl(i\text{-}C_4H_9)_2\right]_2 \xrightarrow{H_2O} CH_3(CH_2)_4CHO$$

70 %

Acid anhydrides have been converted into aldehydes using disodium tetracarbonyl ferrate [21]:

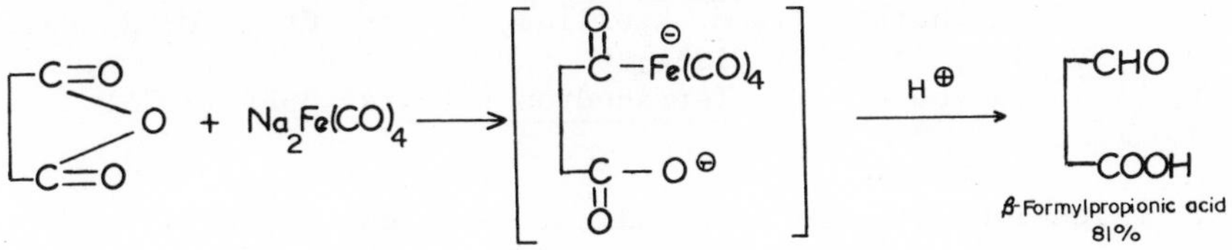

β-Formylpropionic acid
81%

Other yields vary from 27% to 90%. Acyclic acid anhydrides give unsubstituted aldehydes. The reaction is simple, rapid, and occurs at room temperature.

Addenda

B.4. Nitriles with triethoxyoxonium tetrafluoroborate form the ethyl immonium salt, which with triethylsilane followed by hydrolysis gives the aldehyde [J. L. Fry, Chem. Commun., 45 (1974)].

B.9. A new reagent has been utilized in the reduction of esters to aldehydes: Di-(N-methylpiperazine) alane, $HAl[N(CH_2CH_2)_2NMe]_2$ [M. Muraki, T. Mukaiyama, Chem. Lett., 215 (1975)].

References

1. A. I. Rachlin et al., Org. Syn., 51, 8 (1971).
2. J. A. Peters, H. Van Bekkum, Rec. Trav. Chim., 90, 1323 (1971).
3. H. B. White, Jr. et al., J. Lipid Res., 8, 158 (1967)
4. J. E. Siggins et al., Org. Syn., 53, 52 (1973).
5. H. R. Johnson, B. Rickborn, Org. Syn., 51, 11 (1971).
6. Y. Watanabe et al., Bull. Chem. Soc. Jap., 44, 2569 (1971).
7. J. D. Citron, J. Org. Chem., 34, 1977 (1969).
8. W. Silhan, W. Schmidt, Monatsh. Chem., 102, 1481 (1971).
9. J. Malek, M. Cerny, Synthesis, 231 (1972).
10. R. V. Stevens et al., J. Org. Chem., 37, 977 (1972).
11. L. I. Zakharkin et al., Tetrahedron, 25, 5555 (1969).
12. T. Van Es, B. Staskun, Org. Syn., 51, 20 (1971).
13. B. Staskun, O. G. Backeberg, Org. Syn., 50 (Mar. 70-3) (1970).
14. T. Balogh et al., Hung Teljes 2492, August 8, 1971; C. A., 75, 140699 (1971).
15. T. Van Es, B. Staskun, J. Chem. Soc., 5775 (1965).
16. B. Staskun, O. G. Backeberg, J. Chem. Soc., 5880 (1964).
17. W. W. Zajac, Jr. et al., J. Org. Chem., 36, 3539 (1971).
18. M. Fetizon et al., Tetrahedron, 29, 1011 (1973).
19. P. Duhamel et al., Compt. Rend., 276 C519 (1973).
20. L. I. Zakharkin, I. M. Khorlina, J. Gen. Chem. USSR (Engl. transl.), 34, 1021 (1964).
21. Y. Watanabe et al., Tetrahedron Letters, 3535 (1973).

C. Friedel-Crafts and Free Radical Reactions (1, 583)

1. From Arenes, Zinc Cyanide, and Hydrogen Chloride (Gattermann)

$$ArH + Zn(CN)_2 + HCl \xrightarrow{ZnCl_2} ArCH=NH \cdot HCl \xrightarrow{H_2O} ArCHO$$

It is interesting to note that one of the side-products formed from hydrogen cyanide in acidic media, the s-triazene, s-C$_3$H$_3$N$_3$ [1], now has been found to be a reagent that can be substituted for the hydrogen cyanide in the Gattermann reaction [2]. Hydrogen chloride is the catalyst required for substitution in activated nuclei such as furans, pyrroles, and indoles, and some polyhydroxybenzenes (Ex. a), but a Friedel-Crafts catalyst is necessary for the rest.

The Gattermann reaction does not work well in substitution of (or addition to) alkenes, probably because the product, the unsaturated aldehyde, is unstable under acidic conditions [3].

a. Preparation of 5-Methylfurfuraldehyde [2]

Through a salt-ice cooled mixture of 0.15 mol of 2-methyl-furan and 0.05 mol of s-triazine in dry ether, hydrogen chloride was passed for 2 hr or until the aldimine hydrochloride had precipated. The mixture was treated with water, neutralized with sodium carbonate, and extracted with ether. Concentration and distillation gave the aldehyde, boiling point 72-73° at 12 mm, 9.3 g (56%).

2. From Arenes, Carbon Monoxide, and Hydrogen Chloride (Gattermann-Koch) (1, 584)

$$ArH + CO + HCl \xrightarrow{AlCl_3} ArCHO + HCl$$

Beyond the description given in 1, 584, titanium tetra-chloride and boron trifluoride in hydrogen fluoride are alternate catalysts for aluminum chloride [4]. However, this reaction and others in this section have been overshadowed by the Vilsmeier reaction (C.6).

3. From Arenes and Formyl Fluoride (1, 584)

$$ArH + HCOF \xrightarrow{BF_3} ArCHO + HF$$

It is of interest to note that the formyl cation (HCO +) has never been detected [5].

4. From Arenes and Dichloromethyl Alkyl Ethers (1, 585)

$$ArH + Cl_2CHOR \xrightarrow{AlCl_3} ArCHORCl \xrightarrow{H_2O} ArCHO$$

The mesitaldehyde preparation, using a titanium tetra-chloride catalyst (1, 585), is described in Organic Synthesis [6]. The Rieche formylation of benzothiophene with butyl dichloromethyl ether has been carried out to give 65% of the 3-carboxaldehyde [7]. This yield is higher than that of the Vilsmeier-Haack (10%). A formylation of an α-bromoketone by methyl dichloromethyl ether is described (see Aldehydes, G).

5. From Phenols or Active Hydrogen Compounds and Triethyl Orthoformate or Derivatives (Formylation) (<u>1</u>, 585)

$$\text{ArH} + \text{HC(OEt)}_3 \xrightarrow{\text{AlCl}_3} \text{ArCH(OEt)}_2 \xrightarrow{\overset{\oplus}{\text{H}_2\text{O}}} \text{ArCHO}$$

A Friedel-Crafts reaction between an orthoformate and a phenyl alkyl ether follows [8]:

$$\text{RO}\langle\text{—}\rangle + \overset{O}{\underset{O}{\bigcirc}}\text{OEt} \xrightarrow[\text{2)H}_3\text{O}^{\oplus}]{\text{1) Ac}_2\text{O}} \text{RO}\langle\text{—}\rangle\text{CHO}$$

The mixture was refluxed for 24 hr to give after hydrolysis a 25-35% yield of the aldehyde.

A modification has extended formylation to active hydrogen compounds [9]:

$$(\text{EtO})_2\text{CHN}\overset{\oplus}{\langle\text{—}\rangle} \overset{\ominus}{\text{BF}}_4 + \text{YH} \longrightarrow (\text{EtO})_2\text{CHY}$$

The reactants are refluxed for 2 hr in CH_2Cl_2. Y may be $\text{CH(CO}_2\text{Et)}_2$, CH_2NO_2, CH(CN)_2 or permutations.

6. From Arenes, Alkenes, or the Like and Formamides (Vilsmeier) (<u>1</u>, 586)

$$\text{ArH} \xrightarrow[\text{2)H}_3\text{O}^{\oplus}]{\text{1)DMF,POCl}_3} \text{ArCHO}$$

A number of variations of the Vilsmeier reaction, mostly using an electrophilic agent other than POCl_3, have been described. With oxalyl chloride either 1- or 2-adamantylmalonaldehyde may be isolated [10]:

$$\text{DMF} + (\text{ClCO})_2 \xrightarrow[\text{-CO}]{(\text{CH}_2\text{Cl})_2} \text{Me}_2\overset{\oplus}{\text{N}}\!\!=\!\!\text{CHCl} \ \overset{\ominus}{\text{Cl}} \xrightarrow[\text{2)neutralize}]{\overset{\text{1-AdCH=CHOCH}_3}{\text{1)80°}}} \text{1-AdC}\overset{\text{CHO}}{\underset{\text{CHOH}}{\diagdown}}$$

A   20% crude

Indeed the use of acid chlorides probably leads to a different form of A, $\text{Me}_2\overset{+}{\text{N}}=\text{CHOCORX}$, which would be expected to have

different characteristics. Horning and Muchowski [11] claim the
yield of indole-3-carboxaldehyde is higher (85%) by using DMF and
benzoyl chloride; however, a 97% yield has been reported with DMF
and POCl$_3$ [12].

Another acid chloride application involves the use of
phosgene and DMF in the formylation of ketones via the enamine
[13]:

$$DMF + COCl_2 + \text{(cyclohexenyl-morpholine enamine)} \longrightarrow \text{(2-formylcyclohexanone)} CHO$$

The 2-ketoaldehyde was obtained in 52% yield by refluxing the
mixture of the three components in CH$_2$Cl$_2$ for 30 min. If the
usual reagent, DMF-POCl$_3$, is used with trichloroethylene and
cyclohexanone, 2-chloro-1-cyclohexene carboxaldehyde is
obtained (53-74%) [14].

A third substitute for the phosphorus oxychloride is the
bromine adduct of triphenylphosphine [15]. With it and DMF to
form

$$\overset{\overset{\displaystyle Br}{\displaystyle |}}{(C_6H_5)_3\overset{+}{P}OCH\overset{+}{N}(CH_3)_2}Br,$$

the scope of formylation is quite broad in
that (a) styrene gives cinnamaldehyde 25%, (b) phenylacetylene,β-
bromocinnamaldehyde 60%, (c) indole, indole-3-carboxaldehyde 78%,
and (d) dimethylaniline,p-dimethylaminobenzaldehyde 72%. Formylation of alkenes is described here (styrene) and in C.8.

Dimethyl thioformamide,DMTF (Ex. a), as a substitute for
DMF, is said to give improved yields [16].

To obtain an unusual product or perhaps to improve the yield
of pyrrole aldehydes, the product complex is decomposed with
dimethylamine rather than water to give the vinylammonium salt
[17]:

The substrate was treated with POBr$_3$ and DMF followed by
dimethylamine.

The Vilsmeier reaction gives 71% of 2,3,4-trimethoxybenzal-
dehyde from 1,2,3-trimethoxybenzene [18]. A number of furans and
thiophenes have been formylated as well (13 ex.), 33-85% yields
[19]. In addition, acetonitrile has beem formylated to give 32%
of 2-cyano-3-dimethylaminoacrolein, Me$_2$NCH=C(CN)CHO [20].

a.   Preparation of 4-Formyl-2-phenyl-6a-thiathiophthene [16]

To 1 mmol of 2-phenyl-6a-thiathiophthene in 1.5 ml of DMTF was added 3 mmol of POCl$_3$; the mixture was warmed at 40° for 10 min and then shaken with 2 N aqueous NaOH. Extraction with benzene, elution from alumina, and crystallization from acetonitrile gave red needles, 62%. Use of DMF in place of DMTF gave a very low yield of aldehyde.

    8.   From Ethylenic Compounds by Hydroformylation (Oxo
          Process) (1, 588)

$$RCH = CH_2 + CO + H_2 \longrightarrow RCH_2CH_2CHO$$

As a substitute for the cobalt catalyst formerly described, a rhodium hydrido catalyst [(PhCH$_2$)$_3$N]$_2$RhHCO] is used at atmospheric pressure to give 78% of nonaldehyde from 1-octene [21].

    12. From Hexamethylenetetramine and Arenes or the Like

$$ArH + (CH_2)_6N_4 \xrightarrow{CF_3CO_2H} ArCHO$$

Two investigators have adapted the Sommelet reaction to arenes and the like rather than to benzyl halides, as shown in the above equation. One, using trifluoroacetic acid as the solvent, demonstrates some scope to the reaction. The yields were t-butylbenzaldehyde 75%, benzaldehyde 32%, p-phenoxybenzaldehyde 29%, and p,p'diformyldiphenyl ether 25%. The latter two products were obtained from the reaction of diphenyl ether [22]. The other uses 33% aqueous acetic acid as the solvent to synthesize 7-azaindole-3-carboxaldehyde in 50% yield [23].

    13. From Trioxane and Arenes (Free Radical)

The reaction gives poor to moderate yields with heterocycles but is worth considering because of its novelty [24]. The reagents are 30% hydrogen peroxide and a saturated solution of ferrous sulfate, which are added separately to a 3% aqueous sulfuric acid solution containing the trioxane and substrate. Overall yields of aldehydes obtained after hydrolysis were 2-methyl-4-quinolinecarboxaldehyde 21%, 4-methyl-2-quinolinecarboxaldehyde 28%, and 2-quinoxalinecarboxaldehyde 17%.

14. From Nitrites

$$RCH_3 \xrightarrow{NO^{\oplus}} RCH{=}NOH \longrightarrow RCHO$$

This subject is discussed in A.15 of this chapter.

15. From 1,3-Dithianes and Phenols (see Addenda)

Addenda

C.8. Another catalyst for the hydroformylation of alkenes with carbon monoxide is diphenylzirconium chlorohydride [C. A. Bertelo, J. Schwartz, J. Am. Chem. Soc., 97, 228 (1975)].

C.9 $$C_6H_5CH_2CH_2CHO \ + \ CH_2{=}CHOEt \xrightarrow[Ti(OPr)_4]{TiCl_4^-} C_6H_5CH_2CH_2CH{=}CHCHO$$

The combined catalyst must be used for best results [T. Mukaiyama et al., Chem. Lett., 569 (1975); see 1, 589 for similar reactions)].

C.15.

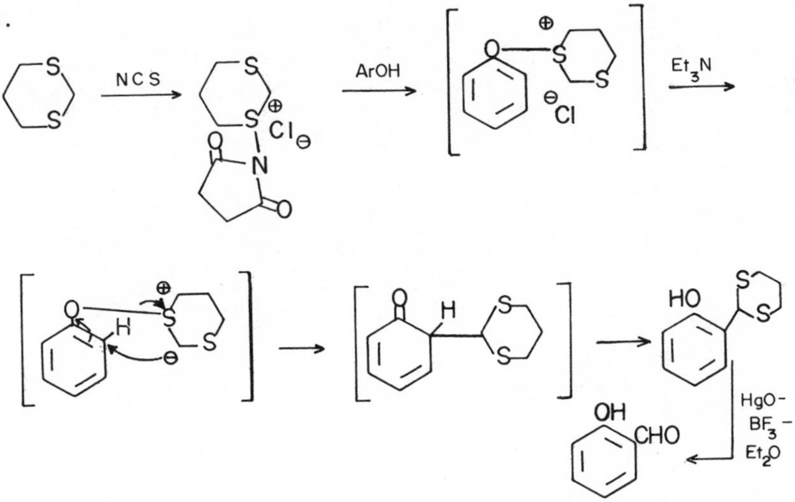

Exclusive ortho formylation is obtained. Yields were 30-46% for o-substitution and 67-79% for the hydrolysis. Oxidation of a methylthiomethylphenol to a salicylaldehyde was also accomplished [P. G. Gassman, D. R. Amick, Tetrahedron Letters, 3463 (1974)].

References

1. D. P. N. Satchell and R. S. Satchell, in S. Patai, The Chemistry of the Carbonyl Group, Interscience, New York, 1966, Vol. I, p. 282.
2. A. Kreutzberger, Arch. Pharm., 302, 828 (1969).
3. Ref. 1, p. 290.
4. Ref. 1, p. 256.
5. G. A. Olah et al., J. Am. Chem. Soc., 94, 4200 (1972).
6. A. Rieche et al., Org. Syn., Coll. Vol., 5, 49 (1973).
7. N. P. Buu-Hoï et al., J. Chem. Soc., C, 339 (1969).
8. G. N. Dorofeenko et al., Khim. Geterotsikl. Soedin, 7, 569 (1971); C. A., 76, 24853 (1972).
9. S. Kabusz, W. Tritschler, Synthesis, 312 (1971).
10. C. Reichardt, E.-U. Würthwein, Synthesis, 604 (1973).
11. D. E. Horning, J. M. Muchowski, Can. J. Chem., 48, 193 (1970).
12. P. N. James, H. R. Snyder, Org. Syn., Coll. Vol., 4, 539 (1963).
13. W. Ziegenbein, Angew. Chem. Intern. Ed. Engl., 4, 358 (1965).
14. L. A. Paquette et al., Org. Syn., Coll. Vol., 5, 215 (1973).
15. H. J. Bestmann et al., Ann. Chem., 718, 24 (1968).
16. D. H. Reid et al., J. Chem. Soc., C 913 (1969).
17. H. Von Dobeneck, T. Messerschmidt, Ann. Chem., 751, 32 (1971).
18. H. Seo, Japanese Patent 32,889, May 2, 1973; C. A., 79, 32098 (1973).
19. G. D. Meakins et al., J. Chem. Soc., Perkin Trans., I, 2327 (1973).
20. C. Reichardt, W.-D. Kermer, Synthesis, 538 (1970).
21. B. Fell, E. Müller, Monatsh. Chem., 103, 1222 (1972); P. Pino, C. Botteghi, Org. Syn., 53, unchecked procedure 1835 (1973).
22. W. E. Smith, J. Org. Chem., 37, 3972 (1972).
23. A. J. Verbiscar, J. Med. Chem., 15, 149 (1972).
24. G. P. Gardini, Tetrahedron Letters, 4113 (1972).

D. Hydrolysis or Hydration

1. From Acetals or Other Aldehyde Derivatives (1, 593)

This subject has been reviewed [1]. The rates of hydrolysis of acetals are exceedingly rapid and show that if diethyl formal is 1, diethyl acetal is 6000, and diethyl cinnamal is 3.7×10^5 [2].

An example of the hydrolysis is that of α-bromoheptaldehyde dimethyl acetal to the aldehyde (90-95%) by concentrated HCl with gentle refluxing and removal of the methanol as it is formed [3]. Dilute acid may be used as well [4] or dilute acid mixed with an immiscible solvent such as benzene [5].

The hydrolysis of the nitrone of o-nitrobenzaldehyde has been conducted recently with 6N H_2SO_4 at 25° with brief stirring by hand [6]:

Quaternary salts of dimethylhydrazones appear to be hydrolyzed readily [7]:

$$R_2C{=}NNMe_2 \xrightarrow{CH_3I} R_2C{=}NNMe_3I \xrightarrow{H_2O} R_2CO$$

Seven examples gave yields of 80-90%. The quaternary salts of thioacetals (or thioketals) hydrolyze readily as well [8].

The malondialdehyde enamine, $Me_2NCH{=}CRCH{=}O$, is readily converted into the crystalline malondialdehyde, $HOCH{=}CRCHO$, by conversion first into the sodium salt with NaOH followed by hydrolysis with aqueous acid [9].

Some efforts have been expended in effecting hydrolysis of aldehyde derivatives by oxidation or reduction of the derivative portion. The technique is useful for those aldehydes sensitive to acidic conditions. One of the most effective of these methods is the hydrolytic reduction of oximes to the corresponding carbonyl compounds by aqueous titanium trichloride [10]. Although chromous acetate has been suggested as a possible reducing agent [11], the titanium trichloride needs no derivative of the oxime, reduces the oxime in 1 hr at room temperature, and serves as its own indicator.

Oximes and semicarbazones have been oxidized rapidly at 0° by cerium ammonium nitrate to give the corresponding carbonyl compounds [12,13]. Yields were 27-90% (camphor oxime gave the lowest yield) and carvone oxime did not racemize in the conversion.

Another method applicable to aldoximes and unhindered ketones is the Meerwein-Ponndorf-Verley reduction [14] with isopropyl alcohol and aluminum isopropoxide. For aldoximes the

acetone formed must be removed as it is formed lest it condense
with the aldehyde.

That oxidation intermediates of thioacetals or thioketals
are hydrolyzed readily is indicated below [15]:

$$(C_6H_5)_2COHCHCHSMe \longrightarrow (C_6H_5)_2COHCHCHO$$

Diphenylglycolic aldehyde (59%) was obtained from the dimethyl
mercaptal S-oxide and concentrated HCl at 25° in 1.5 hr.

2.   From Five-Membered Heterocyclic Rings (Meyers) (1, 593)

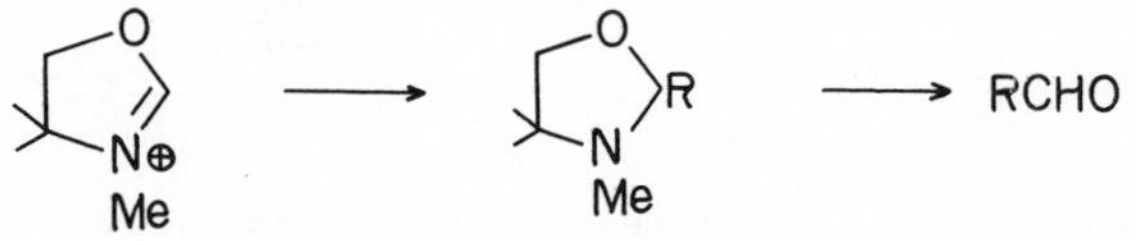

In this growing segment of organic synthesis, created and
developed by A. I. Meyers, the major step is alkylation and the
minor one is simple hydrolysis. This section and the one following take up both these steps in reactions which have been
reviewed [16]. The principle of this work is that a masked ester
or carbonyl group or an equivalent may be used to alkylate, as,
for example, with the masked ester:

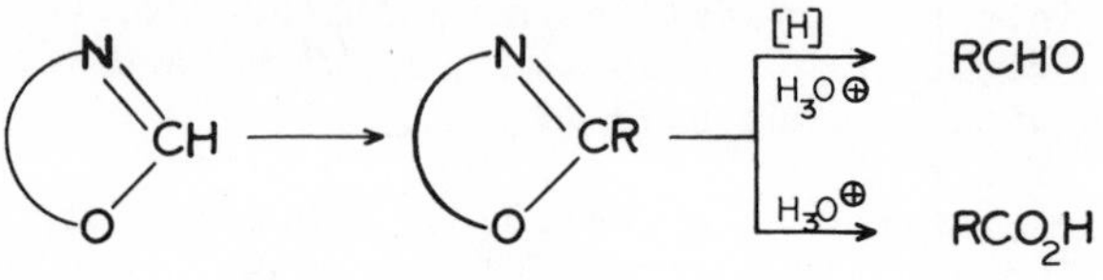

A   B

Structure B, as shown, must be reduced before it can be
hydrolyzed to an aldehyde and A may have all the possible permutations of O, N, or S that are stable.

Other structures similar to A which have an active hydrogen
(underlined) may be utilized as well:

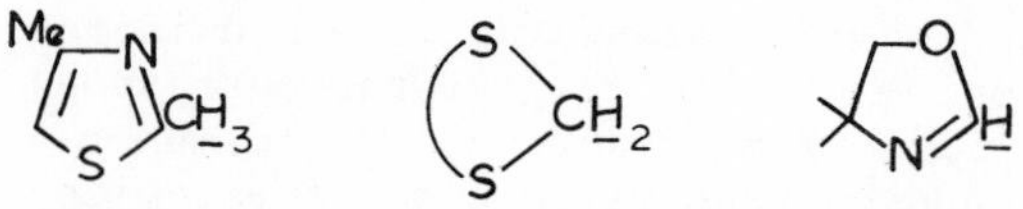

C   D

Structure D is unusual in that the Grignard reagents tend to form
the anion of it, which then opens to form an amino alcohol, but

the Grignard reagent complexed with two molecules of HMPA, $ArMgX \cdot 2HMPA$, simply adds to the methiodide salt as follows [17]:

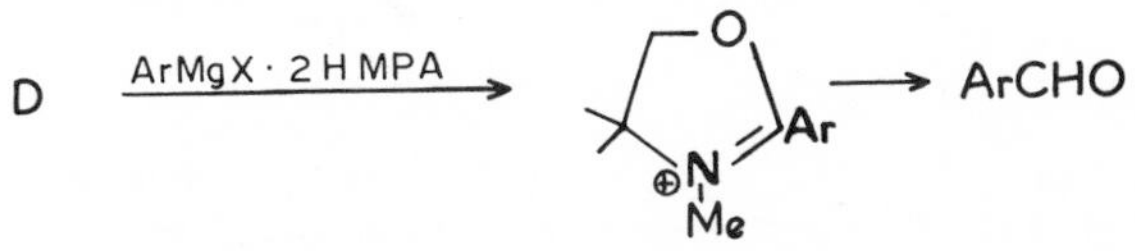

The best hydrolytic conditions in general for aldehydes are steam distillation as they form on the addition of aqueous oxalic acid. Five aromatic Grignard reagents gave 51-90% yields of the aldehyde. For a synthesis using the quaternary salt and D, see the work by Meyers and co-workers [18]. If C is used as the heterocycle, the product must be reduced before hydrolysis [19]:

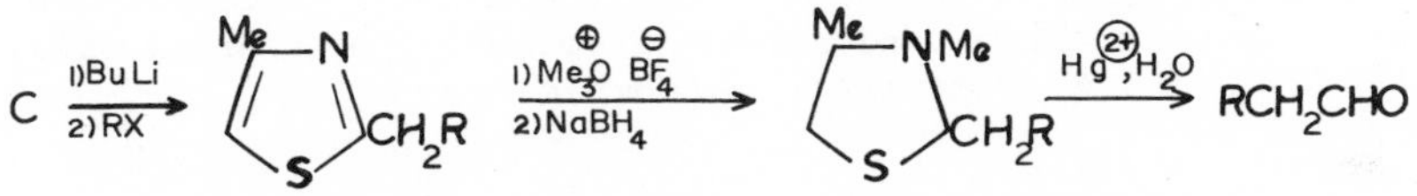

The yield of β-phenylpropionaldehyde ($R=C_6H_5CH_2$) was 52% overall. Meyers [20] used aluminum amalgam in moist ether for the reduction step, a procedure that permits hydrolysis in the absence of any acid. More widely studied synthesis of aldehydes by the Meyers technique is to be found in D.3. Other masked acetals prepared by sigmatropic rearrangement are to be found in H.1.

Not to be disregarded are furan, pyrrole, and thiophene as potential sources of succinaldehyde (1, 593) by hydrolytic splitting; in addition 2,5-disubstituted heterocycles may generate 1,4-diketones (1, 667). It seems to be generally agreed [21] that a route less fraught with polymerization hazards for the hydrolytic splitting of furans proceeds via the 2,5-dimethoxy-furan, E:

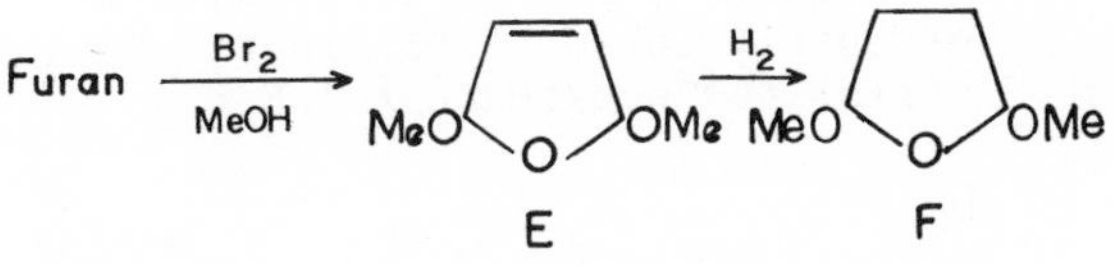

The succinaldehyde generated from F being unstable is used preferably in situ.

3.   From Six-Membered Heterocyclic Rings (Meyers) (1, 594)

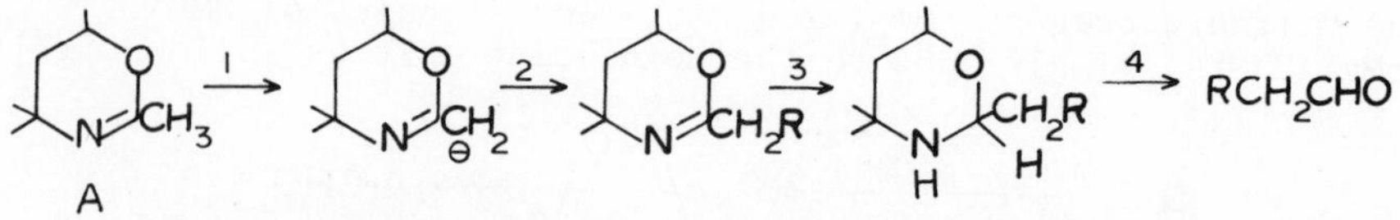

A

The preceding Meyers synthesis of aldehydes is the most studied
of the masked ester or carbonyl reactions [22,23]. The substrate is commercially available, but expensive. Step 1 is
carried out at -78° with BuLi, Step 2 utilizes not only alkyl
halides, but also carbonyl compounds [22,24] and epoxides, Step 3
employs aqueous sodium borohydride for the reduction. If sodium
borodeuteride in D_2O is used in Step 3, the deuterated aldehyde,
RCD=0, is obtained, and Step 4 utilizes aqueous oxalic acid for
the hydrolysis. An extensive series of aldehyde preparations
are listed [25,26] as well as variations on the original substrate, such as the use of the chloromethyl compound to prepare
α-chloro- and α,β-unsaturated carboxaldehydes [25].

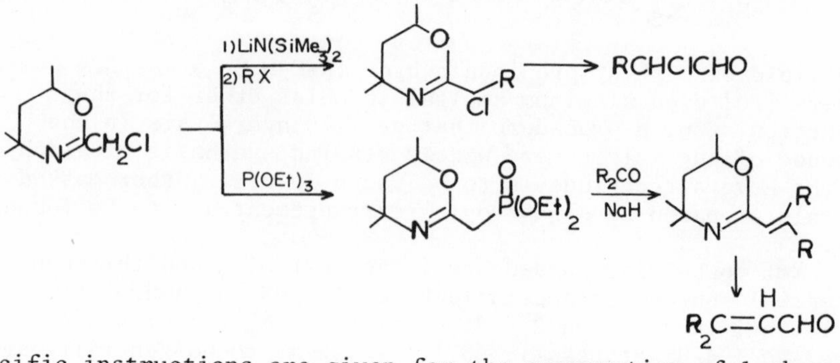

Specific instructions are given for the preparation of 1-phenyl-
cyclopentane-1-carboxaldehyde in 50-55% yield from 2-benzyl-4,6,
6-trimethyldihydrooxazine [26]; α-chloroaldehydes have been prepared from 2-chloromethyl-5,6-dihydro-1,3-oxazines [27].
    The need for butyllithium is eliminated by starting with
the quaternary salt of the substrate A [28]:

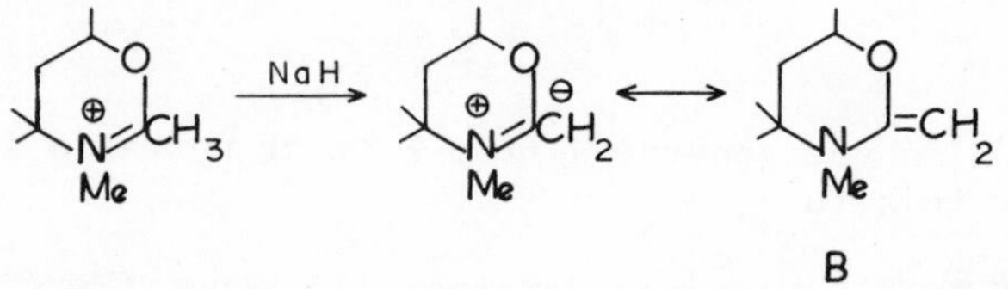

B

Structure B reacts with alkyl halides just as a vinylamine is
expected to react.

An alternate to the Meyers synthesis of aldehydes is the di- or trithiane method. A preparation is given from the tri-thiane [29]:

RCHO
R=C$_{14}$H$_{29}$,

Pentadecaldehyde
47 - 55%, overall

The solvolysis of C is carried out with HgO-HgCl$_2$ in methanol. The dithiane, a preparation of which is available [30], may also be used as a formylating agent. At times 1,3-dithianes do not hydrolyze readily. In such cases it has been found that a mixture of HgO and BF$_3$ is an effective reagent [31].

1,3-Dithianes may also be subjected to an oxidative desul-furization with chloramine-T in methanol-acetone-water to give fair to excellent yields of carbonyl compounds [32]. However the most rapid hydrolysis of dithianes appears to take place with thallium trifluoroacetate. In fact the reaction is complete almost instantaneously at 25° [33]. Similarly, triaryl-s-tri-thianes give good yields of benzaldehydes by treatment with iodine in DMSO [34]. A sigmatropic rearrangement to give an aldehyde with a dithiane is described in H.1.

Of interest as well are the hydrolyses of Reissert compounds [35]:

RCHO + 2-QCOOH

Q=quinoline

Photohydration of pyridine has led to traces of the amino-dienal, H$_2$NCH=CHCH=CHCHO [36]. Other heterocycles also serve as a source of aldehydes [37].

4. From gem-Dihalides (1, 597)

RCHCl$_2$ ⟶ RCHO

The hydrolysis of benzal chlorides to benzaldehydes by 60-80% phosphoric, p-toluenesulfonic, or methanesulfonic acid at 120° has been patented [38]. Yields were about 96%. The dipyridinium salt of benzal bromide may be synthesized and treated with D$_2$O at pD 6.9 to form the deuterated benzaldehyde, C$_6$H$_5$CD=O [39].

In a complex reaction, involving at least both reduction and hydrolysis, 2-trichloromethylpyrazine is converted into 5-methoxy-pyrazine-2-carboxaldehyde [40]:

NaOMe
MeOH
N
N
CCl3
MeO
CHO

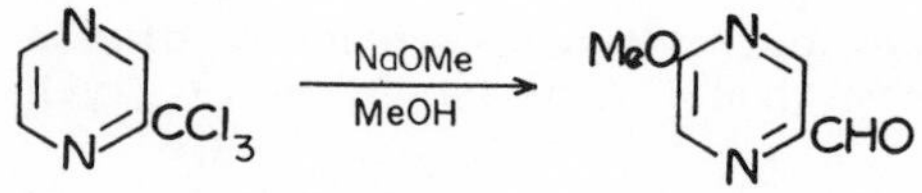

6. Hydration of Some Acetylenes and the Like (1, 599)

$$HC\equiv CH \longrightarrow CH_3CHO$$

It was stated (1, 599) that acetylene was the only alkyne that gave an aldehyde. Elaboration of the acetylene molecule now extends the scope to derivatives as shown [41]:

RC≡CH
Me3SiH
RCH=CHSiMe3
m-ClC6H4CO3H
O
RCH—CHSiMe3
H3O⊕
RCH2CHO

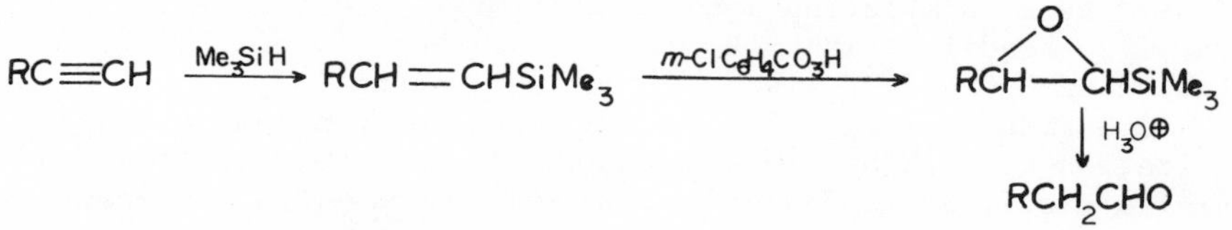

For instance, 1-octyne gave 65% of octanaldehyde as the 2,4-DNPH.

Moreover, 1-adamantanol in 98% sulfuric acid with acetylene gives 90% of 1-adamantylacetaldehyde and 10% of 1-methylhomoadamantan-2-one [42]. The latter is the product of a slower reaction and can be made the predominant product with boron trifluoride etherate in place of sulfuric acid.

In another example an α,β-unsaturated aldehyde has been synthesized from an acid chloride and bistrimethylsilyl acetylene as indicated [42a]:

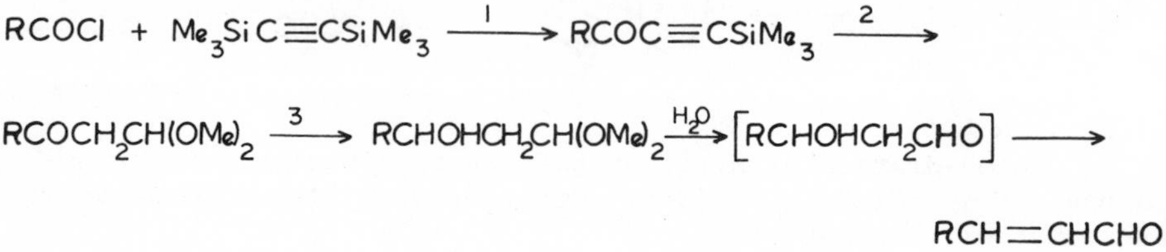

RCOCl + Me3SiC≡CSiMe3
1
RCOC≡CSiMe3
2
RCOCH2CH(OMe)2
3
RCHOHCH2CH(OMe)2
H2O
[RCHOHCH2CHO]
RCH=CHCHO

Step 1 consists of a Friedel-Crafts acylation with $AlCl_3$ in CH_2Cl_2, solvation in Step 2 is accomplished with $NaOCH_3$, reduction in Step 3 is achieved with $NaBH_4$, and Step 4 consists of hydrolysis. If R=biphenyl, the overall yield of 3-biphenyl-2-propenal was 92%.

Allene alkyl ether, derived from propargyl ether, hydrolyzes to give unsaturated aldehydes [43]:

C6H5
R
C=C=CHOCH3
H3O⊕
C=CHCHO
86-87%

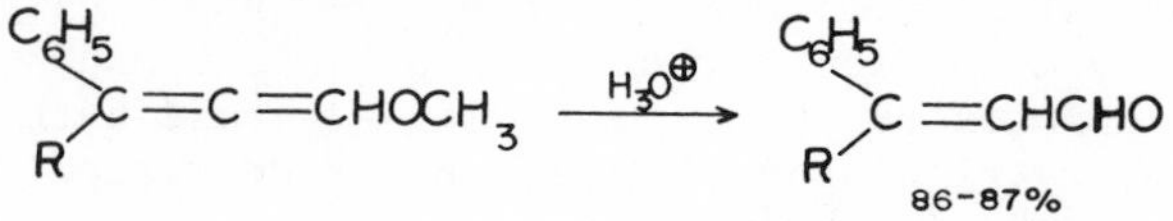

7. From Methyl Methylthiomethyl Sulfoxides and Related
    Types

$$\overset{\ominus}{\text{MeSOCHSMe}} \longrightarrow \text{MeSOCHRSMe} \longrightarrow \text{RCHO}$$

A

Methyl methylthiomethyl sulfoxide, $MeSOCH_2SMe$, is an acetal
derivative that is easily converted into the anion A by NaH.
The anion, which may be alkylated in the usual way, requires
catalytic quantities of mineral acid only for its hydrolysis to
the aldehyde, or ethyl orthoformate may be used to form the
acetal, which is more easily hydrolyzed. The method appears to
have been used first by Ogura [44], who obtained yields of 33-
80% for four aldehydes. Later Ogura succeeded in synthesizing
α-hydroxyaldehydes, which tend to dimerize by the addition of
carbonyl compounds to A [45]. Other elaborations of the method
have been achieved by Herrmann [46] in the synthesis of ketones,
α-hydroxycarbonyl, and α-dicarbonyl systems (see Chapter 11,
D.7).

γ-Methylthiopropenyl methyl sulfide might be called a
vinylogue of a thioacetal. Its anion B may be formed, alkylated,
and the product may be hydrolyzed to the aldehyde by mercuric
chloride in good yield [47].

$$\text{MeSCH}_2\text{CH}=\text{CHSMe} \longrightarrow \overset{\ominus}{\text{MeSCHCH}}=\text{CHSMe} \xrightarrow{\text{C}_5\text{H}_{11}\text{Br}}$$

B

$$\underset{\overset{|}{\text{MeSCHCH}}=\text{CHSMe}}{\overset{\text{C}_5\text{H}_{11}}{}} \longrightarrow \text{C}_5\text{H}_{11}\text{CH}=\text{CHCHO}$$

The reaction is claimed to be shorter and more efficient than
other methods of synthesizing unsaturated aldehydes. A checked
synthesis of trans-4-hydroxy-2-hexenal (60-62%) is available by
this method [48].

Addendum

D.5. The 1,3,5-trithiane anion adds to carbonyl compounds, the
adduct of which when hydrolyzed gives α-hydroxycarboxaldehydes
[D. Seebach et al., Chem. Ber., 107, 367 (1974)].

References

1. P. Salomaa, in S. Patai, The Chemistry of the Carbonyl
    Group, Interscience, New York, 1966, Vol. 1, p. 177.

2. Ref. 1, p. 189.
3. P. Z. Bedoukian, Org. Syn., Coll. Vol., 3, 127 (1955).
4. W. L. Evans et al., Org. Syn., Coll. Vol., 2, 305 (1943).
5. J. H. Billman et al., Syn. Commun., 1, 127 (1971).
6. A. Kalir, Org. Syn., Coll. Vol., 5, 825 (1973).
7. J. Levisalles et al., Chem Commun., 445 (1969).
8. M. Fetizon, M. Jurion, Chem. Commun., 382 (1972).
9. C. Reichardt et al., Tetrahedron Letters, 3979 (1973).
10. G. H. Timms, E. Wildsmith, Tetrahedron Letters, 195 (1971).
11. E. J. Corey, J. E. Richman, J. Am. Chem. Soc., 92, 5276
    (1970).
12. J. W. Bird, D. G. M. Diaper, Can. J. Chem., 47, 145 (1969).
13. T.-L. Ho, Synthesis, 347 (1973).
14. J. K. Suyden, Chem. Ind. (London), 680 (1972).
15. K. Ogura, G. Tsuchihashi, Tetrahedron Letters, 2681 (1972).
16. A. I. Meyers, Heterocycles in Organic Synthesis, Wiley,
    New York, 1974.
17. A. I. Meyers, E. W. Collington, J. Am. Chem. Soc., 92, 6676
    (1970); Ref. 16, p. 186.
18. A. I. Meyers et al., Org. Syn., 54, 42 (1974).
19. L. J. Altman, S. L. Richheimer, Tetrahedron Letters, 4709
    (1971).
20. A. I. Meyers et al., Tetrahedron Letters, 3929 (1972);
    Ref. 16, p. 210.
21. Ref. 16, p. 222.
22. Ref. 16, p. 201.
23. A. I. Meyers et al., J. Org. Chem., 38, 36, 2136 (1973).
24. A. I. Meyers et al., J. Am. Chem. Soc., 91, 764 (1969).
25. Ref. 16, p. 204.
26. I. R. Politzer, A. I. Meyers, Org. Syn., 51, 24 (1971).
27. G. R. Malone, A. I. Meyers, J. Org. Chem., 39, 618 (1974).
28. A. I. Meyers, N. Nozarenko, J. Am. Chem. Soc., 94, 3243
    (1972); Ref. 16, p. 205.
29. D. Seebach, A. K. Beck, Org. Syn., 51, 39 (1971).
30. E. J. Corey, D. Seebach, Org. Syn., 50, 72 (1970).
31. E. Vedejs, P. L. Fuchs, J. Org. Chem., 36, 366 (1970).
32. H. Wynberg et al., Syn. Commun., 2, 7 (1972).
33. T.-L. Ho, C. M. Wong, Can. J. Chem., 50, 3740 (1972).
34. J. B. Chattopadhyaya, A. V. Rama Rao, Tetrahedron Letters,
    3735 (1973).
35. F. D. Popp, Advances in Heterocyclic Chemistry, Vol. 9,
    Academic, New York, 1968, p. 5.
36. K. E. Wilzbach, D. J. Rausch, J. Am. Chem. Soc., 92, 2178
    (1970).
37. Ref. 16, p. 160.
38. H. Coates, W. E. Billingham, German Patent, 2,261,616,
    June 28, 1973; C. A., 79, 66015 (1973).

39. R. A. Olofson, D. M. Zimmerman, J. Am. Chem. Soc., 89, 5057
     (1967).
40. R. J. Tull et al., U.S. Patent, 3,558,625, January 26,
     1971; C. A., 75, 5946 (1971).
41. G. Stork, E. Colvin, J. Am. Chem. Soc., 93, 2080 (1971).
42. D. R. Kell, F. J. McQuillin, J. Chem. Soc., Perkin Trans.,
     I, 2100 (1972).
42a. H. Newman, J. Org. Chem., 38, 2254 (1973).
43. Y. Leroux, R. Mantione, J. Organomet. Chem., 30, 295
     (1971).
44. K. Ogura, G. Tsuchihashi, Tetrahedron Letters, 3151 (1971).
45. K. Ogura, G. Tsuchihashi, Tetrahedron Letters, 2681 (1972).
46. J. L. Herrmann et al., Tetrahedron Letters, 4707 (1973).
47. E. J. Corey et al., J. Am. Chem. Soc., 93, 1724 (1971).
48. B. W. Erickson, Org. Syn., 54, 19 (1974).

E. Rearrangements by Acid Catalysts

   1. From Pinacols or Other 1,2-Disubstituted Compounds
       (1, 600)

The factors influencing the stereochemical course of the
pinacol rearrangement have recently been discussed [1]. A more
detailed treatment of the rearrangement may be found in another
work [2]. For a recent account of the conversion of 1,2-cyclo-
butanediol into cyclopropanecarboxaldehyde (65-80%) by heating
with $BF_3 \cdot Bu_2O$ at 230°, see the work by Conia et al. [3].
Other 1,2-disubstituted types, obtainable from the alkene,
undergo an oxidative rearrangement as shown [4]:

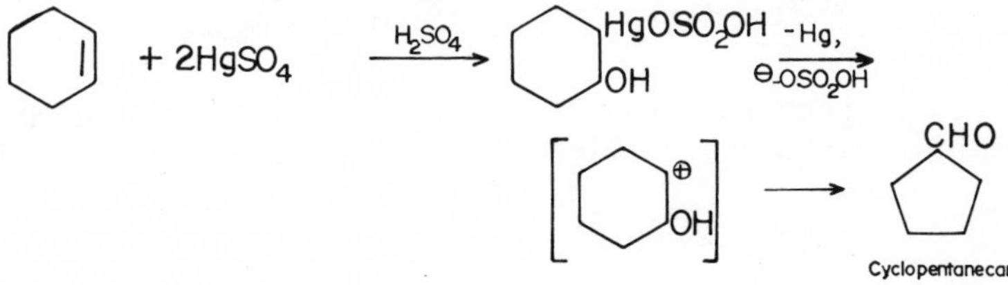

Cyclopentanecarboxaldehyde
46-53%

Although the mechanism of this reaction appears to not have been
determined [4a], a 1,2-addition product is apparently involved.

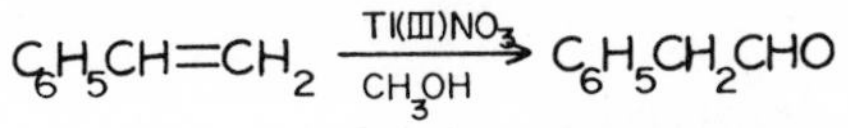

Phenylacetaldehyde
85%

[5]

This procedure, carried out at 25°, is simple and leads to aldehydes and ketones in yields of 75-98% of high purity. The
mechanism has been represented as:

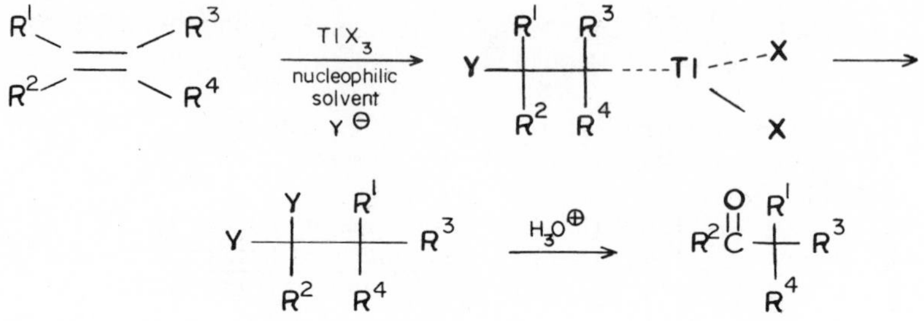

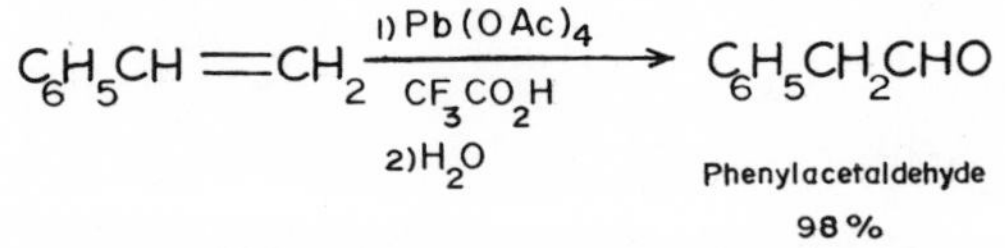

$$C_6H_5CH=CH_2 \xrightarrow[\substack{CF_3CO_2H \\ 2)H_2O}]{1)Pb(OAc)_4} C_6H_5CH_2CHO$$

Phenylacetaldehyde
98%

[6]

Other yields for aldehydes and ketones vary within 5-98%.

2.   From Unsaturated Alcohols (1, 603)

2-Alkylidenecyclobutanols, when heated with 5% H_2SO_4 under
pressure, undergo ring contraction to give 1-alkylcyclopropyl
carbonyl compounds [7]. Thus 2-methylenecyclobutanol gives 1-
methylcyclopropanecarboxaldehyde as indicated:

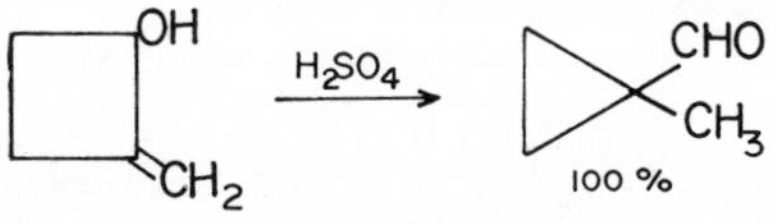

Similarly, in respect to carbonyl formation, 1-vinylcyclopropanol
gives 2-methylcyclobutanone [8]:

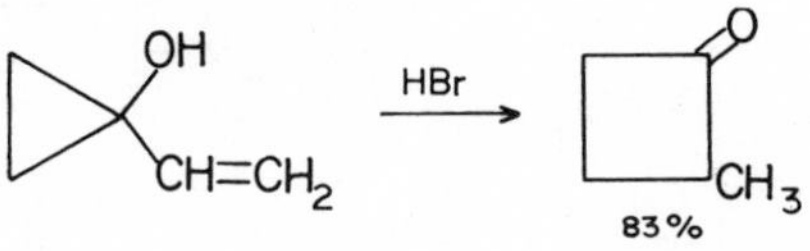

Propargyl alcohols are rearranged via the ketal as follows
[9]:

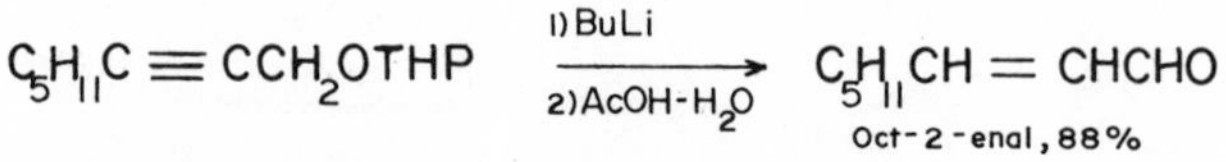

$$C_5H_{11}C\equiv CCH_2OTHP \xrightarrow[\substack{2)AcOH-H_2O}]{1)BuLi} C_5H_{11}CH=CHCHO$$

Oct-2-enal, 88%

Acid stronger than acetic may be needed to rearrange the <u>cis</u> to the <u>trans</u> aldehyde.

3. From Ethylene Oxides ($\underline{1}$, 605)

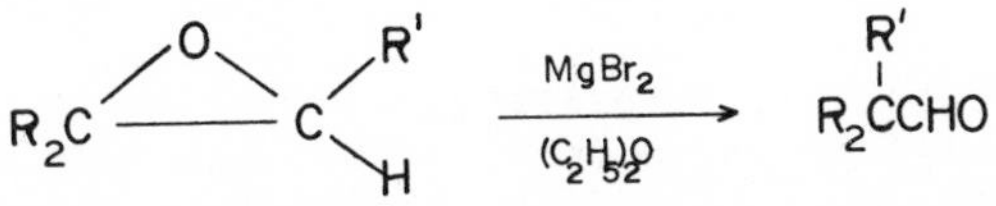

This rearrangement with cyclic epoxides often leads to a change in the size of the ring. Thus cyclooctatetraene epoxide gives phenylacetaldehydes [10] while

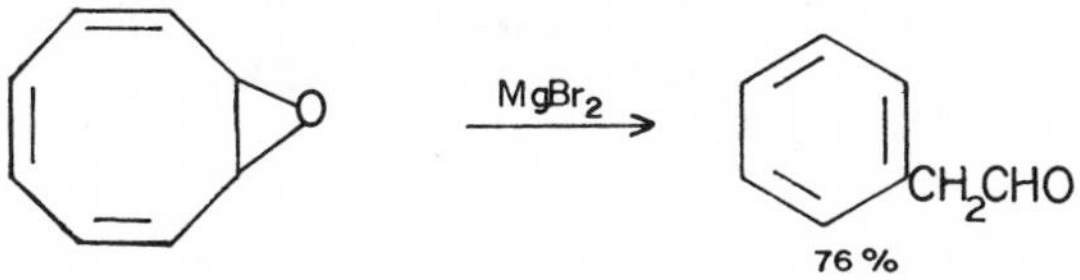

76 %

1-methylcyclohexene epoxide leads to 1-methylcyclopentanecarbox-aldehyde [11]:

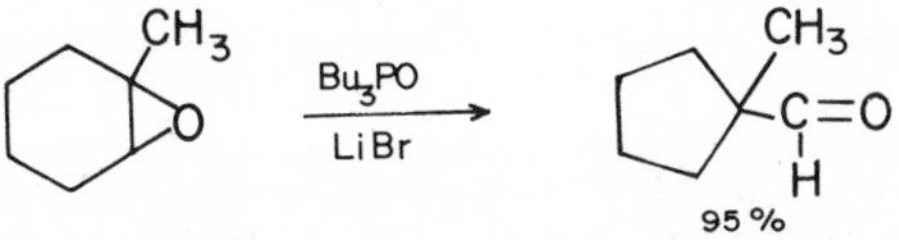

95 %

The phosphine oxide serves as a solubilizing influence on the salt. By contrast the hydroxycyclohexene epoxide leads to cyclopentenecarboxaldehyde [12]:

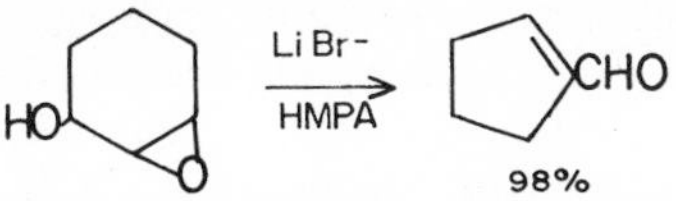

98%

Catalysts and their limitations in this epoxide rearrangement have been listed [11].

For the synthesis of 1,6-hexanedial from cyclohexene epoxide, see A.25.

Heating styrene oxide at 200-300° gives phenylacetaldehyde, while at 500° the main product is toluene [13].

6. From Bicyclic Keto Alcohols

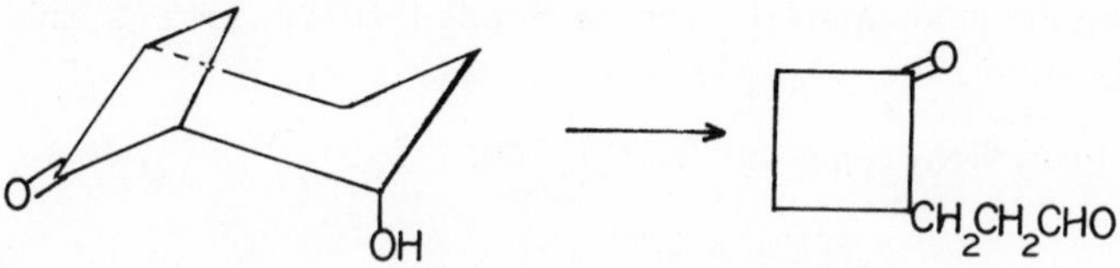

This reverse aldol, surprisingly conducted in acid solution, probably is general provided dehydration, the more common re- action, does not occur [14].   The intermediate appears to be:

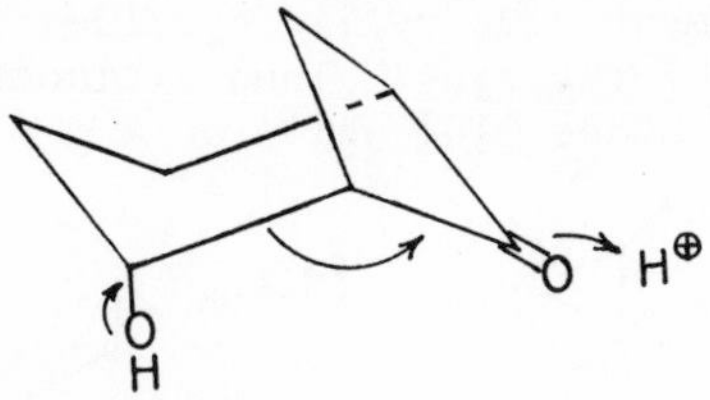

7.   From β-Alkoxyketoximes

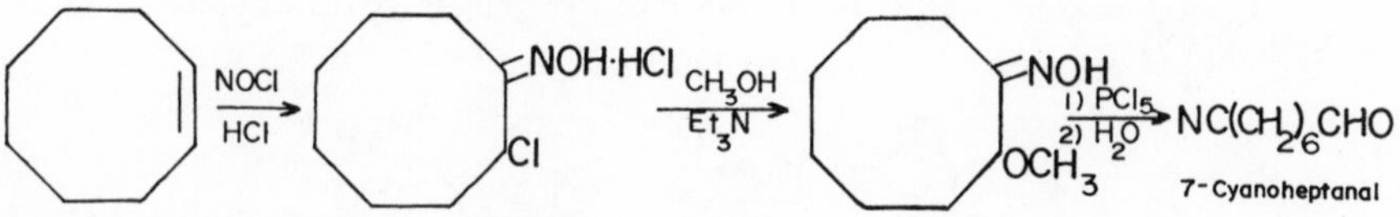

The sequence above represents an easily accessible method for the preparation of ω-cyanoaldehydes [15].   The Beckmann cleavage in the last step occurs in 85.2% yield.

8.   Hydroboration Methods

Although hydroboration is more common in the synthesis of ketones (E.8) than of aldehydes, it may be used for the preparation of the latter.   The synthesis, in which the carbon chain is increased, involves the migration of an alkyl group in the presence of an active hydride reagent, such as lithium trimethoxy-aluminohydride [16]:

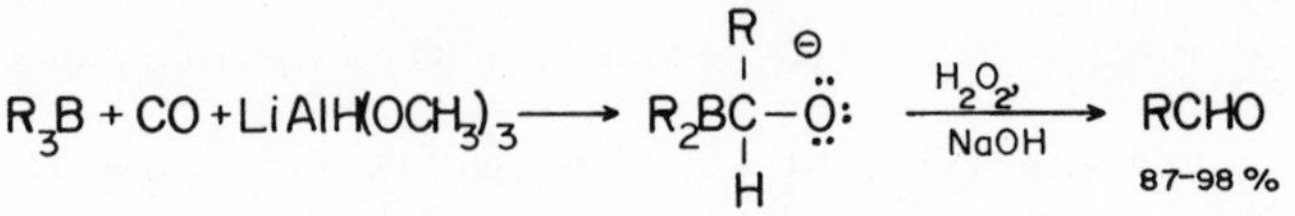

$$R_3B + CO + LiAlH(OCH_3)_3 \longrightarrow R_2BC\overset{R}{\underset{H}{-}}\overset{\ominus}{\ddot{O}}\text{:} \xrightarrow[NaOH]{H_2O_2} RCHO$$

$$87\text{--}98\%$$

To utilize the three R groups in R_3B, 9-BBN is combined with one mol of the alkene to give B-R-9-BBN, in which R is available completely.

Another method for the synthesis of aldehydes involves the

substitution of Ar(R)C≡CH for ⟨◯⟩C≡CCH$_3$ in the second method given in Chapter 11, p. 583.

References

1. B. P. Mundy, R. D. Otzenberger, J. Chem. Educ., 48, 431 (1971).
2. T. S. Stevens, W. E. Watts, Selected Molecular Rearrangements, Van Nostrand Reinhold, New York, 1973, p. 20.
3. J. M. Conia et al., Org. Syn., 53, unchecked procedure 1827 (1973).
4. O. Grummitt et al., Org. Syn., Coll. Vol., 5, 320 (1973).
4a. H. Arzoumanian, J. Metzger, Synthesis, 533 (1971).
5. A. McKillop, E. C. Taylor et al., Tetrahedron Letters, 5275 (1970); J. Am. Chem. Soc., 95, 3635 (1973); C. Lion, J.-E. Dubois, Compt. Rend., 274, C1073 (1972).
6. R. O. C. Norman et al., J. Chem. Soc., Perkin Trans., I, 35 (1973).
7. J. M. Conia et al., Chem. Commun., 103 (1973).
8. H. H. Wasserman et al., J. Am. Chem. Soc., 91, 2375 (1969).
9. E. J. Corey, S. Tereshima, Tetrahedron Letters, 1815 (1972).
10. T. Matsuda, M. Sugishita, Bull. Chem. Soc. Jap., 40, 174 (1967).
11. B. Rickborn, R. M. Gerkin, J. Am. Chem. Soc., 93, 1693 (1971).
12. G. Magnusson, S. Thoren, J. Org. Chem., 38, 1380 (1973).
13. J. M. Watson, B. L. Young, J. Org. Chem., 39, 116 (1974).
14. P. Yates, R. J. Crawford, J. Am. Chem. Soc., 88, 1561 (1966).
15. M. Ohno et al., Org. Syn., 49, 27 (1969).
16. H. C. Brown et al., J. Am. Chem. Soc., 91, 499 (1968); 91, 2144 (1969).

F. Condensation

1. Formylation with Ethyl Formate (Claisen) and the Like (1, 608)

A series of β-oxoaldehydes has been synthesized from ketones and ethyl formate in the presence of sodium methylate or sodium in ether [1]. The sodium salts were isolated and used as such. 3-Keto-Δ^4-steroids with t-BuOK and phenyl formate followed by hydrolysis give 3-hydroxy-4-formyl steroids [2]. Thus 4-cholesten-3-one gave the 3-hydroxy-4-formyl-3,5-cholestadiene as shown:

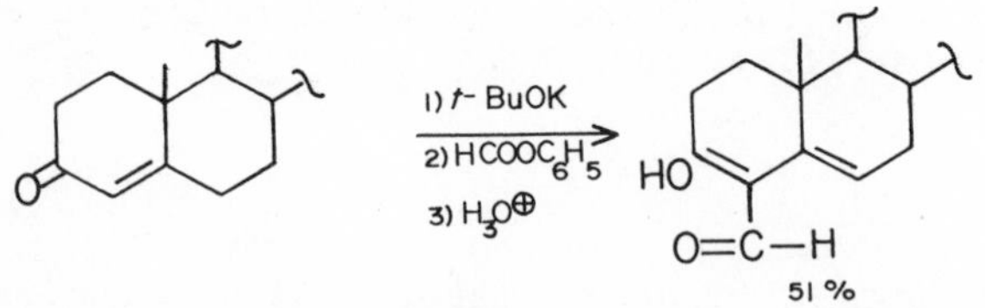

51 %

Two methods are available for obtaining α,β-unsaturated aldehydes by formylation. The first is the Vilsmeier formation given in C.6. The second involves carbon homologation of a ketone, which may be illustrated by the transformation of cyclo-hexanone into 1-cyclohexenecarboxaldehyde [3]:

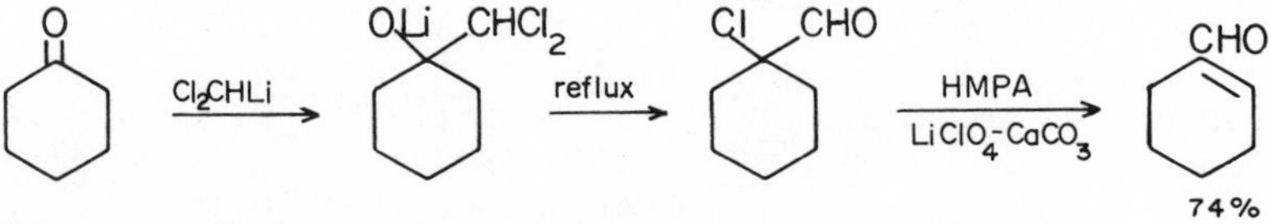

74%

By this procedure 2-octanone gave (E)2-methyl-2-octenal exclusively in 74% yield (glpc).

Another method of homologation involves condensation with the proper Wittig reagent [4]:

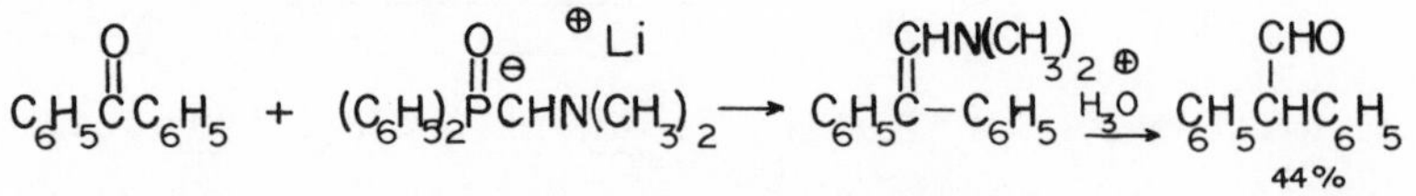

44%

Aldehydes have also been prepared from ketones in the Wittig reaction by the use of 2(cyclohexylimino)vinyl phosphonate [5]:

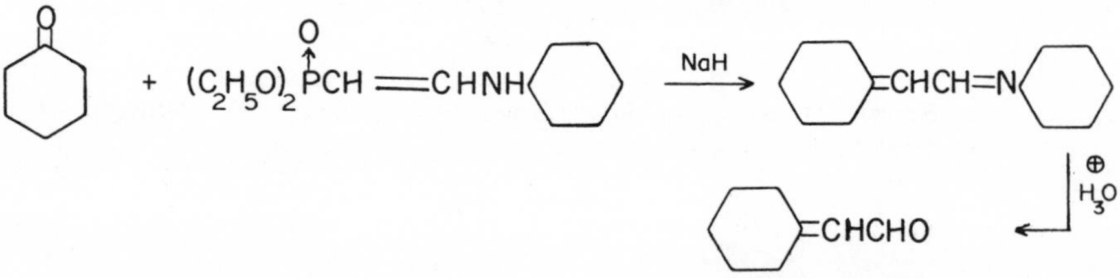

Cyclohexylideneacetaldehyde
83 %

The procedure occurs stereoselectively, when applicable, to afford the <u>trans</u> isomer only.

Finally, a typical enamine alkylation to produce aldehydes from the corresponding ketone has been reported [5a].

2. Reimer-Tiemann Condensation (<u>1</u>, 609)

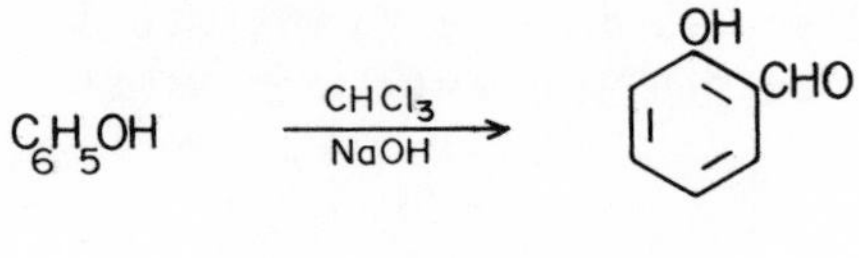

That the Hines mechanism ($\underline{1}$, 610) is correct is supported by evidence for the existence of the carbanion intermediates

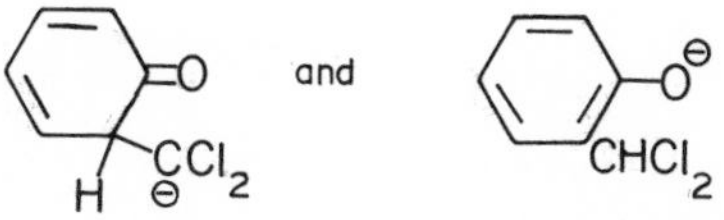

[6]

3. Aldol and Related Mannich and Michael Condensations ($\underline{1}$, 611)

A reaction with some of the characteristics of a reverse benzoin condensation has been utilized in the preparation of benzaldehyde-formyl-d from benzil [7] by a direct procedure:

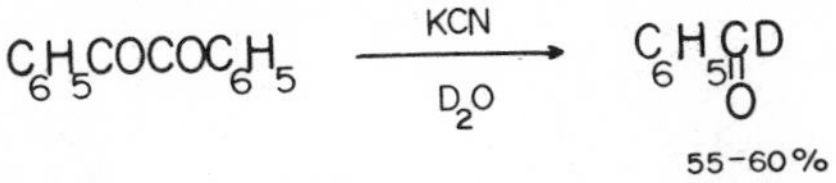

$$C_6H_5COCOC_6H_5 \xrightarrow[D_2O]{KCN} C_6H_5CD$$

55-60%

Myristicinaldehyde has also been synthesized from isomyristicin by what may be regarded as a reverse aldol [8]:

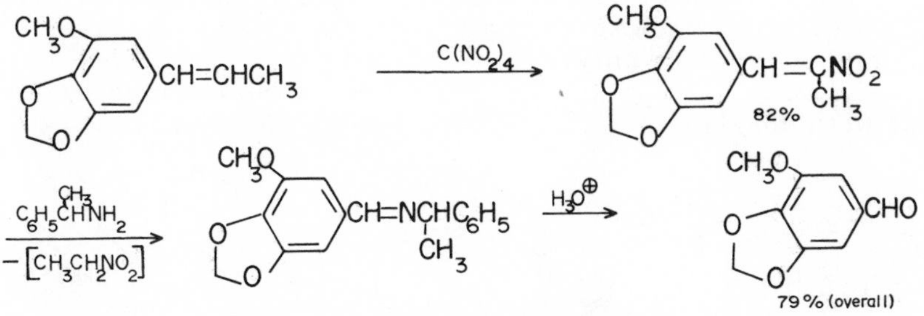

82%

79% (overall)

As is seen, the amine liberates nitroethane with the formation of the aldehyde, which with an excess of the original amine gives the Schiff base from which the aldehyde was recovered by hydrolysis. This reaction may prove to be useful in the conversion of essential oils into benzaldehydes.

A review of the directed aldol condensation is now available [9]. Directed aldol condensations via the preformed lithium enolates have also been reviewed [10]. The synthesis of β,β-diphenylacrolein as described in $\underline{1}$, 611 has now appeared in Organic Synthesis [11].

Ketene thioacetal monoxides have been utilized as Michael receptors (D.7) [12].

A dialdehyde may be converted into a monofunctional derivative by means of a polymeric protective group as shown [12a]:

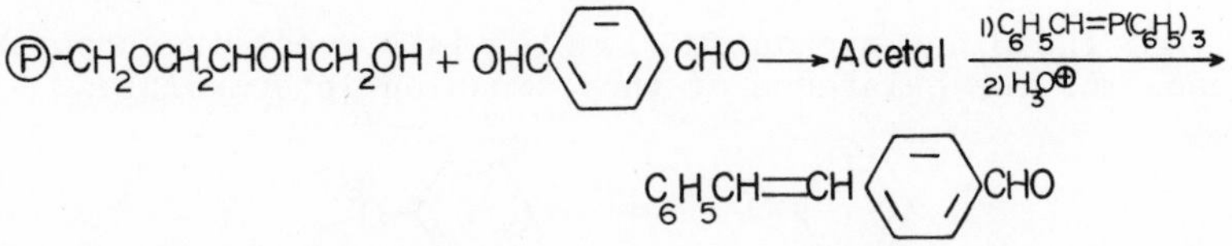

Overall yield of p-styrylbenzaldehyde was 76%.

4. Alkylation of Aldehydes, Mainly via Enamines (1, 614)

Some progress has been achieved in the direct α-alkylation of aldehydes. Thus Dietl and Brannock [13] accomplished this end by treating the aldehyde with the alkyl halide in 50% aqueous NaOH and a catalytic amount of tetrabutylammonium ions (phase transfer):

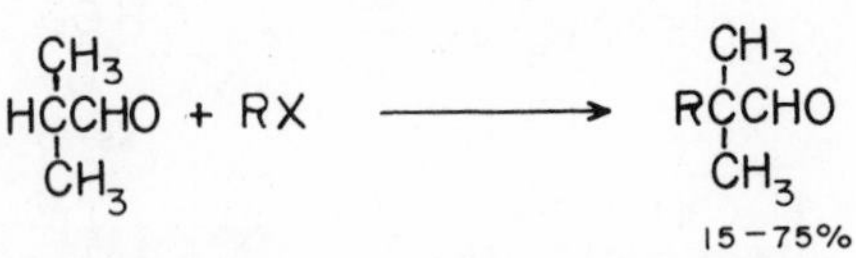

Self-condensation of isobutyraldehyde by the phase transfer method occurs to some extent although the alkylation gives C-alkylation products only.

Salisbury [14] alkylated 9-fluorenecarboxaldehyde by treatment with an ylid:

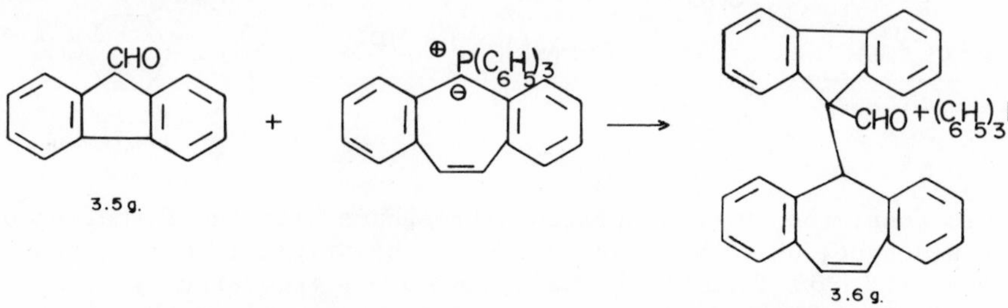

This divergent reaction, which is unprecedented, probably occurs by the carbanion of the aldehyde displacing triphenylphosphine from the dibenzocycloheptatrienyltriphenylphosphonium cation.

An indirect method of alkylation was employed by Odic and Pereyre [15] as shown for the readily available enol acetate of dimethylacetaldehyde:

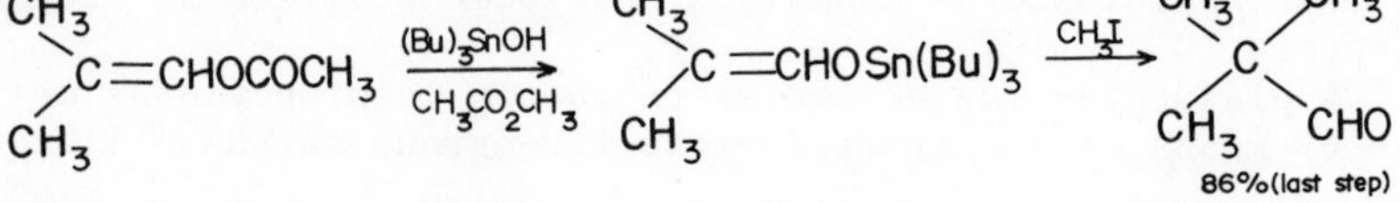

Other last-step yields vary within 27-92%.

The indirect method of Stork and Dowd ($\underline{1}$, 614) via the reaction of the magnesioenamine salt of an aldehyde has now appeared in Organic Synthesis [16].

6. Hydroboration Methods

The principal hydroboration methods of preparing aldehydes are covered under E.8. However, a 1,4-addition of a borane to an unsaturated aldehyde occurs to permit the formation of a saturated aldehyde [17]:

$$R_3B + CH_2=CHCHO \xrightarrow[25°]{THF} RCH_2CH=CHOBR_2 \xrightarrow{H_2O} RCH_2CH_2CHO$$
$$77\text{-}96\% \text{(glpc)}$$

Apparently this reaction is a free radical chain one.

Addenda

F.4.   Alkylation of α,β-unsaturated aldehydes has been accomplished for the first time by treatment with KNH_2 in liquid NH_3 at -60° followed by RX [A. van der Gen et al., Tétrahedron Letters, 1653 (1974)].

F.4.   N-Allylcarbazoles may be alkylated by BuLi followed by RX to form the enamine, $RCH_2CH=CHNHet$ which may be hydrolyzed to the aldehyde, RCH_2CH_2CHO [M. Julia et al., Tetrahedron Letters, 3433 (1974)].

F.7.   Treatment with methoxymethylenetriphenylphosphorane followed by acidic hydrolysis converts the >C=O of ketones, such as androstenolone, into >CHCHO of carboxaldehydes in good yield [S. Danishefsky et al., J. Org. Chem., 40, 1989 (1975)].

References

1.   M. Regitz, F. Menz, Chem. Ber., 101, 2622 (1968).
2.   C. Hunyh, S. Julia, Bull. Soc. Chim. France, 4402 (1971).
3.   H. Tagachi et al., Tetrahedron Letters, 2465 (1973); G. Köbrich, J. Grosser, Tetrahedron Letters, 4117 (1972).
4.   D. J. Peterson, J. Am. Chem. Soc., 93, 4027 (1971).
5.   W. Nagata, T. Wakabayashi, Org. Syn., 53, 104 (1973).
5a. S. F. Martin, R. Gompper, J. Org. Chem., 39, 2814 (1974).
6.   D. S. Kemp, J. Org. Chem., 36, 202 (1971).
7.   A. W. Burgstahler et al., J. Org. Chem., 37, 1272 (1972).

8. A. T. Shulgin, Can. J. Chem., 46, 75 (1968).
9. H. Reiff, W. Foerst (ed.), Newer Methods of Preparative Organic Chemistry, 6, 48 (1971).
10. H. O. House et al., Org. Syn., 54, 49 (1974).
11. G. Wittig, A. Hesse, Org. Syn., 50, 66 (1970).
12. J. L. Herrmann et al., Tetrahedron Letters, 4715 (1973).
12a. C. C. Laznoff, J. Y. Wong, Can. J. Chem., 51, 3756 (1973).
13. H. K. Dietl, K. C. Brannock, Tetrahedron Letters, 1273 (1973).
14. L. Salisbury, J. Org. Chem., 35, 4258 (1970).
15. Y. Odic, M. Pereyre, Compt. Rend., 270, C100 (1970).
16. G. Stork, S. R. Dowd, Org. Syn., 54, 46 (1974).
17. H. C. Brown et al., J. Am. Chem. Soc., 89, 5709 (1967).

G. Organometallic Methods (1, 616)

Essentially, this section involves the alkylation, arylation, or alkylolation of some masked formic acid or formal derivative. The product is then hydrolyzed to the aldehyde. Included as well are other nucleophilic additions where the protected aldehyde group is part of the reagent as in the synthesis of 1-cyclohexenecarboxaldehyde (F.1).

The reaction of alkyllithium compounds with DMF maintains its usefulness (1, 618) as shown [1]:

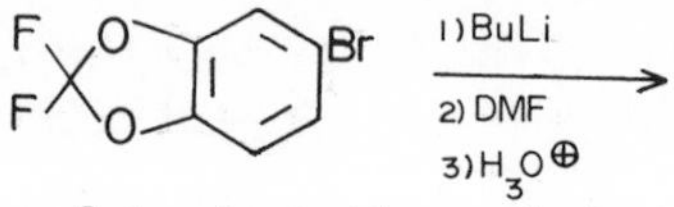

A novel, useful adaptation of the formic acid precurser is illustrated in the action of 1,1,3,3-tetramethylbutylisocyanide on alkyllithium compounds or Grignard reagents [2]:

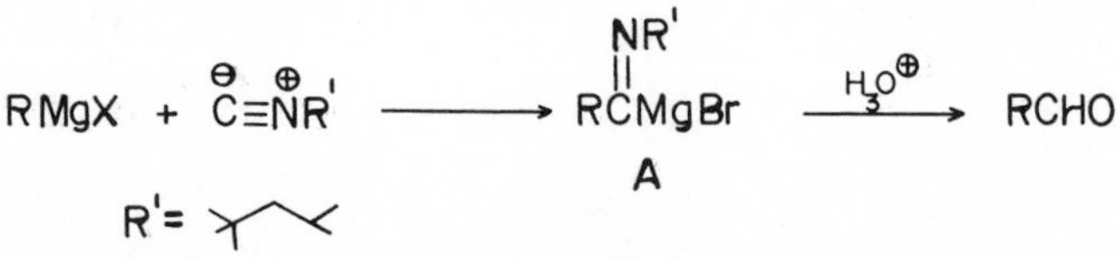

Yields were 48-67% for five examples. Comparable were the yields from alkyllithium compounds that could be used as well to alkylate A to ketones or to treat with D_2O to give eventually the D-aldehyde, RCD=O.

Stetter and Reske claim that phenyl diethyl orthoformate, $PhOCH(OEt)_2$, is better than ethyl orthoformate in reacting with Grignard reagents to form acetals [3]. Eliel and Nader [4] used

2-methoxy-1,3-dioxane to react with Grignard reagents. The
products could have been hydrolyzed to aldehydes,
but the authors were more interested in the
stereo- chemistry of the dioxanes. 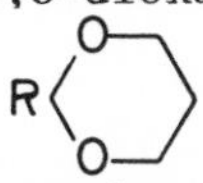
    Still another formic acid derivative used to prepare
aldehydes is 1,1-dichlorodimethyl ether, which with α-bromoketones reacts as indicated [5]:

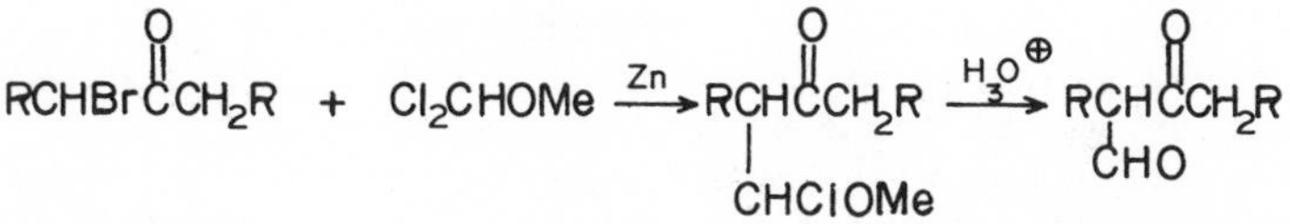

$$RCHBr\overset{O}{\overset{\|}{C}}CH_2R \ + \ Cl_2CHOMe \ \xrightarrow{Zn} RCH\overset{O}{\overset{\|}{C}}CH_2R \ \xrightarrow{H_3O^{\oplus}} RCH\overset{O}{\overset{\|}{C}}CH_2R$$

The ketoaldehydes (36-44%) were purified by vacuum distillation.
    Ethyl formate itself yields carboxyaldehydes with Ivanov
salts since the product conveniently decarboxylates to the
aldehyde [6]:

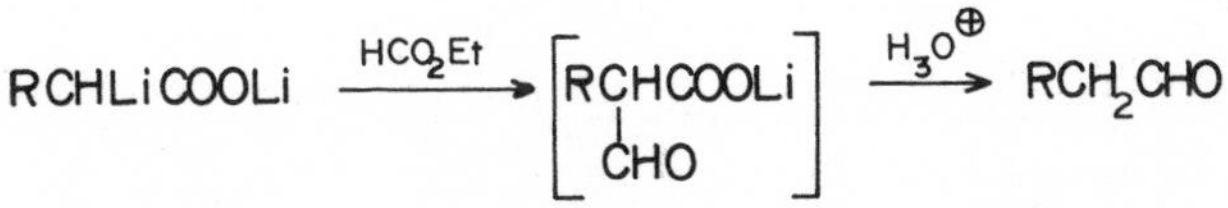

$$RCHLiCOOLi \ \xrightarrow{HCO_2Et} \left[\begin{array}{c} RCHCOOLi \\ | \\ CHO \end{array} \right] \xrightarrow{H_3O^{\oplus}} RCH_2CHO$$

Six examples gave 30-65% yields. The formation of Ivanov salts
has been reviewed [7].
    The protected aldehyde group may be part of the reagent as
shown in the next two equations:

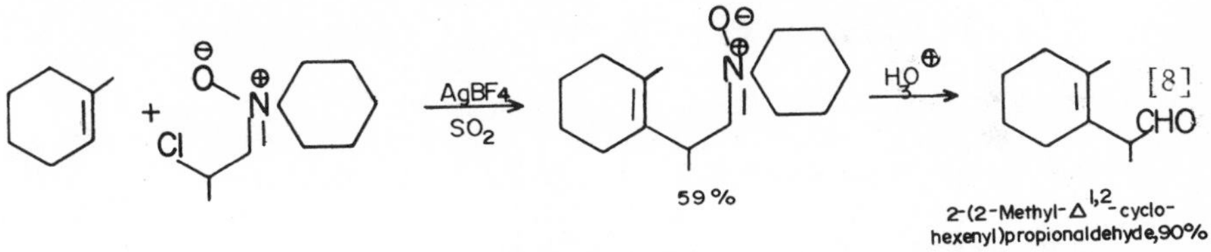

59%   [8]

2-(2-Methyl-Δ^{1,2}-cyclo-
hexenyl)propionaldehyde,90%

BrMgO CHOEt₂   CHO   [9]

EtOCH₂MgBr   HCO₂H
-10°   <80°

78%

For other methods of going from ketones to aldehydes, see F.1.
    Schiff bases may also be alkylated and then hydrolyzed to
aldehydes [10]:

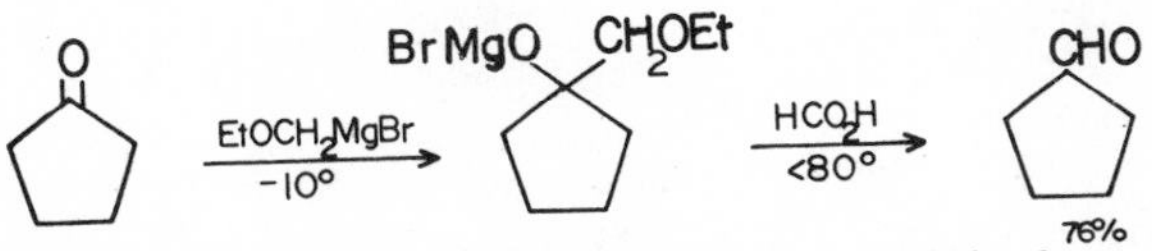

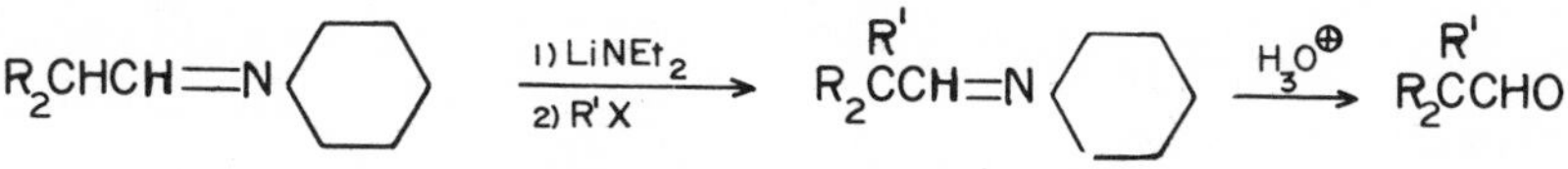

$$R_2CHCH=N\bigcirc \ \xrightarrow[\text{2) R'X}]{\text{1) LiNEt}_2} \ R_2\overset{R'}{\underset{}{C}}CH=N\bigcirc \ \xrightarrow{H_2O^{\oplus}} \ R_2\overset{R'}{\underset{}{C}}CHO$$

Yields for twenty-six examples ranged from 22 to 81%.

Silane derivatives react with carbonyl compounds to give aldehydes as indicated [11]:

$$\text{Me}_3\overset{\ominus}{\text{SiCHSC}_6\text{H}_5}\overset{\oplus}{\text{Li}} \ + \ \text{R}_2\text{C}{=}\text{O} \longrightarrow \text{R}_2\text{C}{=}\text{CHSC}_6\text{H}_5 \xrightarrow{\text{H}_2\text{O}} \text{R}_2\text{CHCHO}$$

Although the first product was not hydrolyzed to the aldehyde, such was possible. Other types as $\text{Me}_3\text{SiCMePO(OEt)}_2$ could be used similarly. The reader is referred to D.1, D.2, D.3, and D.4 for other examples of this nature. Other indirect methods of alkylation may be found in F.4.

Of general, but not synthetic, interest is the breakdown of THF with butyllithium. Butane, ethylene, and the lithium salt of vinyl alcohol are the products, a result of a cycloreversion [12].

References

1. E. I. Stogryn, J. Org. Chem., 37, 673 (1972).
2. H. M. Walborsky et al., J. Am. Chem. Soc., 92, 6675 (1970); Org. Syn., 51, 31 (1971); J. Org. Chem., 37, 187 (1972); D. Hoppe, Angew. Chem. Intern. Ed. Engl., 13, 789 (1974).
3. H. Stetter, E. Reske, Chem. Ber., 103, 643 (1970).
4. E. L. Eliel, F. W. Nader, J, Am, Chem. Soc., 92, 584 (1970).
5. I. I. Lapkin, F. G. Saitkulova, J. Org. Chem. USSR, 6, 450 (1970).
6. P. E. Pfeffer, L. S. Silbert, Tetrahedron Letters, 699 (1970).
7. B. Blagoev, D. Ivanov, Synthesis, 615 (1970).
8. A. Eschenmoser et al., Helv. Chim. Acta, 56, 2961 (1973).
9. M. de Botton, Compt. Rend., 272, C 118, 239 (1971).
10. H. T. Cuvigny, H. Normant, Bull. Soc. Chim. France, 3976 (1970).
11. F. A. Carey, A. S. Court, J. Org. Chem., 37, 939 (1972).
12. R. B. Bates et al., J. Org. Chem., 37, 560 (1972).

H. Electrocyclic and Decarboxylative Reactions

1. Claisen-Cope and Selected Pyrolytic Rearrangements ($\underline{1}$, 619)

The sigmatropic rearrangement, which may be brought about by heat alone, has been reviewed [1]. It is not the purpose here to summarize this review, but to update it and point out trends in extending its scope. One innovation is to use the corresponding sulfide, a procedure that permits alkylation before rearrangement [2]:

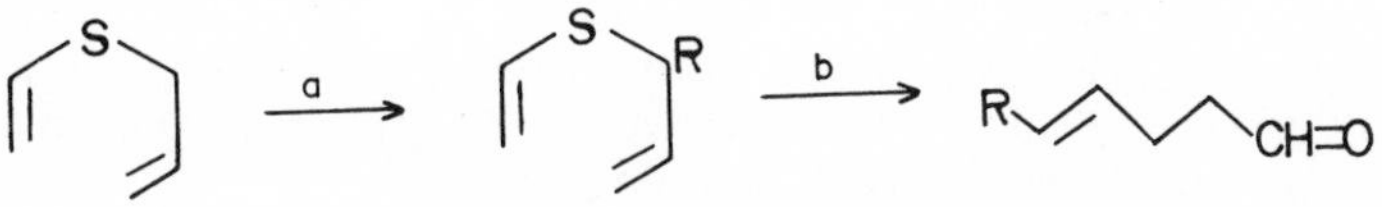

The alkylation (a) is accomplished with BuLi and Dabco followed by treatment with an alkyl halide; the rearrangement and hydrolysis (b) are brought about quite stereospecifically.
     The substrate may be a heterocycle as shown [3]:

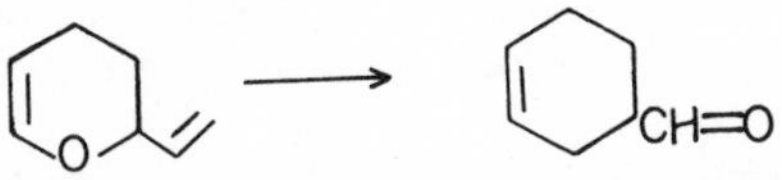

$\Delta^{3,4}$-Cyclohexenecarboxaldehyde was obtained in 68% yield (small scale) by refluxing the substrate in hexane. The substrate may also contain a propargylic rather than an allylic group [4].
     The allyl vinyl ether may be prepared by a Wittig reaction to give the substrate, as indicated, which rearranges at 152° in 45 min to give 9-allyl-9-fluorenecarboxaldehyde (74%) [5]:

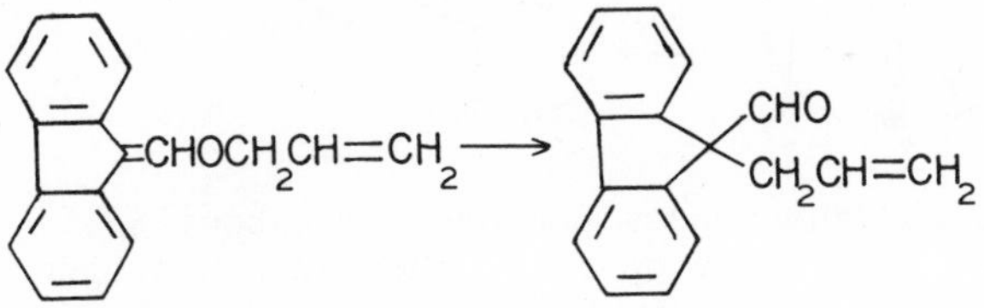

     A simpler technique is to synthesize the allyl vinyl ether without isolation and rearrange it in the same pot [6]. To accomplish the first step in this reaction the vinyl ether ether is mixed with the allyl alcohol and exchange is brought about with mercuric acetate catalyst or phosphoric acid [6a]. Decalin was found to be a good solvent for simultaneous exchange and pyrolysis. Yields of aldehyde as high as 85% were obtained.
     Other means of aligning structural units with Claisen-like behavior is via the acetal as shown in the generalized form [7]:

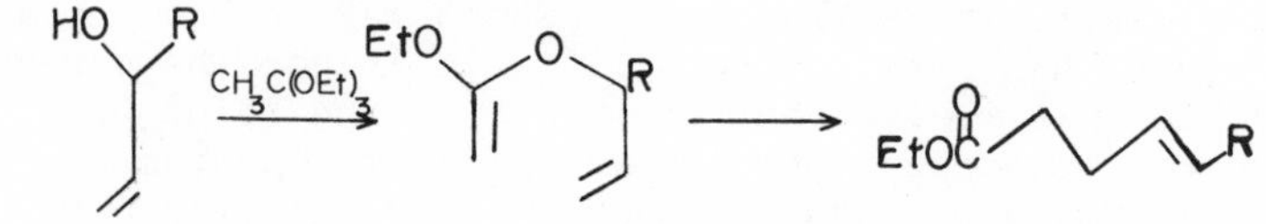

With LiAlH$_4$ the ester may be reduced to the alcohol, which may be oxidized to the aldehyde by Collins reagent. This seemingly long route may give yields as high as 70% with greater stereo-specificity than the normal Claisen route.

The crotyl derivative of heterocyclic compounds of the type conceived by Meyers (D.3) undergoes the Claisen rearrangement, the product of which can be converted into the aldehyde by the Meyers technique [8].

There is a tendency now to vary the electronic centers, particularly the vinyl center, to obtain aldehydes by a Claisen-like rearrangement. Several examples are given to show the scope [9,10]:

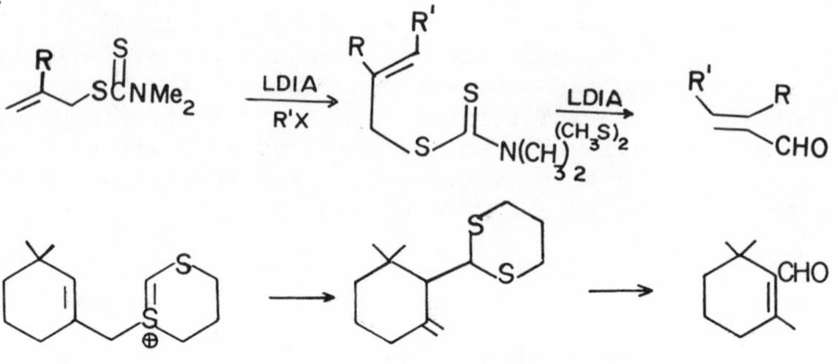

Yields of cyclohexenecarboxaldehydes were about 80%. A third example, which apparently operates through a nitrogen ylid A follows [11]:

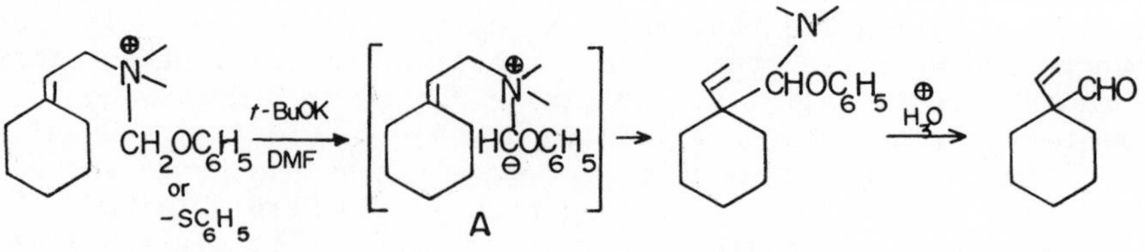

The final example of the section is that of Mander and co-workers [12]:

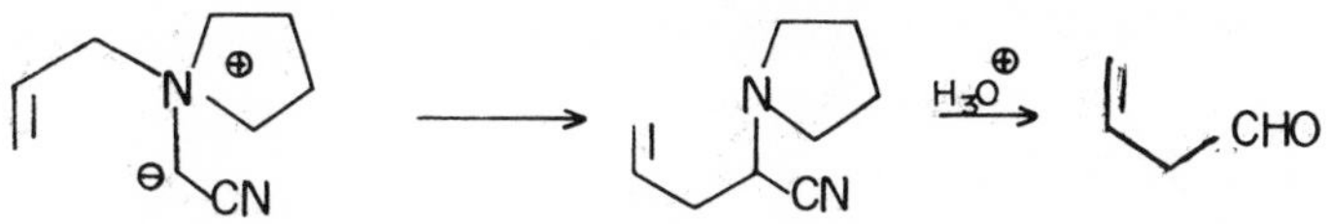

The only carbon atoms in the substrate that become a part of the aldehyde are those in the allyl group and the single anionic carbon atom adjacent to the nitrile group. The attractiveness of the Claisen rearrangement in aldehyde synthesis may be attributed to the relative ease of the reaction and the simplicity of the workup procedure, which is sometimes fractionation.

Addendum

H.1. Steric control in the Cope rearrangement is discussed [D. J. Faulkner, M. R. Peterson, J. Am. Chem. Soc., 95, 553 (1973)].

References

1. S. J. Rhoads, N. R. Raulins, Org. Reactions, 22, 1 (1975); see R. C. Cookson, N. R. Rogers, J. Chem. Soc., Perkin Trans., I, 2741 (1973) for a recent example.
2. H. Yamamoto et al., J. Am. Chem. Soc., 95, 2693 (1973).
3. G. Büchi, J. E. Powell, Jr., J. Am. Chem. Soc., 92, 3126 (1970).
4. R. Rossi, P. Diversi, Synthesis, 34 (1973).
5. E. J. Corey, J. I. Shulman, J. Am. Chem. Soc., 92, 5522 (1970).
6. W. G. Dauben, T. J. Dietsch, J. Org. Chem., 37, 1212 (1972).
6a. R. Marbet, G. Saucy, Org. Syn., 51, unchecked procedure 1752 (1971).
7. W. S. Johnson et al., J. Am. Chem. Soc., 92, 741 (1970).
8. R. E. Ireland, A. K. Willard, J. Org. Chem., 39, 421 (1974).
9. T. Nakai et al., Tetrahedron Letters, 3625 (1974); H. Takahashi et al., J. Am. Chem. Soc., 95, 5803 (1973).
10. E. Hunt, B. Lythgol, Chem. Commun., 757 (1972).
11. S. Julia et al., Bull. Soc. Chim. France, 4057 (1972).
12. L. N. Mander, J. V. Turner, J. Org. Chem., 38, 2915 (1973).

KETONES

The field of ketone synthesis is so tremendous that it staggers
the imagination. It is apparent that great progress is being
made in the discovery of new syntheses, new reagents, and new
conditions for carrying out old syntheses. Only two words
describe the overview: awesome and inspiring.

Although the H.-W. compendium on the synthesis of ketones
offers an extremely thorough treatment, it suffers from the fact
that few references appear beyond 1970.

The Sarett reagent, $CrO_3 \cdot 2C_5H_5N$, so valuable for the preparation of aldehydes, is also useful for the preparation of ketones (A.1), but it has been supplanted, we believe, by the incomparable Corey reagent (see Chapter 10 Addenda, A.1). A competitor for the Jones reagent, $CrO_3/H_2SO_4/CH_3COCH_3$, in oxidizing secondary alcohols to ketones is a heterogeneous mixture of $Na_2Cr_2O_7$, acid and ether (A.1). The alcohol seems to disappear from the ether layer quickly as the ketone accumulates in a protected environment. The number of agents for oxidizing olefins to ketones of the same number of carbon atoms has grown (A.5). A convenient conversion of adamantane into 2-adamantanone by sulfuric acid is given (A.7). Whereas few methods of converting amines into ketones, $>CHNH_2 \rightarrow >C=O$, were given previously, several, including oxidation by tautomerism of a Schiff base (Calo method, A.13) are listed. The latter is applicable to both aldehydes and ketones.

A new method of synthesizing α,β-unsaturated ketones from the ketone via the α-phenylselenium compound appears to be reliable (A.21).

The possibility of conducting the Friedel-Crafts acylation with little or no catalyst is mentioned (C.1). The control of orientation in the Fries rearrangement has been made somewhat more definitive (C.3). As with the Vilsmeier reaction in Chapter 10, the Friedel-Crafts acylation of alkenes, some aliphatic hydrocarbons, and other unsaturated compounds has expanded and seems to have become almost limitless in scope (C.7). It includes a three-component acylation, the Nenitzescu reaction, and methods for preparing cyclopentanones and β-diketones, not to mention simple monoketones and unsaturated ketones.

The Meyers synthesis as applied to the preparation of ketones is discussed in D.1, as well as the hydrolysis of ketals by hydride abstraction with trityl chloride from the glycol part of a dioxolane (D.5). A flurry of papers has appeared on possible ways to hydrolyze 1,3-dithianes or 1,3-dithiolanes since these two compounds are useful precursors of ketones if they are alkylated in the 2-position (D.1).

Wagner-Meerwein rearrangements of epoxides or ketoepoxides have been studied of late perhaps to obtain the rarer types of carbonyl compounds such as spiroketones or cyclobutanones (E.3).

It is of interest to note that an intramolecular Claisen reaction may be achieved with PPA rather than the usual base catalyst (F.2). The reaction is probably limited to the synthesis of nonenolizable dicarbonyl types or ketoesters. However the most important development in the acylation field is the complete fixation of the ethyl acetate anion enolate at -78° with LiICA or similar base (F.2). By the addition of an acylating agent of a structure different from that of the ethyl

acetate enolate, a mixed β-ketoester may be prepared.   Previous
to this development ethyl acetoacetate was a significant by-
product.

A most curious reaction is that of an unsaturated ketone and
hydrazoic acid to produce a diketone rather than the product of
the Schmidt reaction (F.4).   A novel reaction, which in all
probability will find frequent use, consists of the acylation of
an alkene containing electron-withdrawing groups by an aldehyde
in the presence of the cyanide ion (F.7).

The alkylation of ketones is a subject of sufficient volume
to produce a book in itself (G).   Regiospecific alkylation of
unsymmetrical ketones is now possible (G.1), as well as alkylidi-
nation of ketones in the Claisen-Schmidt reaction (G.10).   Not
only are dianions γ-alkylated but also compounds of ylid-anion
combinations, not to mention phase transfer alkylation of
ketones (G.1).   Normal β-alkylation as well as alkylation of
enamines via a cyclopropanation reaction is now possible (G.2).
The field of alkylation of ketones by the Michael reaction (G.3)
is so extensive that adequate coverage was difficult.   Topics
such as annelation, equilibrium enolates versus kinetic enolate
control in the regiospecific alkylation of unsaturated ketones,
dialkyl cuprates as conjugate additive reagents (a burgeoning
field), and unusual alkylating agents have been discussed (G.3).
Conditions have been defined, by the use of a divalent metal
ion, for the isolation of β-hydroxyketones, rather than the
usual α,β-unsaturated ketones, in the aldol reaction (G.10).

The conversion of esters into ketones has been achieved
without the incursion of much tertiary alcohol formation (H.1).
Alkyl- or aryllithium addition to salts of carboxylic acids to
yield ketones has become a most important method of synthesis
(H.2).   Respectable as well is the addition of lithium dialkylcuprates to acid chlorides (H.2).   A new method consists of the
addition of an alkyllithium to an isocyanide (H.3).   In this
reaction the isocyanide supplies only the carbonyl carbon to the
ketone.   Another new method of preparing ketones involves the
alkylation of the anion of ethyl vinyl ether, $CH_2=\bar{C}OEt$ (prepared with t-butyllithium and TMEDA) (H.8).   The latter when
methyl vinyl thioether is employed bids fair to supplant the
Seebach alkylation of dithioketals (D.1).

Electrocyclic reactions to prepare ketones are quite
versatile.   Cyclobutanones are available through the cyclization
of alkenes with dichloroketene or the ketene immonium cation;
with alkynes and enamines, higher-numbered cyclanones result
(I.1).   The 2-methoxyallyl carbonium ion offers some usefulness
as a dienophile in the preparation of ketones (I.1).   An
improvement in the oxy-Cope preparation of ketones is to be
expected since it has been discovered recently that the anion

rearranges much faster than the corresponding alcohol and that
the reaction appears to be accelerated by cluster breaking (I.2).

A. Oxidation

1. From Alcohols (Secondary) (1, 625)

$$\text{\textbackslash CHOH} \longrightarrow \text{\textbackslash C=O}$$

For a detailed discussion of the oxidation of secondary
alcohols to ketones, see H.-W., 7, Pt. 2a, 1973, p. 699, and the
work by Lee [1].

The newer reagents for the oxidation to ketones are listed
under Chapter 10, Aldehydes, A.1. In addition the following
reagents have been employed: Sodium dichromate-H_2SO_4-H_2O in
Et_2O [2], 2,3-dichloro-5,6-dicyano-1,4-benzoquinone (DDQ) [3],
iodobenzene dichloride [4], chromium trioxide-3,5-dimethylpyra-
zole [5], sodium ruthenate [6], Br_2-Et_2O [7], $NaIO_4$-RuO_4 [8], 1-
chlorobenzotriazole [9], CrO_3-$2C_5H_5N$ [10], $DMSO$-O_2 or
$DMSO$-$(CH_3CO)_2O$ [11], Ag_2CO_3 on celite [12], and potassium
ferrate, K_2FeO_4 [13].

For the specific oxidation of 1,2-diols to α-diketones a
combination of benzalacetone and tris(triphenylphosphine)ru-
thenium dichloride is recommended [14].

Comments on the most important of these oxidizing agents
follow.

a. Chromium (VI) Compounds [1]

The most common of these compounds is the Sarett complex,
$CrO_3 \cdot 2C_5H_5N$. In the hands of Ratcliffe and Rodehorst [15] 2-
octanol (6:1-mol ratio of complex and alcohol) in methylene
chloride after 15 min gave 97% of 2-octanone. The reagent is
available commercially (Eastman) or may be prepared without the
fire hazard in methylene chloride.

A method for oxidizing secondary alcohols in diethyl ether
with aqueous chromic acid has been developed by Brown and co-
workers [16]. With unstrained alcohols and an equivalent of
acid at 25° yields were 61-97% (glpc); with strained alcohols
and 100% excess of acid at 0°, yields were 74-99% (glpc).

Oxidation with the Jones reagent (CrO_3-aq H_2SO_4) and isolation of a water-soluble ketone have been reviewed [17].

b. Complexes of Cl_2 with $(CH_3)_2S$ or $(CH_3)_2SO$ and of N-Chloro-
succinimide with $(CH_3)_2S$

For a discussion of these complexes in oxidation of

secondary alcohols, see Chapter 10, A.1. Yields of benzophenone and 4-t-butylcyclohexanone were 98 and 97% (glc) respectively.

c. HNO_3 in Aqueous Glyme

For a discussion of this reagent in converting secondary alcohols to ketones, see Chapter 10, A.1. Dialkylcarbinols gave isolated yields of ketones from 92 to 99%.

d. Ag_2CO_3 on Celite

Although the oxidation of glycols usually results in degradation leading to aldehydes or ketones (A.8), such is not the case with Ag_2CO_3 on celite. With the latter hydroxyketones are often obtained as indicated [18]:

$$CH_3CHOHCH_2CH_2OH \xrightarrow[\;C_6H_6\;]{Ag_2CO_3\, on\, Celite} CH_3COCH_2CH_2OH$$

1-Hydroxybutan-3-one, 80%

Butane 1,4-diol, pentane-1,5-diol, and hexane-1,6-diol give the corresponding γ-, δ-, and ε-lactones.

e. DDQ in Toluene

This oxidizing agent is of interest in that it attacks hindered alcohols much more satisfactorily than unhindered ones [3]. Isoborneol, for example, gives a superior yield of isobornenone (Ex. a):

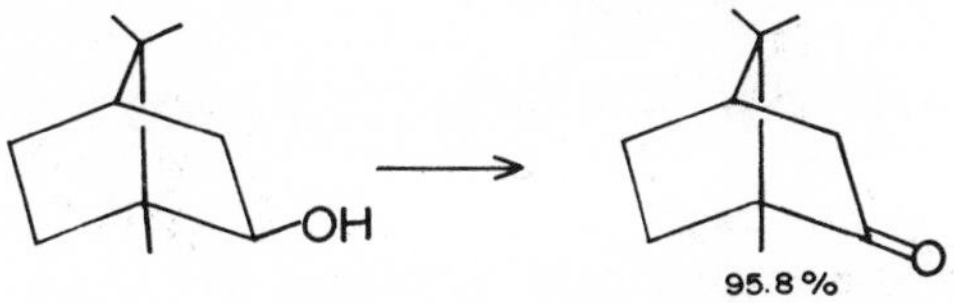

95.8%

On the other hand, β-norborneol does not respond at all to the oxidation:

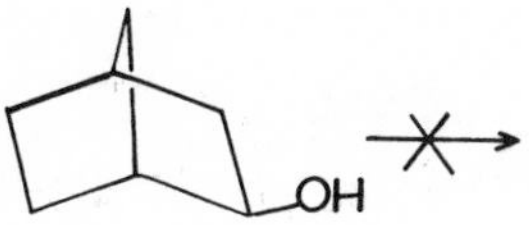

1-Hydroxyadamantane, a tertiary alcohol, when treated with 96% H_2SO_4 rearranges to 2-hydroxyadamantane, a secondary alcohol, which in turn is oxidized by the acid to form adamantanone [19]:

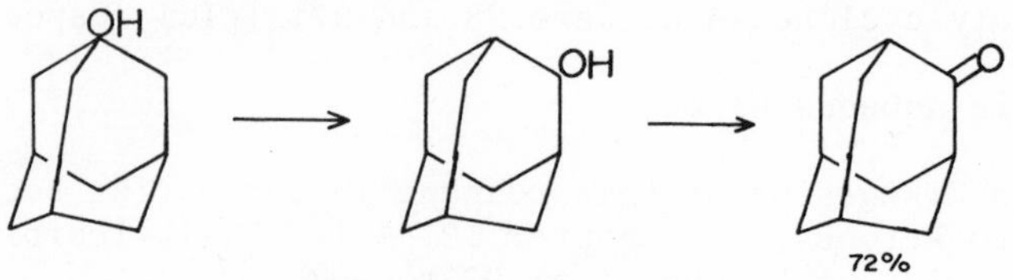

72%

a. Preparation of Isobornenone [3]

Yield was 95.8% from a 1:1 molar ratio of isoborneol and DDQ refluxed in $C_6H_5CH_3$ for 8 hr.

2. From Alcohols (Secondary) and Aluminum t-Butoxide
   (Oppenauer) (1, 628; H.-W., 7, Pt. 2a, 1973, p. 714)

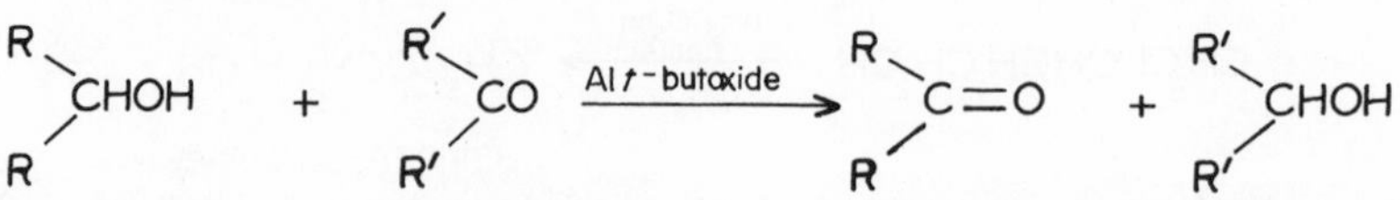

1-Methyl-4-piperidone has been found to be a quite satis-factory hydride acceptor in the Oppenauer oxidation since the excess of the original ketone and the alcohol formed are both readily removed by washing the organic layer with dilute aqueous acid [20]. By this procedure cholesterol was converted into cholesterone in 83% yield by a 4-hr reflux of a toluene solution.

4. From Acyloins or Benzoins (1, 630)

$$(Ar)RCOCHOHR(Ar) \longrightarrow (Ar)RCOCOR(Ar)$$

Potassium permanganate in acetic anhydride has been utilized in converting benzoin into benzil [21]. The same reagent permits the conversion of alkenes into α-diketones (A.5).

Diketone formation also results from the oxidation of the bis-trimethylsilyl derivative of the enolic form of the acyloin [22]:

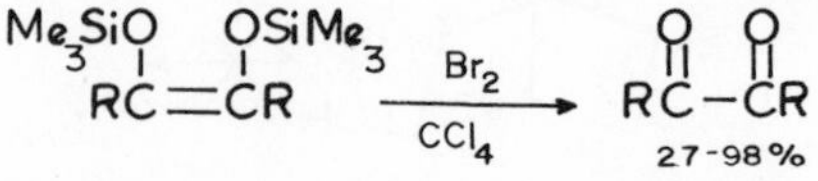

A similar process occurs with the bis-trimethylsilyl derivatives of the enolic form of cyclic acyloins, although in this case ring enlargement may result [23]:

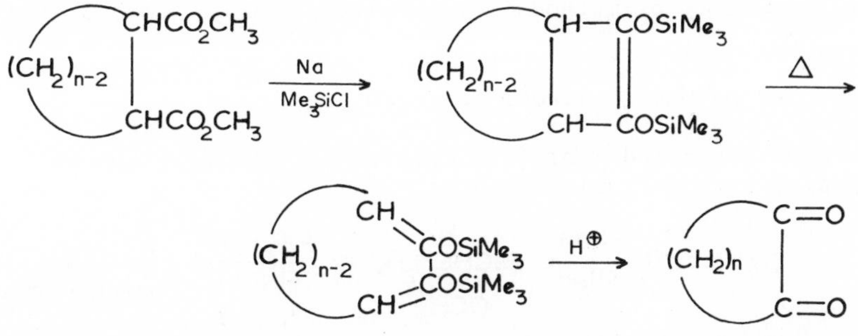

$$71-74\%$$

Similarly, cyclobutoin may be converted into 1,2-cyclobutanedione [24].

Benzoins have been oxidized successfully to benzils by the use of triphenylphosphine dibromide [25]. The α-aminoketone may also be converted into a diketone as indicated [26]:

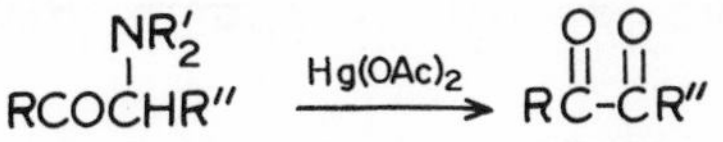

High yields

5. From Olefins via the Ozonide and Related Reactions (1, 631)

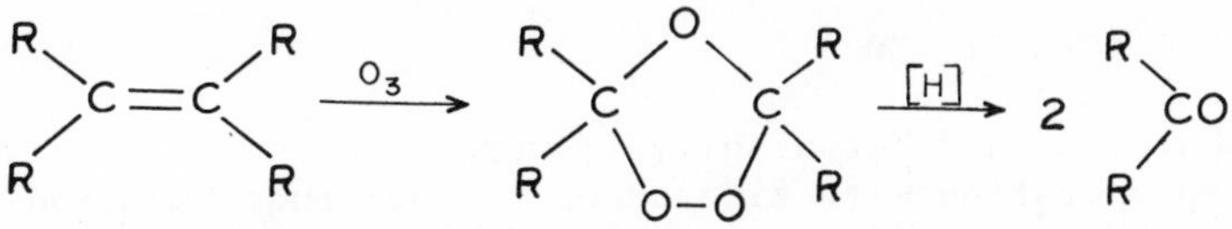

Tetrasubstituted olefins usually on being ozonized give products that cannot be reduced to the ketone. Instead of being the normal ozonide, as represented above, this new product is thought to be a peroxide of the formula $-[CR_2-O-O]_x-$. Nebel [27] showed that this type, formed in alcohol at low temperature, was converted into the ketone by the addition of water. Yields of ketones as the 2,4-dinitrophenylhydrazone varied within 77-99%.

Diketospiroheptane may be obtained by the oxidation of the corresponding dimethylene derivative [28]:

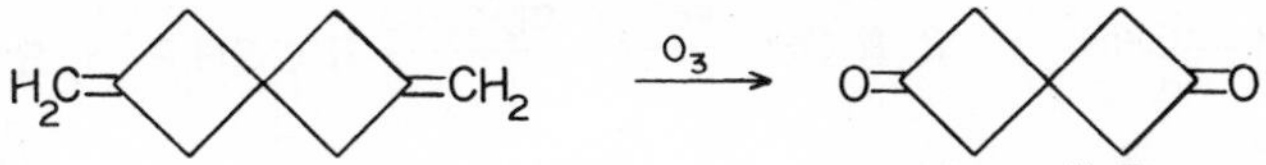

2,6-Dioxospiro [3.3] heptane
61 %

As is true of aldehydes as well, ketones are being prepared

more frequently by methods not involving ozone. Some of these
are summarized.

a. Use of $Hg(OAc)_2$-Li_2PdCl_4 or $HgSO_4$-Base

Rodeheaver and Hunt [29] carried out the reaction as
indicated with mercuric acetate:

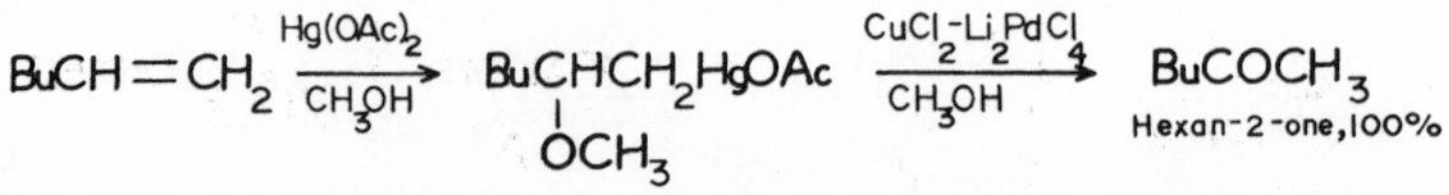

$$BuCH=CH_2 \xrightarrow[CH_3OH]{Hg(OAc)_2} BuCHCH_2HgOAc \xrightarrow[CH_3OH]{CuCl_2-Li_2PdCl_4} BuCOCH_3$$
$$\underset{OCH_3}{|} \qquad \text{Hexan-2-one,100\%}$$

In the reaction the intermediate need not be isolated. A mixture
of the alkene and mercuric salt may be added to the mixture of
the cupric salt and Li_2PdCl_4. Other yields vary from 82 to 100%.
A simple procedure has been developed by Arzoumanian and
co-workers [30] as shown:

$$RCH=CHR + H_2O \xrightarrow[90°]{HgSO_4} RCOCH_2R$$

The suspension of the mercuric salt and olefin is stirred and
then made basic (pH 13) and heated. With propene and 1-butene
yields of the ketone exceed 90%. In some cases the epoxide is
the major product of the reaction.

b. Use of CrO_2Cl_2-Zn

Sharpless and Teranishi [31] were able to effect a similar
one-step operation with zinc added to the crude reaction mixture
of the substrate and reagent, as indicated:

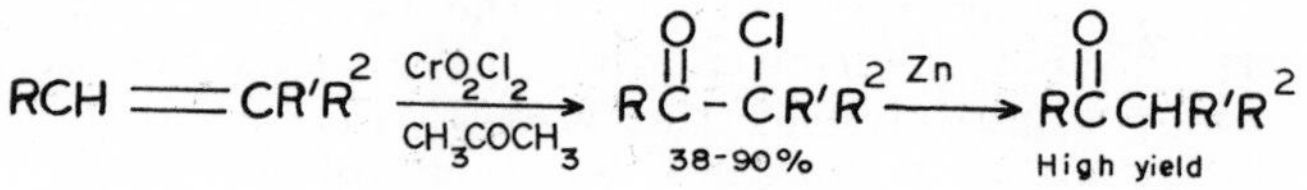

$$RCH=CR'R^2 \xrightarrow[CH_3COCH_3]{CrO_2Cl_2} \underset{38-90\%}{RC-CR'R^2} \xrightarrow{Zn} \underset{\text{High yield}}{RCCHR'R^2}$$

c. Use of Pd(II) Salts

Lloyd and Luberoff [32] found that the oxidation of alkenes
as shown gives the ketone and the acetal or ketal:

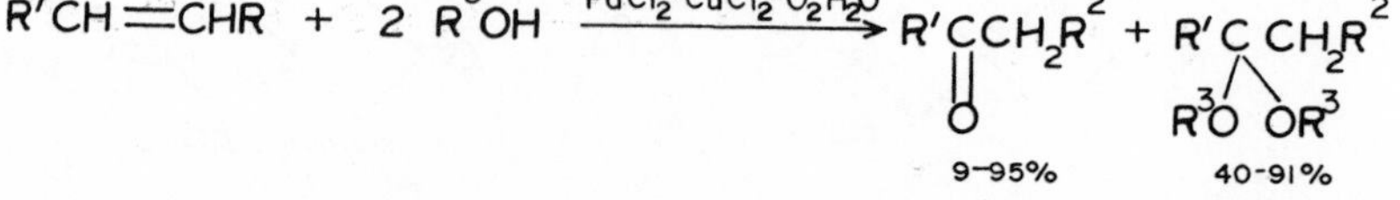

$$R'CH=CHR^2 + 2 R^3OH \xrightarrow{PdCl_2-CuCl_2-O_2H_2O} \underset{9-95\%}{R'CCH_2R^2} + \underset{40-91\%}{R'C CH_2R^2}$$

The oxygen is introduced under pressure and small amounts of
water favor the formation of the ketone.

Cyclohexene gave cyclohexenone, the alcohol, benzaldehyde, and other products when oxygen was passed through a toluene solution of the alkene in the presence of $RhCl(PPh_3)_3$ [33].

Potassium permanganate in acetic anhydride below 10° gives fair yields of 1,2-diketones from olefins [21].

For the use of $HgSO_4$-H_2SO_4, thallium(III) nitrate and lead tetraacetate in trifluoroacetic acid in the conversion of alkenes into ketones, see Chapter 10, E.1.

6. From Methylene or Methinyl Derivatives (1, 633; H.-W., 7, Pt. 2a, 1973, p. 677)

Methylene groups may at times be converted into carbonyl groups by air oxidation as shown for tetralin:

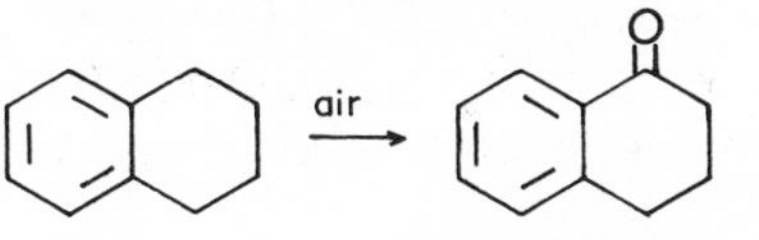

1-Keto-1,2,3,4-tetrahydronaphthalene

Bergman and co-workers [34] obtained this product in 90% yield (48-60% conversion) with $RhCl(PPh_3)_3$, Saratov and co-workers [35] with a trace of manganese (resinate solution) report an 88-90% yield, and Fenton [36] recovered 49 g from 400 ml of the hydrocarbon with $Ir(PPh_3)_3 \cdot 2HCl$ as a catalyst.

Chemical reagents were employed to oxidize other methylene groups as indicated:

$$RC{\equiv}CCH_2R' \xrightarrow[2\,C_5H_5N]{CrO_3^-} RC{\equiv}CCOR' \quad 31\text{-}42\% \qquad [37]$$

$$RCH_2CH{=}CH_2 \xrightarrow[H_2O]{Hg^{2+}} RCHOHCH{=}CH_2 \xrightarrow{Hg^{2+}} RCOCH{=}CH_2 \quad \text{good yields} \qquad [38]$$

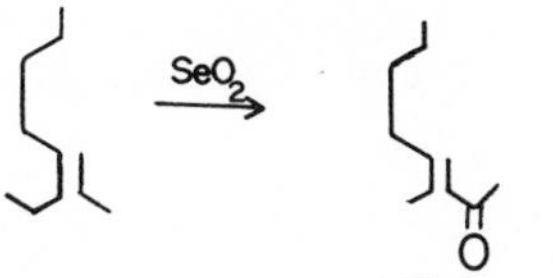

78 % [39]

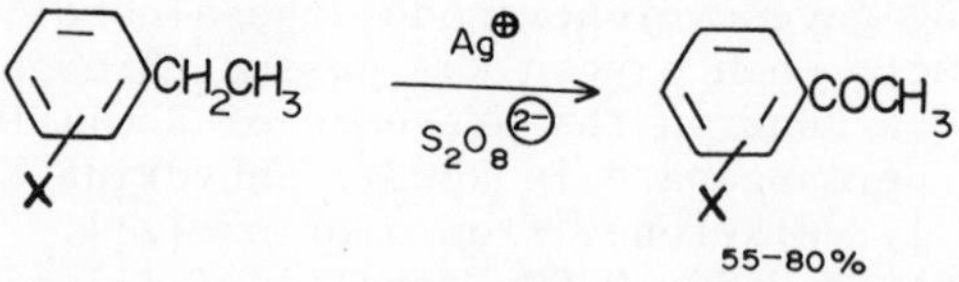

$$[40]$$

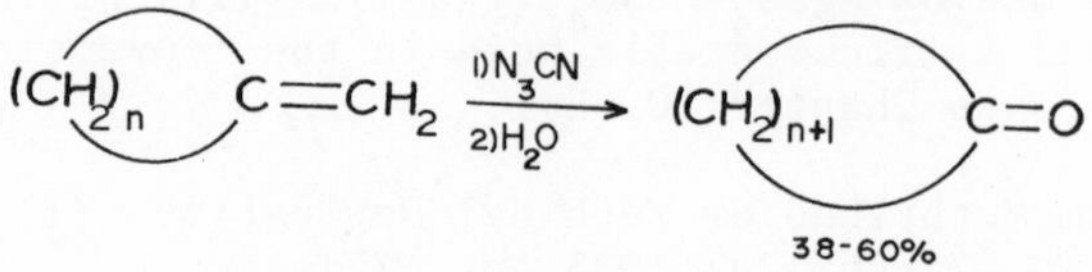

55-80%

$$[41]$$

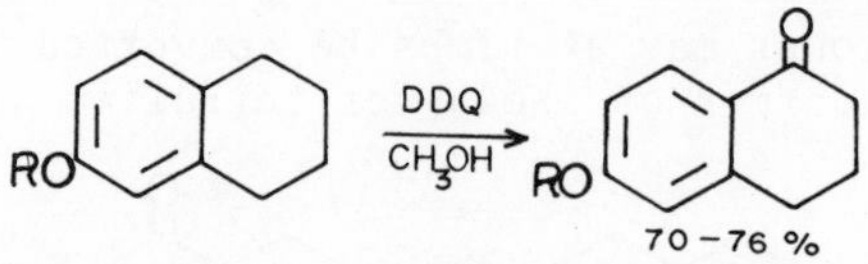

38-60%

$$[42]$$

70-76%

It will be noted that ring expansion occurs in the second to last equation above.

The oxidation of ketones to diketones is illustrated in the equations below:

$$C_6H_5CH_2COC_6H_5 \xrightarrow[\text{HBr}]{\text{DMSO}} C_6H_5COCOC_6H_5 \qquad [43]$$

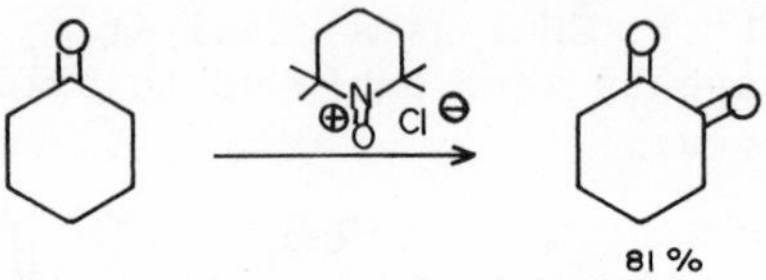

95%

$$[44]$$

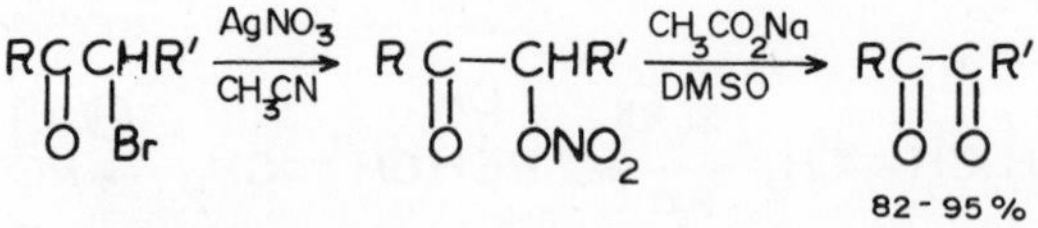

81%

$$[45]$$

82-95%

Aldehydes containing an α-methinyl group, such as isobutyr-aldehyde, are degraded to ketones by treatment with air, 1,4-diazobicyclo[2.2.2]octane, and Cu(OCOCH$_3$)$_2$-2,2'-bipyridyl complex [46]:

$$(CH_3)_2CHCHO \longrightarrow CH_3COCH_3$$

75%

The mechanism appears to be:

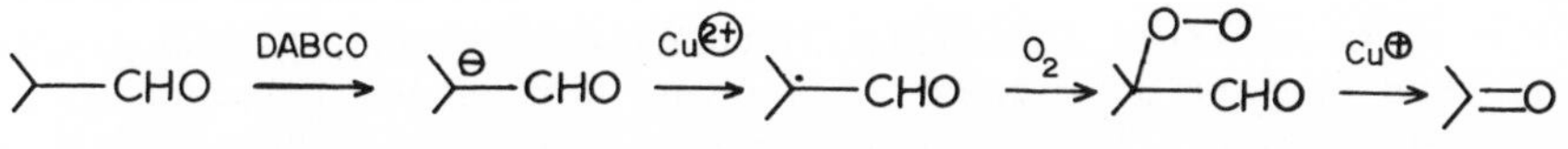

$$\text{>—CHO} \xrightarrow{\text{DABCO}} \text{>}^{\ominus}\text{—CHO} \xrightarrow{\text{Cu}^{\oplus}} \text{>·—CHO} \xrightarrow{O_2} \text{>—CHO} \xrightarrow{\text{Cu}^{\oplus}} \text{>=O}$$

An abnormal oxidation occurs when certain steroidal ketones are treated with mercury (II) acetate [47]:

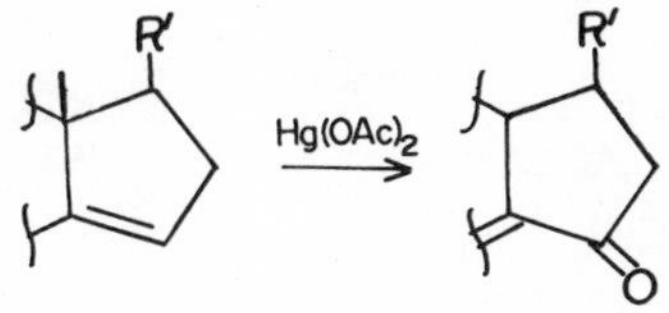

7. From Tertiary Hydrocarbons via Hydroperoxides or Highly Substituted Benzenes and Peroxytrifluoroacetic Acid (1, 637)

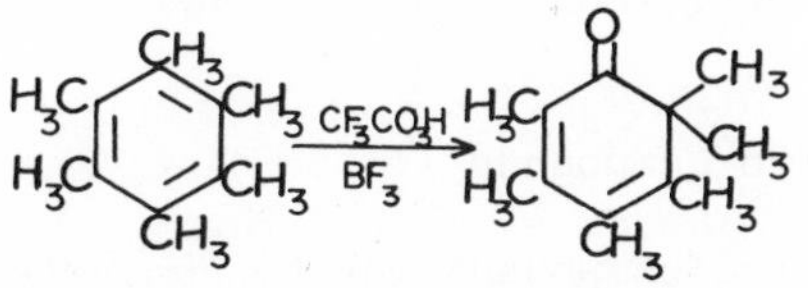

2,3,4,5,6,6-Hexamethyl-2,4-cyclohexadien-1-one has been synthesized as shown above in 82-90% yield [48]. The same oxidative rearrangement may be accomplished in 86% yield by using 30% H_2O_2 to which acetic anhydride and concentrated sulfuric acid have been added [49]. (Care should be taken that highly concentrated hydrogen peroxide does not accumulate in the dehydration medium.) The method is applicable to other highly substituted benzenes.

2-Adamantanone may be synthesized conveniently by stirring the hydrocarbon in 98% sulfuric acid at 80° [50]:

$$\text{AdH} \longrightarrow \text{1-AdOH} \rightleftharpoons \text{2-AdOH} \longrightarrow \underset{47\text{-}48\%}{\text{2-Ad}=\text{O}}$$

Sulfur dioxide is released during the oxidation.

8. From Glycols (1, 638; H.-W., 7, Pt. 2a, 1973, p. 927)

$$\underset{R}{\overset{R}{\diagdown}}\text{COHCHOHR} \longrightarrow \underset{R}{\overset{R}{\diagdown}}\text{CO} + \text{RCHO}$$

Although periodic acid and lead tetraacetate are the most widely used reagents in this oxidation, the use of thallium salts, $Pb(H_2PO_4)_2\text{-}H_2[Pb(H_2PO_4)_2(HPO_4)_2]$, and active precipitated manganese(IV) oxide in forming ketones from glycols is given under Chapter 10, A.7.

It is possible to oxidize both hydroxy groups in the glycol and thus prepare diketones. Regen and Whitesides [51] accomplished this feat for vicinal diols by using benzalacetone as the hydrogen acceptor and tris(triphenylphosphine)ruthenium dichloride as the catalyst.

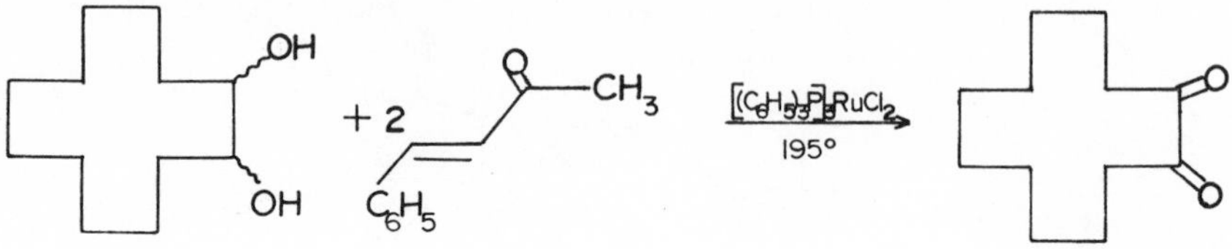

1,2-Cyclododecanediol

1,2-Cyclododecanedione 78-100%(glpc)

With permanganate and acetic anhydride an olefin probably initially forms the glycol, which is then oxidized further to the α-diketone [21].

9. From Nitro Compounds (1, 639)

Secondary nitro compounds may be converted into ketones by a method [52] other than the Nef:

$$R_2CHNO_2 + CH_3(CH_2)_2ONO + NaNO_2 \longrightarrow \underset{67\text{-}88\%}{R_2CO} + CH_3CH_2CH_2OH + NaNO_3 + N_2O$$

For a second method of converting secondary nitro compounds into ketones, see Chapter 10, A.18. Both of these methods are satisfactory for preparing diketones.

N-Nitrososulfonamides on irradiation oxidize the sec-alkyl group attached to nitrogen to the ketone in mediocre yield [53].

10. From Alkenes via Hydroboration (1, 641)

For a discussion of these methods, see E.8. The preparation of ketones from boranes and diazoketones is found in G.9.

13. From Amines (1, 642; H.-W., 7, Pt. 2a, 1973, p. 778)

The methods of Calo, Hyatt, and Andersen (see Chapter 10, A.21) are also applicable to the synthesis of ketones. Other methods for ketone preparation follow.

Corey and Achiwa [54] oxidized primary amines to ketones by a mild method as indicated (Ex. a):

$$\underset{R'}{\overset{R}{\diagdown}}CHNH_2 \xrightarrow[\substack{2)\ \text{DBN-DMSO-THF} \\ 3)\ H_3O^{\oplus}}]{1)\ \text{Mesityl glyoxal }-C_6H_6} \underset{R'}{\overset{R}{\diagdown}}C=O \quad 55-90\%$$

The mesityl glyoxal first forms a Schiff base, $R_2CH-N=CR'_2$, which undergoes prototropic isomerization to $R_2C=NCHR'_2$, after which the latter hydrolyzes to the ketone. 3-Nitromesityl- and 3,5-dinitromesitylglyoxal and 3,5-di-t-butyl-1,2-benzoquinone may also be used as the reagent. Here again the Schiff base is an intermediate and the yields of ketones vary within 84-97%.

An unusual but not too important synthetic method consists of the treatment of the amine with an excess of the alkyllithium followed by hydrolysis [55]. Yields are poor. The steps for α-methylbenzylamine have been represented as:

$$\underset{NH_2}{\overset{}{C_6H_5CHCH_3}} \xrightarrow{2\ RLi} \underset{NLi_2}{\overset{}{C_6H_5CHCH_3}} \xrightarrow{-LiH} \underset{NLi}{\overset{}{C_6H_5CCH_3}} \xrightarrow{H_2O} \underset{O}{\overset{}{C_6H_5CCH_3}}$$

a. Preparation of Cyclododecanone [54]

Cyclododecylamine, 183.3 mg in C_6H_6 was added to 176 mg of mesitylglyoxal in C_6H_6 at 23° and the stirring was continued under N_2 for 30 min. After removal of solvent the Schiff base was dissolved in a solution of 0.1 mmol of DBN in 1 ml of DMSO-THF. After allowing the reaction to proceed for 10.5 hr under N_2, the mixture was hydrolyzed at pH 3 for 2 hr with stirring at 23° after the addition of 8 ml of CH_3OH-THF and crystalline oxalic acid dihydrate. The ketone recovered amounted to 86%.

14. From Halides (1, 643; H.-W., 7, Pt. 2a, 1973, p. 777)

For the synthesis of ketones from halides, see Chapter 10, A.10.

15. From Epoxides (1, 643)

See E.3.

16. From Enamines

Aldehyde enamines, as indicated, absorb O_2 in the presence of Cu(I)Cl to give a ketone and an amide in quantitative yields [56]:

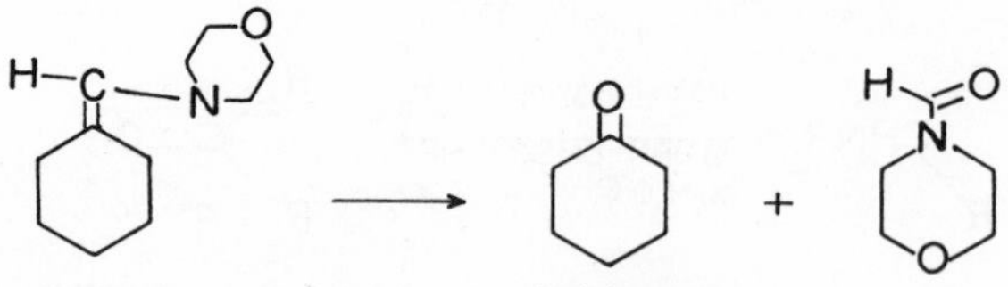

Cyclohexanone

17.   From Phosphoranes

Alkenyl phosphoranes are oxidized by $KMnO_4$ [57] or $NaIO_4$ [58] to diketones as shown:

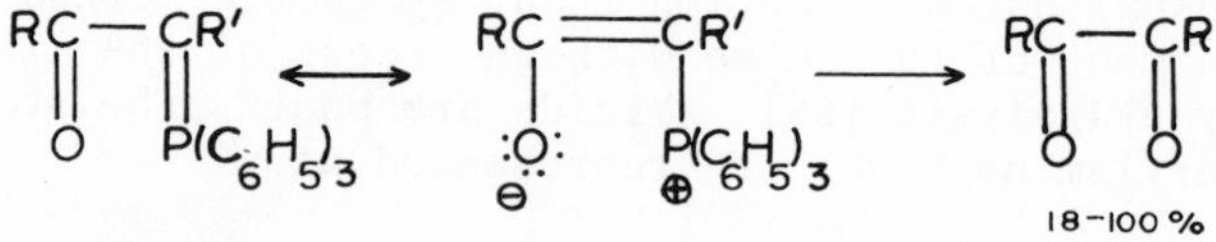

18-100 %

18.   From Ethers

A series of ethers in CCl_4 was oxidized with chromyl chloride, CrO_2Cl_2 [59]. Dibenzhydryl ether responded as indicated:

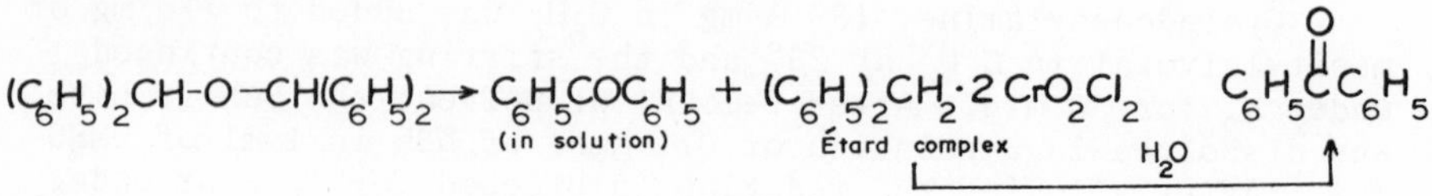

Here the return is quantitative. In most cases little of the ketone was obtained except when one of the alkyl groups was the benzhydryl.

An α,β-unsaturated ketone has been obtained by the irradiation of the trimethylsilyl enol ether of isopropyl phenyl ketone in CCl_4 containing tetraphenyl porphyrin as a sensitizer in the presence of oxygen [60]:

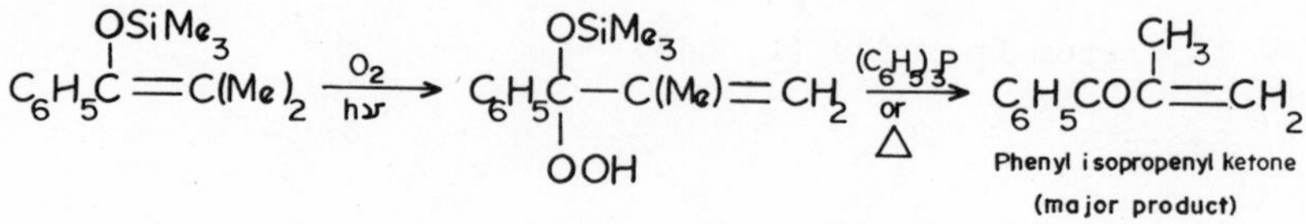

Phenyl isopropenyl ketone
(major product)

19.   From Acetylenes

$$RC \equiv CR' \longrightarrow RCOCOR'$$

Various oxidizing agents have been employed in converting acetylenes into diketones.

N-Bromosuccinimide, NBS, less than 2 molar equiv, and diphenylacetylene in anhydrous DMSO produce a near-quantitative yield of benzil [61].   It is important that the DMSO be anhydrous when employed with NBS.   Terminal acetylenes lead to α-ketoaldehydes.

Ruthenium tetraoxide as the oxidant with NaOCl in CCl_4-H_2O at 0° leads to the diketone and carboxylic acid with nonterminal acetylenes [62].   The highest yield of the diketone obtained (90-98%) was from diphenylacetylene.

Thallium(III) nitrate, 2 equiv, in aqueous acidic glyme converts diarylacetylenes into benzils (60-90%) [63].   With this reagent monoalkylacetylenes give carboxylic acids, dialkylacetylenes give acyloins, and alkylarylacetylenes in methanol give methyl arylacetates.

Terminal acetylenes or acetylenic bromides may be converted into α-keto esters via ozonization as shown for the bromide [64]:

$$RC{\equiv}CBr + O_3 \xrightarrow{MeOH} \left[\begin{array}{c} OH \\ | \\ RC-COMe \\ | \ \ \ \| \\ OMe \ \ O \end{array} \right] \xrightarrow{KI} \underset{40-50\%}{RC\ COOMe}$$

Acetylenes have been transformed into α,β-unsaturated ketone systems as indicated [65]:

$$RC{\equiv}CH + R'CHCH{=}\overset{\oplus}{N}\bigcirc \xrightarrow[\text{basic Al}_2O_3]{\text{aq. BF}_4, \text{SO}_2} \underset{72-82\%}{R\overset{O}{\overset{\|}{C}}CH{=}CH}\diagup^{H}_{R'}$$

20.   From Ketones by Oxidative Coupling

Ketones containing an α-methylene group couple in the presence of $FeCl_3$ as shown [66]:

$$C_6H_5COCH_2CH_3 \longrightarrow [C_6H_5CO\overset{\cdot}{C}HCH_3] \longrightarrow$$

$$\underset{\underset{CH_3}{|}}{C_6H_5COCH\overset{CH_3}{\underset{|}{C}}HCOC_6H_5} + \text{(furan ring)} + C_6H_5CO\overset{Cl}{\underset{|}{C}}HCH_3$$

The reaction was studied under various conditions and it was found that $FeCl_3$-CuO was an effective reagent for oxidative coupling. With this combination, a mmol ratio between diketone, furan, and chloroketone of approximately 12:1:2 was achieved.

21.   From 2-Phenylselenoalkanones to Unsaturated Ketones (Phenylselenious Acid Elimination)

2-Phenylselenocyclohexanone, formed from the enol acetate of cyclohexanone, eliminates as shown [67]:

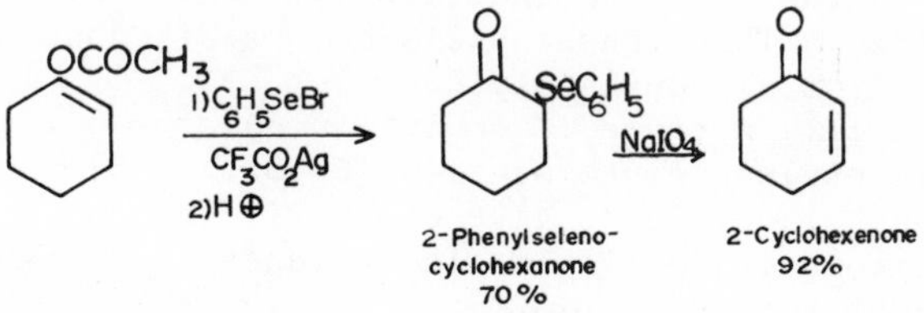

2-Phenylseleno-
cyclohexanone
70%

2-Cyclohexenone
92%

A similar elimination to give unsaturated ketones has been achieved with acyclic ketones [68]:

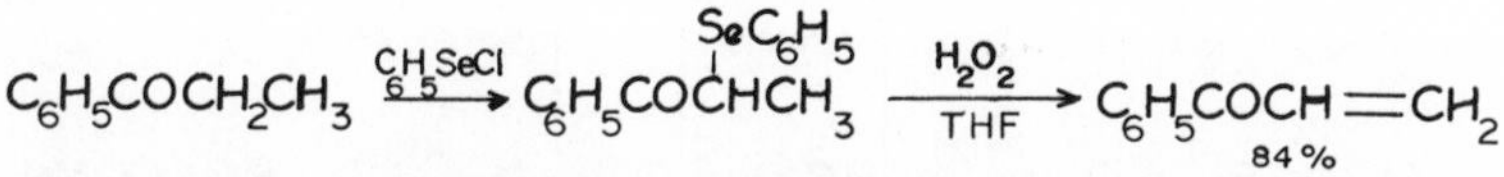

For dialkylated ketones an additional step has been introduced [69]:

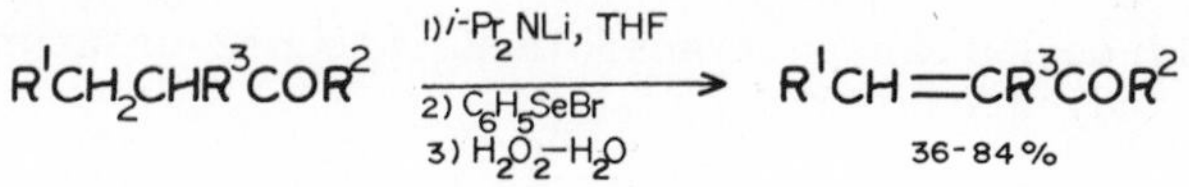

22.   From Secondary Nitriles

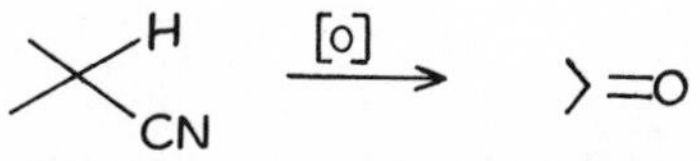

This transformation may be illustrated in the conversion of 2-carbomethoxy-α-bromophenylacetonitrile into ligusticumic acid as shown [70]:

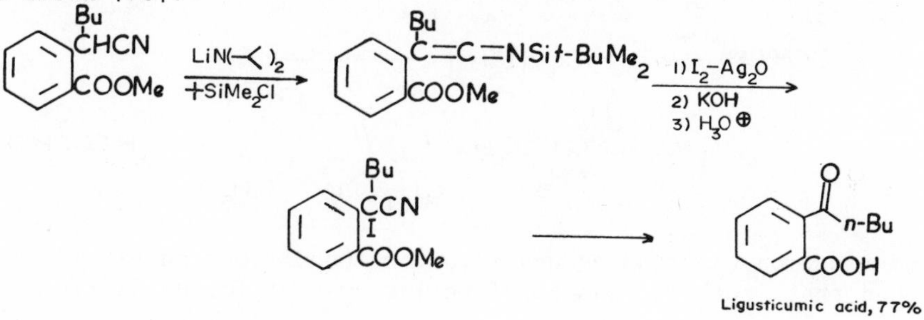

Ligusticumic acid, 77%

23.   From Cyclic Compounds and Singlet Oxygen

A series of cyclic compounds has been oxidized with singlet oxygen in the absence of solvent in a special apparatus and in some cases ketones have been produced [71]. Tetracyclone, for example, gave cis-dibenzoylstilbene:

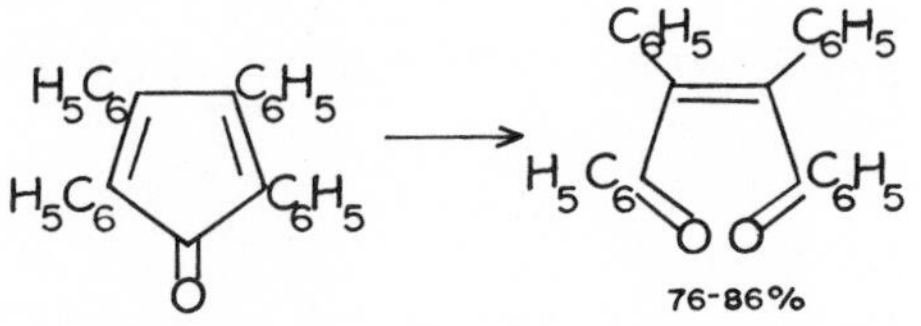

76-86%

24.   From 1,3-Diketones

1,3-Diketones are deacylated when heated in the presence of their copper (II) chelates [72]:

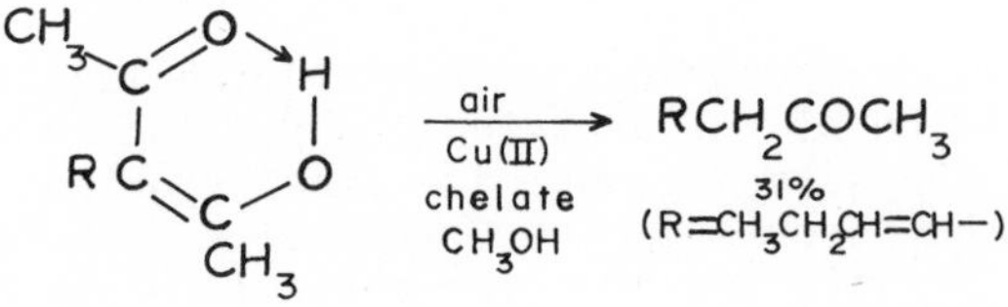

25.   From Disubstituted Malonic Acids (Oxidative Decarboxylation)

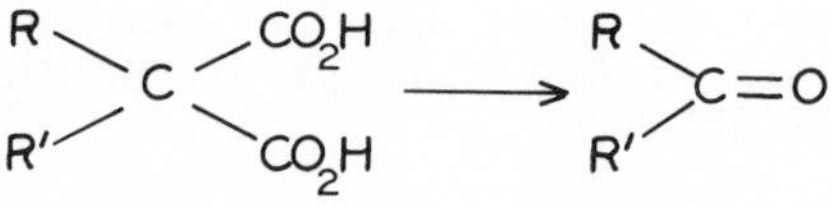

Tufariello and Kissel [73] found that the conversion of disubstituted malonic acids into ketones, which has required five steps, may be accomplished in two steps as indicated:

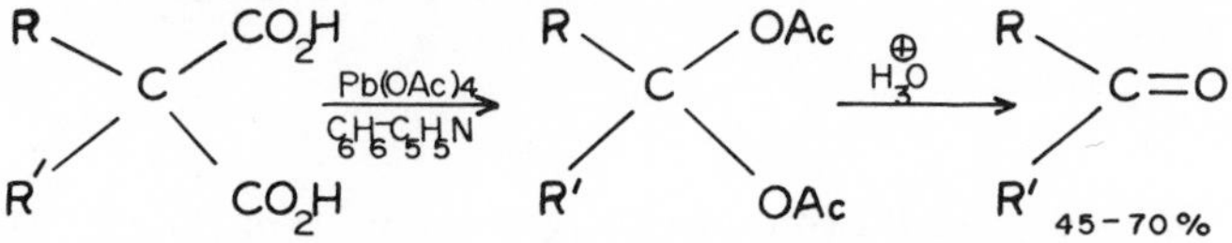

This reaction is related to the Kochi oxidation of acids (see Chapter 2, F.7).

26.   From Active Hydrogen Compounds, the Nitroso Ion, and the Like

Nitrosation to prepare both aldehydes and ketones has been discussed under Chapter 10, A.15.

27. Acetoxylation of Ketones

See Chapter 10, A.26.

28. From p-Alkylsubstituted Phenols and DDQ (see Addenda)

Addenda

A.1. The Moffatt reagent, DMSO-DCC-$\overset{+}{H}$ (1, 625), has been simplified, if not improved; in other words, Me$_2$SOSO$_2$CH$_3$ CH$_3$SO$_3$ in HMPA [J. D. Albright, J. Org. Chem., 39, 1977 (1974)], as well as the triflate salt of DMSO have been used [J. B. Hendrickson, S. M. Schwartzman, Tetrahedron Letters, 273 (1975)]. However, both suffer from the presence of the side product, ROCH$_2$SMe.

A.1. Conventional procedures for preparing 1,4-diketones consist of formation of the acetylenic glycol of an aldehyde, reduction of the alkyne group, and finally oxidation of the glycol with the Sarett reagent. Typical yields for the three steps were 32%, 44%, and 80%, respectively, perhaps that of the first being lower than usual [W. B. Sudweeks, H. S. Broadbent, J. Org. Chem., 40, 1131 (1975)].

A.14. α-Bromoketones are oxidized to α-diketones by DMSO catalyzed by KI-Na$_2$CO$_3$ [D. P. Bauer, R. S. Macomber, J. Org. Chem., 40, 1990 (1975)].

A.28. 6-Hydroxytetralin may be oxidized by 2,3-dichloro-5,6-dicyanobenzoquinone to a quinonemethide, which is oxidized further in the same reaction to 6-hydroxytetralone (78%) [J. W. A. Findlay, A. B. Turner, Chem. Ind. (London), 158 (1970)].

References

1. D. G. Lee, Oxidation, Vol. 1, Dekker, New York, 1969, p. 56.
2. H. C. Brown et al., J. Org. Chem., 36, 387 (1971).
3. J. Iwamura, N. Hirao, Tetrahedron Letters, 2447 (1973).
4. J. Wicha et al., Tetrahedron Letters, 3635 (1973).
5. E. J. Corey, G. W. J. Fleet, Tetrahedron Letters, 4499 (1973).
6. D. G. Lee et al., Can. J. Chem., 50, 3741 (1972).

7.   T. Holm, I. Crossland, Acta Chem. Scand., 25, 59 (1971).
8.   J. A. Caputo, R. Fuchs, Tetrahedron Letters, 4729 (1967).
9.   C. W. Rees, R. C. Storr, J. Chem. Soc., C, 1474 (1969).
10. K. E. Stensiö, Acta Chem. Scand., 25, 1125 (1971).
11. W. H. Clement et al., Chem. Ind. (London), 755 (1969).
12. M. Fétizon et al., Chem. Commun., 1102 (1969).
13. R. J. Audette et al., Tetrahedron Letters, 279 (1971).
14. S. L. Regen, G. M. Whitesides, J. Org. Chem., 37, 1832
     (1972).
15. R. Ratcliffe, R. Rodehorst, J. Org. Chem., 35, 4000 (1970);
     J. C. Collins, W. W. Hess, Org. Syn., 52, 5 (1972).
16. H. C. Brown et al., J. Org. Chem., 36, 387 (1971).
17. R. M. Kanojia, R. E. Adams, Org. Prep. Proced. Int., 4,
     559 (1972).
18. M. Fétizon et al., Chem. Commun., 1102 (1969).
19. H. W. Geluk, J. L. M. A. Schlatmann, Tetrahedron, 24, 5361
     (1968); Chem. Commun., 426 (1967); see Ref. 50.
20. R. Reich, J. F. W. Keana, Syn. Commun., 2, 323 (1972).
21. K. B. Sharpless et al., J. Am. Chem. Soc., 93, 3303 (1971).
22. H. Wynberg et al., Synthesis, 209, 211 (1971); H.-G. Heine,
     Chem. Ber., 104, 2869 (1971).
23. T. Mori et al., Can. J. Chem., 47, 3266 (1969).
24. J. M. Conia, J. M. Denis, Tetrahedron Letters, 2845 (1971).
25. T.-L. Ho, Synthesis, 697 (1972).
26. H. Mohrle, D. Schittenhelm, Chem. Ber., 104, 2475 (1971).
27. C. Nebel, Chem. Commun., 101 (1968).
28. E. Buchta, A. Kröniger, Ann. Chem., 716, 112 (1968).
29. G. T. Rodeheaver, D. F. Hunt, Chem. Commun., 818 (1971).
30. H. Arzoumanian et al., J. Org. Chem., 39, 3445 (1974).
31. K. B. Sharpless, A. Y. Teranishi, J. Org. Chem., 38, 185
     (1973).
32. W. G. Lloyd, B. J. Luberoff, J. Org. Chem., 34, 3949
     (1969).
33. K. Kaneda et al., Bull. Soc. Chem. Jap., 46, 3810 (1973).
34. E. D. Bergmann et al., Tetrahedron Letters, 3665 (1967).
35. I. E. Saratov et al., Neftekhimiya, 8, 895 (1968); C. A.,
     70, 67976 (1969).
36. D. M. Fenton, U.S. Patent 3,422,147, January 14, 1969;
     C. A., 70, 67997 (1969).
37. J. E. Shaw, J. J. Sherry, Tetrahedron Letters, 4379 (1971).
38. H. B. Tinker, J. Organomet. Chem., 32, C25 (1971).
39. W. T. Bhalerao, H. Rapoport, J. Am. Chem. Soc., 93, 4835
     (1971).
40. F. A. Daniher, Org. Prep. Proced., 2, 207 (1970).
41. J. E. McMurry, A. P. Coppolino, J. Org. Chem., 38, 2821
     (1973).
42. J. W. A. Findlay, A. B. Turner, J. Chem. Soc., C23, 547
     (1971).

43. E. Schipper et al., Tetrahedron Letters, 6201 (1968).
44. V. A. Golubev, R. V. Miklyush, J. Org. Chem. (USSR), 8, 1376 (1972).
45. N. Kornblum, H. W. Frazier, J. Am. Chem. Soc., 88, 865 (1966).
46. V. Van Rheenan, Tetrahedron Letters, 985 (1969).
47. E. C. Blossey, P. Kucinski, Chem. Commun., 56 (1973).
48. H. Hart et al., Org. Syn., 48, 87 (1968).
49. H. Hart et al., Synthesis, 195 (1972).
50. H. W. Geluk, V. G. Keizer, Org. Syn., 53, 8 (1973); Ref. 19.
51. S. L. Regen, G. M. Whitesides, J. Org. Chem., 37, 1832 (1972).
52. N. Kornblum, P. A. Wade, J. Org. Chem., 38, 1418 (1973).
53. Th. J. deBoer et al., Rec. Trav. Chim., 90, 901 (1971).
54. E. J. Corey, K. Achiwa, J. Am. Chem. Soc., 91, 1429 (1969).
55. H. G. Richey, Jr. et al., Tetrahedron Letters, 2187 (1971).
56. V. Van Rheenen, Chem. Commun., 314 (1969).
57. E. Zbiral, M. Rasberger, Tetrahedron, 24, 2419 (1968).
58. H.-J. Bestmann et al., Chem. Ber., 102, 2259 (1969).
59. C. D. Nenitzescu et al., Rev. Roumaine Chim., 14, 1553 (1969).
60. G. M. Rubottom, M. I. Lopez Nieves, Tetrahedron Letters, 2423 (1972).
61. S. Wolfe et al., Can. J. Chem., 49, 1099 (1970).
62. H. Gopal, A. J. Gordon, Tetrahedron Letters, 2941 (1971).
63. A. McKillop, E. C. Taylor et al., J. Am. Chem. Soc., 93, 7331 (1971).
64. S. Cacchi et al., J. Org. Chem., 38, 3653 (1973).
65. S. Shatzmiller, A. Eschenmoser, Helv. Chim. Acta, 56, 2975 (1973).
66. H. Inoue et al., Bull. Chem. Soc. Jap., 46, 2211 (1973).
67. D. L. J. Clive, Chem. Commun., 695 (1973).
68. K. B. Sharpless et al., J. Am. Chem. Soc., 95, 6137 (1973).
69. H. J. Reich et al., J. Am. Chem. Soc., 95, 5813 (1973).
70. D. S. Watt, J. Org. Chem., 39, 2799 (1974); S. J. Selikson, D. S. Watt, Tetrahedron Letters, 3029 (1974).
71. J. R. Scheffer, M. D. Ouchi, Tetrahedron Letters, 223 (1970).
72. K. Uehara et al., Bull. Chem. Soc. Jap., 45, 1570 (1972).
73. J. J. Tufariello, W. J. Kissel, Tetrahedron Letters, 6145 (1966).

B. Reduction

3. From Phenols, Phenol Ethers, Ketones, or Ketals (Birch
    and Benkeser Reductions) (1, 647)

A comparison of the Birch (sodium-alcohol-liq. ammonia) and
Benkeser (lithium in low-molecular-weight amines) reductions has
been published [1]. In general the Benkeser is more powerful but
less selective than the Birch reduction; however, the selectivity
of the former may be increased by the proper choice of solvents.
The Birch reduction has been reviewed [2].
    The reduction of anisole gives a variety of products
including ketones as indicated [3]:

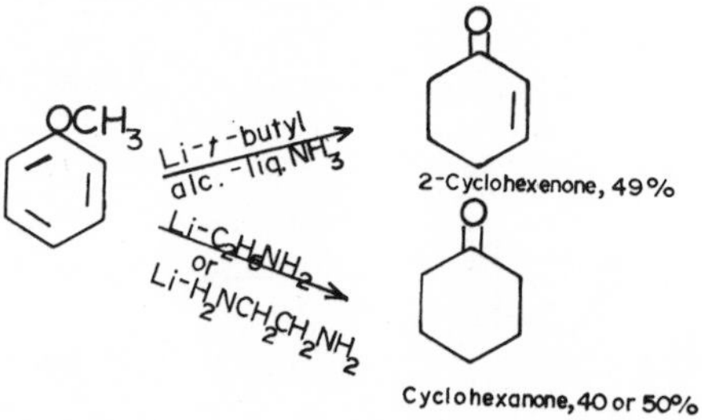

2-Cyclohexenone, 49%

Cyclohexanone, 40 or 50%

The amounts here, which are far from predictable, vary with the
experimental conditions as discussed in 1, 647.
    The return of the unsaturated ketone by the Birch reduction
of 2-methoxynaphthalene, as indicated, offers a likely method of
synthesis [4]:

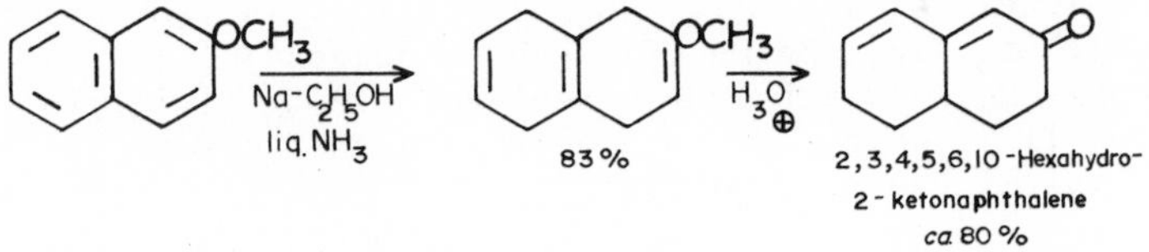

83%

2,3,4,5,6,10-Hexahydro-
2-ketonaphthalene
ca 80%

    In a modified Birch reduction (lithium-n-propylamine-t-butyl
alcohol) Kwart and Conley [5] found that the amounts of the
dihydroanisoles formed from p-substituted anisoles were similar
to those obtained by the Birch reduction. No ketones were isolated in these experiments.
    The reductive alkylation of acetophenone and its aromatic
derivatives under Birch conditions gives rather good yields of
1-alkyl-1-acetocyclohexadienes-2,5 [5a].
    Ketals have been reduced to ketones by lithium in liquid
ammonia followed by acid hydrolysis [6]. When applied to equi-
lenin ethylene ketal, partial hydrogenation of the ring system
occurs as well.

5. From α-Diketones and α-Ketols (<u>1</u>, 649)

Diketones have been reduced to monoketones by hydrogenation in the presence of Pd/BaSO$_4$ in AcOH/HClO$_4$ [7], by aluminum and Na amalgam [8], by Zn-Hg (Clemmensen reagent) [9], by benzpinacol [10], and by LiAlH$_4$ [11].

For reduction of a 1,3-diketone see G.3. The most unusual of these reducing agents is the benzpinacol [10] (Ex. a). With benzil it gives benzoin in 85% yield. Other yields vary from 63 to 91%. Benzpinacol at temperatures above 100° forms the benzhydryl radical, (Ph)$_2$ĊOH, which provides the hydrogen for the reduction.

a. Preparation of 2,2,5,5-Tetramethyl-3-keto-4-hydroxytetrahydrofuran [10]

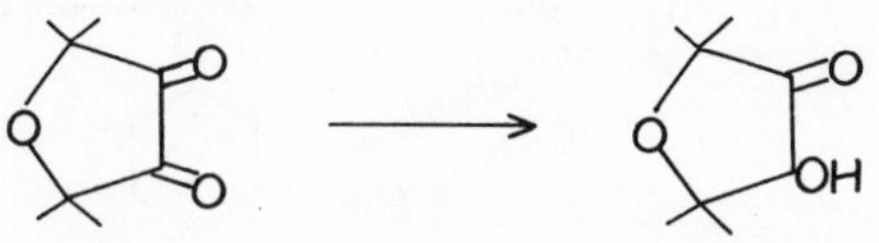

Yield was 73% by heating the diketone and benzpinacol in decalin at 165° for 15 min.

6. From Unsaturated Ketones (<u>1</u>, 650)

Numerous reducing agents have been employed in hydrogenating the double bond in unsaturated ketones. Among those are precipitated nickel catalyst [12], HIrCl$_2$(Me$_2$SO)$_3$ in 2-propanol [13], 5HCOOH·2NEt$_3$ [14], Zn-Hg (Clemmensen reagent) [9], iron pentacarbonyl-OH̄ [15], ethylene glycol-RuCl$_2$[P(C$_6$H$_5$)$_3$]$_3$ [16], and Li-4NH$_3$ [17], which has been useful in the steroid series.

Typical examples follow:

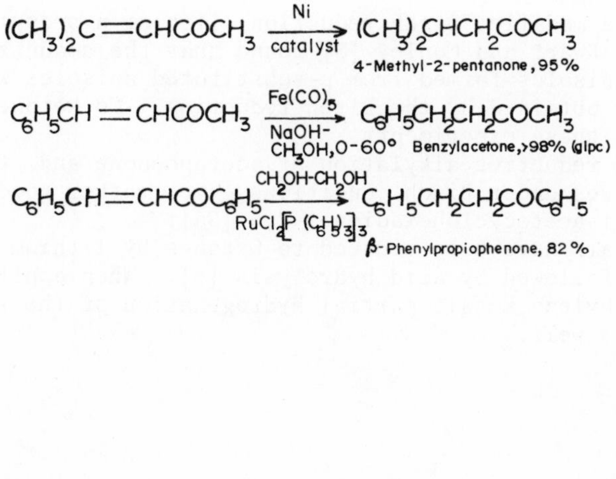

8. From α-Substituted Ketones

$$RCOCHR' \xrightarrow{[H]} RCOCH_2R'$$
with X below the first CH.

Substituents, usually a halogen, attached to the α-carbon may be removed by reagents such as $TiCl_3$ [18] (Ex. a), $LiI-BF_3 \cdot Et_2O$ [19] (Ex. b), CH_3SNa [20], $Me_2SO-Zn/Cu-NaI-$ collidine [21], and $C_6H_5N(CH_3)_2$ [22]. Of these five reagents the first two are preferred in the formation of the unsubstituted ketone. Thus the α-haloketones with $TiCl_3$ give ketones in 84-100% isolated yields, while with $LiI-BF_3 \cdot Et_2O$ the ketones were recovered in 85-98% yields.

a. Preparation of Acetophenone [18]

Yield was 100% from α-bromoacetophenone in CH_3CN added to aqueous 20% $TiCl_3$ and refluxed for 18 hr at 60 torr.

b. Preparation of 2-Methyl-5-methoxy-1-tetralone [19]

Yield was 97% from 2-bromo-2-methyl-5-methoxy-1-tetralone and $BF_3 \cdot Et_2O$ in ether added over 45 min to LiI in ether under N_2 at 25°.

9. From Cyclic Ketones (Cleavage)

Exclusive C_1-C_2 cleavage occurs on the hydrogenation of cyclopropyl methyl ketones [23]:

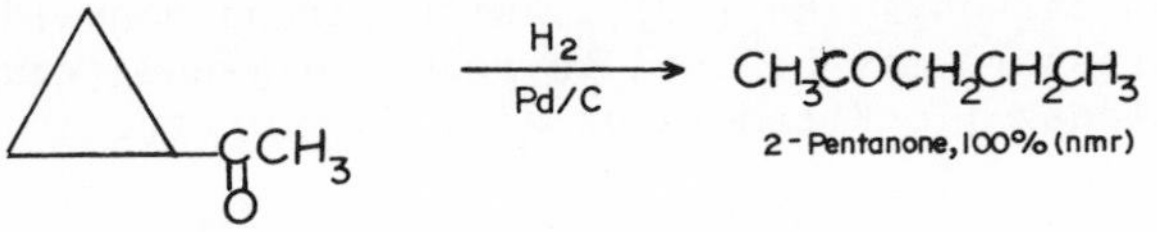

$$\xrightarrow[Pd/C]{H_2} CH_3COCH_2CH_2CH_3$$

2-Pentanone, 100% (nmr)

Similar results are obtained with Bu_3SnH in the presence of a free radical initiator [24], although stereoelectronic factors appear to be involved:

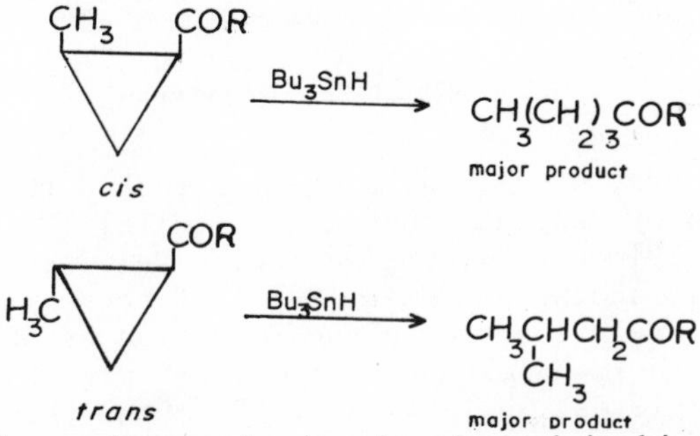

$$\overset{CH_3\quad COR}{\triangledown} \xrightarrow{Bu_3SnH} \underset{major\ product}{CH(CH_3)_2(CH_2)_3COR}$$

cis

$$\underset{trans}{\overset{COR}{H_3C\ \triangledown}} \xrightarrow{Bu_3SnH} \underset{\underset{major\ product}{CH_3}}{CH_3CHCH_2COR}$$

As expected, the cyclopropyl ring is cleaved in bicyclo [4.1.0]heptan-2-one in preference to the cyclohexyl one:

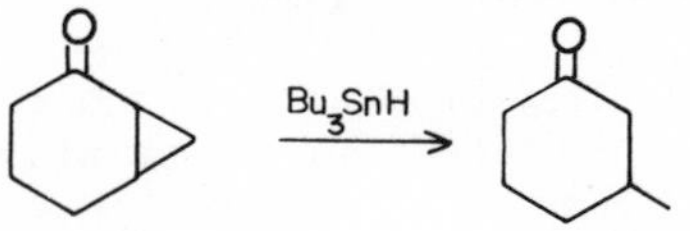

3-Methylcyclohexanone
exclusively

Another transformation of the methylene group of a cyclopropane to a methyl group is found in G.2.

    10.   From Nitro Compounds (see Addenda)

Addenda

B.6.   Rhodium carbonyl, $Rh_6(CO)_{16}$, with carbon monoxide and water under high pressure may be used to reduce α,β-unsaturated to saturated ketones [T. Kitamura et al., Chem. Lett. (3), 203 (1975)].

B.10.   In a reaction of some generality 5-nitro-2-heptanone in THF may be reduced at pH < 1 with aqueous $TiCl_3$ to give 2,5-heptanedione (85%) [J. E. McMurry, Acc. Chem. Res., 7, 281 (1974)].

References

1. E. M. Kaiser, Synthesis, 391 (1972).
2. A. J. Birch, G. Subba Rao, Advances in Organic Chemistry, Vol. 8, 1972, p. 1.
3. Ref. 1, p. 401.
4. A. J. Birch et al., J. Chem. Soc., 1945 (1951).

5.   H. Kwart, R. A. Conley, J. Org. Chem., 38, 2011 (1973).
5a. M. Narisacla, F. Watanabe, J. Org. Chem., 38, 3887 (1973).
6.   D. J. Marshall, R. Deghenghi, Can. J. Chem., 47, 3127
     (1969).
7.   R. Kuhn, I. Butula, Ann. Chem., 718, 50 (1968).
8.   S. Ito et al., Bull. Chem. Soc. Jap., 42, 2068 (1969).
9.   J. G. St. C. Buchanan, P. D. Woodgate, Quart. Rev. (London),
     23, 522 (1969).
10. M. B. Rubin, J. M. Ben-Bassat, Tetrahedron Letters, 3403
     (1971).
11. G. Stork, R. L. Danheiser, J. Org. Chem., 38, 1775 (1973).
12. K. Sakai, K. Watanabe, Bull. Soc. Chem. Jap., 40, 1548
     (1967).
13. J. Trocha-Grimshaw, H. B. Henbest, Chem. Commun., 544
     (1967).
14. M. Sekuja, K. Suzuki, Chem. Pharm. Bull. 18 (8), 1530
     (1970); C. A., 73, 98065 (1970).
15. R. Noyori et al., J. Org. Chem., 37, 1542 (1972).
16. Y. Sasson et al., Synthesis, 359 (1973).
17. V. I. Mel'nikova, K. K. Pivnitskii, J. Org. Chem. USSR
     (Engl. transl.), 6, 2635 (1970).
18. T.-L. Ho, C. M. Wong, Syn. Commun., 3, 237 (1973).
19. J. M. Townsend, T. A. Spencer, Tetrahedron Letters, 137
     (1971).
20. M. Oki et al., Bull. Chem. Soc. Jap., 44, 828 (1971).
21. E. Ghera et al., Chem. Commun., 858 (1973).
22. A. G. Giumanini, Chimia, 464 (1967).
23. A. L. Schultz, J. Org. Chem., 36, 383 (1971); M. Pereye,
     J.-Y. Godet, Tetrahedron Letters, 3653 (1970).
24. J.-Y. Godet, M. Pereyre, Compt. Rend., 273, C1183 (1971).

C. Friedel-Crafts and Related Acylations

The principles that control orientation in acylation were
discussed in 1, 651. An expanded discussion now has been
published [1].

1. From Arenes or Heterocycles and Acylating Agents
    (1, 652)

$$RCOCl + ArH \longrightarrow RCOAr + HCl$$

The capricious character of the nuclei discussed in
acylation reflects the nature of the literature on the subject.
    A review on Friedel-Crafts acylation using little or no
catalyst has been published [2]. When the process does succeed

using very small amounts of $FeCl_3$, $ZnCl_2$, I_2, or Fe at higher temperatures, the simpler conditions and workup seems preferable to the higher yields of the more prosaic procedures. The need for the acylation of alkylbenzenes under Perrier conditions has been reiterated [3].

The checked synthesis of 2-acetyl-6-methoxynaphthalene (45-48%) by the acetylation of 2-methoxynaphthalene has been published [4]. The acetylation of 2-acetylthiophene using more than a molar amount of $AlCl_3$ gives about a 40% conversion to 2,4-diacetylthiophene [5].

Diacylation to give two different acyl groups on mesitylene has been shown to be possible even though acylation for this compound is a reversible process [6]. Either of the two monoacetyl isomers of hemimellitene (1,2,3-trimethylbenzene) has been obtained by the manipulation of conditions, the 5-isomer by Perrier conditions (1, 653), and the 4-isomer by normal acetylation conditions [7].

Naphthalene responds similarly to hemimellitene in that Perrier conditions give the 2-, whereas normal conditions lead to the 1-isomer. In contrast, toluene and tetralin orientation is not changed by either condition. In the diacetylation of naphthalene in CS_2 with $AlCl_3$ at 80°, the 1,6-diacetyl isomer, which was obtained, could be separated from the 1,5-isomer (20%), which was present. A superior synthesis of the 1,6-isomer from 2-acetonaphthone has been published [8]. The acetylation of 2,6-dimethylnaphthalene has been found to occur in the 3-position [9], contrary to previous reports that it was in the 1-position. The acetylation of 2,3-dimethylnaphthalene gave a mixture of 1-, 5-, and 6-monacyl and some diacyl products [10]. The acetylation of chrysene in CH_2Cl_2 gave only 2-acetylchrysene [11].

The chloroacetylation process to form α-chloromethyl ketones is superior on occasion to the chlorination of the related methyl ketone [12].

The acylation of 10-methylphenothiazine gave higher yields with $ZnCl_2$ in $CHCl_3$ rather than with $AlCl_3$ [13].

Intramolecular acylations are common. Thus the sulfide shown gave the corresponding ketone [14] when treated with P_2O_5

4,9-Dihydrothieno-2,3-b-benzothiepin-
4-one, 61%

in refluxing toluene for 6 hr. In the ring closure of 3-methoxy-phenylpropionic acid to 5-methoxyhydrindane, $POCl_3$ added to PPA

raised the yield from 55 to 85% [15]. A study of the yield of
xanthone from o-phenoxybenzoic acid showed that between 1250 and
300 g of PPA/mol of acid, the yield was about 95%, but it dropped
to 80% at 200 g/mol [16].

The rearrangement of benzanilide with PPA at 160-170° gave
about 40% of p-aminobenzophenone [17].

Copper and magnesium benzoate catalyze the benzoylation of
phenyl benzoate to give p-benzoylphenyl benzoate in low yield
[18]. On the other hand, a small amount of trifluoromethylsul-
fonic acid catalyzes the benzoylation of m-xylene [2,19].

It is apparent that acylating agents take a variety of
forms. Thus P_2O_5 in methanesulfonic acid has been recommended
[20]. Another is the mixed anhydride, $RCO_2SO_2CF_3$, which at 60°
acylated arenes as inactive as chlorobenzene [21]. If R were
aliphatic, the anhydride was thermolabile, but if R were phenyl,
the anhydride could be distilled.

The reaction between an aryltrimethylsilane, $XC_6H_4SiMe_3$,
an aroyl chloride, and $AlCl_3$ in CS_2 gives moderate to good
yields of substituted benzophenones, XC_6H_4COAr. In addition,
phenyltrimethylsilane was demonstrated to be about 36 times more
reactive than toluene in electrophilic substitution [22].

For the acylation of heterocycles by acyl free radicals, see
F.5.

2. From Phenols and Nitriles (Hoesch) (1, 654; H.-W., 7,
    Pt. 2a, 1973, p. 389)

$$C_6H_5OH + RC{\equiv}N \xrightarrow{HCl} p\text{-}HOC_6H_4\overset{\overset{\displaystyle NH_2Cl}{\|}}{C}R \xrightarrow{H_2O} p\text{-}HOC_6H_4\overset{\overset{\displaystyle O}{\|}}{C}R$$

This reaction has been of little use of late. Perhaps it
should be mentioned that more than 1 equiv of RCN is needed to
obtain maximum yields because of dimerization or trimerization of
the nitrile [23].

3. From Phenols and Acids or Phenolic Esters (Fries
    Rearrangement) (1, 655; H.-W., 7, Pt. 2a, 1973, p. 285,
    379)

Conditions are still being sought to increase ortho
substitution or the reverse. It should be noted, however, that
the isomers are easily separated because of the intramolecular

hydrogen bonding existing in the ortho isomer. Apparently 1 mol
of ester and 1 mol of AlCl$_3$ in chlorobenzene maximizes the
ortho:para ratio > 1 in the rearrangement of phenyl benzoate
[24]. The fact that one mol each of phenol, AlCl$_3$, and benzoyl
bromide in chlorobenzene gave a lower ortho:para ratio indicates
that the pathways for each reaction differ and that the former
has more intramolecular character. In the rearrangement of 2-
methoxy-5-methylphenyl butyrate, para substitution takes place
with TiCl$_4$, but with no catalyst and irradiation (Photo-Fries)
ortho predominates over para substitution [25].

Some adjustments have been made in Fries conditions, two of
which are restricted to dihydroxybenzenes such as resorcinol.
The first refluxes the organic acid with 0.1 mol of perchloric
acid per mole of phenol and for high boiling acids the temperature is held at 150° for 30 min to give yields of dihydroxy-
acylphenones averaging about 30% [26]. The second fuses the
organic acid with anhydrous ZnCl$_2$ and then treats the mix at
130-140° with the phenol [27]. A third modification utilizes
hydrogen fluoride as both the reagent and solvent [28]. In this
connection it is noteworthy that 3-t-butylphenyl benzoate yields
about 40% of 2-hydroxy-4-t-butylbenzophenone. The t-butyl group
is most frequently lost in an electrophilic reaction of this
nature, as was the case with the Fries rearrangement of 4-t-
butylphenyl benzoate.

5.   From Arenes and N-Substituted Amides (1, 658)

$$RCONMe_2 + ArH \longrightarrow RCOAr$$

Acetanilide has been used to acetylate toluene and chlorobenzene but possesses no important qualifications as a reagent
[29]. However azulene has been acylated usefully by N,N-
dimethylacylamides and phosphorus oxychlorides [30]. In
addition, good yields of p-dialkylaminobenzophenones have been
obtained by the reaction of benzanilides, dialkylanilines, and
phosphorus oxychloride (H.-W., 7, Pt. 2a, p. 277).

7.   From Unsaturated Compounds and Acylating Agents
     (Friedel-Crafts with Aliphatic Compounds, Free Radical
     Reactions) (1, 660; H.-W., 7, Pt. 2a, 1973, p. 427)

$$RCH{=}CH_2 + R'COCl \longrightarrow RCHClCH_2\overset{O}{\overset{\|}{C}}R' \longrightarrow RCH{=}CH\overset{O}{\overset{\|}{C}}R'$$

A review has been published on this reaction [31]. Since
both substrate and products are quite reactive, side reactions
may ensue. Such may be minimized by avoiding an excess of AlCl$_3$

in the following ways: (a) complexing the $AlCl_3$ and the acid chloride in methylene chloride and removing the complex by decantation from the insoluble inorganic halide, and (b) using a nitroalkane solvent which complexes with free $AlCl_3$.   It is also possible to substitute $ZnCl_2$ and an anhydride for $AlCl_3$ and an acid chloride.   Although α,β-unsaturated ketones are the most common product, the β,γ-isomer, which is considered by some to be the kinetically controlled product, is obtained on occasion [32]. Saturated ketones may be obtained by adding a hydride source such as cyclohexane.   Cyclopropenes react with acyl halides and a catalyst to produce open-chain unsaturated ketones [33].

Of interest is the three-component acylation system [34]:

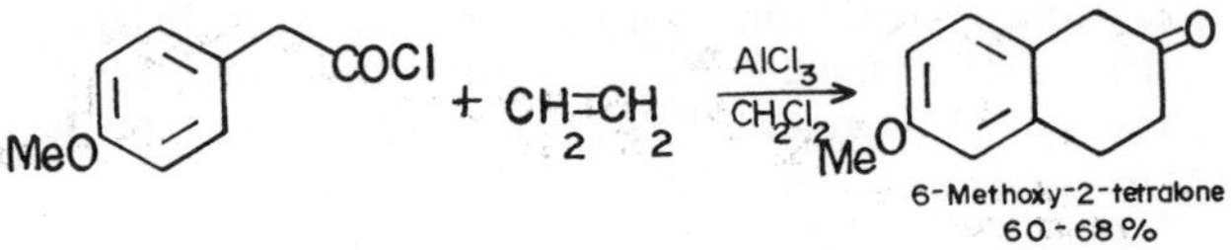

6-Methoxy-2-tetralone
60-68%

Methylene chloride as a solvent is preferred over carbon disulfide.   The tetralone may also be prepared by a free radical pathway [35]:

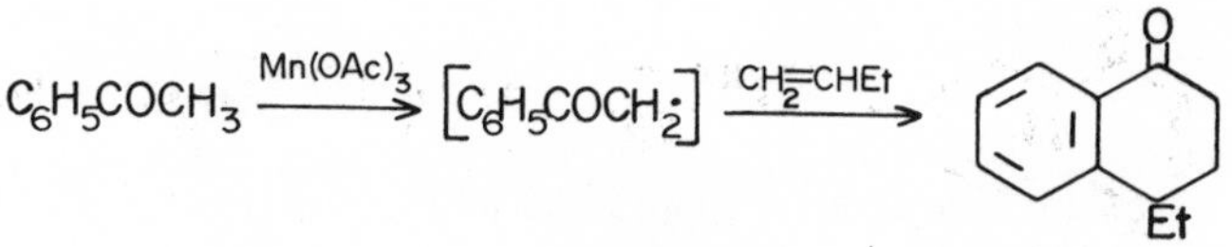

4-Ethyl-1-tetralone was obtained in 49% yield.   If the alkene has more carbon atoms than ethylene and if the reaction is conducted with benzene present, the Nenitsescu reaction, which apparently is quite general, occurs [36,37]:

R=CH$_3$, 62%
R=C$_6$H$_5$, 74%

The phenylhexyl alkyl or aryl ketones are best obtained by adding the suspension of $AlCl_3$ in benzene to the alkene and acyl chloride at $-10°$ to $-15°$.   If benzene is not present, the β,γ-unsaturated ketone may be obtained.   For instance, 1-hexene with acetylium tetrafluoroborate, $CH_3COBF_4^-$, gave 51% of 4-octen-2-one [38].

An unsaturated ester as the alkene and an acylating agent gives a cyclopentenone [39]:

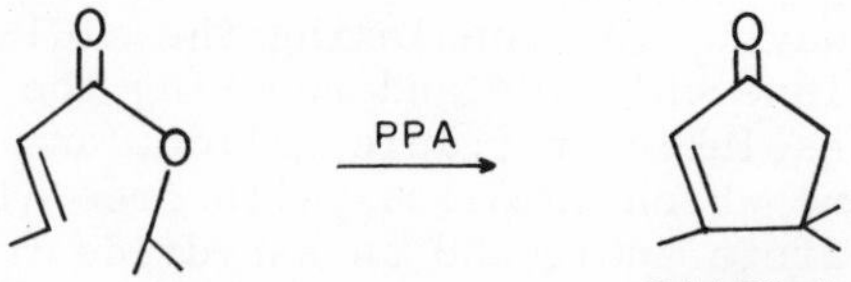

3,4,4-Trimethyl-2-
cyclopentenone
60%

This reaction is like the acylation of propylene except that it
is followed by ring closure.

Three other reactions lead to cyclic mono- or diketones.
One is the succinoylation of isopropenyl acetate to give 2-
acetyl-1,3-cyclopentanedione, 27-30% [40]. This product may be
hydrolyzed to 1,3-cyclopentanedione in 63% yield [40]. The
second is the succinylation of propionyl chloride to form up to
as much as 80% of 2-methyl-1,3-cyclopentanedione if 4 equiv of
propionyl chloride are used [41]. The third is the acylated
alkene obtained from a lactone and PPA [42]:

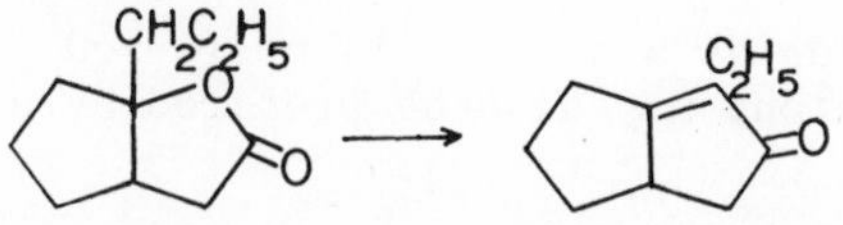

2-Ethylbicyclo[3.3.0]-
octene-1-one-3

Two grams of the substrate gave 0.8 g of the product, 90% pure.
A similar reaction has been carried out by using P_2O_5 dissolved
in methanesulfonic acid [20].

Another acylation is carried out via the enol form of the
ketone [43]:

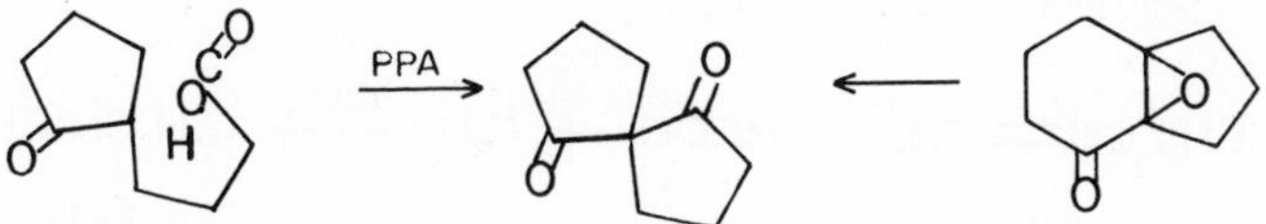

A

Spiro[4.4]nonane-1,6-dione was obtained in 85% yield in the
acid-catalyzed Claisen reaction. The same ketone was obtained
by treatment of A with $BF_3 \cdot Et_2O$.

1-Acetyl-4-chlorocyclooctane (62%) has been prepared from
acetyl chloride and cyclooctene; the product on being dehydrohalogenated by base gave 1-acetylbicyclo[4.2.0]octene [44].
Similarly, benzoylcycloheptatriene (42%) was obtained from cyclo-
heptatriene and benzoyl chloride [45]. Care must be taken with
the product to prevent its rearrangement to desoxybenzoin.

Acylation of acetylenes usually leads to 1,3-diketones. For
example, 1-hexyne and acetylium tetrafluoroborate gave 64% of

valeroyl acetone [46]. Other acylations, some using acid
catalysts, are to be found in F, while unsaturated diketones are
obtained by the gas phase pyrolysis of propargylic esters (see
I.2).

Considerable progress has been made in preparing 1,3-
diketones by acylation with the use of vinyl esters. Thus the
rearrangement of isopropenyl esters with $AlCl_3$ leads to 65-70%
yields of diketones [47]:

$$2 \ RCO_2C\overset{CH_2}{\underset{CH_3}{\diagdown}} \xrightarrow{AlCl_3} (R\overset{O}{\overset{\|}{C}})_2CH_2 \ + \ (CH_3CO)_2CH_2$$

Similarly, cyclohexenyl acetate gave 81% of 2-acetylcyclohexanone
with silver tetrafluoroborate in nitromethane [48]. A vinyloxy-
silane was also used as a substrate for acylation [49]:

$$RC\underset{OSiMe_3}{\overset{\|}{=}}CH_2 \quad \xrightarrow[25°]{R'COCl} \quad RCOCH_2COR'$$

Radical R' was usually a chloromethyl group. With R=$\underline{t}$-butyl and
R'=CCl_3, the yield of diketone was 41%.

Ketoesters were synthesized by using a silane derivative of
a higher oxidation level [50]:

$$CH_2\overset{OEt}{\overset{|}{=}}COSiMe_2CMe_3 + RCH_2COCl \longrightarrow RCH_2\overset{O}{\overset{\|}{C}}CH\overset{OEt}{=}COSiMe_2CMe_3$$

$$\longrightarrow RCH_2COCH_2CO_2Et$$

The reaction is conducted with triethylamine in THF at 25°,
after which the product is hydrolyzed with dilute acid at the
same temperature. Ethyl neohexoylacetate, R=Me_3C, was formed
in 40% (glc) yield, while ethyl crotonylacetate, R=$CH_3CH=CH$, was
isolated in 90% yield, both as dimethylsilyl derivatives.

Typical free radical acylation occurs between acetaldehyde
and camphene [51]:

$$CH_3CHO \ + \ \underset{}{\text{[camphene]}} \ \xrightarrow{(C_6H_5COO)_2} \ \text{[product]}$$

1-(2,2-Dimethyl-3-norbornyl)
propanone, 30-40 %

The substitution of the acyl free radical in heterocycles
has been reviewed [52].

An intramolecular acylation has been carried out with the triphenylphosphine complex of rhodium chlorides [53].

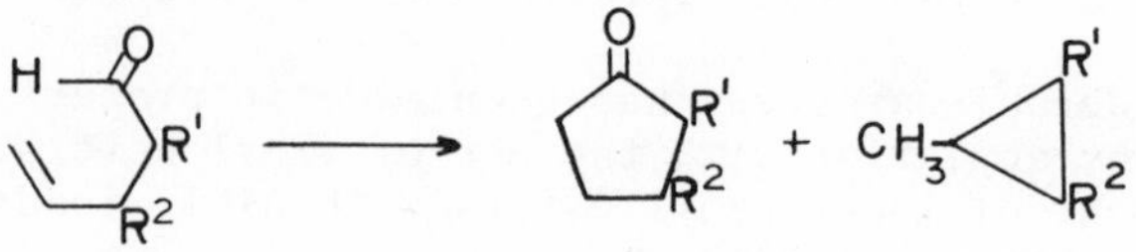

Total yield was 55% (about 15 parts of the cyclopropane and 40 parts of the cyclopentanone).

10.   From Alkenes or Unsaturated Acids and Hydrazoic Acid

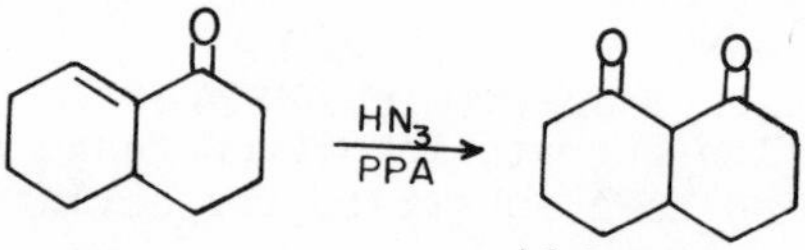

1,8-Decalindione

Instead of the Schmidt reaction, an unusual insertion at the alkene site takes place [54]. It is tempting to believe that a nitrene inserts itself, after which the aziridine tautomerizes to the imino ketone which hydrolyzes to the preceding product.

Addenda

C.1.   Alkaline or alkaline-earth salts, particularly $LiClO_4$, act as a Lewis acid catalyst in the acetylation of anisole. Rates are faster if 1 equiv of salt is used [N. Maigrot-Tournois et al., Compt. Rend., 279, C 911 (1974)].

C.1.   Acylation of 1,3,5-trineopentylbenzene with carbon monoxide in a Friedel-Crafts reaction gives 5,7-dineopentyl-2,2,3-trimethylindanone (60%) [E. Dahlberg et al., Acta Chem. Scand., B28, 1143 (1974)].

C.7.   Hex-5-enoyl chloride with tributylstannane forms the acyl free radical, which adds intramolecularly to the olefin group to give cyclohexanone (36%) [Z. Cekovic, Tetrahedron Letters, 749 (1972)].

C.7.   Treatment of 4-methyl-3-cyclohexenylacetyl chloride with $SnCl_4$ followed by DBN-HMPA gives 4-methylbicyclo [3.2.1] oct-3-en-6-one (50%). Other elimination reagents gave mixtures with the intermediate chloroketone [S. A. Monti, G. L. White, J. Org. Chem., 40, 215 (1975)].

C.10.   The Curtius rearrangement may be invoked with unsaturated acids in sulfuric acid by adding sodium azide,

$$-CH=\overset{|}{C}HCOOH \rightarrow -CH_2-\overset{|}{C}=O$$ [P. Deslongchamps et al., Syn. Commun., 3, 161 (1973); E. W. Garbisch, Jr., J. Wohllebe, J. Org. Chem., 33, 2157 (1968)].

References

1. D. E. Pearson, C. A. Buehler, Synthesis, 455 (1971).
2. D. E. Pearson, C. A. Buehler, Synthesis, 533 (1972).
3. P. G. Olafsson, J. Org. Chem., 35, 4257 (1970).
4. L. Arsenijevic et al., Org. Syn., 53, 5 (1973).
5. V. L. Gol'dfarb et al., J. Org. Chem. USSR, 9, 1975 (1973).
6. P. H. Gore, J. A. Hoskins, J. Chem. Soc., C 517 (1970).
7. L. Friedman, P. J. Honour, J. Am. Chem. Soc., 91, 6344 (1969).
8. D. E. Pearson, C. R. McIntosh, J. Chem. Eng. Data, 9, 245 (1964).
9. P. H. Gore, M. Yusuf, Chem. Commun., 1487 (1969).
10. P. H. Gore et al., J. Chem. Soc., C2502 (1968).
11. W. Carruthers, J. Chem. Soc., 3486 (1953).
12. S. K. Datta, N. K. Bhattacharyya, Syn. Commun., 2, 97 (1972).
13. M. H. Litt et al., J. Org. Chem., 37, 1045 (1972).
14. M. Rajsner et al., Coll. Czech. Chem. Commun., 32, 2854 (1967).
15. A. J. Birch, G. S. R. Subba Rao, Tetrahedron Letters, 2763 (1967).
16. G. Metz, Synthesis, 612 (1972).
17. K Desai, C. M. Desai, J. Indian Chem. Soc., 48, 863 (1971).
18. E. J. Strojny, J. Org. Chem., 34, 3685 (1969).
19. F. Effenberger, G. Epple, Angew. Chem. Intern. Ed. Engl., 11, 300 (1972).
20. P. E. Eaton et al., J. Org. Chem., 38, 4071 (1973).
21. F. Effenberger, G. Epple, Angew. Chem. Intern. Ed. Engl., 11, 299 (1972).
22. C. Eaborn et al., Organomet. Chem. Syn., 1, 151 (1970/71).
23. D. P. N. Satchell, R. S. Satchell, in S. Patai, The Chemistry of the Carbonyl Group, Interscience, New York, 1966, Vol. 1, p. 285.
24. M. J. S. Dewar, L. S. Hart, Tetrahedron, 26, 973 (1970).
25. H. T. J. Chan, J. A. Elix, Australian J. Chem., 26, 1069 (1973).
26. V. V. Mezheritskii, G. N. Dorofeenko, J. Org. Chem. USSR, 5, 502 (1969).

27. H. Schildknecht, H. Schmidt, Z. Naturforsch., 22b, 287
     (1967).
28. J. R. Norell, J. Org. Chem., 38, 1924 (1973).
29. V. I. Minkin, G. N. Dorofeenko, Russ. Chem. Rev., 29, 616
     (1960).
30. K. Hafner, C. Bernhard, Ann. Chem., 625, 108 (1959).
31. J. K. Groves, Chem. Soc. Rev., 1, 73 (1972).
32. J. K. Groves, N. Jones, J. Chem. Soc., C2215 (1968).
33. Ref. 31, p. 91.
34. M. Cadogan et al., Org. Syn., 51, 109 (1971).
35. E. I. Heiba, R. M. Dessau, J. Am. Chem. Soc., 94, 2888
     (1972).
36. A. D. Grebenyuk, N. T. Zaitseva, J. Org. Chem. USSR, 4, 293
     (1968).
37. M. F. Ansell, S. A. Mahmud, Tetrahedron Letters, 4129
     (1971).
38. V. A. Smit et al., Doklady Chem. (Engl. transl.), 203, 272
     (1972).
39. J. M. Conia, M.-L. Leriverend, Bull. Soc. Chim. France,
     2981, 2992 (1970).
40. F. Merényi, M. Nilsson, Org. Syn., 52, 1 (1972).
41. H. Schick, G. Lehmann, J. Prakt. Chem., 38, 391 (1968).
42. J. Ficini, A. Maujean, Bull. Soc. Chim. France, 4392 (1972).
43. H. Gerlach, W. Müller, Helv. Chim. Acta, 55, 2277 (1972);
     T. N. Wheeler et al., J. Org. Chem., 39, 1318 (1974).
44. J. K. Groves, N. Jones, J. Chem. Soc., C2350, 1718 (1969).
45. J. A. Blair, C. J. Tate, J. Chem. Soc., C1592 (1971).
46. V. A. Smit et al., Doklady Chem. (Engl. transl.), 203, 345
     (1972).
47. E. S. Rothman, G. G. Moore, Tetrahedron Letters, 2553
     (1969).
48. N. Ya Grigor'eva et al., Izv. Akad. Nauk Khim. SSSR, Ser.
     Khim., 154 (1973); C. A., 78, 135731 (1973).
49. S. Murai et al., Chem. Commun., 946 (1972).
50. M. W. Rathke, D. F. Sullivan, Tetrahedron Letters, 1297
     (1973).
51. K. Suga, S. Watanabe, Australian J. Chem., 20, 2033 (1967).
52. F. Minisci, Synthesis, 17 (1973).
53. K. Sakai et al., Tetrahedron Letters, 4375 (1972).
54. K. Mitsuhashi et al., Chem. Pharm. Bull. Jap., 17, 1572
     (1969); C. A., 71, 112, 476 (1969).

D. Hydrolysis or Hydration

    1. From Heterocyclic Rings (Meyers) (1, 667)

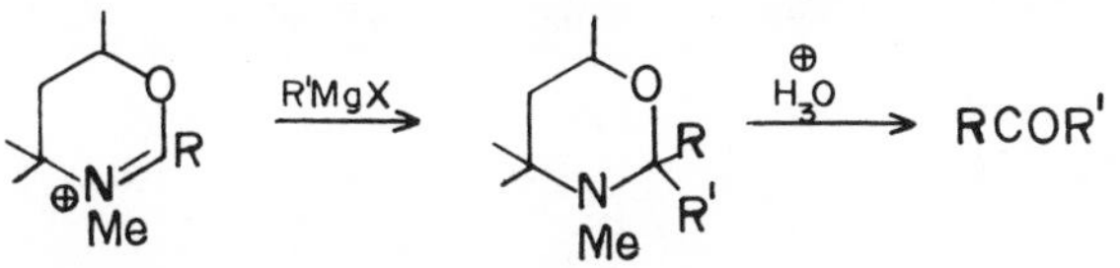

Most of the discussion of the Meyers synthesis is concentrated in Chapter 10, D.2 and D.3, but the preceding reaction has been carried out with 28 examples giving yields of 22-85% [1].   The yield with 2-hexanone, R = Me, and R' = Bu, was 22%.   Highly alkylated ketones may be formed by the following procedure [2,3]:

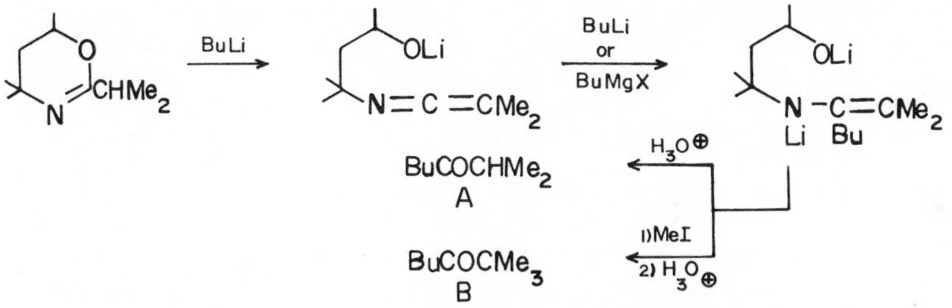

The yield of 2-methylheptanone (A) was 73%, and butyl t-butyl ketone (B) was 60%.   Reference 3 utilizes the five- and six-membered heterocycles.   Still another variation is the 1,4-addition to the heterocycle as follows [4].

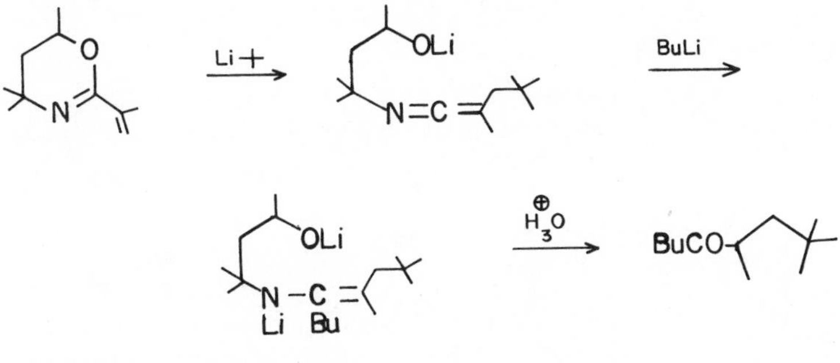

The 2,2,4-trimethyl-4-nonanone was obtained in 77% overall yield with 17 other examples given.

The superior hydrolysis of 2,5-dialkylfurans to diketones as compared to unsubstituted furans in yielding aldehydes is illustrated [5]:

cis-Undec-8-en-2,5-dione was obtained in about 50% yield by hydrolysis with sulfuric in aqueous acetic acid.

One of the most studied reactions of recent times is the cleavage of substituted dithiolane or 1,3-dithiane [6]. Need became apparent for a facile conversion into the carbonyl compound:

The simplest, most economical, and apparently among the most facile reactions is that of C or of oxathiolanes with chloramine T, TsNClNa, in aqueous methanol [7]. A second promising agent, if expense is not considered, is thallium trifluoroacetate, which reacts almost instantaneously with dithianes (6 ex.) (77-95%) (see Chapter 10, D.3). Ceric ammonium nitrate appears to behave somewhat similarly [8]. The new reagents proposed to supplant the older ones, Br_2-CH_2COOH or Hg^{++} with or without BF_3 [7], are Ag_2O in CH_3OH [9], CH_3I followed by hydrolysis [10,11], triethyloxonium tetrafluoroborate-H_2O_2 [12], $CuCl_2$-CuO in 99% aqueous acetone [13], and NBS or NCS [14].

2-Carbethoxy-1,3-dithiane, D, makes a good precursor of carbonyl compounds if lithiumalkyls are to be avoided [15]:

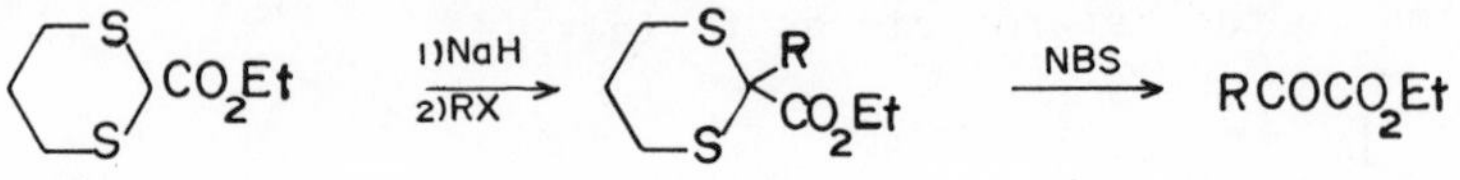

1,3-Dithenium fluoroborate forms Diels-Alder adducts, which are isomerized first to cyclopropanes and then to cyclopentenones as shown [16]:

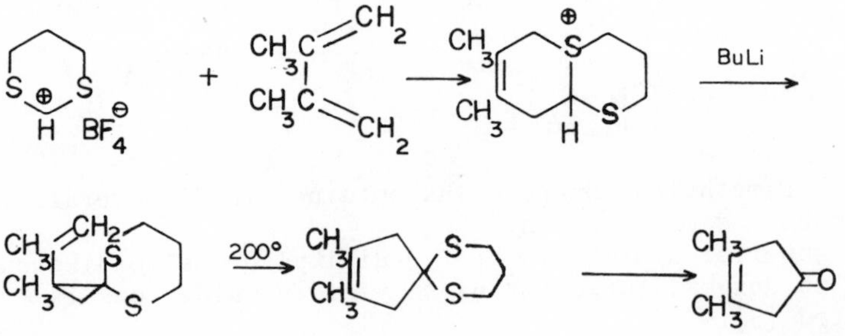

3,4-Dimethylcyclo-pent-3-enone,excellent yield

3. From Vinyl Halides, Ethers, Amines, or the Like
    (1, 669; H.-W., 7, Pt. 2a, 1973, p. 813)

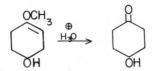

The preceding reaction has become a source of cyclic
ketones. The substrate is derived from the reduction of p-
methoxyphenol with lithium and liquid ammonia [17]. In the
reduction the lithium:phenol ratio is critical at about 11:1,
but the hydrolysis may be carried out easily with aqueous acid
to give an 89% overall yield of 4-hydroxycyclohexanone. A
similar hydrolysis with 10% aqueous H_2SO_4 on 3-ethoxy-2-cyclo-
hexenol gave 2-cyclohexenone in 62-75% yields [18]. α-Methyl-β-
phenylvinyl methyl thioether gave phenylacetone (84%) when
treated with $HgCl_2$ in aqueous acetonitrile [19]:

$$C_6H_5CH=\overset{\overset{\displaystyle Me}{|}}{C}SMe \longrightarrow C_6H_5CH_2COMe$$

An interesting hydrolysis is one used in the preparation of
1-phenyl-4-phosphoranone [20]:

$$\underset{\substack{| \\ C_6H_5}}{\overset{\displaystyle NH_2}{\bigcirc}}\!\!CN \quad\xrightarrow[\text{reflux}]{6\,N\ HCl}\quad \underset{\substack{| \\ C_6H_5 \\ 68-69\%}}{\overset{\displaystyle O}{\bigcirc}}$$

4. From Acetylenes and the Like (Rupe and Meyer-Schuster
    Rearrangements) (1, 670; H.-W., 7, Pt. 2a, 1973, p. 816)

$$RC\equiv CH \longrightarrow RCOCH_3$$

Since acetylenic alcohols are so easily obtained, they
become important sources of ketones by hydrolysis. Fortunately,
the field has benefited by the publication of two papers, one a
review [21] and the other a clarification of the somewhat
variable yields of the types of products [22]. When ethynylcyclohexanol is mixed with 80% formic acid 1:5 ratio and heated
for several hours, 1-acetylcyclohexene (A) is obtained as the main
product with small amounts of ethynylcyclohexene (B) and cyclohexylideneacetaldehyde (C) [22]:

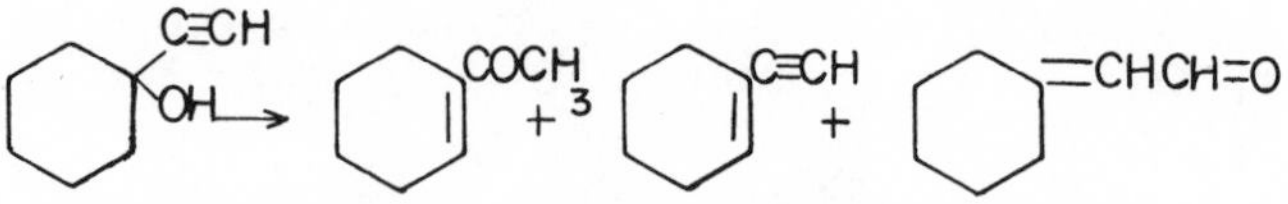

A   B   C

The yields of A, B, and C depend on the experimental conditions. For instance, the yield of A can be raised considerably by refluxing for 2-8 hr.

The acid-catalyzed Rupe rearrangement may be represented by the sequence [23] (Ex. a).

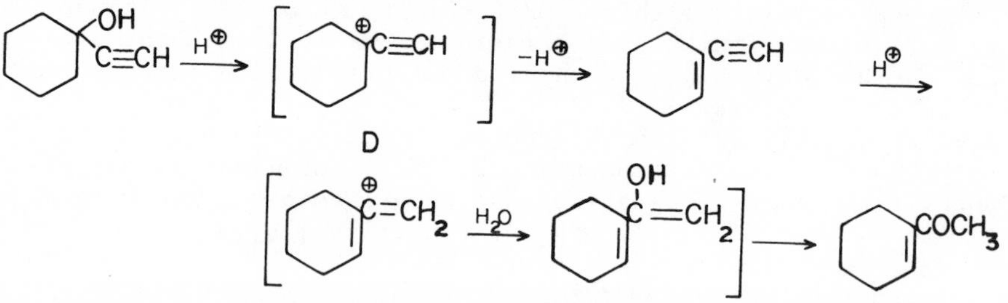

The Meyer-Schuster rearrangement, closely related to the Rupe rearrangement, is an isomerization by a 1,3 shift of secondary and tertiary α-acetylenic alcohols also to α,β-unsaturated ketones. It may be represented by the sequence:

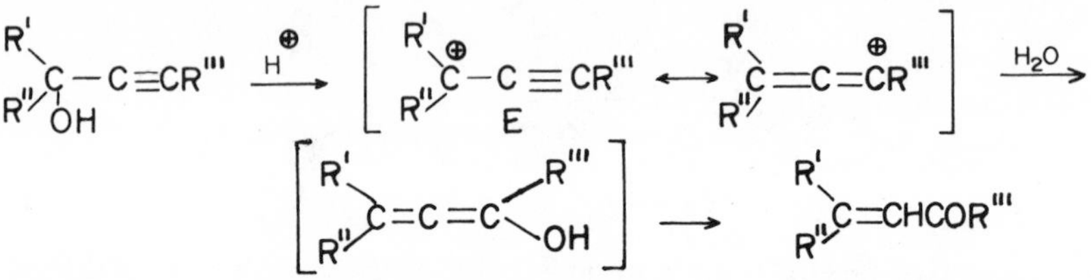

The difference in the R and M-S rearrangements appears to be due to the fact that the alkynyl cation E cannot expel a proton, as in the case of the alkynyl cation D in the R mechanism. Acid catalysts, of which formic acid is the most common, are employed in both rearrangements.

The recent literature appears to concentrate on the synthesis of decalones and octalones as shown by two examples:

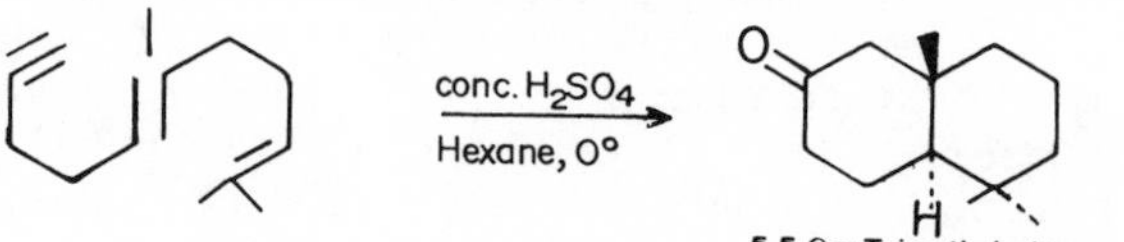

conc. H$_2$SO$_4$
Hexane, 0°

[24]

5,5,9—Trimethyl- *trans*-
2-decalone, 64 %,
93 % pure

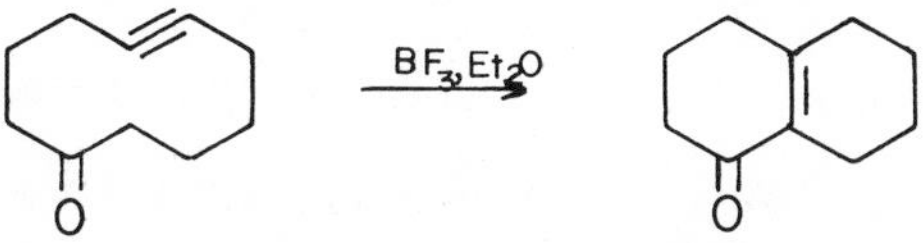

[25]

9,10-Octalone-1
almost quant.

Both solvation and metathesis are accomplished in the synthesis [26]:

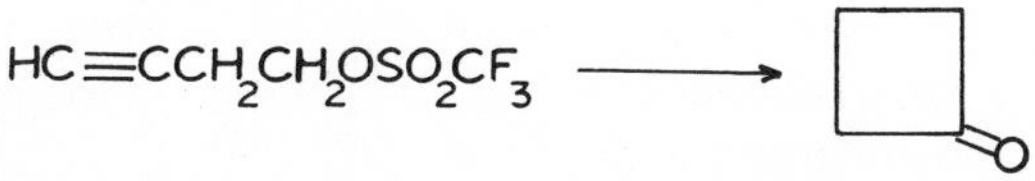

$$HC \equiv CCH_2CH_2OSO_2CF_3 \longrightarrow$$

By using an aqueous solution of trifluoroacetic acid and its salt, cyclobutanone was prepared in 31-36% yield.

An interesting preparation of a dichloromethyl ketone is described in Ex. b [27]:

$$C_4H_9C \equiv CH \xrightarrow[\text{2 MeOH}]{\text{2 NCS}} C_4H_9\overset{(OMe)_2}{\underset{|}{C}}CHCl_2 \xrightarrow{\overset{\oplus}{H_2O}} C_4H_9COCHCl_2$$

For the conversion of acetylenes into α,β-unsaturated carbonyl systems, see A.19.

a. Preparation of 1-Acetylcyclohexene [28]

The yield was 61% from 1-ethynylcyclohexanol and P_2O_5, 85% by the use of an acidic resin, and 49% by the use of 90% formic acid.

b. Preparation of 1,1-Dichloro-2-hexanone [27]

To NCS, 0.2 mol in 300 ml of MeOH, 1-hexyne, 0.1 mol was added in 30 min. After an induction period, the temperature increased to 42° and the mixture was cooled to 25° and stirred for an additional several hr. The methanol was removed by rotary evaporation and replaced by pentane. The succinimide was filtered off, after which the ketal was concentrated and distilled to give an 80% yield. The dichlorohexanone (68%) was recovered from the ketal by stirring for 24 hr in 1:1 concentrated $HCl-H_2O$.

5. From Ketone Derivatives (1, 671; H.-W., 7, Pt. 2a, 1973, p. 799)

$$R_2C = NOH \longrightarrow R_2C = O$$

Section D.1 of Chapter 10 should be consulted for the many methods of converting both aldehyde and ketone derivatives into carbonyl compounds. Those described here are best suited for ketone derivatives. The best of these reagents are thallium(III) nitrate (TTN) in methanol for ketoximes [29] and sodium hypochlorite, ordinary bleach, for tosyl hydrazones [30]. Thallium(III) nitrate appears to react instantly at 25°, may be used in aqueous solution if perchloric acid is added, and is applicable as well to semicarbazones and phenylhydrazones, but not to those derivatives that possess phenolic or aromatic amino groups. Sodium hypochlorite gave 60-85% yields with ketones but poor yields with aldehyde tosylhydrazones. Sodium bisulfite in refluxing aqueous ethanol gives 77-98% yields of ketones from oximes [31]. The hydrolysis of 2,4-dinitrophenylhydrazones with p-toluenesulfonic acid hydrate in chloroform precipitates the 2,4-dinitrophenylhydrazine sulfonic acid salt, a result that permits the ketone to be isolated from the filtrate [32]. Possibly, some 2,4-DNPHs of aldehydes, which do not polymerize in acid, could be hydrolyzed in this way. Although not necessary on every occasion, it is claimed that to obtain pure hydrazones it is best to convert the ketone into the N,N-dimethylhydrazone and then react this derivative with hydrazine [33]. The synthesis of some aromatic imines such as $(C_6H_5)_2C{=}NH$, which are extremely sensitive to hydrolysis, was accomplished by solvolyzing the imine complex from benzonitrile and phenylmagnesium bromide with methanol [34].

Considerable attention has been devoted to the hydrolysis of ketals, for which dilute acid appears to be the preferred reagent. For instance, the ketal of bicyclo[2.2.1]hepten-7-one is hydrolyzed to the ketone with 5% aqueous H_2SO_4 by stirring at room temperature for 20 hr [35]. Sometimes alcohol is added to produce homogeneity. Most unusual methods of deketalization are found in D.1, but some are listed below. The most interesting involves the use of trityl fluoroborate in methylene chloride [36]. Recognition of the mechanism has led to application to the oxidation of sugar ketals:

For instance, the conversion as shown was achieved:

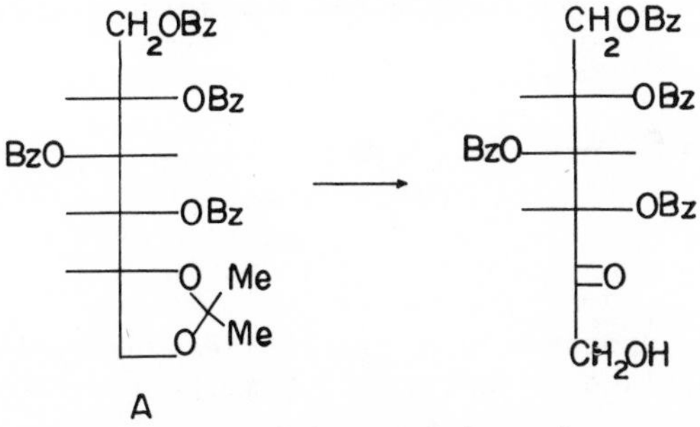

A

The product, 3,4,5,6-tetra-O-benzoyl-L-sorbose, was obtained in 50% yield by stirring A in tritylfluoroborate-CH_2Cl_2 for 5 hr.

A class of spin labels, shown in the case of doxyl, may be converted into ketones by treatment with NO_2 [37]:

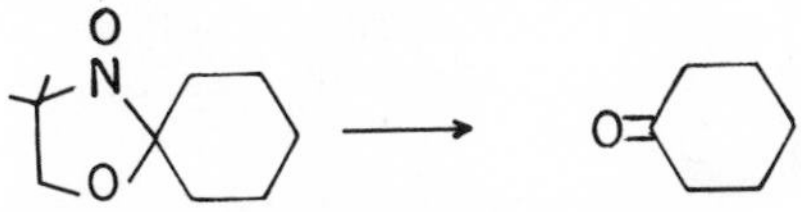

Cyclohexanone was obtained in 95% yield.

6. From β-Ketoesters (Acetoacetic Ester Synthesis) (**1**, 673)

β-Ketoesters have been decarboxylated to the related ketones by lithium iodide in collidine [38].

7. From Methyl Methylthiomethyl Sulfoxides and Related Types

In realization of the fact that higher oxidized states of sulfur in thioacetals or thioketals alkylate and hydrolyze more readily, growing use of thioketal monoxides is made in synthesizing ketones. For example, not only aldehydes (Chapter 10, D.7), but ketones, α-hydroxycarbonyl compounds, and α-dicarbonyl compounds are synthesized as indicated [39]:

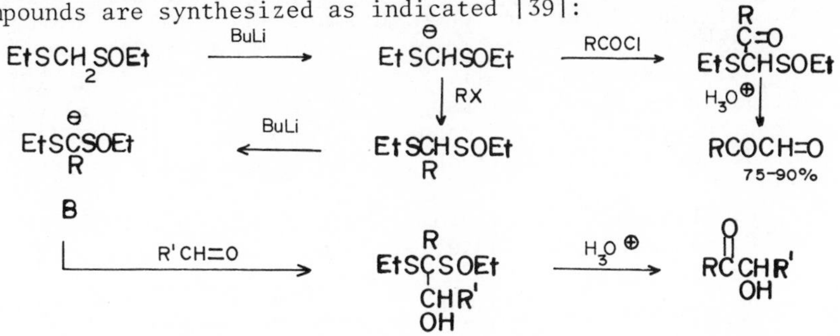

To prepare 1,4-dicarbonyl compounds the anion is treated with an α,β-unsaturated compound:

$$\text{CH}_2\!\!=\!\!\text{CHCOR}' \xrightarrow{\text{B}} \underset{\underset{\text{CH}_2\text{CH}_2\text{COR}'}{|}}{\overset{\overset{\text{R}}{|}}{\text{EtSCSOEt}}} \longrightarrow \text{RCOCH}_2\text{CH}_2\text{COR}'$$

Other Michael condensations may be found in G.3.

It has been proposed that the facile hydrolysis of thioketal monoxides takes place by an oxidative mechanism [40].

Addendum

D.5.   2,4-Dinitrophenylhydrazones are reduced and hydrolyzed to the ketone with titanous trichloride in glyme [J. E. McMurry, M. Silvestri, J. Org. Chem., 40, 1502 (1975)].

References

1. A. I. Meyers, E. M. Smith, J. Org. Chem., 37, 4289 (1972).
2. A. I. Meyers et al., J. Org. Chem., 38, 2129 (1973); J. Am. Chem. Soc., 93, 2314 (1971).
3. J.-E. Dubois, C. Lion, Compt. Rend., 274, C303 (1972); Tetrahedron, 29, 3417 (1973).
4. A. I. Meyers et al., J. Org. Chem., 38, 2136 (1973).
5. G. Buchi, H. Wüest, J. Org. Chem., 31, 977 (1966).
6. D. Seebach, Synthesis, 17 (1969); D. Seebach, A. K. Beck, Org. Syn., 51, 76 (1971).
7. H. Wynberg et al., Tetrahedron Letters, 3445, 3449 (1971); Syn. Commun., 2, 7 (1972).
8. T.-L. Ho et al., Chem. Commun., 791 (1972).
9. D. Gravel et al., Chem. Commun., 1323 (1972).
10. H.-L. W. Chang, Tetrahedron Letters, 1989 (1972).
11. M. Fetizon, M. Jurion, Chem. Commun., 382 (1972).
12. T. Oishi et al., Tetrahedron Letters, 1185 (1972).
13. K. Narasaka et al., Bull. Soc. Chem. Jap., 45, 3724 (1972).
14. E. J. Corey, B. W. Erickson, J. Org. Chem., 36, 3553 (1971).
15. E. L. Eliel, A. A. Hartmann, J. Org. Chem., 37, 505 (1971).
16. E. J. Corey, S. W. Walinsky, J. Am. Chem. Soc., 94, 8932 (1972).
17. P. Radlick, H. T. Crawford, J. Org. Chem., 37, 1669 (1972).
18. W. F. Gannon, H. O. House, Org. Syn., Coll. Vol. 5, 294 (1973).
19. E. J. Corey, J. I. Shulman, J. Org. Chem., 35, 777 (1970).

20. T. E. Snyder, D. L. Morris, Org. Syn., 53, 98 (1973).
21. S. Swaminathan, K. V. Narayanan, Chem. Rev., 71, 429 (1971).
22. R. W. Hasbrouck, A. D. A. Kiessling, J. Org. Chem., 38, 2103 (1973).
23. M. S. Newman et al., J. Am. Chem. Soc., 75, 4740 (1973); J. Org. Chem., 26, 727 (1961).
24. G. Ohloff et al., Helv. Chim. Acta, 56, 1414 (1973).
25. C. E. Harding, M. Hanack, Tetrahedron Letters, 1253 (1971).
26. M. Hanack et al., Org. Syn., 54, 84 (1974).
27. S. F. Reed, Jr., J. Org. Chem., 30, 2195 (1965).
28. R. S. Monson, Advanced Organic Synthesis, Academic, 1971, p. 129.
29. A. McKillop et al., J. Am. Chem. Soc., 93, 4918 (1971).
30. T.-L. Ho, C. M. Wong, J. Org. Chem., 39, 3453 (1974).
31. S. H. Pines et al., J. Org. Chem., 31, 3446 (1966).
32. M. E. N. Nambudiry, G. S. K. Rao, Australian J. Chem., 24, 2183 (1971).
33. G. R. Newkome, D. L. Fishel, Org. Syn., 50, 102 (1970).
34. P. L. Pickard, T. L. Tolbert, Org. Syn., Coll. Vol., 5, 520 (1973).
35. P. G. Gassman, J. L. Marshall, Org. Syn., Coll. Vol., 5, 91 (1973).
36. D. H. R. Barton et al., Chem. Commun., 861 (1971); J. Chem. Soc., Perkin Trans., I, 542 (1972).
37. J. A. Nelson et al., Chem. Commun., 1580 (1971).
38. F. Elsinger, Org. Syn., Coll. Vol., 5, 76 (1973).
39. J. L. Herrmann et al., Tetrahedron Letters, 4707, 4711, 4715 (1973).
40. H. Nieuwenhuyse, R. Louw, Tetrahedron Letters 4141 (1971).

E. Rearrangements

1. From Pinacols (1, 677; H.-W., 7, Pt. 2a, 1973, p. 927)

For a recent discussion, see Chapter 10, E.1.
Two fairly recent examples of the classical pinacol rearrangement are shown [1]:

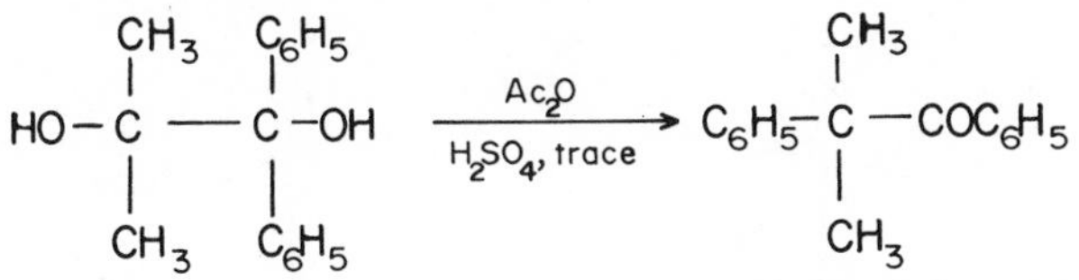

Dimethyldesoxybenzoin, 60%

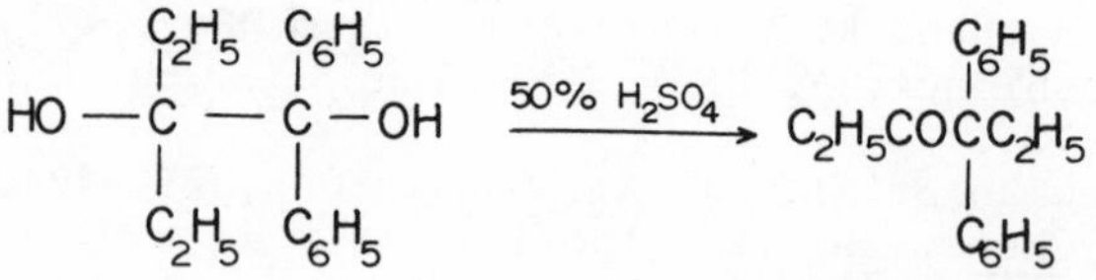

$$HO-\underset{\underset{C_2H_5}{|}}{\overset{\overset{C_2H_5}{|}}{C}}-\underset{\underset{C_6H_5}{|}}{\overset{\overset{C_6H_5}{|}}{C}}-OH \xrightarrow{50\% \ H_2SO_4} C_2H_5CO\underset{\underset{C_6H_5}{|}}{\overset{\overset{C_6H_5}{|}}{C}}C_2H_5$$

4,4-Diphenyl-3-hexanone
77.6%

Some control of the leaving group has been described [1] in (1, 677) and in the following illustration [2]:

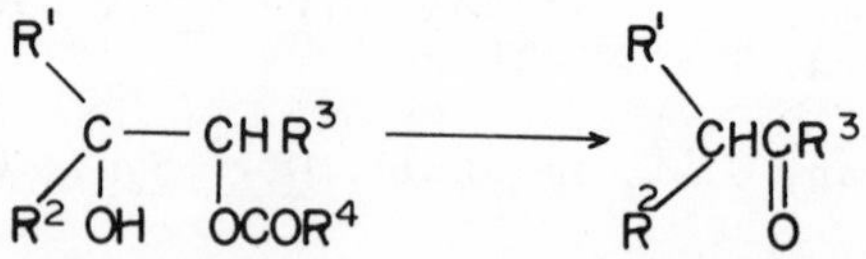

$$\underset{\underset{R^2 \ OH}{}}{\overset{\overset{R^1}{\diagdown}}{C}}-\underset{\underset{OCOR^4}{|}}{CHR^3} \longrightarrow \overset{\overset{R^1}{\diagdown}}{\underset{R^2}{CH}}\underset{\underset{O}{\|}}{CR^3}$$

53-95%

This control occurs except when R^4 is the methyl group, in which case the group transfers to the adjacent hydroxyl group via an orthoacetate. Rearrangement then occurs to give what appears to be an abnormal product but actually is normal if one considers the above-mentioned transfer. The reaction is applicable to acetates, benzoates, and p-nitrobenzoates and if an aryl group is attached to the secondary carbon, some of the aldehyde is also produced.

2. From Allyl Alcohols (1, 678; H.-W., 7, Pt. 2a, 1973, p. 893)

A rearrangement occurs in 1-acetoxy-7-methylene bicyclo[3.2.1]octane when it is treated with THF-HCl [3]:

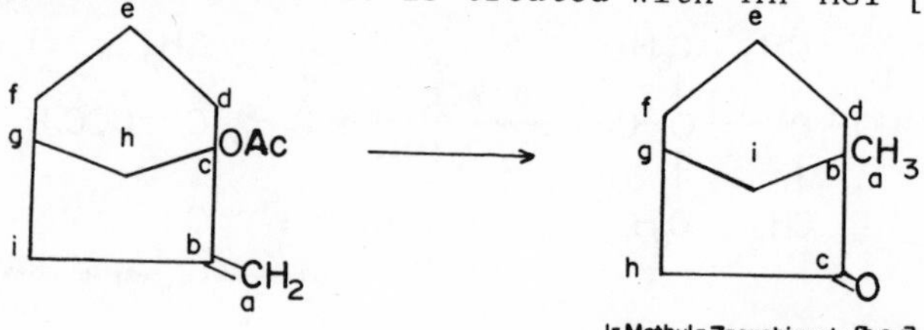

l-Methyl-7-oxobicyclo[3.2.l]octane
unstated yield

3. From Ethylene Oxides (Including the Trost Synthesis)
    (1, 679; H.-W., 7, Pt. 2a, 1973, pp. 950, 1000)

$$ArCH \overset{O}{\underset{}{\diagdown}} CHAr \longrightarrow ArCOCH_2Ar$$

Ring enlargement is a common occurrence in this rearrangement. Thus Conia and co-workers [4] from the oxospiropentane,
in the sequence given, obtained cyclobutanone:

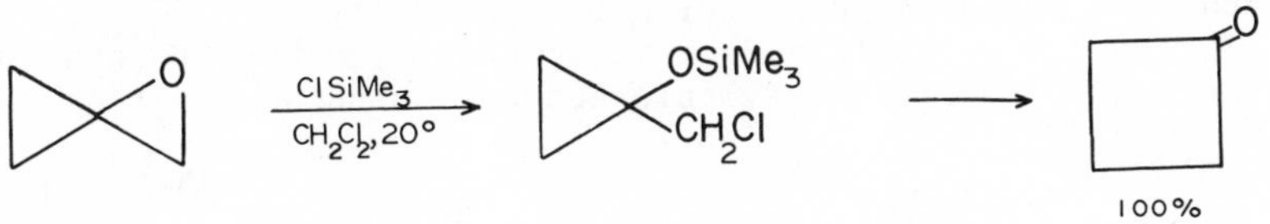

Hawkins and Large [5] in a similar rearrangement converted 1,1'-
epoxy-bicyclohexyl-2-one by treatment with $SbCl_5$-SO_2 with ring
enlargement into the spirodiones shown:

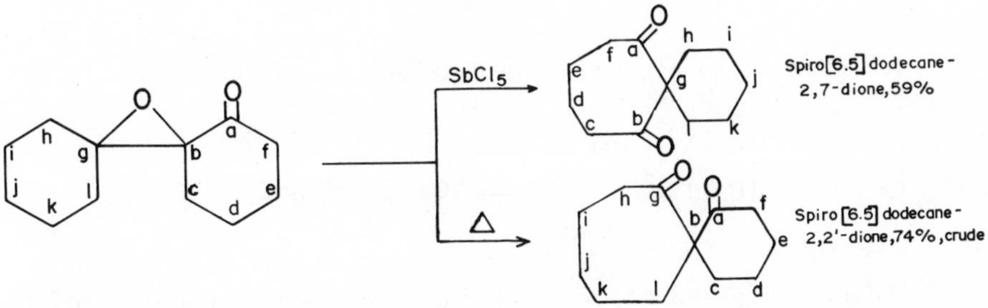

Spiro[6.5]dodecane-
2,7-dione,59%

Spiro[6.5]dodecane-
2,2'-dione,74%,crude

Williams and co-workers [6] obtained an isomeric diketone by
heating.
    The attachment of a spiro ring at the carbonyl carbon has
been accomplished via an ethylene oxide as indicated in the
sequence [7]:

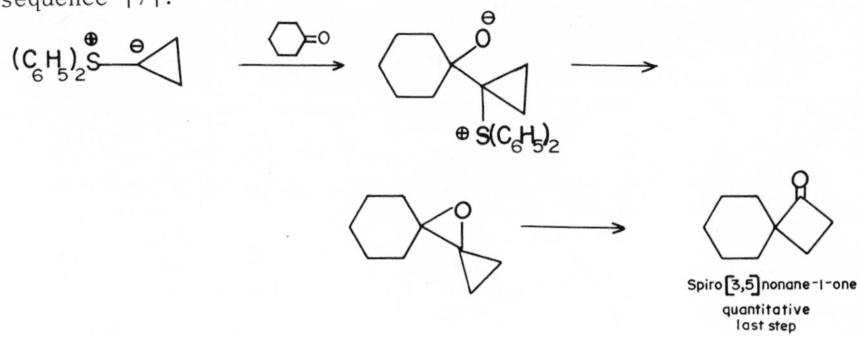

Spiro[3,5]nonane-1-one
quantitative
last step

A number of ketoepoxides have been rearranged to unsaturated ketones [8] by Hart and co-workers.

Epoxidation of a vinylsilane gives the epoxide that hydrolyzes to the ketone [8a].

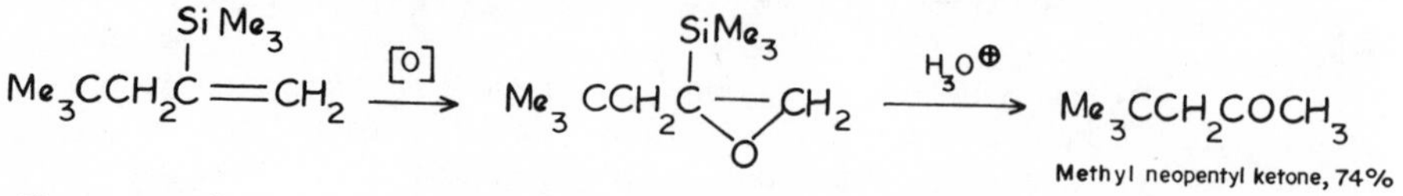

Methyl neopentyl ketone, 74%

The reaction possesses considerable versatility.

3-Keto-4,5-epoxysteroids have been cleaved by p-toluenesulfonylhydrazine to give acetylenic ketones as shown [8b]:

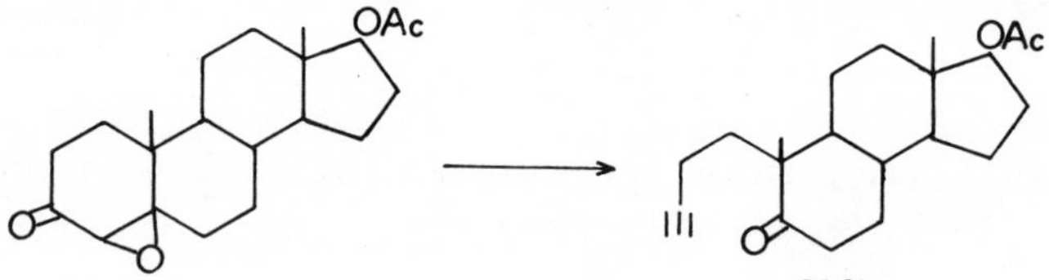

83 %

4.   From Amino Alcohols (or Diamines or Halohydrins)
      (1, 681; H.-W., 7, Pt. 2a, 1973, p. 946)

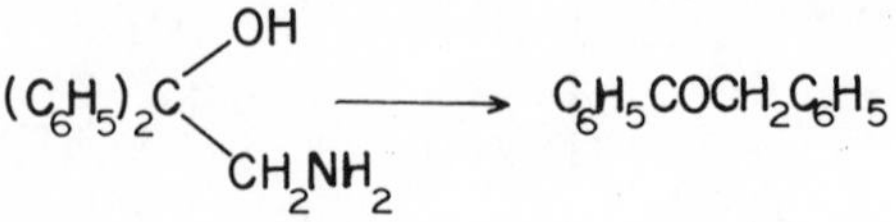

The homologation of cyclic ketones has been accomplished, through the use of the Tiffeneau-Demjanov ring expansion, by the use of isocyanomethyllithium [9]. Thus cyclohexanone was transformed into cycloheptanone by the sequence:

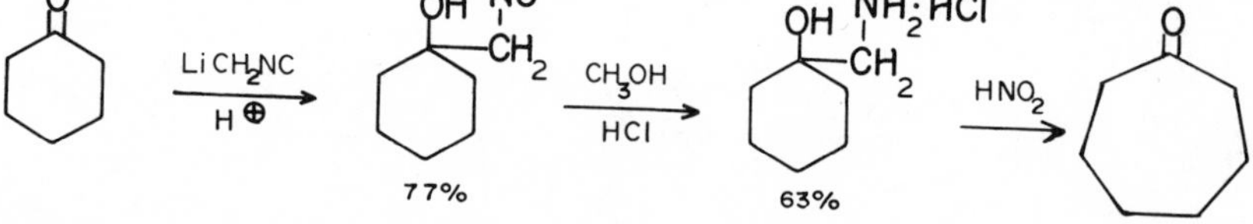

77%   63%   71.5%

An interesting agent for the rearrangement of a bromohydrin is the Grignard reagent [9a]:

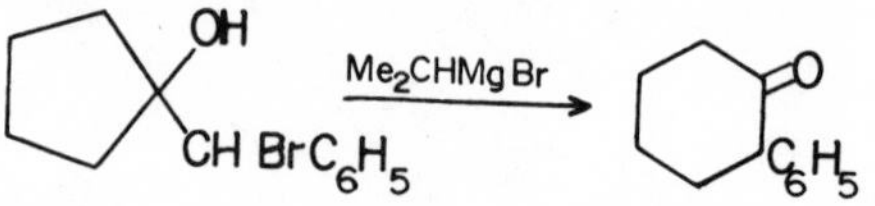

The 2-phenylcyclohexanone was obtained in 80% yield.

6.   From Aldehydes or Ketones (1, 684)

Rearrangements among aldehydes and ketones are common.   For
example, α-acetoxy aldehydes are converted into acetoxy ketones
by treatment with CH_3COOH containing a drop of H_2SO_4 [10]:

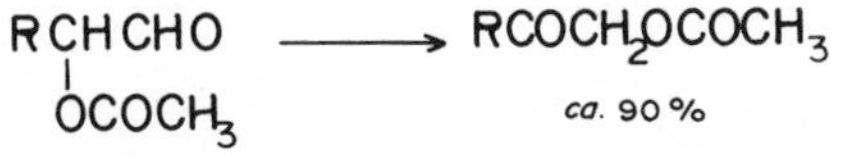

$$RCHCHO \longrightarrow RCOCH_2OCOCH_3$$
$$\underset{OCOCH_3}{|}$$

ca. 90%

The transposition of ketone A into ketone B has been
accomplished by a nonoxidative method as indicated [11]:

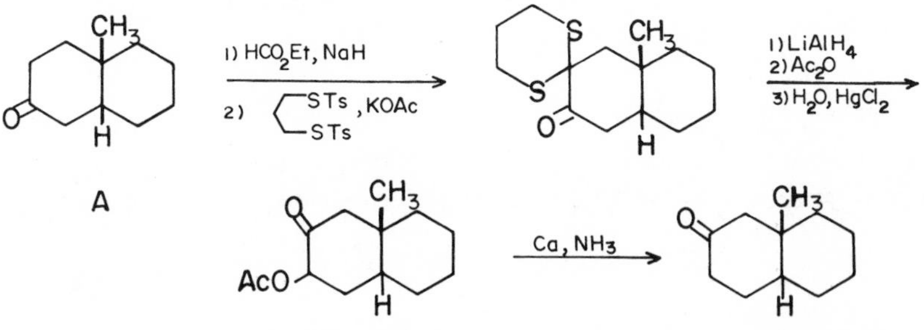

A review of the acid-catalyzed rearrangements of α,β-
unsaturated ketones is available [12].   One illustration from
this review follows:

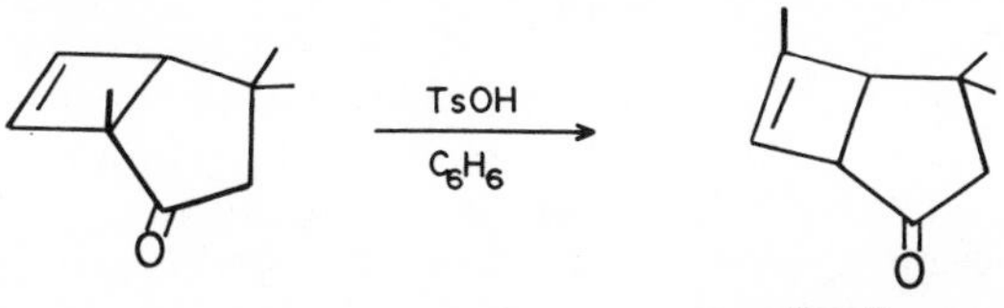

Bicyclo[3.2.0]hept-6-en-2-one
nearly quantitative

In some isomerizations of bicyclic compounds mixtures of
products are formed.   Thus bicyclo[6.1.0]nonanone-4 with
BF_3-Et_2O gave two principal products [13]:

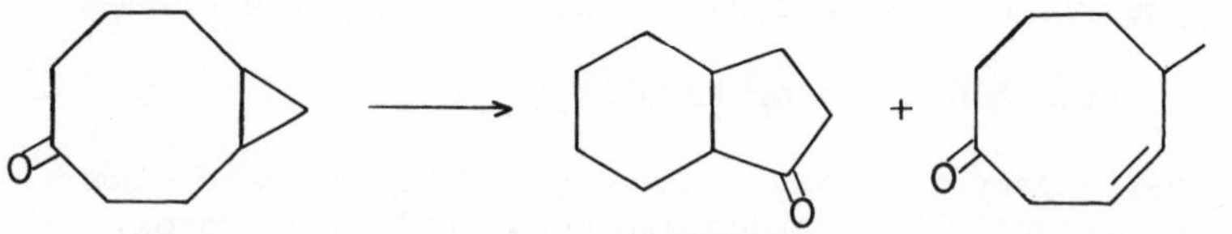

1-Hydrindanone 55 % 5-Methyl-3-cyclooctenone 40 %

7. From Unsaturated Ethers

Allyl phenyl ether with 2 equiv of n-butyllithium in DME-hexane isomerizes to propiophenone [14]:

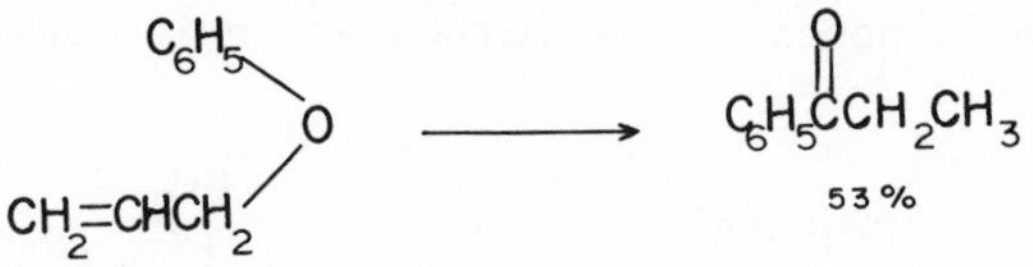
C_6H_5
O
$CH_2{=}CHCH_2$
$C_6H_5CCH_2CH_3$
53 %

The mechanism appears to be:

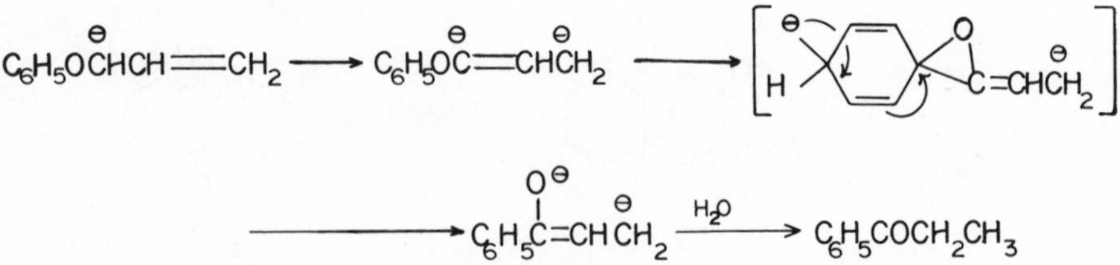

Other canonical forms probably are involved in this Wittig type rearrangement.

8. Hydroboration Methods

The synthesis of ketones via hydroboration has been reviewed in two books [15]. For similar methods to prepare aldehydes, see Chapter 10, E.8. The use of organoboranes in the synthesis of ketones via rearrangement is such an extensive subject that a brief treatment only will be given. Much more detail will be found in other works [15]. Brown's book is particularly important in that the techniques which are essential in many syntheses are described.

Typical syntheses follow:

$$R_3B + CO \xrightarrow{H_2O} \left[\begin{matrix} RB & - & CR_2 \\ | & & | \\ OH & & OH \end{matrix}\right] \xrightarrow[NaOH]{H_2O_2} \underset{80\text{-}90\ \%}{R_2CO} \qquad [16,17]$$

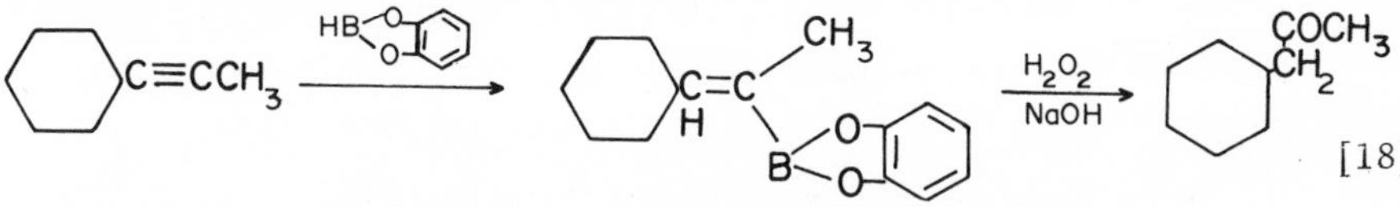

$$\text{(cyclohexyl)}C\equiv CCH_3 \xrightarrow{HB\underset{O}{\overset{O}{<}}\text{(benzene ring)}} \text{(cyclohexyl)}\underset{H}{\overset{CH_3}{C=C}}\underset{B\overset{O}{\underset{O}{<}}}{} \xrightarrow[NaOH]{H_2O_2} \text{(cyclohexyl)}\overset{COCH_3}{\underset{CH_2}{}} \quad [18]$$

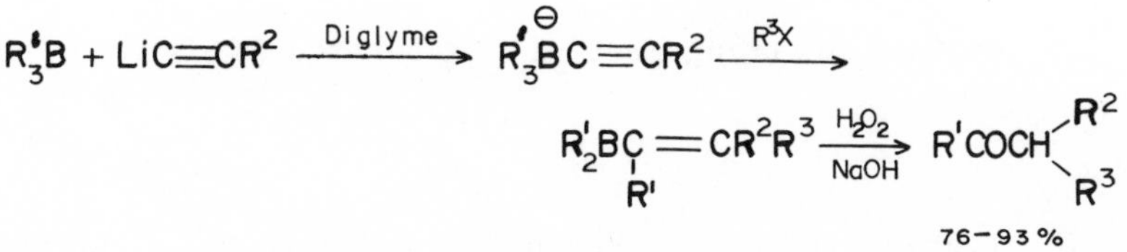

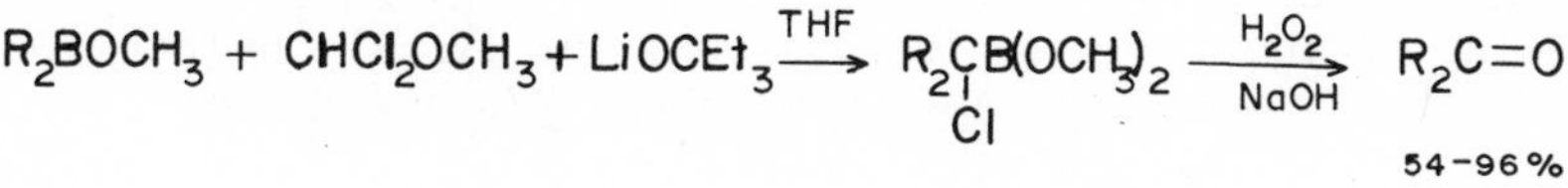

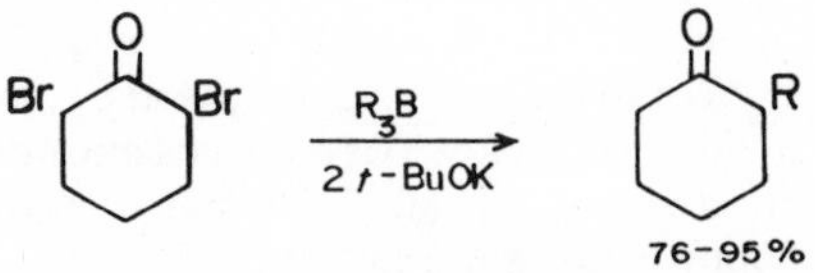

$$R_3'B + LiC\equiv CR^2 \xrightarrow{Diglyme} R_3'\overset{\ominus}{B}C\equiv CR^2 \xrightarrow{R^3X}$$

$$R_2'B\underset{R'}{\overset{}{C}}=CR^2R^3 \xrightarrow[NaOH]{H_2O_2} R'COCH\underset{R^3}{\overset{R^2}{<}} \quad [19]$$

76-93%

$$R_2BOCH_3 + CHCl_2OCH_3 + LiOCEt_3 \xrightarrow{THF} R_2\underset{Cl}{\overset{}{C}}B(OCH_3)_2 \xrightarrow[NaOH]{H_2O_2} R_2C=O \quad [20]$$

54-96%

$$\underset{Br \text{(cyclohexanone)} Br}{} \xrightarrow[2\,t-BuOK]{R_3B} \underset{\text{(cyclohexanone)} R}{} \quad [21]$$

76-95%

$$R_3B + LiC\equiv CR' \longrightarrow R_3\overset{\ominus}{B}C\equiv CR' \xrightarrow{BrCH_2COC_6H_5}$$

$$R_2B\overset{R}{\underset{CH_2COC_6H_5}{C=C}}\overset{R'}{} \xrightarrow{O} RCOCHR'CH_2COC_6H_5 \quad [22]$$

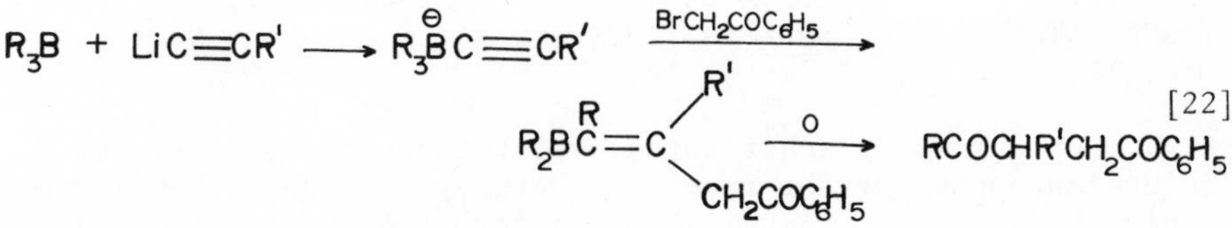

When R and R' = hexyl, the 1,4-diketone yield was 74%.

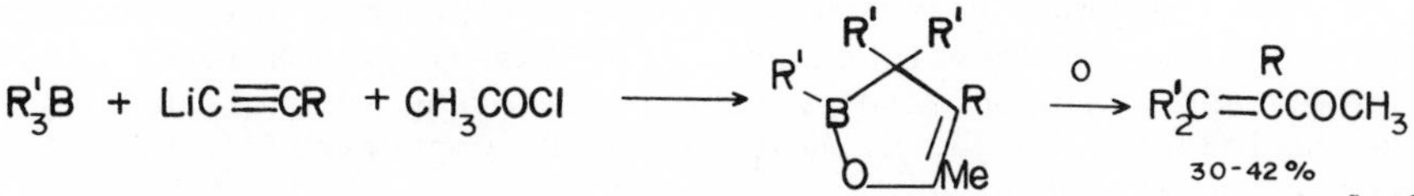

$$R_3'B + LiC\equiv CR + CH_3COCl \longrightarrow R'\overset{R'\ R'}{\underset{O\text{---}Me}{B}}R \xrightarrow{O} R_2'\overset{R}{C}=CCOCH_3 \quad [23]$$

30-42%

The oxidation was accomplished with Jones reagent since hydrogen peroxide and base were unsatisfactory.

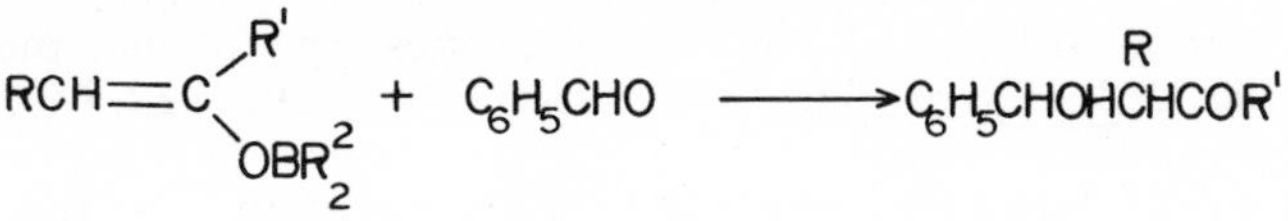

$$RCH=C\underset{OBR_2^2}{\overset{R'}{<}} + C_6H_5CHO \longrightarrow C_6H_5CHOHCHCOR'\overset{R}{} \quad [24]$$

Yield of β-hydroxyketone after hydrolysis was 91% when
$R = CH_3(CH_2)_4$, $R^1 = CH_3$, and $R^2 = CH_3(CH_2)_3$.
See G-9 for the preparation of ketones from boranes and
diazoketones.

9. From Ketones by Irradiation (see Addenda)

10. Curtius Rearrangement of Unsaturated Acids (see C.10)

Addenda

E.3. The transposition of a ketone group, $RCOCH_2R^1 \rightarrow RCH_2COR^1$,
has been accomplished by forming the α-phenylthioketone,
reducing to the alcohol, dehydration, and finally hydrolysis
[B. M. Trost et al., J. Am. Chem. Soc., 97, 438 (1974)].

E.3. The use of sulfur ylids in synthesis has been described
more fully [B. M. Trost, L. S. Melvin, Jr., Sulfur Ylids,
Emerging Synthetic Intermediates, Academic, New York, 1975].
The sequence of producing the fused cyclanone from the $-CH_2C=O$
structure has been described (E.3 of main section). The overall
yield in the cyclopentane annelation (73-85%) is almost
unbelievable [B. M. Trost et al., J. Am. Chem. Soc., 95, 5311
(1973); Tetrahedron Letters, 1929 (1974); Chem. Eng. News.,
Dec. 15, 1975, p. 33.].

E.4. Appropriately substituted cyclobutanes may be contracted
to cyclopropanes or the reverse. Pathways are described [J. M.
Conia, M. J. Robson, Angew. Chem. Intern. Ed. Engl., 14, 473
(1975)].

E.5. The Demjanov rearrangement with ketones may now be carried
out with alkylidenecyclohexanes. For example, methylenecyclohexanes and cyanogen azide give expanded ring ketones such as
the cycloheptanone [J. E. McMurry, A. P. Coppolino, J. Org.
Chem., 38, 2821 (1973)].

E.8. Trialkylboranes react with sodium cyanide and an acid
chloride to form a cyclic intermediate that is converted into a
dialkyl ketone by sodium peroxide. The carbonyl group is furnished by the cyanide and the alkyl groups by the borane [A.
Pelter et al., J. Chem. Soc., Perkin Trans., I, 129 (1975)].

E.9. The field has been reviewed [K. Schaffner, O. Jeger,
Tetrahedron, 30, 1891 (1974)]. Major reactions by irradiation
are α,β- to β,γ-unsaturated ketones, elimination of some

α-substituents, cyclobutane formation of α,β-unsaturated ketones
by dimerization, formation of cyclobutanols, and ring closure.
However, the response to irradiation is not predictable, based
on homolog analogy.

E.10. Curtius Rearrangement of Unsaturated Acids (see C.10).

References

1. T. E. Zalesskaya, I. K. Lavrova, J. Org. Chem. USSR, 4,
    1999 (1968).
2. E. Ghera, J. Org. Chem., 35, 660 (1970).
3. F. E. Ziegler, J. A. Kloek, Tetrahedron Letters, 2201
    (1971).
4. J. M. Conia et al., Tetrahedron Letters, 1747 (1973).
5. E. G. E. Hawkins, R. Large, J. Chem. Soc., Perkin Trans.,
    I, 2169 (1973).
6. J. R. Williams et al., J. Org. Chem., 39, 1028 (1974).
7. B. M. Trost et al., Tetrahedron Letters, 3449 (1970); J.
    Am. Chem. Soc., 95, 5321 (1973).
8. H. Hart, I. Huang, J. Org. Chem., 39, 1005 (1974) and
    earlier papers.
8a. B.-T. Gröbel, D. Seebach, Angew. Chem. Intern. Ed. Engl.,
    13, 82 (1974).
8b. A. Eschenmoser et al., Helv. Chim. Acta, 50, 210 (1967).
9. U. Schöllkopf, P. Böhme, Angew. Chem. Intern. Ed. Engl.,
    10, 491 (1971).
9a. A. J. Sisti, J. Org. Chem., 33, 453 (1968); J. Org. Chem.,
    39, 1182 (1974); D. Seebach et al., Chem. Ber., 108, 2368
    (1975).
10. J.-J. Riehl, A. Fougerousse, Bull. Soc. Chim. France, 4083
    (1968).
11. J. A. Marshall, H. Roebke, J. Org. Chem., 34, 4188 (1969).
12. R. L. Cargill et al., Acct. Chem. Res., 7, 106 (1974).
13. J. L. Gras, M. Bertrand, Bull. Soc. Chim. France, 2024
    (1972).
14. D. R. Dimmel, S. B. Gharpure, J. Am. Chem. Soc., 93, 3991
    (1971).
15. H. C. Brown, Organic Synthesis via Boranes, Wiley, New York,
    1975; G. M. L. Cragg, Organoboranes in Organic Synthesis,
    Dekker, New York, 1973.
16. H. C. Brown, M. W. Rathke, J. Am. Chem. Soc., 89, 2738
    (1967).
17. Ref. 15a, p. 127.
18. H. C. Brown, S. K. Gupta, J. Am. Chem. Soc., 94, 4370
    (1972).
19. A. Pelter et al., Chem. Commun., 544 (1973).

20. B. A. Carlson, H. C. Brown, J. Am. Chem. Soc., 95, 6876 (1973).
21. R. H. Prager, J. M. Tippett, Tetrahedron Letters, 5199 (1973).
22. A. Pelter et al., Tetrahedron Letters, 4491 (1973).
23. M. Naruse et al., Tetrahedron Letters, 795 (1973).
24. T. Mukaiyama et al., J. Am. Chem. Soc., 95, 967 (1973).

F. Acylations

1. From Acids and Anhydrides (Decarboxylation) (1, 686; H.-W., 7, Pt. 2a, 1973, p. 622)

The oxidative decarboxylation of β-carboxypropiophenones, easily obtained from succinic anhydride and arenes, leads to unsaturated ketones as shown [1]:

$$ArCOCH_2CH_2CO_2H \xrightarrow[Cu(II),C_5H_5N]{Pb(OCOCH_3)_4} ArCOCH{=\!=}CH_2$$
$$20-45\%$$

Another example is the decarboxylation of 9-fluorenecarboxylic acid to fluorenone (56%) by heating with 2 equiv of $CuCO_3Cu(OH)_2$ at 265° [2]. With Cu in refluxing quinoline, the acid gives 85% of fluorene.

Fluorenones are produced from benzoic anhydrides by a catalyst such as $RhCl-(PPh_3)_3$ [3]:

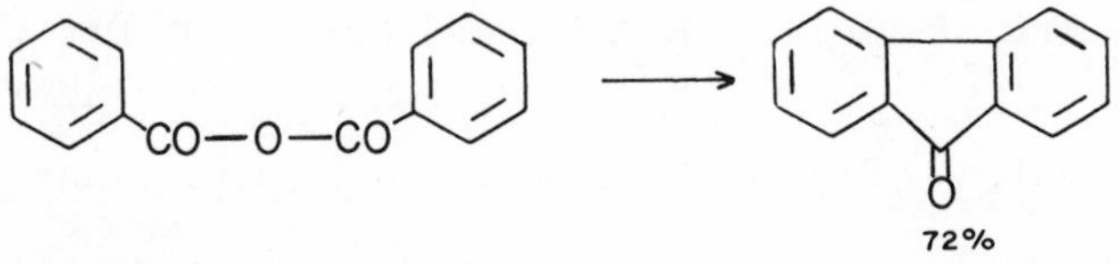

72%

By contrast irradiation of glutaric anhydrides leads to a decrease in ring size [4]:

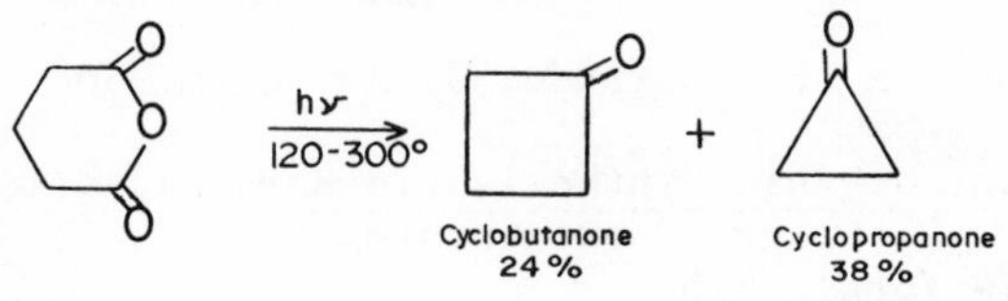

Cyclobutanone
24%

Cyclopropanone
38%

2. From Esters (Claisen, Dieckmann, Reformatsky, etc) (1, 688; H.-W., 7, Pt. 2a, 1973, pp. 492, 504)

The Claisen reaction is discussed under Chapter 14, C.1, the Dieckmann reaction, under C.2 of the same chapter, and the

Reformatsky reactions, under Chapter 4, E.3.

In regard to the Claisen reaction it is now possible to obtain the mixed β-ketoesters; such was not possible previously by any direct procedure [4a].

$$CH_3CO_2Et \xrightarrow[-78°]{LiICA} LiCH_2CO_2Et \xrightarrow{EtCOCl} EtCOCH_2CO_2Et$$

Ethyl propionylacetate was obtained in 60% yield. Evidently, the base is strong enough to convert all of the ethyl acetate into its anion, thus allowing no chance for self-condensation.

If the product is nonenolizable, the Claisen reaction may be conducted by the use of an acid catalyst as shown [5]:

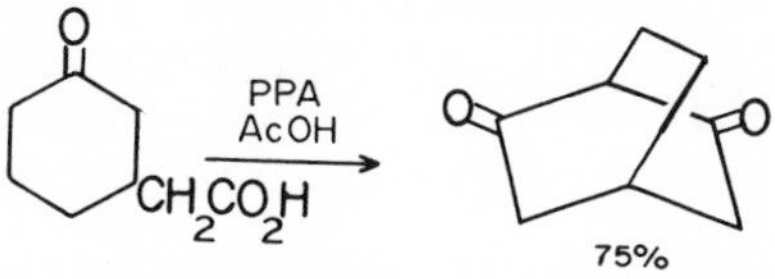

75%

The bicyclo[2.2.2]octane-2,6-dione was obtained in 75% yield by heating for 7 hr in PPA-CH$_3$COOH.

The reversible nature of the Claisen condensation is indicated [6]:

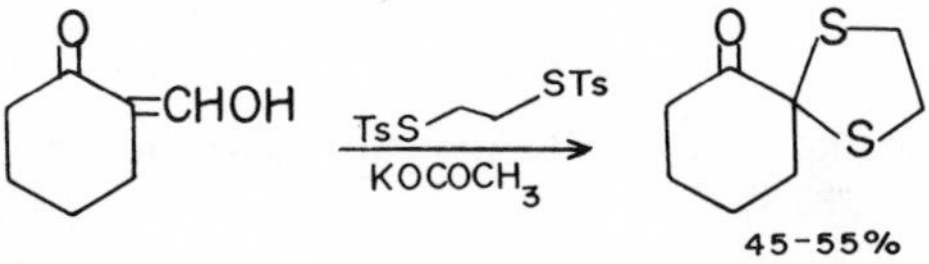

45-55%

The reaction appears to possess more selectivity in forming the dithiolane from the hydroxymethylene ketone than from the ketone itself, although no mention was made of its merits.

A Reformatsky reaction has been accomplished with anhydrides, rather than with carbonyl compounds, to give keto derivatives instead of alcohols [7]:

$$(RCO)_2O \ + \ BrZn\overset{R'}{\underset{R}{C}}CN \longrightarrow RCO\overset{R'}{\underset{R}{C}}CN$$

64-75%

To obtain the β-ketonitrile, R and R[1] must be alkyl groups, not hydrogen.

3. From Malonic and Acetoacetic Esters and β-Diketones
(1, 688)

In the reaction of a symmetrical ketone with an acyl chloride, which accomplishes the same results as a reverse Claisen reaction, probably via the triacylmethane, the mixed diketone is obtained [8]:

$$CH_3COCH_2COCH_3 + C_3H_7COCl \xrightarrow[C_6H_5NO_2]{AlCl_3} CH_3COCH_2COC_3H_7$$

2,4-Heptanedione, 45% (best conditions)

The acylation of the thallium salt of acetoacetic ester at -78° with acetyl chloride gave O-acetylation, but at 25° with acetyl fluoride C-acetylation was obtained [9].

The aroylation of dianions is shown [10]:

1-(p-Methoxyphenyl)-5-phenyl-1,3,5-penta-trione, 77-86%

For more detail on dianion chemistry, see G.1.

Ketene acylates malonic ester and other active hydrogen compounds to give mono or diacyl derivatives [11]:

$$2 \quad CH_2{=}C{=}O \quad + \quad CH_2(CO_2Me)_2 \xrightarrow[130°, 1 hr.]{ClCH_2CO_2Na} (CH_3CO)_2C(CO_2Me)_2$$

Dimethyl diacetomalonate, 95%

4.   From Ketones or Enamines (1, 690; H.-W., 7, Pt. 2a, 1973, p. 473)

A more recent study of the acetylation of ketones by Type II (1, 690) shows that not only direct C-acetylation of the ketone is involved, but also O-acetylation of the ketone and C-acetylation of the resulting ketone enol ester [12]. Thus methyl ethyl ketone responds as indicated:

$$CH_3COCH_2CH_3 \xrightarrow[\substack{fast \\ saturation}]{\overset{(CH_3CO)_2O}{BF_3}} CH_3COCH_2COCH_2CH_3 + CH_3COCHCH_3$$

$$\overset{|}{COCH_3}$$

A   B

14%   86%

However, slow saturation with BF_3 or the use of the BF_3-diacetic acid complex gives 100% of B. It appears that A is produced by direct C-acetylation, while B probably arises by O-acetylation through the enol form:

$$CH_3\overset{OH}{\overset{|}{C}}{=}CHCH_3.$$

On the other hand, it is reported that the enolates of optically active ketones may be acylated without much loss of optical activity [13].

Indirect acylation may be accomplished by forming the α-acylthioester, which is treated with tributylphosphine, triethylamine, and lithium perchlorate to remove the sulfur and form the β-diketone [14].

The acylation of α-bromoketones may be accomplished by the use of zinc and an acid chloride [15]:

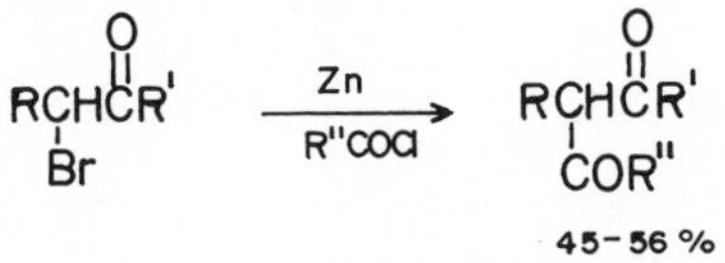

$$\underset{\underset{Br}{|}}{RCH}\overset{\overset{O}{\|}}{C}R' \xrightarrow[R''COCl]{Zn} \underset{\underset{COR''}{|}}{RCH}\overset{\overset{O}{\|}}{C}R'$$

45–56 %

Acylation may also be accomplished through the trimethylsilyl enol form of the keto.   [16]:

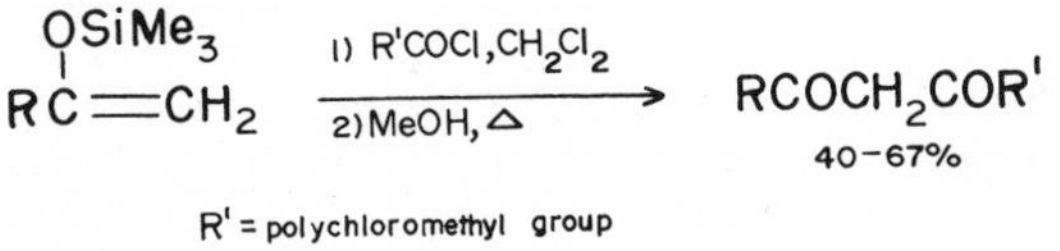

$$\underset{RC}{\overset{OSiMe_3}{|}}=CH_2 \xrightarrow[2)MeOH, \triangle]{1) R'COCl, CH_2Cl_2} RCOCH_2COR'$$

40–67 %

R' = polychloromethyl group

Unsaturated ketones are converted into diketones by treatment with hydrazoic acid-PPA [17]:

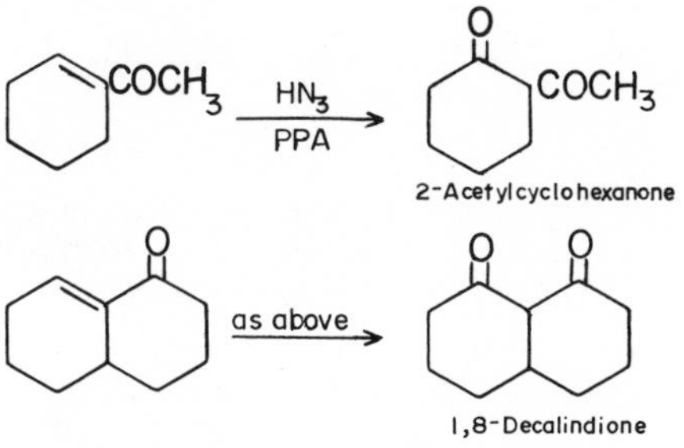

2-Acetylcyclohexanone

1,8-Decalindione

These reactions will occur only if the hydrazoic acid reacts with the olefinic rather than the carbonyl groups.

The acylation of enamines has been reviewed [18]. With an acid chloride and 1-morpholinocyclododecene-1 the reaction follows:

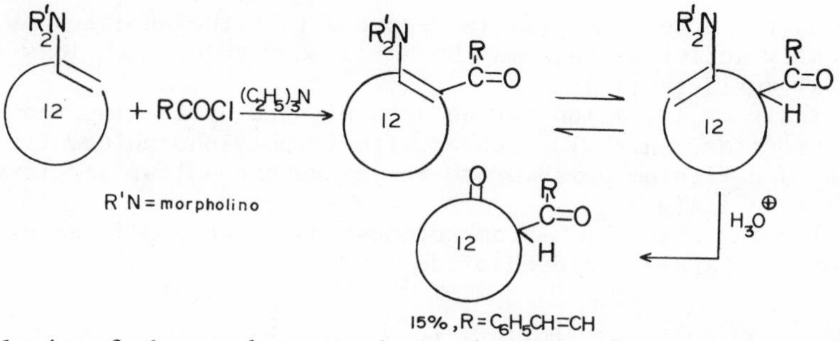

15%, R = C$_6$H$_5$CH=CH

Hydrolysis of the product as shown gives the 2-oxoacylcyclododecane.

The adduct of phenyllithium and pyridine has been acylated at the 3-position of pyridine in poor yield [19].

5.   From Nitriles, Diazoalkanes, Heteroaromatic Bases, and
       the Like (Mostly Free Radical Acylation) (1, 693)

Minisci and co-workers [20] have succeeded in acylating a series of heteroaromatic bases by acyl radicals generated from aldehydes and t-butylhydroperoxide or H$_2$O$_2$ and FeSO$_4$. The method has been applied to quinoline, quinoxaline, and benzo-1, 3-thiazole. The equation with quinoxaline follows:

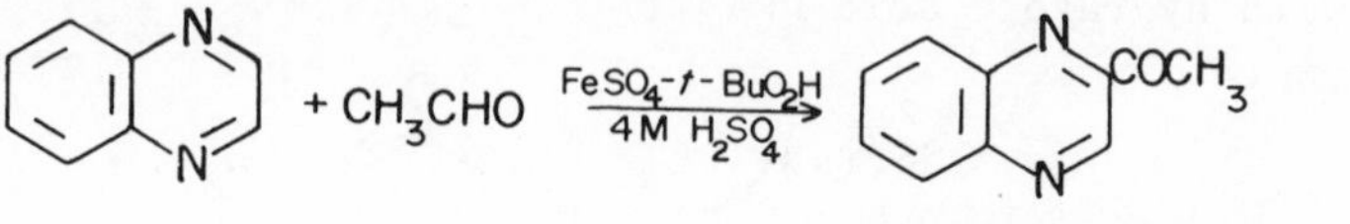

2-Acetylquinoxaline, 70% (isolated)

An improvement in the benzoylation of nucleosides and nucleotides has been achieved by the use of benzoyl cyanide [21]. For example, thymidine with C$_6$H$_5$COCN in CH$_3$CN and N(n-C$_4$H$_9$)$_3$ at 25° gave a 77% yield of the 3',5'-di-O-benzoyl derivative.

6.   From Dimethyl Sulfoxide, Dimethyl Sulfone, N,N-Dimethyl-
       methanesulfonamide, and the Like (1, 693)

Acetylfurans have been produced by the reaction of dimethyl-prop-2-ynylsulfonium bromide on 1,3-diketones or β-ketoesters [22]:

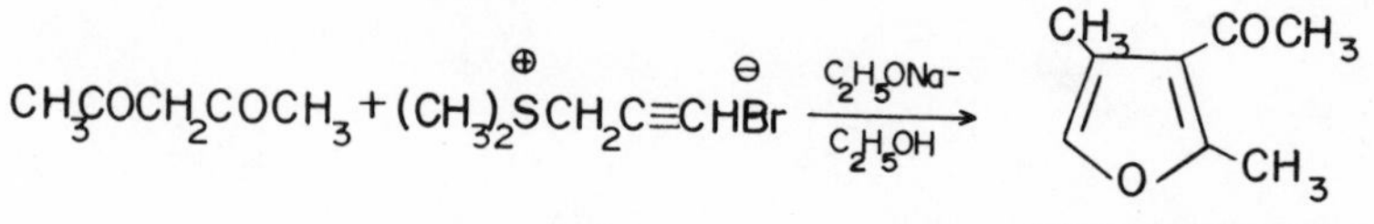

3-Acetyl-2, 4-dimethylfuran
81 %

An allenesulfonium salt, $(CH_3)_2\overset{+}{S}CH{=}C{=}CH_2$, is formed first; this is quite susceptible to anionic addition followed by displacement of dimethyl sulfide. The facile synthesis enlarges the scope of furan preparation.

7. From Aldehydes and Olefins or Biacetyl and Cyclohexane (1, 694; H.-W., Pt. 2a, 1973, p. 462)

$$RCHO + R'CH{=}CH_2 \longrightarrow R\overset{O}{\overset{\|}{C}}CH_2CH_2R'$$

The preceding reaction, catalyzed by the cyanide ion, is anionic in nature. Free radical acylations are described in C.7. In an anionic reaction the olefin must contain electron-withdrawing groups such as NO_2, CN, COOR, or COR. Moreover, since the cyanide ion is basic, aliphatic aldehydes cannot in general be used as acylating agents. However, a different catalyst, as discussed later, can then be used. Stetter has demonstrated the versatility of the reaction in that it occurs with aromatic and heterocyclic aldehydes [23]:

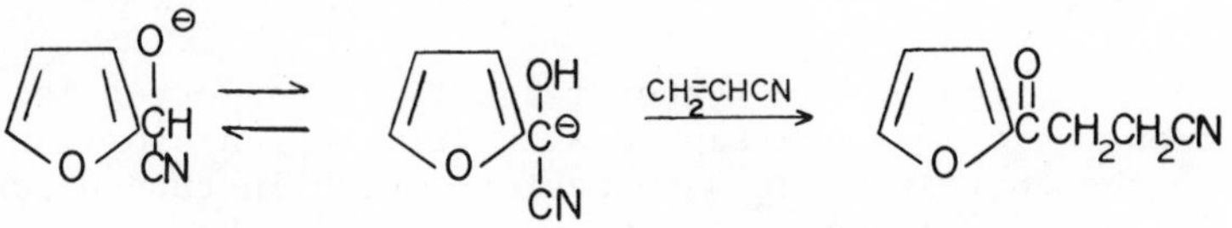

2.β-Cyanopropanoylfuran was obtained in 32% yield.

Two methods have been devised to permit the reaction to encompass aliphatic aldehydes without their undergoing self-condensation. One is to substitute a thiazolium salt (A) for the cyanide catalyst as shown [24]:

$$CH_3CHO \xrightarrow[A]{CH_2{=}CHCOCH_3} CH_3COCH_2CH_2COCH_3$$

A =

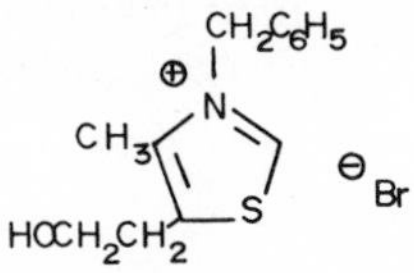

2,5-Hexanedione was obtained in 50% yield by heating in DMF or an alcohol for 24 hr under N_2. The second utilizes as a protective group for the cyanohydrin either the acetal [25a], ketal [25b], or the trimethylsilyl derivative [25c]. An illustration of the former follows:

$$C_5H_{11}\underset{\underset{CN}{|}}{\overset{\overset{OEt}{|}}{\overset{OCHCH_3}{|}}}{CH} \xrightarrow[\text{2) }C_4H_9Br]{\text{1)LDIA}} C_5H_{11}\underset{\underset{CN}{|}}{\overset{\overset{OEt}{|}}{\overset{OCHCH_3}{|}}}{CC_4H_9} \xrightarrow{\overset{\oplus}{H_3O}} C_5H_{11}COC_4H_9$$

5-Decanone, 70%

It is true that alkylation with butyl bromide was pursued rather than addition to an unsaturated carbonyl compound, but either reaction could have been performed. Even secondary alkyl halides give fair yields in the reaction. The dihydropyranyl ketal may be used similarly [25b]. The second method utilizes the silylated cyanohydrin [25c]:

$$Ar\underset{\underset{CN}{|}}{\overset{\overset{OSiMe_3}{|}}{C}}H \xrightarrow[\text{2)RX}]{\text{1) LDIA}} ArCOR'$$

Fifteen examples give 59-98% yields.

9.   From Dianions of Carboxylic Acids (Ivanov Reagents) and Esters

The reaction of the Ivanov reagent with an ester leads to a carbanion, which may be trapped with trimethylchlorosilane (TMCS) to give the corresponding β-ketoester, which in turn undergoes solvolysis under neutral conditions to yield β-keto acids in good yield. These events may be represented as shown [26]:

$$\underset{R}{\overset{R}{>}}CHCOOH \xrightarrow{LiN[CH(CH_3)_2]_2} R_2\overset{\ominus}{C}\overset{\ominus}{CO_2} \xrightarrow{R'CO_2Me} R'COCR_2\overset{\ominus}{CO_2} \xrightarrow{TMCS}$$

$$R'COCR_2CO_2SiMe_3 \xrightarrow{MeOH} R'COCR_2CO_2H$$

The Ivanov reagent is discussed as well under Chapter 4, G.4. Since the latter give ketones quantitatively on being heated, the method permits the formation of highly hindered types.

10.   From Aldehydes

The benzoin and acyloin condensations are discussed under Chapter 4, C.8. An alternate method of some versatility for the preparation of acyloins follows [27]:

$$HCHO + CH_3OCH_2CHO \xrightarrow{\overset{\oplus \quad \ominus}{(C_2H_5)_2NH_2Cl}} CH_2{=}\underset{OCH_3}{\overset{\mid}{C}}CHO \xrightarrow[2)H_3O^{\oplus}]{1)RLi} CH_3\underset{O}{\overset{\parallel}{C}}CHOHR$$

15-86%
(10 examples)

A most unusual reaction involving the synthesis of ketones from aldehydes has been seined from the plankton of synthetic methods [28].

$$Et\,O_2C\overset{\overset{NO}{\mid}}{N}(CH_2)_6\overset{\overset{NO}{\mid}}{N}CO_2Et + 2\,C_6H_5CHO \xrightarrow{base} C_6H_5\overset{O}{\overset{\parallel}{C}}(CH_2)_6\overset{O}{\overset{\parallel}{C}}C_6H_5$$

1,6-Dibenzoylhexane was obtained in 90% yield by allowing the two reagents to stand in alcohol at 5° with solid K_2CO_3 for about a month or until the N_2 gas ceased evolving. The high yield is remarkable if one assumes that a diazo intermediate is formed followed by the attack of the hexamethylene free diradical on the aldehyde. Of course, events could take place stepwise.

11.   From α,β-Unsaturated Ketones and Vinyl Silyl Ethers (see Addenda)

Addenda

F.2.   The aldol reaction between glyoxal and dimethyl β-keto-glutarate gives a mixture of ketoesters, which after hydrolysis and decarboxylation yields bicyclo [3.3.0] octane-3,7-dione-2, 4,6,8-tetramethyl ester, the tetracyclic trione hexaester, as well as other ketoesters, all of which could be separated [U. Weiss et al., Tetrahedron Letters, 3767 (1975)].

F.10.   The acyloin condensation with aldehydes may be carried out heterogeneously with aqueous phosphate buffer containing 10 mol % of N-laurylthiazolium bromide. For hexaldehyde 67% of hexoin was obtained in 5-12 hr of stirring [W. Tagaki, H. Hari, Chem. Commun., 891 (1973)].

F.11.

$$C_6H_5C(OSiMe_3){=}CH_2 + Me_2C{=}CHCOCH_3 \xrightarrow[-78°]{TiCl_4} C_6H_5\underset{O}{\overset{\parallel}{C}}CH_2\underset{Me}{\overset{\overset{\displaystyle Me}{\mid}}{C}}CH_2\underset{O}{\overset{\parallel}{C}}CH_3$$

76 %

The Michael reaction using a weak Lewis acid may also be carried out with an unsaturated ketal [T. Mukaiyama et al., Chem. Lett., 1223 (1974)].

References

1.   P. P. Sane et al., Synthesis, 541 (1973).
2.   B. M. Trost, P. L. Kinson, J. Org. Chem., 37, 1273 (1972),
3.   J. Blum, Z. Lipshes, J. Org. Chem., 34, 3076 (1969).
4.   H. Hiraoka, U.S. Patent, 3,748,242, July 24, 1973; C. A.,
     79, 104,806 (1973).
4a. M. W. Rathke et al., Tetrahedron Letters, 2953 (1971).
5.   H. Gerlach, W. Müller, Angew. Chem. Intern. Ed. Engl., 11,
     1030 (1972).
6.   R. W. Woodward et al., Org. Syn., 54, 37 (1973).
7.   N. Goasdoué, M. Gaudemar, J. Organomet. Chem., 39, 29
     (1972).
8.   M. Motoi et al., Bull. Chem. Soc. Jap., 42, 3359 (1969).
9.   E. C. Taylor et al., J. Am. Chem. Soc., 90, 2421 (1968).
10. C. R. Hauser et al., Org. Syn., Coll. Vol., 5, 718 (1973).
11. H. Eck, H. Prigge, Ann. Chem., 731, 12 (1970).
12. C. R. Hauser et al., J. Org. Chem., 34, 1425 (1969).
13. D. Seebach, V. Ehring, Angew. Chem. Intern. Ed. Engl., 11,
     127 (1972).
14. A. Eschenmoser et al., Helv. Chim. Acta, 54, 710 (1971).
15. I. I. Lapkin, F. G. Saitkulova, J. Org. Chem. USSR, 2586
     (1971).
16. S. Murai et al., Chem. Commun., 946 (1972).
17. K. Mitsuhashi et al., Chem. Pharm. Bull., 17, 1572 (1969).
18. A. G. Cook, Enamines, Dekker, New York, 1969, p. 384; S. F.
     Dyke, Chemistry of Enamines, Cambridge U. P., 1973, p. 22;
     S. Hunig, H. Hoch, Fortsch. Chem. Forsch., 14, 235 (1970).
19. C. S. Giam et al., J. Org. Chem., 39, 3565 (1974).
20. F. Minisci et al., Chem. Commun., 201 (1969); J. Chem.
     Soc., C929 (1970); C1747 (1971); for a review, see
     Synthesis, 1 (1973).
21. A. Holy, M. Soucek, Tetrahedron Letters, 185 (1971).
22. P. D. Hawes, C. J. M. Stirling, Org. Syn., 53, 1 (1973).
23. H. Stetter et al., Angew. Chem. Intern. Ed. Engl., 12, 81
     (1973); 13, 539 (1974); Tetrahedron Letters, 1461 (1973);
     Chem. Ber., 107, 2453 (1974).
24. H. Stetter, H. Kuhlmann, Angew. Chem. Inter. Ed. Engl., 13,
     539 (1974).
25a. G. Stork, L. Maldonado, J. Am. Chem. Soc., 93, 5286 (1971).
25b. A. Kalir, D. Balderman, Synthesis, 358 (1973).
25c. S. Hunig et al., Synthesis, 777 (1973).
26. C. Ainsworth et al., J. Am. Chem. Soc., 93, 6321 (1971);
     B. Angelo, Compt. Rend., C293 (1973).
27. G. A. Russell, M. Ballenegger, Synthesis, 104 (1973).
28. C. M. Samour, J. P. Mason, J. Am. Chem. Soc., 76, 441
     (1954).

G. Alkylations

1. From Ketones (<u>1</u>, 697)

$$RCOCH_3 \xrightarrow[2)R'X]{1)Base} RCOCH_2R'$$

The field of alkylation perhaps is the most active of those
involved in ketone preparations. Matters such as different
bases, the great number of new alkylating agents, the question
of alkylation on the more or less substituted side of the ketone,
enamine alkylation (G.2), and Michael alkylation (G.3) have been
examined in depth.

An organic synthesis preparation, which gives the conditions
to select for the alkylation isomer desired, is available [1].
The formation of two isomers is shown:

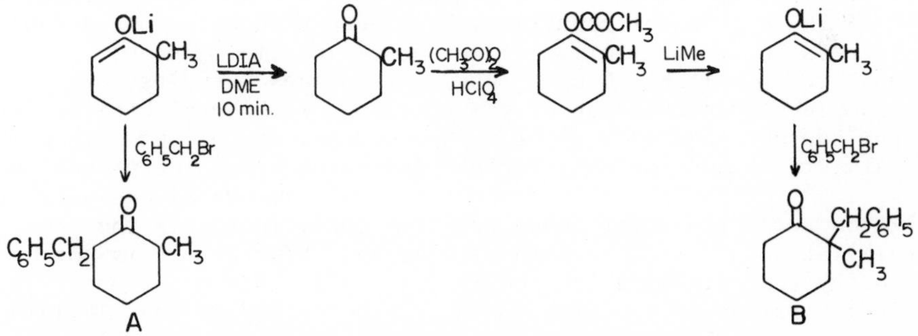

2-Benzyl-6-methylcyclohexanone (A) was obtained in 58-61% crude
yield, 42-45% pure, while the 2-benzyl-2-methyl isomer (B) was
obtained in 54-58% yield. The trimethylsilyl enol ether may be
used to synthesize B as well [2]. In addition the silyl enol
ether may be utilized for annelation [3]:

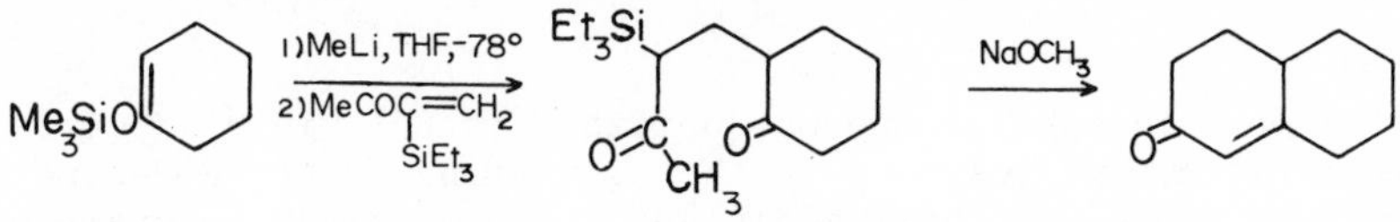

$\Delta^{1,9}$-2-Octalone was obtained in about 80% overall yield. Methyl
vinyl ketone should behave similarly to the α-silylated vinyl
ketone above, but actually it undergoes considerable polymerization under these conditions.

The trimethylsilyl enolate may also be utilized in
methylation via the cyclopropane intermediate [4]:

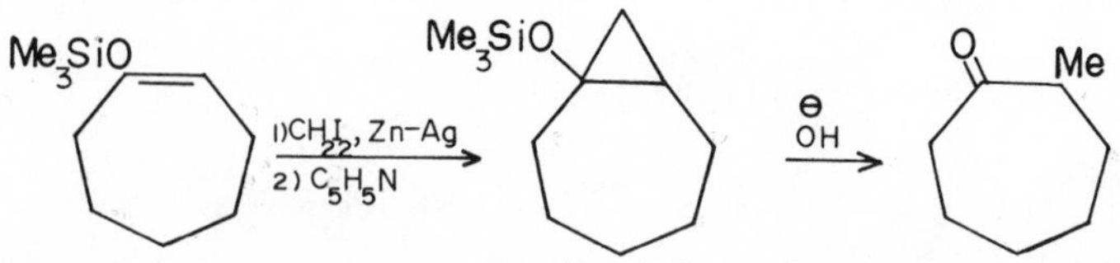

2-Methylcycloheptanone was obtained in 64% overall yield starting from the ketone.

Still another use of the silyl enol ethers is in the synthesis of dichloro unsaturated ketones [5]:

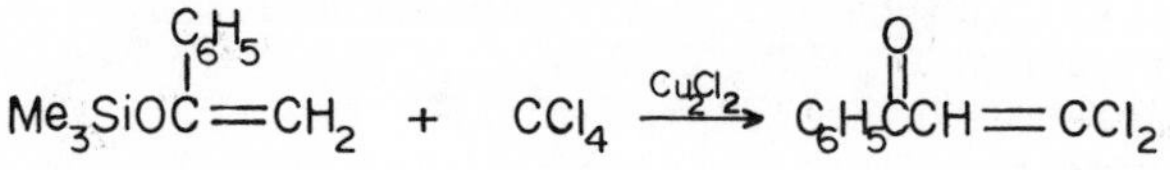

$$Me_3SiO\overset{\underset{|}{C_6H_5}}{C}=CH_2 \ + \ CCl_4 \ \xrightarrow{CuCl_2} \ C_6H_5\overset{\overset{O}{\|}}{C}CH=CCl_2$$

Phenyl β,β-dichlorovinyl ketone was obtained in good yield by refluxing the components in a mixture of DMF and chlorobenzene.

Phase transfer conditions have been used with considerable success in the alkylation of ketones. The importance of the quaternary salt has been pointed out in these studies. For example, the alkylation of phenylacetone with butyl bromide in 50% aqueous NaOH gave 5% of 3-phenyl-2-heptanone, but with a catalytic amount of benzyltriethylammonium hydroxide present as well 90% of the ketone was obtained [6]. α,ω-Dihalides alkylate ketones readily under phase transfer conditions. In the alkylation of desoxybenzoin with benzyl chloride, it is noteworthy that dialkylation occurs [7]. Also if the quaternary salt concentration is too high (3 equiv), the yield of monoalkylated product drops, while that of the dialkylated increases [8].

For hindered ketones, which resist addition and favor enolization, alkylation may be accomplished as follows [9]:

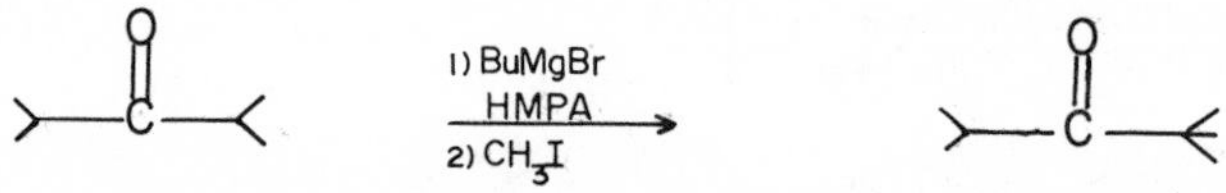

t-Butyl isopropyl ketone was obtained in 87% yield. Similarly by using methyl tosylate as the alkylating agent, 2,2-dimethylcyclohexanone was obtained in 49% yield with 24% recovery of the substrate, 2-methylcyclohexanone. Propargyl bromide may be utilized in alkylation if the active hydrogen of the acetylene group is replaced as shown [10]:

1,2-Cyclohexane- or 1,2-cyclopentanedione is methylated in the 3-position by treating the dione with 2 equiv of lithium diisopropylamide, LDIA, followed by methyl iodide at -78° [11]. Yields were about 70%. Crown ethers used to increase the basicity of the catalyst increase the amount of O-alkylation up to 40-50% [12].

Arylation of acetone is accomplished in high yield by irradiation with an aryl halide [13]. Another way to arylate ketones is to irradiate an iodobenzene in the presence of the anion of the ketone. In this way cephalotaxinone was prepared in 94% yield [14].

Dianions of β-dicarbonyl compounds are very readily alkylated as follows:

$$C_6H_5COCHCOCH_2 \xrightarrow{C_6H_5CH_2Cl} C_6H_5COCH_2COCH_2CH_2C_6H_5$$

The high nucleophilicity of the γ-carbon gives exclusive γ-alkylation. The scope of both alkylation and arylation of dianions has been reviewed [15]. See H.7 and discussion for the use of anionic ylids in alkylation.

The Wittig reagent permits the introduction of the acetonyl group [16]:

$$LiCH_2\overset{O}{\overset{\|}{C}}CH=P(C_6H_5)_3 \xrightarrow{RX} RCH_2\overset{O}{\overset{\|}{C}}CH=P(C_6H_5)_3 \xrightarrow{H_2O} RCH_2\overset{O}{\overset{\|}{C}}CH_3$$

An unusual alkylation is that of benzoin to α-alkylbenzoin, ArCOCOHAr, formed by treatment with an alkyl iodide and sodium hydroxide in DMSO [17]. The iodide, not the bromide, must be used to avoid O-alkylation. Yields ranged within 14-100% for 23 examples.

The thallium salts of β-diketones are crystalline and non-hygroscopic (but poisonous). High yields of alkylated β-diketones are obtained on treatment of the salts with alkyl halides [18].

An unusual alkylation is possible for an alcohol [19]:

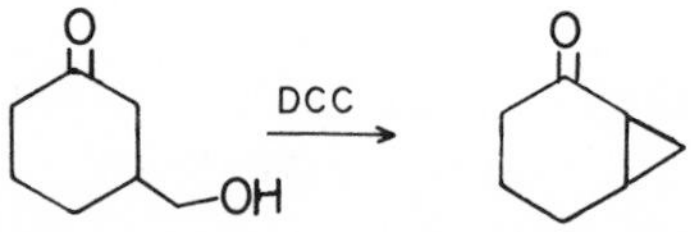

Bicyclo[4.1.0]heptanone was obtained in 60% yield. Similar alkylations are feasible with vinyl ethers [20]:

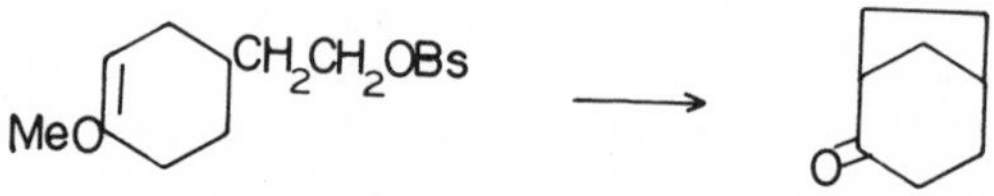

Bicyclo[3.2.1]octanone-2 was obtained in 79% yield on a small scale when the brosylate was heated in CH_3CN and $(C_2H_5)_3N$ at 80°.

As in the Wittig directed aldol synthesis (1, 236, 611), Schiff bases have been alkylated as shown [21]:

$$\mathrm{LiCH_2\overset{CH_3}{C}{=}NC_6H_{11}} \xrightarrow[2)\,H_3O^{\oplus}]{1)\,RX} \mathrm{MeCOCH_2R}$$

Yields were 48-80% with primary, secondary, or allyl halides in conducting the reaction at -20° to -50° for bromides and at 50° for chlorides.

Reductive alkylation occurs when the ketone is treated with formaldehyde followed by reduction of the methylol product with $KHFe(CO)_4$ [22]. Presumably, the Mannich bases could be reduced with other agents; actually, they have been dehydroaminated to the unsaturated ketone [23].

A comprehensive review of reactions of Mannich bases has been published [24]. Among the reactions covered to give substituted ketones are those of deaminomethylation (p. 744), deamination (p. 745), and replacement of the amino group by hydrogen, carbon (p. 750), or other elements or groups. The deaminomethylation is made possible by the equilibrium established between the Mannich base and the ketone as shown in the two equations:

$$\mathrm{R\overset{O}{\overset{\|}{C}}CH_2CH_2NR'_2} \rightleftharpoons \mathrm{R\overset{O}{\overset{\|}{C}}CH{=}CH_2} + \mathrm{R'_2NH} \qquad \text{(p. 745)}$$

$$\mathrm{R\overset{O}{\overset{\|}{C}}CH_2CH_2NR'_2} \overset{H_2O}{\rightleftharpoons} \mathrm{R\overset{O}{\overset{\|}{C}}CH_3} + \mathrm{HOCH_2R'_2} \qquad \text{(p. 744)}$$

The Mannich product of acetoacetic esters was deaminated and decarboxylated in DMF to give unsaturated ketones [24a]. The methiodide of the Mannich base is more useful than the related free base for replacement by hydroxide or alkoxide (p. 756), although one of the authors was unable to duplicate the replacement of the quaternary amino group with hydroxide. The review also includes many cyclization reactions.

2.   From Enamines (1, 703)

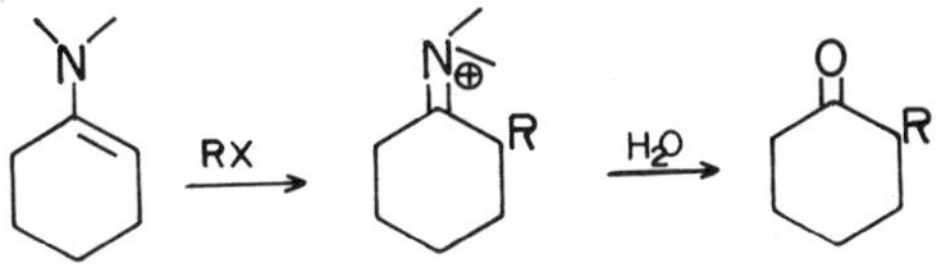

This subject has been reviewed [25]. Preparation of the enamines was discussed previously (1, 703), but two simple syntheses have been described recently. In the first a N-trimethylsilyldialkylamine, preferably of dimethylamine, is treated with the ketone and a trace of p-toluenesulfonic acid to give the enamine in 54-92% yields [26]. The second consists of refluxing cyclohexanone and HMPA for 40 min (no more), cooling, adding the halide, and then hydrolyzing to give 49%, 52%, and 99% yields of 2-allyl-, 2-benzyl-, and 2-carbethoxymethylcyclohexanones, respectively [27]. Longer refluxing gave the redox products, the tertiary amine, and octahydroacridine.

A novel methylation of cycloalkanone enamines takes advantage of the ease of cyclopropanation [28].

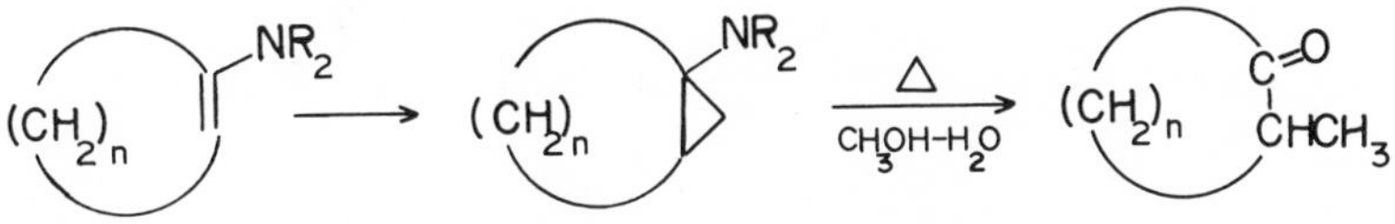

Yields of cyclopropanations for eleven examples, by using either methylene iodide or diazomethane and cuprous iodide as the carbene source, were 32-68%. Ring opening and hydrolysis occurred by heating in 150-170° in aqueous methanol (sealed tubes) or by prolonged refluxing in the presence of Pd(C). Carbethoxymethylation of ketones, via carbethoxycyclopropanation of enamines, has been carried out as well [29].

Epoxides add to enamines to give eventually $\alpha(\beta$-hydroxyethyl) ketones [25,30].

Allyl halides are capable of forming spiro ketones with enamines as shown for the synthesis of methyl spiro[4.5]decane-1-one-4-carboxylate (78%) [31]:

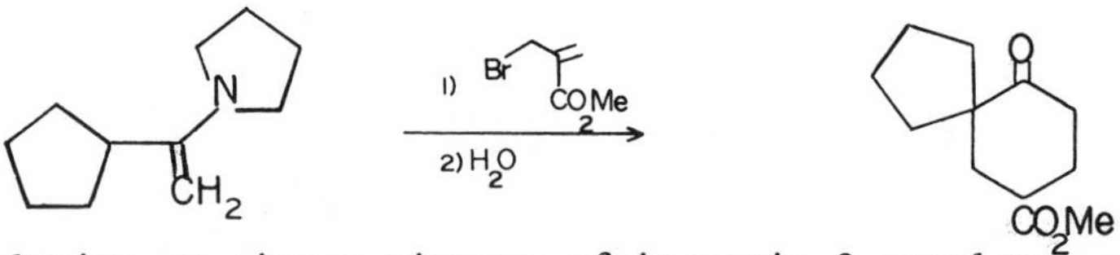

Annelation to give a mixture of isomeric 2-octalones occurs on the addition of methyl vinyl ketone to an enamine [32].

Essentially, a β-alkylation of ketones via enamines may be carried out in several ways [33], one of which is shown:

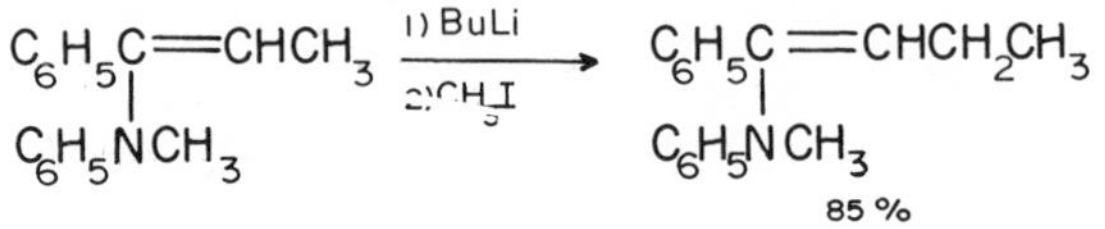

The product was not hydrolyzed to butyrophenone, as would be possible.

Alkylation of ketones via enamines has been conducted with phenyl allyl ether [34] or allyl acetate [35] in the presence of the palladium acetate-triphenylphosphine complex. An interesting alkylation of an enamine occurs with $C_5H_5Fe(CO)_2$-ethylene [36]. The alkylation of a chloroenamine gives not only the alkylated ketone, but also a cyclopropylamine as shown [37]:

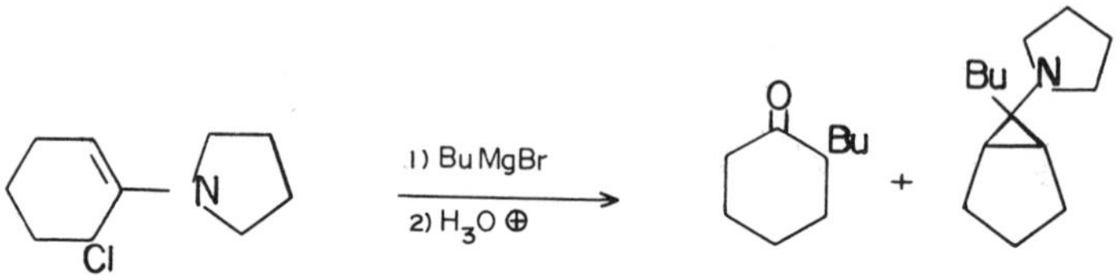

2-Butylcyclohexanone and 6-butyl-6-pyrrolidinobicyclo[3.1.0] hexane were obtained in the ratio of 30:70; the ratio varied with different Grignard reagents.

3. From Unsaturated Carbonyl Compounds and Carbanions (Michael) (1, 706)

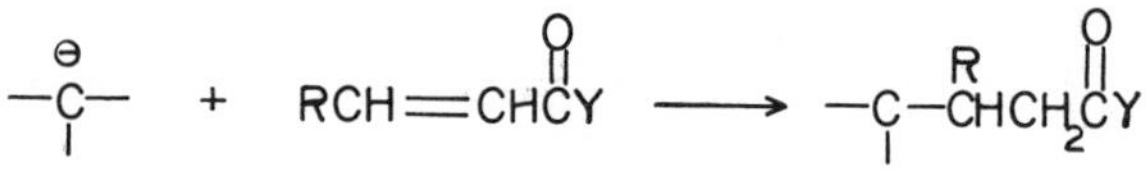

Alkylation of unsaturated ketones has occupied the attention of organic chemists for years, particularly as it applies to annelation and to the altering of the steroid structure. Annelation procedures have been reviewed [38]. An example is shown [39]:

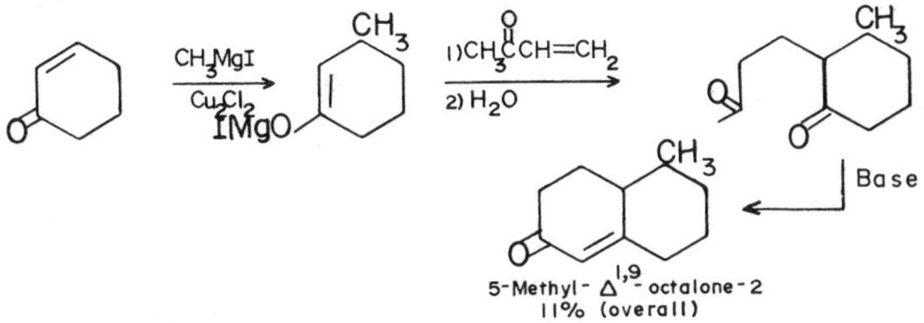

5-Methyl- $\Delta^{1,9}$- octalone-2
11% (overall)

The most direct use of methyl vinyl ketone (MVK) in annelation involves a sulfuric acid catalyst [40]:

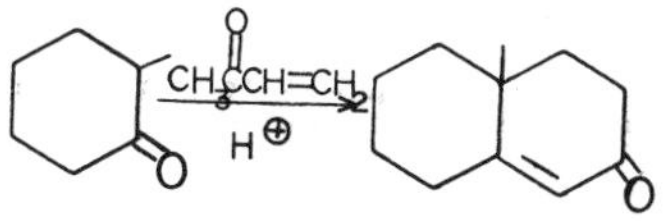

10-Methyl-$\Delta^{1,9}$-octalone-2 was obtained in 49-55% yield by refluxing about 0.5 mol quantities of MVK and the substrate with 0.3 ml of concentrated H_2SO_4. The MVK has been protected in the form of an isoxazole group subsequently used for annelation to prepare 1-methyl-$\Delta^{1,9}$-octalone-2 [41]. Another protective group for the MVK is the trimethylsilyl one in $CH_2COC(SiMe_3)=CH_2$ [42]. Annelation with this reagent gave 52% of 5-methyl-$\Delta^{1,9}$-octalone.

Another promising new annelation procedure is as follows [43]:

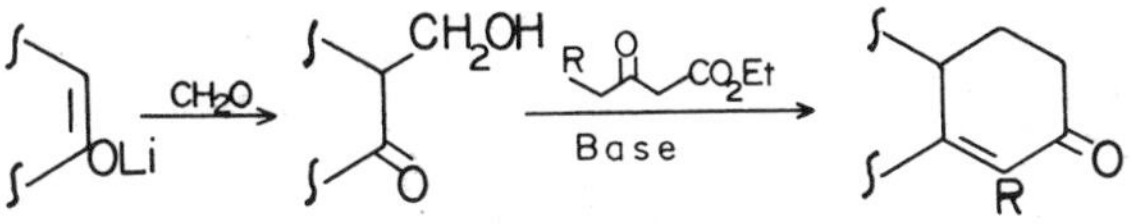

A similar use of the kinetic enolate is also reported [44].
A totally different Michael type of ring closure is accomplished with unsaturated Schiff bases [45]:

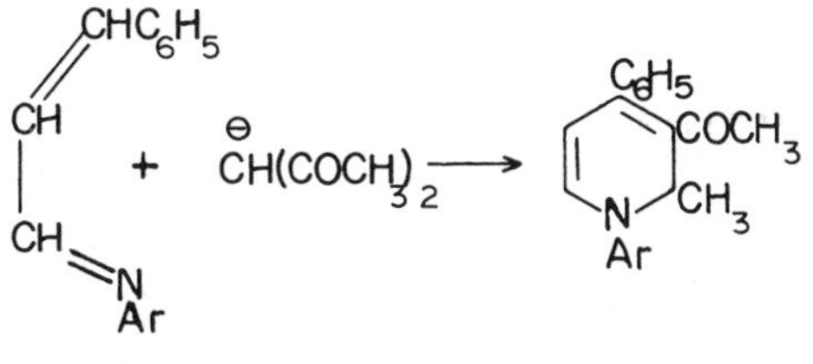

The acetyldihydropyridines were obtained in about 50% yields.
Annelation via Friedel-Crafts acylation has been reviewed [46].

The methylation of the conjugated octalones has turned up an interesting generality. Methylation of octalones or their derivatives with ordinary bases tends to give the equilibrium enolate, which leads to 1-methylation as in A; however, with a powerful base in excess the kinetic enolate, which gives 3-methylation as in B, is obtained [47]:

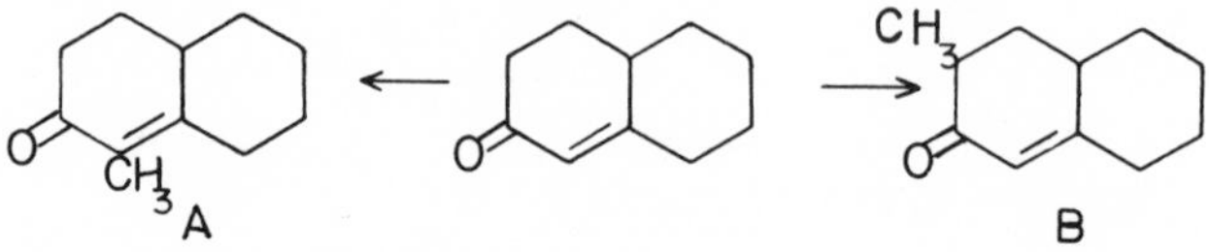

It is important to realize that the kinetic enolate can be trapped only with reactive halides and occasionally with primary iodides. A similar principle has been used to synthesize 4-alkylcyclohexenones [48]:

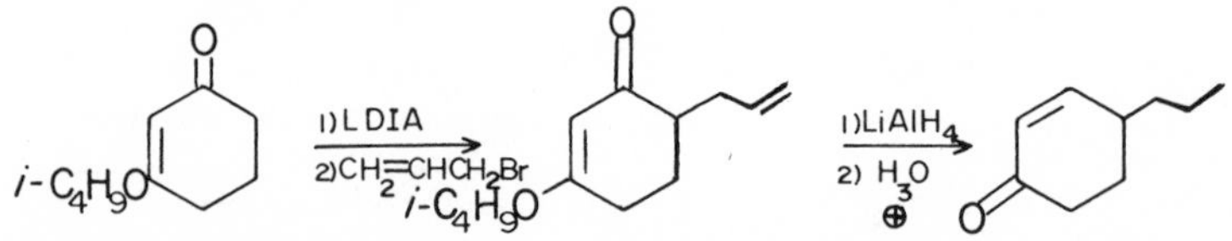

The 4-propyl-$\Delta^{2,3}$-cyclohexenone was obtained in 80% overall yield. The reaction has considerable versatility since it has been used with 1,2- as well as 1,3-diketones [49].

a.   Lithium Dialkylcuprates or Borane Addition

See Chapter 14, C.13, for addition of cuprates or dialkyl-copper reagents to esters.

Some interesting Michael alkylations have been carried out with lithium dialkylcuprate. In this connection 2-dithiomethyl-enecyclohexanone as shown is a useful intermediate [50]:

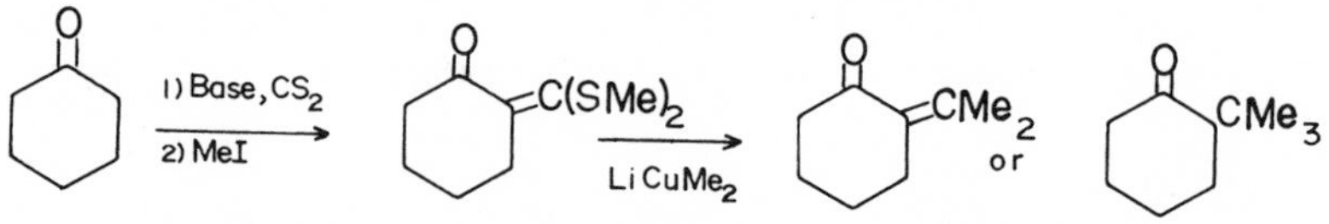

It is essential that the base be lithium 2,6-di-t-butyl-4-methyl-phenolate. Either 2-isopropylidene- or 2-t-butylcyclohexanone may be obtained in high yield, depending on the experimental con-ditions and whether 2 or 3 equiv of lithium dimethylcuprate are added. Tendencies for either conjugate addition of lithium dimethylcuprate or Grignard reagents plus cuprous chloride have been reviewed [51].

Sometimes there is a dramatic increase in yield of the

1,4-alkylation product if the Grignard reagent is added to the unsaturated ester or nitrile rather than in the reverse addition order [51]. The 3-hydroxypropyl group has been introduced through 1,4-addition to cyclopentenone by a cuprate with the hydroxyl group originally protected as an acetal [52]. Tributylphosphine complexed with lithium divinylcuprate when added to cyclohexanone has been found to increase the yield of 3-vinylcyclohexanone dramatically [53], a fact that bids fair to extend the scope of the addition of lithium divinylcuprates. In other cases no complexing agent need be added to produce 1,4-addition of the vinyl group [54].

A study of the complex of divinylcuprates and dimethyl sulfide has been reported [55]. It has been stated that lithium diethynylcuprates or other acetylenic cuprates do not add conjugatively at all [56]. This inactivity has been capitalized on in mixed cuprates such as $LiB\bar{u}CuC\equiv CC_3H_7$, which form 3-butylcyclohexanone in more than 95% yield with cyclohexenone [57]. Not only the butyl, but the methyl and vinyl groups have been added as well [58]. Besides, the acetylenic group may now be added conjugatively to unsaturated ketones. It is accomplished by using a diethylethynylalane, $Et_2AlC\equiv CR$. With 1-acetylcyclohexene this reagent gave yields of the 1,4-addition product as high as 79% [59]. The adducts are used to synthesize cyclopentenones by hydration and ring closure. Alanes also have been found to add more efficiently by a free radical process using light or air [60].

Numerous mixed cuprates have been described. The simplest one is the complex of an alkyllithium with cuprous cyanide, LiRCuCN. High yields of adducts are obtained if the ratio of the copper complex to unsaturated ketone is 7:1 [61], but mediocre if 1:1. For alkyl groups other than methyl or vinyl, the t-butoxycopper complex, t-BuO(R)CuLi is recommended for adding the R group conjugatively [62]. PhS(t-Bu)CuLi appears to be the best reagent to alkylate acid chlorides (see H.2 for a more extensive discussion) and 2 equiv must be used to add conjugatively in reasonable yield to cyclohexenone [62].

$Li[(PhS)_2\overset{Ph}{\underset{|}{C}}]_2Cu$ (A) has been used to prepare 1,4-diketones by conjugate addition [63]:

$$CH_2{=}CHCOCH_3 \xrightarrow{A} (C_6H_5)_2\overset{C_6H_5}{\underset{|}{C}}CH_2CH_2COCH_3 \xrightarrow[CuCl_2]{CuO} C_6H_5COCH_2CH_2COCH_3$$

1-Phenylpentane-1,4-dione was obtained in about 80% overall yield.

A similar addition occurs when conjugated dienones are treated with lithium dimethylcopper [64]:

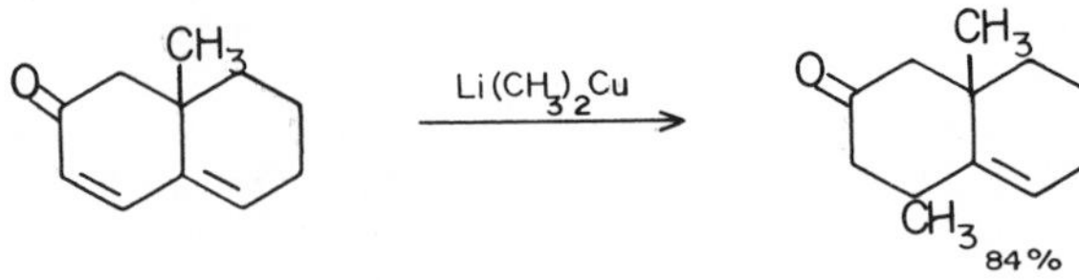

Conjugate ethylation with diethylcadmium occurs as well, with the exception of mesityl oxide and fluorenone types, in which 1,2-addition takes place [65].

Borane also exhibits 1,4-addition as shown for one example [66]:

$$CH_3COCH = CH_2 + Oct_3B \longrightarrow OctCH_2CH = \underset{\underset{OBOct_2}{|}}{C}CH_3 \xrightarrow{H_2O} OctCH_2CH_2COCH_3$$

2-Dodecanone was obtained in about 86% yield with about 14% of the isomer, 3-methyl-2-undecanone, present.

b.   Other Alkylations

Two alkylations were carried out in a single pot as indicated [67]:

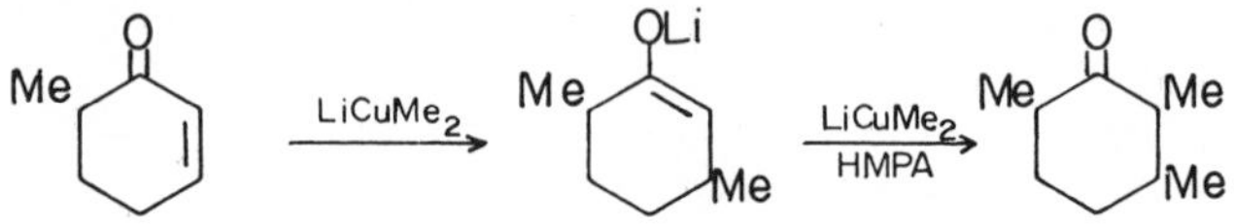

The 2,3,6-trimethylcyclohexanone was obtained in 95% yield.

Among different types of alkylating agents are diethyl-aminopropyne, $Et_2NC \equiv CCH_3$ [68], and Mannich bases, the latter of which is shown [69]:

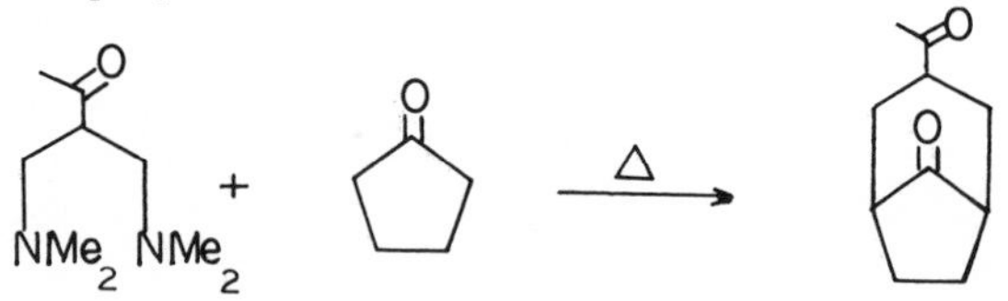

3-Acetylbicyclo[1.2.3]octanone-8 was obtained in 19% yield from a complex mixture. The reaction, which is called a thermal Michael reaction (1, 707), requires no catalyst other than the amino groups present. Other alkylating agents are hexynyldi-ethylaluminum, $BuC \equiv CAlEt_2$, which adds only to cis-enones [59], and dianions [70], the latter of which is shown:

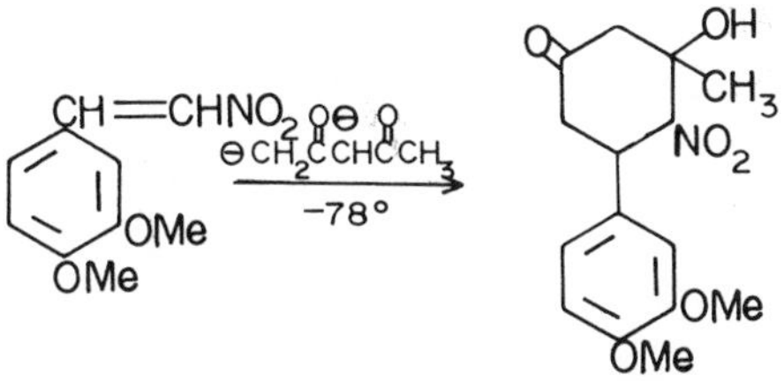

3-Methyl-3-hydroxy-4-nitro-5(3,4-dimethoxyphenyl)cyclohexanone
was obtained in 73% yield.

Cyclopropanone methyl hemiketal is oxidized by the copper
ion to the carbomethoxyethyl free radical, $\cdot CH_2CH_2CO_2Me$, which
adds conjugatively to unsaturated ketones [71].

Anions and nucleophiles add well to methyl 2-methylthio-
acrylate to give various ketoesters (see D.7), and diketones as
shown [72]:

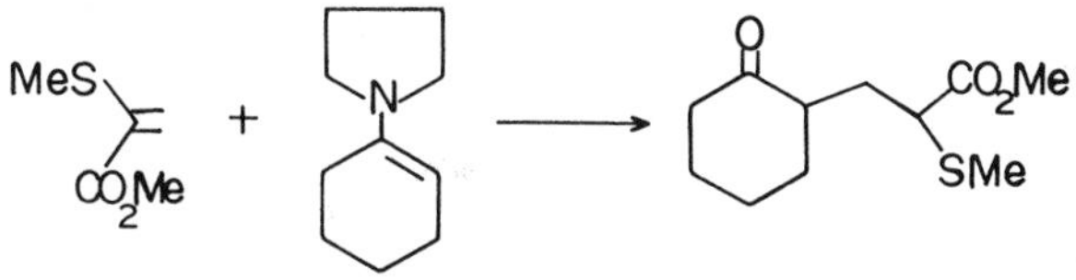

Methyl 2-methylthio-3(2-oxocyclohexyl)propionate was obtained in
95% yield after hydrolysis. Conjugative additions, ring opening,
and closure take place with an unsaturated cyclopropane ester as
indicated [73]:

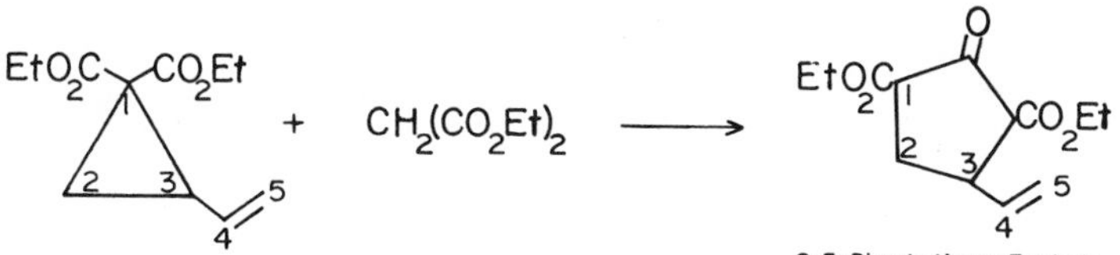

2,5-Dicarbethoxy-3-vinyl-
cyclopentanone

This reaction might be called a "homo Michael addition" [74].

Tributylphosphine has been used as a catalyst in the Michael
reaction [75]. In fact, it works well in the Michael reaction
of nitroalkanes with unsaturated ketones, nitriles, and esters.

Finally, to prepare allenic ketones from β,γ-unsaturated
ketones, the following route has been investigated [76]:

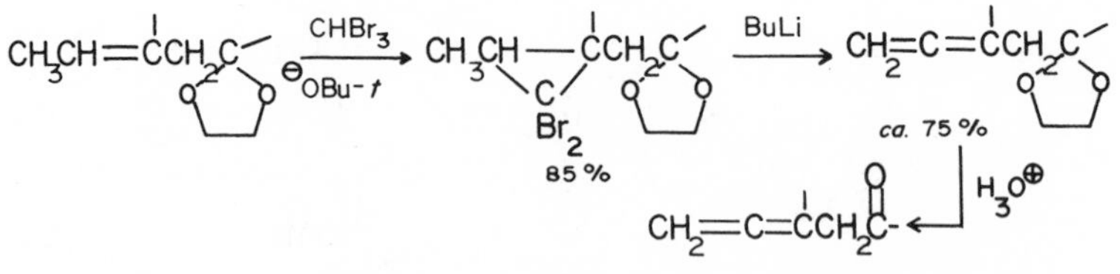

4. From Allylic Rearrangements or Cyclopropane Ring
    Openings (1, 707)

Ketocyclopropanes may be prepared from diazoketones, a
potential source of ketocarbenes, by treatment with a vinyl
acetate as shown [77]:

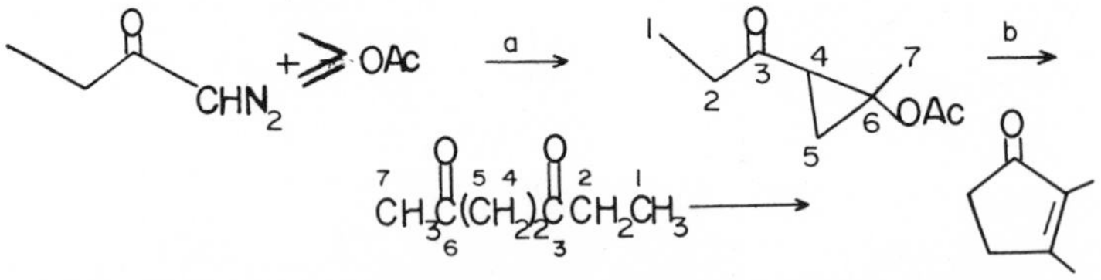

In the first step (a) cuprous acetylacetone is by far the best
catalyst in giving 55% of the cyclopropane. The latter cleaves
to the diketone in the second step (b) by splitting between carbons 4 and 6, with alkali. Then the diketone cyclizes to 2,3-
dimethyl $\Delta^{2,3}$-cyclopentenone in 85% yield.

5. From Aldehydes or Their Derivatives (1, 708)

Discussion of this subject has been diverted to F.7.

9. From Dihalo-, Diazoketones, or Other Substituted Ketones

This reaction is used occasionally to prepare hindered
ketones as, for instance, di-t-butyl ketone from α,α'-dibromo-
diisopropyl ketone [78]. Similarly, lithium t-butoxy(butyl)cu-
prate has been employed to prepare 2-butylcyclohexanone (77%)
from α,α'-dibromocyclohexanone [62]. (See H.2 for the addition
of LiCuROBu-t to conjugated ketones.) α-Chlorothioethers also
alkylate α-bromoketones if the latter are first treated with
zinc to form the Reformatsky-like organometallic intermediate
[79].

Diazoketones are converted best to alkylated ketones by treatment with boranes followed by hydrolysis as shown [80]:

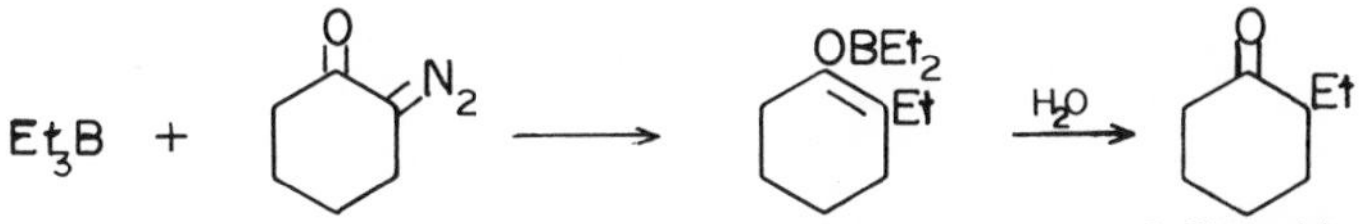

2-Ethylcyclohexanone
98%

Also, the vinyloxyboranes such as A may be treated with bromine to give a regiospecific brominated ketone [81]. Others have treated the vinyloxyboranes with carbonyl compounds to produce β-hydroxyketones [82]:

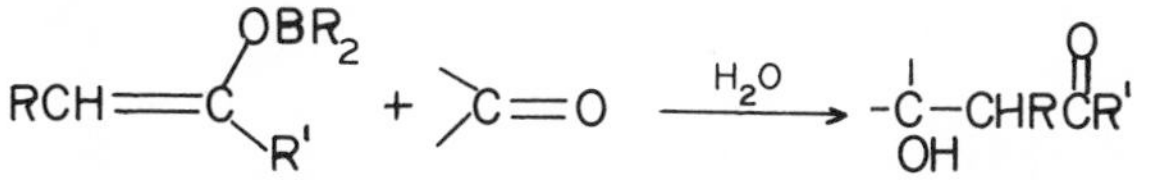

Yields were 42-98% for four examples. The diazoketone may be elaborated before further reaction [83]:

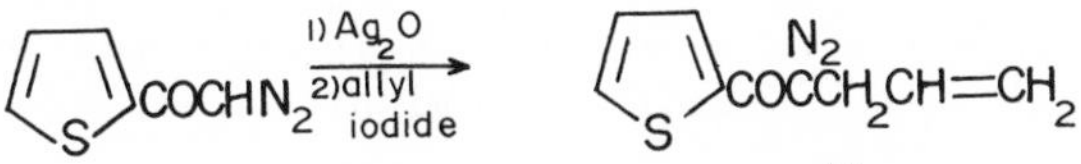

2-(α-Diazo-Δ^{4,5}-pentenoyl)-
thiophene, 52%

A ring closure ensues from an intramolecular alkylation with a diazoketone [84]:

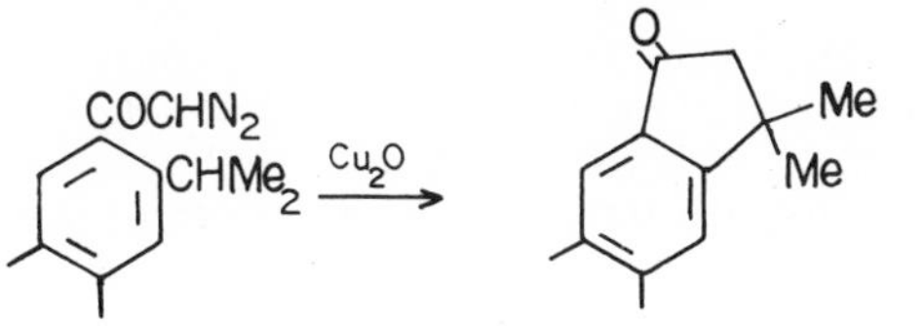

Another group that may be replaced in alkylation or arylation is the formyl, as in β-ketoaldehydes [85]:

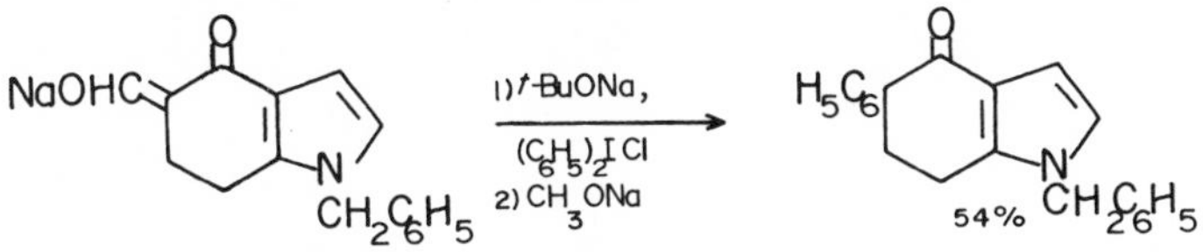

When CH$_3$Br was substituted for Ph$_2$ICI, the yield of the 5-methyl-4-oxotetrahydroindole was 82%.

10.   From Aldehydes and Ketones (Aldolization, Retroaldolization, and the Like

The aldol product usually dehydrates to the unsaturated ketone very readily during workup (see 1, 236), but House [86] found that zinc or other divalent ions stabilized it so that no dehydration occurs:

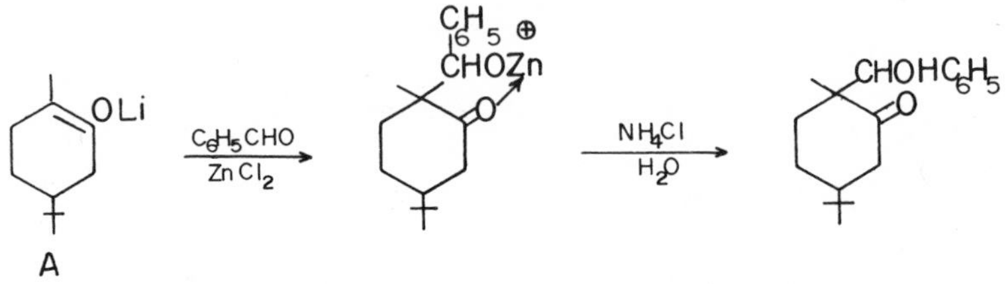

A

5-<u>t</u>-Butyl-2-methyl-2(phenylmethylol)-cyclohexanone was obtained in 85% yield.   The enolate (A) was usually prepared from the ketone with LDIA or from the enol acetate plus MeLi.   Care must be taken in breaking up the zinc complex.

The other most novel advance in the field of aldolization is concerned with the partial success in choosing the product desired in the Claisen-Schmidt reaction as indicated:

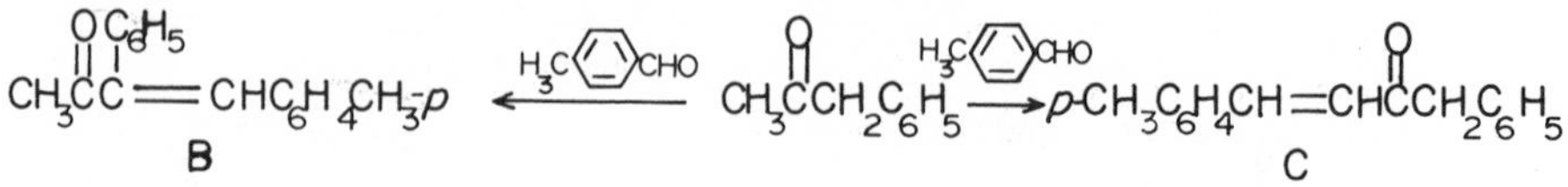

B   **C**

The usual route by using catalysts such as 10% NaOH is condensation at the methyl group to give C.   Fine and Pulaski [87] argue that the C route is the thermodynamically controlled one. A weaker base as a catalyst should give both aldol products at least.   By using piperidine and a Dean-Stark water trap (B) methyl α-phenyl-<u>p</u>-methylstyryl ketone was obtained in 62% yield.

The Claisen-Schmidt reaction is more profound with benzils than with monoketones as shown [88]:

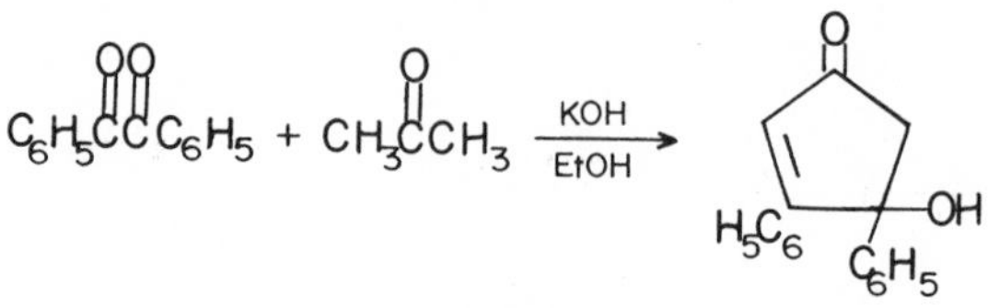

The 3,4-diphenyl-4-hydroxycyclopentenones are obtained in fair yield.   In the synthesis of dypnone, $C_6H_5C = CHCOC_6H_5$,
CH_3

diethylpivaloyloxyboron, $Et_2BOCOMe_3$ [89] has been used as a catalyst. Some of the tricondensation product of acetophenone, $PhCOCH=CPhCH=CPhCH_3$, was obtained from the residue of one of these reactions.

An interesting retroaldol has led to the more stable isomer of the following cyclopentenone system [90]:

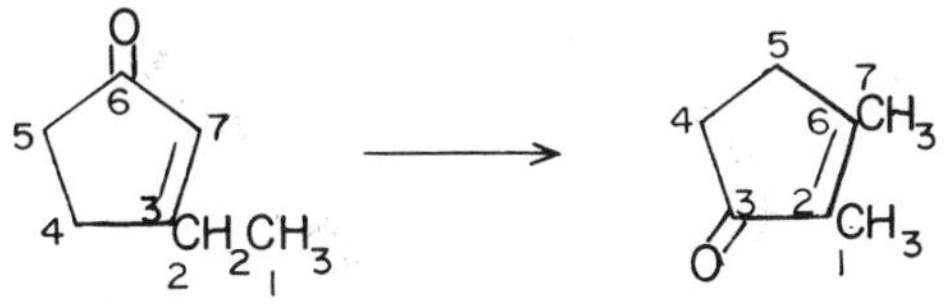

The substrate was refluxed with 3-5% aqueous NaOH for 72 hr to give 2,3-dimethyl-$\Delta^{2,3}$-cyclopentenone. The corresponding reaction with cyclohexenones is much more sluggish.

Addenda

G.1.   The enolate anion seems to be more selective (but gives lower yields) in regiospecific alkylation in the presence of a quaternary ammonium fluoride [I. Kuwajima, E. Nakamura, J. Am. Chem. Soc., 97, 3257 (1975)].

G.1.   A method of controlling the orientation of the double bond in 2-methylcyclohexenone has been described [P. L. Stotter, K. A. Hill, J. Org. Chem., 38, 2576 (1973)].

G.1.   The following examples demonstrate the increased interest in enolate anions, generated by the use of LDIA:

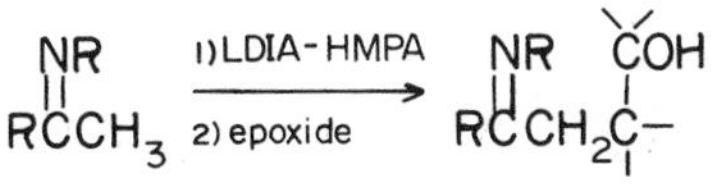

[M. Larcheveque et al., Synthesis, 256 (1975)];

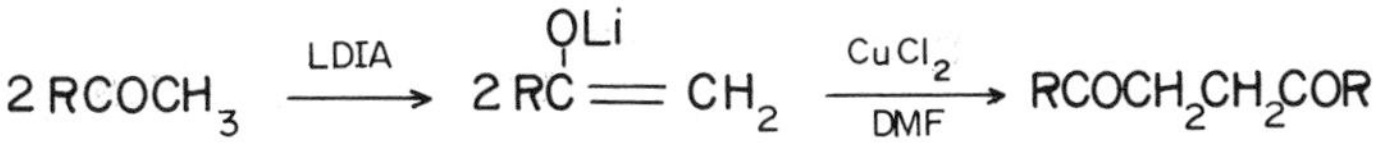

[Y. Ito et al., J. Am. Chem. Soc., 97, 2912 (1975)];

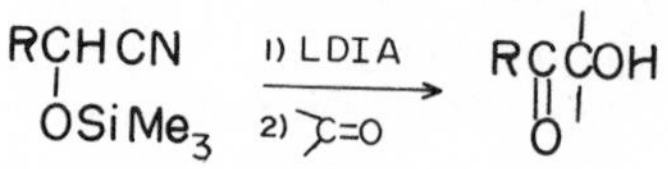

$$\begin{array}{c} RCHCN \\ | \\ OSiMe_3 \end{array} \xrightarrow[2)\; \diagdown C=O]{1)\; LDIA} \begin{array}{c} | \\ RC\!-\!COH \\ \| \quad | \\ O \end{array}$$

[S. Hünig, G. Wehner, Synthesis, 391 (1975)].

G.1. The enolates of many ketones may be prepared at 20° with potassium hydride, which is a much more reactive base than sodium hydride, with no evidence of self-condensation. These conditions offer a contrast to those of LDIA and ketones (-78°) [C. A. Brown, J. Org. Chem., 39, 3913 (1974)].

G.1. Corey has reported a clever α-alkylation of α,β-unsaturated ketones

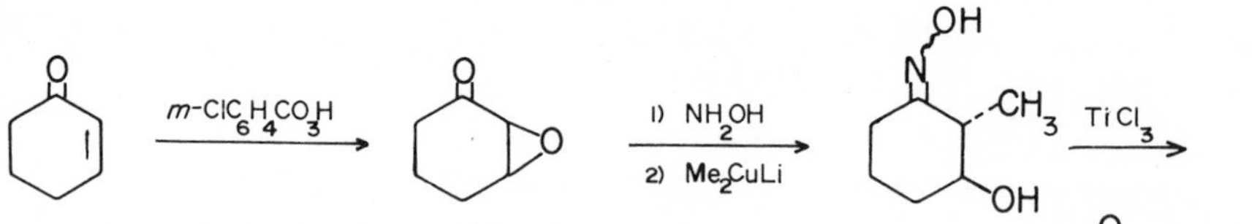

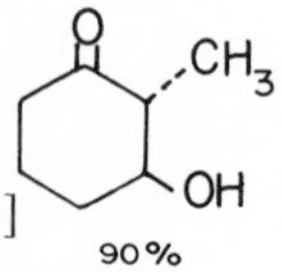

[E. J. Corey et al., Tetrahedron Letters, 3117 (1975)]

90%

G.1. Halobenzenes undergo an intramolecular condensation with ketones as shown

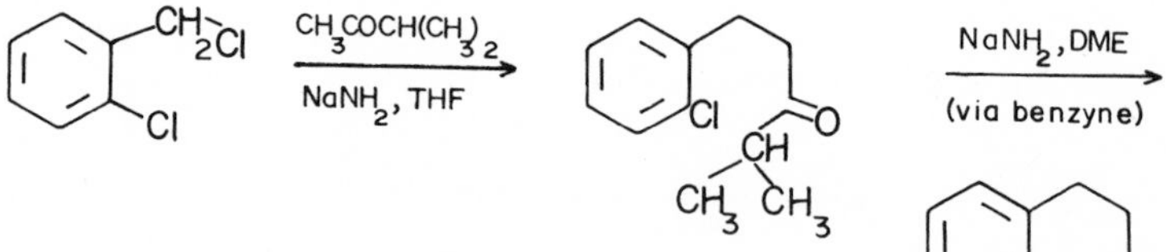

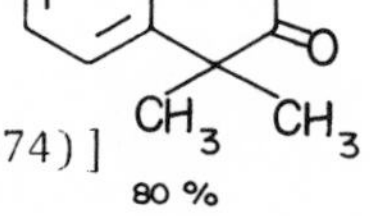

[B. Loubinoux, P. Caubere, Synthesis, 201 (1974)]

80 %

G.3.   The addition of lithium dialkylcuprate to a vinylcyclo-
propyl alkyl ketone to produce α,β-unsaturated ketones has been
described [A. Suzuki et al., Synthesis, 317 (1975)].

G.3.   Both conjugate and nonconjugate addition may be carried out
consecutively as illustrated:

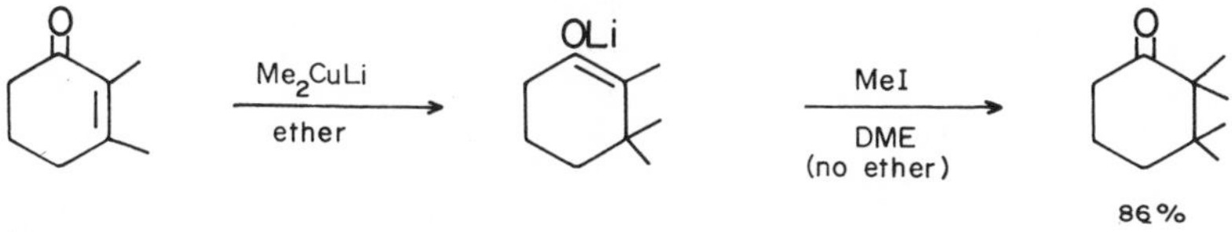

86%

[R. M. Coates, L. O. Sandefur, J. Org. Chem., 39 (1974)]

G.3.   For annelation ICH$_2$CH=C(CH$_3$) SiMe$_3$ with the lithium enolate
of 2-methyl cyclohexane forms

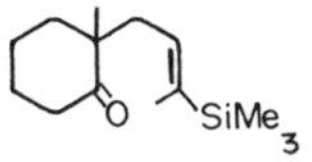

which with m-chloroperbenzoic acid at 0° followed by base gives
[G. Stork, M. E. Jung, J. Am. Chem. Soc., 96,
3682 (1974)].

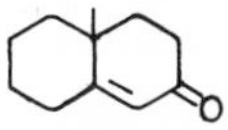

G.3.   Kinetically controlled methylation of an α,β-unsaturated
ketone (on the saturated side) may be accomplished with t-BuOK in
THF at 70° followed by MeI [L. Nedelec et al., Tetrahedron, 30,
3263 (1974)], or by cyclopropanation of the enol of trimethylsilyl ether followed by alkali ring opening [J. M. Conia et al.,
Tetrahedron Letters, 3327, 3329, 3333 (1974)]; however, 1,4-
methylation may be accomplished with AlMe$_3$ - Ni(acac)$_2$ [L. Bagnell
et al., Australian J. Chem., 28, 801 (1975)].

G.3.   A new, superior synthesis of lithium dialkylcuprates is
recorded:

$$CuBr + Me_2S \longrightarrow Me_2SCuBr \xrightarrow{2\ RLi} LiCuR_2$$

G.3.   Diazomethane has been used to conjugatively methylate an
α,β-unsaturated ketone [T. Uyehara et al., Syn. Commun., 3, 365
(1973)].

References

1. M. Gall, H. O. House, Org. Syn., 52, 39 (1972).
2. H. O. House et al., J. Org. Chem., 36, 2361 (1971).
3. G. Stork, B. Ganem, J. Am. Chem. Soc., 95, 6152 (1973).
4. J. M. Conia, G. Girard, Tetrahedron Letters, 2767 (1973).
5. S. Murai et al., Chem. Commun., 741 (1972).
6. M. Makosza et al., Tetrahedron Letters, 1351 (1971).
7. M. Makosza et al., Rocz. Chem., 47, 44 (1973); C. A., 79, 18,305 (1973).
8. A. Brandstrom, W. Junggren, Tetrahedron Letters, 473 (1972).
9. J. Fauvarque, J.-F. Fauvarque, Bull. Soc. Chim. France, 160 (1969).
10. R. B. Miller, Syn. Commun., 2, 267 (1972).
11. A. S. Kende, R. G. Eilerman, Tetrahedron Letters, 697 (1973).
12. A. L. Kurts et al., J. Org. Chem. USSR, 9, 1341 (1973).
13. R. A. Rossi, J. F. Bunnett, J. Org. Chem., 38, 1407 (1973).
14. M. F. Semmelhack et al., Tetrahedron Letters, 4519 (1973).
15. T. M. Harris, C. M. Harris, Org. Reactions, 17, 155 (1969).
16. M. P. Cooke, Jr., J. Org. Chem., 38, 4082 (1973); J. Am. Chem. Soc., 95, 7891 (1973).
17. H. G. Heine, Ann. Chem., 735, 56 (1970).
18. E. C. Taylor et al., J. Am. Chem. Soc., 90, 2421 (1968).
19. C. Alexandre, F. Rouessac, Bull. Soc. Chim. France, 1837 (1971).
20. H. Felkin, C. Lion, Tetrahedron, 27, 1387 (1971).
21. H. Normant et al., Compt. Rend., 277, C511 (1973).
22. G. Cainelli et al., Tetrahedron Letters, 2491 (1973).
23. R. B. Miller, B. F. Smith, Syn. Commun., 3, 129 (1973).
24. M. Tramontini, Synthesis, 703 (1973).
24a. R. B. Miller, B. F. Smith, Tetrahedron Letters, 5037 (1973).
25. A. G. Cooke, Enamines, Dekker, New York, 1969, p. 346.
26. R. W. Franck et al., Tetrahedron Letters, 3107 (1973).
27. R. S. Monson et al., Tetrahedron Letters, 929 (1972).
28. M. E. Kuehne, J. C. King, J. Org. Chem., 38, 304 (1973).
29. S. A. G. de Graff, U. K. Pandit, Tetrahedron, 29, 2141 (1973).
30. A. Z. Britten et al., Tetrahedron, 25, 3157 (1969).
31. D. J. Dunham, R. G. Lawton, J. Am. Chem. Soc., 93, 2074 (1971).
32. R. L. Augustine, J. A. Caputo, Org. Syn., 45, 80 (1965).
33. H. Ahlbrecht, G. Rauchschwalbe, Synthesis, 417 (1973); M. Yoshimoto et al., Tetrahedron Letters, 39 (1973); H. W. Thompson, B. S. Huegi, Chem. Commun., 636 (1973).
34. H. Onoue et al., Tetrahedron Letters, 121 (1973).

35. J. Tsuji, Bull. Chem. Soc. Jap., 46, 1896 (1973).
36. M. Rosenblum et al., J. Am. Chem. Soc., 95, 3062 (1973).
37. D. Cantacuzene et al., Tetrahedron, 29, 4233 (1973).
38. B. P. Mundy, J. Chem. Educ., 37, 4483 (1972).
39. R. A. Kretchmer et al., J. Org. Chem., 37, 4483 (1972).
40. J. F. McMurry et al., Tetrahedron Letters, 4995 (1971).
41. J. E. McMurry, Org. Syn., 53, 70 (1973).
42. R. K. Boeckman, Jr., J. Am. Chem Soc., 95, 6867 (1973); 96, 6179 (1974); G. Stork, B. Ganem, ibid., 95, 6152 (1973); 96, 6180 (1974).
43. G. Stork, J. d'Angelo, J. Am. Chem. Soc., 96, 7114 (1974).
44. M. Kobayashi, T. Matsumato, Chem. Lett., 957 (1973).
45. A. Sammour et al., J. Prakt. Chem., 314, 139 (1972).
46. Y. Hayashi et al., Chem. Lett., 387 (1975).
47. G. Stork, J. Benaim, J. Am. Chem. Soc., 93, 5938 (1970); M. Tanabe, D. F. Crowe, Chem. Commun., 564 (1973); K. P. Dastur, Tetrahedron Letters, 4333 (1973); W. Reusch et al., Tetrahedron Letters, 965 (1973); A. G. Schultz, D. S. Kashdan, J. Org. Chem., 38, 3814 (1973); J. A. Marshall et al., Tetrahedron Letters, 3795 (1971).
48. G. Stork, R. L. Danheiser, J. Org. Chem., 38, 1775 (1973).
49. C. J. Sih et al., J. Am. Chem. Soc., 97, 857 (1975).
50. E. J. Corey, R. H. K. Chen, Tetrahedron Letters, 3817 (1973).
51. G. H. Posner, Org. Reactions, 19, 1 (1972).
52. P. E. Eaton et al., J. Org. Chem., 37, 1947 (1972).
53. J. Hooz, R. B. Layton, Can. J. Chem., 48, 1626 (1970).
54. E. J. Corey, R. L. Carney, J. Am. Chem. Soc., 93, 7318 (1971).
55. H. O. House et al., J. Org. Chem., 40, 1460 (1975).
56. H. O. House et al., J. Org. Chem., 34, 3615 (1969); J. F. Normant, Synthesis, 63 (1972).
57. E. J. Corey, D. J. Beames, J. Am. Chem. Soc., 94, 7210 (1972).
58. H. O. House, M. J. Umen, J. Org. Chem., 38, 3893 (1973).
59. J. Hooz, R. B. Layton, J. Am. Chem. Soc., 93, 7320 (1971).
60. G. W. Kabalka, R. F. Daley, J. Am. Chem. Soc., 95, 4428 (1973).
61. J.-P. Garlier et al., Chem. Commun., 88 (1973).
62. G. H. Posner et al., J. Am. Chem. Soc., 95, 7788 (1973).
63. T. Mukaiyama et al., J. Am. Chem. Soc., 94, 8641 (1972).
64. J. A. Marshall et al., Tetrahedron Letters, 3795 (1971).
65. M. Goemen et al., Bull. Soc. Chim. France, 562 (1973).
66. P. I. Paetzold, H. Grundke, Synthesis, 653 (1973); H. C. Brown et al., J. Am. Chem. Soc., 93, 3777 (1971); 89, 5708 (1967).
67. R. K. Boeckman, Jr., J. Org. Chem., 38, 4450 (1973).

68. J. Ficini, J. d'Angelo, Compt. Rend., 276, C803 (1973);
     J. Ficini, J. P. Genet, Tetrahedron Letters, 1565 (1971).
69. E. M. Austin et al., Tetrahedron, 25, 5517 (1969).
70. D. Seebach, V. Ehrig, Angew. Chem. Intern. Ed. Engl., 13,
     400 (1974).
71. Th. J. de Boer, Tetrahedron Letters, 827 (1973).
72. J. L. Herrmann et al., Tetrahedron Letters, 2603 (1973).
73. N. A. Abraham, Tetrahedron Letters, 451 (1973) and earlier
     papers.
74. P. L. Fuchs, J. Am. Chem. Soc., 96, 1607 (1974).
75. D. A. White, M. M. Baizer, Tetrahedron Letters, 3597
     (1973).
76. M. Santelli, M. Bertrand, Bull. Soc. Chim. France, 2326
     (1973).
77. J. E. McMurry, T. E. Glass, Tetrahedron Letters, 2575
     (1971).
78. J.-E. Dubois et al., Tetrahedron Letters, 177 (1971).
79. I. I. Lapkin, J. Org. Chem. USSR, 9, 1433 (1973).
80. J. Hooz et al., Can. J. Chem., 49, 2371 (1971); H. Kono,
     J. Hooz, Org. Syn., 53, 77 (1973).
81. J. Hooz, J. N. Bridson, Can. J. Chem., 50, 2387 (1972).
82. K. Fromata et al., Bull. Chem. Soc. Jap., 46, 1807 (1973);
     T. Mukaiyama et al., J. Am. Chem. Soc., 95, 967 (1973).
83. U. Schöllkopf, N. Rieber, Chem. Ber., 102, 488 (1969).
84. R. C. Cambie, R. A. Franich, Australian J. Chem., 24, 117
     (1971).
85. W. A. Remers et al., J. Org. Chem., 36, 1232 (1971).
86. H. O. House et al., J. Am. Chem. Soc., 95, 3310 (1973).
87. S. A. Fine, P. D. Pulaski, J. Org. Chem., 38, 1747 (1973).
88. T. J. Clark, J. Org. Chem., 38, 1749 (1973).
89. R. Köster, A.-A. Pourzal, Synthesis, 674 (1973).
90. P. M. McCurry, Jr., R. K. Singh, J. Org. Chem., 39, 2319
     (1974).

H. Organometallic Methods

Organometallic methods have been used generously in the examples provided in Sections F and G and are completed here. Cross references are introduced at appropriate places.

1. From Esters, Lactones, and Polyfunctional Ketones (1, 715)

$$RCO_2R' + R''MgX \longrightarrow RCOR''$$

Earlier it was stated that the preceding reaction was not very useful because of the great tendency of the ketone or ketone complex

to form tertiary alcohols. Apparently, however, the reaction
works with isopropylmagnesium bromide and aliphatic esters, at
least if an equivalent of HMPA is included with the solvent [1].
Under such conditions it is thought that the enolate of the
ketone, which does not add another equivalent of the Grignard
reagent to form the tertiary alcohol, is formed. Yields ranged
from 7.5% ($C_6H_5CH_2COOCH_3$ + C_3H_7MgCl) to 98% ($(CH_3)_3CCOOCH_3$ +
C_3H_7MgCl). Lithium dimethylcuprate also moderates the ester
addition [2]. It gave 85% of acetophenone with ethyl benzoate.
Lithium 2,6-dimethoxyphenyl has been added to pyridinecarboxylic
esters to give low to fair yields of the aroylpyridine [3]. For
other methods of going from acid derivatives to ketones see
Addenda H.2 (two examples).

 2-Thioacyloxypyridine reacts with the Grignard reagent to
form ketones as indicated [4]:

$$RC(O)S\text{-pyridyl} + R'MgBr \longrightarrow RCOR'$$

Seven examples gave yields of 83-97%. Methyl thiobenzoate yields
dibenzoylmethane (64%) by treatment with a powerful base such as
lithium 2,2,6,6-tetramethylpiperidide [5]. The reaction may
occur as shown:

$$C_6H_5COSCH_3 \longrightarrow C_6H_5COS\overset{\ominus}{C}H_2 \longrightarrow C_6H_5COSCH_2\overset{\ominus}{C}OC_6H_5 \longrightarrow$$

$$C_6H_5COSCH(COC_6H_5)_2 \xrightarrow[\text{2) }H_2O]{\text{1) }SCH_3} (C_6H_5CO)_2CH_2$$

 Lithiumvinyl behaves as though it inserts itself into thio-
lactones as indicated [6]:

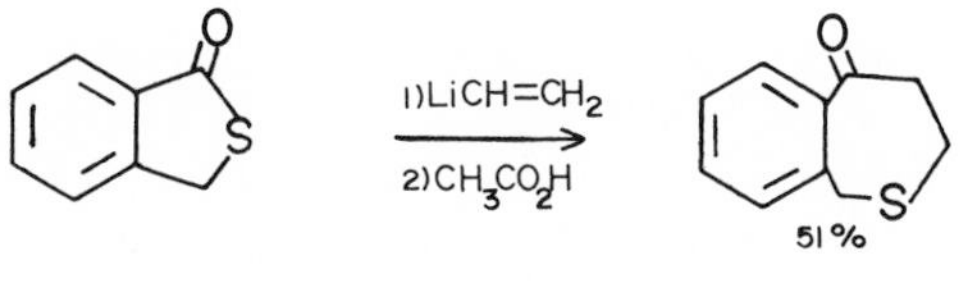

51%

 If one utilizes the theory of vinology, the pseudoester
below behaves normally [7]:

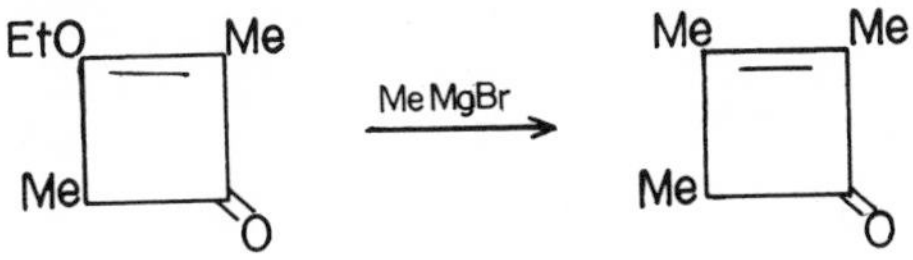

The 2,3,4-trimethylcyclobutenone was obtained in 60% yield.

2. From Acids, Acid Chlorides, and Anhydrides (1, 716)

$$RCOCl + R'MgX \longrightarrow RCOR'$$

Another splendid method of preparing ketones has been added to the already extensive list of ketone preparations. The new method [8] is simple and direct and when it succeeds is the method of choice. It involves adding 2 equiv of an alkyl- or aryllithium to the acid or 1 equiv to the lithium salt of the acid:

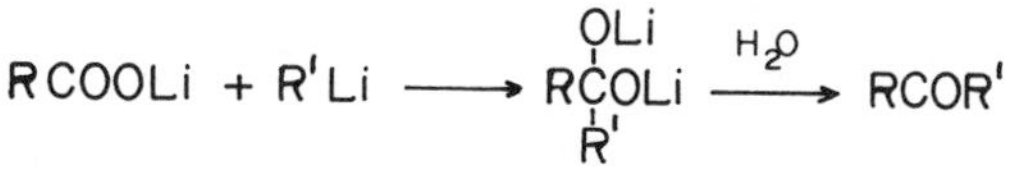

$$RCOOLi + R'Li \longrightarrow \underset{\underset{R'}{|}}{R\overset{\overset{OLi}{|}}{C}OLi} \overset{H_2O}{\longrightarrow} RCOR'$$

In a typical reaction a flocculent precipitate forms on the addition of the first equiv of alkyllithium to the acid in ether and partial or complete solution occurs on the addition of the second equiv. Another way is to prepare the lithium salt from lithium hydride and then add 1 equiv of the alkyllithium. Yields and conditions for synthesizing methyl ketones by a tested procedure have been published [9]. Numerous methods have been compared in the review [8] with the result that the alkyllithium procedure appears to be superior. Under special circumstances it may be desirable to resort to the mixed Claisen reaction (1, 690), to decarboxylation (1, 686), to the organometallic reagent— nitrile (1, 717), or the organometallic reagent-acid chloride (to be discussed shortly) to prepare a specific ketone.

The alkyllithium addition has been extended to include addition to anthranilic acids [10]. In the case of β,γ- unsaturated acids a rearrangement occurs [11]:

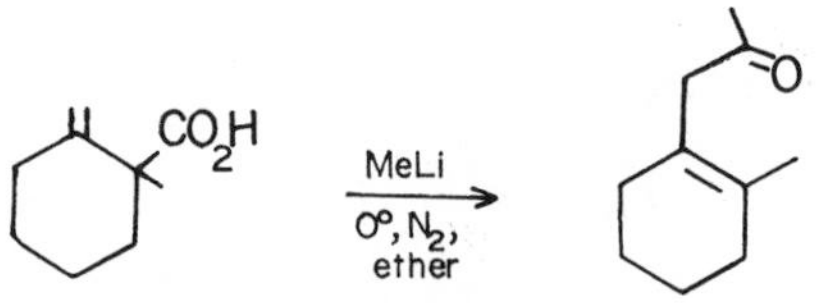

The rearrangement probably occurs via formation of lithium acetate and the anion of 2-methylmethylidenecyclohexane, which recombine to form 2-methylcyclohexenylacetone. If hexane is sub- stituted for the ether (just enough ether to dissolve the compo- nents), no rearrangement occurs and normal 2-methylidene-1-methyl- 1-acetylcyclohexane is obtained.

The lithium salt of the trimethylsilyl ethyl malonate adds to an unsaturated acid chloride as indicated [12]:

$$CH_3CH = CHCOCl + \underset{\underset{CO_2Et}{|}}{LiCHCO_2SiMe_3} \longrightarrow CH_3CH = CH\overset{\overset{O}{\|}}{C}CH_2CO_2Et$$

Ethyl crotonylacetate was obtained in 75% yield by warming the components overnight at 25°.

Lithium dimethylcuprate has been found to be an excellent reagent for addition to acid chlorides [13]:

$$R'COCl + LiCuR_2 \xrightarrow[-78°]{Et_2O} R'COR$$

Good to high yields were obtained for 16 examples. And the reaction has been extended to the preparation of 2-acetylindene (87% at -60°) [14]. At -50° or higher 1-methyl-2-acetylindane is obtained by conjugate addition to the unsaturated ketone. The reaction has been extended to the preparation of α-chloroketones from α-chloroacid chlorides [15] and acetylenic ketones by the addition of lithium alkynyliodocuprate, Li(RC≡CCuI) to acid chlorides [16]. For addition of alkyl groups other than methyl, the compound lithium t-butoxyalkylcuprate, t-BuO(R)CuLi, was found to be a highly effective reagent [17]. The reagent should be used within 1-2 hr if stored at -50°. The yield of pivalophenone (82%) from this reagent (R=t-butyl) and benzoyl chloride exemplifies its effectiveness. Incidentally, yields of 3-alkylketones from the addition of the above reagent to unsaturated ketones were very good except when R=vinyl.

Catalytic amounts of cuprous chloride added to the Grignard reagent and acid chlorides promote free radical reactions. A study of the aging of ethylmagnesium bromide with cuprous chloride shows that the yield of ketone from diisopropylacetyl chloride increases to 80% (40 days) from 60% (10 days) [18]. It is stated (a) that cuprous chloride is indispensable for the reaction of secondary or tertiary alkyl Grignard reagents with acid chlorides, (b) that the Grignard reagent should be added to the acid chloride and cuprous chloride for best yields, and (c) that although no reaction takes place at -78°, it is desirable to begin at this temperature and allow the mixture to warm gradually [19]. However, alkyllithiums do add to acid chlorides at -78°.

The Grignard addition to acid chloride to form ketones has been improved considerably by the addition of 3 equiv of HMPA [20]. This complexing agent also improved yields in the addition of Grignard reagents to DMF to form aldehydes except when t-butyl or styryl Grignard reagents were used. (See H.1 for an explanation of the effectiveness of HMPA.) For that matter, the addition of 3 equiv of phenylmagnesium bromides to thiophthalic anhydride gives o-dibenzoylbenzene in 80% yield, provided an oxidative hydrolysis is used [21].

Unusual couplings with acid chlorides are those with ethylaluminum sesquichloride [22]:

$$\underset{\displaystyle MeO\overset{\textstyle O}{\overset{\|}{C}}(CH_2)_3COCl}{} \xrightarrow[CH_2Cl_2]{Et_3Al_2Cl_3} \underset{\displaystyle MeO\overset{\textstyle O}{\overset{\|}{C}}(CH_2)_3\overset{\textstyle O}{\overset{\|}{C}}Et}{}$$

Methyl 5-oxoheptanoate was obtained in 85% yield. Another is that with vinylmagnesium chloride and cuprous chloride [23]:

$$C_4H_9CO_2H \; + \; CH_2{=}CHMgCl \longrightarrow C_4H_9CO(CH_2)_4COC_4H_9$$

The yield of 1,4-dipentanoylbutane was 53%, but it was lower with other acids or esters. A third is that resulting from the addition of a Grignard reagent to the product of an acid chloride and triethyl phosphite [24]:

$$C_6H_5CO\underset{O}{\overset{\|}{P}}(OEt)_2 \xrightarrow{C_6H_5MgBr} (C_6H_5)_2\overset{OMgBr}{\underset{O}{\overset{|}{C}\overset{\|}{P}}}(OEt)_2 \xrightarrow{H_2O} (C_6H_5)_2C{=}O$$

A fourth involves the reaction of alkyl halides with lithium acylirontetracarbonyl, $LiRCOFe(CO)_4$, to form ketones in 21-67% yields (7 ex.) [25].

3.   From Nitriles or Isocyanides (1, 717)

$$RC{\equiv}N \; + \; R'MgX \longrightarrow R\underset{R'}{\overset{\textstyle}{\overset{|}{C}}}{=}N\overset{\oplus}{M}gX \xrightarrow{H_2O} R\underset{R'}{\overset{|}{C}}{=}O$$

Although the nitrile method is a good general procedure for preparing ketones, acetonitrile produces low yields because of its tendency to form a Grignard reagent complex and to liberate some hydrocarbon from its active hydrogen site [26]. Replacement of the ether with benzene raises the yield somewhat. If a large excess of the Grignard reagent is used, the ketone and the corresponding amine are both obtained [27]. A mixture of two different Grignard reagents reacts with a dinitrile to form the mixed diadduct in low but reasonable yield (22%) [28]. Addition occurs with the Grignard reagent and benzoyl cyanide to give ketones, but reduction is prevalent with acetyl cyanide [29].

Cyanohydrins may be used as substrates for Grignard reagent additions, provided the hydroxyl group is protected by the trimethylsilyl moiety [30]. Protected as the acetal, aldehyde cyanhydrins may be alkylated before Grignard reagent addition

[31]. Grignard reagent addition followed by alkylation may be achieved with aromatic nitriles [32]:

$$C_6H_5CN + CH_3CH_2MgBr \xrightarrow[\text{overnight}]{0-25°} \underset{\overset{\|}{NMgBr}}{C_6H_5\overset{}{C}CH_2CH_3} \xrightarrow[\text{2) H}_3O^{\oplus}]{\text{1)HMPA,}} \underset{\overset{|}{Bu}}{C_6H_5COCHCH_3}$$

Phenyl sec-hexyl ketone was obtained in 63% yield.

An elaboration of the nitrile synthesis results from Reformatsky conditions [33]:

$$\underset{\overset{\|}{O}\ \overset{\|}{O}}{C_6H_5\overset{}{C}CH_2CN} + Me_2CBrCO_2Et \xrightarrow[\text{2)H}_3O^{\oplus}]{\text{1) Zn}} C_6H_5COCH_2COCHMe_2$$

Benzoylisobutyromethane was obtained in 62% yield.

The isonitrile synthesis of Walborsky, which served so well in the synthesis of aldehydes (Chapter 10, G), has now been used to prepare ketones [34]. Triphenylmethylisocyanide, which with 2 equiv of butyllithium gave 59% of 5-nonanone and 9% of valeronitrile, is the reagent used.

4. From Amides, Imidazoles, or the Like (1, 718)

$$RCONR'_2 + R''MgX \longrightarrow RCOR''$$

Dimethylformamide addition to an organometallic compound serves as an excellent method of aldehyde preparation (Chapter 10, G, Ref. 20, H.2). Although the addition of organometallic compounds to amides other than DMF does not work as well, it is useful on occasion. Organolithium compounds have been added to diamides at -78° to produce diketones in 4-76% yields [35]. In addition, a method has been described for producing the alkyllithium in situ; the latter then adds to the amide to form the ketone [36]. With diethylbenzamide yields of ketones were in the 70% range, but with diethylacetamide in the 10% range. Aryllithiums have been added to tetramethyloxamide to form aroylformamides, ArCOCONMe$_2$, in 15-92% yields (11 ex.) [37]. The S-alkylthioamidium iodides are also reagents that react with organometallic compounds to form ketones [38]:

$$\left[\underset{\overset{|}{SMe}}{Me_2\overset{\oplus}{N}=\overset{}{C}C_6H_5} \right]_I^{\ominus} \xrightarrow[\text{2) H}_2O]{\text{1) C}_6H_5MgBr} \underset{97\%}{C_6H_5COC_6H_5}$$

6.   From Unsaturated Ketones ($\underline{1}$, 720).

This material has been incorporated into G.3.

7.   From Phosphorus Ylids and the Like ($\underline{1}$, 721)

$$C_6H_5COCHO + (C_6H_5)_3P=CHCOC_6H_5 \longrightarrow C_6H_5COCH=CHCOC_6H_5$$

The preceding reaction is a typical alkenylation. The ylid
is stable and the reaction, which gives dibenzoylethylene in 85%
yield, is exothermic [39]. Other types of phosphorus-ylid
manipulation have been studied. Those of enaminophosphorus are
sources of α,β-unsaturated ketones [40]:

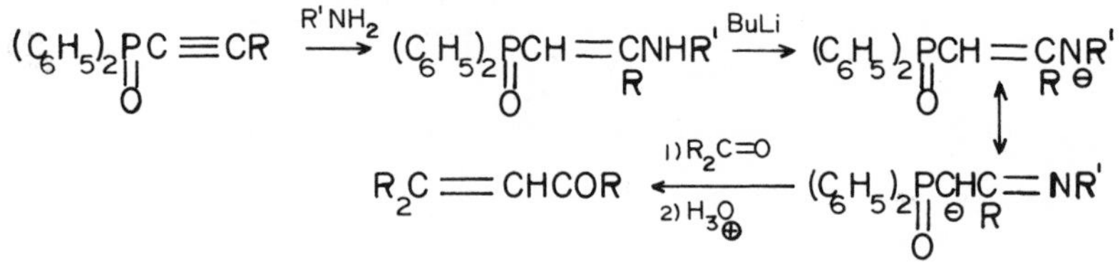

$$(C_6H_5)_2PC \equiv CR \xrightarrow{R'NH_2} (C_6H_5)_2PCH=CNHR' \xrightarrow{BuLi} (C_6H_5)_2PCH=CNR'$$

$$R_2C=CHCOR \xleftarrow[2)\,H_3O^{\oplus}]{1)\,R_2C=O} (C_6H_5)_2PCHC=NR'$$

Yields for 11 examples were 52-70%.

Ylids also add well to nitriles [41]:

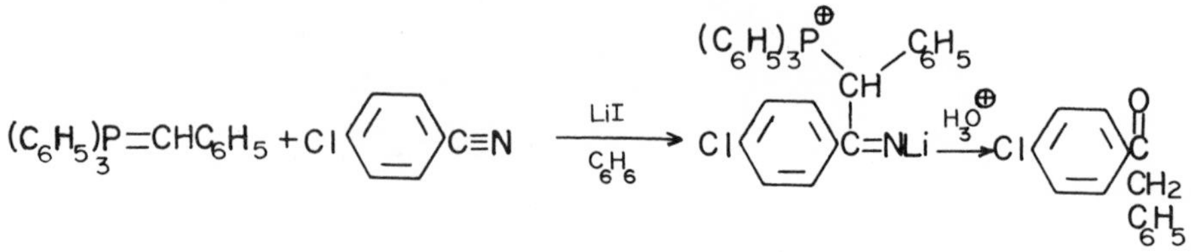

$$(C_6H_5)_3P=CHC_6H_5 + Cl\langle C_6H_4 \rangle C\equiv N \xrightarrow[C_6H_6]{LiI} Cl\langle \rangle C=NLi \xrightarrow[]{H_3O^{\oplus}} Cl\langle \rangle \overset{O}{\underset{\underset{C_6H_5}{CH_2}}{C}}$$

The yield of p-chlorophenyl benzyl ketone was 91% and the lithium
iodide, generated from producing the ylid, was found to facili-
tate the addition to the nitrile.

The reduction of a phosphonium salt is shown [42]:

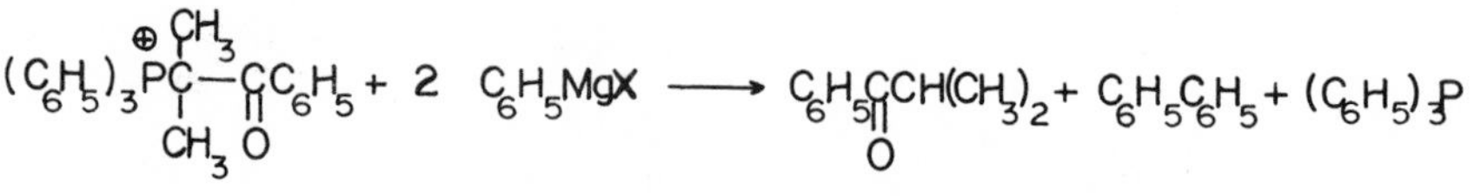

$$(C_6H_5)_3P\overset{\oplus}{\underset{CH_3}{C}}-CC_6H_5 + 2\ C_6H_5MgX \longrightarrow C_6H_5CCH(CH_3)_2 + C_6H_5C_6H_5 + (C_6H_5)_3P$$

The isobutyrophenone was obtained in 57% yield, biphenyl, 65%,
and triphenylphosphine, 66%.

The oxidation of a ylid is indicated [43]:

$$C_6H_5COCl + 2 \ CH_2 = P(C_6H_5)_3 \longrightarrow \underset{\underset{NaIO_4}{\Big\downarrow}}{C_6H_5CO\overset{\overset{P(C_6H_5)_3}{\|}}{CH}} + CH_3\overset{\oplus}{P}(C_6H_5)_3 \quad Cl^{\ominus}$$

$$C_6H_5COCH = O$$

Phenylglyoxal was obtained in nearly 100% yield as the 2,4 DNPH; other examples ranged in yields of 28-100%.

The acetonyl group may be attached by using the lithium salt of acetylmethylenetriphenylphosphine ylid [44]:

$$Li \, CH_2COCH = P(C_6H_5)_3 + C_9H_{19}Br \longrightarrow C_9H_{19}CH_2COCH = P(C_6H_5)_3$$

$$\Big\downarrow C_2H_5OH - H_2O$$

$$C_9H_{19}CH_2COCH_3$$

2-Dodecanone , 93 %

The alkylation (at -78°) was complete when the intense color of the lithium salt disappeared. The hydrolysis is a mild one.

Dianion alkylations have been carried out as well in cases in which one anionic center is the ylid [45]:

$$(MeO)_2\overset{O}{\underset{\|}{P}}CH_2\overset{O}{\underset{\|}{C}}CH_2C_3H_7\text{-}i \xrightarrow{a} (MeO)_2\overset{O}{\underset{\|}{P}}\overset{\ominus}{CH}\overset{O}{\underset{\|}{C}}\overset{\ominus}{CH}C_3H_7\text{-}i \xrightarrow[b]{\substack{1)CH_2C=CHCH_2Cl \\ 2)H_3O^{\oplus}}}$$

$$(MeO)_2\overset{O}{\underset{\|}{P}}CH_2\overset{O}{\underset{\|}{C}}CH\overset{CH_2CH=\overset{Cl}{C}CH_3}{\underset{C_3H_7\text{-}i}{\diagdown}} \xrightarrow{c} (MeO)_2\overset{O}{\underset{\|}{P}}CH_2\overset{O}{\underset{\|}{C}}CH\overset{CH_2CH_2\overset{O}{\underset{\|}{C}}CH_3}{\underset{C_3H_7\text{-}i}{\diagdown}}$$

$$\Big\downarrow d$$

$$CH_3 \text{ — ring with } CHMe_2 \text{ and } =O$$

Dianion or anion-ylid formation (Step a) is brought about by using NaH and then BuLi, the alkylation (Step b) gives a yield of ketophosphoric ester of 80%, the hydrolysis of the vinyl chloride (Step c) was accomplished by good mixing of the ester with methylene chloride and sulfuric acid at 0°, and ring closure (Step d), was achieved in 70% yield with NaH.

8.   From Vinyl Ethers and the Like

$$CH_2=CHOEt \longrightarrow CH_2=\overset{\ominus}{C}OEt \xrightarrow{C_6H_5CHO} C_6H_5\overset{\overset{\ominus}{O}\ OEt}{\underset{H}{C}-C}=CH_2 \xrightarrow[Hg^{\oplus}]{H_2O} C_6H_5\overset{O}{\underset{OH}{CHCCH_3}}$$

A   B

The preceding vinyl anion (A) is available by treating vinyl ethyl ether with t-butyllithium in the presence of TMEDA. A can be added to any type of carbonyl compound to produce mixed acyloins. The preceding reaction gave 43% of the precursors (B) and 80% of phenylacetylcarbinol [46]. However, another group reported 91% yield for this reaction [47]. The latter group also reported the addition of two equiv of methyl vinyl ether lithium (A) to esters to produce 2-hydroxy-1,3-diketones, $CH_3COC(R)OHCOCH_3$. Similarly, the ethyl vinyl thioether lithium salt may be prepared to be used for alkylation as shown [48]:

$$EtSCH=CH_2 \xrightarrow[HMPA]{s\text{-}BuLi} EtS\overset{Li}{C}=CH_2 \xrightarrow{OctBr} EtS\overset{Oct}{C}=CH_2 \xrightarrow[CH_3CN]{Hg^{\oplus}} Oct\overset{O}{C}CH_3$$

$$\downarrow Br(CH_2)_4Br$$

$$\underset{\underset{EtSC=CH_2}{\overset{\displaystyle |}{(CH_2)_4}}}{\overset{\displaystyle |}{EtSC=CH_2}} \xrightarrow[CH_3CN]{Hg^{\oplus}} CH_3CO(CH_2)_4COCH_3$$

2-Decalone was obtained on small scale in 90% yield and octane-2,7-dione, in 60% yield. The authors state that this reaction is simpler than the dithiane method (D.1).

α-Methyleneoxetane reacts with phenyllithium to give β-phenethyl methyl ketone [49]:

$$\xrightarrow{C_6H_5Li} C_6H_5CH_2CH_2COCH_3$$

The reaction was carried out on small scale.

The dianion of a propargyl ether may be dialkylated as indicated [50]:

$$C_6H_5C\equiv CCH_2OMe \xrightarrow{2BuLi} C_6H_5\overset{\ominus}{C}=C=\overset{\ominus}{C}OMe \xrightarrow{2 MeI}$$

$$C_6H_5\overset{Me}{C}=C=\overset{Me}{C}OMe \xrightarrow{H_3O^{\oplus}} C_6H_5\overset{Me}{\underset{|}{C}}=CHCOMe$$

4-Phenylpent-3-enone-2 was obtained in 65% yield. Two different alkyl groups may be added consecutively.

9.   From Carbon Monoxide

$$C_6H_5CH_2I + Ni(CO)_4 \longrightarrow (C_6H_5CH_2)_2C\!=\!O$$

This reaction appears to be superior to some other organometallic ketonizations. The reaction proceeds via the adduct
(A) [51]:

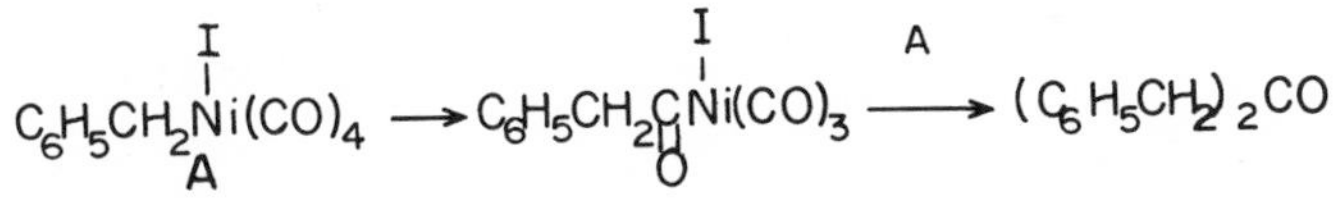

$$C_6H_5CH_2\overset{I}{\underset{A}{Ni(CO)_4}} \longrightarrow C_6H_5CH_2\overset{I}{\underset{O}{CNi(CO)_3}} \xrightarrow{A} (C_6H_5CH_2)_2CO$$

The yield is 95% with DMF as the solvent, but less with nonpolar
ones.
     Mercurials and nickel tetracarbonyl also give the sym-
metrical ketone [52]:

$$2\ RHgCl + Ni(CO)_4 \xrightarrow[N_2, 60-70°]{DMF} RCOR + 2\ Hg + NiCl_2 + 3\ CO$$

Workup of this product is easier than that prepared from a cobalt
carbonyl [53]. The symmetrical ketone is obtained as well from
lithium dialkylcuprate and carbon monoxide at atmospheric
pressure [54]. The Grignard reagent itself does not absorb carbon monoxide; however, if HMPA is added, absorption takes place
at atmospheric pressure, but better at 500 psi to give a ketone
indicative of complex addition [55]:

$$3\ C_3H_7MgX \xrightarrow[2)\,H_3O^\oplus]{1)\,CO,\,HMPA} (C_3H_7)_2CHCOC_3H_7$$

5-Propyl-4-octanone was obtained in 56% yield.
     A mixed ketone is obtained from lithium acyltetracarbonyl
ferrate [56]:

$$Li(C_6H_5COFe(CO)_4) + C_6H_5CH_2Br \longrightarrow C_6H_5COCH_2C_6H_5$$

Desoxybenzoin was obtained in 57% yield by maintaining the
mixture at -40° for 2 hr and then warming to 50°.
     Diacetonylmercury is an interesting acetonylation agent as
shown [57]:

$$C_6F_5CHO + (MeCOCH_2)_2Hg \xrightarrow{Br^\ominus} C_6F_5CHOHCH_2\overset{O}{\overset{\|}{C}}CH_3$$

4-Pentafluorophenyl-4-hydroxy-2-butanone was obtained in 65%
yield as the 2,4-DNPH by holding the above mixture in DME at 25°
for 1 day.

Ketones prepared from boranes are described in E.8.

1,4-Diketones are available by the Michael addition of
RLi-Ni(CO)$_4$ to α,β-unsaturated carbonyl compounds [57a]. Thus
3-phenyl-2,5-hexanedione (82%) was obtained from benzalacetone
and CH$_3$Li-Ni(CO)$_4$ at -78°.

A ring closure occurs with a diene and nickel carbonyl as
indicated [58]:

$$CH_2{=}CH(CH_2)_2CH{=}CH_2 \xrightarrow[HCl]{Ni(CO)_4}$$

The 70% yield consisted of 35% of 2-methylcyclohexanone and 65%
of 2,5-dimethylcyclopentanone.

The coordination and specificity obtained in organometallic
syntheses has been discussed. An example of a complicated case
is given [59]:

$$CH_2{=}CHCH_2Cl + 2\ HC{\equiv}CH + 2\ CO \xrightarrow[HCl]{Ni}$$

Addenda

H.1.   The action of esters at low temperature with LiCHCl$_2$
followed by hydrolysis with 2N HCl gives RCOCHCl$_2$, 30-90%
[J.-F. Normant et al., Compt. Rend,., C929 (1974)].

H.1.   Enolates of ketones and benzyne give cyclobutanols such as

, which may be expanded to 2,3-benzcyclo-
octenone -1 by sodamide in HMPA [P. Caubere,
Acc. Chem. Res., 7, 301 (1974)].

H.1.   The syntheses and chemistry of cyclopropenones are
reviewed [K. T. Potts, J. S. Baum, Chem. Rev., 74, 189 (1974)].
The principal general method of synthesis consists of the
reaction of a dihalocarbene and an acetylene followed by hydrolysis. Among the reactions are the formation of acrylic acids
with a base, of acetylenes and carbon monoxide with a base, and
of cyclo products with the proper olefins or dienes.

H.1.   Lithium tetraalkylborates may be added to acyl halides to
give ketones [E. Negishi et al., J. Org. Chem., 40, 1676 (1975)].

H.2. Phenyllithium added to lithium carboxylates (Ivanov reaction) gives phenyl ketones [R. Levine et al., $\underline{J}$. $\underline{Org}$. $\underline{Chem}$., 40, 1770 (1975).

H.3. Cyanocyclobutanol and methylmagnesium bromide gave 75% of 2-hydroxy-2-methylcyclopentanone [J. d'Angelo, $\underline{Bull}$. $\underline{Soc}$. $\underline{Chim}$. $\underline{France}$, 333 (1975)].

H.8. The γ-thioallenic anion, $CH_2SCR=C=\bar{C}OMe$, is a synthon for obtaining β-dicarbonyl compounds [R. M. Carlson et al., $\underline{Tetrahedron\ Letters}$, 1741 (1975)].

References

1.   F. Huet al., $\underline{Tetrahedron}$, 29, 479 (1973).
2.   R. H. Schlessinger et al., $\underline{Chem}$. $\underline{Commun}$., 1244 (1971).
3.   R. Levine, J. R. Sommers, $\underline{J}$. $\underline{Org}$. $\underline{Chem}$., 39, 3559 (1974).
4.   T. Mukaiyama et al., $\underline{J}$. $\underline{Am}$. $\underline{Chem}$. $\underline{Soc}$., 95, 4763 (1973); $\underline{Bull}$. $\underline{Chem}$. $\underline{Soc}$. $\underline{Jap}$., 47, 1777 (1974).
5.   P. Beak, R. Farney, $\underline{J}$. $\underline{Am}$. $\underline{Chem}$. $\underline{Soc}$., 95, 4771 (1973).
6.   W. C. Lumma, Jr., et al., $\underline{J}$. $\underline{Org}$. $\underline{Chem}$., 35, 3442 (1970).
7.   J. Ficini et al., $\underline{Tetrahedron\ Letters}$, 3357 (1973).
8.   M. J. Jorgensen, $\underline{Org}$. $\underline{Reactions}$, 18, 1 (1970).
9.   T. M. Bare, H. O. House, $\underline{Org}$. $\underline{Syn}$., Coll. Vol. 5, 775 (1973).
10. J. Itier, A. Casadevall, $\underline{Bull}$. $\underline{Soc}$. $\underline{Chim}$. $\underline{France}$, 2342 (1969).
11. J. C. Dalton, H.-F. Chan, $\underline{Tetrahedron\ Letters}$, 3145 (1973).
12. L. Pichat, J.-P. Beaucourt, $\underline{Synthesis}$, 537 (1973).
13. G. H. Posner et al., $\underline{J}$. $\underline{Am}$. $\underline{Chem}$. $\underline{Soc}$., 94, 5106 (1972).
14. J. N. Marx, D. C. Cringle, $\underline{Syn}$. $\underline{Commun}$., 3, 95 (1973).
15. N. T. L. Thi et al., $\underline{Bull}$. $\underline{Soc}$. $\underline{Chim}$. $\underline{France}$, 2102 (1973).
16. M. Bourgain, J.-F. Normant, $\underline{Bull}$. $\underline{Soc}$. $\underline{Chim}$. $\underline{France}$, 2137 (1973).
17. G. H. Posner et al., $\underline{Tetrahedron\ Letters}$, 1815 (1973); $\underline{J}$. $\underline{Am}$. $\underline{Chem}$. $\underline{Soc}$., 95, 7788 (1973).
18. M. Boussu, J.-E. Dubois, $\underline{Compt}$. $\underline{Rend}$., 273 C1270 (1971).
19. J.-E. Dubois et al., $\underline{Tetrahedron\ Letters}$, 829 (1971).
20. J. Fauvarque et al., $\underline{Compt}$. $\underline{Rend}$., 275 C511 (1972).
21. R. H. Schlessinger, I. S. Ponticello, $\underline{Chem}$. $\underline{Commun}$., 1013 (1969).
22. H. Reinheckel, R. Gensike, $\underline{J}$. $\underline{Prakt}$. $\underline{Chem}$., 37, 214 (1968).
23. S. Watanabe et al., $\underline{Can}$. $\underline{J}$. $\underline{Chem}$., 50, 2786 (1972).
24. I. Shahak, E. D. Bergmann, $\underline{Isr}$. $\underline{J}$. $\underline{Chem}$., 4, 225 (1966).
25. Y. Sawa et al., $\underline{Tetrahedron\ Letters}$, 5189 (1969).
26. M. Gordon et al., $\underline{J}$. $\underline{Org}$. $\underline{Chem}$., 37, 3369 (1972).
27. G. Alvernhe, A. Laurant, $\underline{Tetrahedron\ Letters}$, 1057 (1973).

28. S. D. Saraf, F. A. Vingiello, Synthesis, 655 (1970).
29. R. F. Borch et al., J. Org. Chem., 37, 726 (1972).
30. J. C. Gasc, L. Nédélec, Tetrahedron Letters, 2005 (1971).
31. G. Stork, L. Maldonado, J. Am. Chem. Soc., 93, 5286 (1971).
32. T. Cuvigny, H. Normant, Bull. Soc. Chim. France, 4990 (1968).
33. M. Bogavac et al., Bull. Soc. Chim. France, 4437 (1969).
34. M. P. Periasamy, H. M. Walborsky, J. Org. Chem., 39, 611 (1974).
35. D. C. Owsley et al., J. Org. Chem., 38, 901 (1973).
36. N. F. Scilly, Synthesis, 160 (1973).
37. E. Campaigne et al., Syn. Commun., 3, 325 (1973).
38. T. Yamaguchi et al., Chem. Ind. (London), 380 (1972).
39. E. Ritchie, W. C. Taylor, Australian J. Chem., 24, 2137 (1971).
40. A. M. Aguiar et al., Tetrahedron Letters, 1401 (1971); J. Org. Chem., 36, 2892 (1971).
41. R. G. Barnhardt, Jr., W. E. McEwen, J. Am. Chem. Soc., 89, 7009 (1967).
42. T. Mukaiyama et al., Tetrahedron Letters, 23 (1969).
43. H.-J. Bestmann et al., Chem. Ber., 102, 2259 (1969).
44. M. P. Cooke, Jr., J. Org. Chem., 38, 4082 (1973).
45. P. A. Grieco, C. S. Pogonowski, Synthesis, 425 (1973).
46. U. Schöllkopf, P. Hänssle, Ann. Chem., 763, 208 (1972).
47. J. E. Baldwin et al., J. Am. Chem. Soc., 96, 7125 (1974).
48. H. Yamamoto et al., J. Am. Chem. Soc., 95, 2694 (1973).
49. P. F. Hudrlik, A. M. Hudrlik, Tetrahedron Letters, 1361 (1971).
50. Y. Leroux, R. Mantione, J. Organomet. Chem., 30, 295 (1971).
51. E. Yoshisato, S. Tsutsumi, J. Org. Chem., 33, 869 (1968).
52. Y. Hirota, Tetrahedron Letters, 1531 (1971).
53. D. Seyferth, R. J. Spohm, J. Am. Chem. Soc., 91, 3037 (1969).
54. J. Schwartz, Tetrahedron Letters, 2803 (1972).
55. R. Louw et al., Tetrahedron Letters, 3377 (1974).
56. S. Tsutsumi et al., Tetrahedron Letters, 5189 (1969).
57. O. A. Reutov et al., J. Organomet. Chem., 42, C17 (1972).
57a. E. J. Corey, L. S. Hegedus, J. Am. Chem. Soc., 91, 4926 (1969).
58. B. Fell et al., Tetrahedron Letters, 1003 (1968).
59. G. P. Chiusoli, Bull. Soc. Chim. France, 1139 (1969).

I.   Electrocyclic Reactions

This field is expanding rapidly as it brings to light

several new generalizations. At times it is difficult to evaluate certain preparations as most of them are carried out on small scale.

1. Cyclic Products (Including Those of the Diels-Alder)

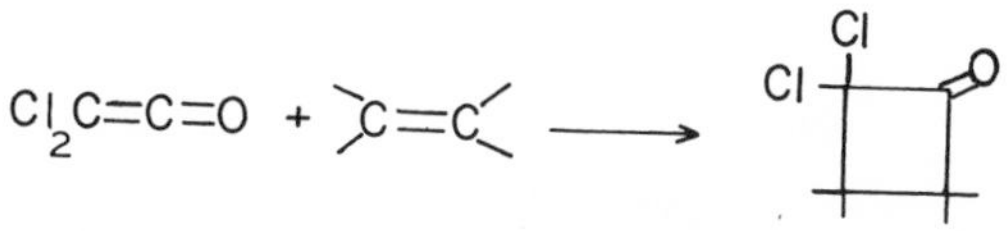

$$Cl_2C=C=O + \; \rangle C=C\langle \; \longrightarrow$$

Ketene itself does not cyclize with alkenes, but the highly reactive dichloroketene does very easily to form cyclobutanones. The general behavior of these ketenes has been reviewed [1]. Not only are the halocyclobutanones available in this manner, but the cyclobutanones are as well by the removal of the halogens from the former with zinc and acetic acid. Cyclopentadiene gives the expected bicyclic dichlorocyclobutanone, which may be hydrolyzed in good yield to tropolone. The dihaloketenes do not react with alkenes that contain electron-withdrawing groups [2]; in fact, the monohaloketenes have difficulty in reacting with any alkene [1].

Substitutions may be made for either the ketene or alkene or both in the formation of cyclo products. Thus acetylene and di(trifluoromethyl)ketene cyclize at 100-150° to form the expected di(trifluoromethyl)cyclobutenones in good yield [3]. In addition, the keteneimmonium cation appears to cyclize more easily than the corresponding ketene [4]:

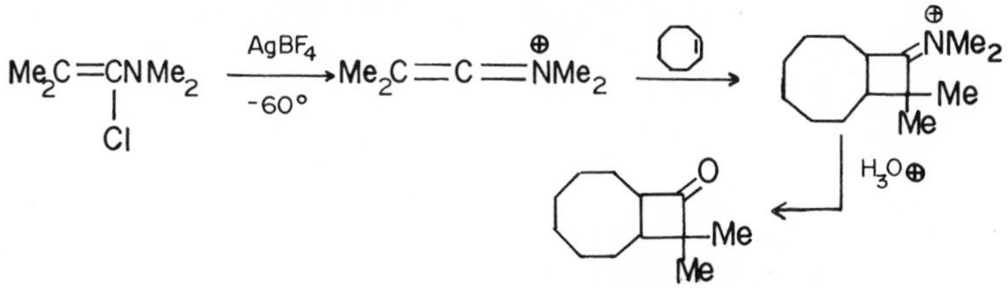

10,10-Dimethylbicyclo[6.2.0]-
decanone-9

The yield of the dimethylcyclobutanone was 87% whereas with cyclohexene 8,8-dimethylbicyclo[4.2.0]octanone-7 was obtained in 89% yield. Enamines, which appear to be better acceptors than alkenes, undergo cyclo reactions with alkynes [5]:

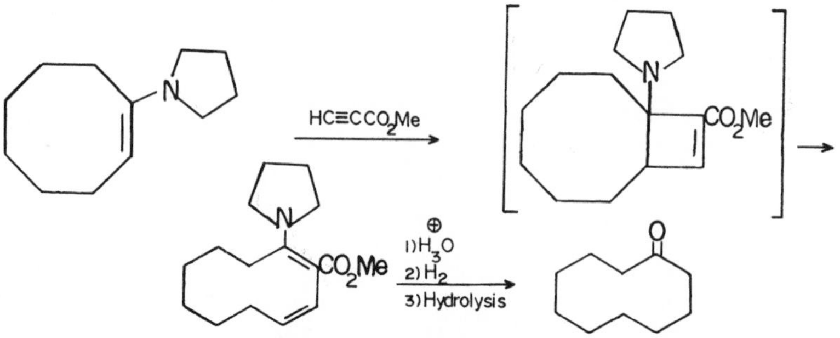

Cyclodecanone was obtained in 44-50% overall yield, which is superior to that obtained by the Dieckmann condensation. The reaction of a dienophile is illustrated [6]:

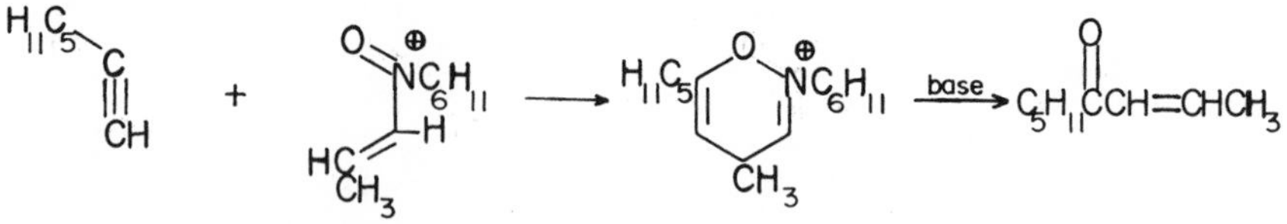

The non-2-en-4-one was obtained in 72% yield. The ketene, from which various types of cyclohexenones may be prepared, may be utilized as the thioacetal [7]:

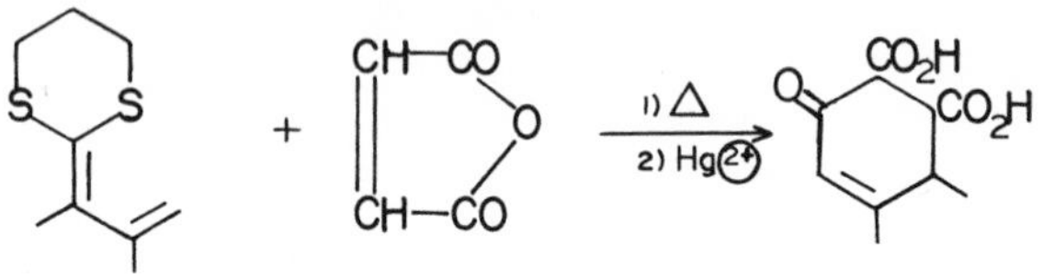

Isolated as the monomethyl ester, the overall yield of the diacid was 58%. Diphenylacetylene, quinone, and diethyl maleate did not undergo the Diels-Alder reaction with the thioacetal.

A dienophile that apparently undergoes Diels-Alder reactions to give a ketone after hydrolysis is the 2-methoxyallyl carbonium ion. It is difficult to work with but if silver ions are added slowly to generate the allylic ion, if no acid is allowed to accumulate, and if the ion is generated in the dark with a vibromixer, yields of as high as 50% are obtained as shown [8]:

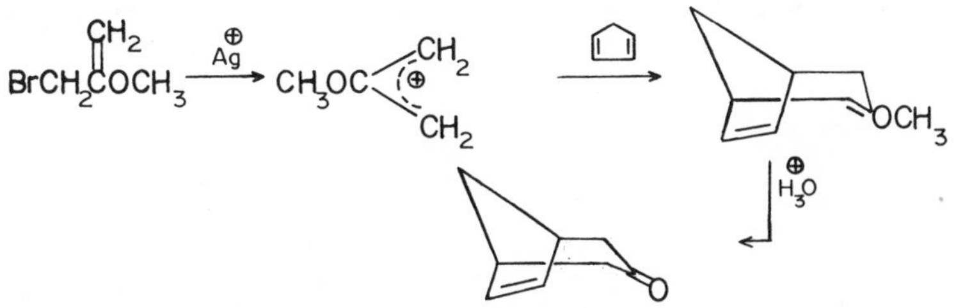

α-Bromoketones behave like allylic ions when they are stabilized by phenyl groups as shown:

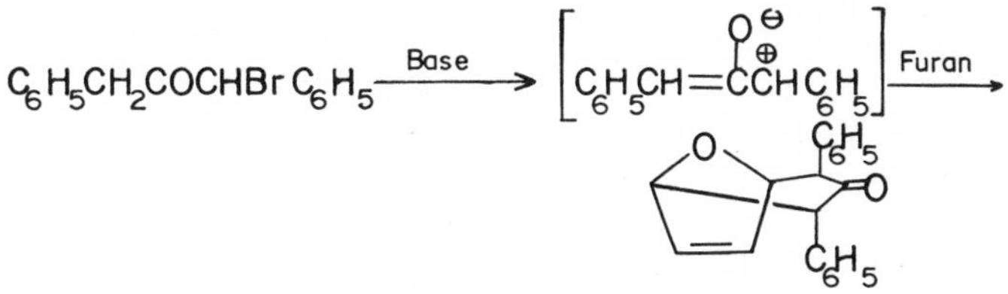

If the ketone is aliphatic, a zinc-copper couple is better than base to form the allylic moiety; iron carbonyl, $[Fe_2(CO)_9]$, may be used as well [9]. Cyclopropanones also behave as though they were oxyallyl moieties as indicated [8]:

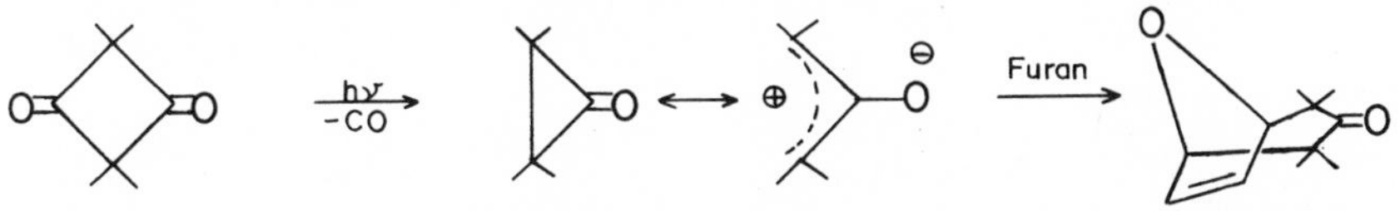

   Perhaps of all dienophiles, cyclopropene is one of the most active since it is the only one known that reacts in a Diels-Alder reaction with tropolone [10].

   The diradical intermediate, often of a singlet nature, is invoked to explain the cyclization of unsaturated ketones [11]:

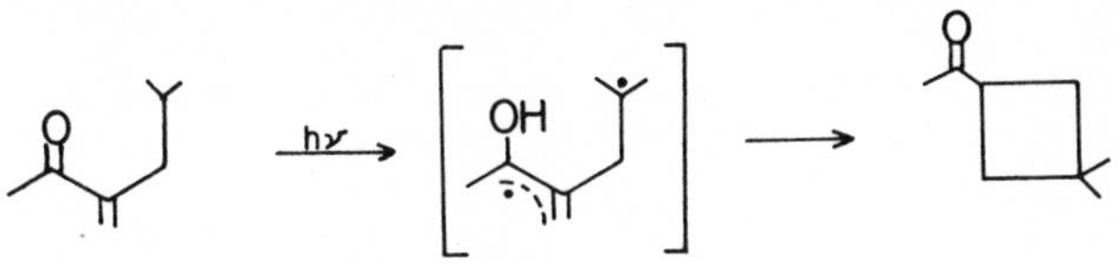

Similarly, 2,3-pentanedione gives 2-hydroxy-2-methylcyclobutanone quantitatively on irradiation [12].

   2.   Cyclic Transition States (Including the Oxy-Cope
        Rearrangement)

In Chapter 10, H.1, the Claisen-Cope rearrangement of allyl vinyl ethers was discussed with some of its modifications in the preparation of aldehydes. In the present chapter the oxy-Cope rearrangement is discussed with illustrations for the synthesis of ketones [13].

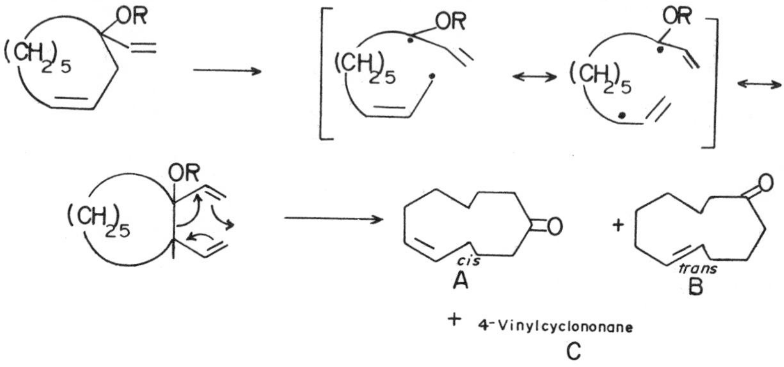

A

B

+ 4-Vinylcyclononane

C

If R=H, then the reaction, at 286° for 12 hr followed by hydrolysis, gives 60% of polymer and 40% of products such as A, B, and C. However, if R=SiMe$_3$, the reaction, known as the siloxy-Cope, conducted as above gives no polymer but 53% of A, 29% of B, and 10% of C. Thus the siloxy-Cope appears to be more promising than the oxy-Cope in that fewer side products are obtained.

It now appears that an even better way to conduct the oxy-Cope is ionically as indicated [14]:

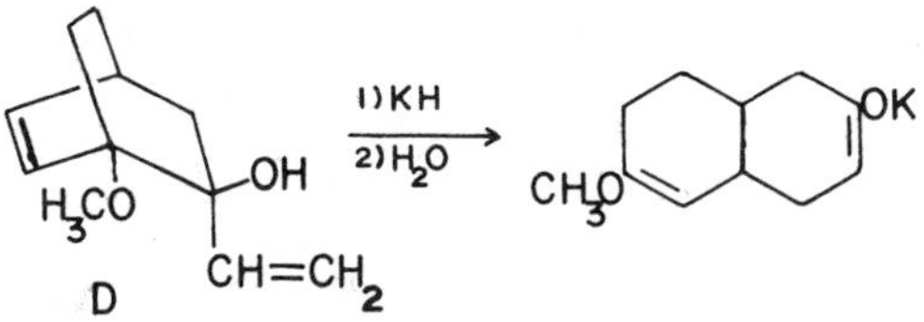

D

The yield of the reaction, which occurs via the alloxy anion of D, was 98% at 66° for several min. Moreover, the reaction is subject to cluster-breaking. If dicyclohexyl-18-crown-6 or HMPA is added, the rate is accelerated 180-fold.

1,2-Divinylcyclohexanols undergo the oxy-Cope [15]:

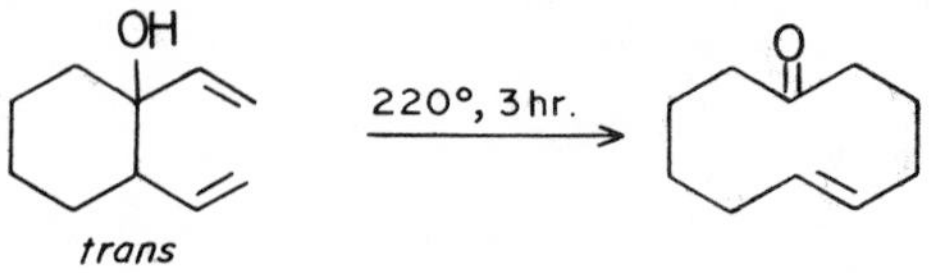

trans

220°, 3 hr.

trans-$\Delta^{4,5}$-Decenone was obtained in 90% yield.

Another type of the Claisen-Cope which gives ketones is indicated [16]:

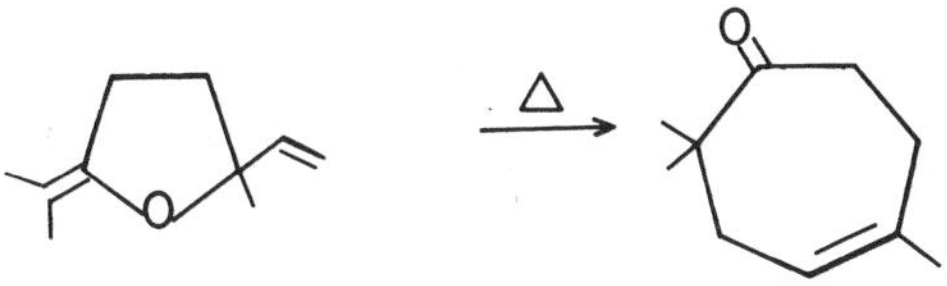

$\Delta^{4,5}$-2,2,5-Trimethylcycloheptenone, one of the constituents of hop oil, was obtained in 62% yield.

Other natural products have been prepared by the oxy-Cope reaction [17]. If no stabilizing features exist, an unsaturated ketone may still be rearranged by brute force, provided decomposition does not take place, as witness [18]:

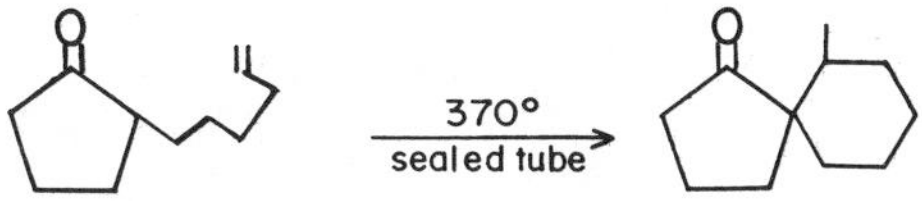

The product consisted of about 60% of the spiro ketone, the remainder being largely a ketone in which alkylation had occurred on both sides of the keto group.

O-Alkylthiobenzoates react with alkenes on irradiation as follows [19]:

$$C_6H_5CH_2CH=CH_2 + C_6H_5\overset{S}{\overset{\|}{C}}OEt \xrightarrow{h\nu} C_6H_5CH_2CH_2COC_6H_5$$
$$E$$

Phenethyl phenyl ketone may arise by addition of the ester to the alkene followed by cyclization and elimination:

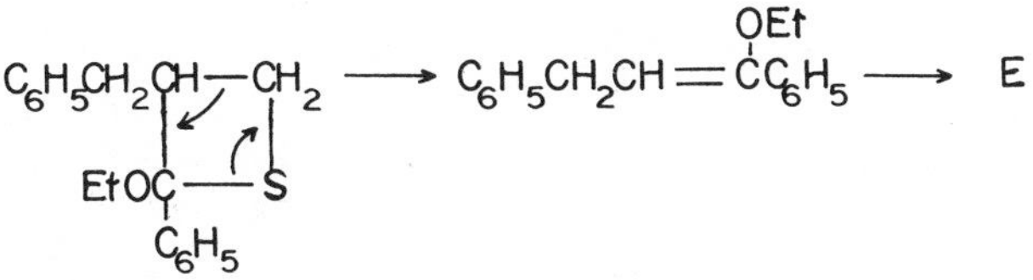

Intramolecular cyclization occurs with propargylic esters at low pressures and high temperatures (650°) [20]:

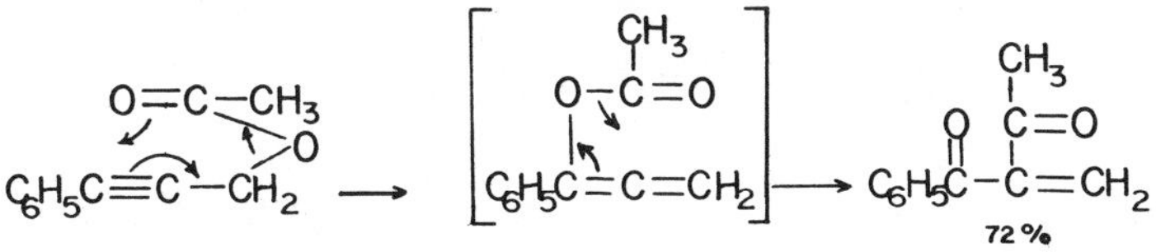

1-Acetyl-1-benzoylethylene was obtained in 72% yield.   Similarly,
2-furylmethyl benzoate gave 2-methylenecyclobutenone in 40% yield
[21]:

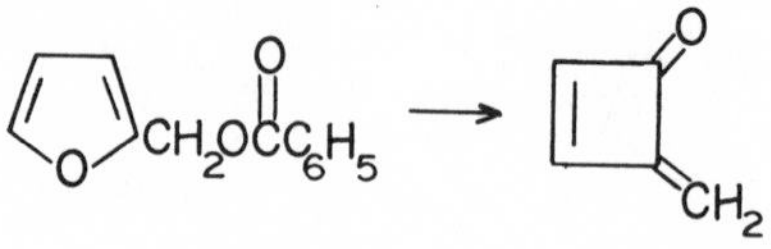

The product must be kept in chloroform below -10° to prevent
polymerization.
     In general the irradiation of ketones with acetylene com-
pounds leads to unsaturated ketones as indicated [22]:

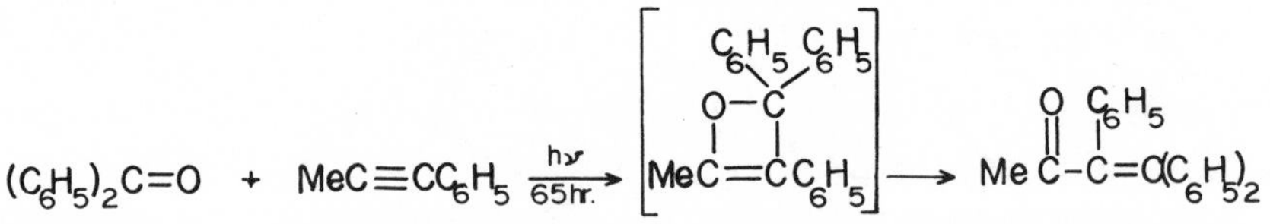

About 25% of methyl α,β,β-triphenylvinyl ketone was isolated.

Addenda

I.1. Photochemical ring expansions via cyclic oxacarbenes occur
with cyclobutanones, some tricyclic ketones, and some cyclopropyl ketones. An illustration of the former follows:

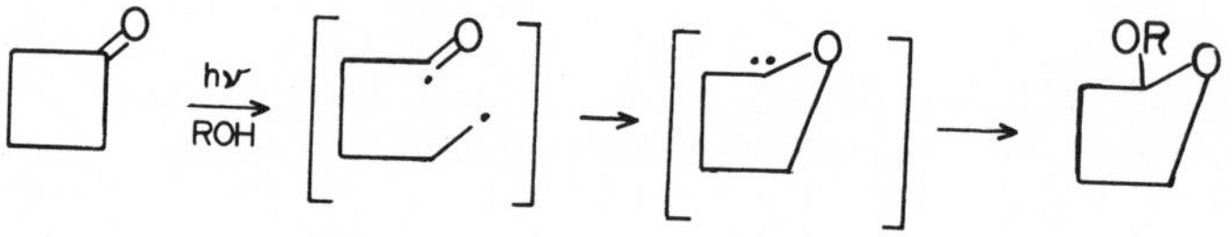

I.1. Thermal cyclizations of unsaturated ketones have been
reviewed [J. M. Conia, P. LePerchec, Synthesis, 1 (1975)]. As
examples, oct-7-ene-2-one cyclized quantitatively at 370° to
2-methylacetylcyclopentanone, and 2-methylacetylcyclobutane gave
3-methylcyclohexanone. Two more transformations are:

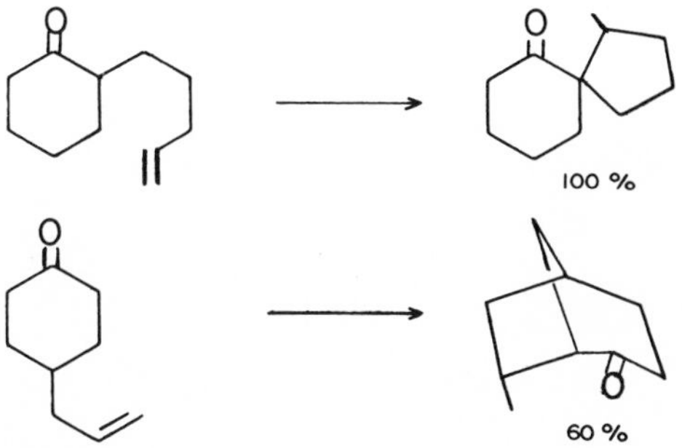

100 %

60 %

I.1. Pentamethylene ketene undergoes a variety of cycloadditions to form cyclic ketones, two of which are illustrated;

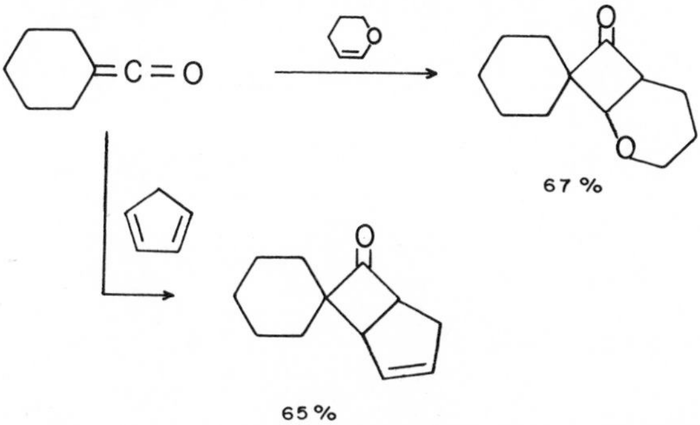

67 %

65 %

[W. T. Brady, P. L. Ting, J. Org. Chem., 39, 763 (1974)]

I.1. Cycloaddition of ketenes to tetramethylallene results in the formation of α,β-unsaturated cyclobutanones:

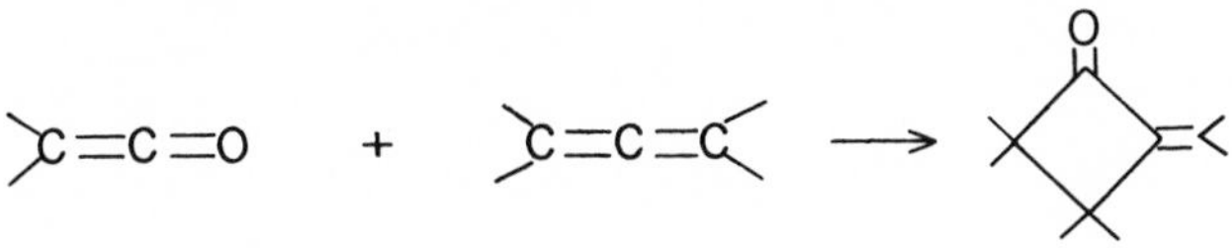

[W. T. Brady et al., J. Org. Chem., 39, 236 (1974)]

References

1. W. T. Brady, Synthesis, 415 (1971).
2. L. Ghosez et al., Tetrahedron, 27, 615 (1971).
3. D. C. England, C. G. Krespan, J. Org. Chem., 35, 3308, 3312, 3322 (1970).
4. J. Marchand-Brynaert, L. Ghosez, J. Am. Chem. Soc., 94, 2870 (1972).
5. R. D. Burpitt, J. G. Thweatt, Org. Syn. Coll. Vol., 5, 277 (1973).
6. S. Shatzmiller, A. Eschenmoser, Helv. Chim. Acta, 56, 2975 (1973).
7. F. A. Carey, A. S. Court, J. Org. Chem., 37, 4474 (1972).
8. H. M. R. Hoffman, Angew. Chem. Intern. Ed. Engl., 12, 819 (1973).
9. R. Noyori et al., J. Am. Chem. Soc., 93, 1272 (1971).
10. Y. Kitahara et al., Chem. Ind. (London), 41 (1973).
11. R. A. Cormier, W. C. Agosta, J. Am. Chem. Soc., 96, 618, 1867 (1974).
12. S. M. Weinreb, R. J. Cvetovich, Tetrahedron Letters, 1233 (1972).
13. R. W. Thies, J. Am. Chem. Soc., 94, 7074 (1972); Chem. Commun., 237 (1971).
14. D. A. Evans, A. M. Golob, J. Am. Chem. Soc., 97, 4765 (1975).
15. E. N. Marvell, W. Whaley, Tetrahedron Letters, 509 (1970).
16. E. Demole, P. Enggist, Helv. Chim. Acta, 54, 456 (1971).
17. R. P. Gregson, R. N. Mirrington, Chem. Commun., 598 (1973).
18. J.-M. Conia et al., Bull. Soc. Chim. France, 963 (1973).
19. A. Ohno et al., Tetrahedron Letters, 4993 (1972).
20. W. S. Trahanovsky, P. W. Mullen, J. Am. Chem. Soc., 94, 5086 (1972).
21. W. S. Trahanovsky, M.-G. Park, J. Am. Chem. Soc., 95, 5412 (1973).
22. H. Polman et al., Rec. Trav. Chim., 92, 845 (1973).

Chapter 12

QUINONES AND RELATED SUBSTANCES

This chapter has been enriched by the recent publication of two good reviews [1, 2] on the synthesis of quinones and a revised edition of Thomson's book has become available [3]. Two reagents now stand out as preferred for the oxidation of phenols to quinones: (a) potassium nitrosodisulfonate, Fremy's salt, for monohydric phenols (A.2) and (b) ceric ammonium nitrate for dihydric phenols (A.4). Although thallium(III) salts may be used in such oxidations, they remove some substituents from the para position of the phenol (A.2). Of the new reagents for the oxidative coupling of phenols to diphenoquinones, paraperiodic acid in DMF appears to be the most satisfactory (A.3). The preparation of dienones or spirodienones has been introduced and more information has been given on benzoquinone monoximes (B.2). An unusual ring closure to form anthraquinones has been made possible by the use of rhodium salt complexes (C.4). An elaboration of the structure of quinones through azidoquinones (B.3) as well as the new synthesis of anthraquinones by ring closure of selected benzophenones offer promise.

A. Oxidation

    1. From Hydrocarbons ($\underline{1}$, 725)

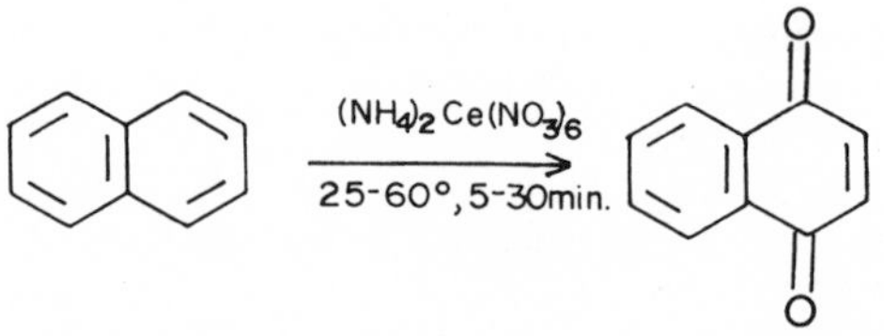

Ceric ammonium nitrate, the capabilities of which are described, has now been used to oxidize polynuclear hydrocarbons (in low to fair yields) to quinones [4]. Naphthoquinone was obtained in 20% yield, anthraquinone in 61%, and 9,10- and 1,4-phenanthraquinones in 27% and 11%, respectively.

2.   From Arylamines or Phenols (1, 726)

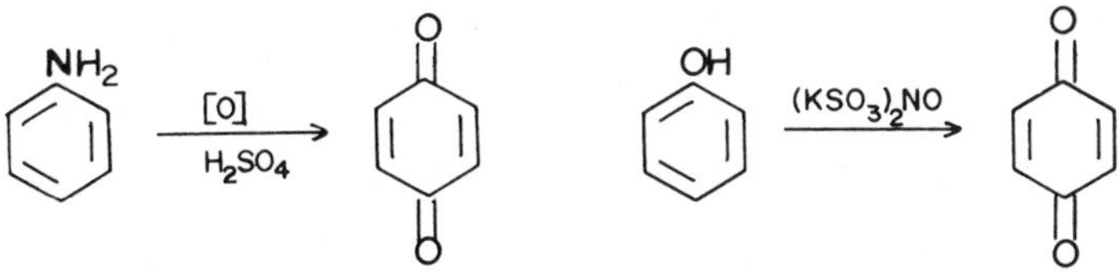

Of particular interest are the o-quinones, which have been produced in large numbers from p-substituted phenols by the use of potassium nitrosodisulfonate [5] as shown:

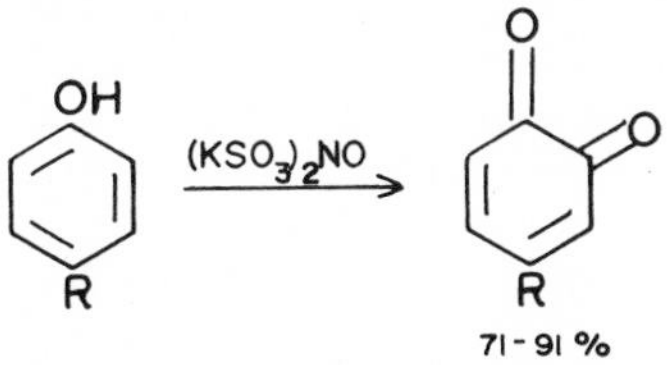

71-91 %

The presence of sodium acetate or phosphate or both as buffers simplifies the procedure.

The oxidation of phenols and amines with potassium nitrosodisulfonate, Fremy's salt, the so-called Teuber reaction has recently been reviewed [2] (Ex. a). This review gives a much better understanding of the scope of the oxidation. Fremy's salt is particularly good for the oxidation of phenols and secondary amines to quinones and iminoquinones, respectively. It has been used to prepare the following rather unusual types of quinones (mostly with other substituents attached):

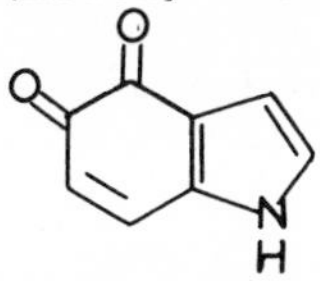

4,5- Indolequinones

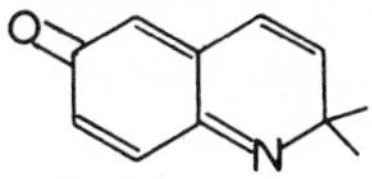

1,2-Dihydro-6-quinolones

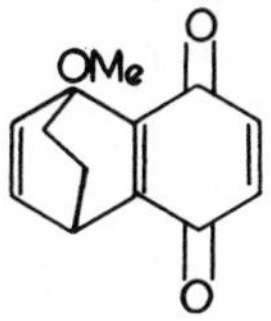

Oxidation product of
Diels-Alder adduct of Benzoquinone and I-Methoxy-I,3-cyclohexadiene

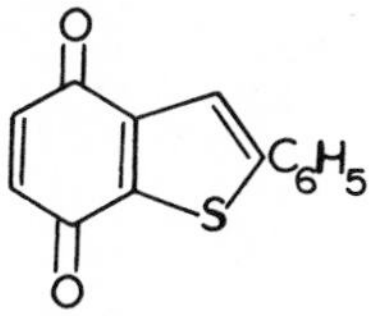

4,7- Dioxobenzothiophene

The synthesis and purification of Fremy's salt is also given in the review [2].

A complex, homogeneous catalyst for the oxidation of phenols with oxygen in chloroform or methanol at room temperature is bis(salicylidene)ethylenediiminocobalt(II) (salcomine) [6]. This reagent of the formula

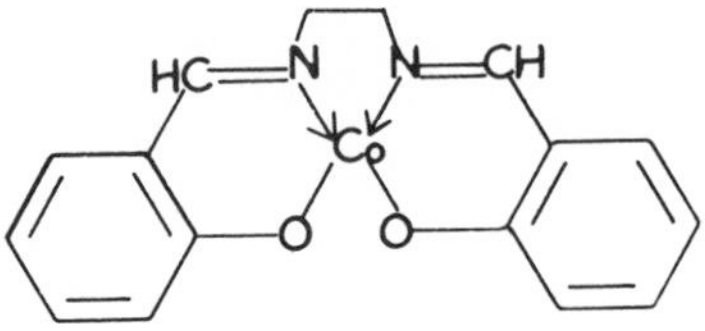

oxidizes a series of substituted phenols with the para position open to p-benzoquinones with yields varying from 14 to 80%.

A more recent method involves the use of thallium(III) tri-fluoroacetate [7], with which phenols containing certain groups in the p-position are converted into p-benzoquinones as indicated:

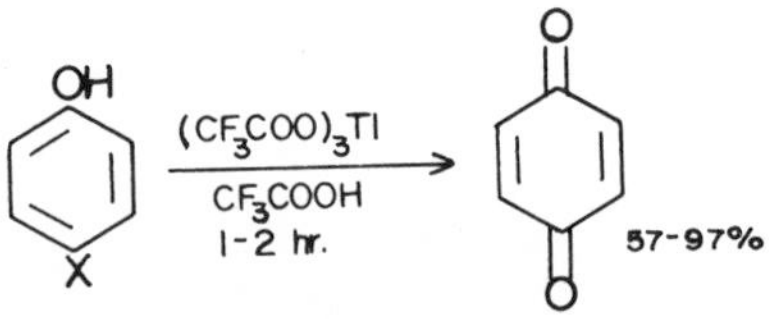

$$X=Cl, Br, I, C(CH_3)_3, AcO$$

Groups such as Cl, Br, I, H, alkyl, or AcO in positions other than 4 are not affected by the reaction.

One of the most unusual oxidations consists of the irradiation of phenols in the presence of air. As alkyl substitution increases in the ring, yields increase from 0.5% for phenol to 75% for 2,3,5,6-tetramethylphenol [7a]. A second method involves the use of vanadium oxychloride in the preparation of the complex perylinequinone [7b]:

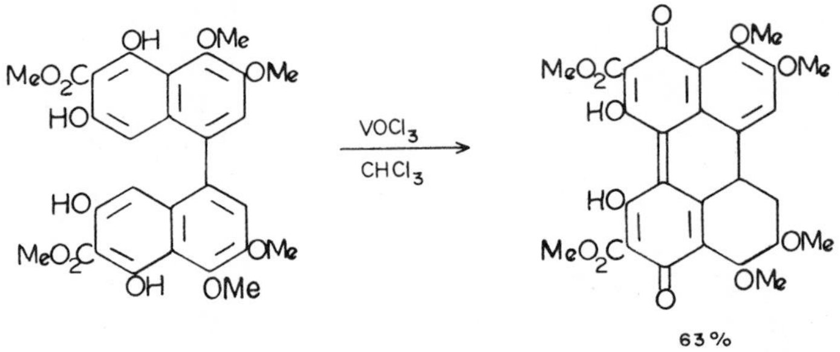

No other oxidizing agent was this satisfactory.

a. Preparation of 4,5-Dimethyl-1,2-benzoquinone [8]

To a solution of 15 g of NaH_2PO_4 in 5 l of distilled water was added 90 g of $(KSO_3)_2NO$ (Fremy's salt) and the mixture was shaken to dissolve the inorganic radical. A solution of 16 g of 3,4-dimethylphenol in 350 ml of ether was then added quickly and the mixture was shaken for 20 min. The o-quinone formed was extracted in 3 parts of $CHCl_3$, total 1.2 l, from which extract there was recovered in the usual manner 8.7-8.9 g (49-50%) of the product.

3. From Phenols (by Oxidative Coupling) (1, 728)

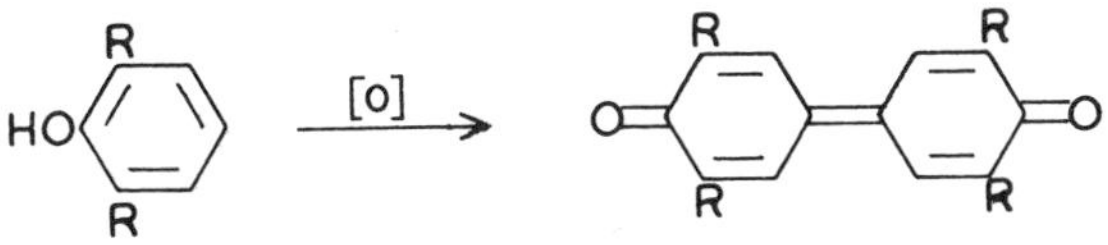

The reagents giving the best phenol couplings have been reviewed [9]. Of these, potassium ferricyanide in alkaline solution has been found to be very effective.

deJonge oxidized phenols (largely 2,6-disubstituted) with lead dioxide in acetic acid to obtain diphenoquinones as shown [10]:

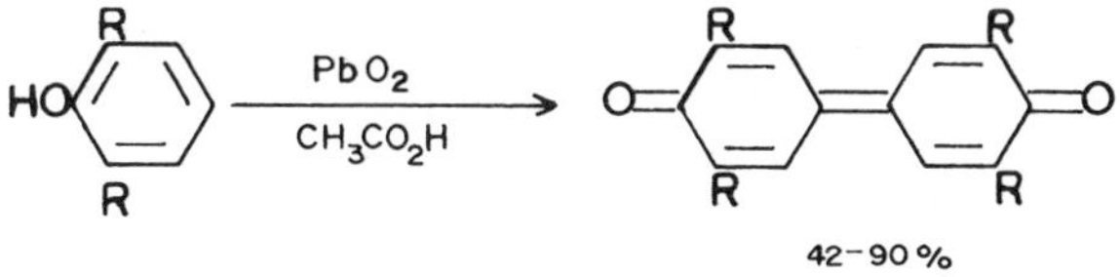

42-90%

Jerussi employed isoamyl nitrite in methylene chloride at 25° for the oxidative coupling of similarly substituted phenols [11]. Yields obtained were 10-65% with the lower yields resulting when sterically hindering or deactivating groups were present in the ortho positions.

A more promising reagent for the formation of diphenoquinones is hydrogen (hexacyanoferrate)(III), $H_3Fe(CN)_6$ [12]. With 2,6-di-t-butylphenol it gives 3,3',5,5'-tetra-t-butyldiphenoquinone as shown:

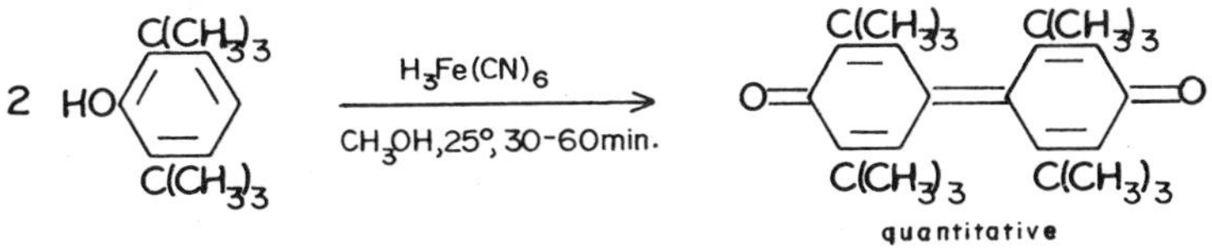

quantitative

Unfortunately, the reagent has been investigated with few 2,6-disubstituted phenols. With 2,6-di-t-butyl-4-methoxyphenol it gives 2,6-di-t-butyl-1,4-benzoquinone (73%). When the 4-position of the phenol is occupied with methyl, 2,6-di-t-butyl-4-methyl-4-methoxy-2,5-cyclohexadienone is obtained in low yield.

The best new coupling agents are silver carbonate on celite in boiling benzene [12a] and paraperiodic acid, H_5IO_6, in DMF [13] (Ex. a). The former gave high yields of the stilbenequinone derived from 2,6-di-t-butyl-4-cresol. With 2,4,6-tri-t-butylphenol it gave the free radical quantitatively. The reagent is particularly useful for the preparation of quinones which are acid or base sensitive and the procedure involving its use "appears to be superior to all known procedures."

A rather obscure oxidative method for β-naphthol has been reported [13a]:

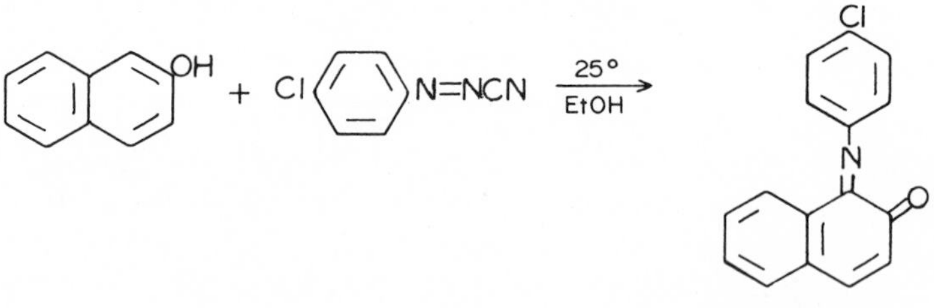

1-*p*-Chlorophenylimino-1,2-naphthoquinone, 83%

a. Preparation of 3,3',5,5'-Tetraalkyldiphenoquinone [13]

The 2,6-dialkylphenol, 10 mmol in 10 ml of DMF, was treated with 4 ml of a 4M solutiion of periodic acid and the solution was stirred at 85-95° for 4-5 min. A colored product partially precipitated. Then at 25° for 5-10 min, the mixture was diluted with 8-10 ml of 50% aqueous methanol and the product was filtered and washed with cold methanol. A second crop of crystals was obtained by dilution of the filtrate with water. Total yield of the diphenoquinone was over 90% except for the product from 2,6-dimethylphenol (60%).

4. From o- or p-Dihydroxybenzenes, Diaminobenzenes, and Related Types (1, 730)

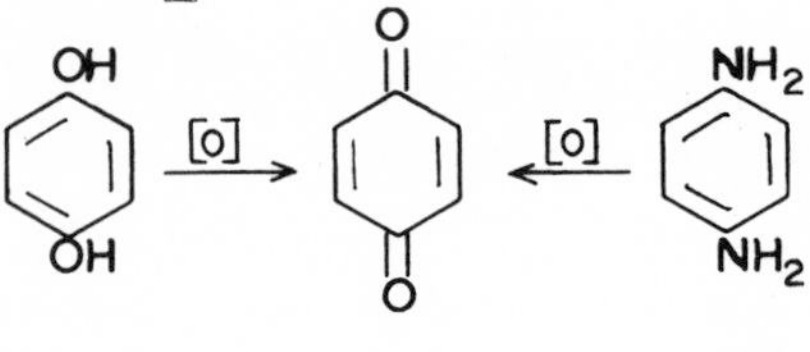

Superior yields of quinones were obtained by the oxidation of dihydric phenols with ceric ammonium nitrate [14] (Ex. a). The method is simple, convenient, and applicable to the production of substituted o- and p-quinones.

The use of thallium trifluoroacetate as the oxidizing agent in the formation of p-benzoquinones (see A.2) has also been applied to dihydric phenols as shown:

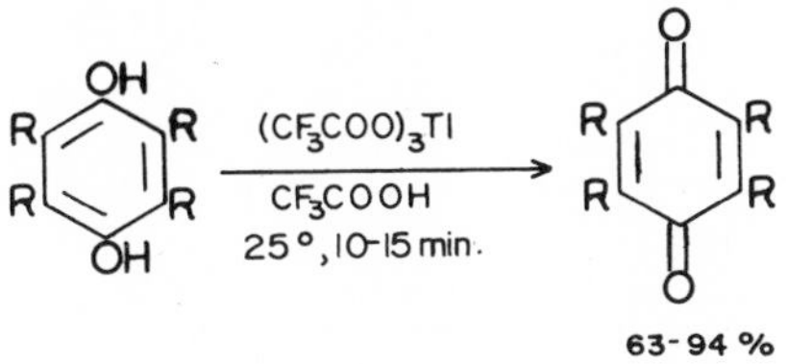

63-94 %

R=H,alkyl,C₆H₅,Br,Cl

Although the most common oxidizing agent for the conversion of catechols into o-quinones consists of silver oxide and sodium sulfate in dry ether, potassium nitrosodisulfonate (see A.2) usually works well with substituted catechols as indicated [15]:

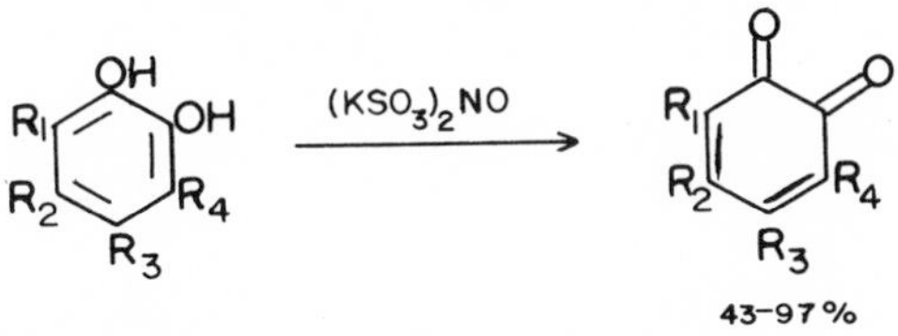

43-97%

The presence of sodium acetate or phosphate or both as a buffer simplifies the procedure.

A second reagent of value in the synthesis of o-quinones is tetrachloro- or tetrabromo-1,2-benzoquinone [16]. The reaction with catechol is as indicated:

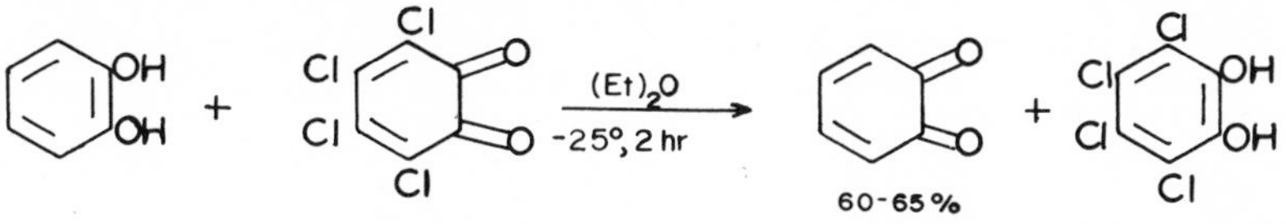

60-65 %

Yields in other cases vary from 20 to 95%. The dihydric phenol produced as a by-product may be readily oxidized to the quinone for future use with HNO_3 in AcOH. One limitation of the method is the fact that the redox potential of the catechol being oxidized must be lower than that of the oxidizing quinone for the equilibrium to lie toward the products. Other reagents for oxidizing catechols are iodosobenzene diacetate [16a] and silver carbonate on celite (A.3).

If amines are present during the oxidation of catechols or hydroquinones containing electron-withdrawing groups, the amino-quinones may be isolated [16b]:

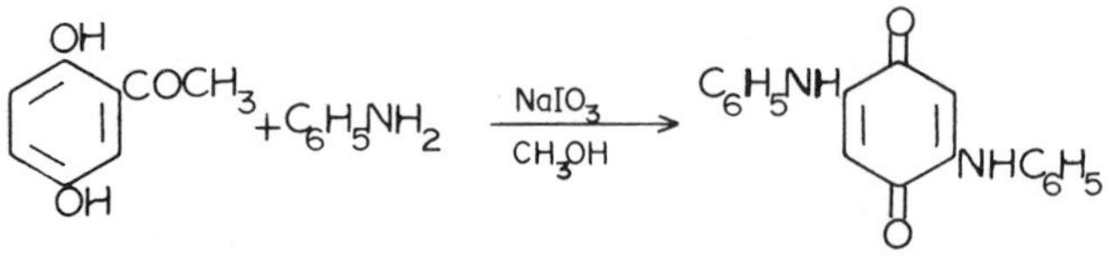

2,5-Dianilino-3-acetobenzoquinone,91%

The 3-position in benzoquinones of the above type is very susceptible to the addition of nucleophiles (see D.2).

An improved method for the synthesis of 1,2-phenanthrene-quinone involves the steps shown [17]:

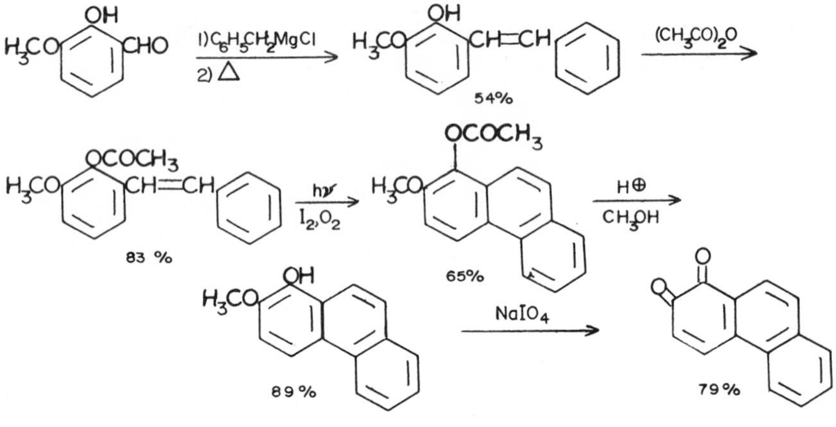

The overall yield from o-vanillin is 25%.

One of the methods described by Norris and Sternhell [18] starting with a p-nitroanisole involves the steps as indicated:

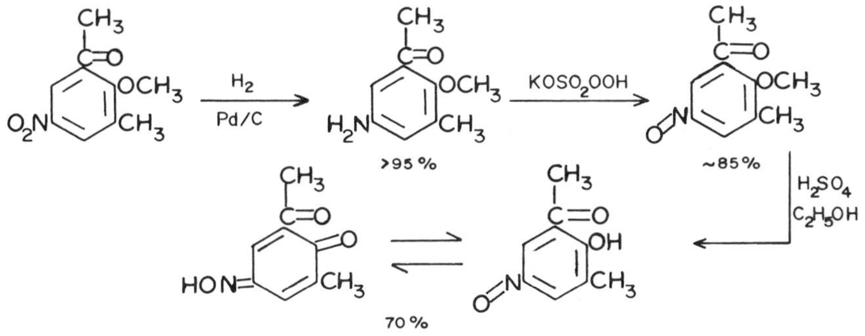

2-Acetyl-6-methyl-4-nitrosophenol (ca. 60% overall)

Newman and Hetzel [19] were successful in synthesizing quinones by the pyrolysis of 4-methylallyloxy-2,3,5,6-tetramethylphenol as indicated:

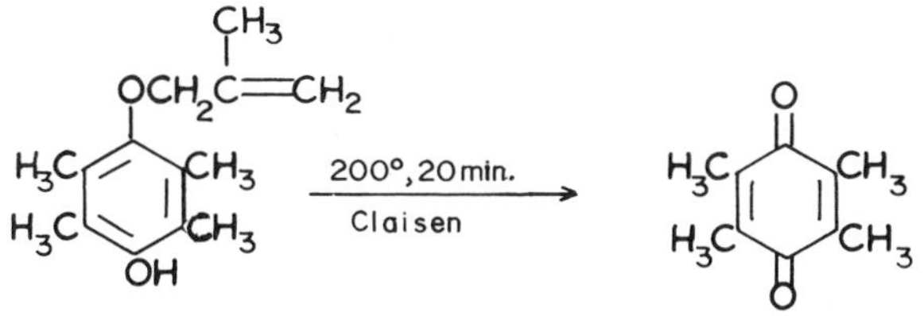

Duraquinone, 68 %

The benzyloxy derivative gave the same quinone in 84% yield. The methallyl group evidently serves as the oxidizing agent in the following way:

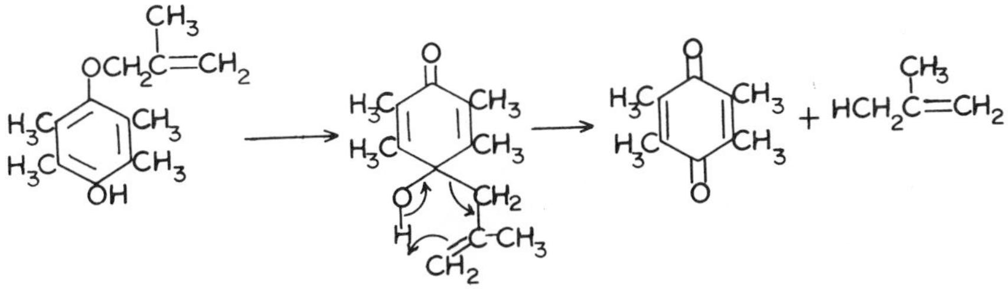

a. Preparation of Benzoquinone [14]

Ceric ammonium nitrate, 1.1 g (2 mmol), was added to a solution of 110 mg (1 mmol) of hydroquinone in a mixture of 3 ml of acetonitrile and 1 ml of water. After 2 min at 25°, the solution was poured into water and extracted with ether. After washing and drying the ether, it was evaporated to yield 90 mg (83%) of benzoquinone.

6.   From p-Substituted Anilines or Phenols (Dienones and
Spirodienones)

The reaction of p-substituted N-chloroanilines with silver
trifluoroacetate in methanol leads to cyclohexa-2,5-dienones [20]
as shown:

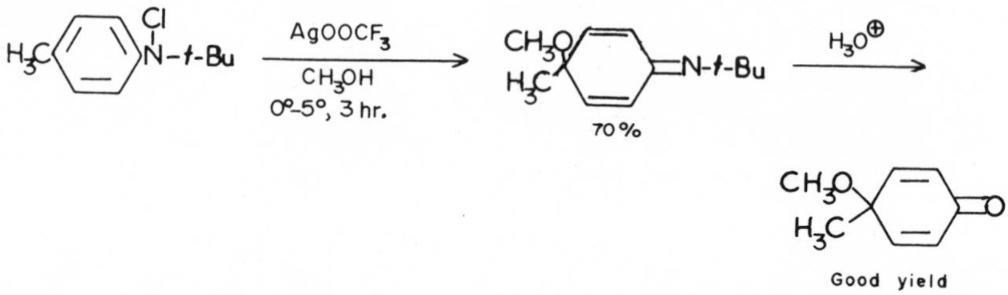

Good yield

The cyclohexa-2,5-dienones, which include the spirodienones, are
valuable intermediates in organic synthesis.

The spirodienones, which are found in nature and which are
more common than originally believed, result in synthesis by
oxidation if properly placed substituents are present. The
general synthesis of a spirodienone and of a dioxaspirodienone
are shown [21]:

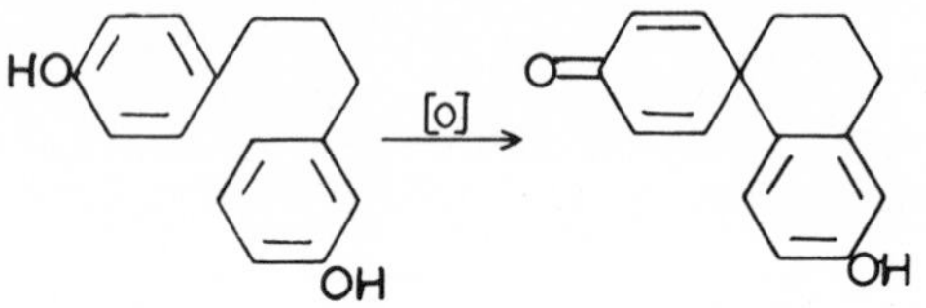

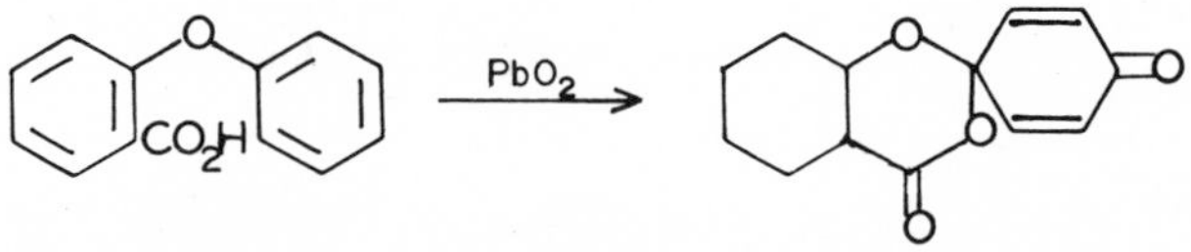

A practical example of the spirodienone synthesis is the
preparation of dehydrogriseofulvin, the oxidized form of the
fungicide, griseofulvin [22]:

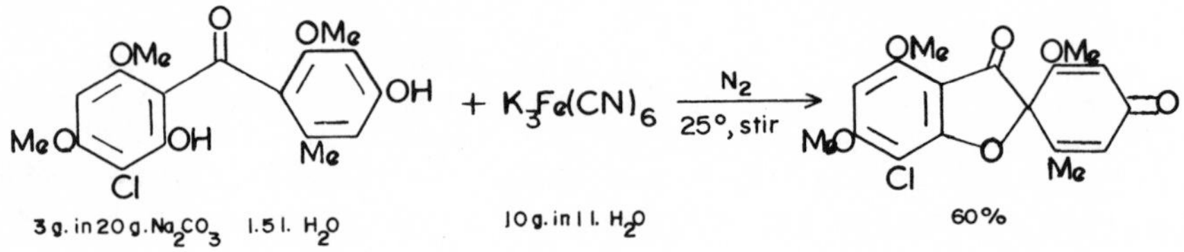

A direct entry into the dienone system is acquired from hindered phenols substituted in the para position [23] as indicated:

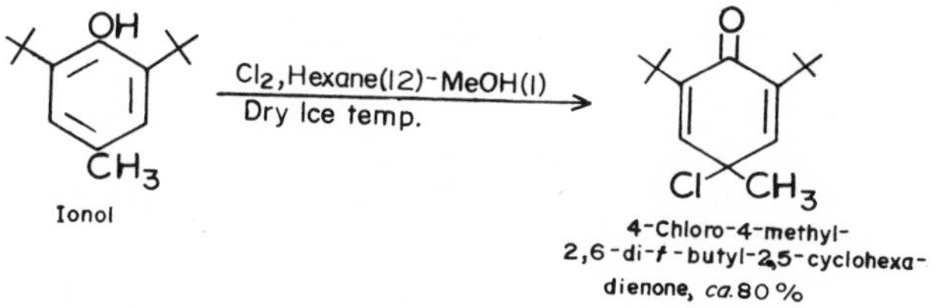

OH
CH_3
Ionol
Cl_2,Hexane(12)-MeOH(1)
Dry Ice temp.
O
Cl
CH_3
4-Chloro-4-methyl-2,6-di-*t*-butyl-2,5-cyclohexadienone, *ca.* 80 %

A large number of compounds of this basic structure are described in the literature.

Addendum

A.2. Ferric chloride and DMF form a complex, $[Fe(DMF)_3Cl_2][FeCl_4]$, which is effective in oxidizing some phenols to quinones [S. Tobinaga, E. Kotani, J. Am. Chem. Soc., 94, 309 (1972)].

References

1. W. M. Horspool, Quart. Rev. (London), 23, 204 (1969).
2. H. Zimmer et al., Chem. Rev., 71, 229 (1971).
3. R. H. Thomson, Naturally Occurring Quinones, 2nd Ed., Academic, New York, 1971.
4. T.-L. Ho et al., Synthesis, 206 (1973).
5. H.-J. Teuber, G. Staiger, Chem. Ber., 88, 802 (1955).
6. H. M. Van Dort, H. J. Geursen, Rec. Trav. Chim., 86. 520 (1967).
7. E. C. Taylor et al., Angew. Chem., 82, 84 (1970); Tetrahedron, 26, 4031 (1970).

7a. K. Pfoertner, D. Böse, Helv. Chim. Acta, 53, 1553 (1970).

7b. K. H. Weisgraber, U. Weiss, J. Chem. Soc., Perkin Trans., I, 83 (1972).

8. H.-J. Teuber, Org. Syn., 52, 88 (1972).
9. W. I. Taylor, A. R. Battersby, Oxidative Coupling of Phenols, Dekker, New York, 1967, p. 81.
10. C. R. H. I. deJonge et al., Tetrahedron Letters, 1881 (1970).
11. R. A. Jerussi, J. Org. Chem., 35, 2105 (1970).
12. L. Taimr, J. Pospišil, Tetrahedron Letters, 2809 (1971).

12a. M. Fetizon et al., J. Org. Chem., 36, 1339 (1971).

13. A. J. Fatiadi, Synthesis, 357 (1973).

13a. O. Simamura et al., J. Chem. Soc., C, 2074 (1971).

14. T.-L. Ho et al., Chem. Ind. (London), 729 (1972).
15. E. Müller et al., Z. Naturforsch., 18B, 1002 (1963).
16. L. Horner, W. Dürckheimer, Z. Naturforsch., 14B, 741 (1959).
16a. A. T. Balaban, Rev. Roumaine Chem., 14, 1281 (1969).
16b. W. Schäfer, A. Aguado, Angew. Chem., 83, 441 (1971).
17. H.-D. Becker, J. Org. Chem., 34, 2026 (1969).
18. R. K. Norris, S. Sternhell, Australian J. Chem., 22, 935 (1969).
19. M. S. Newman, F. W. Hetzel, J. Org. Chem., 34, 1216 (1969).
20. P. G. Gassman, G. A. Campbell, Chem. Commun., 427 (1970).
21. R. S. Ward, Chem. Brit., 9, 444 (1973).
22. A. I. Scott et al., J. Chem. Soc., 4067 (1961).
23. N. N. Kalibabchuk, V. D. Pokhodenko, J. Org. Chem. USSR, 4
     320 (1968).

B. Electrophilic Reactions

1, From o-Aroylbenzoic Acids (1, 737)

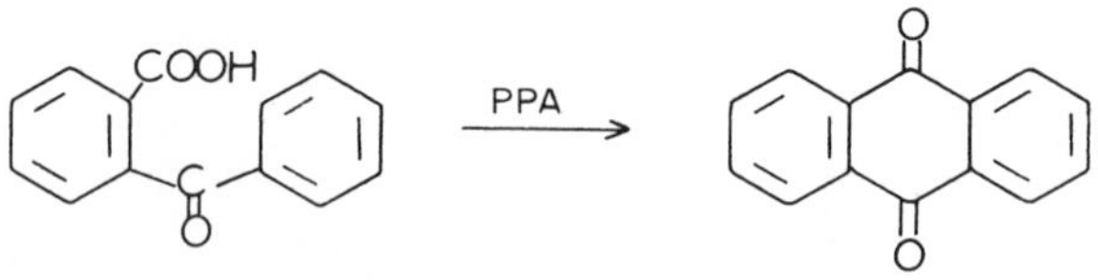

At last an investigator has taken the time and trouble to
determine the most desirable ratio of PPA to substrate for a
maximum yield of quinone. Metz [1] found that 1 mol of o-
aroylbenzoic acid per 350 g of PPA gave a 95% yield of anthraquinone. Less PPA diminishes the yield until with 200 g it drops
to 79%.

2. From Phenols and Nitrous Acid (1, 738)

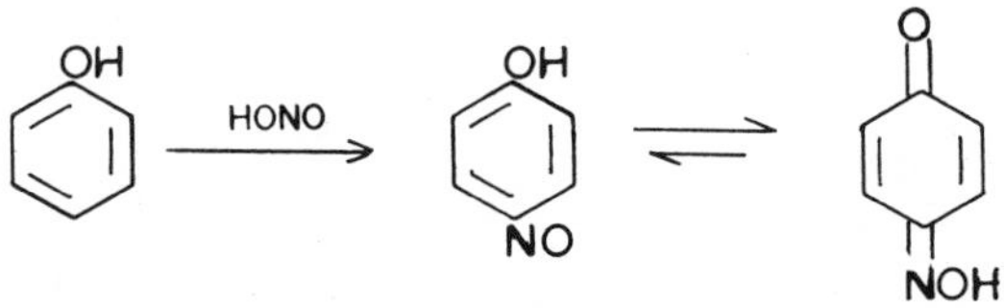

Norris and Sternhell [1a] have described three methods for
preparing quinone monoximes in equilibrium, as already indicated,
with the isomeric nitrosophenol. It appears that the equilibrium
in dioxane lies heavily toward the oxime form unless intramolecular hydrogen bonding between the substituent at C2 (or C6) and
the phenolic hydroxyl group of the nitroso form is possible. The
methods of synthesis for 27 examples are given in detail because
of their critical nature. The first method (see A.4 and D.2 for

the others) consists of the nitrosation of the phenol in the 4-position as indicated (Ex. a):

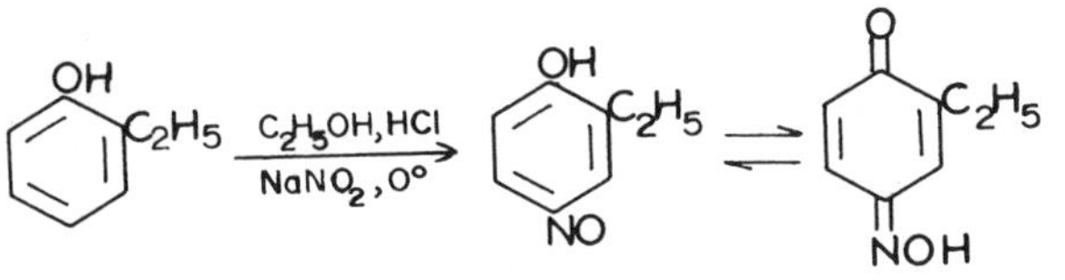

70 %

Considerable spectroscopic data are given on the compounds synthesized.

a. Preparation of 2-Ethyl-1,4-benzoquinone-4-oxime [1]

An ice-cold, stirred solution of 2-ethylphenol, 10 g in 50 ml of 10 N HCl and 50 ml of 95% ethanol was treated with 7.5 g of NaNO$_2$ over 10 min and the solution was stirred for 1 hr at 0°. Addition of 500 ml of water gave a precipitate that was collected and washed with water. Solution in aqueous Na$_2$CO$_3$ followed by acidification with HCl and crystallization from ether-light petroleum gave 8.5 g (70%) of the oxime.

3. From Quinones and Phenols

Substituted arylquinones may be formed by treating the pro-tonated quinone with a phenol, phenol ether, arylamine, or arene followed by oxidation. A specific example follows [2]:

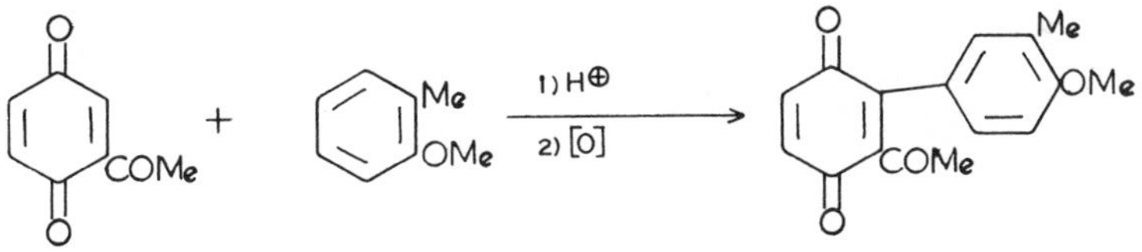

2-Acetyl-3(4-methoxy-3-methylphenyl)-
benzoquinone, 50% overall

4. From Quinones and Trialkylborons

Essentially, this reaction is a method for alkylating quinones via the hydroquinone (see Chapter 5, C.1) as shown [3]:

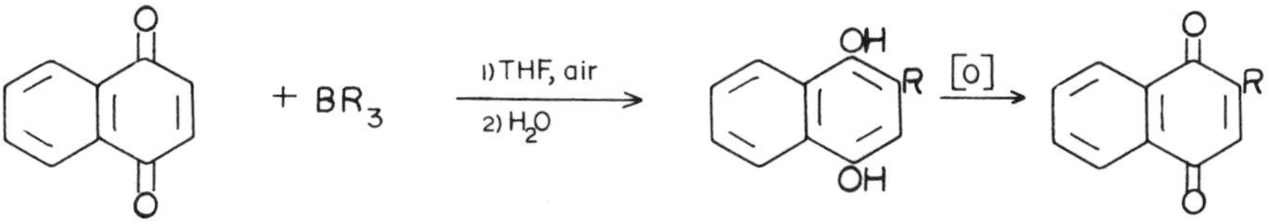

60% (R=C$_6$H$_{11}$)

Addendum

B.1.   Cyclization of some substituted o-benzoylbenzoic acids may lead to spirocyclic intermediates which may give an unexpected anthraquinone as shown [M. S. Newman, Acc. Chem. Research, 5, 354 (1972)]:

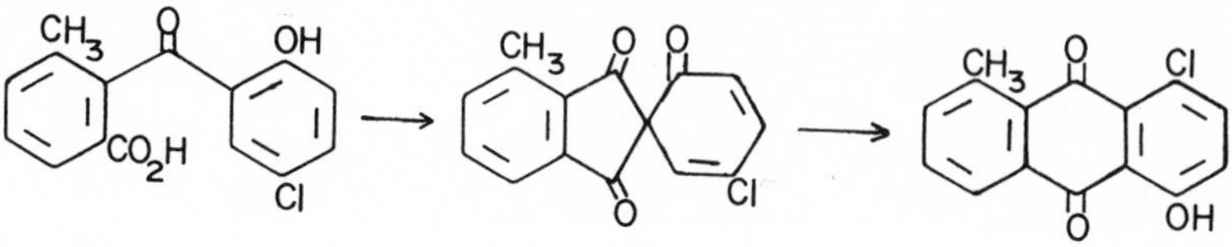

References

1.   G. Metz, Synthesis, 612 (1972).
1a.   R. K. Norris, S. Sternhell, Australian J. Chem., 22, 935 (1969).
2.   P. Kuser et al., Helv. Chim. Acta, 54, 980 (1971).
3.   G. W. Kabalka, J. Organomet. Chem., 33, C25 (1971).

C.   Condensation Reactions

2.   From Aromatic o-Dialdehydes and Glyoxal (1, 740)

The earlier work described the mixed condensation of o-dialdehydes with glyoxal to give a naphthoquinone. The reaction is a type of benzoin condensation, another example of which is now available [1].

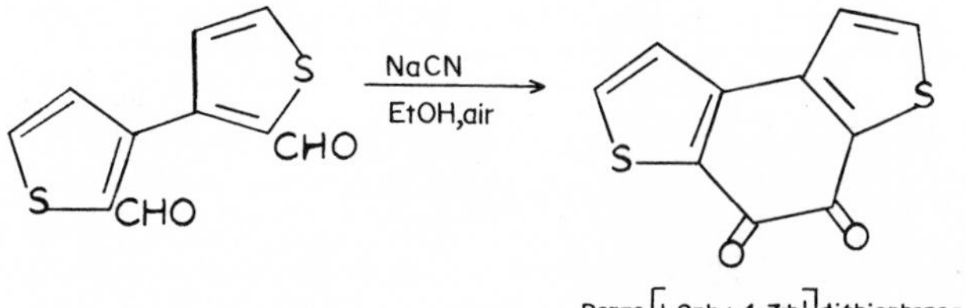

Benzo[1,2-b : 4,3-b']dithiophene-
7,8-quinone, 50 %

Quinones with sulfur atoms in other positions were also synthesized.

4.   From Aromatic o-Dialdehydes and Acetylenic Magnesium Bromide

A series of polycyclic quinones has been synthesized by Müller [1a] from aromatic o-dialdehydes as indicated:

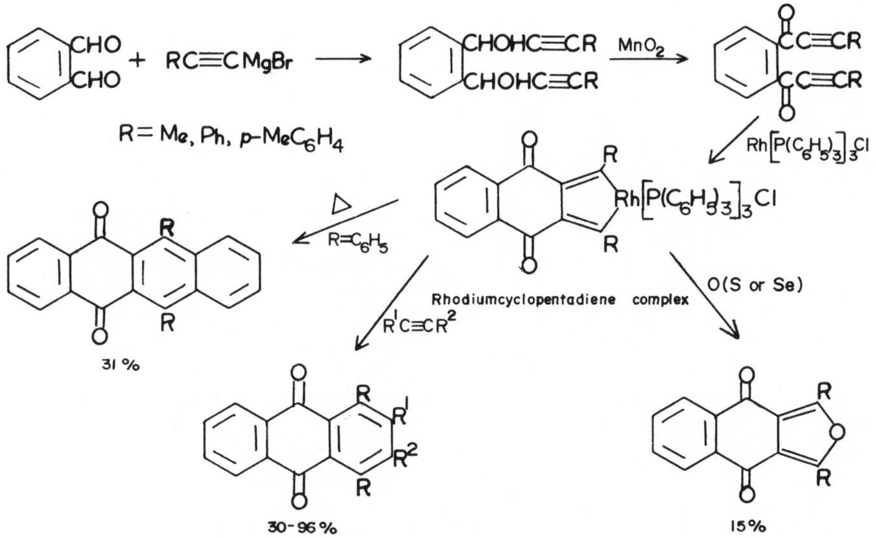

R = Me, Ph, p-MeC₆H₄

31 %

30-96%   15%

5. From Phenols and Carbon Tetrachloride (Zincke-Suhl Reaction)

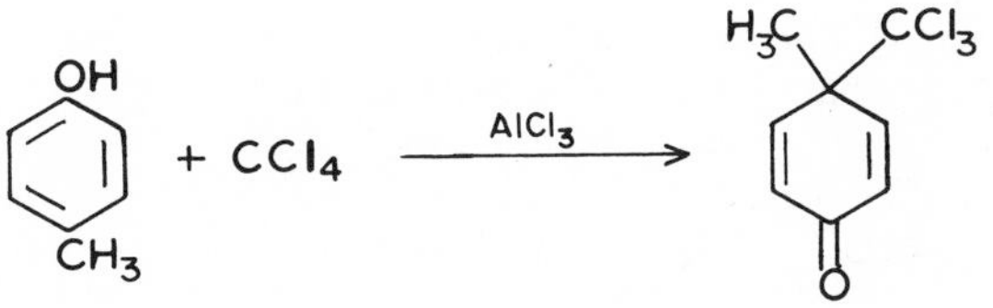

The condensation of carbon tetrachloride with phenols in the presence of aluminum chloride has been investigated by Newman and co-workers [2]. In the case of p-cresol and 3,4-dimethylphenol the yields of the cyclohexadienone were 60% and 72%, respectively. The use of carbon disulfide as a solvent improved the yield of the p-cresol derivative. The generality of the reaction is limited since no comparable results were obtained when methyl iodide, methylene chloride, chloroform, or hexachloroethane was substituted for carbon tetrachloride. With benzotrichloride the principal product (78%) was 2-hydroxy-5-methyl-benzophenone.

References

1. H. Wynberg, H. J. M. Sinnige, Rec. Trav. Chim., 88, 1244
    (1969).
1a. E. Muller et al., Ann. Chem., 754, 64 (1971); Chem.-Ztg.,
    97, 387 (1973); C. A., 79, 92367 (1973).
2. M. S. Newman et al., J. Org. Chem., 19, 978, 985, 992
    (1954); J. Am. Chem. Soc., 81, 6450 (1959).

D. Nucleophilic Reactions

Both 1,2- and 1,4-benzoquinones undergo Michael-type
addition of nucleophiles to give substituted catechols or hydroquinones. If an oxidizing agent is present, this reagent or
even the original quinone tends to reoxidize the reduced product
to a substituted quinone. 1,2-Benzoquinones are more reactive
than the 1,4-isomers in nucleophilic addition and the former may
rearrange to 1,4-benzoquinones. The nucleophiles may be alcohols, amines, halides, or sulfides [1].

1. From Quinones and Alcohols

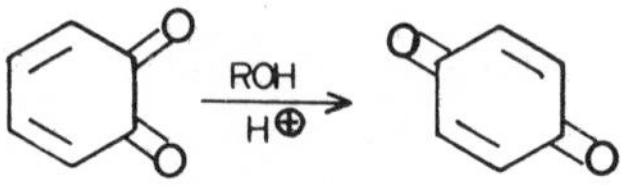

A review [2] deals with this reaction and an example is
given to illustrate the method of addition [3]:

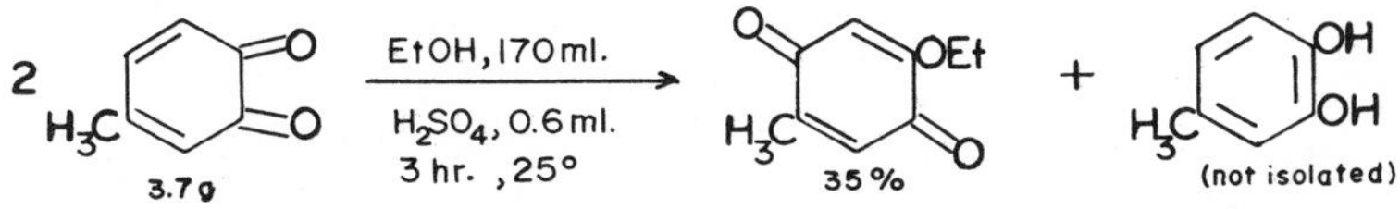

Electron-withdrawing groups attached to the quinone make
the 3-position very sensitive to substitution, as illustrated by
the mild conditions in the example [4]:

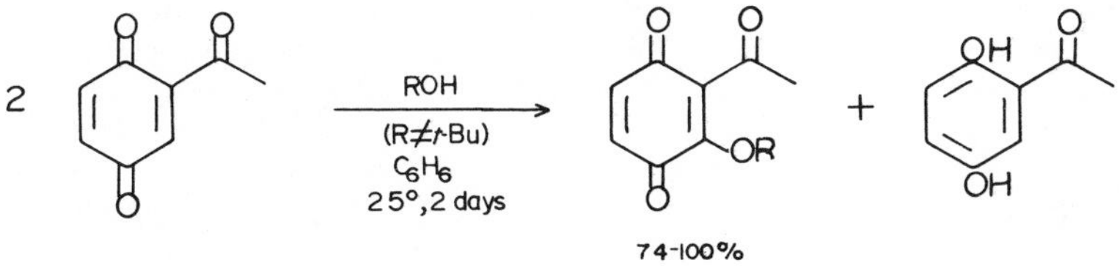

An unusual degradation occurs in the reaction [4a]:

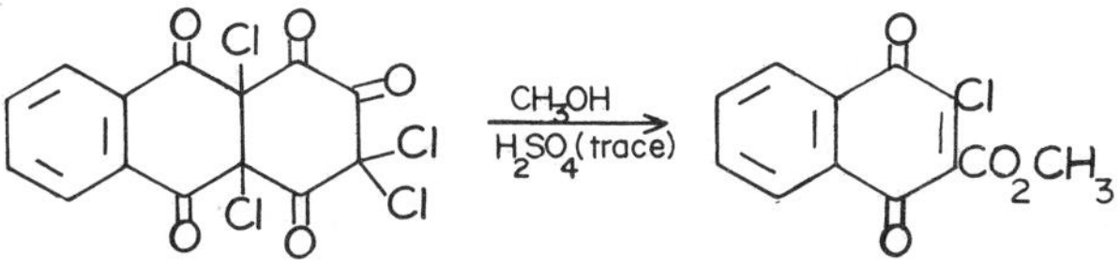

The 2-chloro-3-carbomethoxy-1,4-naphthoquinone was obtained in 89% yield.

2. From Quinones and Amines

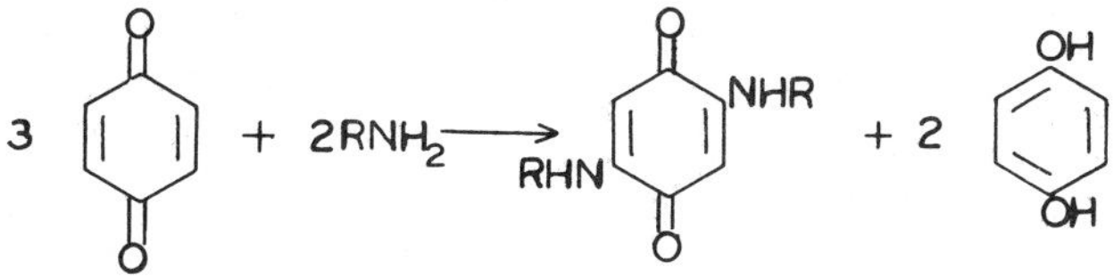

With primary amines rapid addition occurs to give the double substituted product, but with secondary amines the mono-substituted product is predominant [5].

On the other hand reagents which react with carbonyl groups may form a quinone derivative as shown [6] (Ex. a):

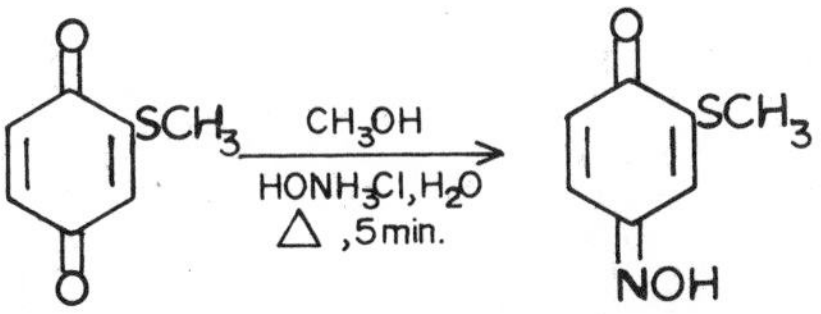

2 g.

1.3 g.
2-Methylthio-1,4-benzoquinone-4-oxime

It is of interest to note that the amino group of quinones may be exchanged quantitatively [7]. Further exchanges in the type are given in [8] and [9]. In the latter a heterocyclic ring is formed:

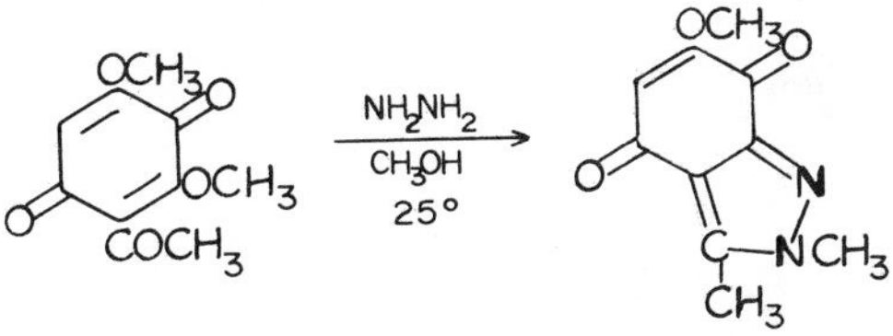

An indazolequinone

a. Preparation of 2-Methylthio-1,4-benzoquinone 4-oxime [6]

Methylthio-1,4-benzoquinone, 2.0 g was suspended in hot methanol and a hot aqueous solution of hydroxyammonium chloride, 3.0 g was added. After the mixture was heated on a steam bath until all the quinone had dissolved and then for an additional 5 min, it was diluted with 100 ml of water and filtered. The ethereal extract of the filtrate was extracted with 10% aqueous Na$_2$CO$_3$ and acidified with 3N HCl to give a crude product that gave 1.3 g of the quinone monoxime by crystallization.

3. From Azidoquinones and Nucleophiles

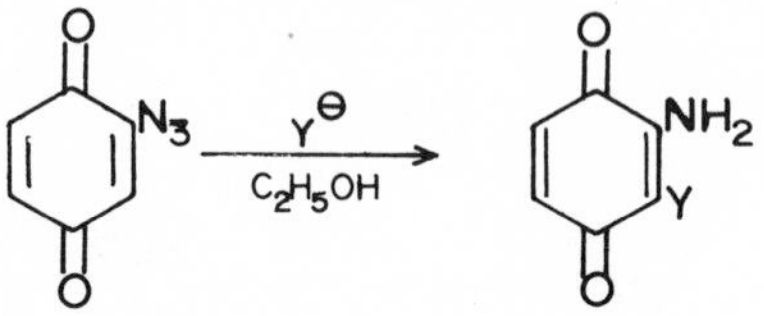

The quinone structure may be elaborated by reaction of the azidoquinone with nucleophiles. Moore and co-workers have devised a method by using the malonic ester anion with a t-butylazidoquinone [10]:

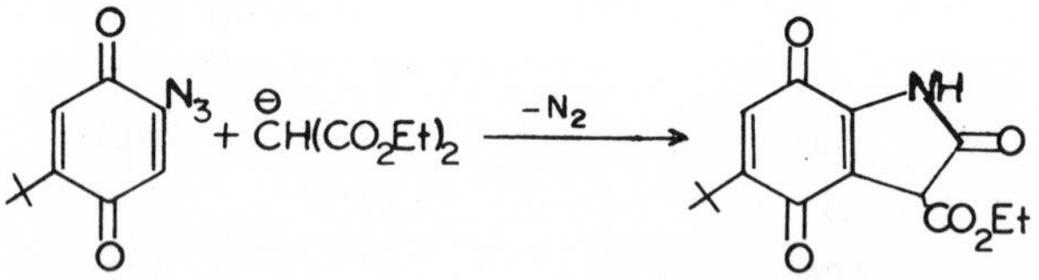

5-t-Butyl-3-carbethoxyindole-
2,4,7-trione, 77 %

Another similar elaboration occurs via 2-amino-3-chloro-benzoquinone, which is readily available from the treatment of aminobenzoquinone with t-butyl hypochlorite [11].

4. From Certain Benzophenones

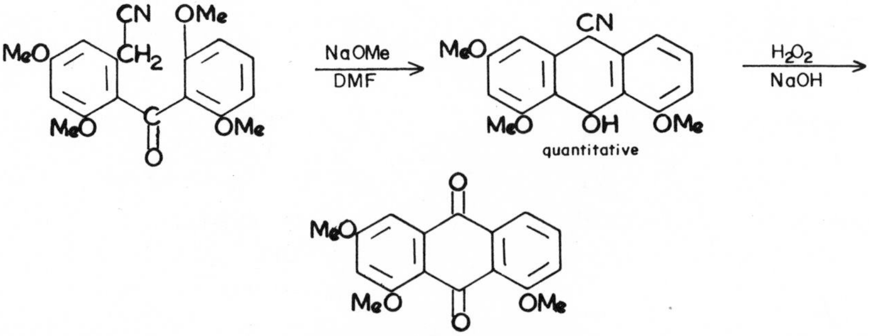

quantitative

1,3,8-Trimethoxyanthraquinone
quantitative

The synthesis is of general applicability in acquiring a number of substituted anthraquinones [12].

5. From Cyclodiketones

An ingenious cyclization has made available a series of compounds that may be considered as benzoquinones if the cyclopropane is considered to have some π character to its bonds. The reaction is as shown [13]:

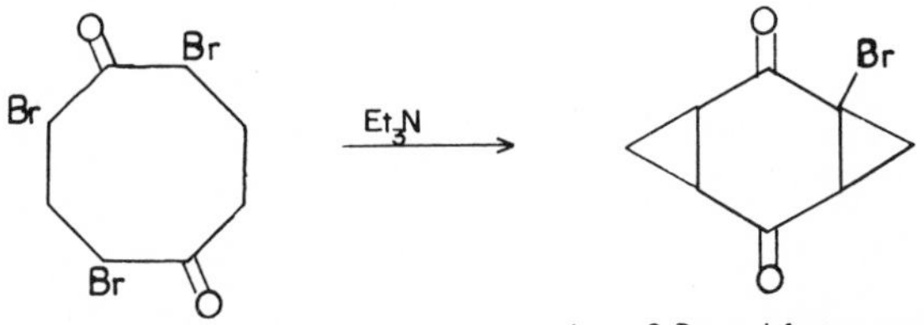

trans-2-Bromo-1,4-bishomoquinone

In the product without bromine, both cis and trans forms have been isolated.

References

1. H. Bosshard, Helv. Chim. Acta, 55, 32 (1972).
2. W. M. Horspool, Quart. Rev. (London), 23, 204 (1969).
3. L. Horner, T. Burger, Ann. Chem., 708 105 (1967).
4. F. Farina, J. Valderrama, Synthesis, 315 (1971).
4a. M. V. Sargent, D. O'N. Smith, Tetrahedron Letters, 2065 (1970).
5. A. Hikosaka, Bull. Chem. Soc. Jap., 43, 3928 (1970).
6. R. K. Norris, S. Sternhell, Australian J. Chem., 22, 935 (1969).
7. W. Schäfer, A. Aguado, Angew. Chem., 83, 441 (1971).
8. F. J. Bullock et al., J. Chem. Soc., C, 1799 (1969).
9. W. Schäfer et al., Angew. Chem. Intern. Ed. Engl., 10, 406 (1971).
10. H. W. Moore et al., Tetrahedron Letters, 4695 (1973).
11. H. W. Moore, G. Cajipe, Synthesis, 49 (1973).
12. C. H. Hassall, B. A. Morgan, J. Chem. Soc., Perkin Trans., I, 2853 (1973).
13. A. S. Dreiding et al., Angew. Chem. Intern. Ed. Engl., 11, 236 (1972) and earlier papers.

E. Cyclo Reactions

A series of substituted anthraquinones, such as emodin, has become available by the cyclo addition of ketene acetals to chloronaphthoquinones [1].

The equivalent of the benzyne intermediate for benzoquinone has now been demonstrated and synthesis as shown has been conducted with this reactive intermediate [2]:

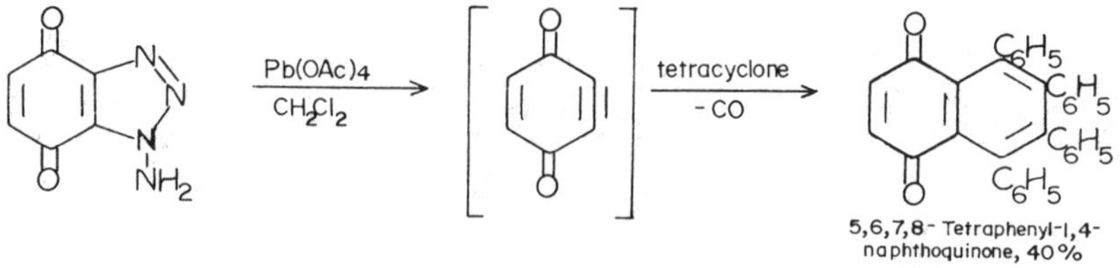

5,6,7,8- Tetraphenyl-1,4-
naphthoquinone, 40%

Finally, the complicated cyclocoupling of a bis (acetoxy-methyl) quinone is given [3]:

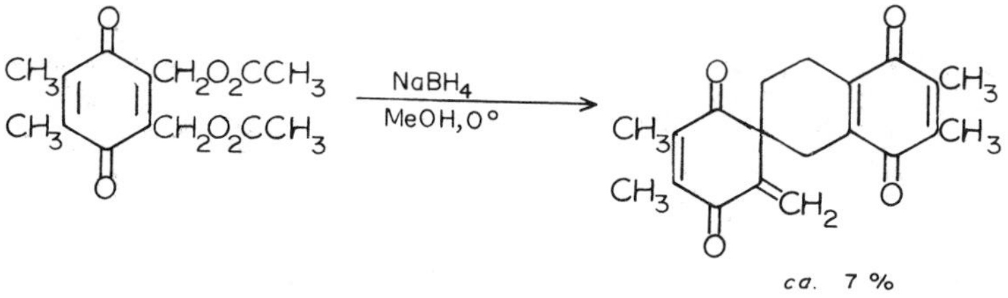

ca. 7%

Addendum

E.   Acetylenes do not react with benzoquinone at an ethylenic bond even on irradiation. However, if the Diels-Alder adduct of benzoquinone and anthracene is treated with a dialkylacetylene under irradiation, coupling occurs to form a fused cyclobutene ring.   A retro Diels-Alder at 220° gives anthracene and the product originally expected from the coupling of benzoquinone and a dialkylacetylene [P. Yates, G. V. Nair, Syn. Commun., 3, 337 (1973)].

Conditions for the Diels-Alder reaction of 1,3-pentadiene and 2,6-dimethylquinone may be so chosen that either of the two possible isomeric adducts may be obtained [Z. Valenta et al., Can. J. Chem., 53, 616 (1975)].

References

1.   J. Banville et al., Can. J. Chem., 52, 80 (1974).
2.   C. W. Rees, D. E. West, J. Chem. Soc., C, 583 (1970).
3.   A. J. Lin, A. C. Sartorelli, J. Org. Chem., 38, 813 (1973).

Chapter 13

CARBOXYLIC ACIDS

A recent review on carboxylic acids [1] covering such diverse
types as orthoacids, isotopically labeled acids, ylid precursors,
peracids, and sulfur-containing acids has become available.   In
this chapter it must be kept in mind that ester syntheses
(Chapter 14) are applicable to acid syntheses since it is usually
easy to convert the former into the latter. We have attempted to
emphasize the direct syntheses of acids with a minimum number
proceeding via the ester.

Although catalytic preparation of acids from various substrates, carbon monoxide, and transition-metal salts is not an
outstanding method, the principles applied are intriguing, far-
ranging, and important enough to warrant good coverage (F.3).
Masking of the potential carboxyl groups in a heterocyclic ring
for protection during nucleophilic operations on other parts of
the molecule has been proved a valuable tool (A.10). The method,
employed even in ylid synthesis (D.3), serves in part to supplement the malonic ester synthesis (A.9). Another supplement to
the malonic ester synthesis, more particularly for some hindered
malonic esters, is the modern adaptation of the Ivanov reagent,
$R_2\bar{C}\ \overline{CO}_2$ (C.2). The Koch-Haaf procedure, as modified by Bott
($R^+ + CH_2 = CCl_2 \rightarrow RCH_2CO_2H$), has now been quite thoroughly
tested (F.4, F.5). The addition of carboxyalkyl free radicals to
alkenes has been reviewed (F.8). Although the scope of the
addition has become extensive and important in specific cases,
the yields maximize with only a large excess of the acid. A
superior saponification agent, $Li\ SCH_2CH_2CH_3$-HMPA, has been
applied successfully to the conversion of hindered esters into
acids (A.2), and DBN also brings about solvolysis to the acid by
O-methyl cleavage of the methyl esters of hindered acids (A.2).
In addition, two superior environments for oxidizing aromatic
hydrocarbons or other substances to their acids have been devised.
Both methods attempt to solubilize the reagent in the organic
layer, one by the use of a crown ether with permanganate and the
other by means of a quaternary salt of high molecular weight in a
heterogeneous mixture of alkali and solvent (B.1).

Some advances have been developed in the Kolbe reaction
(C.3) and in carbonation (C.1). Aromatic acids are readily produced from hydrocarbons by a Friedel-Crafts reaction to obtain an
o-chlorobenzoylarene which is split with a strong base to yield
the arenecarboxylic acid (E.3). A host of other specialized but
useful syntheses, including several with baffling mechanisms,
(e.g., B.12; C.3) are described.

A. Hydrolysis

    2. From Esters (1, 748)

The cleavage of hindered esters under mild conditions has continued to receive the attention of chemists. Thus Bartlett and Johnson [2] utilized lithium n-propyl mercaptide in HMPA at room temperature for the cleavage of hindered methyl esters. The reagent is superior to those previously employed (1, 749-750) in that methyl mesitoate (Ex. a), methyl O-methylpodocarpate, and methyl triisopropylacetate give yields of 100%, 100%, and 99% of the acid, respectively, with the new reagent. Methyl 3β-acetoxy-Δ^5-androsten-17-carboxylate,

which gives only 49-50% of the acetoxy acid with lithium iodide in 2,6-lutidine [3], leads to a 92% yield with the new reagent.

The mercaptides, like most reagents, do possess some disadvantages. Their salts must be protected from air to prevent oxidation to the disulfide. Their odors are unpleasant and they may lead to acyl as well as the desired alkyl cleavage. In addition, their selectivity toward methyl esters in the presence of primary alkyl esters leaves something to be desired. For these reasons McMurry and Wong [4] studied the LiI-DMF reagent in an attempt to drive the reaction to completion at a lower temperature. Although these investigators determined no yields of acids obtained from hindered methyl esters, they succeeded in reducing the temperature of the cleavage of methyl benzoate about 40° by the use of a competing nucleophile such as sodium acetate or sodium cyanide.

Boron trichloride has also served, although less satisfactorily, for the cleavage of hindered esters [5]. With methyl-O-methylpodocarpate it gives a 90% return of O-methylpodocarpic acid; with methyl mesitoate the yield of the free acid is "excellent." Recently 1,5-diazabicyclo[4.3.0] nonene-5 (DBN) has been utilized in the O-alkyl cleavage of hindered methyl esters [6]. This reagent in o-xylene at refluxing temperature for 6 hr gave 91%, 94%, and 94% of the free acids from methyl O-methyl-podocarpate, methyl mesitoate, and methyl triisopropylacetate, respectively. The yield from methyl 3β-acetoxy-Δ^5-etienate was less satisfactory.

a.   Preparation of Mesitoic Acid [2]

Methyl mesitoate, 212.4 mg in 8.5 ml of 0.54M (3.9 equiv)

mercaptide reagent (lithium n-propyl mercaptide in HMPA), kept under N$_2$ for 1.25 hr at 25°, was placed in 150 ml of 1 N HCl. Extraction with ether gave a crude product that was dissolved in NaOH. Washing with ether and reprecipitation with HCl produced 195.1 mg (100%) of colorless acid crystals.

3. From Amides ($\underline{1}$, 751)

$$RCONH_2 \longrightarrow RCOOH$$

This process is conducted normally by hydrolysis in strong acid or in a base. As the amide becomes larger and more compli- cated in structure, it is not so easily hydrolyzed. Ames and Binns overcame this difficulty as shown [6a]:

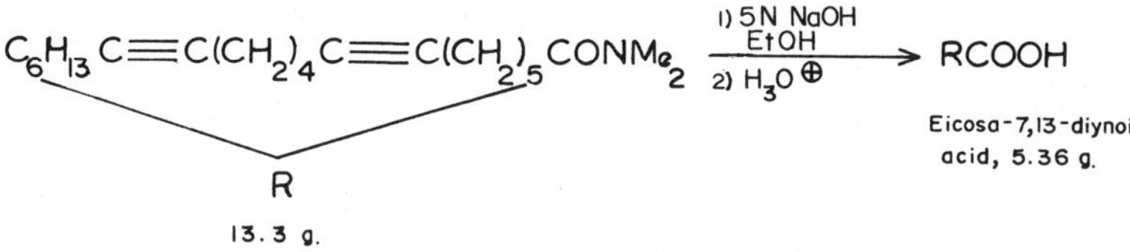

C$_6$H$_{13}$C≡C(CH$_2$)$_4$C≡C(CH$_2$)$_5$CONMe$_2$ $\xrightarrow[\text{2) H}_3\text{O}^\oplus]{\substack{\text{1) 5 N NaOH}\\\text{EtOH}}}$ RCOOH

Eicosa-7,13-diynoic acid, 5.36 g.

R
13.3 g.

The dimethylamide was employed above since it was available by treating 1-iodododec-5-yne with N,N-dimethyloct-7-ynamide and lithium amide.

Nitrosation followed by hydrolysis has been applied as indicated to the conversion of primary and secondary amides into carboxylic acids ($\underline{1}$, 752):

RCONH$_2$ $\xrightarrow[\text{CH}_3\text{I}]{\text{NaH-}}$ RCONHCH$_3$ $\xrightarrow{\text{NOCl}}$ RCON(CH$_3$)(NO) $\xrightarrow[\text{2) H}^\oplus]{\text{1) OH}^\ominus}$ RCOOH

Applied to N-xanthylamide, the results are as shown [6b]. Here the process

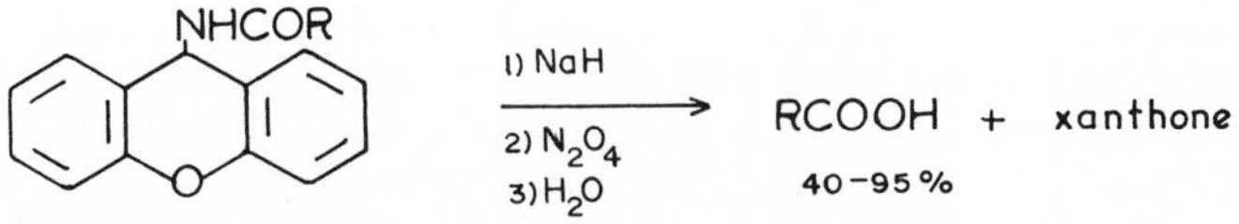

$\xrightarrow[\substack{\text{2) N}_2\text{O}_4\\\text{3) H}_2\text{O}}]{\text{1) NaH}}$ RCOOH + xanthone

40-95%

derives driving force by the loss of xanthone.

α-Aminoxycarboxylic acids have become available similarly via N-hydroxyphthalimide [6c]:

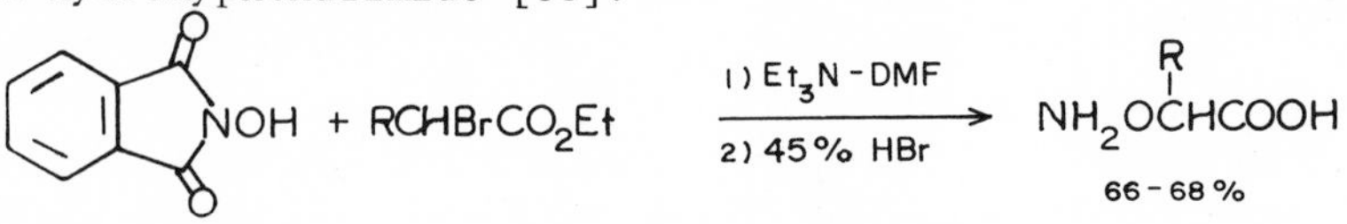

NOH + RCHBrCO$_2$Et $\xrightarrow[\text{2) 45\% HBr}]{\text{1) Et}_3\text{N - DMF}}$ NH$_2$OCH(R)COOH

66-68%

4.   From Nitriles ($\underline{1}$, 752)

$$RCN \longrightarrow RCOOH$$

Nitriles may be hydrolyzed in an acid salt melt containing water at elevated temperatures.   After removing the carboxylic acid, the melt may be restored by heating to remove ammonia.   In this way acetonitrile was converted into acetic acid as shown [7]:

$$CH_3CN + H_2O \xrightarrow[\substack{24 \ : \ 28 \ : \ 1 \\ 180°}]{KHSO_4 - NaHSO_4 - (NH_4)_2SO_4} CH_3COOH$$

78 g.   95 g.   94.8 %

A new industrial process for the manufacture of terephthalic acid from p-xylene, via the terophthalonitrile, has been announced by Lummus [8].   The nitrile is hydrolyzed in three stages to the dicarboxylic acid with controlled pH:

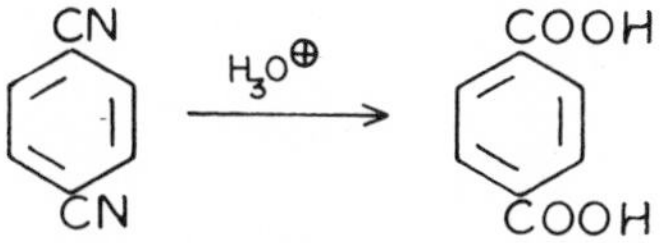

Cyanohydrins have been converted into α-keto acids by the sequence of reactions as indicated [9]:

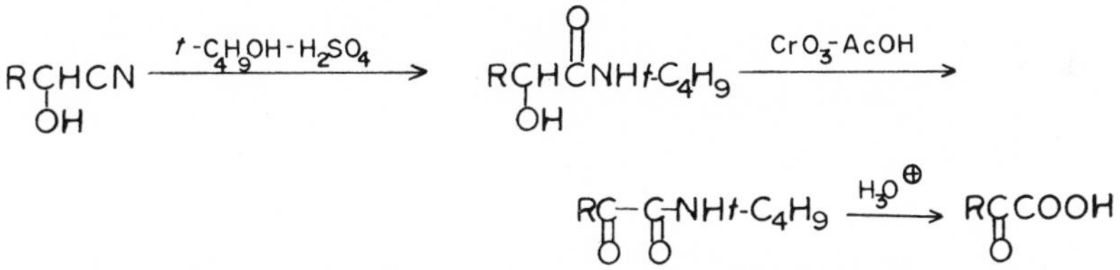

The α-hydroxy-N-t-butylcarboxamides in the first step were obtained in "good yield."   In the second step, the α-ketoacid amide yields varied within 56-90%; the final step led to α-keto acids in yields of 48-89%.

α-Chloroacrylonitrile is hydrolyzed to α-chloroacrylic acid with 60% aqueous sulfuric acid containing some hydroquinone [10].

Tetrachlorodibasic acids have become available by a simple process [11] (see Ex. a).

a.   Preparation of α,α,α',α'-Tetrachloroglutaric Acid

$$NC(CH_2)_3CN + PCl_5 \longrightarrow (Cl_3P\!\!=\!\!NCCl_2CCl_2)_2CH_2 \xrightarrow{\text{HCl}}$$

$$(HO_2CCCl_2)_2CH_2$$

A mixture of 0.1 mol of the dinitrile of glutaric acid and 0.6 mol of PCl$_5$ was refluxed in chlorobenzene until the evolution of hydrogen chloride ceased. Excesses of the phosphorus halide and chlorobenzene were removed by distillation at 10-20 mm, with bath at 40-60°. Concentrated hydrochloric acid, 100 ml was added dropwise to the stirred residue, which mixture was refluxed for 3-4 hr, cooled, and filtered. The yield of acid, after recrystallization from benzene, was 25%. Yields of higher homologs were larger.

b.   Preparation of 3-Chlorocyclobutanecarboxylic Acid [11a]

Yield was 84% by refluxing 3.9 mol of the nitrile in 3 l of concentrated HCl with stirring for 20 hr, extracting with CH$_2$Cl$_2$, evaporating, and distilling at 0.5 mm.

6.   From Trihalides (1, 754)

$$RCX_3 \xrightarrow[\text{H}_2\text{O}]{\text{H}_2\text{SO}_4} RCO_2H$$

This synthesis has been utilized in the preparation of α-alkoxycarboxylic acids as indicated [12]:

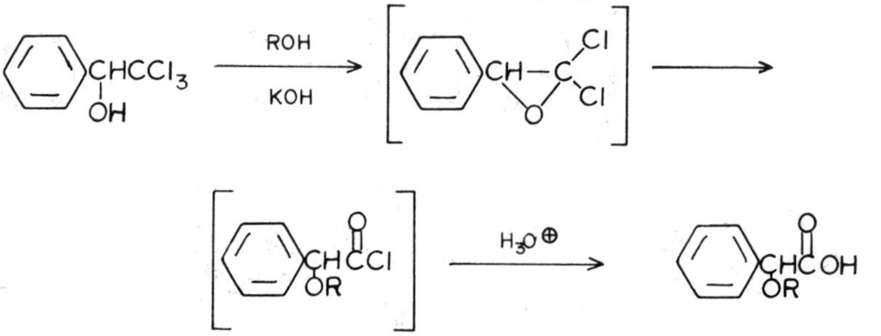

The synthesis of the carbinol has been reviewed [12a] and the mechanism studied [12b].

Reeve [13] prepared the aryl-(trichloromethyl) carbinol in situ from the aromatic aldehyde and either chloroform or bromoform, and the equation of the one operation takes the form (Ex. a):

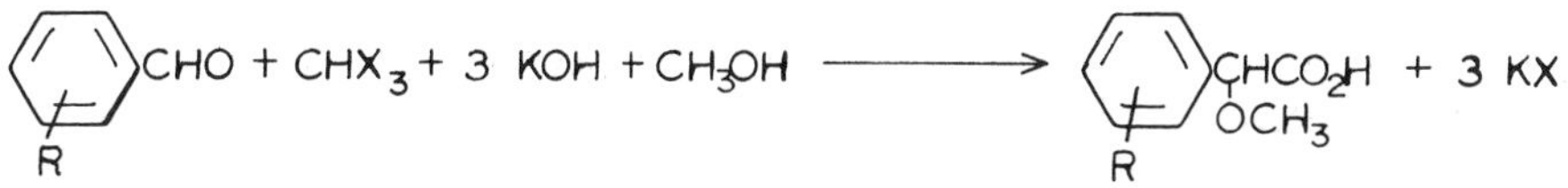

$$\text{R-}C_6H_4\text{-CHO} + CHX_3 + 3\ KOH + CH_3OH \longrightarrow \text{R-}C_6H_4\text{-CH}(OCH_3)CO_2H + 3\ KX$$

As with the Perkin reaction good yields are obtained with benzaldehydes containing electron-withdrawing groups, poor yields
with methylbenzaldehydes, and no products with benzaldehydes
containing strongly electron-donating groups. The amino tri-
chloro compound, $ArCH(NH_2)CCl_3$, may be prepared by treatment of the
hydroxy compound
with KNH_2.

In a similar manner under alkaline conditions acetone,
phenol, and chloroform condense in a one-step reaction to give
phenoxyisobutyric acid [14] as indicated:

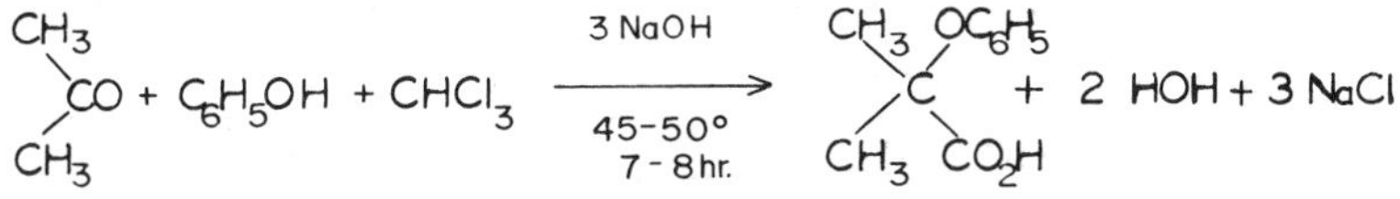

$$(CH_3)_2CO + C_6H_5OH + CHCl_3 \xrightarrow[\substack{45-50° \\ 7-8\,hr.}]{3\ NaOH} (CH_3)_2C(OC_6H_5)(CO_2H) + 2\ HOH + 3\ NaCl$$

It is well known that o- and p-hydroxybenzotrifluorides and
p-aminobenzotrifluoride are hydrolyzed readily to the corresponding benzoic acids with 1 N aqueous sodium hydroxide. By
contrast, conversion to the m-hydroxy- and m-aminobenzoic acids
requires rather drastic heating in concentrated sulfuric acid.
Grinter and co-workers [15] found that m-hydroxy- and m-amino-
benzotrifluorides could be converted readily into the corresponding carboxylic acids by light-induced hydrolysis. Thus the
m-hydroxy isomer was obtained as shown:

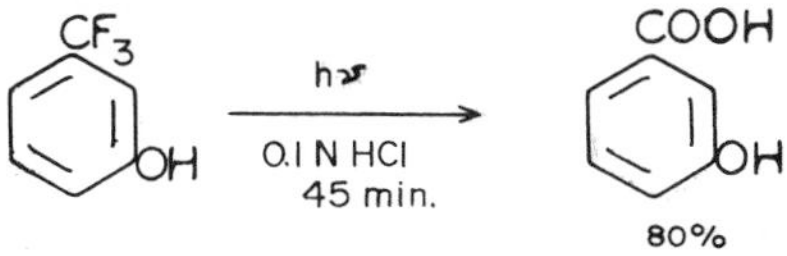

Likewise the m-amino isomer resulted as indicated:

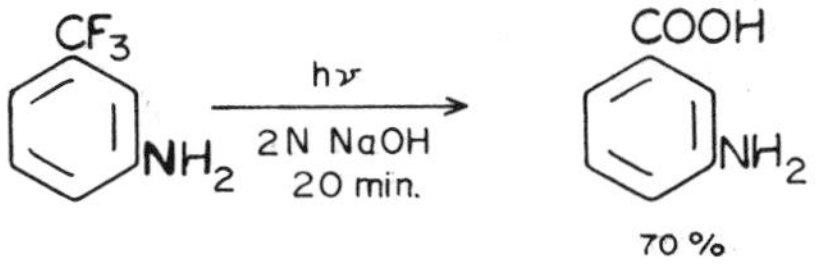

The light-induced hydrolysis of the ortho and para isomers was
less satisfactory.

a.   Preparation of 3-Chloro-α-methoxyphenylacetic Acid [10]

3-Chlorobenzaldehyde, 0.1 mol, bromoform, 0.12 mol, and

50 ml of methanol were mixed, and a solution of KOH, 0.5 mol in 110 ml of commercial methanol was added during 3 hr at 0-5°. After standing overnight, the mixture was treated with 100 ml of water and 300 ml of a half-saturated aqueous salt solution and warmed and filtered. The cooled filtrate at a pH of 3.3 gave sodium hydrogen bis(3-chloro-α-methoxyphenyl acetate), 75%. Excess of hydrochloric acid gave the free acid.

9. From Diethyl Malonate (1, 757)

Takeda and co-workers [16] utilized diethyl malonate with alkyl chloropyruvate in the synthesis of λ-ketocarboxylic acids as shown:

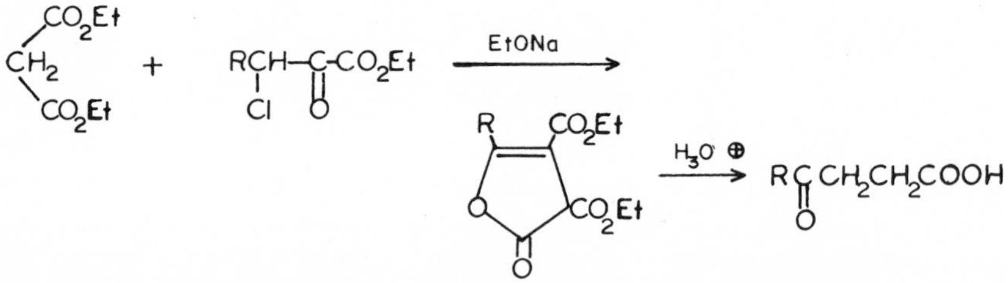

The yields were 32-46% in the first step and 51-74% in the second.

Fleming and Owen [17] employed the diethyl aroylmalonate in the synthesis of acetylenic acids. The two-step reaction follows:

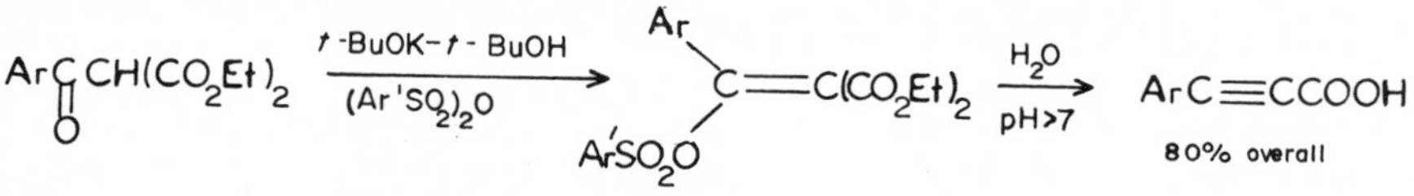

The mechanism appears to be:

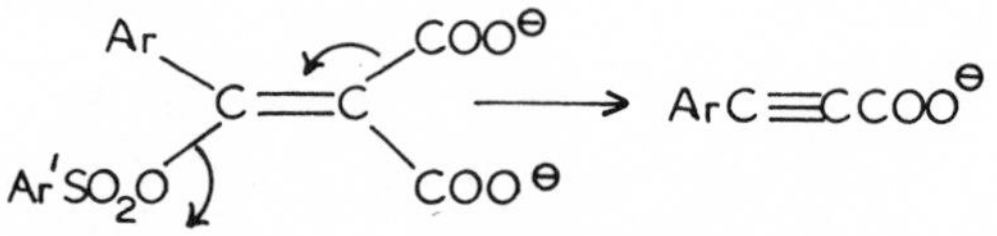

The arenesulfonic anhydride is usually the preferred sulfonating agent, but in some cases the t-butyl esters are preferred.

10. From Hydantoins, 2-Oxazolines, Dihydro-1,3-oxazines, Dithiane Esters, and Orthothioformates (Masked Carboxyl Groups) (1, 758)

   In recent years there has been a trend to incorporate the
potential carboxyl group in some heterocyclic compound that can
be alkylated, acylated, or treated with any type of nucleophilic
agent to give a product capable of hydrolysis to the acid.   The
possibilities of the method are extensive since the heterocycle
may contain O, N, S, or other hetero atoms.
   Meyers and Temple [18] succeeded in synthesizing α-substituted carboxylic acids as their esters via the 2-oxazolines as
shown:

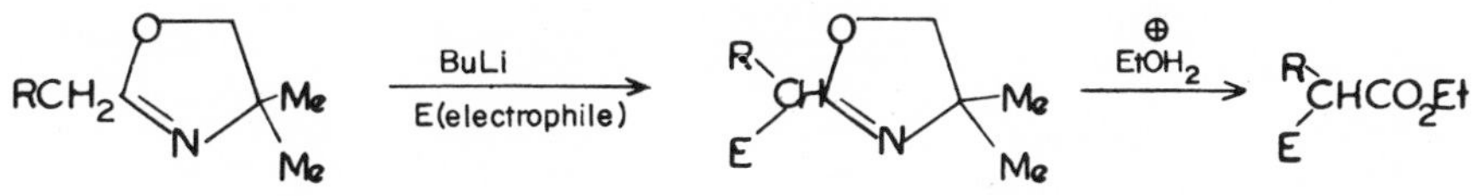

Since the starting material is prepared from readily available
materials

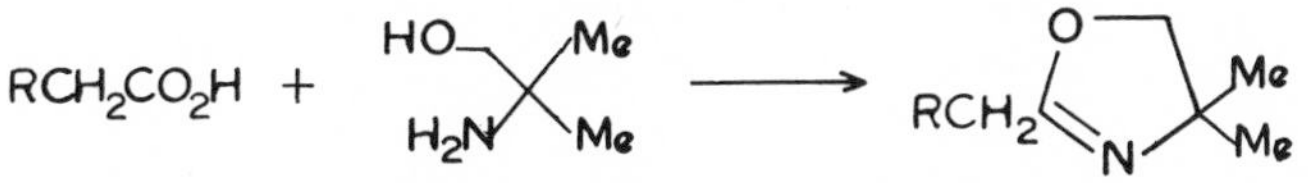

the method actually represents a means of masking the carboxyl
group while attaching various electrophiles to the methylene
group.   The success of the method has been demonstrated by
introducing a variety of electrophiles, some of which contain
active functional groups.   An example follows:

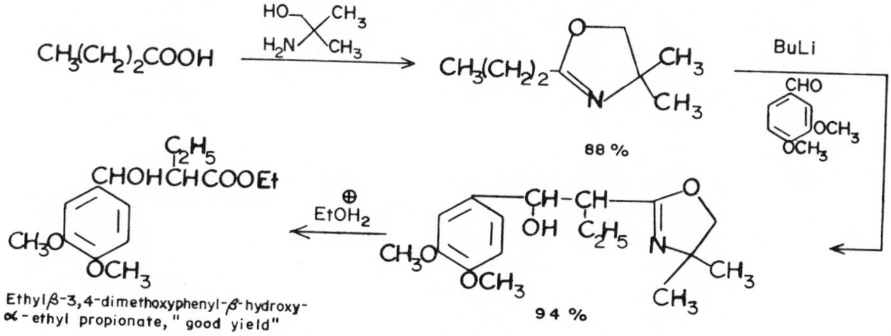

Ethyl β-3,4-dimethoxyphenyl-β-hydroxy-
α-ethyl propionate, "good yield"

   In a similar manner, substituted benzoic acids or esters
have been synthesized via the Grignard reagent of the 2-oxazo-
lines [19], as indicated in the preparation of
p-carboxybenzophenone:

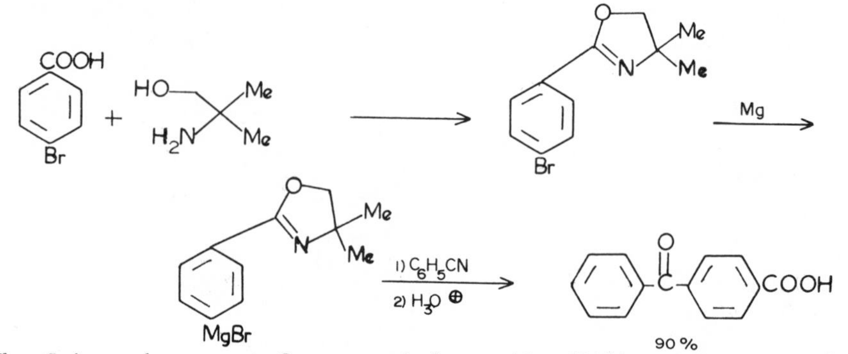

The Grignard reagent forms satisfactorily (90%), although it is necessary for the magnesium to be pure to prevent as much as 25% coupling.

In a similar manner Meyers and co-workers [20] synthesized alkyl and aryl carboxylic acids via dihydro-1,3-oxazines as shown in the preparation of 7-benzoylheptanoic acid:

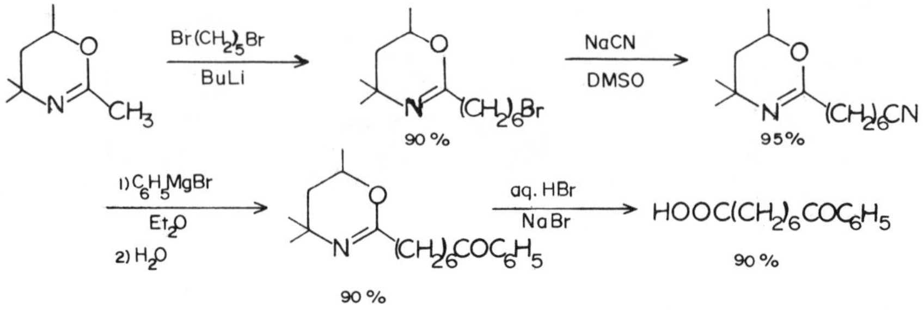

The starting material, 2-methyl-5,6-dihydro-1,3-oxazine, is available commercially. By starting with 2-phenyldihydro-1,3-oxazine aromatic acids may be synthesized.

The oxazole-5-ones from N-acylaminoacids, such as N-benzoylvaline, are capable of alkylation, as has been shown in a one-pot reaction [20a]:

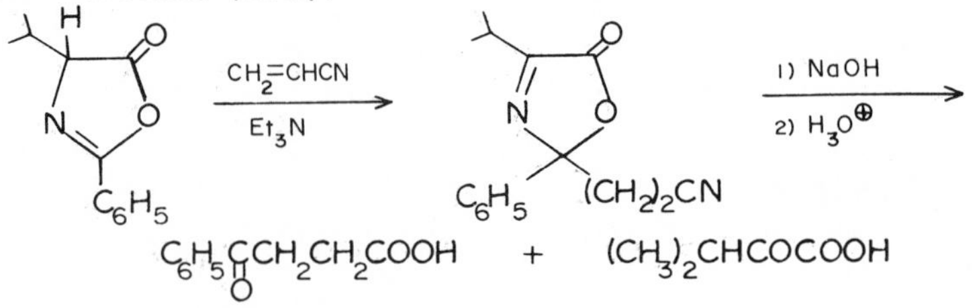

$C_6H_5CCH_2CH_2COOH$ + $(CH_3)_2CHCOCOOH$

4-Oxo-4-phenylbutyric acid
89 %

The heterocycles derived from α-carbethoxycycloalkanones may be used to form 1-carboxycycloalkenes [20b]:

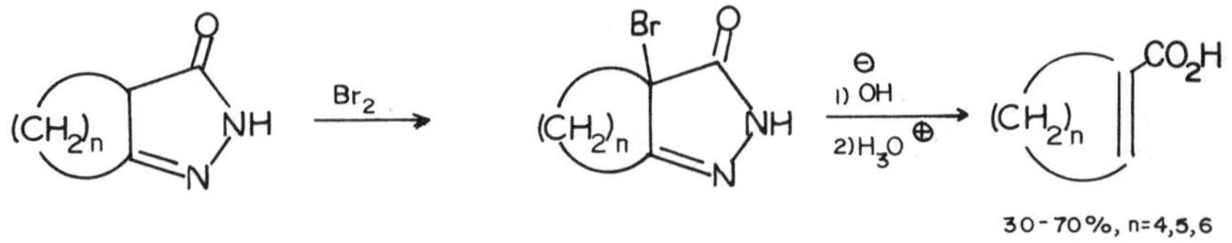

Br_2
1) OH
2) $H_3O^{\oplus}$
30-70%, n=4,5,6

An optically active heterocycle as shown has been dialkylated and hydrolyzed to give an optically active acid of known configuration [20c]:

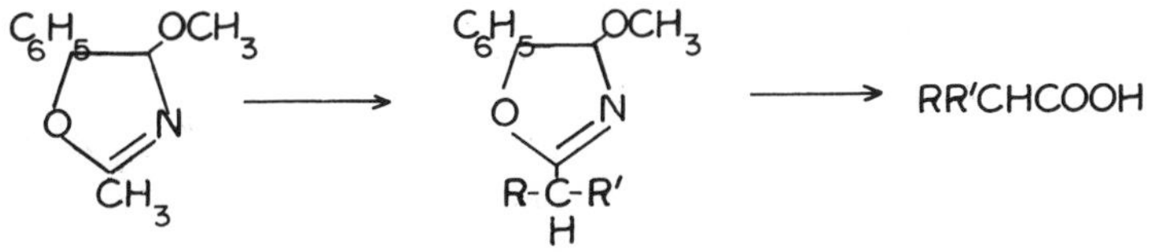

RR'CHCOOH

Orthothioformates are hydrolyzed in the presence of mercuric salts [21] to acids in rather low yield.

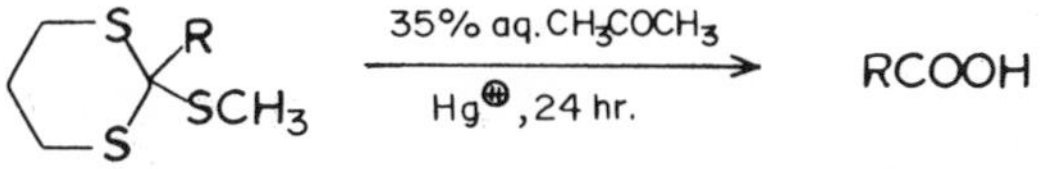

35% aq. CH_3COCH_3
$Hg^{\oplus}$, 24 hr.
RCOOH

Refluxing in alcohol in the presence of Hg^{++} gives the ester in 92-99% yields.

Ethyl 1,3-dithiane-2-carboxylate has also been utilized in the synthesis of esters and α-ketoesters, both of which may be converted into the corresponding acids [22]. Thus the dithiane, readily available from ethyl diethoxyacetate and 1,3-propanedithiol, may be converted stepwise as shown into the ester or α-ketoester:

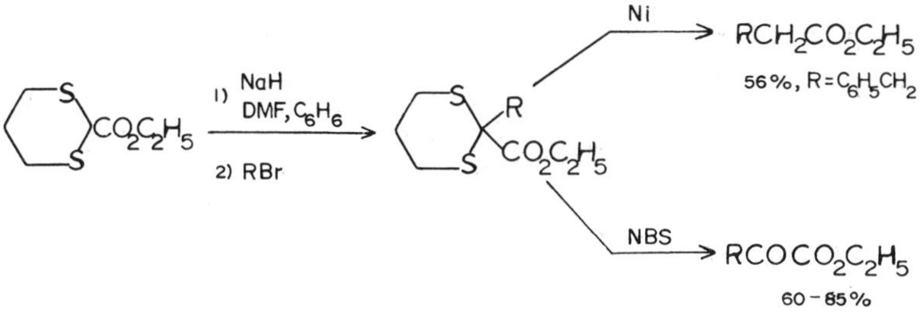

1) NaH DMF, C_6H_6
2) RBr
Ni
$RCH_2CO_2C_2H_5$
56%, R=$C_6H_5CH_2$
NBS
$RCOCO_2C_2H_5$
60-85%

Addenda

A.2. Hindered esters may be solvolyzed by a refluxing mixture of quinoline and acetic acid through which a slow stream of N_2 is passed [G. Aranda, M. Fetizon, Synthesis, 330 (1975)].

A.2. Racemic methyl α-acetoxyesters are hydrolyzed by α-chymotrypsin to optically active acids [I. Tabushi et al., Tetrahedron Letters, 309 (1975)].

A.3. Primary or secondary amides may be hydrolyzed to thiol acids, RCOSH, by treating first with NaH and then CS_2. Conditions are mild [I. Shahak, Y. Sasson, J. Am. Chem. Soc., 95, 3440 (1973)].

References

1. S. Patai, The Chemistry of Carboxylic Acids and Esters, Wiley-Interscience, New York, 1969.
2. P. A. Bartlett, W. S. Johnson, Tetrahedron Letters, 4459 (1970).
3. A. Eschenmoser et al., Helv. Chim. Acta, 43, 113 (1960).
4. J. E. McMurry, G. B. Wong, Syn. Commun., 2, 389 (1972).
5. P. S. Manchand, Chem. Commun., 667 (1971).
6. D. H. Miles, E. J. Parish, Tetrahedron Letters, 3987 (1972).
6a. D. E. Ames, S. H. Binns, J. Chem. Soc., Perkin Trans., I, 255 (1972).
6b. T. B. Patrick, J. G. Dolan, J. Org. Chem., 38, 2828 (1973).
6c. K. S. Suresh, R. K. Malkani, Indian J. Chem., 10, 1068 (1972); Synthesis, 457 (1974).
7. W. Neugebauer, L. Schmidt, U.S. Patent 3,492,345, January 27, 1970; C. A., 72, 89793 (1970).
8. Chem. Eng. News, March 19, 1973, p. 31; April 2, 1973, p. 10.
9. J. Anatol, A. Medete, Synthesis, 538 (1971); Compt. Rend., 272, C, 1157 (1971).
10. B. Raduechel et al., German Patent 2,213,734, September 20, 1973; C. A., 79, 136515 (1973).
11. V. P. Kukhar et al., Synthesis, 545 (1973).
11a. S. C. Cherkofsky et al., J. Am. Chem. Soc., 93, 121 (1971).
12. C. Weizmann et al., J. Am. Chem. Soc., 70, 1153 (1948); E. D. Bergmann et al., ibid., 72, 5012 (1950).
12a. A. B. Galun, A. Kalir, Org. Syn., Coll. Vol. 5, 130 (1973).
12b. W. Reeve et al., J. Org. Chem., 40, 339 (1975).
13. W. Reeve, Synthesis, 131 (1971); W. Reeve et al., J. Am. Chem. Soc., 82, 4062 (1960); 83, 2755 (1961).

14. G. Andrescu, Rom., 52, 775, May 20, 1971; C. A., 78, 58070 (1973).
15. R. Grinter et al., Tetrahedron Letters, 3845 (1968).
16. A. Takeda et al., Bull. Chem. Soc. Jap., 44, 1342 (1971).
17. I. Fleming, C. R. Owen, J. Chem. Soc., C 2013 (1971).
18. A. I. Meyers, D. L. Temple, Jr., J. Am. Chem. Soc., 92, 6644 (1970).
19. A. I. Meyers, D. L. Temple, Jr., J. Am. Chem. Soc., 92, 6646 (1970).
20. A. I. Meyers et al., J. Am. Chem. Soc., 91, 5886 (1969).
20a. W. Steglich, P. Gruber, Angew. Chem. Intern. Ed. Engl., 10. 655 (1971).
20b. L. A. Carpino, E. G. S. Rundberg, Jr., J. Org. Chem., 34, 1717 (1969).
20c. A. I. Meyers, G. Knaus, J. Am. Chem. Soc., 96, 6508 (1974).
21. R. A. Ellison et al., J. Org. Chem., 37, 2757 (1972).
22. E. L. Eliel, A. A. Hartmann, J. Org. Chem., 37, 505 (1972).

B. Oxidation

   1. From Alcohols, Carbonyl Compounds, or Esters (1, 761)

$$RCH_2OH \xrightarrow{[O]} RCHO \xrightarrow{[O]} RCO_2H$$

   Potassium permanganate solubilized in benzene by complexing with dicyclohexyl-18-crown-6 has been proven an effective oxidizing agent for oxidizing olefins, alcohols, aldehydes, and alkyl benzenes to carboxylic acids under mild conditions [1]. The reagent represented as

is prepared by stirring equimolar amounts of dicyclohexyl-18-crown-6 and KMNO$_4$ in benzene at 25°. A recent synthesis of 18-crown-6 types has been described [2]. The product of oxidation, the potassium salt of the acid, may be converted into the free acid in the usual manner.

   By the procedure 1-heptanol gives 70% of heptanoic acid and benzyl alcohol gives 100% of benzoic acid (Ex. a). Efficient and abrasive stirring is needed to sustain solubilization since the manganese dioxide that forms has a tendency to coat the unreacted permanganate.

Quaternary ammonium salts also carry the permanganate ion into the organic phase [3] (Ex. b).

Aldehydes have been oxidized to carboxylic acids by the use of a mixture of sodium chlorite and aminosulfonic acid in high yield as shown [4]:

$$RCHO \xrightarrow[\text{NH}_2\text{SO}_3\text{H}]{\text{NaClO}_2} RCOOH$$

The method is also satisfactory for the oxidation of hydroxybenz-aldehydes. The sulfamic acid, used in equivalent quantities, inhibits the formation of chlorine dioxide, ClO_2, which is less discriminate in attack on the substrate. Sodium chlorite, which has been used in carbohydrate oxidation, serves well with vanillin and similar phenolic aldehydes, provided sulfamic acid is present.

Cyclododecanol has been oxidized to dodecanedioic acid in excellent yield by the use of nitric acid with the catalysts as shown [5]:

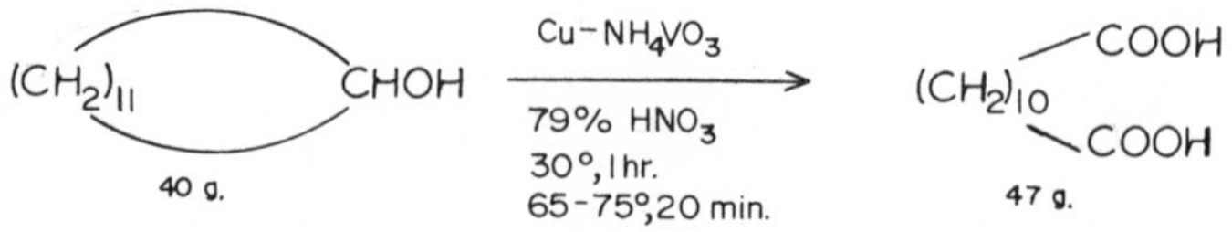

1,2-Cyclanediols with Ag and Ag_2O in alcoholic KOH give acyclic dicarboxylic acids as shown after acidification [6].

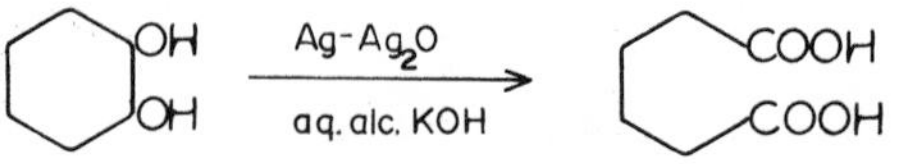

Adipic acid, 89 %

On the other hand, one carbon atom is lost with 1,2-acyclic diols under similar conditions as indicated to give a monocarboxylic acid (Ex. c):

$$C_{10}H_{21}CHOHCH_2OH \xrightarrow[\substack{\text{aq. NaOH} \\ (\text{CH}_3)_2\text{CHOH}}]{\text{Ag-Ag}_2\text{O}} C_{10}H_{21}COOH$$

Undecanoic acid, 98-99%

deVries and Schors obtained similar results in using molecular oxygen in the presence of Co(II) salts [7]. Thus 1,2-dihydroxy-decane gives, as shown, pelargonic acid:.

$$C_8H_{17}CHOHCH_2OH \xrightarrow[\substack{Cd(II) laurate \\ C_6H_5CN \quad 70\%}]{O_2} C_8H_{17}COOH$$

Although cycloalkanones on oxidation usually give dicarboxylic acids (1, 761), such is not the case when cyclopentanone
is oxidized with dimsyl sodium in DMSO [8]:

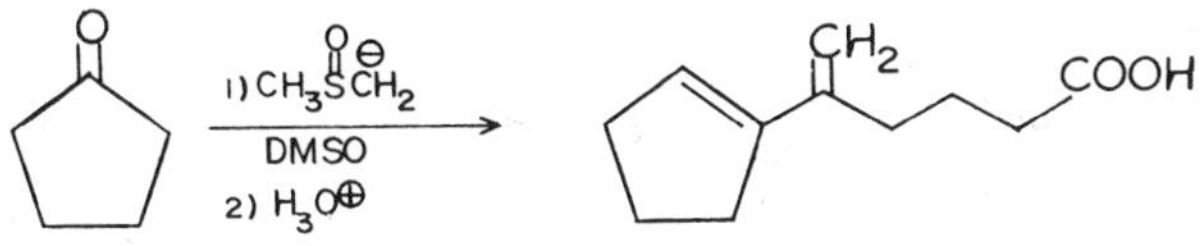

δ-Methylene-δ-(1-cyclopentenyl)pentanoic acid, 59%

Strangely enough, the reaction appears to be limited to
cyclopentanones.

Deno and co-workers [9] have shown that aliphatic ketones
when treated with potassium persulfate, $K_2S_2O_8$, in 50% H_2SO_4
undergo the Baeyer-Villiger oxidation to give carboxylic acids
quantitatively. Thus 2-butanone gives both acetic acid and
ethanol quantitatively as shown:

$$CH_3COCH_2CH_3 \xrightarrow[50\% H_2SO_4]{K_2S_2O_8} CH_3COOH + HOCH_2CH_3$$

Apparently, ethyl migration occurs in the intermediate cation
(1, 829) as indicated:

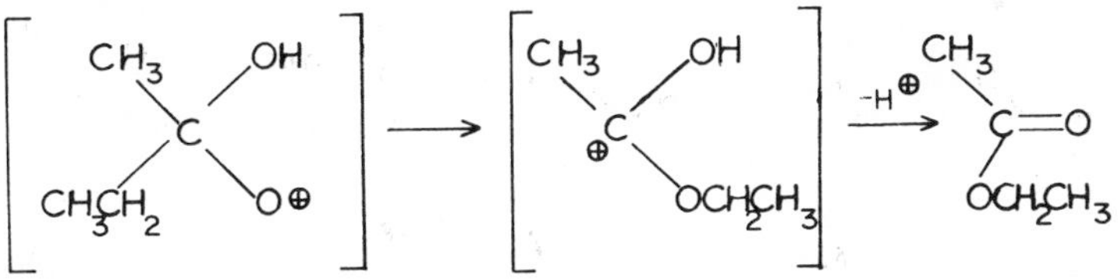

Under the reaction conditions the ester hydrolyzes to give the
acid and alcohol.

The oxidation of esters is the final step in the synthesis
of 2-fluorocarboxylic acids from an alkene. The acids are produced easily and conveniently in three steps as indicated in the
synthesis of 2-fluoroheptanoic acid [10]:

$$CH_3(CH_2)_4CH=CH_2 \xrightarrow[HF]{CH_3CONHBr} CH_3(CH_2)_4CHFCH_2Br \xrightarrow{NaOAc-NaI-DMF}$$

60-77 %

$$CH_3(CH_2)_4CHFCH_2OAc \xrightarrow[AcOH]{HNO_3} CH_3(CH_2)_4CHFCOOH$$

63-78 %   73-84 %

a. Preparation of Benzoic Acid [1]

Benzyl alcohol, 7.3 g (68 mmol), 28.6 g (181 mmol) of $KMnO_4$, 2.4 g (6 mmol) of dicyclohexyl-18-crown-6, and 1.0 l of benzene were rolled in a ball mill for 2 hr. (Caution!) The solid that formed was removed by filtration and dissolved in 5% NaOH solution. After removal of MnO_2 by filtration, the aqueous solution was extracted with ether to remove any crown ether. Acidification gave benzoic acid, 100%.

b. Preparation of Nonanoic Acid [3]

1-Decene, 0.2 mol, was added to a stirred mixture of 50 ml of benzene, 0.01 mol of tricaprylmethylammonium chloride, 0.8 mol of $KMnO_4$, and 100 ml of water at such a rate as to maintain the temperature at 40°. After stirring for an additional 30 min, the excess permanganate was destroyed with Na_2SO_3. The filtrate was acidified with dilute HCl; the benzene solution was extracted with 10% aqueous NaOH; and the acid (91%) was recovered from the alkaline phase by acidification and extraction with ether.

c. Preparation of Undecanoic Acid [6]

1,2-Dodecanediol, 10 g in 200 ml of isopropyl alcohol, was diluted with water, after which 10 ml of 40% NaOH was added, and with stirring at 20-25°, 36 g of freshly precipitated Ag_2O and 10 g of amorphous Ag. After 1 hr, the original black suspension changed to a heavy precipitate. On filtration the filtrate was extracted with ether and the lower layer of isopropyl alcohol was distilled up to 97-98°. The liquid remaining was acidified with HCl, after which 9.0-9.1 g (98-99%) of undecanoic acid was recovered in the usual manner.

6. From Alkenes (1, 766)

$$RCH=CH_2 \xrightarrow{KMnO_4} RCOOH$$

As shown in B.1, potassium permanganate solubilized in benzene by complexing with dicyclohexyl-18-crown-6 oxidizes olefins effectively to carboxylic acids under mild conditions.

Thus <u>trans</u>-stilbene gives benzoic acid and cyclohexene gives adipic acid, both with 100% yield. In all probability quaternary ammonium salts could be substituted for the crown ether [3].

8. From Alkylated Arenes, Heterocycles, or Phenols (<u>1</u>, 769)

$$-CH_3 \xrightarrow{KMnO_4} -COOH$$

Reference has already been made (B.1) to the effectiveness of $KMnO_4$ solubilized in benzene by complexing with dicyclohexy-18-crown-6 or by forming quaternary ammonium permanganate in the oxidation of alkylbenzenes to carboxylic acids under mild conditions. In this manner toluene gives 78% of benzoic acid, while <u>p</u>-xylene produces 100% of the corresponding toluic acid.

The oxidation of alkyl to carboxyl groups in heterocyclic rings is a common procedure. Cooper and Rickard [11], using such a procedure, as indicated, have reported an improved method for the synthesis of 5-nitropyridine-2-carboxylic acid (Ex. a).

a. Preparation of 5-Nitropyridine-2-carboxylic Acid [11]

2-Methyl-5-nitropyridine, 4.7 g, was added slowly with stirring to 35 ml of concentrated H_2SO_4. Chromium(VI) oxide, 10.2 g, was added in small portions while stirring and maintaining the temperature below 70°. Stirring was continued until the temperature dropped to 25° and then the mixture was poured into crushed ice. The solid which separated was crystallized from hot water to give an 80% yield.

12. From Methyl Ketones and β-Diketones (<u>1</u>, 772)

The haloform reaction is a useful method of degradation of an acetyl group to the carboxylic acid group. Carbon tetrachloride in alkali has been found to accomplish the same end as sodium hypochlorite although the mechanism may be different [12]. The reaction is a curious one of limited practicality:

13. From Alkynes

$$RC\equiv CH \xrightarrow{Tl(III)(NO_3)_3} RCOOH$$

McKillop and Taylor [13] have studied the oxidation of acetylenes with thallium(III) nitrate. Although the oxidation products vary with the type and number of substituents in the acetylene, monoalkylacetylenes yield carboxylic acids with the loss of one carbon atom. 1-Octyne, for example, gives heptanoic acid:

$$C_6H_{13}C\equiv CH \xrightarrow{Tl(III)(NO_3)_3} C_6H_{13}COOH$$
80 %

The investigators suggest the following conversion:

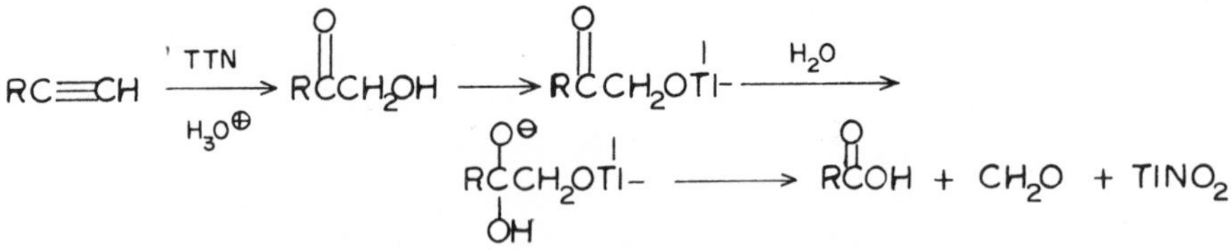

$$RC\equiv CH \xrightarrow[H_3O^\oplus]{TTN} RCCH_2OH \longrightarrow RCCH_2OTl- \xrightarrow{H_2O}$$

$$RCCH_2OTl- \longrightarrow RCOH + CH_2O + TlNO_2$$
         OH

14. From Hydroxamic Acids and Hydrazides

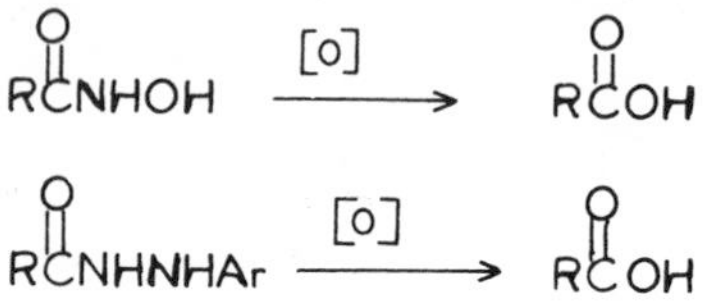

$$RCNHOH \xrightarrow{[o]} RCOH$$

$$RCNHNHAr \xrightarrow{[o]} RCOH$$

Methyl N-hydroxysuccinamate has been oxidized to methyl hydrogen succinate as indicated [14]:

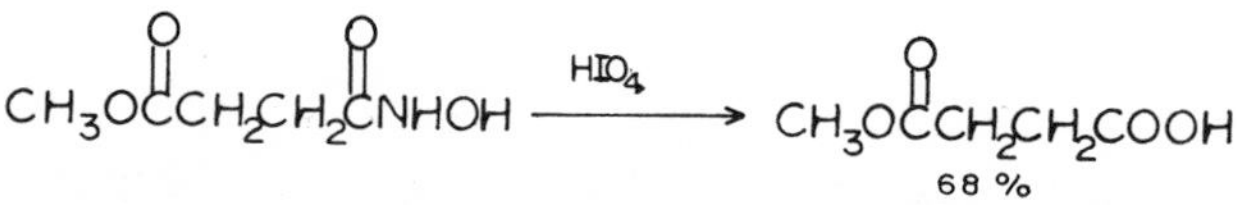

$$CH_3OCCH_2CH_2CNHOH \xrightarrow{HIO_4} CH_3OCCH_2CH_2COOH$$
                                    68 %

The oxidation of γ-phenylhydrazides of N-carbobenzoxy-α-L-glutamylamino acid esters has been accomplished. In the process the phenylhydrazide group is converted into a carboxyl group and the carbobenzoxy and ester groups remain intact (1, 752). The reaction is illustrated in the synthesis of ethyl N-carbobenzoxy-α-L-glutamylglycinate.

The oxidation of aroyl hydrazides has also been conducted very rapidly with ceric ammonium nitrate to give acids in 70-90% yields [14a].

15. From Acids

$$RCHCOONa \xrightarrow[THF]{BuLi} [R_2\overset{\ominus}{C}C\overset{\ominus}{O}O] \xrightarrow[H_2O]{air} R_2\underset{OH}{C}COOH$$

The salt is converted first into the dianion, which is then oxidized at 20° for 5-6 hr [14b]. Yields of α-hydroxyacids were 32-90%.

16. From Benzoins or Like Compounds

$$ArCOCHOHAr \longrightarrow ArCHO + ArCOOH$$

The oxidation is carried out with ceric ammonium nitrate in aqueous acetonitrile to give 81-88% of the aromatic acid (6 ex.) [14a].

In a similar manner Reichstein's compound S has been converted into the hydroxyacid by treatment with an alkaline triphenyltetrazolium chloride (TTC) solution [14c]:

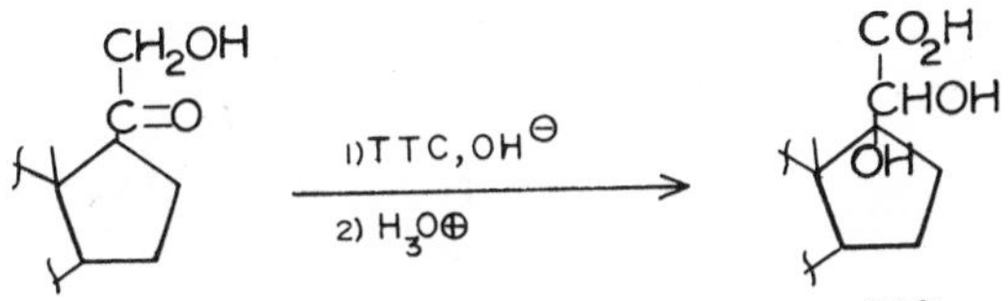

Unsaturated cyclopentanones are cleaved by peroxides to give a δ-oxohexanoic acid [14d]:

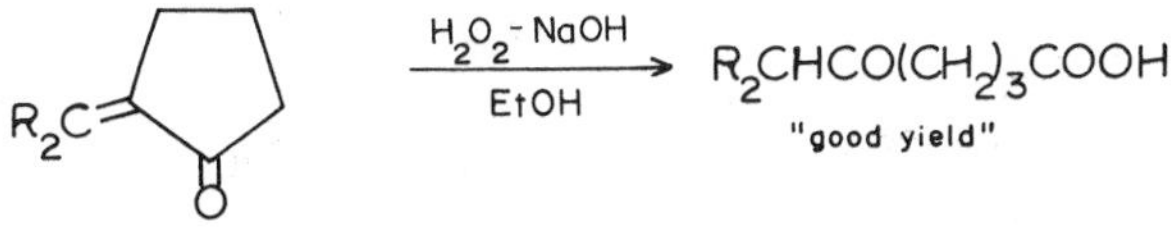

Actually the product was a mixture of acid and ester.

Addendum

B.8. The oxidation of 1,2,4,5-tetramethylbenzene to pyromellitic acid with nitric acid followed by sodium hypochlorite is reviewed as well as the synthesis of most tetracarboxylic acids [B. I. Zapadinskii et al., Russ. Chem. Rev. (Engl. transl.), 42, 939 (1973)]. The oxidation of 1,4-bis(dimethylbenzyl) benzene, A, to 1,4-bis(dicarboxybenzoyl)benzene, B, has been recommended to one of us (D.P.) as suitable for difficult aromatic oxidations: A,

10.6 g, and 132 g of 30% HNO_3 were heated to 200° in a
magnetically stirred type of autoclave. When the pressure
reached 25 kg/cm^2, the autoclave was degassed, closed, and held
at 200° for 4 hr. The acid, B, melting above 300° was then
recovered [S. Nishizaki, Kagyo Kagaku Zasshi, 69, 1069 (1966)].

References

1.   D. J. Sam, H. F. Simmons, J. Am. Chem. Soc., 94, 4024
     (1972).
2.   R. N. Greene, Tetrahedron Letters, 1793 (1972); D. J. Cram
     et al., J. Org. Chem., 39, 2445 (1974).
3.   J. Dockx, Synthesis, 441 (1973).
4.   B. O. Lindgren, T. Nilsson, Acta Chem. Scand., 27, 888
     (1973).
5.   T. Sakurai, M. Takamatsu, Japanese Patent, 7,310,771,
     April 7, 1973; C. A., 79, 18112 (1973).
6.   J. Kubias, Coll. Czech. Chem. Commun., 31, 1666 (1966).
7.   G. deVries, A. Schors, Tetrahedron Letters, 5689 (1968).
8.   W. T. Comer, D. L. Temple, J. Org. Chem., 38, 2121 (1973).
9.   N. C. Deno et al., J. Org. Chem., 35, 3080 (1970).
10. F. H. Dean et al., Org. Syn., 46, 10, 37 (1966).
11. G. H. Cooper, R. L. Rickard, Synthesis, 31 (1971).
12. C. Y. Meyers et al., J. Am. Chem. Soc., 91, 7510 (1969).
13. A. McKillop et al., J. Am. Chem. Soc., 93, 7331 (1971).
14. T. Emery, J. B. Neilands, J. Am. Chem. Soc., 82, 4903
     (1960).
14a. T. L. Ho et al., Synthesis, 562 (1972).
14b. G. W. Moersch, M. L. Zwiesler, Synthesis, 647 (1971).
14c. H. Möhrle, D. Schittenhelm, Arch. Pharm., 303, 771 (1970);
     Synthesis, 225 (1971).
14d. G. Le Guillanton, Bull. Soc. Chim. France, 2871 (1969).

C.   Carbonation and Carboxymethylation of Organometallic
     Compounds; Decarboxylation

     1.   Carbonation (1, 776)

     The carboxylation of α-anions of esters has been described
by Wynberg and co-workers [1]. Thus ethyl isobutyrate when
treated first with lithium diisopropylamide and then with CO_2
gives monoethyl dimethyl malonate as indicated:

$$(CH_3)_2CHCO_2Et \xrightarrow[\text{2) } CO_2]{\text{1) } LiN(i\text{-}pr)_2} (CH_3)_2C\overset{\displaystyle COOH}{\underset{90\%\ \ CO_2C_2H_5}{}}$$

Other esters possessing α hydrogen atoms respond similarly, the
yields being greatest for esters which are crowded around the
anionic carbon atom.   The procedure offers an attractive route to
malonic acid derivatives.

Wynberg and co-workers [2] have also shown that cyclic
monocarboxylic acids may be converted via the ester into dicarboxylic acids.   Thus methyl adamantane-2-carboxylate was converted into the monomethyl ester of the 2,2-dicarboxylic acid as
indicated:

$$\text{(adamantane)}\!-\!COOCH_3 \xrightarrow[\text{2) } CO_2]{\text{1) } LiN(i\text{-}pr)_2} \text{(adamantane)}\!\!<\!\!\begin{array}{l}COOCH_3\\ COOH\end{array}$$
88%

The use of lithium diisopropylamide has been extended to
prepare derivatives of carboxylic acids.   Thus α-amino acids may
be prepared, usually in low yields, as indicated [3] in one
stage:

$$RCH_2CO_2H \xrightarrow[\text{2) } NH_2OCH_3]{\text{1) } LiN(i\text{-}pr)_2} RCH(NH_2)COOH$$

Similarly, β-hydroxy or α,β-unsaturated acids may be obtained as
shown [4].

$$RCH_2COOH \xrightarrow[\text{THF-HMPA}]{\overset{\text{1) } LiN(i\text{-}pr)_2^-}{}} HOCH_2CHRCO_2H \xrightarrow[\triangle]{H^\oplus} CH_2{=}CRCO_2H$$
$$\text{2) } CH_2O \qquad\qquad 80\text{-}93\% \qquad\qquad 90\text{-}94\%$$

The advantage of using HMPA-THF as a mixed solvent in these
two processes is discussed later under C.2.

In the synthesis of β-methylene carboxylic acids, Maercker
[5] formed the Grignard reagent of the β-methylene ether, which
with CO_2 followed by acidification gave the unsaturated acid
desired as indicated:

$$CH_2{=}\overset{R}{\underset{}{C}}CH_2OC_6H_5 \xrightarrow[\text{THF}]{Mg} CH_2{=}\overset{R}{\underset{}{C}}CH_2MgOC_6H_5 \xrightarrow[\text{2) } H^\oplus]{\text{1) } CO_2} CH_2{=}\overset{R}{\underset{}{C}}CH_2COOH + C_6H_5OH$$
$$63\text{-}66\%$$

The results with cinnamyl phenyl ether are noteworthy in that a 90% yield of a 1:1 mixture of 2- and 4-phenylbutenoic acids is obtained:

$$C_6H_5CH=CHCH_2OC_6H_5 \xrightarrow[\text{THF}]{\text{Mg}} C_6H_5CH=CHCH_2MgOC_6H_5 \xrightarrow[\text{2) H}^{\oplus}]{\text{1) CO}_2}$$

$$C_6H_5CH=CHCH_2COOH + C_6H_5CHCH=CH_2$$
$$\qquad\qquad\qquad\qquad\qquad\qquad \underset{COOH}{|}$$

By treatment of ethereal dilithium cyclooctatetraenide with solid CO_2 followed by quenching with water and acidification, Cantrell[6] obtained a product that consisted largely of the acid shown:

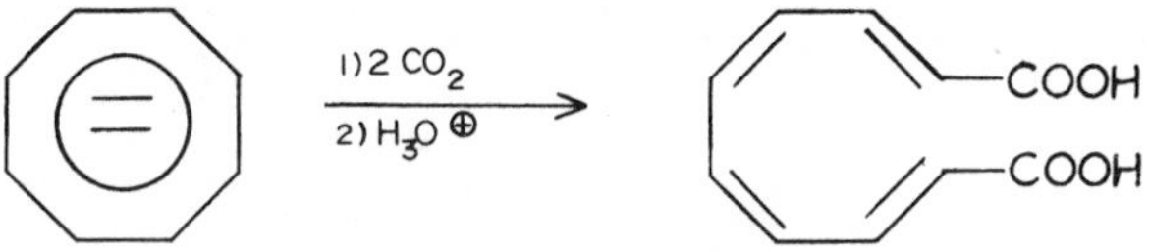

trans,cis,cis,trans-2,4,6,8-
Decatetraene-1,10-dioic acid

The carbonation of the bicyclic hydrocarbon, indene, was accomplished by Patmore and co-workers [7] as shown:

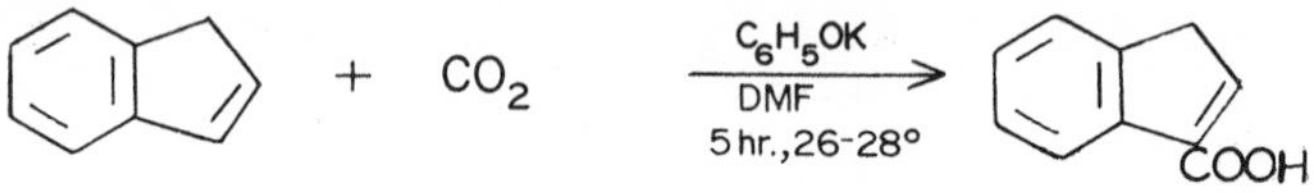

3-Indenecarboxylic acid
37.4%

The use of carbon dioxide with potassium phenolate in DMF was also utilized by Mori and co-workers [8] in the carbonation of cyclohexanone:

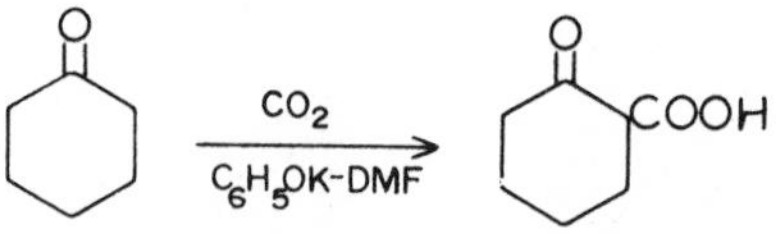

40%

Ketones may also be carbonated by the use of lithium 2,6-t butyl-4-methylphenoxide [9] as indicated:

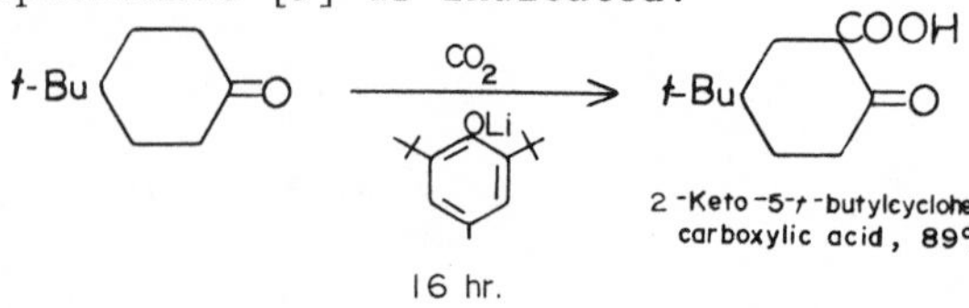

2-Keto-5-*t*-butylcyclohexane-
carboxylic acid, 89%

16 hr.

The phenoxide is ineffective toward less acidic substrates.
    Walborsky and co-workers [10] have utilized the carbonation
reaction in the synthesis of α-keto acids by the sequence shown:

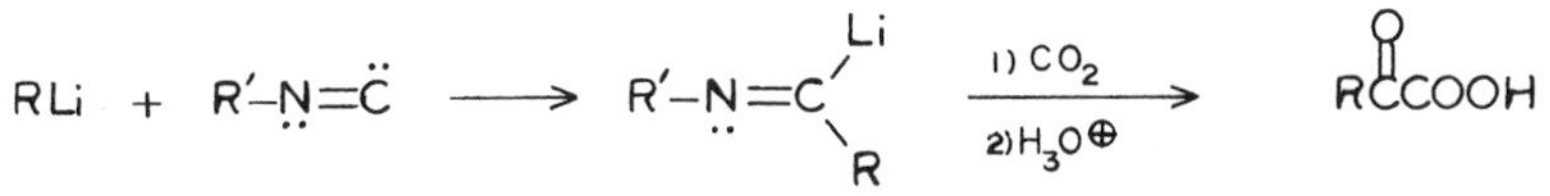

$$RLi \ + \ R'-N\overset{..}{=}\overset{..}{C} \ \longrightarrow \ R'-N=C\overset{Li}{\underset{R}{\diagdown}} \quad \xrightarrow[2)H_3O^{\oplus}]{1)\ CO_2} \quad R\overset{O}{\overset{\|}{C}}COOH$$

This is a simple method, illustrated in the synthesis of 2-keto-
3-methylpentanoic acid (80%) (Ex. a).
    For the preparation of α-amino acids by this procedure, see
[11].
    The carbonation of a variety of organic compounds has been
accomplished by Angelo [12]. The method consists of dissolving
an alkali metal in a solvent, such as THF, containing an aryl
pyridine, such as 2,6-diphenylpyridine (DPP). These solutions,
being strongly basic, are anion free radical reagents that promote the formation of a large variety of other anion free
radicals, which when treated with CO_2 and hydrolyzed give carboxylic acids. Thus as indicated benzophenone gives benzilic
acid (Ex. b).

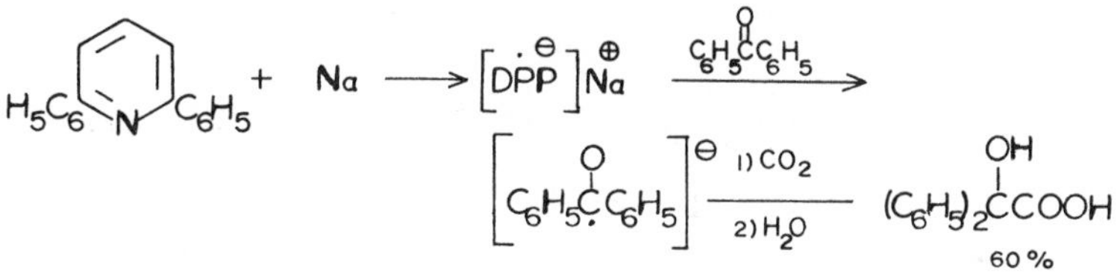

Other carbonations by this method follow:

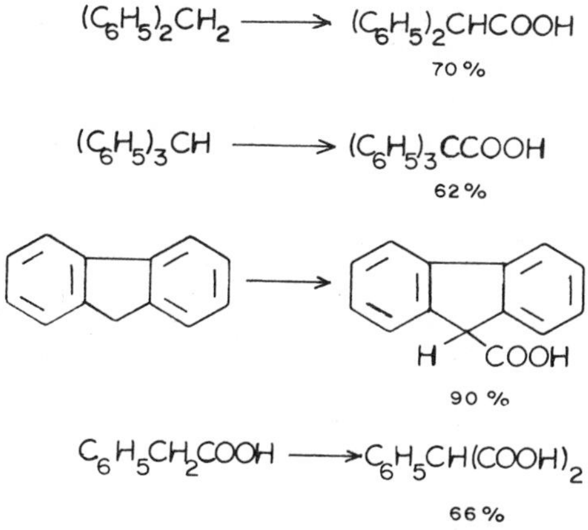

$$(C_6H_5)_2CH_2 \ \longrightarrow \ (C_6H_5)_2CHCOOH$$
$$70\%$$

$$(C_6H_5)_3CH \ \longrightarrow \ (C_6H_5)_3CCOOH$$
$$62\%$$

$$90\%$$

$$C_6H_5CH_2COOH \ \longrightarrow \ C_6H_5CH(COOH)_2$$
$$66\%$$

    The carbonation of lithium acetylides has been accomplished
by Corey and Fuchs [13]. The transformation is the last step of

a sequence starting with the aldehyde as shown:

$$ArCHO \xrightarrow[\text{Zn-CH}_2\text{Cl}_2]{(C_6H_5)_3P,CBr_4} ArCH=CBr_2 \xrightarrow[\text{THF}]{BuLi} ArC\equiv CLi \xrightarrow{CO_2} ArC\equiv CCOOH$$
$$82-90\%$$

The method may be regarded as an analog of the Wittig olefin synthesis since both methods originate with aldehydes.

The carbonation of nitrobenzenes and derivatives has been accomplished via lithiation with phenyllithium to give, in the case of nitrobenzene, o-nitrobenzoic acid [13a]. Yields varied from 87-97%.

Carbonation may be accomplished with stilbene to form meso-2, 3-diphenylsuccinic acid by electrolysis in DMF while passing CO_2 through the mixture [13b].

a.   Preparation of 2-Keto-3-methylpentanoic Acid [10]

To a stirred solution of 0.027 mol of 1,1,3,3-tetramethyl-butyl isonitrile in 27 ml of ether at 0° under N_2 was added rapidly 0.027 mol of sec-butyllithium in hexane. The mixture, after 10 min, was added to an ether slurry of dry ice. After evaporating the solvent, the carbonated imine was refluxed in an oxalic acid solution (7.5 g of acid in 60 ml of water) for 15 min. Extraction with methylene chloride and evaporation of the solvent yielded 2.8 g (80%) of the acid.

b.   Preparation of benzilic acid [12]

2,6-Diphenylpyridine, 2.3 g, 0.46 g of Na, and 40 ml of THF were agitated for 2 hr. Benzophenone, 0.91 g, in 10 ml of THF was then added rapidly. After agitating for 1 hr, passing in CO_2, and hydrolyzing with water, 0.7 g (60%) of the acid was recovered.

c.   Preparation of 2-Butynoic Acid (Tetrolic Acid) [14]

$$CH_3C\equiv CH \xrightarrow{NaNH_2} CH_3C\equiv CNa \xrightarrow{CO_2} CH_3C\equiv CCO_2Na \xrightarrow{H^{\oplus}} CH_3C\equiv CCOOH$$

Yield was 50-59% from propyne and sodamide in liquid ammonia. The ammonia was removed by evaporation, the residue was dissolved in dry THF and ether, and CO_2 was passed through the solution.

2.   Carboxyalkylation from Acids and Salts (Including the Modified Ivanov Reagent) (1, 241, 778)

The Ivanov conception of converting carboxylic acids into
dianions has been improved by treating the sodium salt with a
lithium amide [15] (Ex. a):

$$Me_2CHCO_2Na \quad + \quad LiN(CHMe_2)_2 \longrightarrow Me_2\overset{\ominus}{C}\overset{\ominus}{C}O_2\overset{\oplus}{Li}\overset{\oplus}{Na} \xrightarrow[2)H_3O^{\oplus}]{1)RBr} Me_2\overset{\overset{\displaystyle R}{|}}{C}CO_2H$$
$$70-76\%$$

The mixed salt gives cleaner products largely because it is more
soluble in the medium [4]. However, the disodium salt has been
employed in the preparation of 2,3-diphenylpropionic acid (80-
84%) by benzylation [16].

Pfeffer and Silbert [17] have also found that the addition
of HMPA mitigates against the problem of the insolubility of the
Ivanov salt. Yields (87-93%) of the 2-butylalkanoic acids were
much superior to those obtained when THF-hexane was used (Ex. b),
but the introduction of a second butyl group (to prepare a
tertiary acid) was no more successful in the one solvent than in
the other. However, see Ex. a for the preparation of a tertiary
acid.

a. Preparation of 2,2-Dimethyl-4-phenylbutyric Acid [15]

Diisopropylamine, 7.75 g, 3.68 g of 54% NaH in mineral oil,
and 75 ml of THF over N_2 were stirred and to the mixture, 6.6 g
of isobutyric acid was added in 5 min. Hydrogen evolution was
completed by heating to reflux for 15 min. On cooling to 0°, 52
ml of a standard solution of n-butyllithium in heptane (1.45
mmol/ml; 0.075 mol) was added at a temperature below 10°. After
retaining ice-bath cooling for 15 min, the mixture was heated to
30-35° for 30 min. (2-Bromoethyl) benzene, 13.9 g, was added over
20 min at 0° and after an additional 30 min at this temperature,
the mixture was heated to 30-35° for 1 hr. Water, 100 ml was now
added at a temperature below 15° and from the acidified, aqueous
layer, 10-11 g (70-76%) of the acid was recovered.

b. Preparation of 2-Butylheptanoic Acid [17]

To 4.9 g of diisopropylamine in 35 ml of THF, 1.6 M n-
butyllithium was added to the stirred solution under N_2 at a rate
so as to keep the temperature below 0°. n-Heptanoic acid, 2.95
g, was then added at the same temperature and after 15 min, 9 ml
of HMPA was added as well. The solution was stirred for 20 min at
15° and then 3.3 g of butyl bromide was added at once at 0°.
After 2 hr of additional stirring, followed by acidification and
extraction with petroleum ether, there was recovered 4.06 g (96%)
of the impure acid containing 96.5% of 2-butylheptanoic acid.

3.   Carbonation of Phenol Salts (Kolbe) (1, 778)

The Kolbe reaction produces o- or p-hydroxybenzoic acids from sodium or potassium phenoxide and carbon dioxide. In a modification of this procedure, Yasuhara and Nogi [18] have studied the reaction of the potassium salt with K_2CO_3 under fairly high carbon monoxide pressure and temperature. The product obtained was the dipotassium salt of p-hydroxybenzoic acid as indicated:

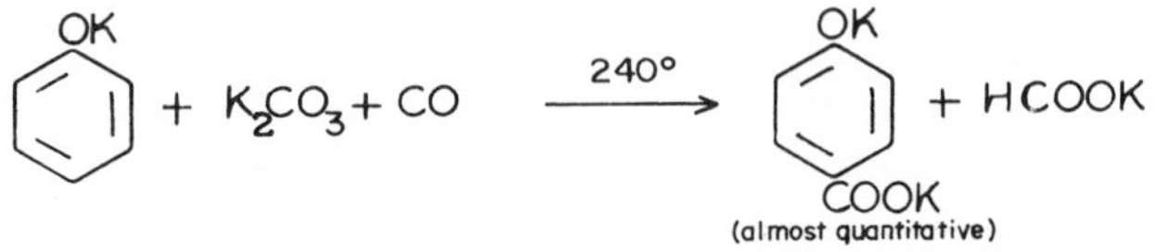

In treating the sodium phenoxide with Na_2CO_3 and sodium formate in a carbon monoxide atmosphere, Yasuhara and Nogi [19] obtained the disodium salt of p-hydroxybenzoic acid (Ex. a):

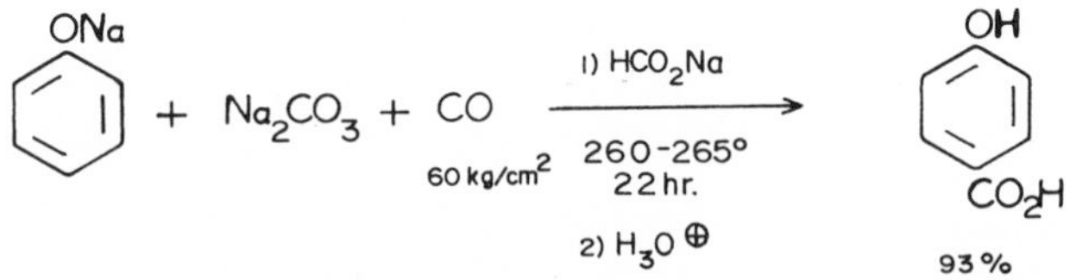

The function of the sodium formate is not known, although it is essential in the process. The reaction is of interest since it permits the synthesis of p-hydroxybenzoic acid from the less costly sodium salts although considerable time is required in the reaction.

Some success has been achieved in the carboxylation of resorcinol by methylmagnesium carbonate, presumably because of the acidic nature of the aromatic hydrogens as shown [20]:

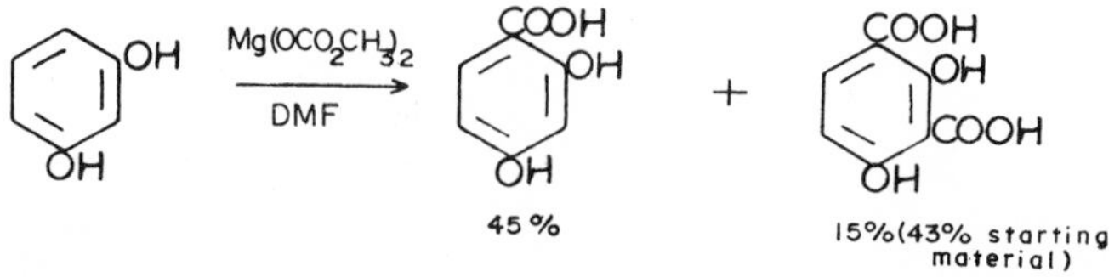

The phenol was heated at 120° for 3 hr in a 2M solution of the carbonate in DMF and the product was recovered from the acidic solution. The method is unsatisfactory when applied to phenol, m- and p-cresol, and hydroquinone, but it has been useful in the carboxylation of cannabidiol.

In a further modification phenols, mostly substituted, were carbonated at atmospheric pressure in the presence of a strong base and an amide solvent [21]. As a rule the structure of the

phenols investigated was such that either only an <u>ortho</u> or <u>para</u>
position was available for the carbonation attack. Examples of
the procedures used follow:

1.   Sodium and ethanol or methanol as the base, DMA as
solvent:

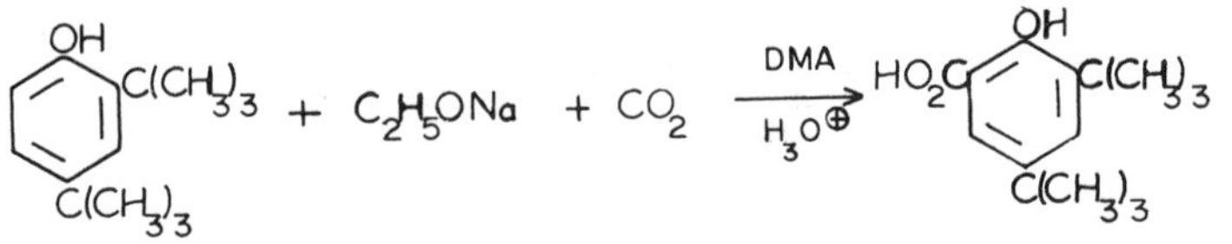

3,5-Di-*t*-butylsalicylic acid
(86% crude)

2.   Sodium methoxide as the base, DMA as solvent:

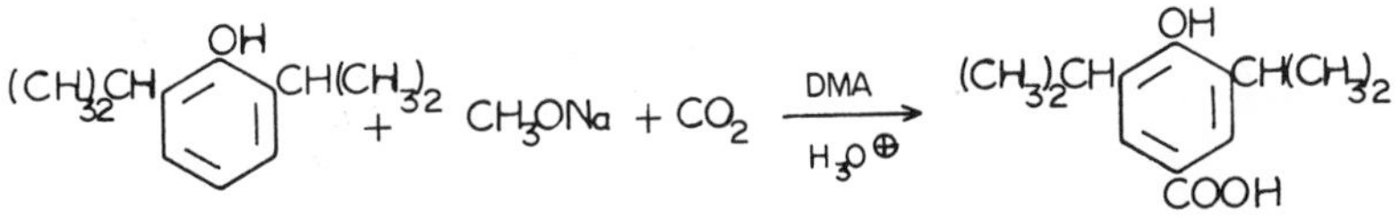

3,5-Diisopropyl-4-hydroxybenzoic acid
(72% crude)

3.   Potassium hydroxide as the base (with subsequent removal
of water), DMF as solvent:

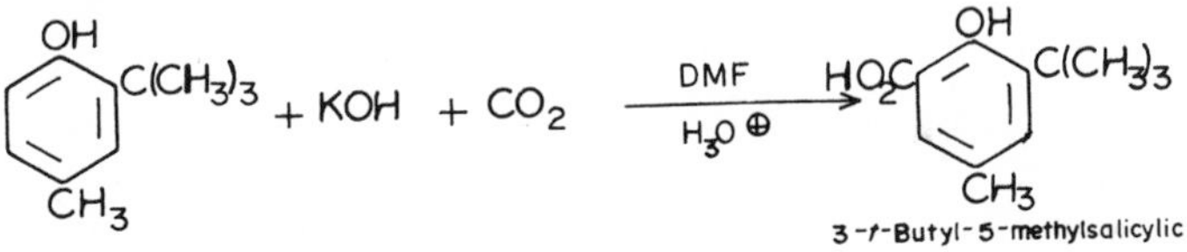

3-*t*-Butyl-5-methylsalicylic acid, 80% (crude)

4.   Potassium hydroxide as the base (with subsequent removal
of water), DMA as solvent:

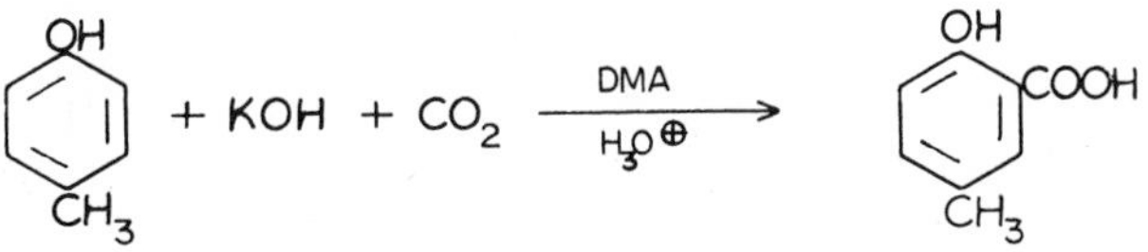

5-Methylsalicylic acid, 82% (crude)

The steric factor appears to be of importance since carboxylation did not occur at sites adjacent to a tertiary butyl or
a dialkyamino group. For example, 3-t-butylphenol gave a 17%
yield of 4-t-butylsalicylic acid and 3,5-di-t-butylphenol gave no
product at all. On the other hand, 4,6-di-t-butylresorcinol gave
a superior yield of 3,5-di-t-butyl-γ-resorcylic acid:

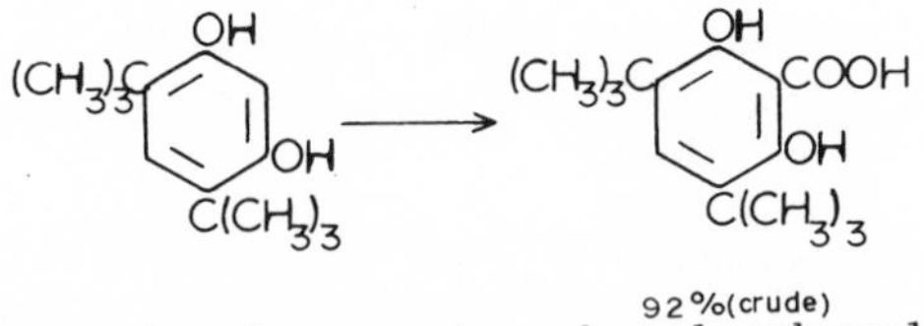

92%(crude)

The older methods of preparing phenolcarboxylic acids have required high temperatures and pressures. Thus 4-hydroxyiso-phthalic acid may be obtained by heating dipotassium salicylate or dipotassium p-hydroxybenzoate under high CO_2 pressure at around 350° [22]. Hydroxytrimesic acid may be obtained when phenol and CO_2 in the presence of K_2CO_3 are heated at about 300° and 1470 atm pressure [23]. Recently, Kito and Hirao [24] have succeeded in preparing hydroxytrimesic acid in high yield from potassium phenolate, potassium n-amylcarbonate, and CO_2 under less severe conditions as indicated:

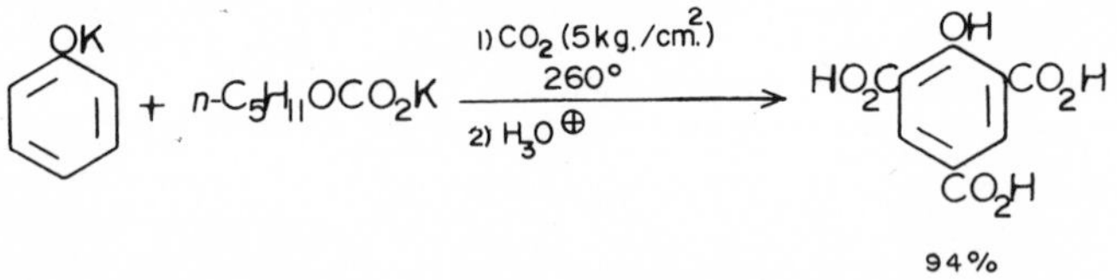

94%

a. Preparation of p-Hydroxybenzoic Acid [19]

Sodium phenoxide, 5.8 g was heated in an autoclave with 5.8 g of Na_2CO_3 and 3.5 g of HCOONa in a CO atmosphere under a pressure of 60 kg/cm² at 260-265° for 22 hr. The crude product in water was neutralized to a pH of 9 and the regenerated phenol (5%) was extracted with ether. Further acidification and extraction with ether gave 6.4 g (93%) of p-hydroxybenzoic acid.

4. Rearrangement of Dicarboxylic Acid Salts (Henkel) (1 779)

Somewhat related to the rearrangement of dicarboxylic acid salts is the formation of potassium terephthalate by the trans-carboxylation of potassium benzoate by means of a potassium naphthoate [25].

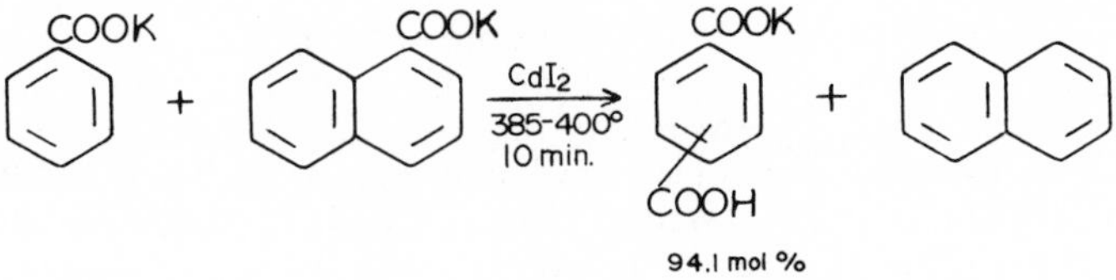

94.1 mol %

Since the crystallizing ability of potassium terephthalate
exceeds that of the other isomers in the reaction mixture, it
separates first, 81 mol % in the above reaction. For best results
the number of COOK groups should be twice the number of benzene
rings. The method is applicable to other carboxylic acid salts;
that is, potassium α-naphthoate carboxylates the salts of pyridine monocarboxylic acids, and the salt of furan-2-carboxylic
acid carboxylates α-naphthalenecarboxylate.

5. Carbonation of Unsaturated Esters

Bottaccio and Chiusoli [26] carbonated α,β- and β,γ-
unsaturated esters as shown:

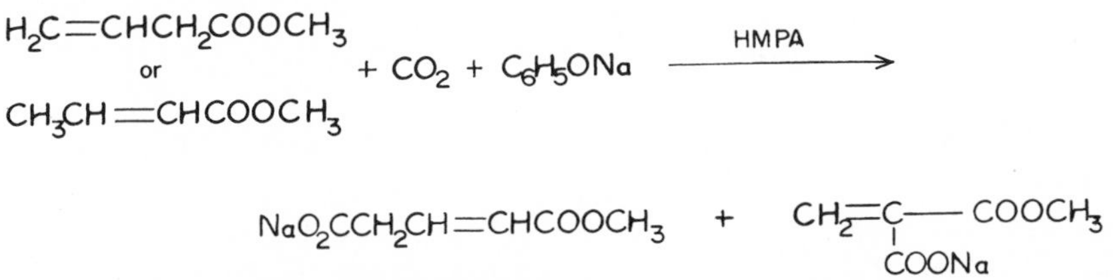

$$H_2C=CHCH_2COOCH_3$$
or
$$CH_3CH=CHCOOCH_3$$ $$+ CO_2 + C_6H_5ONa \xrightarrow{\text{HMPA}}$$

$$NaO_2CCH_2CH=CHCOOCH_3 \quad + \quad CH_2=C-COOCH_3$$
$$\overset{|}{COONa}$$

Sodium methyl glutaconate

Sodium methyl methylenemalonate

From 0.036 mol of the vinyl ester, 0.010 mol of the two products
were obtained, the ratio between the straight and branched chain
products being 11.5. The process might be improved by the
selection of a base such as the salt of a hindered phenol.

6. Decarboxylation

$$R_2C(COOH)_2 \longrightarrow R_2CHCOOH$$

The alkylation of malonic esters was discussed in Chapter
14 (1, 845). Hydrolysis of the β-cyanoesters is more difficult
than that of the β-diesters, as is shown by the need for refluxing the following substrate in concentrated HCl and acetic acid
for 8 hr [27]:

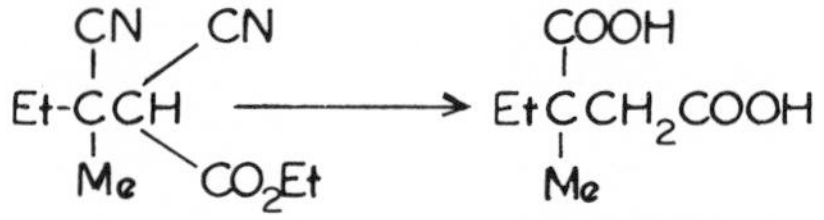

$$Et-\overset{\overset{\displaystyle CN}{|}}{\underset{\underset{\displaystyle Me}{|}}{C}}\overset{\overset{\displaystyle CN}{\diagup}}{\underset{\diagdown CO_2Et}{CH}} \longrightarrow Et\overset{\overset{\displaystyle COOH}{|}}{\underset{\underset{\displaystyle Me}{|}}{C}}CH_2COOH$$

α-Ethyl-α-methylsuccinic
acid, 41-47 %

Another decarboxylation of a diacid is that utilized in
forming an indole-3-alkanoic acid [28].

76-93%
(4 examples)

In this manner the decarboxylation is superior to that resulting from copper powder in boiling quinoline.

Addendum

C.1. Carbonation of active methylene compounds occurs if CO_2 is passed through them mixed with an equivalent amount of DBU and DMSO [E. Haruki et al., Chem. Lett., 427 (1974)].

References

1.   H. Wynberg et al., Tetrahedron Letters, 3001 (1971).
2.   H. Wynberg et al., Tetrahedron Letters, 2339 (1971).
3.   S. Yamada et al., Chem. Commun., 623 (1972).
4.   P. E. Pfeffer et al., J. Org. Chem., 37, 1256 (1972).
5.   A. Maercker, J. Organomet. Chem., 18, 249 (1968).
6.   T. S. Cantrell, Tetrahedron Letters, 5635 (1968).
7.   E. L. Patmore et al., U.S. Patent 3,692,826, September 19, 1972; C. A., 78, 4011 (1973).
8.   H. Mori et al., Chem. Pharm. Bull., 20, 2440 (1972); C. A., 78, 42443 (1973).
9.   E. J. Corey, R. H. K. Chen, J. Org. Chem., 38, 4086 (1973).
10. H. M. Walborsky et al., J. Am. Chem. Soc., 91, 7778 (1969); ibid., 92, 6675 (1970).
11. W. Vaalburg et al., Syn. Comm., 2, 423 (1972).
12. B. Angelo, Bull. Soc. Chim. France, 1710 (1969).
13. E. J. Corey, P. L. Fuchs, Tetrahedron Letters 3769 (1972).
13a. G. Köbrich, P. Buck, Chem. Ber., 103, 1412 (1970).
13b. S. Wawzonek, Synthesis, 294 (1971).
14. J. C. Kauer, M. Brown, Org. Syn., Coll. Vol. 5, 1043 (1973).
15. P. L. Creger, Org. Syn., 50, 58 (1970).
16. C. R. Hauser, W. R. Dunnavant, Org. Syn., Coll. Vol. 5, 526 (1973).
17. P. E. Pfeffer, L. S. Silbert, J. Org. Chem., 35, 262 (1970).
18. Y. Yasuhara, T. Nogi, J. Org. Chem., 33, 4512 (1968).
19. Y. Yasuhara, T. Nogi, Chem. Ind. (London), 77 (1969).
20. R. Mechoulam, Z. Ben-Zvi, Chem. Commun., 343 (1969).

21. W. H. Meek, C. H. Fuchsman, J. Chem. Eng. Data, 14, 388
    (1969); I. Hirao et al., Yuki Gosei Kagaku Kyokaishi, 24,
    1047, 1051 (1966); 25, 66, 412, 417, 577 (1967).
22. J. C. Wygant, U. S. Patent, 3,089,905, May 14, 1963; C. A.,
    59, 11352 (1963).
23. Henkel et Cie, British Patent, 968,829, September 2, 1964;
    C. A., 62, 6435 (1965).
24. T. Kito, I. Hirao, Bull. Chem. Soc. Jap., 44, 3123 (1971).
25. J. Ratusky, Chem. Ind. (London), 1347 (1970).
26. G. Bottaccio, G. P. Chiusoli, Z. Naturforsch., 23B, 1016
    (1968).
27. F. S. Prout et al., Org. Syn., Coll., 5, 572 (1973).
28. R. E. Bowman, P. J. Islip, Chem. Ind. (London), 154 (1971).

D. Condensation

    1. From Aromatic Aldehydes and Anhydrides (Perkin) (1, 780)

    In addition to triethylamine, boric esters have been shown
to be catalysts in the Perkin reaction [1]. Thus cinnamic acid
was prepared in 70% yield in the reaction of benzaldehyde and
acetic anhydride in the presence of 1 ml of $(MeO)_3B$ per 0.1 mol
of reactant. None of the customary sodium or potassium acetate
was present in the reaction mixture.
    In the Perkin reaction the β-hydroxyacid rather than the
unsaturated acid has now been isolated [1a]:

$$C_6H_5CH_2CO_2CH_3 + C_6H_5CHO \xrightarrow[Et_2O,25°]{NaNH_2} C_6H_5CHOHCHCOOH$$
$$\underset{C_6H_5}{|}$$

β-Hydroxy-α,β-diphenylpro-
pionic acid, 10-32 %

The threo form of the product was usually the predominant one.

    3. From Ylids (Wittig) (1, 782)

    1,3-Dithiacyclohexylidine trimethoxyphosphorane and alde-
hydes have been utilized in the synthesis of carboxylic acids
containing an additional carbon atom as shown [2]:

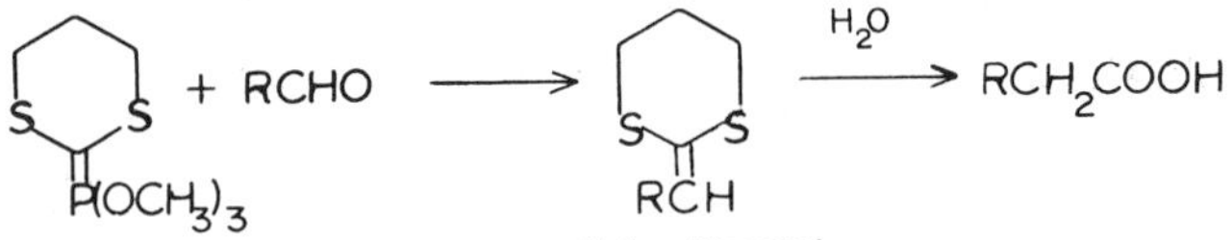

Ketene thioacetal
90%, R=C_6H_5

Since the reaction does not occur with ketones, it may be employed in the selective transformation of aldehydes, RCHO, to acids, RCH$_2$COOH.

Aldehydes have also been converted into carboxylic acids containing one more carbon atom by the use of tetraethyl dimethylaminomethylene diphosphonate, which may be prepared from dimethylformamide acetal and diethyl phosphite [3]. The phosphonate reacts first with sodium hydride and then with an aldehyde to produce the aminoalkenyl phosphonate, which in acid hydrolysis gives the carboxylic acid as indicated:

$$\left[(C_2H_5O)_2 \underset{O}{\overset{}{P}} \right]_2 CHN(CH_3)_2 \xrightarrow[\text{2)RCHO}]{\text{1)NaH}} RCH = \underset{N(CH_3)_2}{\overset{O}{\underset{}{C}}} P(OC_2H_5)_2 \xrightarrow{\overset{\oplus}{H_3O}} RCH_2COOH$$

$$\text{60-75\%} \qquad\qquad \text{75-90\%}$$

Again, the reaction does not occur with ketones.

The various elegant methods of synthesizing unsaturated acids via ylids have been reviewed [4]. One trend noted is the utilization of the phosphonate rather than the phosphorane. In this way the process is sometimes applicable to ketones as well as aldehydes:

$$(EtO)_2 \overset{O}{\underset{}{P}} \overset{\ominus}{CH} CO_2Et + R_2C = O \longrightarrow R_2C = CHCO_2Et + (EtO)_2 \overset{O}{\underset{}{P}} O^{\ominus}$$

4. From Ketones (via the Enamine and Morpholide) (Carboxyethylation) (1, 783)

Hunig and co-workers [5] have utilized the enamine for the synthesis of long-chain carboxylic acids. The simplest case is shown in the steps:

$$\text{75-85\%} \qquad \text{53-57\%} \qquad \text{80-90\%}$$

In this manner propionic acid may be converted into nonanoic acid with an overall yield of 65%. Thus the method succeeds in lengthening the chain of the carboxylic acid by six carbon atoms. Similar β-ketoester cleavages are described in E.1 (1, 784).

Like reactions were conducted with dicarboxylic ester chlorides and with diacyldichlorides (Ex. a) to obtain dicarboxylic acids.

a. Preparation of Docosanedioic Acid [6]

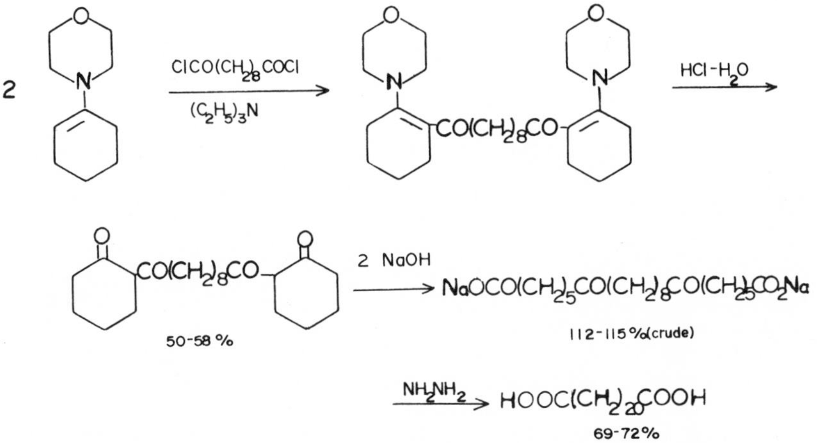

$$2 \quad \text{[morpholino cyclohexene]} \xrightarrow[\text{(C}_2\text{H}_5)_3\text{N}]{\text{ClCO(CH}_2)_{18}\text{COCl}} \text{[bis-morpholino cyclohexenyl } CO(CH_2)_{18}CO] \xrightarrow{\text{HCl-H}_2\text{O}}$$

$$\text{[bis-cyclohexanone } CO(CH_2)_{18}CO]} \xrightarrow{\text{2 NaOH}} \text{NaOCO(CH}_2)_5\text{CO(CH}_2)_{18}\text{CO(CH}_2)_5\text{CO}_2\text{Na}$$

50-58 %   112-115%(crude)

$$\xrightarrow{\text{NH}_2\text{NH}_2} \text{HOOC(CH}_2)_{20}\text{COOH}$$

69-72%

5. From Ketones or Esters and Carboxylic Salt Anions

A direct synthesis of β-hydroxyacids has been accomplished
from ketones and carboxylic acids [7], as shown for cyclohexanone
and acetic acid:

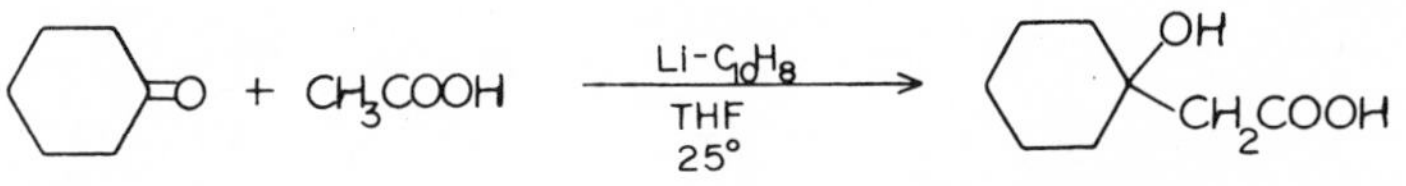

$$\text{[cyclohexanone]} + CH_3COOH \xrightarrow[\substack{\text{THF} \\ 25°}]{\text{Li-C}_{10}\text{H}_8} \text{[1-hydroxycyclohexyl } CH_2COOH]}$$

(1'-Hydroxycyclohexan-1'-yl)acetic acid, 65%

Yields with four other ketones, including acetone and
acetophenone, varied within 30-65%. The Ivanov reagent (1, 241)
is an intermediate. Similar condensations of the salt anion with
ketones have been carried out to give β-hydroxyacids in 0-98%
yields [7a].

A similar reaction occurs with cyclohexanone and crotonic
acid in the presence of lithium diisopropylamide [8]. It is proposed that the intermediate in this case is the dianion as shown:

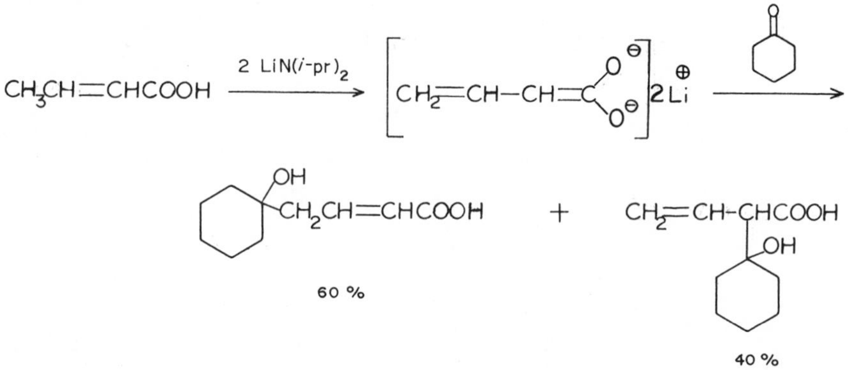

$$CH_3CH=CHCOOH \xrightarrow{2\ LiN(i\text{-}pr)_2} \left[CH_2=CH-CH=C\begin{smallmatrix}O^{\ominus}\\ \\O^{\ominus}\end{smallmatrix} \right] 2\overset{\oplus}{Li} \longrightarrow$$

The percentages of the α- and γ-substituted unsaturated acids vary with the α,β-unsaturated acid. Hexenoic acid gives essentially the corresponding α-substituted acid while methyl 4-bromocrotonate gives exclusively the corresponding γ-product.

Esters have also been condensed with salt anions [8a] to form β-keto acids in good yield:

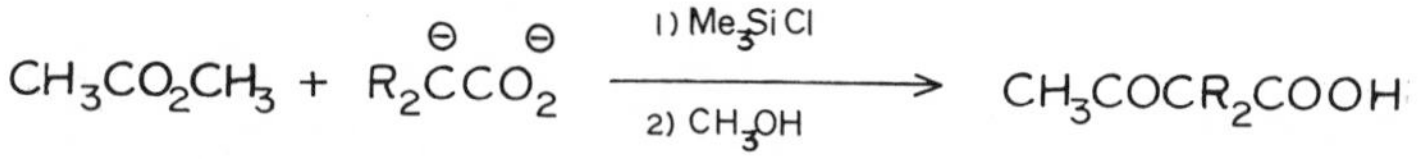

$$CH_3CO_2CH_3 + R_2\overset{\ominus}{C}\overset{\ominus}{CO_2} \xrightarrow[\text{2) CH}_3\text{OH}]{\text{1) Me}_3\text{SiCl}} CH_3COCR_2COOH$$

6. From Unsaturated Esters and Grignard Reagents (Michael)

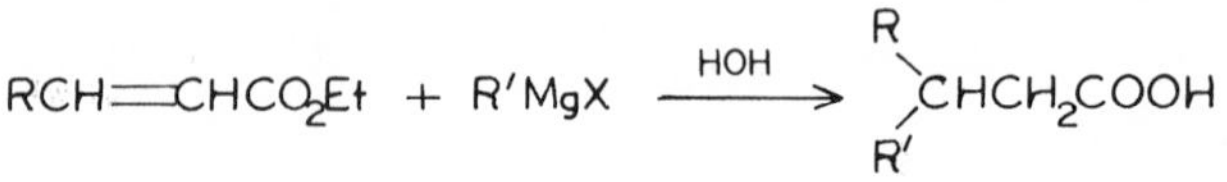

$$RCH=CHCO_2Et + R'MgX \xrightarrow{HOH} \overset{R}{\underset{R'}{\diagdown}}CHCH_2COOH$$

A variety of branched-chain acids are available through this simple Michael reaction. To insure complete 1,4 addition the unsaturated sec-butyl ester is employed.

a. Preparation of 3-Methylheptanoic Acid [9]

Yield was 61-73% overall from sec-butyl crotonate added dropwise to butylmagnesium bromide followed by hydrolysis.

7. From Aldehydes, Glyoxal Bisulfite, and a Cyanide

$$\text{ArCHO} + \text{NaO}_3\overset{\overset{\displaystyle OH}{|}}{S}\text{CH}\text{—}\overset{\overset{\displaystyle OH}{|}}{C}\text{HSO}_3\text{Na} \xrightarrow{\overset{\ominus}{CN}} \text{Ar}\overset{|}{C}\text{HCHOH}\overset{\overset{\displaystyle O}{||}}{C}\text{CN}$$

$$\text{ArCHOHCOCHO} \rightleftarrows \text{ArCHOH}\overset{\overset{\displaystyle}{||}}{C}\text{CHOHCN}$$

ArCHOHCOCHO → (with $B^\ominus$) → ArCH=C=O → (with $OH^\ominus$) → Ar CH₂COOH

"A"

Interest in this relatively obscure reaction seems to be revived, perhaps largely for the acquisition of the dihydroxy-tetronimides (A) which may be isolated. The yields of arylacetic acids are low (about 40% for furfurylacetic and p-hydroxyphenyl-acetic) and lower still (11%) for 4-hydroxy-3-methoxyphenylacetic acid [10].

8.   From Carbonyl Compounds and α-Metalated Isocyanomethyl
      Aryl Sulfones

$$\overset{R_1}{\underset{R_2}{\diagup}}\text{CO} \xrightarrow[\substack{t\text{-BuOK}^- \\ THF}]{C\equiv NCH_2SO_2Ar} \overset{R_1}{\underset{R_2}{\diagup}}C=C\overset{NHCHO}{\underset{SO_2Ar}{\diagdown}} \xrightarrow{\overset{\oplus}{H_3O}} \overset{R_1}{\underset{R_2}{\diagup}}\text{CHCOOH}$$

55-67 %

This conversion of carbonyl compounds into carboxylic acids has been described by Schöllkopf and co-workers [11]. Thus the method represents a procedure for going from a ketone to the carboxylic acid containing one more carbon atom.

9.   From Carbonyl Compounds and Metalated Silyl Thioacetals

$$\overset{R^1}{\underset{R^2}{\diagup}}C=O + \overset{RS}{\underset{RS}{\diagup}}\overset{Li}{\underset{SiMe_3}{\diagup}}C \xrightarrow[-80°]{THF} \overset{RS}{\underset{RS}{\diagup}}C=C\overset{R^1}{\underset{R^2}{\diagdown}} \xrightarrow{H_2O} \overset{R^1}{\underset{R^2}{\diagup}}\text{CHCOOH}$$

Carbonyl compounds react with metalated silyl thioacetals (Peterson olefination) as shown to give ketene thioacetals (30-90%) [12]. The latter on hydrolysis yield carboxylic acids. The

2-lithio-2-trimethyl silyl-1,3-dithiane may be prepared from dithiane, butyllithium, and chlorotrimethylsilane.

10.   From the Cleavage of Lactone-like Compounds

A reductive ring opening has been realized to give the best synthesis of a cis-pentadienoic acid [13]:

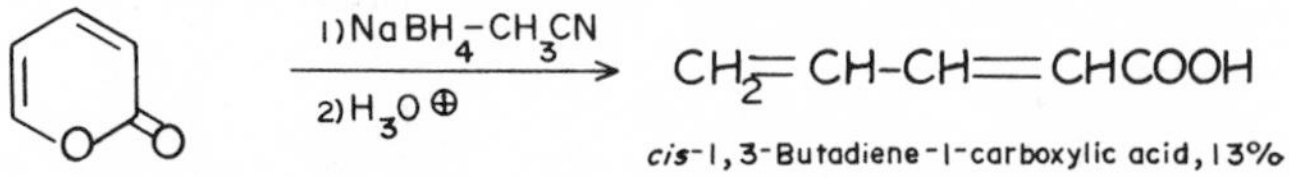

$$CH_2 = CH-CH = CHCOOH$$

cis-1,3-Butadiene-1-carboxylic acid, 13%

A novel ring opening followed by ring closure using an enamine is shown [14].

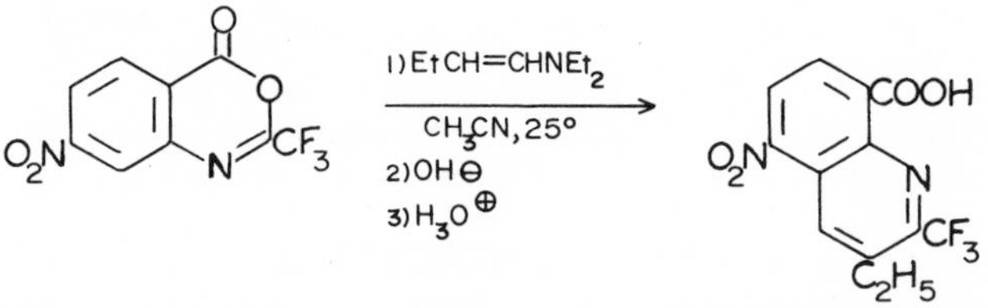

5-Nitro-2-trifluoromethyl-
3-ethyl-8-quinolinecarboxylic acid
67 %

References

1.   V. V. Gertsev, Ya. Ya. Makarov-Zemlyanskii, Zh. Vses. Khim. Obshchest, 17, 598 (1972); C. A., 78, 29658 (1973).
1a. C. G. Kratchanov et al., Synthesis, 317 (1971).
2.   E. J. Corey, G. Markl, Tetrahedron Letters, 3201 (1967).
3.   H. Gross, B. Costisella, Angew. Chem. Intern. Ed. Engl., 7, 391 (1968).
4.   L. D. Bergelson, M. M. Shemyakin, in S. Patai, The Chemistry of Carboxylic Acids and Esters, Wiley-Interscience, New York, 1969, p. 295.
5.   S. Hünig et al., Chem. Ber., 100, 4010 (1967) and preceding papers.
6.   S. Hünig et al., Org. Syn., 43, 34 (1963).
7.   S. Watanabe et al., Chem. Ind. (London), 1811 (1969).
7a. G. W. Moersch, A. R. Burkett, J. Org. Chem., 36, 1149 (1971).
8.   P. E. Pfeffer et al., Tetrahedron Letters, 1163 (1973); S. Watanabe et al., Chem. Ind. (London), 80 (1972).
8a. C. Ainsworth et al., J. Am. Chem. Soc., 93, 6321 (1971).
9.   J. Munch-Petersen, Org. Syn., Coll. Vol. 5, 762 (1973).
10. L. Breen et al., Australian J. Chem., 26, 2221 (1973).
11. U. Schöllkopf et al., Angew. Chem. Int. Ed. Engl., 11, 311 (1972); Ann. Chem., 766, 130 (1972).

12. D. Seebach et al., _Angew. Chem. Int. Ed. Engl._, <u>11</u>, 443 (1972).
13. W. Kirmse, H. Lechte, _Ann. Chem._, <u>739</u>, 235 (1970).
14. W. Steglich, O. Hollitzer, _Angew. Chem._, <u>85</u>, 505 (1973).

E. Alkali Cleavage

3. From Ketones (<u>1</u>, 786)

$$\text{Ar}\overset{\overset{\displaystyle O}{\|}}{C}\text{Ar} + H_2O \xrightarrow[\substack{H_2O \\ Glyme}]{t\text{-BuOK}} \text{Ar}\overset{\overset{\displaystyle O}{\|}}{C}\text{OH} + \text{ArH}$$

Nonenolizable ketones have been hydrolyzed with t-BuOK-H$_2$O, 10:3 equiv, in glyme at room temperature [1]. The method, as indicated, may be utilized in introducing a carboxyl group into an aromatic ring:

$$C_6H_5H \xrightarrow[AlCl_3]{2\text{-ClC}_6H_4COCl} C_6H_5CO \underset{\underline{\quad}}{\langle \overset{Cl}{\bigcirc} \rangle} \xrightarrow[Glyme]{t\text{-BuOK-H}_2O} C_6H_5COOH + \underset{\underline{\quad}}{\langle \overset{Cl}{\bigcirc} \rangle}$$

Of the two possible cleavages of the benzophenone, the preferable one is that between the carbonyl group and the benzene ring containing the 2-chloro substituent. Substituted benzenes give benzophenones (70-87%), which in turn give substituted benzoic acids (80-99%). For example, mesitylene leads to mesitoic acid:

2-Chloro-2',4',6'-tri-
methylbenzophenone, 77%   Mesitoic acid, 96%

Ferrocenecarboxylic acid (88% overall) has been prepared similarly by the o-chlorobenzoylation of ferrocene followed by t-BuOK cleavage [2].

4. From N-(Aroylmethyl)pyridine Salts

Royer and co-workers [3] converted phenacyl chlorides into the N-(aroylmethyl) pyridine salts that were hydrolyzed with aqueous NaOH to give the corresponding aromatic carboxylic acid as shown:

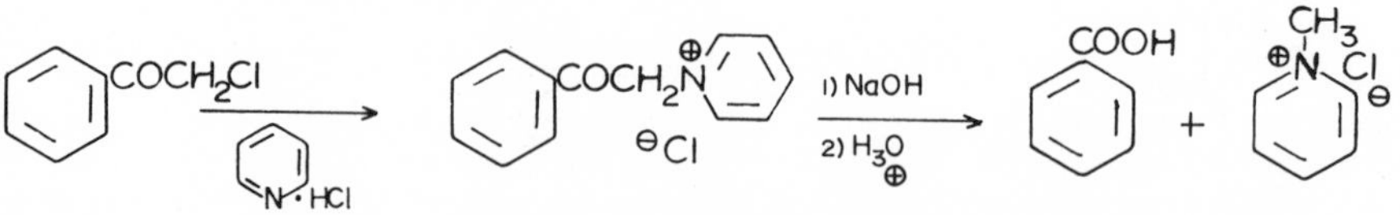

The reaction was successful with substituents such as alkyl, methoxy, and hydroxy in the benzene ring of the original phenacyl chloride. Yields of carboxylic acids based on the phenacyl chloride utilized varied within 38-73%. Similar experiments were successful with substituted 7-chloroacetylbenzofurans.

Although the authors do not discuss the mechanism, an interesting speculation may be offered. If the cleavage is assumed to occur between the aliphatic carbon atoms, two possibilities exist, of which the first appears to be the more reasonable:

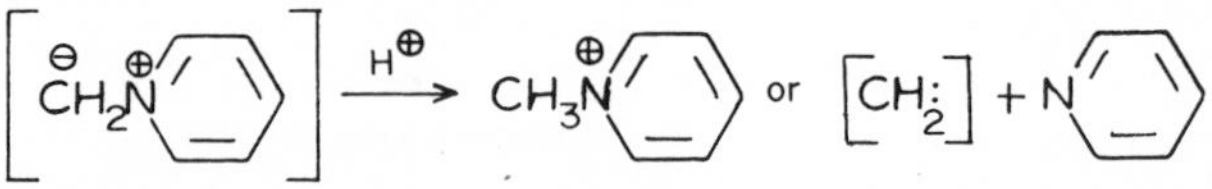

Nonenolizable ketones are best split in a solvent containing 10 equiv t-BuOK to 3 equiv water in an aprotic solvent such as DMSO or ether [4].

References

1. P. Hodge et al., J. Chem. Soc., C455 (1971); Tetrahedron Letters, 3825 (1971).
2. E. R. Biehl, P. C. Reeves, Synthesis, 360 (1973).
3. R. Royer et al., Bull. Soc. Chim. France, 878 (1969).
4. P. G. Gassman et al., J. Am. Chem. Soc., 89, 946 (1967).

F.   Substitution and Addition

1.   From Aromatic Compounds by Acylation or Carboxylation (Friedel-Crafts)

$$C_6H_6 \xrightarrow[AlCl_3]{COCl_2} C_6H_5COCl \xrightarrow{H_2O} C_6H_5COOH$$

$$C_6H_6 \xrightarrow[\substack{AlCl_3 \\ H_3O^{\oplus}}]{ClCOCOCl} C_6H_5COOH$$

See E.3 for the synthesis of nonenolizable ketones via the Friedel-Crafts reaction.   The ketones are then split by alkali to acids.

Gross and co-workers [1], by the use of the dichloromethylene ether of catechol, prepared a series of benzoates which were hydrolyzed to benzoic acids as indicated:

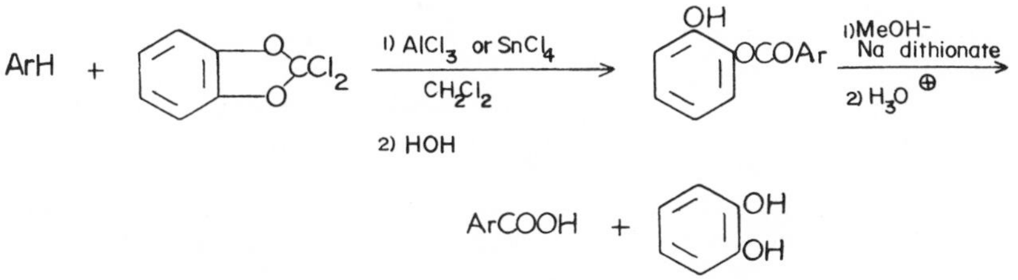

$$ArH \quad + \quad \text{(structure)} \quad \xrightarrow[\text{2) HOH}]{\substack{\text{1) AlCl}_3 \text{ or SnCl}_4 \\ \text{CH}_2\text{Cl}_2}} \quad \text{(structure) OCOAr} \quad \xrightarrow[\text{2) H}_3\text{O}^{\oplus}]{\substack{\text{1)MeOH-} \\ \text{Na dithionate}}}$$

$$ArCOOH \quad + \quad \text{(structure) OH, OH}$$

Various substituents such as alkyl, hydroxy, and methoxy groups may be present in the aromatic hydrocarbon and several aromatic polycyclic and heterocyclic compounds were arylated as well.   The yield of the intermediate esters varied from 51% to quantitative, while the final acid yields, based on the ester, varied from 70% to quantitative.

Hopff and Osman [1a] report a lower yield (24%) in preparing γ-benzoylbutryic acid from benzene and glutaric anhydride than that (80-85%) reported in the Organic Synthesis preparation [1b].

a.   Preparation of Mesitoic Acid [2]

A suspension of 146 g of AlCl$_3$ in 700 ml of CS$_2$ was cooled to 10-15° and 139 g of oxalyl chloride was added with stirring. Mesitylene, 120 g in 200 ml of CS$_2$ was added with stirring at the same temperature.   The mixture was then refluxed for 1 hr and poured cautiously onto a mixture of 2 kg of crushed ice and 300 ml of 12 N HCl.   From the CCl$_4$ extract there was recovered 106-124 g (65-76%) of crude acid.

2.   From Aromatic Compounds by Alkylation (Friedel-Crafts) (1, 791)

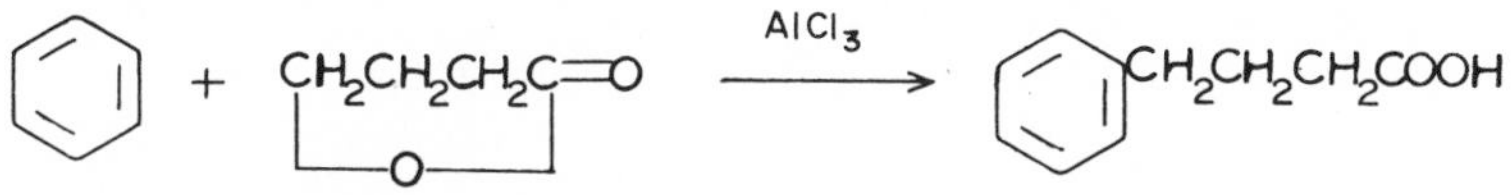

$$\text{(benzene)} \quad + \quad \text{CH}_2\text{CH}_2\text{CH}_2\text{C}=\text{O (glutaric anhydride)} \quad \xrightarrow{\text{AlCl}_3} \quad \text{(benzene)CH}_2\text{CH}_2\text{CH}_2\text{COOH}$$

Three methods have been described for the carboxymethylation of aromatic compounds [3]:

1.   Thermal carboxymethylation with chloroacetic acid, bromoacetic acid, or chloroacetyl polyglycolic acid.   In favorable cases the yields based on the aromatic compound consumed are approximately 70%.   The method has been employed to prepare

1-naphthylacetic acid in one step as shown:

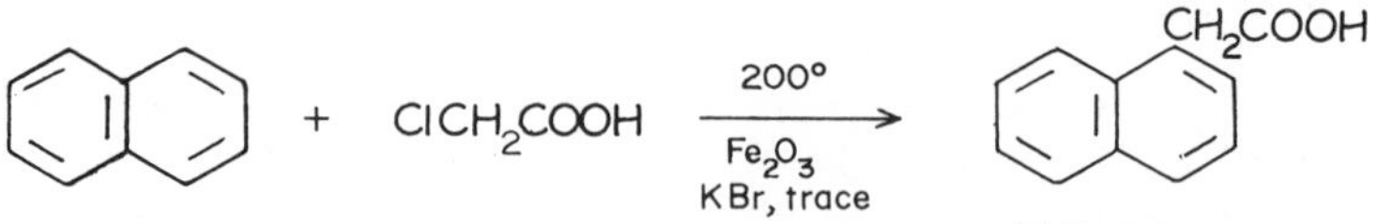

70% crude

The mechanism appears to be a free radical one with $\cdot CH_2COOH$ attacking the aromatic ring.

2. Oxidative carboxymethylation with mixtures containing acetic acid or acetic anhydride and an organic peroxide or oxidizing agent such as potassium permanganate, ceric acetate, or manganic acetate. Yields again are not high. This method has also been employed to prepare 1-naphthylacetic acid as indicated:

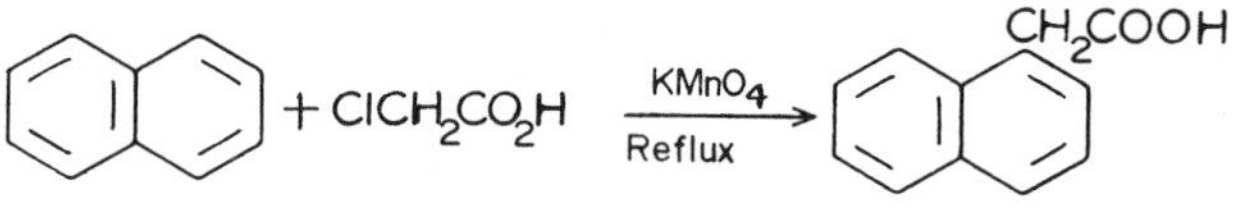

66%

As before, the mechanism appears to be a free-radical one with the attacking particle being $\cdot CH_2COOR$ or $\cdot CH_2COOK$.

3. Photochemical carboxymethylation with thioglycolic acid, iodoacetic acid, or ethyl chloroacetate as reagents. Although this procedure has been utilized in a limited manner, a few syntheses, such as that for phenylacetic acid, as shown, have been reported:

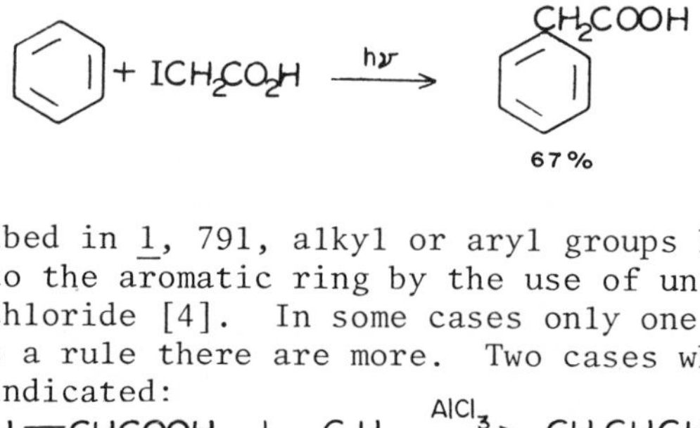

67%

As described in 1, 791, alkyl or aryl groups have been introduced into the aromatic ring by the use of unsaturated acids and aluminum chloride [4]. In some cases only one product is formed, but as a rule there are more. Two cases which give single products are indicated:

$$CH_3CH{=}CHCOOH + C_6H_6 \xrightarrow{AlCl_3} CH_3CH(C_6H_5)CH_2COOH$$

3-Phenylbutyric acid

$$CH_2{=}CHCH_2CH_2COOH + C_6H_6 \xrightarrow{AlCl_3} CH_3CH(C_6H_5)CH_2CH_2COOH$$

4-Phenylpentanoic acid

Unfortunately, no yields are given. It is interesting to note, however, that the phenyl group in the product is never attached to the α- or terminal carbon atom.

A curious arylation of acetic anhydride has been observed recently [5]:

$$(C_6H_5)_3COH + (CH_3CO)_2O \xrightarrow[\substack{4\,hr,\,50°}]{1)\,HBF_4} (C_6H_5)_3CCH_2COOH$$
$$2)\,H_2O \quad \beta,\beta,\beta\text{-Triphenylpropionic acid, 50\%}$$

The enol form of the anhydride no doubt reacts with the carbonium ion of the alcohol.

Propiolactone and phenol at 130° for 12 hr give β-phenoxy-propionic acids in good yield [5a]. In addition, the unsaturated pyrrolidone derived from pyrrole may be used as an alkylating agent with reactive aromatic nuclei [5b].

3. From Carbon Monoxide, Various Substrates, and Transition-metal Salts (Hydrocarboxylation of Olefins and Acetylenes (Reppe) (1, 792)

This subject has now been reviewed [6]. The survey below is based on the facts of the review but does not reflect all of its conclusions. The problem of the synthesis of acids by catalytic carbonylation is complex and its mechanisms are no doubt diverse and as yet not completely understood. The carbon monoxide ligand of a transition metal is attached to the metal atom by a covalent bond that differs electronically from an ordinary metal acyl bond in that it possesses greater stabilization by back (synergistic) bonding from one of the orbitals of the metal, called dπ-pπ bonding [7] (carbonyl frequency in metal

$$-\overset{|}{M} + CO \longrightarrow \overset{\ominus}{M}\overset{\oplus}{CO} \longleftrightarrow \overset{\delta-}{M}\overset{\delta+}{CO}$$
$$\text{A} \qquad \text{B}$$

2000 cm^{-1}, in carbon monoxide, 2200 cm^{-1}). The carbon atom in B although partially positive, is not as positive as in ordinary metal-acyl bonds (e.g., RMgCOR) and thus is more resistant to external reagents such as water or alcohol. However, it is more susceptible to concerted electron movement within a transition-metal complex. The transition-metal ions mostly involved are those of nickel, cobalt, mercury, rhodium, ruthenium, platinum, and palladium. Rhodium salts seem to possess better power to decarbonylate than to carbonylate (1, 350).

One of the most interesting preparations of late is that of dimethyl maleate (90%) from acetylene, carbon monoxide, air, and methanol in the presence of catalytic amounts of palladium chloride complexed with thiourea [6]. A probable mechanism is presented for this reaction since many of the steps are to be found in a number of catalytic sequences involving transition metals. An outstanding source book [8] as well as one of our colleagues [9] has been consulted in these matters. π-Acetylene complex formation is described in the following steps illustrated on p. 699:

1 → 2. The single most important property of the metal catalyst is the presence of a vacant coordination site [10]. It is represented here by the replacement of the ligand, thiourea, by acetylene to form a π-acetylene complex.

2 → 3. 1,2-Addition of Pd-Cl across the triple bond to give the 6-vinyl complex in 3.

3 → 4 or 5. Carbonyl insertion, the crucial step aided by resonance stabilization between the metal and acyl group as shown in 5. The stabilities cannot be much different, however, because the reverse reaction, deacylation, is brought about by higher temperatures and by a change to rhodium metal (Chapter 7, Halides, A.12); 5 is called a chlorine bridged dimer.

5 → 6. This step represents the crucial attack of the solvent, MeOH, on the acyl group. In a catalytic reaction precursors earlier than 5 are not attacked by the solvent.

6 → 7. Oxidation-addition [11]. A common occurrence in transition-metal chemistry.

7 → 8. Carbonyl insertion as in 3 → 4 or 5.

8 → 9. Reductive elimination.

9 → 12. Repetition of previous processes.

The palladium salt accompanying 12 is returned to its original state as follows:

$$L_2 \overset{II}{Pd}(H)(Cl) \xrightarrow{-HCl} L_2 \overset{o}{Pd} \xrightarrow[\substack{2\ HCl \\ CO}]{\frac{1}{2}O_2} L_2 \overset{II}{Pd}(Cl)(CO)^{\oplus} + H_2O + Cl^{\ominus}$$

Because of this recycling by air oxidation, as much as 1 mol of palladium salt produces 60 mol of dimethyl maleate [6]. In many

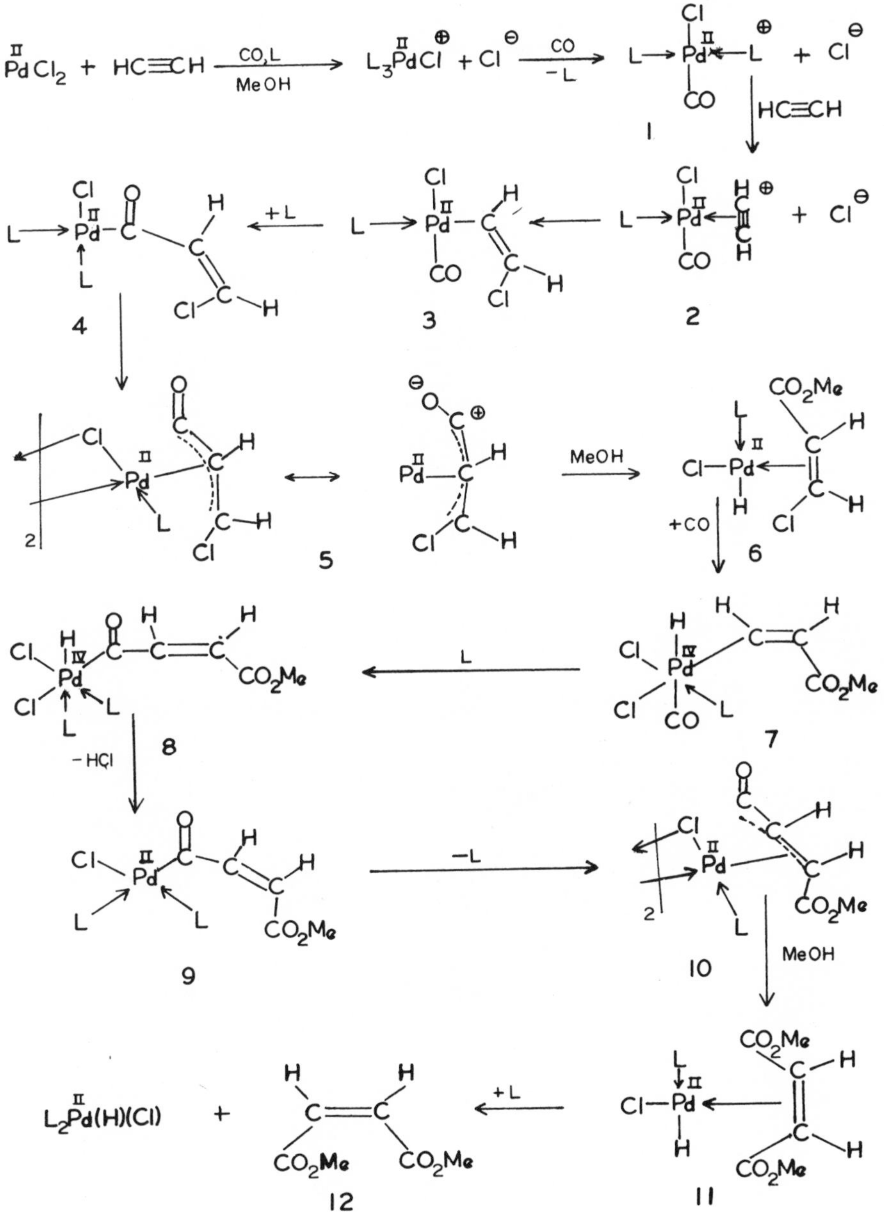

$$\overset{\text{II}}{Pd}Cl_2 + HC\!\equiv\!CH \xrightarrow[\text{MeOH}]{CO,L} L_3\overset{\text{II}}{Pd}Cl^{\oplus} + Cl^{\ominus} \xrightarrow[-L]{CO}$$

1

2

3

4

5

MeOH

6

+CO

8

− HCl

7

9

−L

10

MeOH

11

12

+L

$$L_2\overset{\text{II}}{Pd}(H)(Cl) +$$

CO_2Me

CO_2Me

CO_2Me

CO_2Me

HC≡CH

reactions of this nature some oxidation step is necessary to prevent the formation of metallic palladium [8, p. 77].

With alkenes, carbon monoxide, and palladium chloride at room temperature but under pressure, β-chloropropionic acids are obtained. Similar products are formed with nickel salts. For example, propionyl chloride, carbon monoxide, and nickel carbonyl with some water produce 2,3-butadienoic acid via a π-propargylic system:

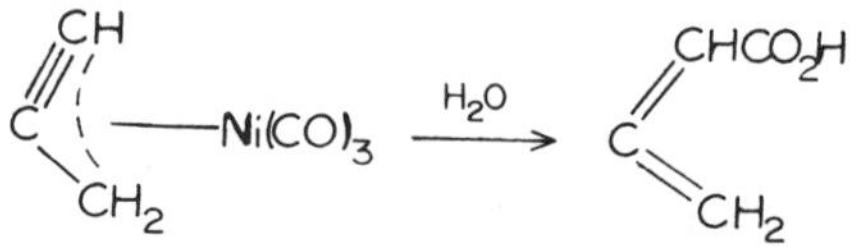

Telomerization occurs in the reaction of acetylene with carbon monoxide, methanol, and nickel carbonyl to form dicarboxylic esters with two or three ethylene units between the functional groups as shown:

$$Me \, O_2CCH_2(CH{=}CH)_{2 \, or \, 3}CH_2CO_2Me$$

Thiourea ligands with nickel salts show increased activity, as was the case with palladium salts. However, clear-cut trends of ligand properties are not to be anticipated.

The carboxylation in the conversion of 1-chloronaphthalene to 1-naphthoic acid (Ex. b) may be considered representative. It should be noted that the reaction occurs at atmospheric pressure. Alkalinity and the use of aprotic solvents contribute to the mild conditions permissible [11a]. The mechanism follows:

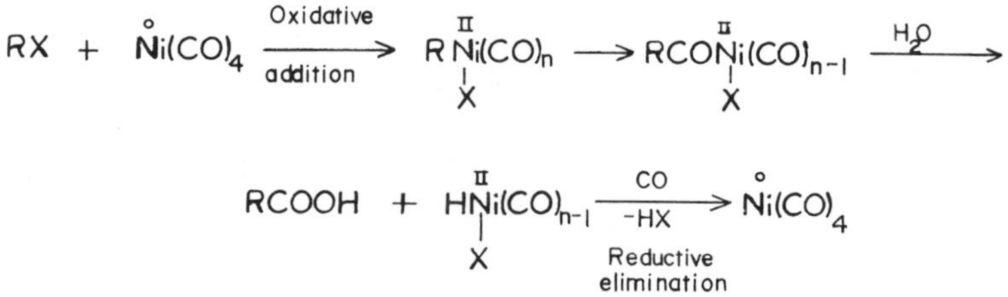

$$RX \; + \; \overset{o}{Ni}(CO)_4 \xrightarrow[\text{addition}]{\text{Oxidative}} R\overset{\mathrm{II}}{Ni}(CO)_n \longrightarrow RCO\overset{\mathrm{II}}{Ni}(CO)_{n-1} \xrightarrow{H_2O}$$

$$RCOOH \; + \; H\overset{\mathrm{II}}{Ni}(CO)_{n-1} \xrightarrow[\substack{-HX \\ \text{Reductive} \\ \text{elimination}}]{CO} \overset{o}{Ni}(CO)_4$$

An alkene with carbon monoxide (200 atm) and water in tne presence of a mixture of H_2PtCl_6 and $SnCl_2$ gives the acid [11b].

One must consult the literature to carry out these reactions since a slight alteration in conditions sometimes leads to remarkable changes in behavior. By all means the catalytic behavior of the transition metal must be maintained.

a. Preparation of 3-Butenoic Acid as its Methyl Ester [6]

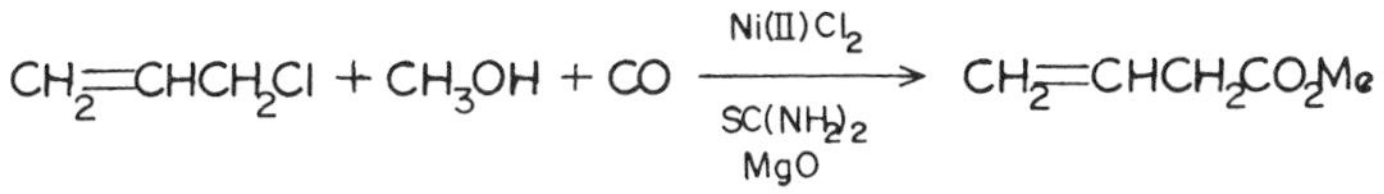

$$CH_2{=}CHCH_2Cl + CH_3OH + CO \xrightarrow[\substack{SC(NH_2)_2 \\ MgO}]{Ni(II)Cl_2} CH_2{=}CHCH_2CO_2Me$$

A stirred mixture of methanol, 50 ml, allyl chloride, 4 ml, nickel(II) chloride, 1 g, thiourea, 0.65 g, and magnesium oxide, 1 g, under a carbon monoxide atmosphere was treated with 0.33 g of powdered manganese iron alloy (80% Mn) (to initiate the formation of nickel carbonyl), whereupon the mixture turned red as CO was absorbed. After 2 hr of stirring, it was filtered and the filtrate was warmed to remove low boiling liquids and then distilled under reduced pressure to give the ester, 63%.

b. Preparation of 1-Naphthoic Acid [6]

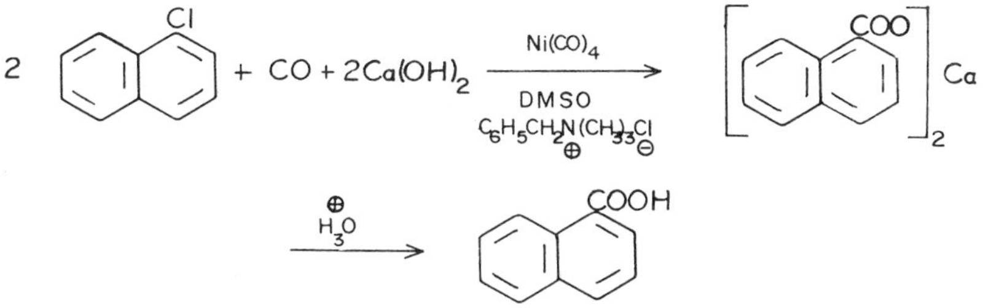

Under a carbon monoxide atmosphere, DMSO, 36 ml, water, 1.5 ml, Ni (CO)$_4$, 1.2 ml, Ca(OH)$_2$, 1.6 g, benzyltrimethylammonium chloride, 2.5 g, and 1-chloronaphthalene, 5.0 g, were mixed and heated at 110° for 7 hr. After 600 ml of CO was absorbed, the mixture was treated with 2 N HCl and extracted with ether. From the ethereal solution there was recovered 4.9 g (92.4%) of the acid.

c. Preparation of Nonanoic Acid [12]

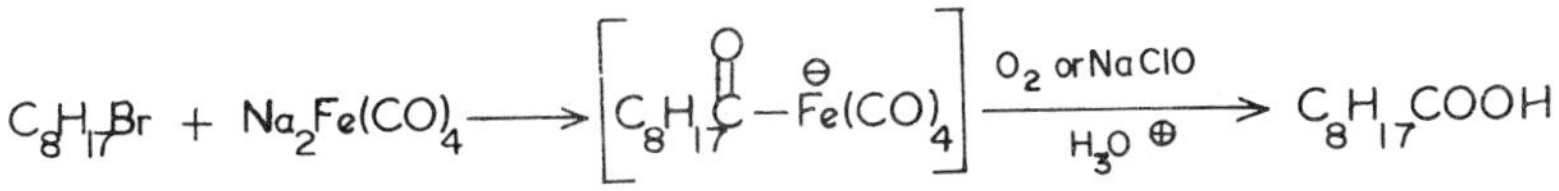

Octyl bromide, 10 mmol, Na$_2$Fe(CO)$_4$, 10+ mmol, in methylpyrrolidone were stirred with CO under 10 psi pressure. The mixture was swept with air or treated with aqueous NaClO to yield about 80% of the acid.

4. From Alcohols, Alkenes, Alkyl Halides, or Esters and Formic-Sulfuric Acid or 1,1-Di- or Trichloroethylene in the Presence of Sulfuric Acid (Koch-Haaf; Bott) (1, 793)

$$>C=C< \xrightarrow[\text{H}_2\text{SO}_4]{\text{HCOOH}} >\underset{\underset{H}{|}}{C}-\underset{\underset{COOH}{|}}{C}<$$

$$>C=C< \ +\text{-CH}=\text{CCl}_2 \xrightarrow[\text{H}_2\text{SO}_4]{\text{BF}_3} >C-\underset{\underset{H}{|}}{C}-\text{CH}_2\text{COOH}$$

A study of the products obtained in the Koch-Haaf reaction with 1-octene showed that the main components were $C_4H_9 \underset{\underset{Et}{|}}{\overset{\overset{Me}{|}}{C}} CO_2H$ and $C_5H_{11} \underset{\underset{Me}{|}}{\overset{\overset{Me}{|}}{C}} CO_2H$ [12a].   The 1-[3.3.0] bicyclooctyl cation

derived from 1,3-cyclooctadiene isomerization (and other rearrangements) is the main species which reacts with carbon monoxide in the Koch-Haaf reaction [12b].
    Souma and co-workers [13] have found that Cu(I) compounds catalyze the Koch-Haaf reaction which is facile enough to be conducted with CO at room temperature and atmospheric pressure. Only t-carboxylic acids were isolated, that is, cyclohexene gave a single acid, 1-methylcyclopentanecarboxylic acid (63%).   The copper catalyst no doubt acts to solubilize CO via $Cu(CO)_3$, but may have a more profound, unknown effect.
    A recent discussion of the synthesis of carboxylic acids with 1,1-dichloroethylene is available [14] (Ex. b):

$$\left[\overset{\oplus}{R}\right] + \text{CH}_2=\text{CCl}_2 \longrightarrow \left[\text{RCH}_2\overset{\oplus}{C}\text{Cl}_2\right] \xrightarrow{\text{H}_2\text{O}} \text{RCH}_2\text{COOH}$$

By using trichloroethylene with carbonium ions and sulfuric acid alone above 80°, Bott [15] produced the corresponding α-chlorocarboxylic acid as shown:

$$\left[\text{R}^{\oplus}\right] \xrightarrow[\substack{\text{H}_2\text{SO}_4 \\ 80\text{-}110°}]{\text{ClCH}=\text{CCl}_2} \left[\text{RCHCl}\overset{\oplus}{C}\text{Cl}_2\right] \xrightarrow{\text{HOH}} \text{RCHClCOOH}$$

Because of the stability required under such conditions, the reaction is restricted to the 1-adamantyl,2-norbornyl, substituted benzyl and benzhydryl cations.   For example, norbornene gives exo-norbornyl-2-chloroacetic acid (66%).

Lansbury and Stewart [16] have employed the Bott reaction in preparing cycloalkanecarboxylic acids by an intramolecular chloroolefin annelation as indicated.

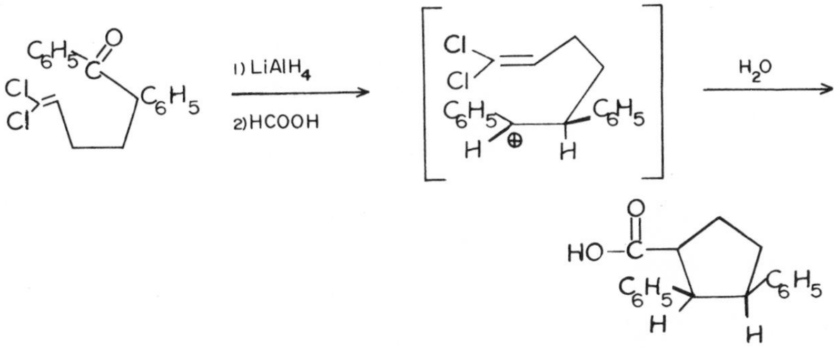

trans-2,3-Diphenylcyclopentanecarboxylic acid
ca. 50%

The starting material is produced from the mesylate of 4,4-dichloro-3-buten-1-ol and desoxybenzoin.

a. Preparation of 1-Methylcyclohexanecarboxylic Acid [17]

Yield was 89-94% from 2-methylcyclohexanol and formic acid added to a mixture of formic acid and 96% H_2SO_4 at 15-20°.

b. Preparation of β,β-Dimethylbutyric Acid [14]

To a vigorously stirred mixture of 90% aqueous H_2SO_4, 200 ml, and boron trifluoride, 8% by weight, at 5-7° was added a solution of 1 mol of t-butyl alcohol and 1.5 mol of 1,1-dichloroethylene over a 2 hr period. The mixture was stirred for an additional 2 hr, poured into ice, and worked up in the usual way for acids, yield 80%.

5. From Tertiary, Saturated Hydrocarbons, Formic Acid, t-Butyl Alcohol, and Sulfuric Acid, or the Tertiary Halide, Formic Acid, Silver Sulfate, and Sulfuric Acid (1, 794)

The Koch-Haaf method has been applied to tertiary halides as well as tertiary hydrocarbons. Thus Holtz and Stock [17a] utilized the method for the conversion of 1-bromo-4-ethylbicyclo-[2.2.2]octane into 1-carboxy-4-ethylbicyclo[2.2.2]octane (49%). In this procedure silver sulfate generates the carbonium ion from the halide, obviating any need for addition of a carbonium ion source such as t-butyl alcohol to abstract a hydride from the hydrocarbon.

Recently, Chapman and co-workers [18] converted both the 4-methyl- and 4-ethylbicyclo[2.2.2]octanes by a modified procedure to the corresponding carboxylic acids as shown:

$R = CH_3, 70\%$
$R = C_2H_5, 75\%$

a. Preparation of Adamantanecarboxylic Acid [19]

Yield was 56-61% from 96% H_2SO_4 and adamantane in carbon tetrachloride to which a mixture of t-butyl alcohol and 98% formic acid was added dropwise.

8. From Alkenes and Carboxyalkyl Free Radicals

$$RCH = CH_2 + [\dot{C}H_2CO_2H] \xrightarrow{CH_3COOH} R(CH_2)_3CO_2H + [\dot{C}H_2COOH]$$

This method has been reviewed [20] and the extensive work done since (to 1970) has been collected in a second review [21]. An excess of acid is essential for high yields, usually at least 6 mol of acid to 0.15 mol of alkene, and the ideal temperature is above 140°. The free radical of the acid may be generated from many sources, including di-t-butylperoxide. Substituted acids with α-hydrogens respond well. α-Chloroacetic acid may be used to add -CHClCOOH to the alkene. α-Bromoacetic acid does not respond. A host of other acids, usually as the esters, have been added to alkenes. Methyl formate is employed to introduce the carbomethoxy group alone. The alkene is usually terminal unless conjugated with some other functional group. Cyclohexene gives poor yields of adducts, that with methyl formate being 36%.
Substitution occurs, as is shown, by using manganese tri-acetate as the free-radical source [21a]:

γ,δ-Unsaturated acid, 38-60%
(3 examples)

a. General Preparation of Carboxylic Acids from Alkenes [21]

The acid, 4 mol, was heated with stirring to 158-162°. Then a mixture of the α-olefin, 0.15 mol, and di-t-butylperoxide, 0.038 mol, in the same acid, 2 mol, was added at 30 drops per min after which the mixture was maintained at the same temperature for 1 additional hour (slower addition improves the yield). After distilling off the excess of acid, the residue was fractionated under reduced pressure. Yields vary but may be in the 50-60% range.

9. From Aromatic Compounds by Carboxyvinylation

Nishimura and co-workers [22] found that cinnamic acid is produced in small yield when benzene, sodium propionate, palladium(II) chloride, propionic anhydride, and propionic acid are heated as indicated:

$$C_6H_6 + CH_3CH_2COONa \xrightarrow[\substack{CH_3CH_2COOH \\ 100^\circ,\ 6\ hr.}]{(CH_3CH_2CO)_2O} \underset{28\%}{C_6H_5CH{=}CHCOOH}$$

Substituted benzenes give cinnamic acids also in low yield. β-methylcinnamic acid (27%) was obtained from benzene, palladium(II) chloride, sodium butyrate, and butyric acid.

10. From Alkenes and Ethyl Diazoacetate (via Carbenes)

2-Phenylcyclopropanecarboxylic acids have been prepared by the sequence of reactions as shown [23] (Ex. a).

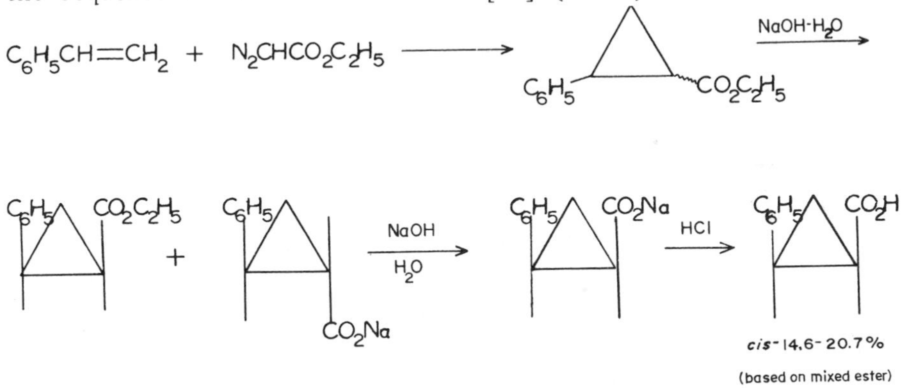

$C_6H_5CH{=}CH_2$ + $N_2CHCO_2C_2H_5$
NaOH-H_2O
C_6H_5
$CO_2C_2H_5$
NaOH
H_2O
CO_2Na
HCl
CO_2H
cis-14.6-20.7%
(based on mixed ester)

The first step produces a mixture of the cis and trans esters, which are separated as a result of the more rapid hydrolysis of the trans due to less steric hindrance.

a. Preparation of cis-2-Phenycyclopropanecarboxylic Acid [23]

To 500 ml of refluxing and stirred xylene a solution of 179 g of ethyl diazoacetate and 163 g of styrene was added in 90 min. After continuing refluxing and stirring for another 90 min, the xylene was removed under pressure and the mixed esters,

boiling point 85-93°/0.5 mm, 155 g, 52%, were recovered.

The mixed esters, 155 g, 200 ml ethanol, 65 ml of water, and 24.5 g of NaOH were refluxed for 5 hr, during which the 200 ml of ethanol was removed and replaced by an equal volume of water. More water, 250 ml, and 150 ml of benzene were then added to the unheated mixture and the total was stirred for 2-3 min. The trans sodium salt was in the aqueous layer, while the cis ester was in the benzene layer.

By a process somewhat similar to the above the cis ester was saponified, after which the mixture was acidified to give 19.5-23.8 g (14.6-20.7%, based on the mixed ester used) of the cis acid.

11.   Reduction of Unsaturated Acids

$$\text{>C} = \overset{|}{\text{CCOOH}} \longrightarrow \text{>CH-CHCOOH}$$

The reduction may be accomplished very easily since there is little tendency for the carboxyl group to be reduced. The reaction may be conducted either chemically or catalytically with hydrogen [23a]. Aromatic acids or amides under Birch conditions give the 1,4-dihydro compound as shown [24]:

$$C_6H_5CO_2H \xrightarrow[NH_3/C_2H_5OH]{Na} \text{(COOH)}$$

1,4-Dihydrobenzoic acid
89-95%

References

1.   H. Gross et al., Chem. Ber., 96, 1382 (1963).
1a.   H. Hopff, M. A. Osman, J. Prakt. Chem., 311, 266 (1969).
1b.   L. F. Somerville, C. F. H. Allen, Org. Syn., Coll. Vol. 2, 81, Note 9 (1943).
2.   P. E. Sokol, Org. Syn., 44, 69 (1964).
3.   P. L. Southwick, Synthesis, 628 (1970).
4.   P. Four, D. Lefort, Tetrahedron Letters, 6143 (1968).
5.   N. C. Deno et al., J. Org. Chem., 35, 278 (1970).
5a.   F. J. Bullock, J. Med. Chem., 11, 419 (1968).
5b.   V. Bocchi, G. P. Gardini, Org. Prep. Proced., 1, 271 (1969).
6.   L. Cassar et al., Synthesis, 509 (1973).
7.   For an illustration see R. E. Harmon et al., Chem. Rev., 73, 27 (1973).
8.   P. M. Maitlis, The Organic Chemistry of Palladium, Vol. 2, Academic, New York, 1971.

9. We are indebted to Dr. Charles M. Lukehart, Vanderbilt University, for assistance.

10. J. P. Collman, Acc. Chem. Res., 1, 136 (1968).

11. J. Halpern, Acc. Chem. Res., 3, 386 (1970).

11a. L. Cassar, M. Foa, J. Organomet. Chem., 51, 381 (1973).

11b. L. J. Kehoe, R. A. Schell, J. Org. Chem., 35, 2846 (1970).

12. J. P. Collman et al., J. Am. Chem. Soc., 95, 249 (1973).

12a. D. R. Kell, F. J. McQuillin, J. Chem. Soc., Perkin Trans., I, 2096 (1972).

12b. M. A. McKervey et al., J. Chem. Soc., C, 2430 (1971).

13. Y. Souma et al., J. Org. Chem., 38, 2016, 3633 (1973).

14. K. Bott, H. Hellmann, Foerst's Newer Methods of Preparative Organic Chemistry, Vol. 6, Academic, London, 1971, p. 67.

15. K. Bott, Chem. Ber., 103, 3850 (1970); 106, 2513 (1973).

16. P. T. Lansbury, R. C. Stewart, Tetrahedron Letters, 1569 (1973).

17. W. Haaf, Org. Syn., 46, 72 (1966).

17a. H. D. Holtz, L. M. Stock, J. Am. Chem. Soc., 86, 5183 (1964).

18. N. B. Chapman et al., J. Org. Chem., 35, 917 (1970).

19. H. Koch, W. Haaf, Org. Syn., Coll. Vol. 5, 20 (1973).

20. G. Sosnovsky, Free Radical Reactions in Preparative Organic Chemistry, Macmillan, New York, 1964, p. 137.

21. H.-H. Vogel, Synthesis, 99 (1970).

21a. M. Okano, Chem. Ind. (London), 423 (1972).

22. S. Nishimura et al., Chem. Commun., 313 (1969).

23. C. Kaiser et al., Org. Syn., 50, 94 (1970).

23a. R. Adams, V. Voorhees, Org. Syn., Coll. Vol. 1, 61 (1941).

24. M. E. Kuehne, B. F. Lambert, Org. Syn., Coll. Vol. 5, 400 (1973).

G. Rearrangements

1. From Diazoketones (Arndt-Eistert and Wolff Rearrangements) (1, 796)

Contrary to earlier reports the Wolff rearrangement of ketones proceeds both by the ketene (A) and the oxirene (B) routes [1]:

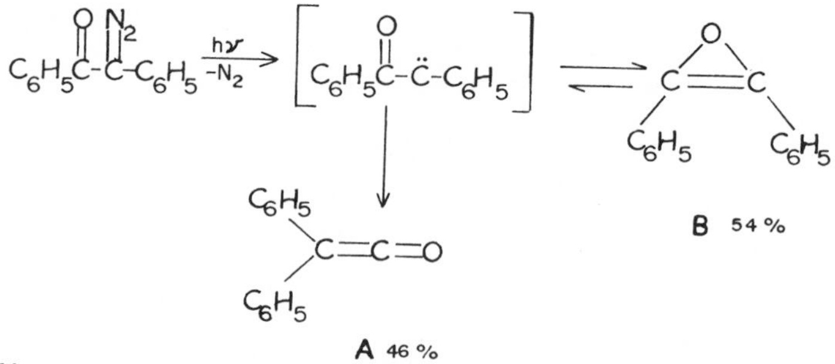

B 54%

A 46%

Other diazocycloalkanones (C_6-C_{12}) gave yields varying within 25-95%.

In the synthesis of a [7] paracyclophane from 4-carboxy [8] paracyclophane, Allinger and Walter [2] employed the Wolff rearrangement in the last step as indicated:

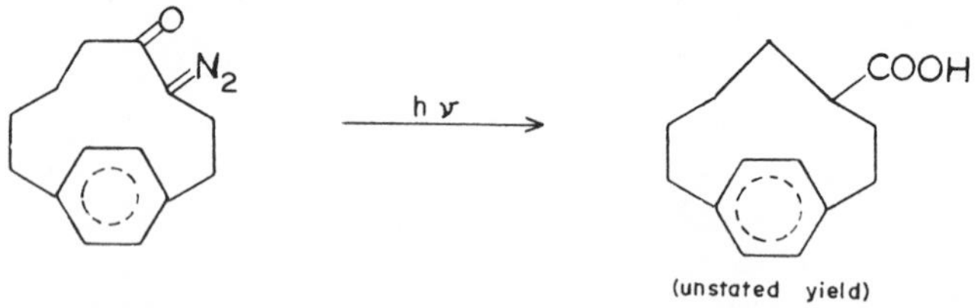

(unstated yield)

The carboxy group in carboxy [7] paracyclophane has been tentatively assigned the 3-position.

2. From Cyclic α-Haloketones (Favorskii) (1, 797)

A review of the Favorskii rearrangement has become available [3]. It lists the chief synthetic uses of the reaction as follows: (a) preparation of branched chain carboxylic acids; (b) stereospecific synthesis of cis-α,β-unsaturated acids; (c) ring contraction in alicyclic and heterocyclic compounds; and (d) modification of steroids.

A reaction resulting in the formation of a branched-chain carboxylic acid follows [4]:

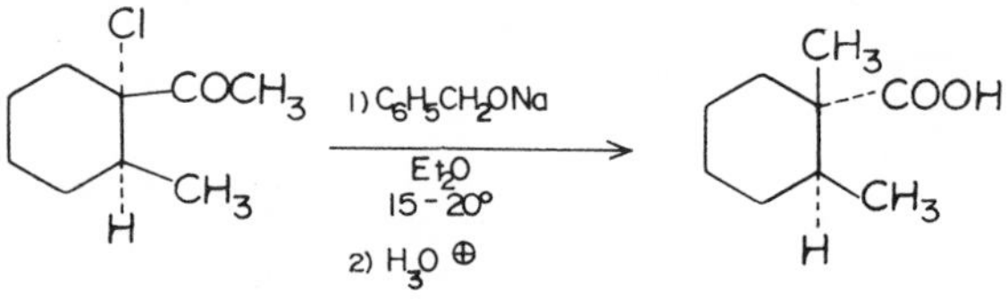

cis -1-Acetyl-1-chloro-2-methyl-
cyclohexane

cis - 1,2-Dimethylcyclohexane-
carboxylic acid, 44%

The variability of the mechanism has been stressed, that is, the cyclopropane intermediate proposed is not always pertinent.

cis-Halogenoacrylic acids have been prepared from 1,1,3-trihalogenoacetones and isocrotonic acid from methyl ethyl ketone as indicated [5, 5a]:

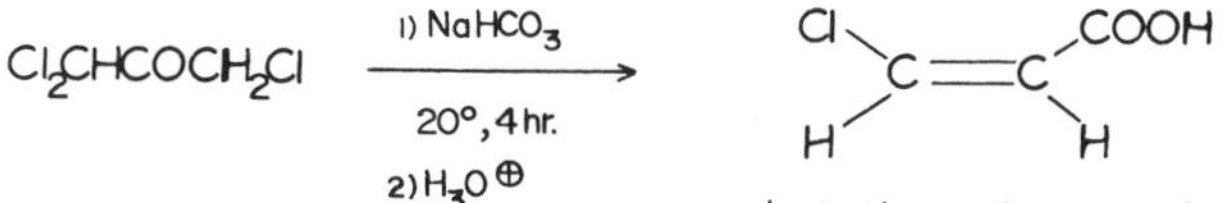
$Cl_2CHCOCH_2Cl$
1) $NaHCO_3$
20°, 4 hr.
2) $H_3O^{\oplus}$
Cl
COOH
H
H

cis-3-Chloroacrylic acid, 74%

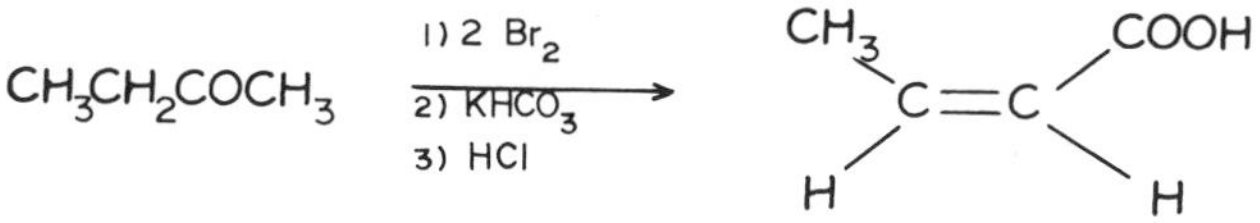
$CH_3CH_2COCH_3$
1) 2 Br_2
2) $KHCO_3$
3) HCl
CH_3
COOH
H
H

*Iso*crotonic acid
69-77% (crude)

An illustration of ring contraction in alicyclic compounds is given in the synthesis of cyclohexanecarboxylic acid (1, 798). Ring contractions in steroids occurs among cholestane derivatives as shown [6]:

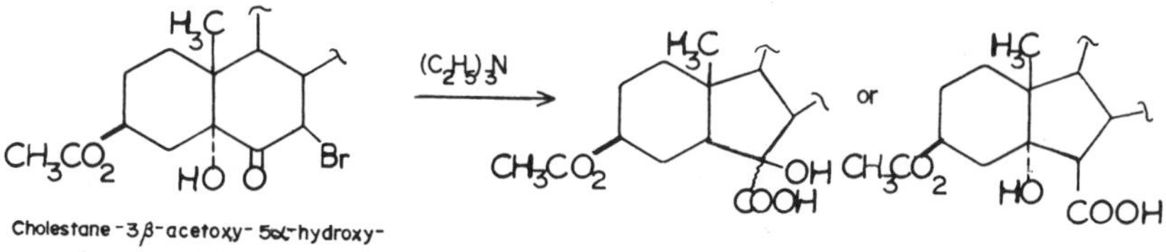
H_3C
CH_3CO_2
HO
Br
$(C_2H_5)_3N$
H_3C
CH_3CO_2
OH
COOH
or
H_3C
CH_3CO_2
HO
COOH

Cholestane-3β-acetoxy-5α-hydroxy-7β-bromo-6-one

The first formula of the product was an original assignment, but later it was suggested that the second is more probable [3].

5. From Sugars

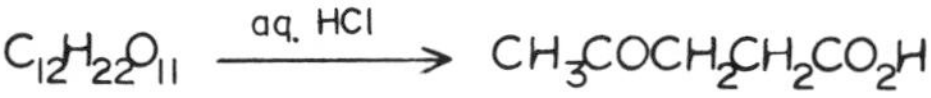
$C_{12}H_{22}O_{11}$
aq. HCl
$CH_3COCH_2CH_2CO_2H$

Saccharose Levulinic acid

A superior preparation (62%) of levulinic acid has been reported [7].

6. From Diketenes

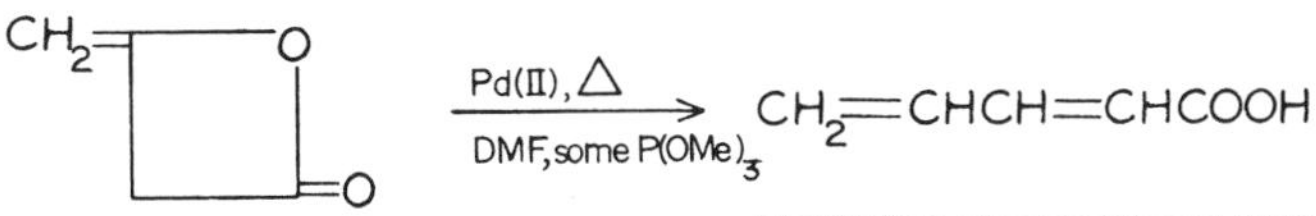
CH_2
O
O
Pd(II), Δ
DMF, some $P(OMe)_3$
$CH_2{=}CHCH{=}CHCOOH$

1,3-Butadiene-4-carboxylic acid, 100%

Heating of diketene itself leads to mixtures of products including the above. However, the palladium catalyst produces a smooth rearrangement to the desired product [8], which should not be heated excessively as decarboxylation may take place, especially without the ligand, trimethyl phosphite.

7.   From Allyl α-Silyloxyvinyl Ether (Claisen) (see Addendum)

Addendum

G.7.

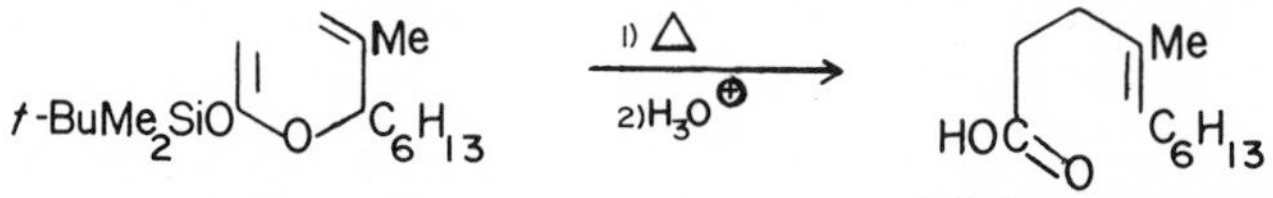

4-Methylundec-4-enoic acid
80%

[J. A. Katzenellenbogen, K. J. Christy, J. Org. Chem., 39, 3315 (1974)]

References

1.   G. Frater, O. P. Strausz, J. Am. Chem. Soc., 92, 6654 (1970).
2.   N. L. Allinger, T. J. Walter, J. Am. Chem. Soc., 94, 9267 (1972).
3.   A. A. Akhrem et al., Russ. Chem. Rev., 39, 732 (1970).
4.   G. Stork, I. J. Borowitz, J. Am. Chem. Soc., 82, 4307 (1960).
5.   C. Rappe, Acta Chem. Scand., 19, 31 (1965).
5a.   C. Rappe, Org. Syn., 53, 123 (1973).
6.   L. F. Fieser, S. Rajagopalan, J. Am. Chem. Soc., 71, 3938 (1949).
7.   J. Dahlmann, Chem. Ber., 101, 4251 (1968).
8.   A. Noels, P. Lefebvre, Tetrahedron Letters, 3035 (1973).

Chapter 14

CARBOXYLIC ESTERS, ORTHOESTERS, AND ORTHOCARBONATES

Although a chemist might be inclined to feel that nothing remains to be accomplished regarding the synthesis of esters from acids and alcohols, the deluge of communications on modifications continues as is described in A.1. Boron trifluoride etherate is recommended as the catalyst for esterification of unstable and unsaturated acids. The combination of boric and sulfuric acids is superior to either acid proper as a catalyst for the esterification of phenols, as is also a combination of polystyrene sulfonated resin plus $CaSO_4$. In the reaction of anhydrides with alcohols 4-dimethylaminopyridine has been found to be a better catalyst than pyridine (A.3). The HMPA-acid-NaOH method with alkyl halides, except tertiary, is a superior method of esterification (A.10), but methods to prepare t-alkyl esters are also described (A.2; A.7).

Two safer methods serve as substitutes for the diazomethane synthesis of methyl esters: the in situ preparation of the diazomethane or the use of 1-methyl-3-p-tolyltriazine (B.2). Lithium isopropylcyclohexylamide (LiICA) appears to be a sufficiently strong base to prevent self-condensation of esters (because the lithium enolate of the ester forms quantitatively) and thus permits acylation with an acyl halide differing in carbon structure from the ester (C.6). In the alkylation of esters the LiICA method seems to be more versatile than the usual malonic ester synthesis (C.6).

The field of acyloxylation has benefited greatly by the publication of a long review in two parts (D.6). Manganese triacetate appears to generate the carboxymethyl free radical ($\cdot CH_2CO_2H$) rather than the expected acyloxy free radical ($CH_3CO_2\cdot$), and thus the formation of lactones results from alkenes (D.10). Another interesting synthesis of lactones from the same source may be found near the end of E.3. Retrocyclo reactions to produce esters different from those used in the original cyclo reaction are discussed in E.9. The carboxylic ester chapter contains so many reactions that others of interest are in all probability described. A short section giving the most important syntheses of orthoesters has been appended at the end of this chapter.

The comprehensive, old review of ester preparations by H.-W. [1] is still applicable. Reviews on the synthesis of esters of hypohalous acid [2a], lactones [2b], β-lactones [2c], and diesters [3] are also available. The isomerization of unsaturated esters has been studied [4] but is not discussed here.

Addendum

General. Reviews of the synthesis of α-methylene lactones are available [T. A. Bryson et al., Syn. Commun., 5, 245 (1975); P. A. Grieco, Synthesis, 67 (1975)].

References

1. H.-W., Methoden der Organischen Chemie, Oxygen Series, III, Vol. 8, G. Thieme Verlag, Stuttgart, 1952, p. 503.
2. (a) H.-W., Methoden der Organischen Chemie, Oxygen Series, I, Vol. 6, part 2, G. Thieme Verlag, 1963, p. 487; (b) p. 561; (c) p. 511.
3. S. Patai, The Chemistry of Carboxylic Acids and Esters, Interscience, New York, 1969, p. 175.
4. S. J. Rhoads, E. E. Waali, J. Org. Chem., 35, 3358 (1970).

The most recent review on the synthesis of carboxylic esters is that of Patai [1].

A. Solvolysis

1. From Carboxylic Acids (Esterification) (1, 802)

$$RCOOH + R'OH \underset{}{\overset{H^{\oplus}}{\rightleftharpoons}} RCOOR' + H_2O$$

The Koch-Haaf reaction for the preparation of carboxylic acids is covered in F.4, Chapter 13 (1, 793). If the reaction mixture is dropped into methanol, the methyl ester may be obtained [2].

The catalysts employed in the customary esterification continue to offer variety. Of these boron trifluoride etherate has found common use. Marshall and co-workers [3] employed it satisfactorily with unstable carboxylic acids such as 1,4-dihydrobenzoic acid as shown (Ex. a):

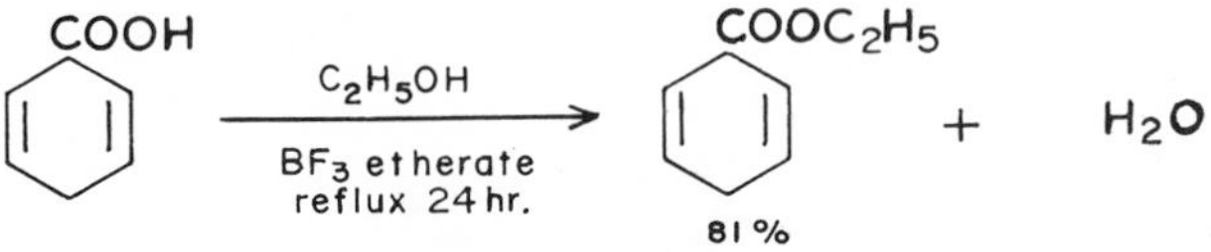

Thus the method is mild and often thorough. Kadaba [4] employed the same reagent in preparing a series of esters of aliphatic and araliphatic unsaturated esters as shown:

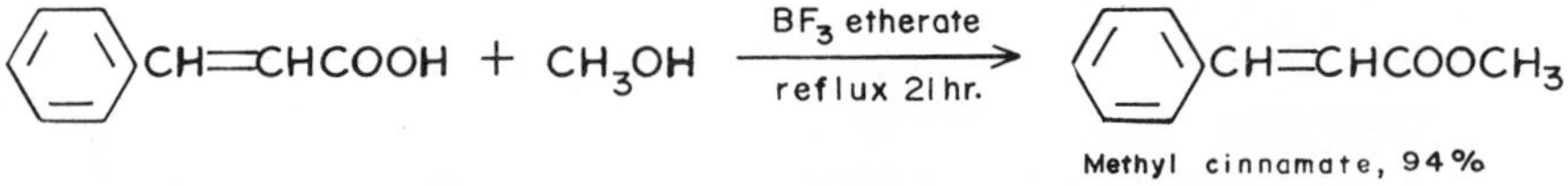

Methyl cinnamate, 94%

The method is applicable as well to the esterification of unsaturated dicarboxylic acids and p-aminobenzoic acid.

Normant and Deshayes [5] prepared cyclohexyl trichloroacetate from the free acid and cyclohexanol as indicated:

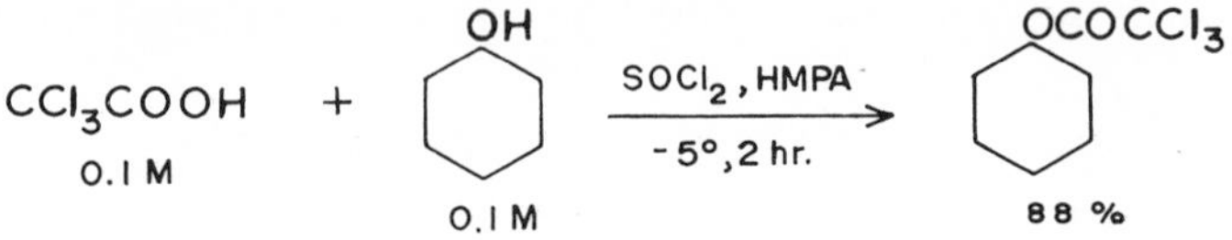

They assumed that the acid chloride is an intermediate and if so, the method should serve in cases where the acid chloride is not readily accessible.

Other reagents that function as catalysts in esterification are β-trichloromethyl-β-propiolactone and β-trichloromethyl-β-ethanesultone [6]. The former gave yields as indicated:

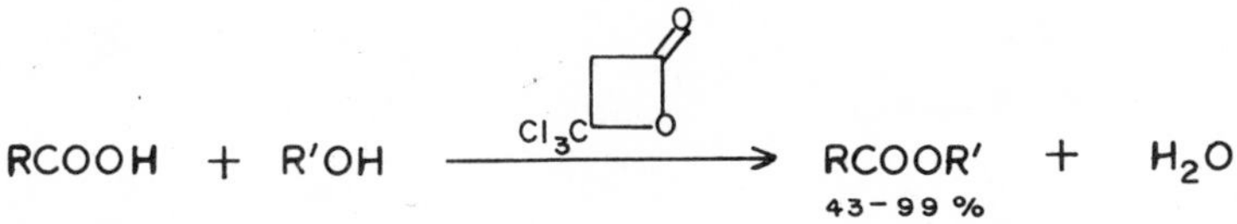

Neckers [7] utilized polymer-protected aluminum chloride, which released aluminum chloride as the resin swelled in the solvent of the reaction:

$$CH_3CH_2COOH \ + \ \textit{n}\text{-BuOH} \ \xrightarrow[]{\substack{\text{polymeric} \\ \text{protected} \\ AlCl_3}} \ CH_3CH_2COOBu$$

Butyl propionate, 96.8%

By substituting phenol for the alcohol, Lowrance [8] prepared a series of aryl esters as shown for phenyl benzoate:

$$C_6H_5COOH \ + \ HO\!\!\!\diagup\!\!\!\bigcirc\!\!\!\diagdown \ \xrightarrow[C_6H_5CH_3, \text{reflux 8 hr.}]{H_3BO_3 \ -H_2SO_4} \ C_6H_5COOC_6H_5$$

94%

The ester was obtained by removing the water by azeotropic distillation from a refluxing toluene solution of phenol, benzoic acid, and catalytic amounts of the two acids. The reaction did not proceed with either of the two acids alone. With terephthalic acid and phenol, diphenyl terephthalate was obtained in 87% yield.

In the synthesis of esters of amino acids Yamazaki and Higashi [9] obtained high yields and high optical purity of esters as indicated:

$$C_6H_5CH_2OCONHCHRCOOH + R'OH \ \xrightarrow[C_5H_5N, 25°, \ 30\,min.]{\overset{\displaystyle HP(OC_6H_5)_2}{\underset{\displaystyle O}{\downarrow}}}$$

$$C_6H_5CH_2OCONHCHRCOOR' \ + \ H_2O$$

Vesley and Stenberg [10] employed a sulfonated polystyrene copolymer with $CaSO_4$, the so-called catalytic dehydrator, in the esterification of acetic acid with methanol and butanol. The acid polymer alone gives quantitative results over a long time period, but the $CaSO_4$ reduces the time required substantially. The apparatus is simple in that the components need only be stirred together at room temperature and the workup is simple as well since the catalyst and catalytic dehydrator are insoluble in the reaction mixture. To cite the results of one experiment, acetic acid with methanol (1:10 ratio by volume) gave the ester in 94 ± 5% yield in 10 min by using the R204 acid polymer (of Fisher Scientific Co.). Longer times were required in the esterification with butanol.

Since a ternary azeotrope forms, Coopersmith and co-workers [11] found that the molecular sieve, Type 3A, was desirable in the esterification of neopentanoic acid with alcohols of low molecular weight by the Fisher method using sulfuric or p-toluenesulfonic acid and an entrainer such as xylene or heptane. By refluxing the mixture through a Soxhlet extractor containing the molecular sieve, ethyl neopentanoate was obtained (100% conversion) in 3.5 hr.

For the esterification of the hindered acid, 1,1'-binaphthyl-8,8'-dicarboxylic acid, Harris and Patel [12] obtained excellent results by using an excess of trimethyl phosphate in alkali:

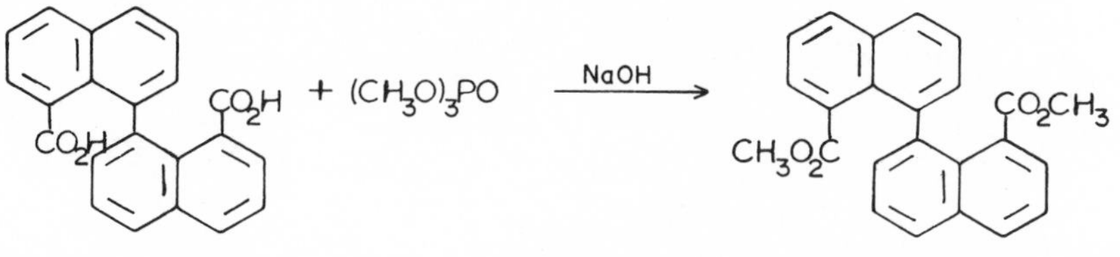

> 90%

In the case of protected nucleotides the reagent, mesitylene-sulfonyl 1-H-1,2,4-triazole appears to give superior yields of the polynucleotides [13].

a.   Preparation of Ethyl 1,4-Dihydrobenzoate [3]

1,4-Dihydrobenzoic acid, 4.11 g, $BF_3 \cdot (C_2H_5)_2O$, 4.9 ml, and 50 ml of anhydrous ethanol were stirred and refluxed for 20 hr. After the addition of 150 ml of water, the cooled mixture was extracted with three portions of ether, from which there was recovered 5.04 g (81%) of the ester.

2.   From Acid Chlorides (1, 807)

$$ROH + R'COCl \longrightarrow R'COOR + HCl$$

In the alcoholysis of carboxylic acid halides containing α-hydrogens with methanol-d in the presence of triethylamine, Truce and Bailey [14] have shown that two mechanisms compete as indicated:

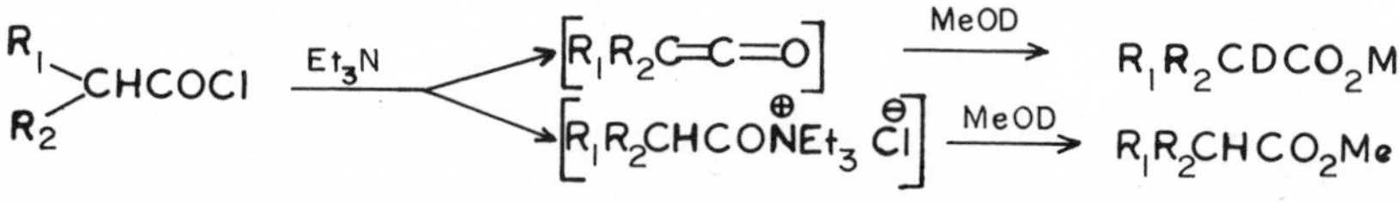

Thus there is an elimination-addition process involving a ketene intermediate that gives a deuterated ester and a substitution process involving an acyl quaternary ammonium intermediate giving an undeuterated ester.

t-Butyl esters have been prepared readily as follows [15].

t-Butyl benzoate
89%

Yields for other t-butyl esters were in the range of 64-94%.

α-Chloroesters are prepared from aldehydes and acetyl chloride with zinc chloride at -5° to 0° or from ketones with aluminum chloride at 25° [16]:

$$R_2C=O + CH_3COCl \longrightarrow R_2CClOCCH_3$$
$$\underset{O}{\|}$$

50-95% (6 examples)

Ethers are split by acid chlorides in the presence of Group VI carbonyls such as Mo(CO)$_6$. For example, THF and acetyl chloride give a 78% return of 4-chlorobutyl acetate [17].

3. From Acid Anhydrides ($\underline{1}$, 809)

$$ROH + (R'CO)_2O \longrightarrow R'COOR + R'COOH$$

Steglich and Höfle [18a] found that 4-N,N-dimethylaminopyridine alone or mixed with triethylamine is an acylation catalyst much superior to pyridine. For example, t-butyl alcohol and 1-methyl-1-cyclohexanol are not acylated by acetic anhydride and pyridine, whereas in the presence of an equivalent amount of the aminopyridine the reaction is complete in 10 hr at room temperature. In preparations it suffices to add a catalytic amount of the aminopyridine and to bind the acid formed with triethylamine (Ex. a) [18b].

Fatiadi [19] on treatment of myo-inositol with acetic anhydride, dimethyl sulfoxide, and pyridine obtained, instead of the acetylated hexaalcohol, the acetylated pentaphenol as shown:

Pentaacetoxybenzene, 46-52%

The yields of the corresponding pentapropionates and pentabutyrates were approximately one-half as great.

a. Preparation of 1-Methyl-1-cyclohexyl Acetate [8]

The yield was 86% from 0.1 mol of 1-methyl-1-cyclohexanol, 0.21 mol of acetic anhydride, 0.15 mol of (C$_2$H$_5$)$_3$N, and 4.1 mmol of N,N-dimethylamino-4-pyridine at 25° for 14 hr [amounts using C$_5$H$_5$N and/or (C$_2$H$_5$)$_3$N are less than 5%)].

4. From Ketenes, Ketene Acetals, and Isocyanates ($\underline{1}$, 811)

$$\text{RNCO} + \text{R'OH} \longrightarrow \text{RNHCOOR'}$$

Beckwith and co-workers [20] prepared a series of carbamic acid esters by starting with the amide and oxidizing with lead tetraacetate in the presence of an alcohol, as shown:

$$\text{RCONH}_2 \xrightarrow{\text{Pb(OAc)}_4} \left[\text{RNCO}\right] \xrightarrow{\text{R'OH}} \underset{33\text{-}96\%}{\text{RNHCOOR'}}$$

The isocyanate was not isolated in the process. The method is satisfactory for primary amides derived from saturated primary, secondary, and tertiary carboxylic acids. In some cases, such as undec-10-enamide, the double bond is unaffected despite the presence of lead tetraacetate.

7. From Esters and Alcohols (Transesterification) ($\underline{1}$, 814)

$$\text{RCOOR'} + \text{R''OH} \underset{\text{OH}^\ominus}{\overset{\text{H}^\oplus}{\rightleftharpoons}} \text{RCOOR''} + \text{R'OH}$$

Pereyre and co-workers [21] carried out the transesterification of a series of saturated and unsaturated aliphatic esters with methyl and ethyl alcohol in the presence of a catalytic amount of tributyltin methoxide or ethoxide as indicated:

$$\underset{0.05\text{ mole}}{\text{RCOOR'}} + \underset{0.5\text{ mole}}{\text{R}^2\text{OH}} \xrightarrow{(\text{C}_4\text{H}_9)_3\text{SnOR}^3} \text{RCOOR}^2 + \text{R'OH}$$

The reaction mixture (0.05 mol ester:0.5 mol alcohol) was heated in a bath at 120° for 40 or 100 hr and the percentage of transesterification, as determined by vpc, varied within 28-72%.

Otsuji and co-workers [22] utilized carbon dioxide as the catalyst in the transesterification of saturated aliphatic esters with monohydric alcohols as shown:

$$\text{RCOOR'} + \text{R}^2\text{OH} \underset{}{\overset{\text{CO}_2}{\rightleftharpoons}} \text{RCOOR}^2 + \text{R'OH}$$

The alcohol was again employed in excess and yields as determined by vpc, with a single exception, varied from 0% to about 30% regardless of whether the CO_2 was simply bubbled through the mixture or applied under pressure. A cyclic mechanism involving R^2OCO_2H is no doubt applicable.

Transesterification of fatty acid esters and amino acid and peptide alkyl esters has been accomplished by the use of a strong anion-exchange resin in methanol or ethanol at room temperature

[23]. The resin was prepared from the chloride form of
BIO-RAD AG I-XS resin by treatment with NaOH.

Transesterification of aliphatic esters was accomplished by
using the mild catalyst, 2% KCN in 95% ethanol [24].

Roelofsen and co-workers [25] carried out ester-interchange
reactions using Union Carbide molecular sieves, types 3A, 4A, and
5A. The sieves were placed in a Soxhlet extractor and as the condensed vapor of the alcohol-ester mixture containing a small
amount of Na or K passed over them, the displaced alcohol was
selectively adsorbed. The conditions recommended for n-aliphatic
alcohols, for secondary and branched primary alcohols, and for
tertiary alcohols are given. Of the many esters prepared by this
procedure, detailed instructions are recorded for n-butyl
methacrylate (isolated yield, 65%), diisopropyl terephthalate
(isolated yield, 87%) and t-butyl benzoate (isolated yield, 83%)
(Ex. a).

a. Preparation of t-Butyl Benzoate [25]

A mixture of methyl benzoate, 20 g, in a solution of 1.2 g
of potassium in 160 g of t-butyl alcohol was refluxed for 70 hr by
heating to 100° while the condensate was passed over 60 g of 5A
pellets in a Soxhlet extractor. The reaction mixture with hexane
washings from the sieve was distilled, the final distillate being
21.7 g (83%) of t-butyl benzoate.

10. From Salts with Alkyl Halides (1, 817)

$$RCO_2\overset{\ominus}{N}\overset{\oplus}{a} + R'X \longrightarrow RCO_2R' + NaX$$

Saegusa and co-workers [26] synthesized a series of carboxylic acid esters by treating the acid with the alkyl halide in
the presence of cuprous oxide and cyclohexyl isocyanide in benzene. It is assumed that cuprous carboxylate is an intermediate.
The reaction is successful with straight and branched-chain
aliphatic, unsaturated aliphatic, and aromatic acids. Branched
halides were less satisfactory than straight-chain ones. In
fact, the esterification with tert-butyl bromide was unsuccessful.
The method may be illustrated by the preparation of butyl phenylacetate (Ex. a).

Lewin and Goldberg earlier studied the cuprous carboxylate-
alkyl halide reaction [27]. These investigators carried out the
reaction in pyridine under an inert atmosphere and anhydrous conditions. Primary, secondary, tertiary aliphatic, allylic,
vinylic, or aryl halides may be employed. There is no rearrangement when neopentyl or neophyl bromides are used and inversion

occurs when optically active neopentyl tosylate-1-d^4 reacts with cuprous benzoate.

Methyl esters were prepared by Mehta [28] by the use of methyl iodide-dimethyl sulfoxide-calcium oxide-calcium sulfate at room temperature. Yields with a series of monocarboxylic and dicarboxylic acids varied from 79-93%. The method is illustrated in Ex. b.

One of the most satisfactory methods of esterification is that of Pfeffer and co-workers [29] involving the use of the alkyl halide with HMPA under alkaline conditions. These investigators showed that the method is usually superior in methylation over the $BF_3 \cdot Et_2O - CH_3OH$ and $(CH_3)_2SO_4 - CH_3OH$ methods. They recommend it particularly for branched-chain aliphatic and mesitoic acids. Methyldipropylacetic acid gave a 99% (glc) conversion of the methyl ester in 30 min, whereas mesitoic acid gave a 96% (glc) conversion of the methyl ester in 20 min. The method is applicable to the synthesis of esters other than methyl.

Shaw and co-workers [30] extended Pfeffer's method to a greater variety of acids. With alkyl iodides the reaction appears to be complete in 1 hr, but with alkyl bromides 20-24 hr at room temperature are necessary. These investigators obtained a 99% absolute yield (glpc) of the methyl ester of mesitoic acid (Ex. c) and 96% absolute yield (glpc) of the methyl ester of a straight-chain aliphatic acid such as octanoic. One experiment with tertiary butyl bromide gave only a trace of the ester.

A second satisfactory method for forming esters from salts is that of Wagenknecht and co-workers [31]. These investigators showed that the use of the quaternary ammonium rather than the metal or tertiary ammonium ion increased the rate dramatically. Thus tetraethylammonium acetate with benzyl chloride gave the ester as indicated (Ex. d).

$$CH_3COO^{\ominus} \overset{\oplus}{N}(CH_3)_4 + C_6H_5CH_2Cl \xrightarrow[25°,1\ hr.]{DMF} CH_3COOCH_2C_6H_5$$
$$100\%$$

Esters of other acids were prepared in yields varying within 60-100%.

In some cases it is not possible to esterify amino acids directly. Maclaren [32] has devised a three-step process, which requires no isolation of intermediates, to accomplish esterification.

$$NH_2CH_2COOK \xrightarrow{CH_3COCH_2CO_2Et} \underset{\underset{CHCOOEt}{\overset{\|}{C}}}{\overset{CH_3}{\overset{|}{C}}}-NHCH_2COOK \xrightarrow{C_6H_5CH_2Br}$$

$$\underset{\underset{CHCOOEt}{\overset{\|}{C}}}{\overset{CH_3}{\overset{|}{C}}}-NHCH_2COOCH_2C_6H_5 \xrightarrow[\text{MeOH}]{HCl} HCl\ NH_2CH_2CO_2CH_2C_6H_5$$

Thus the condensation product with acetoacetic ester blocks the amino group and in the last step the amino ester is obtained. The method, in which yields cover the 59-98% range, is more convenient than former methods for the preparation of 4-methoxybenzyl and 2,4,6-trimethylbenzyl esters.

a. Preparation of Butyl Phenylacetate [26]

Under nitrogen a mixture of phenylacetic acid and butyl bromide, each 20 mmol, Cu_2O, 10 mmol, and cyclohexyl isocyanide, 30 mmol, in 15 ml of benzene was heated at 80° with stirring for 17 hr. The Cu(I) complex was precipitated by the addition of 20 ml of petroleum ether and distillation of the organic layer gave the ester (86%).

b. Preparation of Dimethyl Adipate [28]

Adipic acid, 0.5 mmol, in dry DMSO, 5 ml, and CH_3I, 5 mol, were mixed followed by the addition of anhydrous CaO, 2 g, and drierite (ca. 1 g). After stirring for 8 hr at room temperature, the reaction mixture was filtered and water was added to the filtrate from which 790 mg (91%) of the ester was recovered in the usual manner.

c. Preparation of Methyl Mesitoate [30]

To a solution of mesitoic acid, 10 mmol, in 25 ml of HMPA in a separatory funnel was added 20 mmol of NaOH (as a 25% aqueous solution). After shaking for 5 min, 40 mmol of CH_3I were added, and the solution was shaken for another 5 min. Then 50 ml of 5% HCl in water was added and the solution was extracted with ether, from which extract a 99% yield (glpc) of the ester was recovered in the customary manner.

d. Preparation of Benzyl Acetate [31]

Tetraethylammonium acetate was preformed from the acid and

aqueous or methanolic quaternary tetraethylammonium hydroxide, generally mixed to pH 7. After stripping the solvent from the salt (generally containing a few moles of water), it was dissolved or suspended in DMF (or DMSO) and a stoichiometric amount of the benzyl chloride was added with stirring at 25°. The ester was removed from the salt by extraction to give a yield of 100%.

    11.   From Salts or Acids and Other Alkylating Agents (<u>1</u>, 818)

$$RCOONa \diagup \begin{array}{l} \xrightarrow{(CH_3)_2SO_4} RCOOCH_3 + NaOSO_2OCH_3 \\ \xrightarrow{CH_3OSOCl} RCOOCH_3 + SO_2 + NaCl \end{array}$$

    As is well known, carboxylic acids may be esterified by reagents other than alcohols. Thus Kantlehner and Funke [33] employed the complex formed between DMF and dimethyl or diethyl sulfate as indicated:

$$RCOOH + HC \overset{\oplus}{\underset{N(CH_3)_2}{\overset{OR'}{|}}} R'OS\overset{\ominus}{O_3} \longrightarrow RCOOR' + R'OSO_3H + HCON(CH_3)_2$$

$$R' = CH_3, C_2H_5$$

By heating the acid, 0.25 mol, with an equivalent quantity of the complex for 20 min at 150°, yields mostly in the 90% range were obtained with alphatic (mono- and dicarboxylic) and aromatic acids.
    Vowinkel [34] utilized the O-alkyl-N,N'-dicyclohexylisourea as the esterification reagent as shown:

$$RCOOH + C_6H_{11}NHC\underset{OR'}{\overset{|}{=}}NC_6H_{11} \longrightarrow RCOOR' + \underset{C_6H_{11}NH}{\overset{C_6H_{11}NH}{>}}C=O$$

By heating the carboxylic acid with an equivalent amount of the isourea in a suitable solvent at 60-80° for a few hours, yields usually over 90% were obtained.
    Derevitskaya and co-workers [35] employed an alkyl t-butyl ether as the esterification reagent as shown:

$$RCOOH + R'O \text{-} C_4H_9 \xrightarrow[p\text{-}CH_3C_6H_4SO_3H]{H_2SO_4 \text{ or}} RCOOR' + CH_2=C(CH_3)_2 + H_2O$$

In this reaction 0.1 mol of the ether is mixed with 0.12-0.15 mol of the acid and 0.02-0.05 ml of H_2SO_4 (or 0.05-0.1 g of p-toluenesulfonic acid) and the mixture is heated under reflux until the evolution of isobutylene ceases. Yields vary within 53-94%.

Koganty and co-workers [36] utilized the complex formed between phenyl chloroformate and DMF in the esterification of acetic acid as shown:

$$CH_3COOH + (CH_3)_2\overset{\oplus}{N}=CHOCO_2C_6H_5 \overset{Cl^{\ominus}}{\longrightarrow} CH_3CO_2C_6H_5 + CO_2 + DMF + HCl$$
75%

The ester is obtained by simply adding the complex to the acid at 0°.

Raber and Gariano [37] obtained excellent results in the esterification of hindered acids with triethyloxonium fluoroborate by generating the amine carboxylate salt in situ as indicated:

$$RCOOH + Et_3\overset{\oplus}{O} \ \overset{\ominus}{BF_4} \xrightarrow{EtN(i\text{-}pr)_2} RCOOEt$$
81-95%

The steric effect is minimized since addition of the oxonium salt to the carbonyl group of the acid is not involved; rather displacement of an alkyl group from the salt by the anionic oxygen of the carboxyl group occurs. The method which functions at room temperature in an essentially neutral medium is applicable not only to hindered acids like mesitoic acid (Ex. a) but to unhindered ones as well.

Grundy and co-workers [38] utilized dimethyl sulfate-aqueous NaOH in dioxane or dimethyl sulfate-acetone-potassium carbonate for the esterification of hindered acids such as 2,6-dimethyl-4-methoxybenzoic acid as indicated:

Equally high yields were obtained with mesitoic and triphenyl-acetic acids.

The value of triethyloxonium fluoroborate in the esterification of acyl amino acids and peptides in aqueous sodium bicarbonate at room temperature was demonstrated previously by Yonemitsu and co-workers [39]. Yields of the simpler types ranged within 58-98%.

a.   Preparation of Ethyl Mesitoate [37]

To a solution of 5.7 mol of mesitoic acid and 6.3 mol of triethyloxonium fluoroborate in 75 ml of dichloromethane was

added 5.7 mol of diisopropylethylamine, after which the mixture
was allowed to stand for 24 hr at 25°. Extraction with 3 parts
of N HCl, 3 parts of normal KHCO₃, and 1 part of saturated aqueous
NaCl led to a dichloromethane solution that was dried over Na₂SO₄
and evaporated at reduced pressure to give 1.04 g of crude
product. Bulb-to-bulb distillation yielded 0.99 g (90%) of the
ester.

14.   From Amides ($\underline{1}$, 819)

$$ArCONH_2 \xrightarrow[CH_3OH]{BF_3} ArCOOCH_3$$

Methyl esters have been produced from amides by heating with
methanolic boron trifluoride in sealed tubes for 3 or 16 hr [40].
Thus benzamide, benzanilide, and 2,2-dichloropropionamide gave
quantitative yields of the methyl ester at 105° in 3 hr.

18.   From 2-Oxazolines, Dihydro-1,3-oxazines, and Orthothioformates (Masked Carboxyl Groups)

These methods, applicable to the synthesis of both acids and
esters, are covered in Chapter 13, A.10.

19.   From Acyl Cyanides

$$ROH + C_6H_5COCN \xrightarrow{(C_2H_5)_3N} ROCOC_6H_5 + HCN$$

Holý and Soucek [41] benzoylated a series of nucleosides and
nucleotides by the use of benzoyl cyanide in the presence of a
catalytic amount of a base such as triethylamine or tributylamine. Thus the nucleoside thymidine, 5 mmol, and benzoyl
cyanide, 11 mmol, were stirred in 10 ml of acetonitrile and on
adding tributylamine, 50 μl, rapid solution occurred and after
stirring for 5 min at 25°, 3',5' di-O-benzoylthymidine
crystallized.

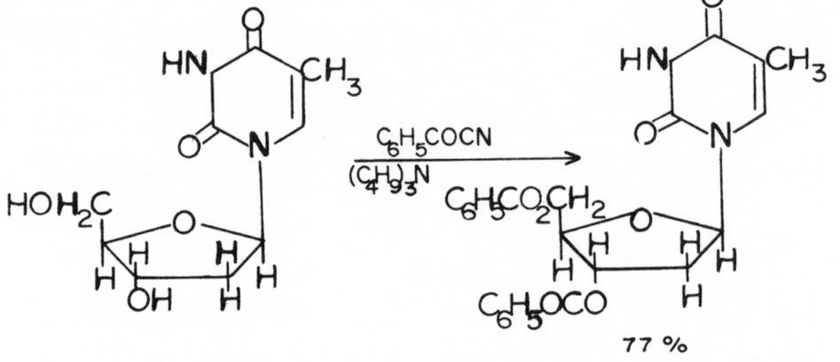

77 %

It should be emphasized that hydrogen cyanide is formed in the reaction. The formation of keto esters from ketocyanides has been given in 1, 813.

20. From Sulfur Compounds

Ogura and Tsuchihashi [42] synthesized phenyl acetates from aldehydes via the 1-methylsulfinyl-1-methylthio-2-arylethylenes as indicated:

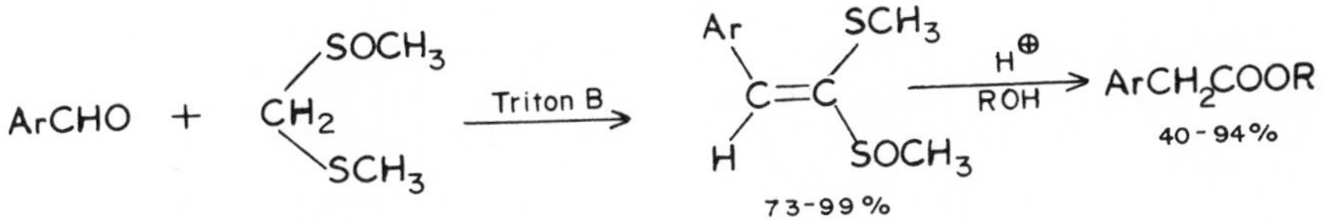

The method is a simple and efficient one for preparing acids such as homopiperonylic and homoveratric, which are key intermediates in the synthesis of isoquinoline alkaloids.

Corey and Beames [43] have sought to protect lactones from reagents that affect other parts of the molecule. It was found that the carbonyl group of the lactone could be protected by conversion into A as follows:

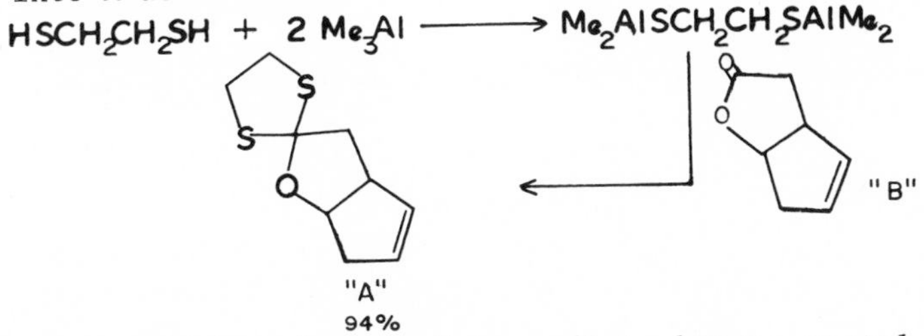

Structure A, available from an intermediate ketene acetal, survived attack by dilute acetic acid, 1.5 equiv KOH in aqueous CH_3OH, or $LiAlH_4$, but could be hydrolyzed readily to B by the reagents as shown:

$$A \xrightarrow[\text{aq. THF}]{\text{HgO-BF}_3\text{etherate}} B$$

21. From Ditosylamines and Salts

In a search for additional methods for replacing C-N bonds with C-O bonds, Andersen and Uh [44] converted the primary amine into the ditosylamine, which in turn was transformed into the acetate as shown:

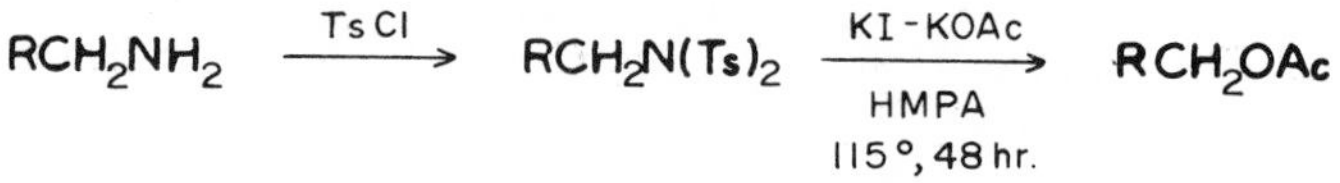

$$RCH_2NH_2 \xrightarrow{\text{Ts Cl}} RCH_2N(Ts)_2 \xrightarrow[\substack{HMPA \\ 115°, 48\,hr.}]{\text{KI-KOAc}} RCH_2OAc$$

Yields obtained are high except for hindered primary amines.

22.   From Acid Derivatives Containing Sulfur

$$RC(SR)_2SMe \longrightarrow RCO_2Et$$

This hydrolysis is conducted by refluxing in aqueous ethanol with a mercuric salt for 4-8 hr [45]. Since the trithio-ester is prepared from a dithioacetal, the method yields essentially an ester from aldehyde:

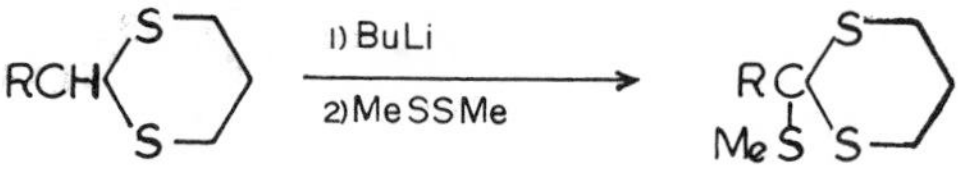

23.   From Cyclobutanones (see Addenda)

Addenda

A.1.   In what are apparently new conditions, t-butyl esters may be prepared in about 90% yields (32 ex.) from the acid, isobutyl-ene, and t-butyl alcohol with H_2SO_4 [S. Pavlov et al., Bull. Soc. Chim. France, 2985 (1974)].

A.2.   Mesylates of alcohols may be prepared from methanesulfonyl chloride and triethylamine via $CH_2=SO_2$ [R. K. Crossland, K. L. Servis, J. Org. Chem., 35, 3195 (1970)]. The carboxylic esters of 1°, 2°, and 3° alcohols may be prepared from the acid chloride and alcohol in the presence of HMPA [J. F. Normant et al., Compt. Rend., 269, C, 1325 (1969)].

A.3.   1-Acyloxy-6-chlorobenzotriazole behaves like an anhydride in that with alcohols it gives good yields of esters [M. Itoh et al., Synthesis, 456 (1975)].

A.4.   A series of t-butyl carbamates (usually 70-90%) were pre-pared by the lead tetraacetate oxidation of the amide [H. E. Baumgarten et al., J. Org. Chem., 40, 3554 (1975)].

A.10.   If potassium acetate and 2-chloro-2-methylcyclohexanone are heated in CH_3CN with a catalytic amount of 18-crown-6, 54% of 3-methyl-2-oxocyclohexyl acetate and 25% of the 1-methyl isomer are obtained [C. L. Liotta et al., Tetrahedron Letters, 2417 (1974)].

A.15.   Chloromesitylene may be trichloromethylated by carbon tetrachloride and AlCl$_3$ and subsequently solvolyzed to the methyl ester by methanol [H. Hart, J. F. Janssen, J. Org. Chem., 35, 3637 (1970)].

A.18.   Epoxides and lithium 2,4,4-trimethyl-2-oxazoline of the Meyers type react to give γ-butyrolactones on hydrolysis [A. I. Meyers et al., J. Org. Chem., 39, 2783 (1974)].

A.22.   Large ring lactones are formed very slowly from ω-hydroxyacids in the presence of 2,2^1-dipyridyl disulfide and triphenylphosphine [E. J. Corey, K. C. Nicolaou, J. Am. Chem. Soc., 96, 5614 (1974)].

A.23.   Cyclobutanones, except the unsubstituted, are cleaved by boiling methanol containing sodium methoxide to form substituted methyl butyrates [B. M. Trost et al., J. Am. Chem. Soc., 97, 2218 (1975)].

References

1. S. Patai, The Chemistry of Carboxylic Acids and Esters, Interscience, New York, 1969.
2. C. A. Grob et al., Helv. Chim. Acta, 55, 2439 (1972).
3. J. L. Marshall et al., Tetrahedron Letters, 4011 (1970).
4. P. K. Kadaba, Synthesis, 316 (1971).
5. J. F. Normant, H. Deshayes, Bull. Soc. Chim. France, 2854 (1972).
6. F. I. Luknitskii, Doklady Akad. Nauk SSSR Engl. Ed., 185, 198 (1969).
7. D. C. Neckers et al., Tetrahedron Letters, 1823 (1973).
8. W. W. Lowrance, Jr., Tetrahedron Letters, 3453 (1971).
9. N. Yamazaki, F. Higashi, Tetrahedron Letters, 5047 (1972).
10. G. F. Vesley, V. I. Stenberg, J. Org. Chem., 36, 2548 (1971).
11. M. Coopersmith et al., Ind. Eng. Chem. Prod. Res. Develop., 5, 46 (1966).
12. M. M. Harris, P. K. Patel, Chem. Ind. (London), 1002 (1973).
13. S. A. Narang et al., Chem. Commun., 325 (1974).
14. W. E. Truce, P. S. Bailey, Jr., J. Org. Chem., 34, 1341 (1969).
15. E. M. Kaiser, R. A. Woodruff, J. Org. Chem., 35, 1198 (1970).
16. R. Kyburz et al., Helv. Chim. Acta, 54, 1037 (1971).
17. H. Alper, C.-C. Huang, J. Org. Chem., 38, 64 (1973).

18. (a) W. Steglich, G. Höfle, Angew. Chem. Intern. Ed. Engl.,
    8, 981 (1969); (b) Synthesis, 619 (1972).
19. A. J. Fatiadi, J. Chem. Eng. Data, 14, 118 (1969).
20. A. L. J. Beckwith et al., Australian J. Chem., 21, 197
    (1968).
21. M. Pereyre et al., Bull. Soc. Chim. France, 262 (1969).
22. Y. Otsuji et al., Bull. Chem. Soc. Jap., 44, 852 (1971).
23. W. Pereira et al., J. Org. Chem., 34, 2032 (1969).
24. K. Mori et al., Synthesis, 790 (1973).
25. D. P. Roelofsen et al., Rec. Trav. Chim., 89, 193 (1970);
    Chem. Ind. (London), 1622 (1966).
26. T. Saegusa et al., Syn. Comm., 2, 1 (1972); J. Org. Chem.,
    38, 1753 (1973).
27. A. H. Lewin, N. L. Goldberg, Tetrahedron Letters, 491 (1972)
    and previous papers.
28. G. Mehta, Synthesis, 262 (1972).
29. P. E. Pfeffer et al., Tetrahedron Letters, 4063 (1972).
30. J. E. Shaw et al., Tetrahedron Letters, 689 (1973).
31. J. H. Wagenknecht et al., Syn. Comm., 2, 215 (1972).
32. J. A. Maclaren, Australian J. Chem., 25, 1293 (1972).
33. W. Kantlehner, B. Funke, Chem. Ber., 104, 3711 (1971).
34. E. Vowinkel, Chem. Ber., 100, 16 (1967).
35. V. A. Derevitskaya et al., Tetrahedron Letters, 4269 (1970).
36. R. R. Koganty et al., Tetrahedron Letters, 4511 (1973).
37. D. J. Raber and P. Gariano, Tetrahedron Letters, 4741
    (1971).
38. J. Grundy et al., Tetrahedron Letters, 757 (1972).
39. O. Yonemitsu et al., Tetrahedron Letters, 1819 (1969).
40. D. J. Hamilton, M. J. Price, Chem. Commun., 414 (1969).
41. A. Holý and M. Souček, Tetrahedron Letters, 185 (1971).
42. K. Ogura, G. Tsuchihashi, Tetrahedron Letters, 1383 (1972).
43. E. J. Corey, D. J. Beames, J. Am. Chem. Soc., 95, 5829
    (1973).
44. N. H. Andersen, H. Uh, Syn. Commun., 2, 297 (1972).
45. R. A. Ellison et al., J. Org. Chem., 37, 2757 (1972).

B.   Electrophilic-type Syntheses

1. From Amines (1, 826)

A second method for the conversion of amines into esters has
been devised by White [1]. The steps involved are as follows:

$$C_6H_5CH_2CH_2NH_2 \xrightarrow[C_5H_5N]{C_6H_5COCl} C_6H_5CH_2CH_2NHCOC_6H_5 \xrightarrow[CH_3COOH]{N_2O_4, CH_3CO_2Na}$$

89-98%

$$\underset{\underset{64\%}{NO}}{C_6H_5CH_2CH_2NCOC_6H_5} \xrightarrow[\underset{CCl_4}{Na_2CO_3}]{\triangle} C_6H_5CH_2CH_2OCOC_6H_5$$

56-59% (last step)

The method is suitable for amides of primary carbinamines; for amides of secondary carbinamines the procedure of White and Aufdermarsh [2] for a trimethylacetamide is recommended.   For nitrosoamides of tertiary carbinamines the method of White and Stuber [3] is suggested.

From the starting amine the preceding reaction generates a diazoalkane which esterifies the acid used to make the amide.

2.   Alkylation of Acids by Diazoalkanes and of Olefins by Diazoesters (1, 826)

$$RCO_2H \; + \; CH_2N_2 \longrightarrow RCO_2CH_3 + N_2$$

Two alternatives to the dangerous but effective method of esterification have been proposed.   One is the in situ preparation of diazomethane followed by its reaction with the acid [4]:

$$O_2N\langle\!\!\!\!\!\!\bigcirc\!\!\!\!\!\!\rangle CO_2H + Et_3N + Me\overset{\overset{NO}{|}}{N}CONH_2 \xrightarrow[\substack{0° \text{ for I hr.} \\ 25° \text{ for 6 hr.}}]{Glyme} O_2N\langle\!\!\!\!\!\!\bigcirc\!\!\!\!\!\!\rangle CO_2Me + N_2$$

Methyl p-nitrobenzoate
quantitative

The method may be modified by adding aqueous KOH dropwise to the acid and nitrosourea at low temperature.

The second alternative involves the use of 1-methyl-3-p-tolyltriazine as the alkylation source [5].   This compound, which is crystalline, stable, and easy to prepare, offers advantages over diazomethane and other diazoalkanes.   With 3,5-dinitrobenzoic acid it reacts as shown (Ex. a):

$$\underset{\underset{NO_2}{\displaystyle|}}{\overset{\overset{NO_2}{\displaystyle|}}{\langle\!\!\!\!\!\!\bigcirc\!\!\!\!\!\!\rangle}}COOH + p\text{-}CH_3C_6H_4N{=}NNHCH_3 \longrightarrow$$

$$\underset{\underset{NO_2}{\displaystyle|}}{\overset{\overset{NO_2}{\displaystyle|}}{\langle\!\!\!\!\!\!\bigcirc\!\!\!\!\!\!\rangle}}COOCH_3 + N_2 + p\text{-}CH_3C_6H_4NH_2$$

Methyl 3,5-dinitrobenzoate
70-90%

a.   Preparation of Methyl 3,5-Dinitrobenzoate [5]

To 7 mmol of 1-methyl-3-p-tolyltriazene in 10 ml of ether,
7.1 mmol of 3,5-dinitrobenzoic acid in 25 ml of ether was added
slowly with occasional swirling of the reaction mixture.   After
N_2 evolution ceased (about 1 hr), the ethereal solution was
washed with 5 N HCl and then with 5% aqueous Na_2CO_3.   Drying was
followed by evaporation to give 1.11-1.42 g (70-90%) of the
ester.

   3.   Rearrangement of Diazoketones (Arndt-Eistert and Wolff
        Rearrangement) (1, 827)

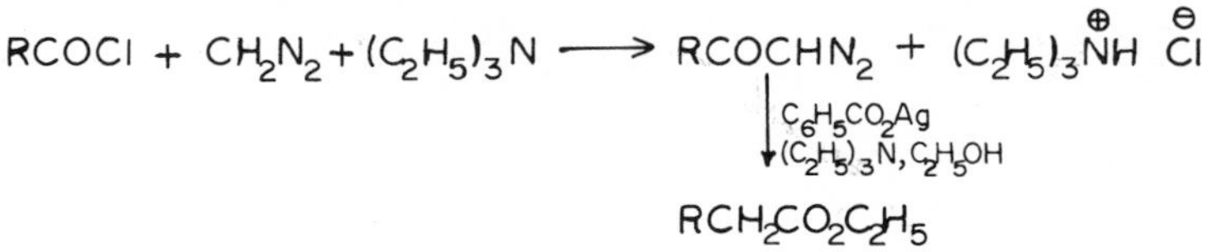

$$RCOCl + CH_2N_2 + (C_2H_5)_3N \longrightarrow RCOCHN_2 + (C_2H_5)_3\overset{\oplus}{N}H \ \overset{\ominus}{Cl}$$

$$\downarrow \begin{matrix} C_6H_5CO_2Ag \\ (C_2H_5)_3N, C_2H_5OH \end{matrix}$$

$$RCH_2CO_2C_2H_5$$

The Newman-Beal modification of the Arndt-Eistert reaction
(1, 828) as shown above has now been utilized [6] in the synthesis of ethyl 1-naphthylacetate.   The method is more reproducible
than the original Arndt-Eistert reaction and permits the
rearrangement to be carried out on larger scale runs.   The use of
triethylamine in the first step permits the use of only one
equivalent of diazomethane.

   4.   Oxidation of Carbonyl Compounds with a Peracid (Baeyer-
        Villiger) (1, 828)

$$\textbf{RCOR} \xrightarrow{\ CH_3CO_2OH\ } \textbf{ROCOR}$$

The synthesis of m-chloroperbenzoic acid, commonly used in
the Baeyer-Villiger synthesis, has now appeared in Organic
Synthesis [7].
   Kawaba and co-workers [8] have prepared p-methoxycarbonyl-
perbenzoic acid, which is a stable and convenient oxidizing agent
suitable for the Baeyer-Villiger oxidation.   It may be prepared
in 80-95% yields by the irradiative oxidation of methyl p-
formylbenzoate in carbon tetrachloride with oxygen.   This reagent,
for example, converts acetophenone into phenyl acetate as shown:

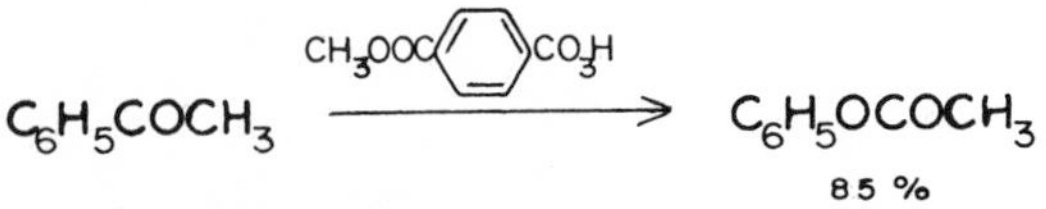

$$C_6H_5COCH_3 \xrightarrow{CH_3OOC-\langle\rangle-CO_3H} C_6H_5OCOCH_3$$

85 %

Soucy and co-workers [9a] oxidized adamantanol in acetonitrile
with ceric ammonium nitrate and obtained the lactone as shown:

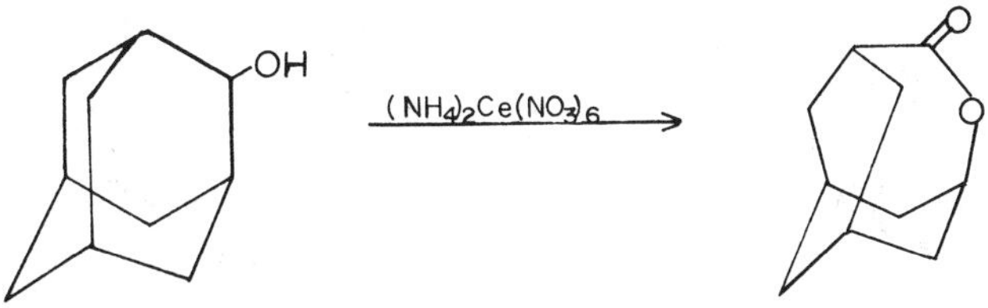

50 %

Ceric ammonium nitrate has also been used to oxidize tetra-cyclone [9b] to the lactone:

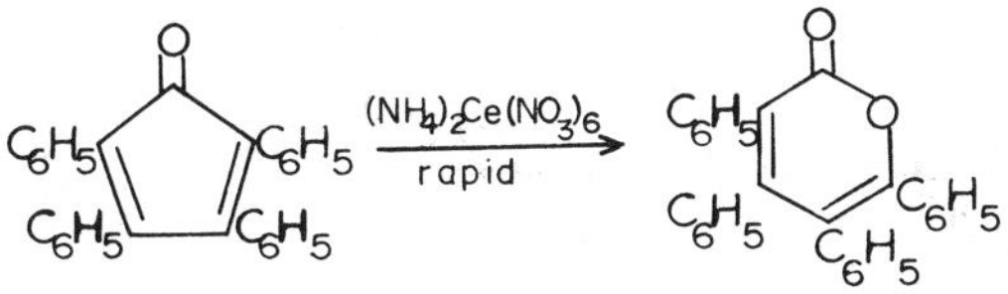

Tetraphenyl-2-pyrone, 77 %

Sodium hypobromite works well in the oxidation of a spiro-butanone [10]:

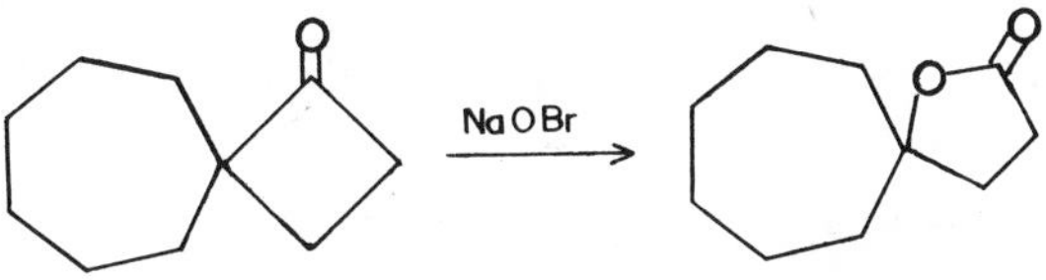

2-Oxa-spiro[4.6]undecan-3-one

94 %

Winnik and Stoute [11] have examined the steric effects in the Baeyer-Villiger reaction of simple ketones. The extent of migration increases, as a rule, with the bulk of the alkyl group. The largest effect, a factor of 13, was found in neopentyl to ethyl migration in 5,5-dimethyl-3-hexanone.

5.   Addition of Carboxylic Acids or Esters to Alkenes, Alkynes or Ethers (1, 830)

Magoon and Slaugh [12] have added acetic acid to bicyclo[2.2.1]hepta-2,5-diene to obtain the acetate as shown:

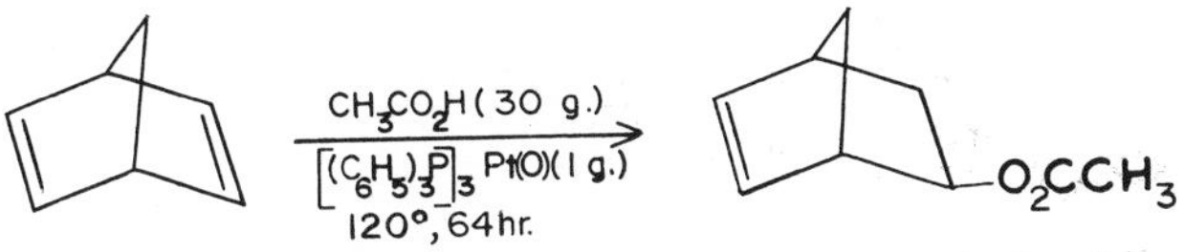

exo-5-Acetoxybicyclo[2.2.1]hept-2-ene, 95 % by analysis

Strong acid added to acetic acid gives the same product as is obtained with the platinum catalyst. However, the modes of addition are quite different. The platinum catalyst with CH_3CO_2D gives the syn-7-deutero compound exclusively while strong acid catalyst gives products with deuterium at various positions. Thus the platinum catalyst provides a unique, stereoselective addition.

Acetic acid has been added to acetylenes, both open-chain and cyclic, by irradiation to give cis- and trans-vinyl acetates [13]. This addition, rather than that yielding ketones, may be enhanced by using acetic anhydride rather than acetic acid, both catalyzed by a mixture of mercuric acetate and boron trifluoride etherate [14].

Nikishin and co-workers [15] have shown that the addition of catalytic amounts of copper(II) acetate completely alters the reaction of diethyl malonate with manganese(III) acetate dihydrate and 1-heptene. In the absence of the catalyst n-heptyl malonate is produced, but in its presence the reaction proceeds as shown to give diethyl hept-2-enylmalonate.

$$CH_2(COOC_2H_5)_2 + (CH_3COO)_3Mn \cdot H_2O + CH_3(CH_2)_5CH{=}CH_2 \xrightarrow{(CH_3COO)_2Cu}$$

$$CH_3(CH_2)_3CH{=}CHCH_2\underset{\underset{\displaystyle C}{}}{\overset{\displaystyle H}{\underset{}{}}}\overset{\displaystyle COOC_2H_5}{\underset{\displaystyle COOC_2H_5}{}}$$

MOH / 40-70% \ $H_3O^{\oplus}$

$$CH_3(CH_2)_3CH{=}CHCH_2CH_2COOM$$

$CH_3(CH_2)_3CH_2$

The latter in acidic solution yields the lactone of γ-hydroxy-nonanoic acid, while in basic solution the salt of a γ,δ-unsaturated acid is produced.

To obtain the vinyl acetate rather than the alkyl acetate, the addition has been carried out with phenylmercuric acetate in the presence of palladium acetate [16]:

$$C_6H_5HgO\overset{O}{\overset{\|}{C}}CH_3 + CH_2{=}CHOCOCH_3 \xrightarrow[CH_3CN]{Pd(OAc)_2}$$

$$\left[C_6H_5CH_2\underset{\underset{\displaystyle OCOCH_3}{}}{CHOCOCH_3}\right] \xrightarrow{-CH_3CO_2H} C_6H_5CH{=}CHOCOCH_3$$

Styryl acetate, cis, 17%; trans, 74%

6. Carbalkoxylation with Carbon Monoxide and a Base ($\underline{1}$, 831)

This topic is almost interchangeable with that utilized in the formation of acids (Chapter 13, F.3). For a review of hydroalkoxycarboxylation see [16a]. Carbalkoxylation has been accomplished by Corey and Hegedus [17] by treating organic halides with nickel carbonyl in a protic medium as shown:

$$RX \xrightarrow[ROH-RO^\ominus]{Ni(CO)_4} RCOOR$$

In this manner a series of methyl and t-butyl esters has been prepared with yields varying within 50-96%. Nickel carbonyl and potassium t-butoxide in t-butyl alcohol is a much more powerful carboxylating system than the methanol-methoxide-nickel carbonyl reagent. The former t-butoxycarboxylates both alkyl iodides and vinyl halides, while the latter is not effective on any alkyl halides.

Esters have also been prepared from alkyl halides by Collman and co-workers [18] by the use of sodium tetracarbonylfer-rate(-II), $Na_2Fe(CO)_4$, which forms alkyltetracarbonyl iron(0) complexes. The latter in the presence of CO forms anionic acyl complexes which in the presence of a halogen and alcohol give esters. The overall reaction may be represented as follows:

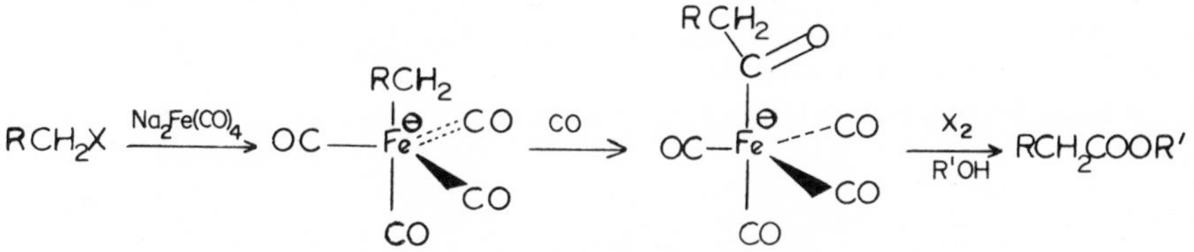

Primary aliphatic halides give isolated yields around 80%. Secondary substrates are less satisfactory.

In a related manner (1, 832) esters may be prepared from alkadienes, carbon monoxide and an alcohol in the presence of palladium-phosphine complexes [19]. Thus from butadiene and carbon monoxide in the presence of a catalytic amount of palladium chloride ethyl 3-pentenoate is obtained as shown:

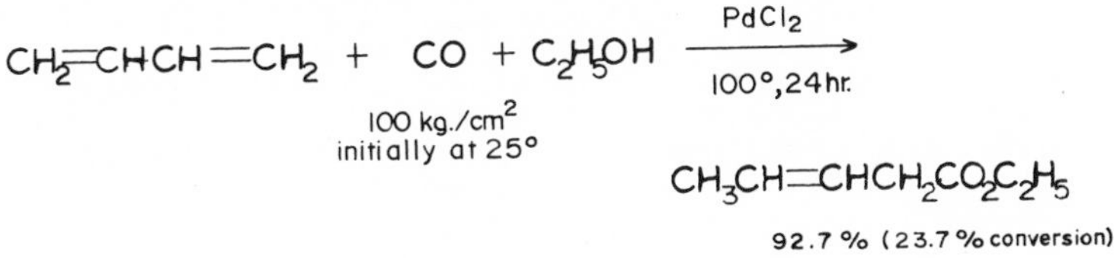

However, if a palladium complex free of the halide is employed, the reaction takes a different course. Thus if butadiene and carbon monoxide in ethanol are treated with palladium(II)

acetate-butylphosphine, phosphine and a base, ethyl 3,8-nonadienoate is obtained as indicated:

$$2\ CH_2{=}CHCH{=}CH_2\ +\ \underset{100\ atm.}{CO}\ +\ \underset{30\ ml.}{C_2H_5OH}\ \xrightarrow[\underset{110°,16\ hr.}{PH_3-OH^{\ominus}}]{Pd(OAc)_2-(C_4H_9)_3P-}$$

20 g.

$$CH_2{=}CH(CH_2)_3CH{=}CHCH_2CO_2C_2H_5$$

31.4 g.

Copolymerization of acrylic esters and butadiene with Fe(acac)$_3$, AlEt$_3$, and the ligand, $(C_6H_5)_3Sb$, gives appreciable yields of diene esters [19a].

Chiusoli and co-workers [20] have developed a method for the carbonylation of allylic halides with the insertion of acetylene at room temperature and atmospheric pressure as shown (Chapter 13, F.3).

$$CH_2{=}CHCH_2Cl\ +\ HC{\equiv}CH\ +\ CO\ +\ MeOH\ \xrightarrow[\underset{NH_2}{S=C\diagdown\ -MgO}]{NiCl_2\cdot 6H_2O}$$

$$CH_2{=}CHCH_2CH{=}CHCO_2Me\ +\ HCl$$

75 %

a.   Preparation of Methyl 4-t-Butyl-1-Cyclohexenecarboxylate [17]

To a solution of NaOCH$_3$, 3.0 mmol, in 5 ml of dry methanol under argon nickel carbonyl (toxic), 6.0 mmol, followed by 1-bromo-4-t-butylcyclohexene, 1.0 mmol, was added and the mixture was heated to 60° and held there for 6 hr. After cooling to 25°, carbon monoxide was bubbled through the mixture for 0.5 hr and the green solution resulting was poured into 50 ml of 0.1 N HCl and 50 ml of ether, from which solution 0.14 g (71%) of the ester was recovered.

7.   Cleavage of Ethers (1, 832)

$$ROR\ \xrightarrow{R'COCl}\ R'COOR\ +\ RCl$$

The use of sulfonic-carboxylic anhydrides in the cleavage of ethers to produce esters has been investigated by Karger and Mazur [21]. The mixed anhydride is more effective than some other cleaving agents. Thus 1,4-dioxane gives an 87% yield of the acetyl tosyl diester of diethylene glycol when treated with acetyl p-toluenesulfonate as shown:

$$\text{(dioxane)} \xrightarrow[80°, 24\ hr.]{CH_3COOSO_2C_6H_4CH_3\text{-}p} CH_3COOCH_2CH_2OCH_2CH_2OSO_2C_6H_4CH_3\text{-}p$$

87%

By contrast, acetyl chloride and stannic chloride with the dioxane for 30 hr at 200° give only a 16% yield of 2-chloroethyl acetate. The new reagent also gives a higher specificity with unsymmetrical ethers.

Alper and Edward [22] have shown that iron pentacarbonyl is an effective catalyst in the cleavage of ethers with acid chlorides. Alkyl ethers are more effectively catalyzed by Lewis acids such as titanium, stannic, or zinc chloride, but at times it may be desirable to employ a slower but milder catalyst. As shown, the pentacarbonyl with butyl ether and the acid chloride gives butyl nonanoate.

$$C_4H_9OC_4H_9 \xrightarrow[\substack{Fe(CO)_5 \\ Reflux}]{C_8H_{17}COCl} C_8H_{17}COOC_4H_9 + C_4H_9Cl$$

75 %

Cyclic ethers were found to submit to ring opening by treatment with trifluoroacetic anhydride [23]. Thus tetrahydrofuran under autogenous conditions in an autoclave gives largely a mixture of esters in which the $O(CH_2)_4$ moiety = 1 and 2:

$$\text{(THF)} + (CF_3CO)_2O \xrightarrow[18\ hr.]{180°} CF_3\overset{O}{\overset{\|}{C}}\Big[O(CH_2)_4\Big]O\overset{O}{\overset{\|}{C}}CF_3 + CF_3\overset{O}{\overset{\|}{C}}\Big[O(CH_2)_4\Big]_2O\overset{O}{\overset{\|}{C}}CF_3$$

0.20mol.   0.20mol.   74% (based on ether)   17.5%(based on ether)

Chlorohydrin acetates have been produced from diols via cyclic orthoacetates [24].

$$CH_3C(OMe)_3 + RCHOHCHOHR \xrightarrow{H^\oplus} \text{(cyclic orthoacetate)} \xrightarrow{(C_6H_5)CCl} RCHClCHR\ (OCOCH_3)$$

38-93% (5 glycols)

8. Alkylation of Esters and Related Processes (1, 833)

$$R_3B + BrCH_2CO_2R' \longrightarrow RCH_2CO_2R' + R_2BBr$$

$$R^\oplus + \,>\!C\!=\!C\!\overset{OH}{\underset{OEt}{\big<}} \longrightarrow R\overset{|}{\underset{|}{C}}CO_2Et + H^\oplus$$

The organoborane process of alkylation has been extended by employing potassium 2,6-di-t-butylphenoxide as the base [25].

The reaction proceeds as with potassium t-butoxide but in this case it is not necessary to protect the product from excess base.

In the carbethoxymethylation of 1-methyl-cyclopentene Brown and co-workers [26] employed B-alkyl-9-borabicyclo [3.3.1] nonane (B-R-9-BBN) since tris (trans-2-methylcyclopentyl) borane failed to react under the usual conditions. The reaction with the cyclopentene is as indicated:

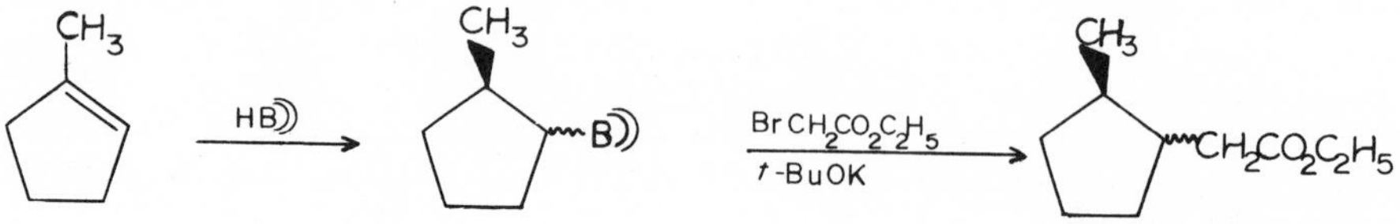

trans-Ethyl(2-methylcyclopentyl) acetate, 60%

To accomplish C-4 homologation Brown and Nambu [27] treated ethyl 4-bromocrotonate as indicated (Ex. a):

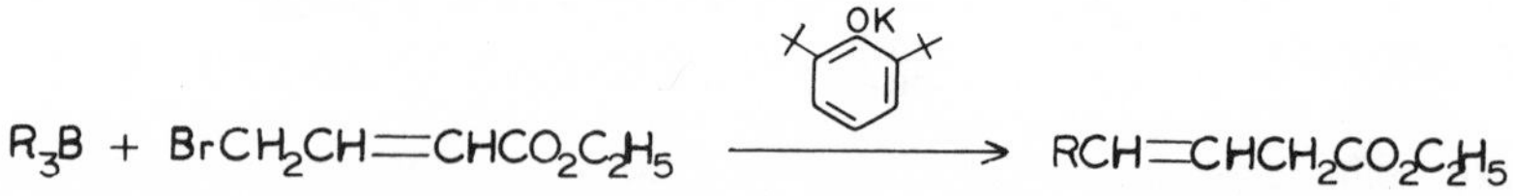

$R_3B + BrCH_2CH=CHCO_2C_2H_5 \longrightarrow RCH=CHCH_2CO_2C_2H_5$

72-89%

It will be noted that in the reaction the double bond migrates from the 2,3- to the 3,4-position.

Similar to the Brown alkylation process for esters is the reaction of R_3B and α-diazoesters (E.1).

The alkylation of Dieckmann esters has been accomplished without prior formation of the enolate by Barco and co-workers [28]. The simple method as indicated consists of heating the ester and alkylating agent with potassium carbonate in acetone (Ex. b).

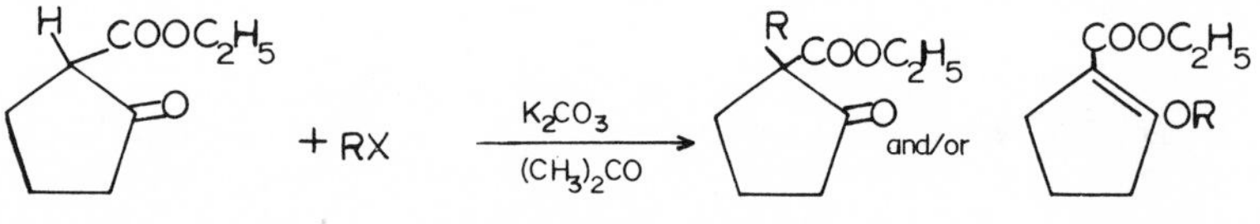

Exclusive C-alkylation (91-100%) occurred with CH_3I, $(H_3CO)_2SO_2$, and $CH_2=CHCH_2Br$; some O-alkylation resulted with i-C_3H_7I, $C_6H_5CH_2OCH_2Cl$, and $ClCOOC_2H_5$; in fact the latter gave 70% O-alkylation.

a. Preparation of Ethyl 3-Hexenoate [27]

Yield was 78% from ethyl 4-bromocrotonate added to a mixture of triethylborane and potassium 2,6-di-<u>t</u>-butylphenoxide in THF.

b. Preparation of Ethyl 1-Methyl-2-oxocyclopentanecarboxylate [28]

Methyl iodide (warning: toxic), 2 mol was added dropwise to a well-stirred suspension of ethyl 2-oxocyclopentanecarboxylate, 1 mol and K_2CO_3, 4 mol in acetone. After refluxing for 1 hr, the mixture was diluted with water, extracted with ether, and by distillation 100% of the alkylation product was recovered.

9. From Carbonyl Compounds and Terephthalic Acid

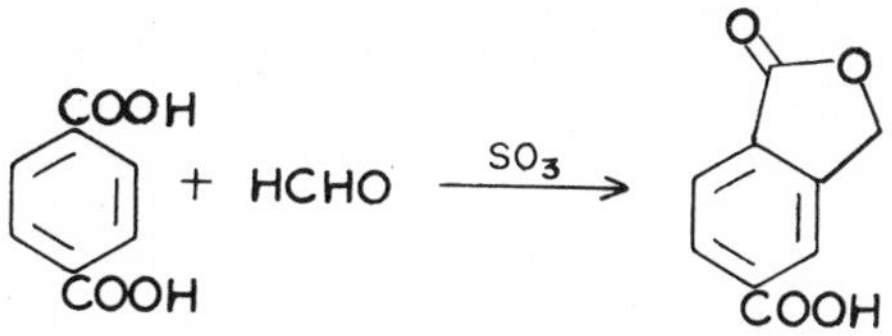

Forney has shown that terephthalic acid reacts with an electron-deficient species, such as formaldehyde, in the presence of sulfur trioxide to produce 5-carboxyphthalide (93%) [29]. Similar reactions occur with benzoic acids containing electron-withdrawing groups. Sulfur trioxide is essential in the reaction.

10. From Carbonyl Compounds and Ketenes or Ketene Acetals

Vuitel and Jacot-Guillarmod [30] synthesized β-hydroxyesters from aliphatic (acyclic, cyclic, and α,β-unsaturated), aromatic, and aliphaticaromatic ketones, ketene, and titanium alkoxides as indicated:

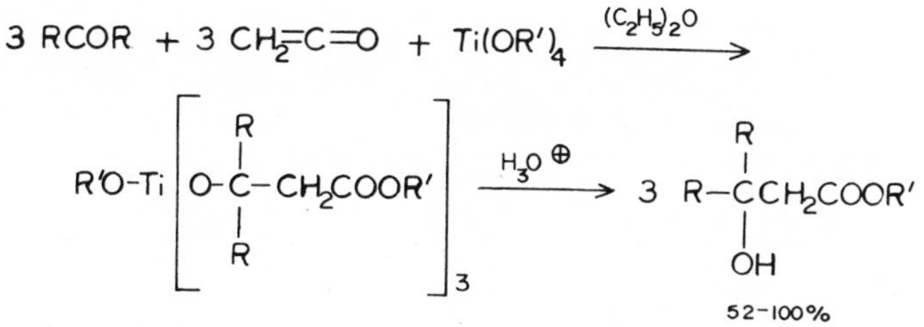

In the one-operation reaction the ketene (from diketene by pyrolysis) is introduced in a stream of nitrogen at room temperature. The product is the same as that obtained in the Reformatsky reaction.

Creger [31] carried out a similar reaction thermally using aromatic aldehydes and silylated ketene acetals as indicated:

$$ArCHO \;+\; \overset{Me}{\underset{R}{\diagdown}}C = C \overset{OSiMe_3}{\underset{OR'}{\diagup}} \quad \xrightarrow[\text{2) } H^{\oplus}]{\text{1) } 150°} \quad Ar\overset{OH}{\underset{CH_3}{\overset{|}{C}H}}\overset{R}{\underset{}{\overset{|}{C}}}COOR'$$

$$61 - 86°$$

11. From Arenes and Ethyl Haloacetate in the Presence of Metallic Halides

Izawa and co-workers [32] have shown that arenes may be photochemically ethoxycarbonylmethylated with ethyl haloacetate in the presence of metallic halides as indicated:

$$C_6H_6 \;+\; ClCH_2CO_2Et \quad \xrightarrow[h\nu]{SbCl_5} \quad C_6H_5CH_2CO_2Et$$

Ethyl phenylacetate, 24%

With other metallic halides the yields were lower.

12. From Cyclodienes and $CO-H_2O$ with $PdCl_2$ (see Addenda)

Addenda

B.3. The Wolff rearrangement of α-diazo carbonyl compounds has been reviewed [H. Meier, K.-P. Zeller, Angew. Chem. Intern. Ed. Engl., 14, 32 (1975)].

B.5. THF is acetylated by acetyl chloride in the presence of catalytic amounts of molybdenum hexacarbonyl to form 4-chloro-butyl acetate (78%) [H. Alper, C.-C. Huang, J. Org. Chem., 38, 64 (1973)].

B.12. 2-Hydroxycyclooct-5-enecarboxylic acid β-lactone was prepared from 1,5-cyclooctadiene, $PdCl_2$, CuCl, and CH_3COONa in water-acetone and carbon monoxide [J. K. Stille, D. E. James, J. Am. Chem. Soc., 97, 674 (1975)].

References

1. E. White, Org. Syn., Coll. Vol. 5, 336 (1973).
2. E. H. White, C. A. Aufdermarsh, Jr., J. Am. Chem. Soc., 83, 1174 (1961).
3. E. H. White, J. E. Stuber, J. Am. Chem. Soc., 85, 2168 (1963).

4.   S. M. Hecht, J. W. Kozarich, Tetrahedron Letters, 1397 (1973).
5.   E. H. White et al., Org. Syn., Coll. Vol. 5, 797 (1973).
6.   V. Lee, M. S. Newman, Org. Syn., 50, 77 (1970).
7.   R. N. McDonald et al., Org. Syn., 50, 15 (1970).
8.   N. Kawabe et al., J. Org. Chem., 37, 4210 (1972).
9.   (a) P. Soucy et al., Can. J. Chem., 50, 2047 (1972); (b) T.-L. Ho, Synthesis, 347 (1973).
10. B. M. Trost et al., Tetrahedron Letters, 923 (1973).
11. M. A. Winnik, V. Stoute, Can. J. Chem., 51, 2788 (1973).
12. E. F. Magoon, L. H. Slaugh, J. Organomet. Chem., 55, 409 (1973).
13. K. Fujita et al., Tetrahedron Letters, 3865 (1973).
14. P. F. Hudrlik, A. M. Hudrlik, J. Org. Chem., 38, 4254 (1973).
15. G. I. Nikishin et al., Chem. Commun., 693 (1973).
16. R. F. Heck, Organomet. Chem. Syn., 1, 455 (1972).
16a. J. Mathieu, J. Weill-Raynal, Formation of C-C Bonds, Georg Thieme, Stuttgart, 1973, p. 336.
17. E. J. Corey, L. S. Hegedus, J. Am. Chem. Soc., 91, 1233 (1969); M. Ryang, S. Tsutsumi, Synthesis, 55 (1971).
18. J. P. Collman et al., J. Am. Chem. Soc., 95, 249 (1973).
19. J. Tsuji et al., Tetrahedron, 27, 3821 (1971); 28, 3721 (1972).
19a. H. Singer et al., Synthesis, 265 (1971).
20. G. P. Chiusoli et al., J. Chem. Soc., C2889 (1968).
21. M. H. Karger, Y. Mazur, J. Am. Chem. Soc., 90, 3878 (1968).
22. H. Alper, J. T. Edward, Can. J. Chem., 48, 1623 (1970).
23. M. A. Wuonola, W. A. Sheppard, J. Org. Chem., 36, 3640 (1971).
24. M. S. Newman, C. H. Chen, J. Am. Chem. Soc., 95, 278 (1973).
25. H. C. Brown et al., J. Am. Chem. Soc., 91, 6855 (1969).
26. H. C. Brown et al., J. Am. Chem. Soc., 91, 2150 (1969).
27. H. C. Brown, H. Nambu, J. Am. Chem. Soc., 92, 1761 (1970).
28. A. Barco et al., Synthesis, 316 (1973).
29. L. S. Forney, J. Org. Chem., 35, 1695 (1970); L. S. Forney, A. T. Jurewicz, ibid., 36, 689 (1971).
30. L. Vuitel, A. Jacot-Guillarmod, Synthesis, 608 (1972).
31. P. L. Creger, Tetrahedron Letters, 79 (1972).
32. Y. Izawa et al., Tetrahedron, 28, 211 (1972).

C. Nucleophilic-type Syntheses

   1. From Esters and Other Active Hydrogen Types with Esters
       (Claisen) (1, 836)

The subject of alkoxycarbonylation, discussed first in this section, has been reviewed recently [1].

A simple high-yield method for the synthesis of β-keto esters has been reported by Krapcho and co-workers [1a]. The preparation of 2-carbethoxycyclooctanone is illustrated (Ex. a):

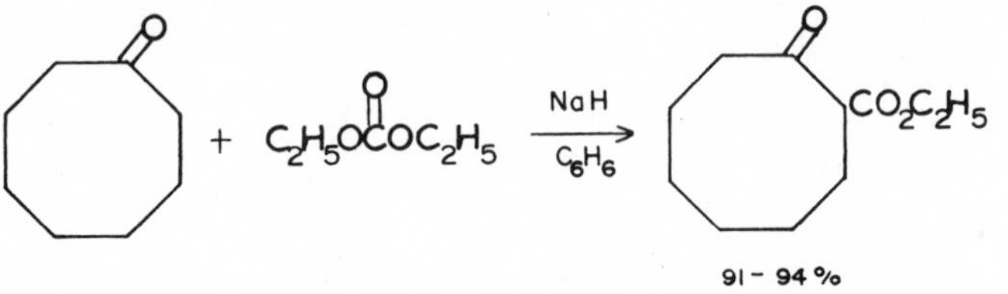

$$91 - 94\%$$

The 2-carbethoxy derivative of cyclononanone, cyclodecanone, and cyclododecanone were obtained in yields of 85%, 95%, and 90% respectively,

To introduce two carbalkoxy groups into a cyclic ketone, Balasubrahmanyam and Balasubramanian [2] recommend the use of the addition product formed between magnesium methoxide and carbon dioxide (Stiles reagent). In this manner, cyclohexanone gave dimethyl cyclohexanone-2,6-dicarboxylate as indicated:

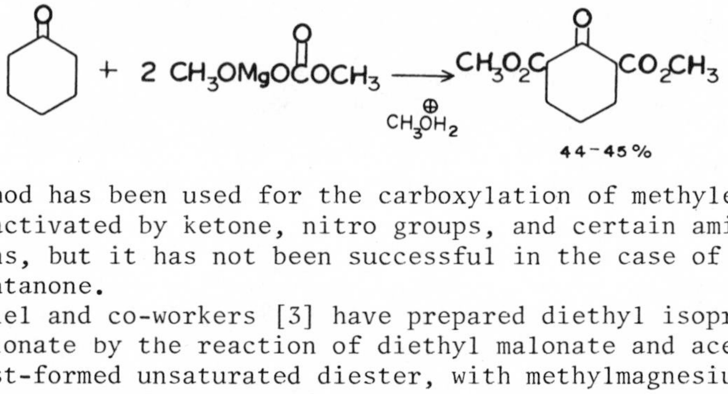

$$44-45\%$$

The method has been used for the carboxylation of methylene groups activated by ketone, nitro groups, and certain amide functions, but it has not been successful in the case of cyclopentanone.

Eliel and co-workers [3] have prepared diethyl isopropyl-idenemalonate by the reaction of diethyl malonate and acetone and the first-formed unsaturated diester, with methylmagnesium iodide in the presence of a trace of copper(I) chloride, gives diethyl t-butylmalonate as indicated (Ex. b):

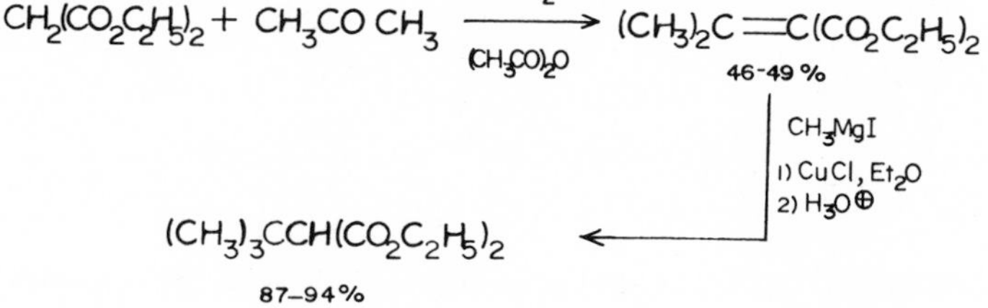

This procedure is preferable over the alkylation of diethyl
sodiomalonate with t-butyl chloride or t-butyl bromide but must
be compared with the malonic ester-t-butyl chloride-AlCl$_3$ method
(1, 834). The method as given illustrates the use of copper(I)
chloride to catalyze the addition reaction in the last step.
High yields are obtained with this catalyst as well as with
lithium dimethylcuprate.

An α-methylene lactone has been synthesized by Harmon and
Hutchinson [4] in the sequence of reactions shown:

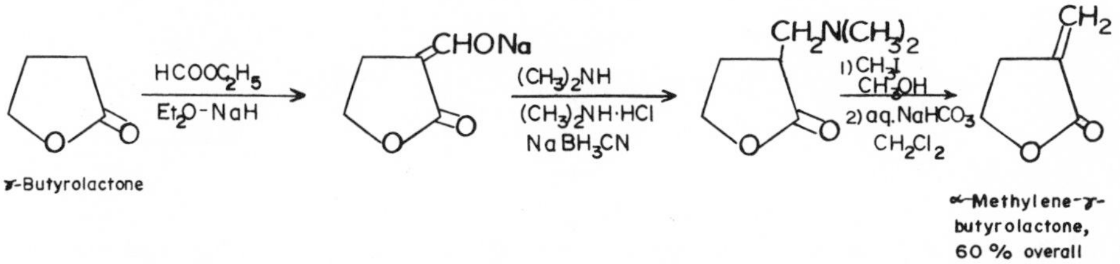

γ-Butyrolactone

α-Methylene-γ-
butyrolactone,
60 % overall

For other routes to the synthesis of α-methylene lactones see
[5], an example of which follows (Ex. c).

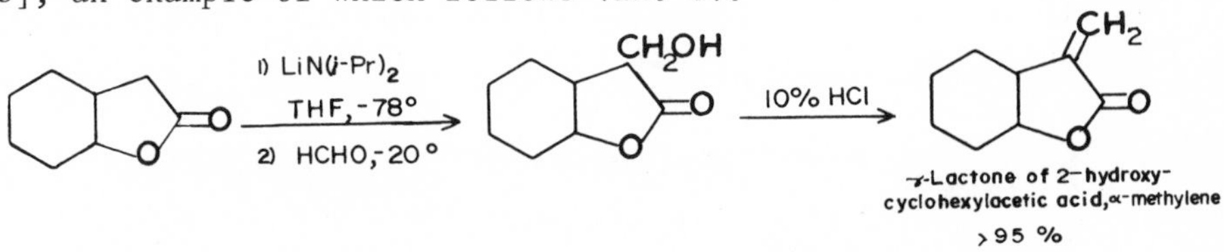

γ-Lactone of 2-hydroxy-
cyclohexylacetic acid,α-methylene
> 95 %

With crotonic esters and a strong base the intermediate
anion is $CH_2=CH-\overline{C}HCO_2Et$, because after quenching with D_2O,
$CH_2=CHCHDCO_2Et$ is obtained as shown [5a]:

Ethyl butene-3-oate d_2 ,87%

a. Preparation of 2-Carbethoxycyclooctanone [1a]

Yield was 91-94% from 35 g of NaH, 400 ml of benzene, 71 g
of diethyl carbonate, and 38 g of cyclooctanone in 100 ml of
benzene.

b. Preparation of Diethyl t-Butylmalonate [3]

Yield was 46-49% of diethyl isopropylidenemalonate from
diethyl malonate, acetone, acetic anhydride and ZnCl$_2$ in reflux
for 20-24 hr; diethyl t-butylmalonate, 87-94% from Mg, CH$_3$I in
$(C_2H_5)_2O$, Cu[I]Cl, diethyl isopropylidenemalonate in $(C_2H_5)_2O$,
and aqueous H$_2$SO$_4$.

c. Preparation of the γ-Lactone of 2-Hydroxycyclohexylacetic
Acid, α-Methylene [5]

To a solution of LiN(i-Pr)$_2$ in THF at -78°, the lactone of
2-hydroxy-1-cyclohexylacetic acid was added and at -20° gaseous
CH$_2$O in N$_2$ was introduced. Quenching with 10% HCl followed by
the workup gave the methylene lactone (>95%).

   3. From Dialkyl Succinates and Homophthalates with Carbonyl
       Compounds (Stobbe) (1, 840)

$$(C_6H_5)_2C{=}O \ + \ (CH_2CO_2C_2H_5)_2 \ \xrightarrow[\text{2) H}^{\oplus}]{\text{1) NaH}} \ (C_6H_5)_2C{=}C{\Large<}^{CO_2C_2H_5}_{CH_2COOH}$$

Hurd and Shah [6] carried out a series of Stobbe condensations with aliphatic aldehydes and dimethyl 3,5-bis(benzyloxy)-
homophthalate using sodium hydride as the base. With butyraldehyde a quantitative yield of the Stobbe half-ester as shown was
obtained if a trace of alcohol is introduced into the reaction
(Ex. a).

With this homophthalate, it should be noted, the reaction proceeds satisfactorily only in the presence of the alcohol.
   The Stobbe reaction is restricted to succinic esters
usually, but glutaric esters have been employed successfully as
the lithium enolate [7] or if the carbonyl reactant has an
appropriately placed ester grouping, such as in methyl α-(1-
oxoindan-2-yl) propionate [8]. It appears that considerable
scope may be added to the preparation by means of lithium
di(trimethylsilyl)amide [9], which has been used so successfully
in preparing β-hydroxy esters (see Chapter 4, E.1).

$$CH_3CO_2C_2H_5 + R_2C{=}O \ \xrightarrow[-78°]{\text{LiN(SiMe}_3)_2} \ R_2COHCH_2CO_2C_2H_5$$

80-93% (4 examples)

a. Preparation of 2,4-Bis(benzyloxy)6-(1-carbomethoxy-1-penten-
1-yl) benzoic Acid [6]

A solution of dimethyl 3,5-bis(benzyloxy)homophthalate, 6.72 g, 53.3% NaH, 0.72 g, and anhydrous ethanol, 12 drops, in 20 ml of dry benzene under N_2, was added in 10 min to butyraldehyde, 1.152 g in 10 ml of dry benzene. Stirring overnight at 25° and the addition of 50 ml of water gave two layers, the aqueous one giving 7.5 g of the Stobbe half-ester when acidified.

4. From Carbonyl Compounds and Esters (Aldol, Knoevenagel, or Doebner Condensation) (1, 841)

An aldol-like reaction has been discovered by Chapman and co-workers [10], who utilized it in the synthesis of α-methylene lactones, a type of current interest (Ex. a):

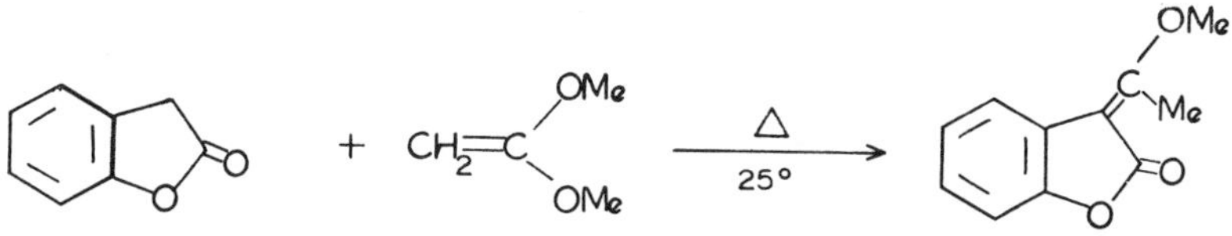

γ-Lactone of 2-hydroxyphenylacetic acid, α-methyl-α-methoxymethylene, 71%

The Knoevenagel reaction has been carried out with ketones and diethyl malonate, ethyl acetoacetate, and ethyl nitroacetate in the presence of titanium tetrachloride and pyridine [11]. Yields of the diethyl unsaturated esters from diethyl malonate vary within 42-96%, an improvement over those obtained by other catalysts. With cyclopentanone the equation is as follows (Ex. b):

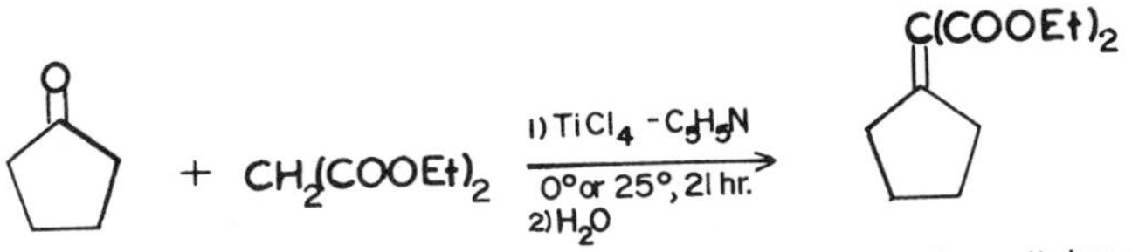

Dicarboethoxymethylene-cyclopentane, 75%

This method is also quite satisfactory for preparing the corresponding methylene derivatives of α-mono-, α,α-, or α,α'-di-, and α,α,α-trihalogeno ketones.

a. Preparation of γ-Lactone of 2-Hydroxy-1-phenylacetic Acid, α-Methyl-α-methoxymethylene [10]

An ethereal solution of the lactone of 2-hydroxyphenylacetic acid, 2.07 g and 1,1-dimethoxyethylene, 5 g at room temperature for 7 days gave, after recrystallization, 1.92 g (71%) of the methylene lactone.

b. Preparation of Dicarbethoxymethylenecyclopentane [11]

Titanium tetrachloride, 0.1 mol in 25 ml absolute CCl_4, was dropped into 200 ml of absolute THF with stirring at 0° while a yellow precipitate formed. Cyclopentanone, 0.05 mol, and diethyl malonate, 0.05 mol, were then added. With further cooling, there was added over 30-60 min dry pyridine, 0.2 mol, in 35 ml absolute THF and the mixture was allowed to react at 0° or at room temperature. After 21 hr, hydrolysis was accomplished with 50 ml of water and a 75% yield of the methylenecyclopentane was recovered from the ethereal extract.

5. From Esters and α,β-Unsaturated Compounds (Michael)
    (1, 844)

Williams and Snyder [12] formed the addition product as shown by adding 9-nitroanthracene to a solution of sodiomalonic ester in DMSO followed by dilution and acidification:

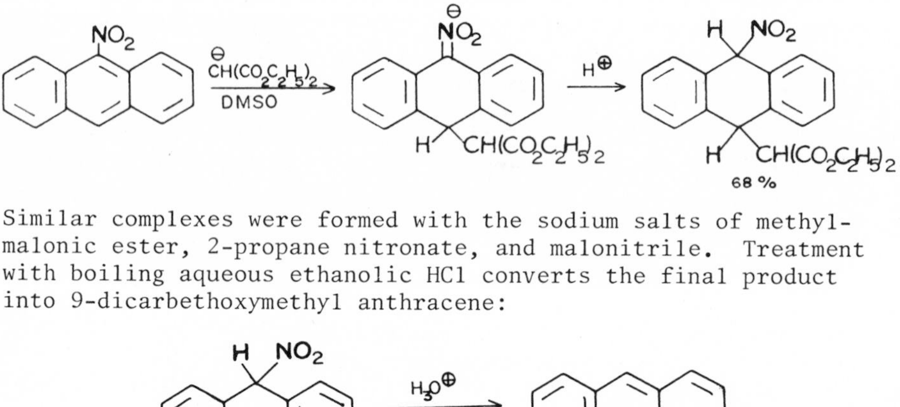

68 %

Similar complexes were formed with the sodium salts of methylmalonic ester, 2-propane nitronate, and malonitrile. Treatment with boiling aqueous ethanolic HCl converts the final product into 9-dicarbethoxymethyl anthracene:

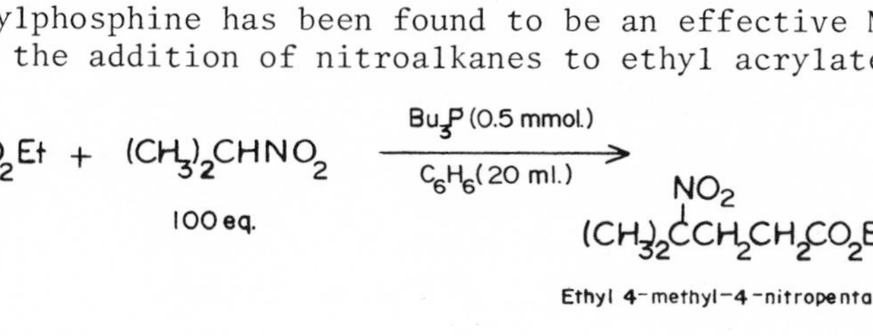

Tributylphosphine has been found to be an effective Michael catalyst in the addition of nitroalkanes to ethyl acrylate [13]:

$CH_2{=}CHCO_2Et$ + $(CH_3)_2CHNO_2$

100 eq.   100 eq.

Bu_3P (0.5 mmol.)

C_6H_6 (20 ml.)

NO_2

$(CH_3)_2CCH_2CH_2CO_2Et$

Ethyl 4-methyl-4-nitropentanoate, 88 %

α-Isocyanoesters add to ethyl acrylate to give N-formylglu-tamates [14]:

$$CNCH_2CO_2C_2H_5 + CH_2=CHCO_2C_2H_5 \xrightarrow[2)H_3O\oplus]{1)NaOC_2H_5} \overset{O}{\overset{\|}{HC}}NHCHCO_2C_2H_5$$
$$\underset{CH_2CH_2CO_2C_2H_5}{|}$$

Tetramethylguanidine was satisfactory as a catalyst in the Michael condensation of nitromethane with 3-methylcrotonate ester while other catalysts failed [15].

Tsuchihashi and co-workers [16] found that a Michael addition occurs between p-tolyl vinyl sulfoxide and diethyl malonate in the presence of sodium ethoxide as shown:

$$p\text{-}C_7H_7\overset{O}{\overset{\|}{S}}CH=CH_2 \xrightarrow[EtONa-EtOH]{CH_2(CO_2Et)_2} p\text{-}C_7H_7\overset{O}{\overset{\|}{S}}(CH_2)_2CH(COOEt)_2$$
61%

By starting with (+)-R-<u>trans</u> β-styryl p-tolyl sulfoxide, these investigators accomplished an asymmetric synthesis as indicated:

$$p\text{-}C_7H_7\overset{O}{\overset{\|}{S}}\underset{H}{\overset{H}{\underset{|}{\overset{|}{C}}}}=C\underset{C_6H_5}{\overset{|}{|}} \xrightarrow[EtONa-EtOH]{CH_2(CO_2Et)_2} p\text{-}C_7H_7\overset{O}{\overset{\|}{S}}CH_2\underset{C_6H_5}{\overset{|}{C}}HCHCH(CO_2Et)_2 \xrightarrow{crystallization}$$

(+)-(R)

82% (diastereoisomers in 8:2 ratio)

$$(+)\text{-}(R) \text{ stereoisomer} \xrightarrow[EtOH]{Raney Ni} C_6H_5\underset{Me}{\overset{|}{C}}HCH(CO_2Et)_2$$

51%

$$[\alpha]_D^{27} +93.1°$$

(c 1.472 CHCl_3)

The method represents a new way for the asymmetric synthesis of 1-phenylethylmalonic ester obtained on treating the (+)-R stereo-isomer with Raney Ni.

6. From Alkyl or Acyl Halides with Esters (1, 845)

$$RCH_2COOR' \xrightarrow[NaOC_2H_5]{R''X} R\underset{R''}{\overset{|}{C}}HCOOR'$$

A theoretical discussion of alkylation with carbanions has been published [17].

Brändström and Junggren [18] have alkylated benzyl cyanide, diethyl malonate, benzyl methyl ketone, methyl cyanoacetate, methyl acetoacetate, and dimethyl benzoylmalonate by ion-pair extraction. Although alkylation is usually carried out in highly polar aprotic solvents such as DMSO, DMF, or HMPA, there are some objections to this procedure [19]: (a) the solvents are expensive and usually not readily recovered, (b) the reactions are often sensitive to traces of water, and (c) in reactions of mesomeric anions there is a tendency for an increase in O-alkylation at the expense of C-alkylation. To overcome these difficulties Brändström and Junggren have formed the tetrabutyl-ammonium salt of methyl cyanoacetate, methyl acetoacetate, or dimethyl benzoylmalonate (from tetrabutylammonium hydroxide and the weak acid) which is soluble in chloroform and may be readily recovered by evaporation. Carbon alkylation of this salt occurs readily, usually in high yield. The steps with methyl acetoacetate are shown (Ex. a).

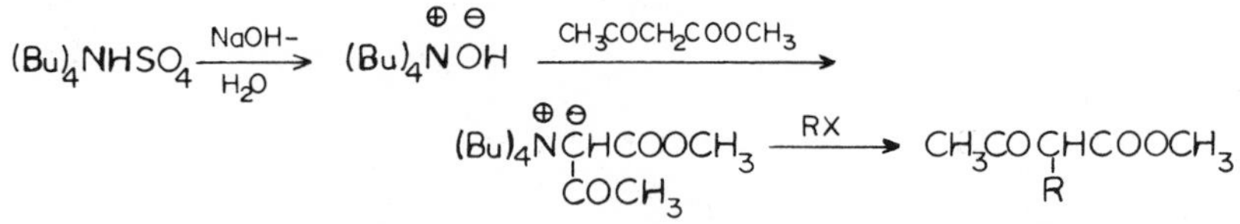

$$(Bu)_4NHSO_4 \xrightarrow[H_2O]{NaOH-} (Bu)_4\overset{\oplus}{N}\overset{\ominus}{O}H \xrightarrow{CH_3COCH_2COOCH_3}$$

$$(Bu)_4\overset{\oplus}{N}\overset{\ominus}{\underset{\underset{COCH_3}{|}}{C}}HCOOCH_3 \xrightarrow{RX} CH_3CO\underset{\underset{R}{|}}{C}HCOOCH_3$$

With dimethyl benzoylmalonate and methyl iodide, the C-alkylated product is obtained in 100% yield, but the percentage of O-alkylated product increases with an increase in size of the alkyl group.

The alkylation and aralkylation of ethyl and t-butyl phenyl-acetate have been accomplished by Hauser and co-workers [20], as indicated for the synthesis of ethyl 2,4-diphenylbutyrate (Ex. c).

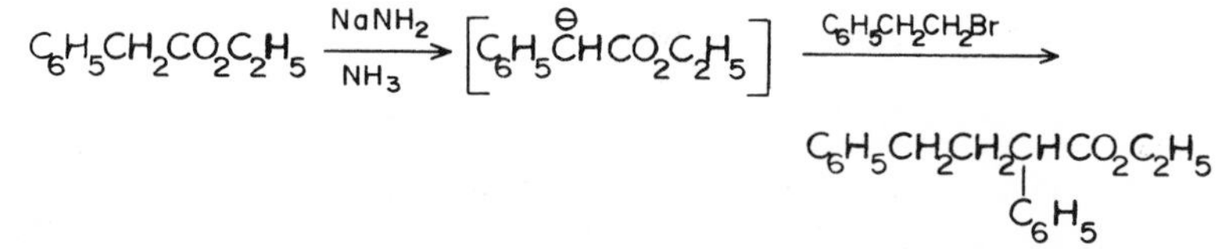

$$C_6H_5CH_2CO_2C_2H_5 \xrightarrow[NH_3]{NaNH_2} \left[C_6H_5\overset{\ominus}{C}HCO_2C_2H_5 \right] \xrightarrow{C_6H_5CH_2CH_2Br}$$

$$C_6H_5CH_2CH_2\underset{\underset{C_6H_5}{|}}{C}HCO_2C_2H_5$$

Ethyl 2,4-diphenylbutyrate, 77-81%

The alkylation of the dianions of α-keto esters has been reviewed [20a].

Rathke and Deitch [21] have synthesized β-keto esters by treatment of the lithium ester enolates with the acid chloride as shown here in Ex. b and in Chapter 11, F.2.

The lithium ester enolates have also been employed in the alkylation of esters [22] as indicated:

$$CH_3COOR \xrightarrow[THF,-78°]{LiICA} LiCH_2COOR \xrightarrow[\substack{20° \\ DMSO-THF}]{R'X} R'CH_2COOR \quad 42-96\%$$

The procedure is simpler and potentially more versatile than the usual malonic ester synthesis; in many cases the yield of alkylation product is superior. Herrmann and co-workers [23] have modified the Rathke process of alkylation of esters in a one-flask procedure under nearly theoretical stoichiometric conditions. The procedure consists of adding 1 equiv of the ester to a 1 molar THF solution containing 1 equiv of LDIA at -78°, of allowing the enolate to form at -78° over 20-40 min, and then adding 1-1.2 equiv of alkylating agent dissolved in 0.3 equiv of HMPA. The equation for the alkylation of methyl butyrate via the enolate follows:

$$CH_3(CH_2)_2COOCH_3 \xrightarrow[-78°]{LDIA} \left[CH_3CH_2CH=\overset{\overset{\ominus}{O}}{C}OCH_3 \right] \xrightarrow[HMPA]{RX} CH_3CH_2\overset{R}{\underset{}{C}}HCO_2CH_3$$

88-98%

Herrmann and co-workers [24] have also utilized the lithium salts in the alkylation of hydracrylic and crotonic esters. Thus methyl hydracrylate responds as follows in the one-operation, two-step sequence:

$$HOCH_2CH_2CO_2CH_3 \xrightarrow[\substack{THF \\ -78°, 40 min.}]{LDIA} \overset{\ominus}{O}CH_2\overset{\ominus}{C}HCO_2CH_3 \xrightarrow[\substack{THF \\ 4°, 5hr.}]{RX} HOCH_2\overset{R}{\underset{}{C}}HCO_2CH_3$$

$$Li^{\oplus} \quad Li^{\oplus}$$

86-95%

This method permits the formation of acrylates as well by dehydration of the product.

When applied to ethyl crotonate, LDIA formed a 1:1 Michael adduct. This difficulty was overcome by utilizing a 1:1 complex of LDIA and HMPA in the alkylation reaction. With this reagent the β,γ-unsaturated ester as shown was produced:

$$CH_3CH=CHCO_2C_2H_5 \xrightarrow[2)\ RX]{1)\ LDIA-HMPA} CH_2=CH\overset{R}{\underset{}{C}}HCO_2C_2H_5$$

90-96%

By repeating the process dialkyl esters were obtained in 88-98% yields.

The alkylation of malonic esters with some α-chloroaldehydes leads to butyrolactones [25].

$$Me_2\overset{}{\underset{Cl}{C}}CHO + CH_2(CO_2Me)_2 \xrightarrow{K_2CO_3} \overset{O}{\overset{\|}{C}}-CHCH(CO_2Me)_2 \xrightarrow{\ominus CH(CO_2Me)_2} $$

In addition to the use of α-lithio salts of esters, it now becomes possible to use the α-copper salt in the synthesis of esters. By employing a low temperature (-110°), ethoxycarbonyl-methylcopper, thought to be the intermediate, was formed satisfactorily [26]:

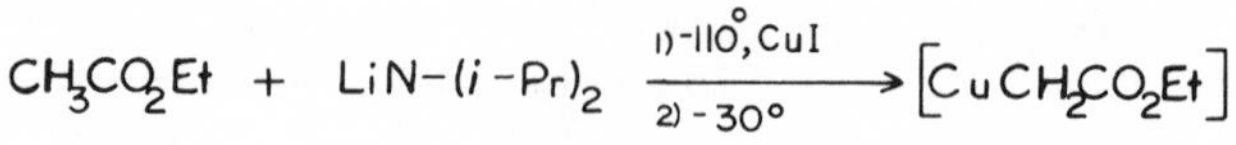

$$CH_3CO_2Et + LiN-(i-Pr)_2 \xrightarrow[\text{2) -30°}]{\text{1) -110°, CuI}} [CuCH_2CO_2Et]$$

The salt is oxidized by the addition of O_2 to diethyl succinate (73%) or is alkylated with allylic halides:

Ethyl cyclohexenylacetate
69 %

Schwartz and co-workers [27] alkylated acetylenic esters by conversion into the vinylrhodium complex which with an alkyl halide gives the corresponding olefinic ester as shown:

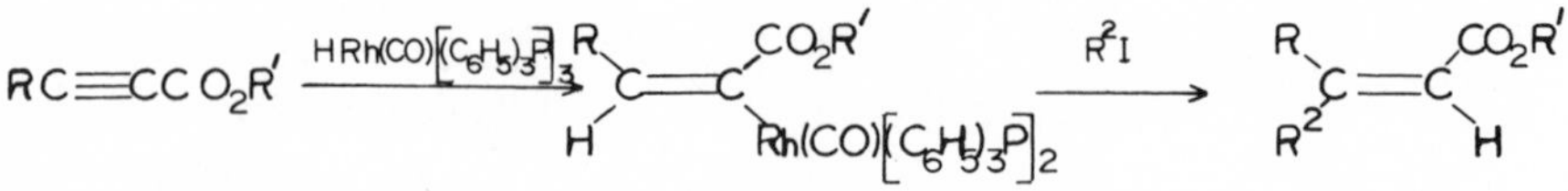

ca. 76%

The reaction is not a free radical one since the cis vinyl rhodium complex gives the cis olefinic ester.

A profound acylation reaction occurs when isopropenyl stearate is refluxed with diethyl malonate in the presence of an acid catalyst [28]:

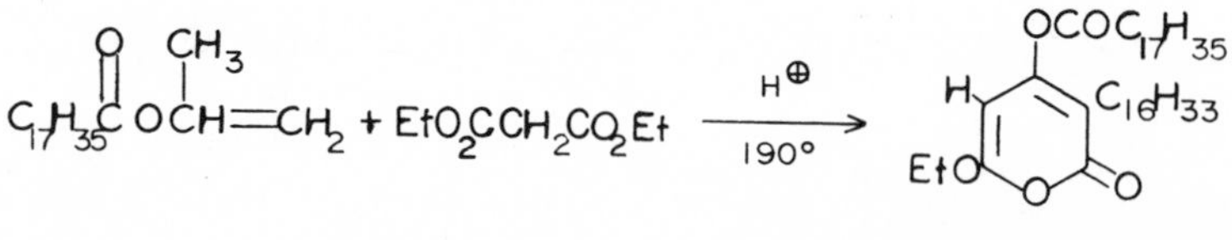

6-Ethoxy-3-hexadecyl-4-stearoyloxy-2-pyrone, 70%

By elimination of hydrogen bromide McDonald and Reitz [29] succeeded in converting α,α-dimethyl dicarboxylates into dimethyl esters of cycloalkenes as indicated:

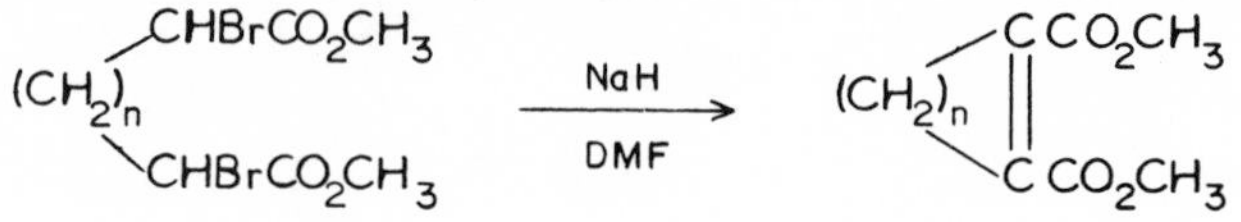

When n = 1 and 6 no unsaturated diester was obtained, but when n = 2,3,4,5 the yields varied within 21-71%.

Cyclopropane formation has been accomplished as shown [30]:

$$ArCH_2-\underset{\underset{CO_2Et}{|}}{\overset{\overset{CO_2Et}{|}}{C}}-CH_2CH_2\overset{\oplus}{N}Me_3 \ \overset{\ominus}{O}Et \xrightarrow{150-200°} EtO_2C \overset{\triangle}{\underset{CH_2Ar}{}}$$

Ethyl 1-arylmethyl-1-
cyclopropanecarboxylate
31-76% (16 examples)

A unique alkylation has been accomplished via the zinc enolate of ethyl dichloroacetate, available under well-defined conditions [31].

$$CCl_2=\underset{OC_2H_5}{\overset{OZnCl}{C}} + ClCH_2OC_2H_5 \longrightarrow C_2H_5OCH_2CCl_2CO_2C_2H_5$$

Ethyl 2,2-dichloro-3-ethoxypropionate
56%

Added to carbonyl compounds the enolate forms Reformatsky products (see Chapter 4, E.3).

Thallium enolates offer no special advantages in promoting exclusive C-alkylation according to Hooz and Smith [32].

The introduction of an acetonyl group into esters may be accomplished via a protective device [33]:

$$\underset{CH_2=CCH_2Br}{\overset{\overset{THP}{\overset{|}{O}}}{}} + (C_6H_5)_2CHCO_2C_2H_5 \xrightarrow[2)H_3O^{\oplus}]{1)NaH-DMF} CH_3COCH_2\underset{\underset{C_6H_5}{|}}{\overset{\overset{C_6H_5}{|}}{C}}-CO_2C_2H_5$$

Ethyl 2,2-diphenyllevulinate, 66-69%

a. Preparation of Methyl α-Butylacetoacetate [18]

Tetrabutylammonium hydrogen sulfate, 0.1 mol, was added to a cooled solution of NaOH, 0.2 mol in 75 ml of water. The mixture was added to a stirred solution of methyl acetoacetate, 0.1 mol and butyl iodide, 0.2 mol in 75 ml of chloroform. In a few minutes the chloroform layer was evaporated and the tetrabutylammonium iodide was precipitated by adding ether to the residue. Filtration removed the iodide and gave a filtrate that yielded 90% of the α-butylacetoacetate on evaporation.

b. Preparation of Ethyl 3-Ketopentanoate [21]

To a solution of LiICA, 50 mmol in 50 ml of THF under

nitrogen in a dry-ice-acetone bath was added ethyl acetate, 25 mmol in 5 min followed after 10 min by propionyl chloride, 25 mmol. The mixture was stirred for an additional 10 min and then quenched with 15 ml of 20% HCl. From the organic layer was recovered by distillation 2.25 g (60%) of the keto ester.

c. Preparation of Ethyl 2,4-Diphenylbutyrate [20]

To a suspension of 0.1 mol of sodium amide in liquid ammonia, 0.1 mol of ethyl phenylacetate in anhydrous ether was added over a 2-min period and the mixture was stirred 20 min. (2-Bromoethyl)benzene in 35 ml of anhydrous ether was added over an 8-min period. After stirring for 3 hr, the mixture was neutralized with 0.1 mol of NH$_4$Cl. Dry ether, 150 ml, was then added, the ammonia was evaporated, and the mixture, cooled to 0°, was hydrolyzed by the addition of 100 ml of 3 N HCl. From the ethereal extract was recovered 20.6-21.8 g (77-81%) of the ester.

7. From Grignard Reagents and the Like and Carbonate Esters (1, 847)

Kaiser and co-workers [34] synthesized ethyl α-cyanophenyl-acetate in the steps shown (Ex. a).

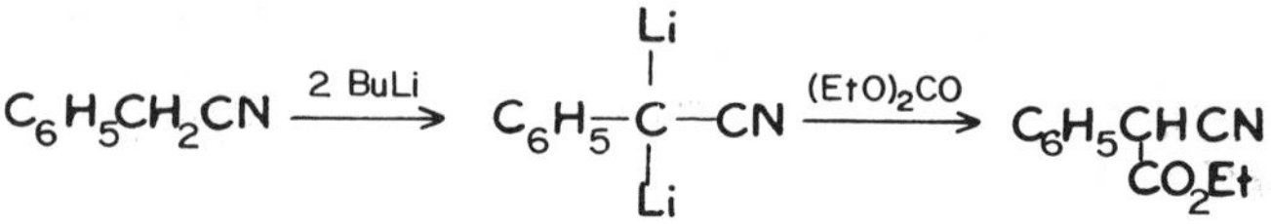

Ethyl α-cyanophenylacetate

79% (last step)

This yield is superior to that available from monolithiophenyl-acetonitrile.

Somewhat similar are the results of De Pasquale [35] where alkylene carbonates were produced from epoxides and carbon dioxide in the presence of nickel(0) complexes as indicated:

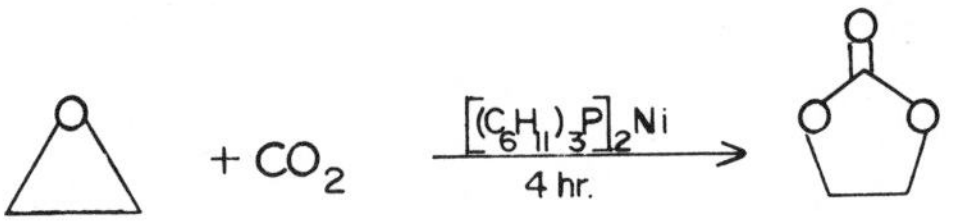

Ethylene carbonate

60% conversion

The rate of carbonate formation is dependent on the structure of both the epoxide and catalyst.

Carbonation of esters has been accomplished by Bottaccio and Chiusoli [36] as shown:

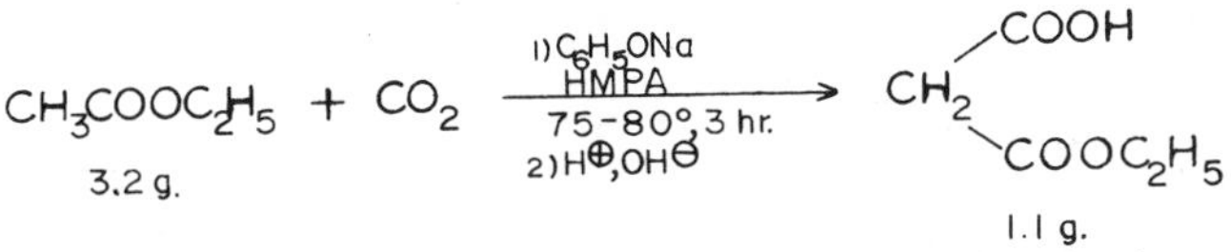

$$CH_3COOC_2H_5 + CO_2 \xrightarrow[\substack{2)H^{\oplus},OH^{\ominus}}]{\substack{1)C_6H_5ONa \\ HMPA \\ 75-80°, 3\ hr.}} CH_2 \begin{array}{l} \diagup COOH \\ \diagdown COOC_2H_5 \end{array}$$

3.2 g.

1.1 g.

Monoethyl malonate

Carbonation of 2,6-dimethylpyridine by Kofron and Baclawski [37] takes the one-operation route indicated:

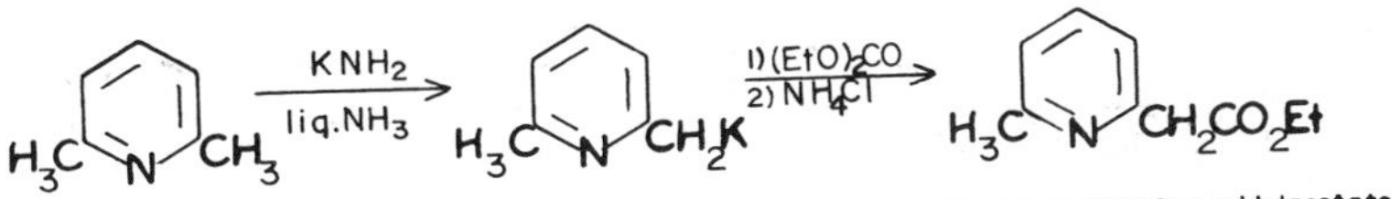

Ethyl 6-methyl-2-pyridylacetate

59-75%

a. Preparation of Ethyl α-Cyanophenylacetate [34]

To 0.11 mol of 1.6M n-butyllithium in hexane at room temperature was added rapidly 70 ml of anhydrous THF followed by a solution of 0.05 mol of phenylacetonitrile in 50 ml of THF in 5 min. After 1 hr 0.05 mol of diethyl carbonate in 50 ml of THF was added to the mixture during 6 min and this mixture was refluxed for 2 hr. Acidification and workup gave 7.45 g (79%) of the cyanoester.

8. Reformatsky, Darzens, and Favorskii Reactions (1, 848)

a. Reformatsky Reaction (see Chapter 4, E.3)

By using a Schiff base rather than an aldehyde in the Reformatsky reaction, either lactams or β-amino esters may be obtained [38]:

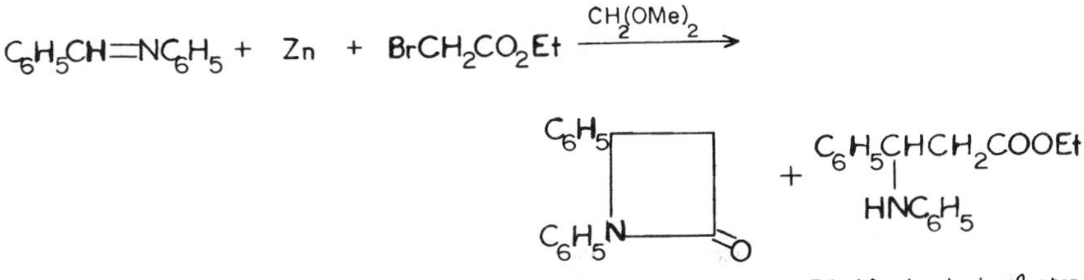

$$C_6H_5CH=NC_6H_5 + Zn + BrCH_2CO_2Et \xrightarrow{CH(OMe)_2}$$

N-β-Diphenylpropiolactam

$$C_6H_5CHCH_2COOEt$$
$$\underset{HNC_6H_5}{|}$$

Ethyl β-phenylamino-β-phenyl-propionate

A   B

At 44°, the lactam A is obtained in 85% yield; at -10° the amino ester B is obtained in 85% yield.

b. Darzens Reaction (see Chapter 6, C.3)

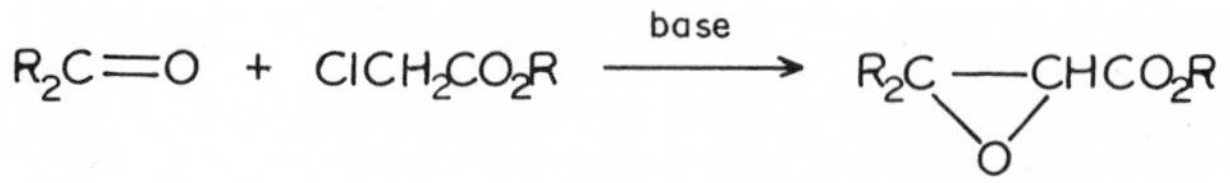

$$R_2C{=}O \ + \ ClCH_2CO_2R \ \xrightarrow{base} \ R_2C\underset{O}{\overset{\triangle}{\text{---}}}CHCO_2R$$

Although the Darzens condensation has been unsuccessful when applied to acetaldehyde and monosubstituted acetaldehyde derivatives, presumably because of competing base-catalyzed self-condensation, Borch [39] was successful in carrying out the reaction by forming the anion from the α-bromoester with lithium bis(trimethylsilyl)amide at -78° and then adding the carbonyl compound at the same temperature as shown:

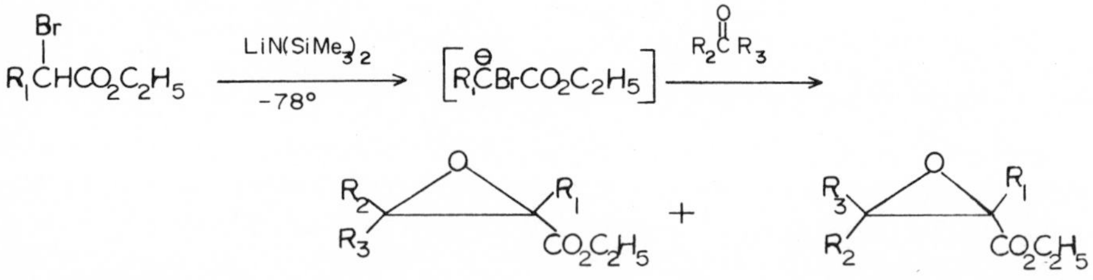

$$R_1\overset{Br}{\underset{}{CHCO_2C_2H_5}} \ \xrightarrow[-78°]{LiN(SiMe_3)_2} \ \left[R_1\overset{\ominus}{C}BrCO_2C_2H_5\right] \ \xrightarrow{R_2\overset{O}{\overset{\parallel}{C}}R_3}$$

A   B

Isolable yields of oxiranes varied within 72-86%, with the A isomer being in excess.

A Darzens-like reaction occurs between aldehydes and thiocyanoacetic esters in the presence of potassium fluoride or carbonate [40]. Thus benzaldehyde reacts as follows:

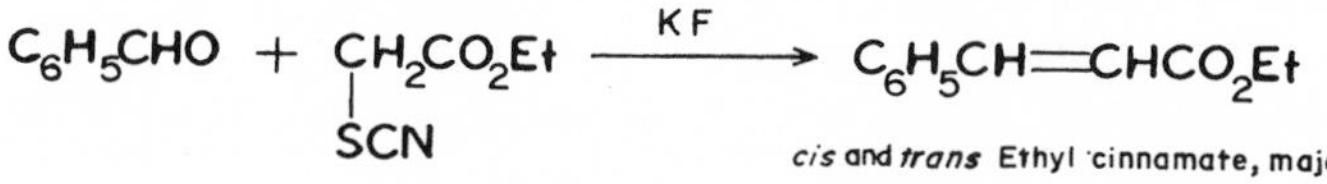

$$C_6H_5CHO \ + \ \underset{SCN}{\overset{}{CH_2CO_2Et}} \ \xrightarrow{KF} \ C_6H_5CH{=}CHCO_2Et$$

cis and trans Ethyl cinnamate, major products

It appears that the episulfide, $C_6H_5CH{-}CH \ CO_2Et$, which loses

sulfur to give the final product, is an intermediate.

c. Favorskii Reaction (see Chapter 13, G.2)

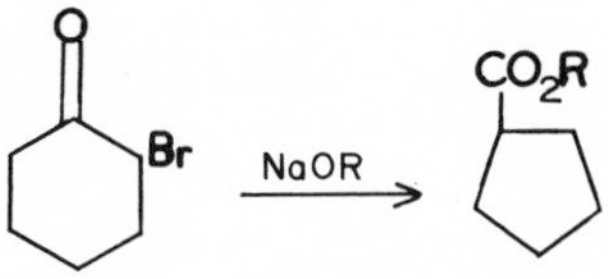

9.   Cleavage of Carbonyl Compounds (1, 848)

Stotter and Hill [41] prepared α-bromoesters from the corresponding α-alkylacetoacetates as indicated:

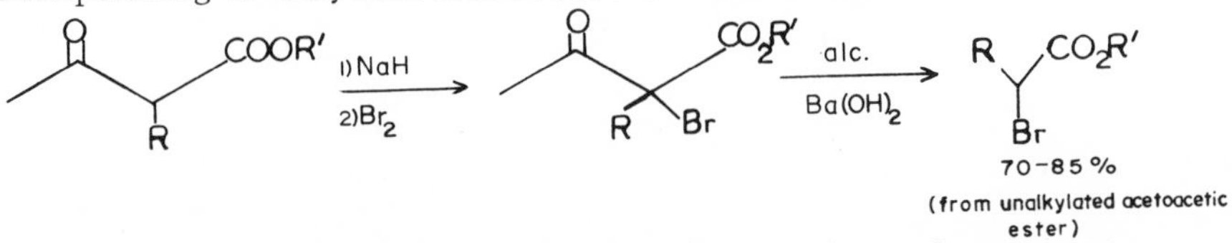

70-85%
(from unalkylated acetoacetic ester)

This method is satisfactory for the preparation of t-butyl α-bromoesters containing oxidation-sensitive and acid/base sensitive functionality.   The mechanism may be represented as:

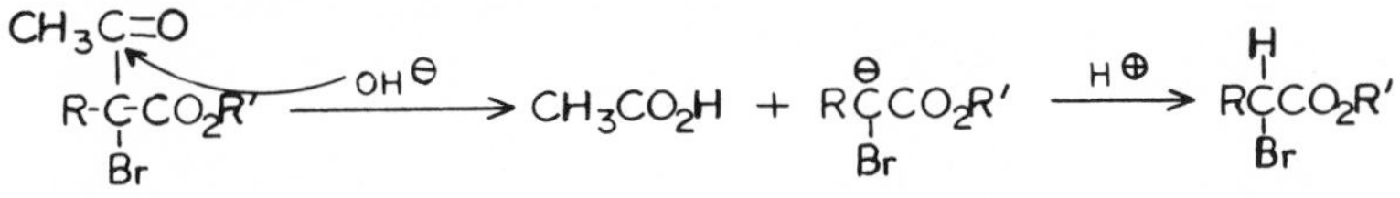

Methyl 3-methyltridecanoate has been obtained by the proper alkylation of 5-methyl-1,3-cyclohexanedione followed by splitting with sodium methoxide [42].

11.   Cleavage of Substituted Malonic Esters (1, 850)

$$R_2C(COOC_2H_5)_2 \xrightarrow{NaOC_2H_5} R_2CHCOOC_2H_5$$

A new method for conversion of malonates to esters of monocarboxylic acids involves heating with sodium and trimethylchlorosilane to form the reactive ketene alkyl silyl acetal, which with an alcohol gives the simple ester as shown [43]:

$$R'R^2C(CO_2Me)_2 + 2 Na + 2 Me_3ClSi \longrightarrow$$

$$R'R^2C=C(OMe)OSiMe_3 + MeOSiMe_3 + CO + 2 NaCl$$

$$\downarrow \text{MeOH}$$

$$R'R^2CH CO_2Me + MeOSiMe_3$$

ca. 85%

α,α-Dichloro-β-ketoesters cleave extraordinarily easily by weak nucleophiles such as sodium acetate [44].

12.   From Carbonyl Compounds or α-Halocarboxylic Acids and Alkylidene Phosphoranes (Wittig)

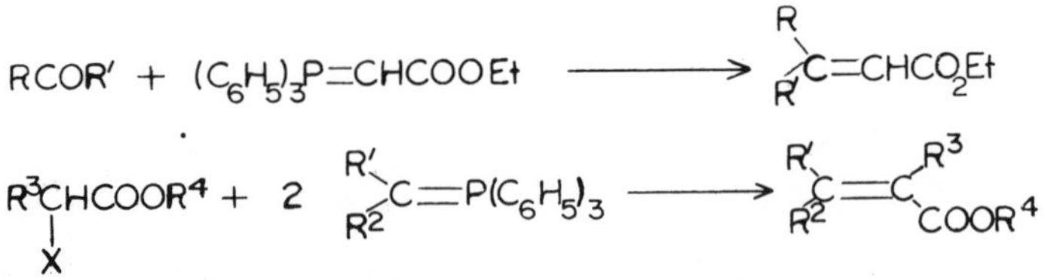

$$RCOR' + (C_6H_5)_3P{=}CHCOOEt \longrightarrow \underset{R'}{\overset{R}{>}}C{=}CHCO_2Et$$

$$\underset{X}{\overset{R^3CHCOOR^4}{|}} + 2 \underset{R^2}{\overset{R'}{>}}C{=}P(C_6H_5)_3 \longrightarrow \underset{R^2}{\overset{R'}{>}}C{=}\underset{COOR^4}{\overset{R^3}{<}}$$

See 1, 141, 1, 782, and the work by Bestmann and co-workers [45] for a discussion of the Wittig reaction.

13.   From Cyclopropanediesters and Lithium Dialkylcuprates

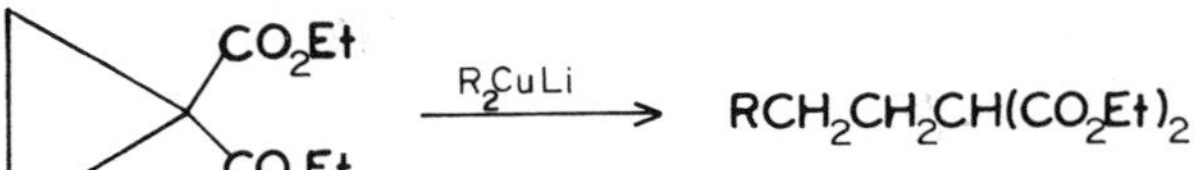

$$\xrightarrow{R_2CuLi} RCH_2CH_2CH(CO_2Et)_2$$
70-85%

Cleavage and alkylation occurs when cyclopropanediesters are treated with lithium dialkylcuprates [46].

14.   From Acetylenic Esters and Vinylcopper Reagents

Corey and co-workers [47] have developed a stereospecific synthesis of 1,3-dienes as shown:

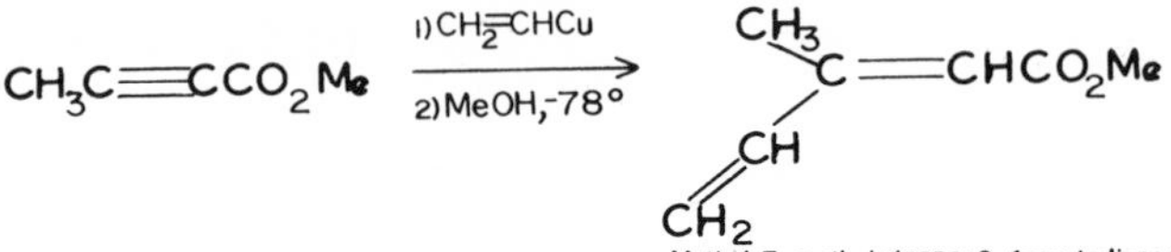

$$CH_3C{\equiv}CCO_2Me \xrightarrow[2)MeOH,-78°]{1)CH_2{=}CHCu} \underset{\underset{CH_2}{\overset{|}{CH}}}{\overset{CH_3}{>}}C{=}CHCO_2Me$$

Methyl 3-methyl-*trans*-2,4-pentadienoate
74%

In a similar way, the use of allylcopper gave stereospecifically 1,4-dienes.

15.   From Vinyl Halides and Olefinic Compounds (see Addenda)

Addenda

C.1.   The Claisen reaction succeeds well with potassium hydride as the base catalyst [C. A. Brown et al., Synthesis, 326 (1975)].

C.1.   Arylmalonic esters may be synthesized from arylacetic esters, ethyl carbonate, and sodium in a simple procedure. The method is applicable to the preparation of aliphatic malonates as well [G. Zvilichovsky, U. Fotodov, <u>Org. Prep. Proced. Int.</u>, <u>6</u>, 5 (1974)].

C.2.   The Dieckmann Reaction occurs as shown:

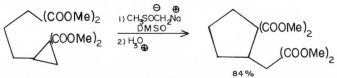

84%

[S. Danishefsky et al., <u>J. Am. Chem. Soc.</u>, <u>96</u>, 1256 (1974)]

C.5.   Ethyl crotonate and diethylmethylmalonic ester condense abnormally in a Michael reaction to give diethyl α-carbethoxy-β,γ-dimethylglutarate. The rearrangement is explained [R. K. Hill, N. D. Ledford, <u>J. Am. Chem. Soc.</u>, <u>97</u>, 666 (1975)].

C.5.   Both β-alkylation (Michael) and α-alkylation occur in the reaction of t-butylmagnesium chloride and ethyl cinnamate [I. Crossland, <u>Acta Chem. Scand.</u>, <u>29</u>, B468 (1975)].

C.6. $(R_2N)_2POCH_3$ with ethyl trichloroacetate gives ethyl 2,2-dichloropropionate (89%). The reagent is a methylating agent prepared from HMPA [J. H. Hargis, W. D. Alley, <u>J. Am. Chem. Soc.</u>, <u>96</u>, 5927 (1974).

C.6.   Lactones may be α-methylated, treated with diphenyl diselenide to form the α-phenyl selenide, and then oxidized with H_2O_2 to give the α-methylene lactone [P. A. Grieco, M. Miyashita, <u>J. Org. Chem.</u>, <u>39</u>, 120 (1974)].

C.6.   The lithium salt of ethyl diazoacetate may be alkylated, acylated, silylated, or subjected to an aldol condensation [M. Regitz, <u>Synthesis</u>, 351 (1972)].

C.11.   Malonic esters are converted into substituted acetic esters by NaCN in DMSO, a quite mild reaction [E. E. van Tamelen, R. J. Anderson, <u>J. Am. Chem. Soc.</u>, <u>94</u>, 8225 (1972)].

C.15.   Ethyl acrylate and 1-iodo-1-hexene combine to form methyl 2,4-nonadienoate (45%) by using palladium acetate in triphenylphosphine as a catalyst [H. A. Dieck, R. F. Heck, <u>J. Org. Chem.</u>, <u>40</u>, 1083 (1975)].

References

1.   J. Mathieu, J. Weill-Raynal, Formation of C-C Bonds, Georg Thieme, Stuttgart, 1973, pp. 325-345.

1a. A. P. Krapcho et al., Org. Syn., Coll. Vol. 5, 198 (1973).

2.   S. N. Balasubrahmanyam, M. Balasubramanian, Org. Syn., Coll. Vol. 5, 439 (1973).

3.   E. L. Eliel et al., Org. Syn., 50, 38 (1970).

4.   A. D. Harmon, C. R. Hutchinson, Tetrahedron Letters, 1293 (1973).

5.   P. A. Grieco, K. Hiroi, Chem. Commun., 1317 (1972).

5a. M. W. Rathke, D. Sullivan, Tetrahedron Letters, 4249 (1972).

6.   R. N. Hurd, D. H. Shah, J. Org. Chem., 38, 607 (1973).

7.   W. H. Puterbaugh, J. Org. Chem., 27, 4010 (1962).

8.   A. Chatterjee et al., J. Chem. Soc., C661 (1971).

9.   M. W. Rathke, J. Am. Chem. Soc., 92, 3222 (1970).

10. O. L. Chapman et al., Chem. Commun., 384 (1971).

11. W. Lehnert, Tetrahedron, 28, 663 (1972); 29, 635 (1973); Tetrahedron Letters, 4723 (1970).

12. R. H. Williams, H. R. Snyder, J. Org. Chem., 36, 2327 (1971).

13. D. A. White, M. M. Baizer, Tetrahedron Letters, 3597 (1973).

14. U. Schöllkopf, K. Hantke, Angew. Chem., 82, 932 (1970).

15. G. P. Pollini et al., Synthesis, 44 (1972).

16. G. Tsuchihashi et al., Tetrahedron Letters, 323 (1973).

17. D. C. Ayres, Chem. Ind. (London), 937 (1973).

18. A. Brändström, U. Junggren, Acta Chem. Scand., 23, 2203, 2204, 2536 (1969); Tetrahedron Letters, 473 (1972).

19. A. Brändström et al., Acta Chem. Scand., 23, 2202 (1969).

20. C. R. Hauser et al., Org. Syn., Coll. Vol. 5, 559 (1973).

20a. S. N. Huckin, L. Weiler, J. Am. Chem. Soc., 96, 1082 (1974).

21. M. W. Rathke, J. Deitch, Tetrahedron Letters, 2953 (1971).

22. M. W. Rathke, A. Lindert, J. Am. Chem. Soc., 93, 2318 (1971).

23. J. L. Herrmann et al., Tetrahedron Letters, 2425 (1973).

24. J. L. Herrmann et al., Tetrahedron Letters, 2429, 2433 (1973).

25. A. Takeda et al., J. Org. Chem., 38, 4148 (1973).

26. I. Kuwajima, Y. Doi, Tetrahedron Letters, 1163 (1972).

27. J. Schwartz et al., J. Am. Chem. Soc., 94, 9269 (1972).

28. E. S. Rothman et al., J. Org. Chem., 38, 2540 (1973).

29. R. N. McDonald, R. R. Reitz, J. Org. Chem., 37, 2418 (1972).

30. C. Kaiser et al., J. Med. Chem., 13, 820 (1970).

31. B. Castro et al., Bull. Soc. Chim. France, 3521 (1969).

32. J. Hooz, J. Smith, J. Org. Chem., 37, 4200 (1972).

33. D. E. Horning et al., Can. J. Chem., 48, 975 (1970).

34. E. M. Kaiser et al., J. Am. Chem. Soc., 93, 4237 (1971).
35. R. J. De Pasquale, Chem. Commun., 157 (1973).
36. G. Bottaccio, G. P. Chiusoli, Z. Naturforsch., 23B, 561 (1968).
37. W. G. Kofron, L. M. Baclawski, Org. Syn., 52, 75 (1972).
38. F. Dardoize et al., Bull. Soc. Chim. France, 3841 (1972).
39. R. F. Borch, Tetrahedron Letters, 3761 (1972).
40. T. Hayashi et al., Bull. Chem. Soc. Jap., 44, 1357 (1971).
41. P. L. Stotter, K. A. Hill, Tetrahedron Letters, 4067 (1972).
42. N. Polgar et al., J. Chem. Soc., C, 870 (1971).
43. J. J. Bloomfield et al., Chem. Commun., 136 (1971).
44. S. K. Gupta, J. Org. Chem., 38, 4081 (1973).
45. H. J. Bestmann et al., Chem. Ztg., 94, 487 (1970); Chem. Ber., 103, 685 (1970).
46. G. Daviaud, P. Miginiac, Tetrahedron Letters, 997 (1972).
47. E. J. Corey et al., J. Am. Chem. Soc., 94, 4395 (1972).

D. Oxidative- and Reductive-type Syntheses

1. Intermolecular Oxidation-Reduction of Aldehydes (Tishchenko) (1, 853)

$$2\ RCHO \longrightarrow RCOOCH_2R$$

It has recently been shown that boric acid is an effective catalyst in converting aldehydes not readily susceptible to aldol condensation or resinification into esters via a Tishchenko reaction [1]. Thus paraformaldehyde in an autoclave gives methyl formate as shown:

$$2\ HCHO \xrightarrow[\substack{C_6H_{12} \\ 250°, 5\ hr.}]{H_3BO_3} HCOOCH_3$$

77%

(100% conversion)

Other aliphatic aldehydes give yields usually of 60-90% with 52-91% conversion.

3. Oxidation of Ethers (1, 855)

$$RCH_2OCH_2R \longrightarrow RCOOCH_2R$$

The spiroether ring in a steroid may be oxidized selectively to a lactone by tert, butyl chromate in carbon tetrachloride [2].

Yields are as high as 48%. Chromium trioxide-acetic acid and ruthenium tetroxide have been used to convert steroidal ethers into lactones while chromic acid in acetone, buffered bromine solutions, and trichloroisocyanuric acid have been found to convert ethers into esters [2].

5.   Electrolysis of Acid Ester Salts (Kolbe), of Acid Esters in the Presence of Olefins, and of α-Acylamino Acid Esters (1, 856) (see Chapter 1, G.7)

$$2\ C_2H_5OOC(CH_2)_nCOOK \xrightarrow{\text{Electrolysis}} C_2H_5OOC(CH_2)_{2n}CO_2C_2H_5$$

Haufe and Beck [3] carried out the Kolbe electrolysis of ω-acyloxyalkanoic acids, as shown for 6-acyloxyhexanoic acid:

$$2\ CH_3COO(CH_2)_5COOH \xrightarrow{\text{Electrolysis}} CH_3COO(CH_2)_{10}OCOCH_3$$

1,10-Decamethylene diacetate, 83.1%

Other acyloxyalkanoic acids, with the exception of the α-acyloxy ones, gave satisfactory yields of the dioldiformates or dioldiacetates.

Schäfer and Pistorius [4], by the electrolysis of ethyl hydrogen oxalate and methyl hydrogen malonate in the presence of mono- or diolefins, have prepared a series of unsaturated esters. Thus ethyl hydrogen oxalate and butadiene on electrolysis, as indicated, gave diethyl 3,7-decadienedioate.

$$A/cm^2 = 0.025$$

$$CH_2{=}CH{-}CH{=}CH_2\ +\ \begin{array}{c} COOH \\ | \\ COOC_2H_5 \end{array} \longrightarrow \begin{array}{c} CH_2CH{=}CHCH_2COOC_2H_5 \\ | \\ CH_2CH{=}CHCH_2COOC_2H_5 \end{array}$$

50% of total 70%

β-Keto esters have been produced by the electrolysis of α-acylamino acid esters [5] as shown:

$$\underset{\underset{NH_2\cdot HCl}{|}}{\overset{\overset{R'}{|}}{RCOCCOOR''}} \xrightarrow[+2H^{\oplus}]{2e^{\ominus}} \underset{\underset{H}{|}}{\overset{\overset{R'}{|}}{RCOCCOOR''}}\ +\ NH_4Cl$$

In the cell a mercury pool served as the cathode, while the anode was platinum. The best yields (80-90%) were obtained for aromatic β-keto esters.

6.   Acyloxylation (1, 856)

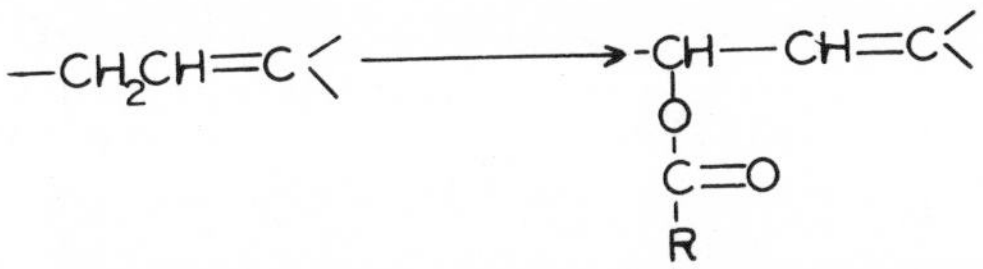

$$-CH_2CH=C\diagdown \longrightarrow -CH-CH=C\diagdown$$

with the -CH carrying an O, then C=O, then R group below.

Pedersen and co-workers have acyloxylated cyclohexene by the use of t-butyl perbenzoate in the presence of a trace of cuprous bromide as indicated (Ex. a) [6]:

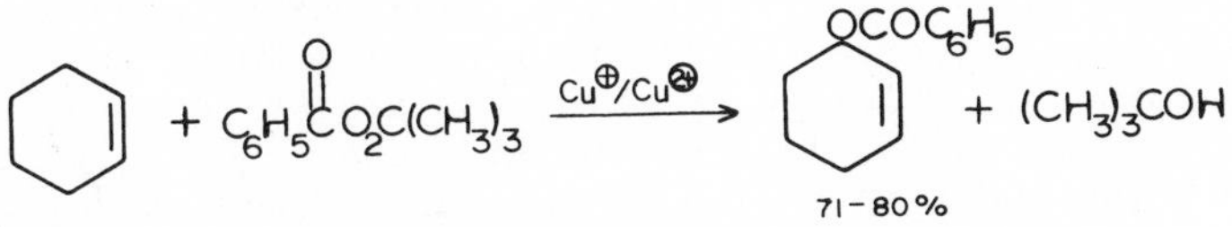

Cyclohexene $+ C_6H_5\overset{O}{\overset{\|}{C}}O_2C(CH_3)_3 \xrightarrow{Cu^{\oplus}/Cu^{\ominus}}$ cyclohexenyl benzoate (OCOC$_6$H$_5$) $+ (CH_3)_3COH$

71-80%

The reaction may be used to prepare benzoyloxy derivatives of saturated hydrocarbons, esters, ethers, sulfides, amides, and certain organosilicon compounds.

Likewise cyclohexene has been oxidized with manganese(III) acetate-potassium bromide to give the cyclohexenyl acetate as indicated [7]:

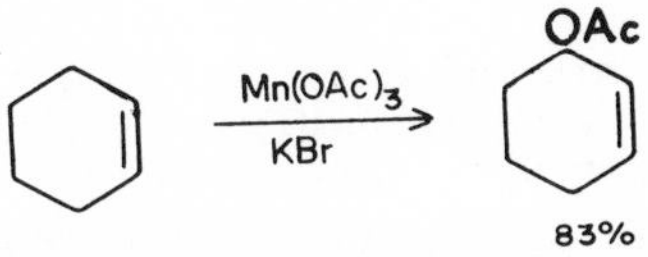

Cyclohexene $\xrightarrow[\text{KBr}]{\text{Mn(OAc)}_3}$ cyclohexenyl acetate (OAc)

83%

2,5-Bis(acyloxy) 2,5-dihydrofurans may be synthesized by the electrolysis of various carboxylic acids and furan using platinum electrodes as shown [8]:

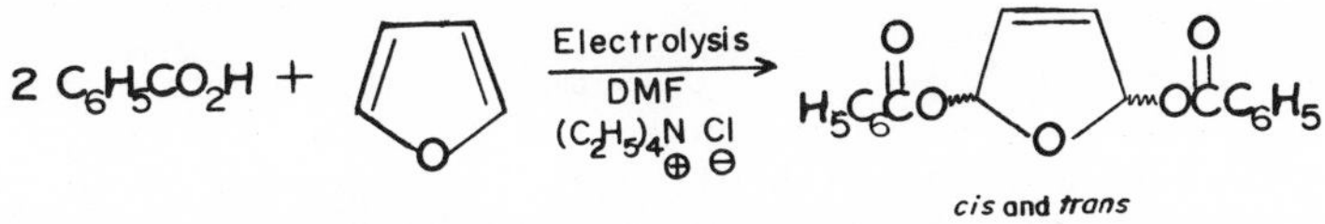

$2 C_6H_5CO_2H +$ furan $\xrightarrow[\substack{\text{DMF} \\ (C_2H_5)_4N\ Cl \\ \oplus\ \ominus}]{\text{Electrolysis}}$ $H_5C_6\overset{O}{\overset{\|}{C}}O\text{---(dihydrofuran ring)---}O\overset{O}{\overset{\|}{C}}C_6H_5$

cis and trans

The products were separated by chromatography.

Acyloxylation at carbon involving peroxides [9] and metal salts [10] has been carefully reviewed. Under the former, both catalyzed and uncatalyzed reactions are discussed. Yields are given for numerous acyloxylations and details are given for the preparation of 1-ethoxyethyl benzoate (77%) (Ex. b), N-acetoxymethyl-N-methylacetamide (67%), 3-acetoxycyclohexene (88%), α-ethyl β-(Δ²-cyclopentyl) camphorate (78%), 4-methylbenzyl acetate (49%), 7-benzoyloxynorbornadiene (36%), p-xylenyl benzoate (24%), 2-benzoyloxycyclohexanone (78%), as well as a general acyloxylation procedure.

The metal salts utilized in acyloxylations are the lead(II), mercury(II), and thallium(III) acetates and palladium salts. In

this review details are given for the preparation of 7-
acetoxycholesteryl acetate (21%), 21-acetoxy-3α-hydroxy-5β-
pregnane-11,20-dione (70%) (Ex. c), benzyl acetate (41%), as well
as a general procedure for the synthesis of α-acyloxycarboxylic
acids.

The acyloxy free radical in reactions involving metal salts
is usually inserted α to O, S, N and vinyl, and with difficulty,
α to carbonyl. The latter insertion is catalyzed by BF_3. In
catalysis with Pd(II), a trace of the metal has been used with a
relatively large amount of Cu(II) to reoxidize Pd(O) to Pd(II).
The Cu(I) formed is oxidized by air. For aralkyl compounds
acylation usually takes place on the side chain, but high temperatures bring about intrusion of nuclear acyloxylation, If for
some reason thermal acyloxylation fails, photochemical acylation
may succeed under milder conditions. The tables of the reviews
[9, 10] illustrate the wide range of reactions investigated.
Acyloxylations of a phenol with benzoyl peroxide and arenes with
lead tetrafluoroacetate have recently been reviewed [11, 12].

a. Preparation of 3-Benzoyloxycyclohexene [6]

Cyclohexene, 0.50 mol and CuBr, 0.00035 mol under N_2 were
heated to 80-82°, after which t-butyl perbenzoate, 0.21 mol was
added with stirring for 1 hr. Stirring and heating were continued for an additional 3 hr and the reaction mixture was then
washed with two 50-ml portions of dilute benzoic acid. From the
organic phase remaining 29-33 g (71-80%) of the acyloxycyclo-
hexene was recovered.

b. Preparation of 1-Ethoxyethyl Benzoate [9]

A mixture of ethyl ether, 1.45 mol t-butyl peroxybenzoate,
0.1 mol and cupric 2-ethylhexanoate, 24 ml of a 1% benzene
solution, was irradiated with ultraviolet light for 96 hr at 30-
32°. After washing with 2 N aqueous Na_2CO_3, the solution was
washed, dried and concentrated at 15 mm pressure. Distillation
in vacuo gave 15.1 g (77%) of the benzoate.

c. Preparation of 21-Acetoxy-3α-hydroxy-5β-pregnane-11,20-dione
[10]

A mixture of the diketone, 1 mol Pb(OCOCH$_3$)$_4$, 1.14 mol benzene, 370 ml and methanol, 20 ml was stirred and BF$_3$(Et)$_2$O, 50 ml was added. After 4 hr at 24° the product was separated with CHCl$_3$. Crystallization from ethyl acetate-light petroleum gave 7.26 g (70%) of acyoxylation product.

7. Reductive Acylation of Carbonyl Compounds or Acetoxy-lation of Quinones (Thiele-Winter) (1, 857)

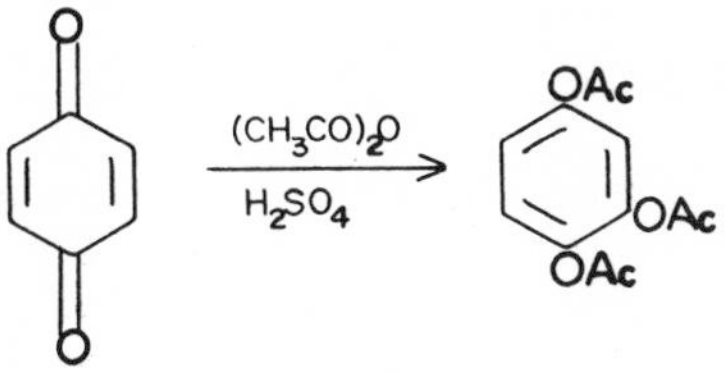

The Thiele-Winter reaction has been reviewed [13]. In this reaction an oxygen atom is introduced into the ring and catalysts such as sulfuric acid, zinc chloride, boron trifluoride etherate, aceticphosphoric anhydride, or perchloric acid are effective. Of the five the latter is probably the most effective. para-Quinones give 1,2,4-triacetoxy compounds, while ortho-quinones give either 1,2,3- or 1,2,4-triacetoxy compounds, although occasionally side-chain acetoxylation occurs. The reaction is a simple one to carry out (Ex. a) and the mechanism is similar to that already described (1, 735):

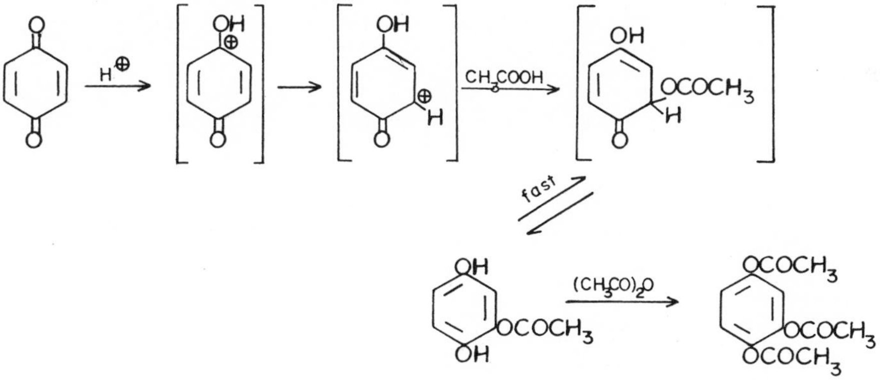

a. Preparation of 1,2,3-Triacetoxy-5-t-butylbenzene [14]

Concentrated H$_2$SO$_4$, 0.3 ml was added to a solution of 15 mmol of 4-t-butyl-1,2-benzoquinone in 30 ml of acetic anhydride at 20°. After 6 min the solution was poured onto 200 g of crushed ice. After stirring for 1 hr, the solid was collected by filtration and crystallized from methanol to give 3.5 g (77%) of the 1,2,3-triacetoxybenzene.

8.   Oxidation of Phenyl Methyl Ketones or Phenyl Alkyl
     Acetylenes

$$ArCOMe \xrightarrow[\text{MeOH-HClO}_4]{\text{Tl(NO}_3)_3} ArCH_2CO_2Me$$

$$ArC{\equiv}CR \xrightarrow[\text{CH}_3\text{OH}]{\text{Tl(NO}_3)_3} ArCHCOOCH_3 \overset{R}{|}$$

A simple process for the conversion of alkyl aryl ketones into ω-arylalkanoic acid derivatives has been devised by Taylor and co-workers [15]. The ketone is treated with thallium(III) nitrate in methanol containing perchloric acid as shown (Ex. a):

$$ArCOCH_3 \xrightarrow[\text{MeOH-HClO}_4]{\text{Tl(NO}_3)_3} ArCH_2CO_2Me$$
$$62-94\%$$

The method is preferred over the Willgerodt-Kindler reaction in that it: (a) succeeds at room temperature, (b) avoids a rather complicated isolation technique, and (c) often gives better yields. Its limitations are that: (a) it is unsuccessful with amino compounds and (b) low yields result in acetophenones in which the aromatic ring is highly deactivated. The mechanism apparently resembles that of the Willgerodt reaction (1, 917).
    In a similar manner, alkylarylacetylenes were oxidized with thallium(III) nitrate as indicated [16]:

$$ArC{\equiv}CR \xrightarrow[\substack{\text{CH}_3\text{OH}\\ \text{Reflux 2 hr.}}]{\text{Tl(NO}_3)_3} ArCHCO_2CH_3 \overset{R}{|}$$
$$92-98\%$$

Again the method is simple and as can be seen the yields of arylacetic esters are excellent.

a.   Preparation of Methyl Phenylacetate [15]

    Acetophenone, 0.01 mol was added to a solution of 0.011 mol of TTN in 25 ml of CH_3OH containing 5 ml of 70% $HClO_4$ and the mixture was stirred at 25° for 5 hr. The $TlNO_3$ precipitate was removed by filtration; the filtrate was diluted with water, and the product was extracted with chloroform. From the chloroform extract was recovered an 84% yield of the ester.

9.   Oxidation-Reduction of Ketals

Mastagli and de Nanteuil [17] treated cyclohexanone ethylene ketal with titanium tetrachloride and found that ring opening and dimerization occurred as indicated:

$$2 \quad \text{[structure]} \xrightarrow[\text{CH}_2\text{Cl}_2]{\text{TiCl}_4} \text{[structure]}(\text{CH}_2)_5\text{COOCH}_2\text{CH}_2\text{Cl}$$

2-Chloroethyl-6(Δ¹-cyclohexenyl)hexanoate

One possible intermediate

does not give as good a yield as the starting material.
In other similar ketals the yields were somewhat lower.

10. Oxidation of Alkenes

$$\text{RCH}=\text{CHR} \xrightarrow[\text{AcOH}]{\text{Mn(OAc)}_3} \text{[structure]}$$

The oxidation of olefins with manganic acetate leads to γ-butyrolactones as shown with yields of 16-79% (not maximized) [18]. The primary thermolysis product appears to be the carboxymethyl ($\cdot\text{CH}_2\text{COOH}$) rather than the acetoxy ($\text{CH}_3\text{CO}_2\cdot$) free radical, a fact which makes the process exceedingly interesting. The procedure is simple in that 0.1M solution of the olefin in glacial acetic acid is refluxed under nitrogen with 2 molar equiv of $\text{Mn(OAc)}_3 \cdot 2\text{H}_2\text{O}$ in the presence of added potassium acetate (300 g/liter of AcOH) until the brown manganic color disappears.

11. Oxidation of Tartrates and α-Acylaminoacrylic Esters

Methyl and ethyl glyoxylates have been prepared conveniently by the oxidation of the appropriate tartrate ester with ethereal periodic acid as indicated [19] (Ex. a).

$$\begin{array}{c} \text{H} \\ | \\ \text{HO}-\text{C}-\text{COOR} \\ | \\ \text{HO}-\text{C}-\text{COOR} \\ | \\ \text{H} \end{array} \xrightarrow[\text{Et}_2\text{O}]{\text{HIO}_6} 2 \ \text{H}-\overset{\text{O}}{\underset{}{\text{C}}}-\text{COOR}$$

76-87% (with 5% H₂O)

α-Acylaminoacrylic esters have been obtained by the oxidation

of β-thio esters with silver carbonate or silver oxide [20].
Thus N-acetylcysteinate gives the α-acetamidoacrylate as shown:

$$
\begin{array}{ccc}
\underset{|}{HSCH_2} & & \underset{||}{CH_2} \\
\underset{|}{CHCO_2Me} & \xrightarrow[CH_3OH]{Ag_2CO_3} & \underset{|}{CCO_2Me} \\
NHCOCH_3 & & NHCOCH_3 \\
& & 78\%
\end{array}
$$

a.   Preparation of Methyl Glyoxylate [19]

To a solution of dimethyl d-tartrate, 0.5 mol in 900 ml of
cold, dry ether under N_2 was added H_5IO_6, 0.5 mol over 1 hr with
stirring. Stirring was continued until a white solid separated,
after which the ether phase was decanted, dried with Linde 4A
molecular sieves, evaporated, and distilled to give 71 g (76%) of
clear methyl glyoxylate containing 5% water. Distillation with
P_2O_5 gives a pure product.

12.   Oxidation of Acetals or Vinyl Silyl Ethers

Deslongchamps and Moreau [21] have shown that acetals may be
oxidized to esters by the use of ozone as indicated:

$$
\underset{\underset{OR'}{|}}{\overset{\overset{OR'}{|}}{RCH}} \xrightarrow[CH_2Cl_2]{O_3} \overset{\overset{O}{||}}{RCOR'}
$$

For example, the dimethylacetal of heptaldehyde gave methyl
heptanoate in 90% yield after 1.5 hr at 25°. Cyclic acetals
react readily at -78° in 2 hr, while acyclic ones respond in
about the same time at room temperature. Methyl 2,3,4,6-tetra-0-
acetyl-β-D-glucopyranoside gives a 95% yield of methyl 2,3,4,5,6-
penta-0-acetylgluconate, but the corresponding anomer is
unaffected.

13.   Reduction of Unsaturated Esters

Moppett and Sutherland [22] have shown that γ-bromo-α,β-
unsaturated esters yield essentially β,γ-unsaturated esters on
treatment with zinc and acetic acid as shown:

$$
BrCH_2CH{=}CHCO_2Et \xrightarrow[AcOH]{Zn} CH_2{=}CHCH_2CO_2Et
$$

Ethyl 4-bromobuten-2-oate   Ethyl buten-3-oate
                                  70% of 95% purity

Similar results were obtained with other γ-bromo unsaturated
esters.

Zurqiyah and Castro [23] reduced conjugated alkenes to alkanes with chromium(II) sulfate. Thus diethyl fumarate responded as indicated (Ex. a):

$$H_5C_2O_2C-CH=CH-CO_2C_2H_5 \xrightarrow[2\ H^\oplus]{2\ Cr(II)^\oplus} H_5C_2O_2CCH_2CH_2CO_2C_2H_5$$

Diethyl succinate, 88-94 %

Chromium(II) sulfate is a mild, versatile reagent that may also be employed to reduce acetylenes to _trans_ olefins, and α,β-unsaturated esters, acids, and nitriles to the corresponding saturated derivatives.

Satoh and co-workers reduced unsaturated esters to saturated ones quantitatively by the use of sodium borohydride-transition-metal salt systems [24]. Thus methyl cinnamate gave methyl β-phenylpropionate quantitatively as shown:

$$C_6H_5CH=CHCOOCH_3 \xrightarrow[\substack{CoCl_2 \cdot 6H_2O-NaBH_4 \\ or \\ CuCl_2 \cdot 2H_2O-NaBH_4}]{\substack{NiCl_2 \cdot 6H_2O-NaBH_4 \\ of}} C_6H_5CH_2CH_2COOCH_3$$

a. Preparation of Diethyl Succinate [23]

To diethyl fumarate, 0.08 mol in 137 ml of DMF over N_2 was added 0.175 mol of 0.55M $CrSO_4$ solution with stirring. After 10 min, 100 ml of water and 30 g of $(NH_4)_2SO_4$ were added. Extraction with ether and distillation gave 12.4-13.2 g (88-94%) of the diester.

14. Coupling of α-Halogenated Esters by Reduction (Wurtz) or Oxidation

α-Bromoesters when treated with zinc (or magnesium) and copper(II) chloride in THF couple as shown [25] (Ex. a).

$$R^2-\underset{\underset{COOR^3}{|}}{\overset{\overset{R^1}{|}}{C}}-Br \xrightarrow[CuCl_2, THF]{Zn} R^2-\underset{\underset{COOR^3}{|}}{\overset{\overset{R^1}{|}}{C}}-\underset{\underset{COOR^3}{|}}{\overset{\overset{R^1}{|}}{C}}-R^2$$

25-40 %

Methyl 2-bromopropanoate couples similarly but the yields of the succinate formed are lower.

The anion of ethyl phenylacetate does not couple with 1,2-dibromides or 1,2-diiodides. Instead it is oxidized and coupled to give the substituted succinic ester as indicated [26]:

$$C_6H_5CHKCO_2Et \xrightarrow{Cl_3CCCl_3} \begin{array}{l} C_6H_5CHCO_2Et \\ | \\ C_6H_5CHCO_2Et \end{array}$$

Diethyl 2,3-diphenylsuccinate
54% *meso* , 24% *dl*

1,2-Dibromides or 1,2-diiodides react similarly.

a.   Preparation of Dimethyl Tetramethylsuccinate [25]

To anhydrous $CuCl_2$, 0.055 mol 65 ml of dry THF was added. Zinc ribbon, 0.055 g-atom methyl 2-bromo-2-methylpropanoate, 0.027 mol and a trace of $HgBr_2$ were added and the mixture was heated to initiate the reaction. After 1 hr refluxing the cooled mixture was decanted, acidified, and extracted with ether from which liquid was recovered 1.0 g (38%) of the succinate.

15.   Reduction of Acid Anhydrides and Acyl Halides

Bailey and Johnson [27] have reported that sodium borohydride reduces cyclic anhydrides to form δ- and γ-lactones in isolated yields of 51-97%. The procedure is more convenient and versatile than if lithium aluminum hydride or lithium tri-t-butyoxyaluminohydride is used as the reducing agent. Usually, the hydride attack occurs at the carbonyl group adjacent to the more highly substituted carbon atom, as illustrated in the case of cis-1-methylcyclohexane-1,2-dicarboxylic acid anhydride:

Lactone of *cis*-2-methyl-
2-hydroxymethylcyclohexane-
carboxylic acid, 65%

Acid fluorides are solvolyzed to esters by triethylsilane as shown [28]:

$$2 \, C_4H_9COF + Et_3SiH \xrightarrow[24\,hr.]{115°} C_4H_9CO_2C_5H_{11}$$

Pentyl pentanoate, 71%

Acid chlorides, on the other hand, in the presence of Pd/C are reduced by the silane to aldehydes.

16.   From Dihydropyrazoles

The dihydropyrazoles obtained from β-ketoacetic esters and hydrazine are readily oxidized to propargylic esters [29].

$$RC{\equiv}CCO_2CH_3$$
67-95%

17.   From Esters to Unsaturated Esters (Use of Phenylselenium
Bromide (see Addenda)

Addenda

D.2.   Another preparation of α-methylenebutyrolactone (see intro-
ductory paragraph of Addenda) is accomplished by the oxidation of
α-methylenetetramethylene glycol with Ag_2CO_3 on celite [M.
Fetizon et al., Tetrahedron, 31, 171 (1975)].

D.8.   A mixture of selenium dioxide and hydrogen peroxide converted cyclobutanone into γ-butyrolactone (95%), but with a trace
of aqueous acid cyclopropanecarboxylic acid (70%) was formed
[J. M. Conia et al., Tetrahedron, 30, 1423 (1974)].

D.12.   In the oxidation of the trimethylsilyl vinyl ether of
cyclopentanones by treatment with ozone followed by $NaBH_4$, even
in the presence of other olefinic groups, valerolactones are
obtained [R. D. Clark, C. H. Heathcock, Tetrahedron Letters, 2027
(1974)].

D.17.   An α-carbethoxymethylenetetrahydrofuran may be converted
into an α-carbethoxymethylenefuran by consecutive treatment with
LDIA, phenylselenium bromide, and hydrogen peroxide [C. H.
Wilson, II, T. A. Bryson, J. Org. Chem., 40, 800 (1975)].   See
A.21 for similar treatment of ketones.

References

1.   P. R. Stapp, J. Org. Chem., 38, 1433 (1973).
2.   G. F. Reynolds et al., Tetrahedron Letters, 5057 (1970).
3.   J. Haufe, F. Beck, Chem. Ing. Tech., 42, 170 (1970).
4.   H. Schäfer, R. Pistorius, Angew. Chem. Intern. Ed. Engl.,
11, 841 (1972).
5.   K. Matsumoto et al., J. Org. Chem., 38, 2731 (1973).
6.   K. Pedersen et al., Org. Syn., Coll. Vol. 5, 70 (1973).
7.   J. R. Gilmore, J. M. Mellor, J. Chem. Soc., C2355 (1971).
8.   S. Arita et al., Kogyo Kagaku Zasshi, 72, 2289 (1969);
C. A., 72, 138926 (1970).

9. D. J. Rawlinson, G. Sosnovsky, Synthesis, 1 (1972).
10. D. J. Rawlinson, G. Sosnovsky, Synthesis, 567 (1973).
11. D. H. R. Barton et al., J. Chem. Soc., C, 2231 (1971).
12. S. Sternhell et al., Tetrahedron Letters, 1763, 5369 (1972).
13. J. F. U. McOmie, J. M. Blatchly, Org. React., 19, 199 (1972).
14. J. M. Blatchly et al., J. Chem. Soc., Perkin Trans., I, 2286 (1972).
15. E. C. Taylor et al., J. Am. Chem. Soc., 93, 4919 (1971).
16. E. C. Taylor et al., J. Am. Chem. Soc., 93, 7331 (1971).
17. P. Mastagli, M. de Nanteuil, Compt. Rend., 268, C1970 (1969).
18. E. I. Heiba et al., J. Am. Chem. Soc., 90, 5905 (1968).
19. T. R. Kelly et al., Synthesis, 544 (1972).
20. D. Gravel et al., Chem. Commun., 1322 (1972).
21. P. Deslongchamps, C. Moreau, Can. J. Chem., 49, 2465 (1971).
22. C. E. Moppett, J. K. Sutherland, J. Chem. Soc., C3040 (1968).
23. A. Zurqiyah, C. E. Castro, Org. Syn., Coll. Vol., 5, 993 (1972).
24. T. Satoh et al., Chem. Pharm. Bull., 19, 817 (1971); C. A., 76, 14037 (1972).
25. C. Fouquey, J. Jacques, Synthesis, 306 (1971).
26. W. G. Kofron, C. R. Hauser, J. Org. Chem., 35, 2085 (1970).
27. D. M. Bailey, R. E. Johnson, J. Org. Chem., 35, 3574 (1970).
28. J. D. Citron, J. Org. Chem., 36, 2547 (1971).
29. E. C. Taylor et al., Angew. Chem. Intern. Ed. Engl., 11, 48 (1972).

E. Irradiative, Carbene, Free Radical, and Cyclo Reactions

1. From Diazo Esters and Trialkylborons or Dialkylchloroboranes

Hooz and Linke [1] prepared esters by the action of the trialkylboron on diazoesters as shown (Ex. a).

$$R_3B + N_2CHCOOC_2H_5 \xrightarrow{H_2O} RCH_2COOC_2H_5$$
$$40-83\ \%$$

The lower yields were obtained in cases in which a steric factor was present. By carrying out the reaction with D_2O rather than H_2O quantitative yields of the deuterated ester are obtained as indicated [2].

$$R_3B + N_2CHCOOC_2H_5 \xrightarrow{D_2O} RCHCOOC_2H_5$$
$$\overset{|}{D}$$

These results are corroborative of the mechanism:

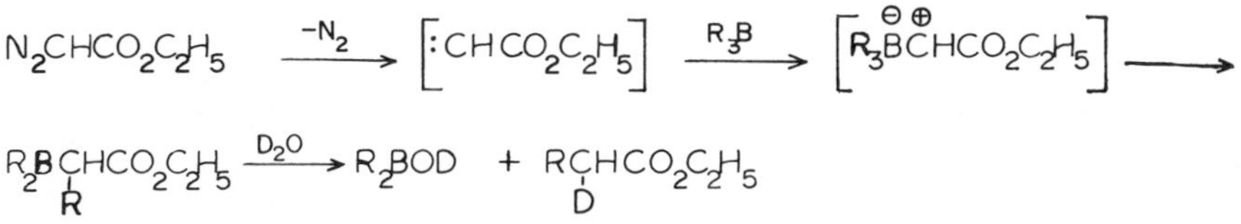

$$N_2CHCO_2C_2H_5 \xrightarrow{-N_2} \left[:CHCO_2C_2H_5 \right] \xrightarrow{R_3B} \left[R_3\overset{\ominus}{B}\overset{\oplus}{C}HCO_2C_2H_5 \right] \longrightarrow$$

$$R_2BCHCO_2C_2H_5 \xrightarrow{D_2O} R_2BOD + RCHCO_2C_2H_5$$
$$\overset{|}{R} \qquad\qquad\qquad\qquad \overset{|}{D}$$

To prepare propargylic esters Hooz and Layton [3] started with the lithium acetylide and proceeded via the trialkynylboron as shown:

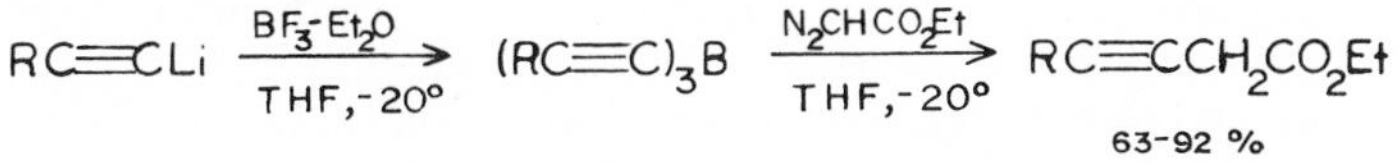

$$RC{\equiv}CLi \xrightarrow[THF,-20°]{BF_3-Et_2O} (RC{\equiv}C)_3B \xrightarrow[THF,-20°]{N_2CHCO_2Et} RC{\equiv}CCH_2CO_2Et$$

63-92 %

Brown and co-workers [4] substituted dialkylchloroboranes for trialkylborons in the reaction with ethyl diazoacetate and found that under mild conditions followed by hydrolysis almost quantitative yields of ethyl esters were obtained. Thus diiso-butylchloroborane gave a 98% yield of ethyl 4-methylpentanoate:

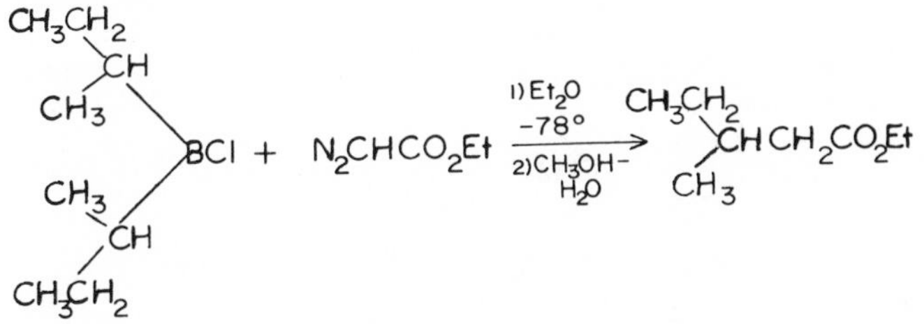

This method has an advantage over that employing the trialkyl-boron in that bulky alkyl groups lead to satisfactory yields.

a. Preparation of Ethyl Octanoate [1]

A solution of ethyl diazoacetate, 20 mmol in 15 ml of THF was added during 20 min to an ice-cooled, stirred solution of trihexylboron, 20 mmol in THF. After keeping the solution at the same temperature for an additional 30 min, it was stirred at room temperature for 2 hr. Water, 5 ml was then added dropwise and the mixture was refluxed for 1 hr. Analysis (glpc) indicated an 83% yield of ester; recovery yielded 2.40 g (70%).

2. From Ethyl Acetoacetate and an Alcohol by Irradiation

Solutions of ethyl acetoacetate and a primary or secondary alcohol under irradiation lead to the formation of a glycol that forms a lactone [5]. Thus ethyl acetoacetate and methyl alcohol respond as shown:

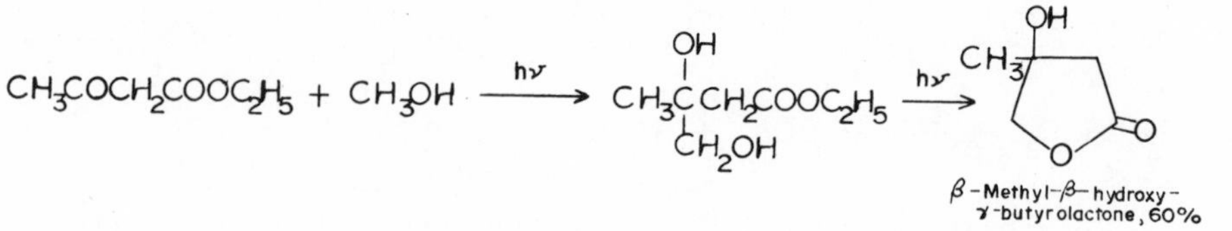

β-Methyl-β-hydroxy-
γ-butyrolactone, 60%

3. Cycloaddition

Boekelheide and Nottke [6] have shown that acetylene derivatives and α-methylene carbonyl compounds react in the presence of a base to form aromatic compounds. Thus phenyl-acetaldehyde with methyl propiolate gave dimethyl 5-phenyliso-phthalate as indicated:

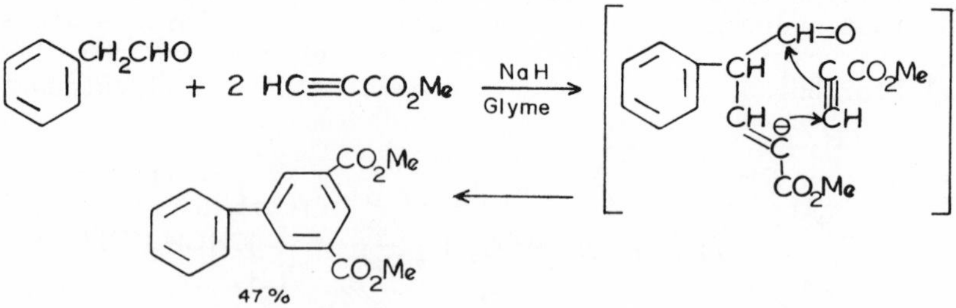

47 %

The reaction is general and of potential value in the synthesis of di- and terphenyl derivatives.

Sample and Hatch [7] have synthesized diethyl trans-Δ^4-tetrahydrophthalate as shown by the Diels-Alder addition of 3-sulfolene and diethyl fumarate (Ex. a).

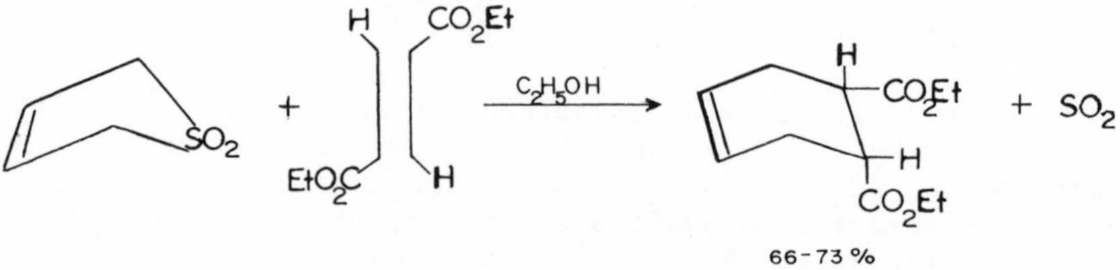

66-73 %

3-Sulfolene is a useful substitute for a diene in the Diels-Alder reaction. Benzene or toluene may be substituted for the alcohol solvent if the latter leads to side reactions. The present reaction with 3-sulfolene rather than with butadiene leads to a higher yield of product at lower reaction temperature and shorter reaction time. 3-Sulfolene also has these advantages

over butadiene: (a) it presents no particular flammability
hazard and is practically nontoxic, (b) it is a nonhygroscopic
solid that remains unchanged on storage, and (c) medium-pressure
reaction vessels can be employed since the autogenous pressures
produced are relatively low.

It should be noted that the use of 3-sulfolene leads to a
trans addition product. Thus the method given is important since
the trans analogs are much more difficult and expensive to prepare and isolate than the corresponding cis isomers.

Furans have been converted into m-hydroxybenzoic esters by
an aluminum chloride catalyzed Diels-Alder reaction [8]:

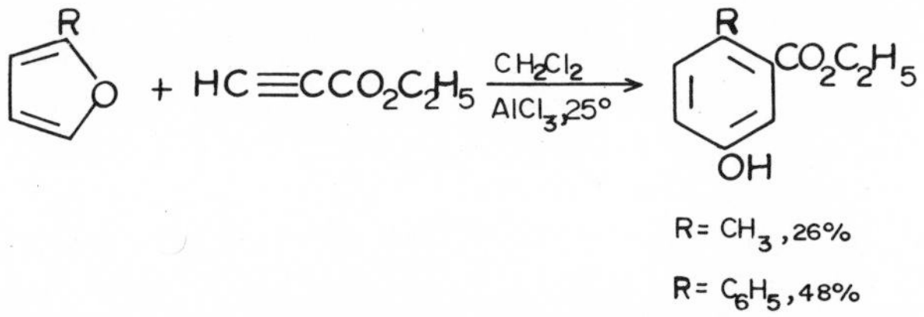

$$R = CH_3, 26\%$$
$$R = C_6H_5, 48\%$$

A number of esters, including cyclopropane ones, have been
obtained by treating the cycloadducts of halogenated ketenes and
alkenes with sodium methoxide in methanol [9].

For additional Diels-Alder reactions leading to esters, see
(1, 123) and the work by Onishchenko [10].

Cadogan and co-workers [11] have produced 1,2,3,4-tetra-
carbomethoxynaphthalenes by the addition of aryl radicals (from
diaroyl peroxides) to dimethyl acetylenedicarboxylate. Thus the
reaction of dibenzoyl peroxide with the acetylenedicarboxylate is
as shown:

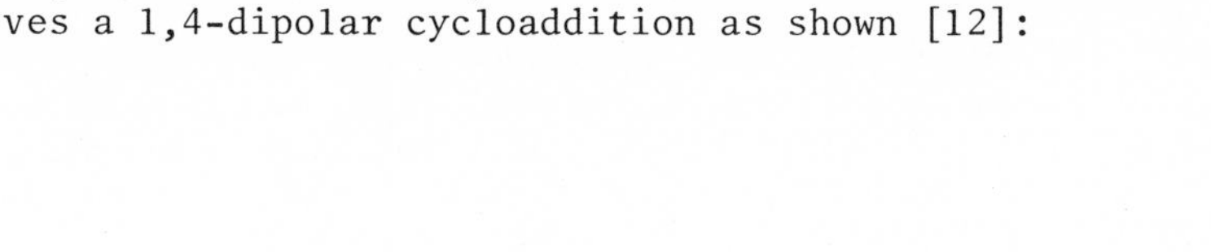

A general but novel synthesis of lactones from alkenes
involves a 1,4-dipolar cycloaddition as shown [12]:

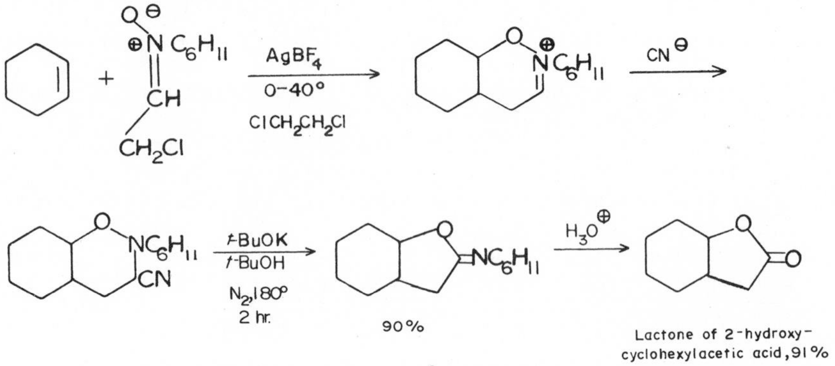

Lactone of 2-hydroxy-
cyclohexylacetic acid, 91%

a. Preparation of Diethyl trans-Δ^4-Tetrahydrophthalate [7]

3-Sulfolene, 0.51 mol diethyl fumarate, 0.50 mol and 1 g of hydroquinone were stirred with 90 ml of absolute ethanol until the solution was almost complete. The mixture was then heated slowly to 105-110° and held there for 8-10 hr. From the reaction mixture made alkaline with Na_2CO_3, the product was extracted with petroleum ether, which upon distillation gave 75-82 g (66-73%) of the ester.

4. Addition of Free Radicals to Heterocycles

Bernardi and co-workers [13] produced free carbethoxy radicals from ethyl pyruvate, hydrogen peroxide, and ferrous sulfate. These radicals add to various heterocycles as is shown for quinoline:

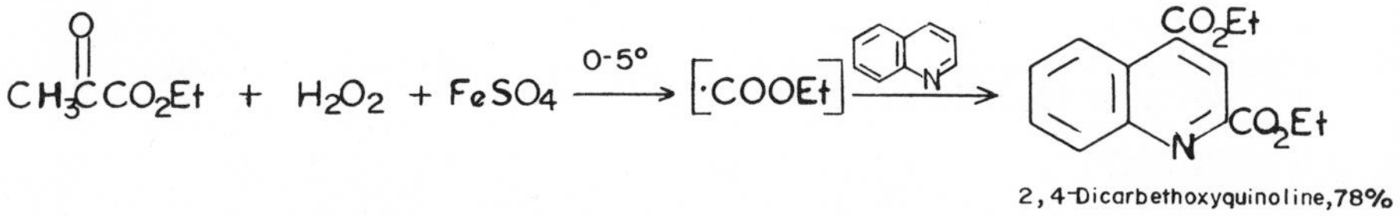

2,4-Dicarbethoxyquinoline, 78%

5. Annelation of Maleates and Fumarates with Dimethylsulfonium 2-Oxotetrahydrofuryl-3-ylid

Trost and Arndt [14] have developed a new annelating agent, dimethylsulfonium-2-oxotetrahydrofuryl-3-ylid, which combines with acrolein, methyl vinyl ketone, acrylonitrile, dimethyl fumarate, diethyl maleate, and benzalacetophenone to give the corresponding cyclopropanes. The reaction with dimethyl maleate is as indicated:

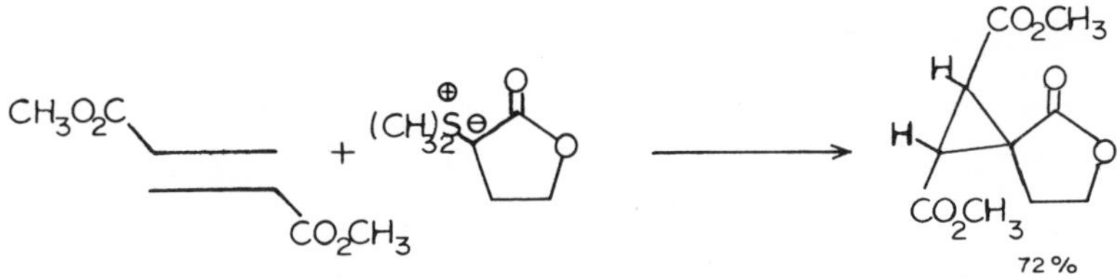

72%

6. From Ketones, Alkyl Diazotates, and Triethyloxonium Fluoroborate

Mock and Hartman [15] have improved on an earlier procedure of homologation in the synthesis of β-keto esters; the method is simple and generally gives good yields. It consists of treating the ketone with ethyl diazoacetate and triethyloxonium fluoroborate and is illustrated for cyclohexanone (Ex. a):

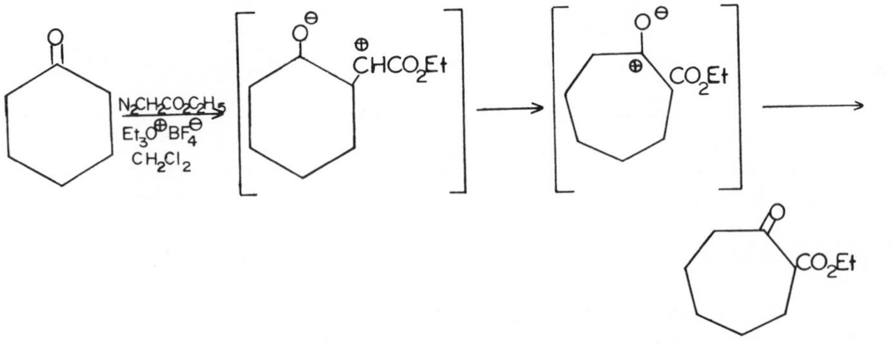

2-Carbethoxycycloheptanone, 90%

a. Preparation of 2-Carbethoxycycloheptanone [15]

To a solution of cyclohexanone, 0.05 mol in methylene chloride at 0° under an inert atmosphere, triethyloxonium fluoroborate, 0.088 mol was added, followed by dropwise addition of ethyl diazoacetate, both with stirring. After the mixture was held at 0° for 3 hr, it was quenched with 150 ml of saturated NaHCO₃ solution. From the methylene chloride layer there was recovered by distillation 8.2 g (90%) of the ketone.

7. Cyclization via the Claisen-Cope Rearrangement

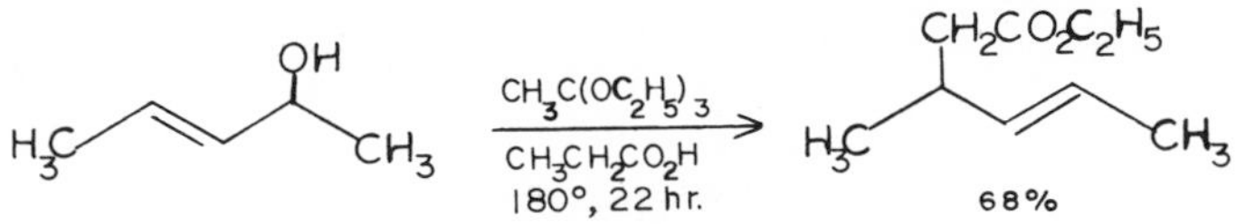

The Claisen-Cope rearrangement has been reviewed [15a].

Hill and co-workers [16] found, as shown, that trans-3-penten-2-ol on treatment with triethyl orthoacetate containing a trace of propionic acid gave trans-ethyl 3-methyl-4-hexenoate. The reaction when applied to optically active alcohols leads to the formation of new carbon-carbon bonds at an asymmetric center with high stereospecificity. The Claisen-Cope rearrangement may be represented as follows:

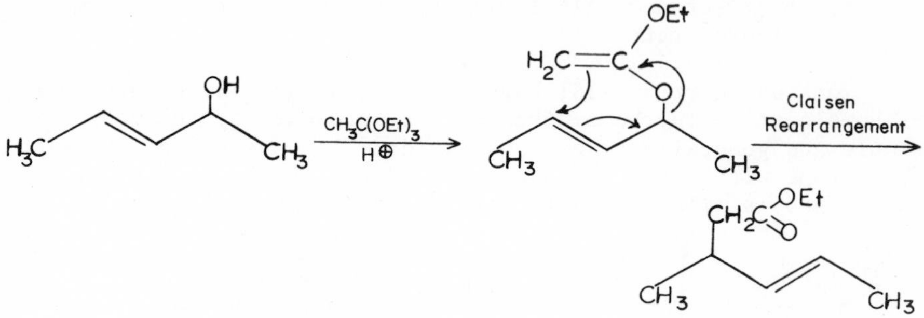

8. From Acrylic Compounds and Maleic and Fumaric Esters

Morita and Kobayashi [17] added acrylic compounds to fumaric and maleic esters in the presence of tervalent phosphorus compounds. This method for the synthesis of 3-butene-1,2,3-tricarboxylic acid 1,2-diethyl-3-methyl ester is shown:

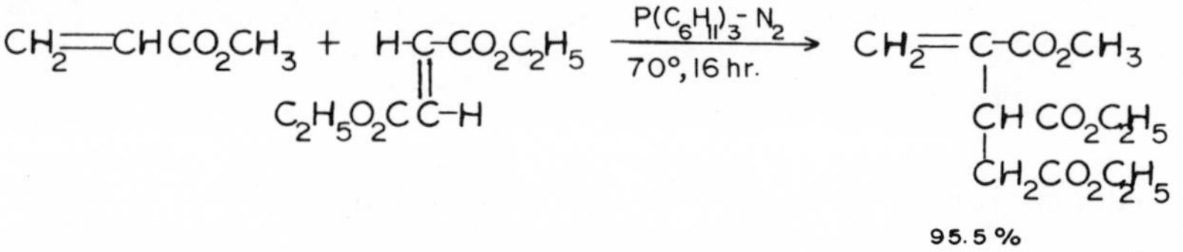

Apparently, the enol form of the acrylic ester is involved in the mechanism as indicated:

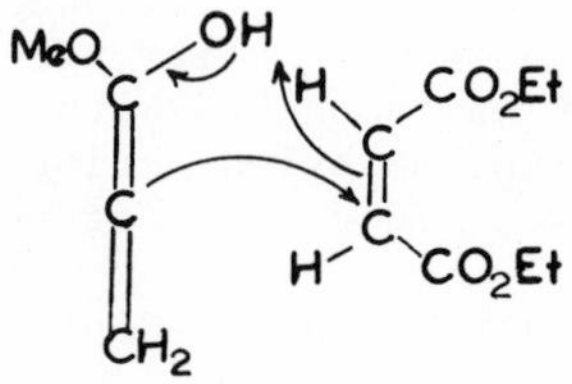

9. Retrocyclo Reactions (see 1, 122)

The objective of this section is to obtain products from a reverse cyclo reaction yielding fragments different from those utilized to form the original cyclo product.

Carlson and Hill [18] synthesized <u>trans,trans</u>-1,4-diacetoxy-1,3-butadiene by the steps which involve a reverse Diels-Alder, as shown:

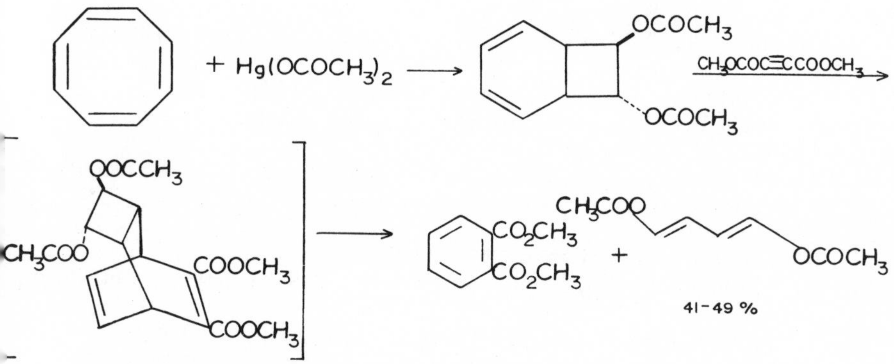

In a second example Jones and co-workers cleaved an oxetane, derived from diethyl maleate, to obtain a methyl crotonate [19].

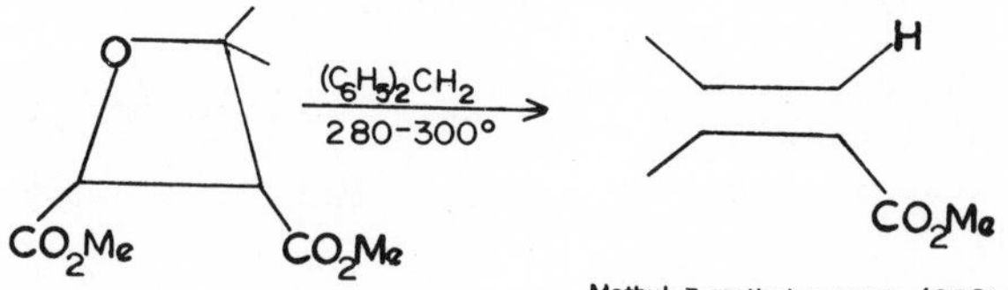

Methyl 3-methylcrotonate,(90% by nmr)

Addenda

E.1.   The carbethoxy free radical, ·CO$_2$Et, which has been generated from ethyl α-oxypropionate, hydrogen peroxide, and iron, has been persuaded to attack heterocycles, usually α or γ to the hetero atom [R. Bernardi et al., <u>Tetrahedron Letters</u>, 645 (1973)].

E.1.   A review on the synthesis and carbene reactions of ethyl α-diazopropionate has been published [M. Jones, Jr., W. von E. Doering, <u>Tetrahedron Letters</u>, 53 (1972).

E.7.   The Cope rearrangement of 1,4-dihydroxy-2-butene by heating with ethyl orthoacetate gives 2-vinylbutyrolactone (91%).   [K. Kondo, F. Mori, <u>Chem</u>. <u>Lett</u>., 741 (1974)].

References

1.   J Hooz, S. Linke, J. Am. Chem. Soc., 90, 6891 (1968).
2.   J. Hooz, D. M. Gunn, J. Am. Chem. Soc., 91, 6195 (1969).
3.   J. Hooz, R. B. Layton, Can. J. Chem., 50, 1105 (1972).
4.   H. C. Brown et al., J. Am. Chem. Soc., 94, 3662 (1972).
5.   S. P. Singh, J. Kagan, Chem. Commun., 1121 (1969).
6.   V. Boekelheide, J. E. Nottke, J. Org. Chem., 34, 4134 (1969).
7.   T. E. Sample, Jr., L. F. Hatch, Org. Syn., 50, 43 (1970).
8.   A. W. McCulloch, A. G. McInnes, Can. J. Chem., 49, 3152 (1971).
9.   W. T. Brady, Synthesis, 415 (1971).
10. A. S. Onishchenko, Diene Synthesis, Daniel Davey, New York, 1964.
11. J. I. G. Cadogan et al., Chem. Commun., 1318 (1972).
12. A. Eschenmoser et al., Helv. Chim. Acta, 55, 2198 (1972).
13. R. Bernardi et al., Tetrahedron Letters, 645 (1973).
14. B. M. Trost, H. C. Arndt, J. Org. Chem., 38, 3140 (1973).
15. W. L. Mock, M. E. Hartman, J. Am. Chem. Soc., 92, 5767 (1970).
15a. S. J. Rhoads, N. R. Raulins, Org. Reactions, 22, 1 (1975).
16. R. K. Hill et al., J. Org. Chem., 37, 3737 (1972).
17. K. Morita, T. Kobayashi, Bull. Chem. Soc. Jap., 42, 2732 (1969).
18. R. M. Carlson, R. K. Hill, Org. Syn., 50, 24 (1970).
19. G. Jones, II et al., Chem. Commun., 374 (1973).

In the following discussion two types, the orthoester, $HC(OR)_3$ or $R'C(OR)_3$ (Section F) and the orthocarbonate, $C(OR)_4$ (Section G), are considered separately. The latest reviews on their synthesis are those by Cordes [1] and DeWolfe [2].

F. Orthoesters

Five methods of synthesis are discussed. The first two appear to be the most common.

1. From Nitriles (Pinner Synthesis) [3]

$$ROH + R'CN \xrightarrow[1]{HCl} \overset{\overset{\displaystyle NH \cdot HCl}{\|}}{R'COR} \xrightarrow[2]{2\,ROH} R'C(OR)_3 + NH_4Cl$$

Yields are quite dependent on anhydrous conditions and are most favorable in the absence of steric effects in either the acid or alcohol portions of the orthoester. Step 1 occurs rapidly at room temperature to form the salt, while Step 2 is best carried out by refluxing in ether with care to keep the temperature below 40° since the hydrochloride may be converted into the amide. A different alcohol may be used for Step 2 to lead to mixed orthoesters, many of which are listed by Post [4].

a. Preparation of Methyl Orthoisobutyrate [5]

Isobutyronitrile, 2 mol methanol, 2.2 mol and 750 ml of dry ether were saturated with hydrogen chloride until 78 g was absorbed. The mixture was allowed to crystallize for 48 hr in a refrigerator, after which the crystalline methyl iminoisobutyrate hydrochloride was filtered, washed with ether, and dried in a vacuum desiccator over solid NaOH for 48 hr to give 270 g (99%). The latter, 275 g, 6 mol of absolute methanol, and 1800 g of dry petroleum ether, boiling point 35°, were stirred for 2 days at 25° and the filtrate was distilled through a 25 cm Heli-Pak column. After removing the methanol and petroleum ether by distillation, the isobutyramide was removed by filtration and the liquid remaining was distilled from 5 g of NaH to yield 208 g (70%) of the orthoester.

2. From Trihalides

$$RCX_3 + NaOR' \longrightarrow RC(OR')_3$$

The reaction is most successful with chloroform, benzotrichlorides, on occasion with 1,1,1-trichloroethane, or with other polyhaloalkanes. Care must be taken in the last two instances to avoid dehydrohalogenation. The reaction is illustrated in Ex. a.

a. Preparation of Tributyl Orthoformate [6]

To dry chloroform, 150 g and butanol, 500 ml, 69 g of metallic sodium was added gradually over 2 hr (initial heating was necessary). The cooled mixture was filtrated and the filtrate was fractionated twice to obtain about 30% of pure orthoformate, boiling point 245-247°.

3. From Orthoesters

$$HC(OEt)_3 + 3\ ROH \longrightarrow HC(OR)_3 + 3\ EtOH$$

This reaction is of interest because commercially available orthoesters, such as triethyl orthoformate, may be exchanged. In the process Lewis acids such as zinc chloride appear to be the best catalysts [7].

4. From Ketene Acetals

$$R_2C{=}C(OEt)_2 + R'OH \xrightarrow{H^{\oplus}} R_2CC(OEt)_2OR'$$

The method was made available by McElvain and co-workers [8].

5. From Ketene Acetals (2,4-Cycloaddition)

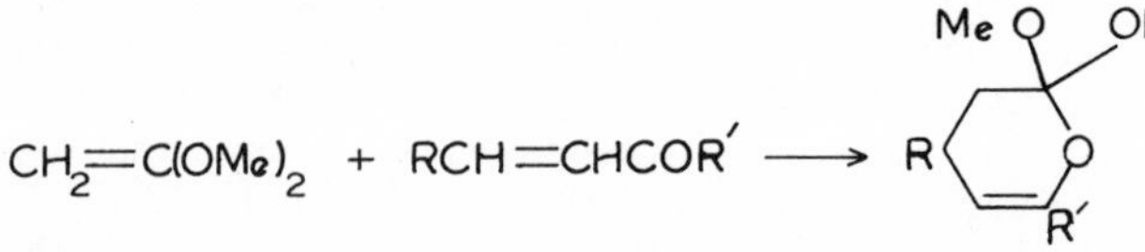
Me O OMe
CH2=C(OMe)2 + RCH=CHCOR' ⟶ R
O
R'

When R and R' are H, an interesting series of conversions is possible [9]:

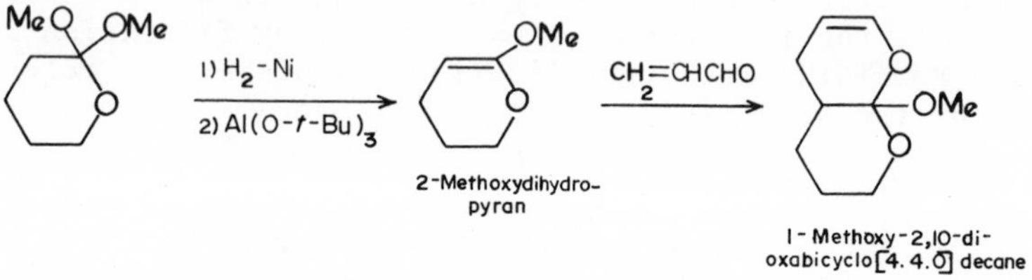
MeO OMe
1) H2-Ni
2) Al(O-t-Bu)3
OMe
2-Methoxydihydro-pyran
CH2=CHCHO
OMe
1-Methoxy-2,10-dioxabicyclo[4.4.0] decane

The preceding are simply illustrations of the possibilities in the synthesis of orthoesters by cycloaddition.

Addendum

F. The application of orthoesters in organic synthesis has been reviewed [V. V. Mezheritskii et al., Russ. Chem. Rev. (Engl. transl.), 42, 392 (1973)].

References

1. E. H. Cordes, in S. Patai, The Chemistry of Carboxylic Acids and Esters, Interscience, New York, 1969, Chapter 13.

2.   (a) R. H. DeWolfe, Synthesis, 153 (1974); (b) Carboxylic
     Ortho Acid Derivatives, Academic, New York, 1970.
3.   Ref. 1, p. 625.
4.   H. W. Post, The Chemistry of the Aliphatic Orthoesters,
     Reinhold, New York, 1943.
5.   S. M. McElvain, C. L. Aldridge, J. Am. Chem. Soc., 75, 3987
     (1953).
6.   P. P. T. Sah, T. S. Ma, J. Am. Chem. Soc., 54, 2964 (1932).
7.   R. P. Narain, R. C. Mehrotra, Proc. Natl. Acad. Sci., India,
     A33, Pt. 1, 45 (1963); C. A., 59, 5018 (1963).
8.   S. M. McElvain et al., J. Am. Chem. Soc., 77, 5601 (1955)
     and earlier papers.
9.   Ref. 2b, p. 46.

G.   Orthocarbonates

   1.   From Chloropicrin

$$4 \ RONa + Cl_3CNO_2 \longrightarrow C(OR)_4 + 3 \ NaCl + NaNO_2$$

This method resembles that described in F.2.   However
carbon tetrachloride is not satisfactory except with fluoroalco-
hols [1].   Instead, chloropicrin works satisfactorily with
sodium alcoholates.

   2.   From Carbon Disulfide

Recently, orthocarbonates have been produced from carbon
disulfide by two schemes:

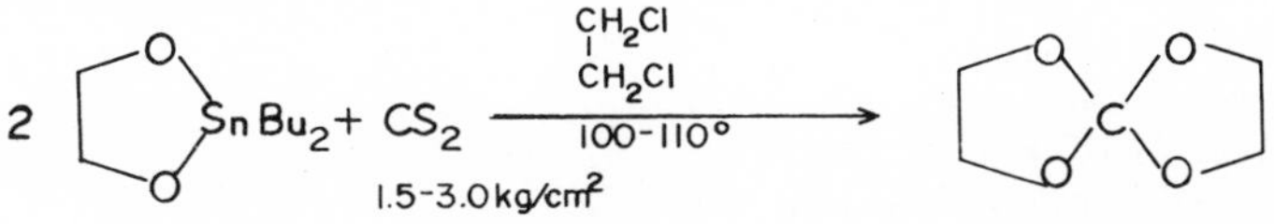

Diethylene-spiro-orthocarbonate
82%   [2]

$$4 \ TlOEt + CS_2 \xrightarrow[4\,hr.,\,25°]{CH_2Cl_2} C(OEt)_4 + 2 \ Tl_2S \qquad [3]$$

Tetraethylorthocarbonate
69%

References

1.   M. E. Hill et al., J. Org. Chem., 30, 411 (1965).
2.   S. Sakai et al., J. Org. Chem., 36, 1176 (1971).
3.   S. Sakai et al., J. Org. Chem., 37, 4198 (1972).

Chapter 15

ACYL HALIDES

The acyl halides should be handled with care. The volatile members are lachrymatory and the nonvolatile ones may likely be skin irritants. All are susceptible to hydrolysis on storage. Evidence now exists that the thionyl chloride preparation of α,β-unsaturated acyl chlorides should be carried out at low temperature either with or without HMPA (A.1). Two new reagents have been developed to give neutral products (rather than SO_2 and HCl with thionyl chloride) (A.2). Triphenylchlorophosphonium chloride appears to be satisfactory for forming acyl chlorides from esters (A.3), while a halosulfonic acid is suitable to form acyl halides from acid anhydrides (A.4). A facile synthesis of the relatively unknown α-chlorosulfonylacyl chlorides is now available (A.1).

An excellent discussion of acyl halides, which gives methods of preparation in considerable detail, is now available [1]. A similar discussion of the synthesis of sulfonyl halides is also available [2].

A. Metathesis

1. From Carboxylic Acids and Inorganic Halides ($\underline{1}$, 860)

$$RCOOH \xrightarrow{PX_3} RCOX$$

Although it is often desirable to use thionyl chloride with a basic material such as pyridine to react with the acid formed in the reaction, such a procedure proved to be undesirable in the synthesis of cis-crotonyl chloride from the cis acid. Hocking [3] found that best results were obtained by using neat thionyl chloride at low temperature as shown:

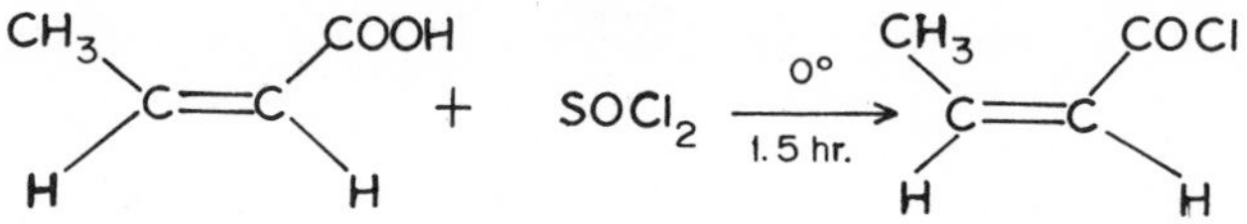

While still cold, the acid chloride was distilled under water-aspirator pressure and then again with a Widmer column (to remove acid) to yield 56% of 97% cis-crotonyl chloride, boiling point <23.5°/10 mm.

Normant and co-workers [4] prepared the acyl chlorides by employing thionyl chloride and HMPA at low temperature as indicated:

$$RCOOH + SOCl_2 \xrightarrow[\substack{-15 \text{ to} \\ -20°}]{\substack{HMPA \\ 60-90\%}} RCOCl + SO_2 + HCl + HMPA$$

As is seen, HMPA absorbs the acid formed and thus protects any acid-sensitive groups that may be present. The $SOCl_2$-HMPA reagent has been used satisfactorily as well in the synthesis of unsaturated acyl chlorides; crotonic acid, for example, gave an 80% yield of crotonyl chloride. It will be recalled (1, 862) that the triphenylphosphine dibromide method for the preparation of acid bromides may be extended to acid chlorides as indicated [5]:

$$RCO_2H + (C_6H_5)_3P + CCl_4 \longrightarrow RCOCl + CHCl_3 + (C_6H_5)_3PO$$

No acid is produced in this reaction. Good yields of aliphatic and benzoic acid chlorides are formed.

The instability of thionyl bromide even at room temperature has precluded or at least minimized its use for the preparation of acid bromides. Saraf and Zaki [6] have found that if the thionyl bromide is added slowly and portionwise to the neat acid, yields of acyl bromides are maximized.

The mixed acyl sulfonyl chlorides have become available by the technique of LeBerre [7] as indicated (Ex. d):

$$RCH_2COOH \xrightarrow[\text{POCl}_3]{\text{ClSO}_3H} RCH \overset{SO_2Cl}{\underset{COCl}{\diagup}}$$

γ-Butyrolacetone, by treatment with a mixture of chlorosulfonic acid and thionyl chloride, gave 2-keto-5-chlorosulfonyltetrahydrofuran (64%)

whereas maleic anhydride with a mixture of sulfur trioxide and thionyl chloride gave α-chlorosulfonylmaleic anhydride (91%):

a. Preparation of Diphenylacetyl Chloride

Yield was 82-94% from the acid, thionyl chloride, and benzene refluxed until gas evolution ceased [8].

b. Preparation of 4-Chloroformylphthalic Anhydride

Yield was 96% from trimellitic anhydride, 0.52 mol thionyl

chloride, 0.85 mol and 100 mg of DMF stirred under reflux until gas evolution ceased [9].

c. Preparation of γ-Chlorobutyryl Chloride [10]

Yield was 62% from 0.55 mol of thionyl chloride and 3 g of anhydrous $ZnCl_2$ to which a stirred solution of 0.5 mol of butyro-lactone was added in a thin stream while the temperature rose to 45°; finally the mixture was heated and stirred at 55° for 22 hr.

d. Preparation of Chlorosulfonylacetyl Chloride [7]

To 1 mol of chlorosulfonic acid and 2 mol of phosphorus oxychloride at 15°, 1 mol of acetic acid was added dropwise and the temperature was brought slowly to 120° and held there for 4 hr. After cooling and decantation from the phosphoric acid formed, the acid chloride layer was distilled under reduced pressure to give a 49% yield.

2. From Carboxylic Acids and Acyl Halides, Haloethers, or Imidoyl Halides (1, 863)

$$RCOOH \xrightarrow[\text{or}]{\substack{COX_2, \\ ArCOX \\ \\ X_2CHOCH_3}} RCOX$$

The use of phosgene in the preparation of acid chlorides has been revised [11]. With esters and phosgene, aluminum chloride has been used as a catalyst for the interchange, while with acids a tertiary amine, DMF, or 1,8-diazabicyclo[5.4.0]undec-7-ene, DBU [12, 13], or with anhydrides ferric chloride has been employed. The easy removal of the excess of phosgene may be advantageous if no ventilation problem exists.

A recent example of the use of phosgene is that of Ulrich and Richter [9] as shown:

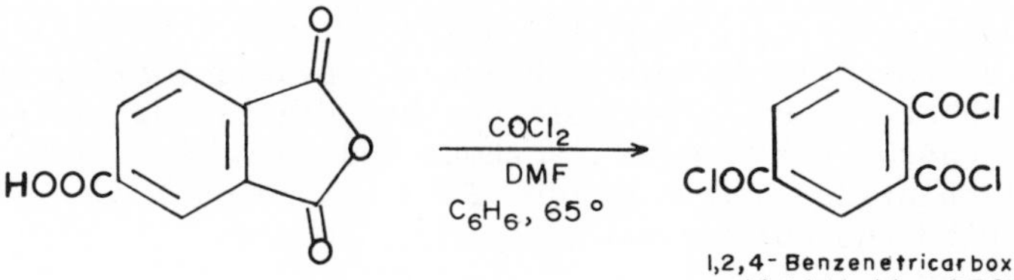

1,2,4-Benzenetricarboxylic acid trichloride, 87%

Phosgene was more satisfactory than thionyl chloride in this preparation.

The newer reagents for the conversion of carboxylic acids into acyl halides are phenyltrimethylacetimidoyl chloride [14] and 2-chlorodioxene [15]. The former, which is conveniently prepared from N-phenyl trimethylacetamide and phosphorus pentachloride, reacts readily with aliphatic or aromatic acids at moderate temperature, as shown below, to give good yields of the acyl chloride:

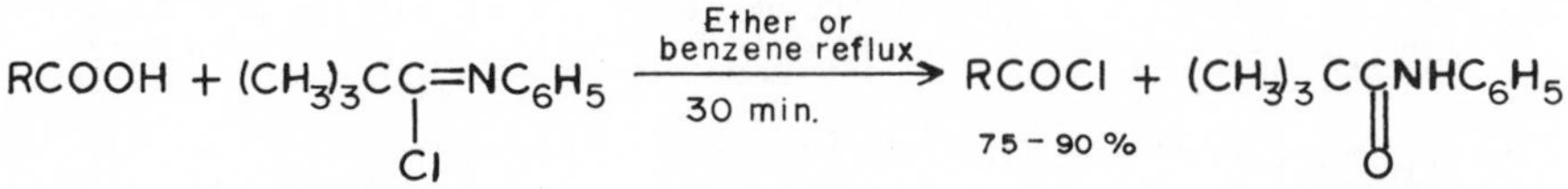

$$RCOOH + (CH_3)_3CC=NC_6H_5 \ \xrightarrow[\substack{\text{30 min.}}]{\substack{\text{Ether or} \\ \text{benzene reflux}}} \ RCOCl + (CH_3)_3CCNHC_6H_5$$

Cl (on left structure)   75 - 90 %

2-Chlorodioxene also gives high yields with the few acids investigated as indicated:

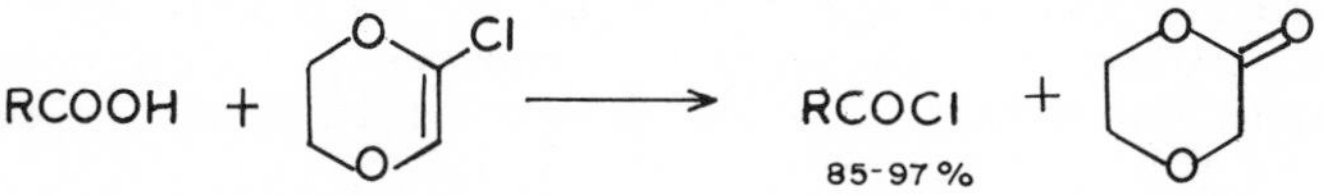

$$RCOOH + \ \ \longrightarrow \ RCOCl \ +$$

85-97 %

An exception in this case was p-nitrobenzoic acid, which gave only 38% of the acyl chloride. Both of these methods, it will be noted, produce no acid as a by-product.

a. Preparation of Methacroyl Chloride [16]

A mixture of 5 mol pyridine and 5 mol methacrylic acid was added to 6 mol of benzenesulfonyl chloride with continuous shaking and cooling. Distillation at 14 cm Hg in a cooled receiver gave a crude product which on distillation at ordinary pressure yielded 450 g (86%) of the acyl chloride.

3. From Esters or Salts (1, 864)

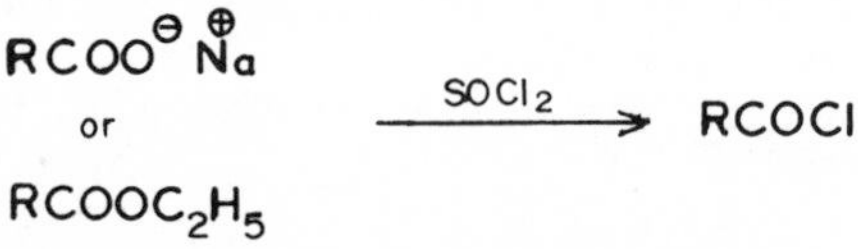

$$\begin{array}{c} RCOO^{\ominus} \ Na^{\oplus} \\ \text{or} \\ RCOOC_2H_5 \end{array} \ \xrightarrow{SOCl_2} \ RCOCl$$

Although esters are usually converted into acyl halides via the salt, by thionyl chloride, phosphorus pentachloride, phosphorus oxychloride, or α,α-dichloromethyl ether, such is not always the case. For example, Rothman and co-workers [17] prepared acyl chlorides and fluorides from isopropenyl esters by treatment with the corresponding acid as shown:

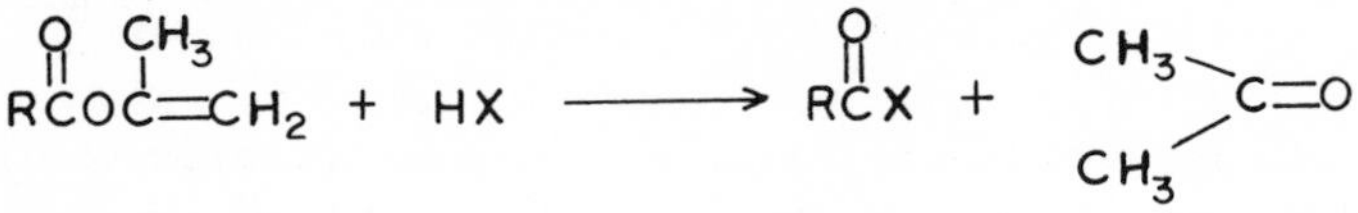

$$RCOC=CH_2 + HX \ \longrightarrow \ RCX \ + \ (CH_3)_2C=O$$

This procedure, which was employed to prepare acetyl fluoride, octanoyl fluoride (50%) (Ex. a), octadecanoyl fluoride, and azelaoyl fluoride (63%), offers some advantage over the acid anhydride-HF method (see 1, 865) in that the latter was only satisfactory for anhydrides derived from C_2 or C_3 acids. With HCl as the reagent, acyl chlorides are obtained in a manner "comparable in simplicity of operation, in yield, and particularly in purity of product" with existing methods.

New reagents for the conversion of esters or lactones into acid chlorides are triphenylchlorophosphonium chloride, triphenylbromophosphonium bromide, and the complex of the former with boron trifluoride, $[Ph_3\overset{+}{P}Cl][BF_3\overline{Cl}]$ [18]. With ethyl cinnamate the reaction with the first reagent is as indicated:

$$C_6H_5CH{=}CHCO_2Et + (C_6H_5)_3PCl_2 \xrightarrow[8\,hr.]{150°}$$

$$C_6H_5CH{=}CHCOCl + (C_6H_5)_3PO + EtCl$$
$$69\%$$

The third reagent gives a 92% yield of cinnamoyl chloride. α-Haloacids respond similarly, but nonhalogenated ones require more strenuous conditions as shown:

Phthalide

$$+ (C_6H_5)_3PCl_2 \xrightarrow[4\,hr.]{180°}$$

2-Chloromethyl-benzoyl chloride
97%

$$+ (C_6H_5)_3PO$$

An unsaturated lactone, 3-hydroxy-2,2,4-trimethyl-3-pentenoic acid β-lactone, has also been converted into an acyl chloride by treatment with the appropriate acid as shown [19]:

$$+ HCl \xrightarrow[10-70°]{ZnCl_2}$$

2,2,4-Trimethyl-3-oxo-valeryl chloride, 88-93%(crude)

Trimethylsilyl esters refluxed with thionyl chloride until sulfur dioxide evolution is complete also give good to high yields of the acid chloride [19a].

a.   Preparation of Octanoyl Fluoride [17]

Isopropenyl octanoate in a Teflon apparatus was heated to

35° for 1 hr while HF was bubbled through the liquid. Extraction with pentane produced a colored layer which was discarded. Distillation of the remainder gave the fluoride, in 50% yield.

4. From Acid Anhydrides (1, 865)

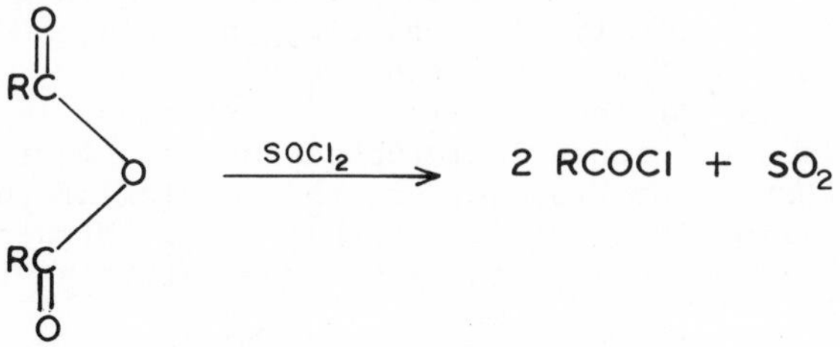

$$\text{RC(O)-O-C(O)R} \xrightarrow{\text{SOCl}_2} 2 \text{ RCOCl} + \text{SO}_2$$

Halosulfonic acids are reagents not previously mentioned in the synthesis of acyl chlorides and acyl fluorides from the appropriate anhydride. They have been utilized by Schmidt and Pichl [20] as shown [Ex. a):

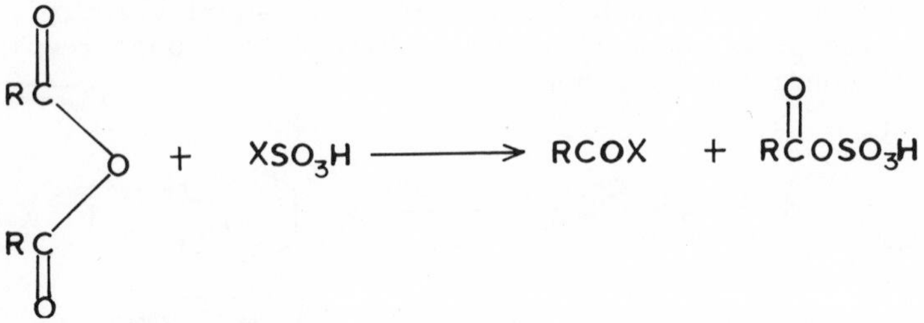

$$\text{(RC(O))}_2\text{O} + \text{XSO}_3\text{H} \longrightarrow \text{RCOX} + \text{RCOSO}_3\text{H}$$

The method is operable with small quantities in simple equipment. Yields of acyl chlorides from the corresponding anhydrides are: (a) $CH_3COCl \sim 95\%$, (b) $C_2H_5COCl \sim 95\%$, (c) $C_3H_7COCl \sim 95\%$, and (d) $C_6H_5COCl \sim 65\%$. The method serves as well in the synthesis of acyl fluorides with yields: (a) $CH_3COF \sim 90\%$, (b) $C_2H_5COF \sim 90\%$, (c) $C_3H_7COF \sim 90\%$, and (d) $C_6H_5COF \sim 75\%$.

a. Preparation of Acetyl Fluoride [20]

Acetic anhydride, 0.25 mol at about -5° was stirred and cooled while 0.2 mol of fluorosulfonic acid was added dropwise. The mixture was warmed slowly to room temperature in vacuo, after which distillation gave the acyl fluoride (90% based on FSO_2OH).

5. From Acyl Halides (Halide Exchange) (1, 866)

$$\text{RCOCl} \xrightarrow{\text{HX}} \text{RCOX} + \text{HCl}$$

This method, widely used for the preparation of acyl
fluorides, has been carried out not only with the free acid or a
normal salt in the presence of acid or tetramethylene sulfone,
but with potassium acid fluoride [21] (Ex. a), sodium fluorosili-
cate [22], potassium fluorosulfinate [23], and the hexafluoro-
acetone-potassium fluoride adduct [24]. Of these the potassium
acid fluoride is perhaps most widely used, although hydrogen
fluoride (1, 866, Ex. a) gives superior yields.

a. Preparation of Fluoroacetyl Fluoride [21]

A mixture of fluoroacetyl chloride, 1 mol and 0.77 mol of
dry KHF$_2$ was heated for 1 hr in a water bath. Distillation gave
the crude acid fluoride, which was fractionated twice, the yield
being 73%.

7. From Tri- or Tetrahalides (1, 866)

Carbonyl chloride fluoride (monofluorophosgene) has been
prepared from trichlorofluoromethane and sulfur trioxide in the
presence of mercury salts [24a].

8. From Carboxylic Acids and Cyanuric Fluoride

The mild reagent, cyanuric fluoride, has been found to
convert acids into acid fluorides [25] (Ex. a). Not only aliphatic and aromatic acids may be used, but also those containing
hydroxyl or unsaturated groups. Yields vary within 54-100%
except in the formation of formyl fluoride (40%). Cyanuric
fluoride was prepared from the chloride by dissolving the
latter in liquid H$_2$F$_2$ and slowly warming to room temperature.
Yield of the fluoride, boiling point 73°, was 76%.

a. Preparation of Benzoyl Fluoride [25]

To 0.12 mol of cyanuric fluoride in 125 ml of acetonitrile
was added in 10 min a solution of 0.3 mol each of benzoic acid
and pyridine in 125 ml of acetonitrile and the mixture was
stirred for an additional 50 min. The mixture was then poured
into water, extracted with ether, dried, evaporated, and distilled to give 36 g (97%), boiling point 156°, of the fluoride.

Addenda

General. A review on the chemistry of phosgene has been published
[H. Babad, A. G. Zeiler, Chem. Rev., 73, 75 (1973)].

A.1. A resin incorporating triphenylphosphine units has been
used with CCl_4 to synthesize acid chlorides [P. Hodge, G.
Richardson, Chem. Commun., 622 (1975)].

A.1. The preparation of acid bromides (40-85%) from the ester
and triphenylphosphine-bromine complex has been described
[A. G. Anderson, Jr., D. H. Kano, Tetrahedron Letters, 5121
(1973)].

References

1. S. Patai, The Chemistry of Acyl Halides, Interscience, New
    York, 1972, p. 35.
2. E. E. Gilbert, Synthesis, 3 (1969).
3. M. B. Hocking, Can. J. Chem., 46, 466 (1968).
4. J. F. Normant et al., Compt. Rend., 269, C 1325 (1969).
5. J. B. Lee, J. Am. Chem. Soc., 88, 3440 (1966).
6. S. D. Saraf, M. Zaki, Synthesis, 612 (1973).
7. A. Le Berre et al., Bull. Soc. Chim. France, 210, 214
    (1973).
8. E. C. Taylor et al., Org. Syn., 52, 36 (1972).
9. H. Ulrich, R. Richter, J. Org. Chem., 38, 2557 (1973).
10. O. P. Goel, R. E. Seamans, Synthesis, 538 (1973).
11. H. Babad, A. G. Zeiler, Chem. Rev., 73, 75 (1973).
12. H. Oediger et al., Synthesis, 591 (1972).
13. C. F. Hauser, German Patent, 1,931,074, January 2, 1970;
    C. A., 72, 78472 (1970).
14. F. Cramer, K. Baer, Chem. Ber., 93, 1231 (1960).
15. M. J. Astle, J. D. Welks, J. Org. Chem., 26, 4325 (1961).
16. J. Heyboer, A. J. Staverman, Rec. Trav. Chim., 69, 787
    (1950).
17. E. S. Rothman et al., J. Org. Chem., 34, 2486 (1969); D. C.
    Hull, United States Patent, 2,475,966, July 12, 1949; C. A.,
    43, 7954 (1949).
18. D. J. Burton, W. M. Loppes, Chem. Commun., 425 (1973); H.-J.
    Bestmann, L. Mott, Ann. Chem., 693, 132 (1966).
19. E. U. Elam et al., Org. Syn., 48, 126 (1968).
19a. H. R. Kricheldorf, Synthesis, 551 (1972).
20. M. Schmidt, K. E. Pichl, Chem. Ber., 98, 1003 (1965).
21. G. Olah et al., Chem. Ber., 89, 862 (1956).
22. J. Dahmlos, Angew. Chem., 71, 274 (1959).

23. F. Seel, J. Langer, Chem. Ber., 91, 2553 (1958).
24. A. G. Pitman, D. L. Sharp, J. Org. Chem., 31, 2316 (1966).
24a. G. Siegemund, Angew. Chem. Intern. Ed. Engl., 12, 918 (1973).
25. G. A. Olah et al., Synthesis, 487 (1973).

B. Oxidation

1. From Aldehydes (1, 869)

$$ArCHO \xrightarrow{Cl_2} ArCOCl + HCl$$

In a rather recent investigation Ol'dekop and Kalinina [1] transformed aromatic aldehydes into aroyl halides by the use of carbon tetrachloride or bromotrichloromethane at high temperatures in a sealed tube as indicated:

$$ArCHO + CCl_4 \xrightarrow{180-210°} ArCOCl + CHCl_3$$

$$ArCHO + CCl_3Br \xrightarrow{145-205°} ArCOBr + CHCl_3$$

The products were recovered by vacuum distillation. By the use of initiators such as a peroxide, the temperature may be lowered somewhat. Yields of acyl chlorides usually vary within 60-94%, whereas the range for acyl bromides is usually 65-88%. The method is not applicable to aliphatic aldehydes, m- and p-nitrobenzaldehydes, p-dimethylbenzaldehyde, cinnamaldehyde, or furfural.

3. From Acid Hydrazides (1, 870)

$$RCONHNH_2 \xrightarrow[HCl]{2 Cl_2} RCOCl + N_2 + 3 HCl$$

An additional reagent for effecting this transformation is thionyl chloride [2]. The hydrazides were simply boiled with thionyl chloride, molar ratio 1:4, in chloroform until HCl, N_2, and SO_2 ceased to be evolved (1-3 days). Yields varied within 68-95%.

Hope and Wiles [2] also prepared the acid chloride by similar treatment from a series of sulfinylhydrazides, obtained from the hydrazide and $SOCl_2$ boiled in $CHCl_3$ for 2 hr, as indicated:

$$RCONHNSO \longrightarrow RCOOH + N_2 + S_2$$
$$\downarrow SOCl_2$$
$$RCOCl$$

In this case the acid chloride yields ranged within 13-80% with only the corresponding carboxylic acid being obtained in one case.

References

1.   Yu. A. Olddekop, A. M. Kalinina, J. Gen. Chem. USSR (Engl. transl.), 34, 3515 (1964).
2.   P. Hope, L. A. Wiles, J. Chem. Soc., 5386 (1965).

C.   Halocarbonylation

2.   By the Use of Carbon Monoxide and a Chlorine Source (1, 872)

$$[\text{R}\cdot] + \text{CO} \longrightarrow [\text{RCO}] \xrightarrow{\text{CCl}_4} \text{RCOCl} + \cdot\text{CCl}_3$$

Dent and co-workers [1] succeeded in the carbonylation of allylic chlorides by heating under pressure with carbon monoxide in the presence of small amounts of π-allylic palladium chloride complexes as shown:

$$\text{CH}_2{=}\text{CHCH}_2\text{Cl} + \text{CO} \xrightarrow[110°, 15\text{min.}]{(\pi\,\text{allyl}-\text{PdCl})_2} \text{CH}_2{=}\text{CHCH}_2\text{COCl}$$

500 atm.   But-3-enoyl chloride, 96%

To obtain this maximum yield there must be 3×10^{-4} mol of the complex per mol of allylic chloride and the time must be short as indicated.

In a Dutch patent [2] aryl halides were carbonylated with CO in the presence of palladium chloride or bromide as a catalyst as shown:

$$\text{C}_6\text{H}_5\text{Cl} + \text{CO} \xrightarrow[160°, 32\text{hr.}]{\text{PdCl}_2, 3.55\text{g.}} \text{C}_6\text{H}_5\text{COCl}$$

127.1 g.   80.5 atm.   80%

Benzoyl bromide was prepared similarly.

The addition of carbon monoxide to ethylene and other terminal olefins to give 4,4,4-trichlorobutanoyl chlorides has been brought about by mixing ethylene, carbon monoxide at 170 atm, carbon tetrachloride, and a catalyst $[\text{C}_5\text{H}_5\text{Mo}(\text{CO})_3]_2$ in an autoclave at 120° for 15 hr [2a].

3.   By the Use of Phosgene

$$C_6H_6 + COCl_2 \xrightarrow{\text{AlCl}_3} C_6H_5COCl + HCl$$

This Friedel-Crafts Reaction is of limited value since the acid chloride has a tendency to react with the original arene to form the ketone [3]. Of the arenes investigated only m-xylene gave an acyl chloride in good yield by its conversion into 2,4-dimethylbenzoic acid [4].

References

1. W. T. Dent et al., J. Chem. Soc., 1588 (1964).
2. Dutch Patent, 6,614,185, April 17, 1967; C. A., 67, 64066 (1967).
2a. M. Ryang, S. Tsutsumi, Synthesis, 55 (1971).
3. G. A. Olah, Friedel-Crafts and Related Reactions, Vol. III, Interscience, New York, 1964, p. 1257.
4. E. Ador, F. Meier, Chem. Ber., 12, 1968 (1879).

Chapter 16

CARBOXYLIC ACID ANHYDRIDES

Although less reactive than the acyl chlorides, the anhydrides
must be protected from the moisture in the air and handled with
caution since irritation of the skin is often encountered.   In
the main, classical methods of synthesis continued to be employed,
but one finds a growing use of carbodiimides in the synthesis of
anhydrides from carboxylic acids (A.1).   Another reagent of some
interest for this purpose is hexamethylphosphorous triamide with
carbon tetrachloride (A.1).   One new development reported is the
preparation of substituted maleic anhydrides by the pyrolysis
of α-alkoxyvinyl α-ketoesters (D.1).   A simple preparation of the
reactive, mixed sulfonic-carboxylic anhydrides is described
(A.10).   This preparation will doubtless lead to more frequent
use of the mixed anhydrides for acylation.

A.   Solvolysis

    1.   From Carboxylic Acids (1, 874)

    One of the most useful reagents for the synthesis of acid
anhydrides from carboxylic acids is the carbodiimide.   The
importance of the reagent is indicated by the new methods
devised for its synthesis [1-3].

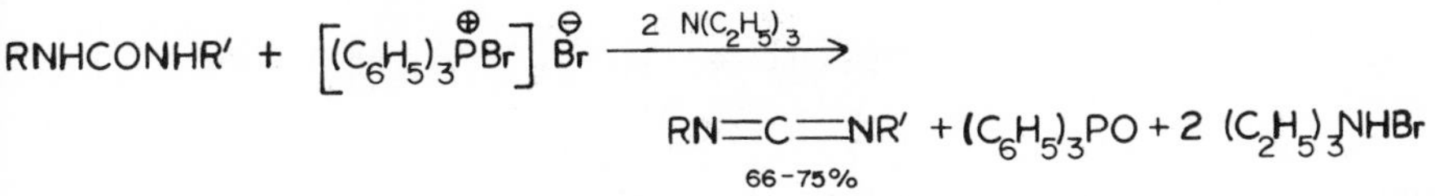

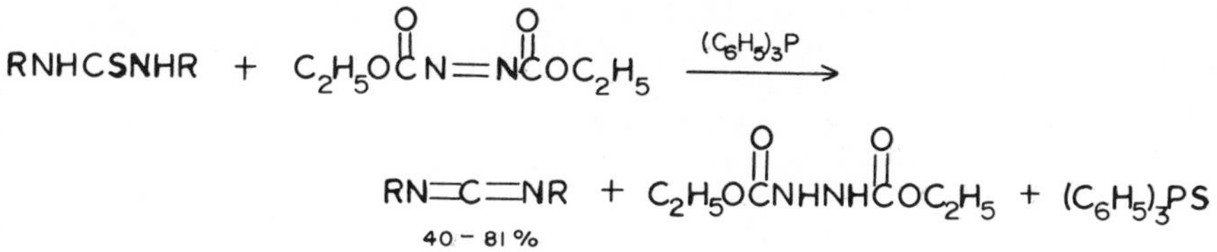

$$RNHCONHR' + \left[(C_6H_5)_3\overset{\oplus}{P}Br\right]\overset{\ominus}{Br} \xrightarrow{2\ N(C_2H_5)_3}$$

$$RN{=}C{=}NR' + (C_6H_5)_3PO + 2\ (C_2H_5)_3NHBr$$

66-75%

$$RNHCSNHR + C_2H_5O\overset{O}{\overset{\|}{C}}N{=}N\overset{O}{\overset{\|}{C}}OC_2H_5 \xrightarrow{(C_6H_5)_3P}$$

$$RN{=}C{=}NR + C_2H_5O\overset{O}{\overset{\|}{C}}NHNH\overset{O}{\overset{\|}{C}}OC_2H_5 + (C_6H_5)_3PS$$

40-81%

$$RNH\overset{\overset{X}{\|}}{C}NHR' + (C_6H_5)_3P + CCl_4 + (C_2H_5)_3N \longrightarrow$$

$$X=O,S \qquad\qquad RN=\!\!=\!\!C=\!\!=\!\!NR' + (C_6H_5)_3PX + CHCl_3 + (C_2H_5)_3NHCl$$

84.5−92%

The synthesis of 1,1'-carbonyldiimidazole, a second reagent
employed in the synthesis of acid anhydrides, has been reviewed
[4]. The diimidazole has a high "transacylation potential" and
is best utilized according to the equation [4a]:

$$3RCOOH + \text{(imidazole structure)} \longrightarrow (RCO)_2O + \text{(imidazolium structure)} \overset{\oplus}{NH}\overset{\ominus}{O_2CR} +$$

$$+ CO_2$$

If the intermediate acylimidazole or other
compounds are insoluble, trifluoroacetylimidazole
may be substituted for the diimidazole as
indicated:

$$2\,RCOOH + \text{(imidazole-COCF}_3\text{)} \longrightarrow (RCO)_2O + \text{(imidazolium)}\overset{\oplus}{NH}\overset{\ominus}{O_2CCF_3}$$

Castro and Dormoy [5] employed hexamethylphosphorous triamide
(trisdimethylaminophosphine) (TDP), with CCl_4 and triethylamine
in the conversion of carboxylic acids into anhydrides as shown
(Ex. a):

$$2RCO_2H + CCl_4 + P[N(CH_3)_2]_3 \xrightarrow[Et_2O]{N(C_2H_5)_3}$$

$$(RCO)_2O + CHCl_3 + [(CH_3)_2N]_3PO + (C_2H_5)_3NHCl$$

65−90%

Applied to formic acid at low temperature (-60° or lower), this
reaction gives a solution that exhibits excellent formylating
properties, a fact which suggests the presence of the unreported
formic anhydride, $(HCO)_2O$.

Another method of preparing the anhydride from dibasic
acids is to heat with acetic anhydride. This procedure gives a
polymeric anhydride mixture from which the monomeric anhydride
may be obtained by distillation at ordinary or reduced pressure.
This method worked particularly well for gem-dialkyl dibasic
acids because of their great tendency to cyclize [5a]. High

molecular weight dianhydrides have been prepared from the tetra-carboxylic acid by sublimation. For example, tetracarboxyhydro-quinone was converted into the dianhydride in this manner at 280° and 15 mm pressure [5b]. Incidentally, the quinone of this compound is so electron-attracting that it forms π complexes with methylbenzenes.

A reagent that forms mixed anhydrides is N,N-diphenylcarba-moyl chloride (Ph_2NCOCl) [5c]. The acid and triethylamine in water is added slowly to a solution of diphenylcarbamoyl chloride in pyridine-water at 0-25°. In about 30 min the mixed anhydride precipitates.

a. Preparation of Benzoic Anhydride [5]

To a stirred solution of 12.2 g of benzoic acid and 30 g of CCl_4 in 250 ml of ether was added 16.3 g of TDP under N_2 at -70°. After 30 min the mixture was hydrolyzed with about 200 ml of water. From the ether extract was recovered 10.2 g (90%) of the anhydride.

2. From Carboxylic Acids or Anhydrides and Acylating Agents, and Related Reactions (1, 876)

Rambacher and Mäke [6] have pointed out that the conversion of the acid chloride in water-pyridine is benefited by the intro-duction of $NaHCO_3$ as shown (Ex. a):

$$2\ RCOCl + 2NaHCO_3 \xrightarrow{C_5H_5N} (RCO)_2O + 2\ NaCl + CO_2 + H_2O$$

The method is simple and gives good yields.

a. Preparation of o-Ethoxybenzoic Anhydride [6]

Pyridine, 4 ml was added to a solution of 16.8 g of $NaHCO_3$ in 300 ml of water. o-Ethoxybenzoyl chloride, 36.9 g was then dropped in during 30 min, after which the mixture was cooled to 20°. With 1 hr of stirring, 29.1 g (92.7%) of the anhydride separated.

3. From the Salts of Carboxylic Acids and Acylating Agents (1, 878)

$$RCOONa + R'COCl \longrightarrow \begin{array}{c} RC{=}O \\ \diagdown \\ O \\ \diagup \\ R'C{=}O \end{array} + NaCl$$

This method is well suited for the synthesis of mixed anhydrides. Recently it has been utilized for the preparation of acetic formic anhydride as indicated [7] (Ex. a):

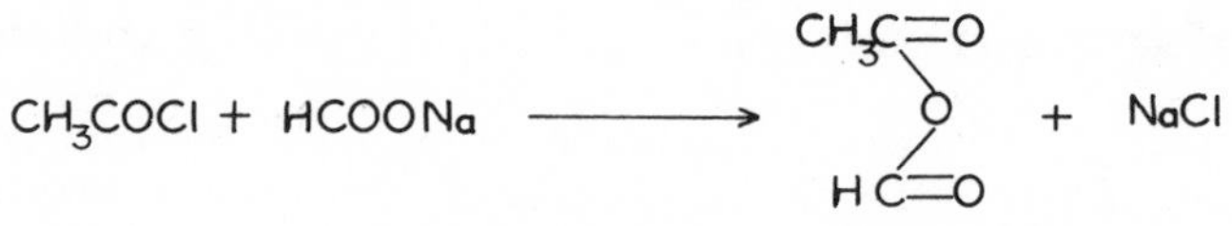

$$CH_3COCl + HCOONa \longrightarrow \begin{matrix} CH_3C{=}O \\ \diagdown \\ O \\ \diagup \\ HC{=}O \end{matrix} + NaCl$$

64%

The method is simple and gives better yields than the other methods employed in preparing this particular mixed anhydride.

a. Preparation of Acetic Formic Anhydride [7]

To sodium formate, 300 g and 250 ml of anhydrous ether was added 294 g of acetyl chloride as rapidly as possible while the temperature was held at 23-27°. After stirring for 5.5 hr at the same temperature, 212 g (64%) of the mixed anhydride was recovered from the ether layer and ether washings of the solid residue by distillation.

7. From Thioesters and a Mercuric Carboxylate

$$\underset{RCOR'}{\overset{S}{\underset{||}{}}} + \quad (R''CO)_2Hg \longrightarrow (R''CO)_2O + RCOOR' + HgS$$

The desulfurization of thioesters was accomplished by heating with a chloroform, methylene chloride, or pyridine solution of a mercuric carboxylate [8] as shown above. The reaction proceeds rapidly at room temperature with yields of 75-96% (Ex. a). In each case an immediate precipitate of mercuric sulfide results.

a. Preparation of Butyric Anhydride [8]

Mercuric butyrate, 0.5 mol in chloroform was mixed with 0.5 mmol of cyclohexyl thioacetate in the same solvent and the mixture was stirred for 10 min. After the unchanged mercury salt was destroyed by passing H_2S through the solution, the mixture was filtered and the filtrate was distilled to remove the excess of solvent. An examination of the residue by glc indicated 84% of the anhydride.

8. From Metal Carboxylates and Various Thio Compounds

$$2 \ RCOOM + C_6H_{11}NH\overset{\overset{\text{S}}{\|}}{C}NHC_6H_{11} \longrightarrow (RCO)_2O + C_6H_{11}NH\overset{\overset{\text{O}}{\|}}{C}NHC_6H_{11} + M_2S$$

M = Ag or Hg(I)

This synthesis proceeds at room temperature in solvents such as acetone, acetonitrile, or chloroform to give yields of the anhydride varying within 68-99% [9]. There is little to choose between the silver and mercurous carboxylates although the latter is preferred to the mercuric carboxylate since the water of crystallization present in some of these mercuric salts is an interfering factor. For the high molecular weight carboxylates practically pure anhydrides were separated by simply removing the black silver or mercury sulfide from the reaction mixture. Dicyclohexylcarbodiimide may be an intermediate in the decomposition of the thiourea.

S-Alkylthioamidium iodides react with mercuric compounds to give acid anhydrides as well [10]. Thus N,N-dimethyl-S-ethyl-thiobenzamidium iodide with mercuric acetate gives acetic anhydride as indicated:

$$\left[\begin{array}{c} (CH_3)_2 N \overset{+}{=} C \cdots C_6H_5 \\ \overset{|}{S}C_2H_5 \end{array} \right]^{\oplus} I^{\ominus} + (CH_3CO_2)_2Hg \longrightarrow$$

$$\begin{array}{c} CH_3C = O \\ \diagup \\ \diagdown O \\ CH_3C = O \end{array} + (CH_3)_2NCC_6H_5 + C_2H_5SHgI$$

$$\underset{85\%}{} \qquad \underset{85.5\%}{\overset{\overset{\text{O}}{\|}}{}}$$

Yields of other anhydrides varied within 25-86%. It will be noted that amides are also obtained, usually in superior yields. The method is a simple one in that to the iodide in acetonitrile the mercuric salt is added and the reaction is complete after stirring for 4 hr at room temperature.

a. Preparation of Succinic Anhydride [9]

N,N'-Dicyclohexylthiourea, 0.02 mol was added to a suspension of mercurous succinate, 0.04 mol in 40 ml of CHCl$_3$ and the mixture was stirred for about 15 hr at 25°. The filtrate and CHCl$_3$ washings of the residue were evaporated to give a 95% yield of succinic anhydride.

9. From Carboxylic Acids and Cyanogen Bromide

$$2\ RCO_2H\ +\ BrCN\ \xrightarrow[\substack{C_6H_6}]{C_5H_5N}\ (RCO)_2O$$

In addition, CO_2, NH_3, and HBr are probably obtained as by-products after workup. The reaction above is exothermic and works well for both aliphatic and aromatic carboxylic acids [11]. The intermediate appears to be RCOOCN.

a.   Preparation of p-Toluic Anhydride [11]

Pyridine, 15 mmol was added dropwise to a stirred solution of 15 mmol of toluic acid and 10 mmol of cyanogen bromide in benzene at 25°. As the reaction mixture rose in temperature, pyridine hydrobromide precipitated. After 0.5 hr the supernatant liquid was decanted, the salt residue was washed twice with dry benzene, and the combined liquids were evaporated to give the anhydride, 78%.

10.   From Acid Chlorides and Sulfonic Acids (Mixed Anhydrides)

$$RCOCl\ +\ R'SO_3H\ \longrightarrow\ RCO_2SO_2R'$$

This preparation is quite simple and usually yields a product (85-100%) not requiring purification if the product is to be employed in some acylation process [12] (Ex. a). The main contaminant present is the sulfonic anhydride that arises by disproportionation, although this process is not very serious unless a temperature of 120° is exceeded during the preparation. These mixed anhydrides are much more reactive than the simple diaroyl anhydrides.

On the other hand, trifluoromethanesulfonic acid and an acyl chloride give only 35-63% of the mixed anhydride [13]. In this case it is better to use the silver salt of the sulfonic acid, which with the acyl halide usually gives yields of 90-100% of mixed anhydrides. According to the authors, the mixed anhydride from the sulfonic acid and aromatic acyl halides may be distilled satisfactorily.

a.   Preparation of p-Toluenesulfonyl Acetyl Anhydride [12]

Anhydrous p-toluenesulfonic acid, 0.5 mol and acetyl chloride, 1.9 mol were refluxed until the hydrogen chloride evolution slowed noticeably. Excess acetyl chloride was removed at room temperature in vacuo. The residue consisting of 97.5% of the crude mixed anhydride solidified, melting point 54-56°. The sulfonic anhydride contaminant could be removed by dissolving

the solid in ether, cooling, and decanting from the deposited
solid sulfonic anhydride.

Addendum

A.3. Mixed anhydrides have been prepared from the sodium salt of
the acid and methyl chlorocarbonate [M. S. Newman et al., J. Am.
Chem. Soc., 90, 747 (1968)].

References

1.   H. J. Bestman et al., Ann. Chem., 718, 24 (1968).
2.   O. Mitsunobu et al., Tetrahedron, 26, 5731 (1970).
3.   R. Appel, Chem. Ber., 104, 1335 (1971).
4.   H. A. Staab, K. Wendel, Org. Syn., 48, 44 (1968).
4a. H. A. Staab, Angew. Chem. Intern. Ed. Engl., 1, 351 (1972).
5.   B. Castro, J.-R. Dormoy, Bull. Soc. Chim. France, 3034
     (1971).
5a. G. Borgen, Acta Chem. Scand., 27, 1840 (1973).
5b. P. R. Hammond, J. Chem. Soc., C1521 (1971).
5c. K. L. Shepard, Chem. Commun., 928 (1971).
6.   P. Rambacher, S. Mäke, Angew. Chem. Intern. Ed. Engl., 7,
     465 (1968).
7.   L. I. Krimen, Org. Syn.. 50. 1 (1970).
8.   J. Ellis et al., Australian J. Chem., 24, 1527 (1971).
9.   T. Hata et al., Bull. Chem. Soc. Jap., 41, 2746 (1968)
10. T. Yamaguchi et al., Bull. Chem. Soc. Jap., 41, 673 (1968).
11. T.-L. Ho, C. M. Wong, Syn. Commun., 3, 63 (1973).
12. M. H. Karger, Y. Mazur, J. Org. Chem., 36, 528 (1971).
13. F. Effenberger, G. Epple, Angew. Chem. Intern. Ed. Engl.,
     11, 299 (1972).

C. Electrophilic Reactions

   3. From N,N'-Dicarbomethoxy-N,N'-diethoxyhydrazine and
       p-Toluenesulfonic Acid

$$\underset{MeO_2CN-NCO_2Me}{\overset{OEt \quad OEt}{|\quad\quad|}} \xrightarrow{\text{TsOH}} TsOCO_2Me$$

The preceding action takes place spontaneously in ether and
gives the product, p-toluenesulfonyl carbomethoxy anhydride, in
38% yield as a crystalline solid [1]. It probably proceeds via

the intermediate carbomethoxy cation. This little known reaction may be worthy of further attention.

Reference

1. R. J. Crawford, R. Raap, J. Org. Chem., 28, 2419 (1963).

D. Cycloaddition (1, 885)

   1. From α-Alkoxyvinyl Pyruvates

    Newman and Stalick [1] have discovered an interesting rearrangement that yields substituted maleic anhydrides.

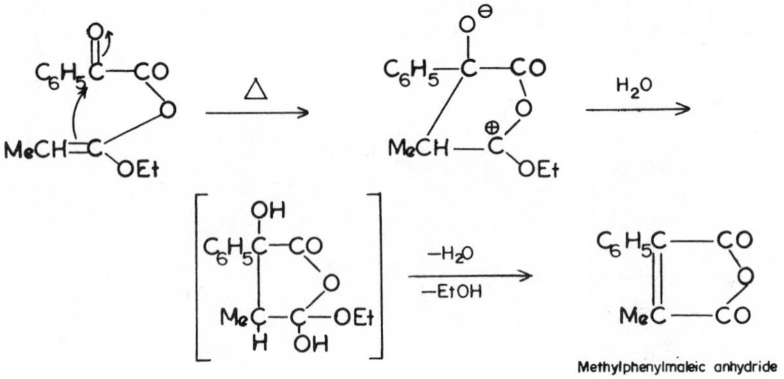

Methylphenylmaleic anhydride
41%

The starting material is available by the addition of the keto acid to 1-ethoxypropyne.

Addendum

D. An elegant way to synthesize anhydrides is from two equivalents of acid and one of the expensive reagent, ethoxy-acetylene. The reaction may be carried out at room temperature to give the anhydride and ethyl acetate by a cyclic mechanism [M. S. Newman, C. Courduvelis, J. Am. Chem. Soc., 88, 781 (1966); H. H. Wasserman, P. S. Wharton, J. Am. Chem. Soc., 82, 1411 (1966).

Reference

1. M. S. Newman, W. M. Stalick, J. Org. Chem., 38, 3386 (1973).

Chapter 17

KETENES,
KETENE DIMERS
AND KETENIMINES

Ketene preparations have been reviewed in the H.-W. series [1].
As mentioned previously (1, 886), the ketenes are highly reactive
and, frequently, toxic substances that cannot be stored well.
Aldoketenes (RCH=C=O) tend to form ß-lactone dimers

and ketoketenes $(R_2C=C=O)$ tend to form 1,3-cyclobutanediones

Those readers interested in the preparation and chemistry of the dimers should consult the H.-W. volume [2]. Because of the activity of the ketenes many are isolated as the acid derivative rather than as the ketene itself. The known types have been tabulated [3].

The unique structure of ketenes limits the number of new syntheses, but two observations reveal general pathways: (a) the presence of a synchronous coupling of electrons (A.1) and (b) the realization that α-carbene ketone, $-\ddot{C}-\overset{\cdot}{C}=O$, is a canonical form of a ketene. Therefore, means of synthesizing carbenes may be applicable to the synthesis of properly substituted ketenes (A.3).

A new synthesis given is the decomposition of alkynyl t-butyl ethers (B.3). A review of the synthesis and reactions of halogenated ketenes has been published [3a].

A.   Pyrolysis or Decomposition

The pyrolytic reactions, which are mostly free radical in nature, have been described (1, 886). At the time of publication of Volume 1, the categories were not well delineated, but seem better so with the following description. All sections except 1 and 3 discuss eliminations and other sections deal with either the radical or diradical type.

1.   From Acids, Anhydrides, Ketones, and Esters (1, 886)

$$CH_3COCH_3 \longrightarrow CH_2=C=O + CH_4$$

In many of the reactions of the type leading to ketenes, the most common intermediate is the acetonyl free radical that decomposes to a methyl free radical and the ketene. An added driving force is brought about by the synchronous coupling of either two unpaired electrons on adjacent carbon atoms or one unpaired electron with another from an adjacent pair. The latter may be illustrated as indicated:

$$\left[CH_3 \colon \overset{\overset{O}{\parallel}}{C} - CH_2 \right] \longrightarrow \left[CH_3 \right] + O=C=CH_2$$

The synchronous coupling of two unpaired electrons is illustrated by the reaction that occurs when cycloalkanones are irradiated [4]:

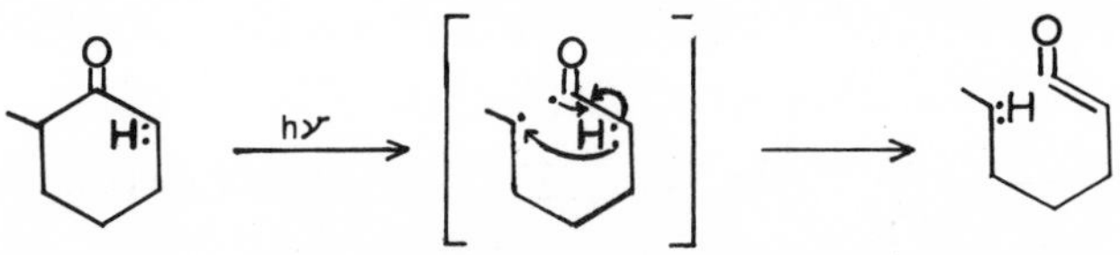

It may be that the hydrogen-atom shift and electron coupling to form the bond are all synchronous. Yields are quite low if attempts are made to isolate the ketene (or its corresponding acid), but may be fair if the ketene is trapped as the amide or ester. For instance, the best yield obtained in the irradiation of a ketosteroid is as shown [4]:

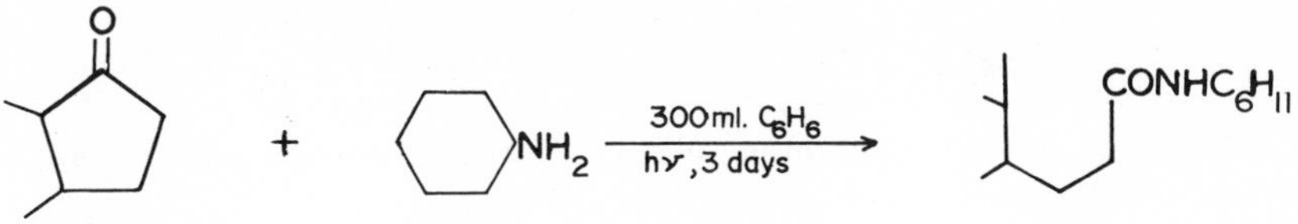

15 g. 10 ml. 3.1 g.

16,16-Dimethylandrostan-5-ol-3β-on-17 methylether

Two other interesting irradiations to produce ketenes are:

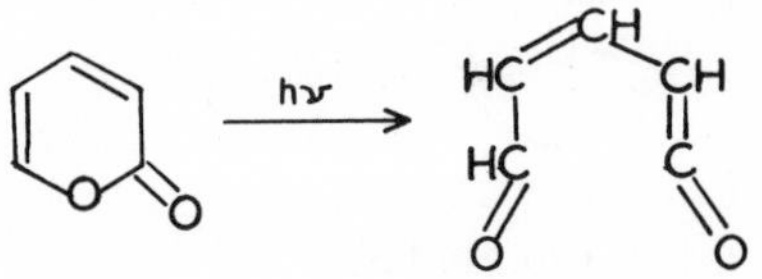

2-Formylvinylketene

[5]

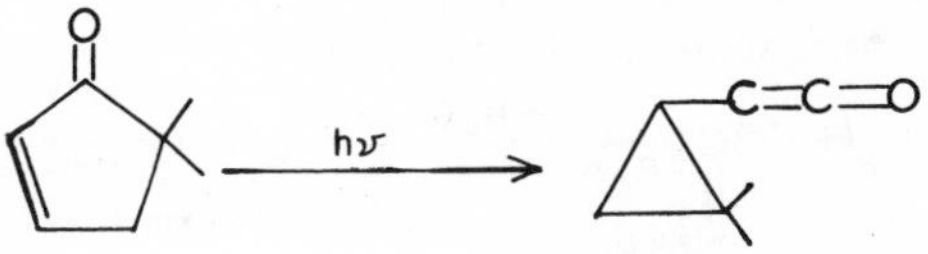

2,2-Dimethylcyclopropylketene
76 % as methyl ester

[6]

The irradiation of 2,2,5,5-tetramethyl-3 ketotetrahydrofuran appears to induce ketene formation [6a]. Products are largely methyl esters

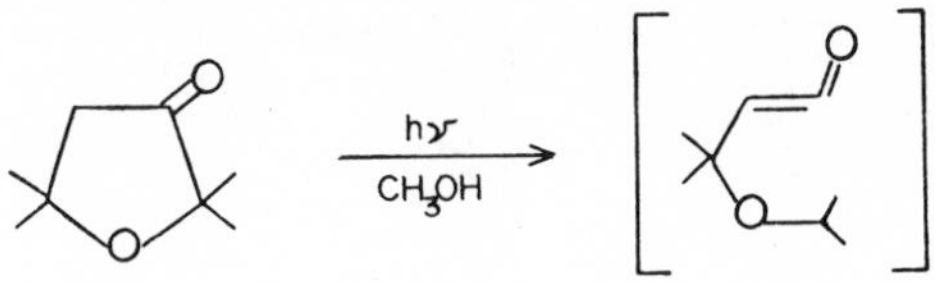

A

derived from A.

In the pyrolysis of a spirohexanone small amounts of vinyl ketene have been produced [7]:

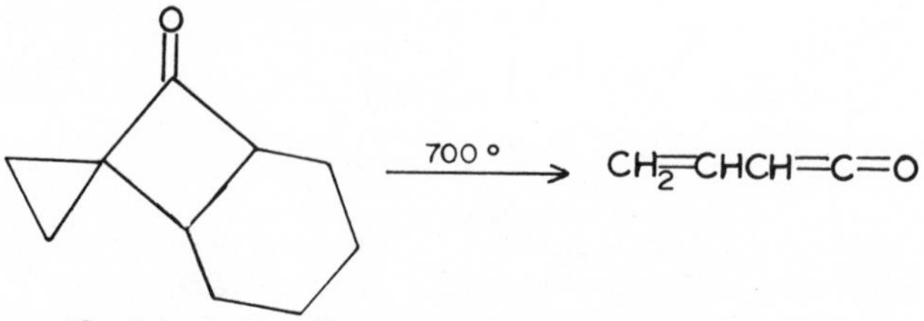

$$\xrightarrow{700°} \quad CH_2{=}CHCH{=}C{=}O$$

Heating of the following unsaturated epoxide enables one to trap propenylketene in CS_2 at $-80°$ [8]:

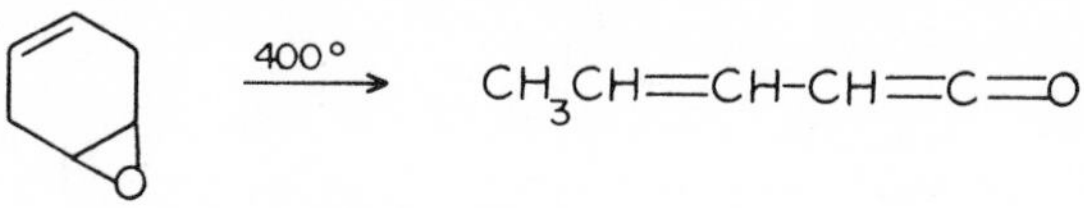

$$\xrightarrow{400°} \quad CH_3CH{=}CH{-}CH{=}C{=}O$$

However, cyclic ketones on irradiation do not always give ketenes [8a].

3. From Ketohydrazones, Diazoketones, or Other Precursors of Ketocarbenes

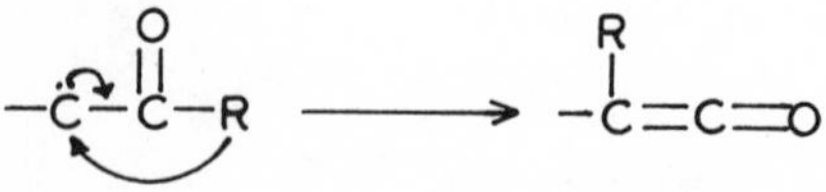

Any method which generates the carbene on the carbon adjacent to a keto group appears to produce the ketene.

One method reported earlier (1, 888) and in the H.-W. series [9] has been improved [10] (Ex. a):

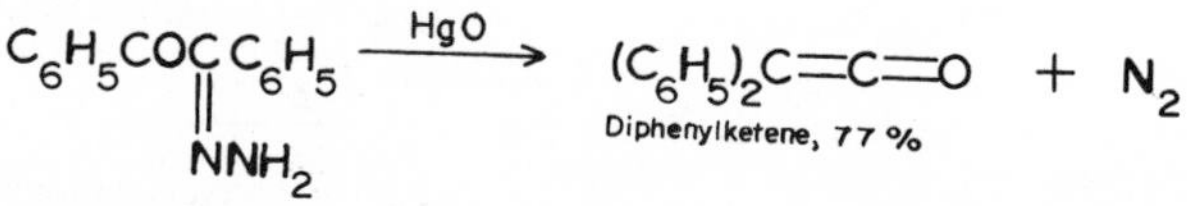

$$C_6H_5COCC_6H_5 \xrightarrow{HgO} (C_6H_5)_2C{=}C{=}O + N_2$$
$$\underset{NNH_2}{\|} \qquad \text{Diphenylketene, 77 \%}$$

Another is a Favorskii rearrangement of α-bromocyclobutanone [11]:

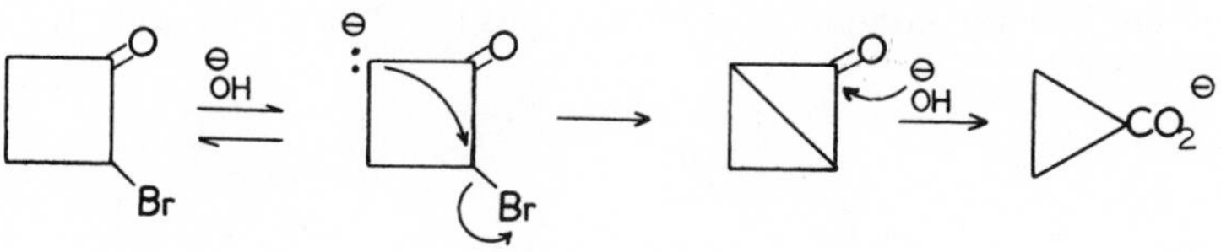

According to DeSelms [12] a cyclobutanedione eliminates carbon monoxide to form a ketene as shown:

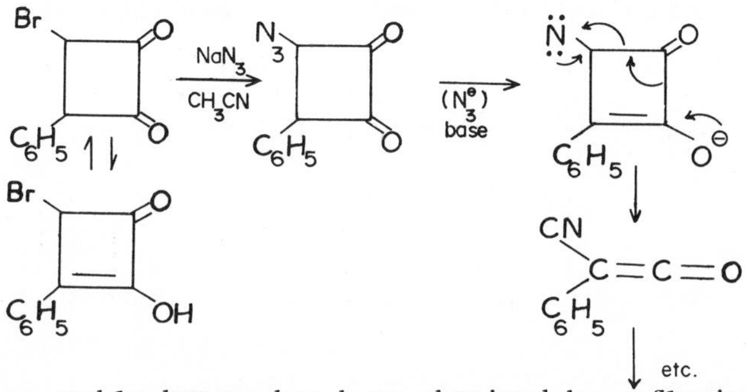

A more stable ketene has been obtained by refluxing 2,5-diazido-3,6-di-t-butyl-1,4-benzoquinone in refluxing benzene as indicated [13]:

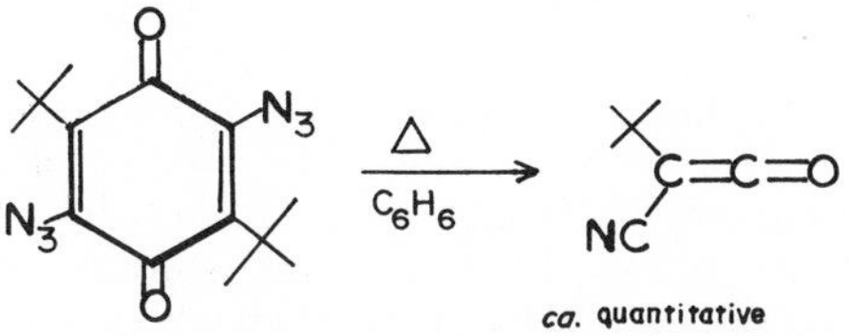

ca. quantitative

The t-butylcyanoketene, which is quite stable in solution, probably results in this case through the dinitrene

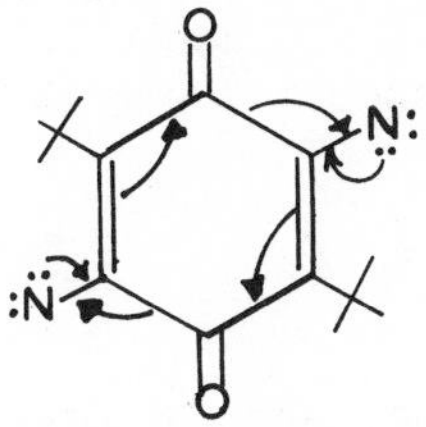

rather than a carbene as an intermediate.

In a similar manner 2-cyano-4-azido-2,5-di-t-butyl-1,3-cyclopentenedione gave the same ketene as shown:

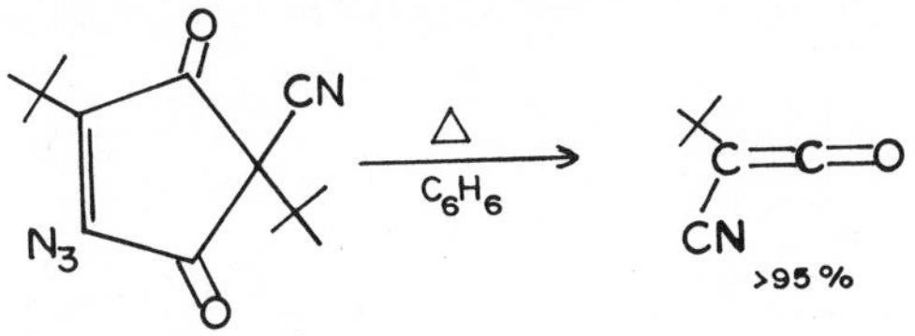

>95%

In fact, the cyclopentenedione is produced from the 1,4-benzoquinone by irradiation.

Deoxygenation is another way of providing the ketocarbene [14]:

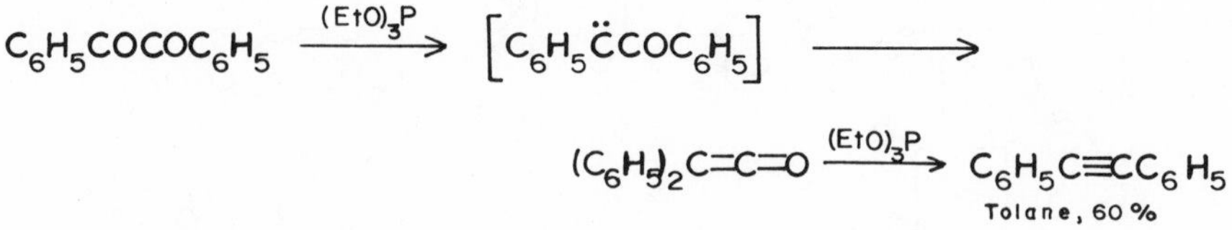

$$C_6H_5COCOC_6H_5 \xrightarrow{(EtO)_3P} \left[C_6H_5\overset{..}{C}COC_6H_5 \right] \longrightarrow$$

$$(C_6H_5)_2C{=}C{=}O \xrightarrow{(EtO)_3P} C_6H_5C{\equiv}CC_6H_5$$
$$\text{Tolane, 60\%}$$

Lastly, a sulfur ylid is the source of a ketocarbene [15]:

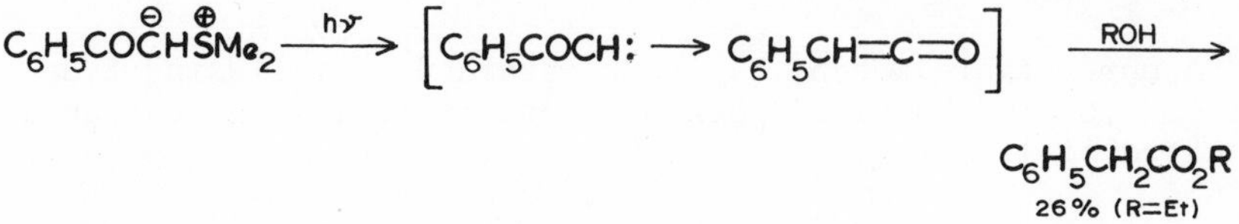

$$C_6H_5CO\overset{\ominus}{C}H\overset{\oplus}{S}Me_2 \xrightarrow{h\nu} \left[C_6H_5COCH: \longrightarrow C_6H_5CH{=}C{=}O \right] \xrightarrow{ROH}$$

$$C_6H_5CH_2CO_2R$$
$$26\% \ (R{=}Et)$$

a.   Preparation of Diphenylketene [10]

Benzilmonohydrazone, 224 g, mercuric oxide, 320 g, and 100 g anhydrous Na_2SO_4 were mixed in 800 ml of anhydrous benzene. The mixture with 1 or 2 ml of methanolic KOH was stirred for 1 hr under ice cooling and filtered. The filtrate was dropped into a special apparatus that possessed a N_2 atmosphere and permitted the intermediate azibenzil to decompose in a condenser containing paraffin oil at 130-135° in its jacket and at the same time allowed the benzene vapor to condense in a second water-jacketed condenser. The crude ketene, 140 g (77%), was recovered from the paraffin-oil condenser.

Addendum

A.2. By modifying the procedure for preparing phenylmalonyl dichloride from phenylmalonic acid and thionyl chloride, more specifically, by refluxing the crude product in toluene under a slow stream of N_2, phenylchlorocarbonyl ketene (92%) may be obtained [S. Nakanishi, K. Butler, Org. Prep. Proced. Int., 7, 155 (1975)].

References

1.   D. Borrmann, in E. Müller, Methoden der Organischen Chemie, Vol. VII, Pt. 4, G. Thieme Verlag, Stuttgart, 1968, pp. 65-286.
2.   Ref. 1, pp. 226, 264.

3.   Ref. 1, p. 108.
3a. W. T. Brady, Synthesis, 415 (1971).
4.   G. Quinkert et al., Chem. Ber., 97, 1799 (1964).
5.   R. G. S. Pong, J. S. Shirk, J. Am. Chem. Soc., 95, 248
     (1973).
6.   W. C. Agosta, A. B. Smith, III, J. Am. Chem. Soc., 93, 5513
     (1971).
6a. P. Yates et al., J. Org. Chem., 35, 3682 (1970).
7.   J.-M. Conia et al., Chem. Commun., 795 (1973).
8.   P. Schiess, P. Radimerski, Angew. Chem. Intern. Ed. Engl.,
     11, 288 (1972).
8a. P. Scribe et al., Tetrahedron Letters, 3441 (1973).
9.   Ref. 1, p. 87.
10. W. Ried, P. Junker, Angew. Chem. Intern. Ed. Engl., 6, 631
     (1967).
11. J.-M. Conia, J.-L. Ripoll, Compt. Rend., 251, 1071 (1960).
12. R. C. De Selms, Tetrahedron Letters, 1179 (1969).
13. H. W. Moore, W. Weyler, Jr., J. Am. Chem. Soc., 92, 4132
     (1970); 93, 2812 (1971).
14. T. Mukaiyama et al., J. Org. Chem., 29, 2243 (1964).
15. B. M. Trost, J. Am. Chem. Soc., 88, 1587 (1966).

B.   Elimination

The two principal elimination methods, B.1 and B.2, as
applied to the formation of halogenated ketenes, have been discussed in the H.-W. volume [1] and in some detail by Brady [2].
As a rule these ketenes polymerize readily, although they may be
utilized in situ in further reactions. Trapping, as has already
been stated, is a common method for determining their presence.

1.   From α-Haloacid Halides (1, 891)

$$\underset{(Ar)R}{\overset{Ar}{\diagdown}}CXCOX \quad \overset{Zn}{\longrightarrow} \quad \underset{(Ar)R}{\overset{Ar}{\diagdown}}C{=}C{=}O \; + \; ZnX_2$$

Preparations are given [1] for methylphenyl-, dichloro-,
dimethyl- and phenylketenes by the use of zinc. An activated
form of this metal may be prepared as shown by Brady [2].

Singhal and Smith [3] substituted copper for zinc in the
formation of the ketene from α-chlorodiphenylacetyl chloride. No
effort was made to isolate the ketene that was trapped with p-
toluidine to give an 80% yield of N(p-tolyl) diphenylacetamide.
The use of copper offers an advantage over zinc in that the
copper halide formed, in contrast to the zinc salt, may be

removed from the benzene solvent by filtration to give a solution of the diphenyl ketene. Some ketenes derived from adamantane have been synthesized [8].

2. From Acid Chlorides [4] or Nitrophenol Esters (1, 891)

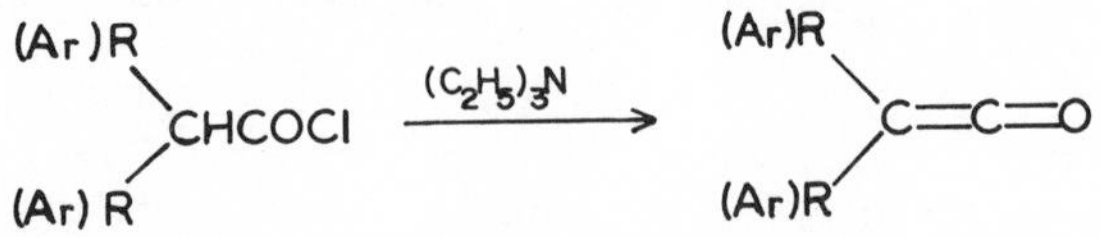

Taylor, McKillop, and Hawks [5] have described in detail the procedure referred to previously (1, 892) for the synthesis of diphenyl ketene from diphenylacetyl chloride by the use of triethylamine. The method, starting with diphenylacetic acid, consists of two simple steps: (a) it involves relatively inexpensive starting materials and (b) it does not require hazardous or toxic chemicals or special apparatus. The ketene, obtained in 73-84% yields, is never exposed to temperatures above 30-35°, except in the final distillation.

In the following example the mechanism of dephenolation resembles that of the E1cB reaction for the synthesis of alkenes [6].

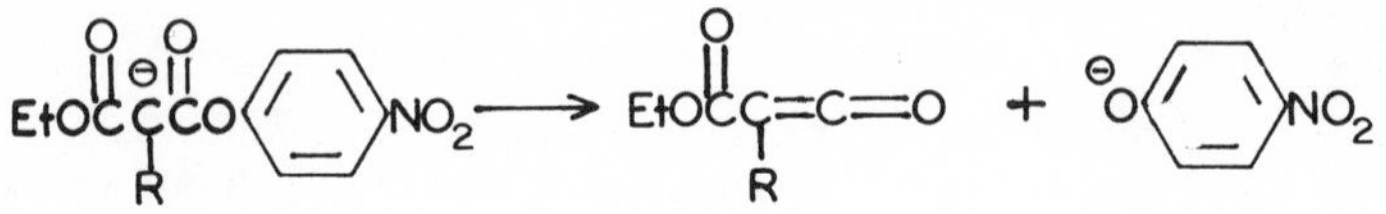

In another example the in situ preparation of a chloroketene is illustrated [7]:

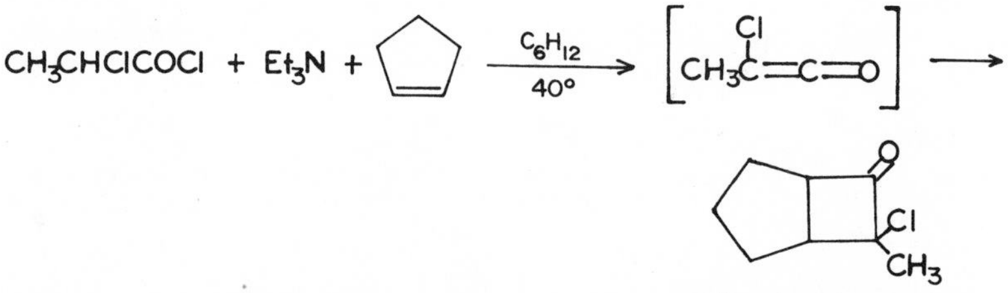

endo- and exo-2-Chloro-
2-methylbicyclo[2.3]heptanone-3
48%

a.   Preparation of 1-Adamantyl Ketene [8]

1-Adamantylacetyl chloride, 10 mmol, and triethylamine, 11 mmol, in 100 ml of dry ether were refluxed for 24 hr, after which the amine hydrochloride was removed by filtration. The ketene may not be isolated from the ethereal solution since a polymer results, but it may be used to prepare acid derivatives that indicate a yield of around 90%.

3. From Acetylenic Ethers

$$RC\equiv COC(Me)_3 \xrightarrow{70°} RCH=C=O \ + \ CH_2=C(Me)_2$$

This elimination occurs with considerable ease [9]. The ethoxytrimethylsilyl acetylene requires a higher temperature:

$$(CH_3)_3Si-C\equiv COC_2H_5 \xrightarrow{120-130°} (CH_3)_3SiCH=C=O \ + \ CH_2=CH_2$$
$$\text{Trimethylsilylketene, 70 %}$$

References

1. D. Borrmann, in E. Müller, Methoden der Organischen Chemie, Vol. VII, Pt. 4, G. Thieme Verlag, Stuttgart, 1968, p. 91.
2. W. T. Brady, Synthesis, 415 (1971).
3. G. H. Singhal, H. Q. Smith, J. Chem. Eng. Data, 14, 408 (1969).
4. Ref. 1, p. 94.
5. E. C. Taylor et al., Org. Syn., 52, 36 (1972).
6. R. F. Pratt, T. C. Bruice, J. Am. Chem. Soc., 92, 5956 (1970).
7. W. T. Brady et al., J. Am. Chem. Soc., 92, 146 (1970).
8. H. Wynberg et al., Rec. Trav. Chim., 89, 23 (1970).
9. Ref. 1, p. 101.

C. Redox Methods

Oxidation or reduction methods may be employed to prepare ketene [1]:

$$CO + H_2 \xrightarrow[200-300°]{ZnO} H_2C=C=O$$

$$HC\equiv CH \xrightarrow[\substack{100°, ZnO-CaO- \\ Ag_2O}]{air} CH_2=C=O \quad 55\%$$

Reference

1. D. Borrmann, in E. Müller, Methoden der Organischen Chemie, Vol. VII, Pt. 4, G. Thieme Verlag, Stuttgart, 1968, pp. 100-102.

D. Ketenimines

The synthesis and chemistry of ketenimines, $R_2C=C=NR'$, have been reviewed [1]. These compounds serve as dehydrating agents in peptide synthesis, as a coreagent in DMSO oxidations, and as substrates for heterocyclic syntheses.

References

1. G. R. Krow, Angew. Chem. Intern. Ed. Engl., 10, 435 (1971).

Chapter 18

CARBOXYLIC ACID AMIDES AND IMIDES

    β-Ketoamides   847
9. From Aryl Halides and Ethyl Malonate via a
    Benzyne   847
10. From α,β-Ynamines   847
11. From Ketones and DMF   848

F. Free Radical Reactions   849

G. Cycloaddition   851

    1. Ketene-Imine Interaction   851
    2. Addition of an Isocyanate to Alkenes   851
    3. Addition of Carbon Dioxide to Ynamines   852
    4. Addition of an Isocyanate to Ynamines   852
    5. From Allyl Alcohols and α-Dimethylaminovinyl
        Ethers, $CH_2=COR(NMe_2)$ (Cope Rearrangement)   853

A new text on the synthesis of amides has become available [1].
The synthesis of β-lactams has also been reviewed recently [2].
Only a few of the β-lactam preparations (A.1, E.1, and G) are
described in this chapter.

The synthesis of amides from acid derivatives appears to be
a relatively simple operation since amide formation is exothermic
with all derivatives except esters and salts. On the other hand,
synthesis from the acid and amine is more difficult because of
the endothermic nature of the process. Interest is high in the
latter process because of its applicability to the synthesis of
peptides; for the most part the reagent here is a phosphine, or a
phosphite, either of which may harness the great ability of phosphorus to remove oxygen (A.1). Of the many reagents attention is
drawn to the simplest one, the combination of diphenyl phosphite
and pyridine. Dicyclohexylcarbodiimide and like compounds are
reagents to be considered in amide formation (A.1). For the synthesis of triacetamide success has been achieved by the use of
2,6-lutidine in conjunction with acetyl chloride (A.2). The rate
of amide formation from esters and amines is increased by using
N-acyloxypypiperidine (A.4), and amide formation from the ester
and amine in general is catalyzed by the sodium salt of 2-
hydroxypyridine (A.4). Very mild hydrolytic conditions have been
developed for the conversion of nitriles into amides (A.5). Triflimides, $RCONHSO_2CF_3$, are unusual reagents in that the triflimide
anion, $N\bar{H}SO_2CF_3$, serves as a leaving group (A.15).

A unique oxidative synthesis consists of the conversion of
an unsaturated or aromatic aldehyde into the amide by the use of

manganese dioxide in the presence of sodium cyanide (C.2). Among electrophilic reactions, noteworthy items are the Friedel-Crafts preparation of amides from arenes and urea (D.1), a Ritter reaction free from strong acid (D.4), and useful amide preparations employing nickel carbonyl or sodium tetracarbonylferrate(-II) (D.8).

Among nucleophilic reactions, the alkylation of phthalimide has been carried out with an alcohol rather than an alkyl halide (E.1) and ynamines have been converted into amides by hydration (E.10). Amide formation from the attack of amines on trihalocarbonyl compounds has attracted some attention as has, for similar reasons, the ring opening of 2-nitrocyclohexanone (E.7). The addition and substitution behavior of the formamide free radical, $\cdot CONH_2$, continue to be studied (F). Most unusual cyclo reactions to produce amides, particularly involving ynamines have been conceived (G).

A. Solvolysis

1. From Carboxylic Acids and Their Ammonium Salts ($\underline{1}$, 895)

The synthesis of amides from carboxylic acids or their salts continues to be used very widely. Employing the common reagent DCC ($\underline{1}$, 896), Thomas [3] synthesized the formamides of amino acids as shown without racemization:

$$RCHCOOCH_2C_6H_5 \quad \xrightarrow[HCOOH,0°]{DCC} \quad RCHCOOCH_2C_6H_5$$
$$\underset{NH_2}{|} \qquad\qquad\qquad\qquad \underset{NHCHO}{|}$$

The N-formyl esters of glycine, β-alanine, and L-leucine were obtained as oils (60-90%). Peptide formation, usually sluggish, between N-benzyloxycarbonylvaline and ethyl phenylalanate was accomplished more readily by first forming the anhydride of the valine. The results using the reagents for anhydride formation as listed are [4]: (a) isopropyl chloroformate 35%, (b) benzoyl chloride 47%, (c) p-methoxybenzoyl chloride 88%, and (d) DCC 59%.

Increasing interest is shown in water-soluble carbodiimides as exemplified [5]:

$$EtN{=}C{=}NCH_2CH_2CH_2NMe_2 + \text{ asparagine} + \text{N-acetylphenylalanine} \xrightarrow{\text{pH 4.75}}$$
N-3-Dimethylaminopropyl-N-ethylcarbodiimide

polypeptide, mol.wt. 5000

Another unusual water-soluble reagent is trimetaphosphate. With glycine (both at 0.1 molar) in water at pH 11, diglycine was formed in 40% yield in 120 hr [6].

Several investigators have employed an alkyl- or arylphos-
phine or hexamethylphosphorous triamide in a carbon tetrahalide
as a reagent. Thus Barstow and Hruby [7] utilized the reagents
as shown (Ex. a):

$$(Ar)RCOOH \ + \ R'NH_2 \ \xrightarrow[\substack{or \\ (C_6H_5)_3P-CBrCl_3 \\ THF}]{(C_6H_5)_3P-CCl_4} \ (Ar)\overset{R}{C}ONHR' \ + \ H_2O$$

Yields for a series of acetamides and benzamides varied
within 61-97%. Yamada and Takeuchi [8] utilized N-protected
amino acids or peptides with amino acid or peptide esters as
indicated:

$$C_6H_5CONHCH_2COOH \ + \ NH_2CH_2COOEt \ \xrightarrow[\substack{P(NMe_2)_3 \\ THF, -20 \, to \, 0°}]{CX_4} $$

$$C_6H_5CONHCH_2CONHCH_2CO_2Et$$

In this manner dipeptides were obtained in 60-85% yields with
little racemization according to the Young racemization test:
Bz-Leu-Gly-OEt, α_{20}^D = -34° for no racemization. Castro and
Dormoy [9] employed the same phosphine, as indicated, to synthesize N-mono- and N,N-disubstituted amides:

$$RCOOH \ + \ R'NH_2 \ \xrightarrow[\substack{CCl_4, THF \\ -70°, 2 \, hr.}]{P(NMe_2)_3} \ \underset{50-96\%}{RCONHR'} \ + \ H_2O$$

In a peptide synthesis Matsueda and co-workers [10] modified the
usual procedure by the addition of a disulfide as a hydrogen
acceptor, a salt to minimize the formation of the thiolester, and
an amine to eliminate the hydrogen halide formed. The equation
for the reaction in simple form becomes:

$$RCOOH + R'NH_2 + (C_6H_5)_3P + (p\text{-}ClC_6H_4S)_2 + CuCl_2 + 2 Et_3N \xrightarrow[3\,hr.]{25°}$$

$$RCONHR' + (C_6H_5)_3PO + (p\text{-}ClC_6H_4S)_2Cu + 2 \ Et_3\overset{\oplus}{N}H \ \overset{\ominus}{Cl}$$

With an N-protected amino acid and a free amino acid ester,
benzyloxycarbonyl-L-phenylalanylglycine ethyl ester was obtained
in 89% yield without racemization by this procedure.

A one-pot reaction has been referred to previously [7]. Also
in a one-step reaction the bis-diethylamide of sulfurous acid,

Et_2NSNEt_2, and the acid gives diethylamides (42-75%) along with sulfur dioxide and diethylamine [11]. In addition, the dichlorophenylphosphine oxide, $C_6H_5POCl_2$, the sodium salt of the acid, and the amine easily yield the amide (70-94%) [12].

The use of silicon tetrachloride as shown gives acceptable yields at room temperature, particularly with aromatic amines and aliphatic and aromatic acids [13]:

$$2\ RCO_2H + 2\ ArNH_2 + SiCl_4 \xrightarrow[\substack{25° \\ 10\ hr.}]{4\ C_5H_5N} 2\ RCONHAr + SiO_2 + 4\ C_5H_5N \cdot HCl + 2\ H_2O$$
$$40-70\%$$

Obviously the separation of the insoluble silica is easily accomplished. The method has been shown to be of less value for peptide synthesis [14].

Thionyl chloride with HMPA has been employed to prepare three benzamides as shown [15]:

$$C_6H_5COOH + C_6H_5NH_2 \xrightarrow[\substack{HMPA \\ 25°, 5\ hr.}]{SOCl_2} C_6H_5CONHC_6H_5 + H_2O$$
$$\text{Benzanilide, 87\%}$$

N-Phosphonium salts of pyridine have been employed in the synthesis of amides as indicated [16]:

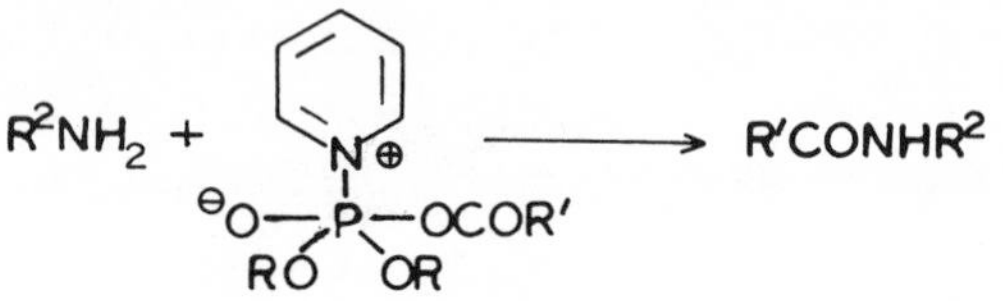

The phosphonium salt, prepared in situ from phosphorous acid oxidized by mercuric chloride in the presence of pyridine, and the carboxylic acid were mixed, after which the mixture was treated with the amine. By the procedure acetanilide was prepared (94%). The method is suitable as well for the synthesis of peptides.

Later Yamazaki and Higashi [17] utilized diphenyl phosphite as the reagent in forming the amides of peptides. For example, benzyloxycarbonyl ethyl glycylglycinate was prepared as shown:

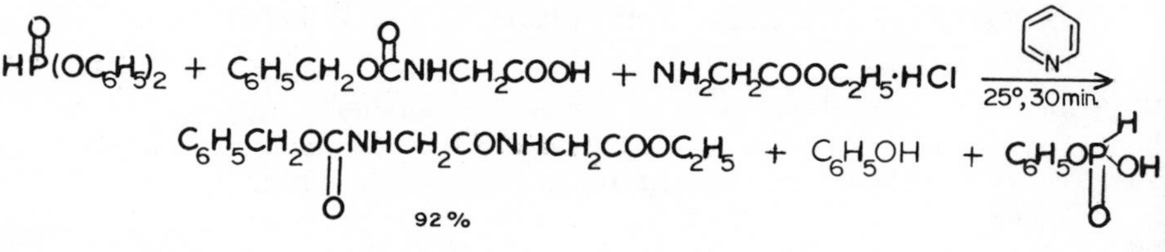

$$HP(OC_6H_5)_2 + C_6H_5CH_2OCNHCH_2COOH + NH_2CH_2COOC_2H_5 \cdot HCl \xrightarrow[25°, 30\ min.]{}$$

$$C_6H_5CH_2OCNHCH_2CONHCH_2COOC_2H_5 + C_6H_5OH + C_6H_5OP\substack{H \\ OH}$$
$$92\%$$

The method in which no oxidizing agent is needed leads to peptides with high optical activity.

β-Trichloromethyl-β-propiolactone and β-trichloromethyl-β-ethane sultone were shown to be effective reagents in amide formation [18]. Thus p-phenetidine and acetic acid with a trace of the sultone boiled for 2 min gave a 93% yield of N-acetyl-p-phenetidine as indicated:

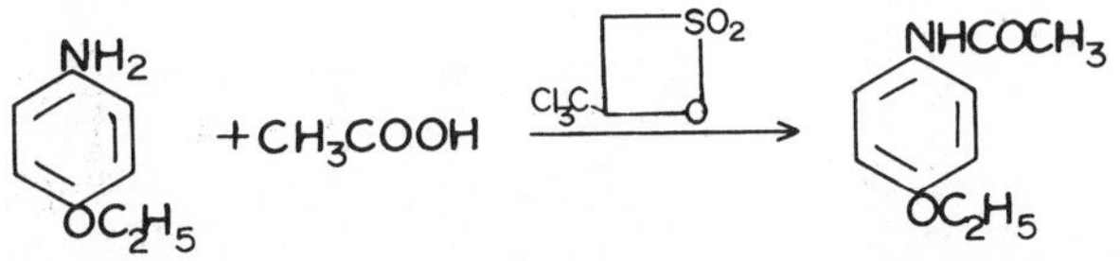

Another unusual reagent is hexachlorocyclotriphosphatriazene, which was employed by Caglioti and co-workers [19] as indicated:

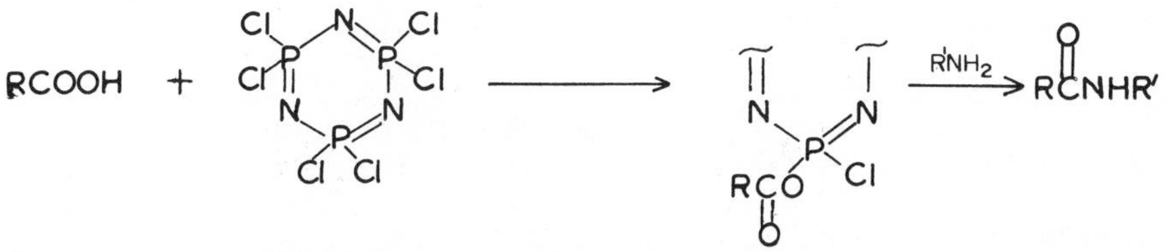

The method, which is applicable to the free acid or its sodium salt, uses an excess of amine, but requires only 20 min for completion. Yields vary within 63-83%.

For compounds containing both the amino and carboxylic acid functions such as 4-aminocyclohexanecarboxylic acid, lactam formation occurs simply by heating with Dowtherm A [20] as indicated (Ex. c):

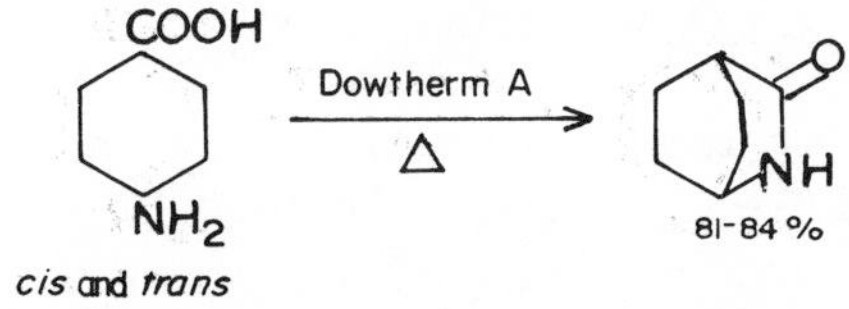

cis and trans

A method of synthesizing β-lactams has been utilized by Ivanov and Dobrev [21] as indicated:

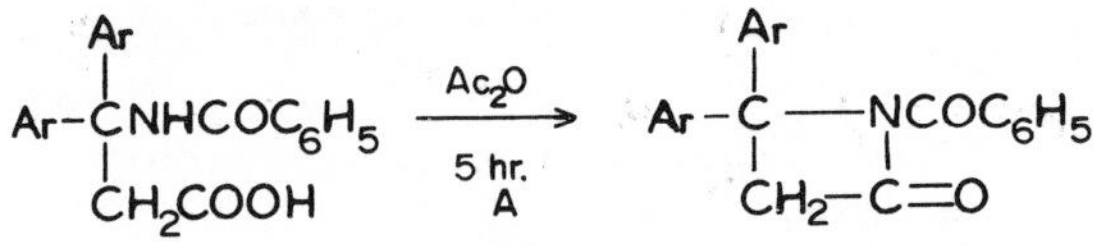

Yields from three N-substituted 3-aminopropionic acids ranged within 95-98%.

The acid may be made susceptible to attack by the amine by complexing it with triphenylphosphine and a sulfenamide as indicated [22]:

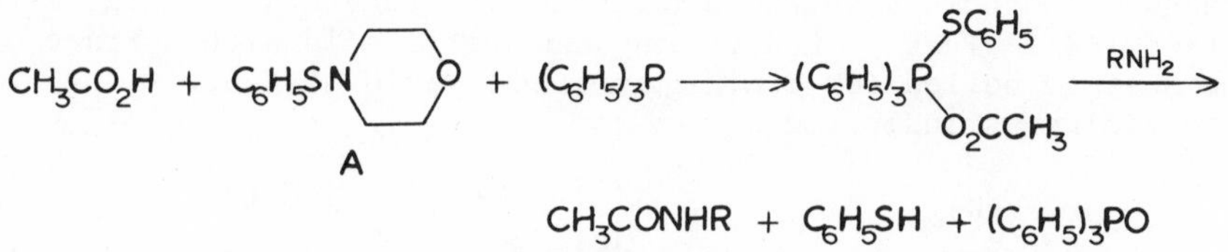

$$CH_3CO_2H + C_6H_5SN \overline{}O + (C_6H_5)_3P \longrightarrow (C_6H_5)_3P\begin{smallmatrix} SC_6H_5 \\ \\ O_2CCH_3 \end{smallmatrix} \xrightarrow{RNH_2}$$

$$A$$

$$CH_3CONHR + C_6H_5SH + (C_6H_5)_3PO$$

Since thiophenol will consume the sulfenamide A, to form the disulfide, it is scavenged by the presence of Cu(II) hexanoate or other organic salts. Yields of amides under the above conditions at 25° ranged within 57-100%.

a. Preparation of N-n-Butylacetamide [7]

Triphenylphosphine, 13.1 g, 50 ml of CCl_4, and 150 ml of THF were refluxed for 30 min, after which 2.85 ml AcOH was added to the solution at 5°. After standing at 5° for 10 min, 9.73 ml of n-butylamine was added and the mixture was refluxed for 45 min. Recovery of the amide amounted to 5.25 g (91%).

b. Preparation of Benzanilide [15]

Benzoic acid, 6.1 g, in 20 ml of HMPA at -10° was treated with 3.6 ml of $SOCl_2$ and 4.65 g of aniline was added. From the mixture, after standing for 5 hr at 25°, 8.3 g (87%) of the anilide was recovered.

c. Preparation of 3-Isoquinuclidone [20]

Yield was 81-84% from a mixture of cis- and trans-4-aminocyclohexane carboxylic acid and Dowtherm A heated at reflux for 20 min.

2. From Acid Chlorides (1, 899)

$$RCOCl \xrightarrow{NH_3} RCONH_2$$

Triacylamides are prepared from the acid chloride and amide preferably with the base, 2,6-dimethylpyridine [23]. Thus triacetamide may be obtained as shown (Ex. a):

$$2 \text{ CH}_3\text{COCl} + \text{CH}_3\text{CONH}_2 \xrightarrow[\text{-40 to 10}^\circ,\text{18 hr.}]{} (\text{CH}_3\text{CO})_3\text{N} + 2 \text{ HCl}$$

Triacetamide, 80-90%

ine yields are dependent on the sequence of mixing of the reagents, the temperature, and whether there are substituents in the α-positions of the acyl chloride and the pyridine.

An alkyl chlorocarbonate has been employed in the formation of the amides of γ,δ-unsaturated acids [24] as indicated:

$$\text{ArCH}_2\text{CHCOOH} \xrightarrow[\text{2)NH}_3]{\substack{\text{1)EtOCOCl-Et}_3\text{N} \\ -30^\circ}} \text{ArCH}_2\text{CHCONH}_2$$

with $\text{CH}_2\text{CH=CH}_2$ substituent groups below each structure

56 - 97 %

A carboxylic amide ketal has been converted into an amide by acetyl chloride [25]:

$$+ \text{ CH}_3\text{COCl} \longrightarrow \text{CH}_3\text{CON}\begin{smallmatrix}\text{CH}_2\text{CH}_2\text{Cl} \\ \\ \text{CH}_2\text{CH}_2\text{OCOCH}_3\end{smallmatrix}$$

N-β-Acetoxyethyl-N-β-chloroethylacetamide

93 %

a.   Preparation of Triacetamide [23]

A cooled methylene chloride solution of 2 equiv of CH_3COCl was treated with slightly more than 2 equiv of 2,6-dimethylpyridine and 1 equiv of acetamide was added, after which the mixture was warmed to approximately 10° for 18 hr. Yield of the amide recovered was 80-90%.

3.   From Acid Anhydrides (1, 900)

$$\text{RC-O-CR} \xrightarrow{\text{NH}_3} \text{RCONH}_2 + \text{RCOOH}$$

Amides have been prepared from anhydrides by treatment with quaternary ammonium salts [26]. Thus butyric anhydride with N,N,N-trimethyl-N-cyclohexylammonium iodide gives N-methyl-N-cyclohexylbutyramide as indicated:

$$\text{C}_4\text{H}_9\text{COCC}_4\text{H}_9 + \text{C}_6\text{H}_{11}\overset{\oplus}{\underset{\text{CH}_3}{\overset{\text{CH}_3}{\text{N}}}}\text{CH}_3 \cdot \text{I}^\ominus \xrightarrow[\text{2)Na}_2\text{CO}_3]{\substack{\text{1) 200}^\circ \\ \text{24 hr.}}} \text{C}_4\text{H}_9\text{CON}\begin{smallmatrix}\text{CH}_3 \\ \\ \text{C}_6\text{H}_{11}\end{smallmatrix} + \begin{smallmatrix}\text{C}_4\text{H}_9\text{CO}_2\text{CH}_3 \\ + \\ \text{CH}_3\text{I}\end{smallmatrix}$$

62 %

The method is satisfactory when at least two alkyl groups are attached to the positive nitrogen atom. If there is only one as in N-methylquinolinium iodide, the free amine is obtained as the product. With aromatic ammonium salts the yields are low. Amide formation is also possible from the anhydride and amine in the presence of NaI, in which case the maximum yield was 62% [26].

Tertiary amines which usually contain at least one benzyl or t-butyl group are degraded to amides by removal of the afore-mentioned groups in refluxing acetic anhydride [27].

4.   From Esters, Lactones, or Phthalides (1, 901)

$$RCOOR' \xrightarrow{NH_3} RCONH_2 + R'OH$$

Phenyl formate has been shown to be effective in the formylation of amines [28]. Thus o-bromoaniline gives the o-bromoformanilide as shown (Ex. a):

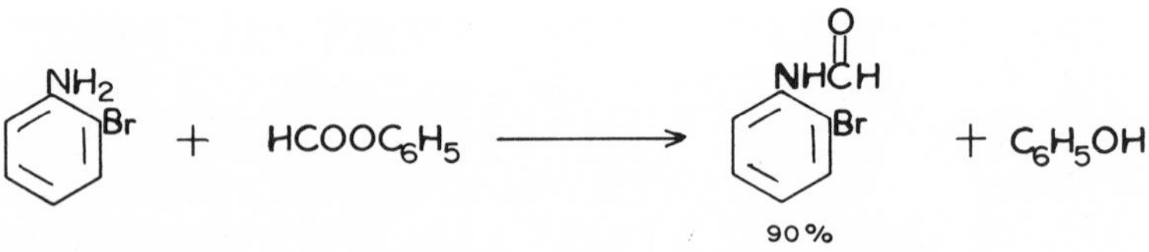

90%

Other yields from primary and secondary amines varied within 60-95%. The amine is simply mixed without heating with a slight excess of the formate and at times the product crystallizes from the reaction mixture.

A unique ester which appears to possess superior driving force in forming the amide is the ester of N-hydroxypiperidine [29]:

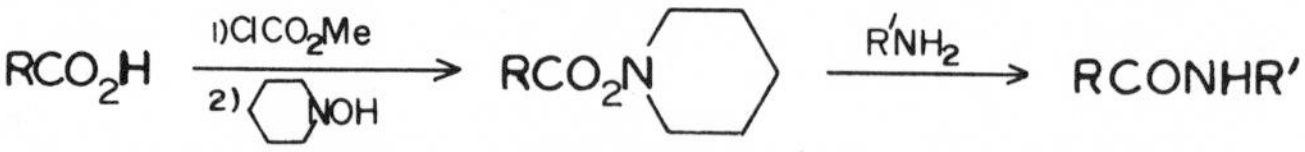

Similarly, the esters of oximes are transformed into amides with an amine at room temperature [30]:

$$>C=NOCOR \xrightarrow{R'NH_2} RCONHR'$$
30-94%

Amides have also been synthesized by the aminolysis of esters by using NaH in DMSO [31]. In this manner ethyl anisate with aniline gives p-ethoxybenzanilide as shown:

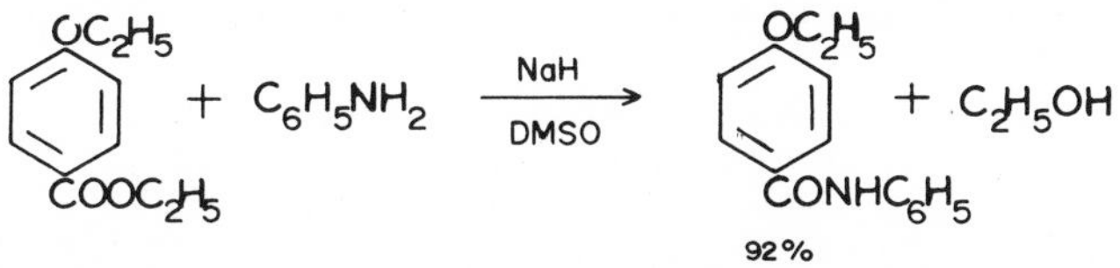

92%

By this method the two components and NaH in the DMSO solvent are simply mixed and stirred overnight at 25°. The yields vary from 68 to 94%.

van Melick and Wolters [32] found isopropenyl formate useful in the formylation of amines as indicated:

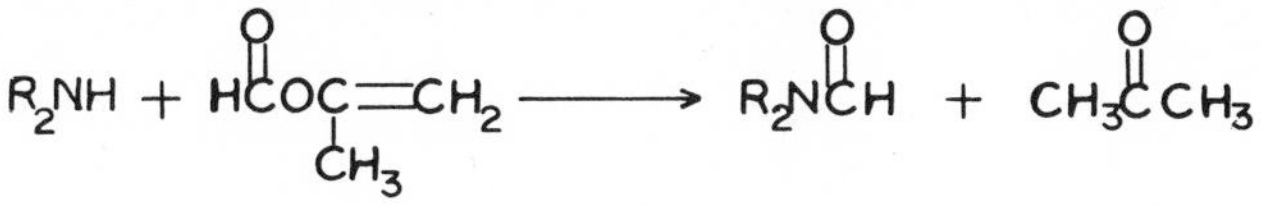

With primary and secondary aliphatic amines the reaction was complete in 5 min. With L-tyrosine methyl ester formylation occurred to give N-formyl-L-tyrosine methyl ester (95%).

Sodium 2- or 4-pyridinolate has been found to be a catalyst in the aminolysis of esters [33]. Thus methyl acetate responded as indicated:

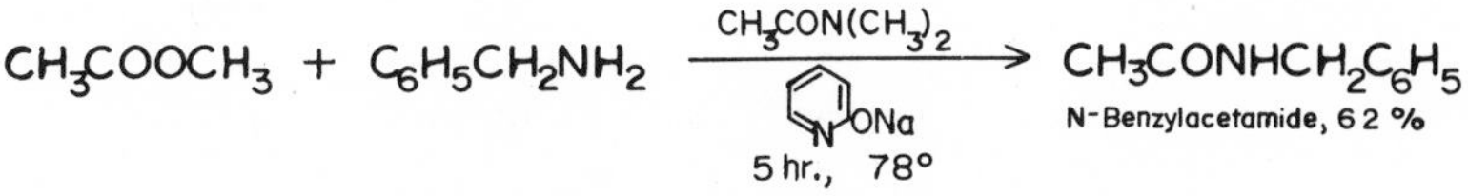

Without the catalyst no amide could be isolated.

Previously, Openshaw and Whittaker [34] showed that 2-hydroxypyridine was an effective catalyst in the aminolysis as indicated:

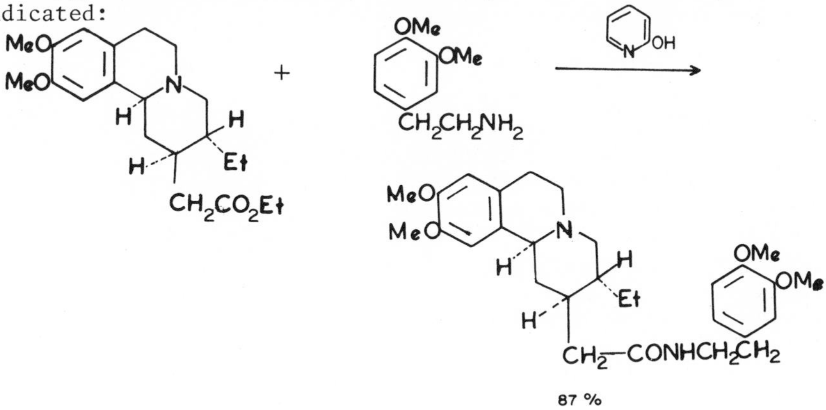

87 %

Without the catalyst the yield of the N-(3,4-dimethoxyphenethyl) amide, was 2%.

An amide in the tetracycline series was prepared by fusing the ester with ammonium formate under nitrogen [35].

5. From Nitriles ($\underline{1}$, 903)

$$RCN \xrightarrow{\text{H}_2\text{O}} RCONH_2$$

A solid-phase catalysis of the hydrolysis of nitriles to amides was accomplished by Cook and co-workers [36] as shown:

$$RCN \xrightarrow[\text{CH}_2\text{Cl}_2, 25°]{\text{MnO}_2} RCONH_2$$

The water necessary apparently comes from that retained by MnO_2, which is employed in great excess. The method, which gives yields of 22-90%, is applicable to aliphatic and aromatic nitriles.

Nickel precipitated on a metal such as zinc has also been used in converting nitriles into amides [37] as indicated:

$$ArCN \xrightarrow[\substack{\text{H}_2\text{O} \\ \text{8hr. reflux}}]{\text{Ni-Zn}} ArCONH_2 \quad \substack{\\ 30-90\%}$$

The method is less satisfactory for aliphatic nitriles.

Other metal catalysts such as copper have been employed. Thus Barber and Fetchin [38] converted acrylonitrile into the amide with copper (prepared from $CuSO_4$, $NaBH_4$, and NaOH in water) as shown:

$$CH_2{=}CHCN \xrightarrow[\text{H}_2\text{O}]{\text{Cu}} CH_2{=}CHCONH_2 \quad \substack{\\ 85.7\%}$$

Raney copper in the same reaction gave a 22.3% yield.

The catalysts $Rh(OH)(CO)(PPh_3)_2$ and $Pt(C_6H_8)(PPh_3)_2$ gave 150 and 58 mol of acetamide per mol of the above catalysts, respectively, when heated in an excess of acetonitrile [39]. $Co(en)_2Br^{2+}$ forms complexes with aminonitriles and this complex with the mercuric ion accelerates the hydration to the amide [40]. Triphenylphosphine formed the monoamide from benzylidenemalonodinitrile [41]:

$$ArCH{=}C(CN)_2 \xrightarrow[\text{2) EtOH reflux}]{\substack{\text{1)} (C_6H_5)_3P{-}CHCl_3 \\ \text{aq. HCl}}} ArCH{=}C\overset{\textstyle CONH_2}{\underset{\textstyle CN}{\Big\langle}}$$

The hydrolysis of nitrile solutions of boron halides produces amide hydrohalides as shown for acetonitrile [42]:

$$CH_3CN \cdot BCl_3 \xrightarrow[\text{1 hr. 25°}]{H_2O} \underset{60\%}{CH_3\overset{\overset{\textstyle O}{\|}}{C}NH_2 \cdot HCl}$$

A like yield from the nitrile saturated with HCl requires 4 days. The result with phenyl cyanide is similar. 2-Acetamidobenzonitrile was converted into 2-acetamidobenzamide as the sole product by irradiation as indicated [43]:

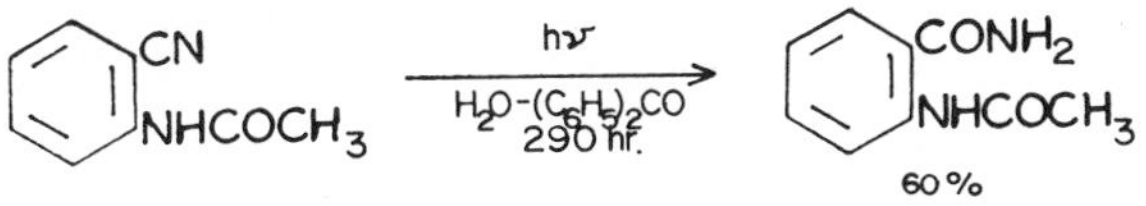

Nitriles may be hydrolyzed to amides by the use of formic acid under pressure [44a]. Thus m-chlorobenzonitrile with an equivalent amount of formic acid in a tantalum- or silver-lined autoclave heated as indicated gave the amide:

$$m\text{-}ClC_6H_4CN + HCOOH \xrightarrow[\text{2 hr.}]{250°} \underset{90\%}{m\text{-}ClC_6H_4CONH_2} + CO$$

Later Becke and co-workers [44b] were able to carry out the reaction at ordinary pressure and temperature by using formic acid alone or in a solvent such as toluene with hydrogen chloride or hydrogen bromide bubbling through the solution. Yields for a series of aliphatic and aromatic nitriles vary within 55-99% and the time required is 2-22 hr.

9.   From Thioamides (1, 906)

$$RCSNH_2 \longrightarrow RCONH_2$$

Thioamides may be prepared from the nitrile and thioacetamide in DMF-HCl at 85° for 30 min [45]. Besides the reagents such as HgO, Ag$_2$O, and litharge (1, 906), trimethyloxonium fluoroborate has been used to convert the thioamide into the amide (60-95%) [46].

11.   From Amides (1, 907)

$$RCONH_2 + R'NH_2 \longrightarrow RCONHR' + NH_3$$

Kraus has extended the study of the use of dimethylformamide in the formylation of aliphatic amines (1, 907-908) [47]. Good

yields were obtained without a catalyst if the time is lengthened.
Thus benzylamine with dimethylformamide responded as shown:

$$C_6H_5CH_2NH_2 + HCON(CH_3)_2 \xrightarrow[\text{30 hr.}]{\text{Reflux}} C_6H_5CH_2NHCH{\overset{O}{\parallel}} + HN(CH_3)_2$$

N-Benzylformamide, 90%

The addition of acid lowers the time required to complete
the reaction. In one case, that of N-decylformamide, a 75% yield
of the amide was obtained by heating 1-aminodecane and dimethylformamide with concentrated sulfuric acid at 120° for 2 hr.

N-Methyl-N-tosylpyrrolidinium perchlorate has been shown to
be selective in tosylating an amino group in the presence of a
hydroxyl group [48]. Thus the bicyclic amino alcohol, A, gave
the N-tosyl derivative, B, as indicated:

A   B 80%

The reagent, prepared from the reaction of N-methylpyrrolidine
and tosyl chloride in the presence of silver perchlorate, tosylates primary and secondary amines with yields of 50-98% (see
B.2 for the solvolysis of $RCONHN=CHCCl_3$ by $R'NH_2$).

The acyl derivatives of chloral hydrazone also exchange with
amines [49]:

$$RCONHN=CHCCl_3 \xrightarrow{R'NH_2} RCONHR'$$

A quite reactive reagent for the synthesis of sulfonimides
is N-sulfinylarenesulfonamide, $O=S=NSO_2Ar$, made from the sulfonamide and refluxing thionyl chloride [50]. With this substituted
sulfonamide and a carboxylic acid in toluene, the carboxylic
imide, $RCONHSO_2Ar$, is formed in yields of 61-84%.

Lastly, phosphoramides of the structure $PO(NHR)_3$ have been
used to form amides from acids. The phosphoramides were tested
when R=aryl or cyclohexyl [51].

12. From Acylazides (1, 909)

Amides may be prepared from acyl azides by treatment with
tributyltin hydride alone or in the presence of azobisisobutyro-
nitrile [52]. Thus benzoyl azide gives benzamide as indicated:

$$C_6H_5CON_3 + (C_4H_9)_3SnH \xrightarrow[\substack{\text{xylene} \\ \text{1.5 hr.reflux}}]{\text{Azobisisobutyronitrile}} C_6H_5CONHSn(C_4H_9)_3 \xrightarrow{H_2O}$$

$$C_6H_5CONH_2 + (C_4H_9)_3SnOH$$
$$\text{91 \%}$$

The intermediate N-(tri-n-butylstannyl) benzamide may be isolated. Without the catalyst more time is required and yields are lower.

Amides have been obtained from acyl azides by irradiation. Thus hexanoyl azide with acetophenone in photosensitized decomposition gave hexanamide as shown [53]:

$$C_5H_{11}CON_3 \xrightarrow[h\nu]{C_6H_5COCH_3} C_5H_{11}CONH_2$$
$$\text{78 \%}$$

13.   From Isocyanates or Isothiocyanates and Precursors (1, 909)

$$RNC{=}O \quad + \quad R'COOH \longrightarrow RNHCOR' + CO_2$$

N-Alkylacetamides have been prepared from amides, lead tetraacetate, and acetic acid as indicated [54]:

$$RCONH_2 \xrightarrow{Pb(OAc)_4} RNCO \xrightarrow{CH_3COOH} RNHCOCH_3$$
$$R \neq aryl \qquad\qquad\qquad\qquad \text{4-78\%}$$

A Hofmann rearrangement has occurred followed by solvolysis of the isocyanate. If instead of acetic acid, an inert solvent such as benzene is used in the reaction, mixtures of N-alkylacetamides and N,N'-dialkylureas result.

15.   From Triflimides

Triflimides contain an unusual but useful leaving group for the preparation of amides in high yield (88-98%) [55].

$$R\overset{O}{\overset{\|}{C}}NHSO_2CF_3 + C_6H_5NH_2 \longrightarrow R\overset{O}{\overset{\|}{C}}NHC_6H_5 + NH_2SO_2CF_3$$

The triflimide is prepared from the anion of the amide:

$$RCONH_2 \xrightarrow[\substack{2) (CF_3SO_2)_2O \\ 5°}]{1) NaH} RCONHSO_2CF_3$$

A ditriflimide apparently is selective in forming amides with primary amines:

$$C_6H_5N(SO_2CF_3)_2 + RNH_2 \longrightarrow CF_3SO_2NHR$$

$$C_6H_5N(SO_2CF_3)_2 + R_2NH \longrightarrow \text{No Reaction}$$

16. From Heterocyclic Compounds (Ring Opening)

Oxazines or oxazoles are likely candidates which may be hydrolyzed to amides or imides. The oxazine obtained from the cycloaddition of N-methylolphthalimide and dichloroethylene behaves in this manner [56]:

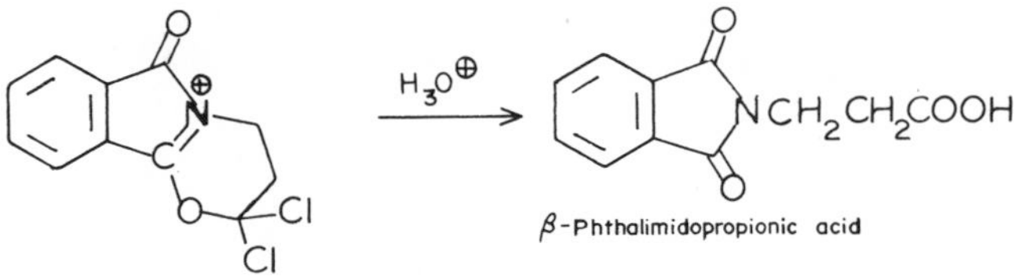
H3O⊕
NCH2CH2COOH
β-Phthalimidopropionic acid

A procedure for obtaining α-acylaminoketones as an alternative to the Dakin-West procedure is as follows [57]:

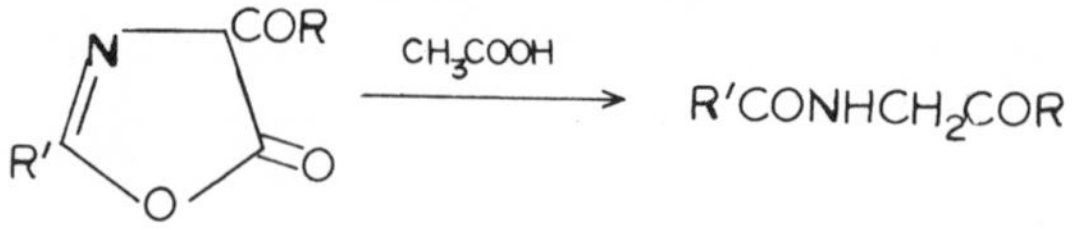
COR
CH3COOH
R'CONHCH2COR

The ring opening of a 2,3-bis-t-butyliminooxetane has been accomplished [58]:

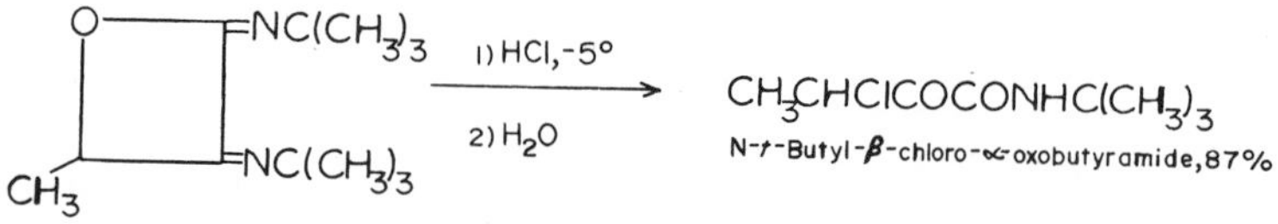
1) HCl,-5°
2) H2O
CH3CHClCOCONHC(CH3)3
N-t-Butyl-β-chloro-α-oxobutyramide,87%

Addenda

A.2. The acid bromide of a phosphite, $(RO)_2\overset{O}{\overset{\|}{P}}H$, may be made in situ by a phase-transfer reaction (with 20% rather than 50% of NaOH) and tetrabromomethane. The former then reacts with dimethylamine to form the amide, $(RO)_2\overset{O}{\overset{\|}{P}}NMe_2$ [A. Zwierzak, Synthesis, 507 (1975)].

A.8. Diketene reacts with primary amines to form acetoacetamides that may be cyclized to 3-acetyl-4-hydroxy-2-pyridones. A review of these syntheses has been published [T. Kato, Acc. Chem. Res., 7, 265 (1974)].

References

1. A. L. J. Beckwith, in Zabicky, The Chemistry of Amides, Interscience, New York, 1970, p. 73.
2. A. K. Mukerjee, R. C. Srivastava, Synthesis, 327 (1973); J. C. Sheehan, E. J. Corey, Org. Reactions, 9, 388 (1957).
3. J. O. Thomas, Tetrahedron Letters, 335 (1967).
4. C. Birr et al., Ann. Chem., 729, 213 (1969).
5. A. Previero et al., French Patent, 2,036,755, February 5, 1971; C. A., 75, 152335 (1971).
6. J. Rabinowitz, Helv. Chim. Acta, 53, 1350 (1970).
7. L. E. Barstow, V. J. Hruby, J. Org. Chem., 36, 1305 (1971).
8. S. Yamada, Y. Takeuchi, Tetrahedron Letters, 3595 (1971).
9. B. Castro, J.-R. Dormoy, Bull. Soc. Chim. France, 3034 (1971).
10. R. Matsueda et al., Bull. Chem. Soc. Jap., 44, 1373 (1971); T. Mukaiyama et al., Tetrahedron Letters, 1901 (1970).
11. G. Rosini et al., Il Farmaco Ed. Sc., 26, 153 (1971); Synthesis, 439 (1972).
12. G. Baccolini, G. Rosini, Chim. Ind. (Milano), 52, 583 (1970); Synthesis, 60 (1972).
13. T. H. Chan, L. T. L. Wong, J. Org. Chem., 34, 2766 (1969).
14. T. H. Chan, L. T. L. Wong, J. Org. Chem. 36, 850 (1971).
15. J. F. Normant, H. Deshayes, Bull. Soc. Chim. France, 2854 (1972).
16. N. Yamazaki, F. Higashi, Tetrahedron Letters, 415 (1972).
17. N. Yamazaki, F. Higashi, Tetrahedron Letters, 5047 (1972).
18. F. I. Luknitskii, Doklady Akad. Nauk SSSR Chem. (Engl. transl.), 184, 198 (1969).
19. L. Caglioti et al., J. Org. Chem., 33, 2979 (1968).
20. W. M. Pearlman, Org. Syn., 49, 75 (1969).
21. C. Ivanov, A. Dobrev, Monatsh. Chem., 96, 1746 (1965).
22. M. Ueki et al., Bull. Chem. Soc. Jap., 44, 1108 (1971).
23. R. T. LaLonde, C. B. Davis, J. Org. Chem., 35, 771 (1970).
24. F. J. McCarty et al., J. Med. Chem., 11, 534 (1968).
25. R. Feinauer, Synthesis, 16 (1971).
26. J. Comin et al., Tetrahedron Letters, 1269 (1973).
27. R. P. Mariella, K. H. Brown, Can. J. Chem., 49, 3348 (1971).
28. H. L. Yale, J. Org. Chem., 36, 3238 (1971).
29. H.-P. Husson et al., Tetrahedron Letters, 2697 (1971).
30. N. C. Bellavista, A. Colonna, Annali di Chimica, 59, 630 (1969); Synthesis, 226 (1971).
31. B. Singh, Tetrahedron Letters, 321 (1971).
32. J. E. W. van Melick, E. T. M. Wolters, Syn. Comm., 2, 83 (1972).
33. N. Nakamizo, Bull. Chem. Soc. Jap., 44, 2006 (1971).
34. H. T. Openshaw, N. Whittaker, J. Chem. Soc., C89 (1969).
35. J. H. Boothe et al., J. Am. Chem. Soc., 81, 1006 (1959).

36. M. J. Cook et al., _Chem. Commun._, 121 (1966).
37. K. Watanabe, _Bull. Chem. Soc. Jap._, _37_, 1325 (1964); ibid., _39_, 8 (1966).
38. W. A. Barber, J. A. Fetchin, German Patent, 2,303,648 August 30, 1973; _C. A._, _79_, 136549 (1973).
39. M. A. Bennett, T. Yoshida, _J. Am. Chem. Soc._, _95_, 3030 (1973).
40. A. M. Sargeson et al., _J. Am. Chem. Soc._, _94_, 8246 (1972).
41. R. L. Powell, C. D. Hall, _J. Chem. Soc._, _C_, 2336 (1971).
42. J. R. Blackborow, _J. Chem. Soc._, _C_, 739 (1969).
43. T. D. Roberts et al., _Tetrahedron Letters_, 1917 (1971).
44. (a) F. Becke et al., _Ann. Chem._, _713_, 212 (1968); (b) _749_, 198 (1971).
45. W. Walter, K.-D. Bode, _Angew. Chem. Intern. Ed. Engl._, _5_, 447 (1966); L. Field, _Synthesis_, 101 (1972).
46. R. Mukherjee, _Chem. Commun._, 1113 (1971).
47. M. A. Kraus, _Synthesis_, 361 (1973).
48. T. Oishi et al., _Chem. Commun._, 1148 (1972).
49. T. Kametani, O. Umezawa, _Chem. Pharm. Bull._, _14_, 369 (1966); _Synthesis_, 45 (1969).
50. F. Bentz, G.-E. Nischk, _Angew. Chem. Intern. Ed. Engl._, _9_, 66 (1970).
51. A. P. Marchenko et al., _Zh. Obshch. Khim._, _44_, 67 (1974); _C. A._, _80_, 108, 141 (1974).
52. M. Frankel et al., _J. Organomet. Chem._, _7_, 518 (1967).
53. I. Brown, O. E. Edwards, _Can. J. Chem._, 45, 2599 (1967).
54. B. Acott et al., _Australian J. Chem._, _21_, 185 (1968); S. S. Simons, Jr., _J. Org. Chem._, _38_, 414 (1973).
55. J. B. Hendrickson, R. Bergeron, _Tetrahedron Letters_, 4607 (1973).
56. R. R. Schmidt, _Synthesis_, 333 (1972).
57. W. Steglich, G. Höfle, _Chem. Ber._, _102_, 899 (1969).
58. H.-J. Kabbe, _Chem. Ber._, _102_, 1410 (1969).

B. Reduction

2. From Mono- and Diacylhydrazides and Related Types (_1_, 914)

$$RCONHNH_2 \xrightarrow[\text{alcohol}]{\text{Raney Ni}} RCONH_2 + NH_3$$

Hydrazide elimination has been accomplished by Kametani and Umezawa [1] by forming the condensation product with chloral followed by treatment with a primary amine as indicated:

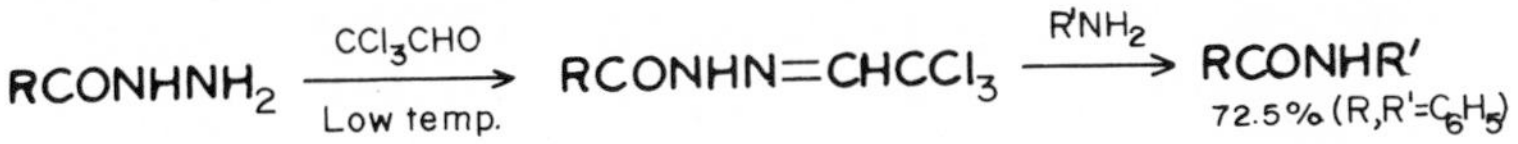

$$RCONHNH_2 \xrightarrow[\text{Low temp.}]{CCl_3CHO} RCONHN=CHCCl_3 \xrightarrow[\text{72.5\% (R,R'=C}_6H_5)]{R'NH_2} RCONHR'$$

This method does not require the isolation of the intermediate.

4. From Heterocycles

Some oxazoles or isooxazoles upon reduction followed by hydrolysis yield amides. One example is given [2]:

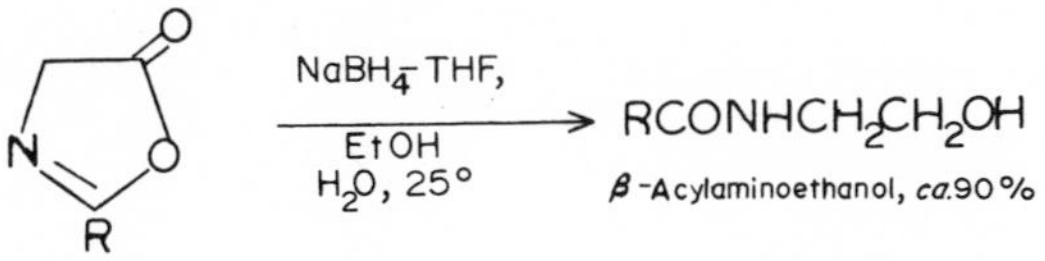

$$\xrightarrow[\substack{EtOH \\ H_2O, 25°}]{NaBH_4 THF,} RCONHCH_2CH_2OH$$

β-Acylaminoethanol, ca.90%

A second example actually represents the preparation of a vinylogous amide [3]:

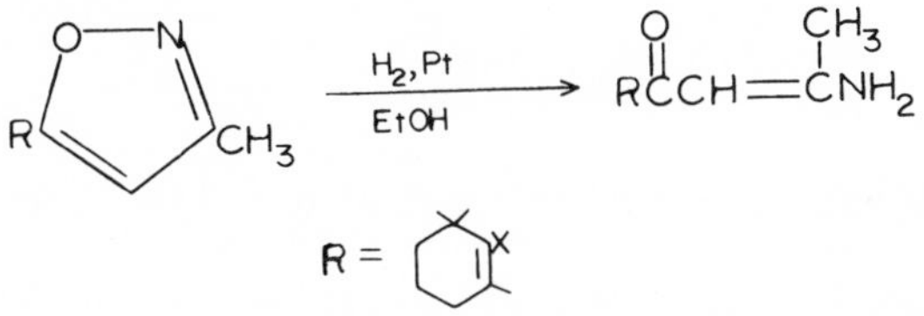

$$\xrightarrow[\text{EtOH}]{H_2, Pt}$$

$$R\overset{O}{\overset{\|}{C}}CH=\overset{\overset{CH_3}{|}}{C}NH_2$$

References

1. T. Kametani, O. Umezawa, Chem. Pharm. Bull. (Tokyo), 14, 369
    (1966); C. A., 65, 5361 (1966).
2. P. Truitt, J. Chakravarty, J. Org. Chem., 35, 864 (1970).
3. G. Büchi, J. C. Vederas, J. Am. Chem. Soc., 94, 9128 (1972).

C. Oxidation

2. From Carbonyl Compounds and the Like (Including
   Willgerodt) (1, 916)

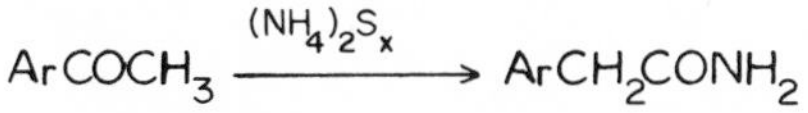

$$ArCOCH_3 \xrightarrow{(NH_4)_2S_x} ArCH_2CONH_2$$

In addition to the Willgerodt reaction, the tetraamino derivatives of glyoxal appear to undergo an internal oxidation-reduction to give an aminoamide [1]:

$$\left[\underset{O}{\overset{\frown}{\underset{\smile}{\bigcirc}}} N \right]_2 CHCH \left[N \underset{O}{\overset{\frown}{\underset{\smile}{\bigcirc}}} \right]_2 \xrightarrow[\text{reflux}]{\underset{H_2O}{CH_3CN}} \underset{O}{\overset{\frown}{\underset{\smile}{\bigcirc}}} NCH_2CON \underset{O}{\overset{\frown}{\underset{\smile}{\bigcirc}}}$$

α—Morpholinoacetylmorpholide, 74%

Aromatic or α,β-unsaturated aldehydes react with secondary amines in the presence of sodium cyanide and manganese dioxide to give high yields of amides, $ArCONR_2$ [2], Schiff bases, ArCH=NAr, may also be oxidized by chromyl chloride in carbon tetrachloride to benzanilides in 60-80% yields [3]. When the sodium salt of N-chlorobenzenesulfonamide is added to acetophenone, the haloform reaction occurs exothermically, but, in place of producing sodium benzoate, the acyl sulfonamide, $C_6H_5CONHSO_2C_6H_5$, is formed on acidification [4]:

$$ArCOCH_3 + 3 \overset{\ominus}{C_6H_5SO_2}\overset{\oplus}{NCl} Na \longrightarrow ArCOCCl_3 + 3 \overset{\ominus}{C_6H_5SO_2}\overset{\oplus}{NH} Na$$

$$ArCONHSO_2C_6H_5 + CHCl_3 \xleftarrow{\quad} \Big\downarrow H_3O^{\oplus}$$

16-82% (for methyl ketones)

A simple way of preparing N,N-dimethylthioformamide (46%) is by heating a mixture of sulfur, formaldehyde, and dimethylamine in alcohol-water at 80° for 5 hr [5].

4. From Amides or Lactams (to Imides)

$$\underset{\underset{O}{\parallel}}{\overset{\overset{R'}{\mid}}{RCH_2NCR^2}} \xrightarrow[Mn^{\oplus}]{CH_3CO_3H} \underset{\underset{O}{\parallel}\;\underset{O}{\parallel}}{\overset{\overset{R'}{\mid}}{RCNCR^2}}$$

Amides are oxidized to imides in good yield by hydroperoxides or peroxy acids in the presence of trace amounts of transition-metal ions such as Co(II), Mn(II), or Mn(III) [6] as shown:

$$\underset{\underset{H}{\mid}}{\overset{\frown}{\underset{N}{\bigcirc}}}\!=\!O \xrightarrow[Mn^{\oplus}]{CH_3COOH} O\!=\!\underset{\underset{H}{\mid}}{\overset{\frown}{\underset{N}{\bigcirc}}}\!=\!O$$

90 %

Similarly, ε-caprolactam gave 40% of adipimide.

The trimethylsilyl derivatives of secondary amides, $RCON(SiMe_3)R'$ may be converted into N-hydroxyamides by molybdenum pentoxide in HMPA at 25° for 3-5 hr. The complex is decomposed with EDTA to yield, for example, 45% of N-hydroxyacetanilide [7].

Addendum

C.2.   The Willgerodt reaction has been reviewed [E. V. Brown, Synthesis, 358 (1975)].

References

1. P. Ferruti et al., J. Chem. Soc., C, 2512 (1970).
2. N. W. Gilman, Chem. Commun., 733 (1971).
3. J. S. Sandhu et al., Chem. Ind. (London), 1297 (1970).
4. M. M. Kremlev, M. T. Plotnikova, J. Org. Chem. USSR, 5, 268 (1969).
5. L. Maier, Helv. Chim. Acta, 53, 1216 (1970).
6. A. R. Doumaux, Jr. et al., J. Am. Chem. Soc., 91, 3992 (1969).
7. S. A. Matlin, P. G. Sammes, Chem. Commun., 1222 (1972).

D.   Electrophilic-type Syntheses

1. From Hydrocarbons and Carbamic Acid Chlorides or Urea (Friedel-Crafts) (1, 919)

Urea has been employed by several investigators as the reagent in attacking hydrocarbon substrates, but the yields of amides produced were very low.   Thus Vingiello and Patel [1] obtained as their best result, in a study involving several poly-cyclic arenes, with naphthalene as shown:

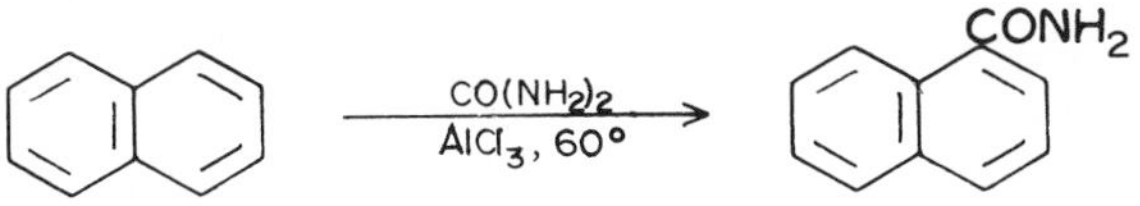

1- Naphthamide, 13%

Wiley and Linn [2] studied benzene and substituted benzenes and obtained the best result as indicated with biphenyl:

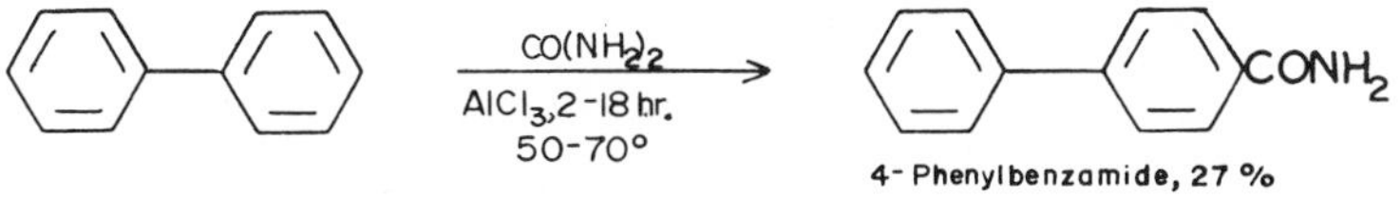

4- Phenylbenzamide, 27 %

With indene and urea Lin and co-workers produced indanecar-boxamide [3]:

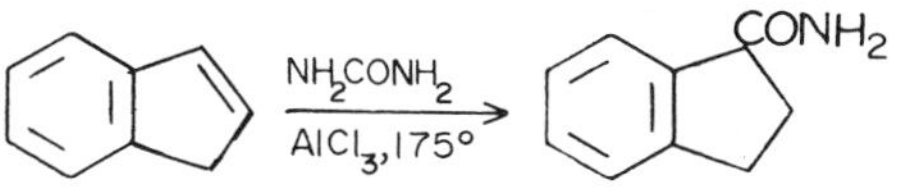

4. From Alkenes and Nitriles (Ritter) ($\underline{1}$, 921) and Alkylation of Amides

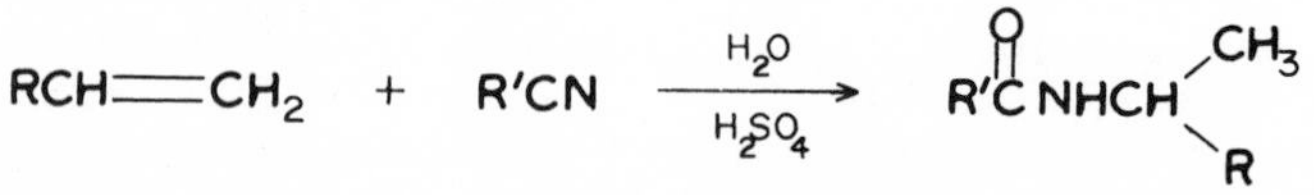

$$RCH={=}CH_2 \quad + \quad R'CN \quad \xrightarrow[H_2SO_4]{H_2O} \quad R'\overset{O}{\overset{\|}{C}}NHCH\overset{CH_3}{\underset{R}{<}}$$

A variety of amides may be produced in the Ritter reaction since carbonium ions, which add to the nitrile, may be produced from different types of organic compounds. Ducker [4] carried out the reaction with cyclohexanone and its cyanohydrin and obtained two amides as indicated:

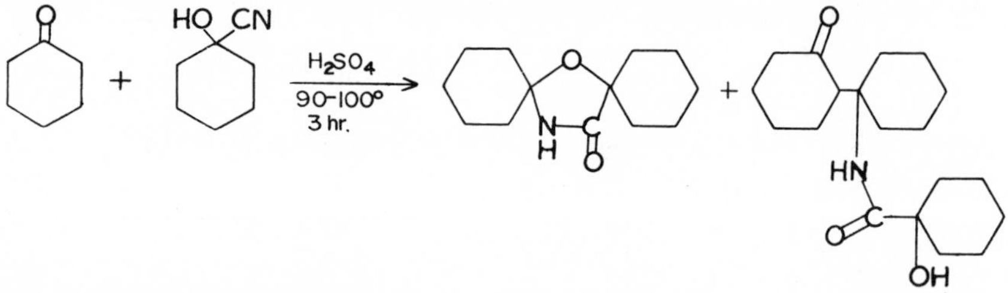

By using milder conditions the percentages of the two amides recovered were reversed.

Brown and Kurek [5] employed the alkene to prepare N-alkylacetamides in a one-operation process free from strong acid as indicated:

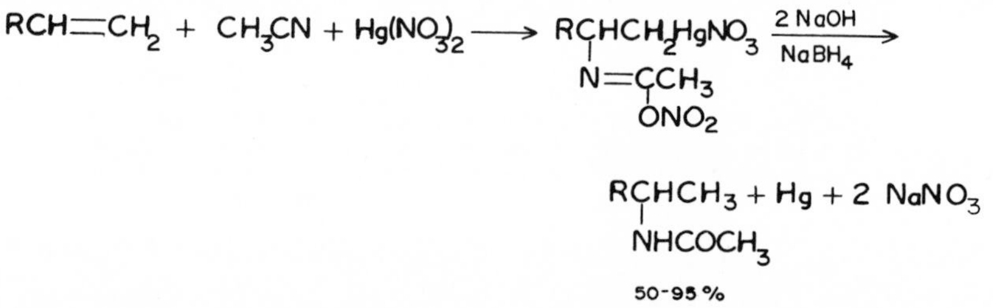

$$RCH={=}CH_2 + CH_3CN + Hg(NO_3)_2 \longrightarrow \underset{\underset{ONO_2}{\overset{|}{N}}=\underset{}{C}CH_3}{R\underset{|}{C}HCH_2HgNO_3} \xrightarrow[NaBH_4]{2\,NaOH}$$

$$\underset{NHCOCH_3}{R\underset{|}{C}HCH_3} + Hg + 2\,NaNO_3$$

50-95 %

Thus solvomercuration-demercuration of olefins offers a convenient technique for the Markovnikov amination of carbon-carbon double bonds.

Propargyl alcohols yield the Ritter product if the reaction is conducted in a mixture of acetic anhydride and sulfuric acid at 0° [6]:

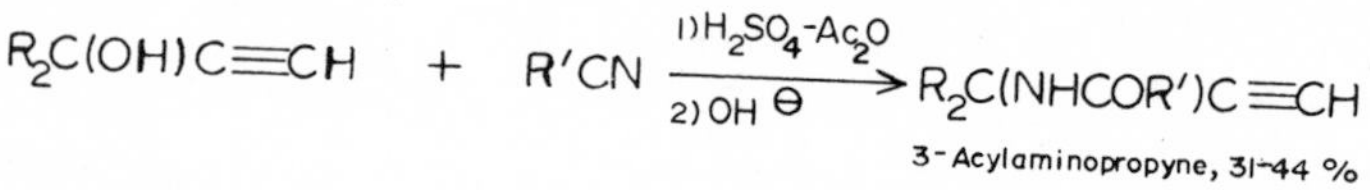

$$R_2C(OH)C{\equiv}CH \quad + \quad R'CN \quad \xrightarrow[2)\,OH^{\ominus}]{1)\,H_2SO_4\text{-}Ac_2O} R_2C(NHCOR')C{\equiv}CH$$

3-Acylaminopropyne, 31-44 %

In addition the chlorodiphenylmethyl antimony hexachloride has been utilized to generate a carbonium ion from borneol [7]:

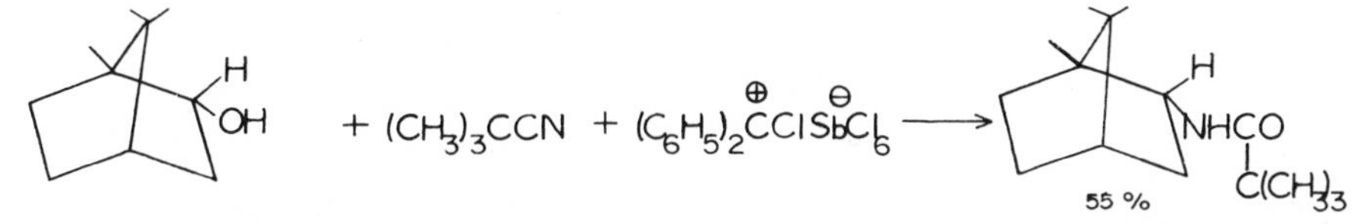

The Ritter reaction has also been studied in liquid hydrogen fluoride [8].

The stereochemistry of the Ritter reaction of bromohydrins has been investigated by Wohl [9] who found that retention of configuration is maintained in the reaction shown. Thus threo-3-bromo-2-butanol gave exclusively threo-2-acetamido-3-bromobutane

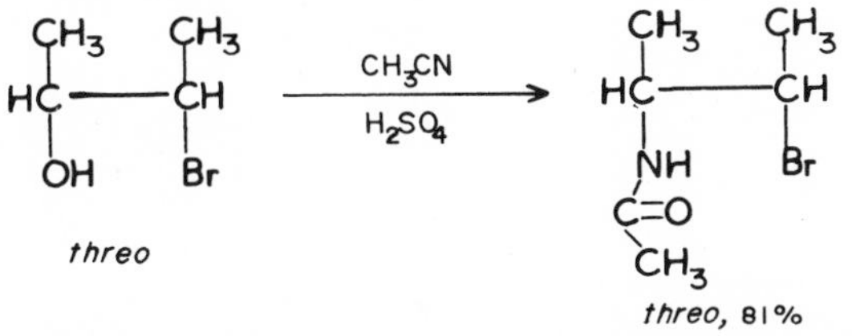

whereas the erythro butanol under similar conditions gave a 73% yield of the erythro amide. In each case the configuration of the product was determined by conversion into the oxazoline, in which case the threo-bromohydrin gave the cis, whereas the erythro-bromohydrin gave the trans-oxazoline.

Anodic acetamidation has been accomplished for adamantane and its 1-halides [10]:

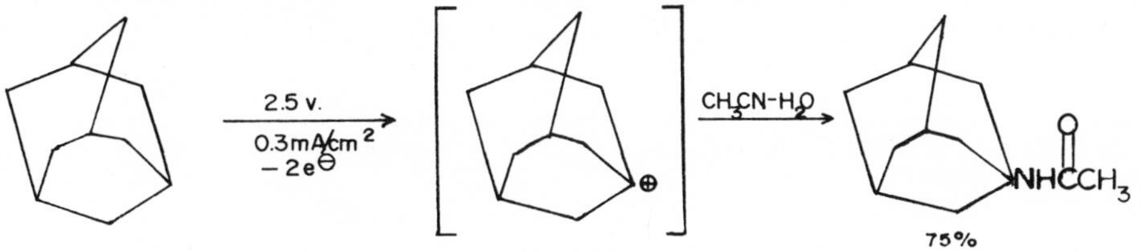

Primary carboxamides may be alkylated by treatment with DMSO-P_2O_5 or DMSO-Ac_2O as shown [11]:

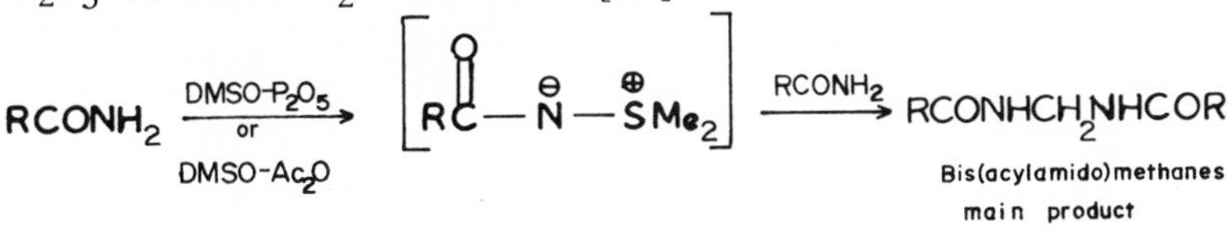

5.   From Oximes (Beckmann Rearrangement) and Related Types
      (1, 922)

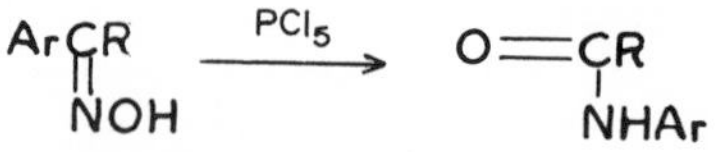

$$ \underset{\underset{NOH}{\shortparallel}}{ArCR} \xrightarrow{PCl_5} \underset{\underset{NHAr}{|}}{O{=}CR} $$

The Beckmann rearrangement usually occurs with various acidic reagents. Kelly and Matthews [12] employed dioxane $\cdot SO_2$ in the rearrangement of cyclohexanone oxime using a so-called flow-through reactor:

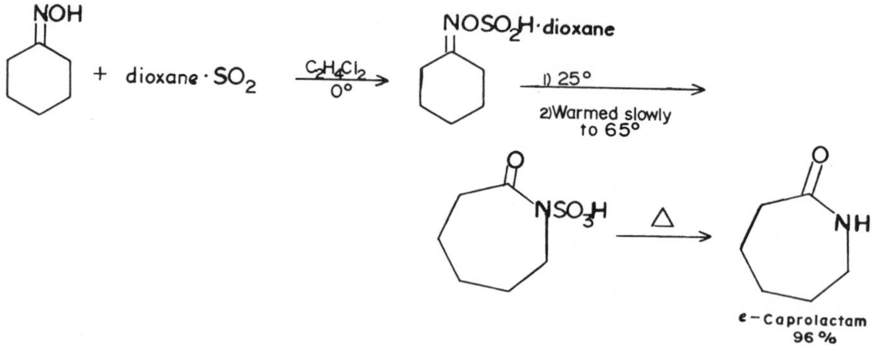

The two intermediates were isolable and in this case pyrolysis was more satisfactory than hydrolysis in the last step.

The substitution of phosphorus pentoxide-methanesulfonic acid for the usual PPA offers certain advantages [13]. The new reagent, prepared simply by dissolving P_2O_5 in methanesulfonic acid, is a mobile, colorless liquid that can be poured and stirred without difficulty. It is inexpensive, readily available, safe to handle, and organic compounds are readily soluble in it. Furthermore, yields compare favorably with those reactions in which PPA is employed. ε-Caprolactam has been synthesized with the new reagent as indicated (Ex. a):

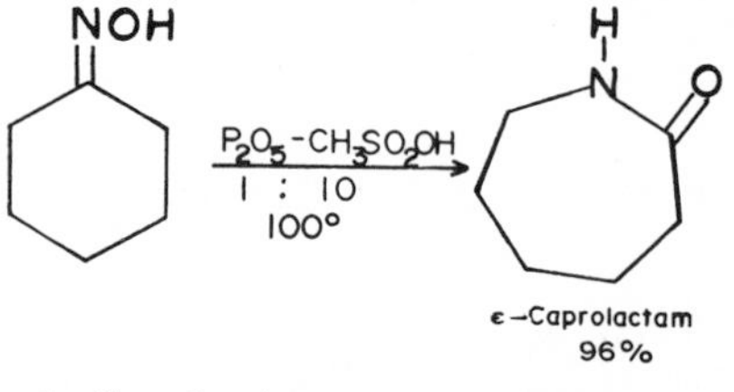

In carrying out the Beckmann rearrangement of adamantane oxime, Sasaki and co-workers employed photolysis in acetic acid as shown [14]:

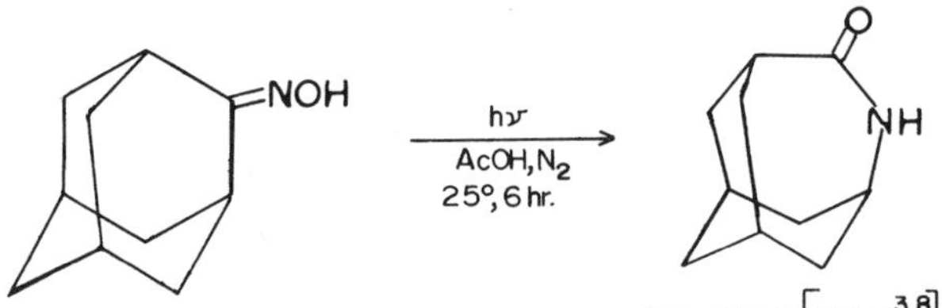

1-Azatricyclo$\left[4.3.3.1^{3,8}\right]$un-
decan-5-one, 89%

Newman and Hung [15] improved the aromatization of α-tetralone oximes to N-(1-naphthyl) acetamides by the use of acetic anhydride and anhydrous phosphoric acid as shown:

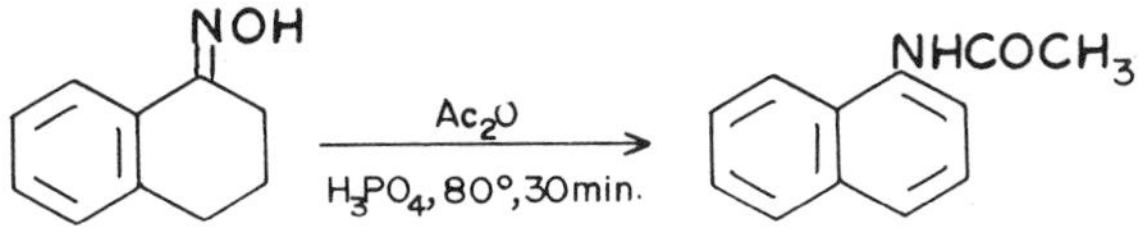

1-Acetylaminonaphthalene, 82%

In the case of anisole, direct amidation occurs as indicated [16]:

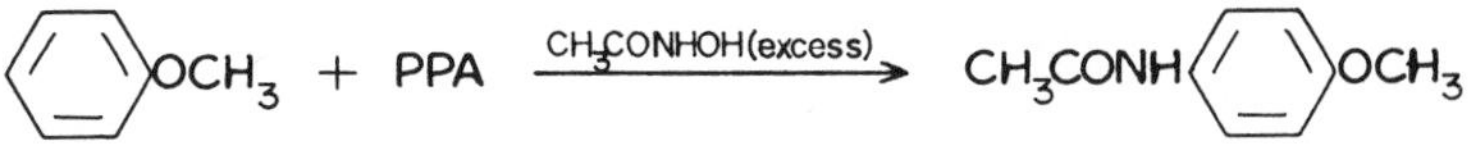

p-Methoxyacetanilide, 57%

The oxime may be either a precursor that rearranges to the amide in the reaction of arenes with acetohydroxamic acid in PPA or perhaps it is the protonated form of the hydroxamic acid that attacks the anisole. It cannot be the isocyanate that attacks because an isocyanate gives a benzamide, not an anilide (1, 919).

The Chapman rearrangement of phenylbenzimidates to benzoyl-diphenylamines in almost quantitative yield has been carried out in boiling tetraglyme [17]:

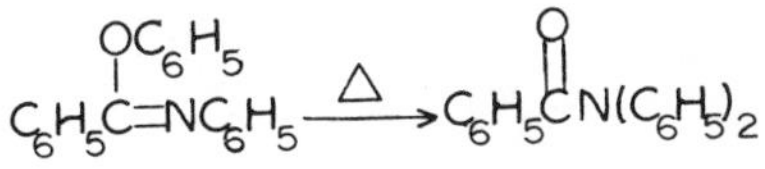

Diphenylbenzamide, 96%

a. Preparation of ε-Caprolactam [13]

Cyclohexanone oxime, 2.0 g was added in small portions to 50 g of rapidly stirred 1:10 P_2O_5-CH_3SO_3H. Each portion of oxime was added only after the previous one had dissolved. The mixture was then heated with stirring to 100°, and 1 hr later it was quenched with aqueous saturated $NaHCO_3$ and extracted with $CHCl_3$. From the extract 1.92 g (96%) of the lactam was recovered.

6.   From Carbonyl Compounds and Hydrazoic Acid or from Alkenes (Schmidt) (1, 924)

$$R_2C=O \xrightarrow{N_3H} RCONHR$$

The Schmidt reaction with ketones has been reviewed recently [18]. The reaction of the carbonyl group with hydrazoic acid takes place selectively in that it does not involve groups such as the carboxy, ester, or nitrile. Aliphatic ketones react the most readily. Cyclic ketones may lead to the formation of lactams (1, 926, Ex. b.3). Alkyl aryl ketones have been investigated the most thoroughly; in fact, substituted acetophenones give acetanilides in high yields. Cyclic alkyl aryl ketones such as benzosuberanone are converted into lactams [19] (ex. a):

3,4,5,6-Tetrahydro-1-benzazocin-2-one
82% crude, as trichloracetic acid adduct

Although benzophenones react less readily than aliphatic or alkyl aryl ketones, satisfactory results may be obtained with sulfuric acid or PPA (1, 926, Ex. b.2) as a catalyst.

a.   Preparation of 3,4,5,6-Tetrahydro-1-benzazocin-2-one [19]

Sodium azide, 1.0 g was added in portions to a solution of 1.6 g of benzosuberan-5-one in 15 g of trichloroacetic acid at 60°. After 4.5 hr, when the evolution of gas almost ceased, an additional 0.1 g of sodium azide was added and the mixture was heated for an additional 1.5 hr. On stirring the pasty mass with ice, a brown oil, which soon solidified, separated. Yield was 2.37 g (82%) of crude product, melting point 95-101°.

8.   From Carbon Monoxide and Metal Carbonyls (1, 927)

The use of carbon monoxide or metal carbonyls has assumed great importance in recent years because of the many laboratory and commercial syntheses with these reagents.
One of the early syntheses of formamides involving carbon monoxide was that of Saegusa and co-workers [20] as shown:

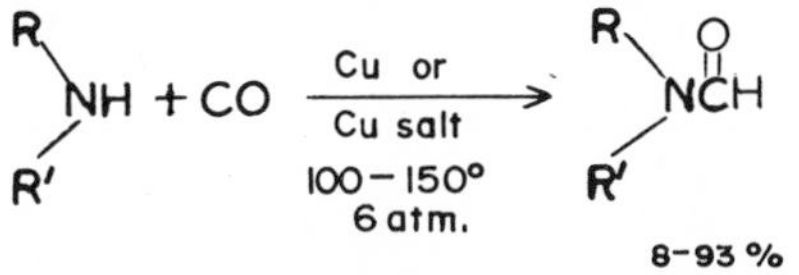

$$R_2NH + CO \xrightarrow[\substack{100-150° \\ 6\,atm.}]{\substack{Cu \text{ or} \\ Cu \text{ salt}}} R_2NCH{=}O$$

8-93 %

The same reaction may be conducted with metal carbonyls [21]. With primary amines at 100° in the presence of sulfur and carbon monoxide, symmetrical ureas are formed [21]:

$$2\ RNH_2 + CO + S \xrightarrow{100°} RNHCONHR + H_2S$$

If aryl amines are used, a catalyst such as a tertiary amine must be present.

A review of two-component carbon monoxide reactions leading to imides and lactams is that of Falbe [22]. Succinimides were prepared as follows (Ex. a) [23]:

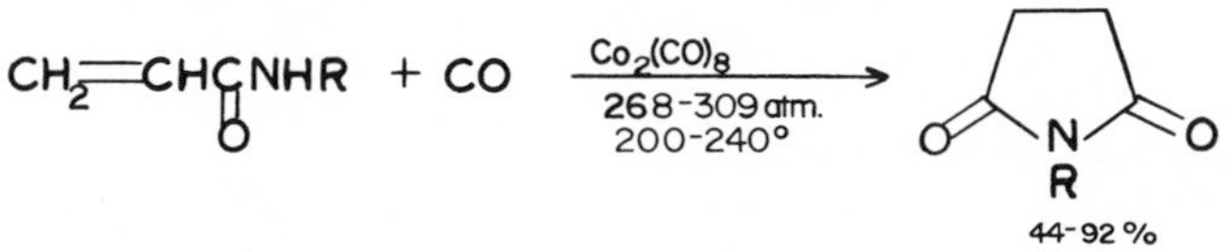

$$CH_2{=}CHCNHR + CO \xrightarrow[\substack{268-309\,atm. \\ 200-240°}]{Co_2(CO)_8}$$

44-92 %

Lactams were prepared [24]:

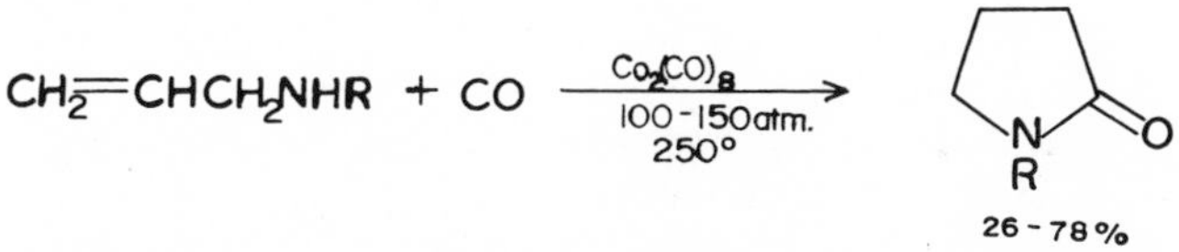

$$CH_2{=}CHCH_2NHR + CO \xrightarrow[\substack{100-150\,atm. \\ 250°}]{Co_2(CO)_8}$$

26 - 78 %

More recently, the carbon monoxide necessary in these reactions has been supplied by metal carbonyls. Thus Corey and Hegedus [25] conducted an aminocarbonylation as shown:

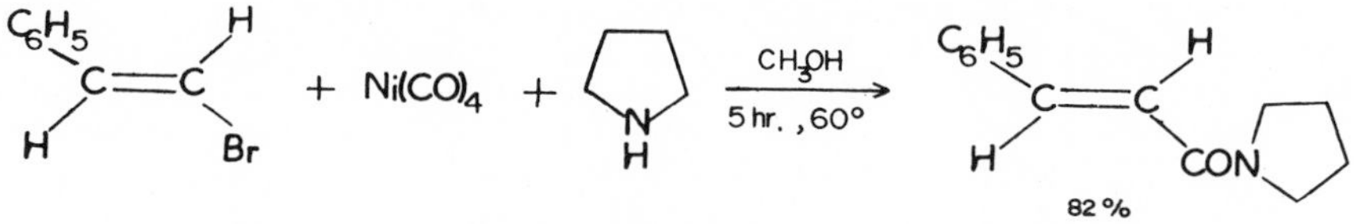

$$\underset{H}{\overset{C_6H_5}{>}}C{=}C\underset{Br}{\overset{H}{<}} + Ni(CO)_4 + \left[\text{pyrrolidine}\right] \xrightarrow[\substack{5\,hr.,60°}]{CH_3OH} \underset{H}{\overset{C_6H_5}{>}}C{=}C\underset{CON}{\overset{H}{<}}$$

82 %

As can be seen, this method involves no carbon monoxide as such and it is carried out at ordinary pressure.

Fukuoka and co-workers [26] carried out a similar reaction from halides or acid chlorides and a complex of lithium dimethyl-amide and nickel carbonyl (Ex. b):

$$C_6H_5CH{=}CHBr + LiN(CH_3)_2{-}Ni(CO)_4 \xrightarrow[\substack{12hr.,33°}]{Et_2O} C_6H_5CH{=}CHCON(CH_3)_2$$

trans   trans - N,N- Dimethylcinnamide, 96%

The complex represented as

$$\begin{array}{c} \text{LiO} \\ \diagdown \\ (CH_3)_2N \diagup \end{array} C \!=\!=\!=\! Ni(CO)_3 \quad \text{also gives}$$

carbamides with benzaldehyde   and
benzophenone.

Collman and co-workers [27] utilized sodium tetracarbonyl-ferrate(-II), $Na_2Fe(CO)_4$, with halides in forming amides as indicated:

$$n\text{-}C_5H_{11}Br \; + \; Na_2Fe(CO)_4 \xrightarrow[\substack{2)(C_2H_5)_2NH \\ 3)I_2-THF}]{1)THF,-15°} n\text{-}C_5H_{11}CON(C_2H_5)_2$$

N,N-Diethylcaproamide, 80% (glpc)

The anionic complex in the reaction is represented as

$$\begin{array}{cc} \begin{array}{c} R \quad CO \\ |\ominus\diagup \\ OC\!-\!Fe\!\!\blacktriangleleft\!\!CO \\ | \\ CO \end{array} & or & \begin{array}{c} R \; O \\ \diagdown C \diagup \\ |\ominus \diagup CO \\ OC\!-\!Fe\!\!\blacktriangleleft \\ |\;\diagdown CO \\ CO \end{array} \end{array}$$

a.   Preparation of N-Methylsuccinimide [23]

N-Methylacrylamide, 54 g in 250 g of C_6H_6 was carbonylated in the presence of 7.5 g of $Co_2(CO)_8$ in a stirred autoclave at 200° under a CO pressure of 300 atm. After 4 hr, the mixture was cooled to 25° and the excess of CO was allowed to escape. On refluxing for 1 hr in a current of N_2, the products were filtered and after distilling off the benzene in a vacuum, the amide, 67.5 g (94%) was recovered.

b.   Preparation of <u>trans</u>-N,N-Dimethylcinnamide [26]

To the ether solution of the complex prepared from 25 mmol of $LiN(CH_3)_2$ and 50 mmol of $Ni(CO)_4$ was added dropwise 50 mmol of <u>trans</u>-β-bromostyrene in 10 ml of ether below 10° and the mixture was stirred for 7 hr at the same temperature. Carbon monoxide was then bubbled through the mixture for 1 hr and then 30 ml of anhydrous ethanol was added, after which the solution was distilled under reduced pressure to give 4.20 g (96%) of the <u>trans</u> amide.

9.   From Quinazolones

$$\text{quinazolone with NBu} \xrightarrow[AlCl_3]{C_6H_5COCl} \text{product with CONHBu and NHCOC}_6H_5$$

N-Benzoyl-N'-butylanthranilamide was prepared in 40% yield as indicated above [28].

   10.   From Enolates and Dichloromethylenedimethylammonium Chloride, $Cl_2C=\overset{+}{N}Me_2\bar{C}l$ (see Addenda)

   11.   From Aromatic Compounds and Acyl α-Hydroxyglycine (see Addenda)

Addenda

D.4.   The Ritter reaction has been extended to anthracene by using acetonitrile-trifluoroacetic anhydride solution under electrolytic conditions to give 9-acetaminoanthracene (82%) [O. Hammerich, V. D. Parker, Chem. Commun., 245 (1974)].

D.5.   The Beckmann rearrangement has been utilized in the total synthesis of perhydrohistrionicotoxin [E. J. Corey et al., J. Am. Chem. Soc., 97, 430 (1975)].

D.10.   This reagent reacts with any type of enolate to produce after hydrolysis an amide.   For example, cyclohexanone gave 2-chloro-1-N,N-dimethylcarbamidocyclohexene 72% (H. G. Viehe et al., Angew. Chem. Intern. Ed. Engl., 10, 575 (1971)].

D.11.   N-Acyl derivatives of aromatic α-amino acids (41-91%) may be prepared from the aromatic compound and acyl α-hydroxyglycine [D. Ben-Ishai et al., Chem. Commun., 349 (1975)].

References

1.   F. A. Vingiello, N. M. Patel, Proc. La. Acad. Sci., 34, 26 (1971); C. A., 79, 115337 (1973).
2.   J. C. Wiley, Jr., C. B. Linn, J. Org. Chem., 35, 2104 (1970).
3.   L.-C. Lin et al., Chin. Chem. Soc., March, 1971, p. 8.
4.   J. W. Ducker, Chem. Ind. (London), 1276 (1968).
5.   H. C. Brown, J. T. Kurek, J. Am. Chem. Soc., 91, 5647 (1969).
6.   N. M. Libman, S. G. Kuznetsov, J. Org. Chem. USSR, 4, 2050 (1968).
7.   D. H. R. Barton et al., Chem. Commun., 331 (1973).
8.   J. R. Norell, J. Org. Chem., 35, 1611 (1970).
9.   R. A. Wohl, J. Org. Chem., 38, 3099 (1973).

10. V. R. Koch, L. L. Miller, Tetrahedron Letters, 693 (1973).
11. J. G. Moffatt, Quart. Rep. Sulfur Chem., 3, 95 (1968).
12. K. K. Kelly, J. S. Matthews, J. Org. Chem., 36, 2159 (1971).
13. P. E. Eaton et al., J. Org. Chem., 38, 4071 (1973).
14. T. Sasaki et al., Chem. Commun., 1239 (1970).
15. M. S. Newman, W. H. Hung, J. Org. Chem., 38, 4073 (1973).
16. F. W. Wassmundt, S. J. Padegimas, J. Am. Chem. Soc., 89, 7131 (1967).
17. O. H. Wheeler et al., Can. J. Chem., 47, 503 (1969).
18. G. I. Koldobskii et al., Russ. Chem. Rev. (Engl. transl.), 40, 835 (1971).
19. P. A. S. Smith, W. L. Berry, J. Org. Chem., 26, 27 (1961).
20. T. Saegusa et al., Tetrahedron Letters, 6125 (1966).
21. M. Ryang, S. Tsutsumi, Synthesis, 55 (1971).
22. J. Falbe, in W. Foerst's Newer Methods of Preparative Organic Chemistry, Vol. III, Academic, New York, 1971, p. 193.
23. J. Falbe, F. Korte, Chem. Ber., 95, 2680 (1962).
24. J. Falbe, F. Korte, Chem. Ber., 98, 1928 (1965).
25. E. J. Corey, L. S. Hegedus, J. Am. Chem. Soc., 91, 1233 (1969).
26. S. Fukuoka et al., J. Org. Chem., 36, 2721 (1971).
27. J. P. Collman et al., J. Am. Chem. Soc., 95, 249 (1973).
28. C. M. Gupta et al., Indian J. Chem., 7, 527 (1969).

E. Nucleophilic-type Syntheses

1. From Amides or Imides by Alkylation or Acylation
    (1, 929)

$$\text{RCONHAr} \xrightarrow[\text{R'I}]{\text{Na}} \underset{\text{R'}}{\text{RCONAr}}$$

Amides were alkylated by Isele and Lüttringhaus [1] as shown (Ex. a):

$$\underset{\text{R'CNHR}^2}{\overset{\text{O}}{\|}} \xrightarrow[\text{2)R}^3\text{X}]{\text{1)KOH-DMSO}} \underset{\underset{\text{R}^3}{\overset{\text{O}}{\|}}}{\overset{\text{R}^2}{\text{R'CN}}}$$

54-90%

The method is applicable to amides and N-alkylamides, but if a long reaction time at higher temperatures is necessary, ethers are formed as a by-product.

N-Alkylphthalimides and N-alkylsuccinimides have been obtained by Mitsunobu and co-workers [2] as indicated:

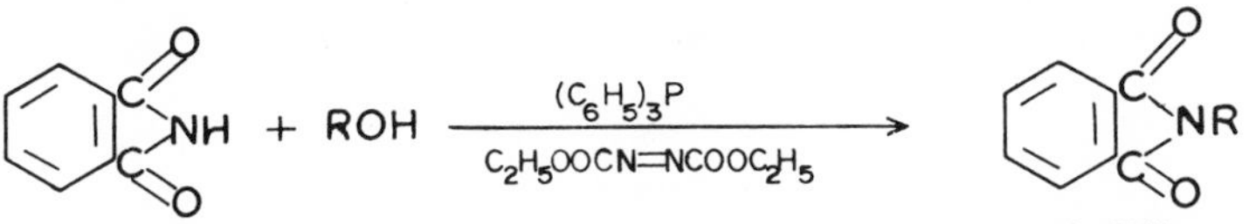

58-93%

Of particular interest in this synthesis is the fact that inversion of the alkyl group occurs:

$$\text{(S)-(+)-2-Octanol} \xrightarrow{\text{as above}} \text{N-2-Octylphthalimide} \xrightarrow[\text{C}_2\text{H}_5\text{OH}]{\text{NH}_2\text{NH}_2\text{H}_2\text{O}} \text{(R)-(-)-2 Octylamine}$$

An α-lactam of surprising stability has been prepared by dehydrohalogenation (or N-alkylation) of 2-bromo-2-t-butyl-N-t-butylpropionamide [3]:

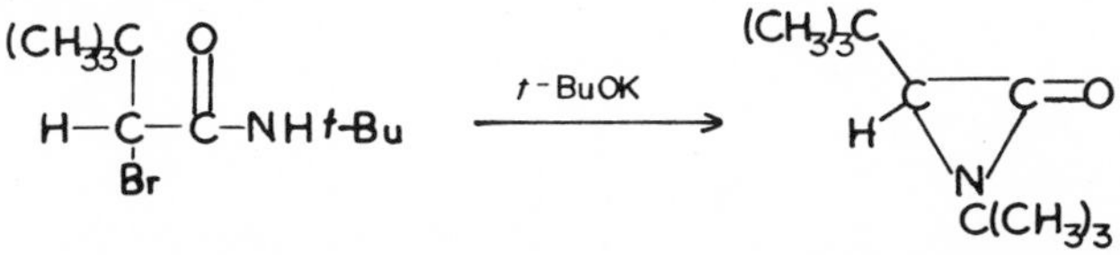

1,3-Di-t-butylaziridinone, ca.100%

In a similar dehydrohalogenation of N-chloroacetyl-N-(4-fluorophenacyl) aniline, a β-lactam was produced [4]:

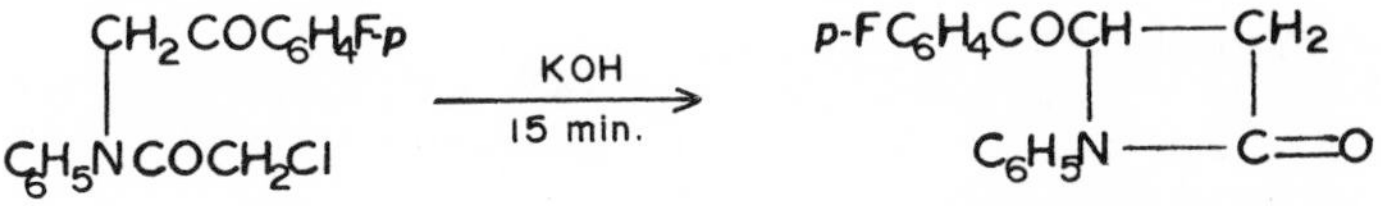

4-(4-Fluorobenzoyl)-2-oxo-1-phenyl-
azetidine, 67 %

Better yields of the β-lactam were obtained by treating the unsaturated amide as shown with piperidine [5]:

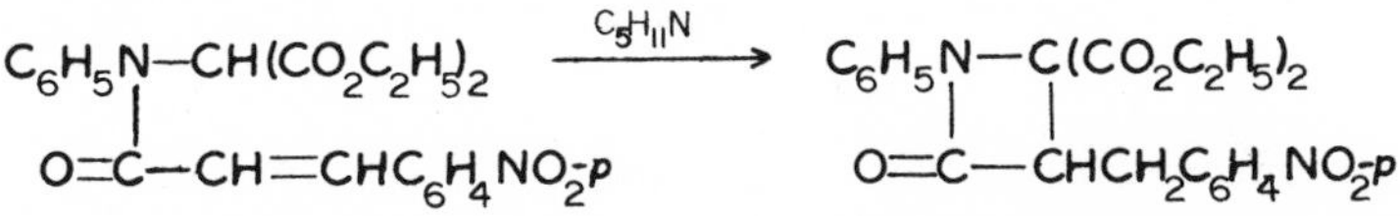

1-Phenyl-3-p-nitrobenzyl-4,4-dicarbethoxy-
2-azetidione, 94 %

Amides have also been vinylated as shown [6]:

$$\text{C}_6\text{H}_5\text{C}{\equiv}\text{CH} + \text{RNHCOR}' \xrightarrow[\text{DMSO}]{\text{NaH}} \text{C}_6\text{H}_5\text{CH}{=}\text{CHNCOR}' \overset{\text{R}}{|}$$

Acylation has been accomplished as well [7]:

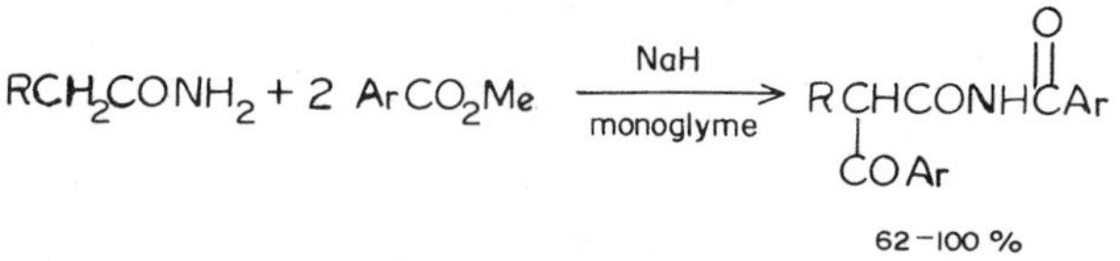

$$RCH_2CONH_2 + 2\ ArCO_2Me \xrightarrow[\text{monoglyme}]{NaH} \underset{\underset{CO\,Ar}{|}}{R\,CHCONHCAr}$$

62-100 %

a. Preparation of N-Butylacetanilide [1]

To 0.25 mol of acetanilide in 150 ml of dry DMSO, 0.3 mol of finely pulverized KOH was added. The mixture was stirred while being cooled with ice water and 0.3 mol of butyl bromide was added in 30 min. After 1 hr at 25°, 5 parts of water were added and from the ethereal extract the acetanilide (88%) was recovered.

2. From Amides or Imides by Hydroxyalkylation or Aminoalkylation ($\underline{1}$, 930)

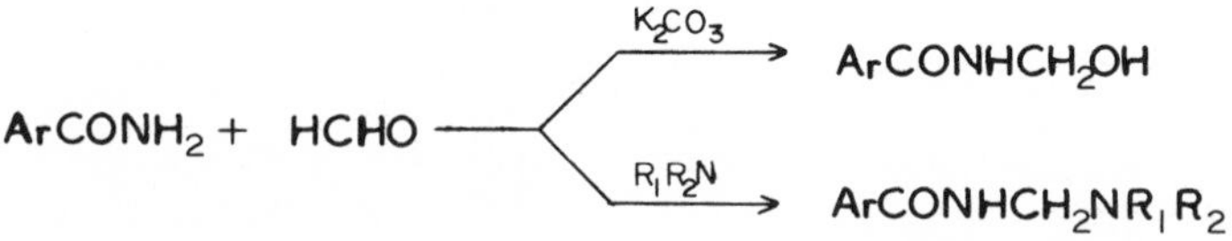

$$ArCONH_2 +\ HCHO \left\{ \begin{array}{l} \xrightarrow{K_2CO_3} ArCONHCH_2OH \\ \\ \xrightarrow{R_1R_2N} ArCONHCH_2NR_1R_2 \end{array} \right.$$

The direct aminomethylation of formanilide has been accomplished by Möhrle and Spillmann [8]. Thus formanilide, diethylamine, and formaldehyde gave the Mannich base, as shown, in unstated yield:

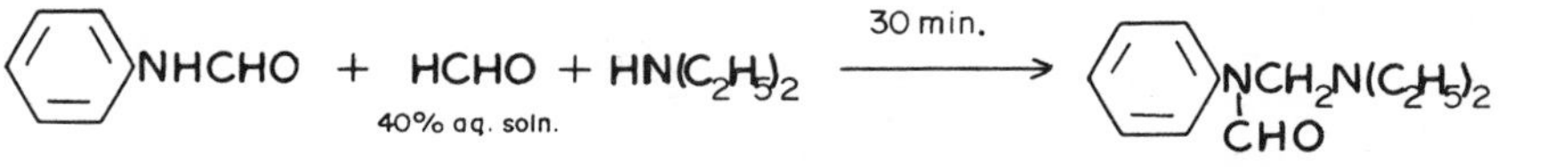

N-Phenyl-N-diethylaminomethylformamide

Since the reaction between formaldehyde, aromatic amines, and amides does not give the corresponding Mannich bases, Stavrovskaya and Drusvyatskaya [9] added triethylamine to acquire a pH value of 8.5-9. In this manner N-methylanilinomethyldichloroacetamide was obtained (Ex. a):

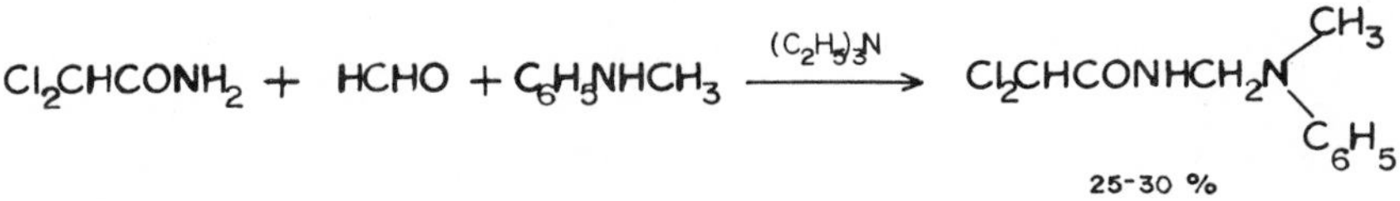

25-30 %

a. Preparation of N-Methylanilinomethyldichloroacetamide [9]

An alcoholic solution of equimolar amounts of dichloro-acetamide, formaldehyde, and methylaniline with an amount of $(C_2H_5)_3N$ necessary to give a pH of 8.5-9 was refluxed for 3 hr. The amide, isolated by chromatography, was recovered in 25-30% yield.

4. From Isocyanates and Similar Compounds by the Addition of the Grignard Reagent or Carbanions (1, 933)

γ-Ketoamides have been prepared from succinimide and a Grignard reagent [10] in one operation:

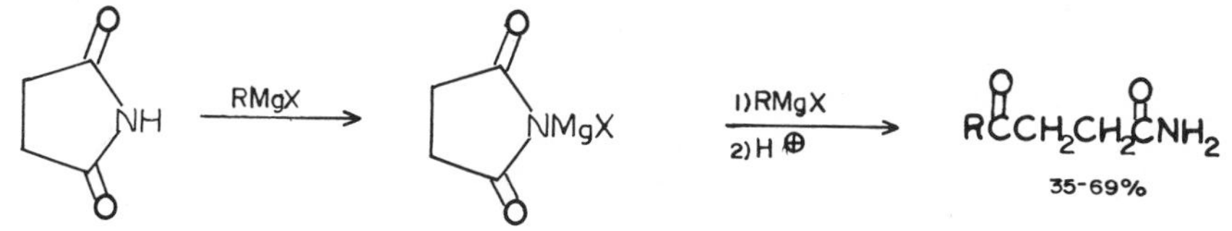

Interest in the isocyanate addition to Grignard reagents to produce amides has been renewed by the reaction [10a]:

$$RC \equiv CMgX \xrightarrow{Me_3SiNCO} RC \equiv CCONH_2$$

6. From Halocarbonyl Compounds by Cleavage with Amines

Trichloroacetaldehyde with an amine gives the amide as shown [11]:

$$CCl_3CHO + RNH_2 \xrightarrow{CHCl_3} HCONHR + CHCl_3$$
$$65-95\%$$

The mechanism appears to be:

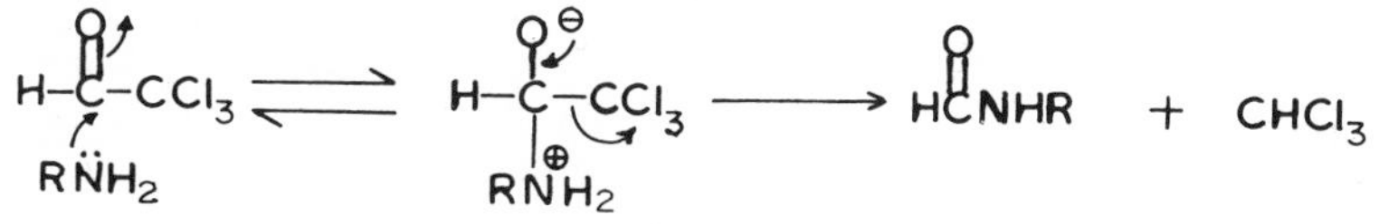

Thus the $C\bar{C}l_3$ carbanion becomes sufficiently stable to function as a leaving group.

Related reactions are the formation of trichloroacetanilide [12] (Ex. a) and α-trifluoroacetylaminocarboxylic acids [13]:

$$CCl_3COCCl_3 + C_6H_5NH_2 \xrightarrow[1.5\ hr.]{C_6H_{14}} C_6H_5NHCOCCl_3 + CHCl_3$$

$$CCl_3COCF_3 + \underset{NH_2}{RCHCOOH} \xrightarrow[25°]{DMSO} \underset{\substack{NHCOCF_3 \\ 20\text{-}100\%}}{RCHCOOH} + CHCl_3$$

In the latter case the reagent may be prepared from the commercially available chloropentafluoroacetone and $AlCl_3$.

a. Preparation of α,α,α-Trichloroacetanilide [12]

Hexachloroacetone, 1 mol in 400 ml of hexane was stirred while 1 mol of aniline was added dropwise over a period of 35-40 min. Then stirring was continued at 65-70° for 45 min, after which the solution was cooled to 0-5°. The solid, crude acetanilide, following crystallization from 90% ethanol, was recovered in 67-69% yield whereas 4-7% more was available from the filtrate.

7. From 2-Nitrocyclohexanone and Amines

Here a nitromethylene group, rather than a trihalomethyl group, as in the preceding section, serves as a leaving group:

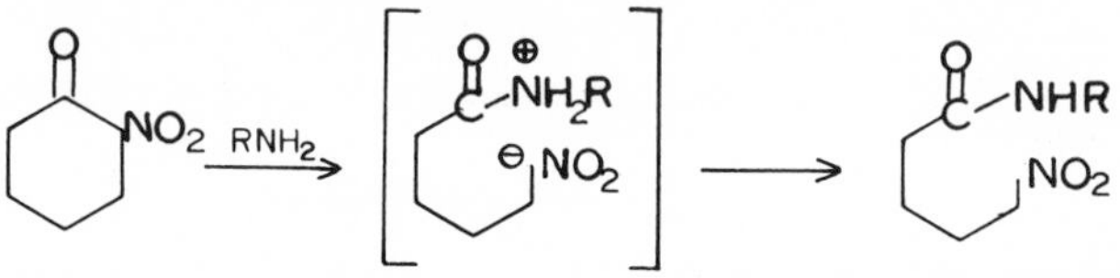
NO2
RNH2
NH2R
NO2
NHR
NO2

Thus with ammonia, ε-nitrocapramide (94%) is produced [14] (Ex. a). With diamines such as hexamethylenediamine the product is bis-1,6-(ε-nitrocapramino) hexane:

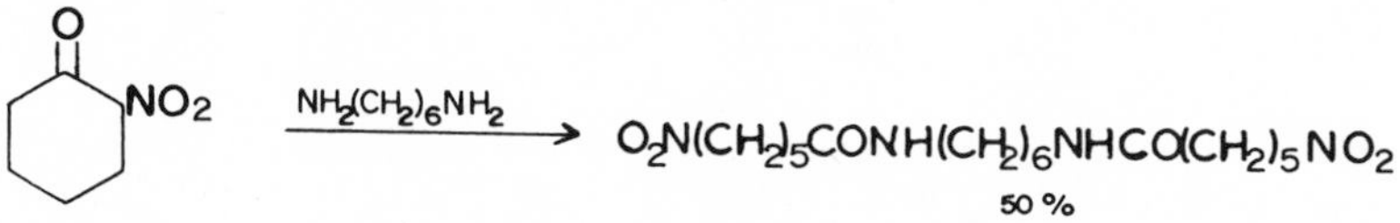
NO2
NH2(CH2)6NH2
O2N(CH2)5CONH(CH2)6NHCO(CH2)5NO2
50 %

a. Preparation of ε-Nitrocapramide [14]

2-Nitrocyclohexanone, 45 g with 150 ml of concentrated ammonia was boiled for 5 min. After standing overnight the amide, 48 g (94%) separated.

8. C-Aroylation of Imides and N-Aroylation of β-Ketoamides

Amides when converted into dianions can be carbon acylated as [15]:

$$C_6H_5CONHCOCH_3 \xrightarrow[\text{NaH} \atop \text{Glyme}]{C_6H_5CO_2CH_3} \left[C_6H_5CO\overset{\ominus}{N}CO\overset{\ominus}{C}H_2 \right] \rightarrow \underset{85\%}{C_6H_5CONHCOCH_2COC_6H_5}$$

The acyl group attaches here to the most nucleophilic C atom. On the other hand, for β-ketoamides the most nucleophilic atom is the N, which under similar conditions [12] reacts as:

$$C_6H_5COCH_2CONH_2 \xrightarrow[\text{NaH} \atop \text{Glyme}]{C_6H_5CO_2CH_3} C_6H_5CO\overset{\ominus}{C}HCO\overset{\ominus}{N}H \rightarrow \underset{91\%}{C_6H_5COCH_2CONHCOC_6H_5}$$

9. From Aryl Halides and Ethyl Malonate via a Benzyne

Aryl halides and ethyl malonate in the presence of sodamide react via a benzyne to give the β-lactam, homophthalimide, as indicated [16]:

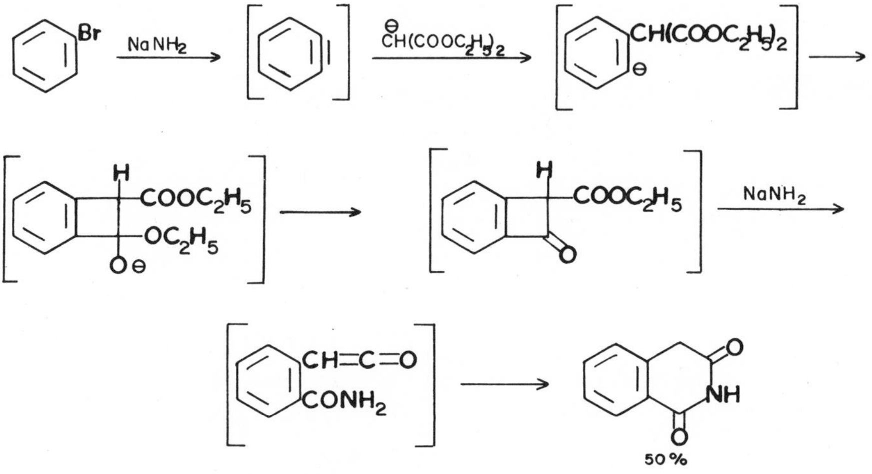

10. From α,β-Ynamines

$$RC\equiv CNR'_2 \xrightarrow{H_2O^{\oplus}} RCH_2CONR'_2$$

If the ynamine is unhindered such as in $CH_3C\equiv CNEt_2$, the hydration may be accomplished with $MgSO_4$ in water [17]. To make the reaction more versatile a hindered ynamine was converted into the amide by the steps:

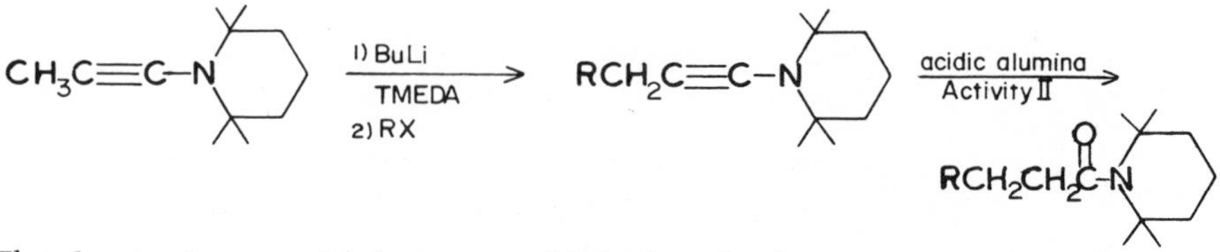

$$CH_3C \equiv C-N \quad \xrightarrow[\substack{TMEDA \\ 2) RX}]{1) BuLi} \quad RCH_2C \equiv C-N \quad \xrightarrow[Activity\,II]{acidic\ alumina}$$

$$RCH_2CH_2\overset{O}{\underset{||}{C}}-N$$

The last step could be accomplished only by passage through a column containing the alumina.

11. From Ketones and DMF

$$R_2C=O \; + \; H\overset{O}{\underset{||}{C}}NMe_2 \quad \xrightarrow[THF, -78°]{LiN(CHMe_2)_2} \quad R_2C(OH)CONMe_2$$

A surprising nucleophilic condensation, as shown above, leads to α-hydroxyamides [18]. When R=phenyl, the yield is 85%; when R_2=pentamethylene, it is 62%. The reaction is surprising in that the DMF anion is unexpectedly the nucleophilic species and self-condensation of the ketone, such as cyclohexanone, does not seem to occur.

Addendum

E.4. Grignard reagents with trimethylsilylisocyanate give amides in about 50% yield [K. A. Parker, E. G. Gibbons, Tetrahedron Letters, 981 (1975)].

References

1. G. L. Isele, A. Lüttringhaus, Synthesis, 266 (1971).
2. O. Mitsunobu et al., J. Am. Chem. Soc., 94, 679 (1972).
3. J. C. Sheehan, J. H. Beeson, J. Am. Chem. Soc., 89, 362 (1967); I. Lengyel, J. C. Sheehan, Angew. Chem. Intern. Ed. Engl., 7, 25 (1968).
4. R. F. Abdulla, Synthesis, 340 (1973).
5. A. K. Bose et al., Tetrahedron, 21, 449 (1965).
6. H. Möhrle, R. Kilian, Tetrahedron, 25, 5745 (1969).
7. J. F. Wolfe, G. B. Trimitsis, J. Org. Chem., 33, 894 (1968).
8. H. Möhrle, P. Spillmann, Tetrahedron, 25, 5595 (1969).
9. V. I. Stavrovskaya, S. K. Drusvyatskaya, J. Org. Chem. USSR (Engl. transl.), 5, 375 (1969).
10. M. Sekiya, Y. Terao, Chem. Pharm. Bull. (Tokyo), 19, 391 (1971).

10a. K. A. Parker, E. G. Gibbons, Tetrahedron Letters, 981 (1975).
11. F. F. Blicke, C.-J. Lu, J. Am. Chem. Soc., 74, 3933 (1952).
12. B. Sukornick, Org. Syn., 40, 103 (1960).
13. C. A. Panetta, T. G. Casanova, J. Org. Chem., 35, 4275
     (1970).
14. C. Bischoff, E. Schröder, J. Prakt. Chem., 314, 891 (1972).
15. C. R. Hauser et al., Can. J. Chem., 46, 2561 (1968).
16. M. Guyot, D. Molho, Tetrahedron Letters, 3433 (1973).
17. E. J. Corey, D. E. Cane, J. Org. Chem., 35, 3405 (1970).
18. B. Bánhidai, U. Schöllkopf, Angew. Chem. Intern. Ed. Engl.,
     12, 836 (1973).

F. Free Radical Reactions (1, 935)

Free radical reactions have become of increasing interest
and at times present possible synthetic methods. They will be
discussed under four headings: (a) unsaturated compounds, (b)
aromatic compounds, (c) some tertiary amines and acid chlorides,
and (d) N-heterocycles.

A review of radical addition to unsaturated compounds [1] is
available. An illustration of formamide formation follows:

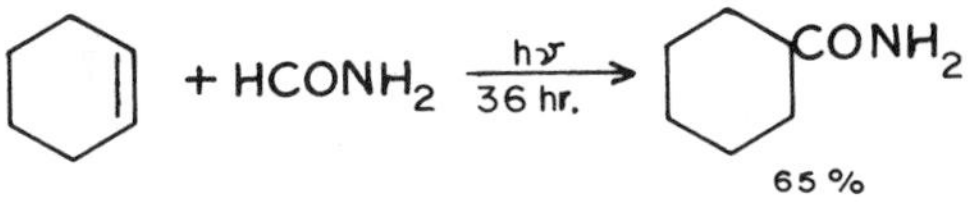

Yields are generally lower with simple acyclic olefins.

The preparation of 1-naphthamide from naphthalene is given
in 1, 936. Phenol and chloroacetamide reacted upon irradiation
in a Friedel-Crafts manner to give a mixture of hydroxyphenyl-
acetamides [2]:

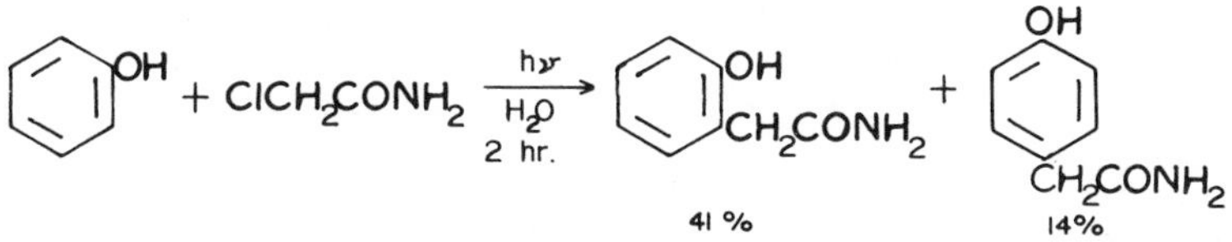

The photo-Fries rearrangement of various kinds of N-aryl
imides was investigated by Katsuhara and co-workers [3]. The
results with N,N-diacetylaniline follows:

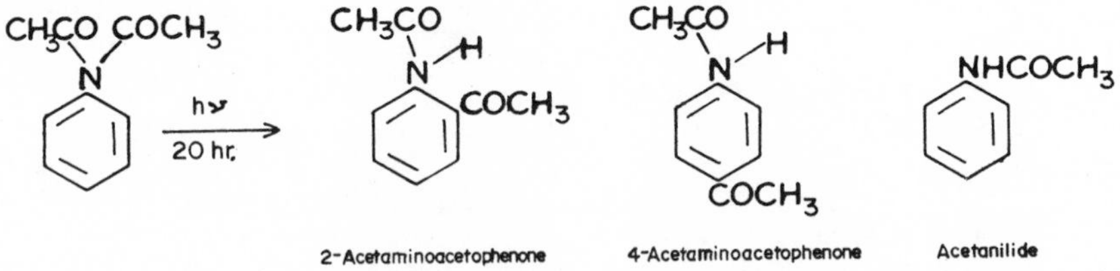

2-Acetaminoacetophenone	4-Acetaminoacetophenone	Acetanilide
30%	25%	45%

In a quite limited reaction, amides may be formed from tertiary amines as indicated [4].

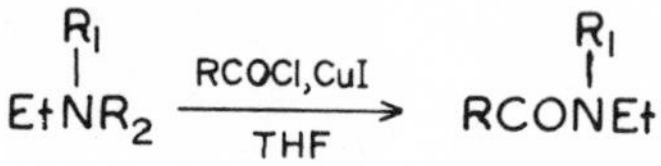

$$EtNR_2 \xrightarrow[\text{THF}]{\text{RCOCl,CuI}} RCONEt$$

(with R_1 on the N of both EtNR₂ and RCONEt)

Yields were 30% when R_2 was benzyl and 80% when it was methallyl. The yield was zero when triethylamine was subjected to the above conditions. The fact that the removable R group must be benzyl, allyl, or other groups stable as free radicals suggests some type of homolytic cleavage.

Various N-heterocycles have been converted into amides by irradiation or treatment with peroxides. Thus Minisci and co-workers obtained 2-quinoxalinecarboxamide from quinoxaline [5]:

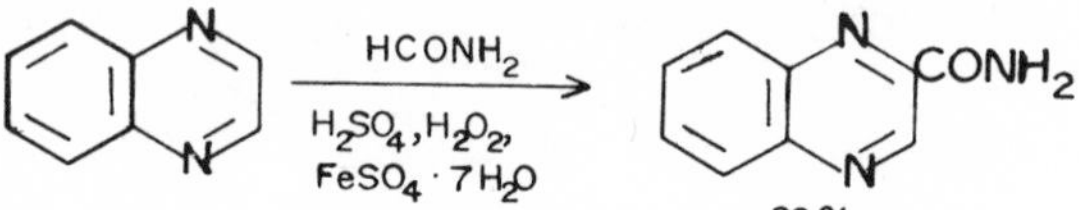

82 %

Similar yields were obtained from several other heteroaromatic bases.

N-Chloroamides with four or more carbon atoms in the acyl group have been converted into γ-chloroamides as shown [6]:

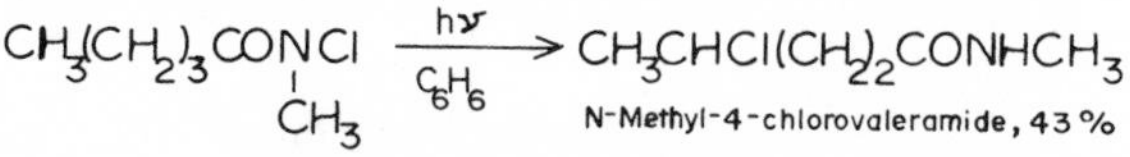

$$CH_3(CH_2)_3CONCl \xrightarrow[C_6H_6]{h\nu} CH_3CHCl(CH_2)_2CONHCH_3$$

(with CH_3 on the N of the starting material)

N-Methyl-4-chlorovaleramide, 43 %

Of more general interest is the decomposition of N-acyl-N'-carboxamidoazo compounds to amides [7]:

$$RCON{=}NCONHR' \xrightarrow[-CO]{-N_2} RCONHR'$$

47 - 82%

R may be ferrocene.

Addendum

F. 1-N-Chloro-N-alkylaminocyclopropanol with $AgNO_2$ forms a
nitrenium ion that rearranges to give a N-alkyl-β-lactam [H. H.
Wassermann et al., J. Am. Chem. Soc., 93, 5586 (1971)].

References

1. H.-H. Vogel, Synthesis, 99 (1970).
2. O. Yonemitsu, S. Naruto, Tetrahedron Letters, 2387 (1969).
3. Y. Katsuhara et al., Tetrahedron Letters, 1323 (1973).
4. P. Caubére, J.-C. Madelmont, Compt. Rend., 275, C, 1305
    (1972).
5. F. Minisci et al., Tetrahedron Letters, 15 (1970); A. Arnone
    et al., Gazz. Chim. Ital., 103, 13 (1973).
6. R. S. Neale, Synthesis, 1 (1971).
7. H.-J. Lorkowski, R. Pannier, J. Prakt. Chem., 311, 936
    (1969).

G. Cycloaddition

     Various cycloaddition reactions have been utilized in the
synthesis of amides.

     1. Ketene-Imine Interaction

     β-Lactams are produced by a ketene-imine interaction [1].

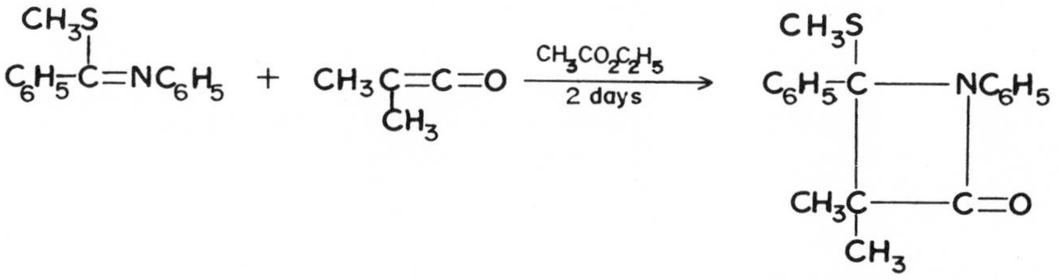

3,3-Dimethyl-1,4-diphenyl-4-methyl-
mercapto-2-azetidinone, 60%

     2. Addition of an Isocyanate to Alkenes

     β-Lactams are also formed in the addition of phenyl
isocyanate to alkenes [2].

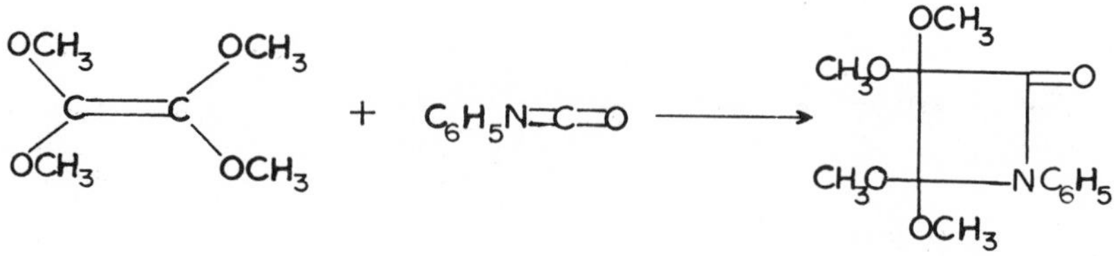

1-Phenyl-3,3,4,4-tetramethoxy-2-azetidinone, 75%

3.   Addition of Carbon Dioxide to Ynamines

Diamides of allene 1,3-dicarboxylic acid are formed by the action of carbon dioxide on ynamines as shown [3]:

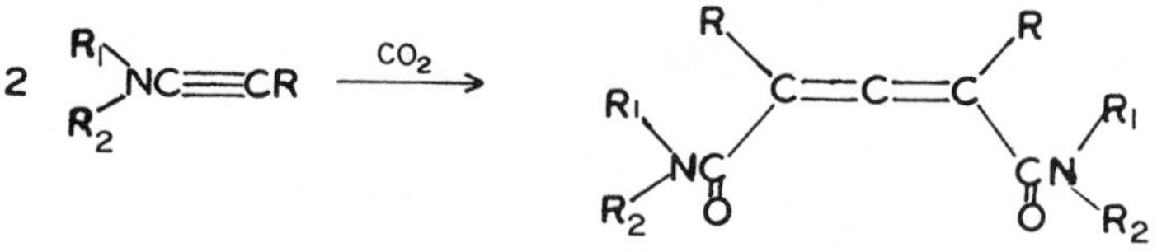

ca. quantitative

If the following mechanism is correct, the entropy of activation must be a very large negative number:

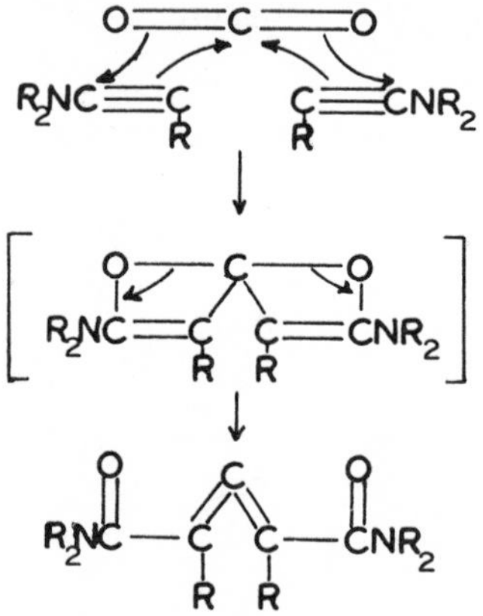

4.   Addition of an Isocyanate to Ynamines

Mixed malonamides are obtained by the reaction of an ynamine with an isocyanate to form the iminoketene [4]:

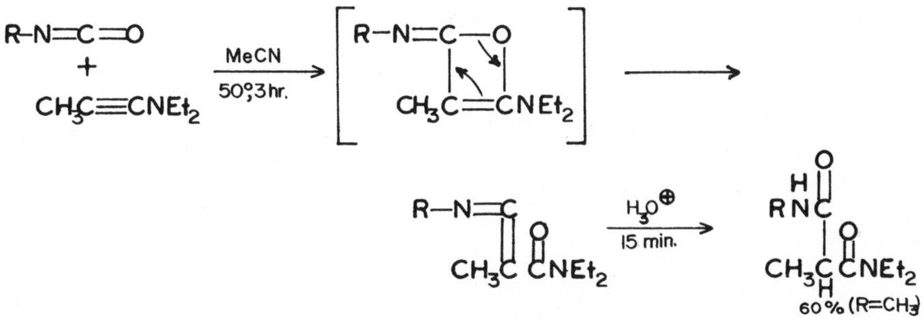

5. From Allyl Alcohols and α-Dimethylaminovinyl Ethers, $CH_2=COR(NMe_2)$ (Cope Rearrangement) (see Addenda)

Addenda

G.2. The cycloaddition reactions of acyl isocyanates to give usually N-acyl-β-lactams has been reviewed [B. A. Arbuzov, N. N. Zobova, Synthesis, 461 (1974)].

G.5. An in situ preparation of the vinyl allyl ether from the two substrates also gives rearrangement to N,N-dimethyl-pent-4-enamide [R. K. Hill et al., J. Org. Chem., 37, 3737 (1972)].

References

1. A. D., R. W. Holley, J. Am. Chem. Soc., 73, 3172 (1951).
2. A. K. Mukerjee, R. C. Srivastava, Synthesis, 327 (1973).
3. J. Ficini, J. Pouliquen, J. Am. Chem. Soc., 93, 3295 (1971).
4. J. Ficini, J. Pouliquen, Tetrahedron Letters, 1139 (1972).

Chapter 19

NITRILES (CYANIDES)

In metathetical preparations of nitriles (RX+C$\bar{\text{N}}$), phase transfer
again has shown a proclivity to increase yields (A.1).   In
elimination reactions from oximes an excess of papers has
appeared describing various new reagents, particularly in view of
the fact that simply refluxing the aldehyde in alcohol with
hydroxylamine hydrochloride and a drop or so of acid returns the
nitrile (C.4).   Moreover the interaction of the aldehyde with
ammonium dibasic phosphate, nitropropane, and acetic acid gives
the nitrile in a one-operation reaction (C.4).   Many mild dehy-
drations of amides are described (C.1).   Innovations have been
made in the cleavage of certain amides (RCONHR') to the nitrile
(RCN) and alcohol (R'OH) (C.2) and in the von Braun reaction to
produce a N-cyanoaminoalcohol (D.6).

    The nitrile-exchange reaction (RCO$_2$H + R'CN $\rightleftharpoons$ RCN + R'COOH)
has been improved by using a dinitrile as the reagent since it
forms an imide that renders reaction irreversible (C.5).
Improvements have been made in the alkylation of nitriles (via
boranes, B.1), in the 1,4-addition of cyanides to α,β-unsaturated
carbonyl compounds (using R$_2$AlCN, D.1), and in the oxidation of
cobalt ammine complexes to nitriles (F.2).   Quite novel preparations are to be found on the formation of a benzonitrile from a
pyrylium salt (B.5), the addition of cyanide to some cyanopoly-
nuclear hydrocarbons (D.1), and the formation of substituted
acrylonitriles (B.4).   It is helpful to recall that nitriles are
sources of acids and their derivatives.

A. Metathesis

    1. From Halides (<u>1</u>, 938)

$$RCH_2X \xrightarrow{NaCN} RCH_2CN + NaX$$

$$ArX \xrightarrow{CuCN} ArCN + CuX$$

a. Aliphatic Halides

Coe and co-workers [1] noted that aliphatic halides with copper(I) cyanide in quinoline react satisfactorily to give the nitrile as indicated:

$$RCl + Cu(I)CN \xrightarrow[\Delta]{quinoline} RCN + Cu(I)Cl$$

Apparently, the ability of the solvent to coordinate readily to copper is important in effecting the reaction. No reaction occurred in a solvent such as DMF.

Starks [2] found that the presence of trace amounts of an organic phase-soluble tetraalkylammonium or tetraalkylphosphonium salt catalyzed displacement reactions of alkyl halides with metallic cyanides in cases in which the alkyl halide is soluble in the organic phase and the inorganic salt is soluble in the aqueous phase. Such a reaction is known by the term "phase transfer." Thus 1-bromooctane gives 1-cyanooctane as shown:

$$CH_3(CH_2)_6CH_2Br \xrightarrow[\substack{(C_{16}H_{33})(C_4H_9)_3 P^{\oplus}Br^{\ominus} \\ 105°, 3hr.}]{aq.\ NaCN} CH_3(CH_2)_6CH_2CN$$
$$91\%$$

The general applicability of phase transfer reactions brought about by quaternary ammonium salts in a two-phase system has been reviewed [3].

Corey and Hegedus [4] have replaced the halide in alkenyl bromides by the use of potassium hexacyanodinickelate(I) as indicated:

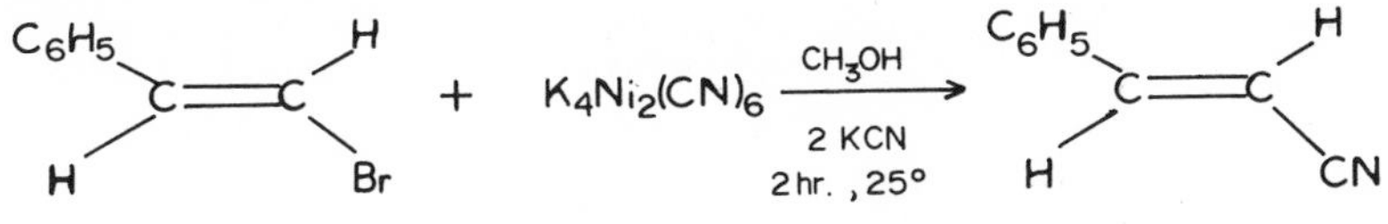

Cinnamic nitrile, 78 %

For allylic bromides Corey and Kuwajima [5] found that cyanomethylcopper as shown was a specific reagent:

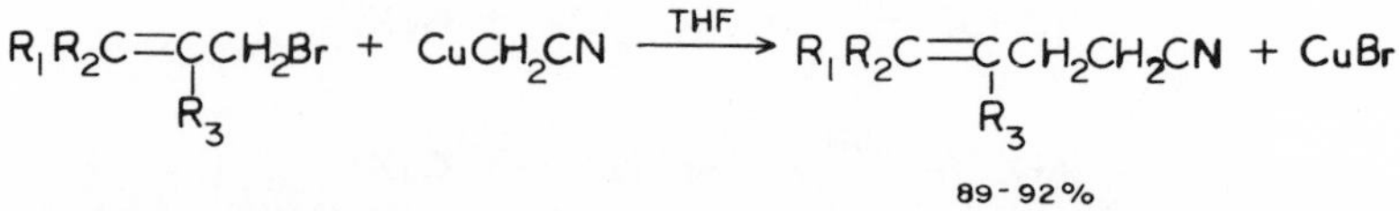
R1R2C=CCH2Br + CuCH2CN
THF
R1R2C=CCH2CH2CN + CuBr
R3
R3
89-92%

By contrast, the reagent did not affect benzyl bromide or unactivated bromides.

Malonitrile has been synthesized by Hashimoto [6] by passing ammonia into a solution of chlorocyanoacetylene as shown:

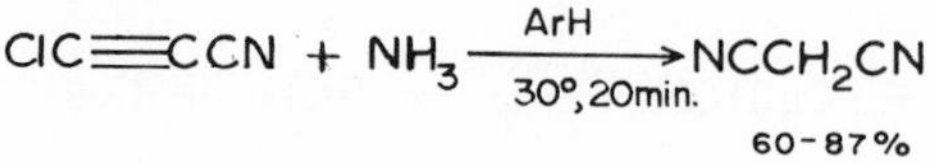
ClC≡CCN + NH3
ArH
30°, 20min.
NCCH2CN
60-87%

The reaction probably occurs by the steps:

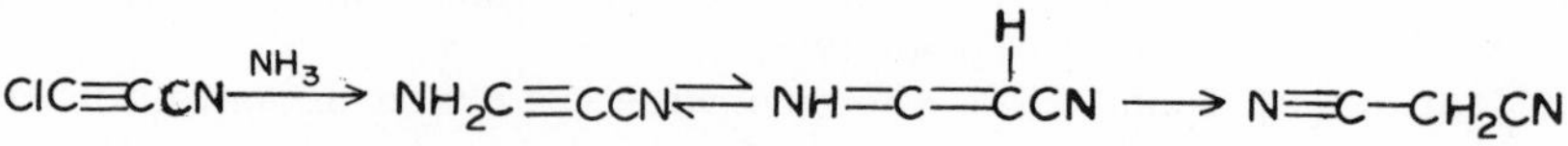
ClC≡CCN
NH3
NH2C≡CCN
NH=C=CCN
H
N≡C—CH2CN

Normant and Piechucki [7] prepared a series of aliphatic and aromatic α-ketonitriles by the action of cuprous cyanide and lithium iodide in ether or a nitrile on the acyl halide as shown:

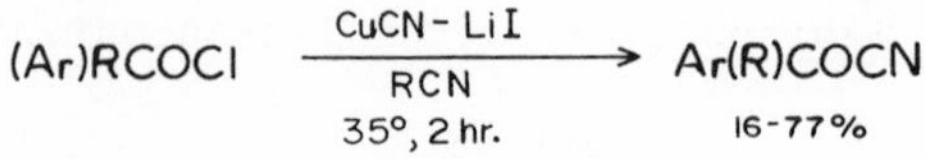
(Ar)RCOCl
CuCN - LiI
RCN
35°, 2 hr.
Ar(R)COCN
16-77%

b. Aromatic Halides

House and Fischer [8] found the use of $NaCu(CN)_2$ in DMF was sometimes of value in the synthesis of cyanides from aromatic halides in that a homogeneous reaction mixture was possible throughout the reaction period. Thus 1-iodonaphthalene gives the cyanide as shown:

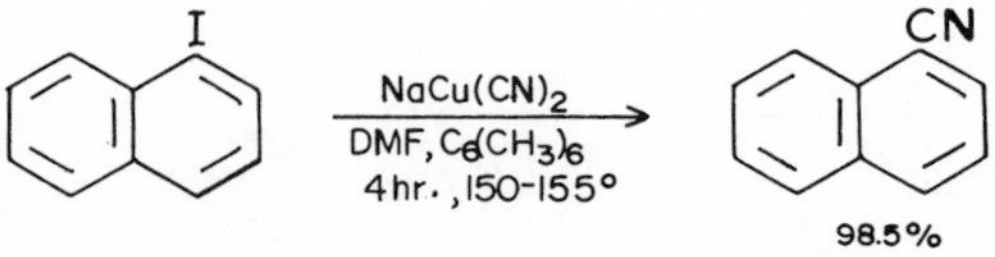
I
NaCu(CN)2
DMF, C6(CH3)6
4hr., 150-155°
CN
98.5%

The same reaction with CuCN in DMF for 4 hr gives a 97% yield of 1-cyanonaphthalene. However, the sodium cyanocuprate shows a greater range of miscibility with organic solvents.

In the synthesis of terephthalodinitrile McNulty and Miller modified the Rosenmund-von Braun procedure by employing KCN and Cu_2CN_2 at 400-425° in a sealed container [9] as indicated:

70 mole %

The product appeared as a sublimate while it formed in the fused salt mixture. Thus a separation problem, common in the synthesis, does not arise.

2. From Esters (Sulfates, Sulfonates, or Thallium Trifluoroacetate) ($\underline{1}$, 941)

$$R_2SO_4 + NaCN \longrightarrow RCN + ROSO_3Na$$

$$ArOSO_2CH_3 + NaCN \longrightarrow ArCN + CH_3SO_3Na$$

$$ArTl(OCOCF_3)_2 + KCN \xrightarrow{h\nu} ArCN + KTl(OCOCF_3)_2$$

Taylor, McKillop, and co-workers [10] have utilized their versatile thallation method, which is obviously not simple metatheses, in the synthesis of nitriles. Starting with the arene there are two steps as shown:

$$ArH \xrightarrow{TTFA} ArTl(OCOCF_3)_2 \xrightarrow[h\nu]{aq.KCN} ArCN$$

Yields in the first step exceed 90% while in the second they vary within 27-80%.

Uemura and co-workers [11] modified Taylor and McKillop's method by using a thallium mixed salt with cupric or cuprous cyanide without irradiation as indicated:

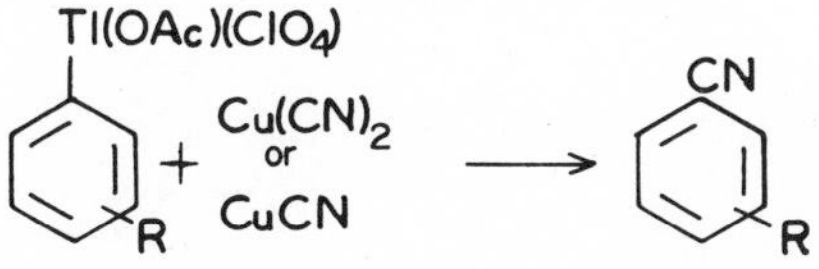

4-90%

As is indicated, the yields are usually modest.

4. From Quaternary Salts and Some Tertiary Amines ($\underline{1}$, 942)

$$-CH_2\overset{\oplus}{N}(CH_3)_3\ \overset{\ominus}{I} \xrightarrow{NaCN} -CH_2CN$$

Although Mannich bases as a rule respond to treatment with NaCN or KCN to give nitriles, such was not the case for the dimethylamino Mannich base of imidazo [1,2-α] pyridine, of its 2-methyl analog, or of the corresponding methiodide [12]. Murmann and co-workers discovered that if the methiodides themselves were employed with the cyanide under anhydrous conditions, satis-factory results as shown, were obtained [12]:

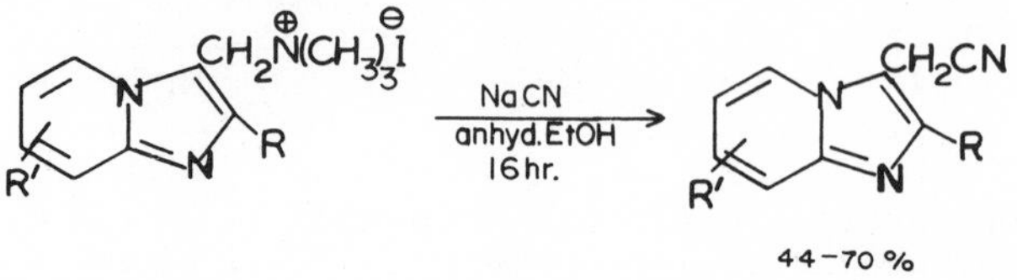

44-70 %

Phenyltrimethylstannane reacts with cyanogen chloride in methylene chloride in the presence of aluminum chloride to give benzonitrile [13]:

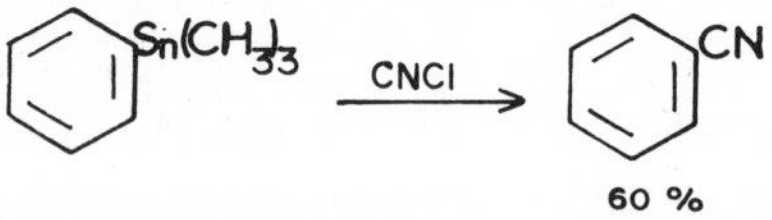

60 %

Similar results were obtained with the three tolunitriles although usually in lower yield.

Cyanylation was accomplished by Brown and Ragault [14] by treating the quaternary ammonium iodide with sodium propanoyl-acetonitrile as indicated:

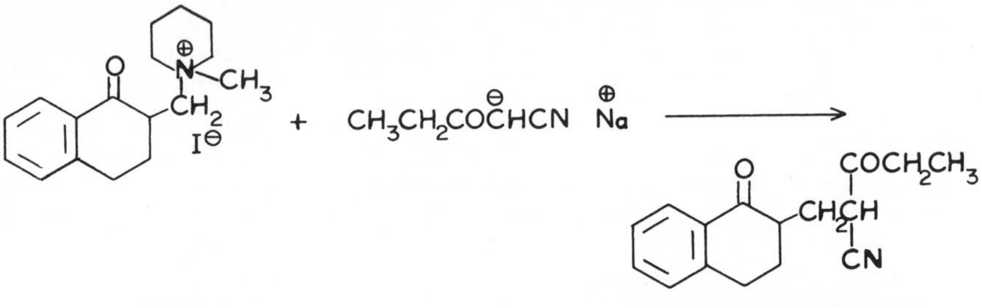

α(1,2,3,4-Tetrahydro-1-naphthyl)-
propanoylacetonitrile, 60 %

6.   From Anions and Cyanogen Compounds or Cyanogen Proper
      (1, 944)

Zweifel and co-workers [15] synthesized trans-α,β-unsaturated nitriles through the sequence as indicated:

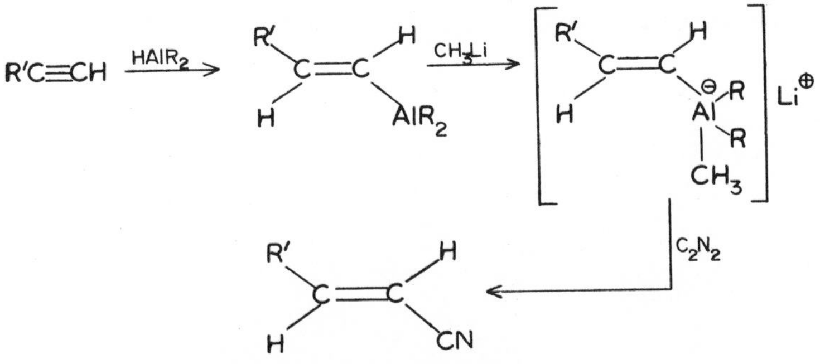

$$R'C\equiv CH \xrightarrow{HAlR_2} \underset{H}{\overset{R'}{>}}C=C\underset{AlR_2}{\overset{H}{<}} \xrightarrow{CH_3Li} \left[\underset{H}{\overset{R'}{>}}C=C\underset{\overset{\ominus}{Al}\underset{CH_3}{\overset{R}{\mid}}R}{\overset{H}{<}} \right] Li^{\oplus}$$

$$\underset{H}{\overset{R'}{>}}C=C\underset{CN}{\overset{H}{<}} \xleftarrow{\quad} \Big\downarrow C_2N_2$$

trans, 62-87% (based on vinyl alanate)

To obtain the cis form of the nitrile, lithium diisobutylmethylaluminum hydride is substituted for diisobutylaluminum hydride
in the first step.

Foulger and Wakefield [16] synthesized aromatic nitriles
(24-34%) by treating the aryllithium with pentachlorobenzonitrile in mesitylene as shown:

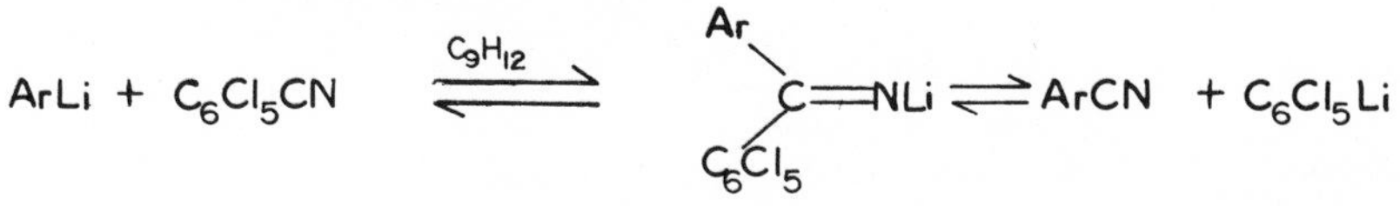

$$ArLi + C_6Cl_5CN \underset{\quad}{\overset{C_9H_{12}}{\rightleftharpoons}} \underset{C_6Cl_5}{\overset{Ar}{>}}C=NLi \rightleftharpoons ArCN + C_6Cl_5Li$$

A

The equilibrium is shifted forward if the temperature is
sufficiently high, since A decomposes to the tetrachlorobenzyne
trapped by the hydrocarbon mesitylene.

7.   From Alcohols

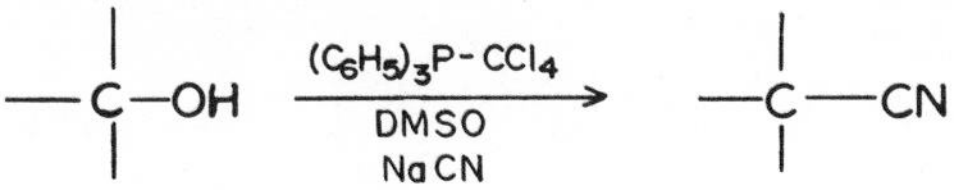

$$-\overset{\mid}{\underset{\mid}{C}}-OH \xrightarrow[\substack{DMSO \\ NaCN}]{(C_6H_5)_3P-CCl_4} -\overset{\mid}{\underset{\mid}{C}}-CN$$

The value of the direct conversion of the hydroxyl group
into the nitrile group is obvious although the reaction doubtless
proceeds via the alkyl chloride.

Brett and co-workers [17] accomplished this feat as
indicated:

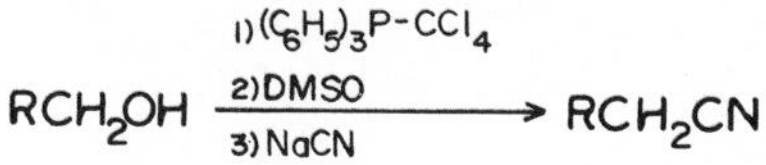

$$RCH_2OH \xrightarrow[\substack{2)DMSO \\ 3)NaCN}]{1)(C_6H_5)_3P-CCl_4} RCH_2CN$$

This method is satisfactory not only for simple monohydric
alcohols, but for carbohydrates as well. 1-Butanol gave an 85%

yield of valeronitrile, while 9.6 g of methyl 2,3,4-tri-O-acetyl-α-D-glucoside gave 4.5 g of the α-D-glucohepturononitrile.

Addenda

A.1. 1,4-Dichlorobutane may be converted into adiponitrile (94%) by refluxing with KCN in acetonitrile containing a catalytic amount of the macrocyclic ether, 18-crown-6. Surprisingly, chlorides respond more rapidly than bromides in the reaction [C. L. Liotta et al., J. Org. Chem., 39, 3416 (1974)].

A.1. Emulsions and removal of the quaternary salt sometimes plague phase transfer reactions. The quaternary salt now has been introduced in the form of a resin, which is easily filtered off at the end of the reaction. 1-Bromooctane was converted into 1-cyanooctane (92%) in this manner [S. L. Regen, J. Am. Chem. Soc., 97, 5956 (1975)].

References

1. P. L. Coe et al., J. Chem. Soc., Perkin Trans., I, 639 (1972).
2. C. M. Starks, J. Am. Chem. Soc., 93, 195 (1971); German Patent, 2,103,547, August 26, 1971; C. A., 75, 117993 (1971).
3. J. Dockx, Synthesis, 441 (1973).
4. E. J. Corey, L. S. Hegedus, J. Am. Chem. Soc., 91, 1234 (1969).
5. E. J. Corey, I. Kuwajima, Tetrahedron Letters, 487 (1972).
6. N. Hashimoto et al., J. Org. Chem., 35, 675 (1970).
7. J. F. Normant, C. Piechucki, Bull. Soc. Chim. France, 2402 (1972).
8. H. O. House, W. F. Fischer, Jr., J. Org. Chem., 34, 3626 (1969).
9. J. S. McNulty, J. F. Miller, Ind. Eng. Chem. Prod. Res. Develop., 8, 96 (1969).
10. E. C. Taylor et al., J. Am. Chem. Soc., 92, 3520 (1970).
11. S. Uemura et al., Tetrahedron, 28, 3025 (1972).
12. W. Murmann et al., J. Med. Chem., 12, 122 (1969).
13. E. H. Bartlett et al., J. Organometal. Chem., 46, 267 (1972).
14. E. Brown, M. Ragault, Compt. Rend., 270, C, 747 (1970).
15. G. Zweifel et al., J. Am. Chem. Soc., 90, 7139 (1968).
16. N. J. Foulger, B. J. Wakefield, Tetrahedron Letters, 4169 (1972).
17. D. Brett et al., J. Org. Chem., 32, 855 (1967).

B.   Nucleophilic Reactions

1.   From Nitriles by Alkylation or Arylation ($\underline{1}$, 946)

$$Ar_2CHCN \xrightarrow[\text{liq. NH}_3]{KNH_2} \left[Ar\overset{\ominus}{C}CN\right] \overset{\oplus}{K} \xrightarrow{ArCH_2Cl} Ar_2\underset{CH_2Ar}{C}CN$$

$$R_3B \;+\; ClCH_2CN \xrightarrow{\quad} RCH_2CN$$

Brändström and Junggren [1] have developed the so-called "extractive alkylation" for weak acids such as benzyl cyanide. If the cyanide in chloroform or methylene chloride is shaken with an aqueous solution or suspension of tetrabutylammonium hydroxide, the carbanion extent is so small that not much of the ion pair is extracted into the organic solvent. However, the reactivity in these solvents is sufficiently high to permit rapid alkylation until one of the components is consumed. In the case of benzyl cyanide the reaction is:

$$C_6H_5CH_2CN \xrightarrow[n\text{-Bu}_4NOH]{RI} C_6H_5\overset{R}{\underset{}{C}}HCN$$

72-90 %

Miyano and Abe [2] alkylated nitriles by the use of an aliphatic primary alcohol or secondary alcohol in the presence of sodium. An ester is introduced as well to prevent the hydrolysis of the product. As is shown in the reaction, the intermediate is the sodium salt of the acetoacetonitrile:

$$C_6H_5CH_2CN \xrightarrow[CH_3COOR]{RONa-} C_6H_5\underset{\underset{CH_3ONa}{\overset{||}{C}}}{C}CN \xrightarrow{ROH} C_6H_5\underset{R}{\overset{}{C}}HCN$$

A

It is best to start with A, which will react with the alcohol if the latter is oxidized to the aldehyde at a high temperature in a manner similar to the Guerbet reaction ($\underline{1}$, 240). Then condensation (a) occurs followed possibly by acyl cleavage (b), dehydration (c), and hydrogenation (d).

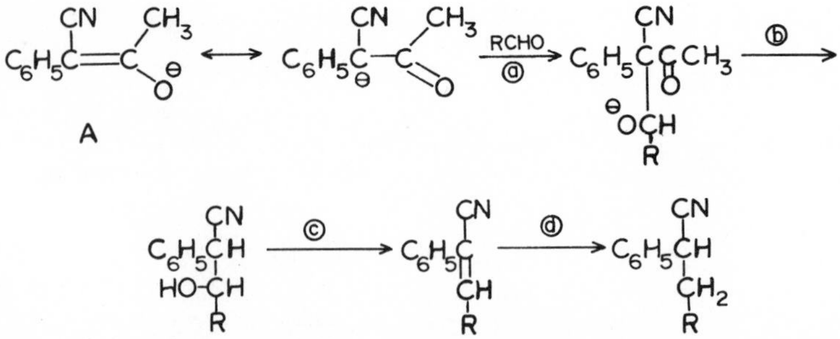

Yields for lower alcohols, which require an autoclave at 210-220° for 1 hr, vary within 34-82%. For higher alcohols no autoclave is necessary; simply heating at 210° for 30 min give yields varying within 33-82%.

A neat alkylative cyclization is possible with nitriles [3]:

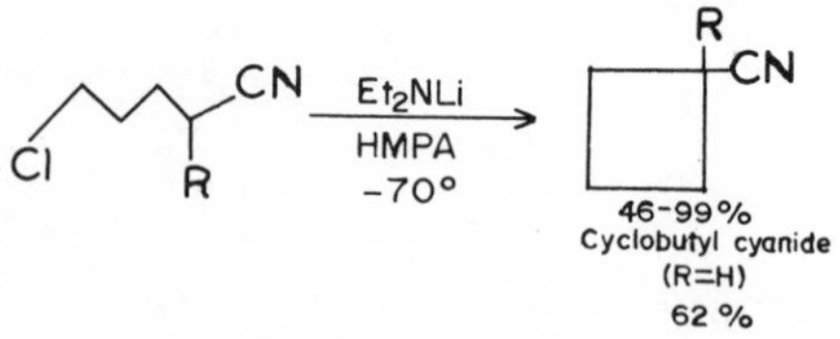

An ingenious method for synthesizing ketocyanohydrins is as follows [4]:

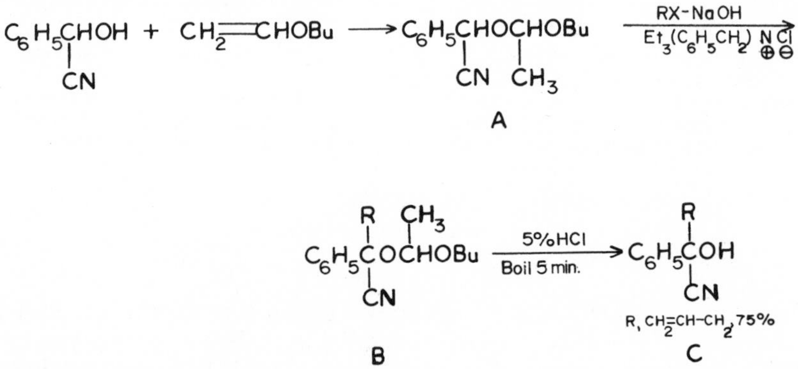

The chemistry which makes this synthesis possible is the stability of A and B to base and of C to acid.

The use of trialkylborons represents a new approach to the alkylation of nitriles. Thus chloroacetonitrile and a trialkyl-boron give the α-alkylated acetonitrile [5]. For success the method requires the influence of a mild base of large steric

requirements such as potassium 2,6-di-t-butylphenoxide.   The
alkylation equation follows (Ex. a, 1):

$$R_3B + ClCH_2CN + \underset{THF}{\overset{0°}{\longrightarrow}} RCH_2CN + R_2BO\langle\text{phenyl}\rangle + KCl$$

If a trialkylboron in which all alkyl groups are utilized is
desired, the β-alkyl 9-borabicyclo [3.3.1] nonane type may be
substituted satisfactorily.   Yields (glpc) for a series of alkyl-
substituted acetonitriles vary within 57-95%.   α-Arylation was
also achieved in that chloroacetonitrile with β-phenyl-9-bora-
bicyclo [3.3.1] nonane gave a 75% (glpc) yield of phenylacetonitrile.

     A similar monoalkylation of a free radical nature has been
achieved by the reaction of the trialkylboron and diazoacetonitrile [6] as shown (Ex. a, 2):

$$R_3B + N_2CHCN \overset{THF}{\longrightarrow} \left[R_2\overset{R}{\underset{|}{B}}CHCN \right] + N_2 \overset{H_2O}{\longrightarrow} RCH_2CN + R_2BOH$$

Yields of a series of alkylated acetonitriles vary within 54-99%
(glpc).
     Dialkylation of phenylacetonitrile has been accomplished by
the use of aqueous NaOH as the condensing agent and DMSO as the
reaction solvent [7] as shown:

$$C_6H_5CH_2CN \xrightarrow[\substack{aq. NaOH \\ 45-50°}]{CH_3Cl-DMSO} C_6H_5C(CH_3)_2CN$$

α,α-Dimethylphenylacetonitrile
93%

     Kaiser and co-workers [8] accomplished dialkylation by
forming first the dianion with butyllithium, after which the
alkyl halide was added as indicated:

$$C_6H_5CH_2CN \xrightarrow[THF]{2\ BuLi} C_6H_5\overset{\ominus}{C}CN \xrightarrow{BuBr} C_6H_5\overset{Bu}{\underset{Bu}{\overset{|}{\underset{|}{C}}}}CN$$

α,α-Dibutylphenylacetonitrile
68%

By the use of ethylene chloride in the second step, 1-cyano-1-
phenylcyclopropane was obtained in 65% yield.
     t-Alkylation of malonitrile was accomplished with t-butyl
halides and AlCl₃ in nitromethane [9] as shown:

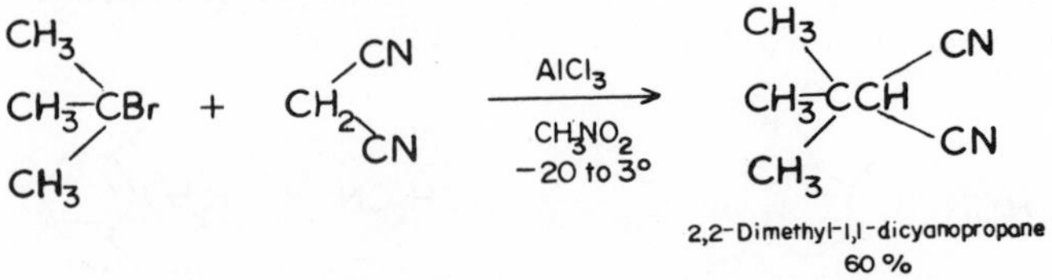

2,2-Dimethyl-1,1-dicyanopropane
60%

The yield decreased with an increase of the branching of the alkyl groups in the β-position of t-butyl bromide.

Reissert compounds with NaH in DMF have been arylated by treatment with 2,4-dinitrofluorobenzene [10] as shown:

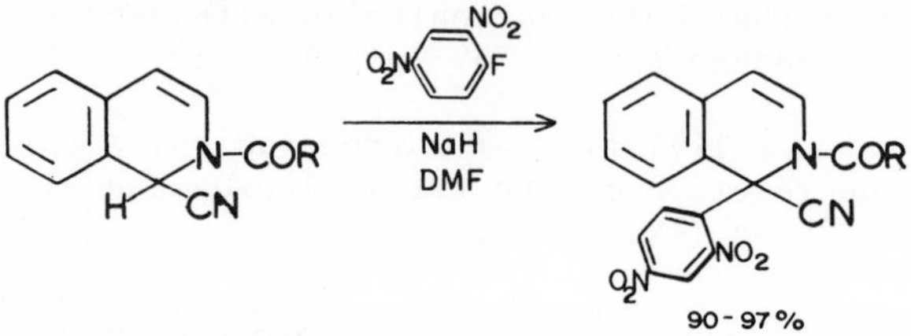

90-97%

On hydrolysis the aryl Reissert compound gives a quantitative yield of 1-arylisoquinoline.

The alkylation of ethyl α-cyanocyclopropanecarboxylate with lithium dimethylcopper results in ring cleavage as shown [11]:

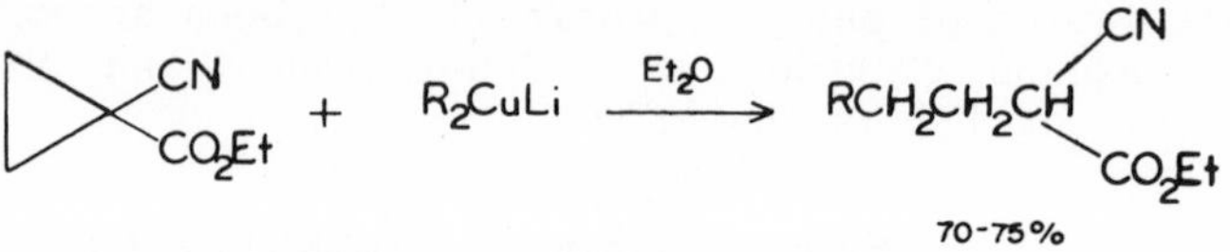

70-75%

a. Preparation of Cyclopentylacetonitrile

1. Brown's method [5]. Potassium 2,6-di-t-butylphenoxide (prepared from 100 mg-atoms of K and 120 mmol of 2,6-di-t-butyl-phenol) in 200 ml of THF at 0° was treated with 100 mmol of tri-cyclopentylborane in THF, after which 7.5 g of chloroacetonitrile in 50 ml of THF was added. Stirring for 1 hr was followed by the addition of n-octane (internal standard). Recovery in the usual manner gave 5.4 g (50%) of the acetonitrile (glpc analysis, 67%).

2. Hooz's method [6]. To an ice-cooled solution of tri-cyclopentylborane was added a solution of diazoacetonitrile, 30 mmol, in 15 ml of THF during 30 min. After stirring for 2 hr at 25° and cooling in an ice bath, 25 ml of a 3 N KOH solution was added. Further stirring for 0.5 hr and the addition of brine followed by recovery of the product from the solvent gave 1.76 g (81%) of the nitrile.

2. From Nitriles by Acylation or Aldolization (<u>1</u>, 947)

$$ArCH_2CN + C_6H_5CH_2OH \xrightarrow[\substack{C_6H_5CH_2OAc}]{Na} ArCH\substack{\diagup CN \\ \diagdown CH_2C_6H_5}$$

54 - 85%

α-Benzylarylacetonitriles (54-85%) were synthesized by
Miyano and co-workers [12] by heating the nitrile with benzyl
alcohol, sodium, and benzyl acetate at 170-180° as shown. The
mechanism of this reaction is undoubtedly that of aldolization
similar to that of the Guerbet reaction (<u>1</u>, 240). The benzyl
acetate was added to protect the nitrile group from hydrolysis.

Malonitrile has been dimerized (for a similar reaction, see
<u>1</u>, 948) by tetrakis (triphenylphosphine) platinum(O) in benzene
as indicated [13]:

$$CH_2\substack{\diagup CN \\ \diagdown CN} \xrightarrow[\substack{C_6H_6,80°,1hr.}]{Pt[(C_6H_5)_3P]_4} \substack{NC \quad CN \\ \diagdown C \diagup \\ \| \\ C-NH_2 \\ CH_2CN}$$

82%

Ethyl cyanoacetate is self-condensed in a similar manner.

3. From Carbonyl Compounds and Malonitrile, Cyanoacetic
    Esters, or Tosylmethyl Isocyanide (1, 949)

$$\substack{O\\ \|} \bigcirc + TsCH_2NC \xrightarrow[\substack{t\text{-BuOH-DME}}]{t\text{-BuOK}} \bigcirc\substack{\diagup CN}$$

Oldenziel and van Leusen [14] have shown that ketones and
tosylmethyl isocyanides give cyanides in a one-step operation, as
indicated, with the yields varying within 36-85%. The method is
simple (Ex. a) and superior to another procedure involving the
ketone as given in C.4. The reaction proceeds as follows:

$$\bigcirc\substack{\diagup OCH \\ \| \\ \diagdown TsCN \\ H} \longrightarrow \bigcirc\substack{=CNCH \\ \| \quad \| \\ Ts \quad O}^H \longrightarrow \bigcirc\substack{\diagdown CN}$$

a. Preparation of Cyclohexyl Cyanide [14]

To a solution of cyclohexanone and tosylmethyl isocyanide in

DME at 0° was added a solution of t-BuOK in t-BuOH-DME. After stirring for 45 min at 0°, the temperature was increased to 20° and stirring was continued for another hr. The cyanide was recovered (80%) by distillation of the pentane extract of the aqueous layer obtained by adding water to the reaction mixture.

    4. From β,β-Dichloroacrylonitrile and Alcohols, Phenols, and the Like

$$Cl_2C{=}CHCN + 2 HY \xrightarrow{base} Y_2C{=}CHCN$$

β,β-Dichloroacrylonitrile reacts with alcohols, phenols, sulfides, and the like to form β,β-disubstituted acrylonitriles [15], as given in the preparation of β,β-dimethoxyacrylonitrile:

$$Cl_2C{=}CHCN + 2 NaOCH_3 \xrightarrow{CH_3OH} (CH_3O)_2C{=}CHCN + 2 NaCl$$

83 %

With phenols, the free phenol was employed.

    5. From Pyrylium Salts and Malonitrile

The reaction of malonitrile, a pyrylium salt, and potassium t-butoxide yields a cyanobenzophenone [16] as shown:

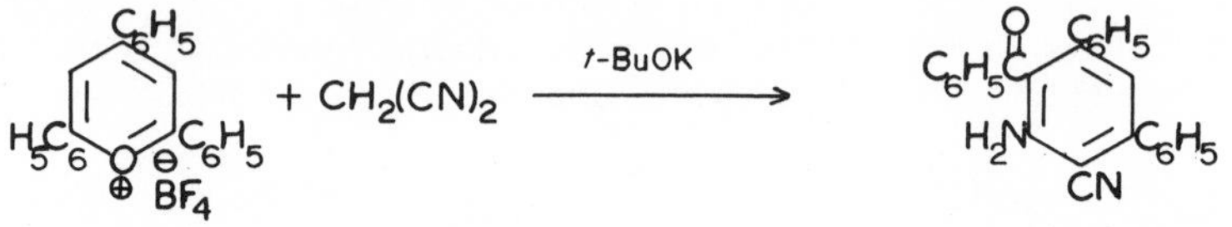

2-Amino-3-cyano-4,6-diphenyl-benzophenone, >75 %

However, the thiapyrylium salt yields the simple cyano compound with loss of CN̄ probably as SCN̄ [17]:

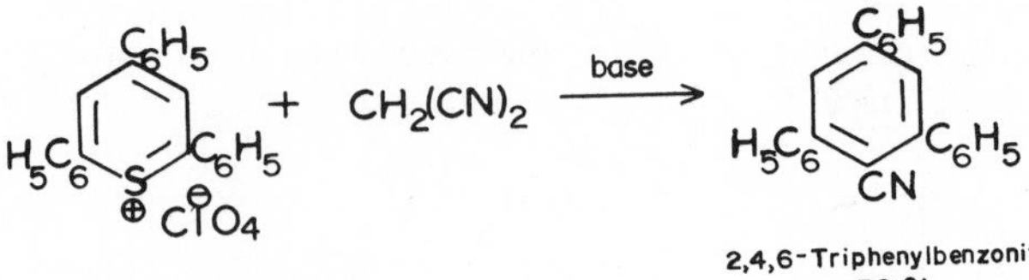

2,4,6-Triphenylbenzonitrile
72 %

The mechanism is discussed.

    6. From Nitrosophosphonium Salts

$$RCH_2\overset{\oplus}{P}(C_6H_5)_3\overset{\ominus}{Cl} \xrightarrow[\text{2)}(CH_3)_2CHONO]{\text{1) HCl-CHCl}_3} RCH(NO)\overset{\oplus}{P}(C_6H_5)_3\overset{\ominus}{Cl} \xrightarrow[\text{EtOH}]{NaOC_2H_5} RCN$$

Nitrosation of phosphonium salts leads to the α-nitroso derivative, which with NaOC$_2$H$_5$-EtOH gives the cyanide as shown [18]. Yields of isonitroso derivatives varied within 66-96%, whereas the nitrile yields in the last step ranged from 44 to 72%. The deoxygenation probably proceeds via the oxime:

$$(C_6H_5)_3\overset{\oplus}{P}\text{---CR} \longrightarrow RCN + (C_6H_5)_3PO + H^{\oplus}$$
$$\underset{HO\text{---N}}{\quad}$$

7. From Carbonyl Compounds and α-Chloronitriles (Darzens)

$$\underset{R_2}{\overset{R_1}{>}}C=O + ClCH_2CN \xrightarrow[\text{TEBA}]{NaOH} \underset{R_2}{\overset{R_1}{>}}\underset{}{C}\overset{O}{\diagdown}CHCN$$

The Darzens reaction between carbonyl compounds and α-chloroesters (1, 316) is also satisfactory if α-chloronitriles are substituted for the α-chloroesters. Thus Jończyk and co-workers [19] conducted the reaction as shown in the presence of aqueous NaOH and triethylbenzylammonium chloride, TEBA. A series of glycidic nitriles were obtained with yields varying within 55-80%. Other solvents such as HMPA, DMF, and DMSO serve almost as well as TEBA.

8. From Nitriles by Alkylidenation

$$\bigcirc\!\!=O + CH_3CN \xrightarrow[\text{C}_6\text{H}_6]{\text{C}_8\text{H}_{17}ONa} \bigcirc\!\!=CHCN$$

Cyclohexylidene -
acetonitrile

Arpe and Leupold [20] heated cyclohexanone with acetonitrile in the presence of sodium n-octyl oxide in benzene and obtained cyclohexylidene nitrile in 70% yield. The nitrile was recovered by azeotropic distillation from water after about 20 hr of refluxing.

References

1. A. Brändström, U. Junggren, Tetrahedron Letters, 473 (1972).
2. S. Miyano, N. Abe, J. Org. Chem., 36, 2948 (1971); 37, 526 (1972).

3. M. Larcheveque et al., J. Organomet. Chem., 57, C33 (1973).
4. M. Makosza, T. Goetzen, Org. Prep. Proc. Int., 5, 203
    (1973).
5. H. C. Brown et al., J. Am. Chem. Soc., 91, 6854 (1969).
6. J. Hooz, S. Linke, J. Am. Chem. Soc., 90, 6891 (1968).
7. L. B. Taranko, R. H. Perry, Jr., J. Org. Chem., 34, 226
    (1969); United States Patent, 3,755,412, August 28, 1973;
    C. A., 79, 104,964 (1973).
8. E. M. Kaiser et al., J. Am. Chem. Soc., 93, 4237 (1971).
9. P. Boldt et al., Ann. Chem., 718, 101 (1968).
10. R. Piccirilli, F. D. Popp, Can. J. Chem., 47, 3261 (1969).
11. E. J. Corey, P. L. Fuchs, J. Am. Chem. Soc., 94, 4014
    (1972).
12. S. Miyano et al., Chem. Pharm. Bull., 18, 550 (1970); C. A.,
    73, 35001 (1970).
13. K. Takahashi et al., Bull. Chem. Soc. Jap., 44, 3484 (1971).
14. O. H. Oldenziel, A. M. van Leusen, Tetrahedron Letters, 1357
    (1973).
15. N. Hashimoto et al., J. Org. Chem., 35, 828 (1970).
16. K. Dimroth, K. H. Wolf, in W. Foerst, Newer Methods of
    Preparative Organic Chemistry, Vol. III, Academic, New York,
    1974, p. 357.
17. G. A. Reynolds, J. A. VanAllan, J. Heterocycl. Chem., 8,
    301 (1971).
18. A. V. Dombrovskii et al., J. Gen. Chem. USSR (Engl. transl.),
    41, 2019 (1971).
19. A. Jończyk et al., Tetrahedron Letters, 2395 (1972).
20. H. J. Arpe and I. Leupold, Angew. Chem. Intern. Ed. Engl.,
    11, 722 (1972).

C. Elimination

1. From Amides (1, 951)

$$RCONH_2 \xrightarrow{P_2O_5} RCN$$

The dehydration of amides continues to be the most widely
used method of synthesizing nitriles. Examples of dehydrating
agents recently utilized are pyrophosphoryl tetrachloride,
Cl_2POPCl_2 [1], phosphonitrilic chloride [2], sodium borohydride
(surprisingly) [3], hexamethylcyclotrisilazane $(HNSiMe_2)_3$ [4],
trimethylsilyl chloride and acetyl or benzoyl chloride [5],
hexamethylphosphoric triamide [6], 2,4,6-trichloro-s-triazine
[cyanuric chloride] in pyridine [7], triphenylphosphine in CCl_4
and THF [8], triphenylphosphine in CCl_4 and Et_3N [9],

titanium tetrachloride in the presence of an organic base [10], ethyl phosphate [11], and chloroform-NaOH-triethylbenzylammonium chloride [12].

Of these dehydrating agents, pyrophosphoryl chloride is effective at mild temperature. It is difficult to state a preference for any one of these reagents, although titanium tetrachloride in an organic base appears to have been tested the most widely and usually gives satisfactory results (Ex. a). The use of phosphonitrilic chloride involves a simple operation and the yields are usually superior. Despite the high cost, the reagent may be competitive since a mixture with higher cyclic and linear material is often satisfactory. The sodium borohydride reagent, in the hands of Borch, by proceeding via the imino ether as shown, has been very satisfactory when applied to primary amides:

$$RCONH_2 + Et_3\overset{\oplus}{O}\overset{\ominus}{BF_4} \longrightarrow \overset{OEt}{RC}\overset{|}{=}\overset{\oplus}{NH_2}\,\overset{\ominus}{BF_4} \xrightarrow{NaBH_4} \left[\overset{OEt}{RC}\overset{|}{=}\overset{\ominus}{N}\right] \longrightarrow \underset{86-100\%}{RCN}$$

The synthesis of nitriles from chlorosulfonamides has been discussed (1, 952). Other excellent yields by this procedure have been obtained by Lohaus [13]. The method is broadly applicable and may be carried out from the acid, usually with no isolation of the intermediate chlorosulfonamide. Mild reaction conditions permit certain functional groups to remain unchanged. Yields obtained in a number of syntheses are reviewed [13b,13c].

a. Preparation of Pivalonitrile [10]

Yield was 81% from pivalamide, $TiCl_4$, and N-methylmorpholine in dry THF at 0° for 20 hr.

b. Preparation of Cinnamonitrile [13b]

$$C_6H_5CH{=}CHCOOH \xrightarrow{ClSO_2NCO} C_6H_5CH{=}CHCONHSO_2Cl \xrightarrow{DMF}$$

$$C_6H_5CH{=}CHCN$$

Yield was 78-87% from cinnamic acid via the chlorosulfonamide treated with DMF.

c. Preparation of Acetonitrile [2]

Acetamide, 12 mol and phosphonitrilic chloride, 1 mol in chlorobenzene was heated at reflux for a few hr. After filtration the nitrile (95%) was recovered from the solvent.

2. From Some Substituted Amides (von Braun) and Some
    Oximes (Second-order Beckmann Rearrangement) (1, 953)

$$RCH_2NHCOR' \xrightarrow{\text{RhCl}[P(C_6H_5)_3]_3} R'CN + RCH_2OH$$

In the von Braun reaction with phosphorus pentachloride as
the reagent, both nitriles and halides are formed. The reaction
has been used to prepare halides (1, 347), but rarely nitriles.
By utilizing $RhCl(PPh_3)_3$ as the reagent, Blum and Fisher [14]
have developed a useful method for synthesizing nitriles from
secondary amides (Ex. a). Yields from 15 secondary amides vary
from a trace to 90%.

a. Preparation of Benzonitrile [14]

A mixture of 4.3 g of N-benzylbenzamide and 225 mg of
$RhCl (PPh_3)_3$ was heated for 2 min at 285° and then for 6 hr at
250°. The distillate, obtained from the reaction mixture under
reduced pressure, gave 1.89 g (90%) of the nitrile by vpc.

4. From Aldehydes via Oximes, Azines, and Related Types
    (1, 956)

$$RCH{=}NOH \longrightarrow RCN$$

Great numbers of reagents for this dehydration are listed in
1, 956. Recent additions include diphenyl hydrogen phosphonate-
triethylamine-carbon tetrachloride [15], p-chlorophenyl
chlorothionoformate, $p\text{-}ClC_6H_4OCCl$ [16], titanium tetrachloride in
the presence of an organic base [17], phenyl chloroformate [18],
and trichloroacetonitrile [19]. Thus for this usually simple
conversion, the flood of communications appears hardly justified.
To illustrate how easily this reaction may be accomplished,
Findlay and Tang [20] have refluxed aliphatic aldehydes, their
trimers, or bisulfite-addition compounds in alcohol with hydroxylamine hydrochloride and a few drops of concentrated hydrochloric
acid for 6 hr to form nitriles in yields of 83-94%.
A very useful one-step conversion of aldehydes into nitriles
has been published [21]:

Indole-3-carbonitrile
48-63%

Mobbs and Suschitzky [22] were successful in converting aryl aldehyde hydrazones into nitriles by heating with mercuric oxide in DME, THF, or diglyme in the presence of a few drops of ethanolic potassium hydroxide as shown:

$$\text{ArCH=NNH}_2 \xrightarrow{\text{HgO}} \underset{\text{10-80\%}}{\text{ArCN}}$$

Tetrazolopyridazines have been converted into cyanocyclopropenes as indicated [23]:

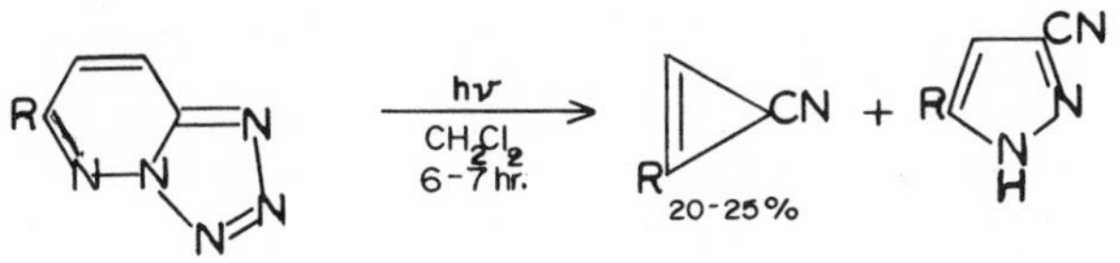

The tosylhydrazone of cyclohexanone on treatment with hydrogen cyanide gave the l-cyanohydrazine, which when heated to 180° yielded cyclohexyl cyanide as indicated [24]:

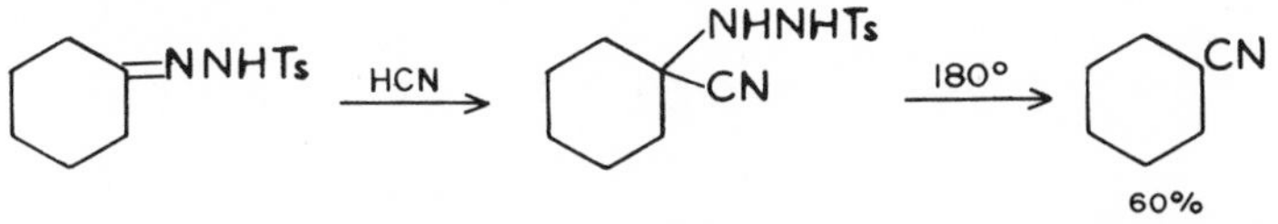

The tosylhydrazone of heptan-4-one responded similarly.

5.   From Carboxylic Acids or Aryl Lithium Compounds and Nitriles (Exchange) (1, 958)

$$\text{ArCOOH} \xrightarrow{m-\text{C}_6\text{H}_4(\text{CN})_2} \text{ArCN}$$

Klein [25] has pointed out that it is desirable to use short-chain dinitriles, such as succinonitrile, glutaronitrile, or α-methylglutaronitrile, in the acid-nitrile exchange since the use of pressure equipment is avoided and the yield, especially in the aliphatic system, is improved. The latter appears to be due to the fact that in the equilibrium as shown

$$\text{RCOOH} + \text{NC(CH}_2)_2\text{CN} \rightleftharpoons \text{RCN} + \text{NC(CH}_2)_2\text{COOH}$$

the cyanocarboxylic acid undergoes an internal cyclization,

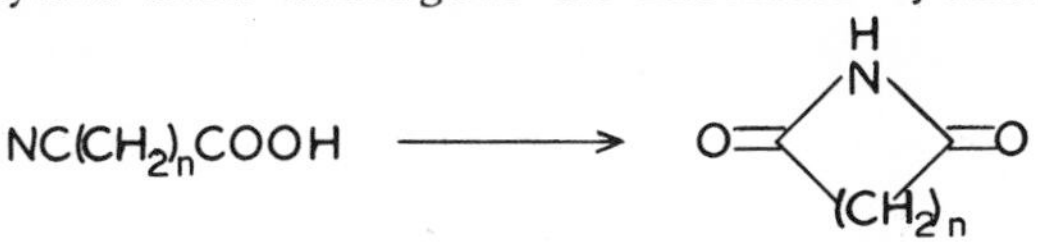

which results in its removal from the equilibrium to lead to a complete reaction. By this method 1,12-dodecanedioic acid with 2 molar equiv of α-methylglutaronitrile in the presence of acid gives a 97% yield of the dinitrile (Ex. a). Pentachlorobenzonitrile has been found to exchange readily with aryllithium [26]:

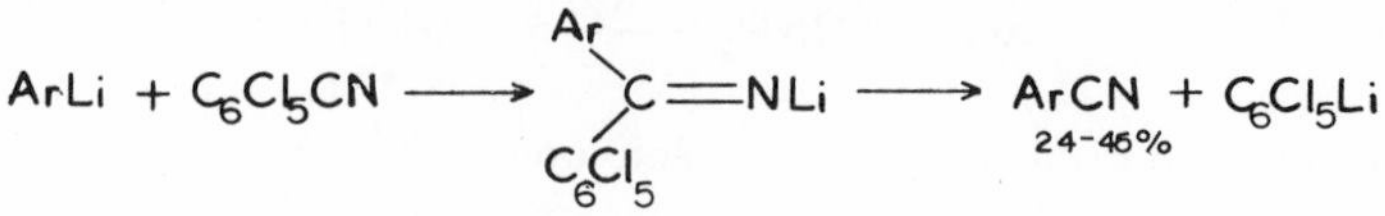

ArLi + C6Cl5CN
Ar
C=NLi
C6Cl5
ArCN + C6Cl5Li
24-46%

a. Preparation of 1,12-Dodecanedinitrile [25]

1,12-Dodecanedioic acid, 5 mol and 10.65 mol of α-methylglutaronitrile with 11.6 g of 85% H_3PO_4 were refluxed for 18 hr. Cooling followed by vacuum distillation gave 929.7 g (97%) of the dinitrile, boiling point 193° (8 mm).

6. From Carboxylic Acids, Sulfonamides, and Phosphorus Pentachloride (Exchange) (1, 959)

$$ArCOOH \xrightarrow[2\ PCl_5]{ArSO_2NH_2} ArCN + ArSO_2Cl + POCl_3 + 3\ HCl$$

The conversion of carboxylic acids into nitriles has been accomplished with urea alone [27]. However, Lücke and Winkler [28] recommend that aminosulfonic acid be added since the reaction with the two reagents proceeds more rapidly at lower temperature. In this manner o-chlorobenzonitrile was prepared as shown:

COOH
Cl
NH2SO2OH
NH2CONH2
200-240°
CN
Cl
87%

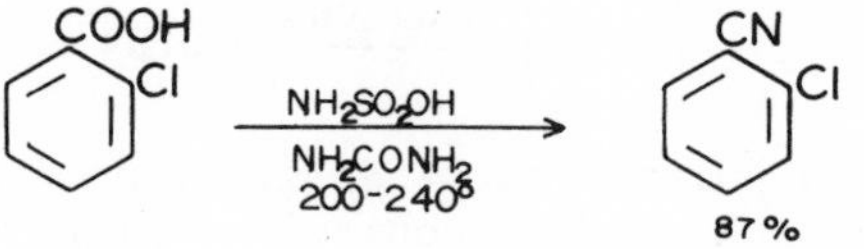

Other carboxylic acids, mostly aromatic, gave yields of 30-90%.

9. From Nitriles by Elimination

$$ClCH_2CHClCN \xrightarrow{\Delta} HC{\equiv}CCN$$

$$RCH_2CH(CN)OPO(OC_2H_5)_2 \xrightarrow{\Delta} RCH{=}CHCN$$

$$\underset{R_2}{\overset{R_1}{\diagdown}}\!\!\underset{CN}{\overset{|}{C}}\!\!N\!\!=\!\!NCO_2Me \xrightarrow[\text{MeOH}]{\text{NaOMe}} \underset{R_2}{\overset{R_1}{\diagdown}}CHCN$$

$$ClCH_2CH_2CN \xrightarrow[\text{N}_2]{[(C_6H_5)_3P]_2Ni(CO)_2} \underset{CH_2CH_2CN}{\overset{CH_2CH_2CN}{\underset{|}{|}}}$$

As is seen, four different eliminations have been employed to produce nitriles. In the first case, cyanoacetylene has been obtained in 40% yield by heating α,β-dichloropropionitrile at 1000° under 20 mm pressure for 62 min [29]. α,α,β-Trichloropropionitrile under somewhat similar conditions gave 78% of chlorocyanoacetylene. In the second case, the steps involved in the sequence starting with the diethyl acylphosphonate are as shown [30]:

$$RCH_2CO\overset{O}{\overset{\|}{P}}\!\!\overset{OC_2H_5}{\underset{OC_2H_5}{\diagdown}} \xrightarrow[\text{2) KCN}]{\text{1) NaHSO}_3} RCH_2\overset{HO}{\underset{CN}{\overset{|}{C}}}\!\!-\!\!\overset{O}{\overset{\|}{P}}\!\!\overset{OC_2H_5}{\underset{OC_2H_5}{\diagdown}} \xrightarrow{\overset{\ominus}{OH}}$$

$$RCH_2CHO\overset{O}{\overset{\|}{P}}\!\!\overset{OC_2H_5}{\underset{OC_2H_5}{\diagdown}} \xrightarrow[\text{3-5 mm.Hg}]{600°} RCH\!\!=\!\!CHCN$$
(in position: CN below first carbon)

Yields in the last step (R=CH₃, C₂H₅, C₄H₉, C₈H₁₇, C₁₀H₂₁) ran 51-71%. In the third case, Ziegler and Wender, because of the difficulty of effecting S_N2 displacements on secondary halides with inorganic cyanides, proceeded from the ketone as shown [31]:

$$\underset{R_2}{\overset{R_1}{\diagdown}}C\!\!=\!\!O \xrightarrow{H_2NNHCO_2CH_3} \underset{R_2}{\overset{R_1}{\diagdown}}C\!\!=\!\!NNHCO_2Me \xrightarrow[\text{MeOH}]{\text{HCN}}$$

$$\underset{R_2}{\overset{R_1}{\diagdown}}\!\!\overset{CN}{\underset{|}{C}}\!\!-\!\!NHNHCO_2Me \xrightarrow[\substack{\text{NaHCO}_3 \\ \text{CH}_2\text{Cl}_2}]{\text{Br}_2} \underset{R_2}{\overset{R_1}{\diagdown}}\!\!\overset{CN}{\underset{|}{C}}\!\!-\!\!N\!\!=\!\!NCO_2Me \xrightarrow[\text{MeOH}]{\text{NaOMe -}} \underset{R_2}{\overset{R_1}{\diagdown}}CHCN$$

The steps to the methyl dialkylcyanodiazene carboxylate are nearly quantitative and in the last step yields varied within 89-97%. To cite a specific case, cyclohexanone was transformed into cyclohexyl cyanide in 80% yield without purification of the intermediates. The mechanism is ionic:

$$\begin{array}{c} R \\ \diagdown \\ \diagup C N = N C \overset{\ominus}{O_2} \\ R \end{array} \xrightarrow[-N_2]{-CO_2} \left[\begin{array}{c} R CN \\ \diagdown \diagup \\ C \ominus \\ \diagup \\ R \end{array} \right] \xrightarrow{H^{\oplus}} \begin{array}{c} R \\ \diagdown \\ CHCN \\ R \end{array}$$

In the fourth case, treatment of β-chloropropionitrile with a phosphine-nickel carbonyl complex yields adiponitrile [32].

A new method of preparing β,β-dichloroacrylonitrile by the pyrolysis of a mixture of acetonitrile and carbon tetrachloride has been reported [33] as shown:

$$CH_3CN \; + \; CCl_4 \xrightarrow[\substack{20-25\,min.}]{\substack{800-1000°}} Cl_2C = CHCN$$
$$\phantom{CH_3CN \; + \; CCl_4 \xrightarrow{800-1000°}} \substack{50-60\,\%}$$

A free radical mechanism, in which in the last step HCl is eliminated from β,β,β-trichloropropionitrile, apparently occurs:

$$CCl_4 \longrightarrow \cdot CCl_3 + \cdot Cl$$

$$Cl\cdot + CH_3CN \longrightarrow \cdot CH_2CN + HCl$$

$$\cdot CCl_3 + \cdot CH_2CN \longrightarrow CCl_3CH_2CN$$

$$CCl_3CH_2CN \longrightarrow Cl_2C = CHCN + HCl$$

10.   From Primary Amines (see Addenda)

Addenda

C.3. 2-Azidobenzoquinones in refluxing benzene or toluene give 2-cyano-4-cyclopentene-1,3-diones (31-95%) (15 ex.) [H. W. Moore et al., J. Am. Chem. Soc., 95, 2603 (1973)]. This appears to be a desirable route to cyclopentenediones.

C.9. A method of synthesizing α-alkylacrylonitrile from methyl cyanoacetate via its Mannich base has been devised [R. B. Miller, B. F. Smith, Syn. Commun., 3, 413 (1973)].

C.10. Primary amines containing an α-methylene group may be brominated with NBS to N,N-dibromoamines, which without isolation may be converted into the nitrile (36-85%) by trimethylamine [W. Gottardi, Monatsh. Chem. 104, 1690 (1973)].

References

1. G. Shaw et al., J. Chem. Soc., C, 2198 (1969).
2. J. C. Graham, D. H. Marr, Can. J. Chem., 50, 3857 (1972).
3. S. E. Ellzey et al., United States Patent, 3,493,567, February 3, 1970; C. A., 72, 100326 (1970); R. F. Borch, Tetrahedron Letters, 61 (1968).
4. W. E. Dennis, J. Org. Chem., 35, 3253 (1970).
5. M. L. Hallensleben, Tetrahedron Letters, 2057 (1972).
6. R. S. Monson, D. N. Priest, Can. J. Chem., 49, 2897 (1971).
7. J. K. Chakrabarti, T. M. Hotten, Chem. Commun., 1226 (1972).
8. E. Yamato, S. Sugasawa, Tetrahedron Letters, 4383 (1970).
9. R. Appel et al., Chem. Ber., 104, 1030 (1971)
10. W. Lehnert, Tetrahedron Letters, 1501 (1971).
11. C. Zinsstag, R. J. Peake, Chimia, 23, 397 (1969).
12. T. Saraie et al., Tetrahedron Letters, 2121 (1973).
13. G. Lohaus, (a) Chem. Ber., 100, 2719 (1967); (b) Org. Syn., 50, 18 (1970); (c) 50, 52 (1970).
14. J. Blum, A. Fisher, Tetrahedron Letters, 1963 (1970).
15. P. J. Foley, Jr., J. Org. Chem., 34, 2805 (1969).
16. D. L. J. Clive, Chem. Commun., 1014 (1970).
17. W. Lehnert, Tetrahedron Letters, 559 (1971).
18. J. M. Prokipcak, P. A. Forte, Can. J. Chem., 49, 1321 (1971).
19. T.-L. Ho, C. M. Wong, J. Org. Chem., 38, 2241 (1973).
20. J. A. Findlay, C. S. Tang, Can. J. Chem., 45, 1014 (1967).
21. H. M. Blatter et al., Org. Syn. Coll., Vol. 5, 656 (1973).
22. D. B. Mobbs, H. Suschitzky, Tetrahedron Letters, 361 (1971).
23. H. Igeta et al., Chem. Commun., 1059 (1972).
24. S. Cacchi et al., Chem. Ind. (London), 213 (1972).
25. D. A. Klein, J. Org. Chem., 36, 3050 (1971).
26. N. J. Foulger, B. J. Wakefield, Tetrahedron Letters, 4169 (1972).
27. B. S. Biggs, W. S. Bishop, Org. Syn., Coll. Vol. 3, 768 (1955).
28. J. Lücke, R. E. Winkler, Chimia, 25, 94 (1971).
29. N. Hashimoto et al., J. Org. Chem., 35, 675 (1970).
30. Y. Okamoto et al., Bull. Chem. Soc. Jap., 42, 543 (1969).
31. F. E. Ziegler, P. A. Wender, J. Am. Chem. Soc., 93, 4318 (1971).
32. O. T. Onsager, German Patent, 2,008,569, September 3, 1970; C. A., 73, 120,124 (1970).
33. N. Hashimoto et al., J. Org. Chem., 35, 828 (1970).

D. Addition Reactions

1. From Unsaturated Compounds and Hydrogen Cyanide (or R_2AlCN) (1, 962)

$$HC\equiv CH \xrightarrow{\text{HCN}} CH_2{=}CHCN \xrightarrow{\text{HCN}} NCCH_2CH_2CN$$

The addition of a cyanide ion to α,β-unsaturated carbonyl compounds has been plagued by reversibility. Nagata and

$$RCOCH{=}CHR' + HCN \rightleftharpoons RCOCH_2\overset{R'}{\underset{CN}{\text{CH}}}$$

co-workers [1] have shown that this difficulty can be overcome either by the addition of $AlEt_3 \cdot HCN$ and THF or Et_2AlCN and THF (<u>1</u>, 963). The dialkyl cyanoalane is prepared by adding the trialkylaluminum to an equivalent amount of hydrogen cyanide with cooling [2]:

$$R_3Al + HCN \longrightarrow R_2AlCN + RH$$

A large number of examples, including those in the steroid family, have been studied [3], as well as the stereochemistry of the products [4].

The cyanation of 9-cyanoanthracene has been accomplished as shown by treatment with NaCN and sodium 9,10-anthraquinone-α-sulfonate (α-SAS) [5] (Ex. a):

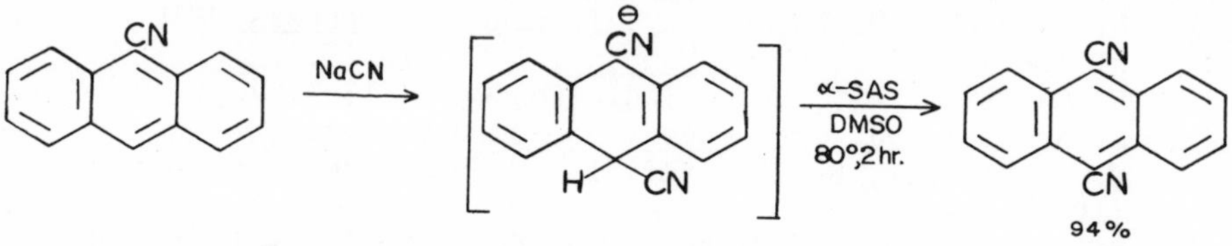

The substrate was obtained by the action of cuprous cyanide on 9-bromoanthracene (A.1), and the carbanion is thought to be an intermediate in the reaction. 9,10-Dicyanophenanthrene was prepared (74%) in a similar manner.

a. Preparation of 9,10-Dicyanoanthracene [5]

An intense magenta color, with the separation of a yellow, crystalline material, developed as 0.008 mol of NaCN was added to 0.003 mol of 9-cyanoanthracene and 0.005 mol of α-SAS in 80 ml of DMSO at 80°. After 1.75 hr, the reaction was worked up to give 0.64 g (94%) of the dicyanoanthracene.

6. From Cyanogen Compounds (<u>1</u>, 969)

The von Braun reaction, in which cyanogen bromide is added to a tertiary amine in an inert solvent to give a haloalkyl-N-cyano-sec-amine, has been modified by Albright and Goldman [6] in that ethanol-chloroform or aqueous THF was used as the solvent. In this way cyclic amines such as yohimbine are transformed into an alkoxy-N-cyano-sec-amine similar to that shown in the next equation. Rönsch [7] introduced a second modification, namely, the use of an inorganic base, especially MgO, which greatly improved the yield of the product formed. For example, tetra-hydropalmitine, as shown, with the use of MgO gives a 94% yield of the hydroxynitrile:

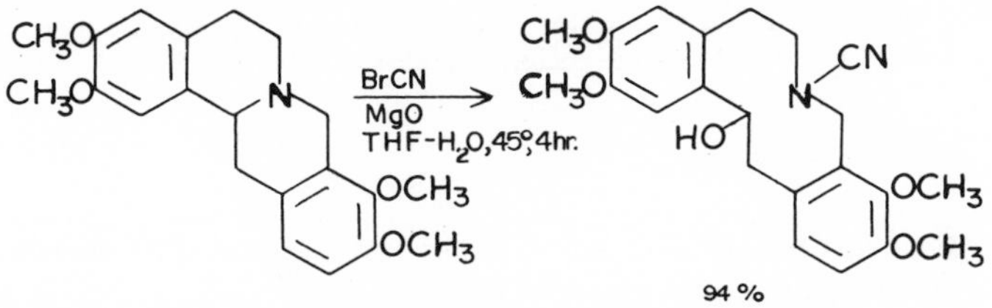

94 %

Yields from two other alkaloids were 87 and 95%.

8. From Unsaturated Nitriles by Hydrogenation

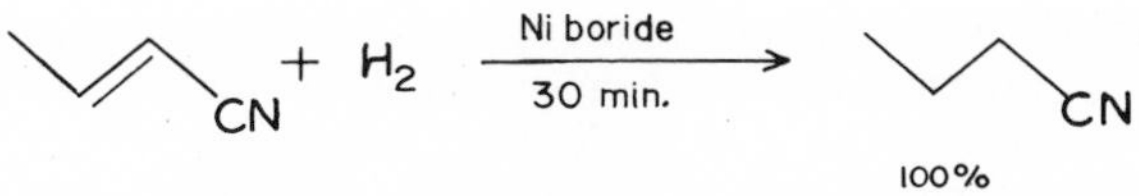

100%

Nickel boride has been proven an effective catalyst in the hydrogenation of C=C bonds in unsaturated compounds containing nitrogen [8]. The catalyst, which is readily prepared from a Ni(II) salt and sodium borohydride, exhibits a high selectivity for the carbon-carbon π bond. As a rule the addition time does not exceed 30 min. Reduction to the amine occurs with longer periods of time.

Addenda

D.1. The synthesis of a cyanohydrin from diethylcyanoaluminum has been checked [W. Nagata et al., Org. Syn., 52, 96 (1972)].

D.1. The Nagata hydrocyanation procedure using HCN-Al(Et)₃ in the 1,4-addition to α,β-unsaturated ketones has now been published in Organic Synthesis [W. Nagata, M. Yoshioka, Org. Syn., 52, 100 (1972)].

References

1. W. Nagata et al., $\underline{J}$. $\underline{Am}$. $\underline{Chem}$. $\underline{Soc}$., $\underline{94}$, 4654 (1972) and previous papers.
2. W. Nagata, M. Yoshioka, $\underline{Org}$. $\underline{Syn}$., $\underline{52}$, 90 (1972).
3. W. Nagata et al., $\underline{Org}$. $\underline{Syn}$., $\underline{52}$, 96, 100 (1972).
4. W. Nagata et al., $\underline{J}$. $\underline{Am}$. $\underline{Chem}$. $\underline{Soc}$., $\underline{94}$, 4672 (1972).
5. K. E. Whitaker, H. R. Snyder, $\underline{J}$. $\underline{Org}$. $\underline{Chem}$., $\underline{35}$, 30 (1970); H. R. Snyder et al., ibid., $\underline{37}$, 314 (1972).
6. J. D. Albright, L. Goldman, $\underline{J}$. $\underline{Am}$. $\underline{Chem}$. $\underline{Soc}$., $\underline{91}$, 4317 (1969).
7. H. Rünsch, $\underline{J}$. $\underline{Prakt}$. $\underline{Chem}$., $\underline{314}$, 382 (1972).
8. T. W. Russell et al., $\underline{J}$. $\underline{Org}$. $\underline{Chem}$., $\underline{37}$, 3552 (1972).

E.   Substitution Reactions

1.   From an Arene and a Cyanylation Source (Including Friedel-Crafts) ($\underline{1}$, 974)

Letsinger and Hautala [1] investigated the effects of substituent groups and solvents on the rates of bimolecular reactions involving nucleophiles and photoexcited nitroaromatic compounds. The results of the reaction of the cyanide ion on 4-nitroanisole, 1-nitronaphthalene, and 4-methoxy-1-nitronaphthalene vary greatly with these two effects. For example, photoexcited 4-nitroanisole with the cyanide ion in aqueous solution gives 2-cyano-4-nitroanisole in high yield as shown:

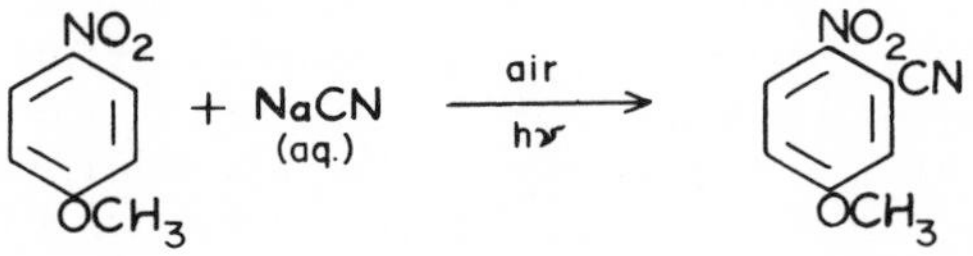

By contrast, in a largely aprotic solvent (95% CH_3CN - 5% H_2O) the reaction of this substrate was greatly retarded, the quantum yield being about 1/50 of that in 90% H_2O - 10% CH_3CN. With the elimination of the methoxy group as in nitrobenzene, no reaction of the photoexcited substrate occurs with common nucleophiles in dilute aqueous solution.

Butler [2] conducted Friedel-Crafts alkylation of benzene with ω-haloalkanenitriles to obtain ω-arylalkanenitriles in superior yield. The reaction for the preparation of 4-phenyl-butyronitrile is as shown:

$$C_6H_5H \; + \; ClCH_2CH_2CH_2CN \xrightarrow{AlCl_3} C_6H_5CH_2CH_2CH_2CN$$

98%

4. From Heterocyclic Compounds and Carbethoxyiminotriphenyl-
    phosphorane

Heterocyclic compounds may be cyanated by treatment with
carbethoxyiminotriphenylphosphorane in the presence of boron trifluoride etherate as shown [3]:

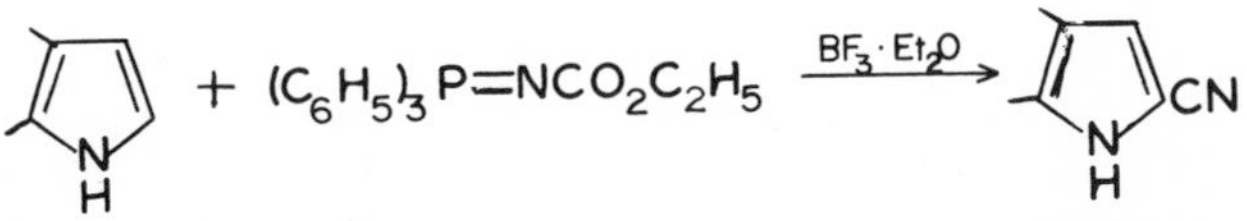

2-Cyanopyrrole
45 %

Indole gave 3-cyanoindole, 65%. It seems possible that acylation
occurs followed by phosphine oxide elimination:

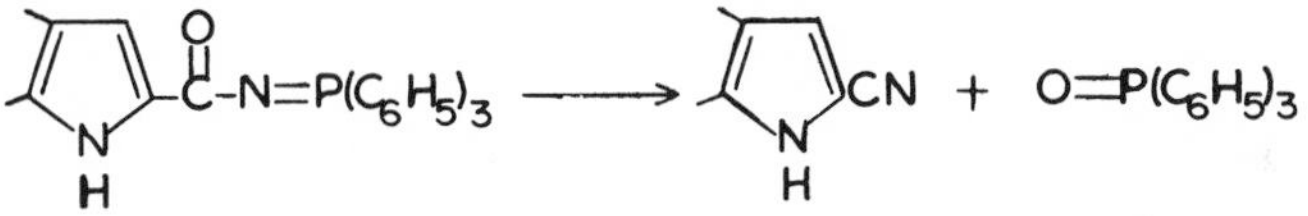

References

1. R. L. Letsinger, R. R. Hautala, Tetrahedron Letters, 4205
    (1969).
2. D. E. Butler, Tetrahedron Letters, 1929 (1972).
3. H. Plieninger et al., Angew. Chem. Intern. Ed. Engl., 7, 377
    (1968).

F. Oxidation

    1. From Amines (Dehydrogenation) (1, 977)

    In applying the lead tetraacetate oxidation to α-
aminoketones, Baumgarten and co-workers [1] found that cleavage
occurred between the carbonyl and carbinamine functions to give,
in the presence of alcohol, moderate yields of the nitrile and
ester. Thus in the case of α-aminovalerophenone the results are:

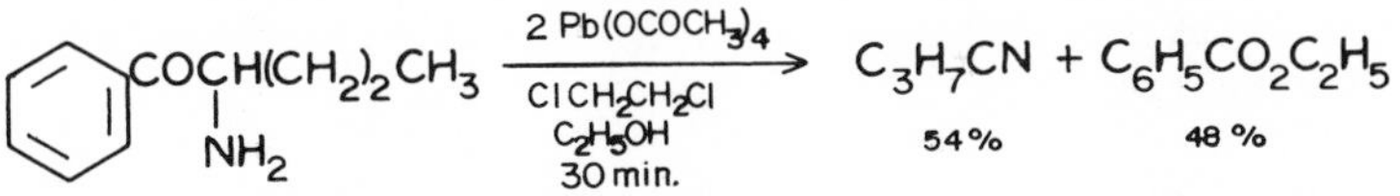

Similarly, ω-aminoacetophenone gave a 67-75% return of benzoyl
cyanide [1a]. The maximum yield of the nitrile (94%) was
obtained in the oxidation of α-amino-p-methylpropiophenone with
iodosobenzene diacetate.

A new method of dehydrogenation consists of passing oxygen through a dry benzene solution of the amine containing cobalt oxide [2]. In this manner, benzylamine with a catalyst:substrate ratio of 7:1 gave 85% benzonitrile in 0.5 hr. The catalyst was prepared by adding 200 ml of 7 N NaOH during 4 hr to 281 g cobalt(II) sulfate heptahydrate in 1.2 l of water. During the process oxygen was passed through the solution and then for an additional 16 hr at room temperature.

A second new process involving cuprous chloride and oxygen has been utilized by Takahashi and co-workers [3] for the preparation of unsaturated dinitriles. The method, starting with o-phenylenediamine, involves cleavage as well as dehydrogenation as shown (Ex. a):

95%

Another synthesis of cis,cis-mucodinitrile has been described (1, 977).

A method known as the Lummus process for preparing terephthalonitrile from p-xylene has been announced [4]. Unfortunately, details of this process, which may be represented as

are lacking.

a. Preparation of cis,cis-Mucodinitrile [3]

In cuprous chloride, 10 mmol in 10 ml of oxygen-free pyridine, oxygen was introduced with stirring. After absorption ceased (ca. 10 min) a solution of o-phenylenediamine, 10 mmol, in pyridine was added slowly. The oxygen absorption was continued during three 10-min periods. After removing the pyridine in vacuo and extracting with ether, 2.97 g (95.2%) of the dinitrile was recovered from the ether layer.

2. From Primary Alcohols or Aldehydes and Ammonia (1, 978)

Although N-unsubstituted imines are unstable, their stability is improved by the formation of cobalt complexes, which may be stored for long periods under an inert gas at 25°. These complexes were utilized by Rhee and co-workers [5] with aldehydes to form nitriles as indicated:

$$C_6H_5CHO + Co(NH_3)_6 X_2 \xrightarrow{CH_3CN} \left[\begin{array}{c} (C_6H_5CH{=}NH)_n \\ \downarrow \\ Co(NH_3)_{6-n} \end{array} \right] X_2 \xrightarrow{Br_2} C_6H_5CN$$

77 %

Another rather simple way of preparing nitriles is by the oxidation of aldehyde-ammonia addition products [6]:

$$RCHO + NH_3 \xrightarrow{MnO_2} RCN$$

The presence of sodium cyanide in the reaction mixture leads to amides (C.5).

References

1. H. E. Baumgarten et al., _J_. _Org_. _Chem_., _36_, 3668 (1971).
1a. Private communication from H. E. Baumgarten, University of Nebraska.
2. J. S. Belew et al., _Chem_. _Commun_., 634 (1970).
3. H. Takahashi et al., _Syn_. _Commun_., _2_, 181 (1972).
4. _Chem_. _Eng_. _News_, April 2, 1973, p. 10.
5. I. Rhee et al., _Tetrahedron Letters_, 3419 (1970).
6. N. W. Gilman, _Chem_, _Commun_., 733 (1971).

G. Reduction and Reductive Dimerization

    1. Reduction of Arylnitromethanes

    In the reduction of phenylnitromethane with a sulfurated borohydride, $NaBH_2S_3$, Lalancette and Brindle [1] obtained benzonitrile as indicated:

$$C_6H_5CH_2NO_2 \xrightarrow{NaBH_2S_3} C_6H_5CN$$

80 %

Only primary aliphatic nitro compounds undergo this change, which likely involves the oxime as an intermediate.

    2. Reductive dimerization

The reductive dimerization of crotonyl nitrile has been accomplished as indicated [2]:

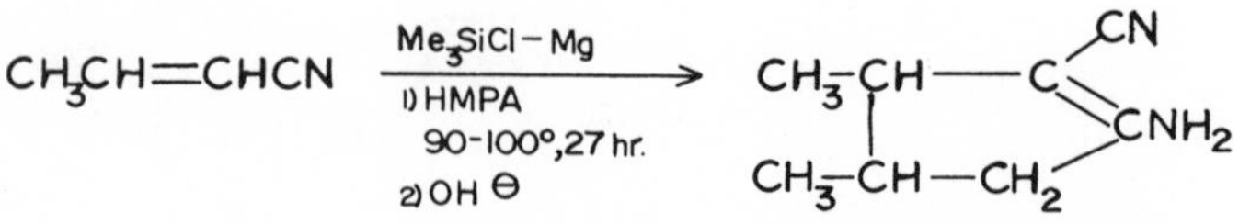

$$CH_3CH{=}CHCN \xrightarrow[\substack{1)\ HMPA \\ 90{-}100°,27\ hr. \\ 2)\ OH^{\ominus}}]{Me_3SiCl-Mg}$$

l-Amino-2-cyano-3,4-dimethyl-
l-cyclopentene, 60%

References

1. J. M. Lalancette, J. R. Brindle, Can. J. Chem., 49, 2990 (1971).
2. M. Bolourtchian et al., J. Organomet. Chem., 33, 303 (1971).

H. Cyclo Reactions

If a diene or a dienophile carries a cyano group, the Diels-Alder product will naturally contain a cyano group. Such compounds as indicated are listed in Onishchenko [1].

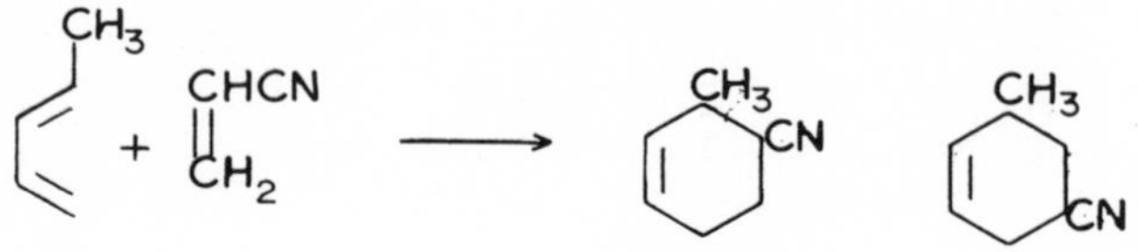

trans

2-Methyl-1,2,5,6-
tetrahydrobenzonitrile
>60%

3-Methyl-12,3,6-
tetrahydrobenzonitrile
trace

Another series was prepared later [2]:

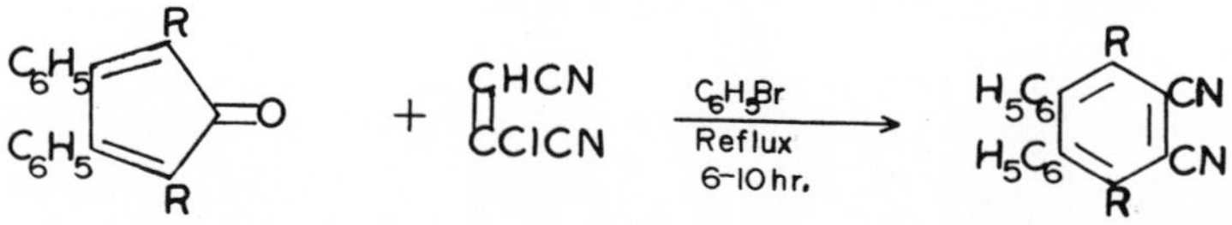

Substituted phthalylnitrile
85-90%

In a more recent example of cycloaddition between a thiophene and dicyanoacetylene, sulfur is eliminated as shown [3] to form a phthalonitrile:

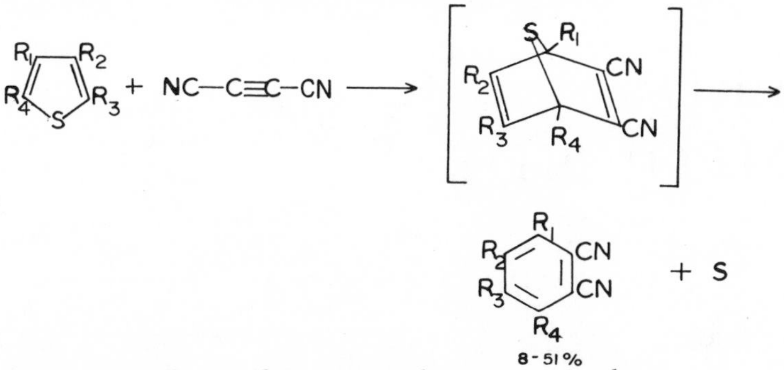

8-51%

Cyclo reactions of an unusual nature involve
2,2-diphenylmethylenecyclopropanes as follows [4]:

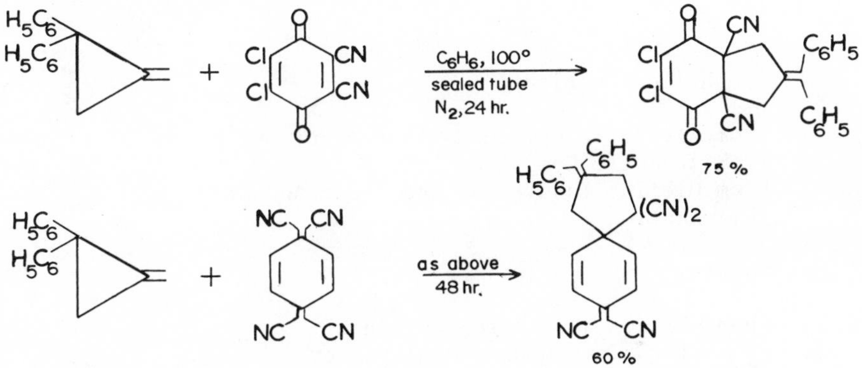

75%

60%

References

1. A. S. Onishchenko, Diene Syntheses, D. Davey, New York,
   1964; R. L. Frank et al., J. Am. Chem. Soc., 69, 2313 (1947).
2. S. A. Mikhalenko, E. A. Luk'yanets, J. Org. Chem. USSR
   (Engl. transl.), 6, 167 (1970).
3. R. Helder, H. Wynberg, Tetrahedron Letters, 605 (1972).
4. R. Noyori et al., Tetrahedron Letters, 2983 (1973).

Chapter 20

NITRO
COMPOUNDS

886

n updated review of the nitro group, including synthesis, is now
vailable [1].   Included in this chapter are just a few words on
ome electrophilic syntheses and the rest is devoted to syntheses
eyond the date of publication of the Feuer book.

Aromatic nitration, including the selection of experimental
onditions, has been discussed in detail [1: Pt. 2].   Here tables
re given containing the mildest type of nitrating agent, that
s, an acyl nitrate or N-nitro-2,6-dimethylpyridinium boron
etrafluoride (with organometallics), to the most drastic type,
amely, the nitronium ion.   Since ordinary nitrating agents
xidize ferrocene, nitroferrocene has been prepared from ferro-
enyllithium and propyl nitrate at -70°.   Cupric nitrate nitrates
rylsilanes, while ordinary reagents tend toward proto-
esilylation.

To avoid dinitration of polyalkylbenzenes, measured equiva-
ents each of methyl nitrate and the arene are treated with boron
rifluoride in nitromethane as indicated [2]:

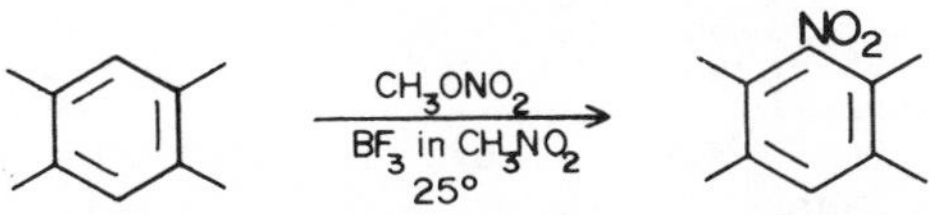

Nitrodurene, 97 %

In our opinion the most interesting developments are those
yntheses of aromatic nitro compounds brought about by Michael
ondensations (C.2).   They are indeed exotic, but perhaps a more
enerally useful synthesis consists of the conversion of short-
hain nitroalkanes into longer ones by a series of simple, con-
ecutive reactions carried out without the isolation of any of the
ntermediates (C.1).   In the same section (C.1) a review is given
escribing the synthesis of cyclic nitro compounds from the nitro-
lkanes and dialdehydes.   Aryl amines are oxidized to nitroarenes
y reagents other than pertrifluoroacetic acid (D.1).

A.   Electrophilic Reactions

1.   From Aliphatic Compounds (Substitution) (1, 980)

Besides the nitration of alkanes in the vapor phase, a similar nitration of aliphatic acids yields nitroalkanes by a nitration-decarboxylation mechanism. Conversions are of the order of 15-62% [3].

The nitration of ketones leads to nitrosonitro derivatives as shown [4]:

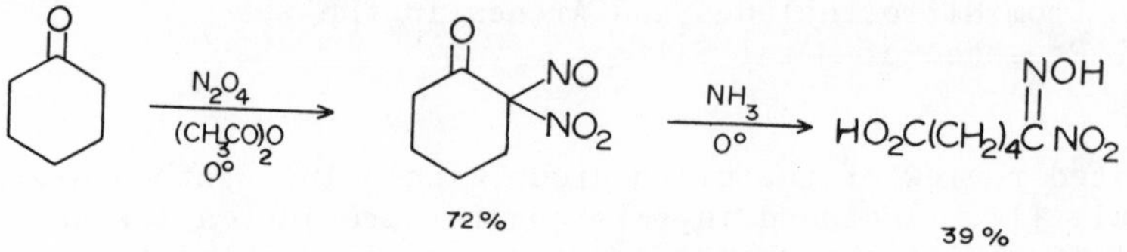

72%   39%

The nitration of α,β-unsaturated esters leads to a mixture of α-nitro unsaturated esters, as well as α-hydroxy-β-nitro saturated esters [5]:

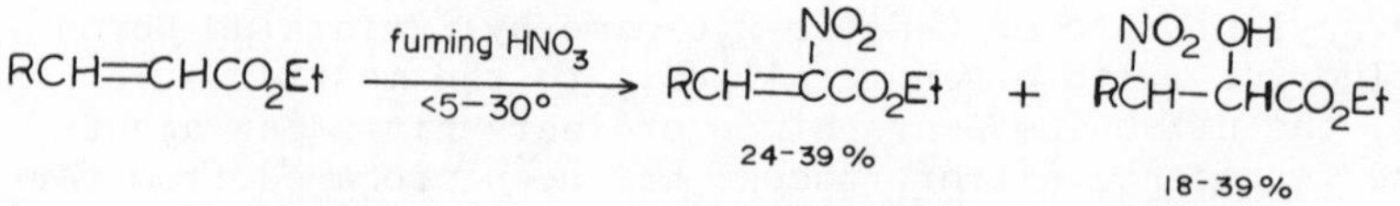

24-39%   18-39%

On the other hand, the nitration of a diazoacetic ester gives first a nitrodiazomethane and then a dinitrodiazomethane [6]:

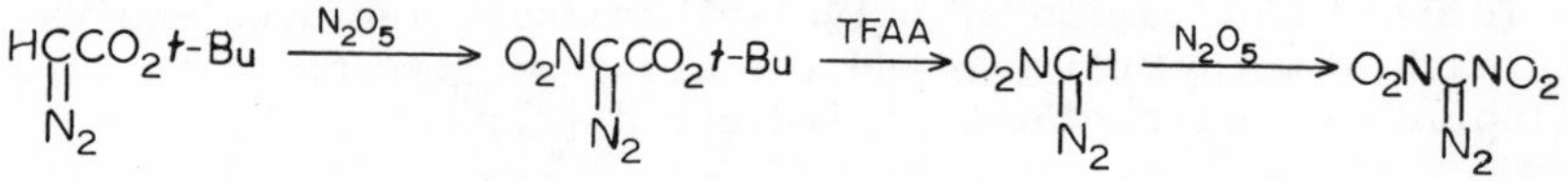

The nitration of an alkylidene diamide gives the corresponding dinitramide [7]:

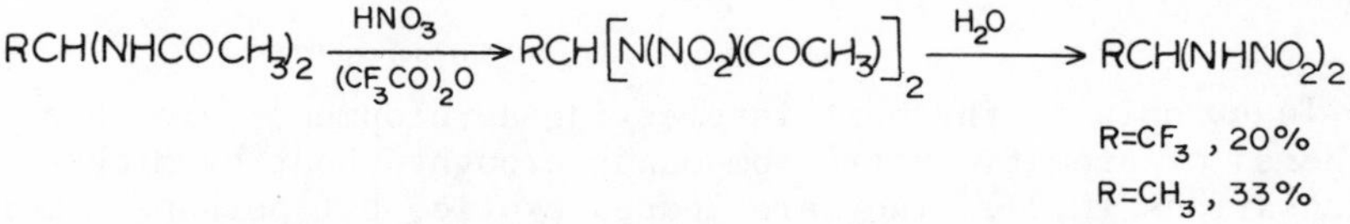

R=CF$_3$, 20%
R=CH$_3$, 33%

2.   From Aromatic Compounds (1, 983)

$$ArH + HNO_3 \longrightarrow ArNO_2 + H_2O$$

Ridd has discussed the mechanism of aromatic nitration in some detail [8]. Briefly, the process may be represented as:

$$HNO_3 \underset{1}{\overset{Step}{\rightleftharpoons}} \overset{\oplus}{NO_2} \underset{2}{\overset{Step}{\rightleftharpoons}} ArHNO_2^{\oplus} \underset{3}{\overset{Step}{\rightleftharpoons}} Ar \overset{NO_2}{\underset{H}{\diagdown}} \underset{4}{\overset{Step}{\rightleftharpoons}} ArNO_2$$

In Step 1 the formation of the intermediate nitronium ion NO_2^+ occurs, Step 2 represents the encounter of the reactants (the formation of the π complex), Step 3 represents the formation of the σ complex, and Step 4 represents the formation of the aromatic nitro compound. All of these four are possible rate-determining steps.

Faith in partial rate factors to assess the susceptibility of each position in arenes to substitution by electrophilic agents has been shaken of late. Substitution at the position of a substituent has been ignored, that is, ortho, meta, and para substitutions are recognized, but ipso or geminal substitution was not, until pointed out by Baciocchi and Illuminati [9]. ipso Substitution may be represented as:

A

If the nitration is carried out with acetyl nitrate and if A is assumed to form the acetate quantitatively, then

Analysis should give the partial rate for ipso substitution. Based on a $C_6H_5CH_3/C_6H_6$ ratio of 25, the partial rate factors are o 44, m 2.1, p 54, and ipso 4.7 [10]. Note that the ipso partial rate factor is twice as great as that of the meta. No doubt more will be heard on this subject in the future.

In the nitration of toluene with nitric acid in the presence of amberlite IR-120 (H^+ form), desiccated by azeotropic distillation, Wright and co-workers found the ratio of o- to p-nitrotoluene to be as shown [11].

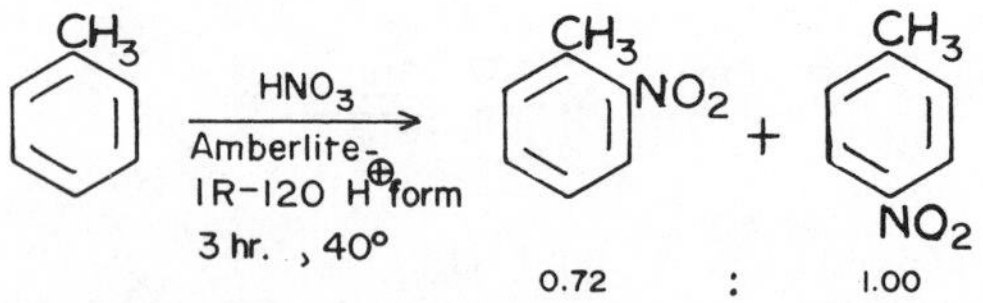

$$0.72 \quad : \quad 1.00$$

In contrast the HNO_3-H_2SO_4 nitration gave a 1.57:1 ratio.

Perylene has been nitrated by Ristagno and Shine [12] by treatment with iodine and silver nitrite in acetonitrile as shown:

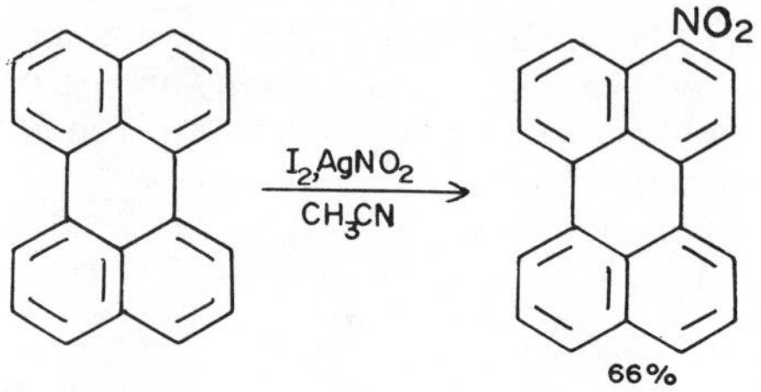

66%

The perylene cation radical, which is readily obtained as a solid, appears to be an intermediate in the reaction since, with sodium nitrite, it also gives 3-nitroperylene. The method is also applicable to the nitration of pyrene, but not as yet to phenanthrene, chrysene, triphenylene, or anthracene.

In the nitration of sodium benzenesulfonate with a mixture of 53% HNO_3, 44% of H_2SO_4, and 3% SO_3, in which there were 10.6 mol of HNO_3 per mol of substrate, Barvinskaya and Spryskov [13] obtained largely the m-nitration product as indicated:

$$C_6H_5SO_2ONa \xrightarrow[-50°]{HNO_3-H_2SO_4-SO_3} NO_2C_6H_4SO_2OH$$

m-84%, o-2.6%, p-14%

In some examples of nitration, nitro groups are not only introduced, but they may replace or be formed from groups already present. Thus in the nitration of 2,3,5,6-tetrafluoro-acetanilide, Belf and Saggers [14] obtained the 1,4 dinitro compound as shown:

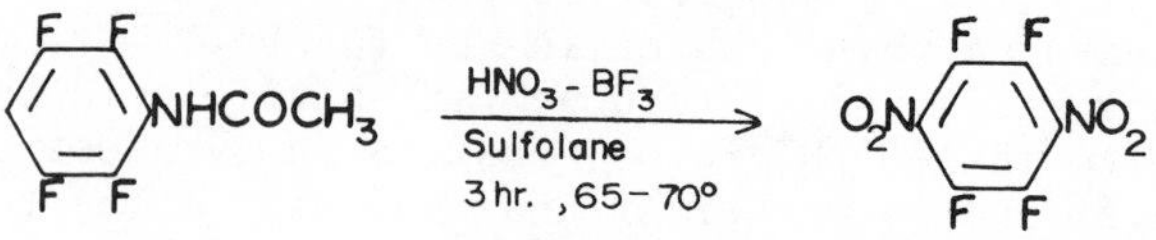

(no yield given)

It is known that nitrodeiodination occurs in part when the substrate to be nitrated possesses hydroxyl or alkoxyl groups ortho or para to the iodine atom.   Recently it has been shown that iodomesitylene and homologs also undergo partial nitro-deiodination [15].

In the nitration of alkyl homologs of benzene, if the nitric acid is added to the hydrocarbon, nitropolyalkylbiphenyls are obtained [16].   Thus o-xylene gives 2-nitro-3',4,4',5-tetra-methylbiphenyl as indicated:

47% of 82% purity

Nitrative coupling is less prominent with the other homologs studied.

Nitronium trifluoromethanesulfonate, $CF_3SO_2ONO_2$, appears to be a faster nitrating agent than nitronium tetrafluoroborate or hexafluorophosphate [17].

The nitration of pentamethylbenzene with NO_2PF_6 in acetonitrile has been found to be complicated in that 20% of the normal nitration product and 38% of N-acetyl-2,3,4,5-tetramethyl-6-nitrobenzylamine are obtained [18].   Likewise, 3,6-dinitro-1,2,4,5-tetramethylbenzene gives with a mixture of HNO_3 and H_2SO_4, 3,6-dinitro-2,4,5-trimethylbenzyl nitrate (63-73%) [19].

Isoquinoline is nitrated in the 4-position with nitric acid in acetic anhydride at 100° for 1 hr [20].   With mixed acid it is nitrated in the benzenoid ring.   The pyrimidine, is nitrated best with potassium nitrate and sul-furic acid in the 5-position (49%) [21].   1-Azulenecarboxylic acid is nitrated in the 3-position (65%) by tetranitromethane in ethanol at 25° [22].

The benzyl onium salts are nitrated with mixed acid as shown in Table 20.1.[23].

Table 20.1.   Nitration of Benzyl Onium Salts

Salt	Y	% meta
$C_6H_5CH_2\overset{+}{Y}(CH_3)_2$	N	88
	S	39
	Se	12

Thus the positive charge on the hetero atom has less effect in directing the substituent to the meta position as the size of the hetero atom increases.

It is of interest to note that if silver nitrate-silicic acid is used as the nitrating agent, no nitration may occur except in the presence of carbon tetrachloride [24].

3. From Olefins (Addition) and Their Adducts (Elimination) (1, 988)

Larkin and Kreuz [25] have developed an improved method for converting 1-alkenes into terminal nitroalkanes. Whereas the old method consisted of converting the olefin into dinitroalkanes, nitro alcohols, nitro nitrites, or mixtures of these compounds, treating the latter with base to form nitro olefins, and then hydrogenating the nitroolefins catalytically, the new procedure consists of transforming the 1-alkene into the β-nitronitrate with nitrogen oxides and oxygen [26] and reducing the latter with sodium borohydride as indicated:

$$RCH{=}CH_2 \xrightarrow[\substack{CCl_4 \\ 0°,7hr.}]{\substack{O_2 \\ N_2O_4}} RCHCH_2NO_2 \xrightarrow[\substack{CCl_4 \\ 0°,50min.}]{NO_2} RCHCH_2NO_2 \xrightarrow{NaBH_4} RCH_2CH_2NO_2$$
$$\underset{O_2NO_2}{} \qquad \underset{ONO_2}{}$$

1-Nitrooctane was obtained in 94% yield from the appropriate β-nitronitrate. The reduction was less satisfactory for more highly branched types. Jäger and Viehe [27] succeeded in converting 3,3-dimethyl-1-butyne into the 1-nitro derivative by the two-step process indicated:

$$CH_3{-}\underset{\underset{CH_3}{|}}{\overset{\overset{CH_3}{|}}{C}}{-}C{\equiv}CH \xrightarrow[I_2,(Et)_2O]{N_2O_4} CH_3\underset{\underset{CH_3}{|}}{\overset{\overset{CH_3}{|}}{C}}{-}CI{=}CHNO_2 \xrightarrow[100°,(vacuum)]{KOH}$$

85%

$$CH_3{-}\underset{\underset{CH_3}{|}}{\overset{\overset{CH_3}{|}}{C}}{-}C{\equiv}CNO_2$$

94%

1-Nitrocyclooctene has been prepared from the cycloalkene as shown by a direct process which involves no isolation of intermediates [28] (Ex. a).

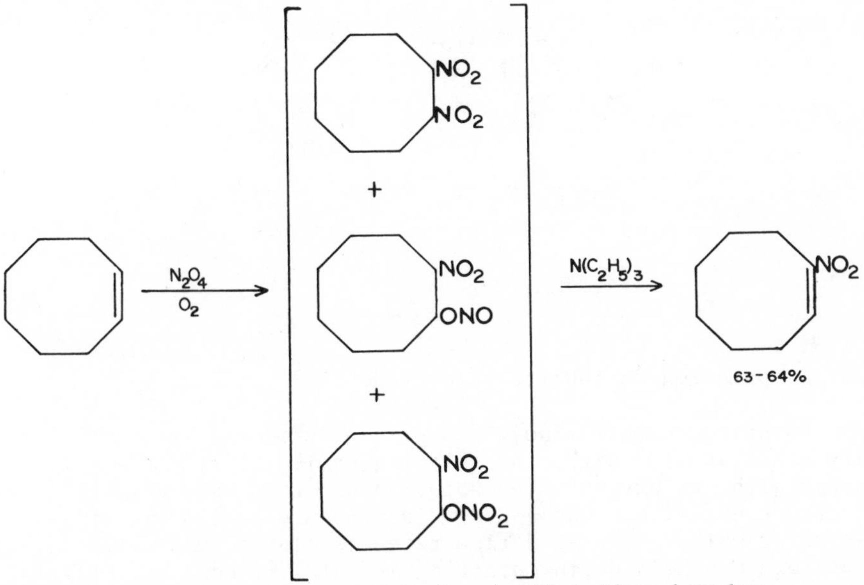

Previous methods are less attractive in that they involve hydrolyzing the addition products to the nitro alcohol from which the cycloalkene was obtained by elimination, or in some cases nitrous and nitric acids were eliminated from dinitro compounds or nitronitrates, respectively. This method has been utilized as well in the synthesis of 1-nitro-1-octadecene from 1-octadecene. The nitronitrate has been reduced to the nitro-alkane as well by prolonged extraction with $NaBH_4$ [25].

A method of proceeding from a dinitrobutene to a dinitrobu-tadiene is as follows [29]:

$$O_2NCH_2CH=CHCH_2NO_2 \xrightarrow[\substack{2)Br_2-CH_3OH \\ 0°}]{1)KOH-CH_3OH} O_2NCH=CH-CH=CHNO_2$$

1,4-Dinitro-1,3-butadiene, 79 %

Nitroform has been added to vinyl acetate [30]:

$$CH_2=CHO_2CCH_3 + HC(NO_2)_3 \xrightarrow{EtOH} CH_3\underset{OEt}{CH}C(NO_2)_3$$

2-Ethoxy-1,1,1-trinitropropane
49 %

An interesting cycloreaction occurs with an enamine and a nitroacetylene [31]:

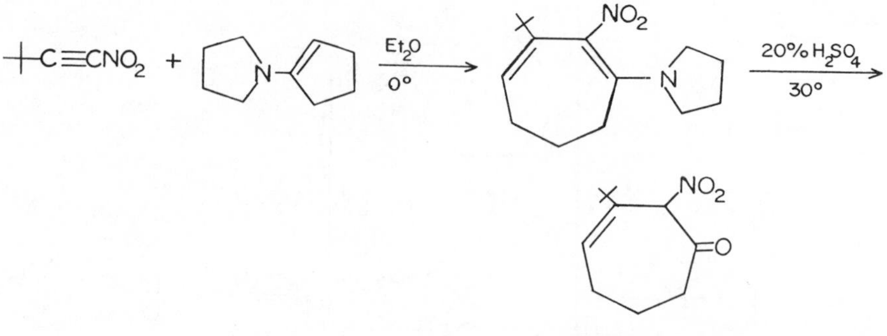

1-*t*-Butyl-2-nitro-7-
cyclohepten-3-one, 76 %

a. Preparation of 1-Nitrocyclooctene [28]

Dinitrogen tetroxide, 39.3 g was introduced into 150 ml of dry ether at -10° with slow stirring while the system was being swept with dry oxygen. The solution was then warmed to 0-5° and cyclooctene, 44.4 g was added over a 30-min period with vigorous stirring while the oxygen flow rate was continued and the temperature was held during stirring at 9-12° by cooling. After stirring for an additional 30 min at 10° with continued oxygen flow, 121 g of triethylamine was added during 12 min with stirring at 4-12°. The mixture was held at 25° for an additional 30 min and then diluted with 150 ml of ether. After neutralizing the excess of triethylamine with dilute acetic acid, the reaction mixture was extracted with ether, from which solution 59-61 g of crude 1-nitrocyclooctene was recovered. Purification gave 39-40 g (63-64%) of pure product.

References

1. H. Feuer, The Chemistry of the Nitro and Nitroso Groups, Pt. 1 and 2, Interscience, New York, 1969, 1970.
2. G. Olah, H. C. Lin, Synthesis, 488 (1973).
3. G. B. Bachman, T. F. Biermann, J. Org. Chem., 35, 4229 (1970).
4. F. Minisci, A. Quilico, Org. Prep. Proced., 1, 5 (1969).
5. C. Shin et al., Bull. Chem. Soc. Jap., 43, 3219 (1970).
6. U. Schöllkopf, P. Markusch, Ann. Chem., 753, 143 (1971).
7. J. A. Young et al., J. Org. Chem., 36, 350 (1971).
8. J. H. Ridd, Acc. Chem. Res., 4, 248 (1971).
9. E. Baciocchi, G. Illuminati, J. Am. Chem. Soc., 89, 4017 (1967).
10. A. Fischer, G. J. Wright, Australian J. Chem., 27, 217 (1974).

11. O. L. Wright et al., Proc. La. Acad. Sci., 31, 134 (1968);
     C. A., 70, 77,478 (1969); J. Org. Chem., 30, 1301 (1965).
12. C. V. Ristagno, H. J. Shine, J. Am. Chem. Soc., 93, 1811
     (1971).
13. I. K. Barvinskaya, A. A. Spryskov, Izv. Vyssh. Ucheb.
     Zaved. Khim. Khim. Tekhnol., 13, 802 (1970); C. A., 73,
     98,522 (1970).
14. L. J. Belf, D. T. Saggers, British Patent, 1,239,552,
     July 21, 1971; C. A., 75, 88,299 (1971).
15. K. Olsson, P. Martinson, Acta Chem. Scand., 26, 3549
     (1972).
16. I. Puskas, E. K. Fields, J. Org. Chem., 31, 4204 (1966).
17. C. L. Coon et al., J. Org. Chem., 38, 4243 (1973).
18. H. Zollinger et al., Helv. Chim. Acta, 54, 2043 (1971).
19. H. Suzuki, K. Nakamura, Synthesis, 606 (1972).
20. J. W. Bunting, W. G. Meatrel, Org. Prep. Proced. Intern.,
     4, 9 (1972).
21. I. Wempen et al., J. Heterocycl. Chem., 6, 593 (1969).
22. P. H. Doukas, T. J. Speaker, J. Pharm. Sci., 60, 184
     (1971); Synthesis, 144 (1972).
23. H. M. Gilow et al., J. Org. Chem., 36, 1745 (1971).
24. J. E. Gordon, J. Org. Chem., 35, 2722 (1970).
25. J. M. Larkin, K. L. Kreuz, J. Org. Chem., 36, 2574 (1971).
26. D. R. Lachowicz, K. L. Kreuz, United States Patent,
     3,282,983, November 1, 1966; C. A., 66, 10577 (1967);
     N. Levy et al., J. Chem. Soc., 52 (1948).
27. V. Jäger, H. G. Viehe, Angew. Chem., 81, 259 (1969).
28. W. K. Seifert, Org. Syn., 50, 84 (1970).
29. G. L. Rowley, M. B. Frankel, J. Org. Chem., 34, 1512 (1969).
30. V. I. Griggs, L. T. Eremenko, Izv. Akad. Nauk SSSR, Ser.
     Khim., 2566 (1969); Synthesis, 597 (1971).
31. V. Jäger, H. G. Viehe, Angew. Chem., 82, 836 (1970).

B. Metathesis

   1. From Halides and Nitro Compounds

   $$RCH_2Br(I) + AgNO_3 \longrightarrow RCH_2NO_2 + AgBr(I)$$

   $$RCHXR' + NaNO_3 \longrightarrow RCH(NO_2)R' + NaX$$

El'tsov and co-workers [1] have shown that p-chloro-, p-
bromo-, and p-iodoaniline, 0.01 M solutions in methanol containing NaNO$_2$ are converted into the corresponding nitro compounds on
being irradiated as shown:

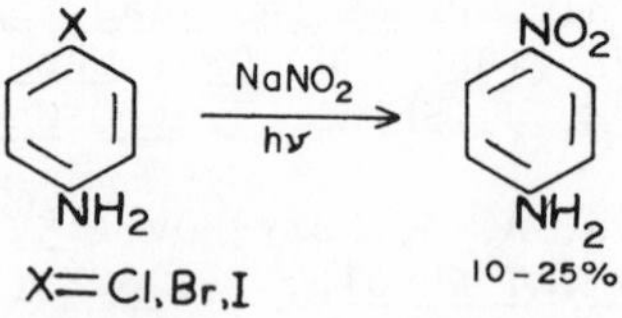

$$X = Cl, Br, I \qquad 10-25\%$$

Since a competitive photoreduction of nitrite ions to ammonia by alcohols occurs, the concentration of the latter should be minimized. It is interesting to note that this replacement occurs only when electron-donating substituents are present in the ring, an effect just the opposite to that occurring under nonirradiating conditions.

Kornblum and co-workers [2] have shown that p-nitrocumyl halides are quite active toward nucleophiles:

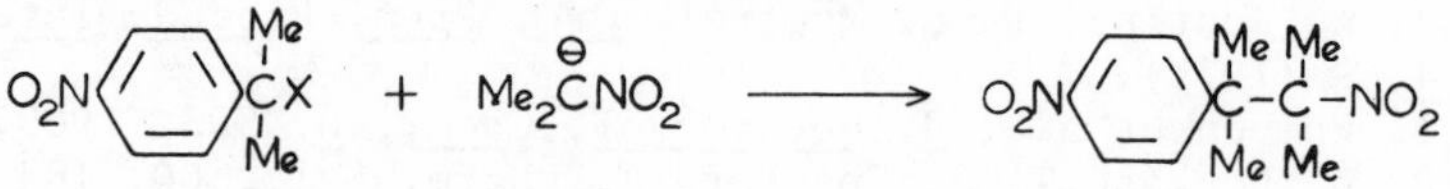

2-Nitro-3-p-nitrophenyl-2,3-dimethyl-butane, 53%

These investigators found the unusual activity to be caused by a radical-anion chain mechanism. Russell and co-workers [3] have encountered a similar situation:

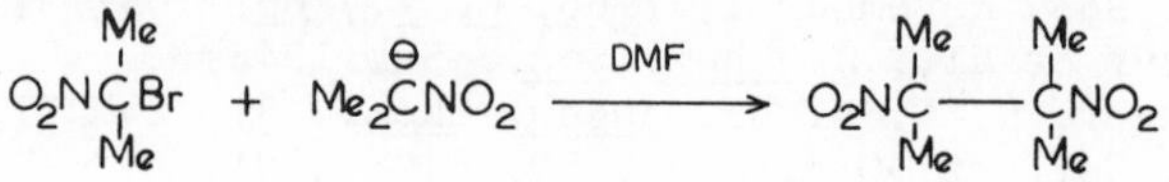

2,3-Dinitro-2,3-dimethyl-butane, ca. 100%

Kornblum also succeeded in replacing nitro groups by anions as indicated [4]:

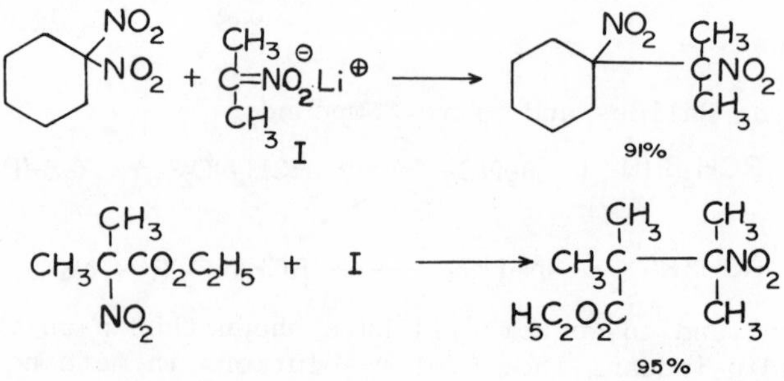

91%

95%

2. From Nitro Compounds and Alkyl- or Aryliodonium Halides or Onium Salts (1, 993)

Displacement occurs when the lithium salt of 2-nitropropane
is treated with anthracene quaternary ammonium salts [5].   Thus
9-anthrylmethyltrimethyl ammonium chloride responds as shown:

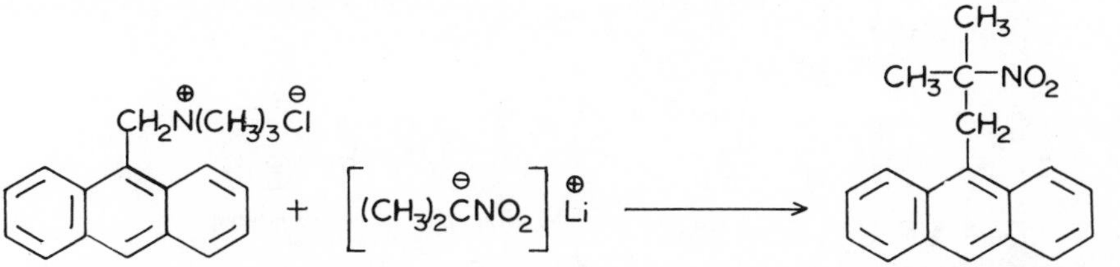

9-(2-Nitro-2-methylpropyl)-
anthracene, 90%

The isomer,

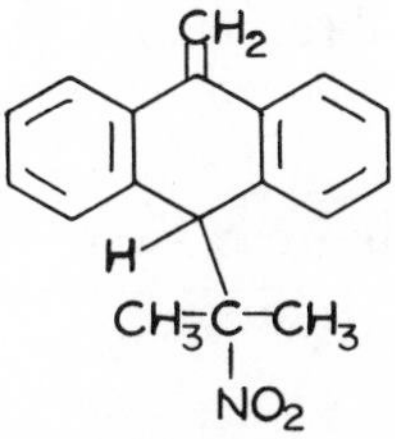

appears to be an intermediate in the reaction.   Presumably, the
reaction occurs well with other polycyclic and heterocyclic
types.

Fluorodenitration occurs when nitroaromatics or nitro-
heterocycles are treated with the fluoride ion in HMPA or N-
methyl-2-pyrrolidone [6].   Thus m-dinitrobenzene gives m-fluoro-
nitrobenzene:

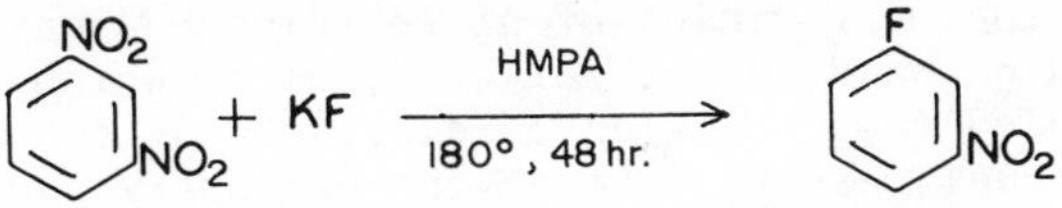

3.   From Carbanions and Nitrate or Nitrite Esters (1, 994)

Continued interest persists in this reaction.   Feuer and
Auerbach [7] have synthesized a new class of compounds, the α-
nitrosulfonates, in this way:

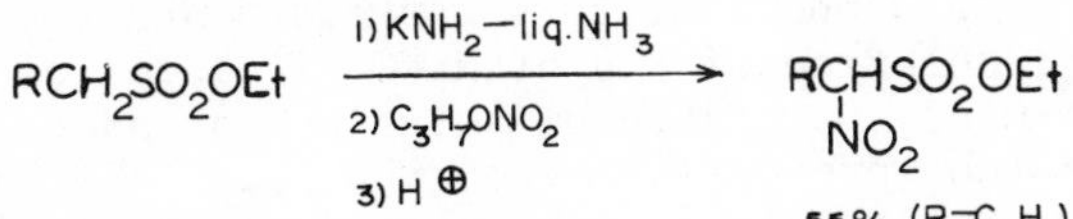

$$RCH_2SO_2OEt \xrightarrow[\substack{2) C_3H_7ONO_2 \\ 3) H^{\oplus}}]{1) KNH_2-liq.NH_3} RCHSO_2OEt$$

$$NO_2$$

55% (R=C_3H_7)

Previously, α-nitrocarboxylic esters were prepared by a similar method [8]. Although the reaction occurs normally when 4-(γ-phenylpropyl)pyridine is treated with sodamide, liquid ammonia, and propyl nitrate [9], it is abnormal for 4-isopropyl-pyridine [10]:

2,3-Di(4-pyridyl)-2,3-dimethyl-butane, 88%

Truce and co-workers have synthesized α-nitrosulfonamides in 25-66% yield [11] and α-nitrosulfones [12] in 8-81% yield by using the anions of the sulfur compounds and ethyl nitrate.

Ivanov salts have also been used to prepare nitroalkanes [13]:

$$RCHCO_2Li \atop Li \quad \xrightarrow[2) H_3O^\oplus]{1) C_3H_7ONO_2, -40°} \quad RCH_2NO_2$$

45-68 %
(3 examples)

4.   From Diazonium Compounds (1, 995)

$$ArN_2^{\oplus}X^{\ominus} \xrightarrow[catalyst]{NaNO_2} ArNO_2 + NaX + N_2$$

For a discussion of the replacement of the diazonium with the nitro group see the work by Feuer [14].

Perhaps the best yields of nitro compounds may be obtained by the method of Ward and co-workers [15] in which the diazonium sulfate is added to a solution of excess $NaNO_2$ containing excess $NaHCO_3$ (the usual base to achieve the necessary neutral or basic media is $CaCO_3$). By the new procedure, o- and p-dinitrobenzenes were synthesized in almost quantitative yield (Ex. a).

a.   Preparation of o-Dinitrobenzene [15]

A solution of 10 g of o-nitroaniline in a mixture of 12 ml of H_2SO_4 and 30 ml of water was added with stirring to 50 ml of ice-cold water. Crushed ice, 20 g was added and then rapidly with stirring a solution of 8 g of $NaNO_2$ in 13 ml of water. After stirring for 5 min, the diazonium solution was added in portions, through a wide tube projecting below the level of the decom-position mixture, to a vigorously stirred solution of 100 g of $NaNO_2$ and 45 g of $NaHCO_3$ in one liter of water, containing a

small amount of silicone "anti-foam" at 60°. Five minutes after
the addition, the solid dinitro compound was filtered, washed with
2 N HCl and then with water, and dried to give 11.8 g (97%).

References

1. A. V. El'tsov et al., J. Org. Chem. USSR, 6, 1955 (1970).
2. N. Kornblum et al., J. Am. Chem. Soc., 92, 5513 (1970).
3. G. A. Russell et al., J. Am. Chem. Soc., 93, 5839 (1971).
4. N. Kornblum et al., J. Am. Chem. Soc., 92, 5783, 5784
    (1970).
5. C. W. Jaeger, N. Kornblum, J. Am. Chem. Soc., 94, 2545
    (1972).
6. G. Bartoli et al., J. Chem. Soc., Perkin Trans., I, 2671
    (1972).
7. H. Feuer, M. Auerbach, J. Org. Chem., 35, 2551 (1970).
8. H. Feuer, R. P. Monter, J. Org. Chem., 34, 991 (1969).
9. H. Feuer, J. P. Lawrence, J. Org. Chem., 37, 3662 (1972).
10. H. Feuer et al., J. Org. Chem., 38, 417 (1973).
11. W. E. Truce, L. W. Christensen, Tetrahedron, 25, 181 (1969).
12. W. E. Truce et al., J. Org. Chem., 34, 3104 (1969).
13. P. E. Pfeffer, L. S. Silbert, Tetrahedron Letters, 699
    (1970).
14. H. Feuer, The Chemistry of the Nitro and Nitroso Groups,
    Pt. 2, Interscience, New York, p. 31.
15. E. R. Ward et al., J. Chem. Soc., 894 (1960).

C. Condensation Reactions

    1. From Aldehydes, Ketones, or Schiff Bases (1, 997)

    A review on the cyclization of dialdehydes with nitromethane,
which includes six preparations, is available [1]. The reaction
offers a general method of cyclization in which the methyl group
of the nitromethane is incorporated into the ring. The reaction
is applicable to 1,4-, 1,5-, and 1,6-dialdehydes and leads to
isomeric mixtures of five-, six-, and seven-membered cyclic
nitrodiols in which the thermodynamically more stable products
predominate.
    With glyoxal and nitromethane in aqueous sodium carbonate a
mixture of isomeric 1,4-dideoxy-1,4-dinitroinositols result.
Because of its insolubility in water, one isomer with the neo-1,4-
configuration, as indicated, may be separated from the fourteen
possible in the reaction mixture:

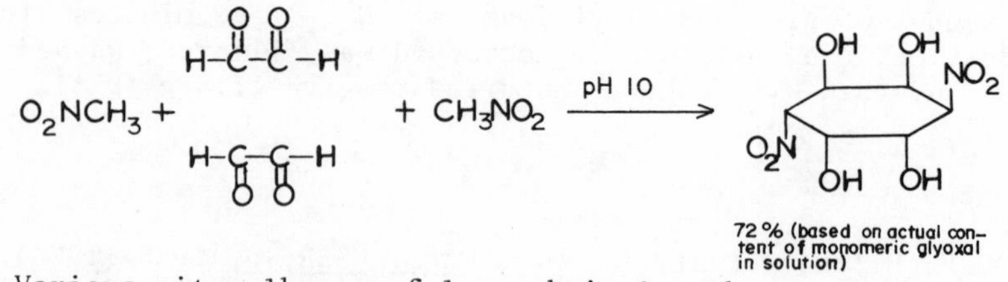

$$O_2NCH_3 + \quad \begin{array}{c} H-\overset{\overset{\displaystyle O}{\|}}{C}-\overset{\overset{\displaystyle O}{\|}}{C}-H \\[4pt] H-\overset{\underset{\displaystyle O}{\|}}{C}-\overset{\underset{\displaystyle O}{\|}}{C}-H \end{array} \quad + CH_3NO_2 \xrightarrow{pH\ 10}$$

72 % (based on actual con-
tent of monomeric glyoxal
in solution)

Various nitroalkanes of long-chain length may now be made from short-chain nitroalkanes. It is a one-pot reaction in which condensation of the nitroalkane with a carbonyl compound, esterification of the nitroalcohol, elimination of the ester, and reduction of the nitroalkene are all carried out consecutively [2]:

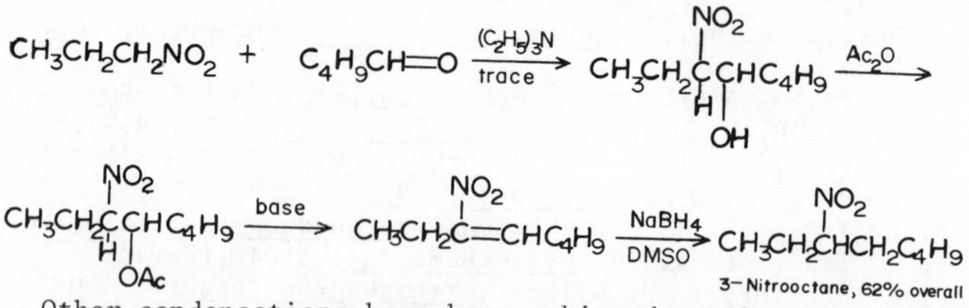

$$CH_3CH_2CH_2NO_2 + C_4H_9CH{=}O \xrightarrow[trace]{(C_2H_5)_3N} CH_3CH_2\overset{\overset{\displaystyle NO_2}{|}}{\underset{\underset{\displaystyle H}{|}}{C}}\overset{}{\underset{\underset{\displaystyle OH}{|}}{C}}HC_4H_9 \xrightarrow{Ac_2O}$$

$$CH_3CH\overset{\overset{\displaystyle NO_2}{|}}{\underset{\underset{\displaystyle OAc}{|}}{\underset{H}{C}}}CHC_4H_9 \xrightarrow{base} CH_3CH_2\overset{\overset{\displaystyle NO_2}{|}}{C}{=}CHC_4H_9 \xrightarrow[DMSO]{NaBH_4} CH_3CH_2\overset{\overset{\displaystyle NO_2}{|}}{C}HCH_2C_4H_9$$

3-Nitrooctane, 62% overall

Other condensations have been achieved, such as the conden-sation of aromatic aldehydes with 4,6-dinitro-1,3-xylene [3] and the Mannich reaction of α,ω-dinitroalkanes [4]. An interesting nitroheterocyclic compound has been produced from the condensa-tion of nitromethane with 2,3-thiophenedicarboxaldehyde [5]:

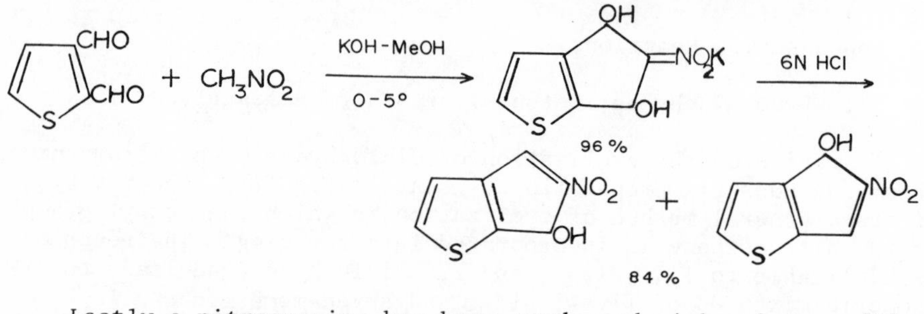

96 %

84 %

Lastly a nitroenamine has been condensed with a ketone [6]:

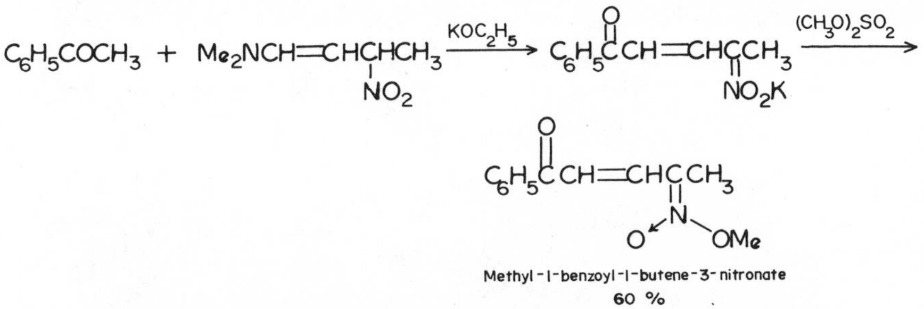

$$C_6H_5COCH_3 + Me_2NCH\!\!=\!\!CHCHCH_3 \xrightarrow{KOC_2H_5} \underset{NO_2}{C_6H_5\overset{O}{\overset{\|}{C}}CH\!\!=\!\!CHCCH_3} \xrightarrow{(CH_3O)_2SO_2}$$

$$\underset{NO_2K}{}$$

$$C_6H_5\overset{O}{\overset{\|}{C}}CH\!\!=\!\!CHCCH_3$$

$$O\overset{N}{\diagup}OMe$$

Methyl-1-benzoyl-1-butene-3-nitronate

60 %

2. From Unsaturated Carbonyl Compounds and the Like (Michael) (1, 999)

The preparation of 2,4,6-triphenylnitrobenzene from the pyrylium tetrafluoroborate (see 1, 1002 for a similar reaction) has appeared in Organic Synthesis [7] as indicated:

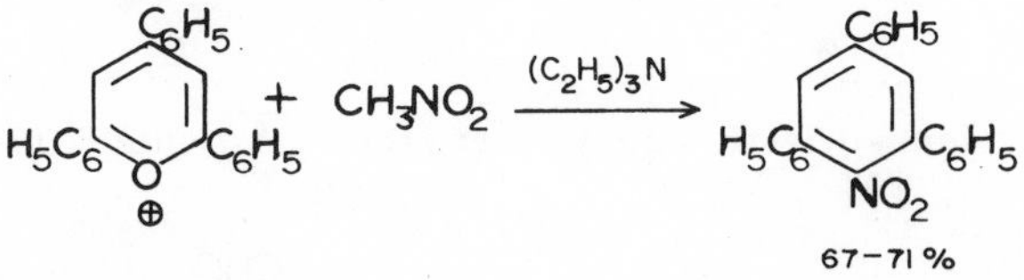

67-71 %

The method possesses an advantage over the preparation by the nitration of 1,3,5-triphenylbenzene since only the one isomer is formed.

Interest in the Meisenheimer complexes (complexes between nitroarenes and a base or nucleophile) has led to some exotic syntheses. 1,3,5-Trinitrobenzene, acetone, and diethylamine, for example, yield p-nitrodiethylaniline probably as follows [8]:

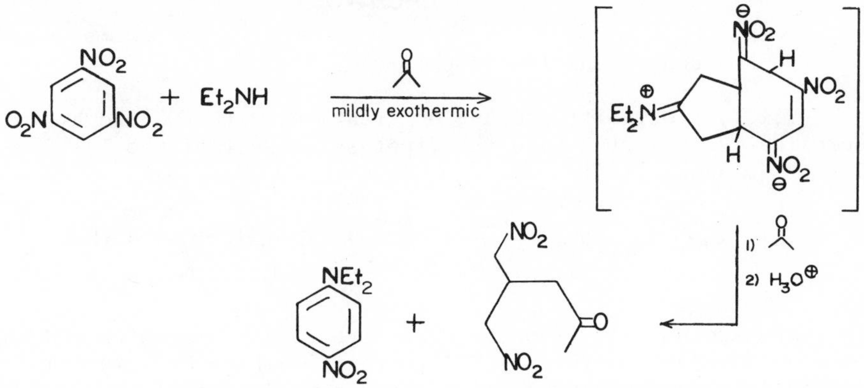

However, other polynitroarenes as 3,5-dinitroacetophenone behave differently:

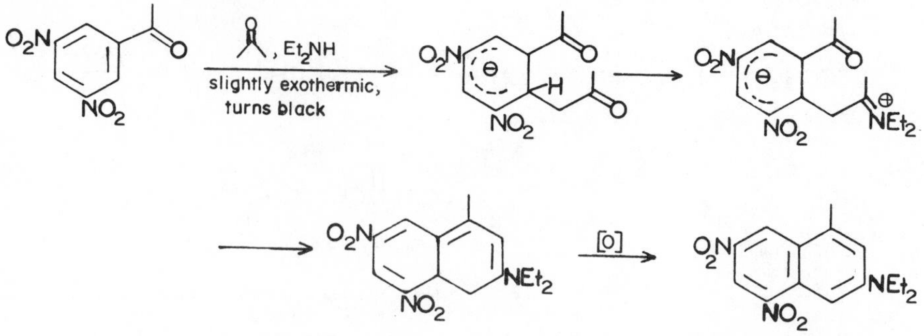

1-Methyl-3-diethylamino-
5,7-dinitronaphthalene, 32%

Another type of Michael addition becomes clear with realization that the hydrolysis of β-nitrovinyldimethylamine, A, leads to β-nitroacetaldehyde:

$$Me_2NCH{=}CHNO_2 \longrightarrow O{=}CHCH_2NO_2$$

A   B

When A is condensed with phenylacetaldehyde, the following reactions occur [9]:

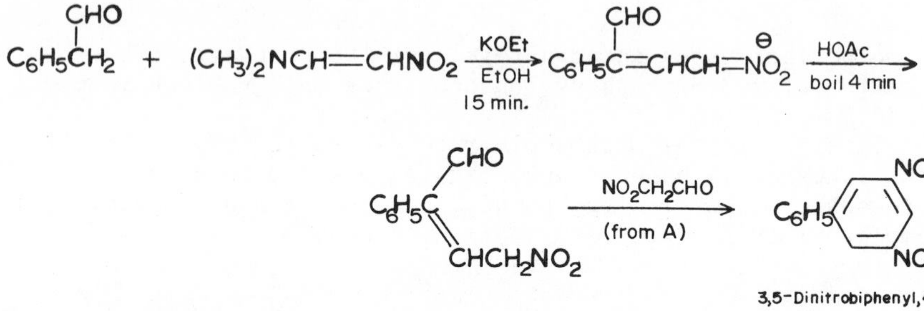

3,5-Dinitrobiphenyl, 40%

3. Condensations (Nucleophilic)

1,1,1,3-Tetranitropropane on treatment with KOH and hydrogen peroxide is converted into the dipotassium salt of the 1,1,3,3-isomer as shown [10]:

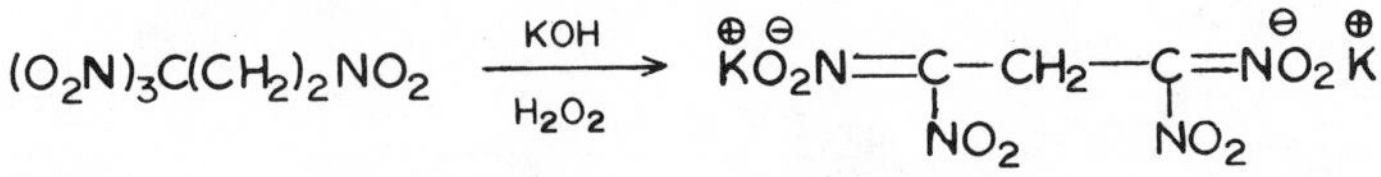

39 %

The dipotassium salt is quite sensitive to a hammer blow and may be transformed into the 1,1,3,3-tetranitroalkane by acid. In all probability the mechanism may be represented as follows:

$$(NO_2)_3C(CH_2)_2NO_2 \xrightarrow{\overset{\ominus}{OH}} (NO_2)_3CCH_2\overset{\ominus}{C}HNO_2 \longrightarrow (NO_2)_2\overset{\ominus}{C}CH_2CH(NO_2)_2 \xrightarrow{H^{\oplus}}$$

$$(NO_2)_2CHCH_2CH(NO_2)_2$$

A review is now available on the synthesis of 1,1,1-trinitroalkanes [11]. The general route is as follows:

$$RCHClNO_2 \xrightarrow{MNO_2} RCH(NO_2)_2 \xrightarrow[base]{C(NO_2)_4} RC(NO_2)_3$$

Yet many others are made via the addition of nitroform to α,β-unsaturated carbonyl compounds or other unsaturated species.

N,N-Dimethylacetamide diethyl ketal condenses with nitromethane to produce an enamine that may be hydrolyzed to α-nitroacetone [12]:

$$Me_2N\overset{\overset{CH_3}{|}}{C}(OEt)_2 + CH_3NO_2 \xrightarrow{10°} Me_2N\overset{\overset{CH_3}{|}}{C}=CHNO_2 \xrightarrow[2)H_3O^{\oplus}]{1)aq.KOH} CH_3COCH_2NO_2$$
$$81\%$$

References

1. F. W. Lichtenthaler, in W. Foerst, Newer Methods of Preparative Organic Chemistry, Vol. IV, Academic, New York, 1968, p. 155.
2. G. B. Bachman, R. J. Maleski, J. Org. Chem., 37, 2810 (1972).
3. R. E. Harmon et al., J. Pharm. Sci., 59, 1356 (1970); Synthesis, 271 (1971).
4. M. Mühlstädt, B. Schulze, J. Prakt. Chem., 313, 205 (1971).
5. J. Skramstad, Acta Chem. Scand., 25, 1287 (1971).
6. T. Severin, H. Kullmer, Chem. Ber., 104, 440 (1971).
7. K. Dimroth et al., Org. Syn., 49, 114 (1969).
8. S. R. Alpha, J. Org. Chem., 38, 3136 (1973).
9. T. Severin et al., Chem. Ber., 104, 2856 (1971).
10. M. J. Kamlet et al., J. Org. Chem., 26, 4881 (1961).
11. L. A. Kaplan, in H. Feuer, The Chemistry of the Nitro and Nitroso Groups, Pt. 2, Interscience, New York, 1970, p. 289.
12. K. K. Babievkii et al., Izv. Akad. Nauk SSSR, Ser. Khim., 1161 (1970); Synthesis, 557 (1971).

D. Oxidation

1. From Amines (1, 1003)

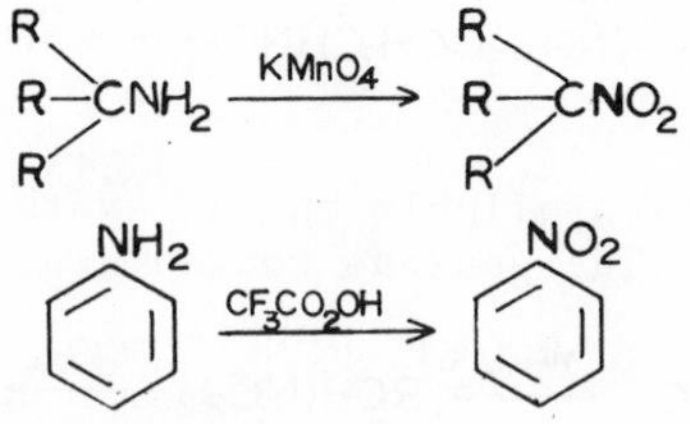

The oxidation of amino compounds to nitro compounds has been discussed by other authors [1].

Pertrifluoroacetic acid is recommended as the oxidizing agent for converting negatively substituted aromatic amines into nitro compounds [2]. With this reagent 2,6-dichloroaniline has been transformed into 1-nitro-2,6-dichlorobenzene as shown (see Ex. a):

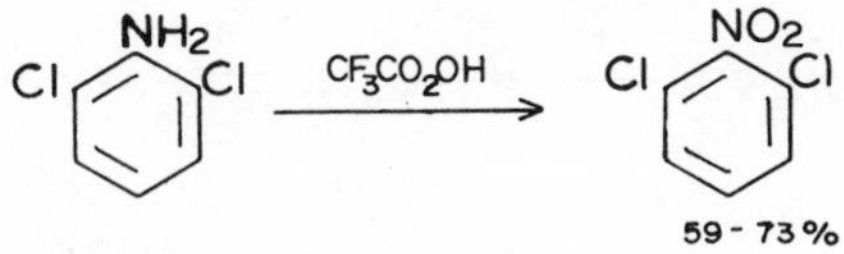

Although the reagent is not as satisfactory for amines such as p-anisidine and β-naphthylamine in which the aromatic nuclei are sensitive to electrophilic attack, it has been employed rather widely with amines because of the direct nature of the transformation. The reaction probably proceeds via the nitroso compound. Boulton [3] has employed another peracid in order to stop at the nitroso stage from which further oxidation as indicated led to the nitro compound:

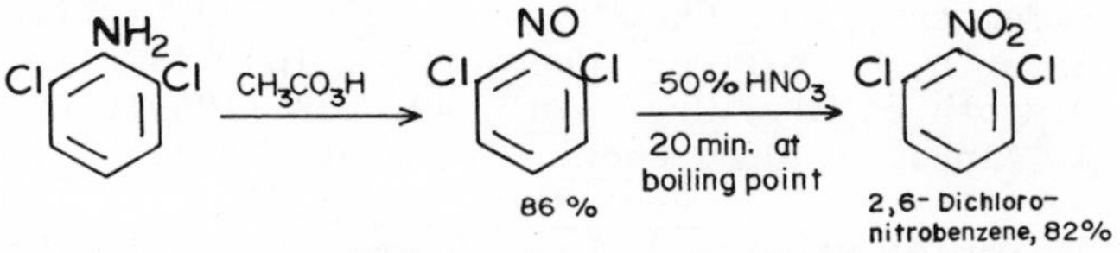

Pozharskii and co-workers [4] discovered an unexpected reaction on the treatment of 1-benzyl-2-aminobenzimidazole with 4 mol of sodium in liquid ammonia. The products were 2-nitrobenzimidazole and 2,2'-azobenzimidazole as indicated:

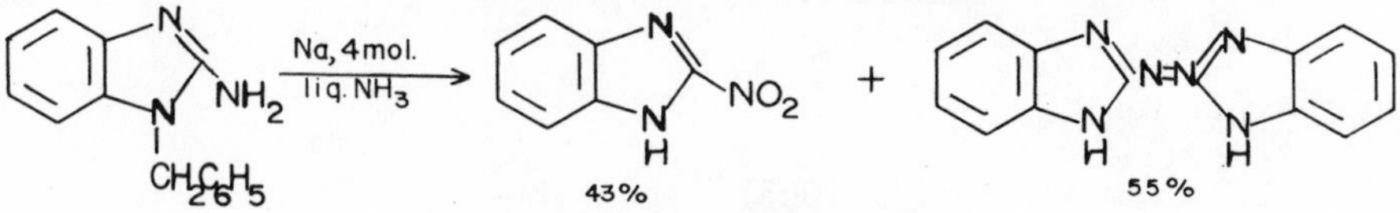

Thus debenzylation results with oxidation of the amino group to the azo or nitro group. It has been suggested that the oxidation occurs from the effect of air on the sodium salts formed from the 2-aminobenzimidazole generated in the reaction.

In the steroid series amino groups have been oxidized to nitro groups by the use of m-chloroperbenzoic acid [5]. Thus 3α-acetoxy-20α-amino-5β-pregnane gave the nitro compound, as indicated:

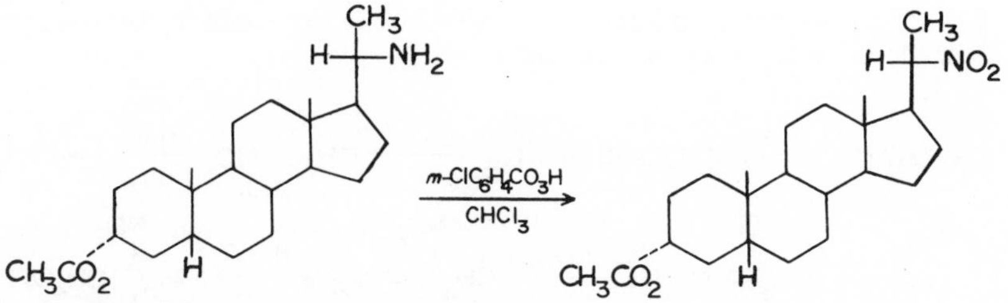

It is interesting to note that retention of configuration at C20 occurs. A large mol ratio of oxidant is necessary to suppress the formation of nitroso dimers. Robinson and co-workers are of the opinion that this direct method is more convenient than that of Bull, which is a stepwise conversion of the oxime to the nitro compound (see D.3).

a. Preparation of 2,6-Dichloronitrobenzene [2]

Hydrogen peroxide, 90%, 5.4 ml in 100 ml of methylene chloride was cooled in ice and the immiscible mixture was stirred. Trifluoroacetic anhydride, 34.0 ml was then added over a 20-min period, after which the ice was removed and the solution was stirred at room temperature for 30 min.

2,6-Dichloroaniline, 8.1 g in 80 ml of methylene chloride was then added over a 30-min period to the prepared peroxytri-fluoroacetic acid reagent while refluxing occurred. The mixture was refluxed for 1 hr, then cooled and poured into 150 ml of cold water. From the organic layer 5,7-7.0 g (59-73%) of the nitro compound was recovered in the usual manner.

2. From Nitro Compounds and Silver or Sodium Nitrite
    (Kaplan-Schechter-ter Meer) ($\underline{1}$, 1004)

A comparison of the Kaplan-Shechter and ter Meer methods has been made in the synthesis of 2,2-dinitropropanol [6]. The procedure involving the Kaplan-Shechter method as modified is as follows:

$$CH_3CH_2NO_2 \xrightarrow{NaOH} CH_3CH=NO_2Na \xrightarrow{AgNO_2} CH_3CH(NO_2)_2 \xrightarrow{NaOH}$$

$$CH_3\overset{NO_2}{\underset{}{C}}=NO_2Na \overset{\ominus}{}\overset{\oplus}{} \xrightarrow[H\oplus]{HCHO} CH_3\overset{NO_2}{\underset{NO_2}{C}}CH_2OH$$

Yields based on nitroethane were 80%. The procedure involving the ter Meer method is as follows:

$$CH_3CH_2NO_2 \xrightarrow{NaOH} CH_3CH=NO_2Na \xrightarrow{Cl_2} CH_3\overset{Cl}{\underset{}{CH}}NO_2 \xrightarrow{NaOH}$$

$$CH_3\overset{Cl}{\underset{}{C}}=NO_2Na \xrightarrow{NaNO_2} CH_3\overset{NO_2}{\underset{}{C}}=NO_2Na \xrightarrow[H\oplus]{HCHO} CH_3\overset{NO_2}{\underset{NO_2}{C}}-CH_2OH$$

Yields based on nitroethane were 65%. Thus the Kaplan-Shechter method gives higher yields although the costs involved in the ter Meer are approximately half as much. On this basis Hamel and co-workers recommend the Kaplan-Shechter procedure for the preparation of small or intermediate quantities and the ter Meer for large-scale production.

3. From Oximes (1, 1006)

$$ArCH=NOH \xrightarrow{CF_3CO_2OH} ArCH_2NO_2$$

Since Bull and co-workers [7] were unsuccessful in oxidizing the oximes of steroids to nitro steroids with peracids, they developed a scheme of oxidation via the pseudo nitrole as indicated:

$$>C=NOH \xrightarrow[\text{fuming}]{HNO_3} >\overset{NO}{\underset{NO_2}{C}} \xrightarrow{H_2O_2-HNO_3} >\overset{NO_2}{\underset{NO_2}{C}} \xrightarrow{H_2-Pt} >\overset{H}{\underset{NO_2}{C}}$$

(pseudo nitrole)

In this manner 3-, 4-, 7-, and 17-nitro steroids were prepared, the overall yield in the latter case being 45%.

Dinitroether oximes have been oxidized to polynitroethers by treatment first with 90% HNO_3 then by adding 30% H_2O_2 to the reaction mixture as shown [8]:

$$\underset{\underset{NO_2}{|}}{\overset{\overset{NO_2}{|}}{F-C}}CH_2OCH_2CH{=}NOH \xrightarrow[\text{2)30\% H}_2O_2]{\text{1)90\% HNO}_3 \ \ 3\text{-}5°} \underset{\underset{NO_2}{|}}{\overset{\overset{NO_2}{|}}{F-C}}CH_2OCH_2CH(NO_2)_2$$

2-Fluoro-2,2-dinitroethyl   2,2-dinitroethyl
ether, 65 %

4.   From Nitro Compounds (Coupling)

$$2 \ \underset{\underset{NO_2}{|}}{\overset{\overset{R^1}{|}}{R^2-C}}{-}H \xrightarrow[\text{AgNO}_3]{\text{KOH}} \underset{\underset{NO_2}{|}}{\overset{\overset{R^1}{|}}{R^2-C}}{-}\underset{\underset{NO_2}{|}}{\overset{\overset{R^1}{|}}{C}}{-}R^2$$

vic-Dinitro compounds may be prepared by the decomposition of the silver salts of secondary nitro compounds at 30° [9]. The unstable silver salts are prepared in situ by adding an aqueous alkaline solution of the nitro compound to silver nitrate, preferably in DMSO. Yields of dinitro compounds vary within 33-84%.

In a similar manner the salts of secondary nitro compounds are oxidized by persulfates in the pH range of 9.5-7.0 at 0-5° to vicinal tertiary dinitro compounds [9] as indicated:

$$2 \ R_2C{=}\overset{\ominus}{N}O_2 \xrightarrow[\text{or} \\ \text{(NH}_4)_2S_2O_8]{\text{Na}_2S_2O_8} \underset{\underset{R}{|}}{\overset{\overset{NO_2}{|}}{R\ C}}{-}\underset{\underset{R}{|}}{\overset{\overset{NO_2}{|}}{C\ R}}$$

14-62%

The principal by-product is the ketone, $R_2C{=}0$.

Coupling also occurs in the nitration of alkylbenzenes if the nitric acid is added to the hydrocarbon [10]. Thus o-xylene gives largely the nitrative coupling product as shown:

2-Nitro-3,4,4,5-tetramethylbiphenyl
20.5 %

The other alkylated benzenes gave a less coupled product.

Coupling has been accomplished by Björklund and Nilsson [11] by heating m-dinitrobenzene with iodobenzene derivatives:

2,6‐Dinitrobiphenyl

Maximum yield was 81% with 1-iodo-4-methoxybenzene. The minimum
temperature to give coupling should be employed to prevent side
reactions. In the coupling perhaps 2,6-dinitrophenylcopper is
formed via copper free radical addition followed by the removal
of a hydrogen atom or via copper anion addition to give
followed by the removal of $H\bar{X}_2$ or a similar species.
The arylcopper then reacts as shown:

$$ArCu + Ar'I \longrightarrow Ar\ Ar' + CuI$$

Coupling of other nitroaryl halides has been accomplished
[12]. Thus Björklund produced 2,2'-dinitrobiphenyls in yields as
high as 52% from 2-nitroaryl halides and CuI in the presence of
pyridine.
    Lastly, the nitronate salts, $Me_2\bar{C}NO_2$, have been coupled
electrolytically to give [13]:

5. From Alkenes

Tetrafluoroethylene has been oxidized to nitrodifluoroacetic
acid in 90-92% yield by treatment with one part of nitric acid
and 10 parts of sulfuric acid [14]. The method was less satisfactory when applied to tetrachloroethylene. To form the nitro-
epoxide Newman and Angier [15] oxidized the nitroethylene with
dilute hydrogen peroxide in a base at 0°.

References

1. H. Feuer, The Chemistry of Nitro and Nitroso Groups, Pt. 2,
    Interscience, New York, 1970, p. 29; M. Hedayatullah, Bull.
    Soc. Chim. France, 2957 (1972).
2. A. S. Pagano, W. D. Emmons, Org. Syn., 49, 47 (1969).
3. A. J. Boulton et al., J. Chem. Soc., B, 1004 (1966).
4. A. F. Pozharskii et al., Tetrahedron Letters, 2219 (1967);
    J. Org. Chem. USSR, 2, 1868 (1966).
5. C. H. Robinson et al., J. Org. Chem., 31, 524 (1966).
6. E. E. Hamel et al., Ind. Eng. Chem. Prod. Res. Dev., 1, 108
    (1962).

7. J. R. Bull· et al., *J*. *Chem*. *Soc*., 2601 (1965).
8. V. Grakauskas, *J*. *Org*. *Chem*., 38, 2999 (1973).
9. A. H. Pagano, H. Shechter, *J*. *Org*. *Chem*., 35, 295 (1970).
10. I. Puskas, E. K. Fields, *J*. *Org*. *Chem*., 31, 4204 (1966).
11. C. Björklund, M. Nilsson, *Acta Chem*. *Scand*., 22, 2338 (1968).
12. C. Björklund, *Acta Chem*. *Scand*., 25, 2825 (1971).
13. S. Wawzonek, *Synthesis*, 285 (1971).
14. I. V. Martynov et al., *J*. *Org*. *Chem*. *USSR*, 5, 420 (1969).
15. H. Newman, R. B. Angier, *Tetrahedron*, 26, 825 (1970).

E. Miscellaneous Reactions

2. From Polynitro Aromatics and Diazomethane (1, 1008)

It has been shown that 1,3,5-trinitrobenzene with diazomethane gives a tricyclic trinitrocompound of the formula

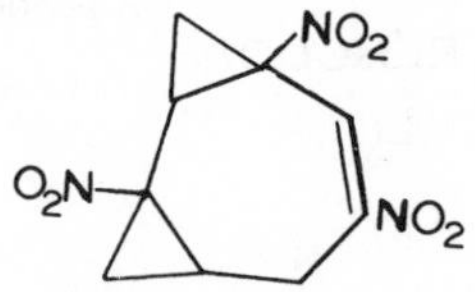

At below 80° an isomer of the above is obtained as shown [1]:

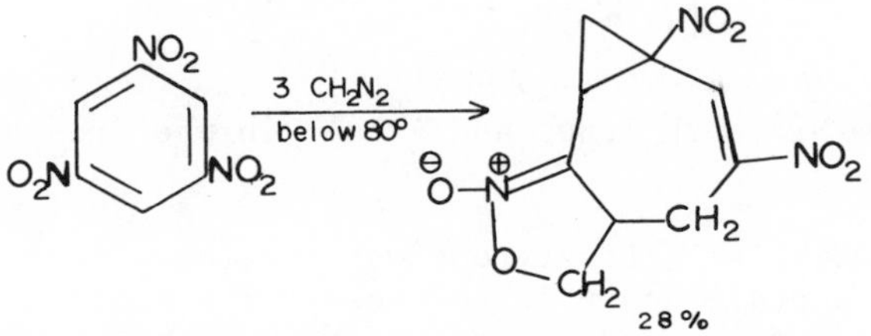

3. From Nitrocarboxylic Acids or Their Derivatives
    (Decarboxylation) (1, 1008)

$$NO_2CHCOOH \longrightarrow NO_2CH_2R + CO_2$$
$$\quad\ \ \underset{R}{|}$$

Bachmann and Biermann [2] converted acid anhydrides into the
acyl nitrate which was mixed with 90% HNO_3 and dropped into a
tube containing glass helices heated to 290°. Decarboxylation
occurs as shown:

$$(RCO)_2O \xrightarrow[\substack{HNO_3 \\ \text{or} \\ N_2O_5}]{N_2O_4,} R\overset{\overset{O}{\parallel}}{C}ONO_2 \xrightarrow{290°} RNO_2 + CO_2$$

The method is most satisfactory in the aliphatic series where conversions as high as 62%, based on the nitrating agent, were obtained. Broadly applicable, the method is the best for the synthesis of many nitroalkanes. It probably proceeds via free radicals as shown:

$$RCO_2{}^{\cdot} + \cdot NO_2 \longrightarrow [R \cdot] \xrightarrow{[NO_2{}^{\cdot}]} RNO_2$$

An indirect procedure for the decarboxylation of carboxylic acids involves the α-anion of the acid [3]. The steps in the process are:

$$RCH_2COOH \xrightarrow[\substack{HMPA - \\ THF}]{LiN(i\text{-}Pr)_2} \underset{\underset{Li}{|}}{RCHCOOLi} \xrightarrow[\substack{LiN(i\text{-}Pr)_2 \\ -40°}]{n\text{-}PrONO_2}$$

$$\left[\underset{\underset{O^{\ominus}}{\overset{}{\underset{}{}}}\; \underset{\overset{\oplus}{N}}{\overset{}{}}\; {}^{\oplus}OLi}{RCCO_2Li} \rightleftharpoons \underset{\overset{|}{NO_2}}{\overset{\overset{Li}{|}}{R}\overset{}{C}\,CO_2Li} \xrightarrow{H^{\oplus}} RCH_2NO_2 + CO_2 \atop 45-68\% \right]$$

4. From Nitroethylenes and Arenes in the Presence of Metal Salts

When β-methyl-β-nitrostyrene was treated with 3 molar equiv of palladium acetate and a large excess of benzene and acetic acid a mixture of cis and trans β-diphenylmethyl-β-nitrostyrenes [4] formed:

$$C_6H_5CH{=}\overset{\overset{NO_2}{|}}{C}Me \xrightarrow[Pd(OAc)_2]{C_6H_6} (C_6H_5)_2CHC{=}\overset{\overset{NO_2}{|}}{C}HC_6H_5$$
$$\textit{cis} \text{ and } \textit{trans, ca. } 60\%$$

Although this reaction resembles a free radical one, the result is similar to an electrophilic substitution.

References

1. Th. J. De Boer et al., Rec. Trav. Chim., 90, 842 (1971).

2. G. B. Bachman, T. F. Biermann, J. Org. Chem., 35, 4229 (1970).
3. P. E. Pfeffer, L. S. Silbert, Tetrahedron Letters, 699 (1970).
4. K. Yamamura et al., Tetrahedron Letters, 2829 (1972).

APPENDIX

A simple purification procedure, which may be helpful to the reader, was recently brought to our attention. It was discovered at the Organic Chemistry Department, Twente University of Technology, Enschede, the Netherlands. We thank Professor Reinhoudt for permission to issue this report as illustrated prior to publication by Reinhoudt and Geevers elsewhere (Note 1).

Purification of Dibenzofuran by Fractional Elution

Dibenzofuran (Aldrich, 29 g) was packed loosely (Note 2) into a glass column, 120 cm x 1 cm, i.e., surrounded by a water-cooling jacket (Note 3); for this experiment, the circulating water was held at 5-10° (Note 4). From a constant-addition Hirshberg funnel (Kontes) ethyl alcohol, 95% (Note 5), was added dropwise to the top of the column at a rate slow enough so that no liquid collected at the top (Note 6). Two hours after the first drop of almost black liquid dripped from the bottom, the black impurity had been eluted completely from the column leaving behind a light tan-colored solid (Note 7). This was recrystallized from aqueous ethanol to give 22 g of off-white crystals, mp 79-82°. This level of purity could not be obtained by one recrystallization alone (Note 8). Other separations accomplished by this procedure were triphenyl phosphine from triphenyl phosphine oxide using methylcyclohexane at 25° and impurities from fluorene using alcohol at 10°.

Possible rationalization. The elution may be a series of solutions and recrystallizations of all components, each action of which enriches the solvent in the more soluble component and impoverishes the solvent in the less soluble component.

Merit of procedure. The process is too new to define its
limits. It has the potential of becoming a fifth parameter of
purification, supplementing the processes of recrystallization,
chromatography, distribution between solvents, and sublimation.
If the scope is extensive, the Reinhoudt-Geevers procedure,
being the simplest and most economical of all five parameters,
should appeal to the industrial segment devoted to organic
synthesis. For a variation in the method see G. J. Arkenbout,
Chem. Tech., 6, 596 (1976).

Notes

1. The principle of the method is outlined in the paper, W. P.
    Trompen, J. Geevers, Rec. Trav. Chim., 95, 106 (1976).
2. If packed too tightly, the solvent may not flow through the
    column.
3. The column was jacketed with two commercial, screw-top outer
    jackets.
4. When the water jacket was held at 25°, a somewhat lower yield
    of product was obtained.
5. The best eluting solvent to choose is one in which the product
    is only slightly soluble at the temperature chosen for the
    water jacket or one from which the product recrystallizes.
6. Liquid collecting at the top stops or slows the flow of the
    liquid.
7. In this example, dibenzofuran began to crystallize from the
    eluting drops as soon as, but not before, the black impurity
    had been eluted.
8. Acknowledgment is made to Karem el Dahan, Cairo University,
    and Villa Mitchell, Vanderbilt University, for performing the
    experiments described here.

SUBJECT INDEX

The Table of Contents at the beginning of each chapter serves as a general index whereas this section is more specific.

REACTION INDEX (VOLUMES 1 AND 2)

Paul F. Hudrlik

Introduction

The purpose of this index is to enable the chemist to find references to ways of converting one organic structure to another. The index is arranged according to the changes taking place at a selected carbon atom that is intimately involved in the transformation from starting material to product. It is based on the structures of the starting materials and products only, without regard to mechanisms or reaction conditions. It is designed to group related reactions together. Thus reactions of acyl anion equivalents are together in one section and most Michael addition reactions are found in another.

How to Use the Index

1. Select a carbon atom (appearing in both starting material and product) that is intimately involved in the reaction or close to the reaction site. For reactions involving the breaking, formation, or rearrangement of carbon-carbon bonds, select a carbon atom directly involved in the reaction.
2. Note whether bonds between the selected carbon atom and other carbon atoms are unchanged, broken, formed, or rearranged. For carbon-carbon bond-forming reactions, determine the length (n) of the newly introduced carbon chain (see Definitions). Locate the appropriate box in the Contents on pp. **1006 and 7.**
3. Note the functionality of the selected carbon atom in both starting material and product (see Definitions). Refer to

said box to find the page number to the appropriate index
section.
       4.   Consult the appropriate section of the index. Within
each section the reactions are arranged according to the following criteria, considered in turn:
       a.   For carbon-carbon bond-forming reactions the
            functionality of C-n in the newly introduced chain,
       b.   The highest atomic number of the elements bonded to
            the selected carbon in the starting material,
       c.   The highest atomic number of the elements bonded to
            the selected carbon in the product.
For easy reference, these atomic numbers in starting material and
product are noted in the left-hand margin.

Definitions

       The functionality of a carbon atom is defined as the sum of
the number of bonds to elements more electronegative than carbon
plus the number of pi bonds to carbon [J. B. Hendrickson, J. Am.
Chem. Soc., 93, 6847 (1971)].   Thus, for example, ketones, enol
ethers, vinyl halides, and acetylenes all have functionality = 2
(at the more functionalized carbon atom).
       Oxidation is defined as an increase in functionality at a
particular carbon.
       Reduction is defined as a decrease in functionality.
       In carbon-carbon bond-forming reactions, if the newly
introduced carbon chain is functionalized on the nth carbon atom,
then n is taken to be the length of the carbon chain. Thus the
following reaction is considered as the addition of a two-carbon
chain.

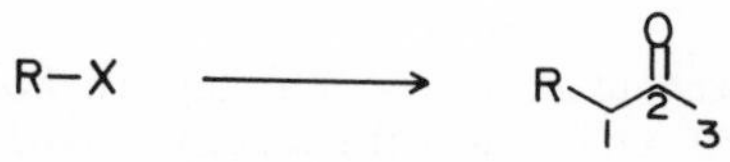

$$R-X \quad \longrightarrow \quad R\underset{1}{\overset{O}{\underset{\quad}{\bigvee}}}\underset{3}{\overset{2}{}}$$

Organization of Index

       The index is separated into the major divisions listed
below:
       I.   Reactions in which the carbon skeleton remains
            unchanged
            A.   Reduction of substrate
            B.   No change in oxidation state of substrate
            C.   Oxidation of substrate
       II. Carbon-carbon bond-breaking reactions

III.   Carbon-carbon bond-forming reactions
    A.   Reduction of substrate
    B.   No change in oxidation state of substrate
    C.   Oxidation of substrate
IV.   Rearrangements of the carbon skeleton

Within the preceding major divisions, the reactions are further subdivided according to each of the following criteria, considered in turn:

1. For carbon-carbon bond-forming reactions only, the length (n) of functional carbon chain added (i.e., the number of bonds from the point of attachment to a functionalized carbon atom in the newly added chain; see examples that follow),

2. The functionality of a selected carbon atom in the starting material,

3. The functionality of the same carbon atom in the product,

4. For carbon-carbon bond-forming reactions only, the functionality of C-n in the newly introduced chain,

5. The highest atomic number of the elements bonded to the selected carbon atom in the starting material,

6. The highest atomic number of the elements bonded to the same carbon atom in the product.

Within these sections, reactions are arranged where possible according to increasing priority of the selected carbon atom in the starting material (or product), according to a modified Cahn-Ingold-Prelog sequence rule [see also P. F. Hudrlik, J. Chem. Doc., 13, 203-206 (1973)].

Arbitrary Conventions and Miscellaneous

Halogens (X) are not normally distinguished from each other and are all considered to have atomic number 17 (except F in some cases).

Alkyl sulfonates (e.g., ROTs) are frequently listed as halides (e.g., RX) when their reactivity is similar.

Reactions involving Se are indexed with those of the corresponding S compounds.

To keep similar reactions together, the atomic numbers of metals (including B, Si, Sn, etc.) in organometallics are disregarded; thus reactions of RMgX are found with those of RLi.

Phenyl groups are normally listed as one-carbon chains except in a few cases where a parafunctional group is present, when they are listed as four-carbon chains). Allyl groups are arbitrarily listed only as three-carbon chains.

The structural formulas in the index are generally not meant to be specific compounds, but compound types. The letter Z refers to an electron-withdrawing group such as CN, CO_2R, NO_2, or SO_2Ar.

Multistep transformations are sometimes listed in several places. On the other hand, reactions are not indexed in every conceivable place. An attempt has been made to list them in places where a synthetic chemist would be most likely to look.

Certain reactions are difficult to place in this index and are generally not included, that is, cycloadditions, electrocyclic reactions, reactions involving both carbon-carbon bond-forming and breaking, reactions involving skeletal change at more than one site in a molecule, complex rearrangements, and some cyclizations and ring expansions and contractions.

Examples

The use of the index is best illustrated by a few examples.

1. Nef Reaction

$$\text{>}\!\!\!-\!NO_2 \longrightarrow \text{>}\!\!\!=\!O$$

In this reaction the carbon skeleton remains unchanged. The carbon atom (*) bearing the nitro group in the starting material undergoes an "oxidation" from functionality = 1 to functionality = 2 (referring to the Contents we see that this section begins on p. 1046 of the Index). The highest atomic number bonded to carbon changes from 7 to 8. The reaction is found on p. 1048 of the Index.

2. Ritter Reaction of a Nitrile with an Olefin

This reaction can be indexed in two places, since either the nitrile or the olefin can be selected as the starting material. In either case the carbon skeleton is unchanged, and no oxidation or reduction takes place.

(a) $R-\overset{*}{C}N \longrightarrow R-\overset{O}{\underset{\|}{C}}-NH-\overset{*}{R}'$

*carbon functionality 3→3 (Index pp. 1038-1042
atomic number 7→8
indexed on p. 1038

(b)

*carbon functionality 1→1 (Index pp. 1016-1027
atomic number 6→7
indexed on p. 1017

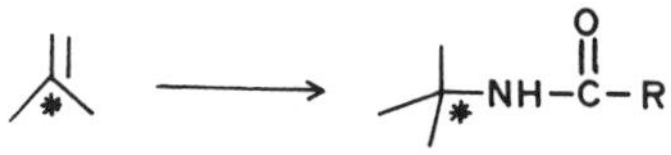

3. Michael Addition of Acetoacetic Ester to MVK

This reaction, as well as most C-C bond-forming reactions, is also listed in two places.

(a)

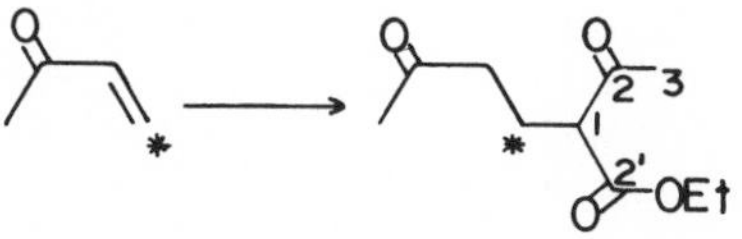

Considering methyl vinyl ketone as the starting material, the selected carbon (*) undergoes reduction (functionality 1→0), and the newly introduced carbon chain is functionalized at C-2 (Index Section III.A.2, pp. 1074-1077). At this point there is again a choice. In the branch containing the keto group, C-2 functionality = 2; this is indexed on p. 1075. In the branch containing the ester group, C-2' functionality = 3; this is indexed on p. 1076.

(b)

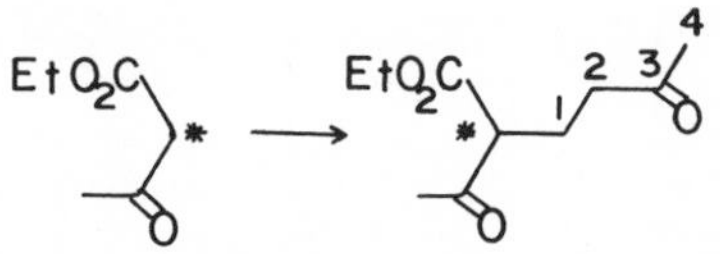

If one considers the acetoacetic ester as the starting material, the selected carbon (*) undergoes no change in oxidation state (functionality 0→0), and the newly introduced carbon chain is functionalized at C-3. This reaction, along with many other Michael addition reactions, is indexed in Section III.B.3, pp. 1098-1099. Carbon-3 functionality = 2; this is indexed on p. 1099.

The final example illustrates why the carbon-carbon bond-forming reactions are subdivided according to the change in oxidation state of the substrate. Most carbon-carbon bond-forming reactions can be viewed as the combination of a nucleophile with an electrophile. Reactions in which the substrate undergoes reduction (III.A) are typically those in which the substrate is the electrophilic partner, whereas reactions in which the substrate undergoes no change in oxidation state (III.B) are typically those in which the substrate is a nucleophile or potential nucleophile. Thus related reactions are found

together; for example, reactions of various substrates with acyl anion equivalents are generally found in Section III.A.1.

I.   REACTIONS IN WHICH THE CARBON
      SKELETON REMAINS UNCHANGED

		Functionality of C in product				
		0	1	2	3	4
Functionality of C in starting material	0	1016	1044	1045	1046	–
	1	1009	1016	1046	1049	–
	2	1010	1011	1028	1050	–
	3	1014	1014	1015	1038	1051
	4	–	1016	–	–	1043

II.   CARBON-CARBON BOND-
       BREAKING REACTIONS

		Functionality of C in product				
		0	1	2	3	4
	0	1051	1052	1054	1054	–
	1	–	1054	1055	1057	–
	2	–	–	1057	1058	1059
	3	–	–	–	1059	–
	4	–	–	–	–	–

III.   CARBON-CARBON BOND-FORMING REACTIONS

0.   Unfunctional Chain

		Functionality of C in product				
		0	1	2	3	4
Functionality of C in starting material	0	1086	–	–	–	–
	1	1060	1087	–	–	–
	2	1061	1061	1088	–	–
	3	1062	1062	1063	1088	
	4	–	1064	–	1064	–

1.   One-Carbon Chain

		Functionality of C in product				
		0	1	2	3	4
	0	1088	1101	–	–	–
	1	1064	1090	1102	–	–
	2	1067	1068	1093	–	–
	3	1072	1072	1073	–	–
	4	–	–	–	–	–

2. Two-Carbon Chain

Functionality of C in starting material	Functionality of C in product				
	0	1	2	3	4
0	1094	1102	–	–	–
1	1074	1096	–	–	–
2	1077	1077	1097	–	–
3	–	1080	1080	–	–
4	–	–	1082	1082	–

3. Three-Carbon Chain

Functionality of C in starting material	Functionality of C in product				
	0	1	2	3	4
0	1098	–	–	–	–
1	1082	1099	–	–	–
2	–	1084	1100	–	–
3	–	–	1084	–	–
4	–	–	–	–	–

4. Four-Carbon Chain

Functionality of C in starting material	Functionality of C in product				
	0	1	2	3	4
0	1100	–	–	–	–
1	1085	1101	–	–	–
2	–	1085	–	–	–
3	–	–	1085	–	–
4	–	–	–	–	–

5. Five- (or Six[*]) Carbon Chain

Functionality of C in starting material	Functionality of C in product				
	0	1	2	3	4
0	1101	–	–	–	–
1	1086	1101	–	–	–
2	–	1086	–	–	–
3	–	–	1086[*]	–	–
4	–	–	–	–	–

IV. REARRANGEMENTS OF THE
CARBON SKELETON

		Functionality of C in product				
		0	1	2	3	4
Functionality of C in starting material	0	1103	1103	–	–	–
	1	1103	1103	1103	–	–
	2	–	1104	1104	1105	–
	3	–	–	1105	–	–
	4	–	–	–	–	–

I. REACTIONS IN WHICH THE CARBON SKELETON REMAINS UNCHANGED

A. REDUCTION OF SUBSTRATE

Functionality 1⟶0

			VOLUME 1		VOLUME 2	
6 ⟶ 6	 R⟍ ⟶ R⟋		I.A.5	11-14	I.A.5 Ad.	11-14; 20
	dienes ⟶ olefins		2.B.2	108	2.B.2	123-124
	 Z⟍ ⟶ Z⟋				II.B.6 Ad. 13.F.6 14.D.13 19.D.8	556; 558; 706; 766-767; 879
	 HO⟍ ⟶ O⟍		10.E.2 11.E.2	603-605; 678-679		
	 RO₂C⟍Br ⟶ RO₂C⟍				14.D.13	766
	⬡ ⟶ ⬡		I.A.5	11-14		
	 ⬡ ⟶ ⬡		2.B.3	108-109	2.B.3 (I.A.10	124-125; 17-18)
	⬡ ⟶ ⬡		I.A.10 II.B.3 (2.B.3 (II.B.2	17-18; 647-648 108-109) 647)	2.B.3 II.B.3 13.F.11 (8.A.8	124-125; 555; 706 407)
	R⟍ ⟶ R⟍B⟍		I.A.5 4.B.2	11-14; 186-188; see also other sections		
7 ⟶ 6	R-N< ⟶ R-H β to keto benzylic		I.A.4 I.A.8 II.G.1 8.A.9	10; 17 700 430-431		
	 R-N⊕< ⟶ R-H		I.A.4 8.A.10	10; 431-432		
	R-NO₂ ⟶ R-H				I.A.13	19
8 ⟶ 6	R-OH ⟶ R-H α to keto		I.A.2 II.B.5	6-8 649	I.A.2	7-9

I.A. 1 ⟶ 0

8 ⟶ 6   R—OR' ⟶ R—H

                epoxides   4.C.2   196–197   4.C.2   233–235
                benzylic   5.C.3   269   4.C.11   241
                 (H$_2$ 4.C.7   203)

R—O—C—R' ⟶ R—H
     ‖
     O
           benzylic   II.F.3   689–690   (Ad.   20)

R—O—SO$_2$R' ⟶ R—H

                                                I.A.3   9–11

16 ⟶ 6   R—SH
                  ⟶ R—H   I.A.7   15–16   I.A.7   14–16
        R—SR'   (I.A.3   10)
                                                 (2.A.6   94)

    SOR ⟶   II.D.6   674–675;   (Ad.   261)
                              II.F.6   693–694

R—SO$_2$Ar ⟶ R—H   I.A.7   16

R—SO$_3$H ⟶ R—H   I.D.6   45

17 ⟶ 6   R—X ⟶ R—H   I.A.3   8–10   I.A.3   9–11
               via RMgX   I.B.1   22
               α-keto   II.B.8   557

Functionality 2 ⟶ 0

6 ⟶ 6   R—≡ ⟶ R⌒
    in presence of double bond   2.B.1   122

7 ⟶ 6   ⟩═N— ⟶ ⟩   (I.A.1   3–6)   (I.A.1   4–7)

8 ⟶ 6   ⟶   II.B.4   648–649

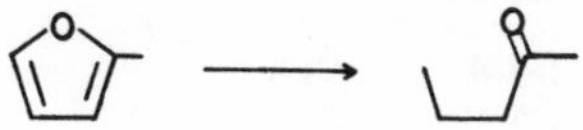

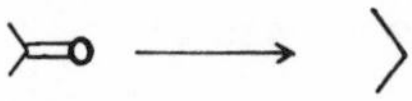

     ═O ⟶ ⟩   I.A.1   3–6   I.A.1   4–7;
                                                   Ad.   20

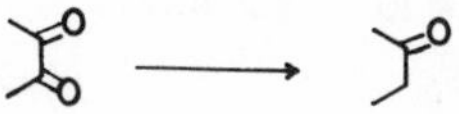

        ⟶   II.B.5   649   II.B.5   556

        ⟶   II.A.9   639–640   II.E.6   581;
                                                   Ad.   584

I.A.   2 → 0

I.A.   2 → 1

7 → 6	$\diagup\hspace{-0.3em}{=}\hspace{-0.3em}N_2$ → $\diagup\hspace{-0.3em}\diagdown$		2.A.14	98–99	2.A.14	104–106
	$\diagup\hspace{-0.3em}{=}\hspace{-0.3em}N-NHTs$ → $\diagup\hspace{-0.3em}\diagdown$		2.A.14	98–99	2.A.14 (2.B.5	104–106 125)
7 → 7	$\diagup\hspace{-0.3em}{-}NO_2$ → $\diagup\hspace{-0.3em}{-}NO_2$		20.C.2	1000	(20.C.1	900)
	$\diagup\hspace{-0.3em}{=}\hspace{-0.3em}NR$ → $\diagup\hspace{-0.3em}{-}N\diagup\hspace{-0.3em}{}^{R}_{Ac}$		18.B.3	914–915		
	$\diagup\hspace{-0.3em}{=}\hspace{-0.3em}NMgX$ → $\diagup\hspace{-0.3em}{-}NH_2$		8.E.2	474		
	pyridine N-oxide → piperidine N–H		8.A.8	429–430		
	$\diagup\hspace{-0.3em}{=}\hspace{-0.3em}N-NH_2$ → $\diagup\hspace{-0.3em}{-}NH_2$		8.A.5	423–424	8.A.5	402–403
	→ $\diagup\hspace{-0.3em}{-}NH-NH_2$				8.A.2	399
	$\diagup\hspace{-0.3em}{=}\hspace{-0.3em}N_2$ → $\diagup\hspace{-0.3em}{-}N$ pyrrolidine				8.D.9	432
	$\diagup\hspace{-0.3em}{=}\hspace{-0.3em}N-OH$ → $\diagup\hspace{-0.3em}{-}NH_2$		8.A.5	423–424	8.A.5	402–403
	→ $\diagup\hspace{-0.3em}{-}NO_2$		20.D.3	1006–7	20.D.3	906–907
	$\diagup\hspace{-0.3em}{<}{}^{NO}_{NO_2}$ → $\diagup\hspace{-0.3em}{-}NHOH$				8.A.2	398
7 → 8	$\diagup\hspace{-0.3em}{=}\hspace{-0.3em}N_2$ → $\diagup\hspace{-0.3em}{-}OH$		4.A.3	179–180		
	→ $\diagup\hspace{-0.3em}{-}OR$		6.B.2	303	6.B.1	313
	→ $\diagup\hspace{-0.3em}{-}OCOR$		10.A.7 14.B.2	555; 826–827		
	→ $\diagup\hspace{-0.3em}{-}X$				7.A.9	353

I.A. 2 → 1

	Reaction				
8 → 6	Ar-OH → Ar-H	I.A.2	6-8	I.A.2	8
	Ar-OR → Ar-H	5.C.3	269		
				2.B.7	126
	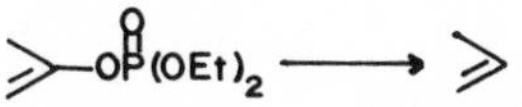			2.A.20	111
	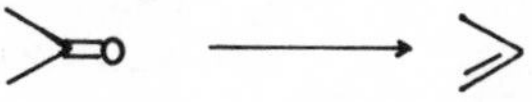			2.A.20 2.A.21	111-113; 113-114
				13.A.10	667
				4.B.2	222
				2.B.5	125
	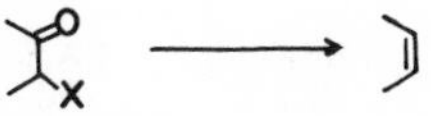	2.B.5 (2.B.4	110 109-110)		
	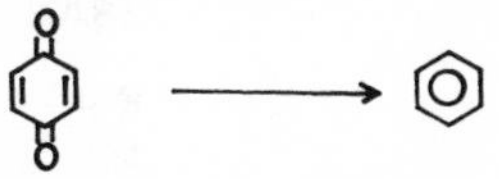	I.A.6	14-15	(Ad.	20)
8 → 7	 via oxime	8.A.5	423-424	8.A.5	402-403
	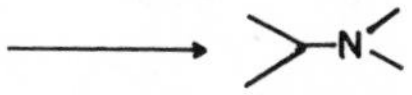	8.A.6,7	424-429	8.A.6,7 (8.A.2 (8.B.1	403-407 399) 412)
8 → 8				Ad.	242
		4.C.1 4.C.3 4.C.5 4.C.6	193-196; 197-198; 200; 201-203	4.C.1 4.C.3 Ad. 4.C.6	228-233; 235-237; 242; 437-438
	(Cannizzaro	4.C.4 (13.B.3	199) 764)		

I.A.   2 → I

8 → 8   (CH₃COCH₃ type ketone) ⟶ (diol with two OH) — Ad.   251

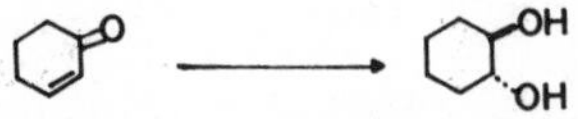

(cyclohexenone) ⟶ (cyclohexane diol) — 4.B.2   223

>=O ⟶ >—OR   6.E.2   325   6.E.2   330

>=O ⟶ >—OCOR   14.D.7   857   (14.D.1   759)
                              (14.D.1   853-855)

>=O ⟶ >—OSiR₃   4.C.1   229-230

(C(OR)₂) ⟶ >—OR   6.E.1   324-325   6.E.1   329-330
                        (10.A.4   550)

16 → 6   ArSO₃H ⟶ Ar—H   I.D.6   44-45

ArSO₃R ⟶ Ar—H   (4.A.1   211)

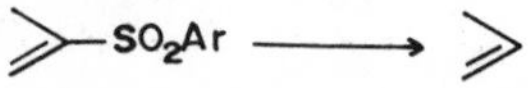

(vinyl)—SO₂Ar ⟶ (alkene)   2.B.7   127

17 → 6   Ar—X ⟶ Ar—H   I.A.3   8-10   I.A.3   9-11;
                          (I.B.1   22)   Ad.   20
                                           (5.C.4   289)
                                           (7.E.1   388)
                                           (8.A.8   408)

(vinyl)—X ⟶ (alkene)   2.B.7   126

17 → 7   (C(X)(NO₂)) ⟶ >—NO₂   20.D.3   1006

17 → 17   (C(X)(X)) ⟶ >—X   7.E.1   408   I.A.3   9-11;
                                              7.E.1   387-388
                                              (7.B.5   369)

Functionality   3 → 0

8 → 6   Ar—CO₂H ⟶ Ar—CH₃   I.A.12   19

17 → 6   R—CX₃ ⟶ R—CH₃   I.A.3   9

Functionality   3 → I

7 → 7   R—CN ⟶ R—CH₂NH₂   8.A.3   419-421   8.A.3   400-401

I.A.   3→1

7 ⟶ 7   R–CN ⟶ R–CH$_2$–NH–R'   (8.A.3   420)

7 ⟶ 8   R–CN ⟶ R–CH$_2$OH   4.C.6   238

8 ⟶ 7   R–$\overset{O}{\overset{\|}{C}}$–N< ⟶ R–CH$_2$–N<   8.A.4   421–423   8.A.4   401–402

R–$\overset{O}{\overset{\|}{C}}$–OH ⟶ R–CH$_2$–NHMe   8.A.12   408–409

8 ⟶ 8   R–$\overset{O}{\overset{\|}{C}}$–N< ⟶ R–CH$_2$–OH   4.C.12   241–242

R–$\overset{O}{\overset{\|}{C}}$–OH ⟶ R–CH$_2$OH   4.C.1   193–196;   4.C.1   228–233;
                                      4.C.5,6   199–201   Ad.   242

R–$\overset{O}{\overset{\|}{C}}$–OR' ⟶ R–CH$_2$OH   4.C.1   193–196;   4.C.1   228–233;
                                       4.C.5,6   199–201   Ad.   242
                                       (2.B.2   108)

R–$\overset{O}{\overset{\|}{C}}$–OR' ⟶ R–CH$_2$–OR'   6.E.3   325–326   6.E.3   330–331

⟶   14.D.15   768

16 ⟶ 8   R–$\overset{O}{\overset{\|}{C}}$–SR' ⟶ R–CH$_2$OH   4.C.9   209   (4.C.1   232)

16 ⟶ 16   R–$\overset{O}{\overset{\|}{C}}$–SH ⟶ R–CH$_2$SH   (4.C.1   232)

17 ⟶ 8   R–$\overset{O}{\overset{\|}{C}}$–X ⟶ R–CH$_2$OH   (4.C.1   193–196)

Functionality   3→2

7 ⟶ 8   R–CN ⟶ R–CHO   10.B.4   572–574;   10.B.4   500;
                        10.B.6–8   575–578   10.B.7   500;
                                             Ad.   502

R– ⟶ R–CHO   10.B.8   577–578

8 ⟶ 6   R$\overset{O}{\overset{\|}{C}}CH_2$$\overset{O}{\overset{\|}{C}}$OMe ⟶ RC≡C$\overset{O}{\overset{\|}{C}}$OMe   14.D.16   768–769

8 ⟶ 8   R–$\overset{NH}{\overset{\|}{C}}$–OR' ⟶ R–CHO   10.B.8   577–578

R–$\overset{NR'}{\overset{\|}{C}}$–OR" ⟶ R–$\overset{NHR'}{\overset{|}{CH}}$–OR"   10.D.3   595

R–$\overset{O}{\overset{\|}{C}}$–N< ⟶ R–CHO   10.B.4,5   572–575   10.B.4   500

R–$\overset{O}{\overset{\|}{C}}$–NH–NH$_2$ ⟶ R–CHO   10.B.8   577–578
                        etc.   (10.B.10   579–580)

I.A.   3 → 2

8 → 8   R—C(=O)—OH ——————→ R—CHO   10.B.11   580-581   10.B.11 501
                                                                  (8.A.12 408-409)

R—C(=O)—OR' ——————→ R—CHO   10.B.9   578-579   10.B.9 501;
                                        (10.B.10 579-580)   Ad.   502

R—C(=O)—O—C(=O)—R' ——————→ R—CHO   10.B.11 501

R—C(OR')$_3$ ——————→ R—CHO   10.B.9   578-579

R—C(OR')$_3$ ——————→ R—CH(OR')$_2$ 9.D.5   538-539   9.D.5 476-477

16 → 8   R—C(=O)—SR' ——————→ R—CHO   10.B.2   571

17 → 6   R—CH=CX$_2$ ——————→ R—C≡CH   3.A.1 173

17 → 8   R—C(=O)—Cl ——————→ R—CHO   10.B.1-3   570-572   10.B.3 499-500
                                        (10.A.7   555-556)   (14.D.15 768)

17 → 17
18.C.3   918

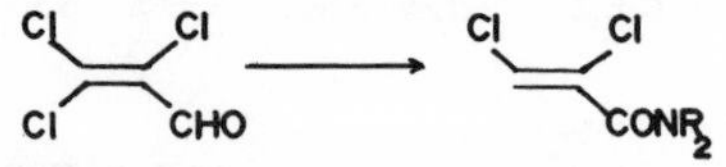

CHX$_3$ ——————→ CH$_2$X$_2$ 7.E.1   408   7.E.1 387-388

Functionality   4 → 1

8 → 7   R—N=C=O ——————→ R—NH—CH$_3$ 8.A.4   421-422

B. NO CHANGE IN OXIDATION STATE OF SUBSTRATE

Functionality   0 → 0

6 → 6   R—B< ——————→ R—H   I.A.5   11-14

R—MgX ——————→ R—H   I.B.1   22

Functionality   1 → 1

6 → 6   olefin isomerization
         via boranes   2.D.1   133-137   2.D.1 147-151
                         2.A.16   100-102   see also 2.A.12
                                                and 2.F.6

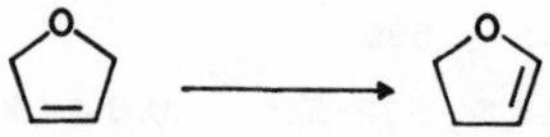

6.C.5   319

dienes ——————→ olefins   2.B.2   108

I.B.   1 → 1

6 → 6

Birch reduction

I.A.10   17-18;
2.B.3   108-109;   2.B.3   124-125;
11.B.3   647-648   11.B.3   555;
                      13.F.11   706

6 → 7   R⌒ ⟶ R⌒NH₂

by hydroboration   8.C.9   457-458

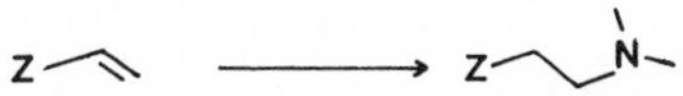

Z⌒ ⟶ Z⌒N⟨   8.D.7   468-470

R⌒ ⟶ R⌒N⟨   8.D.1   460-461   8.D.1   428-429
                      (8.F.1   479)   (8.F.1   437)
                                          (8.F.2   438)

R⌒ ⟶ R⌒N(Ac)

Ritter   18.D.4   921-922   18.D.4   834-835
          (8.F.1   479)   (8.F.1   437)

radical method   18.F   935-936
Michael   18.E.3   931-932

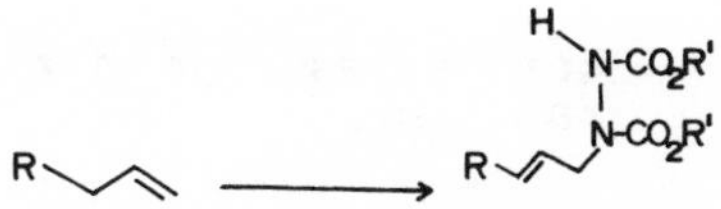

R⌒⌒ ⟶ R⌒⌒N(CO₂R')₂
H, N-CO₂R'

ene   2.C.2   134

Br⌒=⌒ ⟶ ≡⌒NHR   8.C.1   417

R⌒ ⟶ R⟨ aziridine ⟩   8.F.4   487;   8.F.4   440
                       18.D.6   925   (2.A.17   107-108)
                       (19.C.6   970)

R⌒ ⟶ R⟨Cl⟩⟨I⟩N⟨   7.B.3   363-364   8.D.1   429

R⌒ ⟶ R⟨N₃⟩⟨I⟩   7.B.3   363-364;
                  8.F.4   487

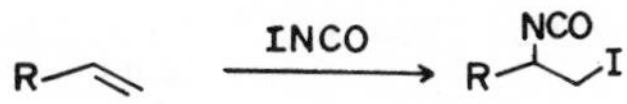

R⌒ —INCO→ R⟨NCO⟩⟨I⟩   2.A.17   107-108

I.B.   I⟶I

6⟶7   R⟍═

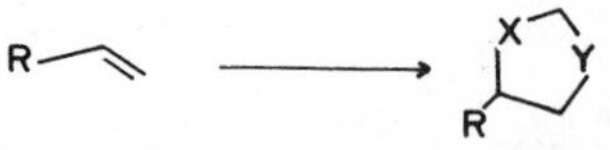

1,3-dipolar cycloadditions   8.F.5   488-490

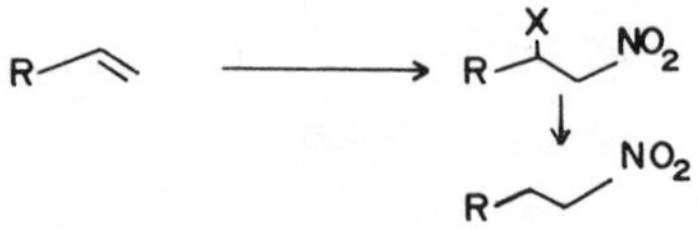

R⟍═ ⟶ R⟍⟋NO₂ (with X)   20.A.3   988-990   20.A.3 892-894

⟶ R⟍⟍NO₂   20.A.3 892

6⟶8   R⟍═ ⟶ R⟍⟍OH

by hydroboration   4.B.2   186-188   4.B.2 219-223

R⟍═ ⟶ R⟍OH   4.B.1   185-186;   4.B.3 223-224
                    4.B.3   188   (4.C.2 235)

R⟍═ ⟶ R⟍OR'   6.B.2   302-303;   6.B.10 316-317
                    6.C.1   312-313   (6.G.1 332)

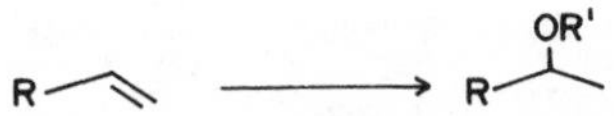

R⟍═ ⟶ R⟍O-CO-R'   14.B.5   830-831   14.B.5 733-734
                        (4.B.3   188)

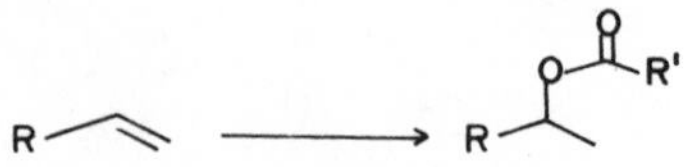

R⟍═ ⟶ R⟍OH-N   (8.A.7 406)

R⟍═ ⟶ R-epoxide   6.D.1   320-321   6.D.1   324-326
                        (4.D.5   219-221)
                        (II.A.5   631-633)
via halohydrin   6.A.4   294-296   6.A.4   308-309

R⟍═ ⟶ R⟍OH,OH   4.D.5   219-221   4.D.5   247-248

R⟍═ ⟶ R-dioxolane   6.B.11 317-318

R⟍═ ⟶ R⟍OH,X   7.B.3   364-365   7.B.3   364
                                            (4.B.1   219)
                                            (6.D.1   326)

I.B. 1→1

6 → 8		7.C.9	389-390		
				2.B.7 4.A.8 Ad.	126; 215; 267
6 → 17		7.B.1 7.C.9	356-359; 389	7.A.18 7.B.1	356-357; 362-363
	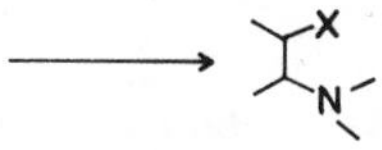	7.B.3 20.A.3	363-364; 988-990		
	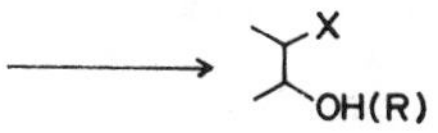	7.B.3 7.C.9	364-365; 389-390	7.B.3 (4.B.1 (6.D.1 (7.B.2	364 219) 325-326) 363)
	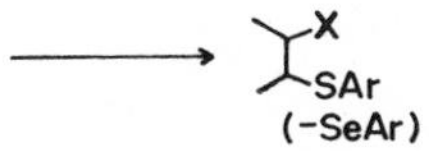	7.B.3	363-364		
		7.B.2	359-362	7.B.2	363-364
				7.A.7	352
7 → 6	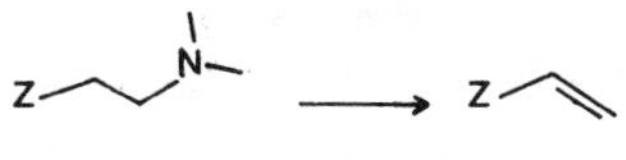	2.A.9 11.G.1 11.G.3 (2.B.6	92-93; 699-700; 707 111)		
	 Hoffmann	2.A.8 (11.G.1	89-92 699-700)	2.A.8	96-98
	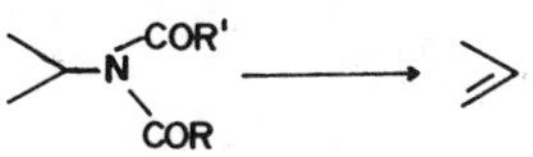	2.A.9	92	2.A.9 Ad.	98-99; 115
	 Cope	2.A.10	93	2.A.10	99

I.B. I⟶I

7 ⟶ 6 NH ⟶   2.A.17 107–108
(2.A.20 III)

NO₂/NO₂ ⟶   2.A.22 114

7 ⟶ 7 R–NH₂ ⟹ R–NH₂
(protection)   14.A.10 723

R–NH₂ ⟶ R–NH–R'   8.A.7   427–429;   8.A.7 405–407;
8.C.1-3 443-450;   8.C.1-3 416-423
8.D.1,2 461-463
(8.A.6   426)   (8.A.6 403-405)
(8.B.5   441-442)
(8.D.4   466)   (8.D.9 432)
(8.D.7   468-470)

R–NH₂ ⟶ R–NH–C≡CH   3.D.1 196

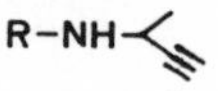

R–NH₂ ⟶ R–NH–Ar   8.C.4   450-452

R–NH₂ ⟶ R–NH–C(=O)–R'   18.A.1-4 895-903   18.A.1-4 816-824
(18.A.6-8,10-12 905-10)   (18.A.11 825-826)
(18.C.3   918)   (18.A.15 827-828)
(18.D.8   927)   (18.D.8 838-840)
(18.E.6 845-846)

R–NH₂ ⟶ R–N(phthalimide)   18.A.1   896-899;
See also other sections above

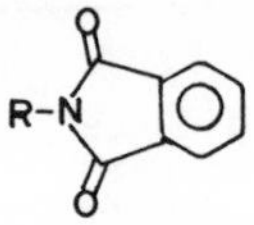

R–NH₂ ⟶ R–NO₂   20.D.1   1003-4   20.D.1 903-905

R–NH₂ ⟶ R–NH–SO₂R'   18.A.15 827-828
(18.A.11 826)

R₃N ⟶ R₂N–H   8.H.1   510-511   8.H.1   453-454
(8.A.9   430-431)
(8.D.7   468-469)

R₃N ⟶(von Braun) R₂N–CN
↓
R₂N–H   19.D.6   969-970   19.D.6   878-879

8.B.3   439-440   8.B.3   414

R₃N ⟶ R₃N→O   (8.A.2 398)

I.B.   I⟶I

7⟶7 R_2N-Ar ⟶ R_2N-H 8.B.6   442

⟶   8.D.5   466-467

$R_4\overset{\oplus}{N}$ ⟶ R_3N 8.A.10   431-432   8.B.5   415-416;
                                       (8.H.3   512)   8.H.3   454

$R-\underset{|}{N}-\overset{O}{\overset{||}{C}}-R'$ ⟶ $R-\underset{|}{N}-H$ 8.B.1,2   436-439;
                                                          8.B.4   440-441

$R-\underset{|}{N}-\overset{O}{\overset{||}{C}}-R'$ ⟶ $R-\underset{|}{N}-CH_2R'$ 8.A.4   421-423

$R-\underset{H}{N}-\overset{O}{\overset{||}{C}}-R'$ ⟶ $R-\underset{R''}{N}-\overset{O}{\overset{||}{C}}-R'$ 18.E.1   929-930   18.E.1   842-844
                                                          (18.E.2   930-931)   (18.E.2   844-845)

R_2N-CN ⟶ R_2N-H 8.B.3   439-440   8.B.3   414

$R-N=CHAr$ ⟶ $R-NH-R'$ 8.B.5   441-442

$R-N=C=O$ ⟶ $R-NH_2$ 8.B.1   436-437
                          (8.G   494-502)

$R-\overset{\oplus}{N}\equiv\overset{\ominus}{C}$ ⟶ $R-NH-CHO$ 18.A.14   910-911

$R-\underset{|}{N}-\underset{|}{N}-$ ⟶ $R-\underset{|}{N}-H$ 8.A.2   417-418;
                                         8.C.3   448-449

$R-N_3$ ⟶ $R-NH_2$ 8.A.2   417-418;   8.A.2   399;
                        8.C.3   448-449   Ad.   409

R_2N-OH ⟶ R_2N-H 8.A.2   417-418

$R_3N\rightarrow O$ ⟶ R_3N 8.A.2   417-418   8.A.2   398-399

$R-NO_2$ ⟶ $R-NH_2$ 8.A.1   413-417   8.A.1   395-398

$R-NH-SO_2Ar$ ⟶ $R-NH_2$ 8.B.4   414-415

7⟶8 $R-NH_2$ ⟶ $R-OH$ 4.A.5   180-181   4.A.5   212
                           (8.H.1   511)

$R-NH_2$ ⟶ $R-O-\overset{O}{\overset{||}{C}}-R'$ 10.A.21   494;
                                       14.A.21   727-728
                          (14.B.1   826)   (14.B.1   730-731)

I.B.   I ⟶ I

7 ⟶ 8   (aziridine NH) ⟶ HO–CH₂CH₂–NH₂ structure   8.E.3   475

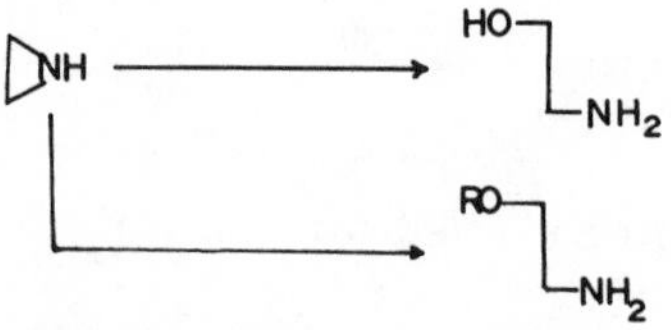

⟶ RO–CH₂CH₂–NH₂ structure   6.B.4   305-306

$R_4\overset{\oplus}{N}$ ⟶ R–OH   4.A.6   182

⟶ R–OR'   6.A.5   296-297

⟶ R–OCOR'   14.A.13   819

7 ⟶ 16   t-Bu–N(thioisoindolinone) ⟶ t-BuSOMe ↓ t-BuOMe   6.H.4   335

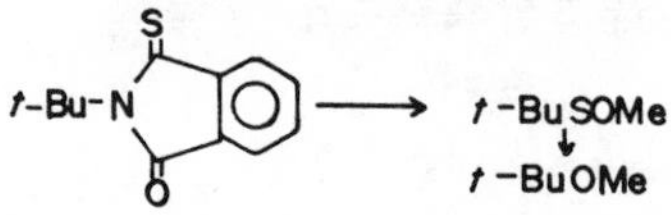

7 ⟶ 17   R–NH₂ ⟶ R–X   (7.A.9   345)

R₃N ⟶ R–X   von Braun   19.D.6   969-970

R–NH–C(=O)–R' ⟶ R–X   7.A.10   347-348   7.A.10   354
(19.C.2   953-954)   (19.C.2   872)

[$R_4\overset{\oplus}{N}$ ⟶ R–CN]   19.A.4   942-943

8 ⟶ 6   (>C–OH) ⟶ (alkene)   2.A.1   71-75   2.A.1   80-83

(>C–OR) ⟶ (alkene)   2.A.13   98   2.A.13   103-104

(oxazoline-Ph) ⟶ Δ ⟶ (alkene)–NHCOPh   2.D.7   154-155

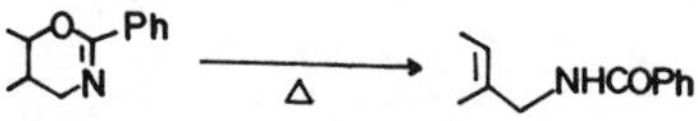

(>C–O–C(=O)–R) ⟶ (alkene)   2.A.6   85-88   2.A.6   92-95

from xanthates   2.A.7   88-89   2.A.7   95-96
from mesylates   2.A.6   85-88   2.A.6   92-95
(2.A.2   75-80)   (2.A.2   83-88)

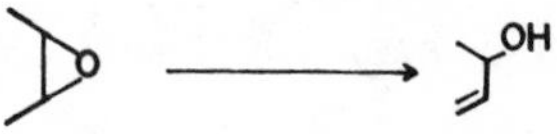

(epoxide) ⟶ (>C–OH allyl)   4.A.7   213-214

I.B.   1→1

8 → 6   epoxide → alkene

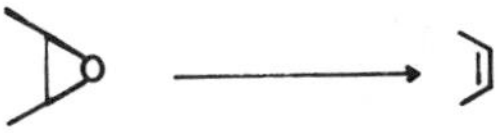

2.A.12   97-98   2.A.12   101-103;
(2.A.5   84-85)   Ad.   115

diol → alkene

2.F.6   148   2.F.6   163-164;
                     Ad.   165

via dimesylate

2.A.6   87   2.A.6   93
                     (2.F.6   163)

β-hydroxysilane → alkene

                     2.A.20   112;
                     Ad.   116
                     (10.G   528)

halohydrin → alkene

2.A.4   82-84;   2.A.4   91;
2.A.5   84-85   2.A.5   91-92

8 → 7   R–OH ⟶ R–N⟨

8.C.7   455   8.C.7   424-425
(8.A.6   425-426)   (8.A.2 399)
                     (8.B.2 413)

via sulfonate

8.C.3   448-450   8.C.3   423

R–OH ⟶ R–N–C–R' (amide, Ritter)

Ritter

18.D.4   921   18.D.4 834-835
(8.F.1   479)

R–OR' ⟶ R–N⟨

8.C.6   453-454

epoxide openings

8.D.5   466-467   (4.A.7 213-214)
(8.D.7   470)

cyclic carbonate ⟶ PhNH–CH₂CH₂–OH

                     4.A.1   211

8 → 8   R–OH ⟶ R–OD

                     4.A.10   217

R–OH ⟶ R–OR'

6.A.1,2   286-292;   6.A.1,2 304-306;
6.A.5   296-297   Ad.   311
                     (6.A.5   309-310)

cyclic

6.A.4   294-296   6.A.4   308-309;
                     6.B.11   318

S$_N$1-type

6.B.1,2   300-304   6.B.1,2 313-314

Michael

6.C.1,2   311-316;
6.E.2   325

R–OH ⟶ R–O–CH=CH₂

6.B.3   305;   6.I   335-342
6.C.1   313
(19.D.7   971)

I.B.   I ⟶ I

8 ⟶ 8   R–OH ⟶ R–O–Ar   6.B.5   306–308;
                                6.C.2   313–316

        R–OH ⟶ R–O–C–OR'   9.A.1,2 514–519;   9A.I   463–467;
                                9.B.1   528–529;   9.B.1,3 470–472
                                9.B.3,5 530–532
                                (9.A.6-8 523–526)
                                see also other parts
                                of Chapter 9

        R–OH ⟶ R–O–C–R'   14.A.1-8 802–816;   14.A.1-7 714–721;
                    ‖   14.D.2   855   Ad.   728
                    O   (14.A.14,15 819–821)   (14.A.14 726)
                                                       (14.A.19 726–727)

        R–OH ⟶ R–O–C–NH₂   14.A.4   811–812;   14.A.4 720
                    ‖   14.A.14 819
                    O

        R–OH ⟶ R–O–C(OR')₂   14.F   778–780

        R–OH ⟶ R–O–C(OR')₃   14.G   781

        R–OH ⟶ R–O–C–X   6.B.7   309

        R–OH ⟶ R–O–NO₂   20.A.I   982

        R–OH ⟶ R–O–SiMe₃   Ad.   311

        R–OH ⟶ R–O–SO₂R'   7.A.5   338   Ad.   728

        R–OR' ⟶ R–OH   (4.A.7   182–183)   4.A.7   212–215;
                                                       4.A.9   215–217
                                                       (4.C.10,11 240–241)
                                                       (4.D.9 250)

        ▷O ⟶ Y⌐OH   4.A.7   182–183;   4.A.7   212–215
                                6.B.4   303–306;   (6.B.3 314)
                    etc.   7.B.7   372–373
                                (4.D.5   219–221)

        ▷·O ⟶ ▷O   6.C.6   323

        ▷O ⟶ ⟨O O⟩X   9.A.3   519–520

        R–O–R' ⟶ R–O–C–R''   14.B.7   832–833   14.B.7 736–737
                     ‖
                     O

I.B.   I→I

8 →8   R-O-CH$_2$R' $\xrightarrow{[O]}$ R-O-$\overset{O}{\overset{\|}{C}}$-R'   14.D.3   855

R-O-$\overset{|}{\underset{R'}{C}}$-OR ⟶ R—OH   *See* other sections of Index
(*e.g.* I.B. 2→2)

⟶ R-O-$\overset{|}{\underset{R'}{C}}$-H   6.E.1   324-325   6.E.1   329-330

⟶ R-O-$\overset{|}{\underset{\underset{=}{R'}}{C}}$-OR   9.A.6   523-524

⟶ R-O-$\overset{|}{\underset{R'}{C}}$-X   6.B.8   310

R-O-$\overset{O}{\overset{\|}{C}}$-R' ⟶ R-OH   4.A.1   176   4.A.1   211
(13.A.2   748-751)   (13.A.2   658-660)

⟶ R-O-CH$_2$R'   6.E.3   325-326   6.E.3   330-331

⟶ R-O-$\overset{O}{\overset{\|}{C}}$-R''   14.A.9   816-817
(14.A.7   814-815)

R-O-$\overset{S}{\overset{\|}{C}}$-SR' ⟶ R-OH   4.A.4   180

R-O-CX$_2$-R' ⟶ R-O-$\overset{O}{\overset{\|}{C}}$-R'   14.A.16   821-822

R-O-O-R' ⟶ R-O-R'   6.E.5   331

R-O-S-t-Bu ⟶ R-O-t-Bu   6.H.4   335

R-O-SO$_2$Ar ⟶ R-OH   4.A.1   211

8 →9   R-OH ⟶ R-F   7.A.14   351   7.A.14   356

8 →16   ▷O ⟶ ▷S   Ad.   312

8 →17   R-OH ⟶ R-X   7.A.1,2   330-334;   7.A.1,2   348-349;
7.A.4-7   336-342   7.A.4,6   350-352;
7.A.16,17   356;
7.A.20,21   358;
Ad.   359-360;
7.B.5   367-368
(7.A.9   354)

R-OR' ⟶ R-X   7.A.8   342-344   7.A.8   352-353

I. B.   I → I

8 → 17

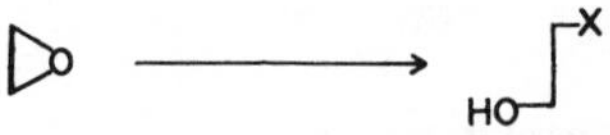

| | 7. B. 7 | 372-373 | 4. A. 7 | 212-215; |
| | | | 7. B. 7 | 370 |

| | | | 4. A. I | 211 |

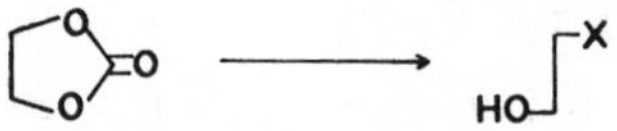

R-O-C(=O)-R' ⟶ R-X   13. A. 2   750   (15. A. 3 786-788)

lactone opening   11. D. I   667   (15. A. I 785)
                                                      (15. A. 3 787)

R-O-SO₂R' ⟶ R-X   7. A. 5   338   7. A. 6   350-352

[R-OH ⟶ R-CN]   19. A. 2   941

16 → 6

sulfoxides and sulfones,   2. A. II   93-97   2. A. II   99-101;
and related Se compounds   4. A. 7   214

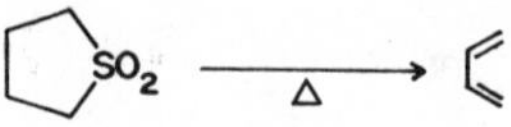

                                                      14. E. 3   772-773

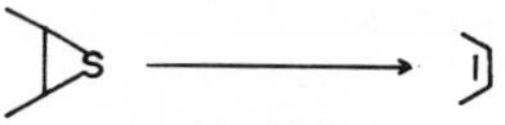

episulfides   2. A. 12   97-98   2. A. 12   101-103
episulfones   2. A. II   93-97
                              (13. G. 2   799)

16 → 7   R-S-H(R) ⟶ R-NH₂   8. C. 6   453-454

16 → 8   R-S-H(R) ⟶ R-OR'   6. H. 4   335

16 → 16 R-SH ⟶ RSO₂Cl   7. C. 8   378

         R-S-R ⟶ R-SCN   19. D. 6   969-970

         R-SO₂N< ⟶ R-SO₃H   8. B. 4   440-441

16 → 17 R-SR' ⟶ R-X   7. A. 16   352;   7. A. 7   352
                                      19. D. 6   969-970

I.B.   1 → 1

17 → 6	$\searrow^X$ (isopropyl-X) ⟶)	2.A.2	75-80	2.A.2	83-88
	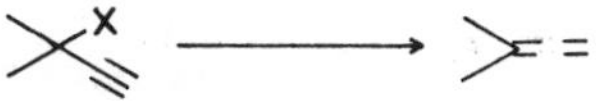 $\times\!\!\!=$ (with X) ⟶ $\searrow=$	3.D.1	171		
	 $\searrow^X_{OH(R)}$ ⟶)	2.A.4,5	82-85	2.A.4,5 (4.A.9	91-92 215-217)
	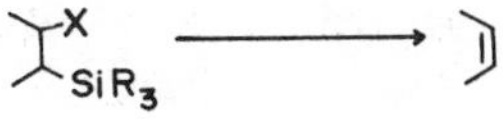 $\searrow^X_{SiR_3}$ ⟶)			2.A.25	115
	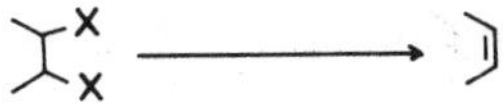 $\searrow^X_X$ ⟶)	2.A.3	80-81	2.A.3 Ad.	88-91; 115

17 → 7	R—X $\xrightarrow[\text{via RMgX}]{\text{Gabriel}}$ R—N$\big\langle$	8.C.1-3 8.B.2 8.C.9 (8.C.5	443-450 437-439 456-458 452-453)	8.C.1-3 8.B.2 8.C.9 (8.B.3	416-423 413-414 425-426 414)
	R—X ⟶ R–N–C– (with =O)	18.E.1 (8.B.2	929-930 437-439)	18.E.1 (8.B.2	842-844 413-414)
	R—X ⟶ R-NHNH$_2$	8.E.3	475		
	R—X ⟶ R-NO$_2$	20.B.1 (20.E.3	991-993 1009)	20.B.1	895-896

17 → 8	R—X $\xrightarrow{\text{via RMgX}}$ R–OH	4.A.2 4.D.2	176-179 214-215		
	R—X $\xrightarrow[\text{$S_N$1-type}]{\text{cyclic}}$ R–OR'	6.A.1,2 6.A.4 6.B.2 (2.A.2	286-292 294-296 302-304 75)	6.A.1,2 6.A.4 (6.B.2	304-306 308-309 314)
	R—X ⟶ R-O-C-R' (with =O)	14.A.10,11	817-818		

17 → 17	R—X ⟶ R—X' Finkelstein	7.A.6	339-342	7.A.6 (7.E.5	350-352 389)
	via RMgX	7.C.9	388-390	7.C.9	378-379

I.B.   2→2

Functionality 2→2

6→6   — ≡ — ⟶ — ≡ —

protection

acetylene isomerization

3.B.2   188
(3.C.1   192-193)
Ad.   199
(3.B.1   185)
(3.D.1   197)
(3.D.3,4 198)

— ≡ — ⇌ /= =

3.D.2   171-172

3.D.1   194-197
(6.I.9   342)

— ≡ —⌐ ⟶ /= =
   X

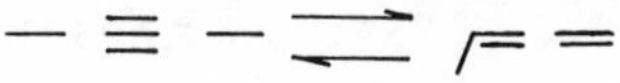

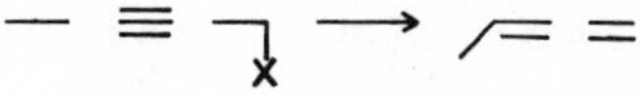

3.E.1   200-201

6→7   ≡ ⟶ =\N—

8.D.1   460-461

18.E.1   843
(20.A.3 892)

6→8   ≡ ⟶ =\OR

6.C.1   311-313
(9.B.4   530-531)

6.I.1   335-336

⟶ =\O-C(=O)R

11.D.5   672;
14.B.5   830-831

14.B.5   734

⟶ ⌄=O

10.D.6   599;
11.D.4   670-671
(9.B.1   528)
(9.B.4   530-531)

10.D.6 514;
11.D.1   571-573;
11.E.8   583
(9.B.1   431)

R/= =\R ⟶ R⌄C(=O)⌄R

11.D.4   671

≡ ⟶ \OR /OR

(9.B.1   528)
(9.B.4   530-531)

— ≡ — ⟶ O⌄⌄O

11.A.19 548-549

R|III ⟶ R\C(OR')(OR')\Cl Cl ⟶ R\C(=O)\Cl Cl

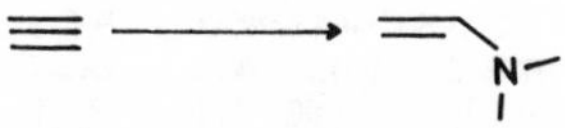

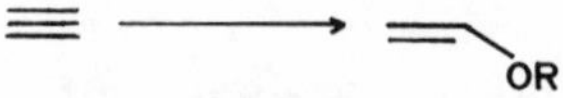

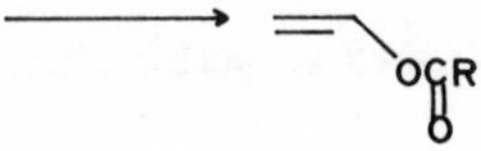

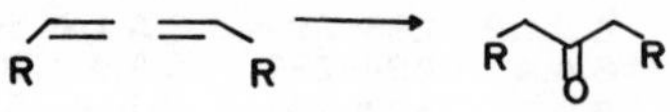

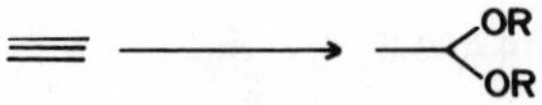

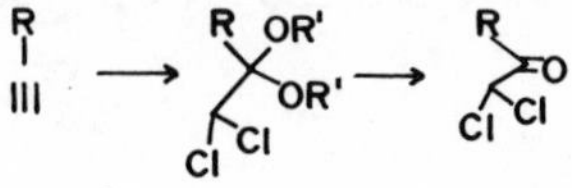

11.D.4   573

I. B.   2 → 2

6 → 17   ≡⟶ ═⟨_X

| | 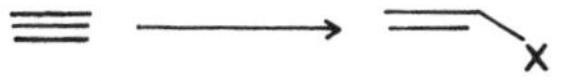 | 2.A.16
7.B.2 | 101;
361-363 | 7.A.18 357
(20.A.3 892) |

×⟨OH≡ ⟶ ⟩═⟨_X

| | 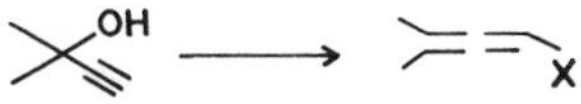 | | | 7.C.3 | 375-376 |

7 → 6   ⟩⟨NR₂ ⟶ ∣∥∣

| | | 3.A.5 | 158-159 | 3.A5 | 179 |

HN–NH
R⟨ ⟩O
or
N–NH
R⟨ ⟩O
⟶ R–≡–CO₂R'

| | 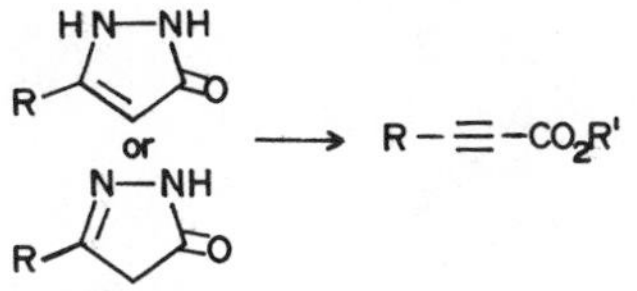 | 3.A.4
13.A.7 | 156;
756 | 3.A.4 | 178 |

⟩N–NH₂ / N–NH₂ ⟶ ∥∥

| | 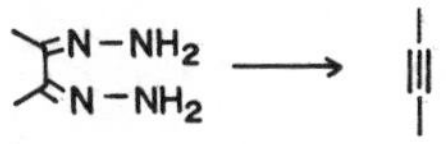 | 3.A.4 | 156-158 | 3.A.4 175-178
(3.A.9 181-182) |

7 → 7   Ar–NH₂ ⟶ Ar–NHR

| | | 8.C.1-3
8.C.8 | 443-450;
456 | 8.C.1-3 416-423 |

see also reactions
of R–NH₂

⟶ Ar–NH–C(=O)–R

| | | 18.A.1-4 | 895-903; |

see also reactions
of R–NH₂

⟶ Ar–N₂⊕

| | | 5.A.6
7.A.9 | 255-257;
344-347 | |

⟶ Ar–NO₂

by oxidation
via Ar–N₂⊕

| | | 20.D.1
20.B.4 | 1003-4
995-996 | 20.D.1 903-905
20.B.4 898-899 |

Ar–NMe₂ —HCOOH→ Ar–N(Me)(CHO)

| | | 18.A.1 | 898 | | |

Ar–NH–C(=O)–R ⟶ Ar–NH₂

| | | 8.B.1,2
8.B.4 | 436-439;
440-441 | |

⟶ Ar–NH–CH₂R

| | | 8.A.4 | 421-423 | |

⟶ Ar–N(R')–C(=O)–R

| | | 18.E.1
(18.E.2 | 929-930
930-931) | 18.E.1 842-844
(18.E.2 844-845) |

Ar–N=C=O ⟶ Ar–NHCH₃

| | | 8.A.4 | 421-422 | |

I.B.   2 → 2

7 → 7						

7 → 7   Ar–N–R ──────→ Ar–NHR   8.A.2   398
        |
        NO

        Ar–N₂⊕ ──────→ Ar–NO₂   20.B.4 995-996   20.B.4 898-899

        Ar–N=N–
                  ⟩──→ Ar–NH₂   8.A.2   417-418   8.A.2 398-399
        Ar–N=O

        Ar–NO₂ ──────→ Ar–NH₂   8.A.1   413-417   8.A.1 395-398
                                       (8.G.6   504)

        ⟩–NO₂ ──────→ ⟩=NOH   11.B.1 646
                                       (11.D.5 672)

        Ar–NHSO₂R ───→ Ar–NH₂   8.B.4   440-441

        ⟩=NR ──────→ [triazine-dione structure]   18.G   935

        ⟩=N–NH₂ ───→ ⟩=N₂   1.A.11 96;
                                       17.A.3 888-889
                                       (3.A.4   156-158)

        [cyclohexene N-OH] ──→ [aniline NH₂]   8.G.1   495-496   (18.D.5 837)

        ⟩=NOH ──────→ ⟩<NO₂   20.D.3 1006-7   20.D.3 906-907
                         NO₂

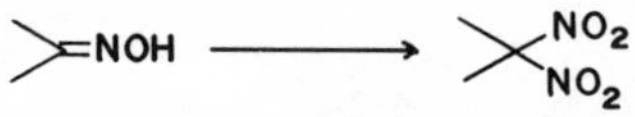

7 → 8   Ar–NH₂ ──────→ Ar–OH   5.A.3   250-253;
                                       5.A.6   255-257   5.A.6   276

        Ar–NR₂ ──────→ Ar–OR'   8.B.6   442   6.A.8   310-311

        Ar–N₂⊕ ──────→ Ar–OH   5.A.6   255-257   5.A.6   276

        Ar–NO₂ ──────→ Ar–OH   5.A.2   249

        Ar–NO₂ ──────→ Ar–OR   6.C.2   315-316

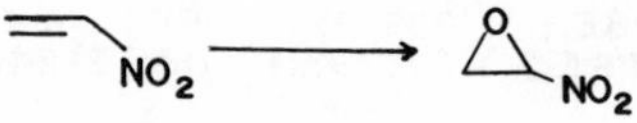

        [CH₂=CH–CH₂–NO₂] ──→ [epoxide NO₂]   20.D.5 908

I.B.   2 ⟶ 2

7 ⟶ 8

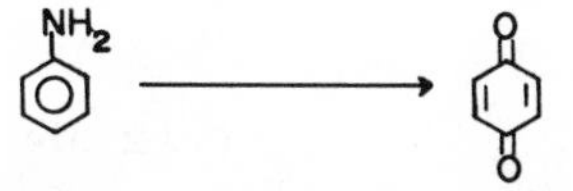

| | | 12.A.2 | 726-728 | 12.A.2 | 637-639 |

| | 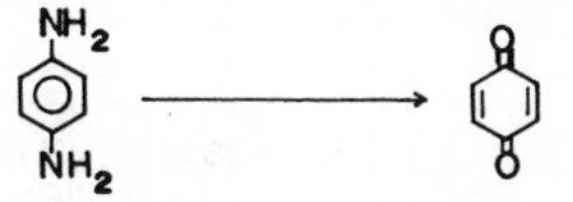 | 12.A.4 | 730-733 | 12.A.4 | 640-643 |

		11.D.3	669-670	11.D.3	571
		(10.D.1	667-668)		
		(11.D.5	673)		
		(11.G.2	703-705)	(11.G.2	599-600)

| | | 11.B.1 | 646 | | |
| | | (11.D.5 | 672) | | |

| | 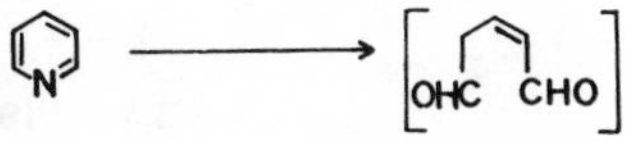 | 10.D.3 | 596-597 | | |

| |  | 8.B.5 | 441-442 | | |

		10.D.1	593;	10.D.1	508-510;
		11.D.5	671-673	11.D.5	573-575;
				Ad.	576

| |  | 9.A.2 | 519 | | |

| | | 10.B.12 | 581; | | |
| | | 11.H.7 | 722 | | |

| |  | 9.D.3 | 538 | 9.D.3 | 476 |

7 ⟶ 9   Ar-NO₂ ⟶ Ar-F

| | | | | 20.B.2 | 897 |

7 ⟶ 17   Ar-N₂⊕ ⟶ Ar-X

| | | 7.A.9 | 344-347 | 7.A.9 | 353-354 |

Ar-NO₂ ⟶ Ar-X

| | | | | 7.A.22 | 358 |

| | | | | 3.A.4 | 177; |
| | | | | 7.A.9 | 354 |

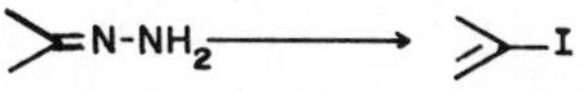

etc.

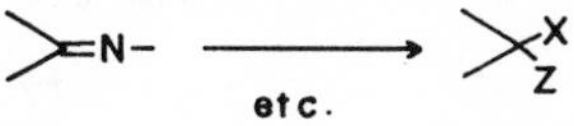

| | | 7.C.5 | 385 | 7.A.9 | 354; |
| | | | | 7.C.6 | 378 |

I.B   2 → 2

8 → 6	R–CH=CH–OR' → R–≡			(3.A2 174)	
	 CH₃–CO–CH₃ → –≡			(3.A1 169-174)(3.A.4 177)(3.A.9 181-182)	
	CH₃–CO–CH₂–COR → =CR–COR			3.E.1 202,205	
	 R–CO–CH₂–CO₂R' → R–≡–CO₂R'			3.A.4 178	
	 CH₃–CO–CO–CH₃ → –≡–	3.A.4	156-158	3.A.4 175-178 (17.A.3 808)	
	 R–CO–CH₂Br → R–≡			3.A.7 180	
8 → 7	Ar–OH → Ar–NH₂	8.C.4 (18.A.8)(18.D.6)	450-452 906 926	8.C.4 423-424 (18.D.5 937)	
	 C=O → C=C–NR₂	11.G.2 (8.D.2)(8.D.6)	703-705 461-463 467-468	11.G.2 599 (8.A.7 407)(8.I 455)	
	 C=O → C=N–Z	(10.D.1)(11.D.5)	593 671-673		
	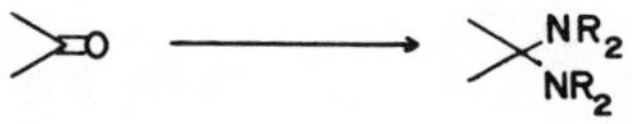 C=O → C(NR₂)(NR₂)	8.D.2	461-463	(8.D.2 429-430)	
	CH₂O → CH₂(NHCOR)₂	18.E.2	930-931	18.E.2 844	
	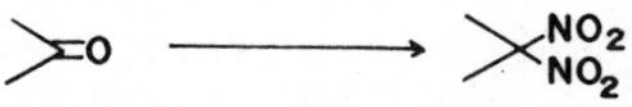 C=O → C(NO₂)(NO₂)	20.D.3	1006-7		
8 → 8	Ar–OH → Ar–OH (protection)			5.A.5 276	
	Ar–OH → Ar–OD			5.A.8 277	

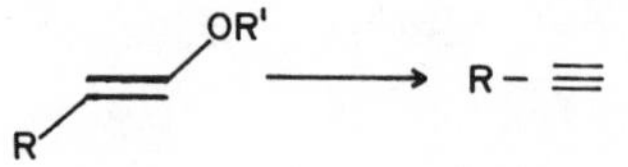

I.B.   2 → 2

8 → 8   Ar–OH ⟶ Ar–OR   6.A.1,2   286-292;   6.A.1,2 304-306
                                6.A.5   296-297;
                                6.B.5   306-308

        ⟶ Ar–O⌒   14.A.17   822

        ⟶ Ar–O–Ar'   6.A.3   292-293   6.A.3   306-308
                                                    (6.H.1 333-334)

                O
                ‖
        ⟶ Ar–O–C–R   14.A   802-816   (14.A.1 717)

>=–OR ⟶ >=–OR'   6.B.3   305   6.I.3   337

Ar–OR ⟶ Ar–OH
            base   5.A.4   253   5.A.4   275
            acid   5.A.5   254-255   5.A.5   275-276;
                                (7.A.8   342-343)   Ad.   277
            reduction   5.C.3   269   5.C.3   288-289

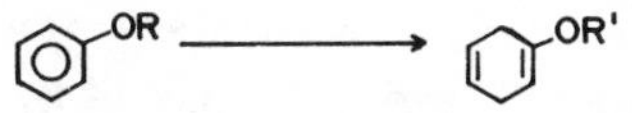

⟨⟩–OR ⟶ ⟨⟩–OR'   11.B.3   647-648   11.B.3 555

⌒OCOR ⟶ ⌒OCOR'   14.B.5   830

        O
        ‖
Ar–O–C–R ⟶ Ar–OH   (5.A.7 277)

⟨⟩–OH ⟶ ⟨⟩=O   11.B.2   647   11.B.3 555

⟨⟩–OR ⟶ ⟨⟩=O   11.B.3   647-648   11.B.3 555

⟨⟩–OH ⟶ O=⟨⟩=O   12.A.2   726-728;   12.A.2 637-639;
                                12.B.2   738-739   12.B.2 646-647
                                                     (12.A.6 644-645)

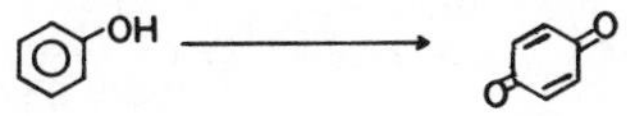

HO–⟨⟩–OH ⟶ O=⟨⟩=O   12.A.4   730-733   12.A.4 640-643

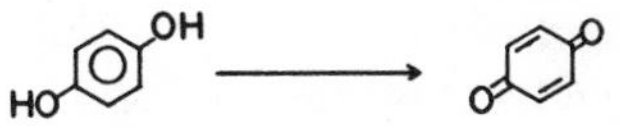

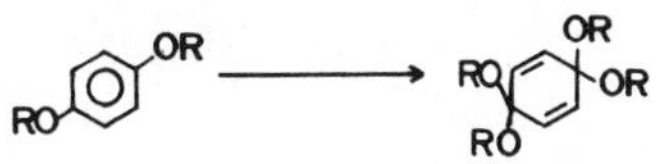

RO–⟨⟩–OR ⟶ RO–⟨⟩(OR)(OR)   6.D.2   322;
                                9.E.3   540-541

I.B.   2→2

8 ⟶ 8

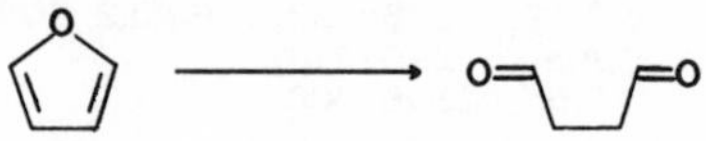

	II.D.1 (II.B.4 (10.D.2	667 648-649) 594)	II.D.1 569-570 (10.D.2 511)
	II.D.3 (10.D.2	669-670 594)	II.D.3 571
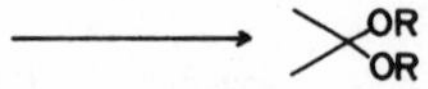	9.B.1	528-529	9.B.1 470-472 (9.B.9 473) (9.D.1 476) (9.E.4 478)
			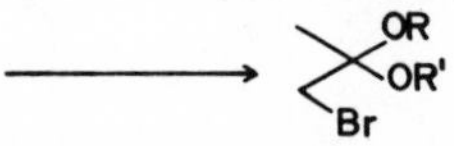 9.B.1 471
	II.A.4	630-631	(II.A.4 540-541)
	(II.D.5	672)	
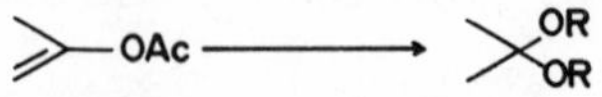	9.B.3	530	9.B.3 472
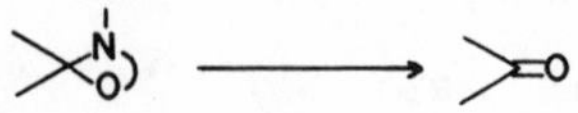	10.D.3	594-596	10.D.2,3 510-513; II.D.1 568-570
	6.B.5	306-308	(6.A.5 310) (6.B.6 314) (6.I.6 339)
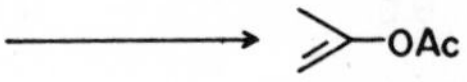	14.A.3,4 (II.F.4	809-812 690-692)	
	5.D.6	275-277	(5.D.10 295)
	5.C.1	267-268	5.C.1 287-288 (4.C.3 237)
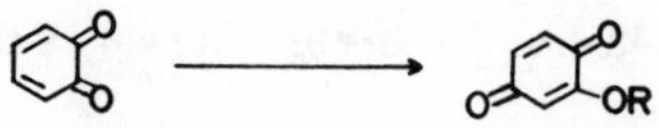	12.A.5	734-735	
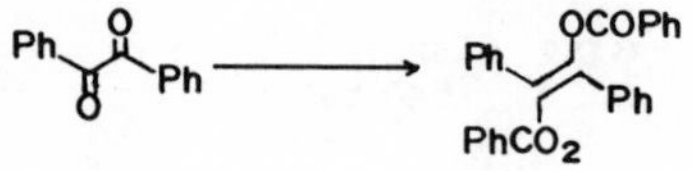	14.D.1	854	

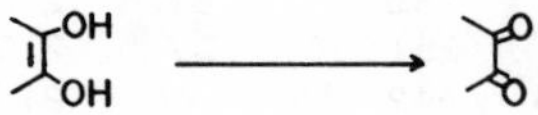

I.B.   2→2

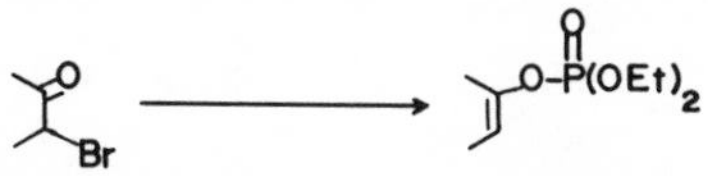

8→8   (structure: CH₂=C(O-P(OEt)₂ with Br)   →   2.A.20 III

CH₂O   →   (structure: —C(OR)(NR₂))   8.D.4   466

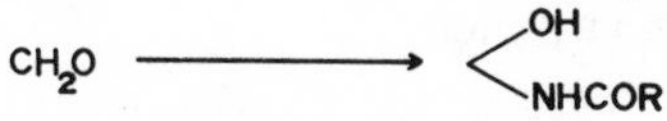

CH₂O   →   (structure: —C(OH)(NHCOR))   18.E.2   930-931   18.E.2   844

(structure: >C=O)   →   (structure: >C(OR)(OR))   9.A.1   514-519;   9.A.1   463-467;
9.A.3-5   519-522   9.A.4   467;
(9.C.1   535)   Ad.   469
(6.B.6   314)
(9.C.1   474-475)
(10.F.3   523-524)

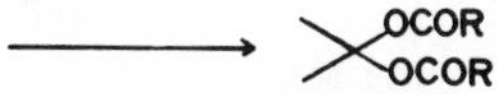

→   (structure: >C(OCOR)(OCOR))   14.A.1   806-807;   14.A.5   812-813

(structure: >C(OR)(OR))   →   (structure: >C—OR)   6.I.2   336-337;
6.I.4   337-338

→   (structure: >C=O)   10.C.5   585-586;   10.D.1,2   508-511;
10.D.2   594;   11.D.5   574-575
11.D.1   667   (9.A.1   463-467)
(11.D.5   672)

→   (structure: >C(OR')(OR'))   9.A.6   523-524   9.A.6   468

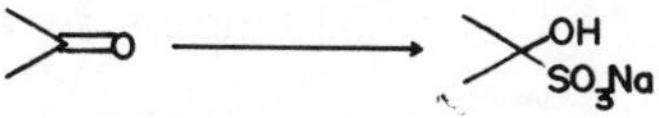

8→16   (structure: >C=O)   →   (structure: >C(OH)(SO₃Na))   4.F.1   232-233

→   (structure: >C(SR)(SR))   1.A.1   3-6   9.A.1   465,467

8→17   Ar—OH   →   Ar—X   (7.A.2   349)

R—CHO   →   R—C(X)(OR')   6.B.7   309

R—CHO   →   R—C(X)(OAc)   14.A.5   812-813

(structure: >C=O)   →   (structure: >C(X)(X))   7.A.3   335-336   7.A.3   350;
(7.A.13   351)   Ad.   359-360

R—C(OR')(OR')   →   R—C(X)(OR')   6.B.8   310

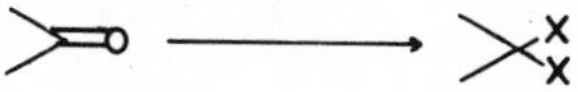

I.B. 2 ⟶ 2

9 ⟶ 7	 (R, F, F alkene) ⟶ R−≡−NR'$_2$			3.A.1	173
15 ⟶ 8	 =PPh$_3$ ⟶ =O			II.A.17	548
16 ⟶ 6	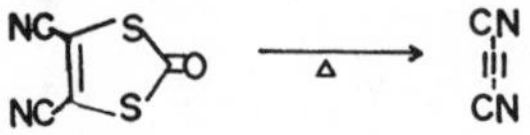 (NC, NC, S, S, O ring) ⟶ CN, CN	3.A.8	161		
	 (R, R, SO$_2$ ring) ⟶ R, R (alkyne)			3.A.8	180
16 ⟶ 8	Ar−SO$_3$H ⟶ Ar−OH	5.A.1	247		
	=SR ⟶ =O			II.D.3	571
	(SR, SR) ⟶ =O	10.D.3	594-596;	II.D.1	570
		II.D.1	667		
	(S(O)−R, SR) ⟶ =O			10.D.1	510;
				10.D.7	515;
				II.D.7	575-576
	 (SR, SR) ⟶ (OR', OR')	9.A.7	524-525	(9.C.1	475)
16 ⟶ 16	Ar−SO$_2$−N< ⟶ Ar−S−S−Ar	8.B.4	440-441		
	Ar−SO$_3$H ⟶ Ar−SO$_2$O−C(=O)−R			16.A.10	800
				(16.C.3	801-802)
16 ⟶ 17	Ar−SO$_2$Cl ⟶ Ar−Cl			7.A.12	355
17 ⟶ 6	=X ⟶ (alkyne)	3.A.1	151-154	3.A.1	169-174
		(3.A.2	154)	(3.A.4	177)
				(6.I.2	337)
				(6.I.5	339)
	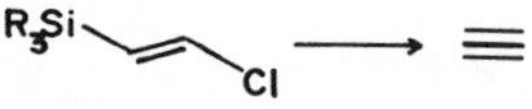 R$_3$Si−CH=CH−Cl ⟶ ≡			3.A.6	179
	(X, X alkene) ⟶ −≡−	3.A.3	155		
	(X, X alkene) ⟶ ==⟍	3.D.1	171	3.E.1	200
		(3.D.3,4	172-173)		

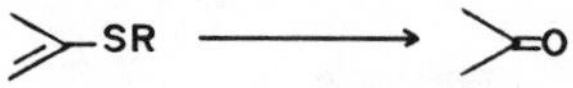

I.B.   2⟶2

17⟶6   ⟶ $-\equiv-$ 3.A.I   151-154   3.A.I   169-174

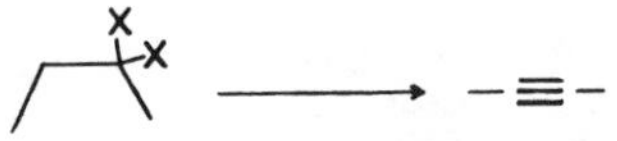

⟶ $-\equiv-$ 3.A.3   154-155

17⟶7   Ar–X ⟶ Ar–N<   8.C.2   446-448   8.C.1,2   418-423

Ar–X ⟶ Ar–NO$_2$ 20.B.I   895-896

Z⌒X ⟶ Z⌒NR$_2$ 8.C.I   417

⟶   3.A.I   153

R–CX–NO$_2$ ⟶ R–C(NO$_2$)–NO$_2$ 20.D.2   1005

17⟶8   Ar–X $\xrightarrow{\text{via ArMgX}}$ Ar–OH   5.A.2   247-250   5.A.2   272-274
                                              5.B.2   260-261   5.B.2   281-282

Ar–X ⟶ Ar–OR   6.C.2   313-316   (6.A.3   308)

Ar–X ⟶ Ar–O–Ar   6.A.3   292-293   6.A.3   306-308
                      (6.A.I   288)

Ar–X ⟶ Ar–O–C(=O)–R   14.A.IO   817-818

>–X ⟶ >–OR   6.I5   338-339

>–X ⟶ >=O   II.D.3   669-670   II.D.3   571

⟶   II.A.14   643

Z⌒X ⟶ Z–CH(OR)–OR   9.B.5   531-532

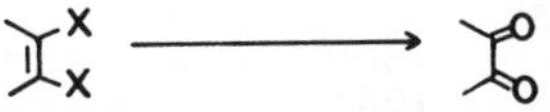

⟶ =⟨OR   6.I.2   336-337

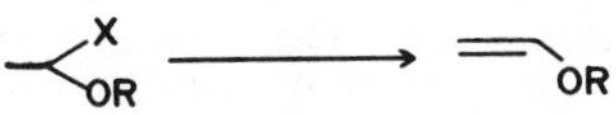

⟶ <(OR)(OR)   9.A.8   468

R–CX–OAc ⟶ R–CHO   IO.D.5   598-599

I.B.   2 → 2

17 → 8  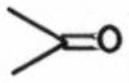	>C(X)(X) ⟶ >C=O	10.D.4	597-598;	10.D.4	513-514
		11.D.2	668		
		(10.F.2	610)		
	⟶ >C(OR)(OR)	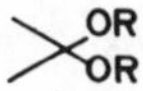 9.A.8	525-526	9.A.8	468
17 → 17	Ar-X ⟶ Ar-X' via ArMgX	7.A.6	339-342		
		7.C.9	388-390		

Functionality   3 → 3

7	>C=C=N- review			17.D	812
7 → 8	R-≡-N< ⟶ R-CH₂-C(O)-N<			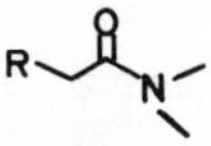 18.E.10	847-848
	R-CN ⟶ R-C(NR')-OR''	8.A.3	420		
		(7.A.7	342)		
	⟶ R-C(O)-NH₂	18.A.5	903-905	18.A.5	824-825
		(7.A.7	342)		
	⟶ R-C(O)-N<	18.D.4	921-922	18.D.4	834-835
		(8.F.1	479)	(8.F.1	437)
	⟶ R-C(O)-OH	13.A.4	752-753	13.A.4	661-662
		(19.C.5	958-959)	(19.C.5	873-874)
	⟶ R-C(O)-OR'	14.A.6	813-814		
	⟶ R-C(OR')₃			14.F.1	778-779
7 → 9	R-CN ⟶ R-CF₃	7.A.13	351		
8 → 7	H-C(O)-NMe₂ ⟶ H-C(NMe₂)₃	8.D.1	462		
	R-C(O)-NH₂ ⟶ R-CN	19.C.1	951-953	19.C.1	870-871
	R-C(O)-NH-R' ⟶ R-CN	7.A.10	347-348;	(7.A.10	354)
		19.C.2	953-954	19.C.2	872
	R-C(O)-NH-X ⟶ R-CN	19.C.8	960		
	R-C(O)-OH ⟶ R-CN	19.C.5,6	958-959	19.C.5,6	873-874
				(19.C.1	871)
8 → 8	R-C≡C-OR' ⟶ R-CH=C=O			17.B.3	811

I.B.   3 ⟶ 3

8 ⟶ 8   R-C≡C-OR' ⟶ R-CH₂-C(=O)-OR'   16.C.1   884-885   (16.D.1   802)

R-C(=N-R')-OR" ⟶ R-C(=O)-OH   13.A.10   664-667

⟶ R-C(=O)-OEt   13.A.10   664-667

R-CH=C=O ⟶ R-CH₂-C(=O)-N⟨   18A.8   905-906
                                   (10.B.4   573)

⟶ R-CH₂-C(=O)-OH   (11.F.5   693)
                                   (13.G.1   796-797)

⟶ R-CH₂-C(=O)-OR'   14.A.4   811-812   14.A.4   720
                                   (14.B.3   827-828)

⟶ R-CH₂-C(=O)-O-C(=O)-R'   16.A.5   880-881

R-C(=O)-NH₂ ⟶ R-C(=O)-NH-NO₂   20.A.1   982

R-C(=O)-NH-R' ⟶ R-C(=O)-NH-R"   18.A.11   907-909   18.A.11   825-826
                                                              (18.A.15 827-828)

⟶ R-C(=O)-N(R')(R")   18.E.1   929-930   18.E.1   842-844
                                   (18.E.2,3 931-932)

R-C(=O)-NH₂ ⟶ (R-C(=O)-)₃N   18.A.2   821

R-C(=O)-NR'₂ ⟶ R-C(=O)-OH   13.A.3   751-752   13.A.3   660
                                         (8.B.2,4 437-441)

⟶ R-C(=O)-OR'   8.B.4   440-441;   14.A.14 726
                                   14.A.14 819-820

R-C(=O)-NH-NH₂ ⟶ R-C(=O)-NH₂   18.B.2   914   18.B.2   830-831

⟶ R-C(=O)-OH   13.B.14 674

R-C(=O)-NH-OH ⟶ R-C(=O)-NH₂   18.B.1   913-914

⟶ R-C(=O)-OH   13.B.14 674

R-C(=O)-N₃ ⟶ R-C(=O)-NH-R'   18.A.12   909   18.A.12 826-827

⟶   14.A.4   812   14.A.4   720

⟶   14.F.4   780

CO ⟶ H-C(=O)-NR₂   18.D.8   927   18.D.8   838-840

I.B.   3 ⟶ 3

8 ⟶ 8   R–C(=O)–OH ⟶ R–(oxazoline) 13.A.10 665-666

R–CH₂–C(=O)–OH ⟶ R–CH=C=O   17.A.1   886-887   17.A.1   804-806

R–C(=O)–OH ⟶ R–C(=O)–N<   18.A.1   895-899   18.A.1   816-820
                                  (18.A.13 909-910)   (18.A.13 827)
                                  (18.B.3   914-915)

R–C(=O)–OH ⟶ R–C(=O)–OR'   14.A.1   802-807;   14.A.1   715-718;
                                  14.A.10,11 817-818;   14.A.10,11 721-726;
                                  14.B.1,2   826-827   14.B.1,2 730-732;
                                  (14.A.9   816-817)   Ad.   728
                                  (14.A.13   819)   (14.A.21 727-728)
                                  (14.B.5   830-831)

R–C(=O)–OH ⟶ R–C(=O)–O–C=C–   14.A.17   822;
                                  14.B.5   830-831   14.B.5   734

R–C(=O)–OH ⟶ R–C(=O)–O–C(=O)–R'   16.A.1-6   874-881;   16.A.1-3 795-798;
                                  16.C.1   884-885   16.A.7-9 798-800;
                                  (18.A.14   910)   Ad.   801

R–CH₂–C(=O)–OR' ⟶ R–CH=C=O   17.A.1   886-887   17.A.1   804-806

R–CH₂–C(=O)–OR' ⟶ R–CH=C(OR'')₂   9.A.1   517

R–C(=O)–OR' ⟶ R–C(=O)–N<   18.A.4   901-903   18.A.4   822-824

R–C(=O)–OR' ⟶ R–C(=O)–OH   13.A.2   748-751   13.A.2   658-660
                                  (4.A.1   176)   (4.A.1   211)
    H₂
Malonic Acids   11.F.3   689-690
                                  13.A.9   757-758

R–C(=O)–OR' ⟶ R–C(=O)–OR'
    (protection)   14.A.20 727

R–C(=O)–OR' ⟶ R–C(=O)–OR''   14.A.7   814-815   14.A.7   720-721
                                  (14.A.8,9 815-816)

R–C(=O)–O–C=C– ⟶ R–C(=O)–OR'   14.A.7   814-815
                                  (14.A.4   811-812)

⟶ R–C(=O)–O–C(=O)–R'   16.A.5   880-881

R–C(=O)–O–C(=O)–R' ⟶ ketene   17.A.1   886-887

⟶ R–C(=O)–N<   18.A.3   900-901   18.A.3   821-822

⟶ R–C(=O)–OH   13.A.1   745-748

I.B. 3⟶3

8⟶8	R-C(=O)-O-C(=O)-R'	⟶	R-C(=O)-OR''	14.A.3 (14.A.5	809-811 812-813)	14.A.3 719
		⟶	R-C(=O)-O-C(=O)-R''	16.A.2 16.A.4	876-878; 880	
	R-C(OR')₃	⟶	R-C(=O)-OR'	9.A.4	520-521	
	R-C(OR')₃	⟶	R-C(OR'')₃			14.F.3 780
8⟶9	R-C(=O)-OH	⟶	R-C(=O)-F			15.A.8 789 (15.A.1-3 783-788)
	R-C(=O)-OH	⟶	R-CF₃	7.A.13	351	7.A.12 355
8⟶16	R-C(=O)-NH₂	⟶	R-C(=O)-SH			Ad. 668
8⟶17	R-C(=O)-N<	⟶	R-C(=O)-X	15.A.6 15.B.3	866-867; 870	15.B.3 791-792
	R-C(=O)-OH	⟶	R-C(=O)-X	15.A.1-3	859-865	15.A.1-3 783-788
	R-C(=O)-OR'	⟶	R-C(=O)-X	15.A.3	864-865	15.A.3 786-788
	R-C(=O)-O-C(=O)-R'	⟶	R-C(=O)-X	15.A.4	865-866	15.A.4 788 (15.A.2 785)
	R-C(=O)-O-SiMe₃	⟶	R-C(=O)-X			(15.A.3 787)
9⟶8	CF₂=CF₂	⟶	CHF₂-C(=O)-NR₂	18.E.3	931-932	
	R-C(=O)-F	⟶	R-C(=O)-OR'			14.D.15 768
16⟶8	R-C(=S)-SH	⟶	R-C(=O)-N<	18.A.10	907	
	R-C(=S)-NH₂	⟶	R-C(=O)-NH₂	18.A.9	906-907	18.A.9 825
	R-C(=S)-OR'	⟶	R-C(=O)-OR'			16.A.7 798
	R-C(SR')₃	⟶	R-C(=O)-OH			13.A.10 667
16⟶17	R-C(=O)-SR'	⟶	R-C(=O)-X	15.B.4	870	
17⟶7	R-C≡C-X	⟶	R-C≡C-NR'₂	8.H.3	512	
	R-C≡C-X	⟶	R-CH₂-CN			(18.A.1 858)
	R-C(=O)-X	⟶	R-CN	19.C.7	959-960	

I.B   3→3

17→8   R–C≡C–X $\longrightarrow$ R–$\overset{O}{\overset{\|}{C}}$–$\overset{O}{\overset{\|}{C}}$–OMe   11.A.19   549

R–$\overset{X}{\overset{\|}{C}}$=N–NHAr $\longrightarrow$ R–$\overset{O}{\overset{\|}{C}}$–NH–NAr$_2$ 18.D.6   926

R–CH$_2$–$\overset{O}{\overset{\|}{C}}$–X $\longrightarrow$ R–CH=C=O   17.A.5   889-890;
                                                 17.B.2   891-892   17.B.2   810-811

R–$\overset{X}{\overset{\|}{C}}$H–$\overset{O}{\overset{\|}{C}}$–X $\longrightarrow$ R–CH=C=O   17.B.1   891   17.B.1   809-810

R–$\overset{O}{\overset{\|}{C}}$–X $\longrightarrow$ R–$\overset{O}{\overset{\|}{C}}$–N<   18.A.2   899-900   18.A.2 820-821

$\longrightarrow$ R–$\overset{O}{\overset{\|}{C}}$–OH   13.A.1   745-748

$\longrightarrow$ R–$\overset{O}{\overset{\|}{C}}$–OR'   14.A.2   807-809;   14.A.2   718-719
                                                  14.D.7   857
                                                  (14.B.7   832-833)

$\longrightarrow$ R–C–O–C–R'   16.A.2-4   876-880   16.A.2-3 797-798

$\longrightarrow$ R–$\overset{O}{\overset{\|}{C}}$–O–SO$_2$R'   16.A.10   800-801

R–CH=CX$_2$ $\longrightarrow$ R–CH=C(OR')$_2$ 9.B.1   470
                                                          (19.B.4   868)

$\longrightarrow$ R–CH$_2$–$\overset{O}{\overset{\|}{C}}$–OH   13.A.7   755-756

R–CX$_2$–OR' $\longrightarrow$ R–$\overset{O}{\overset{\|}{C}}$–OR'   14.A.16   821-822

R–CX$_3$ $\longrightarrow$ R–$\overset{O}{\overset{\|}{C}}$–N<   18.A.7   905

$\longrightarrow$ R–$\overset{O}{\overset{\|}{C}}$–OH   13.A.6   754-755   13.A.6   662-664

$\longrightarrow$ R–$\overset{O}{\overset{\|}{C}}$–OR'   14.A.15   821

$\longrightarrow$ R–C(OR')$_3$ 14.F.2   779

17→17   [pyridine]–X $\longrightarrow$ [pyridine]–X'   7.A.6   339-342

R–$\overset{O}{\overset{\|}{C}}$–X $\longrightarrow$ R–$\overset{O}{\overset{\|}{C}}$–X'   15.A.5   866   15.A.5   788-789

R–CH=CX$_2$ $\longrightarrow$ R–C≡C–X   3.A.1   172-173

R–CX$_3$ $\longrightarrow$ R–$\overset{O}{\overset{\|}{C}}$–X   15.A.7   867

I.B. 4 → 4

Functionality 4 → 4

$$7 \longrightarrow 8 \quad R-N=C=N-R \longrightarrow R-N=\overset{\overset{\displaystyle OPh}{|}}{C}-NHR \qquad 18.A.8 \quad 906$$

$$R-N=C=N-R \longrightarrow R-NH-\overset{\overset{\displaystyle O}{\|}}{C}-NH-R \qquad \begin{array}{ll} 6.A.2 & 291; \\ 16.A.1 & 875; \\ 18.A.1 & 896; \\ 18.A.8 & 906 \end{array}$$

$$R_2N-CN \longrightarrow CO_2 \qquad 8.B.3 \quad 439\text{-}440$$

$$8 \longrightarrow 7 \quad R-NH-\overset{\overset{\displaystyle O}{\|}}{C}-NH-R \longrightarrow R-N=C=N-R \qquad\qquad 16.A.1 \quad 795\text{-}796$$

$$8 \longrightarrow 8 \quad R-N=C=O \longrightarrow R-NH-\overset{\overset{\displaystyle O}{\|}}{C}-OR' \qquad \begin{array}{ll} 14.A.4 & 812; \\ 14.A.14 & 819\text{-}820 \\ (18.A.13 & 909) \end{array} \quad 14.A.4 \quad 720$$

$$R-N=C=O \longrightarrow CO_2 \qquad 8.B.1 \quad 436\text{-}437$$

$$R-NH\overset{\overset{\displaystyle O}{\|}}{C}-NH_2 \longrightarrow R-N=C=O \qquad \begin{array}{ll} 14.A.14 & 819; \\ 18.A.13 & 909 \end{array}$$

$$16 \longrightarrow 7 \quad R-NH-\overset{\overset{\displaystyle S}{\|}}{C}-NH-R \longrightarrow R-N=C=N-R \qquad\qquad 16.A.1 \quad 795\text{-}796$$

$$16 \longrightarrow 8 \quad R-S-\overset{\overset{\displaystyle O}{\|}}{C}-NH-R \longrightarrow CO_2 \qquad 8.B.1 \quad 436\text{-}437$$

$$R-NH-\overset{\overset{\displaystyle S}{\|}}{C}-NH-R \longrightarrow R-NH-\overset{\overset{\displaystyle O}{\|}}{C}-NH-R \qquad\qquad 16.A.8 \quad 799$$

$$CS_2 \longrightarrow C(OR)_4 \qquad\qquad 14.G.2 \quad 781$$

$$\longrightarrow C(OCOR)_4 \qquad 16.A.3 \quad 879$$

$$16 \longrightarrow 16 \quad R_2N-\overset{\overset{\displaystyle S}{\|}}{C}-NR_2 \longrightarrow RO-\overset{\overset{\displaystyle S}{\|}}{C}-OR \qquad 2.F.6 \quad 148$$

$$CS_2 \longrightarrow R-O-\overset{\overset{\displaystyle S}{\|}}{C}-S-R' \qquad 2.A.7 \quad 88\text{-}89$$

$$16 \longrightarrow 17 \quad R-N=C=S \longrightarrow R-N=CX_2 \qquad 7.C.8 \quad 388$$

$$CS_2 \longrightarrow CX_4 \qquad 7.C.8 \quad 388$$

$$17 \longrightarrow 8 \quad R_2N-\overset{\overset{\displaystyle O}{\|}}{C}-X \longrightarrow R_2N-\overset{\overset{\displaystyle O}{\|}}{C}-O-\overset{\overset{\displaystyle O}{\|}}{C}-R \qquad (16.A.1 \quad 797)$$

$$RO-\overset{\overset{\displaystyle O}{\|}}{C}-X \longrightarrow RO-\overset{\overset{\displaystyle O}{\|}}{C}-O-\overset{\overset{\displaystyle O}{\|}}{C}-R' \qquad\qquad Ad. \quad 801$$

$$Cl_3C-NO_2 \longrightarrow C(OR)_4 \qquad\qquad 14.G.1 \quad 781$$

$$17 \longrightarrow 17 \quad FCCl_3 \longrightarrow F-\overset{\overset{\displaystyle O}{\|}}{C}-Cl \qquad\qquad 15.C.7 \quad 789$$

I.C. O→I

C. OXIDATION OF SUBSTRATE

Functionality O→I

6 → 6		I.E 2.D.5	49-52; 139	I.E 2.D.5 (I.C.6 (3.D.3 | 48-49; 153 36-37) 198)

II.A.21 550

6 → 7	R—H ⟶ R—N< α to carbonyl	7.E.3 8.F.4 8.E.3	410; 485-488 475	8.F.2 (18.D.4 13.C.1	437-438 835) 677
	R—H ⟶ R—NO₂	20.A.1 20.B.3	980-983; 994-995	20.A.1 20.B.3	888; 897-898
	R—B< ⟶ R—N<	8.C.9	456-458	(8.F.4	442)
	R—MgX ⟶ R—N<	8.C.9	456-458	8.C.9	425-426

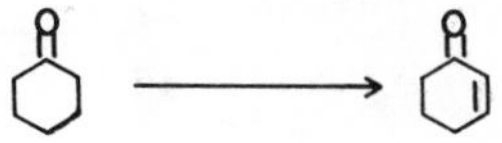

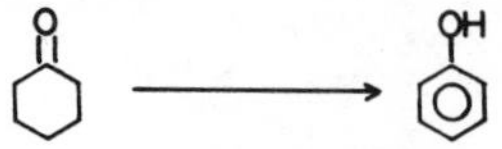

R—NO₂   20.A.1   980-983;   20.A.1   888;
        20.B.3   994-995   20.B.3   897-898

R—B< → R—N<   8.C.9   456-458   (8.F.4   442)

R—MgX → R—N<   8.C.9   456-458   8.C.9   425-426

6 → 8   R—H ⟶ R—OH
        allylic
        α to carbonyl

4.D.1   212-214   4.D.1   245
4.D.4   216-219   4.D.4   247
                  4.D.2   246;
                  4.D.7   249;
                  Ad.   251

R—H ⟶ R—OR'   6.D.2   327-328

R—H ⟶ R-OAc
  allylic   4.D.4   218-219
  α to carbonyl 14.D.6   858-859   14.D.6   760-763;
                                    10.A.26 496

R—H ⟶ R—OOH   5.B.4   263;
                II.A.7   637

R—B< ⟶ R—OH   4.B.2   186-188   4.B.2   219-221;
                                   Ad.   227

R—HgX ⟶ R—OH   Ad.   262

6 → 16 R—CH₂C̈—OH ⟶ R-CH-C̈-Cl   15.A.1   784
                        SO₂Cl

I.C.   0 → I

6 → 16   11.A.21 550

6 → 17   R—H ⟶ R—X

                                 7.C.1   376-380   7.C.1   374-375
                                 (7.C.4   382-383)   (18.F   850)

         allylic   7.C.2   380-382   7.C.2   375
                                   (4.D.4   219)
                                 (3.B.3   166)

     α to carbonyl   7.C.5,6   383-387   7.C.5,6 376-378
                                                    (7.C.9   379)
                                                  (11.A.22 550)

    R—B< ⟶ R—X   7.A.18   356-357
                                                   (7.C.9   379)

Functionality   0 → 2

6 → 7   10.A.15   562;
                                              11.D.5   671-672

           20.A.1 888

6 → 8   10.D.4   597-598;   11.A.6   543-545
                                         11.A.6   633-637;   (1.C.7   38-40)
                                         11.D.2   668   (11.A.7   545)
                                         (10.A.9   557-558)

           11.A.9   639-640   11.E.6   581;
                                         (11.E.6   684-685)   Ad.   584

           10.A.14-15 561-563   (10.A.15 490-492)

       Ar—CH₃ ⟶ Ar—CHO   10.A.11-15 560-563   10.A.11-15 489-492

       Ar—CH₃ ⟶ Ar—CH(OAc)₂   14.D.4   856
                                             (10.A.11   560)

           12.A.1   726-727

6 → 16   9.D.6   477;
                                                11.E.6   581
                                               (11.F.2   587)

6 → 17   (7.C.1   379)
                                              (10.D.4   597)

I.C. 0→3

Functionality 0→3

6 ⟶ 7   Ar–CH₃ ⟶ Ar–CN   19.F.1   882

6 ⟶ 8   Ar–C(=O)–CH₂CH₃ ⟶ Ar–CH₂CH₂–C(=O)–NH₂   18.C.2   916–918   18.C.2   831–832

          Ar–CH₃ ⟶ Ar–C(=O)–OH   13.B.8   769–770   13.B.1   669;
                                        (13.B.10 771)   13.B.8   673

          Ar–C(=O)–CH₃ ⟶ Ar–CH₂–C(=O)–OMe   14.D.8   764

6 ⟶ 17   R–CH₃ ⟶ R–CX₃   7.C.1   378
                                        (13.B.12 772–774)

Functionality 1→2

6 ⟶ 6   R–CH=CH₂ ⟶ R–C≡CH   3.A.1   151–154;   3.A.1   169–174
                                        3.A.6   159–160

6 ⟶ 7   ⟩ ⟶ ⟩–NO₂   20.A.3   988–990   20.A.3   892–893
                                                            (10.F.3   523)

          Ar–H ⟶ Ar–N<   8.D.7   470;   5.D.9   294;
                                        8.F.2   480–482   8.F.8   445;
                                                            18.D.5   837

          Ar–H ⟶ Ar–NO   12.B.2   738–739

          Ar–H ⟶ Ar–NO₂   20.A.2   983–987   20.   887
                                        (5.D.3,4 274)   20.A.2   888–892

          HO–⟨⟩ ⟶ O=⟨⟩=N–Y   12.B.2   738–739   12.B.2   646–647

6 ⟶ 8   Ar–H ⟶ Ar–OH   5.B.1   259–260;   5.B.1   278–281;
                                        5.B.7,8   264–266   5.B.10   286
                                        (5.D.4,5 274–275)   (5.B.2   281–282)
                                                            (8.A.1   396)

          Ar–H ⟶ Ar–OMe   6.G.2   333

          ⟨⟩ ⟶ O=⟨⟩=O   12.A.1   725–726   12.A.1   636
                                        (9.E.3   541)
                                        (12.A.2   726–728)
                                        (12.A.5   734–735)

I.C.   1→2

6 ⟶ 8

		11.A.5	631-633;	10.A.12	490;
		11.A.10,11	641	10.E.1	517-518;
		(11.A.13	642)	11.A.5	542-543
		(11.E.3	680)	(11.C.10	566)
				(11.E.3	579-580)
				(11.F.4	589)

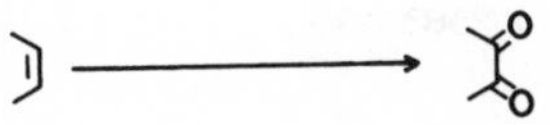

11.A.5   543

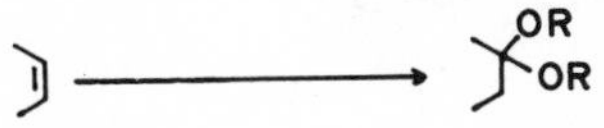

| | | 9.D.1 | 536-537 | 9.D.1 | 475-476 |
| | | | | (11.A.5 | 542) |

		10.D.6	514;
		11.E.3	580;
		Ad.	611

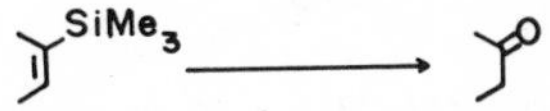

6 ⟶ 16   Ar-H ⟶ Ar-S-Ar   7.D.7   386

⟶ Ar-SO₂-Ph   8.F.3   439

⟶ Ar-SO₃H   5.D.3   274

6 ⟶ 17   Ar-H ⟶ Ar-X

7.D.1-5	392-404	7.D.1-5	380-385
(5.D.2	273)	(5.D.2	292-293)
		(8.F.7	444)

Ar-BR₂
Ar-MgX
Ar-SiMe₃ ⟶ Ar-X   7.C.9   388-389
Ar-SnR₃
Ar-HgX

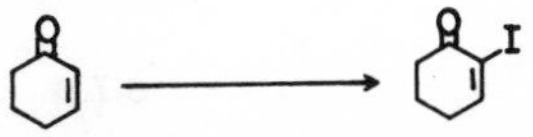

7.B.3   364

7 ⟶ 6   ⟶   3.A.5   158-159

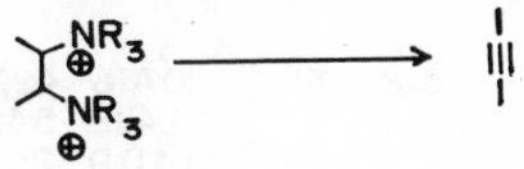

7 ⟶ 7   R-CH₂-NO₂ ⟶ R-CH-NO₂   20.D.2   1004-5   20.D.2   905-906
                              |
                             NO₂

7 ⟶ 8   Ar-N-CH₃ ⟶ Ar-N-CH₂OMe   6.D.2   322
            | |

⟶NH₂ ⟶ ⟶O

(11.A.13	642-643)	10.A.21	493-494;
		11.A.13	546-547
		(10.A.23	495)
		(11.A.4	541)

I. C. 1 → 2

	Reaction		
7 → 8	(spiro oxazolidinone N-NO) → cyclohexylidene-CH-OR		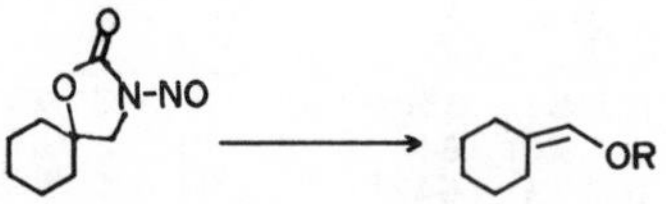 6.I.7 340
	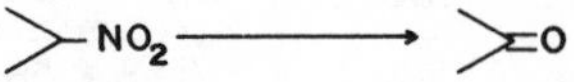 >NO₂ → >=O	10.A.18 565-566; 11.A.9 639-641	10.A.18 492-493; 11.A.9 546; Ad. 558
7 → 17	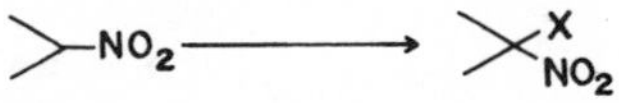 >NO₂ → >C(X)(NO₂)	7.C.5 383-385	
8 → 16	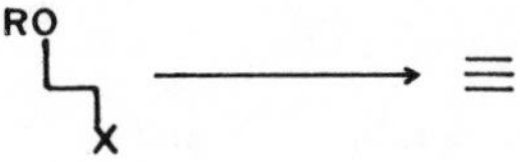 RO-CH-CH₂-X → ≡	3.A.2 154	
8 → 8	 catechol (C₆H₄(OH)₂) → phenol (C₆H₅OH)	5.D.7 277-278	
	 >OH → >=O	10.A.1-3 545-550; 10.A.5 551-552; 11.A.1-4 625-630	10.A.1 484-486; 10.A.5 487; Ad. 496; 11.A.1-4 538-541 (11.A.8 546) (Ad. 552)
	allyl-OH (C=C-C-OH) → ketone	10.E.2 603-605; 11.E.2 678-679	
	diol (HO-C-C-OH) → ketone	10.E.1 600-603; 11.E.1 677-678	11.E.1 577-578 (10.E.1 517-518)
	R—CH₂—OH → R—CH(OR')₂	9.D.4 538	
	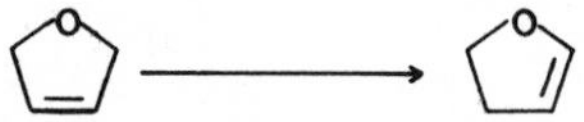 2,3-dihydrofuran → 2,5-dihydrofuran	6.C.5 319	
	≡—OR (propargyl ether) → C=C=C-OR (allene)		 6.I.9 342
	 >CH-OR → >=O	10.A.4 550-551	10.A.16 492; 11.A.18 548 (4.D.9 250)
	>C(OR)(OR) → >C(=O)(OH)		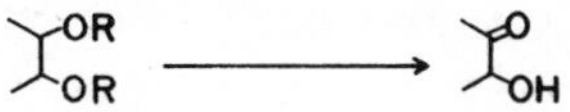 (11.D.5 575)
	epoxide (>C-C<O) → ketone	10.E.3 605-606; 11.E.3 679-681	11.E.3 579-580 (10.E.3 519)

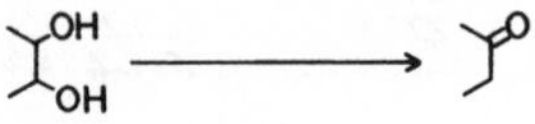

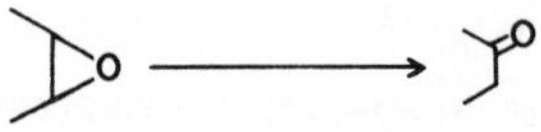

I.C.   1→2

8 → 8   [epoxide structure] → [structure] II.A.15   643

[trimethylsilyl epoxide structure] → [ketone structure]   10.D.6   514;
11.E.3   580;
Ad.   611

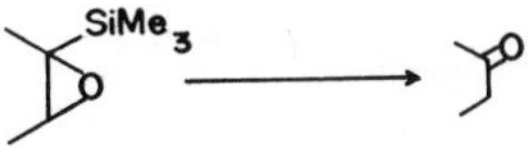

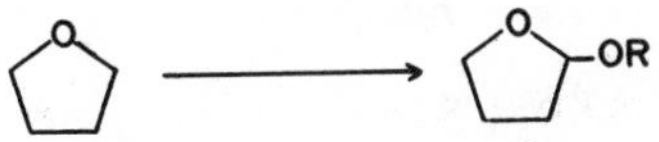

[tetrahydrofuran structure] → [structure]-OR   9.D.2   537-538;
10.A.16 563-564

R-CH(OAc)-CHO → R-C(O)-CH₂OAc   II.E.6   581

8 → 17   R-CH₂-OR' → R-CH(X)-OR'   7.C.4   382-383   7.C.4   376

14 → 8   [SiMe₃ structure] → [ketone structure]   10.D.6   514;
11.E.3   580;
Ad.   611

16 → 7   [S-Me structure] → [N-OH structure]   10.A.15 492

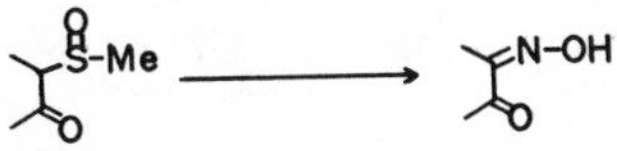

16 → 8   CH₃-S(O)-CH₃ → CH₂(OR)₂   9.D.6   477

16 → 16   R-CH₂-S-Me → R-CH(OH)-SMe   10.A.19 566-567   10.A.19 493

16 → 17   R-CH₂-SR' → R-CH(X)-SR'   7.C.4   382-383

R-CH₂-SO₂-Z → R-CH(NO₂)-SO₂-Z   20.B.3   897-898

17 → 6   [X Y structure] → ≡   3.A.1,2 151-154   3.A.1,2 169-175

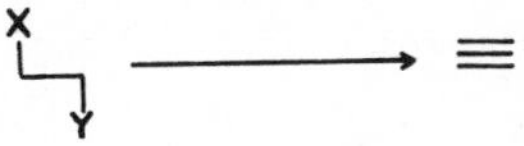

17 → 7   [=X structure] → [=NOH structure]   II.D.5   672

17 → 8   [=X structure] → [=O structure]   10.A.5   551-552;   10.A.5   487;
10.A.9-10 557-560   10.A.10 489
(10.A.19 566-567)   (II.A.6   544)
(II.D.5   672)   (II.A.14 547)
(Ad.   552)

17 → 16   [=X structure] → [OH/SMe structure]   10.A.19 566-567

Functionality 1→3

7 → 7   R-CH₂-NH₂ → R-CN   19.F.1   977-978   19.C.10 876;
19.F.1   881-882

I.C. 1 ⟶ 3

7 ⟶ 7   R—CH₂—NO₂ ⟶ R—CN   19.G.1   883

7 ⟶ 8   R—CH₂—N— ⟶ R—C(=O)—N—   18.C.1   915-916   18.C.4   832

        CH₃—NO₂ ⟶ CO + NH₂OH   8.G.4   501

        R—CH₂—NO₂ ⟶ R—C(=O)—OH   13.G.4   799

8 ⟶ 7   R—CH₂—OH ⟶ R—CN   19.F.2   978

8 ⟶ 8   R—CH₂—OH ⟶ R—C(=O)—OH   13.B.1   760-763   13.B.1   669-672

        ⟶ R—C(=O)—OR'   (14.D.2   855)

        R—CH₂—OR' ⟶ R—C(=O)—OR'   14.D.3   855   14.D.3   759-760

        R—CH₂—OAc ⟶ R—C(=O)—OH   13.B.1   672

17 ⟶ 7   R—CH₂—X ⟶ R—CN   19.B.6   869

17 ⟶ 8   R—C(=O)—CH₂—X ⟶ R—C(=O)—C(=O)—OR'   11.A.6   634

Functionality 2 ⟶ 3

6 ⟶ 7   R—C≡C—H ⟶ R—C≡C—NO₂   20.A.3   892

6 ⟶ 8   (CH₃)₂C(OH)—C≡ ⟶ =CH—CO₂H   13.A.8   756

6 ⟶ 17   R—C≡C—H ⟶ R—C≡C—X   3.B.3   166;   3.B.3   189-190;
                              7.C.3   382   7.C.3   375-376

        R—C≡C—SiR'₃ ⟶ R—C≡C—X   7.C.3   375-376

7 ⟶ 7   (pyridine) ⟶ (2-aminopyridine) NH₂   8.E.4   477-478   (8.F.5   443)

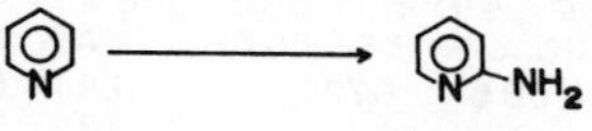

        Ar—CH=N—NH₂ ⟶ Ar—CN   19.C.4   873

        R—CH=N—OH ⟶ R—CN   19.C.4   956-958   19.C.4   872-873
                           (18.D.5   922-924)
                           (19.C.2   954)

7 ⟶ 8   R—CH=N—OH ⟶ R—C(=O)—N<   18.D.5   922-924

I.C. 2 → 3

7 → 8   R—CH≡N—OH ⟶ R—C(=O)—OH   13.A.5   754

8 → 7   R—CHO ⟶ R—CN   19.C.2-4 954-958;   19.C.4 872-873;
                          19.F.2   978   19.F.2 882-883
                          (18.D.6   924)

8 → 8   [furan] ⟶ [RO, OR, RO, OR cyclic]   9.E.2   540

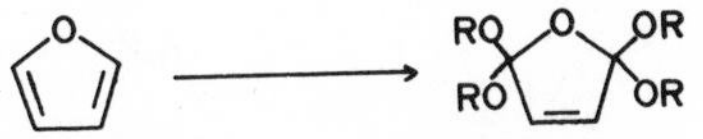

        R—CHO ⟶ R—C(=O)—N<   18.C.3   918   18.C.2 832

        ⟶ R—C(=O)—OH   13.B.1   760-763   13.B.1 669-672
                         (4.C.4   199)
                         (13.B.3   764)

        ⟶ R—C(=O)—OR'   (14.D.1,2 853-855)   14.A.22 728
                                                 (14.D.1   759)

        ⟶ R—C(=O)—O—C(=O)—R'   16.B.3   884

        R—CH(OR')$_2$ ⟶ R—C(=O)—OR'   (11.G.5   709)   14.D.12 766

8 → 17 R—CHO ⟶ R—C(=O)—X   15.B.1   869   15.B.1   791

17 → 7 R—CH(X)—NO$_2$ ⟶ R—CN   19.C.8   960

        ⟶ R—C(NO$_2$)$_3$ 20.C.3 903

17 → 17 R—CHX$_2$ ⟶ R—C(=O)—X   15.B.2 869-870

Functionality   3 → 4

8 → 8   CO ⟶ —N—C(=O)—N—   18.D.8   927   18.D.8   839

16 → 17 R—S—CHO ⟶ R—S—CCl$_3$ 7.A.10   354

II. CARBON-CARBON
     BOND-BREAKING REACTIONS

Functionality 0 → 0

6 → 6   [triangle] —H$_2$→ [open]   I.A.10   I.A.10 17-18;
                                               II.B.9 557-558

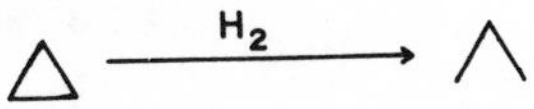

        R—CH$_2$—OH ⟶ R—H   I.A.2   8

        R—CHO ⟶ R—H   I.G.5   59-61

II.   O⟶O

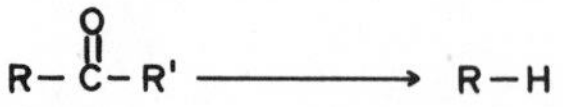

$$R-\overset{\overset{O}{\|}}{C}-R' \longrightarrow R-H$$

I.G.5	61-62;	I.G.5	59-61
13.B.5	765;		
13.E.3	786-789		

$$Z-\overset{|}{\underset{|}{C}}-\overset{\overset{O}{\|}}{C}- \longrightarrow Z-\overset{|}{\underset{|}{C}}-H$$

13.E.1	784-785;	11.A.24	551;
14.C.9	848-849	13.D.4	688;
		14.C.9	755

$$R-CH_2-CN \longrightarrow R-CH_3$$

		I.A.11	18-19

$$Z-\overset{|}{\underset{|}{C}}-\overset{\overset{O}{\|}}{C}-OR \longrightarrow Z-\overset{|}{\underset{|}{C}}-H$$

I.F.1	53;	I.F.1	50-52;
10.D.1	667;	I.F.3	53;
11.D.6	673-675;	11.D.6	575;
11.F.3	668-690;	13.C.6	685-686;
13.A.9	757-758;	14.C.11	755
14.C.11	850-851;		
19.B.1	947		

Functionality   O⟶I

6⟶6

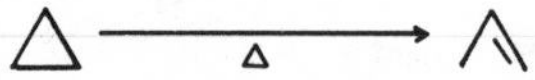

2.D.3	138		
2.D.3	138		
10.H.2	620-621		
10.H.1	620		
2.A.3	81-82	3.E.1	201-202
19.B.1	946	2.A.18	108-109
		4.C.2	235
11.F.1	688	2.F.7	165;
(2.F.1	145)	11.F.1	586
		4.D.8	249-250
(2.F.1	145)	2.A.6	94;
(17.A.2	887-888)	2.A.23	114;
		Ad.	116
		(2.B.7	127)
		(3.E.1	201-202)

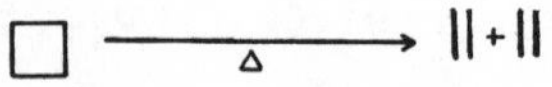

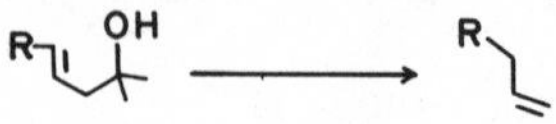

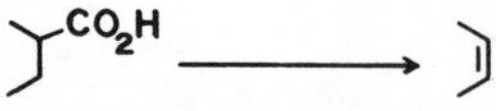

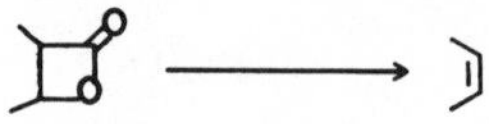

II.   0→1

6→6	 (structure: CO₂H, X) →	2.F.1	145	2.F.7 Ad.	165; 166	
	 (structure: CO₂H, CO₂H) →	2.F.3	146-147	2.F.3 (3.E.1	162-163 202)	
	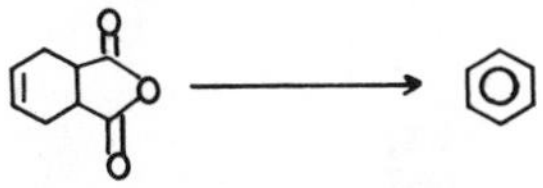 (cyclic anhydride structure) → (benzene)	1.F.2	53-54			
	 (structure: COCl) →	2.F.4	147-148			

6→7

R–CHO	→ R–NH–CHO	19.C.3	954-956		
R–C(=O)–R'	→ R–NH–C(=O)–R'	8.G.3 18.D.5-7 19.C.2	501-502; 922-927; 953-954	8.G.5	449
R–CN	→ R–NO₂	20.B.3	994		
R–C(=O)–OH	→ R–N<	8.G (10.E.5 (14.A.14 (18.D.6 (19.C.3	494-502 606-607) 819-820) 924-926) 954-956)	8.G Ad.	447-449; 452
R–C(=O)–OH	→ R–NO₂			20.B.3 20.E.3	898; 909-910

6→8

(bicyclohexane structure)	→ (cyclohexanol with OH)			4.B.2	222
R–C(=O)–R'	→ R–O–C(=O)–R'	14.B.4	828-830	14.B.4 Ad.	732-733; 769
R–C(=O)–X	→ R–OH			4.H	268

6→17

HO–(cyclopropyl structure)	→ (structure with X)	2.A.2	78		
X X (cyclopropane)	 → (structure with X)	2.A.3	81-82		
Z–C–C(=O)–R	→ Z–C–Br			14.C.9	755
R–C(=O)–OH	→ R–X	7.A.11	348-350	7.A.11	354-355

II.   0 → 2

Functionality   0 → 2

6 → 6 $Ar-\overset{O}{\overset{\|}{C}}-CH(CO_2R)_2 \longrightarrow Ar-C\equiv C-CO_2H$ 13.A.9   664

6 → 7 $\overset{O}{\overset{\|}{\diagup}}CHO \longrightarrow \overset{O}{\overset{\|}{\diagup}}N_2$ 4.A.3   179

   $\rangle-CO_2H \longrightarrow \rangle=N-OH$ 18.D.7   927

6 → 8 $\rangle-Ar \longrightarrow \rangle=O$ II.A.7   637-638
   (5.B.4   263)

   $\rangle-CHO \longrightarrow \rangle=O$ II.A.6   544-545

   $\rangle-CN \longrightarrow \rangle=O$ II.A.22 550

   $\rangle-CO_2H \longrightarrow \rangle=O$ 10.A.20 567;
   II.A.6   634-636;
   II.A.12 642

   $\overset{CO_2H}{\underset{CO_2H}{\rangle\!\!\times}} \longrightarrow \rangle=O$ 8.G.3   498;   II.A.25 551
   10.E.5   606-607

6 → 17 $NC-CH_2CO_2H \longrightarrow NC-CHX_2$ 7.A.15   352

Functionality   0 → 3

6 → 8 $Ar-R \longrightarrow Ar-\overset{O}{\overset{\|}{C}}-OH$ 13.B.8,9   769-770   13.B.8   673
   (16.B.1   882-883)   (13.B.1 669)

   $R-CH_2CH_2-\overset{O}{\overset{\|}{C}}-OH \longrightarrow R-\overset{O}{\overset{\|}{C}}-OH$ 13.E.2   785-786

   $R-CH_2-\overset{O}{\overset{\|}{C}}-OH \longrightarrow R-\overset{O}{\overset{\|}{C}}-OH$ 13.B.6   766;
   19.C.2   954

6 → 17 $CH_3-\overset{O}{\overset{\|}{C}}-R \longrightarrow CHX_3$ 13.B.12   772-774;   (13.B.12 673)
   14.C.10   850

Functionality   1 → 1

6 → 6 $Ar-\overset{O}{\overset{\|}{C}}-Ar \longrightarrow Ar-H$ 13.E.3   768-769;
   18.E.5   933-934

   $Ar-\overset{O}{\overset{\|}{C}}-OH \longrightarrow Ar-H$ 1.F.1,2   52-54   1.F.1   50-52
   (13.E.3 693)

   $\overset{CHO}{\diagup\!\!\diagdown} \longrightarrow \parallel$ 2.A.20 113

II.   1 → 1

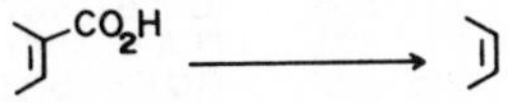

6 → 6	R-CH=C(CO₂H) → (ring)	2.F.1	144-145	2.F.1 162

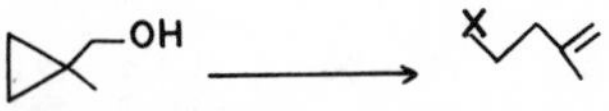

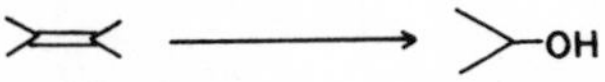

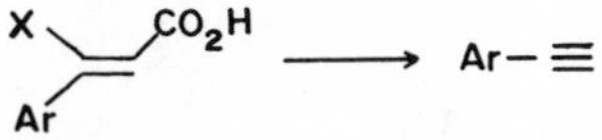

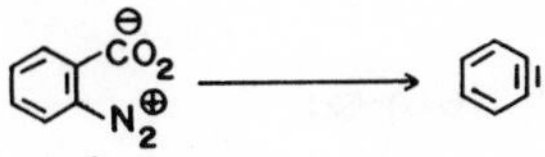

Functionality   1 → 2

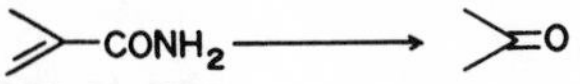

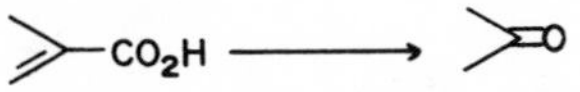

II.   1→2

6 ⟶ 8   〉=〈 ⟶ 〉=O   10.A.6   553-555;   10.A.6   487;
                                  11.A.5   631-633   11.A.5   541
                                  (4.D.3   215-216)   (10.F.3   523)
                                  (13.B.6,7   766-768)

   (enone, R) ⟶ (ketone)   3.A.4   157-158   3.A.4   176;
   (R-≡)   11.E.3   580

6 ⟶ 17   〉=〉-CONH₂ ⟶ 〉=〉-X   7.B.2   359

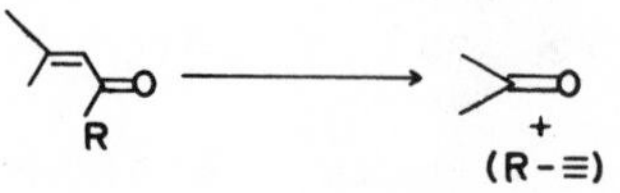

   Ar-C(O)-X ⟶ Ar-X   7.A.12   350

7 ⟶ 7   Ph, NMe₂, CN, Ph ⟶ Ph, NMe₂, Ph   2.A.18   109

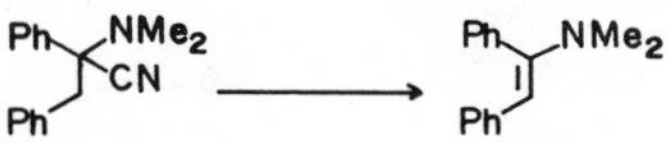

7 ⟶ 8   R-N-X / R,R ⟶ R-N-R / OMe   8.F.2   481

   〉C(NH₂)(CO₂H) ⟶ 〉=O   11.A.13   642

8 ⟶ 8   〉C-OH ⟶ 〉=O   (Δ)   10.H.2   620-621

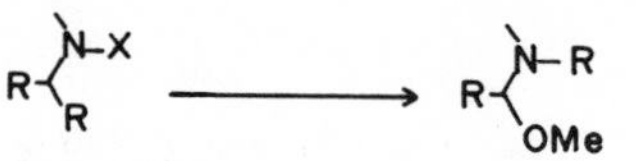

   HO-, X cyclopentane ⟶ enone   2.A.6   86

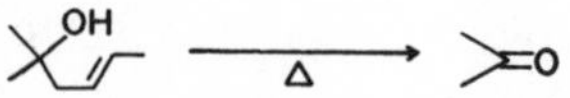

   〉C(OH)-C(O) ⟶ 〉=O   10.E.6   520

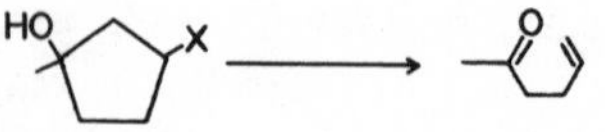

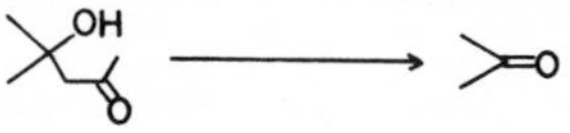

   Ar-CH(OH)-R ⟶ Ar-CHO   10.A.17   492

   OH OH diol ⟶ 〉=O   10.A.7   555-557;   10.A.7   488;
                                  11.A.8   638   11.A.8   545-546
                                                    (14.D.11   765-766)

   [OAc, OAc cyclobutene] ⟶ OAc, OAc diene (Δ)   14.E.9   777

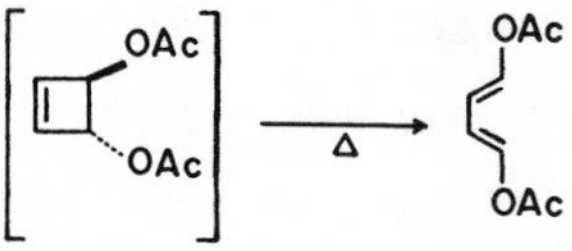

   cyclohexanone-O ⟶ CHO, CHO   10.A.25   495

   R-CH(OH)-CHO ⟶ R-CHO   10.A.8   557

II. 1 → 2

8 → 8   II.A.9   639-640

8.G.3   499;
10.E.5   607;
10.H.3   621-622

9.E.1   539-540

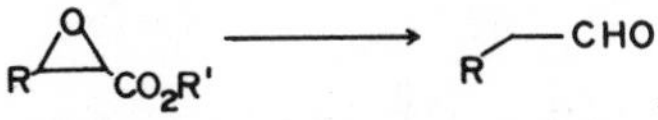

10.H.3   621-622

17 → 8   R–CH–C–OH → R–CHO   10.A.20   567;
10.E.5   607

17 → 17   7.E.2   409-410

Functionality   1 → 3

6 → 7   10.E.7   520

6 → 8   R–CH=C< → R–C–OH   13.B.6,7   766-768   13.B.6   672-673
(11.A.1   626)
(13.E.2   785-786)

9.B.8   534;
19.D.7   971

7 → 7   R–CH–C–Ph → R–CN   19.F.1   881

8 → 8   R–CH–CH₃ → R–C–OH   13.B.12   772-774   (13.B.1   670-671)
(14.C.10   850)

13.B.1   670-671

Functionality   2 → 2

6 → 6   R–C≡C–C–OH → R–C≡C–H   3.B.2   165   3.B.2   189

8 → 8   Ar–C–CH₃ → Ar–CHO   10.E.4   606

Ar–C–CH₂–OH → Ar–CHO   10.E.4   606

Ar–C–C–Ar → Ar–CHO   10.F.3   523

R–C–C–OH → R–CHO   10.H.3   621-622

Ⅱ.   2 ⟶ 3

<u>Functionality 2⟶3</u>

6 ⟶ 8   R–C≡C–H ⟶ R–C(=O)–OH   13.B.13 674

7 ⟶ 7   (cycloheptane)=N–OH / OMe ⟶ (ring)–CN / CHO   10.E.7 520

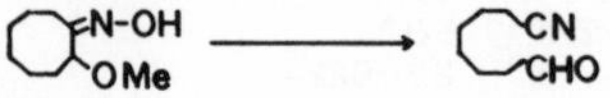

R–C(=NOH)–C(=O)–R' ⟶ R–CN   19.C.2   954

7 ⟶ 8   R–C(=N–OH)–R' ⟶ R–C(=O)–NH–R'   18.D.5   922–924
                                                       (18.D.7   927)

8 ⟶ 7   R₂C(OH)–C(OH)R [R,R–C(OH)(OH)] ⟶ R–CN+(RCO₂H)   19.F.3   979

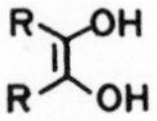

R–C(=O)–R' ⟶ R–CN   19.C.2   953–954
                                                 (13.G.3   799)

8 ⟶ 8   R–(furan) ⟶ R–C(=O)–OH   13.B.9   770

(cyclohexene)–OH/OH ⟶ (ring)–CO₂H/CN   19.F.3   979

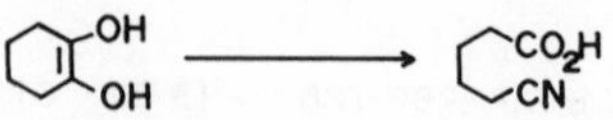

Ar–C(=O)–CH(OH)–Ar ⟶ Ar–C(=O)–OH + (ArCHO)   13.B.16 675

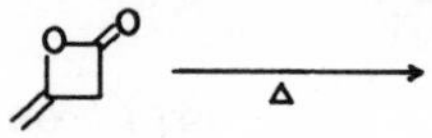

(β-methylene-β-lactone) —Δ→ C=O / =   17.A.4   889

CH₃–C(=O)–CH₃ —Δ→ O=C=O / =   17.A.1 886–887   17.A.1 804–806

R–C(=O)–R' ⟶ R–C(=O)–NH₂   18.E.5   933–934

R–C(=O)–R' ⟶ R–C(=O)–NH–R'   8.G.5   501–502;   18.D.5,6 836–838
                                        18.D.5–7 922–927   (18.C.2 832)
                                        (8.G   494)

R–C(=O)–R' ⟶ R–C(=O)–OH

  oxidation   13.B.1   761   (13.B.1 671)
                                   (18.D.7   927)
  haloform reaction   13.B.12   722–724   13.B.12 673
                                                 (18.C.2 832)
  base cleavage   13.E.3   786–789   13.E.3,4 693–694
  photochemical   13.B.5   765
                                   (14.A.4   811)

II.   2→3

8→8   R–C(=O)–R' ⟶ R–C(=O)–OR'   14.B.4   828-830   14.B4   732-733;
                                                        (14.C.10   850)   Ad.   769
                                                                            (13.B.1   671)

R–C(=O)–C(Z)< ⟶ R–C(=O)–OR'   14.C.9   848-849

R–C(=O)–C(NH₂)< ⟶ R–C(=O)–OR'   19.F.1   881

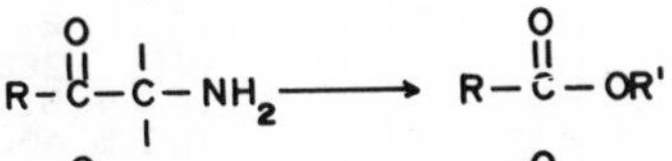

R–C(=O)–C(NO₂)< ⟶ R–C(=O)–NH–R'   18.E.7   846

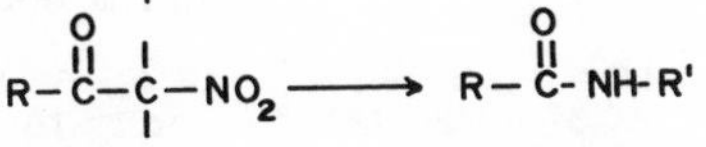

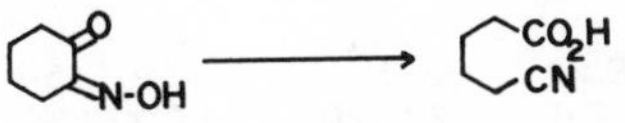

19.C.2   954

R–C(=O)–C(=O)–R ⟶ R–C(=O)–O–C(=O)–R   16.B.2   883-884

Ar–C(=O)–C(=O)–Ar ⟶ Ar–C(=O)–OR   14.C.9   849

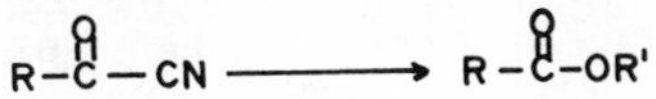

R–C(=O)–CN ⟶ R–C(=O)–N<   18.A.6   904-905

R–C(=O)–CN ⟶ R–C(=O)–OR'   14.A.19   726-727

R–C(=O)–C(=O)–OH ⟶ R–C(=O)–OH   13.B.11   771

Functionality   2→4

7→7   H–C(N₂)–C(=O)–O-t-Bu ⟶ O₂N–C(N₂)–NO₂   20.A.1   888

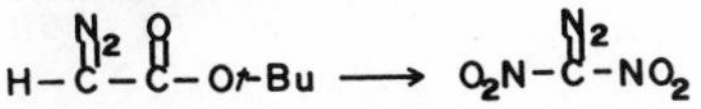

Functionality   3→3

8→8   R–C(=O)–C(=O)–OH ⟶ R–C(=O)–OH   13.B.11   771

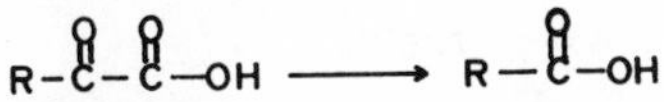

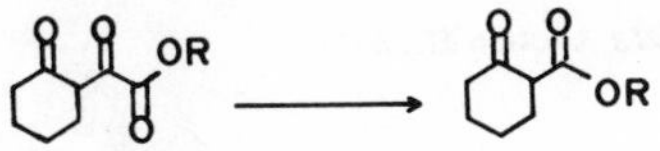

14.C.1   837

III. CARBON – CARBON
      BOND-FORMING REACTIONS

A.   REDUCTION OF SUBSTRATE

      O. Unfunctional Chain

III. A.0.   1→0

Functionality   1→0

| | | 6→6 | structure | | I.G.6 | 62-63 | I.A.10 | 17-18; |

6→6

‖ ⟶ ▷

I.G.6	62-63	I.A.10	17-18;
(6.C.4	318)	I.G.6	61-67;
(II.G.8	712)	Ad.	69
		(I.H	74)
		(2.A.14	106)
		(3.E.1	204)
		(4.H	268)
		(6.C.4	321)
		(8.I.5	458)

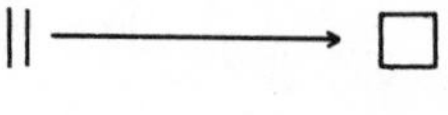

‖ ⟶ ☐

I.G.4	59-61;	I.G.4	56-59;
2.C.3	124-127	Ad.	75
(2.C.1	114)		

⟩ ⟶ ⟩R

I.C.1,2	29-31;	I.C.2	34-36;
2.C.4	127-128	I.C.7	39;
(I.D.4	42-43)	4.E.4	261;
		8.E.5	436

⟩ ⟶ ⟩R–X

| | | 7.B.6 | 370 |

Z⟩ ⟶ Z–⟩R

II.H.6	720-721	II.G.3	600-606;
(14.C.6	847)	13.D.6	690
(18.E.4	933)	(I.A.7	16)

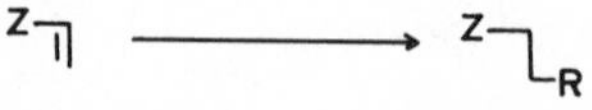

(enone) ⟶ (ketone R)

| II.G.1 | 700-701 | | |

⟶ (ketone R R')

| | | II.G.3 | 604; |
| | | Ad. | 611 |

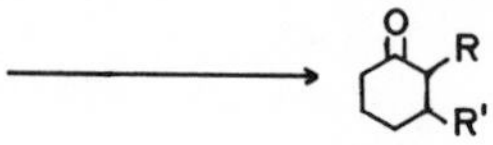

⟶ (ketone R OH)

| | | Ad. | 610 |

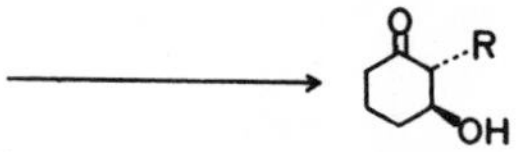

OMe / ⟶ (Δ) ⟶ O Me

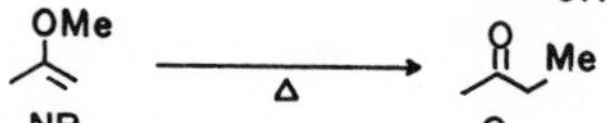

| II.G.4 | 708 | | |

NR₂ / ⟶ O R

See section III. B. 0

7→6

(N=N ring) ⟶ ▱

| | | I.G.4 | 58-59 |
| | | (2.A.14 | 105-106) |

8→6   R–OH ⟶ R–R'

| I.A.9 | 17 | I.B.5 | 31 |

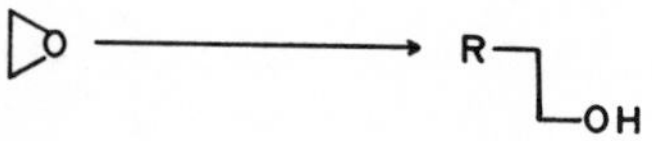

▷O ⟶ R–OH

4.E.2	228-229	4.A.7	214;
		4.E.2	258
		(4.C.2	234)
		(Ad.	610)

III. A. O.   1 ⟶ 0

16 ⟶ 6   SR ⟶ R'   2.B.7   126

R–S⁺–R ⟶ R–R   I.A.7   15

SO₂ ⟶   I.G.4   59

17 ⟶ 6   R–X ⟶ R–R'   I.B.2   22-27   I.B.2   23-27;
                                                  2.C.9   138-139
                                                  (I.B.4   29-30)
                                                  (I.B.6   31-32)
                                                  (Ad.   33)
                                                  (I.C.7   37-40)
                                                  (I.D.4   46-47)

X ⟶ R   II.G.I   701   II.G.9   606-607
                                             (II.H.5   719-720)

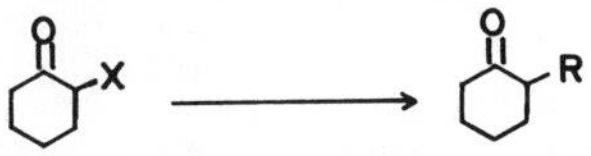

Functionality   2 ⟶ 0

6 ⟶ 6   ≡ OTs ⟶ □   I.B.7   32

7 ⟶ 6   N₂ ⟶ R   II.G.I   701   II.G.9   606-607

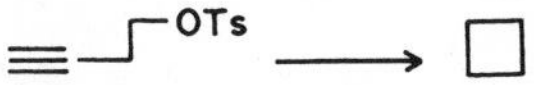

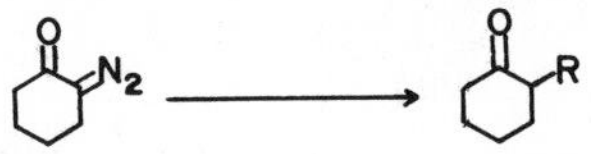

8 ⟶ 6   O ⟶ CH₃/H(R)   I.A.1   6;
                                                              I.A.7   16;
                                                              I.B.5   31

⟶ R/R   I.B.8   32

17 ⟶ 6   X/X ⟶   I.B.2   26

Functionality   2 ⟶ 1

6 ⟶ 6   R–C≡C–H ⟶ R–CH=CH–R'   2.A.16   101   2.B.1   121

R– ≡ ⟶ R–◁   2.A.14   106

R– ≡ –CO₂R' ⟶ CO₂R'   2.C.11   140;
                                                       14.C.6   750
                                                       (14.C.14   756)

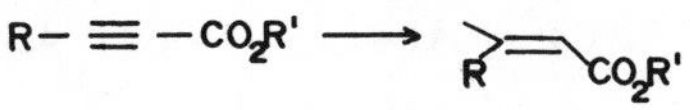

7 ⟶ 7   NR₂ ⟶ NR₂   8.I.5   458;
                                                              II.G.2   599

III. A. O.   2 → I

7 → 7 $\geq$N−R ⟶ $\underset{\text{NHR}}{\overset{\text{R'}}{\diagup}}$ 8.E.3   474-477

$\geq$N−OH ⟶ $\underset{\text{NH}_2}{\overset{\text{R}}{\diagup}}$ 8.A.5   403

8 → 6   R−CH$_2$CHO ⟶ R−CH=CH−R'   2.A.4   82-84

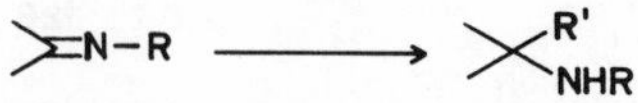

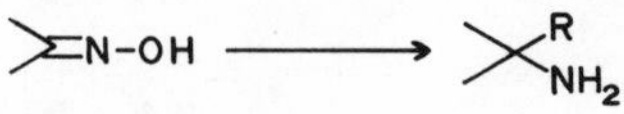

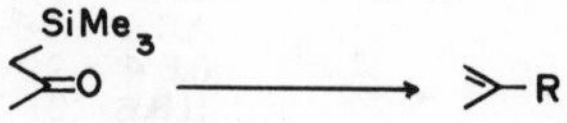

$\overset{\text{SiMe}_3}{\diagup}$=O ⟶ $\diagup$−R   2.A.20   112

8 → 8   (OSiMe$_3$ cyclohexene) ⟶ (cyclohexanol)   4.H   268;
II.G.I   596

$\diagup$=O ⟶ $\underset{\text{OH}}{\overset{\text{R}}{\diagup}}$ 4.E.I   224-227   4.E.I   252-258;
(4.G.5   242)   Ad.   261
(5.C.2   268-269)
(6.A.7   298)
(II.H.I   713-715)

(dioxolane) ⟶ $\underset{\text{O}}{\overset{\text{R}}{\diagup}}$−OH   6.H.2   334

17 → 6   (=/−X) ⟶ (=/−R)   2.C.7   136-137

Ar−X ⟶ Ar−R   I.B.2   25-26   I.B.2   26-27;
I.C.8   40

Ar−X ⟶ Ar−Ar   I.B.3   27-28;   I.B.3   27-30
I.G.3   58-59

17 → 8   R−$\underset{\text{OR'}}{\overset{\text{X}}{\diagup}}$ ⟶ R−$\underset{\text{OR'}}{\overset{\text{R''}}{\diagup}}$ 2.A.4   82-84;
6.A.6   298

Functionality   3 → O

8 → 6   R−$\overset{\text{O}}{\overset{\|}{\text{C}}}$−OH ⟶ R−CMe$_3$ I.B.5   31

Functionality   3 → I

7 → 7   R−CN ⟶ R−$\underset{\text{NH}_2}{\text{CH}}$−R'   8.E.2   474   8.E.2   434-435

8 → 6   (cyclohexanone O) ⟶ (cyclohexylidene cyclohexane)   2.F.5   148

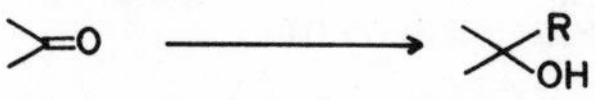

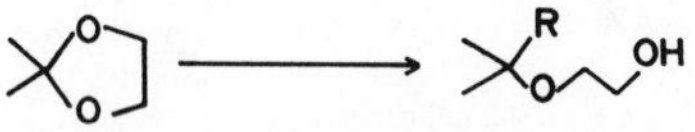

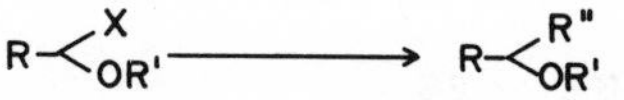

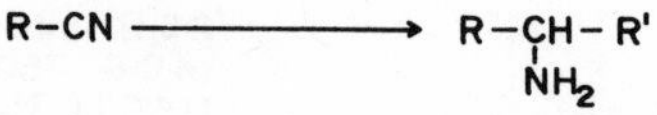

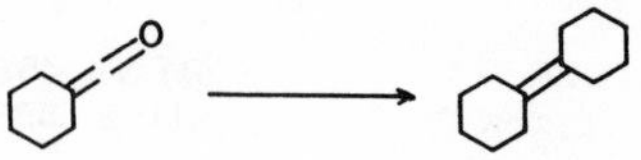

III. A.O.   3 → I

8 → 7	$R_2N{-}CHO$ → $R_2N{-}CHR'_2$	8.E.I	473-474		
8 → 8	$R{-}\overset{O}{\overset{\|}{C}}{-}OR'$ → $R{-}\overset{OH}{\underset{R'}{\overset{\|}{C}}}{-}R''$	4.E.I (II.H.I)	224-227 713-715	4.E.I	252-258
17 → 6	$R{-}{\equiv}{-}X$ → (alkene $R{-}CH{=}CHR'$)			2.B.I	120-121

Functionality   3 → 2

7 → 8	$R{-}CN$ → $R{-}\overset{O}{\overset{\|}{C}}{-}R'$	II.H.3 (8.E.2	717-718 474)	II.H.3 (II.G.7	618-619 620)
8 → 6	$R{-}C{\equiv}C{-}OEt$ → $R{-}C{\equiv}C{-}R'$			3.A.2	174-175
8 → 8	(oxazine ring) → $R{-}\overset{O}{\overset{\|}{C}}{-}R'$			II.D.I	568-570
	$={=}O$ → $\triangleright{=}O$	II.G.8	712		
	$R{-}\overset{O}{\overset{\|}{C}}{-}N{<}$ → $R{-}\overset{O}{\overset{\|}{C}}{-}R'$	10.G II.H.4 (8.E.I	616-619; 718-719 473-474)	II.H.4	619
	$R{-}\overset{O}{\overset{\|}{C}}{-}OH$ → $R{-}\overset{O}{\overset{\|}{C}}{-}R'$	II.F.I (II.H	686-688 713)	II.H.2	616-618
	$R{-}\overset{O}{\overset{\|}{C}}{-}OR''$ → $R{-}\overset{O}{\overset{\|}{C}}{-}R'$	II.F 6 II.H.I	693-694; 713-716	II.H.I	614-615
	(Cl-cyclopropane CO₂Et) → (cyclopropane OEt/OSiMe₃)			9.D.7	477
	$R{-}\overset{O}{\overset{\|}{C}}{-}O{-}\overset{O}{\overset{\|}{C}}{-}R$ → $R{-}\overset{O}{\overset{\|}{C}}{-}R'$	II.H.2	716-717		
	$R{-}C(OR'')_3$ → $R{-}\overset{O}{\overset{\|}{C}}{-}R'$	10.G (9.C.2	616-619 536)		
	→ $R{-}\underset{OR''}{\overset{OR''}{\overset{\|}{\underset{\|}{C}}}}{-}R'$	9.C.2	536		
16 → 6	$R{-}\underset{O}{\overset{\|}{C}}{-}SR'$ → $R{-}\underset{O}{\overset{\|}{C}}{-}R'$			II.H.I	614-615
17 → 6	$R{-}\underset{O}{\overset{\|}{C}}{-}X$ → $R{-}\underset{O}{\overset{\|}{C}}{-}R'$	II.F. 3 II.H.2 II.H.7 (20.E.3	688-690; 716-717; 721-722 1009)	II.H.2	616-618

III. A. O.   4→1

Functionality   4→1

8 →8	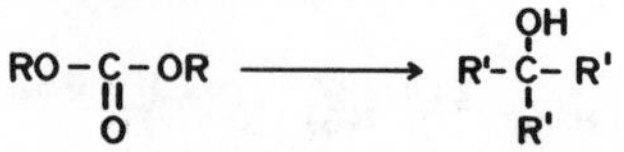 RO−C(=O)−OR → R'−C(OH)(R')−R'	4.E.1 (14.C.7	224-227 847-848	4.E.1	252-258

Functionality   4→3

7→7	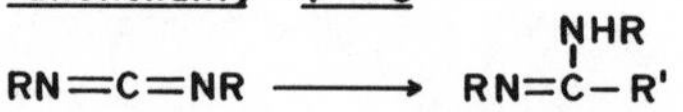 RN=C=NR → RN=C(NHR)−R'	18.E.4	933		
8→8	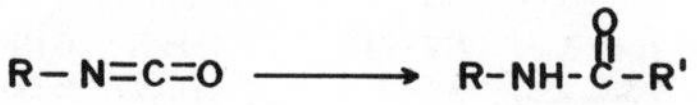 R−N=C=O → R−NH−C(=O)−R'	18.E.4	933		
	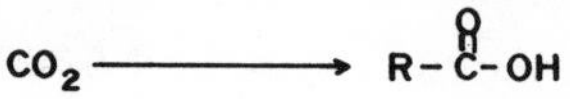 CO₂ → R−C(=O)−OH	13.C.1	776-778	13.C.1	676-680
	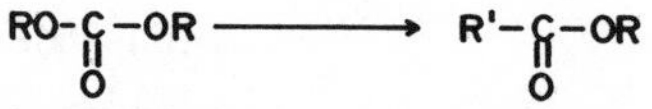 RO−C(=O)−OR → R'−C(=O)−OR	14.C.7 (4.E.1	847-848 224-227)		
16→16	 R−N=C=S → R−NH−C(=S)−R'	18.E.4	933		

I. One-Carbon Chain

Functionality   1→0

(I) C−1 Functionality = 1

6 →6	 ⟩ → NMe₂			8.D.1	429
	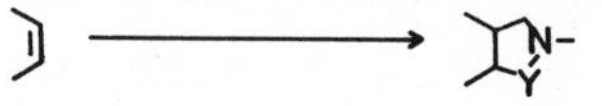 ⟩ → N−Y	8.F.5	488-490		
	Z−|| → Z−NO₂	20.C.2	999-1002	14.C.5	746-747
	 ⟩ → ▷−OH			4.A.9	216
	 ⟩ → ⟩−OH *via borane*	4.B.6	191	4.B.6	225-226
	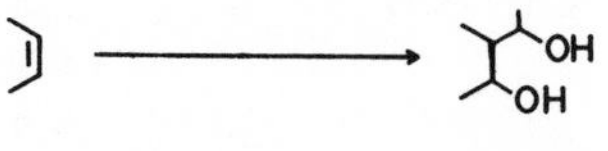 ⟩ → ⟩(−OH)(−OH)	4.B.4 6.B.6 (9.B.7	188-190; 308-309 533)	4.B.4	224-225
	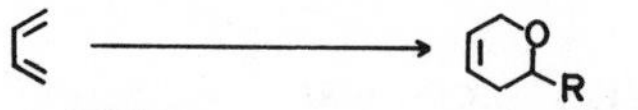 ⟨ → (O ring)−R	6.F	327-328		
	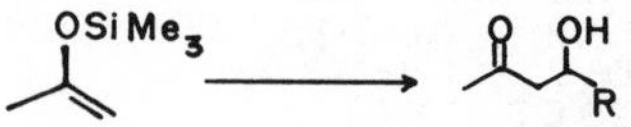 (OSiMe₃) → (O)(OH)−R	*see also* Section III. B.1.		4.B.4	224-225

III. A. 1.   1 ⟶ 0

6 ⟶ 6   )═ ⟶ (CCl₂) >⟨ with Cl, Cl   7.D.6   385

7 ⟶ 6   R–CH₂–N⁺< ⟶ R–CH₂–CH₂–N<   (8.G.8   451)

17 ⟶ 6   R–CH₂–X ⟶ R–CH=CH₂   2.C.6   129   2.E.2   156–158
                                       (2.E.2   141–144)

         R–X ⟶ R–Ar   I.D.1   33–39   I.D.1   42–45

         ⟶ R–C–N<   8.E.1–3   473–477   8.C.9   426

         ⟶ R–CH₂NO₂   20.B.2   993   20.B.1,2 896–897

         ⟶ R–C–OH   4.E.1   224–227   4.E.1   252–258
                                         (4.E.3   229–230)

         ⟶ R–CH₂–X   7.A.16   352

(2)   C – 1   Functionality = 2

6 ⟶ 6   R∕═ ⟶ R∕∕∕CHO   4.B.5   190;
                                       10.C.8   588–589   10.C.8   506

         ⟶ R∕∕C(=O)R'   11.C.7   663   11.C.7   563
                                        (10.C.9   589)
                                        (11.F.7   694–695)
                                        (14.C.6   847)

         Z∕═ ⟶ Z∕∕C(=O)R'   11.D.7   576;
                                                              11.F.7   591–592;
                                                              11.G.3   603

         R∕═ ⟶ R∕C(Cl)∕C(=O)R'   11.C.7   562–566

         ∥< ⟶ (cyclopentenone)=O   11.D.1   570

         Z∕═ ⟶ Z∕∕C(S(=O)R)(SR)   11.D.7   576
                                                              (11.G.3 603)

)═ ⟶ >⟨X,X (dihalocyclopropane)   7.B.5   368–370   7.B.5   366–370

         (cyclohexylidene–X) ⟶ (cyclohexyl with CHO)
                                                              10.H.1   530–531

III. A. I.   I → O

6 → 6   18.G.I   851

N O
7 → 6 R–N–CO$_2$Et ⟶ R–C–Ar   II.F.IO 593

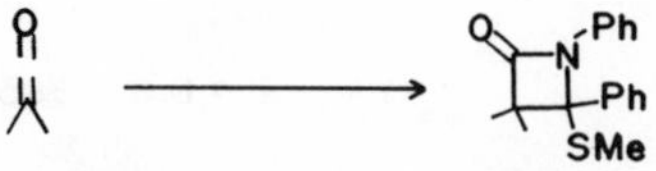

8 → 6 Ar$_2$CH–OH ⟶ Ar$_2$CH–C–R   II.G.6   711

17 → 6 R–X ⟶ R–CHO

via RMgX	10.G	616-619	9.C.2	475;
	(8.E.I	473-474)	10.G	526-528
	(9.C.2	536)	(10.D.2	510-511)
	(10.A.17	564)		
other methods	10.D.3	595	10.D.7	515
			(9.C.I	475)

        O
R–X ⟶ R–C–R'   II.G.5   708-711

via RMgX	II.H.1-4	713-719	II.H.1-4	614-619
other methods	10.D.3	595;	10.D.2,3	510-513;
	II.D.I	667;	II.D.7	575-576
	II.H.7	721-722	II.F.7	591-592;
			II.H.2	618;
			II.H.8,9	622-624
			(II.H.3	619)
			(19.B.I	864)

II.F.4   589

R–X ⟶ R–CH(OR')$_2$ 9.C.2   536   9.C.I,2   475

R–X ⟶ R–CH(SR')$_2$ 10.D.3   595;   9.C.I   475;
                        including   II.D.I   667   10.D.7   515;
                        sulfoxides   II.D.7   575-576
                                                (II.H.8   622)

(3) C–I Functionality = 3

6 → 6 Z⟋ ⟶ Z⟋CN   19.D.I   962-963   19.D.I   877-878;
                                            Ad.   879

R⟋ ⟶ R⟋CONMe$_2$ 18.F   933-936   18.D.8   839;
                                        18.F   849-850

R⟋ ⟶ R⟋CO$_2$H   13.F.3,4   792-794   13.F.4   701-703

CH$_2$=CH$_2$ ⟶ Br·CH$_2$·CH$_2$·CCl$_3$ 7.B.4   365-367   7.B.4   365-366
                                            (13.A.6   754)

III. A.I.   1→0

6 → 6	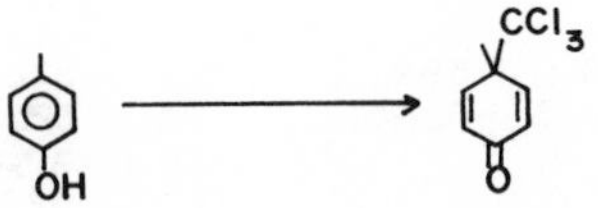 ⟶			12.C.5 649
7 → 6	R–NR₃⁺ ⟶ R–CN	19.A.4	942-943	19.A.4 859-860
8 → 6	R–OH ⟶ R–CN			19.A.2 859; 19.A.7 861-862

R–C≡C–H etc.

Reaction	Refs left	Refs right

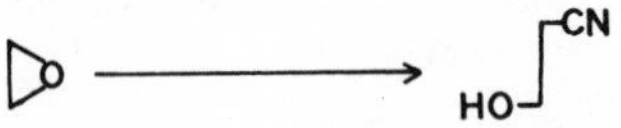

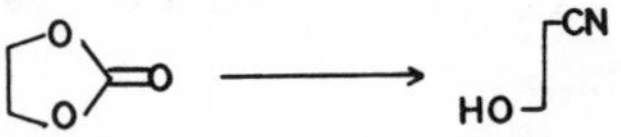

$$R-C\equiv C-H \longrightarrow$$

6 → 6   (epoxide) ⟶ (HO–CH₂–CH₂–CN)   19.D.5   968-969

(cyclic carbonate) ⟶ (HO–CH₂–CH₂–CN)   4.A.I   211

R–OH ⟶ R–C(=O)–OH   13.F.4   793-794   13.F.4,5 702-704

R–O–Ph ⟶ R–C(=O)–OH   13.C.I   677-678

CH₃–O–C(=O)–CH₃ ⟶ CH₃CO·O·CH₃   16.C.2   885

17 → 6   R–X ⟶ R–CN   19.A.1,2 938-941   19.A.1,2 856-859; Ad.   862

⟶ R–C(=O)–N<   18.D.8   840

⟶ R–C(=O)–OH   13.C.I 776-778; 13.F.4 793-794   13.C.I 676-680; 13.F3 701; 13.F.5 703-704 (10.D.2 510-511)

⟶ R–C(=O)–OR'   14.B.6 831-832; 14.C.7 847-848   14.A.22 728; 14.B.6 734-736; 14.C.7 752-753

⟶ R–C(=O)–X   15.C.2 792

Functionality   2 → 0

(1) C–I   Functionality = I

(2) C–I   Functionality = 2

6 → 6   R–C≡C–H ⟶ R–C(=O)–CH₂–C(=O)–R'   II.C.7   564-565

III. A. I.   2 ⟶ O

8 ⟶ 6   ⟩=O ⟶ ⟩—CHO   10.D.2   594   10.F.1   522;
                                     (10.H.3   621-622)   Ad.   525
                                                          (6.C.3   319-321)
                                                          (10.G   527-528)

⟩=O ⟶ ⟩—C(=O)—R   11.B.1   646

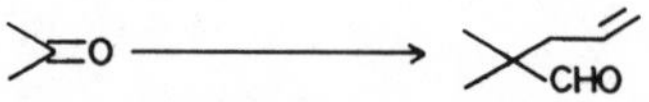

⟩=O ⟶ ⟩⟨—CH=CH₂ / CHO   10.H.1   529

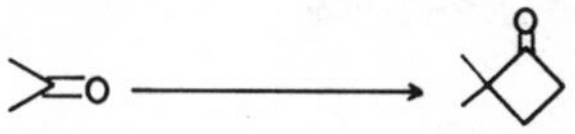

⟩=O ⟶ ⟩⟨□(=O)   11.E.3   579

(3)   C—I   Functionality = 3

8 ⟶ 6   ⟩=O ⟶ ⟩—CN   19.B.3   867-868;
                          19.C.4   873;
                          19.C.9   875-876

R—CHO ⟶ R—CH₂—C(=O)—OH   13.D.3   687-688;
                            13.D.7-9 690-692

R—CHO ⟶ R—CH₂—C(=O)—OR'   14.A.20   727

Functionality   2 ⟶ I

(I)   C—I   Functionality = I

6 ⟶ 6   R—C≡C—H ⟶ R—CH₂—CH=CR'₂   2.B.1   120

R—C≡C—H ⟶ R—C(=O)—CH=CH—R'   11.A.19   549;
                                11.I.1   628
                                (11.I.2   632)

R—C≡C—H ⟶ R—CH=CH—C(—OH)   4.E.1   257

R₂N—≡—R' ⟶ R₂N—C(=O)—C(=CH₂)—R'   8.E.3   436

7 ⟶ 6   Ar—N₂⊕ ⟶ Ar—Ar'   I.G.1   55-57   I.G.1   53-54

⟩=N₂ ⟶ ⟩=⟨   2.A.11   95

III.A.I.   2 ⟶ I

8 ⟶ 6   >=O ⟶ >=⟨R

2.A.II	95-96;	2.A.6	94;
2.E.I,2	140-144	2.A.II	99-I0I;
(12.C.3	742-743)	Ad.	II6;
		2.B.I	I20;
		2.B.7	I27;
		2.E.2	I56-I58;
		2.E.3-5	I58-I60;
		Ad.	I60-I6I
		(2.A.8	97-98)
		(2.A.I3	I04)
		(2.A.20	III-II3)
		(2.C.I2	I4I-I42)

8 ⟶ 7   >=O ⟶ ⟨NH

2.A.20 III

8 ⟶ 8   >=O ⟶ ⟨OH, NH₂

(II.E.4	682)	4.E.I	254;
		8.A.3	40I;
		8.C.2	42I;
		II.E.4	580
		(4.F.2	263)

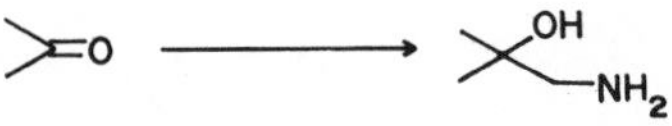

>=O ⟶ ⟨OH, NO₂

20.C.I	997-999	20.C.I	899-90I

>=O ⟶ ⟨OH OH⟩

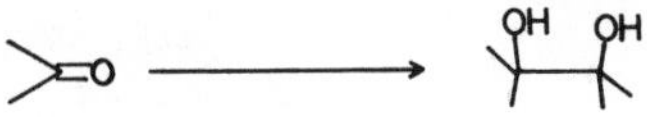

4.C.8	204-205	4.C.8	238-239;
		Ad.	243
		(I4.E.I	77I-772)

>=O ⟶ ⟨OH, OR

I0.E.I	602-603		

>=O ⟶ △O (epoxide)

6.C.4	3I7-3I8	6.A.4	308-309;
(6.E.4	326-327)	6.C.3,4	3I9-322;
		Ad.	323-324

I6 ⟶ 6   (S, N=N ring) ⟶ >=⟨

2.A.I9 II0

I7 ⟶ 6   Ar–X ⟶ Ar–Ar

I.B.3	27-28;	I.B.2	23-27;
I.G.3	58-59	I.B.3	27-30;
		I.G.3	55-56
		(20.D.4	908)

⟨X, NO₂ ⟶ >=<

2.A.22 II4

⟨X, X ⟶ >=<

2.A.3	8I-82	2.A.3	90-9I

Ar–X ⟶ Ar–C–N<

8.C.9 426

Ar–X ⟶ Ar–C–NO₂

20.B.2	993		

I7 ⟶ 8   ⟨X, X ⟶ △O (epoxide)

6.C.3 320

III. A. I.   2 → I

(2) C-I   Functionality = 2

6 → 6   R-≡⌿O⌿R' $\xrightarrow{\Delta}$ R-C(=O)-CH₂-C(=O)-R'   II.I.2   631

7 → 6   Ar-N₂⊕ ⟶ Ar-CHO   10.C.II   591

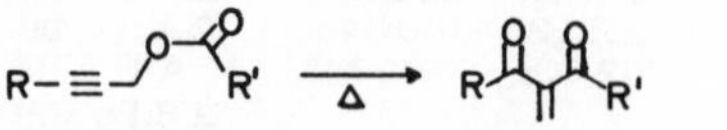

                    ⟶ Ar-C(=O)-R   II.C.9   665

8 → 6   Ar-O-CH₂-CH=CH₂ ⟶ Ar-C(=O)-CH₂CH₃   II.E.7   582

⟩=O ⟶ ⟩=⟨ NR₂   (10.F.I   522)

       ⟶ ⟩=⟨ NO₂   II.B.I   646;   20.C.I   899-901
                            20.C.I   997-999

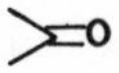

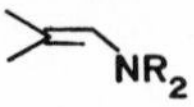

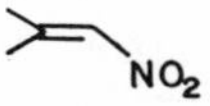

       ⟶ ⟩=⟨ OR   (Ad.   525)
                           (10.H.I   529)

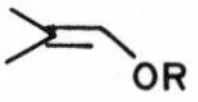

       ⟶ ⟩=⟨ C(=O)R(H)   II.D.4   670   II.D.4   571-573
                                                      (10.F.I   522)

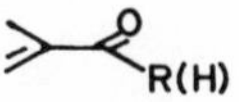

       ⟶ ⟩=⟨ SR   (10.G   528)

       ⟶ ⟩=⟨ SO₂Ar   I.A.7   16

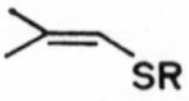

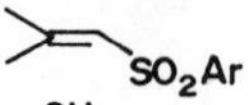

8 → 8   R-CHO ⟶ R-CH(OH)-C(=O)-R'   10.B.4   573   Ad.   515;
                                                                  II.D.7   575-576;
                                                                  Ad.   610;
                                                                  II.H.8   622

         Ar-CHO ⟶ Ar-CH(OH)-C(=O)-Ar   4.C.8   205-207   4.C.8   239;
                                                                       Ad.   243

         R-CHO ⟶ R-◁ SO₂R'   6.C.3   317   6.C.3   320-321
              including sulfides
              and sulfoxides

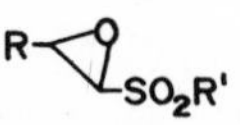

⟩=O ⟶ ⟩⟨ OH / CHCl₂   Ad.   261-262

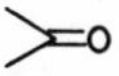

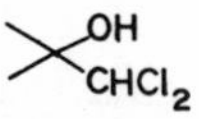

(3) C-I   Functionality = 3

6 → 6   ≡ ⟶ ⟋=⟍ CN   19.D.I   962-963   19.A.6   860-861
                                                        (19.D.I   878)

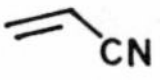

III.A.1.   2 → 1

6 → 6   ≡   ⟶   ⟍⟍CONR$_2$ 18.D.8   927

⟶   ⟍⟍CO$_2$H   13.F.3   792-793   13.F.3   697-701

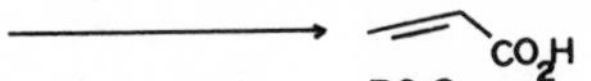

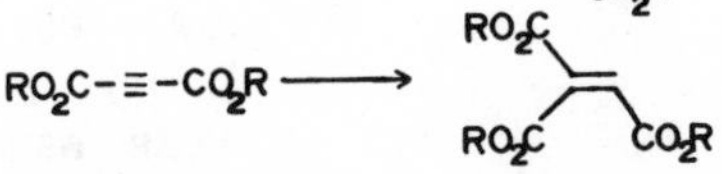

RO$_2$C-≡-CO$_2$R ⟶   RO$_2$C〉=〈CO$_2$R   14.B.6   832

7 → 6   Ar–N$_2$⊕   ⟶   Ar–CN   19.A.5   943
(7.A.9   344-347)

Ar–NH–CHO ⟶ Ar–CN   19.C.1   952

7 → 7   〉=N–R   ⟶   ⟩〈CN NH-R   19.D.4   966-968

⟶   ⟩〈CCl$_3$ NH-R   8.E.3   436

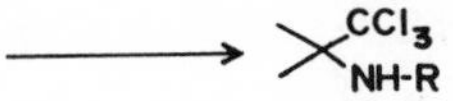

〉=N$_2$ ⟶   ⟍/〈Cl Cl

8 → 6   〉=O   ⟶   ⟩〈X X   2.E.2   142   2.B.7   127;
2.E.2   157;
7.B.5   368
(3.A.1   173)

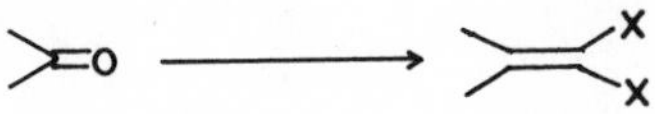

8 → 7   〉=O   ⟶   ⟩〈CN NR$_2$ 19.D.3   965-966
(11.G.5   709)
(13.A.4   753)

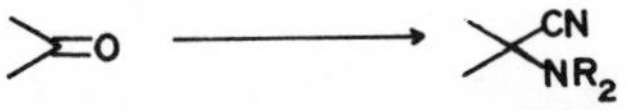

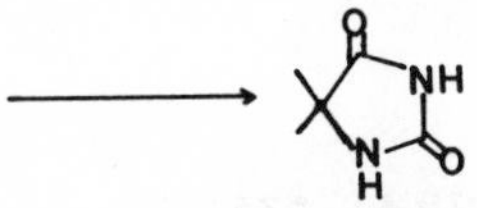

⟶   hydantoin NH O   19.D.3   965

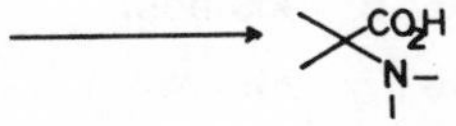

⟶   ⟩〈CO$_2$H N–   13.A.4   753

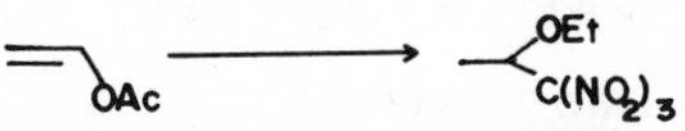

8 → 8   ⟍⟍OAc   ⟶   ⟩〈OEt C(NO$_2$)$_3$ 20.A.3   893

〉=O   ⟶   ⟩〈CN OH   4.F.2   231-233;   4.F.2   263-264;
19.D.2   963-965   Ad.   879
(13.A.4   753)   (8.A.3   401)

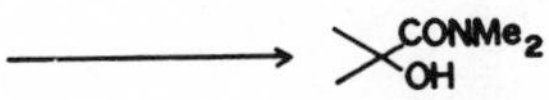

⟶   ⟩〈CONMe$_2$ OH   18.E.11   848

⟶   ⟩〈CO$_2$H OH   7.B.5   368-370;   13.A.6   662-664
13.A.4   753   (13.C.1   679)
(13.A.1   778)

⟶   ⟩〈CX$_3$ OH   7.B.5   368-370

III. A. I.   2⟶I

8⟶17 R-CHO ⟶ R-CH-CN   2.A.20 III
                        |
                        Cl

16⟶6 Ar-SO₃H ⟶ Ar-CN   19.A.3   942

17⟶6 ⟋⟍⟋X ⟶ ⟋⟍⟋CN   19.A.I   857

      ⟋⟍⟋X ⟶ ⟋⟍⟍C-N<   18.D.8   839-840
                          ‖
                          O

      Ar-X ⟶ Ar-CN   19.A.I,2 939-941   19.A.I,2 858-859

           ⟶ Ar-C-OH   13.F.3   701
                 ‖
                 O

           ⟶ Ar-CF₃   7.A.I9   357-358

           ⟶ Ar-C-X   15.C.2   792
                 ‖
                 O

17⟶8 Ar   X   Ar   CO₂H
        \ /   \ /
         X   ⟶   X   13.C.I   777
        / \   / \
      Ar   X   Ar   OH

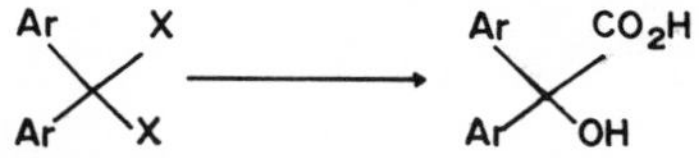

Functionality   3⟶0

(1) C-I Functionality = I

(2) C-I Functionality = 2

(3) C-I Functionality = 3

8⟶6   R-C-OH ⟶ R-CH₂-C-N<   10.B.4   573
          ‖   ‖   (18.A.8 905-906)
          O   O

           ⟶ R-CH₂-C-OH   13.G.I   796-797
                    ‖
                    O

           ⟶ R-CH₂-C-OR' 14.B.3   827-828   14.B.3 732
                    ‖
                    O

      Ar-C-OH ⟶ Ar-CH-C-OH   II.F.5   693
          ‖ | ‖
          O   CH₃ O

Functionality   3⟶I

(1) C-I Functionality = I

7⟶7   R-CN ⟶ R-CH-CH₂-SOAr   8.E.2   434
                  |
                  NH₂

Ⅲ. A. I.   3→1

(2) C—1 Functionality = 2

8 ⟶ 8   R-C(=O)-OR' ⟶ R-CH(OH)-C(=O)-R   4.C.8   207-208   4.C.8   239-240

(3) C—1 Functionality = 3

17⟶6   R-CH₂-C(=O)-X ⟶ R-CH=CH-CN   19.C.9   874-875

17⟶8   R-C(=O)-X ⟶ R-CH(OH)-C(=O)-OH   4.F.2   233

<u>Functionality   3⟶2</u>

(I) C—1 Functionality = I

8 ⟶ 8   CF₃-C(=O)-OEt ⟶ CF₃-C(CHR)-OEt   6.I.10   342

R-C(OEt)(OEt)(NMe₂) ⟶ R-C(=O)-CH₂-NO₂   20.C.3   903

R-C(=O)-OR' ⟶ R-C(=O)-CH₂-S-CH₃   II.F.6   693-694
(10.A.19   566-567)

9 ⟶ 9   CF₂=CF₂ ⟶ CHF₂-CF₂-CH₂OH   4.D.6   222

17⟶6   R-C≡C-X ⟶ R-C≡C-Ar   (3.B.I   185)

17⟶8   R-C(=O)-X ⟶ R-C(=O)-Ar   II.C.1-5   651-659;   II.C.1-5 559-562
see also other
sections, e.g. Ⅲ.B.I

⟶✗ R-C(=O)-CH₂NO₂   (20.C   996)

⟶ R-C(=O)-CH₂OH   4.A.3   179;
II.F.5   693

⟶ R-C(=O)-CH₂-X   II.F.5   693   7.C.6   378

(2) C—1 Functionality = 2

8 ⟶ 8   R-C(=O)-OR' ⟶ Me₃Si—C(OSiMe₃)(R)(R)   4.C.8   207-208   4.C.8   239-240;
II.A.4   540-541

16 ⟶ 6   Ar-C(=S)-S-)₂ ⟶(Ni) Ar-C≡C-Ar   3.A.8   160

III. A. I.   3 → 2

$17 \longrightarrow 6$ $R-\underset{O}{\overset{\|}{C}}-X \longrightarrow R-C\equiv C-Z$ 3.A.7   160   3.A.7   179-180

$17 \longrightarrow 8$ $R-\underset{O}{\overset{\|}{C}}-X \longrightarrow R-\underset{O}{\overset{\|}{C}}-CH=N_2$

II.F.5	693;
13.G.1	796-797;
14.B.3	827-828;
17.A.3	888-889
(4.A.3	179)
(10.A.7	555)
(14.B.2	827)

$\longrightarrow R-\overset{O}{\overset{\|}{C}}-CHO$ II.D.7   575-576

$Ar-\overset{O}{\overset{\|}{C}}-X \longrightarrow Ar-\overset{O}{\overset{\|}{C}}-\overset{O}{\overset{\|}{C}}-Ar$ 4.C.8   208;   10.B.7   650

(3) C – I Functionality = 3

$9 \longrightarrow 9$ $CF_2=CF_2 \longrightarrow ClCF_2-CF_2-CCl_3$ 3.A.3   154

$17 \longrightarrow 8$ $R-\underset{O}{\overset{\|}{C}}-X \longrightarrow R-\underset{O}{\overset{\|}{C}}-CN$ (19.A.1   158)

$\underline{\underline{2}}$. Two-Carbon Chain

Functionality I → 0

(I) C – 2 Functionality = I

$8 \longrightarrow 6$ $R\diagdown\diagup OH \longrightarrow R\diagdown\diagup\diagdown\underset{R}{\diagup}OH$

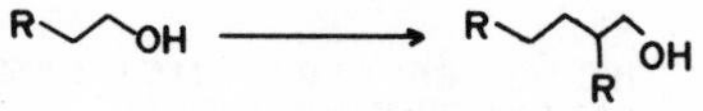

Guerbet   4.G.3   240-241   4.G.3   266

$17 \longrightarrow 6$ $R-X \longrightarrow R-\overset{|}{C}=\overset{|}{C}-$ 2.C.1   130;   2.C.3   135;   2.C.7   136-137

$R-X \longrightarrow R\diagup\underset{R'}{\diagdown}X$ 7.B.6   370

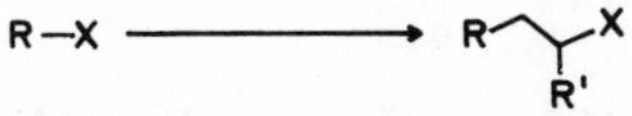

(2) C – 2 Functionality = 2

$6 \longrightarrow 6$ II.G.3   603

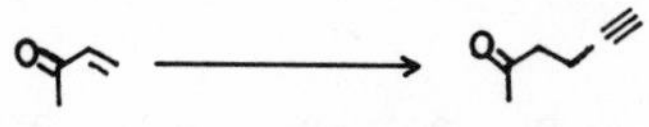

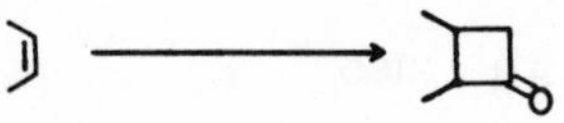

II.I.I   627

III. A. 2.   I ⟶ O

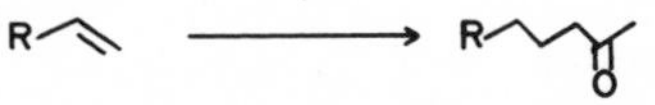

6 ⟶ 6   R⌇ ⟶ R⌇⌇⌇=O   11.G.1   701

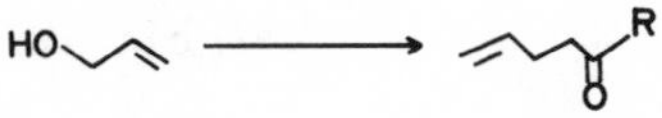

HO⌇ ⟶ ⌇⌇⌇=O R   10.H.1   619-620;   10.H.1   528-531
                           11.G.4   707-708   (2.C.13   142)
                           (3.D.4   170)   (13.G.7   710)
                           (5.F.1   282-284)   (14.E.7   775-776)

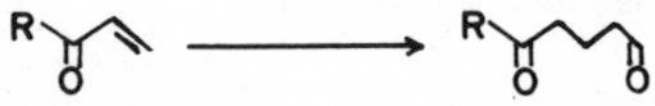

R⌇⌇=O ⟶ R⌇⌇⌇=O ⌇=O   10.D.3   595-596

Z⌇ ⟶ Z⌇⌇⌇=O R   10.F.3   611-614;
                           11.G.3   706-708   11.G.3   600-606
                           (11.G.1,2   699-705)   (5.E.3   297)
                           (14.C.5   844-845)   (Ad.   593)
                           (18.E.3   931-932)

RO⌇ ⟶ RO⌇⌇⌇OR RO OR   (9.E.4   478)

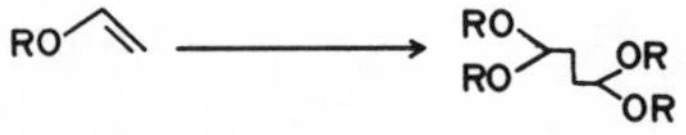

8 ⟶ 6   ⌇=O ⟶ ⌇≡—   OH   3.B.5   190

17 ⟶ 6   R—X ⟶ R—C≡C—R'   3.B.1   162-164   3.B.1   184-187

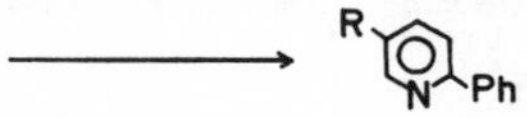

⟶ R⌇—N—Ph   8.C.2   421

⟶ R-CH₂-CHO   10.D.3   595;   10.D.3   511-513;
                           10.F.4   614-615   10.F.4   524-525

⟶ R-CH₂-C-R' O   10.D.3   595;   11.G.1,2   595-600
                           11.D.6   673-675;   (10.D.3   511-513)
                           11.G.1,2   697-705   (8.I.1   455-456)
                           (11.C.7   663)   (11.E.8   583)
                           (11.G.4   708)   (11.H.7   621)

Ph₃C-X ⟶ Ph₃C-CHC-H(R') R O   10.C.10   590;
                           11.C.8   665

(3) C-2 Functionality = 3

6 ⟶ 6   ⌇) ⟶ ⌇⌇CONH₂   18.F   935-936

⟶ ⌇⌇CO₂H   13.F.4   793-794;   13.F.4   701-703;
                           13.F.7   795   13.F.8   704-705

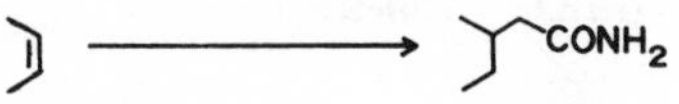

Ⅲ. A. 2.   I ⟶ O

6 ⟶ 6   ⟶   14.C.6   847

⟶   2.C.14   143

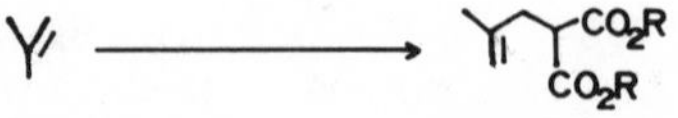

⟶   2.C.2   116-124;   2.C.2   130-134
see also other sections

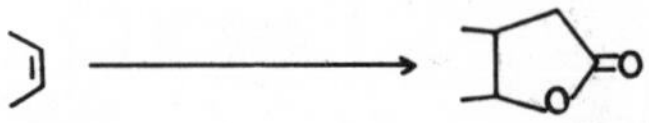

⟶   14.B.8   834   14.D.10 765;
14.E.3   773-774

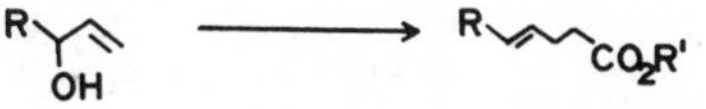

R⟶   2.C.13 142;
OH   10.H.1 530;
13.G.7 710;
14.E.7 775-776

⟶ R⟶CONMe₂   18.G.5   853

R⟶ ⟶ R⟶CO₂R'   (14.F.5   780)

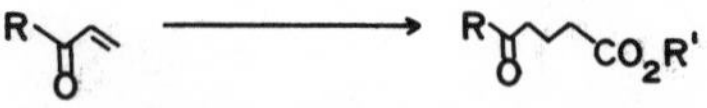

Z⟶ ⟶ Z⟶   10.F.3   611-614;   11.G.3   600-606;
and other   14.C.5   844-845   14.C.5   746-747
Michael additions   (11.G.1-3 699-708) (5.E.3   297)
(18.E.3   931-932) (19.A.4   860)

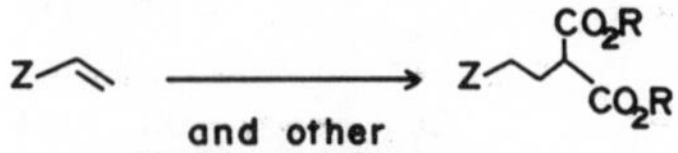

Z⟶ ⟶ Z⟶CO₂Me   14.E.8   776

⟶ CO₂R   14.B.2   827;   13.F.10   705-706
14.C.5   844-845   (14.E.5 774-775)

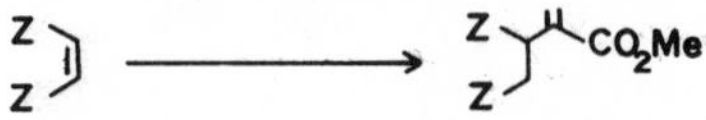

8 ⟶ 6   R–OH   ⟶ R–C–CN   (19.E.3   976)   (19.B.1,2 863-867)

⟶ R–C–C–OH   13.F.6   794-795   13.F.2   697;
13.F.4   702-703

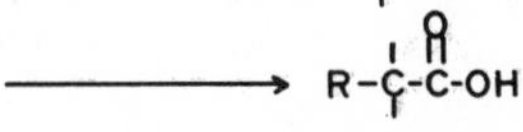

17 ⟶ 6   R–X   ⟶ R–C–CN   19.B.1   946-947;   19.B.1   863-866
(11.G.5   709)   (19.A.1   857-858)

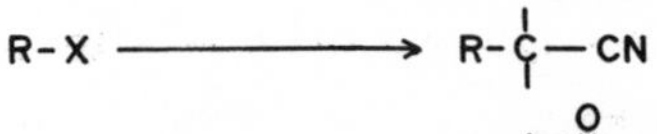

⟶ R–C–C–N<   (18.E.1   929-930)

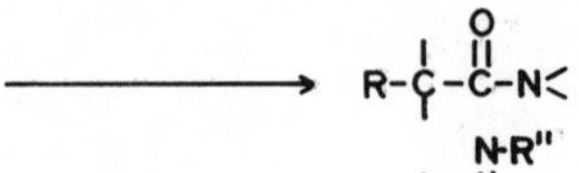

⟶ R–C–C–OR'   10.D.3   595   10.D.3   511-513;
13.A.10 664-667

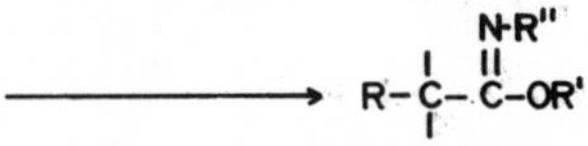

III. A. 2.   1 → 0

17 → 6   R—X ⟶ R—C—C—OH   13.A.9   757–758;   13.B.15   675;
(with O double bond and two open valences)   13.D.3   782;   13.C.2   680–681
                                    13.E.1   784–785   (10.D.3 511–513)
                                                (13.A.10 664–667)

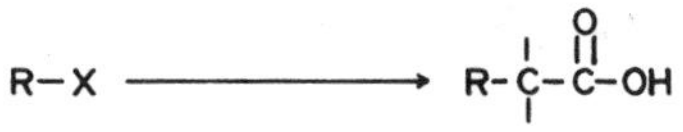

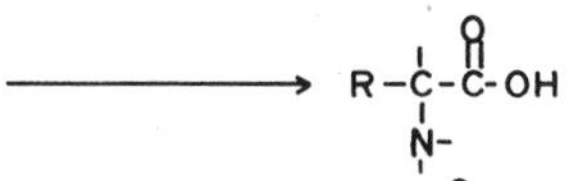

⟶ R—C—C—OH   13.A.9   757–758
(with N)

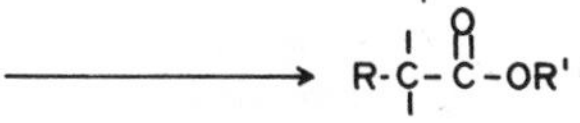

⟶ R—C—C—OR'   14.C.6   845–847   13.A.10   665–667;
                (11.D.6   673–675)   14.C.6   747–752
                (13.D.3   782)   (14.A.18 726)
                (14.B.8   833–834)
                (19.B.1   946–947)

RO₂C—CH₂—X ⟶ RO₂C⌐   (structure with CO₂R)   14.D.14   767

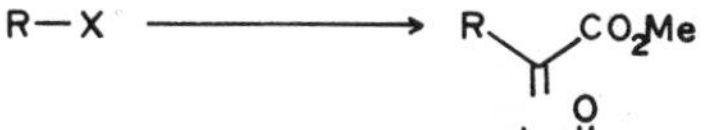

R—X ⟶ R⌐CO₂Me   (structure)   14.C.6   749

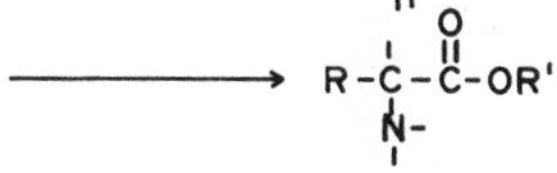

⟶ R—C—C—OR'   (11.G.5   709)   8.B.1   413
(with N)   (13.A.9   757–758)

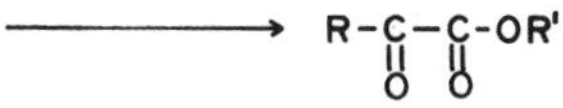

⟶ R—C—C—OR'   11.D.1   570;
(two O double bonds)   13.A.10   667

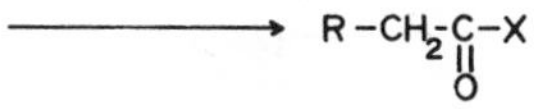

⟶ R—CH₂—C—X   15.D   872–873
(with O)

Functionality   2 → 0

8 → 6   R—CHO ⟶ R⌐(CN)(CN)   19.B.2   949

⟶ R⌐(CO₂H)(NH₂)   13.A.10   758

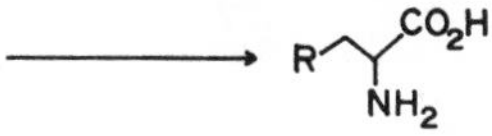

Functionality   2 → 1

(1) C – 2   Functionality = 1

6 → 6   R—≡ ⟶ R⌐═⌐R   2.A.16   101

⟶ R═⌐R   Ad.   144–145

7 → 6   Ar—N₂⁺ ⟶ Ar—CH=CH—Ar'   2.F.2   146   2.C.10   139

8 → 6   ⟩═O ⟶ ⟩═⟨(R)(R')   3.E.1   201

III. A. 2. 2 → 1

8 → 6	R−CHO →	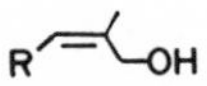			2.E.2 4.E.1	157; 256-257
8 → 8	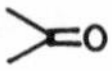 →	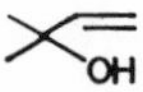			4.E.1	257
	R−CHO → (Prins)	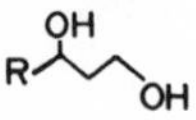	4.B.4 (9.B.7	188-189 533)	4.B.4	224-225
	 →		6.F	327-328		
17 → 6	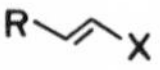 →	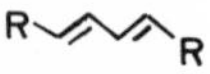	2.C.8	131	2.C.12	140-142
	Ar-X →	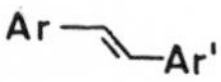			2.C.10	139

(2) C−2 Functionality = 2

7 → 6	$Ar\text{-}N_2^{\oplus}$ →	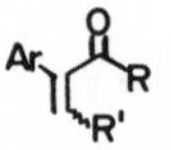	11.C.6	459		
8 → 6	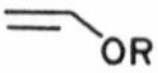 →	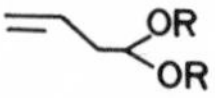	9.B.2	529-530		
	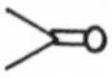 →	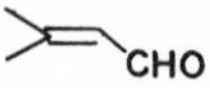	4.G.1 10.F.3 (10.E.2	234-238; 611-614 605)	4.G.1 Ad. 10.F.1,3 (10.D.3 (11.D.1	264-266; 507; 522-523 512) 572)
	→	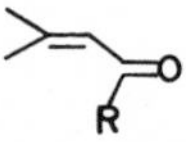	4.G.1 (5.E.1 (14.C.4	234-238 280-282) 841-844)	4.G.1 11.D.1 11.G.10 (5.E.1 (11.H.7	264-266; 572; 608-609 296) 620)
	$R\text{-}CH(OR')_2$ →	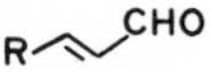	10.C.9	589		
8 → 7	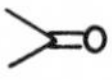 →	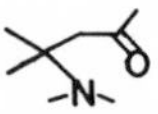	10.D.3	463-465		
8 → 8	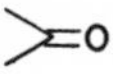 →	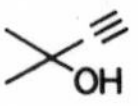	3.B.2 4.E.1	164-166; 226-227	3.B.2 Ad. (4.E.1	187-189; 261 257)
	→	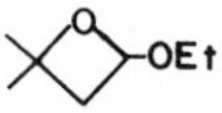			9.F	479

III. A. 2. 2 → 1

8 → 8		4.G.1 10.F.3	234-238; 611-614	4.B.4 4.G.1 10.F.3 11.G.10 (11.E.8 (11.G.9	225-226; 264-266; 523; 608-609 583-584) 607)
	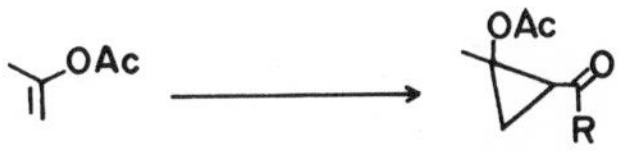			11.G.4	606
	R-CH(OR')$_2$ → 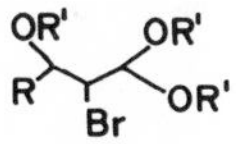			9.B.6	472-473
17 → 6	Ar-X → Ar-C≡C-R			(3.B.1	185)

(3) C-2 Functionality = 3

7 → 7				14.C.8	753
8 → 6				19.B.8	869
				8.E.3	436
		13.D.1-3 (13.A.8	780-783 756)	(13.D.1	687)
	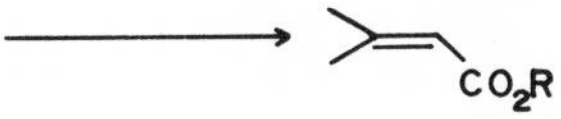	4.G.1 13.D.3 14.C.4	234-238; 782-783; 841-844	13.D.3 (2.A.13 (14.C.8 (14.C.12 (14.E.9	688 104) 754) 756) 777)
	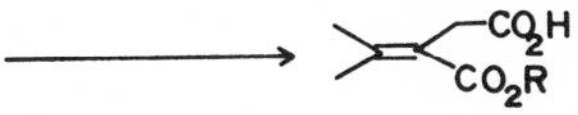	14.C.3	840-841	14.C.3	744-745
	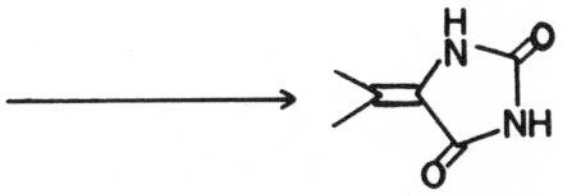	13.A.10	758		
	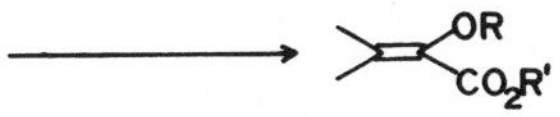	14.C.4	842		
	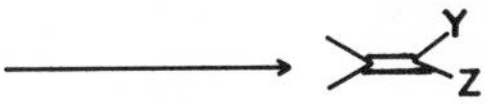 Knoevenagel	14.C.4 19.B.3 (4.G (4.G.8	841-844; 949-950 234-235) 244)	14.C.4 (14.C.1	745-746 742)

III. A. 2.   2 ⟶ 1

8 ⟶ 6   [structure: CH₃-CH(OH)-COOH] ⟶ [butenolide with CN]   14.C.1   837

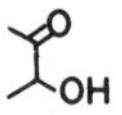

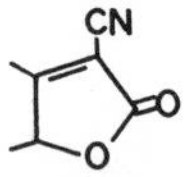

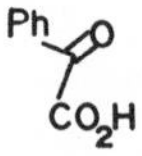

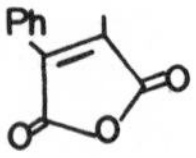

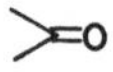

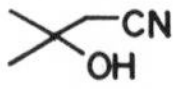

[Ph-CO-CH₂-CO₂H] ⟶ [Ph maleic anhydride]   16.D.1   802

8 ⟶ 8   [ketone C=O] ⟶ [(CH₃)₂C(OH)CN]   19.B.1   946-947

⟶ [epoxide-CN]   19.B.7   869

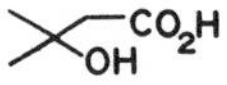

⟶ [(CH₃)₂C(OH)CO₂H]   4.G.4   241-242   13.D.1   687;
                                                        13.D.5   689-690
                                                        (4.E.1   252-253)

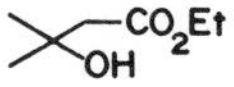

⟶ [(CH₃)₂C(OH)CO₂Et]   4.E.3   229-230;   4.E.3   259-261;
                          4.G.1   234-238   14.B.10 739-740
                          (4.G.8   244)   (4.E.1   252)
                          (14.C.4 841-844)   (13.A.10 664-667)
                          (14.C.8 848)   (14.C.3 744-745)
                                              (14.C.8 753)

Ar-CHO ⟶ [Ar-CH(OH)-C(CO₂Et)=N-CH-Ph]   14.C.4   842

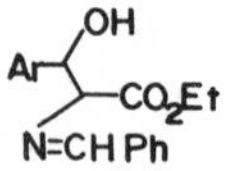

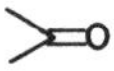

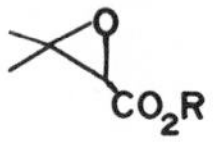

[ketone C=O] ⟶ [glycidic ester epoxide-CO₂R]   6.C.3   316-317   6.C.3   319-321;
                                                  (14.C.8 848)   Ad.   323;
                                                                       14.C.8   754

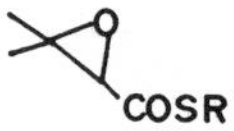

⟶ [epoxide-COSR]   Ad.   323

17 ⟶ 8   RO−CH₂−X ⟶ RO-CH(CH₃)-CO₂Et   6.A.6   310

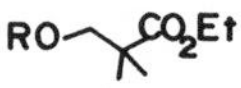

<u>Functionality   3 ⟶ 1</u>

17 ⟶ 6   R−$\overset{\text{O}}{\underset{}{\text{C}}}$−Cl ⟶ R-CH=CH-CHO   10.D.6   514

<u>Functionality   3 ⟶ 2</u>

(1) C − 2   Functionality = 1

17 ⟶ 8   R−$\overset{}{\underset{\text{O}}{\text{C}}}$−X ⟶ R-$\overset{}{\underset{\text{O}}{\text{C}}}$-CH=CH-R'   11.C.7   660-664

III. A. 2.   3 ⟶ 2

17 ⟶ 8   R–C(=O)–X ⟶ R-C(=O)-CH₂CH-R'   II.C. 7   660-664

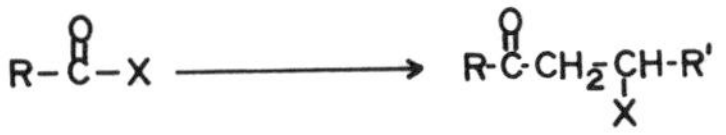

where product is R-C(=O)-CH₂CH(X)-R'

(2) C – 2 Functionality = 2

8 ⟶ 8   R–C(=O)–OR' ⟶ R-C(=O)-CH₂-C(=O)-R'   II.F.4   690-692   (II.F.3 588)

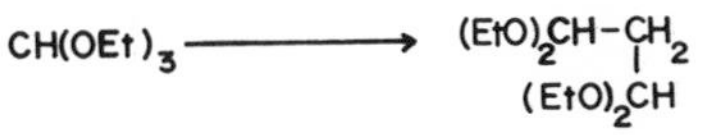

CH(OEt)₃ ⟶ (EtO)₂CH–CH₂–CH(OEt)₂   10.C.9   589

17 ⟶ 6   R–C≡C–X ⟶ RC≡C–C≡C–R'   3.C.1   192-193

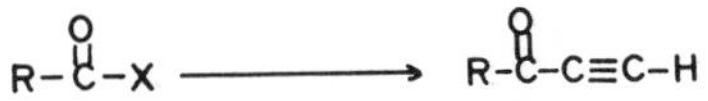

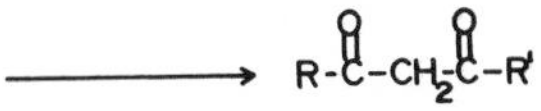

17 ⟶ 8   R–C(=O)–X ⟶ R-C(=O)-C≡C–H   3.B.5   190

⟶ R-C(=O)-CH₂-C(=O)-R'   II.C.7   564-565;
II.F.4   588-590

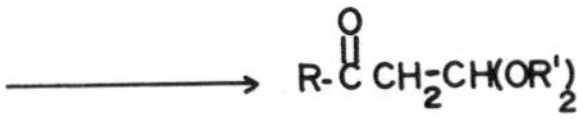

⟶ R-C(=O)-CH₂-CH(OR')₂   9.B.5   531-532

⟶ R-C(=O)-CH=CH-X   7.A.6   370-372
(9.B.5   531-532)
(II.H.7   662)

(3) C – 2 Functionality = 3

7 ⟶ 7   R–CN ⟶ R-C(NH₂)=C-CN   19.B.2   947-949

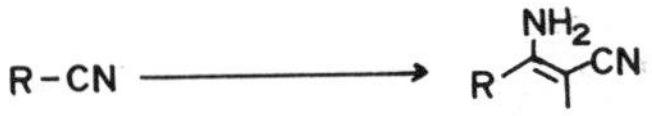

7 ⟶ 8   R–CN ⟶ R-C(=O)-CH(CN)   19.B.2   947-949

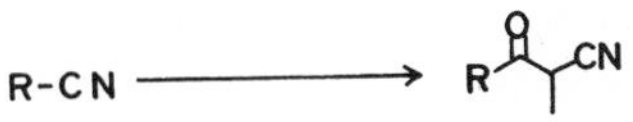

8 ⟶ 8   R–C(=O)-O-C(=O)–R ⟶ R-C(=O)-C-CN   II.F.2   587

(CH₃)₂C=O ⟶ (CH₃)₂C(...)CO₂R   II.F.8   695

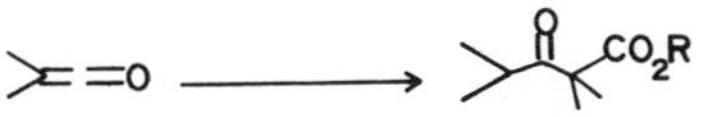

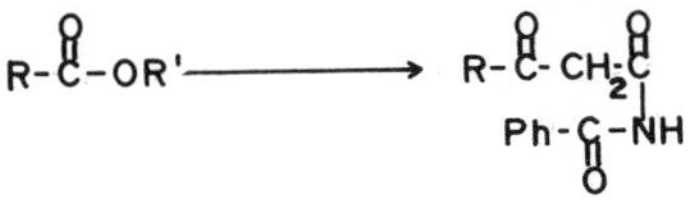

R–C(=O)–OR' ⟶ R-C(=O)-CH₂-C(=O)- ; Ph-C(=O)-NH   18.E.8   847

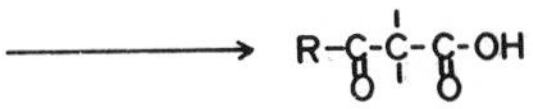

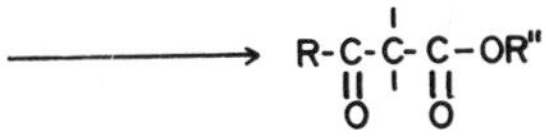

⟶ R-C(=O)-C(=O)-C-OH   II.F.9   592

⟶ R-C(=O)-C-C(=O)-OR"   14.C.1,2   836-840
(14.C.6   845-847)
(19.B.2   947-949)

III. A. 2.   3 → 2

8 → 8   CH(OEt)$_3$ ⟶   (structure: EtO–CH=C(CO$_2$Et)$_2$ type) CO$_2$Et / OEt CO$_2$Et   6.B.6   309

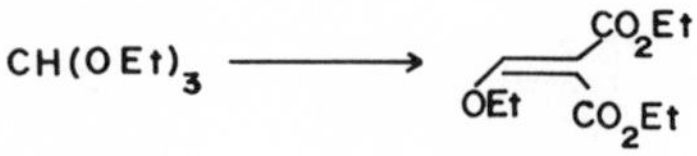

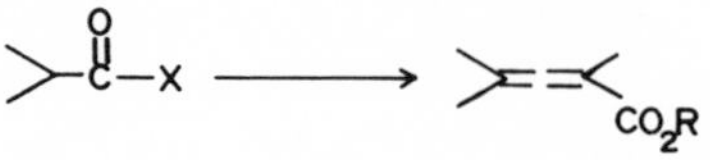

16 → 16   Ph–N=C(Ph)(SMe)   ⟶   (β-lactam structure: Ph–N ... O, Ph, SMe)   18.G.1   851

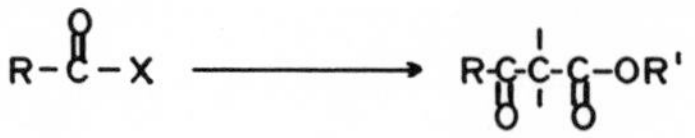

17 → 6   (structure: >C(=O)–X)   ⟶   (structure: >C=C<CO$_2$R)   3.E.1   203

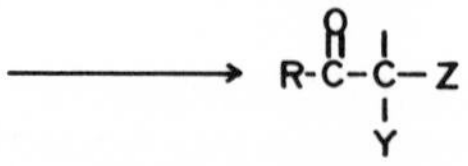

⟶   (structure: >C–C≡C–CO$_2$R)   3.A.7   160

17 → 8   R–C(=O)–X   ⟶   R–C(=O)–C–C(=O)–OR'   II.H.7   722   II.C.7   565;
                                                                          II.F.2   586-587

⟶   R–C(=O)–C(Y)–Z   II.F.3   688-690;   II.H.2   617;
                         14.C.6   845-847;   14.C.6   747-749
                         19.B.2   947-949   (II.F.3   588)

Functionality 4 → 2

8 → 6   CO$_2$ ⟶   (structure: R$_2$NOC–C=C–CONR$_2$)   18.G.3   852

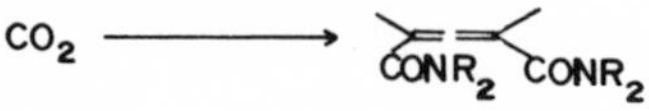

Functionality 4 → 3

8 → 8   R–N=C=O   ⟶   R–NH–C(=O)–C–C(=O) with Et$_2$N   18.G.4   852-853

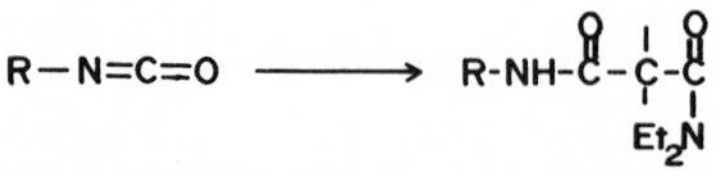

3. Three-Carbon Chain

Functionality 1 → 0

(1) C–3 Functionality = 1

6 → 6   (structure: ene reaction)   ⟶   (product structure)   (2.C.2   117)   2.C.2   134

ene reaction

III. A. 3. 1 ⟶ 0

6 ⟶ 6	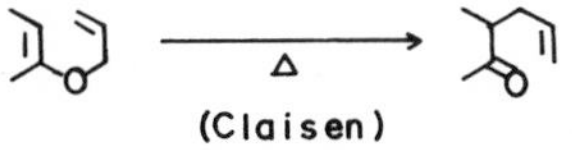	10.H.1 11.G.4 (3.D.4	619-620; 707-708 170)	10.H.1 (2.C.13 (14.E.7	528-531 142) 775-776)
	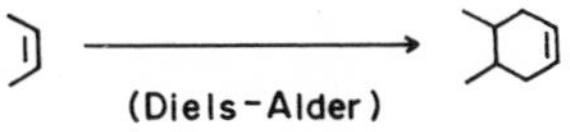	2.C.2 5.F.2 6.F 8.F.6 (2.A.11	116-124; 284; 327; 491-492 97)	2.C.2	130-134
				3.E.1	204
	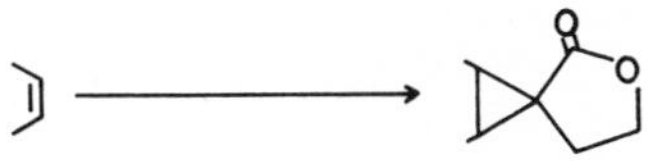			14.E.5	774-775
17 ⟶ 6		2.C.7	129-131	2.B.7 2.C.7 2.C.9	126; 136-137; 138-139
	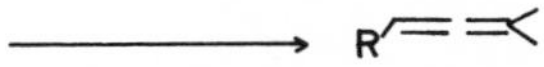			3.B.1	186
				2.B.7 4.A.8 Ad.	126; 215; 267

(2) C-3 Functionality = 2

6 ⟶ 6	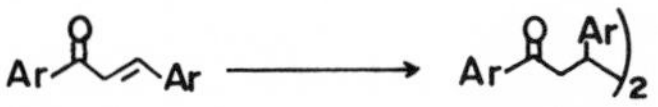	11.B.6	650		
17 ⟶ 6	R—X ⟶ R-CH(Ph)-C≡C-H			3.D.1	196
				11.G.2 (8.I.1	600 456)
	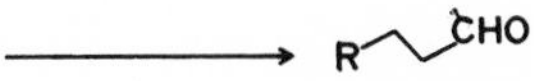			Ad.	525
				10.D.7 10.E.2	515; 518
	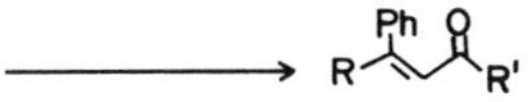			11.H.8	622

(3) C-3 Functionality = 3

6 ⟶ 6	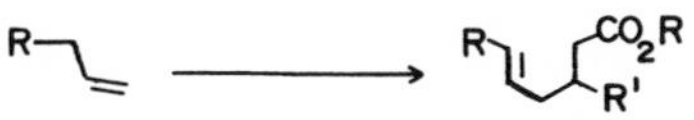			2.C.2 Ad.	134; 144

III. A. 3.　1 ⟶ 0

6 ⟶ 6　　　2.C.2　　116-124　　2.C.2　130-134

17 ⟶ 6　R–X ⟶ R-CH$_2$C≡C-N<　　　　　　18.E.10　847-848

　　　　　R-CH$_2$CH$_2$C-N<　　　　　18.E.10　847-848

　　　　　 ⟶ 　　　　19.C.9　875-876

<u>Functionality　2 ⟶ 1</u>

(1) C - 3　Functionality = 1

8 ⟶ 6　 ⟶ 　　2.C.15　143

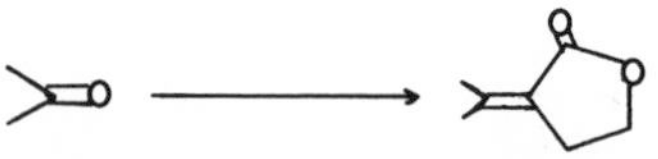

　　　　 ⟶ 　14.C.4　842

8 ⟶ 8　 ⟶ 　　4.E.1　253

17 ⟶ 6　Ar–X ⟶ 　　2.C.7　137

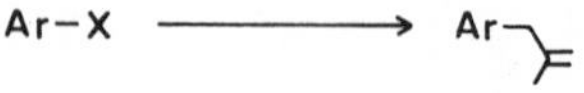

(2) C - 3　Functionality = 2

17 ⟶ 6　Ar–X ⟶ ArC≡CCH(OMe)$_2$　　3.B.1　187

(3) C - 3　Functionality = 3

8 ⟶ 6　 ⟶ 　14.C.3　840-841　14.C.3　744-745

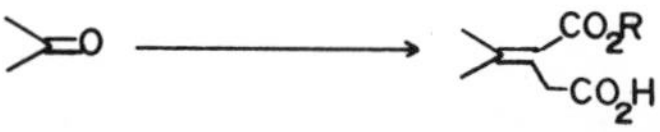

17 ⟶ 6　 ⟶ 　　2.C.8　138;
　　　　　　　　　　　　　　　　　14.C.14　756

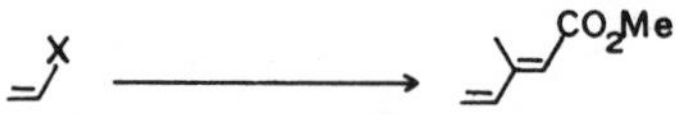

<u>Functionality　3 ⟶ 2</u>

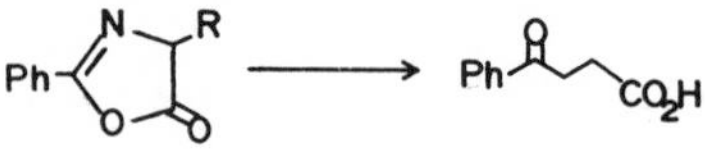

⟶ 　　13.A.10　666

Ⅲ. A. 4.   1⟶0

4. Four-Carbon Chain

　　Functionality   1⟶0

　　(1) C – 4 Functionality = 1

8⟶6　Ar–CH$_2$–OH ⟶ ArOH　　4.B.1　219

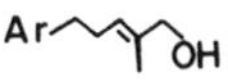

　　(2) C – 4 Functionality = 2

17⟶6　R–X ⟶ R–X　I.D.1　33-39;　I.D.1　42-45
　　　　　　　　　　　　　　　　　　　5.D.1　270-273

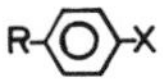

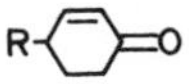

　　　　⟶ R–=O　　　　II.G.3　602

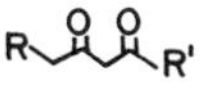

　　　　⟶ RR'　　　　II.G.1　597

　　Functionality   2⟶1

　　(1) C – 4 Functionality = 1

8⟶8　R– CHO ⟶ 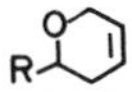 　6.F　327-328

　　(2) C – 4 Functionality = 2

　　(3) C – 4 Functionality = 3

8⟶6　=O ⟶ CO$_2$R　2.C.13　142;
　　　　　　　　　　　　　　　　　　　　　　　10.H.1　530;
　　　　　　　　　　　　　　　　　　　　　　　13.G.7　710;
　　　　　　　　　　　　　　　　　　　　　　　14.E.7　775-776
　　　　　　　　　　　　　　　　　　　　　　　(18.G.5　853)

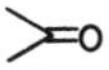

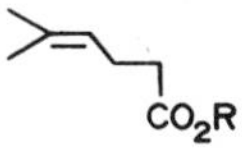

8⟶8　=O ⟶ CO$_2$H　13.D.5　690

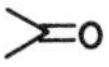

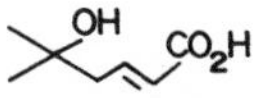

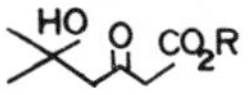

　　　　⟶ CO$_2$R　4.G.1　264

　　Functionality   3⟶2

　　(1) C – 4 Functionality = 1

　　(2) C – 4 Functionality = 2

III. A. 4.   3 ⟶ 2

17 ⟶ 8   R–C–X $\underset{\text{O}}{\overset{\text{||}}{}}$ ⟶ R-C$\overset{\text{O}}{\overset{\text{||}}{}}$⟨O⟩X   II.C.1-5   651-659   II.C.1-5   559-562

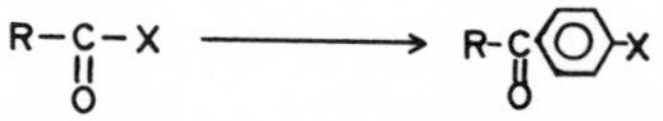

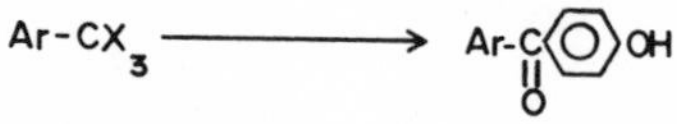

Ar-CX$_3$ ⟶ Ar-C⟨O⟩OH   II.G.7   711-712

5. Five-Carbon Chain

Functionality   1 ⟶ 0

17 ⟶ 6   R–X ⟶ R⤳CHO   10.H.1   529

Functionality   2 ⟶ 1

17 ⟶ 6   ⟋⟍X ⟶ ⤳C=O / OMe   2.C.8   138

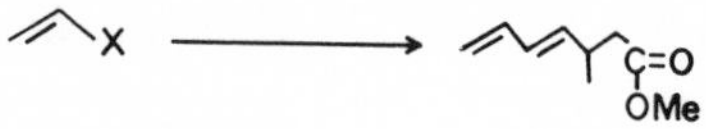

6. Six-Carbon Chain

Functionality   3 ⟶ 2

17 ⟶ 8   R-C-X ⟶ R⤳CO$_2$H   13.D.4   688-689

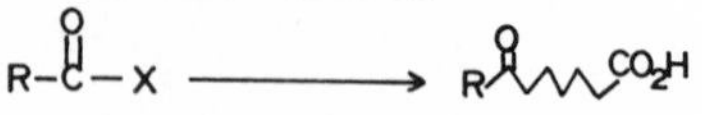

B. NO CHANGE IN OXIDATION STATE OF SUBSTRATE

0. Unfunctional Chain

Functionality   0 ⟶ 0

6 ⟶ 6   R–H $\xrightarrow{\text{:CH}_2}$ R–CH$_3$ 1.G.6   63-64

R–H ⟶ R — R'   1.D.4   42-43

R–MgX ⟶ R — R'   1.B.2   22-27   I.B.2 23-27;
and other   I.B.6 31-32;
organometallics   Ad.   33;
                                             I.D.4 46-47

R–CO$_2$H ⟶ R — R'   1.G.7   65   I.G.7 67-68;
(Kolbe)   14.D.5 760

III. B.O.   O → O

6 ⟶ 6   Z–CH$_3$ $\xrightarrow[\text{reactions}]{\text{alkylation}}$ Z–CH$_2$–R

Z = –Ar	I.C.3	31	I.C.7	37–40
–CHO	10.F.4	614–615	10.F.4 10.G (8.I.1	524–525; 527–528 455–456)
$-\overset{\text{O}}{\underset{}{\text{C}}}-R$	11.G.1,2	697–705	11.G.1,2 (11.G.3	595–600 602)
–CN	19.B.1 (11.G.5	946–947 709)	19.B.1	863–866
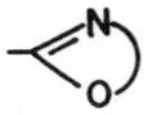 (oxazoline)	10.D.3	595	10.D.3 13.A.10	511–513; 664–667
$-\overset{\text{O}}{\underset{}{\text{C}}}-OH$			13.B.15 13.C.2	675; 680–681
$-\overset{\text{O}}{\underset{}{\text{C}}}-OR'$	14.C.6	844–847	14.C.6	747–752

Z–CH$_2$ ⟶ Z–CH–R
(with Y)

Z–CH$_2$	11.D.6	673–675;
|	13.A.9	757–758;
Y	13.E.1	784–785;
→ Z–CH–R	14.B.8	833–834;
|	14.C.6	844–847
Y	14.B.8	738–739;
	14.C.6	747–752

Functionality   1 ⟶ 1

6 ⟶ 6   (alkene) ⟶ (alkene)^R

	2.C.1	113–116
	(7.B.6	370–372)

Ar–H ⟶ Ar–R

	I.C.5	32;	I.D.1	42–45;
	I.D.1	33–39;	Ad.	47
	20.C.2	1001	(5.D.1	289–292)
	(5.D.1	270–273)	(6.B.9	315–316)
			(8.D.10	433)

(quinone) ⟶ (catechol)^R

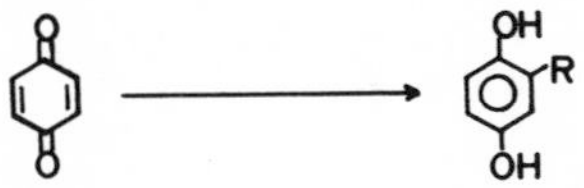	5.C.1	288

7 ⟶ 7   O$_2$N–CH$_3$ ⟶ O$_2$N–CH$_2$–R

	20.B.2	993	20.B.1,2	896–897

8 ⟶ 6   R–(epoxide) $\xrightarrow{R'Li}$ R$\diagdown\diagup$R'

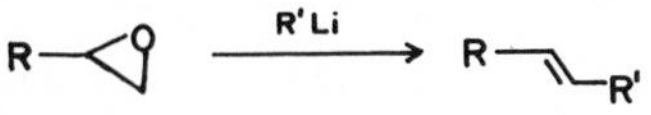

	2.A.15	100

III. B. O.   I ⟶ I

8 ⟶ 8   Ar–CH$_2$O–R ⟶ Ar–CH–OH   4.G.7   243-244   4.G.7   266-267
                             |
                             R

16 ⟶ 16 Ar–S–CH$_3$ ⟶ Ar–S–CH$_2$R   7.A.16   352

Functionality   2 ⟶ 2

6 ⟶ 6   R–C≡C–H ⟶ R–C≡C–R'   3.B.1   162-164   3.B.1   184-187
                                                              (3.C.5 193-194)

7 ⟶ 7   ⟶   8.E.3   476   8.E.3   435-436

8 ⟶ 8   R–C–H ⟶ R–C–R'   10.D.3   595;   10.D.2,3 510-513;
            ‖   ‖   11.D.1   667;   11.F.7   591-592
            O   O   11.G.5   708-711   (11.D.7   575-576)
                                  (10.C.9   589)   (19.B.1   864)
                                  (11.E.5   682-683)
                                  (11.F.7   694-695)

15 ⟶ 15 RO$_2$C   RO$_2$C   R'   13.D.3   782
              \   \   /
               ‖   ⟶   ‖
              PPh$_3$ PPh$_3$

16 ⟶ 16   RS   RS   10.D.3   595;   9.C.1   475;
            \   \   11.D.1   667   10.D.7   515;
             >   ⟶   >–R'   11.D.7   575-576
            /   /   (11.H.8 622)
           RS   RS
           (including sulfoxides)

Functionality   3 ⟶ 3

7 ⟶ 7   HCN ⟶ R–CN   19.A.1,2   938-941;   19.A.1,2 856-859
                                see also other sections

I. One-Carbon Chain

Functionality   0 ⟶ 0

(1) C–1 Functionality = 1

6 ⟶ 6   ⟶   (11.G.1   597)

R–MgX ⟶ R–C–N<   8.E.1-3   473-477   8.C.9   426;
              | 8.I.6   458

III. B. 1. O ⟶ O

	Reaction				
6 ⟶ 6		8.D.3 10.F.3 (11.G.1	463-465; 611-614 700)	8.D.3	430-431
	R-MgX ⟶ R–C–OH	4.E.1 (11.H.1,2	224-227 713-717)	4.E.1	252-258
	⟶ R–C–OR'	2.A.4 6.A.6	82-84; 298	6.A.6	310
	R_3B ⟶ R_3C–OH	4.B.6	191	4.B.6	225-226
		4.G.1,2 10.F.3	234-240; 611-614	4.G.1,2 11.G.10 (4.B4	264-266; 608-609 224-225)
	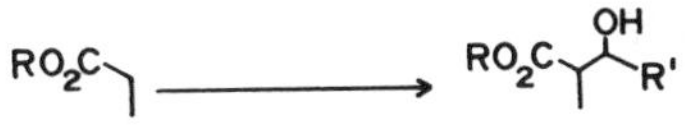	4.G.1 (4.E.3 *see also* section III.A.2	234-238 229-230)	4.E.3 14.B.10	259-261; 739-740

(2) C-1 Functionality = 2

	Reaction				
6 ⟶ 6	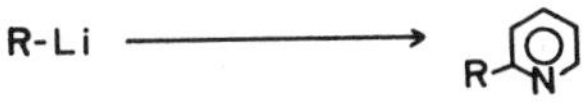	8.E.3	476-477		
	R–H ⟶ R–C(=O)–R'	11.F.7	694		
	R-Li, R-MgX ⟶ R-CHO	10.G (8.E.1 (9.C.2 (10.A.17	616-619 473-474) 536) 564)	10.D.2 10.G (9.C.2	510-511; 526-528 475)
	R-Li, R-MgX ⟶ R–C(=O)–R'	11.H.1-4 (8.E.1,2 (10.G (11.G.5	713-719 473-474) 616-619) 709)	11.D.1 11.H.1-4	568-570; 614-619
	R_3B ⟶ R-CHO			10.E.8	520
	⟶ R–C(=O)–R'	11.C.7	663	11.E.8 Ad.	582-584; 584
	R-MgX ⟶ R-CH(OR')$_2$	9.C.2 (10.G	536 616-619)		
	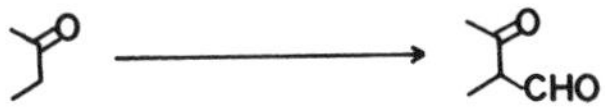	10.F.1 (6.B.6 (10.C.1 (10.C.6 (11.G.1	608-609 309) 584) 586) 698,702)	10.C.5,6 10.F.1	504-506; 521-522

III. B. I.   O⟶O

6 ⟶ 6 ⟶

	II.C.7	663;	II.C.7	565;
	II.F.4	690-692	II.F.4	588-590
			(5.E.2	296)
			(8.I.3	457)

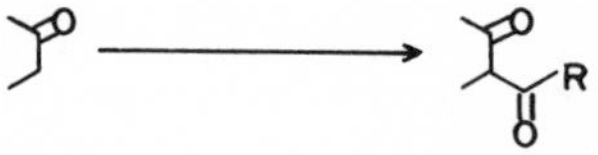

⟶

II.F.5	693;	(II.F.2	587)
19.B.2	947-949		

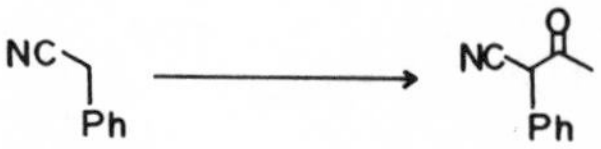

⟶

		II.F.9	592
		(10.G	527)

⟶

II.F.3	688-690;	II.F.2	587;
14.C.1,2	836-840;	II.F.3	587-588;
14.C.6	845-847	14.C.6	747-749

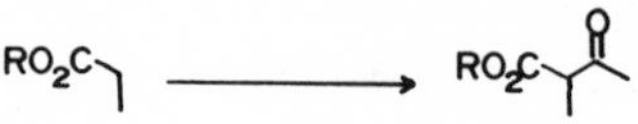

(3) C – I Functionality = 3

R—H ⟶ R—CN	19.E.2	975-976		
R—MgX ⟶ R—CN	19.A.6	944		
R—MgX ⟶ R-C-NH-Ph (N-Ph)	18.E.4	933		
R—MgX ⟶ R-C-NH₂ (O)	18.E.4	933	18.E.4 A d.	845; 848
R—H ⟶ R-C-OH (O)	13.F.5	794	13.F.5	703-705
R—MgX ⟶ R-C-OH (O)	13.C.1	776-778	13.C.1 (10.D.2	676-680 510-511)
R—MgX ⟶ R-C-OR' (O)	14.C.7	847-848	14.C.7	752-753
Z—CH₃ ⟶ Z-CH₂-C-OH (O)			13.C.1 14.C.7	676-680; 752-753
Z—CH₃ ⟶ Z-CH₂-C-OR' (O)	14.C.1	836-839	14.C.1 14.C.7	741-744; 752-753
R—MgX ⟶ R-C-NH-Ph (S)	18.E.4	933		
R—H ⟶ R-C-Cl (O)	15.C	871-872		

Functionality I ⟶ I

(I) C – I Functionality = I

6 ⟶ 6 Ar⌁ ⟶ Ar⌁Ar   2.C.10 139

III.B.I.   I⟶I

6⟶6	Ar–H ⟶ Ar–Ar	I.D.7	45–47	I.B.3 28–29; I.G.I-3 53–56 (5.B.9 284-286) (20.D.4 907-908)	
	Ar–H ⟶ Ar–C–N<	8.D.3 (8.G.8 (18.D.3	463–466 505–507) 920-921)	8.D.3 430-431 (8.G.8 450-451)	
	Ar–MgX ⟶ Ar–C–N<			8.C.9 426	
	Ar–H ⟶ Ar–C–OH	(4.B.7	191-192)		
	Ar–MgX ⟶ Ar–C–OH	4.E.I	224-227	4.E.I 252-258	

$$\underset{CO_2H}{\overset{CO_2H}{\bigodot}} \longrightarrow \bigodot$$

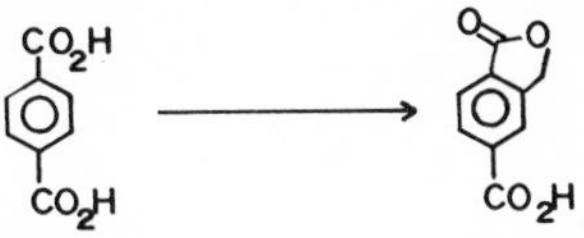

14.B.9   739

	Ar–H ⟶ Ar–CH₂–X	7.D.6	404-406	7.D.6 385	
7⟶6	 >–NO₂ ⟶ ><			2.A.22 114	
7⟶7	 O₂N⟍ ⟶ O₂N⟍⟋N–	20.C.I	998-999		
	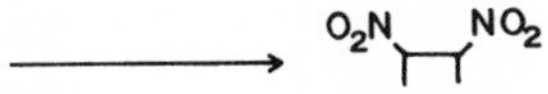 ⟶ O₂N⟍NO₂			20.D.4 907-908	
	 ⟶ O₂N⟍OH	20.C.I	997-999	20.C.I 899-901	
15⟶6	R–CH₂–⁺PPh₃ ⟶ R–CH=C<	2.E.2 (2.A.II	141-144 96)	2.E.2 156-158 (I.C.9 40-41)	
16⟶6	R–CH₂–SR' ⟶ R–CH=C<	2.A.II (13.G.2	95 799)	2.A.II 99-101; 2.A.19,20 110-112; Ad. 115-116 (I.A.7 15)	
17⟶6	R–CH₂–X ⟶ R–CH=C<	2.E.2	141-144	2.C.12 140-141; 2.E.2 156-158 (14.C.12 756)	

(2) C–I Functionality = 2

6⟶6	Ar–H ⟶ Ar–CH=N–R	10.C.7	587-588		

8.I.4   457

III. B. I.   I ⟶ I

6 ⟶ 6   10.H.1   530

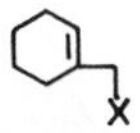

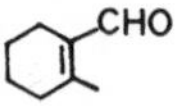

Ar—⟶   Ar⟶CHO   10.C.6   587

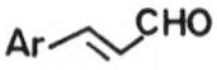

Ar—H ⟶ Ar—CHO   10.C.1-7   583-588   10.C.1-6 502-506;
(10.F.2   609-610)   10.C.12,13 506-507;
Ad   507-508
(10.F.2   522-523)

R—⟶   R—C(=O)—R′   11.C.7   660-664   11.C.7   562-566

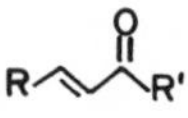

Ar—H ⟶ Ar—C(=O)—R   11.C.1-5   651-659   11.C.1-5 559-562
(5.D.8   278)   (6.B.9   315-316)
(11.G.7   711-712)   (12.B.1   646)
(12.B.1   737-738)
(13.F.1   789-791)

Ar—MgX ⟶ Ar—C(=O)—R   11.H.1-4   713-719   11.H.1-4 614-619

Ar—SiMe₃ ⟶ Ar—C(=O)—R   11.C.1   561

Ar—HgX ⟶ Ar—C(=O)—R   11.C.7   663

Ar—H ⟶ Ar—CH(OR)₂   10.C.5   585-586   10.C.5 503-504

⟶ Ar—CH(X)—OR   10.C.4   503

⟶ Ar—CHX₂   (10.F.2   609-610)   (10.F.2 522-523)

7 ⟶ 7   O₂N—CH₃ ⟶ O₂N—CH₂—C(=O)—R   20.C.3   903

8 ⟶ 8   RO—CH(R′)₂ ⟶ RO—CH(R′)—CHX₂   6.D.2   328

16 ⟶ 16   CH₃—S(=O)—CH₃ ⟶ CH₃S(=O)CH₂C(=O)—R   11.F.6   693-694

(3) C—I Functionality = 3

6 ⟶ 6   Ar—H ⟶ Ar—CN   19.E.1   974-975   19.A.2   859;
(19.D.2,4 964-967)   19.E.1   880
(19.E.4   881)

Ar—Li ⟶ Ar—CN   19.A.6   861;
19.C.5   874

Ar—SnMe₃ ⟶ Ar—CN   (19.A.4   860)

III. B. I.   I ⟶ I

6 ⟶ 6   Ar–H ⟶ Ar–C(=O)–N<   18.D.1,2   919-920   18.D.1   833

Ar–H ⟶ Ar–C(=O)–OH   13.C.3   778-779;   13.C.3   682-684;
13.F.1   789-791   13.F.1   694-695
(13.C.4   779-780)   (13.C.4   684-685)
(13.E.3   693)

Ar–H ⟶ Ar–C(=O)–OR   14.B.8   833-834   14.E.4   774

)(⟶)(–CO₂Me   14.B.6   831

[benzene-CO₂H] ⟶ [cyclic structure NH, S]   13.F.1   791;
18.D.2   920

Ar–H ⟶ Ar–C(=O)–X   15.C.1   871-872   15.C.3   792-793

7 ⟶ 7   O₂N\ ⟶ O₂N\/CO₂H   13.C.1   777

O₂N\ ⟶ O₂N\/CO₂R   20.C.1   997

Functionality 2 ⟶ 2

(1) C–I Functionality = I

6 ⟶ 6   R–C≡C–H ⟶ R–C≡C–Ar   (3.B.1   184-187)

⟶ R–C≡C–C–N<   3.B.4   166-167
(8.B.3   463-465)

⟶ R–C≡C–C–OH   3.B.2   164-166;   3.B.2   187-189
4.E.1   226-227

8 ⟶ 8   R–CHO ⟶ R–C(=O)–C(OH)–   4.C.8   205-207   4.C.8   239;
Ad.   243;
Ad.   610

(2) C–I Functionality = 2

6 ⟶ 6   R–C≡C–H ⟶ R–C≡C–CH(OR')₂   3.B.3   166   3.B.3   189-190
(9.B.6   532)   (9.B.6   472)
(9.C.2   536)

⟶ R–C≡C–C(=O)–R'   3.B.5   190

Ⅲ.B.I.   2 ⟶ 2

7 ⟶ 7	Ph-CH≡N-Ph ⟶			

| | | II.G.5 | 709 | |

| | | | | 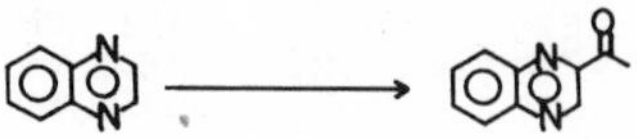 II.F.5 | 590 |

8 ⟶ 6	R-CHO ⟶ R − C≡C − H			3.A.I	173
				(3.A.9	180-181)
				(13.C.I	680)

8 ⟶ 8	Ar-CHO ⟶	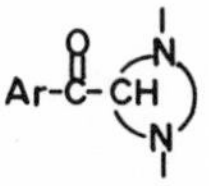 II.G.5	710		

Ar-CHO ⟶ Ar-C=C-Ar   4.C.8   205-207   4.C.8   239;
         HO   OH   Ad.   243
                                          (Ad.   610)

15 ⟶ 15   R   ⟶   R   R'   II.H.7   721-722

(3) C − I Functionality = 3

6 ⟶ 6	R-C≡C-H ⟶ R − C≡C−CN	3.C.2	169		

⟶ R-C≡C-C-N<   3.B.3   166   3.B.3   190;
                        (18.E.4 933)   18.E.4 845
                                        (Ad.   848)

⟶ R-C≡C-C-OH   3.B.2   164-166
                (13.C.I 776-778)   (13.C.I   676-680)

7 ⟶ 7 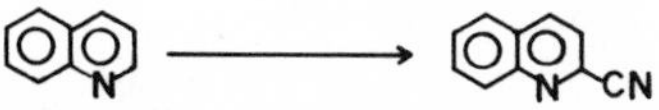	⟶	10.B.I	570	

2. Two-Carbon Chain

Functionality O ⟶ O

(I) C − 2 Functionality = I

6 ⟶ 6   R-Li
         R-MgX   ⟶ R−CH=C<   2.A.4   82-84;   2.A.20 112;
                               2.A.15 100   2.C.7   136-137
                               (2.C.5 128)

R-B< ⟶ R−CH=C<   2.B.I   121

III. B. 2.   O⟶O

II.G.2   703-705

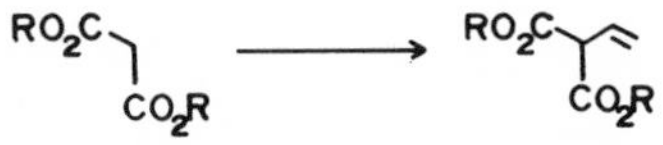

14.B.8   834

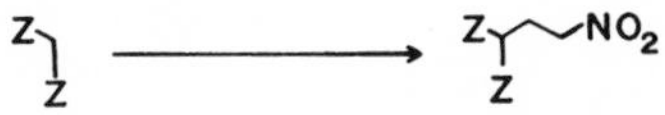

20.C.2   1000-2   (20.C.2 901-902)

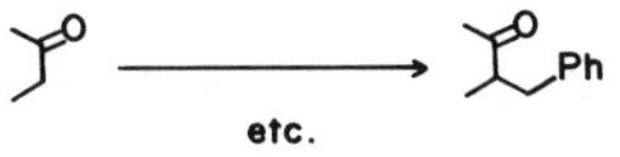

R-Li
R-MgX   ⟶   R-CH$_2$CH$_2$OH   4.E.2   228-229   4.A.7   214;
                                                 4.E.2   258
                                                 (4.C.2 234)
                                                 (4.E.4 261)

Ad.   609;
                       Ad.   729

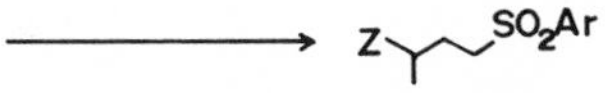

I.A.7   16;
                       14.C.5 747

(2) C-2 Functionality = 2

6⟶6   R-Li   ⟶   R-C≡C-R'   3.A.2   174-175

      R-B⟨   ⟶   R-C≡C-R'   3.C.5   193-194

      R-Li
      R-MgX   ⟶   II.H.5   719-720   II.G.9   606-607

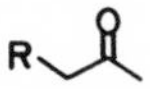

      R-B⟨   ⟶   II.G.1   701   II.G.9   606-607

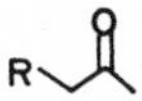

      R-Li   ⟶   II.F.10 592-593

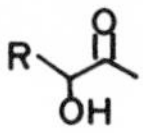

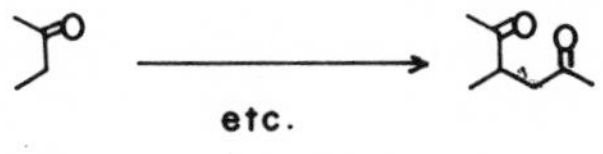

        II.G.2   703-705   10.A.22 495;
                                                11.A.20 549-550;
                                                II.G.4   606;
                                                Ad.   609;
                                                14.C.6   751

(3) C-2 Functionality = 3

6⟶6   R-B⟨   ⟶   R-CH$_2$CN   19.B.1   947   19.B.1   863-866

Ⅲ. B. 2.   O⟶O

6 ⟶ 6	(structure) ⟶ (structure) CONR₂, NR₂			8.D.3 431

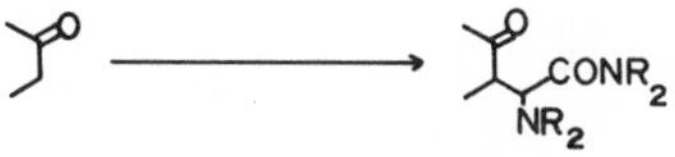

$6 \longrightarrow 6$

Reactant	Product	Col1	Col2	Col3
(acetone)	(structure) CO₂H	13.C.2	778	(10.F.3 523)
R-Li	R-C-C-OH (diketone)			13.C.1 679
(cyclopentadiene)	(structure) CO₂Et			4.B.2 222
(acetone)	(structure) CO₂R	11.G.2	703-705	(14.C.6 750) (14.D.14 768)
R-B<	R-CH₂-C-OEt, O	14.B.8	833-834	14.B.8 737-738; 14.E.1 770-771
R-B<	R-CH-C-OEt, X O	7.C.7 (14.B.8	387-388 833-834)	

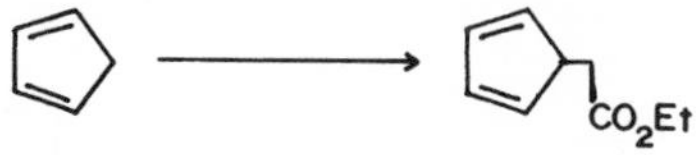

Functionality ⏐⟶⏐

(1) C-2 Functionality=1

	Reactant	Product	Col1	Col2	Col3
6 ⟶ 6	R-CH=CH-CH₂-MgX	R(structure)R	2.C.8	131	
	Ar-H	Ar-CH=C<	(12.C.3	742)	(1.D.2 45) (2.C.1 130)
	Ar-H	Ar-CH₂CH₂OH	4.B.7	191-192	4.B.7 226-227

(2) C-2 Functionality= 2

	Reactant	Product	Col1	Col2	Col3
6 ⟶ 6	(phenylhydrazine)	(indole)	8.G.7	504-505	8.G.7 449-450 (9. 463)
	Ar-HgX	Ar-CH=CHOAc			14.B.5 735
	Ar-PdCl	Ar-C-CHO	10.C.10	590	
	(cyclohexene)	(structure) CHO			10.G 527

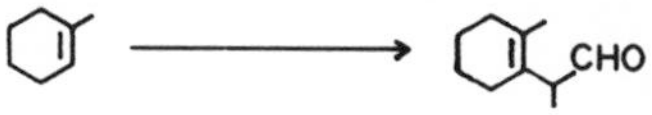

III.B.2.   I ⟶ I

6 ⟶ 6   Ar–H ⟶ Ar–C(=O)–C(–)   II.C.6   659

(3) C – 2 Functionality = 3

6 ⟶ 6   Ar–H ⟶ Ar–CH(Ar)–CN   19.E.I   974-975

Ar–H ⟶ Ar–CH₂–C(=O)–NH₂   18.F   849

Ar–H ⟶ Ar–CH₂–C(=O)–OH   13.F.2   695-697

Ar–H ⟶ Ar–CH₂–C(=O)–OEt   14.B.II   740

Ar–H ⟶ Ar–CH=CCl₂   7.D.6   385

17 ⟶ 6   [Br, CO₂Me / Br, CO₂Me] —NaH→ [CO₂Me / CO₂Me]   2.A.2   87;
14.C.6   750-751

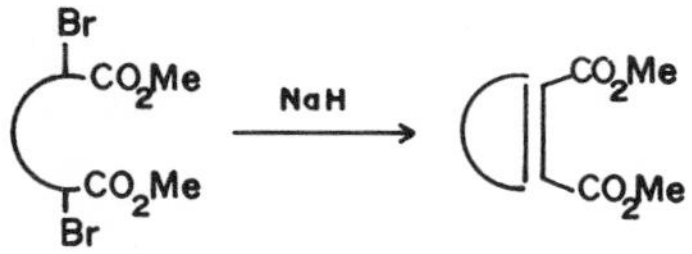

Functionality   2 ⟶ 2

(1) C – 2 Functionality = I

6 ⟶ 6   R–C≡C–H ⟶ RC≡CCH=CHR   3.C.I   192-193

R–C≡C–H ⟶ RC≡CCH₂CH₂OH   3.B.5   190

(2) C – 2 Functionality = 2

6 ⟶ 6   R–C≡C–H ⟶ RC≡CC≡CR   3.C.I   168-169   3.C.I   192-193
(4.D.6   222)

16 ⟶ 16   [S, S ring] ⟶ [S, S ring –CH(OR)(OR)]   10.D.3   595

(3) C – 2 Functionality = 3

6 ⟶ 6   R–C≡C–H ⟶ RC≡CCH₂C(=O)OEt   14.E.I   771

8 ⟶ 6   R–CHO ⟶ RC≡C–C(=O)–OH   13.C.I   680

R–CHO ⟶ RC≡C–C(=O)–OR   3.A.9   181

III.B.3.   O ⟶ O

3. Three-Carbon Chain

Functionality   O ⟶ O

(1) C-3 Functionality = 1

6 ⟶ 6   R—MgX ⟶ R⟋⟍⫝̸   2.C.7   129-131   2.C.7   136-137

R—B⟨ ⟶ R⟋⫝̸   2.A.16   106-107

R—Li ⟶ R⟍⟋   3.E.I   203

⟋⟍O ⟶ ⟍⟋O   10.H.I   619-620;   10.H.I   528-531
                              II.G.2   703-705;
                              II.G.4   707-708

orthoester Claisen   2.C.13   142;
                                                10.H.I   530;
                                                13.G.7   710;
                                                14.E.7   775-776
                                                (18.G.5   853)

R⟍(C=O)⟍NH ⟶ R⟍CN   3.E.I   203

R—MgX ⟶ R⟍⟋⟍OH   4.E.2   228-229

(2) C-3 Functionality = 2

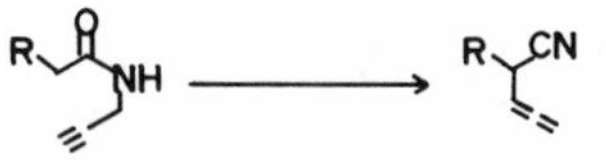

6 ⟶ 6   ⟋⟍O ⟶ ⟍O⟍≡—SiMe₃   II.G.I   597

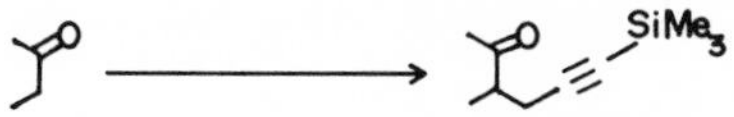

R—B⟨ ⟶ R⟍⟍CHO   10.F.6   525

⟶ R⟍CH(Br)CHO   I.A.5   13

⟶ R⟍(quinone)   12.B.4   647

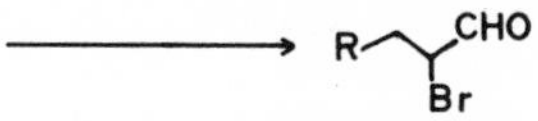

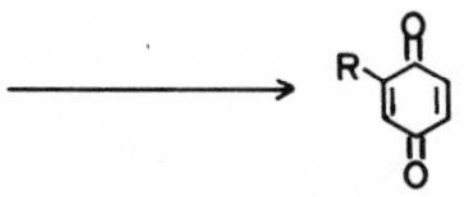

R—MgX
R₂CuLi ⟶ R⟍⟍(C=O)⟍   II.H.6   720-721   II.G.3   600-606
R₃B   (II.D.I   569)
                                                   (II.H.8   622)

III.B.3.   O ⟶ O

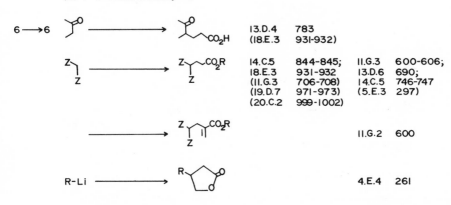

6 ⟶ 6   Z—⟶ Z—C(=O)

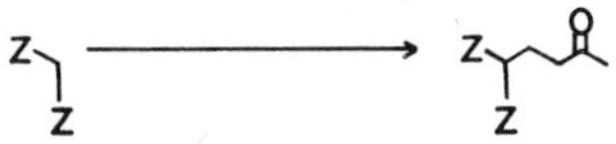

10.F.3	611-614;	Ad.	593;
11.G.3	706-708;	11.G.3	600-606
14.C.5	844-845	(11.G.1	595)
(11.G.1	699-700)		
(11.G.2	703-705)		

(3) C-3 Functionality= 3

6 ⟶ 6

| 13.D.4 | 783 |
| (18.E.3 | 931-932) |

14.C.5	844-845;	11.G.3	600-606;
18.E.3	931-932	13.D.6	690;
(11.G.3	706-708)	14.C.5	746-747
(19.D.7	971-973)	(5.E.3	297)
(20.C.2	999-1002)		

| | | 11.G.2 | 600 |

R-Li ⟶

| | | 4.E.4 | 261 |

Functionality 1 ⟶ 1

(1) C-3 Functionality = 1

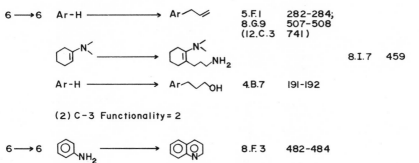

6 ⟶ 6   Ar-H ⟶ Ar—⟍⟍

5.F.1	282-284;
8.G.9	507-508
(12.C.3	741)

| | 8.I.7 | 459 |

Ar-H ⟶ Ar—OH

| 4.B.7 | 191-192 |

(2) C-3 Functionality= 2

6 ⟶ 6

| 8.F.3 | 482-484 |

Ar-H ⟶ Ar—CHO

| 10.C.8 | 587 |

⟶ Ar—

| 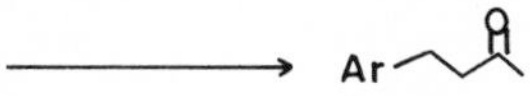 11.C.6 | 659-660 |

⟶

| 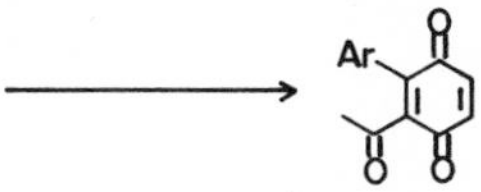 | 12.B.3 | 647 |

III. B. 3.   I ⟶ I

(3) C − 3 Functionality = 3

6 ⟶ 6   Ar–H ⟶ Ar⌒⌒CO₂H   13.F.2   696-697

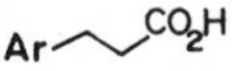

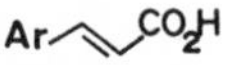

⟶ Ar⌒⌒CO₂H   13.F.9   705

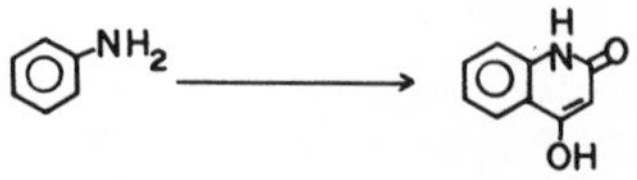

[aniline] ⟶ [quinolinone]   5.D.8   278

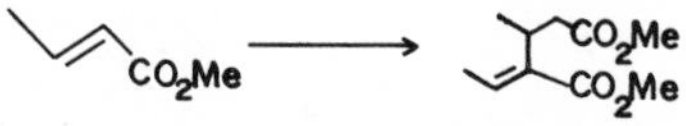

[crotonate CO₂Me] ⟶ [diester CO₂Me / CO₂Me]   2.C.I   129

O₂N⌒⌒ | R ⟶ O₂N⌒⌒CO₂R′ | R   20.C.2   999-1002   14.C.5   746-747
                                                    (11.G.3   605)

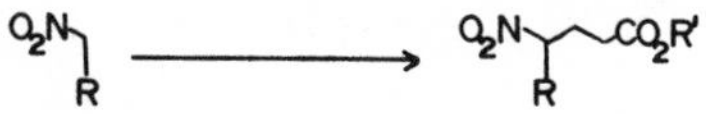

Functionality   2 ⟶ 2

6 ⟶ 6   R–C≡C–H ⟶ R–≡⌒⌒=O   11.G.3   603

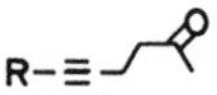

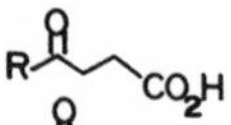

8 ⟶ 8   R–CHO ⟶ R(=O)⌒⌒CO₂H   13.A.7   755

R–CHO ⟶ R(=O)⌒⌒Z   11.F.7   591-592
                                    (11.D.7   576)
                                    (11.G.3   603)

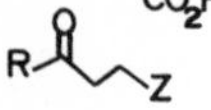

16 ⟶ 16   R⟨SR′/SR′ ⟶ R⟨SR′/SR′⌒Z   11.D.7   576;
                                        11.G.3   603

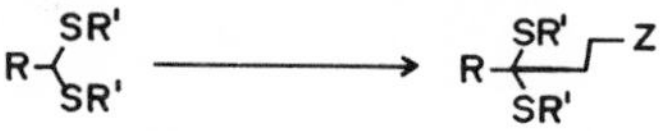

4. Four-Carbon Chain

Functionality   0 ⟶ 0

6 ⟶ 6   R–MgBr ⟶ R⌒≡⌒OH   3.A.2   175

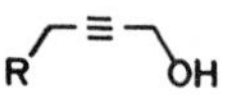

R–B⟨ ⟶ R⌒⌒OH   4.B.2   222

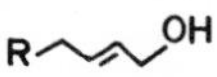

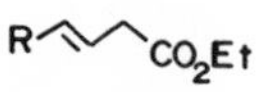

⟶ R⌒⌒CO₂Et   14.B.8   738

R₂CuLi ⟶ R⌒⌒CO₂R / CO₂R   14.C.13   756;
                                    19.B.I   866

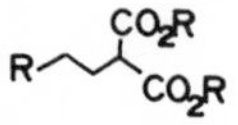

Z⟩Z ⟶ [Z⌒Z / Z⌒Z]   11.G.3   605

R–MgX ⟶ R(=O)⌒⌒CONH₂   18.E.4   845

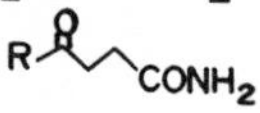

III.B.4.   I⟶I

Functionality   I⟶I

6⟶6   Ar-H ⟶ Ar-Ar'   (I.D.7   45-47)
            benzidine rearrangement   8.G.6   503-504   8.G.6   449
                phenolic coupling   5.B.9   266   5.B.9   284-286;
                                                                 5.D.II   295

                                         12.A.3   728-729   12.A.3   639-640
                                                                 (5.B.9   284-286)

        Ar-H ⟶ Ar⌒⌒CN   19.E.I   880

             ⟶ Ar⌒⌒CO₂H   13.F.2   791-792   13.F.2   695-697

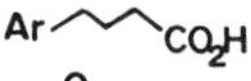

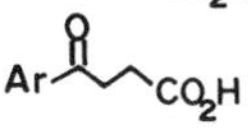

             ⟶ Ar⌒⌒CO₂H   5.D.8   278-279;
                                  II.C.4   657-658
                                  (13.F.I   789-791)

5. Five-Carbon Chain

    Functionality O⟶O

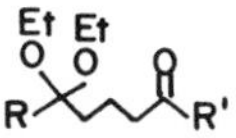

6⟶6   R-MgX ⟶ R⌒⌒R'   9.C.2   475

    Functionality I⟶I

6⟶6   ⌒Cu ⟶ ⌒⌒C=O   2.C.8   138
                        OMe

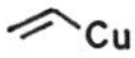

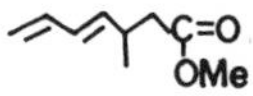

16⟶6   R⌒SH ⟶ R⌒⌒⌒CO₂H   13.G.2   799

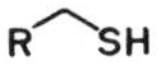

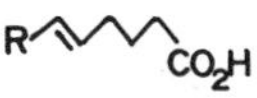

C. OXIDATION OF SUBSTRATE

    I. One-Carbon Chain

        Functionality O⟶I

        (I) C-I Functionality = I

6⟶6   R-CH₂-MgX ⟶ R-CH=C<   2.C.5,6   128-129

                                         Ad.   268

III. C. I.   O → I

6 → 6 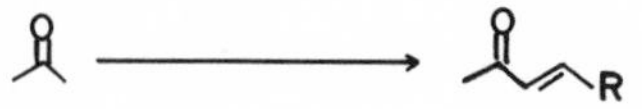   4.G.I   234-238;   4.G.I   264-266;
                                    see also other sections,   II.G.I   598;
                                    e.g. III.A.2   II.G.3   602

13.C.I   677

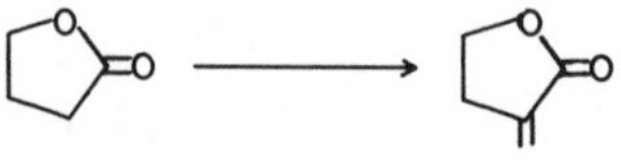

14.C.I   743-744;
                                                              Ad.   715; Reviews
                                                              Ad.   757

(2) C – I   Functionality = 2

6 → 6 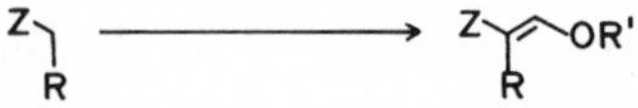   14.C.I   838   6.I.8   341;
                                               14.C.4   745

10.C.7   587

(3) C – I   Functionality = 3

6 → 6 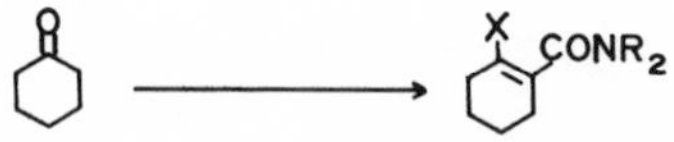   Ad.   841

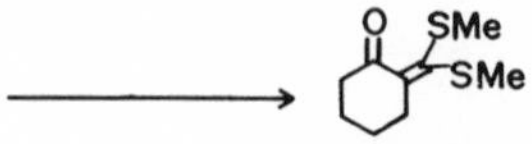

II.G.3   602

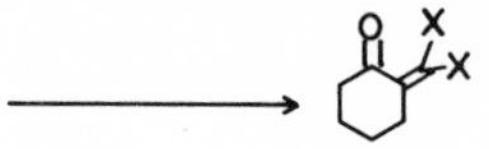

II.G.I   596

Functionality   I → 2

(I) C – I   Functionality = I

7 → 7 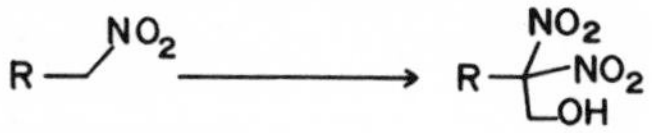   20.D.2   906

2. Two-Carbon Chain

Functionality   O → I

6 → 6 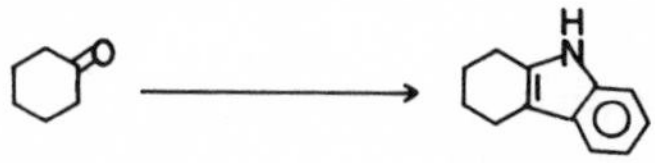   8.G.7   504-505

III.C.2.   O ⟶ I

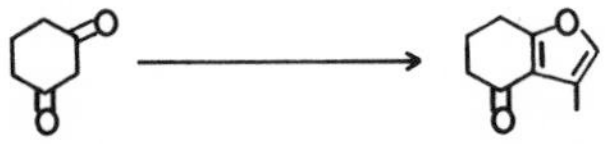

II.F.6   590-591

(2.A.12 103)

IV. REARRANGEMENTS OF THE CARBON SKELETON

Functionality O ⟶ O

hydrocarbon rearrangements	I.D.3	41-42	I.D.3 Ad. I.G.8	45-46; 47; 68-69

Functionality O ⟶ I

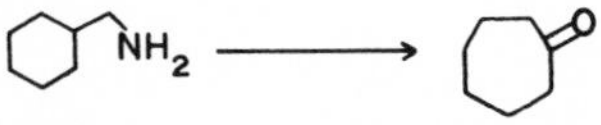

4.A.5   180-181

dienone ⟶ phenol	5.D.6	275-277	5.D.6	293
Cope rearrangement	2.D.1	135	(11.I.2	630-632)

Functionality I ⟶ O   See also O→I

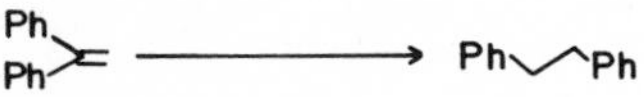

I.A.5   14

Functionality I ⟶ I

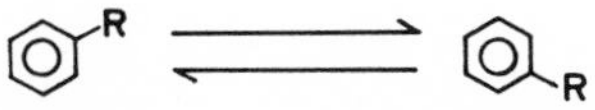

I.D.5   43-44   (5.G.1   300)

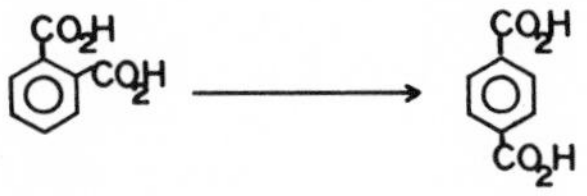

13.C.4   779-780

Functionality I ⟶ 2

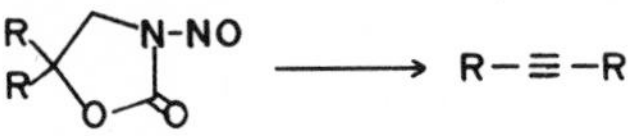

3.A.4   177

IV. 1 → 2

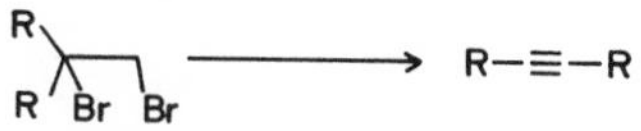

R₂C(Br)CH₂Br → R−≡−R		3.A.1	171

(cyclohexane with exocyclic methylene) → (cycloheptanone)   II.A.6 544; Ad. 584

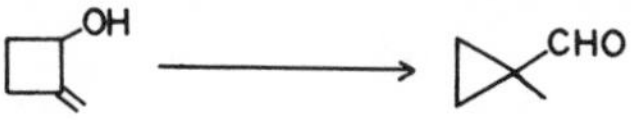

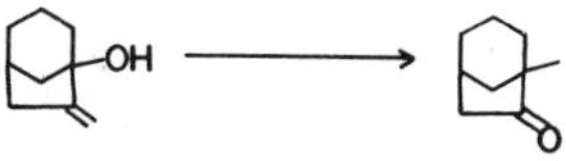

(butadienol) → (cyclopropene-CHO)   II.I.2 630-632

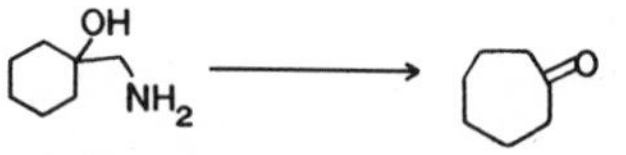

II.E.4 681-682 II.E.4 580
(II.E.5 682-684)

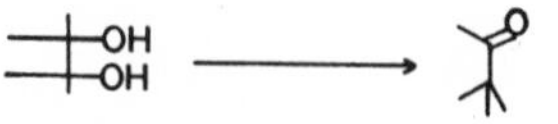

II.E.1 677-678 II.E.1 577-578
(10.E.1 600-603) (10.E.1 517-518) 
(11.A.5 631-632)

10.E.3 519;
II.E.3 679-681 II.E.3 579-580
(5.D.7 293-294)

10.E.2 518

10.E.2 518

II.E.2 578

Functionality 2 → 1

Ph−CO−CO−Ph → Ph₂C(OH)CO₂H   13.B.4 764-765

Functionality 2 → 2

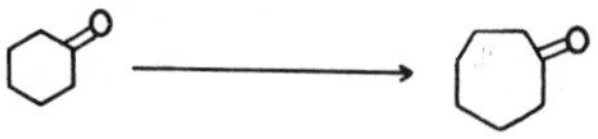

(cyclohexanone) → (cycloheptanone)   II.E.5 682-684 (II.E.4 580)
(20.E.2 1008)

Ar₂CH−CHO → Ar−CH₂−CO−Ar   II.E.6 684-685

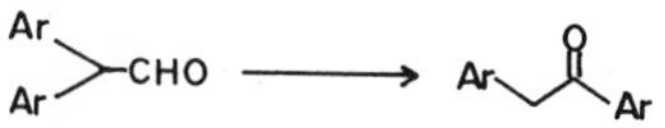

Ⅳ. 2 ⟶ 2

$$\underset{Ar}{\overset{Ar}{>}}C=\underset{}{}X \longrightarrow Ar-\equiv-Ar \qquad 3.A.1 \quad 152 \qquad\qquad 3.A.1 \quad 172$$
$$(3.A.2 \quad 175)$$
$$(3.A.4 \quad 176,178)$$

Functionality   2 ⟶ 3

$$R-\overset{O}{\overset{\|}{C}}-CH=N_2 \longrightarrow R-CH=C=O \qquad
\begin{array}{ll}
13.G.1 & 796\text{-}797; \\
17.A.3 & 888\text{-}889 \\
(10.B.4 & 573) \\
(11.F.5 & 693) \\
(14.B.3 & 827\text{-}828) \\
(18.A.8 & 905\text{-}926)
\end{array}
\qquad
\begin{array}{ll}
13.G.1 & 707\text{-}708; \\
17.A.3 & 806\text{-}808
\end{array}$$

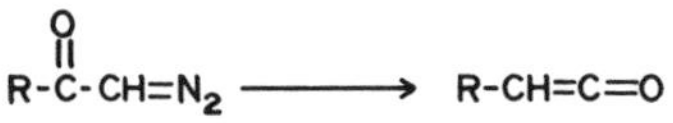

$$\underset{Ar}{\overset{Ar}{>}}\!\!\!\begin{array}{c}O\\ \\O\end{array} \longrightarrow \underset{Ar}{\overset{Ar}{>}}\!\!\!\begin{array}{c}OH\\CO_2H\end{array} \qquad 13.B.4 \quad 764\text{-}765$$

$$\underset{R'}{\overset{R}{>}}\!\!\!\begin{array}{c}O\\ \\X\end{array} \longrightarrow \underset{R}{\overset{R}{>}}\!\!CO_2H \qquad
\begin{array}{ll}
13.G.2 & 797\text{-}799; \\
14.C.8 & 848
\end{array}
\qquad
\begin{array}{ll}
13.G.2 & 708\text{-}709
\end{array}$$

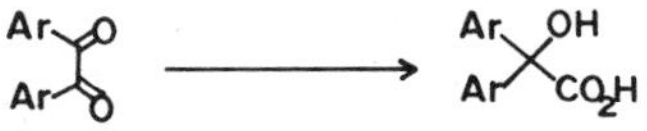

Functionality   3 ⟶ 2

$$\underset{R'}{\overset{R}{>}}\!\!\!\begin{array}{c}X\\ \\X\end{array} \longrightarrow R-\equiv-R' \qquad\qquad\qquad\qquad 3.A.2 \quad 175$$